Mechanical Engineering Series

Frederick F. Ling
Editor-in-Chief

Mechanical Engineering Series

J. Chakrabarty, **Applied Plasticity, Second Edition**

G. Genta, **Vibration Dynamics and Control**

R. Firoozian, **Servo Motors and Industrial Control Theory**

G. Genta and L. Morello, **The Automotive Chassis, Volumes 1 & 2**

F. A. Leckie and D. J. Dal Bello, **Strength and Stiffness of Engineering Systems**

Wodek Gawronski, **Modeling and Control of Antennas and Telescopes**

M. Ohsaki and KiyohiroIkeda, **Stability and Optimization of Structures: Generalized Sensitivity Analysis**

A.C. Fischer-Cripps, **Introduction to Contact Mechanics, 2nd ed.**

W. Cheng and I. Finnie, **Residual Stress Measurement and the Slitting Method**

J. Angeles, **Fundamentals of Robotic Mechanical Systems: Theory, Methods and Algorithms, 3rd ed.**

J. Angeles, **Fundamentals of Robotic Mechanical Systems: Theory, Methods, and Algorithms, 2nd ed.**

P. Basu, C. Kefa, and L. Jestin, **Boilers and Burners: Design and Theory**

J.M. Berthelot, **Composite Materials: Mechanical Behavior and Structural Analysis**

I.J. Busch-Vishniac, **Electromechanical Sensors and Actuators**

J. Chakrabarty, **Applied Plasticity**

K.K. Choi and N.H. Kim, **Structural Sensitivity Analysis and Optimization 1: Linear Systems**

K.K. Choi and N.H. Kim, **Structural Sensitivity Analysis and Optimization 2: Nonlinear Systems and Applications**

G. Chryssolouris, **Laser Machining: Theory and Practice**

V.N. Constantinescu, **Laminar Viscous Flow**

G.A. Costello, **Theory of Wire Rope, 2nd ed.**

K. Czolczynski, **Rotordynamics of Gas-Lubricated Journal Bearing Systems**

M.S. Darlow, **Balancing of High-Speed Machinery**

W. R. DeVries, **Analysis of Material Removal Processes**

J.F. Doyle, **Nonlinear Analysis of Thin-Walled Structures: Statics, Dynamics, and Stability**

J.F. Doyle, **Wave Propagation in Structures: Spectral Analysis Using Fast Discrete Fourier Transforms, 2nd Edition**

P.A. Engel, **Structural Analysis of Printed Circuit Board Systems**

A.C. Fischer-Cripps, **Introduction to Contact Mechanics**

A.C. Fischer-Cripps, **Nanoindentation, 2nd ed.**

J. García de Jalón and E. Bayo, **Kinematic and Dynamic Simulation of Multibody Systems: The Real-Time Challenge**

W.K. Gawronski, **Advanced Structural Dynamics and Active Control of Structures**

W.K. Gawronski, **Dynamics and Control of Structures: A Modal Approach**

G. Genta, **Dynamics of Rotating Systems**

(continued after index)

J. Chakrabarty

Applied Plasticity, Second Edition

Springer

J. Chakrabarty
Visiting Professor
Department of Mechanical Engineering
Florida State University
Tallahassee FL 32303
USA
jchak9@hotmail.com

ISSN 0941-5122
ISBN 978-0-387-77673-6 e-ISBN 978-0-387-77674-3
DOI 10.1007/978-0-387-77674-3
Springer New York Dordrecht Heidelberg London

Library of Congress Control Number: 2009934696

Printed on acid-free paper

Springer is part of Springer Science+Business Media (www.springer.com)

Mechanical Engineering Series

Frederick F. Ling
Editor-in-Chief

The Mechanical Engineering Series features graduate texts and research monographs to address the need for information in contemporary mechanical engineering, including areas of concentration of applied mechanics, biomechanics, computational mechanics, dynamical systems and control, energetics, mechanics of materials, processing, production systems, thermal science, and tribology.

Series Preface

Mechanical engineering, an engineering discipline forged and shaped by the needs of the industrial revolution, is once again asked to do its substantial share in the call for industrial renewal. The general call is urgent as we face profound issues of productivity and competitiveness that require engineering solutions, among others. The Mechanical Engineering Series features graduate texts and research monographs intended to address the need for information in contemporary areas of mechanical engineering.

The series is conceived as a comprehensive one that covers a broad range of concentrations important to mechanical engineering graduate education and research. We are fortunate to have a distinguished roster of consulting editors on the advisory board, each an expert in one of the areas of concentration. The names of the consulting editors are listed on the facing page of this volume. The areas of concentration are applied mechanics, biomechanics, computational mechanics, dynamic systems and control, energetics, mechanics of materials, processing, production systems, thermal science, and tribology.

Austin, Texas

Frederick F. Ling

This book is humbly dedicated to the loving memory of MA INDIRA who continues to be the source of real inspiration to me.

Preface

The past few years have witnessed a growing interest in the application of the mechanics of plastic deformation of metals to a variety of engineering problems associated with structural design and technological forming of metals. Written several years ago to serve as a companion volume to the author's earlier work under the title *Theory of Plasticity*, which comprehensively expounds the fundamentals of plasticity of metals, the present work seems to have stood the test of time and has established itself as a comprehensive reference work that is equally useful for classroom purposes. While the earlier work is mainly concerned with the application of the theory to the solution of elastic/plastic problems, limit analysis of framed structures, and problems in plane plastic strain involving slipline fields, several important areas of plasticity related to the analysis of multidimensional structures and various metal-forming processes had to be left out for obvious reasons. The present text is intended to fill this gap and to make available to the reader in a single volume a detailed account of a wide range of useful results that are scattered in numerous periodicals and other sources.

The fundamentals of the mathematical theory of plasticity are discussed in Chapter 1 with sufficient details, in order to eliminate the need for frequent references to the author's earlier volume. The theory of plane plastic stress and its applications to structural analysis and sheet metal forming are presented in Chapter 2. The axially symmetrical plastic state, as well as a few three-dimensional problems of plasticity, is treated in Chapter 3. The plastic behavior of plates and shells, mainly from the point of view of limit analysis, is discussed with several examples in Chapters 4 and 5. The plasticity of metals with fully developed orthotropic anisotropy and its application to the plastic behavior of anisotropic sheets are presented in Chapter 6. The generalized tangent modulus theory of buckling in the plastic range for columns, plates, and shells is treated in Chapter 7 from the point of view of the bifurcation phenomenon. Chapter 8 deals with a wide range of topics in dynamic plasticity, including the wave propagation, armor penetration, and structural impact in the plastic range. The fundamentals of the rigid/plastic finite element method, with special reference to its application to metal-forming processes, are presented in Chapter 9, where several examples are included for illustration.

The publication of the revised second edition of *Applied Plasticity* is deemed necessary not only for the obvious need for updating the book but also for the purpose of making it more suitable for the teaching of appropriate courses on plasticity at the graduate level. During the preparation of the second edition, several parts of the text have been extensively revised in the light of the recent developments of the subject, and new references to the published literature have been made in appropriate places. The discussion of the finite element method in plasticity, previously relegated to an appendix in the first edition, has now been expanded into a new chapter to permit a more complete treatment of the subject. A new section has been added in Chapter 4 to discuss the yield line theory for plate bending, not only for the derivation of complete solutions but also for the estimation of upper bounds on the limit load. A set of homework problems has been included at the end of each chapter for the benefit of both the student and the instructor, many of these problems having been designed to supplement the text. The references to the published literature have now been collected together and placed at the very end of the book for the sake of the expected convenience of the reader.

The book in its present form would be suitable for teaching advanced graduate level courses on plasticity and metal forming to students of mechanical and manufacturing engineering, as well as on structural plasticity to students of civil and structural engineering. The book will also be found useful for teaching courses on dynamic plasticity to both the mechanical and civil engineering students. Though intended primarily for research workers in the field of plasticity, senior undergraduate students and practicing engineers are also likely to benefit from this book to a large extent.

I take this opportunity to express my gratitude to the late Professor J. M. Alexander, formerly of Imperial College, London, who not only stimulated my interest in plasticity but also encouraged me to undertake the task of writing this book. I am also grateful to Dr. Frederick F. Ling, the Editor-in-Chief of this Series, for his encouragement and support for the publication of the second edition of *Applied Plasticity*. It is a pleasure to offer my sincere thanks to Ms. Jennifer Mirski, the Assistant Engineering Editor of Springer for her helpful cooperation and support during the preparation of the manuscript. Finally, I am deeply indebted to my wife Swati, who gracefully accepted the hardship of many lonely hours to enable me to complete this work in a satisfactory manner.

J. Chakrabarty

Contents

Chapter 1
Fundamental Principles

1.1 The Material Response

1.1.1 Introduction

In a single crystal of many metals, the main mechanism of plastic deformation is simple shear parallel to preferred planes and directions, which at ordinary temperatures coincides with those of the highest atomic density. Slip is initiated along a particular plane and in a given direction when the associated component of the shear stress attains a critical value under increasing external load. The amount of plastic deformation in a single crystal is specified by the glide strain, which is the relative displacement of two parallel slip planes at a unit distance apart. When there are several possible slip directions in a crystal lattice, the displacement of any point in the crystal due to simultaneous shears in the appropriate directions can be found from simple geometry. The mechanism of slip-induced plasticity in single crystals, governed by the glide motion of dislocations along corresponding slip planes, has been the subject of numerous investigations in the past.

The change in shape of a single crystal requires, in general, the operation of five independent slip systems (von Mises, 1928). This is due to the fact that an arbitrary state of strain is specified by the six independent components of the symmetric strain tensor, while the sum of the normal strain components vanishes by the condition of constancy of volume of the plastic material. The existence of five independent slip systems in a single crystal is necessary for the material to be ductile in the polycrystalline form. Face-centered cubic metals, having 12 potential slip systems in each crystal grain, satisfy this requirement and are known to have high degrees of ductility, while hexagonal close-packed metals having relatively low symmetry are noted for limited ductility at room temperatures. The ductility of a polycrystalline metal also requires slip flexibility which enables the five independent slip systems to operate simultaneously within a small volume of the aggregate. Mathematical theories of slip-induced plasticity in single crystals have been developed by Hill (1966), Hill and Rice (1972), and Asaro (1983).

Some attempts have been made in the past to relate the tensile yield stress of polycrystalline metals in terms of the shear yield stress of the corresponding single crystals. Assuming each crystal grain to undergo the same uniform strain as the polycrystalline metal, and by minimizing the sum of the magnitudes of a set of geometrically possible shears, Taylor (1938) determined the uniaxial stress–strain

J. Chakrabarty, *Applied Plasticity, Second Edition*, Mechanical Engineering Series,
DOI 10.1007/978-0-387-77674-3_1,

curve for an aluminum aggregate, in good agreement with the experimental curve. The selection of active slip planes from a very wide range of possible combinations, on the basis of the minimum principle, is not necessarily unique. Taylor's approach has been generalized by Bishop and Hill (1951), who developed a method of deriving upper and lower bounds on the yield function for a polycrystalline metal under any set of combined stresses. The plastic behavior of polycrystalline aggregates in relation to that of single crystals, including the effect of elastic deformation, has been similarly examined by Lin (1957, 1971). A useful review of the recent developments of the micromechanics of polycrystal plasticity has been presented by Khan and Huang (1995).

Taylor's theoretical model ensures the compatibility of strains across the grain boundaries, but fails to satisfy the conditions of equilibrium across these boundaries. In order to satisfy both the conditions of compatibility and equilibrium, a self-consistent model has been proposed by Kröner (1961) and by Budiansky and Wu (1962). These authors approximated each individual crystal grain by a spherical inclusion embedded in an infinitely extended homogeneous elastic matrix. The relationship between the stresses and the strains in the individual grains and those applying to the aggregate has been obtained by an averaging process based on an elastic/plastic analysis for the inclusion problem. Berveiller and Zaoui (1979) modified the theoretical model by introducing a plastic accommodation factor based on the assumption of isotropy of the elastic and plastic responses of both the crystal grain and the aggregate. A self-consistent model in which the anisotropy of the material response is allowed for has been developed by Hill (1965) and Hutchinson (1970), who considered the individual crystal grain as an ellipsoidal inclusion embedded in an elastic/plastic matrix. Another extension of the self-consistent model based on equivalent body forces has been put forward by Lin (1984).

The macroscopic theory of plasticity, which the present volume is concerned with, is based on certain experimental observations on the behavior of ductile metals beyond the elastic limit under relatively simple states of combined stress. The theory is capable of predicting the distribution of stresses and strains in polycrystalline metals, not only in situations where the elastic and plastic strains are of comparable magnitudes but also in situations where the plastic strains are large enough for the elastic strains to be disregarded. The mathematical formulation of the plasticity problem is essentially incremental in nature, requiring due consideration of the complete stress and strain history of the elements that have been deformed in the plastic range. An interesting theoretical model, based on the nucleation of voids in the deforming material including interaction of neighboring voids, has been developed by Gurson (1977) and Tvergaard (1982). A theory of plastic yielding and flow of porous materials has been advanced by Tsuta and Yin (1998). A practical method of predicting the macroscopic behavior of polycrystalline aggregates from microstructural data has been discussed by Lee (1993).

1.1.2 The True Stress–Strain Curve

In order to deal with large plastic deformation of metals, it is necessary to introduce the concept of true stress and true strain occurring in a test specimen. The true stress σ defined as the applied load divided by the current area of cross section of the specimen, can be significantly different from the nominal stress s, which is the load divided by the original area of cross section. If the initial and current lengths of a tensile specimen are denoted l_0 and l, respectively, then the engineering strain e is equal to the ratio $(l - l_0)/l_0$, while the true strain ε is defined in such a way that its increment $d\varepsilon$ is equal to the ratio dl/l, where dl is the corresponding increase in length. It follows that the total true strain produced by a change in length from l_0 to l during the tensile test is

$$\varepsilon = \ln\left(\frac{l}{l_0}\right) = \ln(1 + e). \tag{1.1}$$

Similarly, in the case of simple compression of a specimen, whose height is reduced from h_0 to h during the test, the engineering strain is of magnitude $e = (h_0 - h)/h_0$, while the magnitude of the true strain is

$$\varepsilon = \ln\left(\frac{h_0}{h}\right) = -\ln(1 - e).$$

As the deformation is continued in the plastic range, the true stress becomes increasingly higher than the nominal stress in the case of simple tension and lower than the nominal stress in the case of simple compression. The true strain, on the other hand, is progressively smaller than the engineering strain in simple tension and higher than the engineering strain in simple compression, as the deformation proceeds.

There is sufficient experimental evidence to suggest that the macroscopic stress–strain curve of a polycrystalline metal in simple compression coincides with that in simple tension when the true stress is plotted against the true strain. Figure 1.1(a) shows the true stress–strain curve of a typical engineering material in simple tension. The longitudinal true stress σ existing in a test specimen under an axial load is a monotonically increasing function of the longitudinal true strain ε. The straight line OA represents the linear elastic response with A denoting the proportional limit. The elastic range generally extends slightly beyond this to the yield point B, which marks the beginning of plastic deformation. The strain-hardening property of the material requires the stress to increase with strain in the plastic range, but the slope of the stress–strain curve progressively decreases as the strain is increased until fracture occurs at G.

Since plastic deformation is irreversible, unloading from some point C on the loading curve would make the stress–strain diagram follow the path CE, where E lies on the ε-axis for complete unloading and represents the amount of permanent or plastic strain corresponding to C. On reloading the specimen, the stress–strain

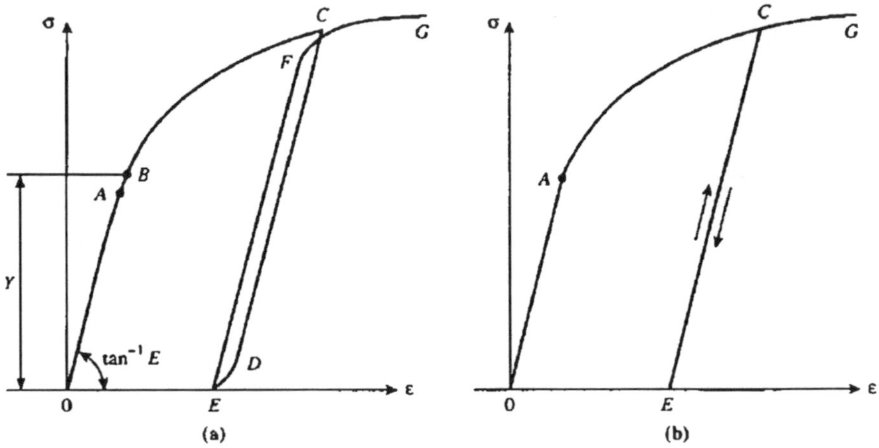

Fig. 1.1 True stress–strain curve in simple tension: (**a**) loading and unloading with reloading and (**b**) idealized stress–strain behavior

curve follows the path *EFG* forming a hysteresis loop of narrow width, where *F* is a new yield point and *FG* is virtually a continuation of *BC*. Following Prandtl (1928), the stress–strain curve may be idealized by neglecting the width of the hysteresis loop and assuming the unloading path to be a straight line parallel to *OA*. The idealized curve, shown in Fig. 1.1(b), implies that the reloading proceeds along the path *ECG*, passing through the point *C* where the previous unloading started. Furthermore, the proportional limit *A* is assumed to coincide with the initial yield point, the corresponding stress being denoted by *Y*.

It follows from the idealized stress–strain curve that the recoverable elastic strain at any point of the curve is equal to σ/E, where *E* is Young's modulus and σ the current stress. Any stress increment $d\sigma$ is associated with an elastic strain increment of amount $d\sigma/E$ and a plastic strain increment of amount $d\sigma/H$, where *H* is the plastic modulus representing the current slope of the curve for the stress against plastic strain. Since the total strain increment is $d\sigma/T$, where *T* is the tangent modulus denoting the slope of the (σ, ε)-curve, we obtain the relationship

$$\frac{1}{T} = \frac{1}{E} + \frac{1}{H}. \tag{1.2}$$

The difference between *T* and *H* decreases rapidly with increasing strain, the elastic strain increment being increasingly small compared to the plastic strain increment. The material is said to be nonhardening when $H = T = 0$, which is approximately satisfied by a material that is heavily prestrained.

Each increment of longitudinal strain $d\varepsilon$ in a tensile specimen is accompanied by a lateral compressive strain increment of magnitude $d\varepsilon'$, the ratio $d\varepsilon'/d\varepsilon$ being known as the contraction ratio denoted by η. Over the elastic range of strains, the contraction ratio has a constant value equal to Poisson's ratio ν, but once the yield point is exceeded, the contraction ratio becomes a function of the magnitude of the strain. For an isotropic material, the elastic and plastic parts of the lateral strain increment have the magnitudes $\nu d\varepsilon^e$ and $d\varepsilon^p/2$, respectively, where $d\varepsilon^p = d\varepsilon - d\sigma/E$ denotes the plastic component of the longitudinal strain increment. Consequently, the total lateral strain increment is given by $2d\varepsilon' = d\varepsilon - (1 - 2\nu)\, d\sigma/E$, and the contraction ratio is therefore expressed as

$$\eta = \frac{1}{2}\left[1 - (1 - 2\nu)\frac{T}{E}\right]. \tag{1.3}$$

Thus, η depends on the current value of T. As the loading is continued in the plastic range, the tangent modulus progressively decreases from its elastic value E and the contraction ratio rapidly increases from ν to approach the fully plastic value of 0.5. It follows that the elastic compressibility of the material becomes negligible when the tangent modulus is reduced to the order of the current yield stress σ. For an incompressible material, the relationship between the nominal stress s and the true stress σ is easily shown to be

$$s = \sigma \, \exp(\mp \varepsilon),$$

where the upper sign corresponds to simple tension and the lower sign to simple compression of the test specimen. The distinction between the behaviors in tension and compression, in relation to the engineering strain, is illustrated in Fig. 1.2.

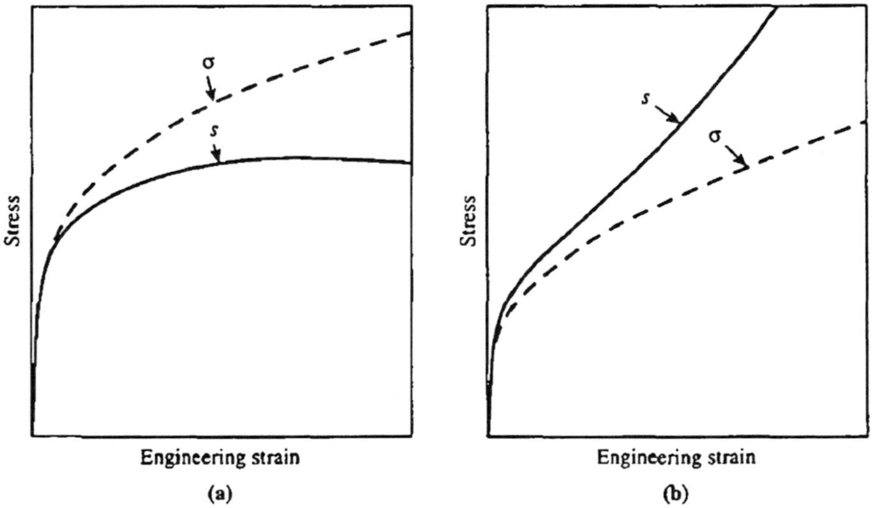

Fig. 1.2 The stress–strain behavior of metals with respect to nominal stress, true stress, and engineering strain: (**a**) simple tension and (**b**) simple compression

The tensile test is unsuitable for obtaining the stress–strain curve up to large values of the strain, since the specimen begins to neck when the rate of work-hardening decreases to a critical value. At this stage, the applied load attains a maximum, and the specimen subsequently extends under decreasing load. Setting the differential of the axial load $P = \sigma A$ to zero, where A is the current cross-sectional area of the specimen, we get $d\sigma/\sigma = -dA/A$. On the other hand, the condition of zero incremental volume change, which holds very closely at the onset necking, gives $-dA/A = dl/l = d\varepsilon$. The condition for plastic instability in simple tension therefore becomes

$$\frac{d\sigma}{de} = \sigma. \tag{1.4}$$

As the extension continues beyond the point of necking, the plastic deformation remains confined in the neck, which grows rapidly under decreasing load leading to fracture across the minimum section. The true stress–strain curve cannot be continued beyond the point of necking without introducing a suitable correction factor (Section 3.3). A neck is also formed in a cylindrical specimen subjected to a uniform fluid pressure on the lateral surface, and the amount of uniform strain at the point of necking is exactly the same as that in uniaxial tension (Chakrabarty, 1972).

The strain-hardening characteristic of materials covering a fairly wide range of strains is most conveniently obtained by the simple compression test, in which a solid cylindrical block is axially compressed between a pair of parallel platens. However, due to the presence of friction between the specimen and the platens, the deformation of material near the regions of contact is constrained, resulting in a barreling of the specimen as the compression proceeds. Since the compression of the cylinder then becomes nonuniform, the true stress–strain curve cannot be derived from the compression test without introducing a suitable correction factor (Section 3.4). Several methods have been proposed in the past to eliminate the effect of friction on the stress–strain curve, but none of them seems to be entirely satisfactory. On the other hand, a state of homogeneous compression is very nearly achieved by inserting thin sheets of *ptfe* (polytetrafluoroethylene) between the specimen and the compression platens. The compressed *ptfe* sheets not only act as an effective lubricant but also help to inhibit the barreling tendency by exerting a radially outward pressure on the material near the periphery. It is necessary to apply the axial load, on an incremental basis, and to replace the deformed *ptfe* sheets with new ones before each application of the load. The true stress and the true strain are obtained at each stage from the measurement of the applied load and the current specimen height, together with the use of the constancy of volume.

Consider an annealed specimen which is loaded in simple tension past the yield point and is subsequently unloaded to zero stress so that there is a certain amount of residual strain left in the specimen. If an axial compressive load is now applied, the specimen will begin to yield under a stress that is somewhat lower than the original yield stress in tension or compression. While the yield stress in tension at the time of unloading is much greater than Y owing to strain hardening of the material, the yield stress in compression is usually found to be lower than Y. A similar lowering of the yield stress is observed if the specimen is loaded plastically in compression and then pulled in tension. This phenomenon is known as the *Bauschinger effect*, which occurs whenever there is a reversal of stress in a plastically deformed element. The phenomenon is generally attributed to residual stresses in the individual crystal grains due to the presence of grain boundaries in a polycrystalline metal.

In some metals, such as annealed mild steel, the load at the elastic limit suddenly drops from an upper yield point to a lower yield point, followed by an elongation of a few percent under approximately constant stress. At the upper yield point, a discrete band of deformed metal, known as *Lüders band*, appears at approximately 45° to the tensile axis at a local stress concentration. During the yield point elongation, several bands usually form at several points of the specimen and propagate to cover the entire length. At this stage, the load begins to rise with further strain and the stress–strain curve then continues in the usual manner as a result of strain hardening. The upper yield point depends on such factors as the rate of straining, eccentricity of the loading, and the rigidity of the testing machine, but its value is usually 10–20% higher than the lower yield point.

1.1.3 Empirical Stress–Strain Equations

In the theoretical treatment of plasticity problems, it is generally convenient to represent the true stress–strain curve of the material by a suitable empirical equation that involves constants to be determined by curve fitting with the experimental curve. For sufficiently large strains, the simplest empirical equation frequently used in the literature is the simple power law proposed by Ludwik (1909), which is

$$\sigma = C\varepsilon^n, \tag{1.5}$$

where C is a constant stress and n is n dimensionless constant, known as the strain-hardening exponent, whose value is generally less than 0.5. Although (1.5) corresponds to an infinite initial slope, it does provide a reasonably good fit with the actual stress–strain curve over a fairly wide range of strains. Since $d\sigma/d\varepsilon$ is equal to $n\sigma/\varepsilon$ according to (1.5), the true strain at the onset of necking in simple tension is $\varepsilon = n$ in view of (1.4). A nonhardening material corresponds to $n = 0$ with C representing the constant yield stress of the material.

When the material is assumed to be rigid/plastic having a distinct initial yield stress Y, the simple power law (1.5) needs to be suitably modified. One such modification, sometimes used in the solution of special problems, is

$$\sigma = Y + K\varepsilon^n,$$

where K has the dimension of stress and n is an exponent. Although this equation predicts a nonzero initial yield stress, it does not provide a better fit with the actual stress–strain curve over the relevant range of strains. The preceding equation includes, as a special case, the linear strain-hardening law ($n = 1$), with K denoting the constant plastic modulus. A more successful empirical equation involving a definite yield point is the modified power law

$$\sigma = C(m + \varepsilon)^n, \tag{1.6}$$

where C, m, and n are constants. The stress–strain curve defined by (1.6), which is due to Swift (1952), is essentially the Ludwik curve (1.5) with the σ-axis moved through a distance m in the direction of the ε-axis. The parameter m therefore represents the amount of initial prestrain with respect to the annealed state. If the same stress–strain curve is fitted by both (1.5) and (1.6), the value of n in the two cases will of course be different. It follows from (1.4) and (1.6) that the magnitude of the true strain at the point of tensile necking is equal to $n - m$ for $m \leq n$ and zero for $n \geq m$. Figure 1.3(a) shows the Swift curves for several values of n based on a typical value of m.

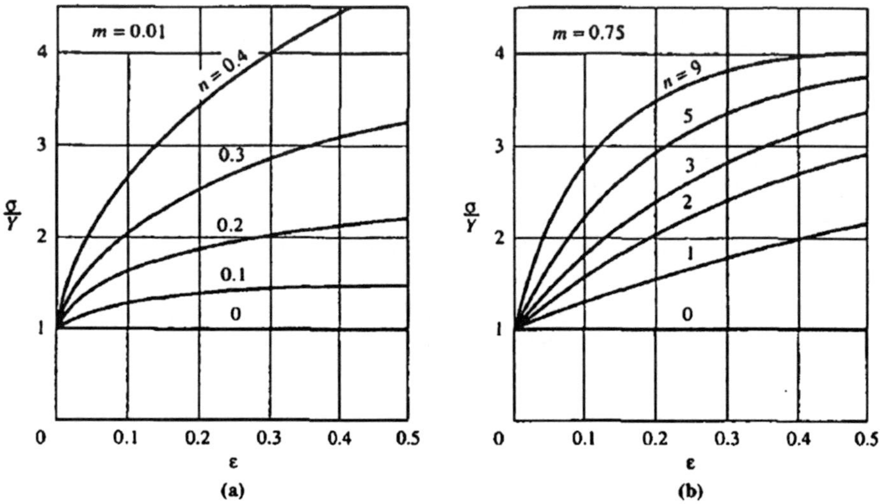

Fig. 1.3 Nature of empirical stress–strain equations. (**a**) Swift equation and (**b**) Voce equation

For certain applications, it is sometimes more convenient to employ a different type of empirical equation proposed by Voce (1948). The Voce equation, which also involves an initial yield stress of the material, may be expressed as

$$\sigma = C(1 - me^{-n\varepsilon}), \tag{1.7}$$

where C, m, and n are material constants, and e is the exponential constant. The stress–strain curve defined by (1.7) exhibits an initial yield stress equal to $(1 - m)C$ and tends to become asymptotic to the saturation stress $\sigma = C$. The slope of the stress–strain curve varies linearly with the stress according to the relation $d\sigma/d\varepsilon - n(C - \sigma)$. Thus, the initial state of hardening of the material is represented by m, while the rapidity of approach to the saturation stress is represented by n. The stress–strain curves defined by (1.7) for a given value of m and several values of n are displayed in Fig. 1.3(b).

The preceding stress–strain equations can be used for elastic/plastic materials, with comparable elastic and plastic strains, provided ε is replaced by the plastic component ε^p. Since the elastic component of the strain ε^e is equal to σ/E, the simple power law of type (1.5) relating the stress to the plastic strain furnishes the total strain in the form

$$\varepsilon = \frac{\sigma}{E}\left\{1 + \alpha\left(\frac{\sigma}{\sigma_0}\right)^{m-1}\right\}, \tag{1.8}$$

where $m = 1/n$, σ_0 is a nominal yield stress, and α is a dimensionless constant. The stress–strain curve defined by (1.8), which is due to Ramberg and Osgood (1943), bends over with an initial slope equal to E, the plastic modulus H associated with any stress a being given by $E/H = m(\sigma/\sigma_0)^{m-1}$. It may be noted that the secant modulus of the stress–strain curve is equal to $E/(1 + \alpha)$ at the nominal yield point $\sigma = \sigma_0$ according to (1.8). Figure 1.4(a) shows several curves of this type for constant values of σ_0/E and α.

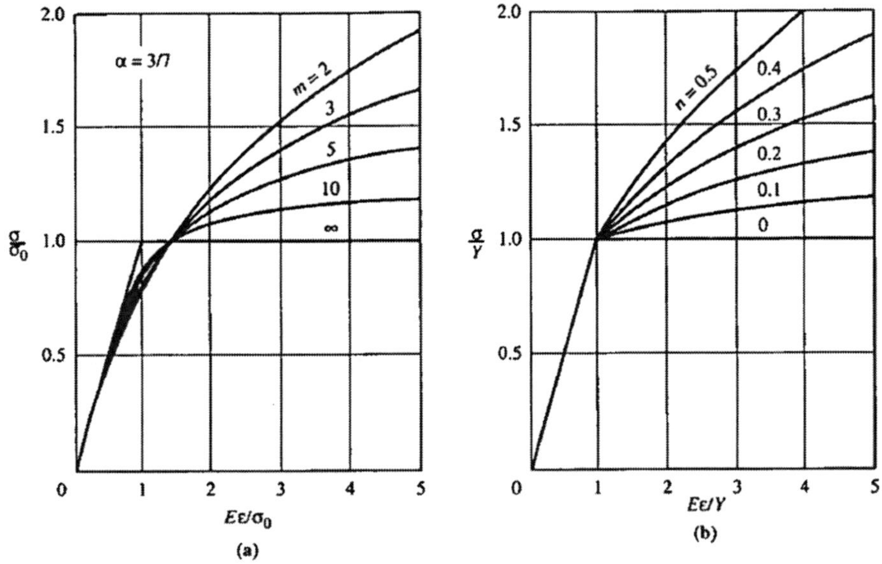

Fig. 1.4 True stress–strain curves for elastic/plastic materials: (**a**) Ramberg–Osgood equation and (**b**) modified Ludwik equation

In some cases, the elastic/plastic analysis can be considerably simplified by using a stress–strain relation that corresponds to the Ludwik curve with its initial part replaced by a chord of slope equal to E. The stress–strain law is then given by the pair of equations

$$\sigma = \begin{cases} E\varepsilon, & \varepsilon \leq Y/E, \\ Y(E\varepsilon/Y)^n, & \varepsilon \geq Y/E. \end{cases} \tag{1.9}$$

The material has a sharp yield point at $\sigma = Y$, and the slope of the stress–strain curve changes discontinuously from E to nE at the yield point, Fig. 1.4(b). The value of the tangent modulus T at any stress $\sigma \geq Y$ is equal to $nE(Y/\sigma)^{m-1}$, where $m = 1/n$. The discontinuity in slope at $\sigma = Y$ can be eliminated, however, by modifying the stress–strain equation in the plastic range as

$$\sigma = Y\left(\frac{E\varepsilon}{nY} - \frac{1-n}{n}\right)^n, \quad \varepsilon \geq Y/E.$$

The tangent modulus according to this equation is easily shown to be $T = E(Y/\sigma)^{m-1}$, where $m = 1/n$ as before. Since $T = \sigma$ at the incipient necking of a bar under simple tension, the magnitude of the uniform true strain exceeds n by the amount $(1 - n)Y/E$, which is negligible for most metals.

The rate of straining has a profound influence on the yield strength of metals, particularly at elevated temperatures. The mechanical response of materials to high strain rates and temperatures is generally established on the basis of uniaxial stress–strain curves obtained under constant values of the strain rate and temperature. For a given temperature, the combined effects of strain ε and strain rate $\dot{\varepsilon}$ can be expressed by the empirical equation

$$\sigma = C\varepsilon^n\dot{\varepsilon}^m, \tag{1.10}$$

where C, m, and n are constants, the parameters m being known as the strain rate sensitivity which is generally less than 0.2 for most metals and alloys. The higher the value of m, the greater is the strain at the onset of tensile necking (Hart, 1967). Since most of the heat generated during a high-speed test remains in the specimen, leading to an adiabatic rise in temperature, the results for a given test must be adjusted appropriately so that they correspond to a constant temperature. In the hot working of metals, the working temperature is high enough for recovery and recrystallization to occur without significant grain growth. Since the rate of work-hardening is then exactly balanced by the rate of thermal softening, the yield stress is practically independent of the strain except for sufficiently small values of the strain.

There are certain metals and alloys known as superplastic materials, which exhibit very large neck-free tensile elongations prior to failure (Backofen et al., 1964). These materials are characterized by high values of the strain rate sensitivity m, which is generally greater than 0.4. The tensile fracture in superplastic materials is caused by the evolution of cavities at grain boundaries rather than by the development of diffuse necks. Conventional superplasticity is observed at relatively low strain rates, usually ranging from 10^{-4}/s to 10^{-3}/s, although recent studies have revealed the existence of superplasticity at considerably higher strain rates in certain alloys and composite materials. The relationship between the flow stress σ and the strain rate $\dot{\varepsilon}$ in superplastic materials is often expressed by the equation

$$\sigma = K\dot{\varepsilon}^m,$$

where m is the strain rate sensitivity and K is generally independent of the strain. The elongation to failure in a tensile specimen of superplastic material increases with increasing values of m. In fine-grained superplastic materials, an m value of about 0.5 is fairly common, and the deformation takes place mainly by grain boundary sliding. Materials which are rendered superplastic by the generation of internal stresses through thermal or pressure cycling can have a strain rate sensitivity as high as unity. The superplasticity of metals and alloys has been discussed at length in the books by Presnyakov (1976), Padmanabhan and Davies (1980), and Nieh et al. (1997).

1.2 Basic Laws of Plasticity

1.2.1 Yield Criteria of Metals

The macroscopic theory of plasticity is based on certain experimental observations regarding the behavior of ductile metals. The theory rests on the assumption that the material is homogeneous and is valid only at temperatures for which thermal phenomena may be neglected. For the present purpose, it is also assumed that the material is isotropic and has identical yield stresses in tension and compression. The plasticity of metals with fully developed states of anisotropy will be discussed in Chapter 6. A further simplific`ation results from the experimental fact that the yielding of metals is unaffected by a moderate hydrostatic pressure (Bridgman, 1945; Crossland, 1954). A law governing the limit of elastic behavior, consistent with the basic assumptions, defines a possible criterion of yielding under any combination of the applied stresses.

The state of stress in any material element may be represented by a point in a nine-dimensional stress space. Around the origin of the stress space, there exists a domain of elastic range representing the totality of elastic states of stress. The external boundary of the elastic domain defines a surface, known as the initial yield surface, which may be expressed in terms of the components of the true stress σ_{ij}, as

$$f(\sigma_{ij}) = \text{constant}.$$

Since the material is initially isotropic, plastic yielding depends only on the magnitudes of the three principal stresses, and not on their directions. This amounts to the fact that the yield criterion is expressible as a function of the three basic invariants of the stress tensor. The yield function is therefore a symmetric function of the principal stresses and is also independent of the hydrostatic stress, which is defined as the mean of the three principal stresses. Plastic yielding therefore depends on the principal components of the deviatoric stress tensor, which is defined as

$$s_{ij} = \sigma_{ij} - \sigma\delta_{ij}, \tag{1.11}$$

where σ denotes the hydrostatic stress, equal to $(\sigma_1 + \sigma_2 + \sigma_3)/3$. Since the sum of the principal deviatoric stresses is $s_{ij} = 0$, the principal components cannot all be independent. It follows that the yield criterion may be expressed as a function of the invariants J_2 and J_3 of the deviatone stress tensor, which are given by (Chakrabarty, 2006)

$$\left.\begin{aligned} J_2 &= -\,(s_1 s_2 + s_2 s_3 + s_2 s_1) = \frac{1}{2}(s_1^2 + s_2^2 + s_3^2) = \frac{1}{2} s_{ij} s_{ij}, \\ J_3 &= s_1 s_2 s_3 = \frac{1}{3}(s_1^3 + s_2^3 + s_3^3) = \frac{1}{3} s_{ij} s_{jk} s_{ki}. \end{aligned}\right\} \quad (1.12).$$

The absence of Bauschinger effect in the initial state implies that yielding is unaffected by the reversal of the sign of the stress components. Since J_3 changes sign with the stresses, an even function of this invariant should appear in the yield criterion.

In a three-dimensional principal stress space, the yield surface is represented by a right cylinder whose axis is equally inclined to the three axes of reference, Fig. 1.5. The generator of the cylinder is therefore perpendicular to the plane $\sigma_1 + \sigma_2 + \sigma_3 = 0$, known as the deviatoric plane. Since $\sigma_1 = \sigma_2 = \sigma_3$, along the geometrical axis of the cylinder, it represents purely hydrostatic states of stress. Points on the generator therefore represent stress states with varying hydrostatic part, which does not have any influence on the yielding. The yield surface is intersected by the deviatoric plane in a closed curve, known as the yield locus, which is assumed to be necessarily convex.

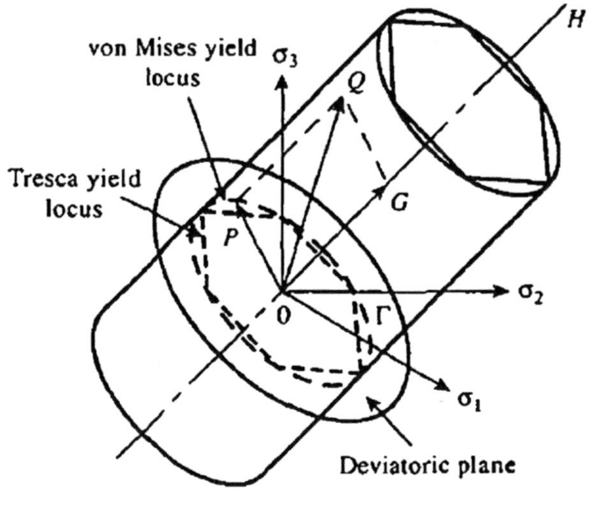

Fig. 1.5 Geometrical representation of yield criteria in the principal stress space

Due to the assumed isotropy and the absence of the Bauschinger effect, the yield locus must possess a six-fold symmetry with respect to the projected stress axes and the lines perpendicular to them, as indicated in Fig. 1.6(b). In an experimental

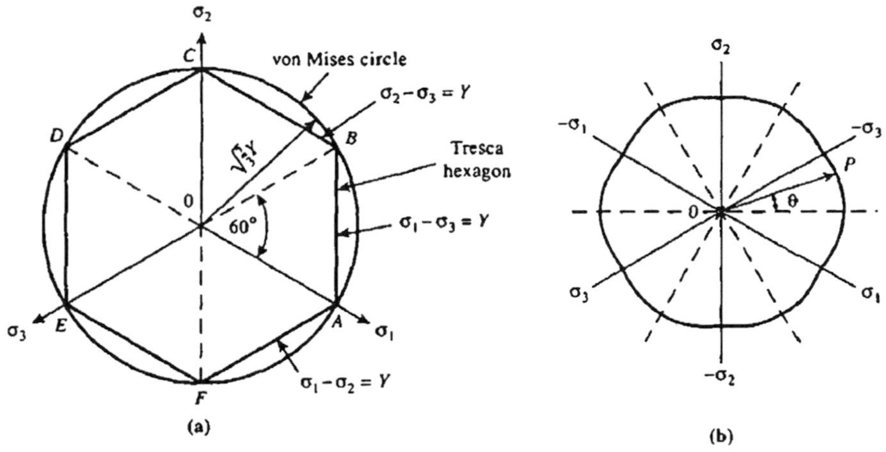

Fig. 1.6 Deviatoric yield locus. (**a**) Tresca hexagon and von Mises circle and (**b**) general shape of the locus

determination of the initial yield locus, it is therefore only necessary to apply stress systems covering a typical 30° segment of the yield locus. This may be achieved by introducing the Lode (1926) parameter μ, which is defined as

$$\mu = \frac{2\sigma_2 - \sigma_3 - \sigma_1}{\sigma_3 - \sigma_1} = -\sqrt{3}\tan\theta, \quad \sigma_1 > \sigma_2 > \sigma_3, \tag{1.13}$$

where θ is the counterclockwise angle made by the deviatoric stress vector with the direction representing pure shear. To obtain the shape of the yield locus, it is only necessary to apply stress systems varying between pure shear ($\mu = \pm$, $\theta = 0$) and uniaxial tension or compression ($\mu = \pm 1$, $\theta = \pm\pi/6$). The yield locus is called regular when it has a unique tangent at each point and singular when it has sharp corners.

The simplest yield criterion expressed in terms of the invariants of the deviatoric stress tensor is $J_2 = k^2$, suggested by von Mises (1913), where k is a constant. The yield function does not therefore involve J_3 at all. In terms of the stress component referred to an arbitrary set of rectangular axes, the von Mises yield criterion may be written as

$$\left.\begin{aligned} s_{ij}s_{ij} &= s_x^2 + s_y^2 + s_z^2 + 2(\tau_{xy}^2 + \tau_{yz}^2 + \tau_{zx}^2) = 2k^2 \\ &\text{or} \\ (\sigma_x - \sigma_y)^2 &+ (\sigma_y - \sigma_z)^2 + (\sigma_z - \sigma_x)^2 + 6(\tau_{xy}^2 + \tau_{yz}^2 + \tau_{zx}^2) = 6k^2. \end{aligned}\right\} \tag{1.14}$$

The second expression in (1.14) follows from the first on subtracting the identically zero term $(s_x + s_y + s_z)^2/3$ and noting the fact that $s_x - s_y = \sigma_x - \sigma_y$, etc. The constant k is actually the yield stress in simple or pure shear, as may be seen by setting $\sigma_x = \sigma$ and $\sigma_y = -\sigma$ as the only nonzero stress components. According to (1.14), the uniaxial yield stress Y is equal to $\sqrt{3}\,k$, which is obtained by considering $\sigma_x = Y$ as the only nonzero stress. The von Mises yield surface is evidently a right circular cylinder having its geometrical axis perpendicular to the deviatoric plane. The principal deviatoric stresses according to the von Mises criterion may be expressed in terms of the deviatoric angle θ as

$$s_1 = \frac{2}{3}Y \cos\left(\frac{\pi}{6} + \theta\right), \quad s_2 = \frac{2}{3}Y \sin\theta, \quad s_3 = -\frac{2}{3}Y \cos\left(\frac{\pi}{6} - \theta\right). \qquad (1.15)$$

In the case of plane stress, the actual principal stresses σ_1 and σ_2 may be expressed in terms of θ using the fact that the sum of these stresses is equal to $-3s_3$ which ensures that σ_3 is identically zero.

On the basis of a series of experiments involving the extrusion of metals through dies of various shapes, Tresca (1864) concluded that yielding occurred when the magnitude of the greatest shear stress attained a certain critical value. In terms of the principal stresses, the Tresca criterion may be written as

$$\sigma_1 - \sigma_3 = 2k, \quad \sigma_1 \geq \sigma_2 \geq \sigma_3, \qquad (1.16)$$

where k is the yield stress in pure shear, the uniaxial yield stress being $Y = 2k$ according to this criterion. All possible values of the principal stresses are taken into account when the Tresca criterion is expressed by a single equation in terms of the invariants J_2 and J_3, but the result is too complicated to have any practical usefulness. For a given uniaxial yield stress Y, the Tresca yield surface is a regular hexagonal cylinder inscribed within the von Mises cylinder.

The Tresca yield surface is not strictly convex, but each face of the surface may be regarded as the limit of a convex surface of vanishingly small curvature. The deviatoric yield loci for the Tresca and von Mises criteria are shown in Fig. 1.6(a). The maximum difference between the two criteria occurs in pure shear, for which the von Mises criterion predicts a yield stress which is $2/\sqrt{3}$ times that given by the Tresca criterion. Experiments have shown that for most metals the test points fall closer to the von Mises yield locus than to the Tresca locus, as indicated in Fig. 1.7. If the latter is adopted for simplicity, the overall accuracy can be improved by replacing $2k$ in (1.16) by mY, where m is an empirical constant lying between 1 and $2/\sqrt{3}$.

1.2.2 Plastic Flow Rules

The following discussion is restricted to an ideally plastic material having a definite yield point and a constant yield stress. The influence of strain hardening will be discussed in the next section. The yield locus for the idealized material retains its size and shape so that the material remains isotropic and free from the Bauschinger effect. Each increment of strain in the plastic range is the sum of an elastic part

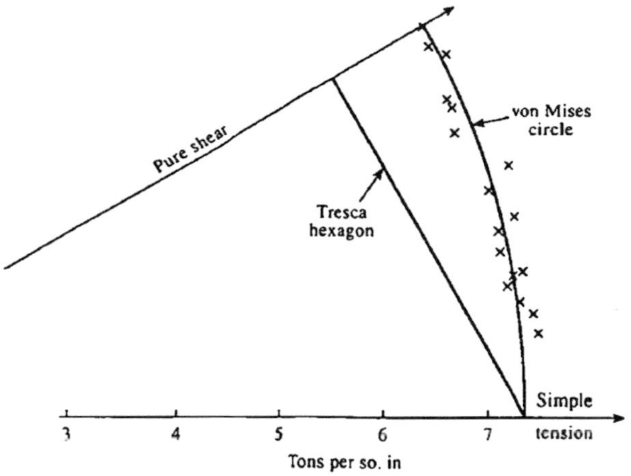

Fig. 1.7 Experimental verification of the yield criterion for commercially pure aluminum (due to Lianis and Ford, 1957)

which may be recovered on unloading, and a plastic part that remains unchanged on unloading. The elastic part of the strain increment is given by the generalized Hooke's law, while the plastic part is governed by what is known as the flow rule.

The consideration of the plastic deformation of a polycrystalline metal in relation to that of the individual crystals leads to the existence of a plastic potential that is identical to the yield function (Bishop and Hill, 1951). The plastic strain increment, regarded as a vector in a nine-dimensional space, is therefore directed along the outward normal to the yield surface at the considered stress point. Denoting the unit vector along the exterior normal by n_{ij}, the associated flow rule for a nonhardening material with a regular yield surface may be written as

$$d\varepsilon_{ij}^{p} = n_{ij}\, d\lambda, \quad n_{ij} d\sigma_{ij} = 0, \tag{1.17}$$

where $d\lambda$ is a positive scalar representing the magnitude of the plastic strain increment vector. The condition $n_{ij} d\sigma_{ij} = 0$ implies that the stress point must remain on the yield surface during an increment of plastic strain. When $n_{ij} d\sigma_{ij} < 0$, indicating unloading from the plastic state, the plastic strain increment is identically zero. Since the components of n_{ij} are proportional to $\partial f/\partial \sigma_{ij}$, where f defines the yield function, of J_2 and J_3, the principal axes of n_{ij} coincide with those of σ_{ij}. The flow rule therefore implies that the principal axes of stress and plastic strain increment coincide for an isotropic solid. The plastic incompressibility condition $d\varepsilon_{ii}^{p} = 0$ is identically satisfied, while the symmetric tensor n_{ij} satisfies the relations

$$n_{ij} n_{ij} = 1, \quad n_{ii} = 0.$$

For a singular yield criterion, (1.17) holds for all regular points of the yield surface. At a singular point of the yield surface, the normal is not uniquely defined, and the plastic strain increment vector may lie anywhere between the normals to the faces meeting at the considered edge. When $n_{ij} d\sigma_{ij} < 0$, the element unloads from the plastic state, and the plastic strain increment vanishes identically. The relation

$$d\sigma_{ij} d\varepsilon_{ij}^p = 0,$$

which holds for both loading and unloading for a regular yield surface, may be assumed to hold even for a singular point of a yield surface when the material is nonhardening (Drucker, 1951).

When the yield criterion is that of von Mises, $f(\sigma_{ij}) = \frac{1}{2} s_{ij} s_{ij}$, which gives $\partial f / \partial \sigma_{ij} = s_{ij}$. The deviatoric stress vector is then of magnitude $\sqrt{2}k$, and the unit normal to the yield surface is $n_{ij} = s_{ij}/\sqrt{2}\,k$. Replacing the quantity $d\lambda/\sqrt{2}\,k$ by $d\lambda$ in (1.17), the associated flow rule may be expressed as

$$d\varepsilon_{ij}^p = s_{ij} d\lambda, \quad s_{ij} ds_{ij} = 0. \tag{1.18}$$

The ratios of the components of the plastic strain increment are therefore identical to those of the deviatoric stress. This relationship was proposed independently by Levy (1870) and von Mises (1913), both of whom used the total strain increment instead of the plastic strain increment alone. It therefore applies to a hypothetical rigid/plastic material whose elastic modulus is infinitely large. The extension of the flow rule to allow for the elastic part of the strain is due to Prandtl (1924) in the case of plane strain and to Reuss (1930) in the case of complete generality. The increment of plastic work per unit volume according to (1.18) and (1.14) is

$$dW^p = \sigma_{ij} d\varepsilon_{ij}^p = s_{ij} s_{ij} d\lambda = 2\,k^2 d\lambda$$

in view of (1.11). Since plastic work must be positive, $d\lambda$ is seen to be necessarily positive for plastic flow. Using the generalized form of Hooke's law, the elastic strain increment may be written as

$$d\varepsilon_{ij}^e = \frac{ds_{ij}}{2G} + \frac{1-2v}{3E} d\sigma_{kk} \delta_{ij}, \tag{1.19}$$

where G is the shear modulus and v is Poisson's ratio. When a stress increment satisfying $s_{ij} ds_{ij} = 0$ is prescribed, the elastic strain increment is known, but the plastic strain increment cannot be found from the flow rule alone.

The flow rule associated with the Tresca criterion furnishes ratios of the components of the plastic strain increment depending on the particular side or corner of the deviatoric yield hexagon. If we consider the side AB of the hexagon, Fig. 1.6(a), the yield criterion is given by (1.16) and the normality rule furnishes

$$d\varepsilon_1^p = -d\varepsilon_3^p > 0, \quad d\varepsilon_2^p = 0.$$

When the stress point is at the corner B of the yield hexagon, defining the equal biaxial state $\sigma_1 = \sigma_2$, the plastic strain increment vector can lie between the normals for the sides meeting at B, giving

$$d\varepsilon_1^p > 0, \quad d\varepsilon_2^p > 0, \quad d\varepsilon_3^p = -(d\varepsilon_1^p + d\varepsilon_2^p).$$

Similar relations hold for the other sides and corners of the yield hexagon. In each case, the rate of plastic work per unit volume is $2k$ times the magnitude of the numerically largest principal plastic strain rate. An interesting feature of Tresca's associated flow rule is that it can be written down in the integrated form whenever the stress point remains on a side, remains at a corner, or moves from a side to a corner, but not when it moves from a corner back to a side.

Let ψ denote the counterclockwise angle made by the plastic strain increment vector with the direction representing pure shear in the deviatoric plane. Evidently, ψ depends on the nature of the plastic potential, which is a closed curve similar to the yield locus. For an experimental verification of the flow rule, it is convenient to introduce the Lode parameter ν (not to be confused with Poisson's ratio), which is defined as

$$\nu = \frac{2d\varepsilon_2^p - d\varepsilon_3^p - d\varepsilon_1^p}{d\varepsilon_3^p - d\varepsilon_1^p} = -\sqrt{3}\tan\psi, \quad d\varepsilon_1^p > d\varepsilon_2^p > d\varepsilon_3^p. \tag{1.20}$$

For a regular yield function and plastic potential, $\nu = 0$ when $\mu = 0$, and $\nu = -1$ when $\mu = -1$. In the case of the von Mises yield criterion and the associated Prandtl–Reuss flow rule, $\mu = \nu$ or $\theta = \psi$ for all plastic states. Tresca's yield criterion and its associated flow rule, on the other hand, correspond to $\nu = 0$ for $0 \geq \mu \geq -1$, and $0 \geq \nu \geq -1$ for $\mu = -1$. The (μ, ν) relations corresponding to the Tresca and von Mises theories are shown in Fig. 1.8. The experimental results of Hundy and Green (1954), included in the figure, clearly support the Prandtl–Reuss rule for the plastic flow of isotropic materials.

Suppose that a plastic strain increment $d\varepsilon_{ij}^p$ is associated with σ_{ij} stress satisfying the yield criterion, while σ_{ij}^* is any other plastic state of stress, so that $f(\sigma_{ij}^*) = f(\sigma_{ij}) =$ constant. The work done by σ_{ij}^* on the given plastic strain increment $d\varepsilon_{ij}^p$ has a stationary value for varying σ_{ij}, when

$$\frac{\partial}{\partial\sigma_{ij}^*}\left\{\sigma_{ij}^* d\varepsilon_{ij}^p - f(\sigma_{ij}^*)d\lambda\right\} = 0,$$

where $d\lambda$ is the Lagrange multiplier (Hill, 1950a). Carrying out the partial differentiation, we have

$$d\varepsilon_{ij}^p = \frac{\partial}{\partial\sigma_{ij}^*}\left\{f(\sigma_{ij}^*)\right\} d\lambda.$$

Since $d\varepsilon_{ij}^p$ is associated with σ_{ij} according to the normality rule, the above equation is satisfied when σ_{ij}^* equals σ_{ij} apart from a hydrostatic stress. The rate of plastic

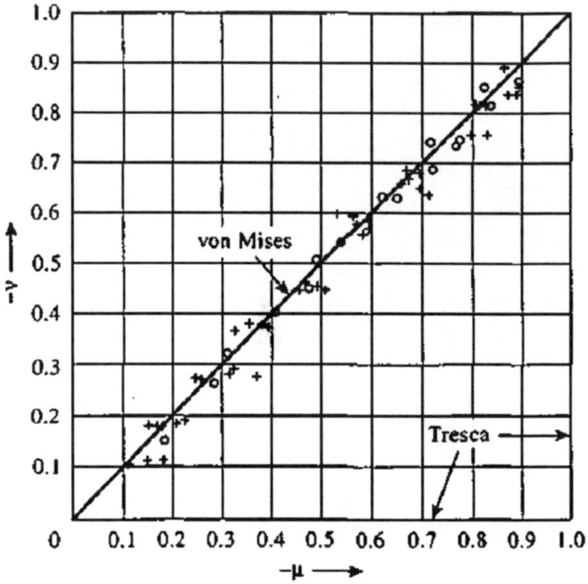

Fig. 1.8 Experimental verification of the (μ, ν) relation due to Hundy and Green (+) and due to Lianis and Ford (o)

work is then a maximum in view of the convexity of the yield function. The maximum work theorem, which is due to von Mises (1928), may therefore be stated as

$$\left(\sigma_{ij} - \sigma_{ij}^*\right) d\varepsilon_{ij}^p \geq 0. \tag{1.21}$$

From the geometrical point of view, the result is evident from the fact that the vector representing the stress difference forms a chord of the yield surface and consequently makes an acute angle with the exterior normal defined at the actual stress point. The identity of the yield function and the plastic potential has a special significance in the mathematical theory of plasticity. Further details regarding the plastic stress–strain relations with and without strain-hardening of the material have been presented by Chen and Han (1988).

1.2.3 Limit Theorems

In an elastic/plastic body subjected to a set of external forces, yielding begins in the most critically stressed element when the load attains a critical value. Under increasing load, a plastic zone continues to spread while the deformation is restricted to the elastic order of magnitude due to the constraint of the nonplastic material. When the plastic region expands to a sufficient extent, the constraint becomes locally ineffective and large plastic strains become possible. For a material whose rate of hardening is of the order of the yield stress, only a slight increase in load can produce an overall distortion of appreciable magnitude. If the material is nonhardening, and the change in geometry is disregarded, the load approaches an asymptotic value which is generally known as the *yield point load*. The basic theorems for the approximate estimation of this load have been obtained by Hill (1951) using a rigid/plastic model and by Drucker et al. (1952) using an elastic/plastic model.

The elastic/plastic asymptotic load, frequently referred to as the collapse load in the context of structural analysis, is very closely attained while the elastic and plastic strains are still comparable in magnitudes. During the collapse, the deformation may therefore be assumed to occur under a constant load while changes in geometry are still negligible. Let $\dot{\sigma}_{ij}$ and v_j denote the stress rate and particle velocity, respectively, under a distribution of boundary traction rate $\dot{T}_j$ at the incipient collapse. If the elastic and plastic components of the associated strain rate are denoted by $\dot{\varepsilon}_{ij}^e$ and $\dot{\varepsilon}_{ij}^p$, respectively, then by the rate form of the principle of virtual work we have

$$\int \dot{T}_j v_j dS = \int \dot{\sigma}_{ij}\,\dot{\varepsilon}_{ij} dV = \int \dot{\sigma}_{ij}\left(\dot{\varepsilon}_{ij}^e + \dot{\varepsilon}_{ij}^p\right) dV, \tag{1.22}$$

where the integrals are taken over the entire surface enclosing a volume V of the considered body. The integral on the left-hand side vanishes at the instant of collapse, since $\dot{T}_j = 0$, while the scalar product $\dot{\sigma}_{ij}\,\dot{\varepsilon}_{ij}^p$ vanishes in an ideally plastic material in view of (1.17). It follows therefore from (1.22) that $\dot{\sigma}_{ij}\,\dot{\varepsilon}_{ij}^e = 0$ at the incipient collapse, which indicates that $\dot{\sigma}_{ij}$ and $\dot{\varepsilon}_{ij}^e$ individually vanish, in view of the elastic stress–strain relation.

A stress field is regarded as statically admissible if it satisfies the equilibrium equations and the stress boundary conditions without violating the yield criterion. Let σ_{ij} and $\dot{\varepsilon}_{ij}$ denote the actual stress and strain rate in the considered body, and σ_{ij}^* any other statically admissible state of stress. If T_j and T_j^* denote the surface tractions corresponding to σ_{ij} and σ_{ij}^*, respectively, then by the virtual work principle,

$$\int \left(T_j - T_j^*\right) v_j dS = \int \left(\sigma_{ij} - \sigma_{ij}^*\right) \dot{\varepsilon}_{ij} dV + \int \left(k - \tau^*\right)[u]\, dS_D \geq 0,$$

where $\tau^* \leq k$ is the magnitude of the shear component of σ_{ij} along a surface S_D that actually involves a tangential velocity discontinuity of magnitude $[u]$. The inequality follows from (1.21) and the fact that the strain rate is purely plastic during the collapse. The assumed stress field may involve stress discontinuities across certain internal surfaces, which are limits of thin elastic regions of rapid but continuous variations of the stress. Since $T_j = T_j^*$ over the part S_F of the surface where the traction is prescribed, the above inequality becomes

$$\int T_j v_j dS_v \geq \int T_j^* v_j dS_v, \tag{1.23}$$

where S_v denotes the part of the surface over which the velocity is prescribed. The above inequality (1.23) constitutes the lower bound theorem, which states that the rate of work done by the actual surface tractions on S_v is greater than or equal to that done by the surface tractions associated with any statically admissible stress field. The theorem provides a lower bound on the load itself at the incipient collapse when the prescribed velocity is uniform over this part of the boundary.

A velocity field is considered as kinematically admissible if it satisfies the plastic incompressibility condition $\dot{\varepsilon}_{ii}^{p} = 0$ and the velocity boundary conditions. Let σ_{ij} and v_j denote the actual stress and velocity, respectively, in a deforming body, and v_j^* any other kinematically admissible velocity producing a strain rate $\dot{\varepsilon}_{ij}^*$. Since the rate of deformation is purely plastic during the collapse, the associated work rate $\sigma_{ij}^*\varepsilon_{ij}^*$ is uniquely defined by the flow rule. By the virtual work principle, we have

$$\int T_j v_j^* dS = \int \sigma_{ij}\,\dot{\varepsilon}_{ij}^* dV + \int \tau\left[u^*\right] dS_D^*,$$

where τ is the magnitude of the actual shear stress and $[u*]$ is the magnitude of the tangential discontinuity in the virtual velocity along a certain surface S_D^*. Since $\tau \leq k$, and $\sigma_{ij}\,\dot{\varepsilon}_{ij}^* \leq \sigma_{ij}\,\dot{\varepsilon}_{ij}^*$ in view of (1.21), the elastic strain rate being zero, the preceding expression furnishes

$$\int T_j v_j dS_v \leq \int \sigma_{ij}\,\dot{\varepsilon}_{ij} dV + \int k\left[u^*\right] dS_D - \int T_j v_j^* dS_F,$$

where use has been made of the fact that $v_j^* = v_j$ on S_v. This result constitutes the upper bound theorem of limit analysis. When the last term of (1.24) is zero, the theorem states that the rate of work done by the actual surface tractions on S_v is less than or equal to the rate of dissipation of internal energy in any kinematically admissible velocity field. When the prescribed velocity is uniform on S_u, the theorem provides an upper bound on the load itself at the instant of collapse.

In a rigid/plastic body, no deformation can occur before the load reaches the yield point value. Over the range of load varying between the elastic limit and the yield point, the body remains entirely rigid, even though partially plastic. Under given surface tractions over a part S_F of the boundary, and given velocities over the remainder S_v, the state of stress at the yield point is uniquely defined in the region where deformation is assumed to occur. On the other hand, the mode of deformation at the incipient collapse is not necessarily unique for an ideally plastic material. When positional changes are disregarded, the physically possible mode compatible with the rate of hardening can be singled out by specifying the traction rate on S_F as an additional requirement (Hill, 1956). If geometry changes are duly taken into account, the deformation mode is found to be unique so long as the rate of work-hardening exceeds a certain critical value (Hill, 1957). The nature of nonuniqueness associated with an ideally plastic body has been illustrated with an example by Hodge et al. (1986).

1.3 Strain-Hardening Plasticity

1.3.1 Isotropic Hardening

The most widely used hypothesis for strain hardening assumes the yield locus to increase in size during continued plastic deformation without change in shape. The yield locus is therefore uniquely defined by the final plastic state of stress regardless of the actual strain path (Hill, 1950a). According to this postulate, the material remains isotropic throughout the deformation, and the Bauschinger effect continues to be absent. The state of hardening at any stage is therefore specified by the current uniaxial yield stress denoted by $\bar{\sigma}$. When the yield criterion is that of von Mises, the current radius of the yield surface is $\sqrt{2/3}$ times $\bar{\sigma}$, and we have

$$\begin{aligned}\bar{\sigma} &= \sqrt{\frac{3}{2}}\left(s_{ij}s_{ij}\right)^{1/2} \\ &= \frac{1}{\sqrt{2}}\left\{\left(\sigma_x-\sigma_y\right)^2+\left(\sigma_y-\sigma_z\right)^2+\left(\sigma_z-\sigma_x\right)^2+6\left(\tau_{xy}^2+\tau_{yz}^2+\tau_{zx}^2\right)\right\}^{1/2}.\end{aligned} \tag{1.25}$$

The quantity $\bar{\sigma}$ is known as the equivalent stress or effective stress, which increases with increasing plastic strain. To complete the hardening rule, it is necessary to relate $\bar{\sigma}$ to an appropriate measure of the plastic deformation. As a first hypothesis, it is natural to suppose that $\bar{\sigma}$ is a function of the total plastic work per unit volume expended in a given element. The work-hardening hypothesis may be therefore stated mathematically as

$$\bar{\sigma} = \phi\left\{\int \sigma_{ij} d\varepsilon_{ij}^p\right\}, \tag{1.26}$$

where the integral is taken along the strain path. Thus, no hardening is produced by the hydrostatic part of the stress which causes only an elastic change in volume. The function ϕ can be determined from the true stress–strain curve in uniaxial tension, where $\bar{\sigma}$ is exactly equal to the applied tensile stress σ, and the incremental plastic work per unit volume is σ times the longitudinal plastic strain increment equal to $d\varepsilon - d\sigma/E$. The argument of the function ϕ is simply the area under the curve for σ plotted against the quantity $\ln(l/l_0) - \sigma/E$, up to the ordinate σ, where l/l_0 is the length ratio at any stage of the extension.

A second hypothesis, frequently used in the literature, assumes $\bar{\sigma}$ to be a function of a suitable measure of the total plastic strain during the deformation. In analogy to the expression for $\bar{\sigma}$, we introduce a positive scalar parameter $\overline{d\varepsilon^p}$, known as the equivalent or effective plastic strain increment, defined as

$$\overline{d\varepsilon^p} = \sqrt{\frac{2}{3}}\left(d\varepsilon_{ij}^p d\varepsilon_{ij}^p\right)^{1/2}$$
$$= \sqrt{\frac{2}{3}}\left\{\left(d\varepsilon_x^p\right)^2 + \left(d\varepsilon_y^p\right)^2 + \left(d\varepsilon_z^p\right)^2 + 2\left(d\gamma_{xy}^p\right)^2 + 2\left(d\gamma_{yz}^p\right)^2 + 2\left(d\gamma_{zx}^p\right)^2\right\}^{1/2}. \tag{1.27}$$

The above definition implies that in the case of a uniaxial tension, $\overline{d\varepsilon^p}$ is equal to the longitudinal plastic strain increment, provided the yield function is regular. The strain-hardening hypothesis may now be stated mathematically as

$$\bar{\sigma} = F\left\{\int \overline{d\varepsilon^p}\right\}, \tag{1.28}$$

where the integral is taken along the strain path of a given element. Thus, the amount of hardening depends on the sum total of all the incremental plastic strains and not merely on the difference between the initial and the final shapes of the element. Both (1.26) and (1.28) imply that the longitudinal tensile stress is the same function of $\ln(l/l_0)$ in uniaxial tension as the compressive stress is of $\ln(h_0/h)$ in simple compression, where h_0/h is the associated height ratio of the specimen.

For a work-hardening Prandtl–Reuss material, the quantity $d\lambda$ appearing in the flow rule (1.18) can be directly related to the equivalent stress and plastic strain increment. Since $s_{ij} = 0$, it follows from (1.18), (1.27), and (1.25) that $\overline{d\varepsilon^p} = (2\bar{\sigma}/3)\,d\lambda$, and consequently,

$$\sigma_{ij} d\varepsilon_{ij}^p = s_{ij} s_{ij} d\lambda = \frac{2}{3}\bar{\sigma}^2 d\lambda = \bar{\sigma}\,\overline{d\varepsilon^p}, \tag{1.29}$$

indicating that in this case the two hypotheses (1.26) and (1.28) are completely equivalent. Inserting the value of $d\lambda$ from (1.29), the Prandtl–Reuss flow rule may be written as

$$d\varepsilon_{ij}^p = \frac{3\overline{d\varepsilon^p}}{2\bar{\sigma}} s_{ij} = \frac{3d\bar{\sigma}}{2H\bar{\sigma}} s_{ij}, \tag{1.30}$$

where $H = d\bar{\sigma}/\overline{d\varepsilon^p}$, representing the current rate of work hardening of the material. Another important result for a Prandtl–Reuss material, which follows from (1.18) and (1.25), is

$$\sigma_{ij} d\varepsilon_{ij}^p = s_{ij} ds_{ij} d\lambda = \frac{2}{3}\bar{\sigma}\, d\bar{\sigma}\, d\lambda = d\bar{\sigma}\,\overline{d\varepsilon^p}, \tag{1.31}$$

where $d\bar{\sigma}$ must be positive for plastic flow. The right-hand side of (1.31) is also equal to $H(\overline{d\varepsilon^p})^2$, where H is a given function of $\int \overline{d\varepsilon^p}$. Adding the elastic and plastic strain increments given by (1.19) and (1.30), we obtain the complete Prandtl–Reuss strain–strain relation in the incremental form.

When yielding occurs according to the Tresca criterion (1.16), it is necessary to replace $2k$ by the current uniaxial yield stress $\bar{\sigma}$. According to Tresca's associated flow rule, the increment of plastic work per unit volume is $\bar{\sigma}$ times the magnitude of the numerically largest principal plastic strain increment denoted by $d\varepsilon^p$. If the work-hardening hypothesis is adopted, it follows that

$$\bar{\sigma} = F\left\{\int \left|d\varepsilon^p\right|\right\},$$

where the integral is taken along the strain path. When the stress point remains on a side, remains at a corner, or moves from a side to a corner, and the principal axes of stress and strain increments do not rotate with respect to the element, the argument of the function F is equal to the magnitude of the numerically greatest principal plastic strain in the element.

1.3.2 Plastic Flow with Hardening

For continued plastic flow of a work-hardening material, the stress increment vector must lie outside the current yield locus, so that $d\bar{\sigma} > 0$. When the yield locus is regular, having a unique normal at each point, the plastic strain increment may be written as

$$d\varepsilon_{ij}^p = h^{-1}\left(n_{kl}d\sigma_{kl}\right)n_{ij}, \quad n_{kl}d\sigma_{kl} \geq 0, \tag{1.32}$$

where n_{ij} is the unit normal to the yield surface in a nine-dimensional stress space, and h (equal to $2H/3$) is a parameter representing the rate of hardening. The equality in (1.32) represents neutral loading since it implies that the stress point remains on the same yield locus. When $n_{kl}d\sigma_{kl} < 0$, the element unloads and no incremental plastic strain is involved. The scalar products of (1.32) with $d\sigma_{ij}$ and $d\varepsilon_{ij}$ in turn furnish the result

$$d\sigma_{ij}d\varepsilon_{ij}^p = h^{-1}\left(d\sigma_{ij}n_{ij}\right)^2 = hd\varepsilon_{ij}^p d\varepsilon_{ij}^p \geq 0. \tag{1.33}$$

The equality holds not only for neutral loading but also for unloading from the plastic state. Since (1.21) holds whether the material work-hardens or not, the basic inequalities for a work-hardening element that is currently in a plastic state may be stated as

$$\left(\sigma_{ij} - \sigma_{ij}^*\right)d\varepsilon_{ij}^p \geq 0, \quad d\sigma_{ij}d\varepsilon_{ij}^p \geq 0. \tag{1.34}$$

In both cases, the equality holds for neutral loading and unloading. If inequalities (1.34) are taken as the basic postulates for the plastic flow of work-hardening materials, the normality rule and the convexity of the yield surface can be easily deduced, as has been shown by Drucker (1951).

Let P_1 and P_2 be two arbitrary stress points located on the two sides of a singular point P, as shown in Fig. 1.9(a). According to the first inequality of (1.34), each of the vectors P_1P and P_2P must make an acute angle with the plastic strain increment vector at P. This condition is evidently satisfied if the direction of the plastic strain increment lies between the normals PN_1 and PN_2 corresponding to the meeting surfaces. A further restriction is imposed by the second inequality of (1.34), which states that the plastic strain increment vector must make an acute angle with the stress increment vector. In Fig. 1.9(b), the vector $d\varepsilon_{ij}^p$ may therefore lie anywhere between the normals PN_1 and PN_2 so long as the vector $d\sigma_{ij}$ lies within the angle T_1PT_2 formed by the tangents at P. If the loading condition is such that $d\sigma_{ij}$ lies outside this angle, the direction of $d\varepsilon_{ij}^p$ coincides with the normal that makes an acute angle with $d\sigma_{ij}$. The flow rule at a singular point has been discussed by Koiter (1953), Bland (1957), and Naghdi (1960).

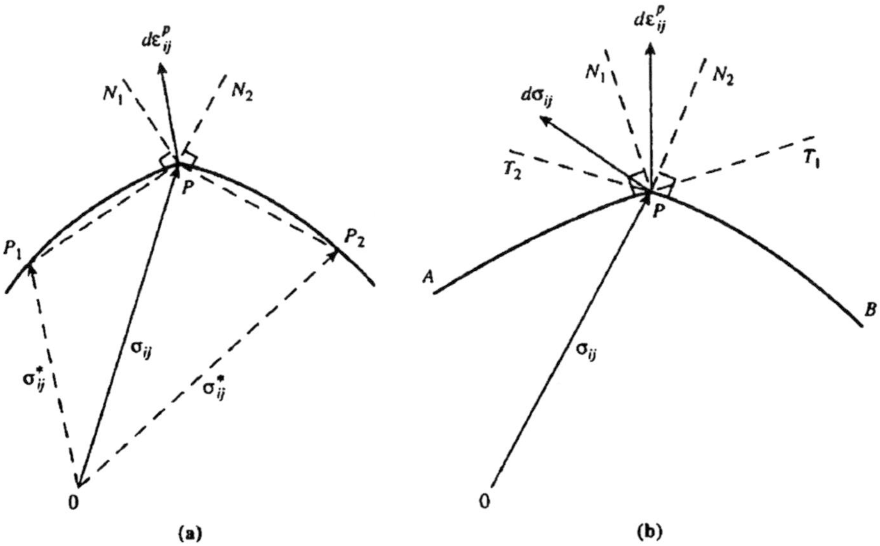

Fig. 1.9 Geometrical representation of the plastic stress–strain relation at a singular point of the yield surface

In the stress–strain relations considered so far, the strain increment $d\varepsilon_{ij}$ must be interpreted as $\dot{\varepsilon}_{ij}\,dt$, where $\dot{\varepsilon}_{ij}$ is the true strain rate and dt is an increment of time scale. The stress increment $d\sigma_{ij}$ is similarly given by a suitable measure of the stress rate, which must be defined in such a way that it vanishes in the event of a rigid-body rotation of the considered element. The most appropriate stress rate in the theory of plasticity is the Jaumann stress rate $\overset{\circ}{\sigma}_{ij}$, which is related to the material rate of change $\dot{\sigma}_{ij}$ of the true stress by the equation

$$\overset{\circ}{\sigma} = \dot{\sigma}_{ij} - \sigma_{ik}\omega_{jk} - \sigma_{jk}\,\omega_{ik}, \qquad (1.35)$$

where ω_{ij} is the rate of rotation of the considered element. The tensors $\dot{\varepsilon}_{ij}$ and ω_{ij} are the symmetric and antisymmetric parts, respectively, of the velocity gradient tensor $\partial v_i/\partial x_j$ and are given by

$$\dot{\varepsilon}_{ij} = \frac{1}{2}\left(\frac{\partial v_i}{\partial x_j} + \frac{\partial v_j}{\partial x_i}\right), \quad \omega_{ij} = \frac{1}{2}\left(\frac{\partial v_i}{\partial x_j} - \frac{\partial v_j}{\partial x_i}\right). \tag{1.36}$$

Equation (1.35), which is originally due to Jaumann (1911), has been rederived by several investigators including Hill (1958) and Prager (1961a). The Jaumann stress rate $\overset{\circ}{\sigma}_{ij}$ is the rate of change of the true stress σ_{ij} referred to a set of axes which participate in the instantaneous rotation of the element. Both σ_{ij}; and $\overset{\circ}{\sigma}_{ij}$ have the same scalar product with any tensor whose principal axes coincide with those of σ_{ij}. For an isotropic material, therefore, the material rate of change of the yield function $f(\sigma_{ij})$ is

$$\dot{f} = \frac{\partial f}{\partial \sigma_{ij}}\dot{\sigma}_{ij} = \frac{\partial f}{\partial \sigma_{ij}}\overset{\circ}{\sigma}_{ij}.$$

Since $\partial f/\partial \sigma_{ij}$ is in the direction of the unit normal n_{ij}, the Jaumann stress rate satisfies the condition that the yield function has a stationary value during the neutral loading of a plastic element. No other definition of the objective stress rate, vanishing in the event of a rigid-body rotation of the element, satisfies this essential requirement.

The constitutive equation for an elastic/plastic solid relates the strain rate to the stress rate, considered in the Jaumann sense. Combining the elastic and the plastic parts of the strain rate, the incremental constitutive equation for an isotropic work-hardening material may be written as

$$d\varepsilon_{ij} = \frac{1}{2G}\left(d\sigma_{ij} - \frac{\nu}{1+\nu}\delta_{ij}d\sigma_{kk}\right) + \frac{3}{2H}n_{ij}n_{kl}d\sigma_{kl} \tag{1.37}$$

for $n_{kl}d\sigma_{kl} \geq 0$ in an element currently stressed to the yield point, the yield surface being considered as regular. The scalar product of (1.37) with n_{ij} furnishes the result

$$n_{ij}d\sigma_{ij} = \frac{2GH}{3G+H}n_{ij}d\varepsilon_{ij}.$$

It follows that $n_{ij}d\sigma_{ij} \geq 0$ for $n_{ij}d\sigma_{ij} \geq 0$ when $H > 0$. Equation (1.37) therefore has the unique inverse

$$d\sigma_{ij} = 2G\left\{d\varepsilon_{ij} + \frac{\nu}{1-2\nu}\delta_{ij}d\varepsilon_{kk} - \frac{3G}{3G+H}n_{ij}n_{kl}d\varepsilon_{kl}\right\} \tag{1.38}$$

whenever $n_{kl}\, d\sigma_{kl} \geq 0$. Equation (1.38) holds equally well for a nonhardening material ($H = 0$), but the magnitude of the last term of (1.37) becomes indeterminate when $H = 0$. When $n_{kl}d\sigma_{kl} < 0$, or $n_{kl}d\varepsilon_{kl} < 0$, implying unloading of the element from the plastic state, the last terms of (1.37) and (1.38) must be omitted. Some computational aspects of the work-hardening Prandtl–Reuss theory of plasticity have been examined by Mukherjee and Liu (2003).

An interesting strain space formulation of the constitutive relations for elastic/plastic solids has been developed by Casey and Naghdi (1981). A generalized constitutive theory for finite elastic/plastic deformation of solids has been developed by Lee (1969) and Mandel (1972) and further discussed by Lubiner (1990). A critical review of the subject of finite plasticity has been made by Naghdi (1990).

A simplified stress–strain relation, proposed by Hencky (1924), assumes each component of the total plastic strain in any element to be proportional to the corresponding deviatoric stress. Although physically unrealistic, the Hencky theory does provide useful approximations when the loading is continuous and the stress path does not deviate appreciably from a radial path. For a work-hardening material, when the yield surface develops a corner at the loading point, the Hencky theory satisfies Drucker's postulates (1.34) over a certain range of nonproportional loading paths, as has been shown by Budiansky (1959) and Kliushnikov (1959). When the material is rigid/plastic, and strain hardens according to the Ludwik power law (1.5), the Hencky theory coincides with the von Mises theory even for nonproportional loading during an infinitesimal deformation of the element, as has been shown by Ilyushin (1946) and Kachanov (1971).

1.3.3 Kinematic Hardening

The simplest hardening rule that predicts the development of anisotropy and the Bauschinger effect, exhibited by real metals, is the kinematic hardening rule proposed by Prager (1956b) and Ishlinsky (1954). It is postulated that the hardening is produced by a pure translation of the yield surface in the stress space without any change in size or shape. If the initial yield surface is represented by $f(\sigma_{ij}) = k^2$ in a nine-dimensional space, where k is a constant, the subsequent yield surfaces may be represented by the equation

$$f\left(\sigma_{ij} - \alpha_{ij}\right) = k^2, \tag{1.39}$$

where α_{ij} is a tensor specifying the total translation of the center of the yield surface at a generic stage, as indicated in Fig. 1.10(b). To complete the hardening rule, it is further assumed that during an increment of plastic strain, the yield surface moves in the direction of the exterior normal to the yield surface at the considered stress point. Following Shield and Ziegler (1958), we therefore write

$$d\alpha_{ij} = c\,d\varepsilon_{ij}^{p}, \tag{1.40}$$

where c is a scalar parameter equal to two-thirds of the current slope of the uniaxial stress–plastic strain curve of the material. When c is a constant, (1.40) reduces to the

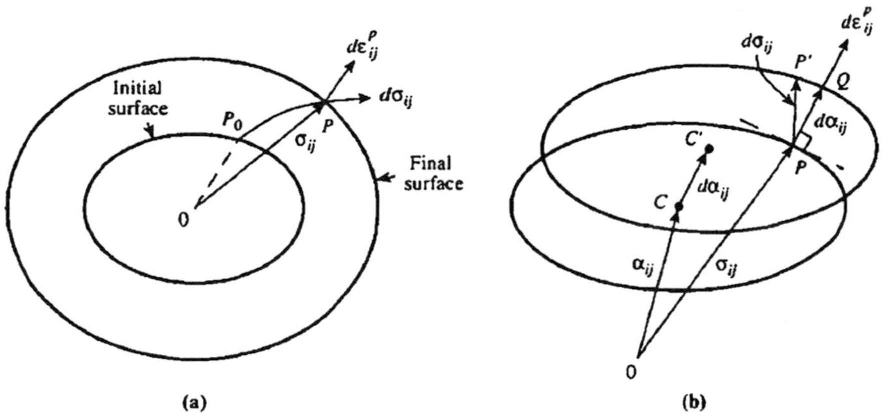

Fig. 1.10 Geometrical representation of the hardening rule considered in the stress space. (**a**) Isotropic hardening and (**b**) kinematic hardening

integrated form $\alpha_{ij} = c\varepsilon_{ij}^{p}$, the deformation being assumed small. In general, c may be regarded as a function of the equivalent plastic strain, the increment of which is defined by (1.27) in terms of the components of the plastic strain increment tensor.

Since the material becomes anisotropic during the hardening process, the principal axes of stress and plastic strain increment do not coincide, unless the principal axes remain fixed in the element as it deforms. The loading condition $df = 0$, which ensures that the stress point remains on the yield surface, furnishes

$$\left(d\sigma_{ij} - d\alpha_{ij}\right)\frac{\partial f}{\partial \sigma_{ij}} = 0 = \left(d\sigma_{ij} - cd\varepsilon_{ij}^{p}\right)d\varepsilon_{ij}^{p} \tag{1.41}$$

in view of (1.39) and (1.40) and the normality rule for the plastic strain increment vector. If the initial yield surface is that of von Mises, $f\left(\sigma_{ij}\right) = \frac{1}{2}s_{ij}s_{j}$, and the yield criterion at any stage of the deformation becomes

$$\left(s_{ij} - \alpha_{ij}\right)\left(s_{ij} - \alpha_{ij}\right) = 2\,k^2, \tag{1.42}$$

where k is the initial yield stress in pure shear. The associated flow rule furnishes the plastic strain increment as

$$d\varepsilon_{ij}^{p} = \frac{\partial f}{\partial \sigma_{ij}}d\lambda = \left(s_{ij} - \alpha_{ij}\right)d\lambda, \tag{1.43}$$

where $d\lambda$ is a positive scalar. Combining (1.43) with (1.41), and using (1.42), it is easily shown that

$$d\lambda = \frac{1}{2ck^2}\left(s_{kl} - \alpha_{kl}\right)d\sigma_{kl}. \tag{1.44}$$

The plastic strain increment for the kinematic hardening is completely defined by (1.43) and (1.44). A modified form of Prager's hardening rule has been proposed by Ziegler (1959), while other types of kinematic hardening have been examined by Baltov and Sawczuk (1965), Phillips and Weng (1975), and Jiang (1993). When the deformation is large, the stress increment entering into the constitutive equation must be carefully defined, and this question has been examined by Lee et al. (1983) and Naghdi (1990).

1.3.4 Combined or Mixed Hardening

The concept of kinematic hardening has been extended to include an expansion of the yield surface along with a translation by Hodge (1957), Kadashevich and Novozhilov (1959), and Mröz et al. (1976). Equation (1.39) is then modified by replacing its right-hand side with a function of the total equivalent plastic strain, whose increment is given by (1.27). Assuming the von Mises yield criterion for the initial state, the combined hardening rule may be stated as

$$\left(s_{ij} - \alpha_{ij}\right)\left(s_{ij} - \alpha_{ij}\right) = \frac{2}{3}\bar{\sigma}^2, \tag{1.45}$$

where $d\alpha_{ij}$ is still given by (1.40). The right-hand side of (1.45) is the square of the current radius of the displaced yield cylinder. The associated plastic strain increment $d\varepsilon_{ij}^p$ is given by (1.43) with $d\lambda = 3\overline{d\varepsilon}^p/2\bar{\sigma}$, as may be seen by substituting (1.43) into (1.27) and using (1.45). The plastic strain increment therefore becomes

$$d\varepsilon_{ij}^p = \left(s_{ij} - \alpha_{ij}\right)\frac{3\overline{d\varepsilon}^p}{2\bar{\sigma}} = \left(s_{ij} - \alpha_{ij}\right)\frac{d\bar{\sigma}}{h\bar{\sigma}}, \tag{1.46}$$

where h is a measure of the isotropic part of the rate of hardening, the anisotropic part being represented by the parameter c. The differentiation of the yield criterion (1.45) gives

$$\frac{2}{3}\bar{\sigma}\,d\bar{\sigma} = \left(s_{kl} - \alpha_{kl}\right)\left(d\sigma_{kl} - c\,d\varepsilon_{kl}^p\right)$$

in view of (1.40) and the fact that $ds_{kk} = 0$. Substituting from (1.46) and using (1.45), we obtain the relation

$$\left(1 + \frac{c}{h}\right)d\bar{\sigma} = \frac{3}{2\bar{\sigma}}\left(s_{kl} - \alpha_{kl}\right)d\sigma_{kl}. \tag{1.47}$$

The flow rule corresponding to the combined hardening process is completely defined by (1.46) and (1.47), the loading condition being specified by $d\bar{\sigma} > 0$ for an element that is currently plastic. Following the early experimental work due to Naghdi et al. (1958), the distortion of the yield surface under continued plastic deformation has been subsequently examined by several investigators.

The resultant strain increment in any element deforming under the combined hardening rule from an initially isotropic state may be written down on the assumption that the material continues to remain elastically isotropic. Then the elastic strain increment, which is given by the generalized Hooke's law, may be written as

$$d\varepsilon_{ij}^e = \frac{1}{2G}\left\{ds_{ij} + \frac{1}{3}\left(\frac{1-2v}{1+v}\right)d\sigma_{kk}\delta_{ij}\right\}, \tag{1.48}$$

where G is the shear modulus and v is Poisson's ratio. Taking the scalar product of the above equation with the tensor $s_{ij} - \alpha_{ij}$ and using (1.47), we have

$$\left(s_{ij} - \alpha_{ij}\right) d\varepsilon_{ij}^e = \left(\frac{c+h}{2G}\right)\bar{\sigma}\overline{d\varepsilon}^p.$$

On the other hand, the scalar product of (1.46) with the same tensor $s_{ij} - \alpha_{ij}$ gives

$$\left(s_{ij} - \alpha_{ij}\right) d\varepsilon_{ij}^p = \bar{\sigma}\overline{d\varepsilon}^p$$

in view of (1.45). The last two equations are added together to obtain the relation

$$\left(s_{ij} - \alpha_{ij}\right) d\varepsilon_{ij} = \left(1 + \frac{H}{3G}\right)\bar{\sigma}\overline{d\varepsilon}^p, \tag{1.49}$$

where H denotes the plastic modulus corresponding to the current state of stress and is defined as

$$H = \frac{3}{2}\left(c + h\right).$$

Thus, H is the slope of the uniaxial stress–plastic strain curve corresponding to a longitudinal plastic strain equal to the total equivalent plastic strain suffered by the given element. Further results related to the mixed hardening rule have been given by Mröz et al. (1976), Rees (1981), and Skrzypek and Hetnarski (1993). A micromechanical model for the development of texture with plastic deformation in polycrystalline metals has been considered by Dafalias (1993).

Consider the special case of proportional loading in which the stress path is a radial line in the deviatoric plane. Let the state of stress at the initial yielding be denoted by s_{ij}^0, satisfying the yield criterion $s_{ij}^0 s_{ij}^0 = 2Y^2/3$. Since the plastic strain increment tensor in this case may be written as

$$d\varepsilon_{ij}^p = \pm\left(\frac{3\overline{d\varepsilon}^p}{2Y}\right)s_{ij}^0,$$

where the upper sign corresponds to continued loading and the lower sign to any subsequent reversed loading in the plastic range, the deviatoric stress increment is

$$ds_{ij} = c\,d\,\varepsilon_{ij}^{p} \pm \left(\frac{d\bar{\sigma}}{Y}\right) s_{ij}^{0} = \pm \left(\frac{3H}{2Y}\right) s_{ij}^{0} \overline{d\,\varepsilon}^{p} \tag{1.50}$$

by the simple geometry of the loading path and the assumption of simultaneous translation and expansion of the yield surface in the stress space. During the unloading of an element from the plastic state, followed by a reversal of the load, the components of the deviatoric stress steadily decrease in magnitude. Plastic yielding would occur under the reversed loading when the vector representing the deviatoric stress changes by a magnitude equal to the current diameter of the yield surface.

1.4 Cyclic Loading of Structures

1.4.1 Cyclic Stress–Strain Curves

The investigations of low-cycle fatigue in mechanical and structural components have resulted in the development of considerable interest in the study of plastic behavior of materials under cyclic loading. In uniaxial states of stress involving symmetric cycles of stress or strain, an annealed material usually undergoes cyclic hardening, and the hysteresis loop approaches a stable limit as shown in Fig. 1.11(a). If the material is sufficiently cold-worked in the initial state, cyclic softening would occur and the hysteresis loop would again stabilize to a limiting state. Based on a family of stable hysteresis loops, obtained by the cyclic loading of a material with different constant values of the strain amplitude, we can derive a cyclic strain-hardening curve, such as that shown in Fig. 1.11(b), which may be compared with the standard strain-hardening curve for the same material (Landgraf, 1970). If the cyclic loading is continued in the plastic range, the stable hysteresis loops are repeated and failure eventually occurs due to low-cycle fatigue. Under certain stress cycles with materials exhibiting cyclic softening, the plastic strain may continue to grow in a unidirectional sense, causing failure by the phenomenon of ratcheting.

Let us suppose that a specimen that is first loaded in tension to a stress equal to σ is subsequently unloaded from the plastic state and then reloaded in compression. It follows from above that yielding would again occur when the magnitude of the applied compressive stress becomes $2\bar{\sigma} - \sigma$, where $\bar{\sigma}$ depends on the magnitude of the previous plastic strain. If we assume the relations

$$h = \frac{2}{3}\beta H, \quad c = \frac{2}{3}(1-\beta) H,$$

where β is a constant less than unity, then $d\bar{\sigma} = \beta d\sigma$, which gives $\bar{\sigma} - Y = \beta(\sigma - Y)$. The initial yield stress σ' in compression during the reversed loading is therefore given by

$$\sigma' - Y = (2\beta - 1)(\sigma - Y).$$

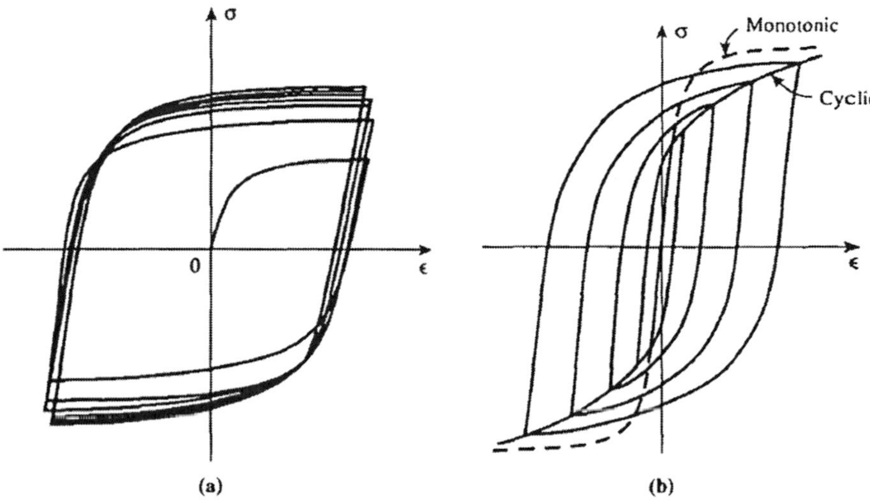

Fig. 1.11 Cyclic loading curves in the plastic: (**a**) constant strain cycles and (**b**) cyclic stress–strain curves

It follows that $\sigma' \gtrless Y$ according as $\beta \gtrless \frac{1}{2}$, irrespective of the rate of hardening. If the specimen is subjected to a complete cycle of loading and unloading with the longitudinal plastic strain varying between the limits $-\varepsilon^*$ to ε^*, then the magnitude of the final stress under a constant plastic modulus H exceeds σ by the amount $4\beta H\varepsilon^*$, which vanishes only when the hardening is purely kinematic.

The shear stress–strain curve of a material under cyclic loading can be derived from the experimental torque–twist curve for a solid cylindrical bar subjected to cyclic torsion in the plastic range. Let T denote the applied torque at any stage of the loading, and let θ be the corresponding angle of twist per unit length. Since the engineering shear strain at any radius r is $\gamma = r\theta$, we have

$$T = 2\pi \int_0^a \tau r^2 dr = \frac{2\pi}{\theta^3} \int_0^{a\theta} \tau \gamma^2 d\gamma,$$

where a denotes the external radius of the bar, and $\tau = \tau(\gamma)$ is the shear stress at any radius. Multiplying both sides of the above equation by θ^3, and differentiating it with respect to θ, it is easy to show that

$$\theta \frac{dT}{d\theta} + 3T = 2\pi a^3 \tau\,(a\theta)\,, \tag{1.51}$$

where $\tau(a\theta)$ is the shear stress at the boundary $r = a$, corresponding to a shear strain equal to $a\theta$. The preceding relation provides a means of obtaining the $(\tau,\ \gamma)$-curve from an experimental (T, θ)-curve during the loading process (Nadai, 1950).

Consider now the unloading and reversed loading of a bar that has been previously twisted in the plastic range by a torque T_0 producing a specific angle of twist

θ_0. For a given value of θ_0, the shear stress acting at any radius r, when the specific angle of twist has decreased to θ may be expressed as

$$\tau\,(r\theta) = \tau_0\,(r\theta_0) + f\,[r\,(\theta - \theta_0)]\,, \tag{1.52}$$

where τ_0 denotes the local shear stress at the moment of unloading. The function f represents the change in shear stress caused by the unloading or reversed loading. The torque acting at any stage is

$$T = T_0 + 2\pi \int_0^a r^2 f\,[r\,(\theta - \theta_0)]dr.$$

Setting $\xi = r(\theta - \theta_0)$, which gives $d\xi = (\theta - \theta_0)\,dr$, the preceding relation can be expressed as

$$T - T_0 = \frac{2\pi}{(\theta - \theta_0)^3} \int_0^{a(\theta - \theta_0)} \xi f\,(\xi)\,d\xi\,.$$

Multiplying both sides of this equation by $(\theta - \theta_0)^3$ and differentiating the resulting expression partially with respect to θ, we have

$$(\theta - \theta_0)\,\frac{\partial T}{\partial \theta} + 3\,(T - T_0) = 2\pi a^3 f\,[a\,(\theta - \theta_0)]\,,$$

since T_0 is a function of θ_0 only. Substituting for $f[a(\theta - \theta_0)]$ from the above equation into (1.52), the shear stress at $r = a$ is finally obtained as

$$\tau\,(a\theta) = \tau_0\,(a\theta_0) + \frac{1}{2\pi a^3}\left\{3\,(T - T_0) + (\theta - \theta_0)\,\frac{\partial T}{\partial \theta}\right\}. \tag{1.53a}$$

Since $\partial T/\partial \theta$ is positive, both terms in the curly brackets of (1.53a) are negative during unloading and reversed loading. The residual shear stress at $r = a$ at the end of the unloading process corresponds to $T = 0$, the corresponding residual shear strain being found directly from the given torque–twist curve.

Suppose that the reversed loading in torsion is terminated when $T = T_1$ and $\theta = \theta_1$, the corresponding shear stress at $r = a$ being denoted by $\tau_1(a\theta_1)$. If the bar is again unloaded, and then reloaded in the same sense as that in the original loading, an analysis similar to the above gives the shear stress at the external radius in the form

$$\tau\,(a\theta) = \tau_1\,(a\theta_1) + \frac{1}{2\pi a^3}\left\{(\theta - \theta_1)\,\frac{\partial T}{\partial \theta} + (T - T_1)\right\}. \tag{1.53b}$$

Equations (1.51) and (1.53) completely define the cyclic shear stress–strain curve based on an experimentally determined cyclic torque–twist curve Wu et al. (1996). The derivative $\partial T/\partial \theta$ is piecewise continuous, involving a jump at each reversal of the applied torque, the correspondence between the various points in the two

Fig. 1.12 The cyclic shear stress–strain curve derived from the cyclic torque–twist curve for a solid cylindrical bar

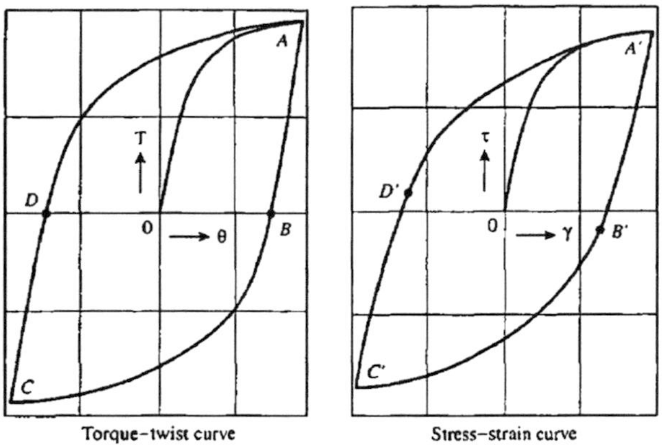

cyclic curves being indicated in Fig. 1.12. It may be noted that (1.53) can be directly obtained from (1.51) if we simply replace T, θ, and τ in this equation by the appropriate differences of the physical quantities.

1.4.2 A Bounding Surface Theory

The anisotropic hardening rule described in the preceding section cannot be applied without modifications to predict the plastic behavior of materials under cyclic loading with relatively complex states of stress (Dafalias and Popov, 1975; Lamba and Sidebottom, 1978). Following an earlier work by Mröz (1967a), various types of theoretical model involving two separate surfaces in the stress space have been widely discussed in the literature, notably by Tseng and Lee (1983), McDowell (1985), Ohno and Kachi (1986), and Hong and Liou (1993), among others. The two-surface model assumes the existence of a bounding surface that encloses the current yield surface throughout the loading history. Both the yield surface and the bounding surface can expand and translate, and possibly also deform in the stress space, as the loading and unloading are continued in the plastic range.

The general features of the two-surface theory are illustrated in Fig. 1.13(a), where the yield surface or the loading surface S and the bounding surface S' are represented by circles with centers C and C', respectively. The current radii of the surfaces S and S' are denoted by $\sqrt{2/3}$ times $\bar{\sigma}$ and $\bar{\tau}$, respectively, while the position vectors of the centers C and C' are denoted by α_{ij} and α'_{ij}, respectively.

The equation for the loading surface is given by (1.35), while that of the bounding surface is expressed as

$$\left(s_{ij} - \alpha'_{ij}\right)\left(s_{ij} - \alpha'_{ij}\right) = \frac{2}{3}\bar{\tau}^2.$$

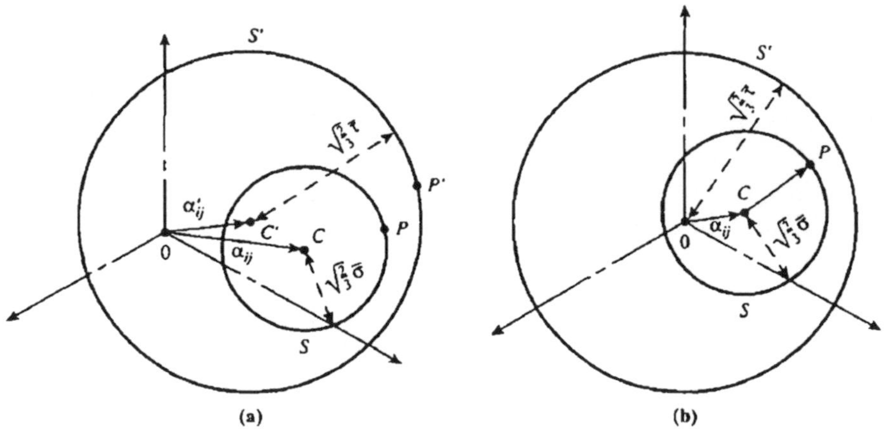

Fig. 1.13 Yield surface S and bounding surface S' in the deviatoric stress space. (**a**) Both surfaces are in translation and (**b**) only the yield surface is in translation

For each point P on the yield surface S, there is a corresponding image point P' on the bounding surface, the distance between the two points P and P', which are defined by the vectors s_{ij} and s'_{ij}, respectively, is an important parameter that enters into the theoretical framework. There are several possible ways of relating the two stress points, the one suggested by Mröz being

$$s'_{ij} - \alpha'_{ij} = (\bar{\tau}/\bar{\sigma})\left(s_{ij} - \alpha_{ij}\right).$$

The variation of α_{ij} and α'_{ij} with continued loading must be defined by appropriate hardening rules, for which there are several possibilities. Since the two-surface theory is not without its limitations in predicting the material response under cyclic loading (Jiang, 1993, 1994), we shall describe in what follows the simplest theoretical model that is consistent with the basic purpose of the theory.

It is assumed, for simplicity, that the bounding surface at each stage is a circular cylinder concentric with the initial yield surface, which is taken as the von Mises cylinder. The radius of the bounding surface S' therefore increases with the amount of plastic deformation following the isotropic hardening rule. The yield surface S, on the other hand, undergoes simultaneous expansion and rotation according to the mixed hardening process, Fig. 1.13(b). At a generic stage of the loading, the deviatoric stresses s_{ij} and s'_{ij} associated with the surfaces S and S', respectively, satisfy the relations

$$\left(s_{ij} - \alpha_{ij}\right)\left(s_{ij} - \alpha_{ij}\right) = \frac{2}{3}\bar{\sigma}^2, \quad s'_{ij}s'_{ij} = \frac{2}{3}\bar{\tau}^2, \tag{1.54}$$

where α_{ij} is the back stress defining the center of the current yield surface. So long as the two surfaces are separated from one another, the translation of the yield surface is assumed to be governed by Prager's kinematic hardening rule, which requires the yield surface to move in the direction of the plastic strain increment. Thus

$$d\alpha_{ij} = c\, d\varepsilon_{ij}^p, \quad \alpha_{ij}\alpha_{ij} \leq \frac{2}{3}(\bar{\tau} - \bar{\sigma})^2, \tag{1.55}$$

where c represents the kinematic part of the rate of hardening of the material. The inequality in (1.55) ensures that the yield surface is not in contact with the bounding surface. The plastic strain increment $d\varepsilon_{ij}^p$ is given by the flow rule (1.46), which applies to the mixed hardening process, the equivalent plastic strain increment $\overline{d\varepsilon^p}$ corresponding to a given strain increment $d\varepsilon_{ij}$ being found from (1.49). The plastic modulus H at any stage is given by the relation

$$H = \frac{3}{2}(h + c),$$

where h represents the isotropic part of the rate of hardening and is two-thirds of the current slope of the curve obtained by plotting $\bar{\sigma}$ against $\bar{\varepsilon}^p$.

In the case of cyclic loading, the parameter c depends not only on the accumulated plastic strain $\bar{\varepsilon}^p$ but also on the distance between the loading point P and its image point P' on the bounding surface. For simplicity, the image point is considered here as the point of intersection of the outward normal to the yield surface at P with the bounding surface, Fig. 1.14(a). If ψ denotes the included angle between the vectors representing the deviatoric stress s_{ij} and the reduced stress $s_{ij} - \alpha_{jj}$, then by the geometry of the triangle OPP', we have

$$\bar{s}^2 + \delta^2 + 2\bar{s}\delta\cos\psi = \bar{\tau}^2,$$

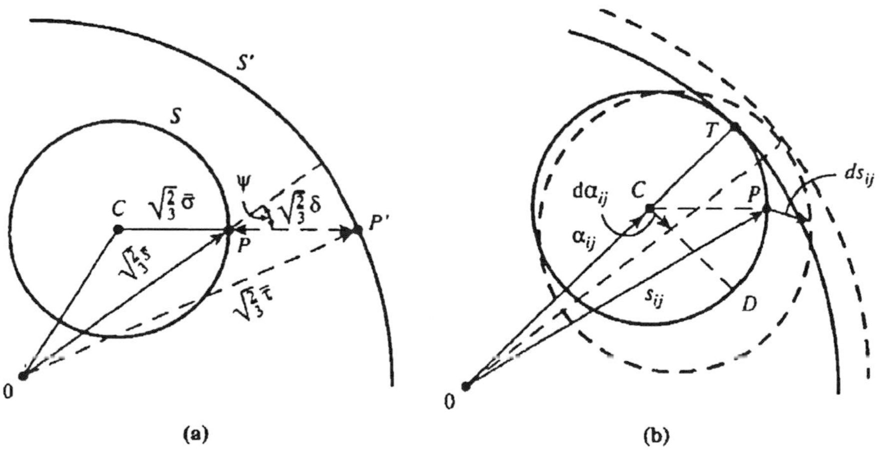

Fig. 1.14 Simplified two-surface model for cyclic plasticity. (**a**) Separate loading and bounding surfaces and (**b**) the two surfaces are in contact

where $\bar{s}$ and δ are $\sqrt{3/2}$ times the lengths of the vectors *OP* and *PP'*, respectively. The above equation immediately gives

$$\delta = -\bar{s}\cos\psi + \sqrt{\bar{\tau}^2 - \bar{s}^2\sin^2\psi}. \tag{1.56}$$

The quantities ψ and $\bar{s}$ appearing in (1.56) can be determined from the relations

$$\cos\psi = \frac{3s_{ij}\left(s_{ij} - \alpha_{ij}\right)}{2\bar{s}\bar{\sigma}}, \quad \bar{s} = \left(\frac{3}{2}s_{ij}s_{ij}\right)^{1/2}. \tag{1.57}$$

The plastic modulus *H* evidently depends on both $\bar{\varepsilon}^p$ and δ. For practical purposes, *H* can be estimated by using the empirical relation

$$H = \left(\frac{d\bar{\tau}}{d\bar{\varepsilon}^p}\right)\exp\left\{\beta\left(\frac{\delta}{\bar{\tau}}\right)^{\gamma}\right\}, \tag{1.58}$$

where β and γ are dimensionless constants to be determined from experimental data on uniaxial stress cycles. Since $\delta/\bar{\tau}$ and $d\bar{\tau}/d\bar{\varepsilon}^p$ monotonically decrease during the process, (1.58) implies a fairly rapid decrease in the value of *H*. When $\delta = 0$, the plastic modulus becomes identical to the slope of the curve for $\bar{\tau}$ against $\bar{\varepsilon}^p$. The quantities $\bar{\sigma}$ and $\bar{\tau}$ are functions of $\bar{\varepsilon}^p$ alone and can be expressed by the empirical equations

$$\left.\begin{aligned} \bar{\sigma} &= \sigma_0\left[1 - m\exp\left(-n\bar{\varepsilon}^p\right)\right], \\ \bar{\tau} &= \tau_0\left[1 - m'\exp\left(-n'\bar{\varepsilon}^p\right)\right], \end{aligned}\right\} \tag{1.59}$$

where σ_0 and τ_0 are saturation stresses, while *m*, *n*, *m'*, and *n'* are appropriate dimensionless constants. The hardening rate parameters *h* and *c* at any given stage follow from (1.58) and (1.59).

Suppose that all the physical quantities have been found for a generic stage of the cyclic loading. During an additional strain increment $d\varepsilon_{ij}$ satisfying the inequality $(s_{ij} - \alpha_{ij})\,d\varepsilon_{ij} > 0$, the equivalent plastic strain increment $d\bar{\varepsilon}^p$ is computed from (1.49), and the associated plastic strain increment tensor $d\varepsilon_{ij}^p$ then follows from (1.46). Since the elastic strain increment $d\varepsilon_{ij}^e$ is equal to $d\varepsilon_{ij} - d\varepsilon_{ij}^p$, the deviatoric stress increment ds_{ij} is obtained from (1.48), where the second term in the curly brackets is equal to $2G\, d\varepsilon_{kk}\delta_{ij}$. The incremental displacement $d\alpha_{ij}$ of the yield surface is determined from (1.55) and the fact that $c = \frac{2}{3}(H - d\bar{\sigma}/d\bar{\varepsilon}^p)$. The new stress tensors s_{ij} and α_{ij}, together with the updated values of $\bar{\sigma}$ and $\bar{\tau}$ obtained from (1.59), enable us to compute the new values of $\delta/\bar{\tau}$ and *H* using (1.58) and (1.59), thereby completing the solution to the incremental problem.

1.4.3 The Two Surfaces in Contact

When the yield surface comes in contact with the bounding surface, the position of the stress point P will generally require the two surfaces to remain in contact during the subsequent loading process. Consider first the situation where P coincides with the point of contact T between the two surfaces, and the coincidence is then maintained following the stress path. Since the back stress α_{ij} in this case is in the direction of the deviatoric stress s_{ij}, it follows from simple geometry that

$$\alpha_{ij} = \left(1 - \frac{\bar{\sigma}}{\bar{\tau}}\right) s_{ij}, \quad s_{ij}s_{ij} = \frac{2}{3}\bar{\tau}^2. \tag{1.60}$$

The yield surface is now assumed to expand at the same rate as the bounding surface so that $d\bar{\sigma} = d\bar{\tau}$ during this loading phase. The plastic modulus H is therefore continuous when the contact begins, in view of (1.58), and the incremental translation of the center of the yield surface is given by the relation

$$d\alpha_{ij} = \left(1 - \frac{\bar{\sigma}}{\bar{\tau}}\right)\left(ds_{ij} - \frac{d\bar{\tau}}{\bar{\tau}} s_{ij}\right). \tag{1.61}$$

This expression implies that the motion of the yield locus consists of a radial expansion of the circle together with a rigid body sliding along the bounding surface.

Suppose now that the stress point P on the yield surface lies between the points T and D, where CD is parallel to the common tangent to the two surfaces in contact, Fig. 1.14(b). This condition, together with the condition of contact can be stated mathematically as

$$\alpha_{ij}\alpha_{ij} = \frac{2}{3}(\bar{\tau} - \bar{\sigma})^2, \quad \frac{2}{3}\bar{\tau}^2 \geq s_{ij}s_{ij} \geq \frac{2}{3}\left[(\bar{\tau} - \bar{\sigma})^2 + \bar{\sigma}^2\right].$$

Assuming $d\bar{\sigma}$ and $d\bar{\tau}$ to be equal to one another as before, thus allowing a slight discontinuity in the plastic modulus as the contact is established, the translation of the center of the yield surface may be written as

$$d\alpha_{ij} = \lambda ds_{ij} - s_{ij}d\mu, \tag{1.62}$$

where λ and $d\mu$ are scalar parameters. The modified hardening rule expressed by (1.62) is consistent with the experimental observation of Phillips and Lee (1979). It implies that the radial expansion of the yield surface is accompanied by its sliding and rolling over the bounding surface. The unknown parameters in (1.62) can be determined from the condition that $d\alpha_{ij}$ is orthogonal to α_{ij}, and the fact that the stress point remains on the yield surface. Thus

$$\alpha_{ij}d\alpha_{ij} = 0, \quad \left(s_{ij} - \alpha_{ij}\right)\left(ds_{ij} - d\alpha_{ij}\right) = \frac{2}{3}\bar{\sigma}\,d\sigma. \tag{1.63}$$

Taking the scalar product of (1.62) with α_{ij}, and using the first relation of (1.63), we get

$$d\mu/\lambda = \left(\alpha_{ij} ds_{ij}\right)/(\alpha_{kl} s_{kl}). \tag{1.64}$$

Using the contact condition, the second condition of (1.63) is easily reduced to

$$s_{ij} d\alpha_{ij} = \left(s_{ij} - \alpha_{ij}\right) ds_{ij} - \frac{2}{3}\bar{\sigma}\, d\bar{\tau}.$$

The scalar product of (1.62) with s_{ij} and the substitution from above lead to the expression

$$\lambda = \frac{\left(s_{ij} - \alpha_{ij}\right) ds_{ij} - \frac{2}{3} H \bar{\sigma} \overline{d\varepsilon}^p}{\left[ds_{kk} - (d\mu/\lambda)\, s_{kl}\right] s_{kl}},$$

where H is the plastic modulus equal to $d\bar{\tau}/\overline{d\varepsilon}^p$ during this phase. Equations (1.64) and (1.65) define the hardening parameters λ and $d\mu$ when $\overline{d\varepsilon}^p$ is known for a given ds_{ij}. In the special case when (1.60) are applicable, we get $\lambda = 1 - \bar{\sigma}/\bar{\tau}$, and $d\mu = \lambda\,(d\bar{\tau}/\bar{\tau})$, and the hardening rule then reduces to (1.61).

Assuming the strain increment $d\varepsilon_{ij}$ to be prescribed, the corresponding value of $\overline{d\varepsilon}^p$ can be approximately estimated from (1.49), which is not strictly valid for the situation considered here. The associated deviatoric stress increment ds_{ij} is then obtained as before, and the parameters λ and $d\mu$ are determined from (1.64) and (1.65). An improved value of $\overline{d\varepsilon}^p$ subsequently follows from the relation

$$\left(1 + \frac{H}{3G}\right) \bar{\sigma} \overline{d\varepsilon}^p = \left(s_{ij} - \alpha_{ij}\right) d\varepsilon_{ij} - \frac{1}{2G} s_{ij} d\alpha_{ij}, \tag{1.66}$$

where $d\alpha_{ij}$ is given by (1.61). Equation (1.66) is obtained by taking the scalar products of (1.46) and (1.48) with the tensor $s_{ij} - \alpha_{ij}$, and using (1.63). The computation may be repeated until the difference between successive values of the effective plastic strain increment becomes negligible. The quantity $\bar{\tau}$ is still given by (1.59), but the value of $\bar{\sigma}$ over this range is obtained from the relation $\bar{\sigma} = \bar{\tau} + (\bar{\sigma}^* - \bar{\tau}^*)$, where the asterisk refers to the instant when the two surfaces first come in contact with one another during the loading.

The theoretical treatment of cyclic plasticity based on a single-surface model has been examined by Eisenberg (1976) and Drucker and Palgen (1981). The constitutive modeling of large strain cyclic plasticity has been discussed by Chaboche (1986), Lemaitre and Chaboche (1989), and Yoshida and Uemori (2003), among other investigators.

The plastic response of materials under cyclic loading has also been discussed in recent years on the basis of an interesting theory of plasticity that does not require the specification of a yield surface. The theory, which is essentially due to Valanis (1975, 1980), who called it the endochronic theory of plasticity, is based on the concept of an intrinsic time that depends on the deformation history, the relationship between the two quantities being regarded as a material property. The theory also introduces an intrinsic time scale which is a function of the intrinsic time, the rate of change of the various physical quantities being considered with respect to this time scale. For further details of the endochronic theory of plasticity, together with some physical applications, the reader is referred to Wu et al. (1995).

1.5 Uniqueness and Stability

1.5.1 Fundamental Relations

Consider the quasi-static deformation of a conventional elastic/plastic body whose plastic potential is identical to the yield function, which is supposed to be regular and convex. The current shape of the body and the internal distribution of stress are assumed to be known. We propose to establish the condition under which the boundary value problem has a unique solution and examine the related problem of stability. When positional changes are taken into account, it is convenient to formulate the boundary condition in terms of the rate of change of the nominal traction, which is based on the configuration at the instant under consideration. When body forces are absent, the equilibrium equation and the stress boundary condition for the rate problem may be written in terms of the nominal stress rate $\dot{t}_{ij}$ and the nominal traction rate $\dot{F}_j$ as

$$\frac{\partial \dot{t}_{ij}}{\partial x_i} = 0, \quad \dot{F}_j = l_i\, \dot{t}_{ij}, \tag{1.67}$$

where x_i denotes the current position of a typical particle and l_i the unit exterior normal to a typical surface element. The relationship between the unsymmetric nominal stress rate $\dot{t}_{ij}$ and the symmetric true stress rate $\dot{\sigma}_{ij}$, referred to a fixed set of rectangular axes (Chakrabarty, 2006), may be written as

$$\dot{t}_{ij} = \dot{\sigma}_{ij} - \sigma_{jk}\frac{\partial v_i}{\partial x_k} + \sigma_{ij}\frac{\partial v_k}{\partial x_k}, \tag{1.68}$$

where v_i denotes the velocity of the particle. The constitutive equations, on the other hand, must involve the Jaumann stress rate $\overset{\circ}{\sigma}_{ij}$, which is the material rate of change of the true stress σ_{ij} with respect to a set of rotating axes, and is given by (1.35). The elimination of $\dot{\sigma}_{ij}$ between (1.35) and (1.68) gives

$$\dot{t}_{ij} = \overset{\circ}{\sigma}_{ij} + \sigma_{ij}\dot{\varepsilon}_{kk} + \sigma_{ik}\omega_{jk} - \sigma_{jk}\dot{\varepsilon}_{ik} \tag{1.69}$$

This equation relates the nominal stress rate $\dot{t}_{ij}$ directly to the Jaumann stress rate $\overset{\circ}{\sigma}_{ij}$. Using the interchangeability of dummy suffixes, the scalar product of (1.69) with $\partial v_j/\partial x_i$ can be expressed as

$$\dot{t}_{ij}\frac{\partial v_j}{\partial x_i} = \left(\begin{smallmatrix}[-4pt]^{\circ} \\ [-6pt]\sigma_{ij}\end{smallmatrix} + \dot{\varepsilon}_{kk}\sigma_{ij}\right)\dot{\varepsilon}_{ij} - \sigma_{ij}\left(2\dot{\varepsilon}_{ik}\,\omega_{jk} + \frac{\partial v_i}{\partial x_k}\frac{\partial v_k}{\partial x_j}\right) \tag{1.70}$$

in view of the symmetry of the tensors σ_{ij} and $\dot{\varepsilon}_{ij}$. This relation is derived here for later use in the analysis for uniqueness and stability.

The constitutive law for the conventional elastic/plastic solid is such that the strain rate is related to the stress rate by two separate linear equations defining the loading and unloading responses. For an isotropic solid, when an element is currently plastic, the constitutive equation for loading may be written down by using the rate form of (1.38). The stress rate is therefore given by

$$\begin{aligned} \overset{\circ}{\sigma}_{ij} &= 2G\left\{\dot{\varepsilon}_{ij} + \frac{v}{1-2v}\dot{\varepsilon}_{kk}\delta_{ij} - \frac{3G}{3G+H}\dot{\varepsilon}_{kl}n_{kl}n_{kl}\right\}, \quad \dot{\varepsilon}_{kl}n_{kl} \geq 0, \\ \overset{\circ}{\sigma}_{ij} &= 2G\left\{\dot{\varepsilon}_{ij} + \frac{v}{1-2v}\dot{\varepsilon}_{kk}\delta_{ij}\right\}, \quad \dot{\varepsilon}_{kl}n_{kl} \geq 0. \end{aligned} \tag{1.71}$$

Consider now a fictitious solid whose constitutive law is given by the first equation of (1.71), whenever an element is currently plastic, regardless of the sign of $\dot{\varepsilon}_{kl}n_{kl}$. Such a solid may be regarded as a linearized elastic/plastic solid, in which the stress rate corresponding to a strain rate $\dot{\varepsilon}_{ij}$ is denoted by $\overset{\circ}{\tau}_{ij}$. The scalar product of (1.71) with $\dot{\varepsilon}_{ij}$ then furnishes

$$\overset{\circ}{\sigma}_{ij} \geq \overset{\circ}{\sigma}_{ij}\,\dot{\varepsilon}_{ij} = 2G\left\{\dot{\varepsilon}_{ij}\,\dot{\varepsilon}_{ij} + \frac{v}{1-2v}\,\dot{\varepsilon}_{kk}^2 - \frac{3G}{3G+H}\left(\dot{\varepsilon}_{ij}n_{ij}\right)^2\right\}, \tag{1.72}$$

where the equality holds only in the loading part of the current plastic region. In contrast to the bilinear elastic/plastic solid, the linearized solid has identical loading and unloading responses for any plastic element.

Let $(\overset{\circ}{\sigma}_{ij}, \dot{\varepsilon}_{ij})$ and $(\overset{\circ}{\sigma}{}^*_{ij}, \dot{\varepsilon}^*_{ij})$ denote two distinct combinations of stress and strain rates in an element of the actual elastic/plastic solid corresponding to a given state of stress. The stress rates for the linearized solid in the two states are $\overset{\circ}{\tau}_{ij}$ and $\overset{\circ}{\tau}{}^*_{ij}$, respectively. If the element is currently plastic, and the two strain rates do not both call for instantaneous unloading, the scalar product of $\dot{\varepsilon}^*_{ij}$ with the appropriate equation of (1.71) shows that

$$\overset{\circ}{\sigma}_{ij}\,\dot{\varepsilon}^*_{ij} \leq \overset{\circ}{\tau}_{ij}\,\dot{\varepsilon}^*_{ij} = 2G\left\{\dot{\varepsilon}_{ij}\,\dot{\varepsilon}^*_{ij} + \frac{v}{1-2v}\dot{\varepsilon}_{ij}\,\dot{\varepsilon}^*_{kk} - \frac{3G}{3G+H}\dot{\varepsilon}_{ij}\,\dot{\varepsilon}^*_{kl}n_{ij}n_{kl}\right\}. \tag{1.73}$$

The equality holds when ε^*_{ij} calls for further loading, whatever the nature of $\dot{\varepsilon}^*_{ij}$. The inequalities satisfied by $\overset{\circ}{\sigma}{}^*_{ij}\,\dot{\varepsilon}^*_{ij}$ and $\overset{\circ}{\sigma}{}^*_{ij}\,\dot{\varepsilon}^*_{ij}$ are similar to (1.72) and (1.73), respectively. Consequently,

$$\left(\overset{\circ}{\sigma}_{ij} - \overset{\circ}{\sigma}{}^*_{ij}\right)\left(\dot{\varepsilon}_{ij} - \dot{\varepsilon}^*_{ij}\right) \geq \left(\overset{\circ}{\tau}_{ij} - \overset{\circ}{\sigma}{}^*_{ij}\right)\left(\dot{\varepsilon}_{ij} - \dot{\varepsilon}^*_{ij}\right). \tag{1.74}$$

If the difference between the unstarred and the starred quantities is denoted by the prefix Δ, then it follows from above that

$$\begin{aligned} \Delta\sigma_{ij}\overset{\circ}{\Delta}\,\dot{\varepsilon}_{ij} &\geq \Delta\tau_{ij}\overset{\circ}{\Delta}\,\dot{\varepsilon}_{ij} \\ &= 2G\left\{\Delta\dot{\varepsilon}_{ij}\Delta\hat{\varepsilon}_{ij} + \frac{v}{1-2v}\left(\Delta\dot{\varepsilon}_{kk}\right)^2\right\} \end{aligned} \tag{1.75}$$

with equality holding for instantaneous loading produced by both $\dot{\varepsilon}_{ij}$ and $\dot{\varepsilon}^*_{ij}$. When both the states call for instantaneous unloading, the relationship between $\Delta\overset{\circ}{\sigma}_{ij}$ and $\Delta\dot{\varepsilon}_{ij}$ is given by the second equation of (1.71), while that between $\Delta\overset{\circ}{\tau}_{ij}$ and $\Delta\dot{\varepsilon}_{ij}$ is given by the first equation of (1.71), leading to the inequality $\Delta\overset{\circ}{\sigma}_{ij}\,\Delta\dot{\varepsilon}_{ij} \geq \Delta\overset{\circ}{\tau}_{ij}\,\Delta\dot{\varepsilon}_{ij}$. For an element that is currently elastic, there is the immediate identity

$$\Delta\overset{\circ}{\sigma}_{ij}\,\Delta\dot{\varepsilon}_{ij} = \Delta\overset{\circ}{\tau}_{ij}\,\Delta\dot{\varepsilon}_{ij} = 2G\left\{\Delta\dot{\varepsilon}_{ij}\,\Delta\dot{\varepsilon}_{ij} + \frac{v}{1-2v}\left(\Delta\dot{\varepsilon}_{kk}\right)^2\right\}. \tag{1.76}$$

It follows, therefore, that the inequality $\Delta\overset{\circ}{\sigma}_{ij}\,\Delta\dot{\varepsilon}_{ij} \geq \Delta\overset{\circ}{\tau}_{ij}\,\Delta\dot{\varepsilon}_{ij}$ holds throughout the elastic/plastic body and under all possible conditions of loading and unloading. This result will now be used for the derivation of the uniqueness criterion.

1.5.2 Uniqueness Criterion

Consider the typical boundary value problem in which the nominal traction rate $\dot{F}_j$ is specified on a part S_F of the current surface of the body, and the velocity v_j on the remainder S_v. Suppose that there could be two distinct solutions to the problem, involving the field equations (1.67), (1.69), and (1.71), together with the prescribed boundary conditions. If the difference between the two possible solutions is denoted by the prefix Δ, then in the absence of body forces, we have

$$\frac{\partial}{\partial x_i}\left(\Delta\dot{t}_{ij}\right) = 0, \quad \Delta\dot{F}_j = l_i\Delta\dot{t}_{ij},$$

in view of (1.67). The application of Green's theorem to integrals involving surface S and volume V gives

$$\int \Delta\dot{F}_j\Delta v_j dS = \int l_i\Delta\dot{t}_{ij}\,\Delta v_j dS = \int \Delta\dot{t}_{ij}\frac{\partial}{\partial x_i}\left(\Delta\dot{t}_{ij}\frac{\partial}{\partial x_i}\left(\Delta v_j\right)\right)dV.$$

The integral on the left-hand side vanishes identically, since $\Delta\dot{F}_j = 0$ on S_F and $\Delta v_j = 0$ on S_v by virtue of the given boundary conditions. The condition for having two possible solutions therefore becomes

$$\int \Delta\dot{t}_{ij}\frac{\partial}{\partial x_i}\left(\Delta v_j\right)dV = 0.$$

The left-hand side of the above equation must be positive for uniqueness (Hill, 1958). Using (1.70), expressed in terms of the difference of the two possible states, a sufficient condition for uniqueness may be written as

$$\int\left\{\Delta\overset{\circ}{\sigma}_{ij}\,\Delta\dot{\varepsilon}_{ij}+\sigma_{ij}\left[\Delta\dot{\varepsilon}_{kk}\,\Delta\dot{\varepsilon}_{ij}-2\Delta\dot{\varepsilon}_{ik}\,\Delta\dot{\omega}_{jk}-\frac{\partial}{\partial x_k}(\Delta v_i)\,\frac{\partial}{\partial x_j}(\Delta v_k)\right]\right\}dV>0. \tag{1.77}$$

for the difference Δv_j of every possible pair of continuous velocity fields taking prescribed values on S_v. For applications to physical problems, it is preferable to replace the above condition by a slightly over-sufficient criterion, using the fact that $\Delta\overset{\circ}{\sigma}_{ij}\,\Delta\dot{\varepsilon}_{ij}\geq\Delta\overset{\circ}{\tau}_{ij}\,\Delta\dot{\varepsilon}_{ij}$ throughout the body. Uniqueness is therefore assured when

$$\int\left\{\Delta\overset{\circ}{\tau}_{ij}\,\Delta\dot{\varepsilon}_{ij}-\sigma_{ij}\left[2\Delta\dot{\varepsilon}_{ik}\,\Delta\omega_{jk}+\frac{\partial}{\partial x_k}(\Delta v_i)\,\frac{\partial}{\partial x_j}(\Delta v_k)\right]\right\}dV>0 \tag{1.78}$$

for all continuous difference fields Δv_j vanishing on S_v. The term in $\Delta\varepsilon_{kk}$ has been neglected, since the contribution made by it is small compared to that arising from a similar term in the quantity $\Delta\overset{\circ}{\tau}_{ij}\,\Delta\dot{\varepsilon}_{ij}$, which is given by (1.75) and (1.76) in the plastic and elastic regions, respectively. It follows from (1.77) and (1.78) that the condition for uniqueness for the linearized elastic/plastic solid also ensures uniqueness for the actual elastic/plastic solid (Hill, 1959). If the constraints are rigid, so that $v_j=0$ on S_v, every difference field Δv_j is a member of the admissible field v_j, and the uniqueness criterion reduces to

$$\int\left\{\overset{\circ}{\tau}_{ij}\,\dot{\varepsilon}_{ij}-\sigma_{ij}\left(2\,\dot{\varepsilon}_{ik}\,\omega_{jk}+\frac{\partial v_i}{\partial x_k}\frac{\partial v_k}{\partial x_j}\right)\right\}dV>0 \tag{1.79}$$

for all continuous differentiable fields v_j vanishing at the constraints. Splitting the tensors $\partial v_i/\partial x_k$ and $\partial v_k/\partial x_j$ into their symmetric and antisymmetric parts, it is easily shown that

$$\sigma_{ij}\left(\frac{\partial v_i}{\partial x_k}\frac{\partial v_k}{\partial x_j}\right)=\sigma_{ij}\left(\dot{\varepsilon}_{ik}\,\dot{\varepsilon}_{jk}-\omega_{ik}\,\omega_{jk}\right).$$

The remaining two triple products cancel one another by the symmetry and antisymmetry properties of their factors. In view of the above identity, the uniqueness criterion (1.79) becomes

$$\int\left[\overset{\circ}{\tau}_{ij}\,\dot{\varepsilon}_{ij}-\sigma_{ij}\left(2\,\dot{\varepsilon}_{ik}\,\omega_{jk}+\dot{\varepsilon}_{ik}\,\dot{\varepsilon}_{jk}-\omega_{ik}\,\omega_{jk}\right)\right]dV>0. \tag{1.80}$$

The leading term in square brackets is given by (1.72) for the current plastic region and by the same equation with the last term omitted for the elastic region. In the treatment of problems involving curvilinear coordinates, it is only necessary to regard the components of the tensors appearing in (1.80) as representing the curvilinear components.

In a number of important physical problems, a part S_f of the boundary is submitted to a uniform fluid pressure p, which is made to vary in a prescribed manner. In this case, the change in the load vector on a given surface element, whose future orientation is not known in advance, cannot be specified. It can be shown that the nominal traction rate for the pressure-type loading is

$$\dot{F}_j = \dot{p}l_j + p\left(l_k\frac{\partial v_k}{\partial x_j} - l_j\frac{\partial v_k}{\partial x_k}\right),$$

where $\dot{p}$ is the instantaneous rate of change of the applied fluid pressure. When the boundary value problem has two distinct solutions under a given pressure rate $\dot{p}$ so that $\Delta\dot{p} = 0$, then the preceding relation gives

$$\Delta\dot{F}_j = p\left\{l_k\frac{\partial}{\partial x_j}(\Delta v_k) - l_j\frac{\partial}{\partial x_k}(\Delta v_k)\right\} \text{ on } S_f. \tag{1.81}$$

It is assumed that the remaining surface area of the body is partly under a prescribed nominal traction rate $\dot{F}_j$ and partly under a prescribed velocity v_j. Since $\Delta\dot{F}_j \neq 0$ on S_f, the bifurcation condition (1.77) must be modified by replacing the right-hand side of this equation with the surface integral

$$p\int\left\{l_k\frac{\partial}{\partial x_j}(\Delta v_k) - l_j\frac{\partial}{\partial x_k}(\Delta v_k)\right\}\Delta v_j dS_f.$$

The uniqueness criterion (1.78) is therefore modified by subtracting the same quantity from the left-hand side of the inequality. In particular, (1.80) is modified to (Chakrabarty, 1969b).

$$\int\left[\overset{\circ}{\tau}_{ij}\dot{\varepsilon}_{ij} - \sigma_{ij}\left(2\dot{\varepsilon}_{ik}\,\omega_{jk} + \hat{\varepsilon}_{ik}\ddot{\varepsilon}_{jk} - \omega_{ik}\,\omega_{jk}\right)\right]dV - p\int\left[l_k\left(\dot{\varepsilon}_{kj} + \omega_{kj}\right) - l_j\dot{\varepsilon}_{kk}\right]v_j dS_f > 0. \tag{1.82}$$

If the functional in (1.82) vanishes for some nonzero field v_j, bifurcation in the linearized solid may occur for any value of the traction rate on S_f and pressure rate on S_f. In the actual elastic/plastic solid, on the other hand, bifurcation will occur only for those traction rates which produce no unloading of the current plastic region. When the material is rigid/plastic, the admissible velocity field is incompressible ($\dot{\varepsilon}_{kk} = 0$), and the scalar product $\overset{\circ}{\tau}_{ij}\dot{\varepsilon}_{ij}$ becomes equal to $\frac{2}{3}H\dot{\varepsilon}_{ij}\,\dot{\varepsilon}_{ij}$, while the triple produce $\sigma_{ij}\,\dot{\varepsilon}_{ik}\,\omega_{jk}$ vanishes due to the coaxiality of the principal axes of stress and strain rate, leading to a considerable simplification of the problem. The condition for uniqueness in rigid/plastic solids has been discussed by Hill (1957), Chakrabarty (1969a), and Miles (1969).

1.5.3 Stability Criterion

Consider an elastic/plastic body which is rigidly constrained over a part S_v of its external surface, while constant nominal tractions are maintained over the remainder S_F. The deformation of the body will be stable if the internal energy dissipated in any geometrically possible small displacement from the position of equilibrium exceeds the work done by the external forces. Since these two quantities are equal to one another when evaluated to the first order, it is necessary to consider second-order quantities for the investigation of stability. The stress and velocity distributions throughout the body are supposed to be given in the current state, which is taken as the initial reference state for the stability analysis. At any instant during a small virtual displacement of a typical particle, its velocity is denoted by ω_j and the associated true stress by s_{ij}. Then the instantaneous rate of dissipation of internal energy per unit mass of material in the neighborhood of the particle is $(s_{ij}/\rho)(\partial\omega_i/\partial z_j)$, where z_j is the instantaneous position and ρ the current density. The rate of change of this quantity following the particle is

$$\left(\frac{\partial}{\partial t} + w_k \frac{\partial}{\partial z_k}\right)\left(\frac{s_{ij}}{\rho}\frac{\partial w_i}{\partial z_j}\right) = \frac{1}{\rho}\left\{\left(\dot{s}_{ij} - \frac{\dot{\rho}}{\rho}s_{ij}\right)\frac{\partial w_i}{\partial z_j} + s_{ij}\left(\frac{\partial \dot{w}_i}{\partial z_j} - \frac{\partial w_i}{\partial z_k}\frac{\partial w_k}{\partial z_j}\right)\right\},$$

where $\dot{w}_j$ is the instantaneous acceleration of the considered particle (Chakrabarty, 1969a). The operator appearing in the first parenthesis represents the material rate of change and may be denoted by D/Dt for compactness. If the initial true stress is σ_{ij}, the initial velocity v_j, and the initial stress rate $\dot{\sigma}_{ij}$, the above expression considered in the initial state furnishes

$$\rho_0\left\{\frac{D}{Dt}\left(\frac{s_{ij}}{\rho}\frac{\partial w_i}{\partial z_j}\right)\right\}_{t=0} = \dot{\sigma}_{ij}\frac{\partial v_i}{\partial x_j} + \sigma_{ij}\left(\frac{\partial \dot{v}_i}{\partial x_j} + \frac{\partial v_k}{\partial x_k}\frac{\partial v_i}{\partial x_j} - \frac{\partial v_i}{\partial x_k}\frac{\partial v_k}{\partial x_j}\right) \tag{1.83}$$

in view of the compressibility condition $\dot{\rho} = -\rho\,(\partial v_k/\partial x_k)$ in the initial state. Since the rate of dissipation of internal energy per unit volume in the initial state is $\sigma_{ij}(\partial v_i/\partial x_j)$, the internal energy dissipated per unit initial volume during an interval of time δt required by the additional displacement δuj may be written as

$$\delta U = \sigma_{ij}\frac{\partial v_i}{\partial x_j}\delta t + \frac{1}{2}\rho_0\left\{\frac{D}{Dt}\left(\frac{s_{ij}}{\rho}\frac{\partial w_i}{\partial z_j}\right)\right\}_{t=0}(\delta t)^2, \tag{1.84}$$

which is correct to the second order irrespective of the strain path. If the nominal traction and its rate of change in the initial state are denoted by F_j, and $\dot{F}_j$, respectively, the work done by the surface forces during the virtual displacement is

$$\delta W = \int\left(F_j + \frac{1}{2}\dot{F}_j\delta t\right)\delta u_j dS = \int F_j \delta u_j dS$$

in view of the boundary conditions $\dot{F}_j = 0$ on S_F and $v_j = 0$ on S_v. Expressing δu_j in terms of the initial velocity v_j and the initial acceleration $\dot{v}_j$, we have

$$\delta W = \delta t\int F_j v_j dS + \frac{1}{2}(\delta t)^2\int F_j \dot{v}_j dS \tag{1.85}$$

to second order. Since the total internal energy dissipated during the additional displacement must exceed the work done by the external forces, a sufficient condition for stability is

$$\int \delta U dV - \delta W > 0.$$

Substituting from (1.83), (1.84), and (1.85), the left-hand side of this inequality can be written entirely as a volume integral, since

$$\int F_f v_j dS = \int \sigma_{ij} \frac{\partial v_i}{\partial x_j} dV, \int F_f \dot{v}_j dS \int \sigma_{ij} \frac{\partial \dot{v}_i}{\partial x_j} dV,$$

by the principle of virtual work, the nominal traction being the same as the actual traction in the initial state. We therefore have

$$\int \left\{ \dot{\sigma}_{ij} \frac{\partial v_i}{\partial x_j} + \sigma_{ij} \left(\frac{\partial v_k}{\partial x_k} \frac{\partial v_i}{\partial x_j} - \frac{\partial v_i}{\partial x_k} \frac{\partial v_k}{\partial x_j} \right) \right\} dV > 0$$

as the required condition for stability. Substituting for $\dot{\sigma}_{ij}$, using (1.68), and introducing the true strain rate $\dot{\varepsilon}_{ij}$ and the spin tensor ω_{ij} in the initial state, the stability criterion is finally obtained as

$$\int \left\{ \overset{\circ}{\sigma}_{ij} \dot{\varepsilon}_{ij} + \sigma_{ij} \left(\dot{\varepsilon}_{kk} \dot{\varepsilon}_{ij} - 2\dot{\varepsilon}_{ik} \omega_{jk} - \frac{\partial v_i}{\partial x_k} \frac{\partial v_k}{\partial x_j} \right) \right\} dV > 0 \qquad (1.86)$$

for all continuous differentiable velocity fields v_j vanishing at the constraints. Since the expression in the curly brackets of (1.86) is equal to $\dot{t}_{ij} \left(\partial v_j / \partial x_i \right)$ in view of (1.70), the surface integral of the scalar product $\dot{F}_j v_j$ must be positive for the stability of the elastic/plastic solid (Hill, 1958). In the special case of rigid/plastic solids, an analysis similar to that presented above has been given by Chakrabarty (1969a).

The stability functional (1.86) may be compared with the uniqueness functional (1.77), which involves the difference field Δv_j instead of the velocity field vj. Due to the nonlinearity of the material response, the difference between two possible solutions is not necessarily a solution itself, and consequently (1.86) is always satisfied when (1.77) is. It follows that a partially plastic state in which the boundary value problem has a unique solution is certainly stable. When the solution is no longer unique, the partially plastic state may still be stable, and a point of bifurcation is therefore possible before an actual loss of stability. At such a stable bifurcation, the load must continue to increase with further deformation in the elastic/plastic range.

Problems

1.1 For a certain application involving an elastic/plastic material, the stress–strain curve in the plastic range needs to be replaced by a straight line defined by $\sigma = Y + T\,\varepsilon$. The actual strain-hardening curve can be represented by $\sigma = Y\,(E\varepsilon/Y)\,.^{n.}$ If the linear strain-hardening law predicts the same area under the stress–strain curve as that given by the power law curve, over the range $0 \leq \varepsilon \leq \varepsilon_0$, when both the hardening laws are extended backward to $\varepsilon = 0$, show that

$$\frac{E\varepsilon_0}{Y} = \left(\frac{1+n}{1-n}\right)^{1/n}, \quad \frac{T}{E} = \frac{2n}{i-n}\left(\frac{1-n}{1+n}\right)^{1/n}.$$

1.2 For an element of work-hardening material yielding according to the von Mises yield criterion under biaxial compression, show that the principal stresses can be expressed in terms of the polar angle θ of the deviatoric stress vector as

$$\sigma_x = -\frac{2\bar{\sigma}}{\sqrt{3}}\cos\theta, \quad \sigma_y = -\frac{2\bar{\sigma}}{\sqrt{3}}\sin\left(\frac{\pi}{6}-\theta\right),$$

where $\bar{\sigma}$ is the equivalent stress. Show also that the components of the plastic strain increment, according to the Prandtl–Reuss flow rule, can be expressed in terms of the angle θ and the current plastic modulus H as

$$d\varepsilon_x^p = -\cos\left(\frac{\pi}{6}+\theta\right)\left(\frac{d\bar{\sigma}}{H}\right), \quad d\varepsilon_y{}^p = \sin\theta\left(\frac{d\bar{\sigma}}{H}\right).$$

1.3 For an element of von Miss material deforming under a plane a strain tension in the x-direction and a stress-free state in the y-direction, show that the applied stress and the deviatoric angle at the initial yielding are given by $\sigma_e = Y/\sqrt{c}$ and $2\cos\,\theta_e = \sqrt{3/c}$, where $c = 1 - \nu + \nu^2$. If a prismatic beam made of such a material having a depth 2 h is bent to an elastic/plastic curvature, so that the depth of the elastic core becomes 2c, prove that the bending couple M is given by

$$\frac{M}{M_e} = \frac{a^2}{h^2} + 2\sqrt{3}c\int_{c/h}^{1}(\cos\theta)\,\xi\,d\xi, \quad \xi = \frac{y}{h},$$

where M_e is the bending moment at the elastic limit. Assuming a mean value of $\cos\theta$, equal to $\sqrt{\cos\theta_e}$, obtain the moment–curvature relationship in the dimensionless form

$$\frac{M}{M_e} = m - (m-1)\left(\frac{\kappa_e}{\kappa}\right)^2, \quad m = \sqrt{\frac{3}{2}}\,(3c)^{1/4}.$$

1.4 An ideally plastic bar of circular cross section is rendered completely plastic by the combined action of an axial force N and a twisting moment T. If the ratio of the rate of extension to the rate of twist at the yield point is denoted by $a\alpha/3$, show that the normal and shear stress distributions over the cross section of the bar are given by

$$\frac{\sigma}{Y} = \frac{\alpha}{\sqrt{\alpha^2 + 3r^2/a^2}}, \quad \frac{\tau}{Y} = \frac{r/a}{\sqrt{\alpha^2 + 3r^2/a^2}}.$$

Denoting the fully plastic values of the axial force and twisting moment by N_0 and T_0, respectively, and setting $\lambda = \sqrt{3+\alpha^2}$, obtain the interaction relationship in the parametric form

$$\frac{N}{N_0} = \frac{2}{3}\alpha\,(\lambda - \alpha)\,, \quad \sqrt{3}\frac{T}{T_0} = \lambda - \frac{2}{3}\alpha^2\,(\lambda - \alpha)\,.$$

1.5 A block of isotropic material is compressed in the x-direction by a pair of rigid smooth dies, while the deformation in the y-direction is completely suppressed by constraints. If the material strain hardens linearly with a constant tangent modulus T, show that the polar equation of the stress path in the deviatoric plane is given by

$$\frac{\bar{\sigma}}{Y} = \left\{\frac{\sqrt{3}\sin\theta - (1-2\nu)\,(T/E)\cos\theta_e}{\sqrt{3}\sin\theta_e - (1-2\nu)\,(T/E)\cos\theta}\right\}^{\alpha T/E} \exp\left[(\theta_e - \theta)\,\frac{\beta T}{E}\right],$$

where $\bar{\sigma}$ is the equivalent stress, and θ is the deviatoric angle having a value θ_e at the initial yielding, while α and β are dimensionless parameters defined as

$$\alpha = \frac{3 + (1-2\nu)^2\,(T/E)^2}{3 + (1-2\nu)^2\,(T/E)} \qquad \beta = \frac{\sqrt{3}\,(1-2\nu)\,(1-T/E)}{3 + (1-2\nu)^2\,(T/E)}.$$

1.6 The plastic modulus of a certain kinematically hardening material varies with the equivalent plastic strain according to the relation

$$H = Kn\,\exp\left(-n\bar{\varepsilon}^p\right),$$

where K and n are material constants. A specimen of this material is first pulled in tension until the longitudinal stress is equal to σ_0 and is then subjected to a complete loading cycle which involves a plastic strain amplitude of amount $\varepsilon*$. Show that the longitudinal stress at the end of the loading cycle exceeds σ_0 by the amount

$$\Delta\sigma = -K\left[1 - \exp(-4n\varepsilon^*)\right]\,\exp\left(-n\varepsilon^*\right).$$

1.7 For a material that hardens according to the combined hardening rule, the isotropic and kinematic parts of the plastic modulus H are assumed to be in the ration $\beta/(1-\beta)$, where β is a constant. Assumie the plastic modulus to be given by

$$H = Kn\,\exp\left(-n\bar{\varepsilon}^p\right)$$

Considering a complete loading cycle of a specimen involving a constant strain amplitude of amount ε^*, following a stress equal to σ_0 applied by simple tension, show that the tensile stress at the end of the cycle exceeds σ_0 by the amount

$$\Delta\sigma = K\left[1 + \exp(-2n\varepsilon^*)\right]\left[(2\beta - 1) + \exp\left(-2n\varepsilon^*\right)\right]\,\exp\left(-n\varepsilon^*\right).$$

1.8 The plastic yielding and flow of a certain isotropic material can be predicted with sufficient accuracy by modifying the von Mises yield criterion in the form

$$J_2\left(1 - \frac{9J_3^2}{4J_2^3}\right)^{1/3} = k^2$$

where k is the yield stress in pure shear. Show that the uniaxial yield stress according to this criterion is $Y = 1.853\ k$. Considering a state of plane stress defined by $\sigma_3 = 0$ and denoting σ_2/σ_1 by α, prove that the ratio of the two in-plane plastic strain increments according to the associated flow rule is given by

$$\frac{d\varepsilon_2^p}{d\varepsilon_1^p} = \left(\frac{2\alpha - 1}{2 - \alpha}\right)\left\{\frac{6\left(1 - \alpha + \alpha^2\right)^2 + (1 + \alpha)(2 - \alpha)\left(2\alpha^2 - 2\alpha - 1\right)}{6\left(1 - \alpha + \alpha^2\right)^2 + (1 + \alpha)(2\alpha - 1)\left(2 - 2\alpha - \alpha^2\right)}\right\}$$

Chapter 2
Problems in Plane Stress

In many problems of practical interest, it is a reasonable approximation to disregard the elastic component of strain in the theoretical analysis, even when the body is only partially plastic. In effect, we are then dealing with a hypothetical material which is rigid when stressed below the elastic limit, the modulus of elasticity being considered as infinitely large. If the plastically stressed material has the freedom to flow in some direction, the distribution of stress in the deforming zone of the assumed rigid/plastic body would approximate that in an elastic/plastic body, except in a transition region near the elastic/plastic interface where the deformation is restricted to elastic order of magnitude. The assumption of rigid/plastic material is generally adequate not only for the analysis of technological forming processes, where the plastic part of the strain dominates over the elastic part, but also for the estimation of the yield point load when the rate of work-hardening is sufficiently small (Section 1.2). In this chapter, we shall be concerned with problems in plane stress involving rigid/plastic bodies which are loaded beyond the range of contained plastic deformation.

2.1 Formulation of the Problem

A plate of small uniform thickness is loaded along its boundary by forces acting parallel to the plane of the plate and distributed uniformly through the thickness. The stress components σ_z, τ_{xz}, and τ_{yz} are zero throughout the plate, where the z-axis is considered perpendicular to the plane. The state of stress is therefore specified by the three remaining components σ_x, σ_y, and τ_{xy}, which are functions of the rectangular coordinates x and y only. During the plastic deformation, the thickness does not generally remain constant, but the stress state may still be regarded as approximately plane provided the thickness gradient remains small compared to unity.

2.1.1 Characteristics in Plane Stress

For greater generality, we consider a nonuniform plate of small thickness gradient, subjected to a state of generalized plane stress in which σ_x, σ_y, and τ_{xy} denote the

J. Chakrabarty, *Applied Plasticity, Second Edition*, Mechanical Engineering Series, DOI 10.1007/978-0-387-77674-3_2, © Springer Science+Business Media, LLC 2010 stress components averaged through the thickness of the plate. In the absence of body forces, the equations of equilibrium are

$$\frac{\partial}{\partial x}(h\sigma_x) + \frac{\partial}{\partial y}(h\tau_{xy}) = 0, \quad \frac{\partial}{\partial x}(h\tau_{xy}) + \frac{\partial}{\partial y}(h\sigma_y) = 0, \tag{2.1}$$

where h is the local thickness of the plate. If yielding occurs according to the von Mises criterion, the stresses must also satisfy the equation

$$\sigma_x^2 - \sigma_x\sigma_y + \sigma_y^2 + 3\tau_{xy}^2 = \sigma_1^2 - \sigma_1\sigma_2 + \sigma_2^2 = 3k^2, \tag{2.2}$$

where σ_1 and σ_2 are the principal stresses in the xy-plane, and k is the yield stress in pure shear equal to $1/\sqrt{3}$ times the uniaxial yield stress Y. In the (σ_1, σ_2)-plane, (2.2) represents an ellipse having its major and minor axes bisecting the axes of reference.

Suppose that the stresses are given along some curve C in the plastically deforming region. The thickness h is regarded as a known function of x and y at the instant under consideration. Through a generic point P on C, consider the rectangular axes (X, y) along the normal and tangent, respectively, to C. If we rule out the possibility of a stress discontinuity across C, the tangential derivatives $\partial\sigma_x/\partial y$,$\partial\sigma_y/\partial y$, and $\partial\tau_{xy}/\partial y$ must be continuous. Since $\partial h/\partial x$ and $\partial h/\partial y$ are generally continuous, the equilibrium equation (2.1) indicates that $\partial\sigma_x/\partial x$ and $\partial\tau_{xy}/\partial x$are also continuous across c. For given values of these derivatives, the remaining normal derivative $\partial\sigma_y/\partial x$ can be uniquely determined from the relation

$$\left(2\sigma_x - \sigma_y\right)\frac{\partial\sigma_x}{\partial x} + \left(2\sigma_y - \sigma_x\right)\frac{\partial\sigma_y}{\partial x} + 6\tau_{xy}\frac{\partial\tau_{xy}}{\partial x} = 0 \tag{2.3}$$

obtained by differentiating the yield criterion (2.2), unless the coefficient $2\sigma_y - \sigma_x$ is zero. When $\sigma_x = 2\sigma_y$, the above equation gives no information about $\partial\sigma_y/\partial x$, which may therefore be discontinuous across C. It follows that S is a characteristic when the stress components are given by

$$\sigma_x = \sigma, \quad \sigma_y = \frac{1}{2}\sigma, \quad \tau_{xy} = \tau,$$

where σ is the normal stress, and τ the shear stress transmitted across C. In view of the yield criterion (2.2), the relationship between σ and τ is

$$\sigma^2 + 4\tau^2 = 4k^2 \tag{2.4}$$

which defines an ellipse E in the (σ, τ)-plane, as shown in Fig. 2.1. If ψ denotes the angle of inclination of the tangent to E with the σ-axis, reckoned positive when τ decreases in magnitude with increasing σ, then

$$\tan\psi = \mp\frac{d\tau}{d\sigma} = \pm\frac{\sigma}{4\tau} = \pm\left(\frac{\sigma_x - \sigma_y}{2\tau_{xy}}\right). \tag{2.5}$$

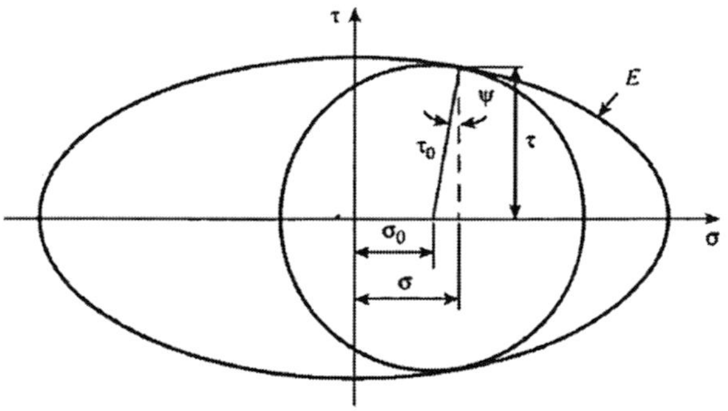

Fig. 2.1 Yield envelope for an element in a state of plane plastic stress including the associated Mohr circle for stress

The expression in the parenthesis is the cotangent of twice the counterclockwise angle made by either of the two principal stresses with the x-axis. It follows that there are two characteristic directions at each point, inclined at an angle $\pi/4 + \psi/2$ on either side of the algebraically greater principal stress direction. The two characteristics are identified as α- and β-lines following the convention indicated in Fig. 2.2, when $\sigma_1 > \sigma_2$. The two characteristics coincide when $\psi = \pm\pi/2$.

The normal and shear stresses acting across the characteristics are defined by the points of contact of the envelope E with Mohr's circle for the considered state of stress. The locus of the highest and lowest points of the circle is an ellipse representing the yield criterion (2.2), which may be written as

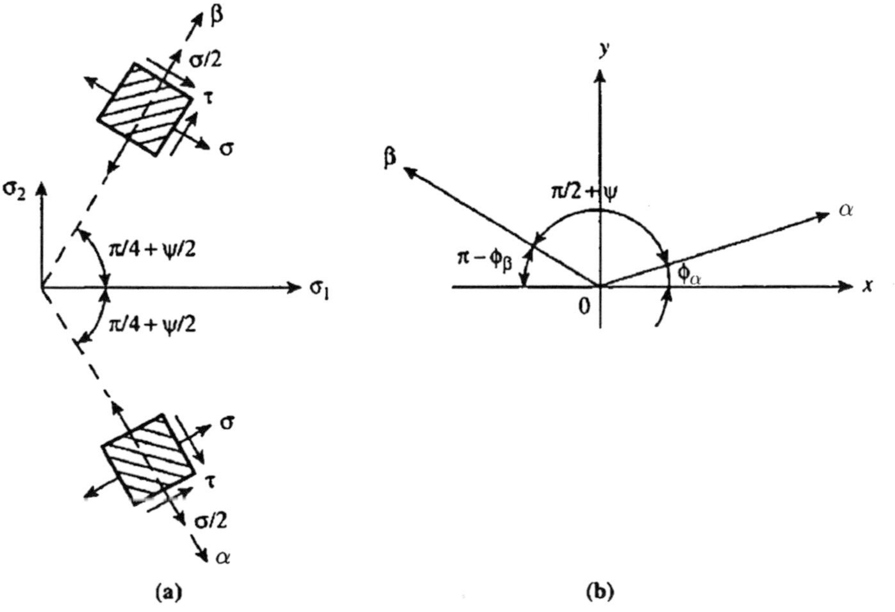

Fig. 2.2 Characteristics directions in plane stress ($\sigma_1 > \sigma_2 > 0$), designated by α- and β-lines

$$F(\sigma_0,\tau_0) = \sigma_0^2 + 3\tau_0^2 = 3k^2,$$

where (2.6)

$$\sigma_0 = \frac{1}{2}(\sigma_1 + \sigma_2) = \frac{3}{4}\sigma, \quad \tau_0 = \frac{1}{2}(\sigma_1 - \sigma_2).$$

Thus, τ_0 is numerically equal to the maximum shear stress and σ_0 the mean normal stress in the plane. Since σ cannot exceed $2k$ in magnitude, the material being ideally plastic, σ_0 be numerically less than $3k/2$ and τ_0 numerically greater than $k/2$ for the characteristics to be real. Since $d\tau/d\sigma_0$ is equal to $-r\sin\psi$ in view of (2.6) and Fig. 2.1, the acute angle which the tangent to the yield locus makes with the σ-axis must be less than $\pi/4$ for the stress equations to be hyperbolic. When τ_0 is numerically less than $k/2$, there is no real contact between the Mohr circle and the envelope E, and the stress equations then become elliptic.

In the limiting case of $|\tau_0| = k/2$, the characteristics are coincident with the axis of the numerically lesser principal stress, the values of the principal stresses being $\pm(2k, k)$ in the limiting state.

2.1.2 Relations Along the Characteristics

When the equations are hyperbolic, it is convenient to establish equations giving the variation of the stresses along the characteristics. Following Hill (1950a), we take the axes of reference along the normal and tangent to a typical characteristic C at a generic point P as before. Setting $2\sigma_y = \sigma_x \tan\psi$ in (2.3), we get

$$\frac{\partial \tau_{xy}}{\partial x} \pm \tan\psi \frac{\partial \sigma_x}{\partial x} = 0,$$

where the upper sign corresponds to the β-line and the lower sign to the α-line, respectively. The substitution of $\partial\sigma_x/\partial x$ from above into the first equation of (2.1) gives

$$h\frac{\partial \tau_{xy}}{\partial x} \pm \tan\psi \frac{\partial \tau_{xy}}{\partial y} = \pm\left(\sigma_x \frac{\partial h}{\partial x} + \tau_{xy}\frac{\partial h}{\partial y}\right)\tan\psi.$$

Regarding the curve C as a β-line, we observe that the space derivatives along the σ and β-lines at P are

$$\frac{\partial}{\partial s_\alpha} = \cos\psi\frac{\partial}{\partial x} - \sin\psi\frac{\partial}{\partial y}, \quad \frac{\partial}{\partial s_\beta} = \frac{\partial}{\partial y},$$

and the preceding equation therefore reduces to

$$h\frac{\partial \tau_{xy}}{\partial s_\alpha} = \frac{\partial h}{\partial s_\alpha}\tan\psi + (\sigma\tan\psi + \tau)\sin\psi\frac{\partial h}{\partial s_\beta} \tag{2.7}$$

since $\sigma_x = \sigma$ and $\tau_{xy} = \tau$ at the considered point. To obtain the derivative $\partial \tau_{xy}/\partial s_\alpha$ at P, let ϕ_β denote the counterclockwise angle made by the tangent to C with an arbitrary fixed direction. If the x-axis is now taken in this direction, then

$$\tau_{xy} = -\frac{1}{4}\sigma \sin 2\phi_\beta - \tau \cos 2\phi_\beta.$$

Differentiating this expression partially with respect to s_α and then setting $\phi_\beta = \pi/2$. we obtain

$$\frac{\partial \tau_{xy}}{\partial s_\alpha} = \frac{1}{2}\sigma \frac{\partial \phi_\beta}{\partial s_\alpha} - \frac{\partial \tau}{\partial s_\alpha} = \tan \psi \left(2\tau \frac{\partial \phi_\beta}{\partial s_\alpha} - \frac{\partial \sigma}{\partial s_\alpha} \right)$$

in view of (2.5). Inserting in (2.7) and rearranging, the result can be expressed in the form

$$d\,(h\sigma) - 2\,h\tau d\phi_\beta = -\,(\sigma \sin \psi + \tau \cos \psi)\,\frac{\partial h}{\partial s_\beta} ds_\alpha \tag{2.8}$$

along an α-line. Similarly, considering the curve C to be an α-line, it can be shown that

$$d\,(h\sigma) + 2\,h\tau d\phi_\alpha = -\,(\sigma \sin \psi + \tau \cos \psi)\,\frac{\partial h}{\partial s_\alpha} ds_\beta \tag{2.9}$$

along a β-line, where ϕ_α is the counterclockwise orientation of the α-line with respect to the same fixed direction. Evidently, $d\phi_\beta$ - $d\phi_\alpha = d\psi$. If the thickness distribution is given, (2.8) and (2.9) in conjunction with (2.4) would enable us to determine the stress distribution and the characteristic directions in the hyperbolic part of the plastic region. When the thickness is uniform, (2.8) and (2.9) reduce to the relations

$$\left.\begin{aligned} d\sigma - 2\tau d\phi_\beta &= 0 \quad \text{along an } \alpha - \text{line}, \\ d\sigma + 2\tau d\phi_\alpha &= 0 \quad \text{along an } \beta - \text{line}, \end{aligned}\right\} \tag{2.10}$$

which are analogous to the well-known Hencky equations in plane strain. In the solution of physical problems, it is usually convenient to express the yield criterion parametrically through the angle ψ, or a related angle θ such that

$$\tan \theta = \sqrt{3} \sin \psi, \quad -\frac{\pi}{2} \leq \psi \leq \frac{\pi}{2}, \quad -\frac{\pi}{3} \leq \theta \leq \frac{\pi}{3}. \tag{2.11}$$

The angle θ denotes the orientation of the stress vector in the deviatoric plane with respect to the direction representing pure shear in the plane of the plate. Indeed, it follows from (2.4), (2.5), and (2.11) that

$$\left.\begin{aligned}\frac{\sigma}{k} &= \frac{4\sin\psi}{\sqrt{1+3\sin^2\psi}} = \frac{4}{\sqrt{3}}\sin\theta,\\ \frac{\tau}{k} &= \frac{\cos\psi}{\sqrt{1+3\sin^2\psi}} = \sqrt{1-\frac{4}{3}\sin^2\theta}.\end{aligned}\right\} \tag{2.12}$$

Since $2(\sigma_1+\sigma_2)=3\sigma$ and $2(\sigma_1-\sigma_2)=\sqrt{\sigma^2+16\tau^2}$, we immediately get $\sigma_1+\sigma_2=\sqrt{3}k\sin\,\theta$ and $\sigma_1-\sigma_2=2k\cos\theta$, and the principal stresses become

$$\sigma_1 = 2k\sin\left(\theta+\frac{\pi}{6}\right), \quad \sigma_2 = 2k\sin\left(\theta-\frac{\pi}{6}\right) \tag{2.13}$$

It is also convenient at this stage to introduce a parameter λ which is defined in the incremental form

$$d\lambda = \frac{1}{2}\left(\frac{d\sigma}{\tau} - d\psi\right) = 2\frac{d\sigma}{\sigma}\tan\psi - \frac{1}{2}d\psi$$

which is assumed to vanish with ψ or θ. Substituting from (2.12) and integrating, we get

$$\begin{aligned}\lambda &= \tan^{-1}(2\tan\psi) - \frac{1}{2}\psi\\ &= \sin^{-1}\left(\frac{2}{\sqrt{3}}\sin\theta\right) - \frac{1}{2}\sin^{-1}\left(\frac{1}{\sqrt{3}}\tan\theta\right).\end{aligned} \tag{2.14}$$

Evidently, $-\pi/4 \leq \lambda \leq \pi/4$ over the relevant range. If ω denotes the counter-clockwise angle made by a principal stress axis with respect to a fixed axis, then

$$d\omega = d\left(\phi_\alpha + \frac{1}{2}\psi\right) = d\left(\phi_\beta + \frac{1}{2}\psi\right).$$

Dividing (2.8) and (2.9) by $h\sigma$ throughout, and substituting for $d\sigma/\sigma$, σ/τ, $d\phi_\alpha$, and $d\phi_\beta$, we finally obtain the relations

$$\left.\begin{aligned}d(\lambda-\omega) + 2\tan\psi\frac{dh}{h} &= -\left(\frac{1+3\sin^2\psi}{2h\cos\psi}\right)\frac{\partial h}{\partial s_\beta}ds_\alpha,\\ d(\lambda-\omega) + 2\tan\psi\frac{dh}{h} &= -\left(\frac{1+3\sin^2\psi}{2h\cos\psi}\right)\frac{\partial h}{\partial s_\alpha}ds_\beta.\end{aligned}\right\} \tag{2.15}$$

along the α- and β-lines, respectively. Similar equations may be written in terms of θ using (2.11). Numerical values of λ are given in Table 2.1 for the whole range of values of ψ and θ. When h is a constant, (2.15) reduces to

Table 2.1 Parameters for plane strees characteristics

Ψ degrees	Ψ Radians	λ radians	θ degrees	Ψ degrees	λ radians
0	0	0	0	0	0
10	0.17453	0.25177	10	5.843	0.15089
20	0.34907	0.4570	20	12.130	0.30013
30	0.52360	0.59527	30	19.471	0.44556
40	0.89813	0.68435	35	23.845	0.51581
50	0.87266	0.73722	40	28.977	0.58352
60	1.04720	0.76616	45	35.264	0.64757
70	1.22173	0.77992	50	55.542	0.75558
80	1.39626	0.78473	55	55.542	0.75558
90	1.57080	0.78540	60	90.000	0.78540

$$\left.\begin{aligned} \lambda - \omega &= \text{constant along an } \alpha\text{-line,} \\ \lambda - \omega &= \text{constant along a } \beta\text{-line,.} \end{aligned}\right\} \tag{2.16}$$

In analogy with Hencky's first theorem, we can state that the difference in values of both λ *and* ω between a pair of points, where two given characteristics of one family are cut by a characteristic of the other family, is the same for all intersecting characteristics. It follows that if a segment of one characteristic is straight, then so are the corresponding segments of the other members of the same family, the values of λ and ω being constant along each straight segment. If both families of characteristics are straight in a certain portion of the plastic region, the state of stress is uniform throughout this region.

2.1.3 The Velocity Equations

Let (v_x, v_y) denote the rectangular components of the velocity averaged through the thickness of the plate. The material is assumed as rigid/plastic, obeying the von Mises yield criterion and the Lévy–Mises flow rule. In terms of the stresses σ_x, σ_y, and τ_{xy}, the flow rule may be written as

$$\frac{\partial v_x/\partial x}{2\sigma_x - \sigma_y} = \frac{\partial v_y/\partial x}{2\sigma_y - \sigma_x} = \frac{\partial v_x/\partial y + \partial v_y/\partial x}{6\tau_{xy}}. \tag{2.17}$$

Consider a curve C along which the stress and velocity components are given, and let the x- and y-axes be taken along the normal and tangent, respectively, to C at a typical point P. Assuming the velocity to be continuous across C, the tangential derivatives dv_x/dy and dv_y/dy are immediately seen to be continuous. From (2.17), the normal derivatives dv_x/dx and dv_y/dx can be uniquely determined unless $2\sigma_y - \sigma_x = 0$, which corresponds to $dv_y/dy = 0$. Thus, C is a characteristic for the velocity field if it coincides with a direction of zero rate of extension. There are two such directions at each point and they are identical to those of the stress characteristics.

Since the characteristics are inclined at an angle $\pi/4+\psi/2$ to the direction of the algebraically greater principal strain rate $\dot{\varepsilon}_1$, the condition $\dot{\varepsilon} = 0$ along a characteristic gives

$$\sin\psi = \frac{\dot{\varepsilon}_1 + \dot{\varepsilon}_2}{\dot{\varepsilon}_1 - \dot{\varepsilon}_2} = \frac{1}{3}\left(\frac{\sigma_1+\sigma_2}{\sigma_1-\sigma_2}\right) = \frac{1}{\sqrt{3}}\tan\theta. \tag{2.18}$$

The range of plastic states for which the characteristics are real corresponds to $|\sigma_0| \leq 3|\tau_0|$ and are represented by the arcs *AB* and *CD* of the von Mises ellipse shown in Fig. 2.3. The numerically lesser principal strain rate vanishes in the limiting states, represented by the extremities of these arcs, where the tangents make an acute angle of $\pi/4$ with the σ_0-axis.

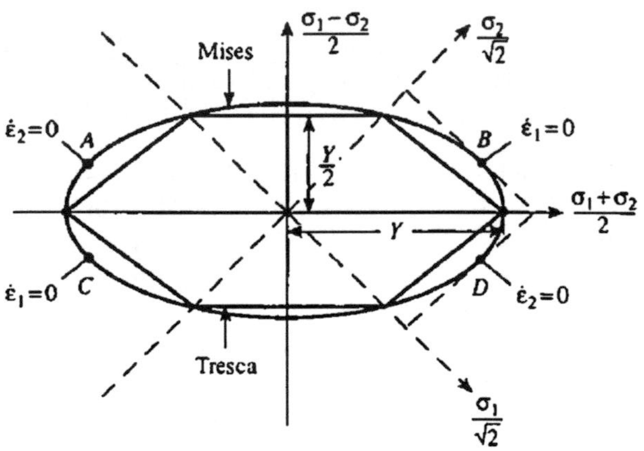

Fig. 2.3 Plane stress yield loci according to Tresca and von Mises for a material with a uniaxial yield stress *Y*

The velocity of a typical particle is the resultant of its rectangular components v_x and v_y. The resolved components of the velocity vector along the α and β-lines, denoted by u and v, respectively, are related to the rectangular components as

$$u = v_x\cos\phi_\alpha + v_y\sin(\psi+\phi_\alpha), \quad v = -v_x\sin\phi_\alpha + v_y\cos(\psi+\phi_\alpha),$$

where ϕ_α denotes the counterclockwise angle made by the α-line with the x-axis. The preceding relations are easily inverted to give

$$\left.\begin{aligned} v_x &= [u\cos(\psi+\phi_\alpha) - v\sin\phi_\alpha]\sec\psi, \\ v_y &= [u\sin(\psi+\phi_\alpha) - v\cos\phi_\alpha]\sec\psi. \end{aligned}\right\} \tag{2.19}$$

Differentiating v_x partially with respect to x, and using the fact that $dv_x/dx = 0$ when $\phi_\alpha = -\psi$, we obtain

$$\frac{\partial v}{\partial s_\alpha} - (u\tan\psi + v\sec\psi)\frac{\partial\phi_\alpha}{\partial s_\alpha} = 0.$$

Similarly, equating dv_y/dy to zero after setting $\phi_\alpha = -\psi$ in the expression for the partial derivative of v_y with respect to y, we get

$$\frac{\partial v}{\partial s_\beta} + (u \sec \psi + v \tan \psi) \frac{\partial \phi_\beta}{\partial s_\beta} = 0$$

in view of the relation $d\phi_\beta - d\phi_\alpha = d\psi$ The velocity relations along the characteristics therefore become

$$\left. \begin{aligned} du - (u \tan \psi + v \sec \psi)\, d\phi_\alpha &= 0 \text{ along an } \alpha\text{-line}, \\ dv + (u \sec \psi + v \tan \psi)\, d\phi_\beta &= 0 \text{ along an } \beta\text{-line}, \end{aligned} \right\} \quad (2.20)$$

When the characteristic directions are known at each point of the field, the velocity distribution can be determined from (2.20). For $\psi = 0$, these equations reduce to the well-known Geiringer equations in plane plastic strain.

Since the thickness strain rate has the same sign as that of $-(\sigma + \sigma_2)$, a thinning of the sheet corresponds to $\sigma_0 > 0$ and a thickening to $\sigma_0 < 0$. If the rate of change of thickness following the element is denoted by h, it follows from the associated flow rule that

$$\frac{\dot{h}}{h} = \frac{1}{h}\left(\frac{\partial h}{\partial t} + \omega \frac{\partial h}{\partial s}\right) = -\left(\frac{\sigma_{\rm n} + \sigma_1}{2\sigma_1 - \sigma_{\rm n}}\right)\frac{\partial \omega}{\partial s}, \quad (2.21)$$

where w is the speed of a typical particle, s is the arc length along the momentary flow line, and (σ_n, σ_t) are the normal stress components along the normal and tangent, respectively, to the flow line. The change in thickness during a small interval can be computed from (2.21). Evidently, whenever there is a discontinuity in the velocity gradient, there is also a corresponding jump in $\partial h/\partial t$, leading to a discontinuity in the surface slope of an initially uniform sheet.

2.1.4 Basic Relations for a Tresca Material

If the material yields according to Tresca's yield criterion with a given uniaxial yield stress $Y = 2k$, the yield locus is a hexagon inscribed in the von Mises ellipse. When the principal stresses σ_1 and σ_2 have

opposite signs, the greatest shear stress occurs in the plane of the sheet, and the yield criterion becomes

$$(\sigma_1 - \sigma_2)^2 = \left(\sigma_x - \sigma_y\right)^2 + 4\tau_{xy}^2 = 4k^2, \quad \left|\sigma_x + \sigma_y\right| \le 2k. \quad (2.22)$$

As in the case of plane strain, the stress equations are hyperbolic, and the characteristics are sliplines bisecting the angles between the principal stress axes. Since the shear stresses acting across the characteristics are of magnitude k, the envelope of the Mohr's circles then coincides with the yield locus. The variation of the normal stress σ along the characteristics is obtained from (2.8) and (2.9) by setting $\tau = k$, $\psi = 0$, and $d\phi_\alpha = d\phi_\beta = d\phi$, the expression in each parenthesis being then equal to k.

For a uniform sheet, these relations reduce to the well-known Hencky equations in plane strain. Since the thickness strain rate vanishes by the associated flow rule, the velocity equations are also hyperbolic and the characteristics are again the sliplines, the velocity relations along the characteristics being given by the familiar Geiringer equations in plane strain.

When the principal stresses σ_1 and σ_2 have the same sign, the greatest shear stress occurs out of the plane of the applied stresses, and the numerically greater principal stress must be of magnitude $2k$ for yielding to occur. On the (σ_0, τ_0)-plane, the yield criterion is defined by the straight lines $\sigma_0 \pm \tau_0 = \pm 2k$. In terms of the principal stresses, the yield criterion may be written as

$$(\sigma_1 + 2k)(\sigma_2 \pm 2k) = 0, \quad \sigma_1\sigma_2 \geq 0.$$

This equation can be expressed in terms of the (x, y) components of the stress, using the fact that $\sigma_1 + \sigma_2 = \sigma_x + \sigma_y$ and $\sigma_1\sigma_2 = \sigma_x\sigma_y - \tau^2_{xy}$ the result being

$$\tau_{xy}^2 - \sigma_x\sigma_y + 2k\left|\sigma_x + \sigma_y\right| = 4k^2, \; 2k \leq \left|\sigma_x + \sigma_y\right| \leq 4k. \tag{2.23}$$

The partial differentiation of (2.23) with respect to x reveals that the stress derivatives are uniquely determined unless $\sigma_x = \pm 2k$. Hence, there is a single characteristic across which the normal stress is of magnitude $2k$. In other words, the stress equations are parabolic with the characteristic coinciding with the direction of the numerically lesser principal stress, whose magnitude is denoted by σ. When the x- and y-axes are taken along the normal and tangent, respectively, to the characteristic, we have

$$\sigma_x = \pm 2k \text{ and } \sigma_y = \pm\sigma, \; \tau_{xy} = \mp\left(k - \frac{\sigma}{2}\right)\sin 2\omega. \tag{2.24}$$

where ω denotes the counterclockwise angle made by the characteristic with a fixed direction which is temporarily considered as the x-axis. Inserting (2.24) into the equilibrium equation (2.1), and setting $\omega = \pi/2$ after the differentiation, we get

$$\left(1 - \frac{\sigma}{2k}\right)\frac{\partial\omega}{\partial s} = -\frac{1}{h}\frac{\partial h}{\partial n}, \; \frac{\partial}{\partial s}\left(\frac{h\sigma}{2k}\right) = -\left(1 - \frac{\sigma}{2k}\right)h\frac{\partial\omega}{\partial n}, \tag{2.25}$$

where ds and dn are the line elements along the characteristic and its orthogonal trajectory, forming a right-handed pair of curvilinear axes.

When the thickness is uniform, ω is constant along each characteristic, which is therefore a straight line defined by $y = x\tan\omega + f(\omega)$, where $f(\omega)$ is a function of ω to be determined from the stress boundary condition. The curvature of the numerically greater principal stress trajectory is

$$\frac{\partial \omega}{\partial n} = \frac{\partial \omega}{\partial x} \sin \omega - \frac{\partial \omega}{\partial y} \cos \omega = -\frac{\cos \omega}{x + f'(\omega) \cos^2 \omega}. \tag{2.26}$$

Inserting from (2.26) into the second equation of (2.25), and using the fact that $ds = sec\ \omega\ dx$ along a characteristic, the above equation is integrated to give (Sokolovsky, 1969),

$$1 - \frac{\sigma}{2k} = \frac{g(\omega)}{x + f'(\omega) \cos^2 \omega}, \tag{2.27}$$

where $g(\omega)$ is another function to be determined from the boundary condition. Since the numerically lesser principal strain rate vanishes according to the associated flow rule, it follows that the velocity equations are also parabolic, and the characteristic direction coincides with the direction of zero rate of extension. The tangential velocity v remains constant along the characteristic, and the normal velocity u follows from the condition of zero ate of shear associated with these two orthogonal directions.

2.2 Discontinuities and Necking

2.2.1 Velocity Discontinuities

In a nonhardening rigid/plastic solid, the velocity may be tangentially discontinuous across certain surfaces where the shear stress attains its greatest magnitude k. For a thin sheet of metal, it is also necessary to admit a necking type of discontinuity involving both the tangential and normal components of velocity. To be consistent with the theory of generalized plane stress, the strain rate is considered uniform through the thickness of the neck, whose width b is comparable to the sheet thickness h. Since plastic deformation is confined in the neck, the rate of extension vanishes along its length, and the neck therefore coincides with a characteristic. The relative velocity vector across the neck must be perpendicular to the other characteristic in order that the velocity becomes continuous across it. Localized necking cannot occur, however, when the stress state is elliptic.

Let v denote the magnitude of the relative velocity vector which is inclined at an angle ψ to the direction of the neck, Fig. 2.4a. The rate of extension in the direction perpendicular to the neck is $(v/b)\ sin\ \psi$, and the rate of shear across the neck is $(v/2b)\ cos\ \psi$. The condition of the zero rate of extension along the neck therefore gives the principal strain rates as (Hill, 1952).

$$\dot{\varepsilon}_1 = \frac{v}{2b}(1 + \sin \psi), \quad \dot{\varepsilon}_2 = -\frac{v}{2b}(1 - \sin \psi), \quad \dot{\varepsilon}_3 = -\frac{v}{b} \sin \psi, \tag{2.28}$$

irrespective of the flow rule. These relations imply that the axis of $\dot{\varepsilon}_1$ is inclined at an angle $\pi/4 + \psi/2$ to the neck. For a von Mises material, ψ is related to the stress according to (2.18), while for a Tresca material, $\psi = 0$ when the stress state

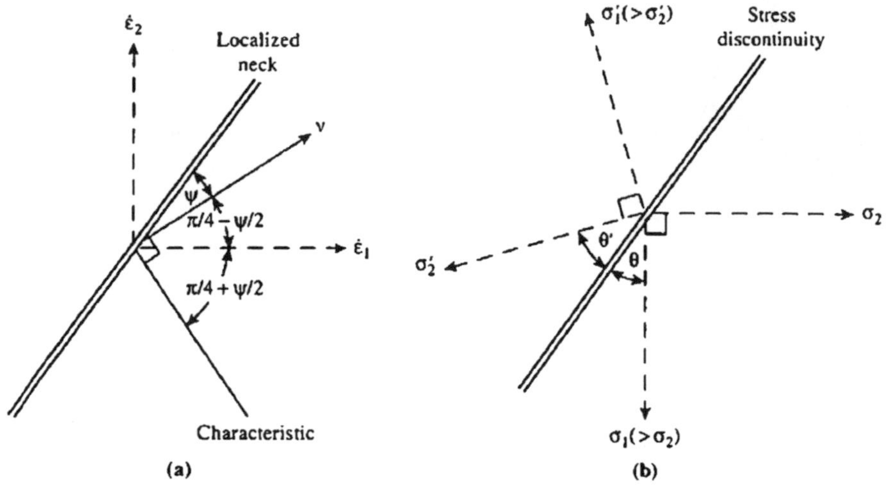

Fig. 2.4 Velocity and stress discontinuities in plane stress including the associated principal directions

is hyperbolic. In the case of a uniaxial tension, $\dot{\varepsilon}_1 = -2\dot{\varepsilon}_2$ for any regular yield function, giving $\psi = \sin^{-1} 1/3 \approx 19.47°$, the angle of inclination of the neck to the direction of tension being $\beta = \pi/4 + \psi/2 = \tan^{-1}\sqrt{2} \approx 54.74°$. For a Tresca material, on the other hand, ψ can have any value between 0 and $\pi/2$ under a uniaxial state of stress.

When the material work-hardens, a localized neck is able to develop only if the rate of hardening is small enough to allow an incremental deformation to remain confined in the incipient neck. For a critical value of the rate of hardening, the deformation is just able to continue in the neck, while the stresses elsewhere remain momentarily unchanged. Since the force transmitted across the neck remains momentarily constant, we have

$$-\frac{dh}{h} = \frac{d\sigma}{\sigma} = \frac{dY}{Y} = \frac{H}{Y} d\varepsilon,$$

where σ is the normal stress across the neck, Y the current yield stress, H the rate of hardening, and $d\varepsilon$ the equivalent strain increment. If the Lévy–Mises flow rule is adopted, then

$$-\frac{dh}{h} = \left(\frac{\sigma_1 + \sigma_2}{2Y}\right) d\varepsilon = \frac{3\sigma}{4Y} d\varepsilon.$$

The last two equations reveal that the critical rate of hardening is equal to $3\sigma/4$ or $(\sigma_1 + \sigma_2)/2$. Using (2.12), the condition for localized necking to occur may therefore be expressed as

$$\frac{H}{Y} \leq \frac{\sigma_1 + \sigma_2}{2Y} = \frac{\sqrt{3}\sin\psi}{\sqrt{1+3\sin^2\psi}}. \tag{2.29}$$

It is also necessary to have $-1 \leq \sigma_1/\sigma_2 \leq 2$ for the equations to be hyperbolic. As ψ increases from 0 to $\pi/2$, the critical value of H/Y increases from 0 to $\sqrt{3}/2$. In the case of a uniaxial tension ($3 \sin \psi = 1$), localized necking can occur only if $H/Y \leq 0.5$. Thus, for a sheet of metal with a rounded stress–strain curve, a gradually increasing uniaxial tensile stress a produces a diffuse neck when $H = \sigma$, and eventually a localized neck when $H = \sigma/2$. A microstructural model for the shear band type of strain localization has been examined by Lee and Chan (1991).

2.2.2 *Tension of a Grooved Sheet*

Consider a uniform rectangular sheet of metal whose thickness is locally reduced by cutting a pair of opposed grooves in an oblique direction across the width, Fig. 2.5a. The grooves are deep enough to ensure that plastic deformation is localized there when the sheet is pulled longitudinally in tension. The width of the sheet is large compared to the groove width b, which is slightly greater than the local sheet thickness h so that a uniform state of plane stress exists in the grooves. The material in the grooves is prevented from extending along its length by the constraint of the adjacent nonplastic material. The principal strain rates in the grooves are therefore given by (2.28) in terms of the angle of inclination of the relative velocity with which the sides of the grooves move apart. If the material is isotropic, the directions of the principal stresses σ_1 and σ_2 are inclined at angles $\pi/4 + \psi/2$ and $\pi/4 - \psi/2$, respectively, to the direction of the grooves, where $\sigma_1 > \sigma_2$, the principal axes of stress and strain rate being coincident.

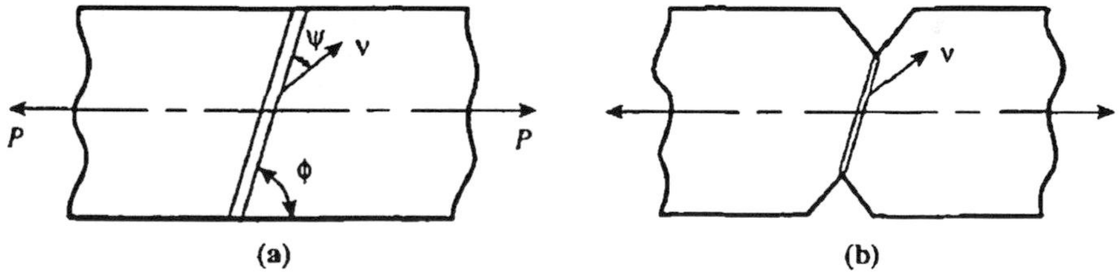

Fig. 2.5 Necking of grooved and notched metal strips under longitudinal tension

Let ϕ denote the counterclockwise angle made by the prepared groove with the direction of the applied tensile force P. Then the normal and shear stresses acting across the grove are $P \sin\phi/lh$ and $P \cos \phi/lh$, respectively, where l is the length of the groove. By the transformation relations for the stress, we have

$$(\sigma_1 + \sigma_2) + (\sigma_1 - \sigma_2)\sin\psi = (2P/lh)\sin\phi,$$
$$(\sigma_1 - \sigma_2)\cos\psi = (2P/lh)\cos\phi.$$

These equations can be solved for the principal stresses in terms of the angles ϕ and ψ, giving

$$\sigma_1 = \frac{P\,[\sin(\phi-\psi)+\cos\phi]}{hl\cos\psi}, \quad \sigma_2 = \frac{P\,[\sin(\phi-\psi)-\cos\phi]}{hl\cos\psi}. \tag{2.30}$$

From (2.28) and (2.30), Lode's well-known stress and strain parameters are obtained as

$$\left.\begin{aligned} \mu &= \frac{2\sigma_2-\sigma_1-\sigma_3}{\sigma_1-\sigma_3} = -\frac{3\cos\phi-\sin(\phi-\psi)}{\sin(\phi-\psi)+\cos\phi}, \\ \nu &= \frac{2\dot{\varepsilon}_2-\dot{\varepsilon}_1-\dot{\varepsilon}_3}{\dot{\varepsilon}_1-\dot{\varepsilon}_3} = -\frac{3\,(1-\sin\psi)}{(1+3\sin\psi)}. \end{aligned}\right\} \tag{2.31}$$

Equation (2.31) form the basis for establishing the (μ, ν) relationship for a material using the measured values of ϕ and ψ. The ends of the sheet must be supported in such a way that they are free to rotate in their plane to accommodate the relative movement necessary to permit the localized deformation to occur. The shape of the deviatoric yield locus may be derived from the fact that the length of the deviatoric stress vector is

$$s = \sqrt{\frac{2}{3}}\left(\sigma_1^2-\sigma_1\sigma_2+\sigma_2^2\right)^{1/2} = \frac{P}{hl}\,\frac{\sqrt{2\left[\sin^2(\phi-\psi)+3\cos^2\psi\right]}}{\sqrt{3}\cos\psi}, \tag{2.32}$$

and the angle made by the stress vector is $\theta = \tan^{-1}(-\mu/\sqrt{3})$ with the direction representing pure shear. If the yield locus and plastic potential have a sixfold symmetry required by the isotropy and the absence of the Bauschinger effect, it is only necessary to cover a 30° segment defined by the direction of pure shear ($\mu = \nu = 0$) and that of uniaxial tension ($\mu = \nu = -1$). This is accomplished by varying ϕ between 90° and $\tan^{-1}\sqrt{2} \approx 54.7°$, and measuring ψ for each selected value of ϕ. It may be noted that according to the Lévy–Mises flow rule ($\mu = \nu$), the relationship $\tan\phi = 4\tan\psi$ always holds.

The preceding analysis, due to Hill (1953), is equally applicable to the localized necking caused by the tension of a sheet provided with a pair of asymmetrical notches as shown in Fig. 2.5b. If the notches are deep and sharp, and the rate of work-hardening is sufficiently low, plastic deformation is localized in a narrow neck joining the notch roots. This method may be used for the determination of the yield criterion and the plastic potential for materials with sufficient degrees of pre-strain, provided it is reasonably isotropic. The method has been tried with careful experiments by Hundy and Green (1954), and by Lianis and Ford (1957), using specimens which can be effectively tested to ensure that they are actually isotropic. These investigations have confirmed the validity of the von Mises yield criterion and the associated plastic potential for several engineering materials, as indicated in Fig. 1.8.

2.2.3 Stress Discontinuities

We begin by considering the normal and shear stresses acting over a surface which coincides with a characteristic. When the state of stress is hyperbolic, there is only one stress circle that can be drawn through the given point on the Mohr envelope without violating the yield criterion. Since the stress states on both sides of the characteristic are represented by the same circle, all components of the stress are continuous. When the stress state is parabolic, and the yield criterion is that of Tresca, the principal stress acting along the tangent to the characteristic can have any value between 0 and ±2 k, permitting a discontinuity in the numerically lesser principal stress. When the considered surface is not a characteristic, a stress discontinuity is always possible with two distinct plastic states separated by a line of stress discontinuity.

Let σ_1, σ_2 be the principal stresses on one side of the discontinuity ($\sigma_1 \geq \sigma_2$), and σ'_1, σ'_2, those on the other side ($\sigma'_1 \geq \sigma'_2$) . The angles of inclination of σ_1 and σ'_2 with the line of discontinuity are denoted by θ and θ', respectively, reckoned positive as shown in Fig. 2.4b. Since the normal and shear stresses across the line of discontinuity must be continuous for equilibrium, we have

$$\left.\begin{aligned}(\sigma_1 + \sigma_2) - (\sigma_1 - \sigma_2)\cos 2\theta &= \left(\sigma_1' + \sigma_2'\right) + \left(\sigma_1' - \sigma_2'\right)\cos 2\theta',\\ (\sigma_1 - \sigma_2)\sin 2\theta &= \left(\sigma_1' - \sigma_2'\right)\sin 2\theta'.\end{aligned}\right\} \tag{2.33}$$

If the von Mises yield criterion is adopted, the stresses on each side of the discontinuity must satisfy (2.2). Considering the stress components along the normal and tangent to the discontinuity, specified by n and t, respectively, and using the continuity conditions $\sigma_n = \sigma'_n$ and $\tau_{nt} = r'_{nl}$, it is easily shown from (2.2) that $\sigma'_n - \sigma'_t = \sigma_t$ giving

$$2\cot 2\theta' = \cot 2\theta + \left(\frac{\sigma_1 + \sigma_2}{\sigma_1 - \sigma_2}\right)\mathrm{cosec}2\theta. \tag{2.34}$$

The elimination of $\sigma'_1 - \sigma'_2$ between the two equations of (2.33), and the substitution from (2.34), lead to the relation

$$2\left(\sigma_1' - \sigma_2'\right) = (\sigma_1 + \sigma_2) - 3(\sigma_1 - \sigma_2)\cos 2\theta. \tag{2.35}$$

Since $\sigma'_1 - \sigma'_2$ is then given by the yield criterion, the principal stresses and their directions are known on one side of the discontinuity when the corresponding quantities on the other side are given.

Considering the Tresca criterion for yielding, suppose that (σ_1, σ_2) represents a hyperbolic state, so that $\sigma_1 - \sigma_2 = 2k$ and $|\sigma_1 + \sigma_2| \leq 2k$. If the ($\sigma'_1, \sigma'_2$) state is also hyperbolic, then $\sigma'_1 - \sigma'_2 = 2\text{k}$ by the yield criterion, and the continuity conditions (2.33) furnish $\theta' = \theta$, and

$$\sigma_1' = \sigma_1 - 2k\cos 2\theta \geq 0, \quad \sigma_2' = \sigma_2 - 2k\cos 2\theta \leq 0.$$

If the (σ'_1, σ'_2) state is parabolic, the yield criterion is either $\sigma'_1 = 2k(\sigma'_2 \geq 0)$ or $\sigma'_2 = -2k(\sigma'_1 \leq 0)$. In the first case, (2.33) gives

$$\left.\begin{aligned} \tan\theta' &= \left(1 + \cos 2\theta - \frac{\sigma_2}{k}\right)\operatorname{co\,sec} 2\theta, \\ \frac{\sigma_2'}{2k} &= -\frac{\sigma_2}{k}\left(\frac{\sigma_2}{2k} - \cos 2\theta\right)\Big/\left(1 + \cos 2\theta - \frac{\sigma_2}{k}\right), \end{aligned}\right\} \frac{\sigma_2}{2k} \leq \cos 2\theta. \tag{2.36}$$

The corresponding results for the second case are obtained from (2.36) by replacing σ_2 and σ'_2 with $-\sigma_1$ and σ'_l, respectively, tan θ with cos θ, and reversing the sign of cos 2θ. Exceptionally, when the stress normal to the discontinuity is $\pm 2k$, the other principal stress on either side can have any value between 0 and $\pm 2k$. Such a discontinuity may be considered as the limit of a narrow zone of a continuous sequence of plastic states.

When yielding occurs according to the von Mises criterion, all the stress components must be continuous across a line of velocity discontinuity. This is evident for a necking type of discontinuity (since the neck must coincide with a characteristic), across which the stress is necessarily continuous. For a shearing discontinuity, the shear stress across the line of discontinuity is of magnitude k, and since the normal stress is continuous for equilibrium, the remaining stress must also be continuous in view of the yield criterion. As a consequence of this restriction, the velocity must be continuous across a line of stress discontinuity. The rate of extension along a line of stress discontinuity, which is the derivative of the tangential velocity, is evidently continuous. Since the flow rule predicts opposite signs for this component of the strain rate on the two sides of the stress discontinuity, the rate of extension must vanish along its length. The discontinuity may therefore be regarded as the limit of a narrow zone of elastic material through which the stress varies in a continuous manner.

2.2.4 *Diffuse and Localized Necking*

It is well known that the deformation of a bar subjected to a longitudinal tension ceases to be homogeneous when the rate of work-hardening of the material is less than the applied tensile stress. At the critical rate of hardening, the load attains its maximum, and the subsequent extension of the bar takes place under a steadily decreasing load. Plastic instabilities of this sort, leading to diffuse local necking, also occur when a flat sheet is subjected to biaxial tension in its plane. The strain-hardening characteristic of the material is defined by the equivalent stress $\overline{\sigma}$, and the equivalent total strain $\overline{\varepsilon}$, related to one another by the true stress–strain curve in uniaxial tension. Using the Lévy–Mises flow rule in the form

$$\frac{d\varepsilon_1}{2\sigma_1 - \sigma_2} = \frac{d\varepsilon_2}{2\sigma_2 - \sigma_1} = -\frac{d\varepsilon_3}{\sigma_1 + \sigma_2} = \frac{\overline{d\varepsilon}}{2\overline{\sigma}}, \tag{2.37}$$

and the differential form of the von Mises yield criterion (2.2) where $3k^2$ is replaced by $\overline{\sigma}^2$, it is easily shown that the stress and strain increments in any element must satisfy the relation

$$d\sigma_1 d\varepsilon_1 + d\sigma_2 d\varepsilon_2 = d\overline{\sigma} d\overline{\varepsilon}.$$

Consider a rectangular sheet whose current dimensions are b_1 and b_2 along the directions of σ_1 and σ_2, respectively. If the applied loads $hb_2\sigma_1$ and $hb_2\sigma_2$ attain stationary values at the onset of instability, where h denotes the current thickness, then

$$\frac{d\sigma_1}{\sigma_1} = \frac{db_1}{b_1} = d\varepsilon_1, \quad \frac{d\sigma_2}{\sigma_2} = \frac{db_2}{b_2} = d\varepsilon_2,$$

in view of the constancy of the volume hb_lb_2 of the sheet material. Combining the preceding two relations, we have

$$\frac{d\overline{\sigma}}{\overline{d\varepsilon}} = \sigma_1 \left(\frac{d\varepsilon_1}{\overline{d\varepsilon}}\right)^2 + \sigma_2 \left(\frac{d\varepsilon_2}{\overline{d\varepsilon}}\right)^2.$$

Substituting from (2.37), and using the expression for $\overline{\sigma}^2$, and setting $\sigma_2/\sigma_1 = p$, the condition for plastic instability is obtained as (Swift, 1952; Hillier, 1966)

$$\frac{H}{\overline{\sigma}} = \frac{(1+\rho)\left(4 - 7\rho + 4\rho^2\right)}{4\left(1 - \rho + \rho^2\right)^{3/2}}, \tag{2.38}$$

where $H = d\overline{\sigma}/d\overline{\varepsilon}$ denotes the critical rate of hardening. The quantity on the left-hand side of (2.38) is the reciprocal of the subtangent to the generalized stress–strain curve. The variation of the critical subtangent with stress ratio ρ is shown in Fig. 2.6 for both localized and diffuse necking. If the stress ratio is maintained constant throughout the loading, the total equivalent strain at instability is obtained directly from (2.38), if we adopt the simple power law $\overline{\sigma} = C\,\overline{\varepsilon}^n$, which gives $H/\overline{\sigma} = n/\overline{\varepsilon}$. In the case of variable stress ratio, the instability strain will evidently depend on the prescribed loading path.

In the biaxial tension of sheet metal, failure usually occurs by strain localization in a narrow neck following the onset of instability. The phenomenon can be explained by considering the development of a pointed vertex on the yield locus, which allows the necessary freedom of flow of the plastic material in the neck. When both the principal strains are positive, experiments seem to indicate that the neck coincides with the direction of the minimum principal strain ε_2 in the plane of the sheet. The incremental form of the Hencky stress–strain relations may be assumed to hold at the incipient neck, where the subsequent deformation remains

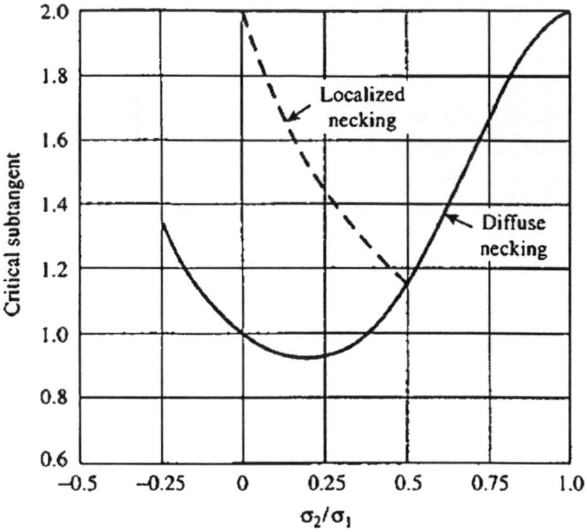

Fig. 2.6 Critical subtangent to the effective stress–strain curve as a function of the stress ratio

confined (Stören and Rice, 1975). The principal surface strains in the neck may therefore be written as

$$\varepsilon_1 = \overline{\varepsilon}\left(\frac{2\sigma_1 - \sigma_2}{2\overline{\sigma}}\right), \quad \varepsilon_2 = \overline{\varepsilon}\left(\frac{2\sigma_2 - \sigma_1}{2\overline{\sigma}}\right),$$

where $\overline{\sigma}$ and $\overline{\varepsilon}$ are the equivalent stress and total strain, respectively. The elimination of σ_2 and σ 1 in turn between these two relations gives the principal stresses

$$\sigma_1 = (2\varepsilon_1 + \varepsilon_2)\,\frac{2\overline{\sigma}}{3\overline{\varepsilon}}, \quad \sigma_2 = (2\varepsilon_2 + \varepsilon_1)\,\frac{2\overline{\sigma}}{3\overline{\varepsilon}}.$$

Assuming the power law $\overline{\sigma} = C\,\overline{\varepsilon}^n$ for the generalized stress–strain curve, the incremental form of the first equation above at the inception of the neck is found as

$$\frac{d\sigma_1}{\sigma_1} = \frac{2d\,\varepsilon_1 + d\,\varepsilon_2}{(2+\alpha)\,\varepsilon_1} - (1-n)\,\frac{d\,\overline{\varepsilon}}{\overline{\varepsilon}}, \tag{2.39}$$

where $\alpha \geq 0$ denotes the constant strain ratio $\varepsilon_2/\varepsilon_1$, prior to the onset of necking. The quantity de is the increment of the equivalent total strain $\overline{\varepsilon}$. Since $\overline{\varepsilon}^2 = \frac{4}{3}\left(\varepsilon_1^2 + \varepsilon_1\varepsilon_2 + \varepsilon_2^2\right)$ according to the Hencky theory, we get

$$\frac{d\,\overline{\varepsilon}}{\overline{\varepsilon}} = \frac{(2+\alpha)\,d\,\varepsilon_1 + (1+2\alpha)\,d\,\varepsilon_2}{2\left(1+\alpha+\alpha^2\right)\varepsilon_1}.$$

The neck is characterized by a discontinuity in $d\varepsilon_1$, but $d\varepsilon_2$ must be regarded as continuous across the neck. Since the material outside the neck undergoes neutral loading at its inception, we set $d\varepsilon_2 = 0$. Combining the last two equations, and using the fact that $d\sigma_1/\sigma_1 = d\varepsilon_1$ for the load across the neck to be stationary, the condition for localized necking is obtained as

$$\left\{\varepsilon_1 + \frac{(1-n)(2+\alpha)}{2\left(1+\alpha+\alpha^2\right)} - \frac{2}{2+\alpha}\right\}\frac{d\varepsilon_1}{\varepsilon_1} = 0.$$

Since $d\varepsilon_1 \neq 0$ for the development of the neck, the expression in the curly brackets must vanish, giving the limit strain over the range $0 < \alpha < 1$ in the form

$$\varepsilon_1 = \frac{3\alpha^2 + n(2+\alpha)^2}{2(2+\alpha)\left(1+\alpha+\alpha^2\right)} = \frac{(2-\rho)\left[(2\rho-1)^2+3n\right]}{6\left(1-\rho+\rho^2\right)}, \qquad (2.40)$$

where ρ is the stress ratio σ_2/σ_1, equal to $(1 + 2\alpha)/(2 + \alpha)$. The value of ε_1 given by (2.40) may be compared with that predicted by (2.38) for a material with a given value of n. The limit strain is seen to be higher than the instability strain except when $\alpha = 0$, for which both the conditions predict $\varepsilon_1 = n$.

For $\alpha < 0$, the neck forms along the line of zero extension, which is inclined at angle $\tan^{-1}\sqrt{-\alpha}$ to the direction of the minimum principal strain in the plane of the sheet. The strain ratio then remains constant during the incremental deformation, and the onset of necking is given by (2.29), with Y = a, the limit strain being easily shown to be

$$\varepsilon_1 = n\left(\frac{2-\rho}{1+\rho}\right) = \frac{n}{1+\alpha}, \quad -\tfrac{1}{2} \leq \alpha \leq 0.$$

The curve obtained by plotting ε_1 against ε_2 corresponding to localized necking in a given material is called the forming limit diagram, which represents the failure curve in sheet stretching. This will be discussed in Section 6.5 on the basis of a different physical model including anisotropy of the sheet metal.

2.3 Yielding of Notched Strips

2.3.1 V-Notched Strips in Tension

Consider the longitudinal extension of a rectangular strip having a pair of symmetrical V-notches of included angle 2α in the plane of the strip. The material is assumed to be uniformly hardened, obeying the von Mises yield criterion and the associated Lévy–Mises flow rule. When the load attains the yield point value, the region of incipient plastic flow extends over the characteristic field shown in Fig. 2.7. The triangular region *OAB* is under a uniaxial tension $\sqrt{3}k$ parallel to the notch face, and the characteristics are straight lines inclined at an angle $\beta = \tan^{-1}\sqrt{2}$ to the notch face. Within the fan *OBC*, one family of characteristics are straight lines passing through the notch root, and the state of stress expressed in polar coordinates (r, ϕ) is given by

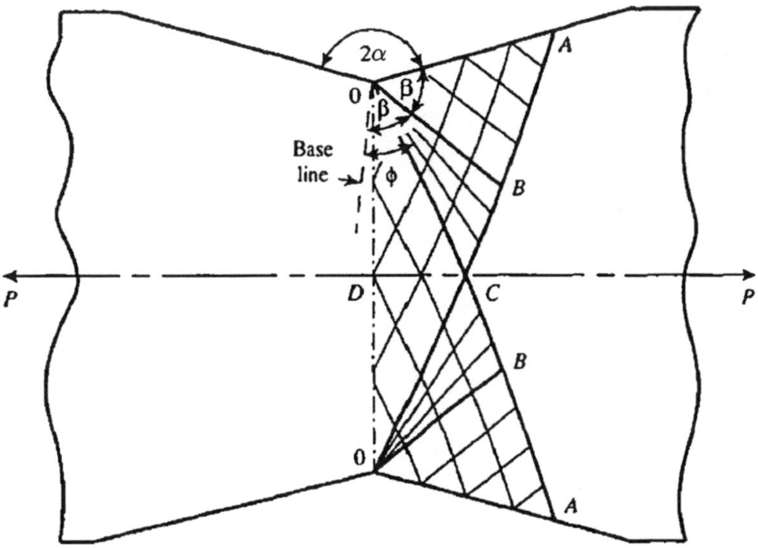

Fig. 2.7 Characteristic field in a sharply notched metal strip under longitudinal tension (α ≥ 70.53)

$$\sigma_r = k \cos \phi, \ \sigma_\phi = 2k \cos \phi, \ \tau_{r\phi} = k \sin \phi, \tag{2.41}$$

so that the equilibrium equations and the yield criterion are identically satisfied. It follows from (2.5) that $\cos \psi = 2\tan \phi$ within the fan. Along the line *OB*, $\psi = \sin^{-1} \frac{1}{3}$, giving $\phi = \beta = \tan^{-1} \sqrt{2}$. The baseline from which ϕ is measured therefore makes an angle 2β with the notch face. The curved characteristics in *OBC* are given by

$$r\,(d\phi/dr) = -\cos\psi = -2\tan\phi, \ \ or \ \ r^2 \sin\phi = \text{constant}.$$

The curved characteristics approach the baseline asymptotically, if continued, and are inflected where $\phi = \pi/2 - \beta$. The region *OCO* is uniformly stressed, and the principal stress axes coincide with the axes of symmetry. From geometry, angle *COD* is equal to $\phi_0 - (2\beta + \alpha - \pi)$, where ϕ_0 is the value of ϕ along *OC*, the corresponding value of ψ being denoted by ψ_0. Since the algebraically lesser principal stress direction in *OCO* is parallel to *00*, angle *COD* is also equal to $\pi/4 - \psi_0/2$. The relation $\cot \psi 0 = 2\tan \phi_0$ therefore gives

$$2\tan\phi_0 + \tan 2\,(\alpha + 2\beta - \phi_0) = 0, \tag{2.42}$$

which can be solved for ϕ_0 when α lies between $\pi - 2\beta$ and $\pi/2$, the limiting values of ϕ_0 being 0 and β, respectively. Since $\sigma_1 + \sigma_2 = 3k \cos \phi_0$ and $\sigma_1 - \sigma_2 = k\sqrt{1 + 3\sin^2\phi_0}$ within *OCO* in view of (2.41), the constraint factor is

$$f = \frac{\sigma_1}{\sqrt{3}k} = \frac{3\cos\phi_0 + \sqrt{1 + 3\sin^2\phi_0}}{2\sqrt{3}}, \quad \pi - 2\beta \leq \alpha \leq \pi/2. \tag{2.43}$$

The field of Fig. 2.7, which is due to Hill (1952), has been extended by Bishop (1953) in a statically admissible manner to show that the solution is in fact complete. As α decreases from $\pi/2$, the constraint factor increases from unity to reach its highest value of $2/\sqrt{3}$ when $\alpha = \pi - 2\beta \approx 70.53°$. The field in this case shrinks to a coincident pair of characteristics along the transverse axis of symmetry. In general, the constraint factor is closely approximated by the empirical formula

$$f = 1 + 0.155\sin\left\{4.62\left(\frac{\pi}{2} - \alpha\right)\right\}, \quad \pi - 2\beta \leq \alpha \leq \pi/2. \tag{2.44}$$

For all sharper notches, the characteristic field and the constraint factor are identical to those for $\alpha = \pi - 2\beta$. Indeed, by the maximum work principle, the constraint factor cannot decrease when material is added to reduce the notch angle, while the value of f certainly cannot exceed $2/\sqrt{3}$ since no stress component can exceed $2k$ in magnitude.

The yield point load can be associated with a deformation mode consisting of localized necking along both characteristics through the center D of the minimum section. If the ends of the strip are moved longitudinally with a unit speed relative to D, the particles on the transverse axis of symmetry must move inward with a speed equal to $\tan(\pi/4 - \psi/2)$. The vector representing the relative velocity of particles across the neck is then perpendicular to the other characteristic as required. Strictly speaking, there is no opportunity for the deformation to occur outside the localized necks.

2.3.2 Solution for Circular Notches

Consider, now, the longitudinal tension of a strip with symmetrical circular notches of radius c, the roots of the notch being at a distance 2α apart. For sufficiently small values of the ratio a/c, the characteristic field is radially symmetric as shown in Fig. 2.8, and the stress distribution is defined by (2.13) where σ_2 and σ_1 represent the radial and circumferential stresses denoted by σ_r and $\sigma\phi$, respectively. The substitution into the equation of radial equilibrium gives

$$\frac{d\sigma_r}{dr} = \frac{\sigma_\phi - \sigma_r}{r} = \frac{2k\cos\theta}{r}, \quad \text{or} \quad r\frac{d\theta}{dr} = \frac{2}{\sqrt{3} + \tan\theta},$$

where r is the radius of a generic point in the field. The boundary condition $\sigma_r = 0$ at the notch surface is equivalent to $\theta = \pi/6$ at $r = c$, and the integration of the above equation results in

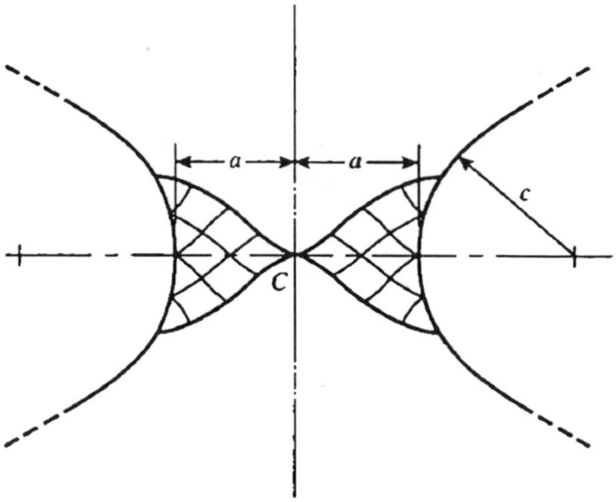

Fig. 2.8 Characteristic field in a circularly notched strip under longitudinal tension ($a/c \leq 1.071$)

$$\frac{r^2}{c^2} = \frac{\sqrt{3}}{2} \sec\theta \exp\left[\sqrt{3}\left(0 - \frac{\pi}{6}\right)\right], \quad \frac{\pi}{6} \leq \theta \leq \frac{\pi}{3}. \tag{2.45}$$

The characteristics coincide when $\theta = \pi/3$, and this corresponds to $r/c \approx 2.071$. The angular span of the circular root covered by the field in this limiting case is $4\beta - \pi \approx 38.96°$, which is obtained from (2.14) as twice the difference between the values of λ corresponding to $\theta = \pi/3$ and $\theta = \pi/6$. The characteristic field is easily constructed using (2.45) and (2.14), and the fact that the polar angle ϕ measured from the transverse axis is, by (2.16), equal to the decrease in the value of λ from that on the transverse axis. The resultant longitudinal force per unit thickness across the minimum section is

$$P = 2\int_c^{a+c} \sigma_\phi dr = 2\int_c^{a+c} \frac{d}{dr}(r\sigma_r)\, dr = 4k(a+c)\sin\left(\theta_0 - \frac{\pi}{6}\right),$$

where θ_0 is the value of θ at the center of the minimum section where $r = a + c$, and is directly given by (2.45). The constraint factor is

$$f = \frac{P}{2\sqrt{3}ka} = \frac{2}{\sqrt{3}}\left(1 + \frac{c}{a}\right)\sin\left(\theta_0 - \frac{\pi}{6}\right), \quad 0 \leq \frac{a}{c} \leq 1.071. \tag{2.46}$$

The value of f computed from (2.46) exceeds unity by the amount $0.226a/(a + c)$ to a close approximation. For higher values of a/c, the characteristics coincide along a central part of the transverse axis, and the longitudinal force per unit thickness is $4k(a - 1.071c)$ over the central part, and $2k(2.071c)$ over the remainder of the minimum section, giving the constraint factor

$$f = \frac{1}{\sqrt{3}}\left(2 - 0.071\frac{c}{a}\right), \quad \frac{a}{c} \geq 1.071.$$

As a/c increases, f approaches its asymptotic value of $2/\sqrt{3}$, which is the ratio of the maximum shear stresses in pure shear and simple tension. Localized necking would occur at the yield point along the characteristics through C if the rate of work-hardening of the material is not greater than the uniaxial yield stress of the material.

Consider, now, the solution for Tresca's yield criterion and its associated flow rule. Since no stress can exceed $2k$, which is now equal to the uniaxial yield stress, the constraint factor f cannot exceed unity. On the other hand, f is unity for a strip of width $2a$. Hence, the actual constraint factor is $f = 1$, whatever the shape of the notch. A localized neck forms directly across the minimum section when the yield point is attained, provided the rate of work-hardening is not greater than $2k$.

2.3.3 Solution for Shallow Notches

The preceding solutions hold only when the notches are sufficiently deep. In the case of shallow notches, the deformation originating at the notch roots spreads across to the longitudinal free edges. Considering a sharply notched bar with an included angle 2α, a good approximation to the critical notch depth may be obtained by extending the characteristic field further into the specimen. With reference to Fig. 2.9, which shows one-quarter of the construction, the solution involves the extension *ABCEFG* of the basic field *OABC*. The extended field is bounded by a stress-free boundary *AG* generated from a point on the notch face, the material lying beyond *AG* being assumed unstressed. The construction begins with the consideration of the curvilinear triangle *CBE*, which is defined by the β-line *CB* and the conditions of symmetry along *CE*. Since $\cot\phi = 2 \tan \psi$ along *CB*, the boundary conditions are

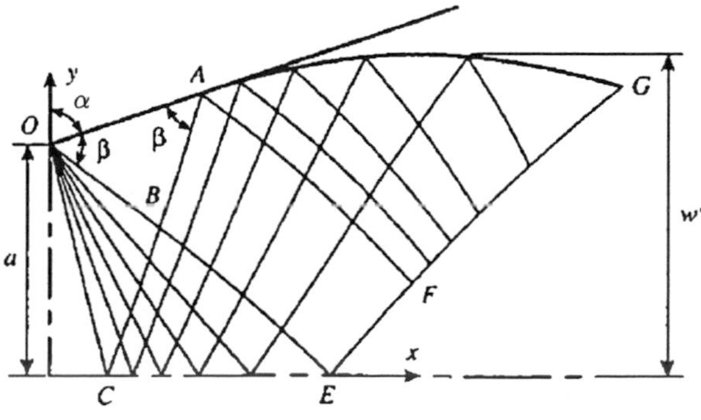

Fig. 2.9 Critical width of a V-notched strip subjected to longitudinal tension

$$\lambda = \frac{\pi}{2} - \phi - \tfrac{1}{2}\tan^{-1}\left(\tfrac{1}{2}\tan\phi\right), \quad \omega = -\gamma + \phi + \tfrac{1}{2}\tan^{-1}\left(\tfrac{1}{2}\tan\phi\right),$$

along *CB* in view of (2.14), where ω denotes the counterclockwise angle made by the algebraically greater principal stress with the longitudinal axis of symmetry, while $\gamma = \sigma + 2\beta - 3\,\pi/4$. Since $\omega = 0$ along *CE* by virtue of symmetry, the values of λ and ω are easily obtained throughout the field *CBE* using the characteristic relations (2.16). Starting with the known coordinates of the nodal points along CB, and using the fact that

$$\phi_\alpha = \omega - \left(\frac{\pi}{4} + \frac{\psi}{2}\right), \quad \phi_\beta = \omega + \left(\frac{\pi}{4} + \frac{\psi}{2}\right), \tag{2.47}$$

where ψ is obtained from (2.14), the coordinates of each point of the field can be determined numerically by the mean slope approximation for small arcs considered along the characteristics,

Since *AB* is a straight characteristic, all the β-lines in the field *ABEF* are also straight, though not of equal lengths. The angles ψ, ω, and λ. along *AF* are therefore identical to those along *BE*. The known values of ϕ_β and ϕ_α along *BE* and *AF* furnish the coordinates of the nodal points of *AF* by simple geometry and the tangent approximation. Since all characteristics meet the stress-free boundary *OAG* at a constant angle $\beta = 54.74°$, we have the boundary conditions $\psi = 19.47°$ and $\lambda = 25.53°$ along *AG*. Starting from point *A*, where $\omega = \pi/2 - \alpha$, and using (2.16), the values of λ and ω throughout the field *AFG* are easily determined. The angles ϕ_α and ϕ_β at the nodal points of the field are then computed from (2.47), and the rectangular coordinates are finally obtained by the mean slope approximation. The stress-free boundary *AG* generated as a part of the construction has a maximum height ω^*, which is the critical semiwidth of the strip. The numerical computation carried out by Ewing and Spurr (1974) suggests the empirical formula

$$\frac{a}{\omega^*} = 1 - 0.286\sin\left\{4.62\left(\frac{\pi}{2} - \alpha\right)\right\}, \tag{2.48}$$

which is correct to within 0.5% over the range $70.5° \leq \alpha \leq 90°$. This formula actually provides an upper bound on the critical semiwidth, since all specimens wide enough to contain the extended field are definitely not overstressed, as may be shown by arguments similar to those used for the corresponding plane strain problem (Chakrabarty, 2006).

When the semiwidth w of the notched bar is less than w^*, we can find an angle $\alpha^* > \alpha$ such that $w^*(\alpha^*) = w$. Then, the corresponding constraint factor $f(\alpha^*)$ calculated from (2.46) would provide a lower bound on the yield point load. Indeed, the yield point load cannot be lowered by the addition of material required to reduce the notch angle from $2\alpha^*$ to the actual value 2α. The constraint factor for subcritical widths is closely approximated by the lower bound value, which may be expressed empirically as

$$f = 1 + 0.54\left(1 - \frac{a}{w}\right), \quad a \leq w \leq w^*.$$

In the case of an unnotched bar, the above formula reduces to $f = 1$. The deformation mode then consists of a localized neck inclined at an angle $\beta = 54.74°$ to the tension axis. Such a neck is also produced in a tensile strip with either a single notch or a symmetric central hole (Fig. 2.10). The tension of single-notched strips has been investigated by Ewing and Richards (1973), who also produced some experimental evidence in support of their theoretical prediction.

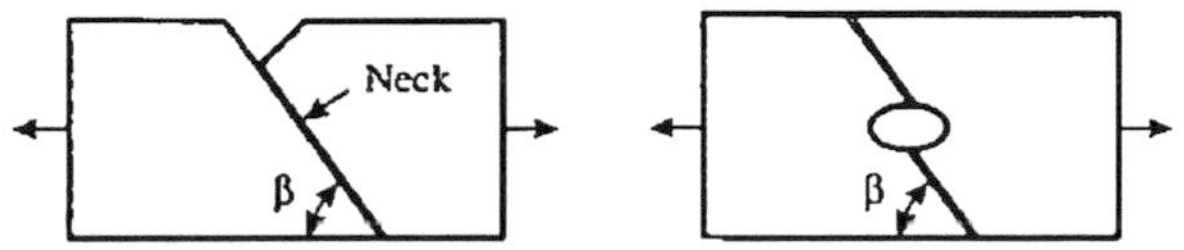

Fig. 2.10 Initiation of a localized neck in a flat sheet inclined at an angle β to the direction of tension

2.4 Bending of Prismatic Beams

2.4.1 Strongly Supported Cantilever

A uniform cantilever of narrow rectangular cross section carries a load kF per unit thickness at the free end, just sufficient to cause plastic collapse, the weight of the cantilever itself being negligible. The material is assumed to obey the von Mises yield criterion and the Lévy–Mises flow rule. Consider first the situation where the cantilever is rigidly held at the built-in end. If the ratio of the length l of the beam to its depth d is not too large, the characteristic field in the yield point state will be that shown in Fig. 2.11. The deformation mode at the incipient collapse consists of rotation of the rigid material about a center C on the longitudinal axis of symmetry. The solution involves localized necking along *EN* and localized bulging along *NF*, together with a simple shear occurring at *N*. The vector representing the relative velocity of the material is inclined to *EF* at an angle ψ which varies along the discontinuity. Although a local bulging can only occur in a strain-softening material, the solution may be accepted as a satisfactory upper bound on the collapse load (Green, 1954a) for ideally plastic materials.

In the triangular regions *ABD* and *GHK*, the state of stress is a uniform longitudinal tension and compression, respectively, of magnitude $\sqrt{3}\,k$, the characteristics being straight lines inclined at an angle $\beta = \tan^{-1}\sqrt{2}$ to the free edge. The region *ADE* is an extension of the constant stress field round the singularity *A*, the corresponding stresses being given by (2.41), where the polar angle ϕ is measured from a baseline that is inclined at an angle 2β to the free edge *AB*. The curved characteristics in *ADE* have the equation $r^2 \sin\phi = \text{constant}$, while the relation $\cot\psi = 2\tan\phi$ holds for the characteristic angle ψ. The normal stresses vanish at *N*, and the stress components along the curve *ENF* in plane polar coordinates are

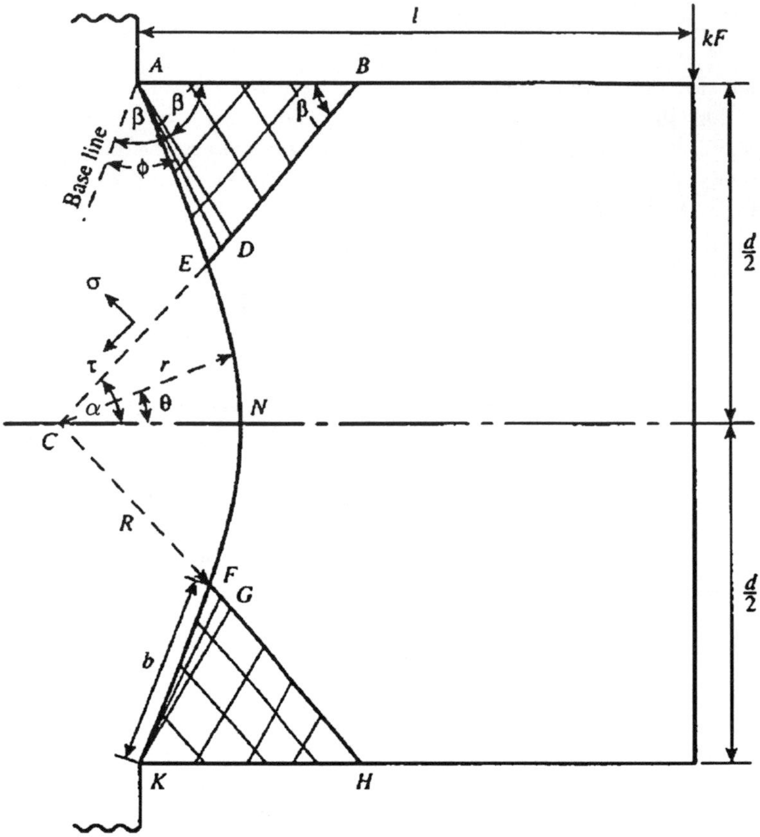

Fig. 2.11 Characteristic fields for an end-loaded cantilever with strong support ($l/d \leq 5.65$)

$$\sigma_r = k\sin\theta,\ \sigma_\theta = 2k\sin\theta,\ \tau_{r\theta} = -k\cos\theta, \tag{2.49}$$

so that the equilibrium equations and the yield criterion are identically satisfied, the polar angled being measured counterclockwise with respect to the longitudinal axis. The relation cot $\psi = 2$ cot θ immediately follows from (2.5) and (2.49). The polar equation to the curve *ENF*, referred to *C* taken as the origin, is given by

$$r\,(d\theta/dr) = \cot\psi = 2\cot\theta, \quad \text{or} \quad r^2\cos\theta = \text{constant}$$

Since the characteristic directions are everywhere continuous, the value of ψ at *E* may be written as

$$\psi_0 = \tan^{-1}\left(\frac{1}{2}\cot\phi_0\right) = \tan^{-1}\left(\frac{1}{2}\tan\alpha\right),$$

where ϕ_0 and α are the values of ϕ and θ, respectively, at E. It follows that $\phi_0 = \pi/2 - \alpha$. From geometry, angle AEC is $2\beta - \phi_0 + \alpha$, which must be equal to $\pi/2 + \psi_0$, and substituting for ϕ_0 and ψ_0 in terms of α, we obtain

$$2\alpha - \tan^{-1}\left(\frac{1}{2}\tan\alpha\right) = \pi - 2\beta. \tag{2.50}$$

The solution to this transcendental equation is $\alpha \approx 51.20°$, which gives $\psi_0 \approx 31.88°$. The fan angle EAD is $\delta = \alpha + \beta - \pi/2 = \psi_0/2$. It is interesting to note that the state of stress in the deforming region in plane stress bending varies from pure tension at the upper edge to pure shear at the center, whereas in plane strain bending the stress state is pure shear throughout the region of deformation.

The geometry of the field is completely defined by the dimensions b and R, representing the lengths AE and CE, respectively. The ratios b/d and R/d depend on the given ratio l/d, and are determined in terms of F/d from the conditions:

(a) the sum of the vertical projections of AE and CE is equal to $d/2$; and
(b) the resultant vertical force transmitted across $AENFK$ per unit thickness is equal to kF.

The resultant force acting across ENF is most conveniently obtained by regarding $CENF$ as a fully plastic region, the normal and shear stresses across CE and CF being of magnitudes σ and τ directed as shown. The pair of conditions (a) and (b) furnishes the relations

$$\left.\begin{aligned} &R\sin\alpha + b\cos\lambda = d/2, \\ &R(\sigma\cos\alpha - \tau\sin\alpha) + b(\tau\sin\lambda - \sigma\cos\lambda) = kF/2, \end{aligned}\right\} \tag{2.51}$$

where $\sigma = 2k\sin\alpha$ and $\tau = k\cos\alpha$ in view of (2.49), while $\lambda = \alpha + 2\beta - \pi/2 \approx 70.68°$. Substituting for σ and τ, and inserting the values of σ and λ, equations of (2.51) are easily solved for b/d and R/d as

$$\frac{b}{d} = 0.6075 - 0.9696\frac{F}{d}, \quad \frac{R}{d} = -0.0941 + 1.1741\frac{F}{d}. \tag{2.52}$$

The ratio F/d at the yield point is finally determined from the condition that the resultant moment of the forces acting on $AENFK$ about the center of rotation C is equal to the moment of the applied force about the same point. Thus

$$KF(l + R\cos\alpha - b\cos\lambda) = \sigma\left(R^2 + b^2 + 2Rb\sin\psi_0\right) + 2\tau Rb\cos\psi_0.$$

Substituting for σ and τ, using the values of σ, λ, and ψ_0 inserting the expressions for b/d and R/d from (2.52), the above equation may be rearranged into the quadratic

$$0.4342 - \left(\frac{1}{d} - 0.2600\right)\frac{F}{d} - 0.5288\frac{F^2}{d^2} = 0, \tag{2.53}$$

when F/d has been calculated from (2.53) for a given value of l/d, the ratios b/d and R/d follow from (2.52). Since $R = 0$ when $F/d \approx 0.080$, the proposed field applies only for $l/d \leq 5.65$. For higher l/d ratios, the characteristic field is modified in the same way as that in the corresponding plane strain problem, the collapse load being closely approximated by the empirical formula

$$\frac{d}{F} = 0.20 + 2.18\frac{1}{d}, \quad \frac{1}{d} \geq 5.65.$$

The values of F/d, b/d, and R/d corresponding to a set of values of $l/d < 5.65$ are given in Table 2.2. The plane stress values of F/d are found to be about 14% lower than the corresponding plane strain values (Chakrabarty, 2006) over the

Table 2.2 Results for an end-loaded cantilever

Strong support				Work support			
l/d	F/d	b/d	R/d	l/d	F/d	b/d	δ
1.33	0.346	0.272	0.313	0.328	0.492	0.536	54.74
1.62	0.287	0.329	0.243	0.275	0.482	0.531	49.06
2.00	0.233	0.382	0.180	0.255	0.477	0.527	43.26
2.55	0.182	0.431	0.120	0.177	0.477	0.523	37.53
3.36	0.137	0.475	0.067	0.134	0.482	0.518	31.82
4.72	0.086	0.514	0.019	0.095	0.492	0.515	26.07
5.65	0.80	0.530	0	0.079	0.498	0.512	23.49

whole range of values of l/d. The yield point load for a tapered cantilever has been discussed by Ranshi et al. (1974), while the influence of an axial force has been examined by Johnson et al. (1974).

Let M denote the bending moment at the built-in end under the collapse load ktF, where t is the thickness of the cantilever. Since the fully plastic moment under pure bending is $M_0 = \sqrt{3}\,ktd^2$, the ratio M/M_0 is equal to $4Fl/\sqrt{3}\,d^2$, and (2.53) may be written in the form

$$\frac{M}{M_0} \approx 1 + 1.23\frac{F}{d}\left(0.49 - \frac{F}{d}\right) \tag{2.54}$$

which is correct to within 0.3% for $F/d \leq 0.62$. The elementary theory of bending assumes $M/M_0 \approx 1$ irrespective of the shearing force. The results for the end-loaded cantilever are directly applicable to a uniformly loaded cantilever if we neglect the effect of surface pressure on the region of deformation. Since the resultant vertical load now acts halfway along the beam, the collapse load for a uniformly loaded cantilever of length $2l$ is identical to that for an end-loaded cantilever of length l.

2.4.2 Weakly Supported Cantilever

Consider a uniform cantilever of depth d, which fits into a horizontal-slot in a rigid vertical support. The top edge of the beam is clear of the support, so that the adjacent plastic region is able to spread into the slot under the action of a load kF per unit thickness at the free end. The length of the cantilever is l, measured from the point where the bottom edge is strongly held. The characteristic field, due to Green (1954b), is shown in Fig. 2.12. It consists of a pair of triangles ABN and CEF, under uniaxial tension and compression, respectively, and a singular field CEN where one family of characteristics is straight lines passing through C. The characteristics in the uniformly stressed triangles are inclined at an angle $\beta - \tan^{-1}\sqrt{2}$ to the respective free edges. The curved characteristics in CEN are given by the polar equation $r^2 \sin\phi = \text{constant}$, where ϕ is measured from a datum making an angle 2β to the bottom edge CF. The stresses in this region are given by (2.41) with an overall reversal of sign.

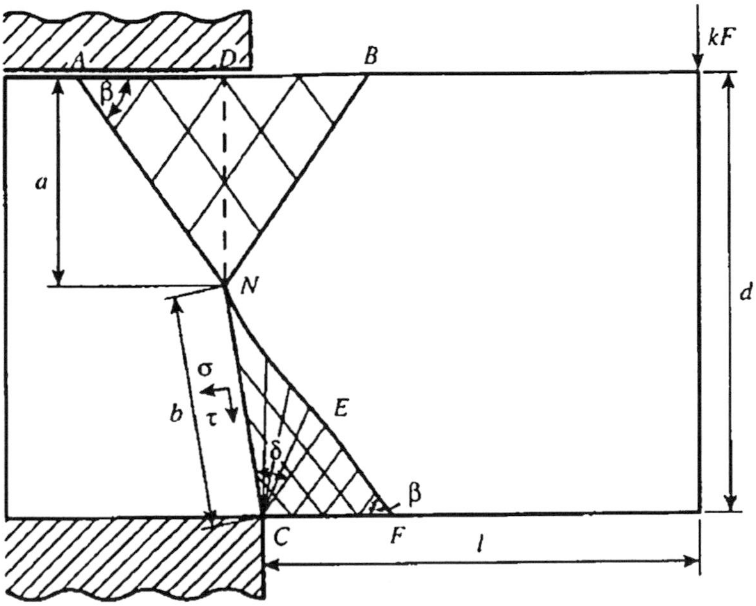

Fig. 2.12 Characteristic field for an end-loaded cantilever with weak support ($l/d \geq 1.33$)

For a given depth of the cantilever, the field is defined by the angle δ at the stress singularity, and the length b of the characteristic CN. The height of the triangle ABN is

$$a = d - b\sin(\beta + \delta)\,. \tag{2.55}$$

The stress is discontinuous across the neutral point N, about which the cantilever rotates as a rigid body at the incipient collapse. The normal and shear stresses over CN are

$$\sigma = -2k\cos(\beta - \delta), \quad \tau = \sin(\beta - \delta).$$

Since the tensile stress across the vertical plane *DN* is $\sqrt{3}\,k$ parallel to *AB*, the condition of zero resultant horizontal force across *CND* and the fact that the resultant vertical force per unit thickness across this boundary is equal to *kF* furnish the relations

$$b\,[\tau\,\cos(\beta + \delta) - \sigma\sin(\beta + \delta)] = \sqrt{3}\,ka,$$
$$b\,[\tau\,\sin(\beta + \delta) + \sigma\cos(\beta + \delta)] = kF.$$

Substituting for σ and τ, and using the value of β, these relations can be simplified to

$$\frac{F}{d} = \sin^2\delta, \quad \sqrt{3}\frac{a}{b} = \frac{2\sqrt{2}}{3} + \left(\frac{1}{3} - \frac{F}{b}\right)\tan(\beta + \delta) \tag{2.56}$$

When *F/b* and *a/b* have been calculated from (2.56) for a selected value of δ, the ratio *d/b* follows from (2.55), while *l/b* is obtained from the condition of overall moment equilibrium. Taking moment about *N* of the applied shearing force *kF* and also of the tractions acting over *CND*, we get

$$\frac{l}{b} = \cos(\beta + \delta) + \frac{b}{F}\left[\cos(\beta + \delta) + \frac{\sqrt{3}a^2}{2b^2}\right]. \tag{2.57}$$

Numerical values of *l/d, F/d, b/d*, and *a/d* for various values of δ are given in Table 2.2. As *l/d* decreases, δ increases to approach the limiting value equal to β. The angle between the two characteristics *CN* and *EN* decreases with decreasing *l/d*, becoming zero in the limit when the triangle *CEF* shrinks to nothing. The ratios *F/d* and *l/d* attain the values 0.328 and 1.332, respectively, in the limiting state.

The ratio M/M_0 at the built-in section (through *C*), which is equal to $4Fl/\sqrt{3}\mathrm{d}^2$, can be calculated from the tabulated values of *l/d* and *F/d*. The results can be expressed by the empirical formula

$$\frac{M}{M_0} \approx 1 + 1.45\frac{F}{d}\left(0.34 - \frac{F}{d}\right), \tag{2.58}$$

which is correct to within 0.5% for $F/d < 0.33$. Equations (2.54) and (2.58) are represented by solid curves in Fig. 2.13, the corresponding relations for plane strain bending being shown by broken curves. Evidently, M exceeds M_0 over the whole practical range, indicating that the constraining effect of the built-in condition outweighs the weakening effect of the shear except for very short cantilevers. In the case of a strong support, the maximum value of M/M_0 is about 1.121 in plane strain and 1.074 in plane stress, corresponding to *F/d* equal to about 0.28 and 0.25, respectively.

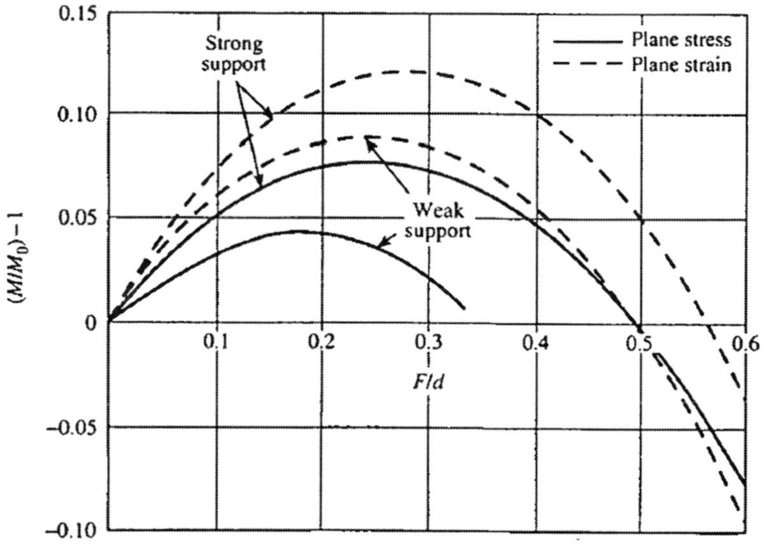

Fig. 2.13 Influence of transverse shear on the yield moment in the plane stress and plane strain bending of beams

2.4.3 Bending of I-Section Beams

Consider an I-beam, shown in Fig. 2.14a, whose transverse section has an area A_ω for the web and A_f for each flange including the fillets. The depth of the web is denoted by d, and the distance between the centroids of the flanges is denoted by h. It is assumed that the flanges yield in simple tension or compression, while the

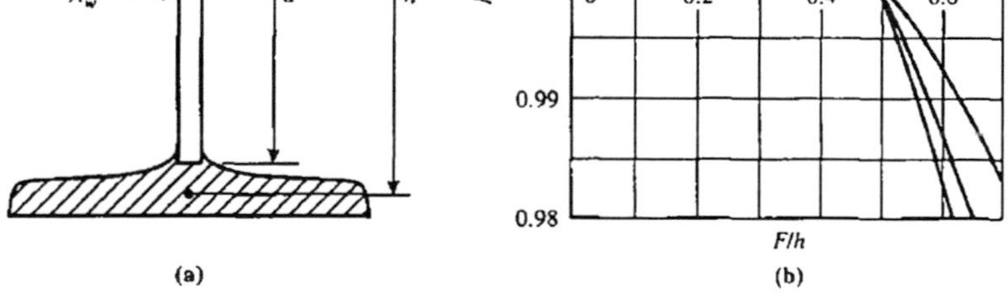

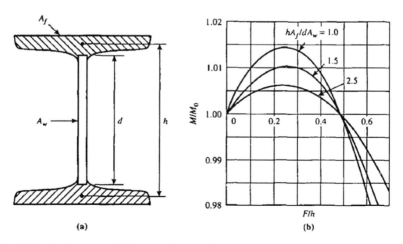

Fig. 2.14 Yield point states for I-section beams subjected to combined bending and shear

web yields under combined bending and shear. The bending moment carried by the flanges is $M_f = \sqrt{3}\, khA_f$, since one flange is in tension and the other in compression with a stress of magnitude $\sqrt{3}\, k$. For a uniform cantilever of length l, carrying a load ktF at the free end, where t is the web thickness, the bending moment existing at the fixed end is $M = ktFl$. Hence the bending moment shared by the web is

$$M_\omega = M - M_f = k\left(tFl - \sqrt{3}hA_f\right)$$

giving

$$\frac{M_\omega}{M_0'} = \frac{4Fl}{\sqrt{3}d^2} - \frac{4\,h}{d}\left(\frac{A_f}{A_w}\right),$$

where M'_0 is the fully plastic moment of the web under pure bending, equal to $\sqrt{3}\, kd^2t/4 = \sqrt{3}\, kA\omega d/4$. When the cantilever is strongly supported at the built-in end, $M\omega/M'_0$ is given by the right-hand side of (2.54), and the collapse load is given by

$$1.23\frac{F^2}{d^2} + \left(2.31\frac{l}{d} - 0.60\right)\frac{F}{d} - \frac{4h}{d}\left(\frac{A_f}{A_w}\right) - 1 = 0. \qquad (2.59a)$$

For a weak end support, $M\omega/M'_0$ is given by the right-hand side of (2.58), and the equation for the collapse load becomes

$$1.45\frac{F^2}{d^2} + \left(2.31\frac{l}{d} - 0.49\right)\frac{F}{d} - \frac{4h}{d}\left(\frac{A_f}{A_w}\right) - 1 = 0. \qquad (2.59b)$$

Equation (2.59), due to Green (1954b), is certainly valid over the practical range of values of l/d for standard I-beams. Since the flanges do not carry any shearing load, $M_f = M_0 - M'_0$, where M_0 is the fully plastic moment of the I-beam under pure bending. The relation $M_f = M - M_w$ therefore gives

$$\frac{M}{M_0} = 1 + \left(\frac{M_w}{M_0'} - 1\right)\frac{M_0'}{M_0}, \quad \frac{M_0}{M_0'} = 1 + \frac{4h}{d}\left(\frac{A_f}{A_w}\right).$$

The relationship between the bending moment and the shearing force at the yield point state of an I-beam is now obtained on substitution from (2.54) and (2.58), where M/M_0 is replaced by M_w/M'_0. Considering the strong support, for instance, we have

$$\frac{M}{M_0} = 1 + \left\{1.23\frac{F}{d}\left(0.49 - \frac{F}{d}\right)\right\}\bigg/\left(1 + \frac{4h}{d}\frac{A_f}{A_\omega}\right). \qquad (2.60)$$

The variation of M/M_0 with F/d is displayed in Fig. 2.14b. A satisfactory lower bound solution for M/M_0 at the yield point of an I-beam has been derived by Neal

(1961). The influence of axial force on the interaction relation, based on the characteristic field in plane stress, has been investigated by Ranshi et al. (1976).

2.5 Limit Analysis of a Hollow Plate

2.5.1 Equal Biaxial Tension

A uniform square plate, whose sides are of length 2α, has a circular hole of radius c at its center. The plate is subjected to a uniform normal stress σ along the edges in the plane of the plate. As the loading is continued into the plastic range, plastic zones spread symmetrically outward from the edge of the hole and eventually meet the outer edges of the plate when the yield point is reached. We begin by considering the von Mises yield criterion and its associated Lévy–Mises flow rule. For a certain range of values of c/a, the characteristic field would be that shown in Fig. 2.15a. The stress distribution within the field is radially symmetrical with the radial and circumferential stresses given by (2.13), where $\sigma_1 = \sigma_\phi$ and $\sigma_2 = \sigma_r$, the spatial distribution of the angle θ being given by (2.45). Assuming $\theta = \alpha$ at the external boundary $r = a$, where $a_r = a$, we obtain

$$\frac{\sigma}{Y} = \frac{2}{\sqrt{3}} \sin\left(\alpha - \frac{\pi}{6}\right), \quad \frac{c^2}{a^2} = \frac{2}{\sqrt{3}} \cos\alpha \exp\left\{-\sqrt{3}\left(\alpha - \frac{\pi}{6}\right)\right\}'. \tag{2.61}$$

The deformation mode at the incipient collapse consists of localized necking along the characteristics through A, permitting the rigid corners to move diagonally

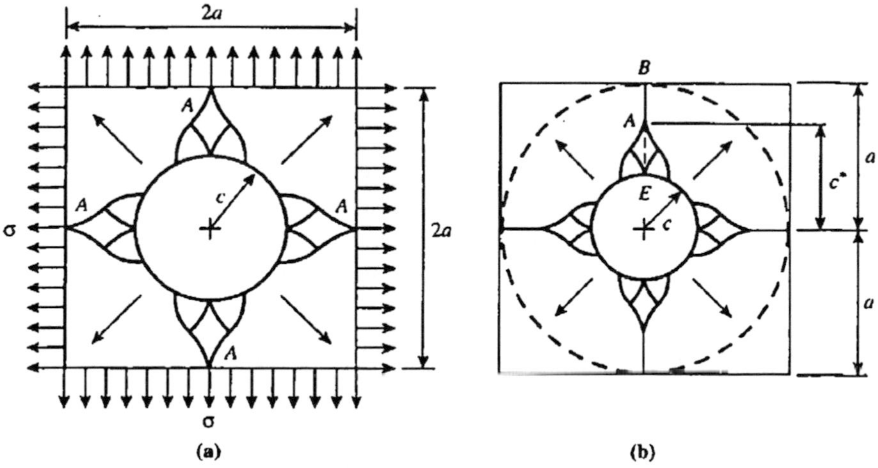

Fig. 2.15 Equal biaxial tension of a square plate with a central circular hole. (**a**) $0.483 \leq c/a \leq$ and (**b**)$0.143 \leq c/a \leq 0.483$

outward. Since the characteristics of stress and velocity exist only over the range $\pi/6 \leq \alpha \leq \pi/3$, the solution is strictly valid for $0.483 \leq c/a \leq 1$. When $\alpha = \pi/3$, the two characteristics coincide at A, and σ attains the value $0.577\,Y$.

For $c/a \leq 0.483$, (2.61) provides a lower bound on the yield point load, since the associated stress distribution is statically admissible in the annular region between the hole and the broken circle of radius a, shown in Fig. 2.15b. The remainder of the plate is assumed to be stressed below the yield limit under balanced biaxial stresses of magnitude σ, a discontinuity in the circumferential stress being allowed across the broken circle. On the other hand, an upper bound solution is obtained by extending the localized necks as straight lines from $r = c^* = 2.071c$ to $r = a$, permitting the same mode of collapse as in (a). Since $\sigma\phi = 2\sigma_r = 2k$ along the straight part of the neck, the longitudinal force per unit thickness across BAE is

$$a\sigma = \int_c^a \sigma_\phi dr = \int_c^{c^*} \frac{d}{dr}(r\sigma_r)dr + 2k\left(a - c^*\right) = \frac{2Y}{\sqrt{3}}\left(a - \frac{c^*}{2}\right),$$

where the second step follows from the equation of stress equilibrium. The upper bound therefore becomes

$$\sigma = \frac{2Y}{\sqrt{3}}\left(1 - 1.035\frac{c}{a}\right), \quad 0.143 \leq \frac{c}{a} \leq 0.483. \tag{2.62a}$$

For $c/a \leq 0.143$, a better upper bound is provided by the assumption of a homogeneous deformation mode in which the rate of plastic work per unit volume is $2U/a$, where U is the normal velocity of each side of the square. Since the rate of external work per unit plate thickness is $8a\sigma U$, we obtain the upper bound

$$\sigma = Y\left(1 - \frac{\pi c^2}{4a^2}\right), \quad 0 \leq \frac{c}{a} \leq 0.143. \tag{2.62b}$$

The difference between the lower and upper bounds, given by (2.61) and (2.62), respectively, is found to be less than 3% over the whole range of values of c/a.

When the material yields according to the Tresca criterion, a lower bound solution is obtained from the stress distribution $\sigma_r = Y(1 - c/r)$, $\sigma_\phi = Y$ in the annulus $c \leq r \leq a$, and $\sigma_r = \sigma_\phi = \sigma$ in the region r = a. The continuity of the radial stress across r = a gives the lower bound a = $Y(1 - c/a)$. To obtain an upper bound, we assume localized necking along the axes of symmetry normal to the sides of the square, involving a diagonally outward motion of the four rigid corners with a relative velocity v perpendicular to the necks. For a unit plate thickness, the rate of internal work in the necks is $4(a - c)vY$, while the rate of external work is $4a\sigma v$, giving the upper bound $\sigma = Y(1 - c/a)$. Since the upper and lower bounds coincide, it is in fact the exact solution for the yield point stress.

2.5.2 Uniaxial Tension: Lower Bounds

Suppose that the plate is brought to the yield point by the application of a uniform normal stress σ over a pair of opposite sides of the square. The effect of the circular cutout is to weaken the plate so that the yield point value of σ is lower than the uniaxial yield stress Y. A lower bound solution for the yield point stress may be obtained from the stress discontinuity pattern of Fig. 2.16a, consisting of four uniformly stressed regions separated by straight lines, across which the tangential stress is discontinuous. The material between the circular hole of radius ρa the inner square of side ρa is assumed stress free. The conditions of continuity of the normal and shear stresses across each discontinuity are given by (2.33). If the principal stresses are denoted by the symbols s and t where $s \leq t$, the stress boundary conditions require

$$t_1 = \sigma, \ \ s_3 = 0, \ \ t_4 = 0,$$

where the subscripts correspond to the numbers used for identifying the regions of uniform stress. Let α denote the counterclockwise angle which the direction of the algebraically lesser principal stress in region 2 makes with the vertical. Then the clockwise angles made by this principal axis with the discontinuities bordering regions 1, 3, and 4 are

$$\theta_1' = \alpha - \theta_1, \ \theta_3' = \frac{\pi}{2} + \alpha - \theta_3, \ \theta_4' = -\left(\frac{\pi}{4} - \alpha\right).$$

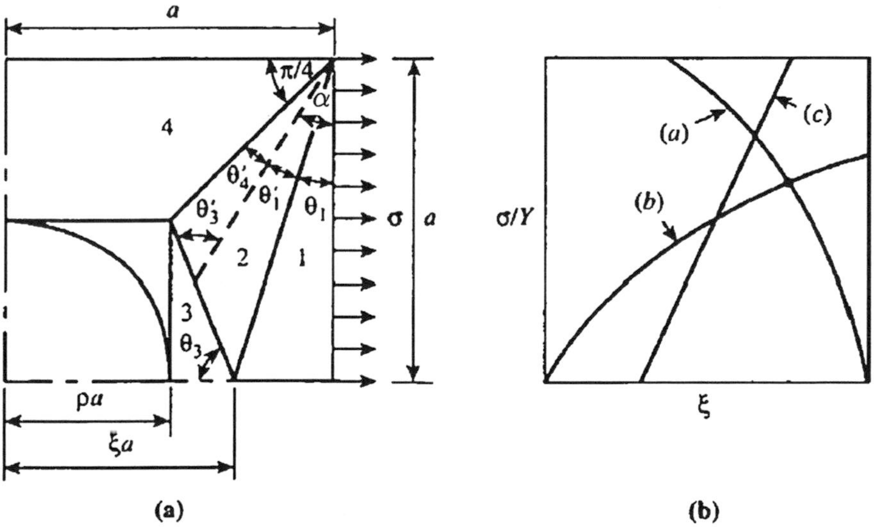

Fig. 2.16 Uniaxial tension of a square plate with a circular hole. (**a**) Stress discontinuities and (**b**) graphical representation of yields equalities

From geometry, the counterclockwise angles made by the algebraically greater principal stress axis with the line of discontinuity in regions 1, 3, and 4 are given by

$$\tan\theta_1 = 1 - \xi, \quad \tan\theta_3 = \rho/(\xi - \rho), \quad \theta_4 = -\frac{\pi}{4}.$$

Considering the discontinuity between regions 4 and 2, and using (2.33), the quantities $(s_2 + t_2)/s_4$ and $(s_2 - t_2)/s_4$ can be expressed as functions of α. The consideration of the discontinuity between regions 3 and 2 then furnishes α and the ratio t_3/s_4. Finally, the continuity conditions across the boundary between regions 1 and 2 furnish s_4 and s_1 in terms of the applied stress σ. The results may be summarized in the form

$$\left.\begin{aligned} s_1 &= \frac{\rho\sigma}{(1-\rho)(1-\xi)}, \quad \tan 2\alpha = \frac{2\rho}{\xi}, \\ s_2 + t_2 &= \frac{\sigma(\xi - 2\rho)}{\xi(1-\rho)}, \quad s_2 - t_2 = \frac{\sigma\sqrt{\xi^2 + 4\rho^2}}{\xi(1-\rho)}, \\ -t_3 &= \frac{\rho\sigma}{(1-\rho)(\xi-\rho)}, \quad s_4 = \frac{\sigma}{1-\rho}. \end{aligned}\right\} \tag{2.63}$$

Since regions 3 and 4 are in uniaxial states of stress, the magnitudes of t_3 and s_4 must not exceed the yield stress Y. For the von Mises criterion, the required inequalities in regions 1 and 2 follow from (2.2) and (2.63). When ρ is sufficiently small, the greatest admissible value of a is that for which region 4 is at the yield limit, giving the lower bound

$$\sigma = Y(1-\rho), \quad 0 \le \rho \le 0.204.$$

For higher values of ρ, region 4 is not critical, and we need to examine the following yield inequalities for the estimation of the lower bound:

$$\frac{\sigma}{Y} \le \frac{(1-\rho)(1-\xi)}{\left[(1-\rho)^2(1-\xi)^2 - \rho(1-\rho)(1-\xi) + \rho^2\right]^{1/2}}, \tag{2.64a}$$

$$\frac{\sigma}{Y} \le \frac{\xi(1-\rho)}{\left(\xi^2 - \rho\xi + 4\rho^2\right)^{1/2}}, \quad \frac{\sigma}{Y} \le \frac{(1-\rho)(\xi-\rho)}{\rho}, \tag{2.64b,c}$$

The parameter ξ must be chosen in the interval $0 \le \xi \le 1$ such that the inequalities (2.64) admit the greatest value of σ. If the right-hand sides of these inequalities are plotted as functions of ξ for a given ρ, the greatest admissible value of σ/Y is the largest ordinate of the region below all the curves, as indicated in Fig. 2.16b. Thus, a/Y is given by the point of intersection of the curves (a) and (b), if this point is below the line (c), and by the point of intersection of (a) and (c), if curve (b) passes above this point. It turns out that the former arises for $0.204 \le \rho \le 0.412$, and the latter for $0.412 \le \mathrm{p} \le 1$.

When the Tresca criterion is adopted, the inequalities corresponding to regions 1 and 2 only are modified, the stress distribution being statically admissible if

$$s_1 \leq Y, \; s_2 - t_2 \leq Y, \quad -t_3 \leq Y, \; s_4 \leq Y.$$

Since $s_2 - t_2$ is greater than s_4 over the whole range in view of (2.63), it is only necessary to consider the first three of the above inequalities. To obtain the best lower bound, the first two conditions should be taken as equalities for relatively small values of ρ, while the first and third conditions should be taken as equalities for relatively large values of ρ. Using (2.63), the results may be expressed as

$$\left.\begin{aligned} \rho &= \frac{\xi\,(1-\xi)}{(3\xi-2)\sqrt{2-\xi}}, \quad \frac{\sigma}{Y} = \frac{(1-\xi)(1-\rho)}{\rho}, \quad 0 \leq \rho \leq 0.44, \\ \xi &= \frac{1+\rho}{2}, \quad \frac{\sigma}{Y} = \frac{(1-\rho)^2}{2\rho}, \quad 0.44 \leq \rho \leq 1. \end{aligned}\right\} \tag{2.65}$$

The lower and upper bound solutions given in this section are essentially due to Gaydon and McCrum (1954), Gaydon (1954), and Hodge (1981). All the bounds discussed here apply equally well to the uniaxial tension of a square plate of side $2a$, containing a central square hole with side $2c$, where $c = \rho a$.

2.5.3 Uniaxial Tension: Upper Bounds

An upper bound on the yield point stress is derived from the velocity field involving straight localized necks which run from the edge of the hole to the stress-free edges of the plate, as shown in Fig. 2.17a. The rigid triangles between the two pairs of neck move away vertically toward each other, while the rigid halves of the remainder of the plate move horizontally at the incipient collapse. Let v denote the velocity of one side of the neck relative to the other, and ψ the angle of inclination of the neck to the relative velocity vector. For a von Mises material, the rate of plastic work per unit volume in the neck is $kv\sqrt{1+3\sin^2\psi}/b$ in view of (2.28), where b is the width

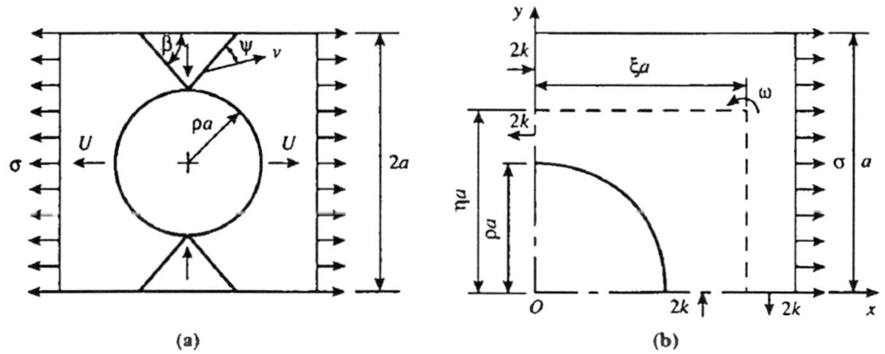

Fig. 2.17 Velocity discontinuity patterns for the plastic collapse of a uniaxially loaded square plate with a circular hole

of the neck. Since the length of each neck is $a(1 - \rho)$ cosec β, where β is the angle of inclination of the neck to the free edge of the plate, the rate of internal work in the necks per unit plate thickness is

$$W = 4kva\,(1 - \rho)\cos\mathrm{ec}\beta\sqrt{1 + 3\sin^2\psi}. \tag{2.66}$$

Equating this to the rate of external work, which is equal to $4\sigma aU = 4\sigma avcos$ $(\beta - \psi)$ per unit thickness, we obtain the upper bound

$$\frac{\sigma}{Y} = \frac{(1 - \rho)\cos\mathrm{ec}\beta\sqrt{1 + 3\sin^2\psi}}{\sqrt{3}\cos(\beta - \psi)}.$$

Minimizing σ with respect to β and ψ, it is found that the best upper bound corresponds to $\beta = \pi/4 + \psi/2$ and $\psi = \sin^{-1}\frac{1}{3}$, giving

$$\sigma = Y\,(1 - \rho)\,. \tag{2.67}$$

Since the upper bound coincides with the lower bound in the range $0 \leq \rho \leq 0.204$, the exact yield point stress is $Y(1 - \rho)$ for a von Mises material over this range.

For a Tresca material, the rate of plastic work per unit volume in a neck is $kv(1 + sin\,\psi)/b$ in view of (2.28), whatever the state of stress in the neck. The upper bound solution is therefore modified to

$$\frac{\sigma}{Y} = \frac{(1 - \rho)\cos\mathrm{ec}\beta\,(1 + 3\sin\psi)}{2\cos(\beta - \psi)}.$$

This has a minimum value for $\beta = \psi = \pi/2$, and the best upper bound is precisely that given by (2.67). The necks coincide with the vertical axis of symmetry, and the two halves of the plate move apart as rigid bodies at the incipient collapse.

For relatively large values of ρ, a better upper bound is obtained by assuming that each quarter of the plate rotates as a rigid body with an angular velocity ω about a point defined by the distances ξa and ηa as shown in Fig. 2.17b. The deformation mode involves localized necking and bulging, with normal stresses of magnitude $2k$ acting along the horizontal and vertical axes of symmetry. Since the normal component of the relative velocity vector is of magnitude $\omega|x - \xi\, a|$ along the x-axis, and to $\omega|y - \eta a|$ along the y-axis, the rate of internal work per unit thickness of the quarter plate is

$$W = 2\,k\omega\left\{\int_{\rho a}^{a}|x - \xi\,\mathrm{a}\,|dx + |\int_{\rho a}^{a}|y - \eta \mathrm{a}|\,dy\right\}.$$

Carrying out the integration, the result may be expressed as

$$W = 2\,k\omega a^2\left\{\left(1 + \rho^2\right) - (1 + \rho)\,(\xi + \eta) + \xi^2 + \eta^2\right\}. \tag{2.68}$$

The rate of external work on the quarter plate is σaU, where $U = a\omega(\eta - \frac{1}{2})$ is the normal component of the velocity of the center of the loaded side. Equating the rates of external and internal work done, and setting to zero the partial derivatives of σ with respect to ξ and η, we get

$$2\xi = 1 + \rho, \quad 2\eta = (1 + \rho) + \frac{\sigma}{2k},$$

for the upper bound to be a minimum. The best upper bound on a is therefore given by

$$\left(\frac{\sigma}{2k}\right)^2 + 2\rho\left(\frac{\sigma}{2k}\right) - 2(1 - \rho)^2 = 0.$$

It is easily verified that the conditions for a minimum σ are the same as those required for equilibrium of the quarter plate under the tractions acting along the neck and the external boundaries. The solution to the above quadratic is

$$\frac{\sigma}{2k} = -\rho + \sqrt{2 - 4\rho + 3\rho^2}, \tag{2.69}$$

where $k = Y/\sqrt{3}$ for a von Mises material, and $k = Y/2$ for a Tresca material. Evidently, (2.67) should be used in the range $0 \leq \rho \leq 0.42$, and (2.69) in the range of $0.42 \leq \rho \leq 1$ for a von Mises material. The ranges of applicability of (2.67) and (2.69) for a Tresca material are modified to $0 \leq \rho \leq \frac{1}{3}$ and $\frac{1}{3} \leq \rho \leq 1$, respectively. The upper and lower bound solutions are compared with one another in Fig. 2.18a and b, which correspond to the von Mises and Tresca materials, respectively.

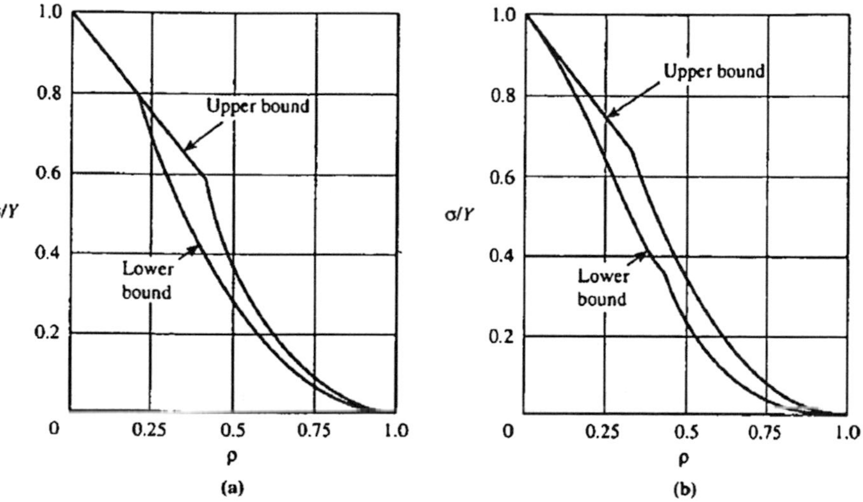

Fig. 2.18 Bounds on the collapse load for a square plate with a circular hole. (**a**) von Mises material and (**b**) Tresca material

2.5.4 Arbitrary Biaxial Tension

Suppose that a uniform normal stress $\lambda\sigma$ $(0 \leq \lambda \leq 1)$ is applied to the horizontal edges of the plate, in addition to the stress σ acting on the vertical edges. For a given value of λ, let σ be increased uniformly to its yield point value. A graphical plot of $\lambda\sigma$ against σ at the yield point state defines a closed interaction curve such that points inside the curve represent safe states of loading. Since such a curve must be convex, a simple lower bound may be constructed by drawing a straight line joining the points representing the lower bounds corresponding to uniaxial and equal biaxial tensions acting on the plate.

Upper bounds on the yield point stress for an arbitrary λ can be derived on the basis of the velocity discontinuity patterns of Fig. 2.17. Considering mode (a), the rate of external work done by the stress $\lambda\sigma$ is obtained from the fact that the speed of the rigid triangle is equal to $-v\sin(\beta - \psi)$, the length of its base being $2a(1 - \rho)\cot\beta$. Equating the total external work rate to the internal work rate given by (2.66), we get

$$\frac{\sigma}{Y} = \frac{(1-\rho)\sqrt{1+\sin^2\psi}}{\sin\beta\cos(\beta-\psi) - \lambda(1-\rho)\cos\beta\sin(\beta-\psi)}$$

for a von Mises material. The upper bound has a minimum value when $\beta = \pi/4 + \psi/2$ and $3\sin\psi = (1+z)/(1-z)$, where $z = \lambda(1-\rho)$, the best upper bound for the considered deformation mode being

$$\frac{\sigma}{Y} = \frac{1-\rho}{\sqrt{(1-\lambda)(1-\rho)+\lambda^2(1-\rho)^2}}, \quad \lambda(1-\rho) \leq 0.5. \tag{2.70}$$

When $\lambda(1-\rho) \geq 0.5$, the best configuration requires $\beta = \pi/2$, giving $\sigma = 2Y(1-\rho)/\sqrt{3}$ as the upper bound. For a Tresca material, the quantity $\sqrt{1+3\sin^2\psi}$ in (2.66) must be replaced by $(1+\sin\psi)$, and the upper bound is then found to be $\sigma = Y(1-\rho)$, on setting $\beta = \psi = \pi/2$ for all values of λ.

If the rotational mode (b) is considered for collapse, the rate of external work per unit thickness is $\sigma a^2\omega(\eta - \frac{1}{2})$ due to the horizontal stress, and $-\lambda\sigma a^2\omega(\xi - \frac{1}{2})$ due to the vertical stress. Equating the total external work rate to the internal work rate given by (2.68), and minimizing $\sigma/2k$ with respect of ξ and η, we get

$$2\xi = (1+\rho) - \frac{\lambda\sigma}{2k}, \quad 2\eta = (1+\rho) + \frac{\sigma}{2k},$$

and the best upper bound on the yield point stress is then given by

$$\left(1+\lambda^2\right)\left(\frac{\sigma}{2k}\right)^2 + 2\rho(1-\lambda)\left(\frac{\sigma}{2k}\right) - 2(1-\rho)^2 = 0 \tag{2.71}$$

for both the von Mises and Tresca materials with the appropriate value of k. When $\lambda = 0$, E (2.70) and (2.71) coincide with (2.67) and (2.69), respectively. When

$\lambda = 1$, these bounds do not differ appreciably from those obtained earlier for the von Mises material, while coinciding with the exact solution for the Tresca material. For sufficiently small values of ρ, a better bound for the von Mises material is obtained by dividing the right-hand side of (2.62b) by the quantity $\sqrt{1 - \lambda + \lambda^2}$, which is appropriate for an arbitrary λ. The bounds on the collapse load for a square plate with reinforced cutouts have been discussed by Weiss et al. (1952), Hodge and Perrone (1957), and Hodge (1981).

2.6 Hole Expansion in Infinite Plates

2.6.1 Initial Stages of the Process

An infinite plate of uniform thickness contains a circular hole of radius a, and a gradually increasing radial pressure p is applied round the edge of the hole, Fig. 2.19. While the plate is completely elastic, the radial and circumferential stresses are the same as those in a hollow circular plate of infinite external radius and are given by

$$\sigma_r = -\frac{pa^2}{r^2}, \quad \sigma_\theta = \frac{pa^2}{r^2}.$$

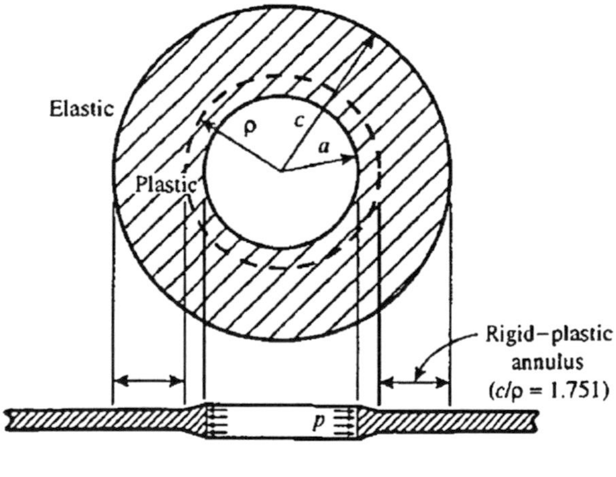

Fig. 2.19 Elastic and plastic regions around a finitely expanded circular hole in an infinite plate

Each element is therefore in a state of pure shear, the magnitude of which is the greatest at $r = a$. Yielding therefore begins at the edge of the hole when p is equal to the shear yield stress k. If the pressure is increased further, the plate is rendered plastic within some radius c, the stresses in the nonplastic region being

$$\sigma_r = -\frac{kc^2}{r^2}, \quad \sigma_\theta = \frac{kc^2}{r^2}, \quad r \geq c.$$

Since the stresses have opposite signs, the velocity equations must be hyperbolic in a plastic region near the boundary $r = c$, with the characteristics inclined at an angle $\pi/4$ to the plastic boundary. Over a range of values of c/a, the plastic material would be entirely rigid. The equation of equilibrium is

$$\frac{\partial \sigma_r}{\partial r} = \frac{\sigma_\theta - \sigma_r}{r}.$$

If the von Mises criterion is adopted, the stresses may be expressed in terms of the deviatoric angle ϕ as

$$\sigma_r = -2k \sin\left(\frac{\pi}{6} + \phi\right), \quad \sigma_\theta = 2k \sin\left(\frac{\pi}{6} - \phi\right), \tag{2.72}$$

where $\phi = 0$ at $r = c$. Inserting the relation $\sigma_\theta - \sigma_r > = 2k \cos\phi$ into the equilibrium equation, we have

$$\cos\left(\frac{\pi}{6} + \phi\right) \frac{\partial \phi}{\partial r} = -\frac{\cos\phi}{r}.$$

Using the boundary condition at $r = c$, this equation is readily integrated to give the radial distribution of ϕ as

$$\frac{c^2}{r^2} = e^{\sqrt{3}\phi} \cos\phi.$$

The relationship between the applied pressure ρ and the plastic boundary radius c is given parametrically in the form

$$\rho = 2k \sin\left(\frac{\pi}{6} + \alpha\right), \quad \frac{c^2}{a^2} = e^{\sqrt{3}\alpha} \cos\alpha, \tag{2.73}$$

where α is the value of ϕ at $r = a$. As α increases from zero, ρ increases from k. The pressure attains its greatest value $2k$ when $\alpha = \pi/3$, giving

$$\frac{c}{a} = \left(\frac{1}{2} e^{\pi/\sqrt{3}}\right)^{1/2} \approx 1.751.$$

The characteristics at this stage envelop the edge of the hole, which coincides with the direction of the numerically lesser principal stress, equal to $-k$. If the hole is further expanded, the plastic boundary continues to move outward, and the inner radius p of the rigid part of the plastic material is such that $c/\rho = 1.751$ throughout the expansion. Since an ideally plastic material cannot sustain a stress greater than $2k$ in magnitude, the plate must thicken to support the load which must increase for continued expansion.

If the material yields according to the Tresca criterion, $\sigma_\theta - \sigma_r = 2k$ in a plastic annulus adjacent to the boundary $r = c$. Substituting in the equilibrium equation and integrating, we obtain the stress distribution

$$\sigma_r = -k\left(1 + 2\text{In}\frac{c}{r}\right), \quad \sigma_\theta = k\left(1 - 2\text{In}\frac{c}{r}\right). \tag{2.74}$$

in view of the continuity of the stresses across $r = c$. The applied pressure attains its greatest value $2k$ when $c/a = \sqrt{e} \approx 1.649$, and the corresponding circumferential stress vanishes at the edge of the hole. A further expansion of the hole must involve thickening of the plate, while a rigid annulus of plastic material exists over the region $\rho \leq r \leq c$, where $c/p = 1.649$ at all stages of the continued expansion.

2.6.2 Finite Expansion Without Hardening

A solution for the finite expansion of the hole will now be carried out on the basis of the Tresca criterion, neglecting work-hardening. The circumferential stress then changes discontinuously across $r = \rho$ from zero to $-k$, the value required by the condition of the zero circumferential strain rate in the presence of thickening. The angle of inclination of the velocity characteristics to the circumferential direction changes discontinuously from $\cot^{-1}\sqrt{2}$ to zero across $r = \rho$. In the plastic region defined by $a \leq r \leq p$, the equations defining the stress equilibrium and the Lévy–Mises flow rule are

$$\left.\begin{aligned} &\frac{\partial}{\partial r}(h\sigma_r) = \frac{h}{r}(\sigma_\theta - \sigma_r), \quad \frac{\partial v}{\partial r} = \left(\frac{2\sigma_r - \sigma_\theta}{2\sigma_\theta - \sigma_r}\right)\frac{v}{r}, \\ &\frac{1}{h}\left(\frac{\partial h}{\partial \rho} + v\frac{\partial h}{\partial r}\right) = -\left(\frac{\sigma_r + \sigma_\theta}{2\sigma_\theta - \sigma_r}\right)\frac{v}{r}, \end{aligned}\right\} \tag{2.75}$$

where h is the local thickness and v the radial velocity with ρ taken as the time scale. These equations must be supplemented by the yield criterion which becomes $\sigma_r = -2k$ in the region $r \leq p$. The set of equations (2.75) is hyperbolic with characteristics $d\rho = 0$ and $dr - vd\rho = 0$ in the (r, ρ)-plane.

The solution for a hole expanded from a finite radius may be obtained from that expanded from zero radius by discarding the part of the solution which is not required. Indeed, it is immaterial whether the pressure at any radius is applied by an external agency or by the displacement of an inner annulus (Hill, 1949). Since the plate is infinite, the stress and velocity at any point must be functions of a single parameter $\xi = r/p$, so that

$$\rho\frac{\partial}{\partial r} = \frac{d}{d\xi}, \quad \rho\frac{\partial}{\partial \rho} = -\xi\frac{d}{d\xi}.$$

Setting $\sigma_r = -2k$ and $\sigma_\theta = -2ks$ in (2.75), where s is a dimensionless stress variable; they are reduced to the set of ordinary differential equations

$$\frac{dh}{d\xi} = (s-1)\frac{h}{\xi}, \quad \frac{dv}{d\xi} = -\left(\frac{2-s}{1-2s}\right)\frac{v}{\xi}, \quad \left(1 - \frac{\xi}{v}\right)\frac{dh}{d\xi} = \left(\frac{1+s}{1-2s}\right)\frac{h}{\xi}. \tag{2.76}$$

The elimination of $dh/d\xi$ between the first and third equations of (2.76) leads to

$$\frac{v}{\xi} = \frac{(1-s)(1-2s)}{2\left(1-s+s^2\right)}, \quad \frac{dv}{d\xi} = -\frac{(1-s)(2-s)}{2\left(1-s+s^2\right)}.$$

Eliminating v between the above pair of equations, we have

$$\xi \frac{d}{d\xi}\left(\frac{2s-s^2}{1-s+s^2}\right) = \frac{3(1-s)^2}{1-s+s^2}. \tag{2.77}$$

In view of the boundary condition $s = \frac{1}{2}$ when $\xi = 1$, the integration of (2.77) results in

$$\text{In}\left(\frac{1}{\xi}\right) = -\frac{1}{3}\left(\frac{1-2s}{1-s}\right) + \frac{1}{2}\text{In}\left\{\frac{3(1-s)^2}{1-s+s^2}\right\} + \frac{1}{\sqrt{3}}\tan^{-1}\left(\frac{1-2s}{\sqrt{3}}\right). \tag{2.78}$$

The elimination of ξ between (2.77) and the first equation of (2.76) gives

$$h\frac{d}{dh}\left(\frac{2s-s^2}{1-s+s^2}\right) = -\frac{3(1-s)}{1-s+s^2}.$$

If the initial thickness of the plate is denoted by h_0, then $h = h_0$ when $s = \frac{1}{2}$, and the above equation is integrated to

$$\frac{h}{h_0} = [2(1-s)]^{-1/3}\exp\left\{\frac{2}{\sqrt{3}}\tan^{1}\left(\frac{1-2s}{\sqrt{3}}\right)\right\}. \tag{2.79}$$

If r_0 denotes the initial radius to a typical particle, the incompressibility of the plastic material requires $h_0 r_0 dr_0 = hr\,dr$ at any given stage of the expansion. Since $r_0 = \rho$ when $r = p$, we obtain the relation

$$1 - \frac{r^2}{\rho^2} = 2\int_{\xi}^{1}\left(\frac{h}{h_0}\right)\xi\,d\xi, \tag{2.80}$$

where the integration is carried out numerically using the relations (2.78) and (2.79). As ξ decreases from unity, s decreases from 0.5 and becomes zero at $r = \rho^*$, which is given by

$$\text{In}\frac{\rho}{\rho^*} = -\frac{1}{3} + \frac{1}{2}(\text{In}\,3) + \frac{\pi}{6\sqrt{3}} \approx 0.518, \quad \frac{c}{\rho^{ast}} \approx 2.768,$$

in view of (2.78). The thickness ratio at $r = p^*$ is $h^*/h_0 \approx 1.453$ in view of (2.79). The solution is therefore complete for $c/a \leq 2.768$, where a is the current radius of the hole. For larger expansions of the hole, σ_θ becomes positive when $r/c \leq 0.361$, and the yield criterion reverts to $\sigma_\theta - \sigma_r = 2k$. Equation (2.76) is then modified in such a way that an analytical solution is no longer possible. A numerical solution furnishes $a/c \approx 0.280$ when the hole is expanded from zero radius, the value of h/h_0 at the edge of the hole being 3.84 approximately. This is in close agreement with the value obtained experimentally by Taylor (1948a).

2.6.3 Work-Hardening von Mises Material

When the material work-hardens and obeys the von Mises yield criterion with the Lévy–Mises flow rule, (2.75) must be supplemented by the yield criterion which is written parametrically through an auxiliary angle ϕ as

$$\sigma_r = -\frac{2\sigma}{\sqrt{3}}\sin\left(\frac{\pi}{6}+\phi\right), \qquad \sigma_\theta = -\frac{2\sigma}{\sqrt{3}}\sin\left(\frac{\pi}{6}-\phi\right), \tag{2.81}$$

where σ is the current yield stress in uniaxial tension or compression. We suppose that the uniaxial stress–strain curve is represented by the equation

$$\sigma = \sigma_0\left(1 - me^{-ne}\right), \tag{2.82}$$

where σ_0 and n are the empirical constants. The initial yield stress is $Y = (1 - m)\sigma_0$, and the current slope of the stress–strain curve is $H = n(\sigma_0 - \sigma)$, which decreases linearly with increasing stress.

The material work-hardens only in the region $r \leq \rho$, where $\rho = 0.571c$, since the plastic material beyond this radius remains rigid. It is convenient to define the dimensionless quantities

$$\xi = \frac{r}{\rho}, \quad \eta = \frac{h}{h_0}, \quad s = \frac{\sigma}{\sigma_0}.$$

We consider the expansion of a hole from zero radius, so that the stresses and strains in any element depend only on ξ. On substitution from (2.81), the last two equations of (2.75) become

$$\frac{dv}{d\xi} = -\left(\frac{\sqrt{3}+\tan\phi}{\sqrt{3}-\tan\phi}\right)\frac{v}{\xi}, \quad \left(\frac{\xi - v}{\eta}\right)\frac{d\eta}{d\xi} = -\left(\frac{2\tan\phi}{\sqrt{3}-\tan\phi}\right)\frac{v}{\xi}. \tag{2.83}$$

To obtain the differential equation for s, we write the expression for the circumferential strain rate in terms of the material derivative by using the Lévy–Mises flow rule. Thus

$$\frac{v}{r} = \frac{2\sigma_\theta - \sigma_r}{2H\sigma}\left(\frac{\partial\sigma}{\partial\rho} + v\frac{\partial\sigma}{\partial r}\right).$$

Substituting for σ_r, σ_θ, and H, and introducing the parameter $\xi = r/\rho$, this equation is reduced to

$$\left(\frac{\xi - v}{1 - s}\right)\frac{ds}{d\xi} = -\left(\frac{2n\sec\phi}{\sqrt{3}-\tan\phi}\right)\frac{v}{\xi}. \tag{2.84}$$

Inserting the expressions for σ_r and σ_θ from (2.81) into the first equation of (2.75) gives

$$\left(\frac{1}{\eta}\frac{d\eta}{d\xi}+\frac{1}{s}\frac{ds}{d\eta}\right)\tan\left(\frac{\pi}{6}+\phi\right)+\xi\frac{d\phi}{d\xi}=-\frac{2}{\sqrt{3}-\tan\phi}.$$

Eliminating $d\eta/d\xi$ and $ds/d\xi$ from the above equation by means of (2.83) and (2.84), we finally obtain the differential equation for ϕ in the form

$$\begin{aligned}&(\xi-\nu)\left(\sqrt{3}-\tan\phi\right)\frac{d\phi}{d\xi}\\&=\frac{\nu}{\xi}\left\{\sqrt{3}\sec\phi+n\left(\frac{1-s}{s}\right)\left(1+\sqrt{3}\tan\phi\right)\right\}\sec\left(\frac{\pi}{6}+\phi\right)-2.\end{aligned} \tag{2.85}$$

Equations (2.84), (2.85), and the first equation of (2.83) must be solved simultaneously for the three unknowns ν, s, and ϕ, using the boundary conditions $\nu=0$, $s=1-\mathrm{m}$, and $\phi=\pi/3$ when $\xi=1$. The second equation of (2.83) can be subsequently solved for η under the boundary condition $\eta=1$ when $\xi=1$. The expressions for all the derivatives given by (2.83) to (2.85) become indeterminate at $\xi=1$, but the application of L'Hospital's rule furnishes

$$\frac{d\phi}{d\xi}=\frac{\sqrt{3}}{2},\quad \frac{d\nu}{d\xi}=\frac{d\eta}{d\xi}=-\frac{1}{2\,(1+\lambda)},\quad \frac{ds}{d\xi}=-\frac{\lambda}{2\,(1+\lambda)},\ \xi=1,$$

where $\lambda=(2/\sqrt{3})mn/(1-\mathrm{m})$. The first derivatives of all the physical quantities are therefore discontinuous across $\xi=1$. For a nonhardening material, it is easy to see that the stress gradients $\partial\sigma_r/\partial \mathrm{r}$ and $\partial\sigma_\theta/\partial_r$ have the values zero and $-3\kappa/2\rho$, respectively, just inside the radius $r=\rho$.

Once the thickness distribution has been found, the initial radius ratio r_0/ρ to a typical element can be calculated from (2.80) for any assumed value of ξ. The current radius σ for a hole expanded from an initial radius σ_0 is obtained from the relation

$$\int_{a/\rho}^{1}\eta\xi\,d\xi=\frac{1}{2}\left(1-\frac{a_0^2}{\rho^2}\right).$$

The stress distribution is shown graphically in Fig. 2.20 for a material with $m=0.60$ and $n=9.0$. The thickness distribution is displayed in Fig. 2.21 for both work-hardening and nonhardening materials. The selected material is similar to that used by Alexander and Ford (1954), who analyzed the corresponding elastic/plastic problem using the Prandtl–Reuss theory. A rigid/plastic analysis for a plate of variable thickness has been given by Chern and Nemat-Nasser (1969).

From the known variations of r_0/ρ, s, and ϕ with ξ, the internal pressure necessary to expand a circular hole from an internal radius a_0 to a final radius a is found by using the first equation of (2.81). On releasing an amount q of the expanding pressure, the plate is left with a certain distribution of residual stresses. This is obtained by adding the quantities $q(a^2/r^2)$ and $-q(a^2/r^2)$ to the values of σ_r, and σ_θ, respectively, at the end of the expansion, so long as the unloading is elastic. For sufficiently large values of q, secondary yielding would occur on unloading, and the

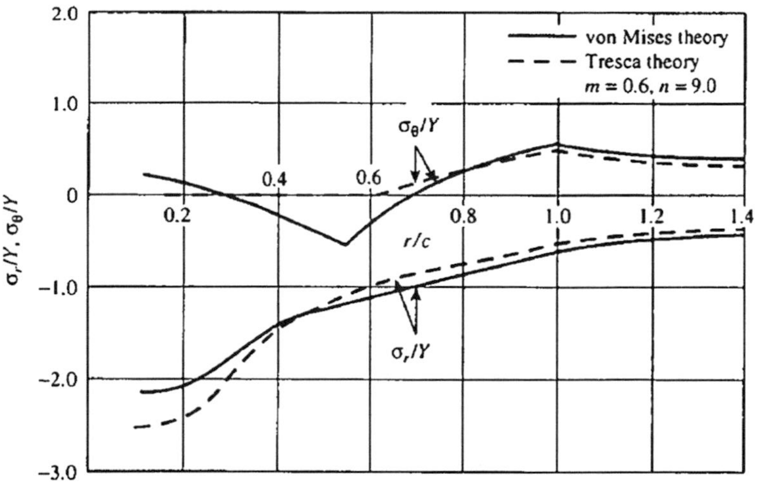

Fig. 2.20 Stress distribution in the finite expansion of a circular hole in an infinite plate of work-hardening material

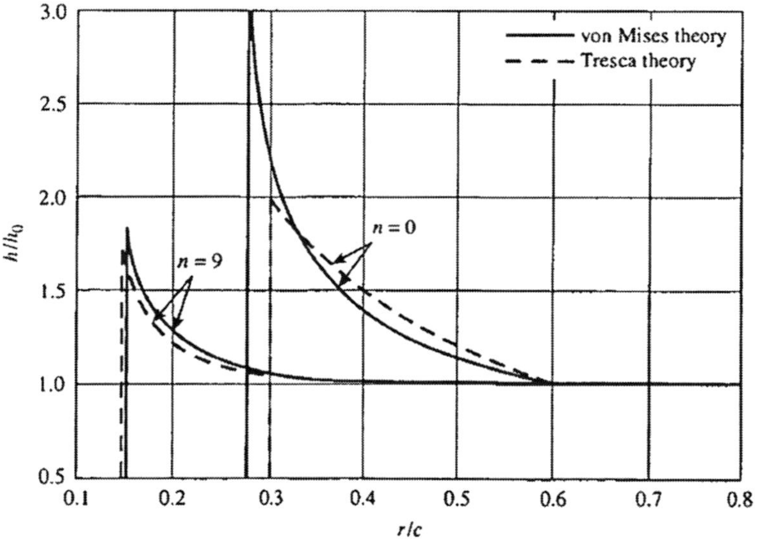

Fig. 2.21 Thickness variation in an infinite plate containing a finitely expanded circular hole

analysis becomes quite involved. A complete analysis of the unloading process for $c/a \leq 1.751$, taking secondary yielding into account, has been given by Alexander and Ford (1954) and by Chakrabarty (2006).

2.6.4 Work-Hardening Tresca Material

Suppose, now, that the material obeys Tresca's yield criterion and its associated flow rule, the stress–strain curve of the material is given by (2.82). For $\rho \leq r \leq c$, where $p = 0.607c$, the stress distribution is still given by (2.74), where $k = Y/2$, since there is no strain hardening in this region. Indeed, the flow rule corresponding to the yield condition $\sigma_\theta - \sigma_r - Y$ implies the thickness strain to be zero, and the velocity vanishes identically in view of the incompressibility condition and the boundary condition. For $r \leq \rho$, the stresses over a certain finite region would be $\sigma_r = -\sigma$ and $\sigma_\theta = 0$, corresponding to a corner of the yield hexagon. The associated flow rule gives

$$\dot{\varepsilon}_r < 0, \quad \dot{\varepsilon}_\theta > 0, \quad \dot{\varepsilon}_z > 0.$$

The sum of the three strain rates must vanish by the condition of plastic incompressibility. The rate of plastic work per unit volume is $\sigma\,\dot{\varepsilon}$, where $\varepsilon = -\,\varepsilon_r$, indicating that the relationship between σ and ε is the same as that in uniaxial tension or compression. It follows from the plastic incompressibility equation written in the integrated form that

$$\frac{hr}{h_0 r_0} = e^{\varepsilon} = \left(\frac{1-s}{m}\right)^{-1/n} \tag{2.86}$$

in view of (2.82), with $s = \sigma/\sigma_0$. Since $\sigma_\theta = 0$, the first equation of (2.75) reveals that $hr\sigma = h_0 \rho Y$ in view of the boundary conditions at $r = \rho$. Equation (2.86) therefore gives

$$\frac{r_0}{\rho} = \frac{Y}{\sigma} e^{-\varepsilon} = \left(\frac{1-m}{s}\right)\left(\frac{1-s}{m}\right)^{1/n}. \tag{2.87}$$

Differentiating (2.87) partially with respect to r, and noting the fact that $\varepsilon = ln(\partial r_0/\partial r)$, we have

$$-\left(\frac{1-m}{s}\right)\left\{\frac{1}{s} + \frac{1}{n\,(1-s)}\right\}\left(\frac{1-s}{m}\right)^{2/n} \frac{\partial s}{\partial r} = \frac{1}{\rho}.$$

Integration of this equation under the boundary condition $s = 1 - m$ at $r = \rho$, and using the integration by parts, furnishes the result

$$\xi = \left(\frac{1-m}{s}\right)\left(\frac{1-s}{m}\right)^{2/n} + \frac{1-m}{mn} \int_{1-m}^{s} \left(\frac{1-s}{m}\right)^{2/n-1} \frac{ds}{s} \tag{2.88}$$

where $\xi = r/p$. The integral can be evaluated exactly for $n = 2$ and $n = 4$, but the numerical integration for an arbitrary value of n is straightforward. The thickness change can be calculated from the relation

$$\frac{h}{h_0} = \frac{\rho Y}{r\sigma} = \frac{1-m}{\xi s}. \tag{2.89}$$

This solution will be valid so long as the thickness strain rate is positive. Since the strain is a function of ξ only, this condition is equivalent to $dh/d\xi < 0$, which gives $-ds/d\xi < s/\xi$ in view of (2.89). Using the expression for $\partial_s/\partial r$, the condition for the validity of the solution may be written as

$$\xi \leq \left(\frac{1-m}{s}\right)\left\{1 + \frac{s}{n\,(1-s)}\right\}\left(\frac{1-s}{m}\right)^{2/n}.$$

This condition will be satisfied for most engineering materials for all values of $r_0/\rho \geq 0$. For a nonhardening material, $h_0/h = \xi$ and $r_0/\rho = \sqrt{2\xi - 1}$, giving $a/c = 0.303$ for a hole expanded from zero radius. For a work-hardening material with $m = 0.6$ and $n = 9.0$, it is found that $h/h_0 \approx 1.69$ and $r/c \approx 0.143$ at the edge of the hole when its initial radius is zero. The computed results for the Tresca theory are plotted as broken curves in Figs. 2.20 and 2.21, which provide a visual comparison with the results corresponding to the von Mises theory.

The Tresca theory for a hypothetical material with an exponentially rising stress–strain curve has been discussed by Prager (1953), Hodge and Sankaranarayanan (1958), and Nemat-Nasser (1968). A rigid/plastic analysis for the hole expansion under combined radial pressure and twisting moment has been given by Nordgren and Naghdi (1963). An elastic/plastic analysis for the finite expansion of a hole in a nonhardening plate of variable thickness has been presented by Rogers (1967). An elastic/plastic small strain analysis for a linearly work-hardening Tresca material has been given by Chakrabarty (1971).

2.7 Stretch Forming of Sheet Metals

2.7.1 Hydrostatic Bulging of a Diaphragm

A uniform plane sheet is placed over a die with a circular aperture and is firmly clamped around the periphery. A gradually increasing fluid pressure is applied on one side of the blank to make it bulge through the aperture. If the material is isotropic in the plane of the sheet, the bulge forms a surface of revolution, and the radius of curvature at the pole can be estimated at any stage from the measurement of the length of the chord to a neighboring point and its corresponding sagitta. The polar hoop strain can be estimated from the radial expansion of a circle drawn from the center of the original blank. The stress–strain curve of the material under balanced biaxial tension obtained in this way is capable of being continued up to fairly large strains before instability. The process has been investigated by Hill (1950b), Mellor (1954), Ross and Prager (1954), Weil and Newmark (1955), Woo (1964), Storakers (1966), Wang and Shammamy (1969), Chakrabarty and Alexander (1970), Ilahi et al. (1981), and Kim and Yang (1985a), among others.

Let r denote the current radius to a typical particle, and r_0 the initial radius, with respect to the vertical axis of symmetry. The local thickness of the bulged sheet is denoted by t, and the inclination of the local surface normal to the vertical by ϕ, as shown in Fig. 2.22. The ratio of the initial blank thickness t_0 to the blank radius a is assumed small enough for the bending stress to be disregarded. The circumferential and meridional stresses, denoted by σ_θ and σ_ϕ, respectively, must satisfy the equations of equilibrium which may be written in the form

$$\frac{\partial}{\partial r}\left(rt\sigma_\phi\right) = t\sigma_\theta, \quad \sigma_\phi \sin\phi = \frac{pr}{2t}, \tag{2.90}$$

where ρ is the applied fluid pressure. If the meridional and circumferential radii of curvature are denoted by ρ_ϕ and ρ_θ, respectively, the equations of normal and tangential equilibrium may be expressed as

$$\frac{\sigma_\theta}{\rho_\theta} + \frac{\sigma_\phi}{\rho_\phi} = \frac{p}{t}, \quad \frac{\sigma_\phi}{\rho_\theta} = \frac{P}{2t}, \tag{2.90a}$$

where

$$\rho\theta = r \cos\text{ec}\phi, \quad \rho_\phi = \frac{\partial r}{\partial \phi}\sec\phi.$$

The elimination of p/t between the two equations of (2.90a) immediately furnishes

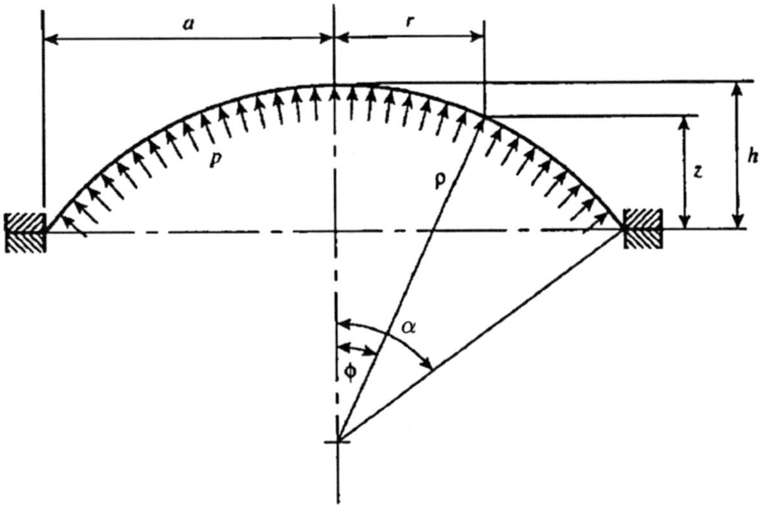

Fig. 2.22 Bulging of a circular diaphragm by the application of a uniform fluid pressure

$$\frac{\sigma_\theta}{\sigma_\phi} = 2 - \frac{\rho_\theta}{\rho_\phi}.$$

This equation indicates that $\sigma\theta \lesseqgtr \sigma\phi$ for $\rho_\theta \gtreqless \rho_\theta$. The principal surface strains ε_θ,ε_ϕ and the thickness strain ε_t at any stage are

$$\varepsilon_\theta = \ln\left(\frac{r}{r_0}\right), \; \varepsilon_\phi = \ln\left(\frac{\partial r}{\partial r_0}\sec\phi\right), \; \varepsilon_t = \ln\left(\frac{t}{t_0}\right). \tag{2.91}$$

The condition for incompressibility requires $\varepsilon_t = -(\varepsilon_\theta + \varepsilon_\phi)$. If the radial velocity is denoted by v the components of the strain rates may be expressed as

$$\dot{\varepsilon}_\theta = \frac{v}{r}, \; \dot{\varepsilon}_\phi = \frac{\partial v}{\partial r} + \dot{\phi}\tan\phi, \; \dot{\varepsilon}_t = \frac{\dot{t}}{t},$$

where the dot denotes rate of change following the particle. Eliminating v between the first two of the above relations, we obtain the equation of strain rate compatibility

$$\frac{\partial}{\partial r}\left(r\dot{\varepsilon}_\theta\right) = \dot{\varepsilon}_\phi - \dot{\phi}\tan\phi. \tag{2.92}$$

It will be convenient to take the initial radius r_0 as the independent space variable, and the polar compressive thickness strain ε_0 as the time scale, to carry out the analysis.

Introducing an auxiliary angle ψ, representing the angle made by the deviatone stress vector with the direction representing pure shear, the von Mises yield criterion and the associated Lévy–Mises flow rule can be simultaneously satisfied by writing

$$\left.\begin{aligned} \sigma_\theta &= \tfrac{2}{\sqrt{3}}\sigma\sin\left(\tfrac{\pi}{6}+\psi\right), \quad \sigma_\phi = \tfrac{2}{\sqrt{3}}\sigma\cos\psi, \\ \dot{\varepsilon}_\theta &= \dot{\varepsilon}\sin\psi, \; \dot{\varepsilon}_\phi = \dot{\varepsilon}\cos\left(\tfrac{\pi}{6}+\psi\right), \end{aligned}\right\} \tag{2.93}$$

where σ and $\dot{\varepsilon}$ are the equivalent stress and strain rate, respectively. Introducing dimensionless variables

$$\xi = \frac{r_0}{a}, \; s = \frac{\sigma}{c}, \; q = \frac{pa}{t_0 C},$$

where C is a constant stress, and using the fact that $\partial r/\partial r_0 = \cos\phi\,\exp(\varepsilon_\phi)$, $r = r_\phi\,\exp(\varepsilon_\phi)$, and $t = t_0\,\exp(\varepsilon_1)$ in view of (2.91), (2.90) and (2.92) can be combined with (2.93) to obtain the set of differential equations

$$\frac{\partial}{\partial\xi}\left[\xi s\cos\psi\,\exp\left(-\varepsilon_\phi\right)\right] = s\cos\phi\sin\left(\frac{\pi}{6}+\psi\right)\exp\left(-\varepsilon_\theta\right),$$

$$s\sin\phi = \frac{\sqrt{3}}{4}q\xi\sec\psi\exp\left(2\varepsilon_\theta+\varepsilon_\phi\right), \tag{2.95}$$

$$\frac{\partial}{\partial \xi}\left[\xi \dot{\varepsilon} \sin \psi \exp\left(\varepsilon_\theta\right)\right] = \left[\dot{\varepsilon} \cos \phi \cos\left(\frac{\pi}{6} + \psi\right) - \dot{\phi} \sin \phi\right] \exp\left(-\varepsilon_\phi\right), \quad (2.96)$$

These equations must be supplemented by the strain-hardening law $\sigma = Cf(\varepsilon)$, where ε is the equivalent total strain. Since $\sigma_\theta = \sigma_\phi = \sigma$ at the pole ($\xi = 0$), while $\dot{\varepsilon} = 0$ at the clamped edge ($\xi = 1$), the boundary conditions may be written as

$$\psi = \frac{\pi}{6} \text{ and } \dot{\varepsilon} = 1 \text{ at } \xi = 0; \quad \psi = 0 \text{ at } \xi = 1.$$

When the distributions of the relevant physical parameters have been found for any given polar strain ε_0, the shape of the bulge can be determined by the integration of the equation

$$\frac{\partial z}{\partial r_0} = -\tan \phi \frac{\partial r}{\partial r_0} = -\sin \phi \exp\left(\varepsilon_\phi\right).$$

Using (2.95), this equation may be written in the more convenient form

$$\frac{\partial}{\partial \xi}\left(\frac{z}{a}\right) = -\frac{\sqrt{3}}{4}\left(\frac{q}{s}\right) \xi \sec \psi \exp\left(2\varepsilon_\theta + 2\varepsilon_\phi\right), \quad (2.97)$$

which must be solved numerically under the boundary condition $z = 0$ at $\xi = 1$. The polar height h is finally obtained as the value of z at $\xi = 0$. The polar radius of curvature ρ is given by

$$\frac{\rho}{a} = \left(\frac{2\sigma_0 t_0}{pa}\right) \exp\left(-\varepsilon_0\right) = \left(\frac{2s_0}{q}\right) \exp\left(-\varepsilon_0\right),$$

where σ_0 is the value of σ_θ or σ_ϕ the pole $\xi = 0$. Plastic instability occurs when the pressure attains a maximum, and this corresponds to $d\rho/\rho$ being equal to $d\sigma_0/\sigma_0 - d\varepsilon_0$ during an incremental deformation of the bulge. This condition can be used to establish the point of tensile instability in the bulging process.

Suppose that the values of s, ψ, and ϕ are known at each point for the mth stage of the bulge. In order to continue the solution, we must find the corresponding distribution of $\dot{\varepsilon}$. Using the boundary condition $\dot{\varepsilon} = 1$ at $\xi = 0$, (2.96) is therefore solved numerically for $\dot{\varepsilon}$, the quantity $\dot{\phi}$ being found from the values of ϕ in the previous and current stages of the bulge. The values of ε_θ and ε_ϕ for the $(m + 1)$th stage are then obtained from their increments, using (2.93) and an assigned change in ε_0. It is convenient to adopt the power law of hardening

$$\sigma = C\varepsilon^n.$$

Since s is a known function of ε, (2.94) can be solved for ψ, assuming a value of q for the $(m + 1)$th stage, and using (2.95). The correct value of q is obtained when the boundary conditions $\psi = 0$ at $\xi = 1$ and $\psi = \pi/6$ at $\xi = 0$ are both satisfied.

If, for a certain stage of the bulge, $\dot{\varepsilon}$ is found to vanish at $\xi = 1$, indicating neutral loading of the clamped edge, the condition $\dot{\varepsilon} = 0$ at $\xi = 1$ must be satisfied at all subsequent stages. The material rate of change of (2.95) gives

$$\dot{\phi} = \left\{\frac{\dot{q}}{q} + \left[\sqrt{3}\sin\left(\frac{\pi}{6} + \psi\right) - \frac{n}{\varepsilon}\right]\dot{\varepsilon} + \dot{\psi}\tan\psi\right\}\tan\phi$$

in view of (2.93). The ratio $\dot{q}/q$ in this case should be found by substituting the above expression for ϕ into (2.96), and integrating it under the boundary conditions $\dot{\varepsilon} = 1$ at $\xi = 0$, and $\dot{\varepsilon} = 0$ at $\xi = 1$, the quantity $\dot{\psi}$ being given by the previous and current values of ψ.

Initially, however, it is reasonable to assume $\sqrt{3}\tan\psi \approx 1 - n\xi^2$ as a first approximation, which is appropriate over the whole bulge except at $\xi = 1$ (when $n \neq 0$). Since changes in geometry are negligible, (2.94) reduces in this case to

$$\frac{\partial}{\partial \xi}(s\cos\psi) + \frac{1}{2\xi}\left(1 - \sqrt{3}\tan\psi\right)(s\cos\psi) = 0.$$

This equation is readily integrated under the boundary condition $s = s_0$ at $\xi = 0$, resulting in

$$s\cos\psi = \frac{\sqrt{3}}{2}s_0\exp\left(-\frac{n}{4}\xi^2\right), \quad \phi = \frac{q\xi}{2s_0}\exp\left(\frac{n}{4}\xi^2\right), \tag{2.98}$$

in view of (2.95). The power law of hardening permits the Lévy–Mises flow rule to be replaced by the Hencky relations, so that the strain rates in (2.93) are replaced by the strains themselves. Substituting for the strain ratio $\varepsilon_\phi/\varepsilon_\theta$ into the equation of strain compatibility, obtained by eliminating r between the first two relations of (2.91), we get

$$\frac{\partial \varepsilon_\theta}{\partial \xi} - \frac{\sqrt{3}}{2\xi}\left(\cot\psi - \sqrt{3}\right)\varepsilon_\theta = -\frac{\phi^2}{2\xi} \tag{2.99}$$

to a sufficient accuracy. Inserting the expressions for $\cot\psi$ and ϕ, this equation can be integrated under the boundary condition $\varepsilon_\theta = 0$ at $\xi = 1$. Since $\varepsilon_\theta = \varepsilon_\theta/2$ at $\xi = 0$, we obtain

$$\frac{q}{s_0} = 2\sqrt{2\varepsilon_0}\left\{\int_0^1 (1 - nx)^{3/4}\exp\left(\frac{nx}{2}\right)dx\right\}^{-1/2} \approx \sqrt{[8]}\frac{\varepsilon_0}{8 - n}.$$

Using the above expression for q, the hoop strain $\varepsilon_\theta = \varepsilon\sin\psi$ can be expressed as a function of ε_θ and ξ. It is sufficiently accurate to put the result in the form

$$\varepsilon\sin\psi \approx \tfrac{1}{2}\varepsilon_0\left(1 - \xi^2\right)\left(1 - \frac{n\xi^2}{8 - n}\right)\left(1 - n\xi^2\right)^{-3/4}. \tag{2.100}$$

Since $s/s_0 = (\varepsilon/\varepsilon_0)^n$, the distributions of s, ε, ψ, and ϕ can be determined from (2.98) and (2.100) for any small value of $\varepsilon 0$. The shape of the bulge and the polar height are obtained from the integration of the equation $\partial z/\partial \xi = \text{-a}\,\phi$, the result being

$$\frac{z}{a} \approx \frac{q}{4s_0}\left(1-\xi^2\right)\left\{1+\frac{n}{32}\left(1+\xi^2\right)\right\}, \quad \frac{h}{a} \approx \frac{1}{2}\left\{\left(\frac{16+n}{8-n}\right)\varepsilon_0\right\}^{1/2}.$$

The initial shape of the bulge is therefore approximately parabolic for all values of n. Since ϕ changes at the rate $\phi = \phi/2\varepsilon_0$ in view of (2.98), the solution for the von Mises material can proceed by integrating (2.96) as explained earlier. Figure 2.23 shows the surface strain distribution for various values of ε_0 in a material with $n = 0.2$, the variation of ε_0 with the polar height being displayed in Fig. 2.24. The theoretical predictions are found to agree reasonably well with available experimental results on hydrostatic bulging. A detailed analysis for the initial deformation of the diaphragm has been given by Hill and Storakers (1980).

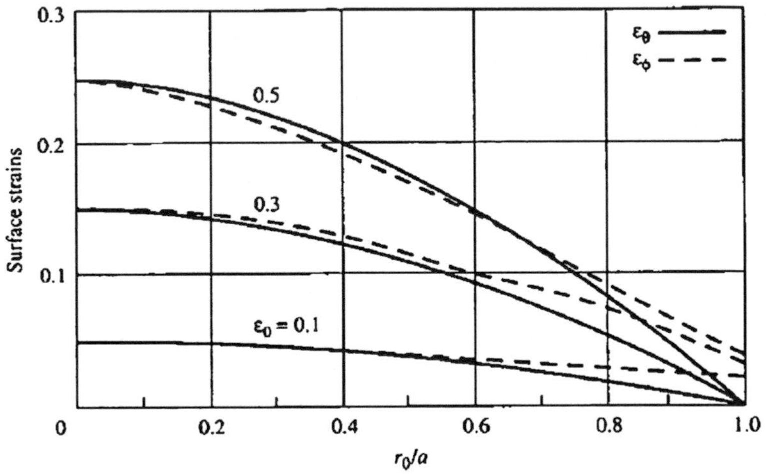

Fig. 2.23 Distribution of circumferential and meridional true strains in the hydrostatic bulging process when $n = 0.2$ (after Wang and Shammamy, 1969)

2.7.2 Stretch Forming Over a Rigid Punch

A flat sheet of metal of uniform thickness t_0 is clamped over a die with a circular aperture of radius a, and the material is deformed by forcing a rigid punch with a hemispherical head. The axis of the punch passes through the center of the aperture and is normal to the plane of the sheet. The deformed sheet forms a surface of revolution with its axis coinciding with that of the punch. Due to the presence of friction between the sheet and the punch, the greatest thinning does not occur at the

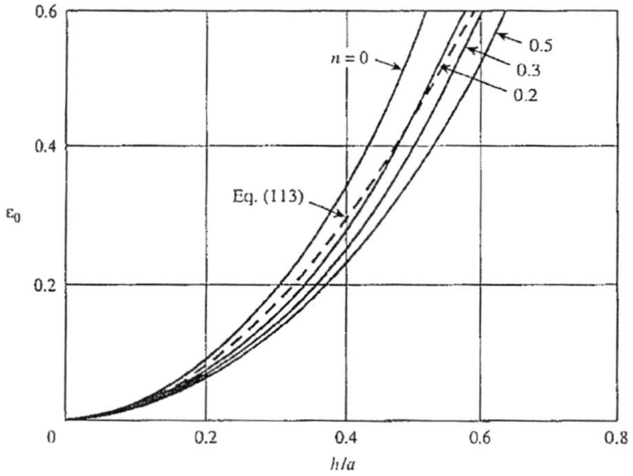

Fig. 2.24 Variation of the polar thickness strain with polar height during the bulging of a circular diaphragm

pole but at some distance away from it, and fracture eventually occurs at this site. In a typical cupping test, known as the Erichsen test, a hardened steel ball is used as the punch head, and the height of the cup when the specimen splits is regarded as the Erichsen number, which is an indication of the formability of the sheet metal. The process has been investigated experimentally by Keeler and Backofen (1963) and theoretically by Woo (1968), Chakrabarty (1970a), Kaftanoglu and Alexander (1970), and Wang (1970), among others. Finite element methods for the analysis of the stretch-forming process have been discussed by Wifi (1976), Kim and Kobayashi (1978), and Wang and Budiansky (1978).

In the theoretical analysis of the forming process, the coefficient of friction μ will be taken as constant over the entire surface of contact. The radius of the punch head, denoted by R, is somewhat smaller than the radius of the die aperture, Fig. 2.25. Let t denote the local thickness of an element currently at a radius r and at an angular distance ϕ from the pole, the initial radius to the element being denoted by r_0.

Over the region of contact, the equations of tangential and normal equilibrium are

$$\frac{\partial}{\partial r}\left(rt\sigma_\phi\right) = t\sigma_\theta + \mu pR \tan\phi, \quad t\left(\sigma_\theta + \sigma_\phi\right) = pR, \tag{2.101}$$

where p is the normal pressure exerted by the punch, and $r = R \sin\phi$. The elimination of pR between the above equations gives

$$\frac{\partial}{\partial r}\left(rt\sigma_\phi\right) = t\sigma_\theta\left(1 + \mu\tan\phi\right) + \mu t\sigma_\phi \tan\phi.$$

If the material obeys the von Mises yield criterion, the stresses are given by (2.93). In terms of the dimensionless variables $\xi = r_0/a$ and $s = \sigma/C$, where C is a constant stress, the above equation becomes

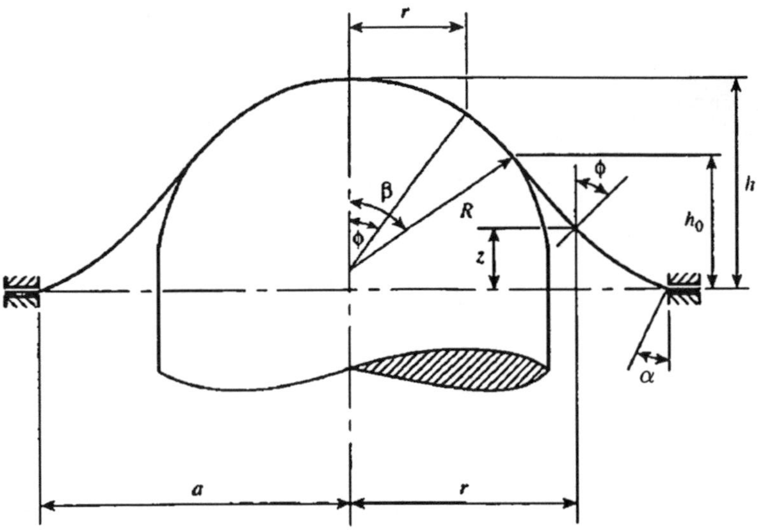

Fig. 2.25 Stretch forming of a circular blank of sheet metal over a hemispherical-headed punch

$$\frac{\partial}{\partial \xi}\left[\xi s \cos\psi \exp\left(-\varepsilon_\phi\right)\right]$$
$$= s\left[(\cos\phi + \mu\sin\phi)\sin\left(\frac{\pi}{6}+\psi\right) + \mu\sin\phi\cos\psi\right]\exp\left(-\varepsilon_\theta\right),$$

where $\sin\phi = (a/R)\xi\exp(\varepsilon_\theta)$. In view of (2.93) and the second equation of (2.101), the pressure distribution over the punch head is given by

$$\frac{pR}{t_0 C} = s\left(\sin\psi + \sqrt{3}\cos\psi\right)\exp\left(-\varepsilon_\theta - \varepsilon_\phi\right). \tag{2.103}$$

The geometrical relation $r = R\sin\phi$ furnishes $\dot{\phi} = \dot{\varepsilon}_\theta \tan\phi$ over the contact region. Substituting into the compatibility equation (2.92), and using (2.93) for the components of the strain rate, we get

$$\frac{\partial}{\partial \xi}\left[\xi\dot{\varepsilon}\sin\psi\exp\left(\varepsilon_\theta\right)\right] = \dot{\varepsilon}\cos\phi\left[\cos\left(\frac{\pi}{6}+\psi\right) - \tan^2\phi\sin\psi\right]\exp\left(\varepsilon_\phi\right).$$

When the stresses and strains are known at the mth stage of the process, (2.104) can be solved using the condition $\dot{\varepsilon} = 1$ at $\xi = 0$, the polar compressive thickness strain ε_0 being taken as the time scale. The computed distribution of $\dot{\varepsilon}$ and an assigned increment of ε_0 furnish the quantities ε_θ, ε_ϕ, and ϕ, while s follows from the given stress–strain curve. Equation (2.102) is then solved for ψ at the $(m + 1)$th stage under the boundary condition $\psi = \pi/6$ at $\xi = 0$, and the stresses are finally obtained from (2.93).

Over the unsupported surface of unknown geometry, the equation of meridional equilibrium is given by (2.101) with $p = 0$. Since the circumferential and meridional curvatures at any point are $\sin\phi/r$ and $(\partial\phi/\partial r)\cos\phi$, respectively, the equation of normal equilibrium is

$$\frac{\sigma_\theta}{r}\sin\phi + \frac{\partial\phi}{\partial r}\sigma_\phi\cos\phi = 0.$$

Using the relations (2.91) and (2.93), this may be rewritten as

$$r_0 \operatorname{cosec}\phi \frac{\partial\phi}{\partial r_0} = -\frac{1}{2}\left(1 + \sqrt{3}\tan\psi\right)\exp\left(\varepsilon_\phi - \varepsilon_\theta\right).$$

If the angle of contact is denoted by β, then $\xi = \xi^* = (R/a)\sin\beta\exp(-\varepsilon_\theta^*)$ at $\phi = \beta$, where the asterisk refers to the contact boundary. The integration of the above equation results in

$$\ln\left\{\frac{\tan(\phi/2)}{\tan(\beta/2)}\right\} = -\frac{1}{2}\int_{\xi^*}^{\xi}\left(1 + \sqrt{3}\tan\psi\right)\exp\left(\varepsilon_\phi - \varepsilon_\theta\right)\frac{d\xi}{\xi}. \tag{2.105}$$

The remaining equilibrium equation and the compatibility equation in the dimensionless form are identical to (2.94) and (2.96), respectively.

To continue the solution from a known value of β at the mth stage, and the corresponding distributions of ψ and s, (2.96) is solved numerically for ε using the condition of continuity across $\xi = \xi^*$, the distribution of $\dot\phi$ being obtained from the previous values of ϕ. The distribution of ϕ for the $(m + 1)$th stage is then obtained from (2.105), assuming a value of β and the previous distribution of ψ. The correct value of β is that for which the continuity condition for ψ across $\xi = \xi^*$ is satisfied, when (2.94) is solved for ψ with the boundary condition $\psi = 0$ at $\xi = 1$. The total penetration h of the punch at any stage can be computed from the formula

$$\frac{h}{a} = \frac{R}{a}(1 - \cos\beta) + \int_{\xi^*}^{1}\sin\phi\exp\left(\varepsilon_\phi\right)d\xi. \tag{2.106}$$

in view of the relation $\partial z/\partial\xi = -a\tan\phi$, where ϕ is given by (2.105) over the unsupported region. The resultant punch load is $P = 2\pi Rt^*\sigma_\phi^*\sigma £\sin^2\beta$, and the substitution for t^* and σ_ϕ^*, furnishes

$$\frac{P}{2\pi Rt_0C} - s^*\sin^2\beta\sin\left(\frac{\pi}{6} + \psi^*\right)\exp\left[-\left(\varepsilon_\theta^* + \varepsilon_\phi^*\right)\right]. \tag{2.107}$$

Equations (2.106) and (2.107) define the load–penetration relation parametrically through β. When the load attains a critical value, a local neck is formed at the thinnest section (leading to fracture) due to some kind of instability of the biaxial stretching.

If the strain hardening is expressed by the power law $s = \varepsilon^{\mathrm{n}}$, the Hencky theory may be used for the solution of the initial problem. Assuming $\sqrt{3}\tan\psi \approx 1 - \mathrm{n}\xi^2$ as a first approximation, and omitting the negligible friction terms in (2.102), it is found that s is given by the first equation of (2.98) throughout the deformed sheet. Furthermore, $\phi = \mathrm{a}\xi/R$ for $0 \leq \xi \leq \xi^*$, and

$$\phi = \frac{a\xi^{*2}}{\xi R} \exp\left\{\frac{n}{4}\left(\xi^2 - \xi^{*2}\right)\right\}, \quad \xi^* \leq \xi \leq 1,$$

in view of (2.105). The strain compatibility equation (2.99), which is the same as that for the stretch-forming process, gives

$$\frac{\partial}{\partial\xi}\left[\varepsilon\left(1 - n\xi^2\right)^{3/4} \sin\psi\right] = -\frac{\phi^2}{2\xi}\left(1 - n\xi^2\right)^{3/4}. \tag{2.108}$$

Substituting for ϕ, and using the conditions of continuity of ε and ψ across $\xi = \xi^*$, we obtain the expression for the polar thickness strain as

$$\varepsilon_0 = \frac{\beta^2}{2}\left\{1 + \int_{x^*}^{1}\left(\frac{\xi^*}{x}\right)^2 (1 - nx)^{3/4} \exp\left[\frac{n}{2}\left(x - x^*\right)\right] dx\right\}, \tag{2.109}$$

where $x = \xi^2$ and $x^* = \xi^{*2} = (RB/a)^2$. The distribution of ψ is now obtained by the integration of (2.108). By (2.106), the polar height is

$$h = \tfrac{1}{2}R\beta^2\left\{1 + \int_{x^*}^{1} \exp\left[n\left(\frac{x - x^*}{4}\right)\right]\frac{dx}{x}\right\}, \tag{2.110}$$

while the punch load is easily found from (2.107). Starting with a sufficiently small value of ε_0, for which a complete solution has just been derived, the analysis can be continued by considering (2.104) and (2.96) as explained before. The distribution of thickness strain and the load–penetration relationship are shown graphically in Figs. 2.26 and 2.27 for a material with $n = 0.2$, assuming $R = a$ and $\mu = 0.2$.

The theory seems to be well supported by available experimental results. It is found that the punch load required for a given depth of penetration is affected only slightly by the coefficient of friction.

2.7.3 Solutions for a Special Material

The solution to the stretch-forming problem becomes remarkably simple when the material is assumed to have a special strain-hardening characteristic. From the practical point of view, such a solution is extremely useful in understanding the physical behavior of the forming process and in predicting certain physical quantities with reasonable accuracy. The stress–strain curve for the special material is represented by

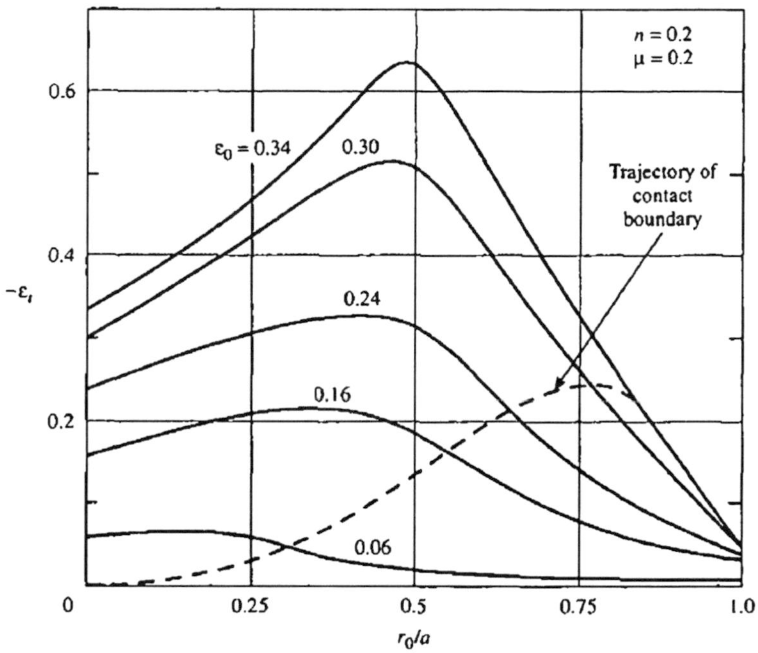

Fig. 2.26 Distribution of thickness strain in the stretch-forming process using a hemispherical punch with $R = a$ (after N.M. Wang, 1970)

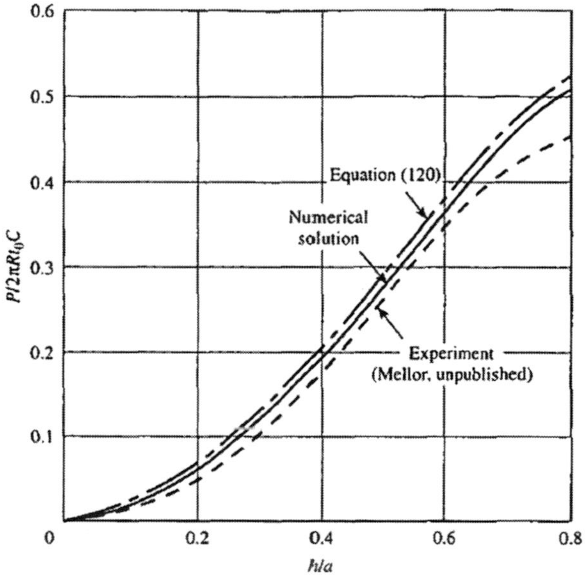

Fig. 2.27 Dimensionless load–penetration behavior in the punch stretching process using $R = a$ ($n = 0.2$, $\mu = 0.2$)

$$\sigma = Y \exp(\varepsilon),$$

where Y is the initial yield stress. The stress–strain curve is unlike that of any real metal, but the solution based on it should provide a good approximation for sufficiently prestrained metals (Hill, 1950b).

Considering the hydrostatic bulging process, it is easy to see that the assumed strain-hardening law requires the bulge to be a spherical cap having a radius of curvature ρ, which is given by simple geometry as

$$\rho = \frac{h^2 + a^2}{2h} = a \operatorname{cosec} \alpha, \tag{2.111}$$

where α is the semiangle of the cap. Indeed, it follows from (2.90a) that $\sigma_\theta = \sigma_\phi = \sigma = p\rho/2t$ when $\rho_\theta = \rho_\phi = \rho$, indicating that $t\sigma$ is a constant at each stage. Since $t = t_0 \exp(-\varepsilon)$, we recover the assumed stress–strain curve. Substituting from (2.91) into the relation $\varepsilon_\theta = \varepsilon_\phi = \varepsilon/2$ given by the flow rule, and using the fact that $r = \rho \sin\phi$, we get

$$r_0 \frac{\partial \phi}{\partial r_0} = \sin\phi \quad \text{or} \quad \frac{r_0}{a} = \frac{\tan(\phi/2)}{\tan(\alpha/2)}$$

in view of the condition $r_0 = a$ at $\phi = \alpha$. The strain distribution over the bulge is therefore given by

$$\varepsilon = 2 \ln \frac{r}{r_0} = 2 \ln\left(\frac{1 + \cos\phi}{1 + \cos\alpha}\right) = 2 \ln\left(1 + \frac{hz}{a^2}\right). \tag{2.112}$$

It follows from (2.112) that $\varepsilon = 0$ at $\phi = \alpha$, indicating that there is no straining at the clamped edge, which merely rotates to allow the increase in bulge height. The magnitude of the polar thickness strain is

$$\varepsilon_0 = 4 \ln \sec\frac{\alpha}{2} = 2 \ln\left(1 + \frac{h^2}{a^2}\right). \tag{2.113}$$

This relation is displayed by a broken curve in Fig. 2.24 for comparison. To obtain the velocity distribution, we consider the rate of change of (2.112) following the particle, as well as that of the geometrical relation $r = a(\sin\phi/\sin\alpha)$, taking α as the time scale. The resulting pair of equations for v and $\dot{\phi}$ may be solved to give

$$\frac{v}{r} = \frac{\cos\phi - \cos\alpha}{\sin\alpha} = \frac{z}{a}, \quad \dot{\phi} = \frac{\sin\phi}{\sin\alpha} = \frac{r}{a}.$$

The rate of change of the above expression for z and the substitution for $\dot{\phi}$ furnish the result $\dot{z} = v \cot\phi$, which shows that the resultant velocity of each particle is along the outward normal to the momentary profile of the bulge.

The relationship between the polar strain and the polar radius of curvature obtained for the special material provides a good approximation for a wide variety of metals. From (2.111) and (2.113), it is easily shown that

$$\frac{a}{\rho} \approx \sqrt{2\varepsilon_0}\exp\left(-\tfrac{3}{8}\varepsilon_0\right).$$

to a close approximation. When the pressure attains a maximum, the parameter $t\sigma/\rho$ has a stationary value at the pole, giving

$$\frac{1}{\sigma_0}\frac{d\sigma_0}{d\varepsilon_0} = 1 + \frac{1}{\rho}\frac{d\rho}{d\varepsilon_0} \approx \frac{11}{8} - \frac{1}{2\varepsilon_0}$$

in view of the preceding expression for ρ as a function of ε_θ. For the simple power law $\sigma_0 = C\varepsilon_0^n$, the polar strain at the onset of instability therefore becomes

$$\varepsilon_0 = \tfrac{4}{11}\,(+2n)\,.$$

The instability strain is thus equal to $\frac{4}{11}$ for a nonhardening material ($n = 0$). This explains the usefulness of the bulge test as a means of obtaining the stress–strain curve of metals for large plastic strains.

The special material is also useful in deriving an analytical solution for stretch forming over a hemispherical punch head, provided friction is neglected (Chakrabarty, 1970). The stress–strain curve is then consistent with the assumption of a balanced biaxial state of stress throughout the deforming surface. We begin with the unsupported region, for which the equilibrium in the normal direction requires

$$\frac{\sigma_\phi}{\sigma_\theta} = -\frac{\rho_\phi}{\rho_\theta} = -\frac{1}{r}\frac{\partial r}{\partial \phi}\tan\phi$$

in view of the first equation of (2.90a) with $p = 0$. The assumption $\sigma_\theta = \sigma_\phi$ therefore implies $\rho_\theta = -\rho\phi$ over the unsupported region, giving

$$\frac{\partial r}{\partial \phi} = -r\cot\phi \quad \text{or} \quad \frac{r}{a} = \frac{\sin\alpha}{\sin\phi}, \tag{2.114}$$

in view of the boundary condition $\phi = \alpha$ at $r = a$. Since $r = R\sin\beta$ at $\phi = \beta$, the angles β and α are related to one another by the equation

$$\sin\alpha = \frac{R}{a}\sin^2\beta \tag{2.115}$$

It may be noted that the meridional radius of curvature changes discontinuously from $-R$ to R across $\phi = \beta$. The shape of the unsupported surface is given by the differential equation

$$\frac{\partial z}{\partial \phi} = -\tan\phi \frac{\partial r}{\partial \phi} = a\left(\frac{\sin\alpha}{\sin\phi}\right),$$

which is integrated under the boundary condition $z = 0$ at $\phi = \alpha$ to obtain

$$\frac{z}{a} = \sin\alpha \ln\left\{\frac{\tan(\phi/2)}{\tan(\alpha/2)}\right\}. \qquad (2.116)$$

The unsupported surface is actually a minimal surface since the mean curvature vanishes at each point. It is known from the geometry of surfaces that the only minimal surface of revolution is the catenoid. Indeed, the elimination of ϕ between (2.114) and (2.116) leads to the geometrical relation

$$\frac{r}{a} = \sin\alpha \cosh\left(\frac{z}{a}\operatorname{cosec}\alpha + \ln\tan\frac{\alpha}{2}\right)$$

which is the equation of a catenoid. The flow rule requires $\varepsilon_\theta = \varepsilon_\phi = \varepsilon/2$, and the substitution from (2.91) and (2.114) gives

$$r_0 \frac{\partial \phi}{\partial r_0} = -\sin\phi \quad \text{or} \quad \frac{r_0}{a} = \frac{\tan(\alpha/2)}{\tan(\phi/2)},$$

in view of the boundary condition $r_0 = a$ at $\phi = \alpha$. The expressions for r and r_0 furnish the compressive thickness strain as

$$\varepsilon = 2\ln\frac{r}{r_0} = 2\ln\left(\frac{1+\cos\alpha}{1+\cos\phi}\right), \quad R\sin\beta \le r \le a. \qquad (2.117)$$

Since $t\sigma$ is constant at each stage due to the assumed strain-hardening law, and the fact that $\varepsilon = \ln(t_0/t)$, the first equation of (2.90) is identically satisfied.

Over the region of contact, $\rho_\theta = \rho_\phi = R$, and equilibrium requires $\sigma_\theta = \sigma_\phi = \sigma$ in the absence of friction, giving $p = 2\sigma t/R$. In view of the relations $\varepsilon_\theta = \varepsilon_\phi$ and $r = R\sin\phi$, the initial radius r_0 to a typical particle is given by

$$r_0 \frac{\partial \phi}{\partial r} = \sin\phi \quad \text{or} \quad \frac{r_0}{a} = \frac{\tan(\alpha/2)\tan(\phi/2)}{\tan^2(\beta/2)},$$

in view of the condition of continuity across $\phi = \beta$. The compressive thickness strain therefore becomes

$$\varepsilon = 2\ln\frac{r}{r_0} = 2\ln\left\{\frac{(1+\cos\alpha)(1+\cos\phi)}{(1+\cos\beta)^2}\right\}, \quad 0 \le r \le R\sin\beta, \qquad (2.118)$$

The continuity of the strains evidently ensures the continuity of the stresses across the contact boundary. The thickness has a minimum value at the pole when there is no friction between the material and the punch head.

The total penetration of the punch at any stage is obtained from (2.116) and the fact that the height of the pole above the contact boundary is equal to $R(1–\cos\beta)$.

Hence

$$\frac{h}{R} = (1 - \cos\beta) + \sin^2\beta \ln\left\{\frac{\tan(\beta/2)}{\tan(\alpha/2)}\right\} \tag{2.119}$$

in view of (2.115). Available experimental results indicate that the relationship between h/R and β is practically independent of the material properties when the punch is well lubricated. It is therefore a good approximation to assume that the relationship between h/R and $P/2Rt_0C$ is independent of the strain-hardening characteristic. Then, for the simple power law $\sigma = C\varepsilon^n$, the load–penetration relationship is given parametrically through β by (2.119) and the formula

$$\frac{P}{2\pi Rt_0C} = \left(\frac{1+\cos\beta}{1+\cos\alpha}\right)^2 \sin^2\beta\left\{2\ln\left(\frac{1+\cos\alpha}{1+\cos\beta}\right)\right\}^n. \tag{2.120}$$

This expression is obtained from the fact that $t^*\sigma^*$ is equal to $t_0C(\varepsilon^*)^n \exp(-\varepsilon^*)$, where ε^* is given by (2.117) with $\phi = \beta$. It is easily shown that the punch load given by (2.120) does not have a stationary value in the interval $0 < \beta < \pi/2$, a result that is in agreement with experiment. A graphical plot of (2.120) is included in Fig. 2.27 for comparison. An application of the preceding results to the hydrodynamic lubrication in stretch forming has been discussed by Wilson and Hector (1991).

2.8 Deep Drawing of Cylindrical Cups

2.8.1 Introduction

In the simplest deep-drawing operation, a fiat circular blank of sheet metal is formed into a cylindrical cup having a flat or hemispherical base. The blank is placed over a die with a circular aperture, Fig. 2.28, and a blank holder is generally used to prevent wrinkling of the flange. A cylindrical punch is forced on the blank, either hydraulically or mechanically, to draw the outer annulus of the blank forming the cup wall. The blank-holding pressure is provided either by spring loading or by bolting down to a fixed clearance. Due to the elasticity of the apparatus, the latter method effectively generates a spring loading of very high intensity. In commercial deep drawing, the clearance between the punch and the die is slightly greater than the metal thickness, so that the thickness change that occurs through the die is unconstrained.

As the punch penetration increases from zero, the material in a central part of the blank is wrapped around the punch head and is therefore subjected to bending under tension. The deformation in the outer portion of the blank remains at the elastic order of magnitude until the punch load is sufficient to render the flange completely plastic. The outer annulus is then drawn radially inward, and its thickness progressively increases so long as the element remains on the flat part of the die. As the material passes over the die lip, it is first bent and then straightened under tension, resulting

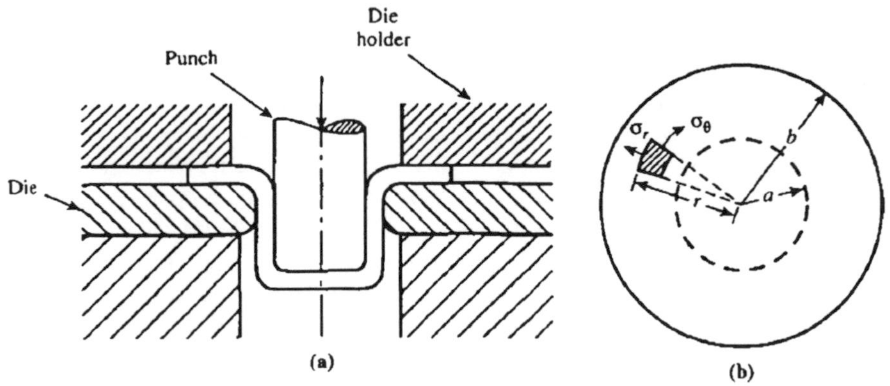

Fig. 2.28 Deep drawing of a cylindrical cup from a circular blank of sheet metal. (**a**) Schematic diagram and (**b**) stresses in an element

in a decrease in thickness. There is a narrow band of relatively thicker metal that has never been in contact with either the die or the punch. In the case of a flat-headed punch, the greatest thinning occurs over the punch profile radius due to bending and friction under biaxial tensile stresses. The load attains a maximum at some intermediate stage of the drawing and falls to zero at the end of the process (Johnson and Mellor 1983). An experimental method of estimating the die profile friction in deep drawing has been discussed by Tsutsumi and Kato (1994).

A complete solution for the deep-drawing process, taking into account the effects of bending and friction over the die and punch, does not seem to have been carried out with sufficient rigor. The early theoretical and experimental results on deep drawing have been reported by Chung and Swift (1951) and reviewed by Alexander (1960a) and Alexander et al. (1987). A numerical solution for deep drawing with a hemispherical headed punch has been obtained by Woo (1968) and Wifi (1976), neglecting the bending effects. Numerical solutions based on the finite element method have been presented by Kobayashi et al. (1989) and Saran et al. (1990). The main contribution to the drawing load is due to the radial drawing of the flange, which will be considered in what follows.

2.8.2 Solution for Nonhardening Materials

Since the blank thickens most at the outer radius, all the blank-holding force is exerted at the rim, but its effect is generally small and may be neglected as a first approximation. The material therefore deforms effectively under conditions of plane stress, the nonzero principal stresses being σ_r and σ_θ in the radial and circumferential directions, respectively. If the local thickness of the blank is denoted by h, the equation of radial equilibrium is

$$\frac{\partial}{\partial r}(h\sigma_r) = \frac{h}{r}(\sigma_\theta - \sigma_r), \qquad (2.121)$$

where σ_r is tensile and σ_θ compressive. We begin with a nonhardening material with a uniaxial yield stress Y, and approximate the von Mises yield criterion by the modified Tresca criterion

$$\sigma_r - \sigma_\theta = mY,$$

where m is a constant, equal to 1.1 approximately. It is found that the thickness variation across the blank at any stage can be neglected in the equilibrium equation without significant errors. Equation (2.121) is then immediately integrated on substitution from the yield criterion to obtain the stress distribution

$$\sigma_r = mY \ln\frac{b}{r}, \quad \sigma_0 = -mY\left(1 - \ln\frac{b}{r}\right), \qquad (2.122)$$

in view of the boundary condition $\sigma_r = 0$ at the current external radius $r = b$, the initial radius of the blank being b_0.

If v denotes the radial velocity of a typical particle, with respect to the rim displacement $b_0 - b$ taken as the time scale, the relationship between the radial and circumferential strain rates according to the Lévy–Mises flow rule may be written as

$$\frac{\partial v}{\partial r} = \left(\frac{2\sigma_r - \sigma_\theta}{2\sigma_\theta - \sigma_r}\right)\frac{v}{r} = -\left\{\frac{1 + \ln(b/r)}{2 - \ln(b/r)}\right\}\frac{v}{r}$$

in view of (2.122). Using the boundary condition $v = -1$ at $r = b$, the above equation is readily integrated to

$$v = -\frac{dr}{db} = -\frac{r}{b}\left(1 - \tfrac{1}{2}\ln\frac{b}{r}\right)^{-3}. \qquad (2.123)$$

This equation can be further integrated with the substitution $\ln(b/r) = \rho$, which on differentiation following the particle gives

$$\frac{dr}{r} = -\frac{d\rho}{1 - (1 - \rho/2)^3} \approx -\frac{4d\rho}{3\rho(2 - \rho)}$$

in view of (2.123), the result being a close approximation over the relevant range. If r_0 denotes the initial radius to a typical particle, the initial condition is $r = r_0$ when $b = b_0$, and the integration of the above equation results in

$$\frac{r_0}{r} = \left\{\frac{\ln(b/r)\left[2 - \ln(b_0/r_0)\right]}{\ln(b_0/r_0)\left[2 - \ln(b/r)\right]}\right\}^{2/3}. \qquad (2.124)$$

Equation (2.124) furnishes b/r as a function of b_0/r_0 for any given value of b/b_0. The relationship between r/b_0 and r_0/b_0 is thus obtained for a selected value of b/b_0. The flow rule gives the incremental thickness change in any element as

$$\frac{dh}{h} = -\left(\frac{\sigma_r + \sigma_\theta}{2\sigma_\theta - \sigma_r}\right)\frac{dr}{r} = -\left\{\frac{1 - 2\ln(b/r)}{2 - \ln(b/r)}\right\}\frac{dr}{r}.$$

Inserting the approximate expression for dr/r, and using the initial condition $h = h_0$ when $b = b_0$, the formula for the thickness variation is obtained as

$$\frac{h}{h_0} = \sqrt{\frac{b_0}{b}}\exp\left\{-\frac{[\ln(b/r)]^2}{4 - 2\ln(b/r)} + \frac{[\ln(b_0/r_0)]^2}{4 - 2\ln(b_0/r_0)}\right\}. \quad (2.125)$$

Equations (2.124) and (2.125) enable us to determine the thickness contour at any given stage of the drawing. Since the external radius $r = b$ is always stressed in simple compression, the thickness of the outside edge is $h_0\sqrt{b_0/b}$ irrespective of the stress–strain curve of the metal. The initial radius to the element that is currently at the die throat $r = a$ is given by (2.124). For a given element, the ratio b/r steadily increases from its initial value b_0/r_0.

A closed-form solution is also possible when the von Mises yield criterion is used, provided the thickness variation in the equilibrium equation is disregarded. It is convenient to express the yield criterion parametrically through an angle ψ as

$$\sigma_r = \frac{2Y}{\sqrt{3}}\sin\psi, \quad \sigma_\theta = -\frac{2Y}{\sqrt{3}}\cos\left(\frac{\pi}{6} + \psi\right), \quad (2.126)$$

where ψ lies between 0 and $\pi/3$. Substituting into the equilibrium equation (2.121), where h is omitted, we get

$$\cos\psi\frac{\partial\psi}{\partial r} = -\frac{1}{2r}\left(\sin\psi + \sqrt{3}\cos\psi\right).$$

The integration of this equation under the boundary condition $\psi = 0$ at $r = b$ results in

$$\frac{r^2}{b^2} = \frac{\sqrt{3}}{2}\sec\left(\frac{\pi}{6} - \psi\right)\exp\left(-\sqrt{3}\psi\right). \quad (2.127)$$

It is interesting to note that ψ depends only on the ratio r/b. Turning now to the Lévy–Mises flow rule, we have the velocity equation

$$\frac{\partial v}{\partial r} = \left(\frac{2\sigma_r - \sigma_\theta}{2\sigma_\theta - \sigma_r}\right)\frac{v}{r} = -\left(1 + \sqrt{3}\tan\psi\right)\frac{v}{2r},$$

where the external displacement $b_0 - b$ is taken as the time scale. Using the expression for $\partial\psi/\partial r$, the above equation is reduced to

$$\frac{\partial v}{\partial \psi} = \left(\frac{1 + \sqrt{3}\tan\psi}{\sqrt{3} + \tan\psi}\right) v.$$

In view of the boundary condition $v = -1$ at $r = b$, this equation is integrated to

$$v^2 = \frac{\sqrt{3}}{2} \sec\left(\frac{\pi}{6} - \psi\right) \exp\left(\sqrt{3}\psi\right).$$

The quantities ψ and v^2 increase as r decreases. It follows from (2.127) and the above equation for v^2 that

$$v = -\frac{dr}{db} = -\frac{r}{b} \exp\left(\sqrt{3}\psi\right). \tag{2.128}$$

Considering the rate of change of (2.127) following the particle, and substituting from above, we obtain an equation for $d\psi/db$, which is integrated to give

$$\int_{\psi_0}^{\psi} \left\{\frac{\sqrt{3} + \tan(\pi/6 - \psi)}{\exp\left(\sqrt{3}\psi\right) - 1}\right\} d\psi = 2\ln\left(\frac{b_0}{b}\right), \tag{2.129}$$

where ψ_0 is the initial value of ψ for a typical particle. The integral must be evaluated numerically to determine ψ for a given ψ_0. For small values of ψ, it is sufficiently accurate to use the approximate expression

$$\frac{\psi}{\psi_0}\left(\frac{\sqrt{3} + \psi}{\sqrt{3} + \psi_0}\right)^2 \left(\frac{2 + \sqrt{3}\psi_0}{2 + \sqrt{3}\psi}\right)^3 = \left(\frac{b_0}{b}\right)^{3/2}, \ \psi < 0.3.$$

The ratio r_0^2/b_0^2 is given by the right-hand side of (2.127) with ψ_0 written for ψ. Consequently,

$$\frac{r_0}{r} = \frac{b_0}{b}\sqrt{\frac{\cos(\pi/6 - \psi)}{\cos(\pi/6 - \psi_0)}} \exp\left[\frac{\sqrt{3}}{2}(\psi - \psi_0)\right]. \tag{2.130}$$

This expression furnishes the total hoop strain whose magnitude at any stage is $\ln(r_0/r)$. The thickness change is given by the differential equation

$$\frac{dh}{h} = -\left(\frac{\sigma_r + \sigma_\theta}{2\sigma_\theta - \sigma_r}\right)\frac{dr}{r} = -2\frac{dr}{r} - \frac{\sqrt{3}\exp\left(\sqrt{3}\psi\right) d\psi}{\exp\left(\sqrt{3}\psi\right) - 1}$$

in view of (2.126) and (2.128), and the fact that $-2db/b$ is equal to the expression in the integral of (2.129). Using the initial condition $h = h_0$ when $r = r_0$ and $\psi = \psi_0$, the above equation is integrated to

$$\frac{h}{h_0} = \left(\frac{b_0}{b}\right)^2 \left\{\frac{1 - \exp\left(-\sqrt{3}\psi_0\right)}{1 - \exp\left(-\sqrt{3}\psi\right)}\right\} \left\{\frac{\cos\left(\pi/6 - \psi\right)}{\cos\left(\pi/6 - \psi_0\right)}\right\}. \tag{2.131}$$

The solution is relevant only over the range of $0 \leq \psi \leq \psi_a$, where ψ_a corresponds to the die throat $r = a$ and is obtainable from (2.127) at any stage of the drawing. The solutions based on the von Mises and modified Tresca criteria are compared with one another in Figs. 2.29 and 2.30, which display the positions of particles and thickness profiles at various stages of the process for a drawing ratio $b_0/a = 2.0$.

The blank will neck at $r = a$ if the radial stress there is equal to Y at the beginning of the drawing. Setting $\psi = \pi/3$ in (2.127), the corresponding drawing ratio is found to be

$$\frac{b_0}{a} = \exp\left(\frac{\pi}{2\sqrt{3}}\right) \approx 2.477.$$

In actual practice, failure would occur over the punch head, where bending and frictional effects are responsible for lowering the limiting drawing ratio to somewhat smaller values. A detailed analysis for the limiting drawing ratio will be given in Section 6.6, including the effect of anisotropy of the sheet metal.

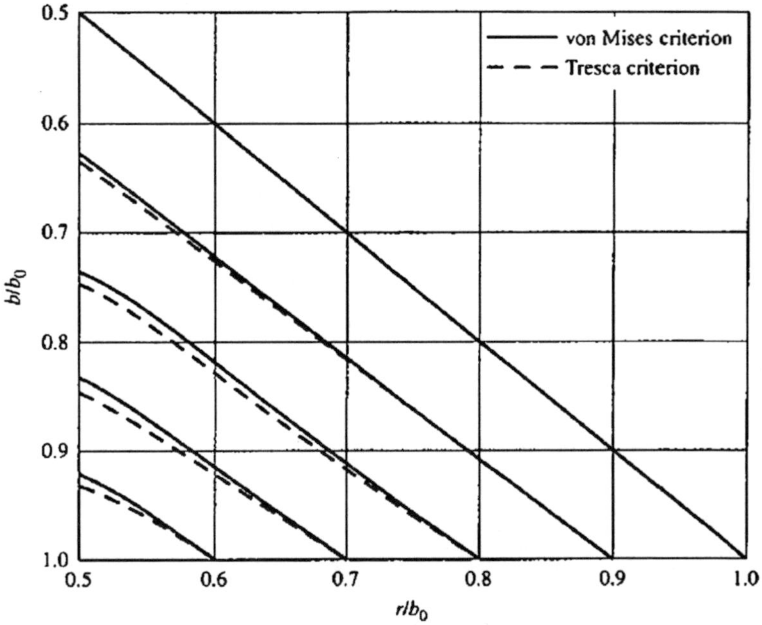

Fig. 2.29 Trajectories of the paths of particles during the progressive radial drawing of a circular blank (no work-hardening)

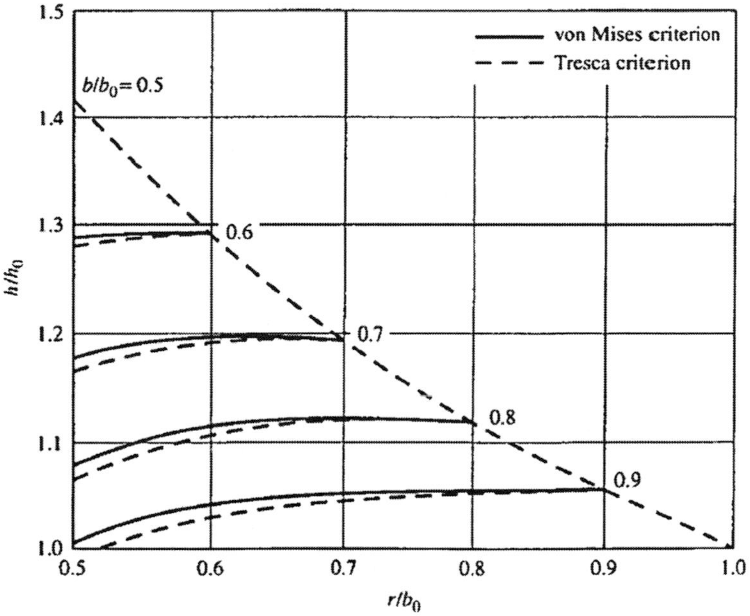

Fig. 2.30 Thickness profile at different stages of radial drawing of a circular blank (no work-hardening)

2.8.3 *Influence of Work-Hardening*

The overall effect of work-hardening is to increase the drawing load and reduce the variation of thickness across the blank at any given stage of the drawing. When the rate of work-hardening is sufficiently high, the drawing stress and the drawing load attain their maximum values at different stages of the process. The paths of the particles are, however, found to be only marginally affected by the work-hardening characteristic of the material. It is therefore a useful approximation to assume that the total equivalent strain at any stage is independent of the rate of work-hardening. By the Lévy–Mises flow rule, the increment of equivalent strain ε in the absence of hardening is

$$d\varepsilon = \left(\frac{2mY}{2\sigma_\theta - \sigma_r}\right) d\varepsilon_\theta = -\left\{\frac{2mY}{2 - \ln(b/r)}\right\}\frac{dr}{r} \approx \frac{8d\rho}{3\rho\,(2-\rho)^2},$$

where $\rho = \ln(b/r)$. Since $\varepsilon = 0$ in the initial state $\rho = \rho_0$, the integration of the above equation furnishes

$$\varepsilon = \frac{2}{3}\left\{\ln\left[\frac{\rho(2-\rho_0)}{\rho_0(2-\rho)}\right] + \frac{2(\rho-\rho_0)}{(2-\rho)\,(2-\rho_0)}\right\}. \qquad (2.132)$$

Equations (2.124), (2.125), and (2.132) indicate that the relationship between the equivalent strain, hoop strain, and thickness strain is

$$\varepsilon = \frac{2}{3}\left(2\ln\frac{r_0}{r} - \ln\frac{h}{h_0}\right).$$

The equivalent strain is therefore significantly higher than the magnitude of the local hoop strain, equal to $\ln(r_0/r)$, in a region near the die throat $r = a$.

The relationship between b/r and b_0/r_0 at any stage may be obtained, to a close approximation, on the basis of a uniform flange thickness equal to $h_0\sqrt{b_0/b}$ at each stage of the drawing. Then by the condition of constancy of volume, we get

$$b_0^2 - r_0^2 = \sqrt{\frac{b_0}{b}}\left(b^2 - r^2\right),$$

this relation furnishes r_0/r explicitly as a function of b/r and b_0/b, the result being

$$\frac{r_0}{r} = \left(\frac{b_0}{b}\right)^{1/4}\left\{1 + \left[\left(\frac{b_0}{b}\right)^{3/2} - 1\right]\frac{b^2}{r^2}\right\}^{1/2}, \qquad (2.133)$$

from which the ratio b_0/r_0 can be found for any given values of b/r and b_0/b, in view of the identity $b_0/r_0 = (b_0/b)(b/r)(r/r_0)$. The ratio b_0/b can be expressed in terms of b/r and $b_0 r_0$ as

$$\frac{b_0}{b} = \left\{\frac{\left(b_0^2/r_0^2\right)\left(b^2/r^2 - 1\right)}{\left(b^2/r^2\right)\left(b_0^2/r_0^2 - 1\right)}\right\}^{2/3}$$

in view of (2.133). The ratio r_0/r can be similarly expressed in terms of b/r and b_0/r_0 using the preceding relation, to establish the variation of the hoop strain in a given element as the drawing proceeds.

Consider now the stress and velocity distributions in a work-hardening blank. If the current uniaxial yield stress of the material is denoted by σ, the modified Tresca criterion in the presence of hardening may be written as

$$\sigma_r - \sigma_\theta = m\sigma = mF(\varepsilon).$$

The function F is defined by the stress–strain curve of the material in simple compression. The substitution in the equilibrium equation (2.121), with the thickness variation disregarded, gives the radial stress as

$$\sigma_r = m\int_r^b \sigma\frac{dr}{r} = m\int_0^\rho F(\varepsilon)\,d\rho. \qquad (2.134)$$

The integration can be carried out numerically for any given stage specified by the ratio b/b_0, the integrand being a known function of ρ in view of (2.132) and

(2.133). Initially, however, $\sigma_r = mY\rho$, as in the case of a nonhardening material. The radial velocity v of a typical particle is governed by the differential equation

$$\frac{1}{v}\frac{\partial v}{\partial \rho} = -\frac{2\sigma_r - \sigma_\theta}{2\sigma_\theta - \sigma_r} = \frac{m\sigma + \sigma_r}{2m\sigma - \sigma_r}.$$

In view of the boundary condition $v = -1$ at $\rho = 0$, the solution to the above equation may be expressed in the form

$$v = -\exp\left\{\int_0^\rho \left(\frac{m\sigma + \sigma_r}{2m\sigma - \sigma_r}\right) d\rho\right\}, \tag{2.135}$$

which furnishes the velocity distribution at any given stage of the drawing. Since $v = -dr/db$, the variation of ρ following the particle is

$$d\rho = \frac{db}{b} = \frac{dr}{r} = -\left(1 + \frac{r}{bv}\right)\frac{dr}{r}.$$

The change in thickness of a given element is finally obtained by the integration of the equation

$$\frac{dh}{h} = -\left(\frac{\sigma_r + \sigma_\theta}{2\sigma_\theta + \sigma_r}\right)\frac{dr}{r} = -\left(\frac{m\sigma - 2\sigma_r}{2m\sigma - \sigma_r}\right)\frac{dr}{r}$$

along the path of the particle. Substituting for dr/r, and using the initial condition $h = h_0$ at $\rho = \rho_0$, we obtain the thickness strain as

$$\ln\left(\frac{h}{h_0}\right) = \int_{\rho_0}^{\rho}\left(\frac{m\sigma - 2\sigma_r}{2m\sigma - \sigma_r}\right)\frac{v\,d\rho}{v + e^{-\rho}}, \tag{2.136}$$

where the integrand is a known function of ρ and ρ_0. The numerical integration is based on a fixed value of ρ_0 and varying values of ρ, which increases with increasing b_0/b. Figure 2.31 shows the computed results for the drawing stress and flange thickness at $r = a$, for a drawing ratio of 2, based on the power law $F(\varepsilon) = \sigma_0\varepsilon^n$ for the strain hardening.

A numerical analysis of the radial drawing process for a work-hardening blank, based on the von Mises yield criterion, can be carried out in a manner similar to that for the stretch-forming process. In terms of an auxiliary angle ψ, the stresses and strains rates may be expressed as

$$\left.\begin{aligned}
\sigma_r &= \frac{2\sigma}{\sqrt{3}}\sin\psi, & \sigma_\theta &= -\frac{2\sigma}{\sqrt{3}}\cos\left(\frac{\pi}{6} + \psi\right),\\
\dot{\varepsilon}_r &= \dot{\varepsilon}\sin\left(\frac{\pi}{6} + \psi\right), & \dot{\varepsilon}_\theta &= -\dot{\varepsilon}\cos\psi,
\end{aligned}\right\} \tag{2.137}$$

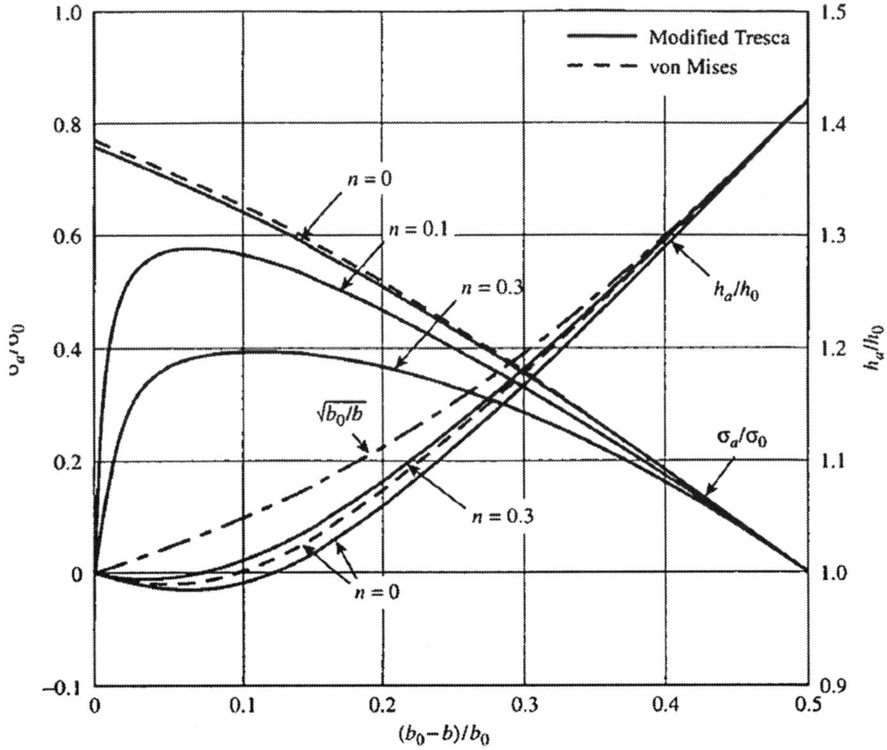

Fig. 2.31 Influence of work-hardening on the drawing stress and flange thickness at the die throat during radial drawing ($b_0/a = 2$)

where σ and $\dot{\varepsilon}$ are the equivalent stress and strain rate, respectively. The relevant differential equations are the equilibrium equation (2.121) and the compatibility equation

$$\frac{\partial \dot{\varepsilon}_\theta}{\partial r} = \frac{\dot{\varepsilon}_r - \dot{\varepsilon}_\theta}{r} \tag{2.138}$$

obtained by eliminating v between the expressions $\dot{\varepsilon}_r = \partial v/\partial r$ and $\dot{\varepsilon}_\theta = v/r$. The substitution from (2.137) into (2.121) and (2.138) results in

$$\left.\begin{aligned} \frac{\partial}{\partial \xi}\left[s \sin\psi \exp(\varepsilon_z)\right] &= -\frac{s}{\xi}\cos\left(\frac{\pi}{6} - \psi\right)\exp(-2\varepsilon_\theta), \\ \frac{\partial}{\partial \xi}(\dot{\varepsilon}\cos\psi) &= -\frac{\sqrt{3}\dot{\varepsilon}}{\xi}\cos\left(\frac{\pi}{6} - \psi\right)\exp(\varepsilon_r - \varepsilon_\theta), \end{aligned}\right\} \tag{2.139}$$

where ε_z is the thickness strain equal to $-(\varepsilon_r + \varepsilon_\theta)$, while ξ and s are dimensionless variables defined as

$$\xi = \frac{r_0}{b_0}, \quad s = \frac{\sigma}{\sigma_0} = f(\varepsilon),$$

with σ_0 denoting a constant stress. The boundary conditions are $\psi = 0, s = f\left[\ln(b_0/b)\right]$ and $\dot{\varepsilon} = 1/b$ at $\xi = 1$. The solution at any stage is relevant only over the range $\xi^* \leq \xi \leq 1$, where $\xi^* = (a/b_0)\exp\left(-\varepsilon_\theta^*\right)$, with the asterisk specifying quantities at the die throat $r = a$.

In order to start the solution, an initial distribution of ψ and s must be found. If we adopt the power law $s = \varepsilon^n$ for strain hardening, the Hencky theory may be used for the solution of the initial problem. Then $\dot{\varepsilon}$ may be replaced by ε in the preceding relations, and (2.139) becomes

$$\frac{n}{\varepsilon}\frac{\partial \varepsilon}{\partial \xi} + \cot\psi \frac{\partial \psi}{\partial \xi} = -\frac{1}{2\xi}\left(1 + \sqrt{3}\cot\psi\right),$$
$$\frac{1}{\varepsilon}\frac{\partial \varepsilon}{\partial \xi} - \tan\psi \frac{\partial \psi}{\partial \xi} = -\frac{\sqrt{3}}{2\xi}\left(\sqrt{3} + \tan\psi\right),$$

to a sufficient accuracy. This is a pair of simultaneous differential equations for ε and ψ, and the elimination of $\partial\varepsilon/\partial\xi$ and $\partial\psi/\partial\xi$ in turn leads to

$$\frac{\partial \psi}{\partial \xi} = -\frac{\left(1 - \sqrt{3}n\tan\psi\right)\left(\sqrt{3} + \tan\psi\right)}{2\xi\left(1 + n\tan^2\psi\right)}, \quad \frac{\partial \varepsilon}{\partial \xi} = -\frac{\varepsilon\left(\sqrt{3} + \tan\psi\right)^2}{2\xi\left(1 + n\tan^2\psi\right)}. \tag{2.140}$$

The first equation of (2.140) is readily integrated to obtain ξ as a function of ψ, using the boundary condition $\psi = 0$ at $\xi = 1$, the result being

$$\xi^2 = \frac{\sqrt{3}}{2}\sec\left(\frac{\pi}{6} - \psi\right)\left(\cos\psi - \sqrt{3}n\sin\psi\right)^{4n/(1+3n^2)}\exp\left\{-\sqrt{3}\left(\frac{1-n^2}{1+3n^2}\right)\psi\right\}. \tag{2.141}$$

The elimination of ξ between the two equations of (2.140) furnishes the differential equation for ε as

$$\frac{\partial \varepsilon}{\partial \psi} = \varepsilon\left(\frac{\sqrt{3}\cos\psi + \sin\psi}{\cos\psi - \sqrt{3}n\sin\psi}\right).$$

Integrating and using the boundary condition $\varepsilon = \ln(b_0/b)$ at $\psi = 0$, we obtain the solution

$$\varepsilon = \left(\ln\frac{b_0}{b}\right)\left\{\frac{\exp\left[\sqrt{3}\left(1 - n^2\right)\psi/\left(1 + 3n^2\right)\right]}{\cos\psi - \sqrt{3}n\sin\psi}\right\}^{(1+3n)/(1+3n^2)} \tag{2.142}$$

Equations (2.141) and (2.142) provide the complete solution for the stresses and strains for an arbitrary small value of $\ln(b_0/b)$. To continue the solution, the second equation of (2.139) is solved for $\dot{\varepsilon}$ using the boundary condition $\dot{\varepsilon} = 1/b$ at $\xi = 1$.

The last two equations of (2.137) then furnish the components of strain increment for an assumed small change in b. Using the hardening law $s = \varepsilon^n$, the first equation of (2.139) is next solved to obtain the corresponding distribution of ψ. The distribution of the radial stress at different stages and the variation of the drawing load with the amount of drawing for $n = 0.4$ are displayed in Fig. 2.32.

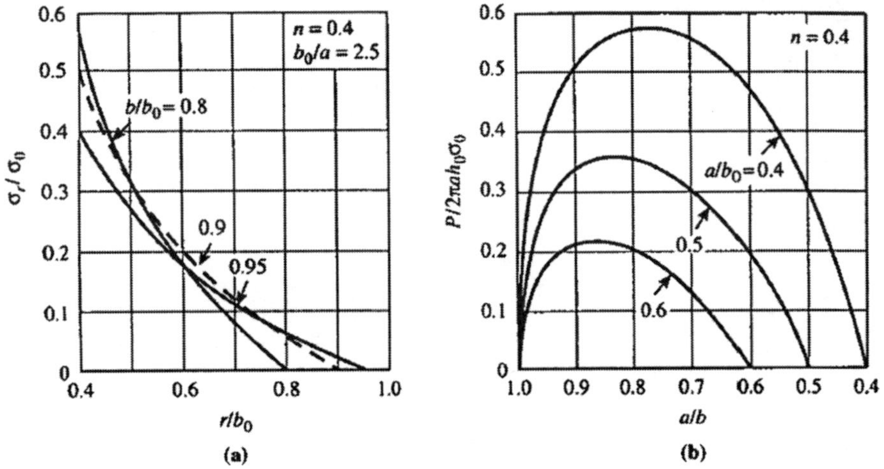

Fig. 2.32 Results for radial drawing of a work-hardening circular blank. (**a**) Distribution of radial stress at different stages and (**b**) variation of drawing load with amount of drawing (after Chiang and Kobayashi, 1966)

Experiments have shown that the limiting drawing ratio in conventional deep drawing with a flat-headed punch is practically independent of the strain-hardening characteristic of the material (Wallace, 1960). The theoretical limiting drawing ratio using a flat-headed punch is therefore about 2.48 for isotropic metals. This ideal value can be approximately realized by lubricating the die and making the punch rough enough to prevent sliding of the material over its surface. Work-hardening has a significant effect, however, on the limiting drawing ratio in the case of pressure-assisted deep drawing, where a high-pressure fluid holds the material against the punch and inhibits fracture over the punch head, as has been demonstrated by Kasuga and Tsutsumi (1965) and El-Sebaie and Mellor (1973). The hydrodynamic deep-drawing process has been further investigated by Yossifon and Tirosh (1985).

2.8.4 Punch Load and Punch Travel

We begin by considering a nonhardening material for which the punch load has its greatest value at the beginning of the drawing. If Tresca's yield criterion is adopted, the radial velocity v_a at $r = a$ is given by (2.123). Since the material entering the die cavity is of length $v_a\, db$ during a small change in the flange radius b, the amount of punch travel z at any stage of the drawing is given by

$$\frac{z}{a} = \int_{b_0}^{b} \left(\frac{va}{a}\right) db = \frac{4\ln(b_0/b)\left[4 - \ln\left(b_0 b/a^2\right)\right]}{\left[2 - \ln(b_0/a)\right]^2 \left[2 - \ln(b/a)\right]^2}, \qquad (2.143)$$

which should be sufficiently accurate for the practical range of values of b_0/a. Setting $b = a$ in (2.143), we obtain the length l of the drawn cup as

$$\frac{l}{a} = \frac{\left[4 - \ln(b_0/a)\right]\ln(b_0/a)}{\left[2 - \ln(b_0/a)^2\right]}.$$

As representative values, l/a is equal to 1.342 when $b_0/a = 2$ and equal to 2.406 when $b_0/a = 2.5$. These lengths are of the same order as those experimentally observed. If the effect of bending over the die profile is disregarded, the punch load P acting at any stage of the drawing is equal to the radial drawing load $2\pi a h_a \sigma_a$. Using (2.122) and (2.125), the punch load can be expressed as

$$\frac{\rho}{2\pi a h_0 Y} = m\sqrt{\frac{b_0}{b}}\ln\left(\frac{b}{a}\right)\exp\left\{\frac{\left[\ln(b_0/a_0)\right]^2}{4 - 2\ln(b_0/a_0)} - \frac{\left[\ln(b/a)\right]^2}{4 - 2\ln(b/a)}\right\}, \qquad (2.144)$$

where a_0 is the initial radius to the element that has currently reached the die throat and can be found by setting $r = a$ and $r_0 = a_0$ in (2.133). The right-hand side of (2.144) progressively decreases from its initial value of $\ln(b_0/a)$ and becomes zero at the end of the drawing process. The thickness distribution in the drawn cup is obtained from the fact that the wall thickness is equal to h_a at a distance z from the bottom of the cup.

When the rate of hardening is sufficiently high, the punch load increases with the punch travel until the load attains a maximum. There is a critical rate of hardening below which the load has its greatest value at the beginning of the drawing (Chakrabarty and Mellor, 1968). The punch load at any stage is given by

$$P = 2\pi a h_a \sigma_a = 2\pi a m \int_a^b hF(\varepsilon)\,\frac{dr}{r} \qquad (2.145)$$

in view of the equilibrium equation and the modified Tresca criterion. Considering the rate of change of the above equation with respect to $b_0 - b$, and then setting $b = b_0$, the initial rate of change of P is obtained as

$$\frac{\dot{P}}{2\pi a m} = F(0)\left\{-\frac{h_0}{b_0} + \int_a^{b0} \dot{h}\frac{dr}{r}\right\} + h_0 F'(0)\int_a^{b0} \dot{\varepsilon}\frac{dr}{r},$$

since $\partial h/\partial r = 0$ in the initial stage. The condition for $\dot{P}$ to be initially positive may therefore be written as

$$b_0 F'(0)\int_a^{b_0} \dot{\varepsilon}\frac{dr}{r} - F(0)\left\{1 - \frac{b_0}{h_0}\int_a^b \dot{h}\frac{dr}{r}\right\} > 0.$$

The expressions for $\dot{h}$ and $\dot{\varepsilon}$ in the initial stage are identical to those for a non-hardening material with a uniaxial yield stress $Y = F(0)$. Using (2.122) and (2.123), and the Lévy–Mises flow rule, it is easily shown that

$$\dot{\varepsilon} = \frac{1}{b_0}\left(1 - \frac{1}{2}\ln\frac{b_0}{r}\right)^{-4}, \quad \dot{h} = h_0\left(\frac{1}{2} - \ln\frac{b_0}{r}\right)\dot{\varepsilon},$$

when $b = b_0$. Inserting these expressions into the preceding integrals and integrating, we obtain the required condition as

$$\frac{F'(0)}{F(0)} > \frac{4 - \ln\alpha\,(2 - \ln\alpha)(4 - \ln\alpha)}{2\ln\alpha\,(2 - \ln\alpha) + (\ln\alpha)^3/3}, \tag{2.146}$$

where α is the drawing ratio b_0/a. The right-hand side of (2.146) decreases as α increases and approaches infinity as α tends to unity. When the strain hardening is represented by the power law $F(\varepsilon) = \sigma_0\varepsilon^n$, which will be assumed in what follows, the above inequality is satisfied for all values of α.

During the continued drawing of a work-hardening blank, the stretching of the flange material that enters the die cavity is usually small and may be disregarded. However, the contribution to the punch travel, made by the stretching of the sheet material over the punch head, must be taken into consideration. Assuming a flat-headed punch, and neglecting the effects of bending and friction, the state of stress over the punch head may be taken as one of balanced biaxial tension of increasing intensity. Since the thickness of the sheet decreases uniformly from h_0 to $h_0\exp(-\varepsilon)$, where ε is the compressive thickness strain at a generic stage, the contribution to the cup length due to the biaxial stretching of the material is

$$d = \frac{a}{2}\left(e^{\varepsilon} - 1\right), \quad \varepsilon^n e^{-\varepsilon} = \left(\frac{h_a}{h_o}\right)\left(\frac{\sigma_a}{\sigma_o}\right). \tag{2.147}$$

The first relation of (2.147) results from the constancy of volume of the material over the punch head, while the second relation follows from the condition of overall longitudinal equilibrium of the partially formed cup. Equation (2.147) holds only for $b \geq b^*$, where b^* denotes the flange radius when the punch load attains its maximum.

During a small time interval represented by db, an annular element of the flange at $r = a$ enters the die cavity to form an elemental ring of length $v_a\,db$. The amount of punch travel due to the combined effects of metal flow into the die, and stretching over the punch, is

$$z = d + \int_{b_0}^{b} v_a db = \frac{a}{2}\left(e^{\varepsilon} - 1\right) + \int_{b_0}^{b} v_a db, \quad b^* \leq b \leq b_0, \tag{2.148}$$

where v_a is given by (2.135) with $\rho = \ln(b/a)$ for the upper limit of integration. Once the punch load has reached its maximum, there is no further stretching of material over the punch head, and the punch travel is then given by

$$z = \frac{a}{2}\left(e^{\varepsilon^*} - 1\right) + \int_{b_0}^{b} v_a db, \quad a \le b \le b^*.$$

The punch load at any stage can be determined approximately from the relation $P = 2\pi a h_a \sigma_a$, or more accurately from (2.145) using (2.136). In Fig. 2.33, the punch load is plotted against the punch travel in dimensionless form for a flat-headed punch, using $b_0/a = 2.0$ and $m = 1.1$. The computed results for the maximum drawing load P^* may be expressed with reasonable accuracy by the empirical equation

$$\frac{P^*}{2\pi a h_0 \sigma_0} = \frac{m}{1+n}\left(\frac{1}{\sqrt{3}}\right)^n \left(\ln\frac{b_0}{a}\right)^{1+\sqrt{n}}, \quad 1 \le \frac{b_0}{a} < 2.5 \tag{2.149}$$

where $m = 1.10$. Due to the effects of friction and bending, which are neglected here, the actual load will be somewhat higher than that predicted by the theory.

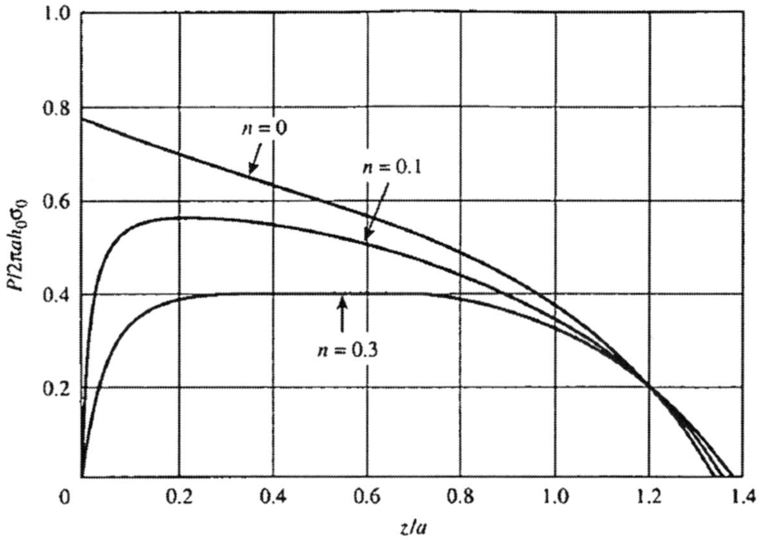

Fig. 2.33 Variation of punch load with punch travel in the deep-drawing process using a flat-headed punch ($b_0/a = 2$)

In the industrial stamping of sheet metal, drawbeads are often used to control the metal flow into the die cavity, thereby suppressing or minimizing the possibility of wrinkling. A drawbead generally consists of a semicylindrical projection on the inner face of the blank holder, designed to fit into a groove provided on the die face, the blank holding force being sufficient to bend the sheet plastically before it is drawn into the die. The restraining force produced by the drawbead is due partly to the bending and unbending of the sheet as it moves over the drawbead, and partly to the friction between the material and the surfaces of contact. The details of the method and its analysis have been discussed by Nine (1978), Wang (1982), Triantafyllidis et al. (1986), and Sanchez and Weinmann (1996).

2.9 Ironing and Flange Wrinkling

2.9.1 Ironing of Cylindrical Cups

In the ironing process, the wall thickness of a deep-drawn cup is reduced by restricting the clearance between the die and the punch. We consider the situation where the internal diameter of the cup is held constant during the process (Hill, 1950a), and the reduction in thickness is achieved by forcing the cup axially through a die as shown in Fig. 2.34a. The circumferential strain imparted to the cup is zero at the punch surface and is negligible elsewhere when the initial wall thickness H is sufficiently small in relation to the cup diameter d. Since the cup is elongated during the ironing, frictional forces arise not only between die and cup but also between cup and punch. If the latter is neglected, the problem is identical to the plane strain drawing of a sheet between rough tapered dies, the surface of the punch being regarded as the center line of the sheet, Fig. 2.34b. The longitudinal force exerted by the punch on the cup wall is

$$P = \pi h (d + h)\, t',$$

where h is the final wall thickness, and t' the mean tensile stress in the ironed wall. By the plane strain theory of sheet drawing (Chakrabarty, 2006), we have

$$t' = t\left\{1 + \left(1 - \frac{1}{2}\ln\frac{1}{1-r}\right)\mu\cot\alpha\right\}$$

to a close approximation, where t is the mean drawing stress for frictionless drawing, μ the coefficient of friction, α the die angle, and $r = (H - h)/H$ the fractional

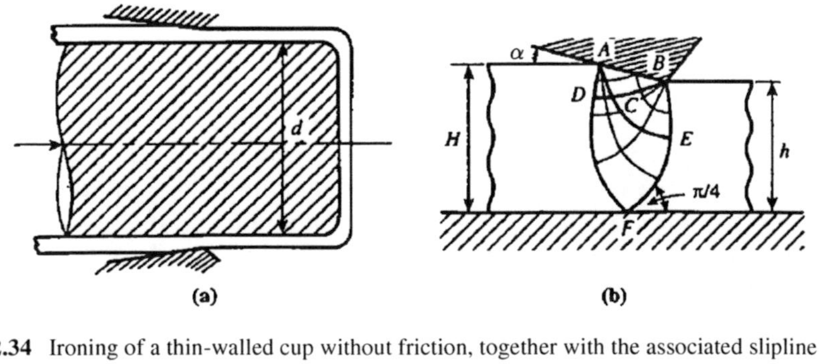

Fig. 2.34 Ironing of a thin-walled cup without friction, together with the associated slipline field

reduction in thickness. Neglecting h/d in comparison with unity, the punch load may be expressed in the nondimensional form

$$\frac{P}{2\pi kdH} = (1-r)\left\{1 + \left(1 - \frac{1}{2}\ln\frac{1}{1-r}\right)\mu\cot\alpha\right\}\frac{t}{2k}, \tag{2.150}$$

where k is the yield stress in pure shear. In terms of a parameter $\lambda = (2/r - 1)\alpha$, the ratio $t/2k$ for a nonhardening material is given by the empirical formulas

$$\left.\begin{aligned} \frac{t}{2k} &= \left(0.93 + 0.07\beta\lambda^2\right)\ln\left(\frac{1}{1-r}\right), & 1.0 \le \lambda \le 2.5, \\ \frac{t}{2k} &= 0.85\lambda^{0.52}\ln\left(\frac{1}{1-r}\right), & 2.5 \le \lambda \le 8.0. \end{aligned}\right\} \tag{2.151}$$

Figure 2.35 shows the variation of the dimensionless punch force with the fractional reduction for different values of α, when $\mu = 0$. Each solid curve passes through a maximum and approaches zero load. The lower broken curve corresponds to $t/2k = -\ln(1-r)$, representing the ideal drawing stress. This curve has a maximum ordinate of $1/e \cong 0.368$ at a fractional reduction of $1 - e^{-1} \cong 0.632$. The upper broken curve on the left represents the bulge limit and is given by the relation

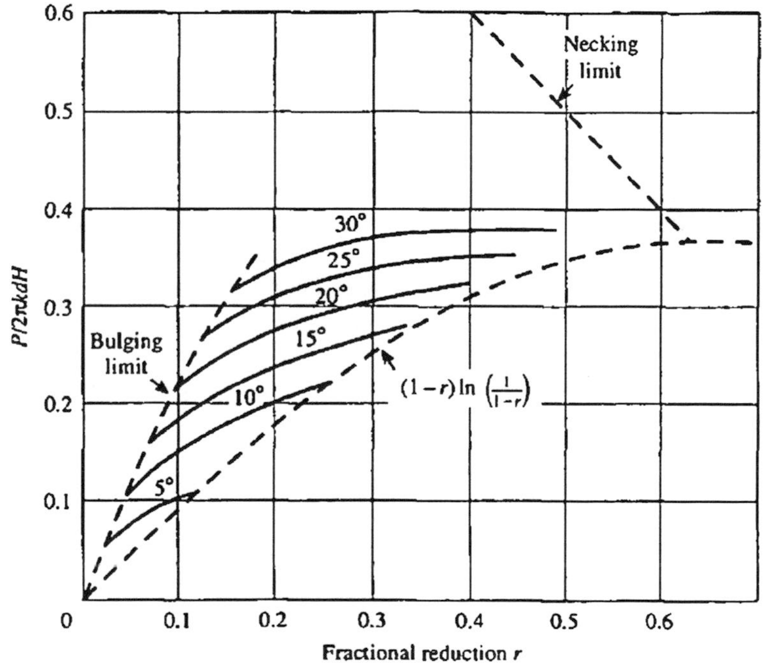

Fig. 2.35 Variation of dimensionless ironing load with fractional reduction in thickness of the cup wall

$r = \alpha\,(0.23 + \alpha/9)$ to a close approximation. The broken straight line on the right-hand side represents $t = 2k$, which corresponds to necking of the ironed wall. The influence of the friction factor in (2.150) is to make the solid curves cross due to the existence of an optimum die angle for which the punch load is a minimum with a given reduction in thickness. The optimum die angle is

$$\alpha_0 \approx 1.49\left(\frac{\mu r^2}{2-r}\right)^{1/3} - 0.08\mu\,(2-r) \tag{2.152}$$

to a close approximation. The influence of work-hardening on the ironing load has been investigated by Fukui and Hansson (1970). An analysis for the redrawing of deep-drawn cups through conical dies has been given by Fogg (1968). The redrawing operation in one or more stages is necessary for producing cups of larger drawing ratio than can be achieved in a single drawing operation.

When the punch is rough, frictional stresses are induced in the cup wall to the left of point F due to the relative motion between the wall and the punch. The frictional stresses are obviously absent over the ironed wall which is carried along with the punch. The effect of punch friction may be approximately estimated by considering the frictional force to be equivalent to a negative back pull, the punch speed being higher than that of the sliding wall. The resultant tension in the ironed wall is thereby reduced by the amount $(1 - \eta)F$, where F denotes the frictional force, and η is the back-pull factor equal to $r(1 + \mu \cot \alpha)$ approximately. The punch load is therefore increased by the amount ηF, where F is an unknown to be determined by experiment.

2.9.2 Flange Wrinkling in Deep Drawing

During the deep-drawing operation, the compressive hoop stress induced in the flange may exceed a certain critical value causing the flange to collapse into waves or wrinkles. When no blank holder is employed to provide lateral support, the critical blank size is generally too small for practical purposes. The development of wrinkles of large amplitude can be prevented by using a blank holder which exerts a lateral pressure on the blank. The effect of the constraint around the inner boundary may be simulated by assuming the flange to be clamped at this radius and subjected to a line load along the circumference of the mean radius $c = (b+a)/2$. An analysis for the associated elastic bending problem reveals that the load per unit circumference is equal to $\beta E\omega$, where ω is the lateral deflection at the mean radius, and

$$\beta \approx \frac{8I}{(b-a)^4} = \frac{2}{3}\left(\frac{h}{b-a}\right)^3, \quad I = \frac{1}{12}\,(b-a)\,h^3, \tag{2.153}$$

over the relevant range, where E is Young's modulus, and I is the second moment of area of the flange section about a radial line through the center. The analysis of the flange wrinkling process in the plastic range is most conveniently carried out by the energy method commonly used in the elastic buckling of columns, if we replace Young's modulus E by a mean value of the reduced modulus

$$E_r = \frac{4ET}{\left(\sqrt{E}+\sqrt{T}\right)^2},$$

where T is the tangent modulus at the current state of hardening. The analysis given below is essentially due to Senior (1956), who extended an earlier treatment by Geckeler (1928), for the flange wrinkling process.

The number of waves m in the wrinkled flange is assumed large enough for each half-wave of the segment to be considered as having a length $l = \pi c/m$ measured along the mean radius $r = c$. For simplicity, the transverse displacement w is assumed to be independent of r, the wave form being expressed as

$$\omega = \delta \sin\frac{\pi s}{l} = \delta \sin\frac{ms}{c}, \tag{2.154}$$

where δ is the amplitude of the half-wave, and s is the distance along the mean circumference. If the radial displacement is assumed to vanish during the formation of wrinkles, the strain energy of bending of a typical flange segment into a half-wave is given by

$$U_1 = \frac{1}{2}E_r I \int_0^l \left(\frac{d^2w}{ds^2}\right)^2 ds = \pi E_r I\left(\frac{\delta^2 m^3}{4c^3}\right),$$

where I is given by (2.153), the current blank thickness h being taken as uniform. For a spring-type blank holder having a stiffness ka, the lateral force acting at the crown of each half-wave is $ka\delta^2/2m$, giving an energy component equal to $ka\delta^2/4m$. In view of the additional lateral force $\beta E_r w$ per unit circumference acting at the mean radius c, the energy stored in the flange segment due to the lateral loading is

$$U_2 = \frac{ka\delta^2}{4m} + \frac{\beta E_r}{2}\int_0^1 w^2 ds = \left(\beta c + \frac{ka}{\pi E_r}\right)\frac{\delta^2}{4m},$$

where β is given by (2.153). Since there are $2m$ half-waves in the wrinkled flange, the total energy of distortion is

$$U = 2m\left(U_1 + U_2\right) = \frac{\pi}{3}E_r h^3\delta^2\left\{\frac{(b-a)\,m^4}{(b+a)^2} + \frac{b+a}{2\,(b-a)^3}\right\} + \frac{1}{2}ka\delta^2. \tag{2.155}$$

It may be noted in passing that the pressure exerted by the blank holder is confined near the periphery $r = b$, where the flange thickness is actually the greatest.

Since w does not depend on r according to (2.154), no work is done by the radial stress distribution during the wrinkling. To obtain the work done by the hoop stress σ_θ due to the circumferential shortening, it is convenient to take the hoop stress distribution approximately in the form

$$\sigma_\theta = -\sigma\left(1 - \ln\frac{b}{r}\right), \quad a \le r \le b,$$

where σ denotes the mean uniaxial yield stress at the considered stage of the flange drawing. The required work is therefore given by

$$W = -\frac{1}{2}\iint h\sigma_\theta\left(\frac{1}{r}\frac{dw}{d\theta}\right)^2 r dr\, d\theta.$$

Substituting for σ_θ, and noting that $w = \delta \sin m\theta$ according to (2.154), it is easily shown that

$$W = \frac{\pi}{4}\sigma h m^2 \delta^2\left(2 - \ln\frac{b}{r}\right)\ln\frac{b}{r}.$$

The critical condition for the wrinkling of the flange is $W = U$. In view of (2.155) and the preceding relation for W, this condition furnishes the critical stress in the form

$$\frac{\sigma}{E_r}\left(2 - \ln\frac{b}{a}\right)\ln\frac{b}{a} = \frac{2h^2}{3a^2}\left(\frac{b-a}{b+a}\right)\left\{2\left(\frac{ma}{b+a}\right)^2 + \left(\frac{a}{b-a}\right)^4\left(\frac{b+a}{ma}\right)^2\right\} + \frac{2k}{\pi E_r}\left(\frac{a}{hm^2}\right). \tag{2.156}$$

This is the required relationship between the various physical parameters affecting the wrinkling process. The value of m that makes σ/E_r a minimum, the remaining quantities being given, is easily shown to be

$$m = 0.841\left(\frac{b+a}{b-a}\right)\left\{1 + \frac{3k}{\pi E_r}\left(\frac{a}{b+a}\right)\left(\frac{b-a}{h}\right)^3\right\}^{1/4}. \tag{2.157}$$

It should be noted that E_r is a function of σ and is defined by the stress–strain curve of the material. When the blank holder is absent ($k = 0$), the preceding results reduce to

$$m = 0.841\left(\frac{b+a}{b-a}\right), \quad \frac{E_r h^2}{\sigma a^2} = 0.531\left(\frac{b^2}{a^2} - 1\right)\left(2 - \ln\frac{b}{a}\right)\ln\frac{b}{a}. \tag{2.158}$$

For a given drawing ratio b_0/a, and a given blank holding pressure defined by k, the critical value of σ/E can be calculated from (2.156) for any stage of the drawing using the relation $h = h_0\sqrt{b_0/b}$. The failure by wrinkling will not occur if the corresponding critical value of σ is greater than the actual mean yield stress in the flange, which may be taken as the arithmetic mean of the yield stresses at $r = a$

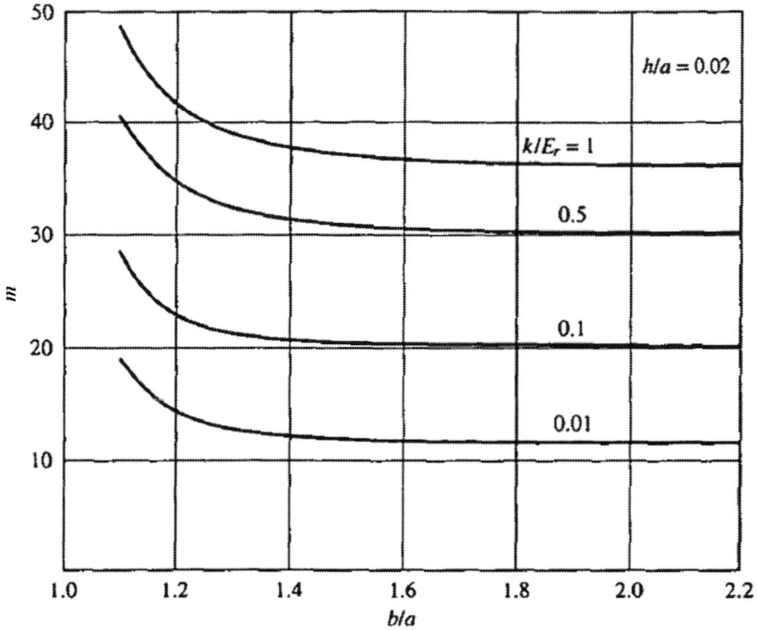

Fig. 2.36 Number of waves in a wrinkled flange as a function of the current radius ratio for different vales of the stiffness parameter

and $r = b$. Figure 2.36 shows the variation of the number of waves of the wrinkled flange with the radius ratio of the flange for various values of the stiffness parameter k/E_r, using a representative value of h/a. The theoretical results are consistent with the observed wrinkling behavior.

The overall effect of the blank holding pressure is to increase the number of waves in the wrinkled flange while reducing their amplitude. In the deep-drawing operation, the number of waves developed by wrinkling should be fairly large so that the wave amplitude is small enough to permit ironing out of the wave tips without unduly raising the punch load. An analysis for flange wrinkling that includes a variation of w with r has been given by Yu and Johnson (1982). A solution based on the consideration of the elastic/plastic bifurcation has been discussed by Chu and Xu (2001). The problem has been treated both analytically and numerically by Carreia et al. (2003). The wrinkling in sheet metal forming has been investigated by Wang and Lee (1989), Hassani and Neale (1991), and Kim et al. (2000).

Problems

2.1 A plane sheet of metal is subjected to biaxial tensile loading in its plane, the strain-hardening exponent of the material being n. Show that the effective strain at the onset of plastic instability, which occurs when the applied loads attain their maximum, is given by

$$\bar{\varepsilon} = \frac{2n\left(1+3\alpha^2\right)^{3/2}}{1+15\alpha^2}, \quad \alpha = \frac{\sigma_1-\sigma_2}{\sigma_1+\sigma_2}$$

Suppose that the load ratio is held constant during the loading, which makes α vary from an initial value α_0. Setting $\lambda_0 = \left(1+3\alpha_0^2\right)^{1/2}$, and assuming a state of plane strain at instability, show that

$$\left(\frac{\lambda_0-1}{\lambda_0+1}\right)\left(\frac{2+\lambda_0}{2-\lambda_0}\right)^2 = \exp\left(-2\sqrt{3}n\right)$$

Compute the numerical values of λ_0 and α_0 in the special case when $n = 0.3$.

2.2 A square plate, whose sides are of length $2a$, has a central circular hole of radius ρa, and is brought to the yield point state by the application of tensile and compressive stresses, each of magnitude σ, in orthogonal directions normal to the edges. Using the stress discontinuity pattern shown in Fig. A, in which the stresses in regions 4, 5, and 6 are equal in magnitude but opposite in sign to those in regions 1, 2, and 3 respectively, and assuming $\rho + \xi > 1$, obtain the lower bounds for a Tresca material as

$$\frac{\sigma}{Y} = \frac{(1-\rho)(1-\xi)}{\rho+\xi-1}, \qquad \frac{\xi}{1-\xi} = \frac{2\sqrt{\rho^2+\xi^2}}{\rho+\xi-1}, \ \rho \le 0.48$$

$$\frac{\sigma}{Y} = -\rho + \sqrt{1-2\rho+2\rho^2}, \ 2\xi = 1+\sqrt{1-2\rho+2\rho^2}, \ \rho \ge 0.48$$

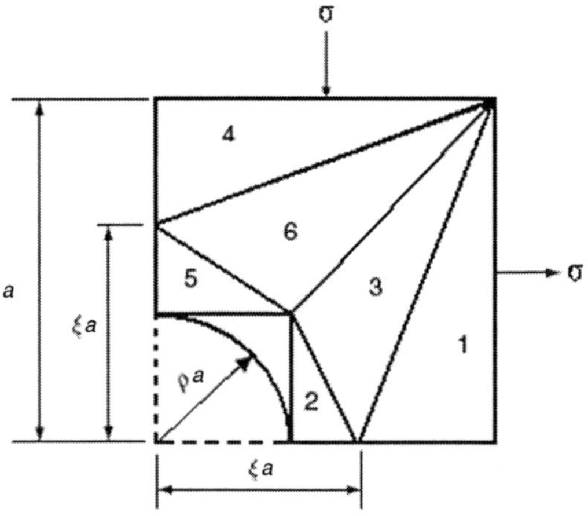

Fig. A

Fig. B

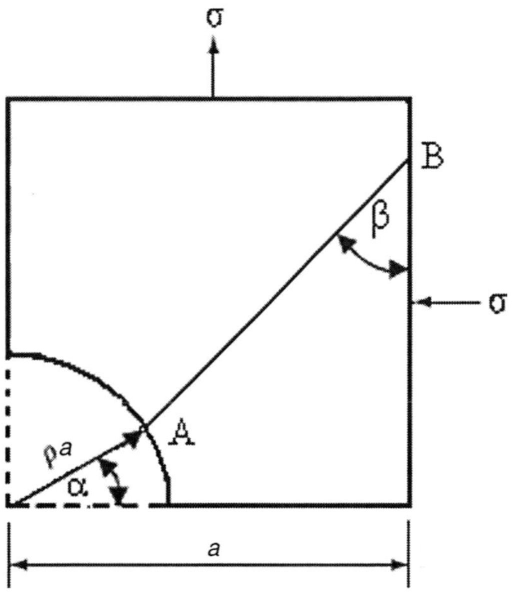

2.3 For sufficiently small values of ρ, a fairly close upper bound solution for the yield point state of the square plate of the preceding problem is obtained by using the velocity discontinuity pattern indicated in Fig. B, where the tangential velocity is discontinuous across the neck *AB*. Assuming the Tresca criterion, show that the best upper bound is given by

$$\frac{2\sigma}{Y} = \frac{1 - \rho\cos\alpha}{\sin 2\beta - \rho\sin\beta\cos(\alpha+\beta)}, \qquad \tan\beta = \frac{\sin\alpha}{\cos\alpha - \rho}.$$

$$4\rho\cos^3\alpha - 4\left(1+\rho^2\right)\cos^2\alpha + \rho\left(1+\rho^2\right)\cos\alpha + 2 = 0$$

Verify that for $\rho \geq 0.37$, an improved upper bound is obtained by setting $\lambda = -1$ in (2.71), which coincides with the preceding lower bound for $\rho \geq 0.48$.

2.4 A circular hole in an infinite plate is expanded from zero radius by the application of a uniform radial pressure, the stress–strain curve of the material being given by

$$\sigma = \sigma_0\left[1 - m\exp(-n\varepsilon)\right]$$

where σ_0, *m,* and *n* are constants. Show that the work done in expanding the hole to any radius *a* for a Tresca material is equal to $\pi\ a\ \rho\ h_0\ Y$, where *Y* is the initial yield stress, h_0 is the initial plate thickness, and ρ is the radius to the rigid/plastic boundary. Using (2.88), show that the formula

$$\frac{a}{\rho} = \frac{1-m}{nm^{2/n}} \ln \left\{ \frac{2+(n-2)\,m^{2/n}}{2\left(1-m^{2/n}\right)} \right\}, \quad 2 \leq n \leq 5$$

is exact for $n = 2$ and $n = 4$ but can be used as a good approximation for other values of n.

2.5 An infinite plate of uniform initial thickness h_0, and containing a circular hole of initial radius a_0, is expanded by the application of an increasing radial pressure p. The material obeys Tresca's yield criterion and its associated flow rule, and strain-hardens according to the law $\sigma = Y\exp\,(n\varepsilon)$, where n is a constant. Show that the current thickness h in the deformed plate varies with the initial radial distance r_0 according to the formula

$$\frac{h_0}{h} = \frac{1}{2+n} \left\{ \left(\frac{\rho}{r_0}\right)^{n/(1+n)} + (1+n)\left(\frac{r_0}{\rho}\right)^{2/(1+n)} \right\}, \quad a \leq r \leq \rho$$

where ρ is the radius to the rigid/plastic boundary. Denoting the current radius of the hole by a, derive the expression

$$\frac{a}{\rho} = \frac{1}{2+n} \left\{ 1 + (1+n)\left(\frac{a_0}{\rho}\right)^{(2+n)/(1+n)} \right\}, \quad \frac{1}{2} \leq \frac{a}{\rho} \leq 1$$

2.6 In the hydrostatic bulging of circular diaphragm using a die aperture of radius a, let the shape of the bulge at each stage be assumed as a spherical cap of angular span equal to 2α. Show that this assumption is consistent with Tresca's associated flow rule for an arbitrary strain-hardening characteristic of the material when the thickness of the bulge is uniform at each stage. Obtain the associated strain distribution in the form

$$\varepsilon_\theta = \frac{1}{2} \ln \left(\frac{1+\cos\phi}{1+\cos\alpha} \right) \quad \varepsilon_\phi = \frac{1}{2} \ln \left\{ \left(\frac{2}{1+\cos\alpha} \right) \left(\frac{2}{1+\cos\phi} \right) \right\}$$

Verify that the magnitude of the polar thickness strain is exactly one-half of that given by (2.125). Derive the expressions for the associated strain rates at a generic stage of the bulge, using the angle α as the time scale.

2.7 In the radial drawing of a circular blank, it is generally a good approximation to assume the total effective strain ε at any radius as equal to the magnitude of the hoop strain, which is given by equation (2.133). If the material yields according to the modified Tresca criterion, and strain-hardens according to the power law $\sigma = \sigma_0 \varepsilon^n$, show that the radial drawing stress at any stage is given by

$$\frac{\sigma_a}{\sigma_0} = m \int_{\varepsilon_b}^{\varepsilon_a} \frac{\varepsilon^n \, d\varepsilon}{1 - \sqrt{\beta}\exp\,(-2\varepsilon)}, \quad \beta = \frac{b_0}{b}$$

where $\varepsilon_b = \ln\,(b_0/b)$, and ε_a follows from (2.133). Obtain a graphical plot of the ratio σ_a/σ_0 against $1 - b/b_0$ when $b_0/a = 2$, using $m = 1.1$ and $n = 0.1, 0.3$, and 0.5.

2.8 For determining the paths of the particles in the radial drawing of a circular blank, a good overall approximation is achieved by writing $dr/r = -\,d\rho/\rho$, where $\rho = \ln\,(b/r)$. Show that the thickness variation based on this approximation for the Tresca criterion is given by

$$\frac{h}{h_0} = \left(\frac{\rho}{\rho_0}\right)^{1/2} \left(\frac{2-\rho}{2-\rho_0}\right)^{3/2}, \quad \frac{r_0^2}{b_0^2} = 1 - \left(\frac{b}{b_0}\right)^{3/2} \left(1 - \frac{r^2}{b^2}\right)$$

where $\rho_0 = \ln(b_0/r_0)$, the last equation being based on the assumption of a uniform blank thickness at each stage. Taking $b_0/a = 2$, make a graphical comparison of the variation of h_a/h_0 with $1 - b/b_0$ given by the preceding relations with that predicted by the more exact solution based on the Tresca criterion presented in Section 2.7 (ii).

2.9 Consider a thin plate in the form of a ring sector defined by concentric circular arcs of radii a and b, where $b > a$. The plate is brought to the yield point state by the application of terminal couples that tend to increase the curvature. The radial and hoop stresses may be expressed in the form

$$\sigma_r = 2k \sin\alpha. \quad \sigma_\theta = \pm 2k \cos\left(\frac{\pi}{6} \mp \alpha\right)$$

which satisfies the von Mises yield criterion parametrically through α. The upper sign applies to $r \le c$, and the lower sign to $r \ge c$, where c is the radius to the neutral surface which corresponds to $\alpha = \alpha_0$. Using the equilibrium equation, show that

$$\frac{b^2}{c^2} = \frac{2}{\sqrt{3}} \cos\left(\frac{\pi}{6} - \alpha_0\right) \exp\left(\sqrt{3}\alpha_0\right), \quad \frac{c^2}{ab} = \left(1 - \frac{4}{3}\sin^2\alpha_0\right)^{-1/2}$$

2.10 For usual values of the b/a ratio in the preceding problem, the auxiliary angle α is fairly small, so that α^2 is generally negligible compared to unity. Show that the value of α at $r = c$, and the associated yield couple M_0, is given approximately by

$$\alpha_0 = \frac{\sqrt{3}}{4} \ln\left(\frac{b}{a}\right), \quad \frac{c^2}{ab} \approx 1 + \frac{1}{8}\left(\ln\frac{b}{a}\right)^2, \quad M_0 \equiv \frac{\sqrt{3}}{4} kh(b-a)^2$$

The last expression may be obtained by taking the hoop stress distribution approximately as

$$\sigma_\theta = k\left(\sqrt{3} + \alpha\right), \quad a \le r \le c; \quad \sigma_\theta = -k\left(\sqrt{3} - \alpha\right), \quad c \le r \le b$$

where α in the two regions is expressed in the way similar to that for α_0 .

Chapter 3
Axisymmetric and Related Problems

Many physically important problems of plasticity involve solids of revolution which are loaded symmetrically about their geometrical axes. Typical examples of axially symmetrical distribution of stress and strain are provided by the expansion of circular cylindrical tubes, upsetting of cylindrical blocks, extrusion of cylindrical billets, and the axial drawing of wires and tubes. In the theoretical analysis of such problems, which involve large plastic strains, it is customary to assume the material to be rigid/plastic, for which no deformation can occur until the load attains the yield point value. The rigid/plastic assumption is adequate for the estimation of the yield point load itself, and also for the determination of stresses and strains in the plastically deforming zone during the continued loading. In this chapter, we shall deal with several examples in which the stress distribution is axially symmetrical, together with a few related problems involving three-dimensional states of stress.

3.1 Basic Theory and Exact Solutions

3.1.1 Fundamental Relations

An axially symmetrical state of stress is defined as that in which the only nonzero stress components are σ_{r}, and σ_θ, σ_z, and τ_{rz}, referred to the cylindrical coordinates (r, θ, z), where the z-axis coincides with the axis of symmetry. Since the deformation is confined to the meridian plane, the nonzero velocity components are u and w in the radial and axial directions, respectively. The symmetry requires the stress and velocity components to be independent of the angle θ measured round the z-axis. When body forces are neglected, the equations of equilibrium become

$$\left.\begin{aligned} \frac{\partial \sigma_r}{\partial r} + \frac{\partial \tau_{rz}}{\partial z} + \frac{\sigma_r - \sigma_\theta}{r} &= 0, \\ \frac{\partial \sigma_z}{\partial_z} + \frac{\partial \tau_{rz}}{\partial r} + \frac{\tau_{rz}}{r} &= 0. \end{aligned}\right\} \tag{3.1}$$

J. Chakrabarty, *Applied Plasticity, Second Edition*, Mechanical Engineering Series,

DOI 10.1007/978-0-387-77674-3_3,

Suppose that the material is nonhardening and obeys the von Mises yield criterion and the Lévy–Mises flow rule. Then, in the region where the material is stressed to the yield limit, the stresses must satisfy the equation

$$(\sigma_r - \sigma_\theta)^2 + (\sigma_\theta - \sigma_z)^2 + (\sigma_z - \sigma_r)^2 + 6\tau_{rz}^2 = 2Y^2, \tag{3.2}$$

where Y is the uniaxial yield stress. Solutions based on the Tresca theory will be discussed in the next section. If the plastic region contains a finite portion of the axis of symmetry, $\tau_{rz} = 0$ and $\sigma_r = \sigma_\theta$ along the axis for the stresses to be finite. The components of strain rate for an axially symmetrical deformation mode are given by

$$\dot{\varepsilon}_r = \frac{\partial u}{\partial r}, \quad \dot{\varepsilon}_\theta = \frac{u}{r}, \quad \dot{\varepsilon}_z = \frac{\partial w}{\partial z}, \quad 2\dot{\gamma}_{rz} = \frac{\partial u}{\partial z} + \frac{\partial w}{\partial r}. \tag{3.3}$$

Since the axis of symmetry is a streamline, $u = 0$, along $r = 0$. The sum of the three rates of extension must vanish by the condition of plastic impressibility, giving

$$\frac{\partial u}{\partial r} + \frac{u}{r} + \frac{\partial w}{\partial z} = 0. \tag{3.4}$$

In terms of a positive scalar factor of proportionality $\dot{\lambda}$, the Lévy–Mises flow rule may be written as

$$\left.\begin{aligned} \dot{\varepsilon}_r &= \dot{\lambda}\,(2\sigma_r - \sigma_\theta - \sigma_z), & \dot{\varepsilon}_\theta &= \dot{\lambda}\,(2\sigma_\theta - \sigma_z - \sigma_r), \\ \dot{\varepsilon}_z &= \dot{\lambda}\,(2\sigma_z - \sigma_r - \sigma_\theta), & \dot{\gamma}_{rz} &= 3\dot{\lambda}\tau_{rz}. \end{aligned}\right\} \tag{3.5}$$

Since there are only three equations involving the four unknown stress components, the velocity equations must be considered along with the stress equations to obtain the solution under prescribed boundary conditions.

Let $-p$ denote the mean normal stress in the meridian plane, and τ the magnitude of the maximum shear stress in the same plane, so that the principal stresses in this plane are equal to $-p \pm \tau$. If the remaining principal stress σ_θ is denoted by σ, the von Mises yield criterion (3.2), expressed in terms of the principal components, becomes

$$(p + \sigma)^2 + 3\tau^2 = Y^2. \tag{3.6}$$

The two orthogonal directions of maximum shear stress in the meridian plane are designated by α and β such that the axis of the algebraically greater principal stress in this plane makes a counterclockwise angle of $\pi/4$ with the α-direction. The stresses acting on a curvilinear element bounded by neighboring α-and β-lines are shown in Fig. 3.1. If ψ denotes the counterclockwise angle made by a typical flow line with the local α-direction, the normal components of the stress, referred to a pair of axes (ξ, η) taken along and perpendicular to the flow direction, are

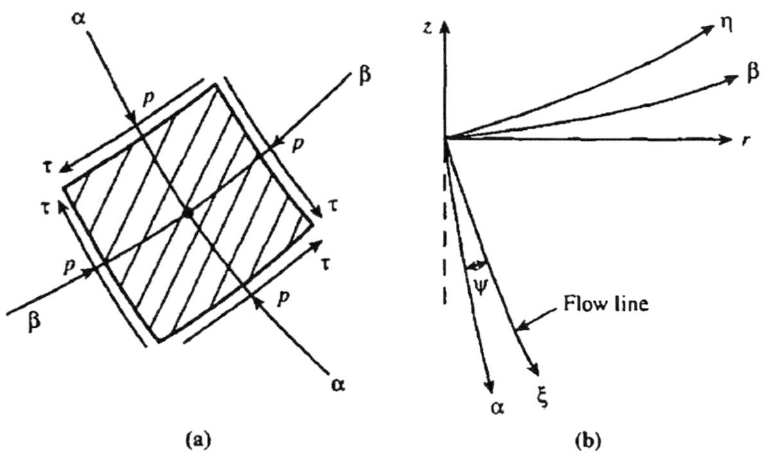

Fig. 3.1 Curvilinear element with maximum shear directions in the meridian plane

$$\sigma_\xi = -p + \tau \sin 2\psi, \quad \sigma_\eta = -p - \tau \sin 2\psi.$$

If the resultant velocity of the particle is denoted by v, its radial component is $u = \mathrm{v}(\partial r/\partial s)$, where s is the arc length along the flow line. The plastic flow rule then furnishes the rates of extension as

$$\dot{\varepsilon}_\xi = \frac{\partial \mathrm{v}}{\partial s} = \dot{\lambda}\left(2\sigma_\xi - \sigma_\eta - \sigma\right),$$

$$\dot{\varepsilon}_\theta = \frac{\mathrm{v}}{r}\frac{\partial r}{\partial s} = \dot{\lambda}\left(2\sigma - \sigma_\xi - \sigma_\eta\right).$$

Substituting for σ_ξ and σ_η into the above relations, and eliminating $\dot{\lambda}$ between the resulting equations, we obtain the differential relation

$$2\frac{d\mathrm{v}}{\mathrm{v}} + \left(1 - \frac{3}{p+\sigma}\sin 2\psi\right)\frac{dr}{r} = 0, \tag{3.7}$$

which holds along the flow lines. When the flow field is given, the maximum shear directions at each point are obtained from the fact that the principal axes of stress and strain rate coincide in an isotropic material. Equations (3.6) and (3.7) then furnish the quantities $p + \sigma$ and τ throughout the field.

Suppose that the stress and velocity components are given along a curve C in the meridian plane. Assuming the stresses to be continuous, the condition of continuity of the velocity requires $\dot{\lambda}$ to be continuous in view of the second equations of (3.3) and (3.5). It follows from the remaining equations of (3.3) and (3.5) that the velocity gradients are, in general, uniquely determined by the given stress and velocity distributions along C. The problem of axial symmetry is therefore not hyperbolic for a von Mises material.

3.1.2 Swaging in a Contracting Cylinder

A solid cylindrical bar is held in a closely fitting container over a part of its length. The swaging action is simulated by the hypothetical process in which the material is squeezed out by a radial contraction of the cylindrical cavity. The frictional stress is assumed to have a constant magnitude mk $(0 \leq m \leq 1)$ along the length of the die, where k is the yield stress in shear. In analogy to the solution for the plane strain compression, we assume the shear stress τ_{rz} to be independent of z. Taking the plane $z = 0$ through one end of the container, Fig. 3.2, we write

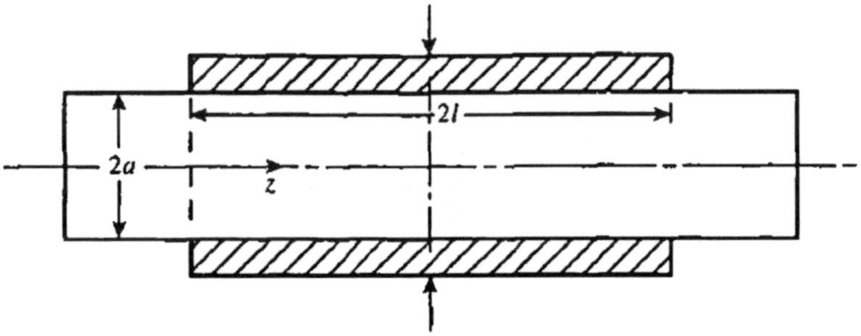

Fig. 3.2 Plastic flow from a radially contracting rough cylindrical container

$$\tau_{rz} = mk\left(\frac{r}{a}\right), \quad u = -U\left(\frac{r}{a}\right), \tag{3.8}$$

where a denotes the current radius of the bar, and U the radial speed of the container. Since $\dot{\varepsilon}_r = \dot{\varepsilon}_\theta$ in view of the assumed expression for u, the flow rule gives $\sigma_r = \sigma_\theta$ and the yield criterion (3.2) then furnishes

$$\sigma_z - \sigma_r = Y\sqrt{1 - \frac{m^2 r^2}{a^2}}.$$

Substituting for τ_{rz} into the equilibrium equations (3.1), and using the fact that σ_r and σ_θ are equal to one another, we obtain the stress distribution

$$\left.\begin{aligned} \frac{\sigma_r}{Y} &= \frac{\sigma_\theta}{Y} = -2\left(\frac{mz}{\sqrt{3}a} + A\right), \\ \frac{\sigma_z}{Y} &= -2\left(\frac{mz}{\sqrt{3}a} + A\right) + \sqrt{1 - \frac{m^2 r^2}{a^2}}, \end{aligned}\right\} \tag{3.9}$$

in view of the yield condition, where A is a constant (Hill, 1950a). Since the resultant axial force across a transverse section must balance the frictional resistance, we have

$$2\pi \int_0^a r\sigma_z dr = -2\pi amkz \text{ or } A = \frac{1-\left(1-m^2\right)^{3/2}}{3m^2}.$$

Substituting from (3.8) into the incompressibility equation (3.4), and integrating, we obtain the axial velocity in the form

$$w = 2U\left[\frac{z}{a} + f(r)\right].$$

The function $f(r)$ can be determined by considering the ratio $\dot{\gamma}_{rz}/\dot{\varepsilon}_\theta$ given by the flow rule (3.5). Thus

$$\frac{\partial w}{\partial r} = \left(\frac{6\tau_{rz}}{\sigma_r - \sigma_z}\right)\frac{u}{r} = \frac{2\sqrt{3}Ur}{a^2}\left(1 - \frac{m^2r^2}{a^2}\right)^{-1/2}.$$

Inserting the preceding expression for w, and integrating, we obtain the function $f(r)$, and the axial velocity is finally given by

$$\frac{w}{U} = \frac{2z}{a} - \frac{2\sqrt{3}}{m}\left(1 - \frac{m^2r^2}{a^2}\right)^{1/2} + B, \tag{3.10}$$

where B is another constant. The rate of axial flow of material across a transverse plane must equal the rate of displacement of material from the container, which requires

$$2\pi \int_o^a wrdr = -2\pi a\,(l - z)\,U,$$

where $2\,l$ denotes the length of the container. The substitution for w into the above relation gives

$$B = -\frac{2\,l}{a} + \frac{4}{\sqrt{3}m^3}\left[1 - \left(1 - m^2\right)^{3/2}\right].$$

The velocity solution is not valid near the central section, where a rigid material must exist at all stages. It follows from the assumed expression for u that the ratio r/a remains constant following the motion of a particle. Indeed, the rate of change of r/a is

$$\frac{d}{dt}\left(\frac{r}{a}\right) = \frac{1}{a}\frac{dr}{dt} - \frac{r}{a^2}\frac{da}{dt} = \frac{u}{a} + \frac{rU}{a^2} = 0.$$

Let ξ denote the axial distance by which a typical particle is in advance of the boundary particle originally situated on the same transverse section. Since the difference between the axial velocities of the two particles is $d\xi/dt$, (3.10) gives

$$-\frac{d\xi}{da} = \frac{1}{U}\frac{d\xi}{dt} = \frac{2\xi}{a} + \frac{2\sqrt{3}}{m}\left\{\sqrt{1 - \frac{m^2 r^2}{a^2}} - \sqrt{1 - m^2}\right\}$$

for $z < l$. Since r/a is constant during the motion, the above equation is readily integrated to

$$\left\{\frac{\sqrt{3}ma^2\xi}{2\left({a_0}^3 - a^3\right)} + \sqrt{1 = m^2}\right\}^2 + \frac{m^2 r^2}{a^2} = 1, \tag{3.11}$$

where a_0 is the initial radius of the bar. An original transverse section is therefore distorted into an ellipsoid of revolution with semimajor and semiminor axes of lengths a/m and $\left(a_0^3 - a^3\right)/\sqrt{3}ma^2$, respectively.

3.1.3 Fully Plastic State in a Cylindrical Tube

A hollow cylinder of internal radius a and external radius b is rendered completely plastic by the combined action of an internal pressure p and an axial tensile force P. If the tube is sufficiently long, the distribution of stress and strain should be independent of z (except near the ends), and transverse plane sections should remain plane during the loading. The axial strain rate, denoted by $\dot{\varepsilon}$, is therefore a constant at each instant. Then the incompressibility equation is immediately integrated, and the velocity field is given by

$$u = \frac{1}{2}\left(\dot{\lambda}\frac{a^2}{r} - \dot{\varepsilon}r\right), \quad w = \dot{\varepsilon}z, \tag{3.12}$$

where $\dot{\lambda}$ is a constant, and the plane $z = 0$ is assumed to coincide with the central section of the tube. Since the shear strain rate $\dot{\gamma}_{rz}$ vanishes in view of (3.12), the coordinate axes are in the principal directions, and the principal strain rates are

$$\dot{\varepsilon}_r = \frac{1}{2}\left(\dot{\lambda}\frac{a^2}{r^2} + \dot{\varepsilon}\right), \quad \dot{\varepsilon}_\theta = \frac{1}{2}\left(\dot{\lambda}\frac{a^2}{r^2} - \dot{\varepsilon}\right), \quad \dot{\varepsilon}_z = \dot{\varepsilon}.$$

Using the above relations and the Lévy–Mises flow rule, the ratios of the principal stress differences are obtained as

$$\frac{\sigma_z - \sigma_r}{\sigma_\theta - \sigma_r} = \frac{1}{2}\left(1 + \frac{3\dot{\varepsilon}r^2}{\dot{\lambda}a^2}\right), \quad \frac{\sigma_\theta - \sigma_z}{\sigma_\theta - \sigma_r} = \frac{1}{2}\left(1 - \frac{3\dot{\varepsilon}r^2}{\dot{\lambda}a^2}\right). \tag{3.13}$$

It follows from (3.13) and the third equation of (3.5) that $\sigma_\theta - \sigma_r$ has the same sign as that of $\dot{\lambda}$, which represents the fractional rate of change of the internal volume of the tube. The substitution for $\sigma_z - \sigma_r$ and $\sigma_\theta - \sigma_z$ from (3.13) into the yield criterion (3.2) furnishes

$$\sigma_\theta - \sigma_r = 2\,k\left(1 + \frac{3\dot{\varepsilon}^2 r^4}{\dot{\lambda}^2 a^4}\right)^{-1/2}. \tag{3.14}$$

The sign of the radical must be positive since $\sigma_r < \sigma_\theta$. Inserting into the first equation of (3.1), and noting that $\tau_{rz} = 0$, this equation is integrated under the boundary condition $\sigma_r = 0$ at $r = b$ to give

$$\frac{\sigma_r}{k} = \sinh^{-1}\left(\frac{\dot{\lambda}a^2}{\sqrt{3}\dot{\varepsilon}b^2}\right) - \sinh^{-1}\left(\frac{\dot{\lambda}a^2}{\sqrt{3}\dot{\varepsilon}r^2}\right) \tag{3.15}$$

The remaining equilibrium equation is identically satisfied. Since $\sigma_r = -p$ at $r = a$, the internal pressure is obtained from (3.15) as

$$\frac{p}{k} = \sinh^{-1}\left(\frac{\dot{\lambda}}{\sqrt{3}\dot{\varepsilon}}\right) - \sinh^{-1}\left(\frac{\dot{\lambda}a^2}{\sqrt{3}\dot{\varepsilon}b^2}\right). \tag{3.16}$$

For a given radius ratio b/a, (3.16) gives $\dot{\lambda}/\dot{\varepsilon}$ as a function of p/k. The axial force is also a function of $\dot{\lambda}/\dot{\varepsilon}$ and is obtained from the fact that

$$\sigma_z = \frac{1}{2}\left(\sigma_\theta + \sigma_r\right) + \frac{3\dot{\varepsilon}r^2}{2\dot{\lambda}a^2}\left(\sigma_\theta - \sigma_r\right)$$

in view of (3.13). The distribution of the axial stress σ_z has a resultant equal to P. Since $r(\sigma_\theta + \sigma_r) = \partial(r^2\sigma_r)/\partial r$ in view of the first equation of (3.1), we have

$$p = 2\pi\int_a^b r\sigma_z dr = \pi a^2 p + \pi a^2 k\int_a^b \frac{6r^3}{a^4}\left(\frac{\dot{\lambda}^2}{\dot{\varepsilon}^2} + \frac{3r^4}{a^4}\right)^{-1/2} dr$$

in view of (3.14) and the boundary conditions for σ_r. Carrying out the integration, and setting $P = \pi a^2\sigma$, it is easily shown that

$$\frac{\sigma - p}{k} = \sqrt{\frac{\dot{\lambda}^2}{\dot{\varepsilon}^2} + \frac{3b^4}{a^4}} - \sqrt{\frac{\dot{\lambda}^2}{\dot{\varepsilon}^2} + 3}. \tag{3.17}$$

For a tube subjected to an axial tension only, $p = 0$ and $\dot{\lambda} = 0$, giving $\sigma = \sqrt{3}\,k\left(b^2/a^2 - 1\right)$. For a closed-ended tube under internal pressure alone, $\sigma = p$ and $\dot{\varepsilon} = 0$, giving the result $p = 2\,k\,\ln(b/a)$. In general, the fully plastic state of the tube is characterized by an interaction curve obtained by plotting $(\sigma - p)/k$ against p/k as shown in Fig. 3.3. The interaction locus is convex, and the vector $(\dot{\lambda}, \dot{\varepsilon})$ is parallel to the outward normal to the locus at any point $(p, \sigma - p)$. Indeed, the rate of plastic work per unit internal volume, which is easily shown to be equal to $p\dot{\lambda} + (\sigma - p)\,\dot{\varepsilon}$, has a maximum value when $(\dot{\lambda}, \dot{\varepsilon})$ is associated with $(p, \sigma - p)$ through the normality rule of plastic flow.

When the tube is subjected to an external pressure q, along with an internal pressure p and an axial load $P = \pi a^2\sigma$, the problem is equivalent to a uniform

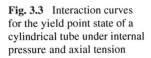

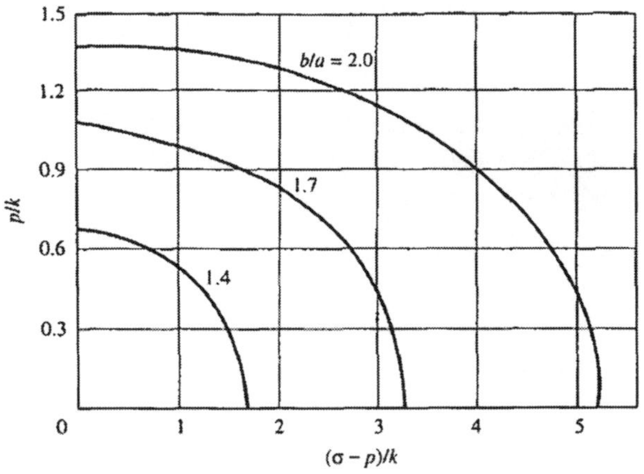

Fig. 3.3 Interaction curves for the yield point state of a cylindrical tube under internal pressure and axial tension

hydrostatic pressure of amount q superimposed on the stress distribution produced by an internal pressure $p - q$ and an axial load $P + \pi(b^2 - a^2)q$. The interaction curve is then defined by (3.16) and (3.17) with the left-hand sides replaced by $(p - q)/k$ and $(\sigma - p)/k + b^2q/a^2k$, respectively. The vector $(\dot{\lambda}, \dot{\varepsilon})$ is then directed along the exterior normal to the locus obtained by plotting $\sigma - p + (b^2/a^2)q$ against $p - q$, as may be verified by computing the coefficients of $\dot{\lambda}$ and $\dot{\varepsilon}$ in the modified expression for the plastic work rate. For a given wall ratio b/a, the interaction locus defined in this way is identical to that without the external pressure (Hill, 1976).

Consider now the related process in which a uniform radial, pressure q is applied to external surface $r = b$, while a smooth rigid mandrel of radius a closely fits into the bore of the cylinder. The circumferential strain rate vanishes at the bore, giving $\dot{\lambda} = \dot{\varepsilon}$, while the condition of zero axial force gives $\sigma = 0$. The modified forms of (3.16) and (3.17) furnish the applied external pressure q as a function of b/a, the result being

$$\left(\frac{b^2}{a^2} - 1\right)\frac{q}{k} = \sqrt{1 + \frac{3b^4}{a^4}} - 2 - \sinh^{-1}\left(\frac{a^2}{\sqrt{3}b^2}\right) + \sinh^{-1}\left(\frac{1}{\sqrt{3}}\right). \qquad (3.18)$$

Since the radial velocity at $r = b$ is of magnitude $\dot{\varepsilon}\left(b^2 - a^2\right)/2b$, the rate of external work done is

$$\dot{W} = \pi q\dot{l}\left(b^2 - a^2\right) = \pi ql\left(b^2 - a^2\right)\dot{\varepsilon} = 2qb\dot{b}V/\left(b^2 - a^2\right)$$

in view of the constancy of volume $V = \pi(b^2 - a^2)l$, where l is the length of the tube. Substituting from (3.18), and integrating, we may obtain the total work done per unit volume in deforming the tube from an initial external radius b_0 to a final radius b. The result is relevant in predicting the ideal die pressure for the extrusion of axisymmetric tubes (Haddow, 1962).

3.1.4 Plastic Flow Through a Conical Channel

Consider the flow of a rigid/plastic material through a conical channel, the direction of flow being toward the virtual apex 0, Fig. 3.4. The frictional stress exerted by the channel is assumed to have a constant value *mk,* where $0 \leq m \leq 1$. The channel is assumed long enough for the flow to be steady and the end effects negligible (Shield, 1955a). It is convenient to introduce spherical polar coordinates (r, ϕ, θ), where ϕ is measured from the axis of symmetry, the boundary of the channel being defined by $\phi = \pm\alpha$ If the flow is everywhere radial, the nonzero components of the strain rate are given by

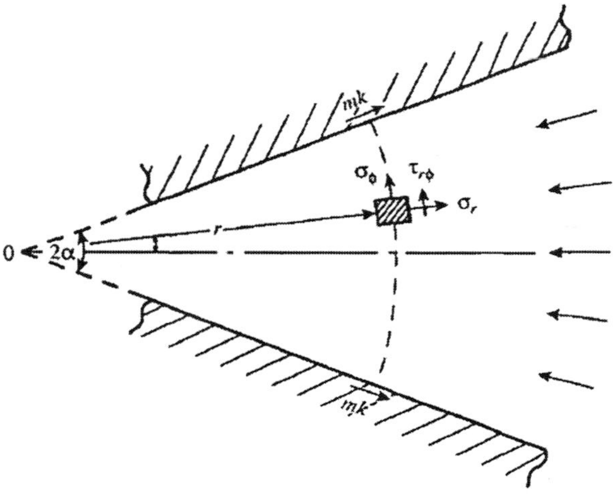

Fig. 3.4 Converging flow of plastic material through a rough conical channel

$$\dot{\varepsilon}_r = \frac{\partial u}{\partial r}, \quad \dot{\varepsilon}_\phi = \dot{\varepsilon}_\theta = \frac{u}{r}, \quad \dot{\gamma}_{r\phi} = \frac{1}{2r}\frac{\partial u}{\partial \phi}, \tag{3.19}$$

where u denotes the outward radial velocity of a typical particle. The nonzero components of the stress are σ_r, σ_ϕ, σ_θ, and $\tau_{r\phi}$, where $\sigma_\theta = \sigma_\phi$ by the Lévy–Mises flow rule. The equations of equilibrium in spherical coordinates become

$$\left.\begin{aligned} &r\frac{\partial \sigma_r}{\partial r} + \frac{\partial \tau_{r\phi}}{\partial \phi} + 2\left(\sigma_r - \sigma_\phi\right) + \tau_{r\phi}\cot\phi = 0,\\ &r\frac{\partial \tau_{r\phi}}{\partial r} + \frac{\partial \sigma_\phi}{\partial \phi} + 3\tau_{r\phi} = 0. \end{aligned}\right\} \tag{3.20}$$

Due to the axial symmetry, the stress and velocity components are independent of θ. The condition $\sigma_\theta = \sigma_\phi$ reduces the von Mises yield criterion to

$$\left(\sigma_r - \sigma_\phi\right)^2 + 3\tau_{r\phi}^2 = 3k^2,$$

which is satisfied by the introduction of a nondimensional stress $S = S(\phi)$, such that

$$\tau_{r\phi} = ks, \quad \sigma_r - \sigma_\phi = \sqrt{3}k\left(1 - s^2\right)^{1/2}. \tag{3.21}$$

Since the friction opposes the relative motion of the plastic material, s has the same sign as ϕ. Substituting for $\tau_{r\phi}$ into the second equation of (3.20), and integrating, we have

$$\sigma_\phi = kf(r) - 3k\int s\, d\phi$$

Combining this with the first equation of (3.20), and using (3.21), we obtain the relation

$$rf'(r) + s'(\phi) + s\cot\phi + 2\sqrt{3}\left(1 - s^2\right)^{1/2} = 0.$$

The first term on the left-hand side is a function of r only while the remaining terms are functions of ϕ only. Consequently,

$$rf'(r) = -c, s'(\phi) + s\cot\phi + 2\sqrt{3}\left(1 - s^2\right)^{1/2} = c,$$

where c is a constant. The first of these equations is integrated to $f(r) = c\ln(a/r)$, where a is a constant radius. The second equation must be solved under the boundary conditions $s(0) = 0$ and $s(\alpha) = m$. When $s(\phi)$ has been determined, the stress distribution is obtained from (3.21) and the relation

$$\sigma_\phi = kc\ln(a/r) - 3k\int s(\phi)\, d\phi.$$

The relationship between s and ϕ is most conveniently obtained by rewriting the above differential equation in the alternative form

$$\left[c - s\cot\phi - 2\sqrt{3}\left(1 - s^2\right)^{1/2}\right]\phi'(s) = 1, \tag{3.22}$$

the boundary conditions being $\phi(0) = 0$ and $\phi(m) = \alpha$. It follows from (3.22) that c must be greater than $2\sqrt{3}$, which is the value required for a smooth channel. For $m > 1$, (3.22) may be integrated numerically with a selected value of c, using the condition $\phi(0) = 0$, to find $\phi(s)$ in the interval $0 \le s \le 1$. The value of α for any given value of m, corresponding to a chosen value of c, is obtained from the remaining boundary condition $\phi(m) = \alpha$.

The radial velocity u must be determined from the incompressibility condition and the isotropy condition, regardless of the particular flow rule. In terms of the components of stress and true strain rate, these conditions can be expressed as

$$\dot{\varepsilon}_r + 2\dot{\varepsilon}_\phi = 0, \quad \frac{\dot{\gamma}_{r\phi}}{\dot{\varepsilon}_r - \dot{\varepsilon}_\phi} = \frac{\tau_{r\phi}}{\sigma_r - \sigma_\phi}.$$

Substituting from (3.19), the first of these equations is readily integrated, and the result is then inserted into the second equation to give

$$u = -\left(\frac{a}{r}\right)^2 g(\phi), \quad g'(\phi) = -\frac{2\sqrt{3}s}{\sqrt{1-s^2}} g(\phi),$$

in view of (3.21). The integration of the second of the above equations furnishes

$$g(\phi) = U \exp\left\{-2\sqrt{3} \int s\left(1 - s^2\right)^{-1/2} d\phi\right\} \tag{3.23}$$

where U is a positive velocity. The solution, which is now formally complete, may be applied to the wire-drawing process on the assumption that the redundant shear at the die entry and exit may be disregarded. The influence of elastic deformation on the conical flow has been examined by Danyluk and Haddow (1965).

An exact solution involving a radial flow field also exists when the material, having an initial shear yield stress k, work-hardens linearly with a plastic modulus H. Introducing dimensionless parameters β and λ, and using the flow rule along with the expression $u = (a/r)^2 g(\phi)$ for the radial velocity, we write

$$\sigma_r - \sigma_\phi = \sqrt{3}k\lambda, \quad \tau_{r\phi} = k\beta\lambda, \quad g'(\phi) = -2\sqrt{3}\beta g(\phi). \tag{3.24}$$

Evidently, β is a function of ϕ only. Since the plastic flow is in a steady state, the effective strain rate £ is equal to $u(\partial e/\partial r)$. If the effective stress is denoted by σ, then

$$\frac{\partial \sigma}{\partial r} = H\frac{\partial \varepsilon}{\partial r} = H\frac{\dot{\varepsilon}}{u} = -\frac{2H}{r}\sqrt{1+\beta^2}$$

The integration of this equation results in

$$\sigma = \sqrt{1+\beta^2}\left(2H \ln\frac{a}{r} + F\right),$$

where a is a reference radius, and F is an unknown function of ϕ. The substitution of (3.24) into the von Mises yield criterion, where $\sqrt{3}k$ must be replaced by σ, then furnishes

$$\lambda = \left(\sigma/\sqrt{3}k\right)\left(1+\beta^2\right)^{1/2} = 2h \ln\frac{u}{r} + F(\phi), \tag{3.25}$$

where $h = H/\sqrt{3}k\ h = H/y/3k$. Inserting (3.24) in the equilibrium equation (3.20), we obtain a pair of differential equations for β and F, which can be integrated using the boundary conditions

$$\beta = 0 \text{ at } \phi = 0, \ \beta \left(1 + \beta^2\right)^{-1/2} = m \text{ at } \phi = \alpha$$

for the parameter β, the shear stress $\tau_{r\phi}$, being considered as m times the local shear yield stress $\sigma/\sqrt{3}$ along the conical surface. The solution has been given in detail by Durban (1979). The steady-state plastic flow field around a rigid cone penetrating in an infinitely extended medium has been investigated by Durban and Fleck (1992).

3.2 Slipline Fields and Indentations

3.2.1 Relations Along the Sliplines

It is assumed that the material obeys Tresca's yield criterion and the associated flow rule with a constant shear yield stress k. In the principal stress space, the plastic state of stress is represented by a point on the surface of a prism, whose cross section is a regular hexagon, and whose axis is equally inclined to the σ_1-, σ_2-, and σ_3-axes, where $\sigma_3, = \sigma_\theta$. A section of the prism made by a plane perpendicular to the σ_3-axis is also a hexagon whose center is located at $(\sigma_\theta, \sigma_\theta)$ with respect to the σ_1- and σ_2-axes, Fig. 3.5a. The stress states corresponding to the corners A, C, D, and F are such that the circumferential stress σ_θ is equal to one of the principal stresses in the meridian plane. This condition, which is originally due to Haar and von Karman (1909), has been examined by Shield (1955b).

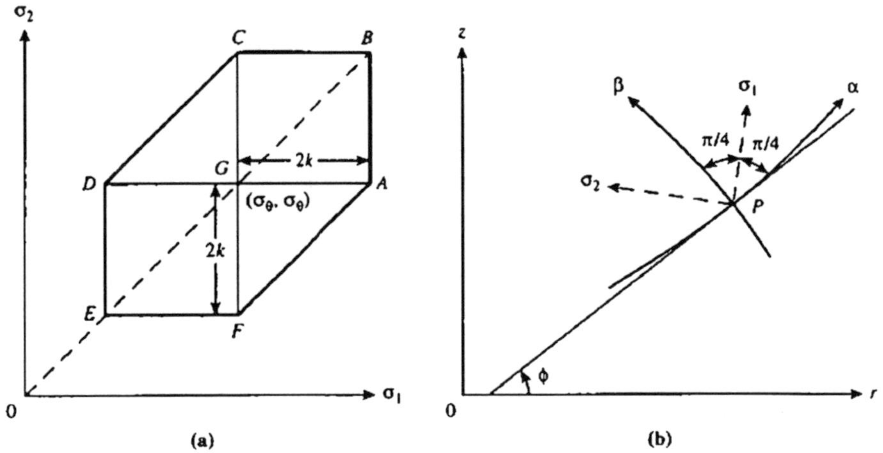

Fig. 3.5 Tresca's yield locus and the orientation of sliplines in an axially symmetrical plastic state of stress

The sliplines, which are the lines of maximum shear stress in the meridian plane, are designated by α and β with the convention that the direction of the algebraically greater principal stress is obtained by a counterclockwise rotation of π/4 from that of the α-line. Consider the corner F of the Tresca hexagon, where σ_θ is equal to the algebraically greater principal stress σ_1, the yield criterion being $\sigma_1 - \sigma_2 = 2\,k$. We denote by ϕ the angle of inclination of the α-line measured away from the r-axis toward the z-axis, Fig. 3.5b. The stresses acting on a curvilinear element formed by neighboring α- and β-lines are those shown in Fig. 3.1a, where $\tau = k$ and $p = -(\sigma_r + \sigma_\theta)/2$. In terms of p and ϕ, the stress components in cylindrical coordinates (r, θ, z) can be expressed as

$$\left.\begin{aligned} \sigma_r = -P - k\sin 2\phi, \quad & \sigma_z = -p + k\sin 2\phi, \\ \tau_{rz} = k\cos 2\phi, \quad & \sigma_\theta = -p + k. \end{aligned}\right\} \tag{3.26}$$

By analogy with the plane strain problem, the stress equations are hyperbolic and the characteristics coincide with the sliplines. The substitution from (3.26) into the equilibrium equations (3.1) results in

$$\frac{\partial p}{\partial r} + 2k\left(\cos 2\phi \frac{\partial \phi}{\partial r} + \sin 2\phi \frac{\partial \phi}{\partial z}\right) + \frac{k}{r}(1 + \sin 2\phi) = 0,$$

$$\frac{\partial p}{\partial r} + 2k\left(\sin 2\phi \frac{\partial \phi}{\partial r} - \cos 2\phi \frac{\partial \phi}{\partial z}\right) - \frac{k}{r}\cos 2\phi = 0.$$

From geometry, the space derivatives along the α- and β-lines, forming a right-handed pair of curvilinear axes, are given by

$$\left.\begin{aligned} \frac{\partial}{\partial s_\alpha} &= \cos\phi \frac{\partial}{\partial r} + \sin\phi \frac{\partial}{\partial z}, \\ \frac{\partial}{\partial s_\beta} &= \cos\phi \frac{\partial}{\partial z} - \sin\phi \frac{\partial}{\partial r}. \end{aligned}\right\} \tag{3.27}$$

Using (3.27), the preceding pair of differential equations can be suitably combined to give

$$\frac{\partial p}{\partial s_\alpha} + 2k\frac{\partial \phi}{\partial s_\alpha} + \frac{k}{r}(\cos\phi + \sin\phi) = 0,$$

$$\frac{\partial p}{\partial s_\beta} + 2k\frac{\partial \phi}{\partial s_\beta} - \frac{k}{r}(\cos\phi + \sin\phi) = 0.$$

These are the equilibrium equations considered along the sliplines. Since $dr = ds_\alpha \cos\phi$ along an α-line and $dr = -ds_\beta \sin\phi$ along a β-line, the above equations may be written in the differential form

$$\left.\begin{aligned} dp - 2kd\phi + k(1 + \tan\phi)\frac{dr}{r} &= 0 \quad \text{along an } \alpha\text{-line}, \\ dp - 2kd\phi + k(1 + \cot\phi)\frac{dr}{r} &= 0 \quad \text{along a } \beta\text{-line}. \end{aligned}\right\} \tag{3.28}$$

These equations differ from Hencky's plane strain equations only in the terms containing r. For numerical computations, the differential relations (3.27) may be replaced by the corresponding finite difference equations, and the values of p, ϕ, r, and z throughout the field can be determined by an iterative method.

The velocity field can be determined by using the incompressibility equation (3.4) and the isotropy condition which requires the principal axes of stress and strain rate to coincide. In terms of the components of the velocity, the isotropy condition may be written as

$$\left(\frac{\partial u}{\partial z}+\frac{\partial w}{\partial r}\right)\sin 2\phi+\left(\frac{\partial u}{\partial r}-\frac{\partial w}{\partial z}\right)\cos 2\phi=0. \tag{3.29}$$

The system of equations (3.4) and (3.29) is hyperbolic, and the characteristics are again the sliplines. If the velocity components along the α- and β-lines are denoted by

U and W, respectively, then

$$u=U\cos\phi-W\sin\phi,\quad w=U\sin\phi+W\cos\phi.$$

Substituting from above into the field equations (3.4) and (3.29), and using (3.27), it is easy to show that

$$\frac{\partial U}{\partial s_\alpha}+\frac{\partial W}{\partial s_\beta}-W\frac{\partial \phi}{\partial s_\beta}+\frac{U\cos\phi-W\sin\phi}{r}=0,$$
$$\frac{\partial U}{\partial s_\alpha}-\frac{\partial W}{\partial s_\beta}-W\frac{\partial \phi}{\partial s_\alpha}-U\frac{\partial \phi}{\partial s_\beta}=0.$$

Adding these two equations and subtracting one from the other, the derivatives along the α- and β-lines are separated from one another, and the velocity equations are then expressed in the differential form

$$\left.\begin{aligned} dU-Wd\phi+(U-W\tan\phi)\frac{dr}{2r}&=0 \quad \text{along an }\alpha\text{-line},\\ dW-Ud\phi+(W-U\cot\phi)\frac{dr}{2r}&=0 \quad \text{along a }\beta\text{-line}\end{aligned}\right\} \tag{3.30}$$

These relations differ from Geiringer's equations in plane strain only in the terms in r. Equation (3.30) can be integrated numerically along the characteristics using the finite difference approximation and an iterative method.

Let $\dot\lambda$ denote the magnitude of the maximum engineering shear strain rate in the meridian plane. It is the shear strain rate associated with the α- and β-directions, taken as a right-handed pair of curvilinear axes, and is given by

$$\dot\gamma=\frac{\partial U}{\partial s_\beta}+\frac{\partial W}{\partial s_\alpha}+U\frac{\partial \phi}{\partial s_\alpha}-W\frac{\partial \phi}{\partial s_\beta}. \tag{3.31}$$

The assumption that the stress state is represented by the corner F of the yield locus requires $\dot\varepsilon_1\geq 0$ and $\dot\varepsilon_3\geq 0$. These restrictions are equivalent to $\dot\varepsilon_r+\dot\varepsilon_z+\dot\gamma\geq 0$

and $\dot{\varepsilon}_\theta \geq 0$. Since the material is incompressible, the condition for validity of the initial assumption becomes $\dot{\gamma} \geq u/r \geq 0$.

For a von Mises material, the sliplines are not characteristics, but the velocity relations (3.30) are still applicable. An additional relation involving the velocity components is furnished by (3.7), where $v = \sqrt{U^2 + W^2}$ and $\psi = tan^{-1}(W/U)$. The stress components σ_r, σ_z, and τ_{rz} are given by (3.26) with k replaced by τ, which is the maximum shear stress in the meridian plane. The substitution into the equilibrium equations (3.1) and the use of (3.27) lead to the differential relations

$$\left.\begin{aligned} dp + 2\tau d\phi + (\sigma + p + \tau \tan\phi)\frac{dr}{r} &= \frac{\partial \tau}{\partial s_\beta} ds_\alpha, \\ dp - 2\tau d\phi + (\sigma + p + \tau \cot\phi)\frac{dr}{r} &= \frac{\partial \tau}{\partial s_\alpha} ds_\beta, \end{aligned}\right\} \quad (3.32)$$

along the α- and β-lines, respectively. These equations are complicated by the presence of the derivatives of τ on the right-hand side (Hill, 1950a). Equations (3.6), (3.7), (3.30), and (3.32) form a set of six equations for the six unknowns p, ϕ, τ, σ, U, and W, and the solution must be carried out numerically using the appropriate boundary conditions.

When the state of stress is represented by the sides *AB, BC, DE,* and *EF* of the Tresca hexagon, the numerically lesser principal strain rate in the meridian plane vanishes by the associated flow rule, and the circumferential strain rate is therefore equal in magnitude to the numerically greater principal strain rate. Considering the side *AB,* for instance, the yield criterion and the flow rule may be written as

$$\sigma_1 - \sigma_3 = 2k, \quad \sigma_1 > \sigma_2 > \sigma_3, \quad \dot{\varepsilon}_2 = 0, \dot{\varepsilon}_1 = -\dot{\varepsilon}_3 = -u/r > 0.$$

Let ψ denote the angle of inclination of the first principal axis with respect to the r-axis, measured toward the z-axis. The components of the strain rate in the meridian plane can be expressed in terms of u/r and ψ as

$$\frac{\partial u}{\partial r} = -\frac{u}{r}\cos^2\psi, \quad \frac{\partial w}{\partial z} = -\frac{u}{r}\sin^2\psi, \quad \frac{\partial u}{\partial z} + \frac{\partial w}{\partial r} = -\frac{u}{r}\sin^2\psi.$$

This is a system of three equations for the three unknowns, u, w, and ψ, defining a kinematically determined problem. The equations governing the three principal stresses are obtained by substituting the expressions

$$\sigma_r = \sigma_1\cos^2\psi + \sigma_2\sin^2\psi, \quad \sigma_z = \sigma_1\sin^2\psi + \sigma_2\cos^2\psi,$$
$$\tau_{rz} = (\sigma_1 - \sigma_2)\sin\psi\cos\psi,$$

into the equilibrium equations (3.1), which are supplemented by the yield criterion. The two systems of equations for the stress and velocity distributions can be shown to be hyperbolic with the principal stress trajectories as characteristics. The relations holding along the principal lines have been discussed by Lippmann (1962, 1965), and further examined by Desdo (1971).

3.2.2 Indentation by a Flat Punch

The plane surface of a semi-infinite medium is indented by a flat-ended smooth punch of circular cross section. The medium occupies the region $z \geq 0$, and the origin of coordinates is taken at the center O of the punch whose radius is equal to OA. The slipline field shown in Fig. 3.6, which is due to Shield (1955b), is based on the assumption that the plastic stress field throughout the deforming zone is represented by the corner F of the yield hexagon (Fig. 3.5). The field in the region ABC is generated by the stress-free surface AB, where the sliplines meet at an angle of $\pi/4$. It follows from the nature of the problem that AC is an α-line and BC a β-line. The characteristic AC generates the fan ACD of angle $\pi/2$, where A is a singular point. The region OAD is defined by the characteristic AD and the condition that the sliplines meet OA at an angle of $\pi/4$.

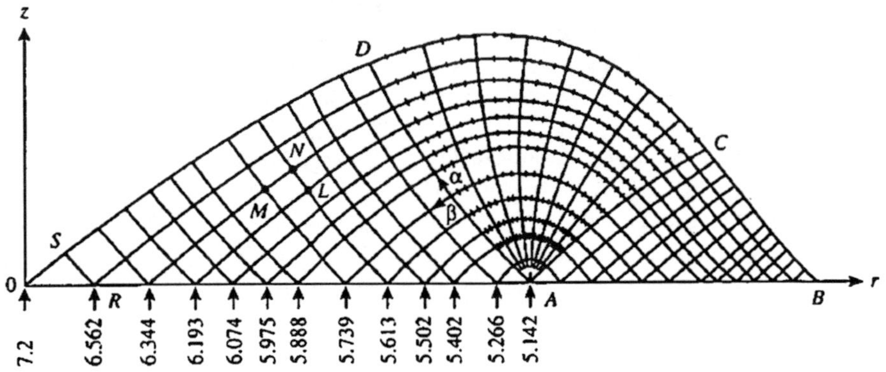

Fig. 3.6 Slipline field due to indentation of a semi-infinite medium by a flat-based cylindrical punch (after R. T. Shield, 1955b)

The values of p and ϕ at a generic point of the field ABC depend only on the angle ψ made by the radius vector with the r-axis. Since $\tan \psi = z/r$, we obtain

$$r\frac{\partial}{\partial r} = -\sin\psi\cos\psi\frac{d}{d\psi}, \quad r\frac{\partial}{\partial z} = -\cos^2\psi\frac{d}{d\psi}.$$

Using these expressions for the partial derivatives, the equilibrium equations (3.1) are reduced to the differential relations

$$p'\sin\psi - 2k\sin(2\phi - \psi)\,\phi' - k(1 + \sin 2\phi)\sec\psi = 0,$$
$$P'\sin\psi - 2k\sin(2\phi - \psi)\,\phi' - k(1 + \sin 2\phi)\sec\psi = 0,$$

where the prime denotes differentiations with respect to ψ. The elimination of p' between the above relations and the substitution $2(\phi - \psi) = \lambda$ result in

$$\lambda'\sin\lambda + 3\sin\lambda + \cos\lambda\tan\psi = -1. \tag{3.33}$$

When $\lambda(\psi)$ has been determined by the numerical integration of (3.33) under the boundary condition $\lambda(0) = \pi/2$, the quantity $p(\psi)$ can be found from the equation

$$p'/k = \lambda' \cos\lambda + (1 - \sin\lambda)\tan\psi + 3\cos\lambda,$$

which is obtained by combining (3.33) with one of the preceding relations. The integration of the above equation results in

$$p/k = \lambda \sin\lambda + \ln\sec\psi + \int_0^{\psi} (3\cos\lambda - \sin\lambda\tan\psi)\, d\psi, \tag{3.34}$$

in view of the boundary condition $p(0) = k$. If ρ denotes the length of the radius vector from 0 to a generic point on a typical slipline, then $d\rho/d\psi = \rho\cot(\phi - \psi)$ along an α-line and $d\rho/d\psi = -\rho\tan(\phi - \psi)$ along a β-line, giving

$$\left.\begin{aligned} \rho &= \rho_1 \exp\left(\int_0^{\psi} \cot(\lambda/2) d\psi\right) && \text{along an } \alpha\text{-line} \\ \rho &= \rho_2 \exp\left(-\int_0^{\psi} \tan(\lambda/2) d\psi\right) && \text{along a } \beta\text{-line} \end{aligned}\right\} \tag{3.35}$$

where ρ_1 and ρ_2 are the values of ρ at the boundary $\psi = 0$ corresponding to the α-and β-lines, respectively. Equating the two expressions for ρ at point C, we have

$$OB = OA \exp\left\{2\int_0^{\psi_0} \mathrm{cosec}\lambda \mathrm{d}\psi\right\},$$

where ψ_0 is the value of ψ at C. It is determined from the condition that the β-line through B passes through O. The geometry of the field ABC is determined from (3.35) when λ. is a known function of ψ.

The distribution of p and ϕ in the region $OACD$ must be found by the numerical integration of (3.28). Let L and M denote two neighboring diagonal points of the slipline net, and N the point of intersection of the α-line through L and the β-line through M. It is required to locate N and to find the values of p and ϕ at the point N, when those at L and M are known. The finite difference form of (3.28) is

$$\left.\begin{aligned} \frac{P_N - P_L}{2k} + (\phi_N - \phi_L) &= -\frac{(r_N - r_L) + (Z_N - Z_L)}{r_N + r_L}, \\ \frac{P_N - P_M}{2k} + (\phi_N - \phi_M) &= -\frac{(r_N - r_M) - (Z_N - Z_M)}{r_N + r_M}. \end{aligned}\right\} \tag{3.36}$$

Using the familiar small arc approximation for constructing the slipline field, N is taken to be the point of intersection of two straight lines from L and M whose slopes are the mean of those of the α-lines at L and N and the β-lines at M and N. Consequently,

$$
\begin{aligned}
r_N - r_L &= \tfrac{1}{2}\left(\cot\phi_L + \cot\phi_N\right)\left(z_N - z_L\right), \\
z_N - z_M &= \tfrac{1}{2}\left(\cot\phi_M + \cot\phi_N\right)\left(r_M - r_N\right).
\end{aligned}
\tag{3.37}
$$

The solution is most conveniently obtained by setting $\phi_N \approx \phi_L$ in the first equation and $\phi_N \approx \phi_M$ in the second equation of (3.37), and substituting the corresponding values of r_N and z_N into (3.36) to obtain the second approximation to ϕ_N. With the help of the modified values of r_N and z_N calculated from (3.37), the values of ϕ and $p/2\,k$ at N are finally computed from (3.36). The distribution of the dimensionless punch pressure q/k, which is equal to $1 + p/k$ along OA, is indicated in Fig. 3.6. The average punch pressure is found to be $5.69\,k$, and the distance of point B from the origin is found to be 1.58 times the punch radius OA.

The boundary conditions for the velocity field require the normal velocity to be constant over the area of contact OA, and zero along the boundary $ODCB$ of the region of incipient deformation. The condition $U = 0$ along $ODCB$ gives $W = 0$ along this slipline, by the second equation of (3.30), in order to avoid an infinite velocity at the origin O. If the speed of the punch is assumed to be unity, $W - U = \sqrt{2}$ along OA. The velocity field in a small region ORS near the origin is obtained with sufficient accuracy on the assumption that the sliplines are inclined at a constant angle $\pi/4$ with the coordinate axes. The velocity components (u, w) in this region can be directly obtained from the condition of incompressibility and the fact that $\dot{\gamma}_{rz} = 0$ throughout ORS. Since the velocity in this region depends only on the angle ψ, we have

$$
\begin{aligned}
-\sin\psi \frac{du}{d\psi} + u \sec\psi + \cos\psi \frac{dw}{d\psi} &= 0, \\
\cos\psi \frac{du}{d\psi} - \sin\psi \frac{dw}{d\psi} &= 0,
\end{aligned}
$$

in view of (3.4) and the last equation of (3.3). The elimination of $dw/d\psi$ and $du/d\psi$ between the above relations gives

$$
\cos 2\psi \frac{du}{d\psi} + u \tan\psi = 0, \quad \cos 2\psi \frac{dw}{d\psi} + u = 0,
$$

Using the boundary conditions $u = w = 0$ along $\psi = \pi/4$, and $w = 1$ along $\psi = 0$, the solutions of these equations are found to be

$$
u = \frac{2}{\pi}\sqrt{1 - \tan^2\psi}, \quad w = \frac{2}{\pi}\cos^{-1}(\tan\psi) \tag{3.38}
$$

The magnitude of the maximum engineering shear rate in the (r, z) -plane is easily shown to be $\dot{\gamma} = (u/r)\sec 2\psi$, and the restriction $\dot{\gamma} \geq u\,/\,r \geq 0$ is therefore satisfied.

The velocity is evidently continuous across the boundary $ODCB$. Since the velocity is known on RS, the boundary conditions $U = W = 0$ along SD and $W - U = \sqrt{2}$ along RA suffice to determine the velocity field in the region $RSDA$. The known velocity on AD and the condition of zero velocity on DCB furnish the velocity distribution in the remaining part $ABCD$ of the deforming zone. The point A is a singularity of the velocity field. The numerical computation is based on (3.30), whose finite difference form is

$$\left.\begin{aligned} U_N - U_L &= \frac{1}{2}(\phi_N - \phi_L)(W_N + W_L) + \frac{1}{2}(U_L \cot\phi_L + U_N \cot\phi_N - W_L - W_N)(z_L - z_N)/(r_L + r_N), \\ W_N - W_M &= \frac{1}{2}(\phi_M - \phi_N)(U_M + U_N) + \frac{1}{2}(U_M \cot\phi_M + U_N \cot\phi_N - W_M - W_N)(r_N - r_M)/(r_N + r_M), \end{aligned}\right\} \tag{3.39}$$

where L and M are two diagonal points of a slipline mesh, and N is the neighboring nodal point where the velocity is required. The solution is obtained by setting on the right-hand side $U_N \approx U_L$, $W_N \approx W_L$ in the first equation, and $U_N \approx U_M$, $W_N \approx W_M$ in the second equation of (3.39), as a first approximation, and then finding the second approximation to compute the final values of U_N and W_N. The values of $\dot{\gamma}$ and u/r obtained numerically from the calculated values of U and W are found to satisfy the inequalities $\dot{\gamma} \geq u/r \geq 0$. The slipline field can be extended into the rigid region in a statically dmissible manner (Shield, 1955b), indicating that the solution is complete. The indentation of a semi-infinite medium using a rough punch has been discussed by Eason and Shield (1960).

3.2.3 Indentation by a Rigid Cone

Consider the indentation of a semi-infinite medium by a smooth conical indenter whose axis is perpendicular to the plane free surface. The problem corresponds to the Rockwell hardness test of a sufficiently prestrained metal. During the indentation process, the free surface of the indented material is forced out to form a raised lip adjacent to the indenter. Since the problem does not involve any characteristic length, the slipline field must be geometrically similar at each stage of the indentation. Since the flat punch is the limiting case of a cone, it is natural to assume that the state of stress in the deformation zone is represented by the corner F of the yield hexagon (Fig. 3.5).

Referring to the slipline field of Fig. 3.7, let $z = f(r)$ represent the shape of the raised lip BE at any stage of the indentation. The incompressibility of the plastic material implies that the volume of material lying below the original free surface (shown broken) must equal the volume of the cone above that surface. If a denotes the radial distance of point B from the axis of the cone, whose semiangle is ψ, the incompressibility condition may be written as

$$z_0 r_0^2 - 2\int_a^{r_0} rf(r)\,dr = \frac{1}{3}a^3 \cot\psi, \tag{3.40}$$

where (r_0, z_0) denote the coordinates of point E referred to the chosen pair of axes. Indeed, the volume of revolution of the area $OBEF$ is equal to π times the left-hand side of (3.40), while that of the area AOB is π times the right-hand side of (3.40).

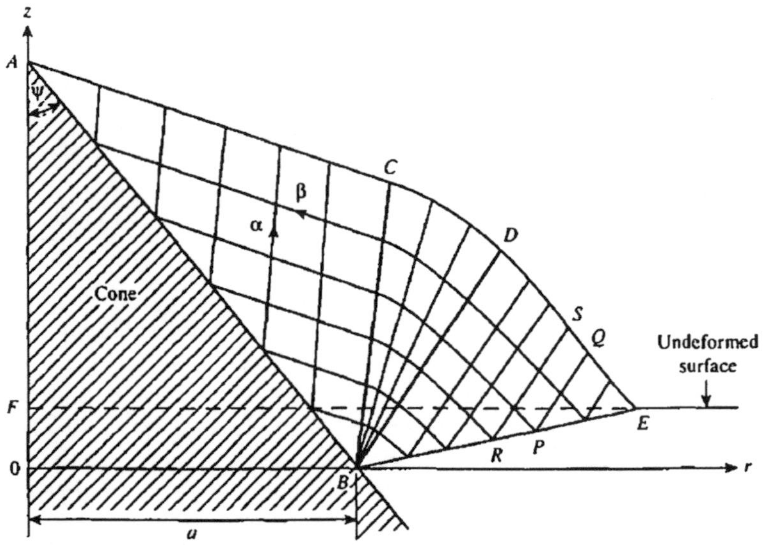

Fig. 3.7 Slipline field due to indentation of a semi-infinite block by a rigid cone (after F. J. Lockett, 1963)

The nature of the problem indicates that *BD* and *BC* are α-lines and *ACDE* is a β-line. The absence of friction between the material and the cone requires $\phi = \pi/4 + \psi$ along *AB,* while the stress-free condition along the raised lip requires

$$p = k, \ \phi = \frac{\pi}{4} + \theta \text{ on } BE,$$

where θ denotes the quantity $\tan^{-1}$ *(df/dr).* Since the material lying outside *ACDE* is assumed rigid, $U = W = 0$ along this interface, the velocity being assumed continuous. The condition of geometrical similarity requires that the normal component of velocity at any point on the surfaces *AB* and *BE* should be proportional to the normal distance of the respective surface element from the stationary point *F.* Taking the factor of proportionality as $1/\sqrt{2}$ without loss of generality, the restrictions on the velocity components *U* and *W* may be written from simple geometry as

$$\begin{aligned} W + U &= r\sin\phi + (z_0 - f)\cos\theta \quad \text{on BE} \\ W - U &= (a\cot\psi - z_0)\sin\psi \quad \text{on AB} \end{aligned}$$

The velocity boundary conditions on *AB* and *ACDE* indicate that there is a velocity singularity at *A.* The velocity components at *E* will be single valued only if $\theta = 0$ at *E,* implying that there should be no discontinuity in the surface slope at *E.* The fan angle at *B* is the value of $\psi - \theta$ at *B.*

For computational purposes, it is convenient to determine the velocity field simultaneously with the slipline field, starting from the point *E,* whose position relative to *0* is adjusted to make the slipline *ACDE* meet the cone at its vertex *A.* Suppose that the complete solution is known along an α-line such as PQ (Fig. 3.7). Then the values of r, z, p, and ϕ can be calculated at the nodal points of RS starting from R, and the velocity components U, W obtained at the same points starting from S. When the region BDE has been found, the singular domain BCD can be determined from the known values of the field quantities along BD, the velocity distribution being subsequently obtained starting from the boundary CD. The remaining field ABC and the associated velocity distribution are defined by the slipline BC, the frictionless condition along AB, and the prescribed velocities along AC and BC. Due to the presence of singularity of the basic equations at $r = 0$, the numerical procedure is least accurate near A. The values of the computed quantities at points away from A are, however, not significantly affected by such inaccuracies.

The numerical computation carried out by Lockett (1963) reveals that the fan angle at B decreases rapidly as ψ decreases from 90° and becomes zero in the limiting case of $\psi \approx 52.5°$. The calculated values of r_0/a, z_0/a, p_0/k, and q/k, where $\pi a^2 q$ represents the total axial load applied to the indenter, are given in Table 3.1 for four different values of the cone semiangle ψ. In all cases, the velocity condition along BE is used in the construction of the field, while that along AB is found to be satisfied within 1%. The incompressibility condition (3.40), used as a final check, is also satisfied to a similar accuracy. The inequalities $\dot{\gamma} \geq \dot{\varepsilon}_\theta \geq 0$, required for the validity of the assumed plastic state of stress, are found to be satisfied at all points of the field. The solution to the indentation problem for $\psi < 52.5°$ is not known.

Table 3.1 Numerical data for cone indentation

ψ	r_0 / a	z_0 / a	Θ	p_0 / k	q / k	Fan angle
52.5°	1.51	0.185	52.5°	1.00	3.88	0°
60°	1.51	0.137	32.2°	1.95	4.28	27.8°
70°	1.52	0.085	18.0°	2.80	4.76	52.0°
80°	1.54	0.041	8.2°	3.50	5.20	71.8°
90°	1.58	0	0	4.15	5.68	90.0°

Useful experimental results on the cone indentation process for several engineering materials have been reported by Dugdale (1954). A theoretical treatment for the indentation of a prepared conical hole in a semi-infinite medium has been presented by Haddow and Danyluk (1964). The stress distribution around a spherical indenter pressed in an existing cavity has been discussed earlier by Ishlinsky (1944), who obtained the value 2.85 Y for the mean indentation pressure.

3.2.4 The Hardness of Metals

In the hardness test of metals, using a cone or a pyramid, geometrical similarity implies that the resistive pressure q, defined as the load per unit projected area of impression, has a constant value that depends on the vertex angle 2ψ, and to some extent on the coefficient of friction μ. When the material is heavily prestrained, we may write $q = cY$, where Y is the uniaxial yield stress of the material, and c is a constant which is found to lie between 2.5 and 3.0. For an annealed or a slightly prestrained metal, the resistive pressure increases with the depth of penetration, but it is still possible to write $q = c\sigma$ approximately, where σ is a mean uniaxial yield stress of the plastic material around the indenter.

In the Brinell hardness test, which uses a hardened steel ball as the indenter, the resistive pressure q, which denotes the hardness number, is considered as the load divided by the projected area of the impression. The pressure q is a function of the ratio a / D, where a is the radius of the impression and D is the ball diameter, Fig. 3.8a. Careful experimental investigations by Meyer (1908) and Tabor (1951) revealed that

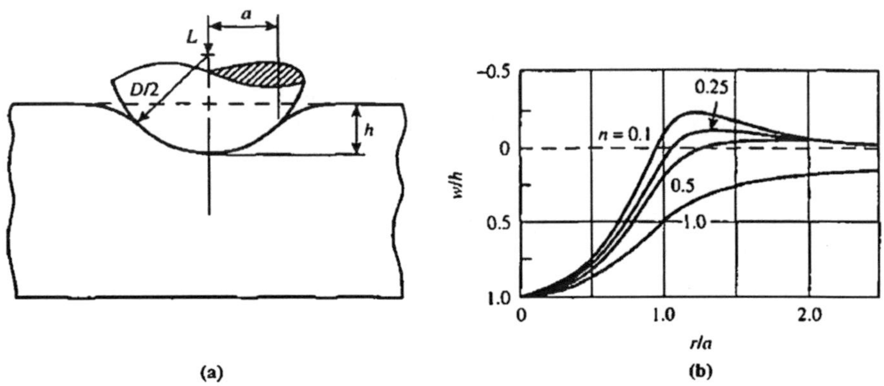

Fig. 3.8 The Brinell hardness test (**a**) geometry of the indentation process and (**b**) normalized displacement profile of the original plane surface

q generally increases with a / D according to the simple power law

$$q = \frac{L}{\pi a^2} = k\left(\frac{a}{D}\right)^n, \tag{3.41}$$

where L is the applied load, k is a constant having the dimension of stress, and n is a dimensionless exponent usually less than 0.5. The range of values of a / D over which (3.41) can be validated empirically is given by $0.01 \leq a / D \leq 0.5$. Equation (3.41) can be rearranged in the form

$$q = \frac{L}{\pi a^2} = k\left(\frac{L}{\pi D^2 K}\right)^{n/(2+n)},$$

which indicates that the hardness number increases with the load for a given ball diameter. Subsequent investigations have shown that the exponent n in (3.41) is approximately the same as that in the Ludwik power law $\sigma = A\varepsilon^n$ for the uniaxial true stress–strain curve of the material. The constant stress A appearing in the stress–strain law, according to Tabor (1951), is related to k as

$$k = A\alpha\beta^n,$$

where α and β are universal constants equal to 2.8 and 0.4, respectively. The substitution of the above expression into (3.41) results in

$$q = \frac{L}{\pi a^2} = A\alpha \left(\frac{\beta a}{D}\right)^n. \tag{3.42}$$

This equation states that the hardness number is directly proportional to the yield stress of the material corresponding to a true strain equal to $\beta a/D$, which may be regarded as a global representative strain in the indentation process.

As the load is increased from zero, the deformation is initially elastic, and the resistive pressure q and the depth of impression h are given by Hertz's theory of elastic contact, the results being expressed as

$$q = \frac{16\,Ga}{3\pi\,(1-v)\,D}, \quad h = \frac{2a^2}{D},$$

where G is the shear modulus and v is Poisson's ratio. If (3.42) is assumed to hold for an elastic material ($n = 1$), when it is incompressible ($v = 0.5$), we have $A = 3G$ and $q = 32\,Ga/3\pi D$, giving $\alpha\beta = 32/9\pi \approx 1.132$, which is close to the value obtained from Tabor's choice for α and β. Plastic yielding begins at a point directly below the center of the ball, at a depth of about $0.5a$ from the free surface, when q becomes approximately equal to $1.1Y$. For an ideally plastic material, there is a fairly well-defined yield point load under which the material is displaced sideways to produce a raised coronet near the periphery of the impression. Work-hardening materials with relatively low values of n exhibit a piling-up mode of deformation, while those with relatively high values of n exhibit a sinking-in mode of deformation around the indenter. The deformation mode can be characterized by a dimensionless parameter m such that

$$\frac{h}{D} = \left(\frac{a}{mD}\right)^2 \quad \text{or} \quad m^2 = \frac{a^2}{hD}.$$

Norbury and Samuel (1928) observed that m is an invariant that depends only on the value of n. The piling-up mode corresponds to $m > 1$ and the sinking-in mode to $m < 1$, the transition between the two modes being given by $m = 1$.

It is possible to establish certain useful results involving the total expenditure of work W in the indentation process without resorting to computational details. If r denotes the radius of the impression and z the depth of penetration at a generic stage of the penetration when the applied load is P, then by (3.41), we have

$$\frac{P}{L} = \left(\frac{r}{a}\right)^{2+n}, \quad \frac{z}{h} = \left(\frac{r}{a}\right)^2.$$

Consequently, the external work done in forming the required impression of depth h is found as

$$W = \int_0^h P d_z = 2Lh \int_0^1 \left(\frac{r}{a}\right)^{3+n} d\left(\frac{r}{a}\right) = \frac{2Lh}{4+n}. \qquad (3.43)$$

In the absence of friction, the external work W is also equal to the total internal energy of the deformed material. Adopting the Hencky theory for small strains, we get

$$W = \int \left(\frac{\sigma \dot{\varepsilon}}{1+n}\right) dV = \frac{1}{1+n} \int \sigma_{ij}\varepsilon_{ij} dV = \frac{2\pi}{1+n} \int_0^a pwr\, dr$$

by Gauss's divergence theorem, with p denoting the normal pressure on the indenter and w the vertical displacement at any radius r. Since $w = h - r^2/D$ by simple geometry when the ball is rigid, the above equation gives

$$(1+n)\, W = Lh - \frac{2\pi}{D} \int_0^a pr^3 dr, \ \mathrm{L} = 2\pi \int_0^{\mathrm{a}} pr\, dr.$$

Combining this expression for W with that given by (3.43) furnishes the result

$$L = \frac{2\pi}{hD} \left(\frac{4+n}{2-n}\right) \int pr^3 dr,$$

which provides a second expression for the load in terms of the pressure distribution over the indenter. In terms of the invariant m, the above equation can be written as

$$\left(\frac{4+n}{2-n}\right) m^2 = \left(\int_0^1 p\xi d\xi\right) / \left(\int_0^1 p\xi^3 d\xi\right), \qquad (3.44)$$

where $\xi = r/a$. When the material is elastic ($n = 1$), the pressure distribution is elliptical by the Hertz theory of contact, and the right-hand side of (3.44) is equal to $\frac{5}{2}$ giving $m^2 = \frac{1}{2}$, which is in agreement with the Hertz theory.

According to a numerical treatment of the ball indentation problem reported by Hill et al. (1989), the pressure distribution depends only slightly on the value of n, and it may be expressed with sufficient accuracy by the empirical relation

$$p \approx p_0 \left(1 - \frac{r^2}{a^2}\right)^{(2+n^2)/(5+n^2)}, \quad 0 \leq n \leq 1$$

where p_0 is the maximum pressure that occurs at $r = 0$. The parameter m^2 is readily evaluated by integration on substituting for p in (3.44), the first integral appearing in (3.44) being equal to half the mean pressure q. Thus

$$m^2 = 3\left(\frac{2-n}{4+n}\right)\left(\frac{4+n^2}{5+n^2}\right), \quad q = \left(\frac{5+n^2}{7+2n^2}\right)p_0. \tag{3.45}$$

It follows from the first relation of (3.45) that the piling-up mode corresponds to $n < 0.24$ and the sinking-in mode to $n > 0.24$ approximately. The downward displacement of the contact perimeter is $(1 - m^2)h$, while a small annular bulge exists over the range $n \leq 0.5$. The distribution of the normalized displacement w/h of the original plane surface of the indented block, obtained numerically by Hill et al. (1989), is shown in Fig. 3.8b for several values of n. Finite element solutions for the ball indentation problem have been reported by Sinclair et al. (1985). Upper and lower bound solutions for the ball indentation problem have been discussed by Tirosh et al. (2008).

3.3 Necking of a Cylindrical Bar

3.3.1 Stress Distribution in the Neck

A cylindrical specimen subjected to axial tension begins to neck when the load attains a maximum. After the neck has formed, the stress distribution in the bar is no longer uniform, and the ratio of the axial load to the minimum cross-sectional area does not represent the yield stress. There is sufficient experimental evidence to indicate that the state of strain is approximately uniform over the minimum section which is therefore uniformly hardened to a current uniaxial yield stress Y. The uniformity of the radial strain rate requires the radial velocity u to be proportional to r, and hence the circumferential strain rate u/r is equal to the radial strain rate. It follows that $\sigma_r = \sigma_\theta$ over the minimum section for any regular isotropic yield surface. Since the state of stress at each point of the minimum section is then an axial tension together with a varying hydrostatic stress equal to σ_r, the yield criterion becomes

$$\sigma_z - \sigma_r = Y$$

across $z = 0$. Combining it with the first equation of equilibrium (3.1) and setting $\sigma_r = \sigma_\theta$, we get

$$\frac{\partial \sigma_z}{\partial r} + \frac{\partial \tau_{rz}}{\partial_z} = 0 \text{ when } z = 0.$$

The second equation of equilibrium (3.1) states that $\partial\sigma_z/\partial z = 0$ when $z = 0$, since τ_{rz} vanishes on this plane due to symmetry.

At a typical point in the neck, let ψ denote the angle of inclination of the major principal stress direction in the meridian plane to the z-axis (Fig. 3.9). For a point sufficiently close to the plane $z = 0$, the transformation of the stress components furnishes

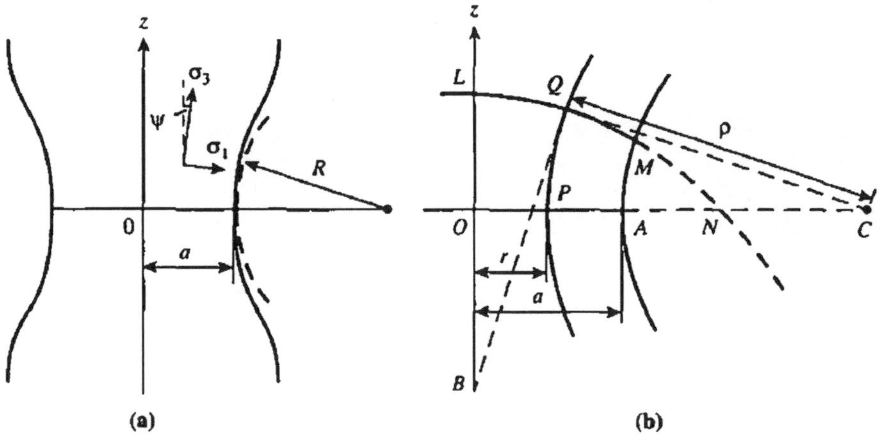

Fig. 3.9 Geometry of neck in a cylindrical tension bar and the principal stress trajectories in the meridian plane

$$\sigma_r \approx \sigma_1, \ \sigma_z \approx \sigma_3, \ \tau_{rz} \approx (\sigma_3 - \sigma_1)\,\psi \approx \mathrm{Y}\psi,$$

where σ_1 and σ_3 are the principal stresses in the meridian plane. Consequently,

$$\left(\frac{\partial \tau_{rz}}{\partial_z}\right)_{z=0} = Y\left(\frac{\partial \psi}{\partial z}\right)_{z=0} = \frac{Y}{\rho},$$

where ρ is the radius of curvature of the major principal stress trajectory where it crosses the plane $z = 0$. The preceding equation of equilibrium then reduces to

$$\frac{\partial \sigma_z}{\partial r} + \frac{Y}{\rho} = 0 \ \text{ when } z = 0. \tag{3.46}$$

It is evident that ρ increases inward from a value R at the surface to infinity on the longitudinal axis. The axial stress σ_z, therefore, increases inward from the value Y at the surface of the neck. The yield criterion then indicates that σ_r and σ_θ also increase inward and are always tensile across the minimum section. The initiation of fracture at the center of the neck is caused by the existence of a state of triaxial tension.

In order to complete the solution, a reasonable assumption must be made for the variation of ρ with r. Obviously, the simplest assumption is to take ρ as inversely proportional to r, as has been done by Davidenkov and Spirodonova (1946). A more accurate solution has been obtained by Bridgman (1944), who assumed that the transverse trajectories of principal stress near the minimum section are circular arcs orthogonal to the longitudinal trajectories. Let LM be such a trajectory, intersecting the minimum section OA externally at N, and having its center of curvature at B. If C is the center of curvature of the longitudinal trajectory at a generic point P, then

$$\rho^2 = CQ^2 = BC^2 - BN^2 = OC^2 - ON^2 = (r + \rho)^2 - ON^2,$$

where r is the radial distance of P from the longitudinal axis. Consequently,

$$r^2 + 2r\rho = a^2 + 2aR = \text{constant},$$

where a is the radius of the bar at the minimum section. The relationship between r and ρ is therefore obtained as

$$\rho = \left(a^2 + 2aR - r^2\right)/2r$$

Inserting this expression in (3.46), and integrating it under the boundary condition $\sigma_z = Y$ at $r = a$, where $\sigma_r = 0$, we obtain the axial stress distribution

$$\frac{\sigma_z}{Y} = 1 + \ln\left(1 + \frac{a^2 - r^2}{2aR}\right)$$

across the minimum section. The expression for $\sigma_r = \sigma_\theta$ is given by the second term on the right-hand side of the above relation. The mean longitudinal stress a across the minimum section is given by

$$\frac{\sigma}{Y} = \frac{2}{a^2}\int_0^a \sigma_z r dr = \left(1 + \frac{2R}{a}\right)\ln\left(1 + \frac{a}{2R}\right). \qquad (3.47)$$

If, on the other hand, ρ is assumed equal to $(a/r)R$, a straightforward integration of (3.46) leads to the mean stress $\sigma = Y(1 + a/4R)$, which predicts an increasingly higher value of the yield stress than that given by (3.47) as a/R increases.

The Bridgman formula (3.47) provides a correction factor for continuing the tensile stress–strain curve of a metal beyond the point of necking. The method involves an accurate measurement of the root radius R of the formed neck at each stage, in addition to that of the applied load and the radius of the minimum section. The stress–strain curve is obtained by plotting the calculated value of Y against the quantity $2\ln(a_0/a)$, where a_0 is the initial radius of the tensile specimen. For details, the reader is referred to Marshall and Shaw (1952). A modification of the Bridgman correction factor, based on a large strain finite element analysis, has been proposed by Celentano et al. (2005).

3.3.2 Initiation of Necking

The bifurcation problem that arises at the onset of necking of a circular cylindrical bar under axial tension may be analyzed by treating the associated rate problem with appropriate boundary conditions. For greater generality, the cylindrical surface of the bar is assumed to be under a uniform fluid pressure of varying intensity, while the ends of the bar are moved apart with a prescribed relative velocity in such a way that both ends are free of tangential traction. The material is assumed to be isotropic and rigid/plastic, obeying Tresca's yield criterion and the associated flow rule. Since the principal axes of stress and strain rate coincide in such a situation, the material derivative of the true stress may be used in formulating the associated rate problem. Due to the uniformity of the current state of stress, the rate equation of equilibrium is $\partial\dot{\sigma}_{ij} / \partial x_i = 0$, where $\dot{\sigma}_{ij}$ is the true stress rate which is nonuniform at the onset of necking. Under conditions of axial symmetry, the rate equations of equilibrium are

$$\left.\begin{aligned} \frac{\partial\dot{\sigma}_r}{\partial r} + \frac{\dot{\sigma}_r - \dot{\sigma}_\theta}{r} + \frac{\partial\dot{\tau}_{rz}}{\partial z} = 0, \\ \frac{\partial\dot{\tau}_{rz}}{\partial z} + \frac{\dot{\tau}_{rz}}{r} + \frac{\partial\dot{\sigma}_z}{\partial z} = 0. \end{aligned}\right\} \tag{3.48}$$

The principal stresses at the onset of necking are $\sigma_r = \sigma_\theta = 0$ and $\sigma_z = \sigma$, while the stress rates $\dot{\sigma}_r$ and $\dot{\sigma}_\theta$ are assumed to be equal to one another. The incompressibility condition and the isotropy condition to be satisfied by the velocity field are

$$\frac{\partial u}{\partial r} + \frac{u}{r} + \frac{\partial w}{\partial z} = 0, \quad \frac{\partial u}{\partial z} + \frac{\partial w}{\partial r} = 0. \tag{3.49}$$

The second equation of (3.48) and the first equation of (3.49) are identically satisfied by introducing functions $\phi\,(r, z)$ and $0(r, z)$ such that

$$\dot{\tau}_{rz} = -\frac{1}{r}\frac{\partial\phi}{\partial z}, \quad \dot{\sigma}_z = \frac{1}{r}\frac{\partial\phi}{\partial r},$$
$$u = -\frac{1}{r}\frac{\partial\psi}{\partial z}, \quad w = \frac{1}{r}\frac{\partial\psi}{\partial r}.$$

Substituting from above into the remaining two equations of (3.48) and (3.49), and using the yield criterion in the rate form

$$\dot{\sigma}_z - \dot{\sigma}_r = H\dot{\varepsilon}_z = H\frac{\partial w}{\partial z},$$

where H is the current slope of the uniaxial stress–strain curve of the material, we obtain the pair of partial differential equations

$$\left.\begin{aligned} r\frac{\partial}{\partial r}\left(\frac{1}{r}\frac{\partial\psi}{\partial r}\right) - \frac{\partial^2\psi}{\partial z^2} = 0, \\ r\frac{\partial}{\partial r}\left(\frac{1}{r}\frac{\partial\phi}{\partial r}\right) - \frac{\partial^2\phi}{\partial z^2} = H\frac{\partial^3\psi}{\partial z^3}. \end{aligned}\right\} \tag{3.50}$$

The origin of coordinates is taken at the geometrical center of the bar whose current radius is denoted by a and the current length by $2l$. The applied fluid pressure p is assumed to remain temporarily constant at the onset of necking, so that the nominal traction rate $\dot{F}_j$ is equal to $pl_k(\partial v_k/\partial x_j)$ along the cylindrical surface whose outward drawn unit normal is denoted by l_i In terms of the true stress rate $\dot{\sigma}_{ij}$, the boundary condition on the lateral surface can be written as

$$l_i\left(\dot{\sigma}_{ij} - \sigma_{jk}\frac{\partial v_i}{\partial x_k}\right) = pl_k\frac{\partial v_k}{\partial x_j} \text{ along r} = \text{a}.$$

The condition of zero tangential traction along the ends of the bar requires $\dot{\tau}_{rz}$ to vanish on $z = \pm$l. The boundary conditions for the eigenvalue problem therefore become

$$\dot{\sigma}_r = 0, \quad \dot{\tau}_{rz} = (\sigma + p)\frac{\partial u}{\partial z} \text{ along } r = a,$$
$$\dot{\tau}_{rz} = 0, \quad w = 0 \text{ along } z = \pm l,$$

where σ denotes the axial stress at the incipient necking. In terms of the functions ϕ and ψ, the above conditions are equivalent to

$$\left.\begin{aligned} &\frac{\partial\phi}{\partial r} - H\frac{\partial^2\psi}{\partial r\partial z} = 0, \quad \frac{\partial\phi}{\partial z} - (\sigma + p)\frac{\partial^2\psi}{\partial z^2} = 0 \text{ on } r = a, \\ &\quad\frac{\partial\phi}{\partial z} = 0, \qquad\qquad \frac{\partial\psi}{\partial r} = 0 \text{ on } z = \pm l. \end{aligned}\right\} \tag{3.51}$$

The deformation mode at the incipient necking, which is by no means unique, is given by a linear combination of the eigenmode and that corresponding to a homogeneous extension of the bar.

In view of the symmetry of the assumed deformation mode at bifurcation, the solution to the governing equations for the eigenvalue problem may be written in the form

$$\phi = HrF(r)\cos\frac{\pi z}{l}, \quad \psi = \frac{1}{\pi}rG(r)\sin\frac{\pi z}{l}, \tag{3.52}$$

so that the boundary conditions on $z = \pm l$ are identically satisfied. The substitution of (3.52) into (3.50) leads to the pair of ordinary differential equations

$$\frac{d}{dr}\left[\frac{1}{r}\frac{d}{dr}(rG)\right] + \frac{\pi^2}{l^2}G = 0,$$
$$\frac{d}{dr}\left[\frac{1}{r}\frac{d}{dr}(rF)\right] + \frac{\pi^2}{l^2}(F + G) = 0.$$

The solution to these equations can be written down in terms of Bessel functions of the first kind and of orders 0 and 1, the result being

$$G(r) = \alpha J_1\left(\frac{\pi r}{l}\right), \quad F(r) = \beta J_1 + \alpha\left(\frac{\pi r}{2l}\right)J_0\left(\frac{\pi r}{l}\right), \tag{3.53}$$

where α and β are constants. The second expression follows from the first if we observe that $\alpha(\pi r/2l)J_0(\pi r/l)$ is a particular integral for the differential equation for F. In view of (3.52) and (3.53), the first two conditions of (3.51) give

$$\alpha\left(\frac{\pi a}{2l}\right)J_1\left(\frac{\pi a}{l}\right) - \beta J_0\left(\frac{\pi a}{l}\right) = 0,$$
$$\alpha\left\{\frac{\pi a}{2l}J_0\left(\frac{\pi a}{l}\right) - \frac{\sigma + p}{H}J_1\left(\frac{\pi a}{l}\right)\right\} + \beta J_1\left(\frac{\pi a}{l}\right) = 0,$$

in view of the well-known relation $(d/dx)[xJ_1(x)] = xJ_0(x)$. For α and β to have nontrivial solutions, the determinant of their coefficients in the above equations must vanish. Since $\sigma + p$ is equal to the equivalent stress $\bar{\sigma}$, the condition for bifurcation may be expressed as

$$\frac{\bar{\sigma}}{H} = \frac{\pi a}{2l}\left\{\frac{J_0(\pi a/l)}{J_1(\pi a/l)} + \frac{J_1(\pi a/l)}{J_0(\pi a/l)}\right\} \tag{3.54}$$

This condition is independent of the applied fluid pressure. The incipient deformation mode at bifurcation, obtained by the superposition of the eigenfield on the homogeneous field given by $u = -Ur/2l$, $w = Uz/l$, where U is the axial speed of each end of the bar, will nowhere involve unloading if $\alpha \leq U/\pi$.

For normal specimen sizes, we may express the right-hand side of (3.54) in ascending powers of $\pi a/l$, and neglect all terms of order higher than $(\pi a/l)^2$. On the generalized stress–strain curve, the condition for necking then becomes

$$\frac{1}{\bar{\sigma}}\frac{d\bar{\sigma}}{d\bar{\varepsilon}} \approx 1 - \frac{1}{8}\left(\frac{\pi a}{l}\right)^2.$$

For a sufficiently long specimen, the condition for bifurcation reduces to the well-known result $d\bar{\sigma}/d\bar{\varepsilon} = \bar{\sigma}$, which corresponds to a maximum load. It can be shown that a cylindrical bar subjected to lateral pressure alone begins to neck at essentially the same value of the true strain (Chakrabarty, 1972), although the boundary conditions at the ends of the bar are somewhat different in this case from those considered above. The bifurcation in an elastic/plastic bar under an axial tension has been investigated by Cheng et al. (1971), Needleman (1972), and Hutchinson and Miles (1974). A theoretical analysis for the cup-and-cone fracture in a tensile specimen has been presented by Tvergaard and Needleman (1984).

3.4 Compression of Short Cylinders

3.4.1 Compression of Solid Cylinders

A solid cylindrical bar of radius a and height h is axially compressed between a pair of overlapping parallel dies. The coefficient of friction μ between the cylinder and the dies is assumed constant, and the material is assumed rigid/plastic with a uniaxial yield stress Y. At the incipient compression of the cylinder, the shear stress

τ_{rz} is equal to μq at the lower die and to $-\mu q$ at the upper die, where q denotes the normal pressure on the die. We consider an approximate analysis, due to Siebel (1923) and Bishop (1958), in which the variation of the normal stress components in the axial direction is disregarded. Then, multiplying the first equation of equilibrium (3.1) by dz, and integrating between $z = 0$ and $z = h$, we have

$$\frac{d\sigma_r}{dr} + \frac{\sigma_r - \sigma_\theta}{r} = \frac{2\mu q}{h}. \tag{3.55}$$

For sufficiently small values of μ, it is a good approximation to set $\sigma_r = \sigma_\theta$ throughout the cylinder. The yield criterion of both Tresca and von Mises then reduces to the form

$$\sigma_r + q = Y,$$

which implies that the z-component of the deviatoric normal stress is equal to $-2Y/3$. Combining this with (3.55), the condition for equilibrium is finally reduced to

$$\frac{dq}{dr} = -\frac{2\mu q}{h}.$$

Using the boundary conditions $q = Y$ at $r = a$ (where $\sigma_r = 0$), the above equation is integrated to give

$$q = Y \exp\left[2\mu(a - r)\,/\,h\right]. \tag{3.56}$$

The pressure therefore rises exponentially from Y at the periphery to reach a peak value of $Y\,exp(\mu a/h)$ at the center. The mean die pressure $\bar{q}$ is given by

$$\frac{\bar{q}}{Y} = \frac{2}{Ya^2}\int_0^a qr\,dr = \frac{h^2}{2\mu^2 a^2}\left\{\exp\left(\frac{2\mu a}{h}\right) - 1 - \left(\frac{2\mu a}{h}\right)\right\}. \tag{3.57}$$

When $\mu a/h$ is small compared to unity, $\bar{q} \approx Y(1 + 2\mu a\,/\,3\,h)$ to a close approximation. It follows that the applied load in a compression test must be divided by $\pi a^2(1 + 2\mu a/3\,h)$ to obtain the current yield stress, so long as no appreciable barreling has taken place.

For sufficiently large values of a/h, the frictional stress on the die face will be equal to the yield stress in shear over an inner circle of radius r_0. Using Tresca's yield criterion, we set $\mu q = Y/2$ at $r = r_0$ to obtain

$$\frac{r_0}{h} = \frac{a}{h} \quad \frac{1}{2\mu} \ln \frac{1}{2\mu}.$$

The pressure distribution is therefore given by (3.56) only over the outer annulus ($r_0 \leq r \leq a$). Over the inner circle ($0 \leq r \leq r_0$), where the frictional stress μq equals $Y/2$, the previous yield condition no longer holds, but it is still a good approximation to take $d\sigma_r/dr \approx -dq/dr$ as before (Bishop, 1955). The equilibrium equation (3.55) then becomes

$$\frac{dq}{dr} = -\frac{Y}{h}, \quad 0 \le r \le r_0.$$

Integrating, and using the continuity conditions $q = Y/2\mu$ at $r = r_0$, the solution is obtained as

$$q = Y\left(\frac{1}{2\mu} + \frac{r_0 - r}{h}\right), \quad 0 \le r \le r_0.$$

Using the die pressure distributions over the two regions $r \ge r_0$ and $r \le r_0$, and integrating, the mean die pressure is easily shown to be given by

$$\frac{\bar{q}}{Y} = \frac{h^2}{2\mu^2 a^2}\left(\frac{1}{2\mu} + \frac{r_0}{h} - 1\right) - \frac{h}{\mu a} + \frac{r_0^2}{a^2}\left(\frac{1}{2\mu} + \frac{r_0}{3\,h}\right), \tag{3.58}$$

which holds for $r_0 \ge 0$. When $h/a \ge 2$, the deformation mode would involve a pair of conical zones having their bases in contact with the dies, and the mean compressive stress is then equal to Y for all values of μ. In Fig. 3.10, the dimensionless mean pressure given by (3.57) and (3.58) is plotted against a/h for various values of μ. The theory is found to be in good agreement with experimental results reported by MacDonald et al. (1960), Unksov (1961), and Lange (1985). An extension of the analysis for blocks of noncircular cross section has been considered by

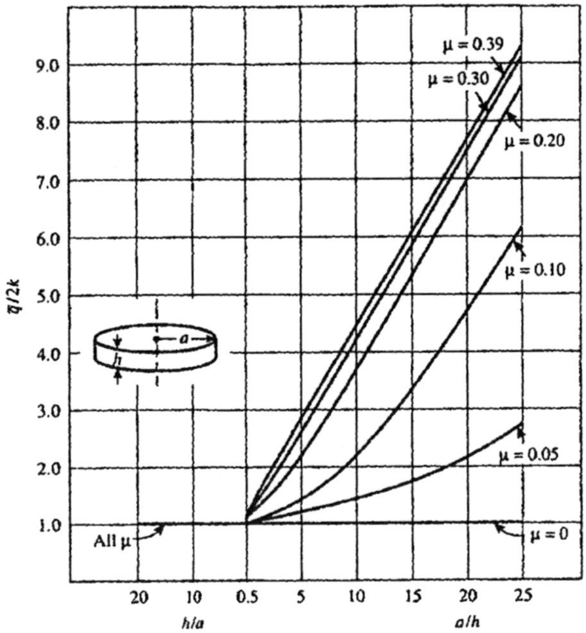

Fig. 3.10 Mean die *pressure* for the incipient compression of a *circular cylinder* in the presence of Coulomb friction

Johnson et al. (1966). Various *upper bound* solutions for the compression problem have been given by Kudo (I960), Haddow and Johnson (1962), Collins (1968), Samanta (1968), and Avitzur (1980). A slipline field solution to this problem has been presented by Kwasczynska and Mröz (1967).

Consider now the axial compression of a prismatic *block,* of arbitrary cross-sectional area A and height h, between rigid parallel dies having a small coefficient of friction μ. The *origin* of rectangular coordinates (x, y, z) is taken at the center of the block *with* the z-axis along the direction of compression. Since the *state* of stress in the *block* does not differ appreciably from that of a uniaxial compression, it is a good approximation to write the yield criterion in the form

$$s_z = \sigma_z - \sigma = -\frac{2}{3}Y,$$

where σ is the hydrostatic part of the stress. Multiplying this *equation by* the volume element dV, and integrating through half the specimen $z \geq 0$, the applied load P is obtained as

$$\frac{P}{A} = -\frac{2}{Ah}\int \sigma_z dV = \frac{2Y}{3} - \frac{2}{Ah}\int \sigma dV.$$

Now, in any self-equilibrated stress field, the average value of the hydrostatic stress is directly obtained from the surface tractions. Indeed, if the position vector to a typical surface element is denoted by x_j, then by the application of Green's theorem, we have

$$\int \sigma_{ij}\delta_{ij}dV = \int \sigma_{ij}\frac{\partial x_j}{\partial x_i}dV = \int \frac{\partial}{\partial x_i}\left(\sigma_{ij}x_j\right) dV = \int F_j x_j dA,$$

where δ_{ij} is the Kronecker delta, F_j is the boundary traction, and use has been made of the equilibrium equation $\partial\sigma_{ij} / \partial x_j = 0$. The above relation immediately gives

$$3\int \sigma dV = -\left(\frac{1}{2}hP + \int r\tau dA\right),$$

where the right-hand integral is taken over the whole area of die contact, and τ denotes the magnitude of the radial component of the frictional stress at a distance r from the centroid of the area. The preceding relations furnish the mean die pressure as

$$\bar{q} = \frac{P}{A} \approx Y\left(1 + \frac{\mu}{Ah}\int r\, d\, A\right), \tag{3.59}$$

since $\tau \approx \mu Y$ within the present order of approximation (Hill, 1963). When the cross section of the block is an ellipse with major axis $2a$ and minor axis $2b$, giving an eccentricity$e = \sqrt{1 - b^2 / a^2}$, Equation (3.59) furnishes

$$\bar{q} = Y\left\{1 + \frac{\mu}{\pi abh}\iint r\,dx\,dy\right\} = Y\left\{1 + \frac{4\mu a}{3\pi h}E(e)\right\},$$

where E is the complete elliptic integral of the second kind. This formula should be sufficiently accurate for small values of μ and for small to moderate values of e. For a circular cylinder of radius a, the second term in the curly brackets reduces to $2\mu a/3\,h$ as expected.

3.4.2 Estimation of Incipient Barreling

The effect of die friction is to produce barreling of the compressed cylinder due to a longitudinal nonuniformity of the deformation mode. The analysis will be based on a general method proposed by Hill (1963) for treating metal-forming problems. In a deformable body, consider a certain distribution of stress σ_{ij} in a region occupied by volume V, together with a distribution of traction t_j defined over its surface S. If the stress field is in equilibrium, and is compatible with the applied surface tractions, then by the principle of virtual work rate, we have

$$\int \sigma_{ij}\frac{\partial w_i}{\partial x_j}dV = \int t_j w_j dS \tag{3.60}$$

for any virtual velocity field w_j which is continuously differentiable and is consistent with the conditions of constraint. If we choose a sufficiently wide class of virtual velocity field w_j, which includes an approximation to the actual field, then (3.60) furnishes a suitable condition of overall equilibrium, although the distribution of σ_{ij} corresponding to the approximating field is not necessary in equilibrium. Since the surface traction t_j is not completely stated at all points of S, the traction corresponding to the approximate stress field may be used along interfaces and in regions of slipping friction. In the language of function space, the considered family of virtual motion may be regarded as orthogonal to the divergence between the approximate and actual stress fields.

Consider the compression of a short circular cylinder of height $2h$ and radius a. In view of the symmetry of the incipient barreling mode about the mid-plane, whose geometrical center is taken as the origin of cylindrical coordinates (r, θ, z), we choose a family of incompressible virtual velocity field in the form

$$w_r = \frac{1}{2}rf'(z), \quad w_z = -f(z).$$

The nonzero velocity components are thus expressed in terms of an arbitrary continuous function of z satisfying the relation $f(-z) = -f(z)$. The substitution from above into (3.60) gives

$$\int \left\{ \int_0^h \left(3s_z f' - r\tau_{rz} f''\right) dz \right\} dA = f'(h) \int r\tau dA - 2Pf(h),$$

where τ is the magnitude of the frictional stress on the die face. The yield criterion may be written as $s_z = -2Y/3$ to a close approximation. Then, integrating by parts, and using the conditions $\tau_{rz} = 0$ at $z = 0$ and $\tau_{rz} = -\tau$ at $z = h$, the above expression is reduced to

$$\int_0^h \left\{ \int \left(2Y - r\frac{\partial \tau_{rz}}{\partial z}\right) dA - 2P \right\} f'(z) dz = 0.$$

Since $f'(z)$ is completely arbitrary, the expression in the curly brackets must vanish, giving

$$\int r\frac{\partial \tau_{rz}}{\partial z} dA = 2\left(AY - P\right).$$

Multiplying both sides of this equation by dz, and integrating between the limits 0 and z, we finally get

$$\int r\tau_{rz} dA = 2\left(AY - P\right) z. \tag{3.61}$$

This equation is in fact exact, apart from the slight approximation in the yield criterion. It can be shown to hold equally well for a cylinder of noncircular cross section with τ_{rz} denoting the outward radial component of the shear stress.

Since the actual mode of deformation at the yield point can differ from homogeneous compression only in first-order quantities involving μ, the velocity field for the incipient barreling corresponding to a unit speed of the die may be taken approximately in the form (except near $r = 0$)

$$v_r = \frac{r}{2h} + \mu\phi'(z), \quad v_z = -\frac{z}{h} + \frac{\mu}{r}\phi(z), \tag{3.62}$$

so that the incompressibility condition is identically satisfied. The quantity ϕ must be an odd function of z, vanishing on the die face $z = h$, and having continuous first and second derivations. By the Lévy–Mises flow rule,

$$\tau_{rz} = \frac{1}{2}\left(\frac{\partial v_r}{\partial z} + \frac{\partial v_z}{\partial r}\right) s_z \Big/ \frac{\partial v_z}{\partial z} = \frac{h}{3}\mu Y \left(\phi'' + \frac{\phi}{r^2}\right).$$

Substituting in (3.61), and using the first approximation $P \approx AY(1 + \mu a/3h)$, we obtain the differential equation for ϕ as

$$\frac{\phi''}{3} + \frac{\phi}{a^2} + \frac{z}{h^2} = 0.$$

The solution of this equation, satisfying the conditions $\phi\ (0) = \phi\ (h) = 0$, is easily found as

$$\phi\ (z) = \frac{a^2}{h}\left\{\frac{\sin\left(\sqrt{3}z\,/\,a\right)}{\sin\left(\sqrt{3}h\,/\,a\right)} - \frac{z}{h}\right\}.$$

The distribution of the radial velocity u of particles on the free surface of the cylinder $(r = a)$ is obtained from (3.62) along with the above relation as

$$u = \frac{a}{2h} + \frac{\mu a}{h}\left\{\sqrt{3}\frac{\cos\left(\sqrt{3}z\,/\,a\right)}{\sin\left(\sqrt{3}h\,/\,a\right)} - \frac{a}{h}\right\}. \tag{3.63}$$

The incipient barreling of the free surface, over and above a uniform radial expansion, is given by the term in μ. The predicted shape of the barreled cylinder, according to (3.63), is in broad agreement with observation for short cylinders at least when $h\,/\,a\ <\ \pi\ /2\sqrt{3}$ approximately. For a finite element solution of the compression problem, including barreling, the reader is referred to Lee and Kobayashi (1971).

3.4.3 Compression of a Hollow Cylinder

A short hollow cylinder of internal radius a and external radius b, and of height h, is axially compressed between a pair of rigid parallel dies. Under frictionless conditions, the deformation of the cylinder is homogeneous, and every material particle moves radially outward with a speed proportional to its radial distance from the axis of symmetry. With increasing friction between the cylinder and the dies, the rate of expansion of the hole decreases, and that of the external diameter correspondingly increases. When the friction is sufficiently high, the compressive hoop stress developed at the edge of the hole is large enough to cause the hole to contract (Kunogi, 1954). As the compression proceeds, the continued change in geometry of the cylinder causes the frictional constraint to progressively increase, so that an initial expansion of the hole may be followed by a subsequent contraction.

The experimental investigation of van Rooyen and Backofen (1960) has revealed that the frictional stress over the interface between the cylinder and the platens may be approximately constant under certain conditions of lubrication. Following Hawkyard and Johnson (1967), a constant frictional stress mk $(0 \leq m \leq 1)$ will therefore be assumed in the theoretical analysis. The incipient barreling of the cylinder, that causes inhomogeneity of the deformation, will be disregarded. If c denotes the radius to the neutral surface, which may be either in the bore or in the material of the cylinder, the velocity field for the incipient deformation may be written as

$$u = -\frac{\dot{h}}{2h}\left(r - \frac{c^2}{r}\right), \quad w = \frac{\dot{h}z}{h},$$

where the z-plane is assumed to coincide with the stationary lower platen, the upper platen being moved downward with a speed$-\dot{h}$. The nonzero components of the strain rate corresponding to this velocity field are

$$\dot{\varepsilon}_r = -\frac{\dot{h}}{2h}\left(1 + \frac{c^2}{r^2}\right), \quad \dot{\varepsilon}_\theta = -\frac{\dot{h}}{2h}\left(1 - \frac{c^2}{r^2}\right), \quad \dot{\varepsilon}_z = -\frac{\dot{h}}{h} \tag{3.64}$$

The incompressibility condition is identically satisfied, and the shear rate $\dot{\gamma}_{rz}$ is found to be identically zero. The assumed velocity field therefore implies the approximation that σ_r, σ_θ, and σ_z are the principal stresses at each point of the cylinder. By the Lévy–Mises flow rule, we have

$$\sigma_r - \sigma_\theta = \frac{\dot{\varepsilon}_r - \dot{\varepsilon}_\theta}{\dot{\varepsilon}_r - \dot{\varepsilon}_z}(\sigma_r - \sigma_z) = \left(\frac{2c^2}{c^2 + 3r^2}\right)(\sigma_r - \sigma_z),$$

$$\sigma_r - \sigma_\theta = \frac{\dot{\varepsilon}_\theta - \dot{\varepsilon}_z}{\dot{\varepsilon}_r - \dot{\varepsilon}_z}(\sigma_r - \sigma_z) = \left(\frac{c^2 - 3r^2}{c^2 + 3r^2}\right)(\sigma_r - \sigma_z),$$

in view of (3.64). The stresses must satisfy the von Mises yield criterion, which is written in terms of the shear yield stress k as

$$(\sigma_r - \sigma_\theta)^2 + (\sigma_\theta - \sigma_z)^2 + (\sigma_z - \sigma_r)^2 = 6k^2.$$

Substituting from the preceding relations, and solving for $\sigma_r - \sigma_z$, we obtain the normal stress differences

$$\sigma_r - \sigma_z = \frac{k\left(c^2 + 3r^2\right)}{\sqrt{c^4 + 3r^4}}, \quad \sigma_r - \sigma_\theta = \frac{2kc^2}{\sqrt{c^4 + 3r^4}}. \tag{3.65}$$

The stresses are also required to satisfy the equation of equilibrium, obtained from (3.55) by replacing μq with $\pm mk$. Thus

$$\frac{d\sigma_r}{dr} + \frac{\sigma_r - \sigma_\theta}{r} = \pm\frac{2mk}{h},$$

where the upper sign holds in the region of outward flow and the lower sign to the region of inward flow. Substituting from (3.65), and setting $\sigma_z = -q$, where q is the normal die pressure, the above equation is reduced to

$$\frac{dq}{dr} = 2k\left(\frac{c^2 + 3r^2}{r\sqrt{c^4 + 3r^4}} \mp \frac{m}{h}\right) \quad . \tag{3.66}$$

Since σ_r vanishes at the radii $r = a$ and $r = b$, the values of q at these two boundaries are

$$q_a = \frac{k\left(c^2 + 3a^2\right)}{\sqrt{c^4 + 3a^4}}, \quad q_b = \frac{k\left(c^2 + 3b^2\right)}{\sqrt{c^4 + 3b^4}},$$

in view of (3.65). The pressure distribution can be determined by the integration of (3.66) using the appropriate sign, and employing the boundary conditions $q = q_a$ at $r = a$ and $q = q_b$, at $r = b$.

Consider the situation $0 \leq c \leq a$, for which the frictional constraint is small enough to permit an outward flow to occur throughout the cylinder. Then the upper sign in (3.66) holds for the whole cylinder, and the integration of this equation under the boundary condition $q = q_b$, at $r = b$ gives

$$\frac{q}{k} = 2m\left(\frac{b-r}{h}\right) + \frac{c^2 + 3r^2}{\sqrt{c^4 + 3r^4}} + \ln\left\{\frac{b^2\left(\sqrt{c^4 + 3r^4} - c^2\right)}{r^2\left(\sqrt{c^4 + 3b^4} - c^2\right)}\right\}. \tag{3.67}$$

The radius to the neutral surface is obtained from the remaining boundary condition $q = q_a$ at $r = a$. Using the preceding expression for q_n, we get

$$\ln\left\{\frac{a^2\left(\sqrt{c^4 + 3b^4} - c^2\right)}{b^2\left(\sqrt{c^4 + 3a^4} - c^2\right)}\right\} = 2\left(\frac{b}{a} - 1\right)\frac{ma}{h}, \quad 0 \leq c \leq a. \tag{3.68a}$$

For a given b/a ratio, the greatest value of ma/h for which (3.67) gives the pressure distribution over the whole cylinder corresponds to $c = a$. For higher values of ma/h, the deformation of the cylinder consists of an inner region of inward flow and an outer region of outward flow, characterized by $a \leq c \leq b$. Considering the lower sign in (3.66), and integrating it under the boundary condition $q = q_a$ at $r = a$, we obtain the solution

$$\frac{q}{k} = 2m\left(\frac{r-a}{h}\right) + \frac{c^2 + 3r^2}{\sqrt{c^4 + 3r^4}} + \ln\left\{\frac{a^2\left(\sqrt{c^4 + 3r^4} - c^2\right)}{b^2\left(\sqrt{c^4 + 3a^4} - c^2\right)}\right\}, \tag{3.69}$$

which gives the pressure distribution in the inner region $a \leq r \leq c$. The pressure distribution in the outer region $c \leq r \leq b$ is still given by (3.67). The condition of continuity of the pressure at $r = c$ furnishes the relation

$$\ln\left\{\frac{a^2\left(\sqrt{c^4 + 3b^4} - c^2\right)}{b^2\left(\sqrt{c^4 + 3a^4} - c^2\right)}\right\} = 2\left(\frac{b}{a} + 1 - \frac{2c}{a}\right)\frac{ma}{h}, \quad a \leq c \leq b, \tag{3.68b}$$

which defines the position of the neutral surface. As the parameter ma/h is increased in this range for a given b/a, the ratio c/a progressively increases, but c cannot exceed the mean radius of the cylinder as will be evident from (3.68b). Equation (3.68) is displayed graphically in Fig. 3.11, where b/a is plotted against ma/h for constant

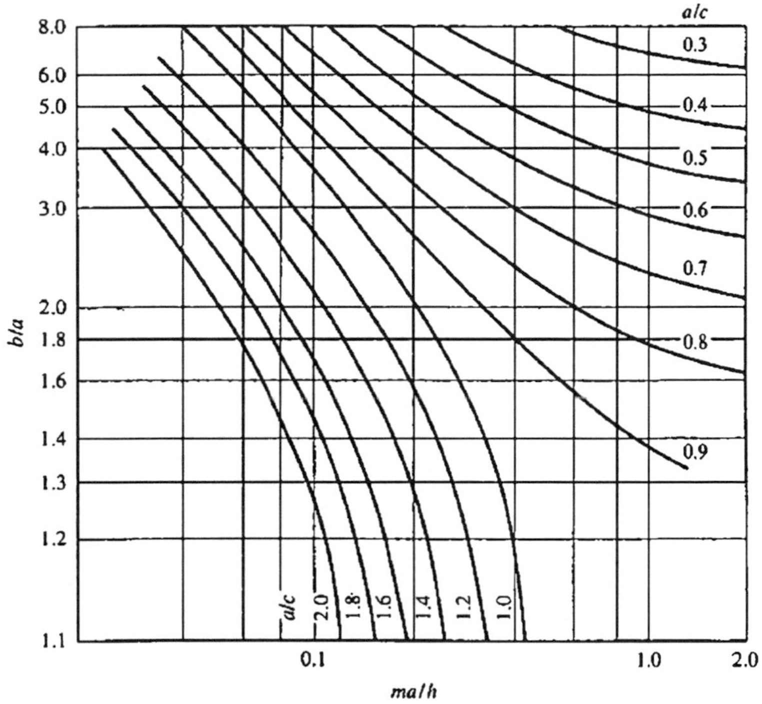

Fig. 3.11 Position of the neutral surface in the axial compression of a hollow cylindrical block

values of *a/c*. By interpolation, the radius to the neutral surface can be determined from these graphs for any given cylinder geometry and friction factor.

The solution for the continued compression of a hollow cylinder of initial radii a_0 and b_0, and initial height h_0, may be obtained from the preceding solution with *a, b,* and *h* defining the current geometry of the cylinder. One of the relations between the initial and current dimensions of the cylinder is given by the overall incompressibility condition $\left(b^2 - a^2\right) h = \left(b_0^2 - a_0^2\right) h_0$, while a second relation is obtained from (3.64) considered at $r = a$. Thus

$$\frac{b^2}{a^2} = 1 + \frac{h_0 a_0^2}{h a^2}\left(\frac{b_0^2}{a_0^2} - 1\right), \quad \frac{da}{dh} = -\left(1 - \frac{c^2}{a^2}\right)\frac{a}{2\,h}, \tag{3.70}$$

where *c/a* is given by (3.68). The solution may be obtained in a step-by-step manner using an iterative procedure. A convenient method is to assume $\Delta a = 0$ as a first approximation corresponding a selected small value of Δh at any given stage. The values of *b/a, c/a,* and *da / Dh* at the end of the incremental deformation are then computed from (3.70) and (3.68). A modified value of Δa is then obtained from the second equation of (3.70), using the trapezoidal rule of integration.

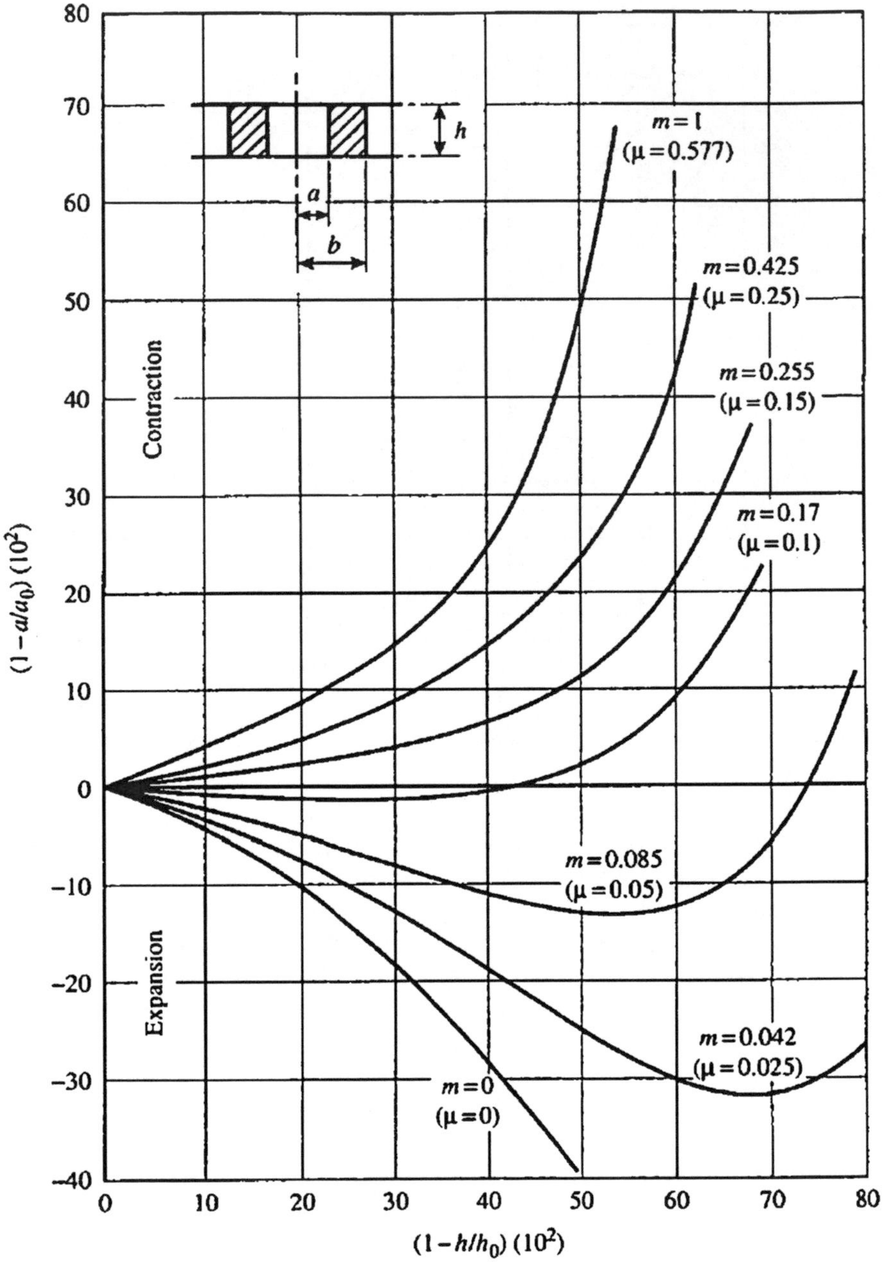

Fig. 3.12 Changes in geometry of an axially compressed hollow cylinder under various frictional condition (after Hawkyard and Johnson, 1967)

The results of the computation are displayed in Fig. 3.12, where the percentage change in bore diameter is plotted against the percentage reduction in height for various values of *m*, assuming $b_0/a_0 = 2$ and $a_0/h_0 = 1.5$. The curves follow the same trend as those found experimentally by Male and Cockroft (1964) and Male and DePierre (1970), using a number of metals and a range of frictional conditions over the die face. Upper bound solutions to this problem have been discussed by Kudo (1960) and Avitzur (1980). An approximate solution for the ring compression with barreling has been presented by Lahoti and Kobayashi (1974). A finite element solution based on a velocity-dependent frictional stress has been discussed by Kobayashi et al. (1989).

3.5 Sinking of Thin-Walled Tubes

A cylindrical tube of uniform thickness is drawn through a conical die so that its external diameter is reduced while improving its mechanical properties and surface finish. Sometimes, a mandrel or plug is used on the exit side to control the thickness of the drawn tube. The hollow-sinking process illustrated in Fig. 3.13a allows a greater reduction in diameter, and it has been treated by Swift (1949) and Hill (1950a), and subsequently by other investigators. The initial tube thickness is assumed small compared to the mean diameter, so that the variation of stress through the thickness can be neglected in the analysis. If the tube is sufficiently long, the problem may be considered as one of steady state for the calculation of stress and strains during the drawing.

3.5.1 Solution Without Strain Hardening

Let t denote the wall thickness of a typical element of the tube within the die, q the mean compressive hoop stress, and σ the mean tensile stress parallel to the die face. The frictional stress along the die face is of magnitude μp, where μ is the coefficient of friction and p the normal pressure between the die and the wall. Consider the stresses acting across the faces of an element formed by meridian planes with an included angle $d\theta$ and normal sections at mean radii r and $r + dr$. Referring to Fig. 3.13b, and resolving the forces perpendicular to the die face, we have

$$qt\left(\frac{dr}{\sin\alpha}\right)\cos\,\alpha\,d\,\theta = p\left(\frac{dr}{\sin\alpha}\right)rd\theta,$$

where α denotes the semiangle of the die. The above relation immediately furnishes

$$p = (qt\,/\,r)\cos\alpha.$$

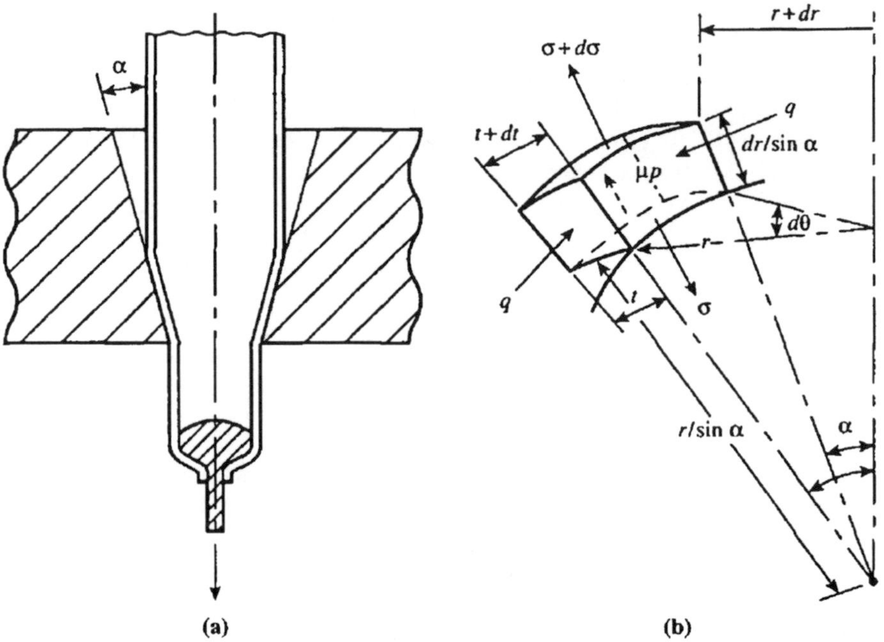

Fig. 3.13 Tube drawing through a conical die. (**a**) The hollow-sinking process and (**b**) stresses acting on a typical tube element

Thus, p is small compared to q when t/r is small. Resolving the elemental forces parallel to the die, we have

$$d\,(tr\sigma)\,d\,\theta + qt\left(\frac{dr}{\sin\alpha}\right)\sin\alpha\;d\,\theta + \mu p\left(\frac{dr}{\sin\alpha}\right)rd\theta = 0.$$

Substituting for p, and simplifying, this relation is reduced to the differential equation

$$\frac{d}{dr}\,(rt\sigma) + tq\,(1 + \mu\cot\alpha) = 0. \tag{3.71}$$

The circumferential and thickness strain increments of any element are dr/r and dt/t, respectively, and by the Lévy–Mises flow rule, the ratio of the two strain increments is

$$\frac{r}{t}\frac{dt}{dr} = \frac{\sigma - q}{\sigma + 2q}. \tag{3.72}$$

Eliminating dt/dr between (3.71) and (3.72), and using the fact that $\sigma\,(2\sigma + q) + q(2q + \sigma) = 2Y^2$ by the von Mises yield criterion, we get

$$r\frac{d\sigma}{dr} + \frac{2Y^2 + q\,(\sigma + 2q)\,\mu\cot\alpha}{\sigma + 2q} = 0. \tag{3.73}$$

This may be regarded as a differential equation for σ, which must be solved under the boundary condition $\sigma = 0$, at $r = b$. Combining (3.73) with (3.72), we obtain the differential equation for the thickness change as

$$t\frac{d\sigma}{dr} + \frac{2Y^2 + q\,(\sigma + 2q)\,\mu\cot\alpha}{\sigma - q} = 0. \tag{3.74}$$

The boundary condition is $t = t_0$ (the initial wall thickness) at $r = b$, where $\sigma = 0$. Equations (3.73) and (3.74) can be solved numerically, using the yield criterion, to obtain the variation of σ and t with r during the steady-state operation.

The numerical solution is greatly facilitated by representing the von Mises yield criterion parametrically through an auxiliary angle ϕ. It is easily verified that the expressions

$$\sigma = \frac{2Y}{\sqrt{3}}\sin\phi,\ q = \frac{2Y}{\sqrt{3}}\cos\left(\frac{\pi}{6} + \phi\right), \tag{3.75}$$

satisfy the von Mises criterion identically. The substitution of (3.75) into (3.73) and (3.74) and the use of boundary conditions $\phi = 0$ and $t = t_0$ when $r = b$ lead to the solution expressed in the form

$$\left.\begin{aligned} \ln\frac{b}{r} &= \int_0^{\phi} \frac{2d\phi}{\sqrt{3}\sec^2\phi + \mu\cot\alpha\left(\sqrt{3} - \tan\phi\right)}, \\ \ln\frac{t}{t_0} &= \int_0^{\phi} \frac{\left(1 - \sqrt{3}\tan\phi\right)d\phi}{\sqrt{3}\sec^2\phi + \mu\cot\alpha\left(\sqrt{3} - \tan\phi\right)}. \end{aligned}\right\} \tag{3.76}$$

The integrals can be evaluated numerically for any given value of μ cot α. The values of ϕ and t at the die exit $r = a$, denoted by ϕ_a and t_a, respectively, can be computed from (3.76). The resultant longitudinal force required for drawing the tube is

$$P = 2\pi a t_a \sigma_a \cos\alpha = \frac{4\pi}{\sqrt{3}} Y a t_a \sin\phi_a \cos\alpha,$$

where σ_a denotes the drawing stress, which is the value of σ at $r = a$. The solid curves in Fig. 3.14 represent the dimensionless drawing stress σ_a/Y, which is plotted against a/b for three different values of μ cot α. The thickness ratio t_a/t_0 of the drawn tube is represented by the lower solid curves in Fig. 3.15. According to the present theory, the load corresponding to a given b/a ratio is proportional to the initial wall thickness, which agrees with experiment when a/t_0 is greater than about 5.

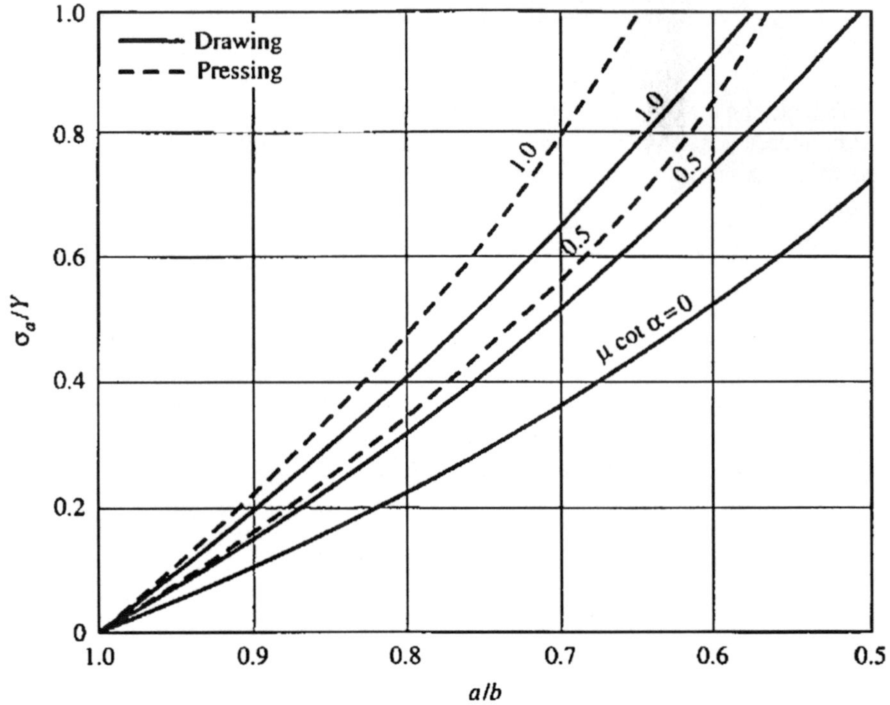

Fig. 3.14 Relation between the drawing stress and the reduction in diameter due to drawing and pressing of thin tubes

A closed-form solution for the tube-drawing process is possible if Tresca's yield criterion is used as an approximation to the von Mises criterion. Since σ and q are both positive in the drawing process, the maximum shear stress criterion of Tresca becomes

$$\sigma + q = Y.$$

Using this relation to eliminate q from (3.73) and (3.74), and integrating under the boundary conditions $\sigma = 0$ and $t = t_0$ at $r = b$, we obtain the solution in the form

$$\ln \frac{b}{r} = \frac{F + G}{2\mu \cot \alpha}, \quad \ln \frac{t}{t_0} = \frac{F - 2G}{\mu \cot \alpha}, \tag{3.77}$$

where F and G are functions of σ only for any given values of μ and α and are easily shown to be given by

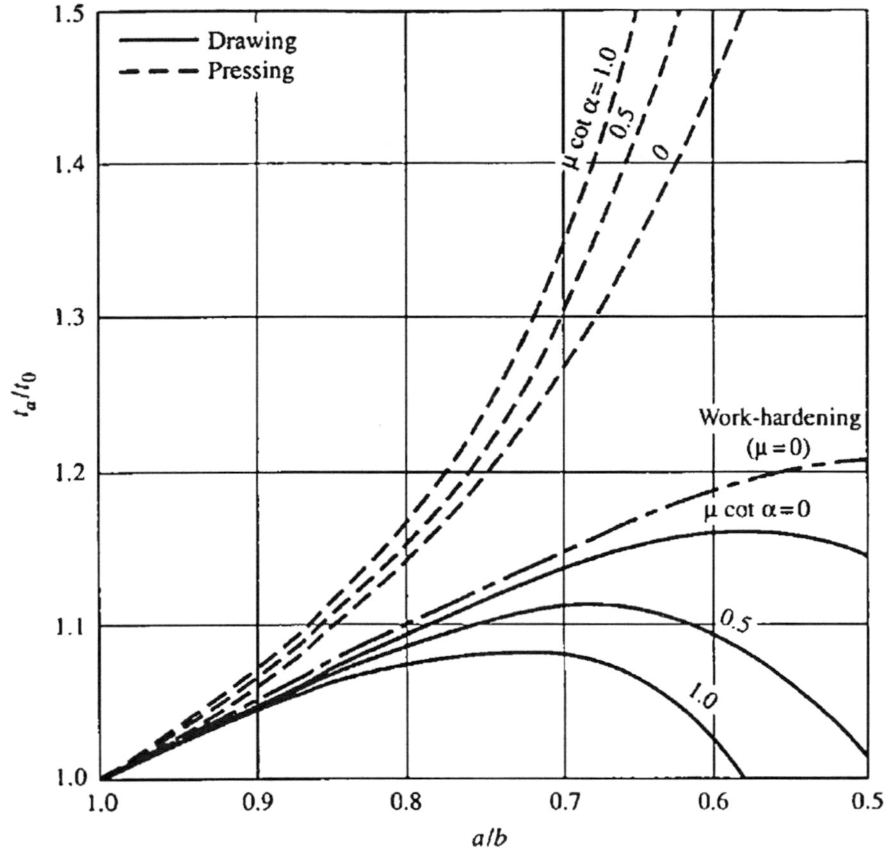

Fig. 3.15 Relation between thickness change and reduction in area due to drawing and presenting of thin tubes

$$\left.\begin{aligned} F(\sigma) &= -\ln\left\{1-\left(\frac{\mu\cot\alpha}{1+\mu\cot\alpha}\right)\frac{\sigma}{2Y}\left(3-\frac{\sigma}{Y}\right)\right\}, \\ G(\sigma) &= 2\sqrt{m}\left\{\tan^{-1}\left(3\sqrt{m}\right)-\tan^{-1}\left[\sqrt{m}\right]\left(3-\frac{2\sigma}{Y}\right)\right\}, \end{aligned}\right\} \tag{3.78}$$

with m denoting the parameter μ cot α/(8 – μ cot α). When $\mu = 0$, the right-hand side of the first equation of (3.78) reduces to $(\sigma/Y)(l - \sigma/4Y)$ and that of the second equation to $(\sigma/2Y)(1 - \sigma/Y)$. The thickness ratio at $r = a$, predicted by (3.77) and (3.78), does not differ significantly from that shown in Fig. 3.15, which indicates the accuracy of the simplified analysis. For an analysis of tube drawing with an internal mandrel, the reader is referred to Blazynski and Cole (1960) and Moore and Wallace (1967). An application of Hill's variational method to the tube-drawing process has been discussed by Lahoti and Kobayashi (1974). A slipline field solution for tube drawing has been given by Collins and Williams (1985).

The tube-pressing process is similar to the tube-drawing process, except that the tube is pushed instead of being pulled through the die by a longitudinal force on the entry side. In this case, σ is negative, and the boundary condition is modified to $\sigma = 0$ at $r = a$. The governing equations (3.73) and (3.74) remain unaffected, and the stresses are still given by (3.75), where ϕ vanishes at $r = a$ and has negative values elsewhere. In terms of ϕ, the von Mises solution for tube pressing is therefore given by (3.77) and (3.78) with b replaced by a. The broken curves of Fig. 3.14 show the variation of $-\sigma/Y$ at $r = b$ with the radius ratio a/b in tube pressing. The thickness ratio t_a/t_0 in tube pressing is represented by the upper set of curves of Fig. 3.15. Since σ is negative, the tube thickens continuously during the pressing. In contrast, the thickness of a tube drawn through the die may decrease near the exit if the reduction in diameter is sufficiently large. The resultant change in thickness is always considerably smaller in tube drawing than in tube pressing.

3.5.2 Influence of Strain Hardening

When the material strain-hardens, the constant yield stress Y in (3.73) and (3.74) must be replaced by the variable equivalent stress $\bar{\sigma}$, which is a function of the total equivalent strain $\bar{\varepsilon}$. The von Mises yield criterion is then expressed by (3.75) with Y replaced by $\bar{\sigma}$, and the substitution into (3.73) and (3.74) leads to the differential relations

$$\left.\begin{aligned} \frac{dr}{r} &= -\frac{2\,(1+\lambda\tan\phi)\,d\phi}{\sqrt{3}\sec^2\phi + \mu\cot\alpha\left(\sqrt{3}-\tan\phi\right)}, \\ \frac{dt}{t} &= -\frac{\left(1-\sqrt{3}\lambda\tan\phi\right)\left(1+\sqrt{3}\lambda\tan\phi\right)d\phi}{\sqrt{3}\sec^2\phi + \mu\cot\alpha\left(\sqrt{3}-\tan\phi\right)}, \end{aligned}\right\} \tag{3.79}$$

where $\lambda = (H/\bar{\sigma})\left(\overline{d\varepsilon}/d\phi\right)$, with H denoting the plastic modulus $d\bar{\sigma}/\overline{d\varepsilon}$. The equivalent strain increment $\overline{d\varepsilon}$ can be expressed in terms of the circumferential and thickness strain increments as

$$\overline{d\varepsilon} = \frac{2}{\sqrt{3}}\left\{\left(\frac{dr}{r}\right)^2 + \left(\frac{dr}{r}\right)\left(\frac{dt}{t}\right) + \left(\frac{dt}{t}\right)^2\right\}^{1/2}$$

For a given stress–strain curve, $\bar{\sigma}$ and H are known functions of $\bar{\varepsilon}$. Equation (3.79) in conjunction with the preceding relation is therefore sufficient to determine r, t, and $\bar{\varepsilon}$ as functions of ϕ for any given value of the friction parameter $\mu \cot a$. For most metals, it is a reasonable first approximation to calculate $\overline{d\varepsilon}$ on the assumption that dr/r and dt/t are those corresponding to $H = 0$ for a given $d\phi$. Equation (3.79) then furnishes modified values of dr/r and dt/t, and a second approximation to $\overline{d\varepsilon}$ then follows. The solution is easily carried out in a stepwise manner to any desired accuracy using the appropriate boundary conditions.

The thickness change for frictionless tube drawing is shown graphically in Fig. 3.15, considering a linear strain hardening with $H = Y$. The general effect of strain hardening in tube drawing is to increase the wall thickness at any given radius and is therefore opposite to that of die friction. When the free edge of the drawn tube reaches the die entry, the process becomes unsteady, since the stresses and strains then depend not only on r but also on the radius of the free edge. The final unsteady state of the tube-drawing process is similar to the deep-drawing process and has been analyzed by Szczepinski (1962) by the numerical integration of the basic equations, which are easily shown to be hyperbolic.

The solution is greatly simplified if the thickness variation in the equilibrium equation is disregarded, the error involved in this approximation for the calculated stresses being usually insignificant. If, in addition, Tresca's yield criterion is adopted, and the material is assumed to strain-harden linearly, a closed-form solution can be found for the tube-drawing process by taking the relationship between σ and q as

$$\sigma + q = Y\left(1 + n\ln\frac{b}{r}\right),$$

where nY represents the constant slope of the uniaxial stress–strain curve. The above expression involves the further assumption that the total equivalent strain in any given element is numerically equal to the total circumferential strain, which is found to be a good approximation over the normal range of reductions. Eliminating q from (3.71) by means of the above relation, and neglecting the variation of t with r, we obtain the simplified differential equation

$$r\frac{d\sigma}{dr} - (\mu\cot\alpha)\,\sigma = -Y(1 + \mu\cot\alpha)\left(1 + n\ln\frac{b}{r}\right).$$

Integrating, and using the boundary condition $\sigma = 0$ at $r = b$, the solution to the above equation is obtained as

$$\frac{\sigma}{Y} = \left(1 + \frac{1-n}{\mu\cot\alpha}\right)\left\{1 - \left(\frac{r}{b}\right)^{\mu\cot\alpha}\right\} + n\ln\frac{b}{r}. \tag{3.80}$$

The thickness change in the tube can be calculated by the integration of (3.72). Eliminating q with the help of the yield condition, this equation is reduced to

$$\frac{dt}{t} = \left\{\frac{2\sigma - Y\left[1 + n\ln(b/r)\right]}{2Y\left[1 + n\ln(b/r) - \sigma\right]}\right\}\frac{dr}{r}.$$

Substituting from (3.80), and using the boundary condition at $t = t_0$ at $r = b$, the solution may be expressed as

$$\ln\frac{t}{t_0} = \int_0^{\ln(b/r)}\left\{\frac{(2 + m - 2n) + mn\xi - 2(1 + m - n)\exp(-m\xi)}{(1 - m - n) - mn\xi - 2(1 + m - n)\exp(-m\xi)}\right\}d\xi, \tag{3.81}$$

where $m = \mu \cot\alpha$, and $\xi = \ln(b/r)$. The integral is easily evaluated numerically for any given values of m and n to obtain the variation of t/t_0 with r/b. The integration can be carried out explicitly for a nonhardening material ($n = 0$), and the result can be expressed in the compact form

$$\ln\frac{t}{t_0} = \frac{2+m}{1-m}\ln\frac{b}{r} + \frac{3}{1-m}\ln\left(1 - \frac{\sigma}{2Y}\right),$$

which is essentially the same as that given by Johnson and Mellor (1983). When the right-hand side of the above equation becomes indeterminate, an independent analysis gives $\ln(t/t_0) = 2\ln(b/r) - 1.5(b/r - 1)$. Approximate solutions for the sinking of relatively thick tubes have been presented by Chung and Swift (1952) and by Moore and Wallace (1967). The distribution of residual stresses present in cold-drawn tubes due to unloading from the plastic state has been discussed by Denton and Alexander (1963).

3.6 Extrusion of Cylindrical Billets

In the process of extrusion, a block of metal is reduced in cross section by forcing it through a rigid die having an aperture of given shape. In the case of direct extrusion, shown in Fig. 3.16a, the container wall is stationary, and the billet is pushed through the die by a ram moving in the same direction. The inverted extrusion, on the other hand, involves a hollow ram fixed to the die, which is forced into the metal to make the extruded billet move in the opposite direction. Since there is no relative motion between the billet and the container wall in inverted extrusion, frictional losses are kept down to a minimum. Except for the initial and final stages, the extrusion process may be considered as one of steady state, and the following treatment is concerned only with the steady-state process.

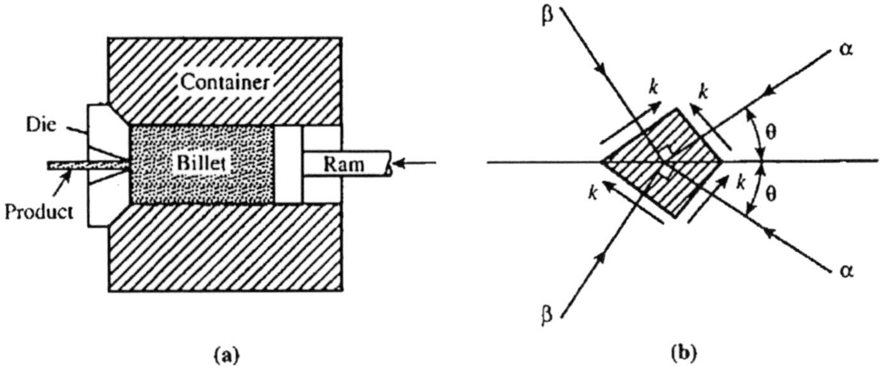

Fig. 3.16 Extrusion of a cylindrical billet. (**a**) The direct extrusion process and (**b**) basic discontinuities in stress

3.6.1 The Basis for an Approximation

A rigorous solution to the extrusion of cylindrical billets requires involved numerical analysis even when friction and work-hardening are disregarded. From the practical point of view, however, solutions based on the consideration of equilibrium, in which the yield limit is exceeded in a limited zone of plastic deformation, should provide fairly realistic estimates of the extrusion load (Chakrabarty, 1993b). Consider an equilibrium distribution of stress which is assumed to be discontinuous across certain surfaces separating regions where the maximum shear stress k, which occurs in the meridian plane, has constant directions. If the material is assumed to be ideally plastic obeying Tresca's yield criterion, the stresses in a typical plastic element are given by

$$\sigma_r = -p - k\sin 2\phi, \quad \sigma_z = -p + k\sin 2\phi,$$
$$\tau_{rz} = k\cos 2\phi, \qquad \sigma_\theta = -p + \lambda k,$$

where p is the mean compressive stress in the meridian plane, and ϕ is the counterclockwise angle made by the α-lines with the r-axis. The yield condition will not be violated if the parameter λ lies between -1 and 1. The substitution from above into the equilibrium equations (3.1) gives

$$\frac{\partial p}{\partial r} + \frac{k}{r}(\sin 2\phi + \lambda) = 0, \quad \frac{\partial p}{\partial z} - \frac{k}{r}\cos 2\phi = 0,$$

since ϕ is assumed to have a constant value. Let $\psi = \tan^{-1}(z/r)$ denote the angle which the radius vector to a generic point of the field makes with the r-axis. The preceding equations then furnish, on integration, the expressions for p and λ as

$$\left.\begin{aligned} p &= k\left(f + \cos 2\phi \tan\psi\right), \\ \lambda &= -\left(rf' + \sin 2\phi\right) + \cos 2\phi \tan\psi, \end{aligned}\right\} \qquad (3.82)$$

where f is a function of r to be found from the boundary condition. Across a surface of stress discontinuity, the normal and shear stresses must be continuous, but p, ϕ, and λ are all discontinuous. As in the case of plane strain, the sliplines are reflected at a stress discontinuity, separating two plastic regions, the jump in the value of p across the discontinuity being of magnitude $2k \sin 2\theta$, where θ is the acute angle made by the α-lines with the discontinuity. When the α- and β-lines are oriented as shown in Fig. 3.16b, the jump is positive from the lower to the upper side of the discontinuity.

Friction between the billet and the container wall is eliminated in hydrostatic extrusion in which the ram is separated from the billet by a pressurized fluid. An improved lubrication between the billet and die is also achieved in hydrostatic extrusion, particularly for large reductions using conical dies (Duffill and Mellor, 1969). In conventional extrusion through conical dies, the lubrication sometimes breaks down as the extrusion proceeds. Moreover, the billet is usually square-ended in conventional extrusion, and there is usually a pressure peak at the beginning of extrusion with conical dies. In hydrostatic extrusion, on the other hand, the billet is chamfered to match the die, and the pressure peak is absent. The technology of hydrostatic extrusion has been described in detail by Pugh (1970) and by Alexander and Lengyel (1971).

3.6.2 Extrusion Through Conical Dies

When the radius ratio b/a is moderate, in relation to the die semiangle α, the stress discontinuity pattern of Fig. 3.17a is appropriate. The included angles at C and B are right angles, while AC is inclined at an angle $\pi/4 - \alpha/2$ with the z-axis. This satisfies frictionless conditions along the die face and the container wall and the condition of symmetry of the stress about the z-axis. The solution begins with the triangle AEC, where $\phi = \pi/4$, $p = k$, and $\lambda = -1$. Since $\theta = \alpha/2$ along AC, the value of p immediately above AC is $k(1 + 2\sin\alpha)$. The reflection of the sliplines in AC means that $\phi = \pi/4 + \alpha$ in the region $CABD$, and the boundary condition for p along AC, where $\psi = \pi/4+\alpha/2$, gives $f = 2(1+2\sin\alpha) - \cos 2\alpha$ by (3.82). Hence the value of p at any point of the region $CABD$ is

$$p_1 = 2k\,(1 + 2\sin\alpha) - k\,(\cos 2\alpha + \sin 2\alpha \tan\psi)\,. \qquad (3.83)$$

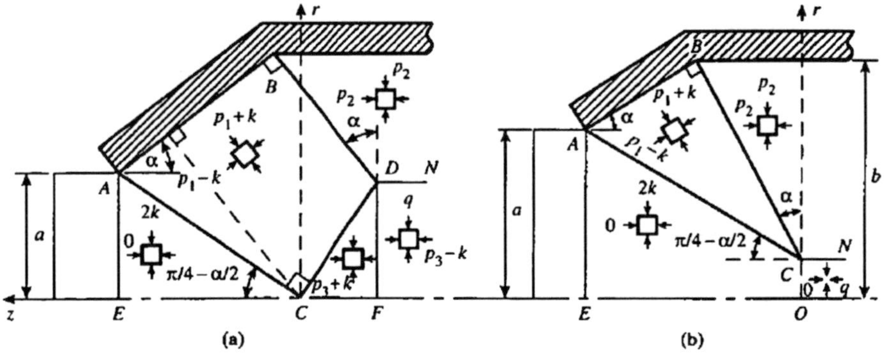

Fig. 3.17 Stress discontinuity pattern for the analysis of extrusion through smooth conical dies. (**a**) $b/a \geq 1+\sin\alpha$ and (**b**) $b/a \leq 1+\sin\alpha$

Setting $\psi = -(\pi/4 - \alpha/2)$ in the first equation of (3.82), the value of p immediately above CD is found to be $k(1 + 6\sin\alpha)$. On crossing the discontinuity CD, where $\theta = \pi/2 - \alpha/2$, the pressure p suffers a jump by an amount $2k\sin\alpha$, giving

$$p_3 = k\,(1 + 8\sin\alpha)\,,$$

which is the value of p throughout the triangle CDF, where $\theta = \pi/4$ and $\lambda = -1$. The material above BDN is assumed to be in a stress state defined by $\sigma_r = \sigma_z = -p_2$, which depends only on r. Evidently, p_2 is equal to the value of $p_1 - k$ along the line BD, whose equation is $z = r\tan\alpha - c\sec\alpha$, where c is the radius to the line DN. It follows from (3.83) that

$$p_2 = 2k\left(2 + \frac{c}{r}\right)\sin\alpha, \quad \frac{c}{b}\sin\alpha = 1 - (1 + \sin\alpha)\,\frac{a}{b},$$

where the second relation is obtained from the geometry of the discontinuity pattern. The condition of continuity requires the radial compressive stress in the region to the right of DF to be $q = 6\,k\sin\alpha$. The circumferential stress in this region is equal to $-q$ for equilibrium. If the extrusion pressure is denoted by p_e, which is the mean pressure exerted by the ram, then

$$b^2 p_e = 2\int_0^c (p_3 - k)\,r\,dr + 2\int_c^b p_2 r\,dr.$$

Substituting for p_2 and p_3, and integrating, we obtain the formula for the extrusion pressure. In view of the above expression for $(c/b)\sin\alpha$, the result becomes

$$\frac{P_e}{Y} = 2\left(1 - \frac{a}{b}\right)(1 + \sin\alpha), \tag{3.84}$$

where Y denotes the uniaxial yield stress of the material. The smallest value of b/a for which the field applies corresponds to $c = 0$, giving $b/a = 1 + \sin\alpha$ as the limit.

For $b/a \geq 1 + \sin\alpha$, the stress discontinuity pattern of Fig. 3.17b is applicable, the angle ACB being again equal to $\pi/4 - \alpha/2$. The expression for p in the triangular region ABC will take the boundary value $k(1 + 2\sin\alpha)$ along AC if

$$f = 2(1 + 2\sin\alpha) - \cos 2\alpha - \frac{2c}{r}\sin\alpha\,(1 + \sin\alpha),$$

where c is the radius to the apex C. The result follows from the first equation of (3.82) and the fact that $z = (r - c)\tan(\pi/4 + \alpha/2)$ along AC. The expression for p in this region then becomes

$$\begin{aligned} p_1 &= 2k(1 + 2\sin\alpha) - k(\cos 2\alpha + 2\sin\alpha\tan\psi) \\ &\quad - \frac{2kc}{r}\sin\alpha\,(1 + \sin\alpha). \end{aligned} \tag{3.85}$$

The state of stress in the region above BCN is given by $\sigma_r = \sigma_z = -p_2$, where p_2 is the value of $p_1 - k$ along the line BC, whose equation is $z = (r - c)\tan\alpha$. Consequently,

$$p_2 = 2k\left(2 - \frac{c}{r}\right)\sin\alpha, \quad \frac{c}{b}\sin\alpha = -1 + \frac{a}{b}(1 + \sin\alpha). \tag{3.86}$$

The stress is uniform in the material to the right of *OC,* the radial compressive stress being $q = 2\,k \sin \alpha$, which is equal to the circumferential compressive stress. The resultant force exerted by the ram furnishes the same extrusion pressure as that given by (3.84). The fractional reduction in area is $R = 1 - (a/b)^2$, which gives $a/b = \sqrt{1-R}$, the area ratio b^2/a^2 being known as the extrusion ratio. Due to the redundant shearing at entry and exit, the extrusion pressure is considerably higher than the ideal pressure $2Y\ln(b/a)$, corresponding to homogeneous compression.

To check the validity of the preceding solution, it is only necessary to examine the circumferential stress in the material in contact with the container wall and the die face. Substituting $\sigma_r = -p_2$ and $\tau_{rz} = 0$ in the first equation of equilibrium (3.1), we find that $\sigma_\theta = -4\,k \sin \alpha$ in the region adjacent to the wall for both the discontinuity patterns. The yield criterion is therefore not violated in this region if $\alpha \leq \pi/6$. Considering the region adjacent to the die, where $\phi = \pi/4 + \alpha$, and inserting the appropriate expression for f in the second equation of (3.82), we get

$$\lambda = -\,(\cos 2\alpha + \sin 2\alpha \tan \psi) - \frac{2c\varepsilon}{r} \sin \alpha\,(1 + \sin \alpha)\,,$$

where $\varepsilon = 0$ for field (a) and $\varepsilon = 1$ for field (b). It is easily verified that $\lambda \leq -1$ throughout the region *ABC* in field (b) and in the triangular region between *CA* and the radial line parallel to *DB* in field (a). The minimum value of λ is $-(1 + \sin 2\alpha)$, occurring at *C* in field (b), and along *CA* in field (a). Despite the violation of the yield criterion, which occurs in a limited region, the solution is expected to provide a good approximation for moderate reductions. An approximate slipline field solution for extrusion through conical dies in due to Chenot et al. (1978).

The dimensionless extrusion pressure given by (3.84) is plotted as solid curves in Fig. 3.18 as a function of the fractional reduction in area for different die angles. Each solid curve terminates at a point on the lower chain-dotted curve, representing the ideal extrusion pressure $2Y\ln(b/a)$ corresponding to homogeneous work. The broken curves have been obtained by Lambert and Kobayashi (1969) on the basis of a kinematic approximation leading to an upper bound. The empirical formula

$$\frac{Pe}{Y} = 0.5 \sin \alpha + \left(0.91 + 0.12 \sin^2 \alpha\right) \ln \left(\frac{1}{1-R}\right) \qquad (3.87)$$

represents approximately the mean of the static and kinematic solutions for small and moderate reductions, while approximately coinciding with the kinematic solution for large reductions. Equation (3.87) may be regarded as a close approximation to the actual extrusion pressure in the range of $0.1 \leq R \leq 0.9$ under frictionless conditions. The presence of a small friction over the die face, defined by a constant friction coefficient μ, may be approximately allowed for by multiplying the right-hand side of (3.87) by the factor $1 + (\mu \cot \alpha)/\sqrt{1 - 0.5R}$, which is found to provide a good approximation for plane strain extrusion.

The influence of work-hardening can be approximately estimated by assuming the mean effective strain to be the same as that occurring in a nonhardening material.

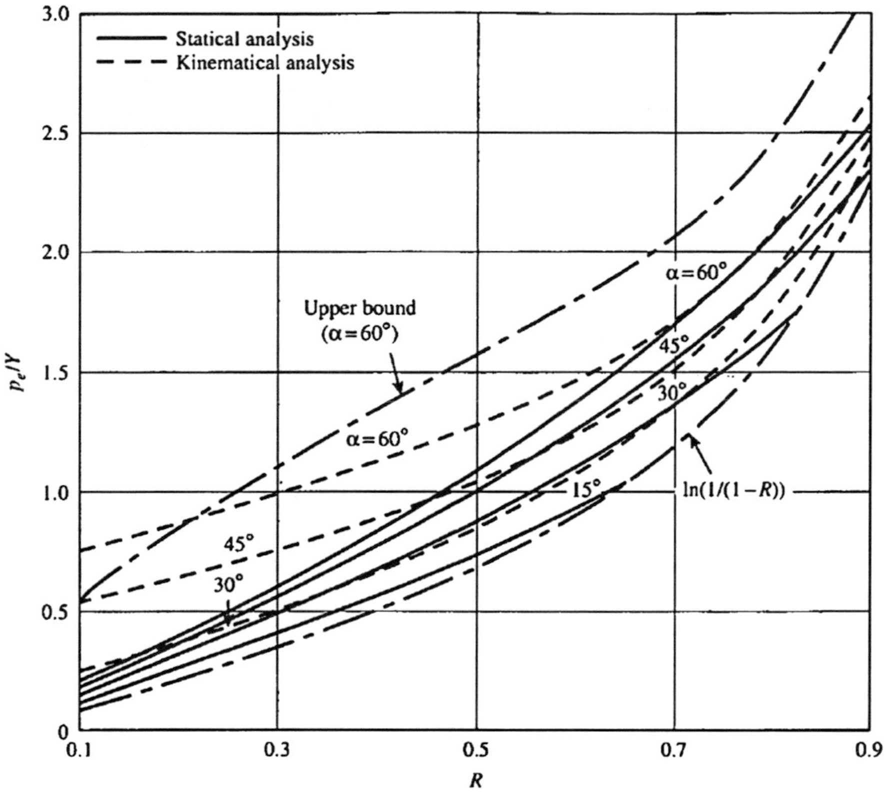

Fig. 3.18 Variation of extrusion pressure with fractional reduction in the area for extrusion through smooth conical dies

Since the extrusion pressure is equal to the work done per unit volume of the material, the mean effective strain ε is given by the right-hand side of (3.87), whether the material work-hardens or not. To obtain the extrusion pressure for a work-hardening material, it is therefore only necessary to replace Y in (3.87) by a mean yield stress $\bar{Y}$, such that $\bar{Y}\varepsilon$ is equal to the area under the uniaxial stress–strain curve up to a strain of ε. When friction and work-hardening are duly taken into account, the predicted extrusion pressure is found to agree with available experimental results on extrusion.

3.6.3 Extrusion Through Square Dies

A statistical analysis for frictionless extrusion through a square-faced die will be carried on the basis of the same discontinuity patterns as those used for satisfactory lower bound solutions in plane strain extrusion (Chakrabarty, 2006). When the reduction in area is sufficiently small, the stress discontinuity pattern is that shown

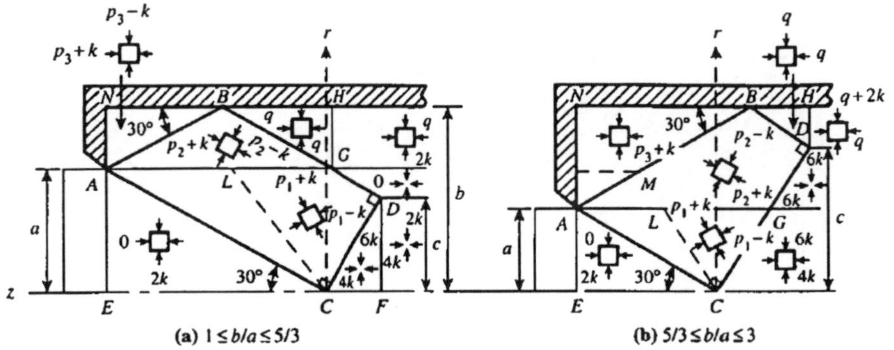

Fig. 3.19 Stress discontinuity patterns for the extrusion of cylindrical billets through square-faced dies

in Fig. 3.19a. The discontinuities *AB, AC,* and *BD* are each inclined at an angle of 30° to the axis of symmetry, while *CD* is perpendicular to both *AC* and *BD.* The stress boundary conditions are satisfied as a result of successive reflection of sli-plines in *AC, CD, AG,* and *AB,*the acute angle made by the α-lines being 15° with the first discontinuity and 75° with the others. The jump in the value of p across each of these discontinuities is therefore equal to k. The stress-free condition on the exit side is satisfied by taking $p = k$ and $\lambda = -1$ in the region *AEC.* Since $\phi = 75°$ in the region *ACDG,* while $p = 2\,k$ along *AC* where $\psi = 60°$, the expressions for p and λ in this region are

$$p_1 = 3k + \frac{k}{2}\left(1 - \sqrt{3}\tan\psi\right), \quad \lambda_1 = -\frac{1}{2}\left(1 + \sqrt{3}\tan\psi\right),$$

in view of (3.82). The condition $\psi = -30°$ along the discontinuity *CD* gives $p = 5\,k$ and $\lambda = -1$ in the region *CDF.* Due to the constant jump in the value of p across *AG,* the expressions for p and λ in the region *AGB,* where $\phi = -75°$, become

$$p_2 = 4k + \frac{k}{2}\left(1 - \sqrt{3}\tan\psi\right), \quad \lambda_2 = \frac{1}{2}\left(1 - \sqrt{3}\tan\psi\right),$$

By simple geometry, $\sqrt{3}\,\tan\psi = 3\,(2a/r - 1)$ along *AB*. The jump condition across *AB* and the equations of equilibrium are therefore satisfied by taking

$$p_3 = k\left(7 - \frac{3a}{r}\right), \quad \lambda_3 = 1 - \frac{3a}{r}, \tag{3.88}$$

as the values of p and λ in the region to the left of *AB.* Since *BG* coincides with a principal stress direction in the region *AGB,* the equilibrium equations and the boundary conditions are satisfied by assuming region *BGH* to be in a state of hydrostatic compression q, which is equal to the value of $p_2 - k$ along *AG* for the same value of r. The fact that $\sqrt{3}\tan\psi = 3 - 4c/r$ along BD, where c is the radius to point D, furnishes the result

$$q = 2k\left(1 + \frac{c}{r}\right), \quad \frac{c}{b} = \frac{3}{2}\left(1 - \frac{a}{b}\right).$$

The material to the right of GH is nonplastic with $\sigma_r = \sigma_\theta = -2\,k$ and $\sigma_z = -q$. Since BD is parallel to AC; the state of stress in the material to the right of GD is identical to that in AEC. Statically admissible states of stress in the regions to the right of the discontinuities DF and GH are as shown. Since the resultant longitudinal thrust exerted by the die must balance the total ram load $\pi b^2 p_e$ for equilibrium, we have

$$b^2 p_e = 2\int_a^b (p_3 + k) r dr = 2k\,(b - a)\,(4b + a)\,.$$

In terms of the uniaxial yield stress Y of the assumed Tresca material, the extrusion pressure is therefore given by

$$\frac{p_e}{Y} = \left(1 - \frac{a}{b}\right)\left(4 + \frac{a}{b}\right), \quad 1 \le \frac{b}{a} \le 3. \tag{3.89}$$

For geometrical reasons, the greatest reduction for which the field of Fig. 3.19a is applicable corresponds to $c = a$, giving $b/a = \frac{5}{3}$. The state of stress is statically admissible in region AGB since λ_2 lies between -1 and $2c/a - 1$, but the yield criterion is violated in the region ABN since $\lambda_3 \le -1$ for $r/a \le \frac{3}{2}$. The yield condition is also violated in the region ACL, where CL makes an angle of 30° with CA, since λ_1 varies between -2 and -1 within this region.

When $b/a \ge \frac{5}{3}$, the stress discontinuity pattern is modified to that of Fig. 3.19b, where point D is situated at a radius $c \ge a$. The principal compressive stress in the different regions separated by the discontinuities are as shown, the quantities p_1, p_2, p_3, and q being the same as before. The material to the right of DH is plastic and has a circumferential compression equal to $4\,k$. The material to the right of DG is under a uniform hydrostatic compression of amount $6\,k$, which is the value of $p_2 + k$ along DG where $\psi = -30°$. Since the expression for the die pressure is unchanged, the extrusion pressure is still given by (3.89). The greatest radius ratio for which the field is applicable corresponds to $b/a = 3$, for which points B and D coincide $(c = b)$. The yield criterion is violated over the region $a \le r \le 1.5a$ within ABN, as well as over the region $\pi/6$ ψ $\pi/3$ within ACG, the extent of the overstressed regions being indicated by broken lines.

When $b/a \ge 3$, the region BDH disappears to form a triangle ACB, where B is situated at a radial distance $b - 3a$ from the container wall. The radial and axial discontinuities BH and NBM are considered through B, where H is on the container wall and N is on the die face. The values of p and λ in the material on the exit side of BH, denoted by p_4 and λ_4, respectively, are obtained from (3.82) by setting $\phi = \pi/4$ and $f(r) = 3 + 2\ln(r/3a)$. Thus

$$p_4 = 2k\left(3 + \ln\frac{r}{3a}\right), \quad \lambda_4 = -1, r \leq 3a. \tag{3.90}$$

Only the circumferential stress is therefore discontinuous across *BN*. The material to the right of *BH* is assumed plastic under the principal stresses

$$\sigma_r = -p_4, \sigma_z = -(p_4 + k), \quad \sigma_\theta = -(p_4 + 2k),$$

which satisfy the equilibrium equation and Tresca's yield criterion, the necessary continuity conditions across *BH* and *BM* being identically satisfied. The state of stress in the material lying above the discontinuity $r = 3a$ is therefore statically admissible. The condition of overall longitudinal equilibrium gives

$$b^2 p_e = 2\int_a^{3a} (p_3 + k)\, r\, dr + 2\int_{3a}^{b} (p_4 + k)\, r\, dr.$$

Substituting for p_3 and p_4 from (3.88) and (3.90), respectively, and integrating, the dimensionless extrusion pressure is obtained as

$$\frac{p_e}{Y} = 3 - \frac{a^2}{b^2} + \ln\frac{b}{3a}, \quad \frac{b}{a} \geq 3. \tag{3.91}$$

The lower solid curve in Fig. 3.20 represents the extrusion pressure in frictionless extrusion plotted as a function of the fractional reduction $R = 1 - (a/b)^2$. The ideal extrusion pressure $2Y \ln(b/a)$, represented by the lower broken curve, indicates how the relative amount of redundant work decreases as the reduction is increased. When the reduction in area is fairly large, a better approximation is provided by the consideration of a single triangular stress discontinuity pattern (Chakrabarty, 1993), which gives the extrusion pressure as

$$\frac{p_e}{Y} = 4\left(\frac{b^2}{a^2} - 1\right) / \left(\frac{b^2}{a^2} + 1\right)$$

The upper solid curve in Fig. 3.20 represents the extrusion pressure under conditions of sticking friction and is obtained by multiplying the frictionless pressure by the same factor as that in plane strain extrusion with a given reduction. For a closer approximation over the range $R \leq 0.3$, the value of p_e/Y for the axisymmetric extrusion is taken to be identical to the plane strain value of $p_e/2k$. The upper solid curve may be expressed by the empirical equation

$$\frac{p_e}{Y} = 0.75 + 1.5\ln\left(\frac{1}{1-R}\right), \tag{3.92}$$

which is correct to within 5% over the range $0.1 \leq R \leq 0.9$. This is the extrusion pressure at the end of the steady-state process when the frictional drag between the billet and the container wall falls to zero. In the conventional extrusion through

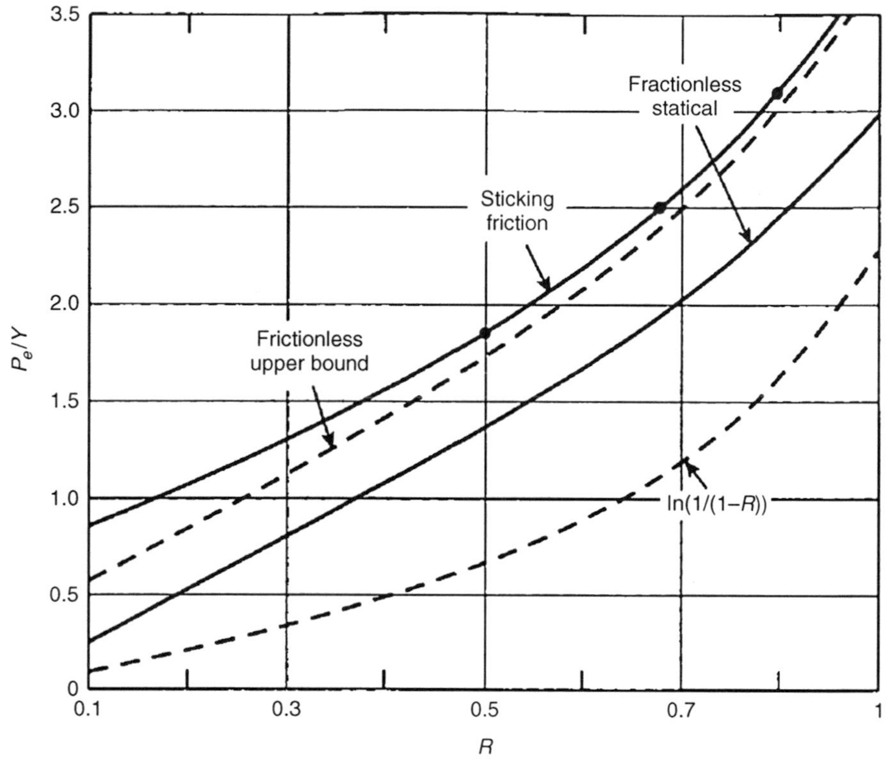

Fig. 3.20 Variation of extrusion pressure with fractional reduction in the area for axis symmetric extrusion through square-faced dies

square-faced dies, the lubricant film usually breaks down, and the extrusion pressure approaches that for conditions of sticking friction. Some experimental results due to Johnson (1957), plotted in Fig. 3.20 for comparison, confirms the usefulness of the above formula. For a work-hardening material, the quantity Y in (3.92) should be replaced by a mean yield stress $\bar{Y}$, as in the case of conical dies. A solution for the axisymmetric tube extrusion, using a perturbation method, has been given by Spencer (1964).

3.6.4 Upper Bound Solution for Square Dies

By the kinematic theorem of limit analysis (Section 1.2), any fictitious velocity field that satisfies the incompressibility condition and the prescribed boundary conditions provides an overestimate on the extrusion pressure. Consider first the extrusion of a cylindrical billet through a square-faced die, the assumed velocity discontinuity pattern being that shown in Fig. 3.21a. The material approaches the die with a unit speed and leaves the deformation zone $ABDC$ with a speed b^2/a^2 in the axial

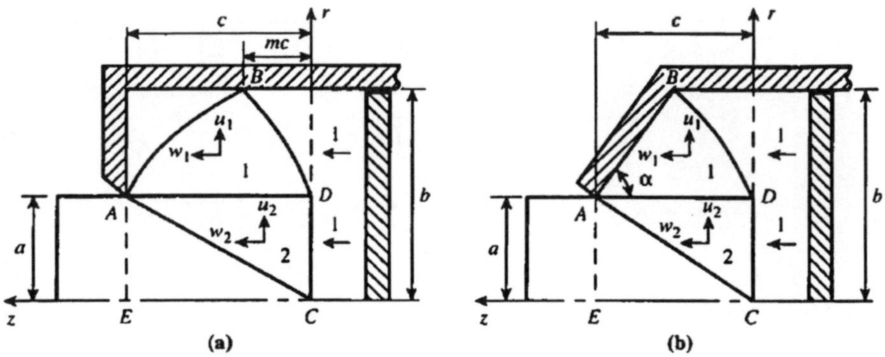

Fig 3.21 Velocity discontinuity patterns in axisymmetric extrusion. (**a**)Square-faced dies and (**b**) conical dies

direction. The components of the velocity vector in regions *ABD* and *ACD*, referred to the pair of axes considered through *C*, are denoted by (u_1, w_1) and (u_2, w_2), respectively, where

$$\left.\begin{aligned} u_1 &= -\frac{b^2-a^2}{2cr}, & w_1 &= 1-m,\\ u_2 &= -\frac{b^2-a^2}{2ca}, & w_2 &= 1+\left(\frac{b^2-a^2}{2ca}\right)\frac{z}{r}, \end{aligned}\right\} \tag{3.93}$$

with *c* and *m* as constants. The incompressibility condition is identically satisfied in each region, and the normal component of velocity is made continuous across the straight discontinuities *AD, DC,* and *AC.* The curved discontinuities *AB* and *BD* meet on the container wall at *B,* their shapes being given by the condition of continuity of the normal component of velocity across these lines. Thus

$$w_1 - u_1\frac{dz}{dr} = 0 \text{ along } AB, \quad w_1 - u_1\frac{dz}{dr} = 1 \text{ along } BD. \tag{3.94}$$

These equations are readily integrated using (3.93), and the boundary conditions are $z = c$ at $r = a$ for *AB* and $z = 0$ at $r = a$ for *BD,* the result being

$$\left.\begin{aligned} z &= c\left\{1-(1-m)\left(\frac{r^2-a^2}{b^2-a^2}\right) \text{ along } AB,\right\}\\ z &= mc\left(\frac{r^2-a^2}{b^2-a^2}\right) \text{ along } BD. \end{aligned}\right\} \tag{3.95}$$

Both these relations furnish $z = mc$ at $r = b,$ defining the location of point *B*. The components of the rate of deformation corresponding to the assumed velocity field (3.93) are easily found as

$$\dot{\varepsilon}_r = -\dot{\varepsilon}_\theta = \frac{b^2 - a^2}{2cr^2}, \quad \dot{\varepsilon}_z = \dot{\gamma}_{rz} = 0 \quad in \quad ABD$$

$$\dot{\varepsilon}_r = 0, \quad \dot{\varepsilon}_z = -\dot{\varepsilon}_\theta = \frac{b^2 - a^2}{2car}, \quad \dot{\gamma}_{rz} = \frac{b^2 - a^2}{4car}\left(\frac{z}{r}\right) \quad in \quad ACD$$

The rates of plastic work per unit volume in the deforming regions ABD and ACD for a von Mises material are $\sqrt{3}k\,\dot{\bar{\varepsilon}}_1$ and $\sqrt{3}k\,\dot{\bar{\varepsilon}}_2$, respectively, where $\dot{\bar{\varepsilon}}_1$ and $\dot{\bar{\varepsilon}}_2$ denote the equivalent strain rates in these regions. By (1.27), we have

$$\dot{\bar{\varepsilon}}_1 = \frac{b^2 - a^2}{\sqrt{3}cr^2}, \quad \dot{\bar{\varepsilon}}_2 = \frac{b^2 - a^2}{\sqrt{3}car}\left(1 + \frac{z^2}{4r^2}\right)^{1/2} \tag{3.96}$$

The rate of dissipation of the total internal energy due to the deformation of the material is therefore given by

$$E = 2\pi\sqrt{3}k\left\{\int_a^b \int_{z_1}^{z_2} \dot{\bar{\varepsilon}}_1 r\,dz\,dr + \int_0^a \int_0^{rc/a} \dot{\bar{\varepsilon}}_2 r\,dzdr\right\}$$

where z_1 and z_2 are the values of z on the curves DB and AB, respectively, for a given r. It follows from (3.95) that $z_2 - z_1$ is independent of m and is equal to $c(b^2 - r^2)/(b^2 - a^2)$. It is convenient to express the energy rate as $E = \pi kb^2 e$, where e is a dimensionless parameter. Carrying out the integration using (3.96), the result may be put in the form

$$e = 2\ln\frac{b}{a} + \left(1 - \frac{a^2}{b^2}\right)\left\{\frac{a}{c}\ln\left[\frac{c}{2a} + \sqrt{1 + \frac{c^2}{4a^2}}\right] + \frac{1}{2}\sqrt{1 + \frac{c^2}{4a^2}} - 1\right\}. \tag{3.97}$$

For a given reduction in area, the quantity e depends only on the axial length c of the deformation zone, but the contribution to the energy rate from the expression in the curly brackets is usually small.

Plastic energy is also dissipated by the shearing across the surfaces of velocity discontinuity, where the shear stress is of magnitude k. For the plane surfaces CD, AD, and AC, the rate of energy dissipation is easily obtained from the fact that the magnitudes of the velocity discontinuity across them are equal to v, m + vz/a, and v$\sqrt{1 + c^2/a^2}$, respectively, where v $= (b^2 - a^2)/2ac$.

Along the curved discontinuities AB and DB, the rates of energy dissipation due to shearing are the line integrals of the quantities $-2\pi kr(u_I dr + w_I\,dz)$ and $2\pi kr(-u_I dr + mdz)$, respectively, which can be evaluated with the help of (3.93) and (3.94). If the rate of dissipation of the total internal energy due to shearing is denoted by $\pi kb^2 f$, where f is a second dimensionless parameter given by

$$f = \frac{2ac}{b^2}\left\{m + \left(1 - 2m + 2m^2\right)\frac{2}{3a}\left(\frac{b^3 - a^3}{b^2 - a^2}\right)\right\} + \left(1 - \frac{a^2}{b^2}\right)\left(\frac{2b - a}{c} + \frac{c}{a}\right). \tag{3.98}$$

The rate of external work done in the steady-state extrusion process is $\pi b^2 p_e$ for a smooth container wall, and equating it to the rate of internal energy dissipation, which is $\pi k b^2 (e + f)$, furnishes an upper bound on the extrusion pressure in the form

$$\sqrt{3} p_e = Y(e + f)$$

which is independent of the frictional condition along the die face. To obtain the best upper bound, the constants m and c should be such that the extrusion pressure is minimized. The minimum conditions $\partial p_e/\partial m = 0$ and $\partial p_e/\partial c = 0$ yield the relations

$$m = \frac{1}{2}\left\{1 - \frac{3a}{4}\left(\frac{b^2 - a^2}{b^3 - a^3}\right)\right\}, \tag{3.99a}$$

$$\begin{aligned}\frac{c^2}{a^2}\left\{1 + \frac{2ma^2}{b^2} + \frac{8m^2 a}{3b^2}\left(\frac{b^3 - a^3}{b^2 - a^2}\right)\right\} = \left(1 - \frac{a^2}{b^2}\right)\left\{\left(\frac{2b}{a} - 1\right)\right. \\ \left. + \ln\left(\frac{c}{2a} + \sqrt{1 + \frac{c^2}{4a^2}}\right) - \frac{c}{2a}\sqrt{1 + \frac{c^2}{4a^2}}\right\}.\end{aligned} \tag{3.99b}$$

Equation (3.99b) can be solved iteratively for c/a, starting with a trial value obtained by neglecting the last two terms on the right-hand side. The computed upper bound for the extrusion pressure is shown by the upper broken curve in Fig. 3.20. For large reductions in area, an improved upper bound is given by the solution for the conical die with $\alpha = \pi/2$. Upper bound solutions based on similar velocity fields have been discussed by Kudo (1960, 1961) and by Kobayashi and Thomsen (1965).

3.6.5 Upper Bound Solution for Conical Dies

A convenient upper bound solution for extrusion through a smooth conical die may be obtained by assuming the material to slide along the die face, as shown in Fig. 3.21b. A curved velocity discontinuity BD is considered through the corner B, the resultant velocity in the region ABD being taken as everywhere parallel to the die face. The assumed velocity distribution in the deforming regions ABD and ACD are

$$\left.\begin{aligned} u_1 &= -\frac{b^2 - a^2}{2cr}, \quad w_1 = \left(\frac{b^2 - a^2}{2cr}\right)\cot\alpha, \\ u_2 &= -\frac{b^2 - a^2}{2ca}, \quad w_2 = 1 + \left(\frac{b^2 - a^2}{2ca}\right)\frac{z}{r}, \end{aligned}\right\} \tag{3.100}$$

where α is the semiangle of the conical die. It may be noted that the velocity vector in region ACD is constant in direction and magnitude along any straight line passing through C. The continuity of the normal component of velocity across the curve BD requires

$$w_1 - u_1 \frac{dz}{dr} = 1 \quad \text{along } BD.$$

The shape of the curved discontinuity is obtained by integrating this equation under the boundary condition $z = 0$ at $r = a$, the result being

$$z = c\left(\frac{r^2 - a^2}{b^2 - a^2}\right) - (r - a)\cot\alpha.$$

The difference between the values of z along BD and AB for a given r is equal to $c\,(b^2 - r^2)/(b^2 - a^2)$. The equivalent strain rate in region 2 is again given by (3.96), but that in region 1 is multiplied by a factor $\lambda = \sqrt{1 + \cot^2\alpha/4}$. The rate of dissipation of internal energy due to the deformation in regions ABD and ACD is equal to $\pi kb^2 e$, where

$$e = 2\lambda \ln\frac{b}{a} + \left(1 - \frac{a^2}{b^2}\right)\left\{\frac{a}{c}\ln\left(\frac{c}{2a} + \sqrt{1 + \frac{c^2}{4a^2}}\right) + \frac{1}{2}\sqrt{1 + \frac{c^2}{4a^2}} - \lambda\right\} \tag{3.97a}$$

The magnitudes of the velocity discontinuity across DC and AC are the same as before, the rate of energy dissipation due to shearing along them being $\pi k(b^2 - a^2)$ times $a/2c$ and $(a^2 + c^2)/2ac$, respectively. The velocity discontinuity across AD gives rise to the rate of internal energy dissipation

$$2\pi ka \int_0^c (w_2 - w_1)\,dz = \pi k\left\{2ac + \left(b^2 - a^2\right)\left(\frac{c}{2a} - \cot\alpha\right)\right\}.$$

Across the curved discontinuity BD, where $dz/dr = -(1 - w_1)/u_1$, the rate of energy dissipation due to shearing is found to be

$$2\pi k \int [-u_1 r\,dr + (1 - w_1)\,r\,dz]$$
$$= \pi k\left\{\frac{4c}{3}\left(\frac{b^3 - a^3}{b^2 - a^2}\right) + \left(b^2 - a^2\right)\left(\frac{b - a}{c}\operatorname{cosec}^2\alpha - 2\cot\alpha\right)\right\}.$$

The total rate of internal energy dissipation due to shearing across all the discontinuities taken together may be written as $\pi kb^2 f$, where

$$f = \frac{2ac}{b^2}\left\{1 + \frac{2}{3a}\left(\frac{b^3 - a^3}{b^2 - a^2}\right)\right\} + \left(1 - \frac{a^2}{b^2}\right)\left\{\frac{a^2 + c^2}{ac} + \left(\frac{b - a}{c}\right)\operatorname{cosec}^2\alpha - 3\cot\alpha\right\} \tag{3.101}$$

The frictionless extrusion pressure is finally obtained on the substitution from (3.97a) and (3.101) into the relation $p_e = k\,(\,e + f\,)$. The ratio c/a is obtained by minimizing the upper bound, given by $dp_e/dc = 0$, which corresponds to

$$\frac{c^2}{a^2}\left\{1 + \frac{4a}{3}\left(\frac{b^3 - a^3}{b^4 - a^4}\right)\right\} = \left(\frac{b^2 - a^2}{b^2 + a^2}\right)\left\{1 + \left(\frac{b}{a} - 1\right)\cos ec^2\alpha\right\} + \left(\frac{b^2 - a^2}{b^2 + a^2}\right)\left\{\ln\left(\frac{c}{2a} + \sqrt{1 + \frac{c^2}{4a^2}}\right) - \frac{c}{2a}\sqrt{1 + \frac{c^2}{4a^2}}\right\} \tag{3.102}$$

This equation can be solved for c/a by an iterative procedure starting with a suitable first approximation. When the die is rough, and the frictional stress along the die is denoted by mk, where $0 \leq m \leq 1$, it is necessary to multiply the second term in the second curly brackets of (3.101) by the factor $1 + m\,(b + a)/2a$ and to modify (3.102) accordingly. For sufficiently small reductions, the preceding solution for the square-faced die is expected to provide a better upper bound. The frictionless upper bound pressure for $\alpha = 60°$ is depicted by the upper chain-dotted curve in Fig. 3.18.

Various upper bound solutions based on radially converging flow fields have been presented by Avitzur (1965, 1968), Hailing and Mitchell (1965), Tirosh (1971), Osakada and Nimi (1975), and Calladine (1985). The upper bound solutions to other kinds of extrusion process have been discussed by Adie and Alexander (1967), Chen and Ling (1968), Yang and Han (1987), and Gordon et al. (2000). The phenomenon of central bursting, which often occurs in extrusion and wire drawing, has been treated by Avitzur (1980) and Reddy et al. (1996). Finite element solutions for the extrusion process have been discussed by Zienkiewicz et al. (1978) and Chen et al. (1979). A slipline field solution for extrusion through conical dies has been presented by Seweryn (1992). Extrusion through rotating conical dies has been investigated by Durban et al. (2001) and Ma et al. (2004).

3.7 Mechanics of Wire Drawing

Wires of small diameter are produced by successive drawing through tapered dies so that its cross section is progressively reduced to the specified size, Fig. 3.22a. The material within the die is rendered plastic by the combined action of the applied longitudinal pull and the pressure developed between the wire and the die. In the absence of friction and back pull, the work done per unit volume of the wire is equal to the mean drawing stress, which is the applied longitudinal force per unit area of

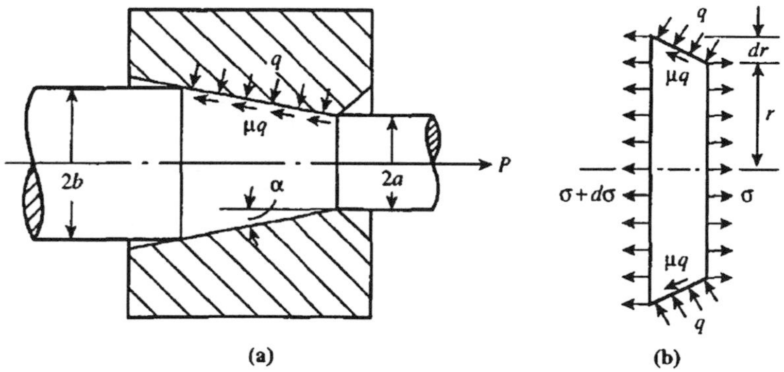

Fig. 3.22 Wire drawing through a conical die. (**a**) Geometry of the process and (**b**) stresses acting on a transverse slice

cross section. If the deformation is assumed to be homogeneous, the plastic work per unit volume is simply the area under the uniaxial stress–strain curve of the material corresponding to the total longitudinal strain in the drawn wire. For a nonhardening material, the ideal drawing stress is therefore equal to $2Y\, ln(b/a)$, where b and a denote the initial and final radii of the wire, respectively. Due to the presence of redundant work and friction, the actual drawing stress can be considerably in excess of the ideal value.

3.7.1 Solution for a Nonhardening Material

When friction is absent, the mean drawing stress t in the wire-drawing process is identical to the mean extrusion pressure during extrusion through the same die. This conclusion follows from the fact that the superposition of a uniform hydrostatic tension of magnitude equal to the mean extrusion pressure p_e leaves the plastic state of stress in the die unaffected. Since the normal pressure on the die is reduced by the amount p_e, due to the superposed tension, the mean die pressure in drawing is $(1 - R)$ times that in extrusion for a given reduction in area R. Indeed, it follows from the condition of overall longitudinal equilibrium of the material in the die that the mean die pressure in wire drawing is equal to $t(1 - R)/R$ when the die is perfectly smooth.

An elementary analysis for the mechanics of wire drawing through a rough conical die has been given by Siebel (1947), on the assumption that redundant work and strain hardening are negligible. Consider the longitudinal equilibrium of a thin slice of the plastic material formed by plane transverse sections of radii r and $r + dr$, as shown in Fig. 3.22b. The axial stress σ is assumed uniform over the transverse section, and the coefficient of friction μ between the material and the die is assumed constant. Resolving axially the forces acting on the element, it is easy to obtain the equilibrium equation

$$r\frac{d\sigma}{dr} + 2\sigma + 2q\,(1 + \mu \cot \sigma) = 0, \tag{3.103}$$

where q denotes the local die pressure. The boundary condition along the conical die surface requires the radial stress σ_r to be equal to $-q(1 + \mu \tan \alpha)$. Since $\mu \tan \alpha$ is small compared to unity over the practical range, it is a good approximation to write $\sigma_r = \sigma_\theta \approx -q$ throughout the element. If the effect of the small shear stress on the plastic yielding is disregarded, the von Mises criterion reduces to

$$\sigma + q = Y.$$

The elimination of q from the equilibrium equation (3.103), with the help of the yield criterion, gives

$$r\frac{d\sigma}{dr} - 2\mu\sigma \cot \alpha = -2Y\,(1 + \mu \cot \alpha)\,. \tag{3.104}$$

For drawing without a back pull, the boundary conditions are $\sigma = 0$ at $r = b$, and $\sigma = t$ at $r = a$. The integration of the above equation then furnishes

$$\frac{t}{Y} = \left(1 + \frac{\tan \alpha}{\mu}\right)\left\{1 - \left(\frac{a}{b}\right)^{2\mu \cot \alpha}\right\}. \tag{3.105}$$

For practical purposes, it is convenient to expand the second term in the curly brackets of (3.105) in ascending powers of μ and neglect all terms of order higher than μ^2. The mean drawing stress is then approximately given by

$$\frac{t}{Y} \approx \left\{1 + \left(1 - \frac{1}{2}\ln\frac{1}{1-R}\right)\mu \cot \alpha\right\}\ln\frac{1}{1-R}.$$

The expression in the curly brackets represents the friction factor in wire drawing for small values μ, which is usually much less than 0.1 when the die is well lubricated. A slipline field solution for the wire drawing process has been discussed by Chenot et al. (1978).

A useful expression for the mean drawing stress, including friction and redundant work, may be obtained on the assumption that the latter is independent of friction. Adopting the right-hand side of (3.87) for the dimensionless drawing stress without friction, the mean drawing stress with friction may be written in terms of the same friction factor as that in the preceding solution, the result being

$$\begin{aligned}\frac{t}{Y} = \left\{1 + \left(1 - 0.5\ln\frac{1}{1-R}\right)\mu \cot \alpha\right\}&\left\{0.5 \sin \alpha\right.\\ &\left.+ \left(0.91 + 0.12 \sin^2 \alpha\right)\ln\frac{1}{1-R}\right\},\end{aligned} \tag{3.106}$$

which is found to be in good agreement with experiment for $\alpha \leq 30°$. By the condition of overall longitudinal equilibrium, the mean die pressure is easily shown to be $\bar{q} = t(1 - R)/R\,(1 + \mu \cot\alpha)$, which decreases with increasing coefficient of friction.

The mean drawing stress and the mean die pressure for frictionless wire drawing are plotted in Figs. 3.23 and 3.24, respectively, as functions of the fractional reduction in area for $\alpha = 10°$, $20°$, and $30°$. The drawing stress for $\alpha = 10°$ and $\mu = 0.05$ is represented by the broken curve in Fig. 3.23. The solution is not valid when $R < 0.3\alpha$ approximately, for which the wire bulges at the die entry before being drawn through it. The greatest reduction for which the solution holds corresponds to $t = Y$, which is the condition for necking of the drawn wire.

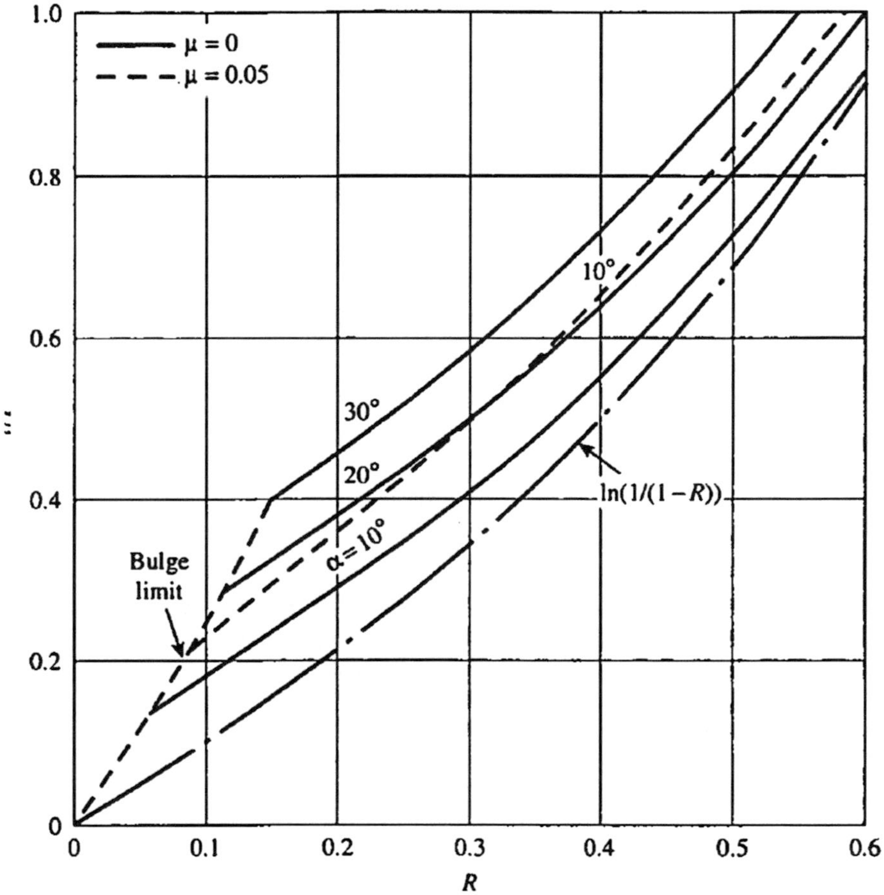

Fig. 2.23 Variation of mean drawing stress with fractional reduction in area for wire drawing through smooth conical dies

For a given reduction in area, the mean drawing stress t has a minimum value for a critical combination of μ and α. Equation (3.106) is not suitable for the computation of the optimum die angle due to the magnification of error in the empirical formula for the frictionless drawing stress. A more appropriate formula for this purpose is

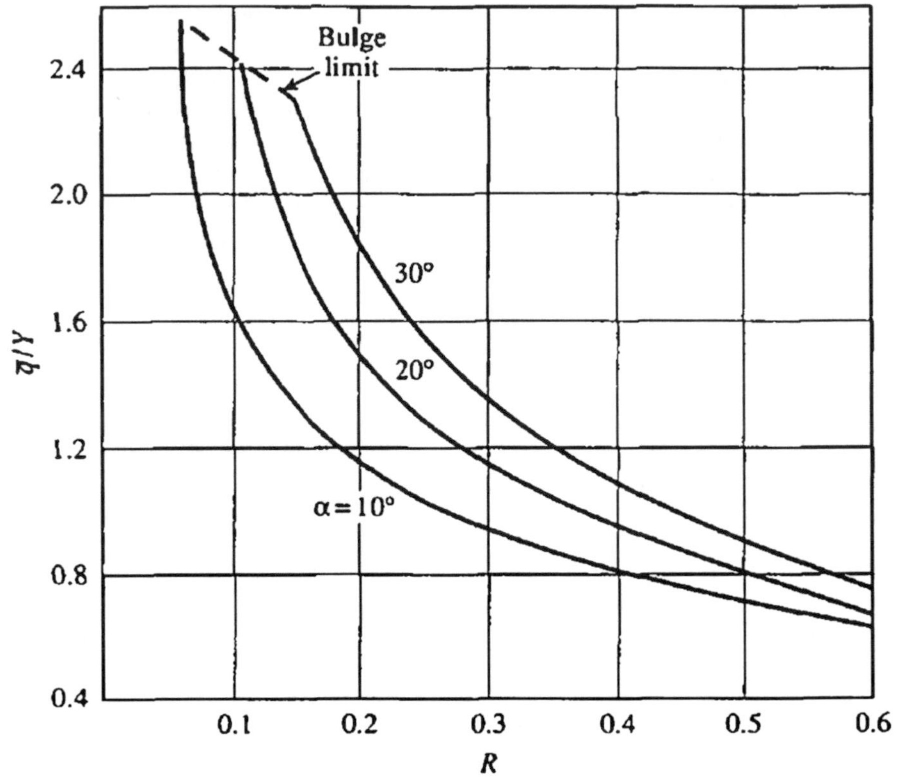

Fig. 3.24 Variation of mean die pressure with fractional reduction in area for wire drawing through smooth conical dies

obtained by assuming the redundant work factor for a given reduction in area to be similar to that in plane strain drawing with low die angles (Chakrabarty, 1987). Over the relevant range of values of R, μ, and α, it is then a good approximation to write

$$\frac{t}{Y} = \left\{1 + \left(\frac{2-R}{2}\right)\frac{\mu}{\alpha}\right\}\left\{0.93 + 0.13\alpha^2\left(\frac{2-R}{2}\right)^2\right\}\ln\frac{1}{1-R}. \qquad (3.107)$$

The expression in the second curly brackets of (3.107) approximately agrees with that corresponding to (3.106) for small α. The minimization of the drawing stress (3.107) with respect to α gives

$$\alpha^3 + \left(\frac{2-R}{2}\right)\alpha^2 - 1.79\left(\frac{\mu R^2}{2-R}\right) = 0.$$

Since the optimum die angle is quite small, the second term on the left-hand side may be neglected as a first approximation. The optimum die semiangle α_0, is then obtained with sufficient accuracy from the second approximation, the result being

$$\alpha_0 \approx 1.21\left(\frac{\mu R^2}{2-R}\right)^{1/3} - 0.08\mu\,(2-R)\,, \tag{3.108}$$

which is found to be in good agreement with experiment. For given values of μ and R, the optimum die angle in wire drawing is somewhat lower than that in sheet drawing, owing to the higher redundant work involved in the wire-drawing process. This has been demonstrated experimentally by Johnson and Rowe (1968) and by Caddell and Atkins (1969). A slipline field solution for the axisymmetric drawing of tubes has been given by Collins and Williams (1985).

The coefficient of friction in wire drawing can be determined with reasonable accuracy by using an experimental setup in which the die is split in halves longitudinally, and the force Q that tends to separate each half during the drawing is measured concurrently with the drawing force P. It follows from the conditions of overall longitudinal and transverse equilibrium of each half of the die that

$$P = \pi q\left(b^2 - a^2\right)(1 + \mu \cot\alpha)\,, \quad Q = q\left(b^2 - a^2\right)(\cot\alpha - \mu)\,.$$

The elimination of the mean die pressure q between these two relations furnishes

$$\mu = \frac{(P/\pi Q) - \tan\alpha}{1 - (P/\pi Q) - \tan\alpha} = \tan\left\{\tan^{-1}\left(\frac{P}{\pi Q}\right) - \alpha\right\}. \tag{3.109}$$

In practice, the two halves of the die are pressed together with a force S which is progressively reduced from $S > Q$ to $S < Q$ during the drawing. The condition $S = Q$ corresponds to the instant when separation of the two halves is about to occur (Maclellan, 1952; Wistreich, 1955).

In the case of extrusion in the presence of Coulomb friction, an analysis for the friction factor can be carried out by solving (3.103) under the appropriate boundary conditions. The solution is directly obtained from (3.105) by interchanging the radii a and b and replacing the drawing stress t by the negative extrusion pressure $-p_e$-. For a given friction coefficient, the friction factor in extrusion is found to be somewhat higher than that in wire drawing. For sufficiently small coefficients of friction, the dimensionless extrusion pressure in the presence of Coulomb friction is given by (3.106) with a change in sign of the logarithmic term appearing before the parameter $\mu \cot \alpha$.

3.7.2 Influence of Back Pull and Work-Hardening

In order to reduce the die pressure, wire drawing is sometimes carried out under a longitudinal tensile force, known as back pull, applied on the entry side of the die. The lowering of die pressure is of course achieved at the expense of an increased drawing force caused by a back pull F, which decreases the axial component of the die load by an amount ηF, where η is a dimensionless quantity called the back-pull factor. If T_0 is the drawing force without the back pull, and T denotes that with a back pull, then the condition of overall longitudinal equilibrium may be written as

$$T = T_0 + (1 - \eta) F.$$

A good approximation to the back-pull factor for a nonhardening material can be obtained by solving (3.103) under the modified boundary conditions $\sigma = F/\pi b^2$ when $r = b$, and $\sigma = T/\pi a^2$ when $r = a$. The result is easily shown to be

$$T = T_0 + F \left(\frac{a^2}{b^2} \right)^{1+\mu \cot \alpha}$$

in view of (3.105) which gives the quantity $T_0 / \pi a^2$. A comparison between the preceding two equations indicates that the back-pull factor is

$$\eta = 1 - (1 - R)^{1+\mu \cot \alpha}. \tag{3.110}$$

As the coefficient of friction increases from zero, the back-pull factor increases from a value equal to the fractional reduction R. For sufficiently small values of μ, cot α, the back-pull factor for a given reduction in area varies almost linearly with the parameter μ cot α.

The simplest way of allowing for strain hardening in the theoretical framework is to replace the constant yield stress Y by a mean yield stress $\bar{Y}$. It is assumed that the overall effective strain imposed in a given drawing operation is independent of the strain-hardening characteristic of the material. Since the mean drawing stress is equal to the work done per unit volume of the material in the absence of friction and back pull, the overall effective strain ε in wire drawing is given by the second factor on the right-hand side of (3.106). The mean yield stress is then obtained from the fact that $\bar{Y}\varepsilon$ is equal to the area under the uniaxial stress–strain curve up to a strain of ε. If the stress–strain curve is represented by the power law $\sigma = C\varepsilon^n$, where C and n are empirical constants, we have

$$\bar{Y}\varepsilon = C \int_0^{\varepsilon} \varepsilon^n d\varepsilon = \left(\frac{C}{1+n} \right) \varepsilon^{1+n}.$$

Substituting for ε, the mean yield stress of the wire being drawn through the die is obtained as

$$\bar{Y} = \left(\frac{C}{1+n} \right) \left\{ 0.5 \sin \alpha + \left(0.91 + 0.12 \sin^2 \alpha \right) \ln \left(\frac{1}{1-R} \right) \right\}^n.$$

The influence of strain hardening on the redundant work factor may be estimated by considering frictionless wire drawing without back pull. If the redundant work factor for a nonhardening material is denoted by ϕ, the second factor on the right-hand side of (3.106) must be equal to $-\phi \ln(1 - R)$. The mean drawing stress for a work-hardening material may therefore be written as

$$t = \frac{C}{1+n}\left\{\phi \ln\left(\frac{1}{1-R}\right)\right\}^{1+n}\left\{1+\left(1-0.5\ln\frac{1}{1-R}\right)\mu\cot\alpha\right\}, \qquad (3.111)$$

which follows from (3.106) with Y replaced by $\bar{Y}$ Since the ideal drawing stress corresponds to $\phi = 1$, it follows from the above relation that the redundant work factor for a work-hardening material is equal to ϕ^{1+n}. The effect of strain hardening is therefore to augment the redundant work by an amount that increases with increasing values of n (Atkins and Caddell, 1968). For given values of μ and R, the optimum die angle for a work-hardening material is somewhat smaller than that for a nonhardening material. A finite element analysis of the wire-drawing process has been given by Chen et al. (1979).

3.7.3 Ideal Wire-Drawing Dies

The reduction of area in wire drawing can be accomplished with perfect efficiency if the profile of the die is such that the direction of the algebraically greater principal stress coincides with the streamline at each point of the deforming region. Since the die profile itself is a streamline, the surface of an ideal wire-drawing die must be perfectly smooth. The velocity of the plastic material is everywhere continuous, and the material leaving the die is uniformly strained, although the deformation within the die is not homogeneous. With efficient lubrication, a streamlined die can be of considerable use in actual wire-drawing operations where friction cannot be eliminated. As in the case of plane strain, the ideal die profile is by no means unique, and the shortest profile should be the best choice for die design. The problem has been investigated by Richmond (1965) and Richmond and Morrison (1967), while a three-dimensional treatment has been given by Hill (1967).

It is assumed that the material obeys Tresca's yield criterion and the associated flow rule, with the circumferential stress σ_θ equal to the algebraically lesser principal stress in the meridian plane. Thus $\sigma_\theta = -(p + k)$, where p denotes the mean compressive stress in the meridian plane, the remaining stress components in cylindrical coordinates being the same as those in (3.26). The characteristic equations for the stress are given by (3.28) with a change in sign of the leading terms in parentheses. The characteristic equations for the velocity are given by (3.30) with $U = W$ throughout the deforming region. Since $dz/dr = \tan\phi$ along an α-line and $dz/dr = -\cot\phi$ along a β-line, where ϕ is the angle of inclination of the α-line with the r-axis measured toward the z-axis, we have

$$\left.\begin{aligned} dp \pm 2k d\phi - \frac{k}{r}(dr \mp dz) - 0, \\ dU \mp U d\phi - \frac{U}{r}(dr \mp dz) = 0, \end{aligned}\right\} \qquad (3.112)$$

along the sliplines. The upper sign applies when the slipline is an α-line, and the lower sign when it is a β-line. It follows from the above relations that

$$dp + 2k\frac{dU}{U} = dp + 2k\frac{dW}{W} = 0$$

along each slipline. The integrated form of these relations may be written as

$$U = W = c\exp\left(-\frac{p}{2k}\right) \tag{3.113}$$

throughout the deforming region, where C is a constant to be determined from the boundary condition. Since the velocity is everywhere continuous, (3.113) can be satisfied only if the entry and exit sliplines are straight, making an angle of $\pi/4$ with the axis of symmetry ($\phi = \pi/4$). Since $p = k$ along the entry slipline, and the speed of the material on the exit side is b^2/a^2 times that on the entry side, (3.113) indicates that $p = k - 4k\ln(b/a)$ along the exit slipline, giving the drawing stress

$$t = 4k\ln\left(\frac{b}{a}\right) = Y\ln\left(\frac{1}{1-R}\right),$$

which is identical to that for homogeneous extension. Since the die profile coincides with a trajectory of the algebraically greater principal stress, the tangent to the profile is inclined at an angle $\pi/4 + \phi$ to the r-axis, giving $dz/dr = \tan(\pi/4 + \phi)$ along the die surface. The greatest permissible reduction in area in wire drawing is 63.2%, which corresponds to $t = Y$ specifying the condition of necking of the drawn wire.

A streamlined flow field for a sigmoidal die can be constructed by assuming a suitable variation of U along the center line AB, Fig. 3.25a. In particular, the assumption $U = c^2/z^2$ along AB corresponds to a simple radial flow field in the region ABC, where c denotes the distance OB. Thus $U = c^2/(r^2 + z^2)$ throughout ABC, the resultant velocity at each point being directed toward the center O. The sliplines in this region, meeting the radial flow lines at a constant angle of $\pi/4$, are therefore logarithmic spirals with pole at O. The condition $U = b^2/a^2$ at A is satisfied by taking $OA = c(a/b)$. If ψ denotes the angle turned through by each of the sliplines AC and BC, then $OB/OA = e^{2\psi}$, giving

$$2\psi = \ln\left(\frac{b}{a}\right) = \ln\left(\frac{1}{1-R}\right).$$

On the entry side of BC, the α-lines are taken as straight and the β-lines as their orthogonal trajectories. On the exit side of AC, the field is completed by taking the β-lines as straight and the α-lines as their orthogonal trajectories. A suitable profile of the sigmoidal die is defined by the major principal stress trajectory passing through C. The slipline field and the die profile in the meridian plane are identical to those for the plane strain drawing of a sheet whose exit and entry thicknesses are $2a$ and $2b$, respectively (Chakrabarty, 1987). The relation $c\psi = b$ therefore holds, giving the unknown quantity c. The stress and velocity distributions for the axisymmetric drawing can be calculated from (3.112), using the appropriate boundary conditions along BD and AE.

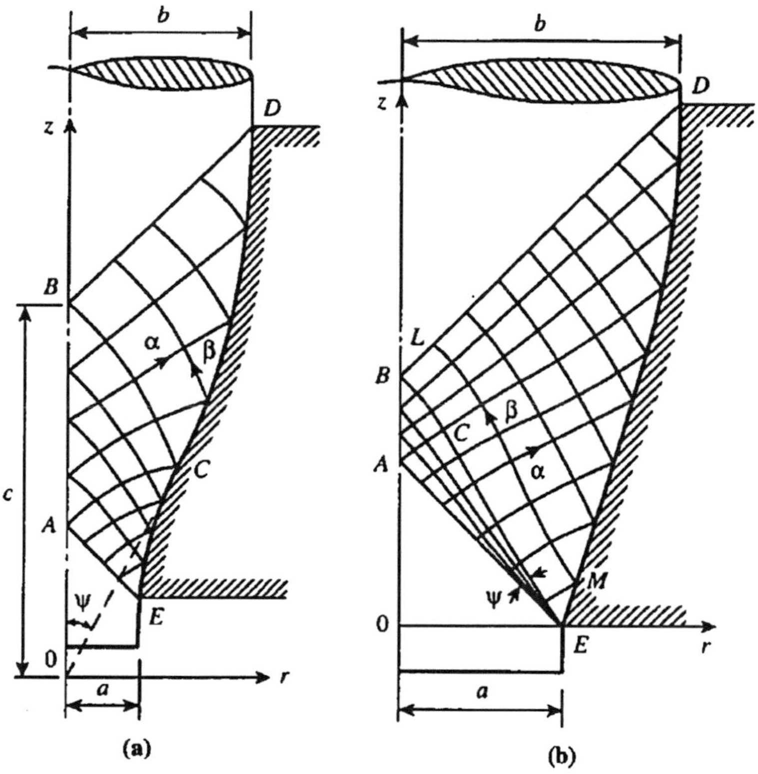

Fig. 3.25 Streamlined dies for the wire drawing process. (**a**) Sigmoidal die and (**b**) minimum length die

The particular distribution of U along AB that gives the shortest possible streamlined die is the one involving a singularity at the exit point of the die, Fig. 3.25b. It is convenient to start with an assumed fan angle ψ for the singular field EAC, which is taken as an equiangular net. Using the boundary conditions $\phi = \pi/4$ and $p = k[l - 4\ ln(b/a)]$ along AE, the values of p and ϕ at the nodal points of the fan are computed from the finite difference form of the first equation of (3.112). The boundary condition $\phi = \pi/4$ along AB and the computed data on AC enable us to carry out similar calculations for the region ABC. If the computed value of p at B differs from the required value k, the procedure is repeated with a new value of ψ until the agreement is obtained. Numerical calculations suggest that $\psi \approx 0.41\ R^2$ is a close approximation over the relevant range.

The boundary conditions $p = k$ and $\phi - \pi/4$ along BD, and the available values of p and ϕ along the known characteristic ECB, are sufficient to determine the field DBE uniquely. The construction begins by guessing the position of L where the first β-line LM meets the entry slipline BD, and finding the location of successive nodal points on LM until the boundary point M is reached. For this point to be acceptable, the boundary condition $dz/dr = \tan(\pi/4+\phi)$ must be satisfied. When

L is located by trial and error, the process is repeated for each successive β-line until the entire field is covered, to obtain the complete die profile DE. The velocity distribution is subsequently obtained from (3.113) using the computed nodal values of p. Numerical results for the die profile, corresponding to several values of ψ, are given in Table 3.2. A useful discussion on the design of optimum die profiles in axisymmetric extrusion are found in Sortais and Kobayashi (1968) and Samanta (1971).

Table 3.2. Coordinates of minimum length die profile

$\psi = 1°(R = 0.197)$		$\psi = 2°(R = 0.293)$		$\psi = 4°(R = 0.421)$		$\psi = 6°(R = 0.511)$	
r/a	z/a	r/a	z/a	r/a	z/a	r/a	z/a
1.000	0	1.000	0	1.000	0	1.000	0
1.019	0.400	1.032	0.400	1.053	0.398	1.072	0.395
1.044	0.797	1.073	0.793	1.117	0.786	1.184	0.903
1.081	1.324	1.129	1.317	1.226	1.433	1.324	1.551
1.104	1.731	1.167	1.725	1.284	1.869	1.404	2.082
1.115	2.056	1.188	2.119	1.312	2.280	1.428	2.460
1.116	2.174	1.190	2.302	1.315	2.531	1.432	2.748

3.8 Some Three-Dimensional Problems

3.8.1 Indentation by a Rectangular Punch

Consider the indentation of the plane surface of a semi-infinite block of metal by a rigid flat punch of rectangular cross section. An exact solution does not seem to exist for this three-dimensional problem, but reasonable bounds on the yield point load can be found by using the limit theorems based on the Tresca criterion. To obtain a lower bound solution, the stress discontinuity pattern of Fig. 3.26a, which is a generalization of the associated plane strain pattern, may be used to construct a statically admissible stress field in the indented block. The rectangle $ABCD$ defines the area of indentation, and the line EF is parallel to the longer sides of the rectangle. The volume $ABCDEF$ has two triangular faces ABE and CDF and two trapezoidal faces $ADFE$ and $BCFE$, all inclined at an angle of 60° to the plane of the rectangle. A state of uniaxial compression of amount $2k$ is assumed in the material occupied by four prisms, extending from the inclined faces of the region $ABCDEF$ and having their axes inclined at an angle of 60° to the plane of the rectangle. The prisms extend infinitely into the material of the block, only two of them being shown in the diagram for clarity. The uniaxial stress state in the prisms is compatible with a triaxial stress state in the region $ABCDEF$ where the principal compressive stresses are $(k, k, 3k)$, the remainder of the material being stress free. Indeed, by considering vertical sections through the mid-points of opposite sides of the rectangle, the

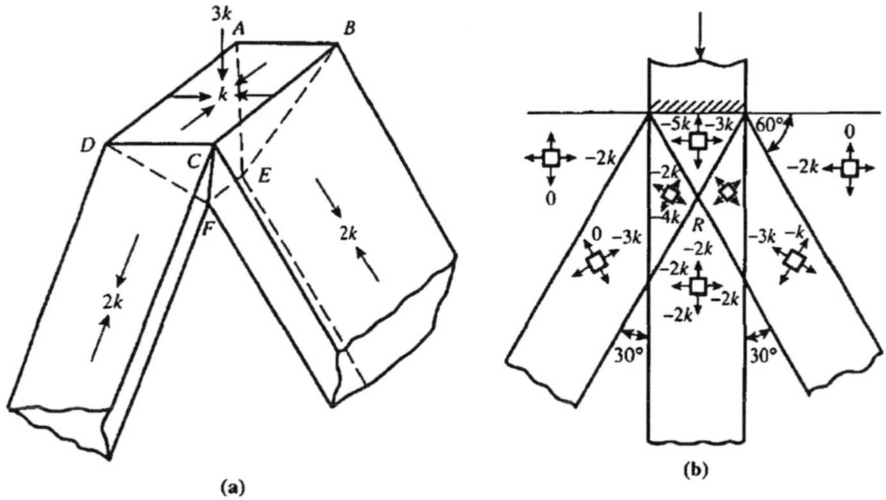

Fig. 3.26 Statically admissible stress field for the indentation of a plane surface by a rectangular flat punch

normal and shear stresses are easily shown to be continuous across the faces of each prism.

If we now superimpose a uniform hydrostatic compression of amount $2\ k$ in the material vertically below the punch and an all-round horizontal compression of amount $2\ k$ in the remainder of the block, Tresea's yield criterion is nowhere violated in the resulting stress field, while the pressure on the punch is increased from $3\ k$ to $5\ k$. The stress is discontinuous not only across the faces of the inclined prisms but also across the faces of a central vertical prism whose cross section is the given rectangle. The final discontinuous state of stress, as it appears on a vertical section through the mid-points of the longer sides of the rectangle, is shown in Fig. 3.26b. It may be noted that the directions of the principal stresses in the material lying between the vertical and inclined discontinuities are modified in such a way that the normal and shear stresses become continuous across these lines. Since the stress field is statically admissible, the pressure $5\ k$ on the rectangle is a lower bound on the punch pressure. Using a modification of the stress field of Fig. 3.26, it can be shown that the lower bound of $5\ k$ for the mean punch pressure applies to any convex area of indentation by a rigid punch (Shield and Drucker, 1953).

An upper bound on the yield point load will now be obtained on the basis of a kinematically admissible velocity field which is symmetrical about the vertical planes through the mid-points of the opposite sides of the rectangle. It is only necessary to consider the part of the velocity field which corresponds to the downward speed U of the quarter $OABC$ of the rectangle, whose longer and shorter sides are of lengths $2a$ and $2b$, respectively, Fig. 3.27a. The incipient deformation of the block is confined in three regions of plane plastic flow, involving the triangles PBS and PBG,

Fig. 3.27 Deformation mode for indentation by a rectangular punch involving surface of velocity discontinuity

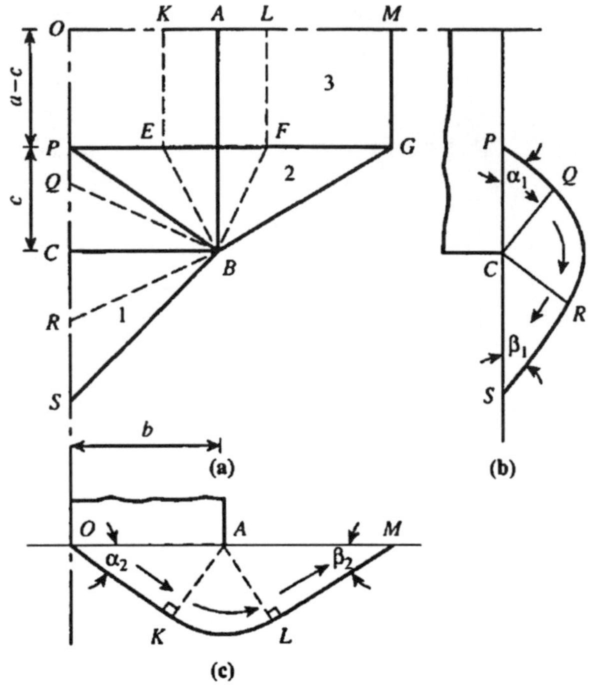

and the rectangle *OPGM*, appearing as projections in the plan view. The flow lines are assumed to be parallel to *OS* in region 1, and parallel to *OM* in regions 2 and 3. The vertical sections through *OS* and *OM*, shown in Fig. 3.27b,c, indicate that the flow is everywhere parallel to the respective rigid/plastic interface, the speed of flow in the two cases being of constant magnitudes U cosec α_1 and U cosec α_2, respectively. The surfaces of velocity discontinuity across the stationary rigid material are indicated by the broken lines in the figure. The flow patterns over typical vertical sections through regions 1 and 2, parallel to *OS* and *OM*, respectively, are similar to those shown in the figure.

Plastic energy is dissipated not only due to shearing across the surfaces of velocity discontinuity but also due to straining in the cylindrical zone *EFLK* in region 3, and to the conical zones *BQR* and *BEF* in regions 1 and 2. Each element of material in these zones is in a state of instantaneous simple shear parallel to the interface, and the engineering shear rate is of magnitude (U/r) cosec α, where r is the radius to a typical element from the center of curvature of the circular interface. The rate of dissipation of total internal energy in region 3 is easily found to be

$$E_3 = k\,(a-c)\,bU\,[2\,(\alpha_2+\beta_2)+\cot\alpha_2+\cot\beta_2]\,.$$

To obtain the energy rates in regions 1 and 2, consider the volume of the deforming material lying between the vertical planes at distances x and $x + dx$ from the apex B, measured along the appropriate side of the rectangle. Plastic energy in this elemental volume of material is dissipated at the rate

$$dE = kU\,[(\alpha+\beta)+\sec\phi\,(\alpha+\beta+\cot\alpha+\cot\beta)]\,\xi dx,$$

where ϕ is the semiangle of the cone, and ξ sin α is the local radius of the circular boundary. By simple geometry, $\phi = \tan^{-1}$ $(c \sin\alpha_1 /b)$, $\xi = xc/b$ in region 1, and $\phi = \tan^{-1}(b \sin\alpha_2/c)$, $\xi = xb/c$ in region 2. The integration of the above equation between the appropriate limits gives the energy rates in regions 1 and 2 as

$$E_1 = \frac{1}{2}kUcb\left\{(\alpha_1+\beta_1)+\left(1+\frac{c^2}{b^2}\sin^2\alpha_1\right)^{1/2}(\alpha_1+\beta_1+\cot\alpha_1+\cot\beta_1)\right\},$$

$$E_2 = \frac{1}{2}kUcb\left\{(\alpha_2+\beta_2)+\left(1+\frac{b^2}{c^2}\sin^2\alpha_2\right)^{1/2}(\alpha_2+\beta_2+\cot\alpha_2+\cot\beta_2)\right\}. \qquad (3.114)$$

Equating the rate of external work $qabU$, where q is the mean punch pressure, to the total rate of dissipation internal energy $E_I + E_2 + E_3$, we obtain

$$q = k\left\{\left(1-\frac{c}{a}\right)[2\,(\alpha_2+\beta_2)+\cot\alpha_2+\cot\beta_2]+\frac{c}{2a}\,(e_1+e_2)\right\}, \qquad (3.115)$$

as the required upper bound on the mean punch pressure, with e_I and e_2 denoting the expressions in curly brackets in the expressions for E_I and E_2, respectively. Equation (3.115) is essentially due to Shield and Drucker (1953). *For* $b/a \geq 0.53$, the best upper bound corresponds to $c = b\sqrt{2}$, $\alpha_I = 46.3°$, $\beta_I = 46.3°$, $\alpha_2 = 34°$, $\beta^2 = 47.1°$, while $b/a \leq 0.53$, requires $c = b\sqrt{2}$, $\alpha_I = 39°$, $\beta_1 = 46.3°$, $\alpha_2 = \beta_2 = 45°$. The substitution of these values into (3.115) furnishes the upper bound formulas

$$\left.\begin{aligned} q &= k\left(5.14+0.66\frac{b}{a}\right), \quad 0<\frac{b}{a}\leq 0.53,\\ q &= k\left(5.24+0.47\frac{b}{a}\right), \quad 0.53\leq\frac{b}{a}\leq 1. \end{aligned}\right\} \qquad (3.116)$$

The first expression of (3.116) reduces *to* $q = 5.14\,k$ for rectangles whose lengths are large compared to their breadths, in agreement with the well-known Prandtl value for the plane strain indentation pressure. In the case of a square punch ($b = a$), the second equation of (3.116) reduces to $q = 5.7lk$, which is about 14% higher than the lower bound $q = 5k$. The three-dimensional indentation problem for a layer of material resting on a rigid foundation has been treated by Shield (1955c). The indentation of a semi-infinite block by a pyramid indenter has been discussed by Haddow and Johnson (1962).

3.8.2 Flat Tool Forging of a Bar

A long rectangular bar of width $2w$ and thickness $2\ h$ is compressed between a pair of lubricated flat tools which overlap the bar width, Fig. 3.28. The width of the dies is denoted by $2b$, and the bite ratio b/w is assumed to be small or moderate (less than 1) so that the frictional contribution to the lateral spread is negligible. The variation of stress and strain through the thickness of the bar will also be neglected in the analysis (Hill, 1963). Taking the origin of coordinates at the geometric center of the deforming zone, with the x- and y-axes considered along the length and width of the bar, respectively, the simplest velocity field for a unit speed of compression may be written as

$$\mathrm{v}_x = x - \int_0^x \phi(x)\,dx, \mathrm{v}_y = y\phi(x), \mathrm{v}_z = -z, \tag{3.117}$$

where $\phi(x)$ is a differentiable continuous even function of x satisfying the condition $\phi(b) = 0$. The x- and y-components of velocity have been made continuous across the rigid/plastic interfaces $x = \pm b$. The plastic incompressibility condition is identically satisfied, and the predicted strain rates are independent of z. A comparable family of virtual orthogonalizing fields may be taken as

$$u_x = -\int_o^x \psi(x)\,dx, \quad u_y = y\psi(x), \quad u_z = 0,$$

where $\psi(x)$ is a continuous even function of x vanishing at $x = \pm b$. Using the variational method characterized by (3.60), with the virtual velocity W_j replaced by u_j, we get

$$\int \left[(\sigma_y - \sigma_x)\,\psi(x) + y\tau_{xy}\psi'(x)\right] dV = 2\int \sigma_x dS.$$

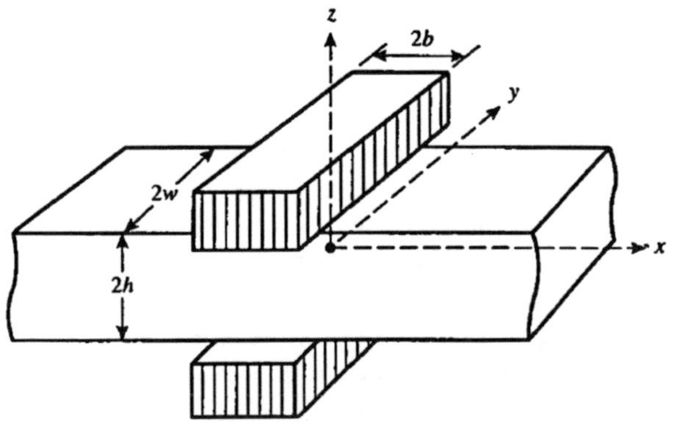

Fig. 3.28 Compression of a rectangular block of metal involving lateral spread of material

The surface integral on the right-hand side, taken over the interfaces $x = \pm b$, is identically zero since the ends of the bar are free to move longitudinally. Integrating by parts, and using the condition $\psi(b) = 0$, the volume integral can be reduced to

$$\iint \left(\sigma_x - \sigma_y + y \frac{\partial \tau_{xy}}{\partial x} \right) dydz = 0 \tag{3.118}$$

for all x, since $\psi(x)$ is an arbitrary function of x. This is the equation of equilibrium exactly suited to the kinematically admissible field (3.117). The usefulness of (3.60) lies in the fact that it supplies the appropriate result automatically.

It would be sufficiently accurate to write the yield criterion as $s_z = -k$, where k is a constant equal to the yield stress in pure shear when the bite ratio is small ($b/t <$ 0.5). For moderate-to-large bite ratios, it is more appropriate to take $k = 2Y/3$, where Y is the uniaxial yield stress. Adopting the Lévy–Mises flow rule together with the admissible velocity field (3.117), we have

$$\sigma_x - \sigma_y = s_z \left(\dot{\varepsilon}_x - \dot{\varepsilon}_y \right) / \dot{\varepsilon}_z = k \left[1 - 2\phi(x) \right],$$

$$\tau_{xy} = s_z \left(\dot{\gamma}_{xy} / \dot{\varepsilon}_z \right) = \frac{1}{2} \dot{k} y \phi'(x), \quad \tau_{xz} = 0.$$

It may be noted that only the mean value of τ_{xy} vanishes along the free sides $y = \pm w$. The local error should have no significant effects, however, on the estimated load and the lateral spread. The substitution from above into (3.118) and integration between the appropriate limits lead to the differential equation

$$\frac{d^2\phi}{dx^2} + \frac{6}{w^2}(1 - 2\phi) = 0.$$

Using the boundary conditions $\phi'(0) = 0$ and $\phi(b) = 0$, the solution is found to be

$$\phi(x) = \frac{1}{2} \left\{ 1 - \frac{\cosh\left(2\sqrt{3}x/w\right)}{\cosh\left(2\sqrt{b}/w\right)} \right\}, \tag{3.119}$$

which defines the shape of the lateral spread during a small reduction in thickness. The spread coefficient c is defined as the ratio of the average lateral strain to the average thickness strain in the compressed material. Thus

$$c = \frac{1}{b} \int_0^b \phi(x)\, dx = \frac{1}{2} \left\{ 1 - \frac{w}{2\sqrt{3}b} \tanh \frac{2\sqrt{3}b}{w} \right\}. \tag{3.120}$$

It follows from (3.120) that $c \to 0$ when $b/w \to 0$ and $c \to 0.5$ when $b/w \to \infty$, indicating that the expected limiting behaviors are correctly predicted by the above spread formula. The forging load P is most conveniently obtained from the relation

$$2hp = \int (\sigma_x - \sigma_z)\, dV,$$

where the integral is taken over the entire volume of the plastic material. The above relation follows from the fact that

$$\iint \sigma_x dydz = 0, \quad P = -\iint \sigma_x dxdy,$$

by the conditions of overall longitudinal and vertical equilibrium. Since

$$\sigma_x - \sigma_z = s_z(\dot{\varepsilon}_x - \dot{\varepsilon}_z) / \dot{\varepsilon}_z = k\,(2 - \phi)$$

by the Lévy–Mises flow rule, the substitution into the preceding volume integral furnishes the forging load as

$$P = kA\left\{2 - \frac{1}{b}\int_0^b \phi\,(x)\,dx\right\} = kA\,(2 - c)\,, \tag{3.121}$$

where A is the contact area between the bar and the tool, and c is given by (3.120). This formula reduces to $P = 2kA$ in plane strain ($c = 0$) and to $P = YA$ in simple compression ($c = 0.5$), with k and Y denoting the yield stresses in pure shear and uniaxial compression, respectively. For large bite ratios ($b/t > 1.5$), friction acts to increase the spread significantly due to a progressive build-up of longitudinal compressive strain in the material between the dies (Hill, 1950c). The problem of lateral spread in flat rolling has been approximately treated by Kummerling and Lippmann (1975) and by Oh and Kobayashi (1975).

3.8.3 Bar Drawing Through Curved Dies

A bar of rectangular cross section is drawn through a pair of smooth curved dies under a front tension T_f and a back-pull T_0, so that its thickness is reduced from $2h_0$ to $2\,h_f$, and its width is increased from $2w_0$ to $2w_f$. The projected length l of the arcs of contact is assumed to be significantly greater than the mean thickness of the bar, so that the variation of stress and strain through the thickness can be disregarded. To analyze this problem, we choose a set of rectangular axes with origin at the centroid of the entry section, as shown in Fig. 3.29. It is assumed at the outset that plastic deformation occurs in the entire volume of material between the planes of entry and exit and that plane transverse sections remain plane during the drawing.

Let a generic section of the material between the two dies be of thickness $2\,h$ and width $2w$. The simplest type of approximating velocity field, satisfying the incompressibility equation and the boundary conditions, has components (Hill, 1963)

$$\mathrm{v}_x = \frac{h_0 w_0}{hw} \quad \mathrm{v}_y = -\frac{y h_0}{h}\frac{d}{dx}\left(\frac{w_0}{w}\right), \quad \mathrm{v}_z = -\frac{z w_o}{w}\frac{d}{dx}\left(\frac{h_o}{h}\right). \tag{3.122}$$

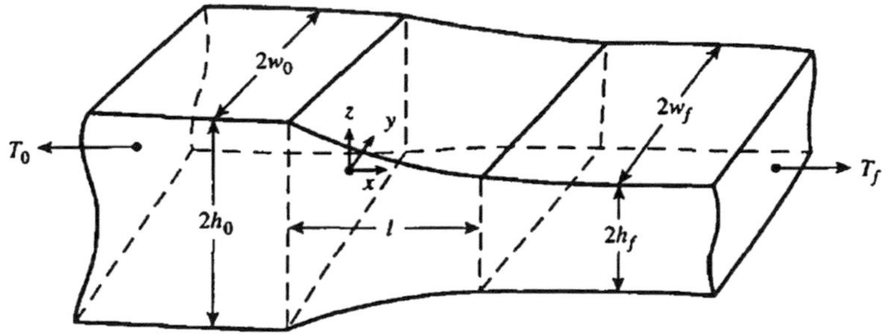

Fig. 3.29 Drawing of bar of rectangular cross section through a smooth curved die

It follows from (3.122) that $v_z/v_x = \pm dh/dx$ along the die faces $z = \pm h$ and $v_y/v_x = \pm dw/dx$ along the sides $y = \pm w$, as required. The material enters the zone of deformation with a unit speed and leaves it longitudinally with a speed equal to h_0w_0/h_fw_f. The y-component of velocity is made continuous across the planes of entry and exit by requiring dw/dx to vanish at $x = 0$ and $x = l$. The z-component of velocity is, however, discontinuous unless the dies have zero entry and exit angles. The nonzero components of strain rate in the zone of deformation are found to be

$$\dot{\varepsilon}_x = -\frac{h_0w_0}{hw}\left(\frac{h'}{h} + \frac{w'}{w}\right), \quad \dot{\gamma}_{xy} = -\frac{yh_0w_0}{2h}\frac{d}{dx}\left(\frac{w'}{hw^2}\right),$$

$$\ddot{\varepsilon}_y = \frac{h_0w_0w'}{hw^2}, \quad \dot{\varepsilon}_z = \frac{h_0w_0h'}{wh^2}, \quad \dot{\gamma}_{xz} = -\frac{zh_0w_0}{2w}\frac{d}{dx}\left(\frac{h'}{wh^2}\right),$$

where the prime denotes derivative with respect to x. Adopting the same yield criterion as that in the preceding problem, and substituting for the strain rates into the Lévy–Mises flow rule, the associated normal stress differences and shear stresses of the approximating field are obtained as

$$\left.\begin{aligned} \frac{\sigma_x - \sigma_y}{k} &= 1 + \frac{2h}{w}\frac{dw}{dh}, \quad \frac{\sigma_x - \sigma_y}{k} = 2 + \frac{h}{w}\frac{dw}{dh}, \\ \frac{2\tau_{xy}}{k} &= -ywh^2\frac{d}{dh}\left(\frac{w'}{hw^2}\right), \quad \frac{2\tau_{xy}}{k} = -zwh^2\frac{d}{dh}\left(\frac{w'}{hw^2}\right), \end{aligned}\right\} \tag{3.123}$$

where k is a constant that lies between $Y/\sqrt{3}$ and $2Y/3$, the material being ideally plastic. For a work-hardening material, k must be an assigned function of x, determined from the uniaxial stress–strain curve of the material.

For the application of the variational principle, a virtual orthogonalizing field similar to the approximating field (3.122) may be written in terms of an arbitrary function $\phi(x)$ in the form

$$u_x = \frac{h_0}{h}\phi(x), \quad u_y = -\frac{yh_0}{h}\phi'(x), \quad u_z = \frac{zh_0}{h^2}h'\phi(x), \tag{3.124}$$

between the planes $x = 0$ and $x = l$. The incompressibility equation and the boundary condition on $z = \pm h$ are identically satisfied. The only restriction to be imposed on the function $\phi(x)$ is that its derivative $\phi'(x)$ must vanish at $x = 0$ and $x = l$ so that u_y becomes continuous at these sections. Substituting in (3.60), where w_j is replaced by u_j given by (3.124), and ignoring the effect of the velocity discontinuity, we get

$$4\int_0^1 \left\{ w\left[(\sigma_x - \sigma_y)\phi' - (\sigma_x - \sigma_z)\frac{h''\phi}{h}\right] - \left(h\int_0^w y\tau_{xy}dy\right) \right.$$
$$\left. \frac{d}{dx}\left(\frac{\phi'}{h}\right) + \left(w\int_0^h z\tau_{xy}dz\right)\frac{d}{dx}\left(\frac{h'\phi}{h^2}\right)\right\} dx = \frac{T_f}{h_f}\phi(i) - \frac{T_0}{h_0}\phi(0),$$

since $\sigma_x - \sigma_y$ and $\sigma_x - \sigma_z$ are functions of x only, while τ_{xy} and τ_{xz} are independent of z and y, respectively. Integrating by parts the terms involving derivations of ϕ, and using the stress boundary conditions $\sigma_x = T_0/4w_0h_0$ at $x = 0$, $\sigma_x = Tf/4w_fh_f$ at $x = l$, and $\sigma_y = \tau_{xy} = \tau_{xz} = 0$ at $x = 0$ and $x = l$, it is easily shown that

$$h\frac{dF}{dh} + G = 0, \quad 0 \le x \le 1, \tag{3.125}$$

where F and G are functions of x only and are defined by the equations

$$\left.\begin{aligned} kF(x) &= \frac{1}{h}\frac{d}{dx}\left(h\int_0^w y\tau_{xy}dy\right) + w(\sigma_x - \sigma_y), \\ kG(x) &= \frac{1}{h}\frac{d}{dx}\left(w\int_0^h z\tau_{xy}dz\right) + w(\sigma_x - \sigma_z). \end{aligned}\right\} \tag{3.126}$$

It follows from (3.126) that $F(0) = T_0/4kh_0$ and $F(l) = T_f/4kh_f$. The integrated form of (3.125) therefore becomes

$$\frac{T_f}{kh_f} = \frac{T_0}{kh_0} + 4\int_{h_f}^{h_0} G\frac{dh}{h} \tag{3.127}$$

giving the relationship between T_0 and T_f when $G(x)$ is known. The substitution of (3.123) into (3.126) results in

$$\left.\begin{aligned} -\frac{1}{3h}\frac{d}{dx}\left\{h^3w^4\frac{d}{dh}\left(\frac{w'}{hw^2}\right)\right\} + 2h\frac{dw}{dh} + w = F, \\ -\frac{1}{3h}\frac{d}{dx}\left\{h^5w^2\frac{d}{dh}\left(\frac{h'}{wh^2}\right)\right\} + h\frac{dw}{dh} + 2w = G. \end{aligned}\right\} \tag{3.128}$$

Equations (3.125) and (3.128) taken together represent a fourth-order nonlinear differential equation for the width function $w(x)$, the four boundary conditions needed for its solution being

$$w(0) = w_0, \quad w'(0) = w'(i) = 0, \quad F(0) = T_0/4kh_0.$$

It is convenient to solve the problem by an iterative method, starting with an approximate expression for w satisfying the three boundary conditions and involving the exit semiwidth w_f. Since h is a given function of x, the function $G(x)$ can be determined from the second equation of (3.128). A reasonable guess for T_f/kh_f then enables us to calculate w_f from (3.127). A first approximation for $F(x)$ follows from the integration of (3.125) with the boundary condition $F(0) = T_0/4kh_0$, and the first equation of (3.128) is subsequently solved for w to obtain the second approximation. The process can be repeated until the desired accuracy is achieved.

The numerical solution may also be considered as that of an initial value problem requiring a suitable guessed value of $w''(0)$. For a given back tension T_0, which specifies the value of $F(0)$, and hence provides a relationship between $w''(0)$ and $w''(0)$ in view of (3.128), the differential equation for w can integrated by the Runge–Kutta method. By adjusting the value of $w''(0)$ with the help of successive iterations, the remaining boundary condition $w'(l) = 0$ can be satisfied to a close approximation. The results of the computation carried out in this way by Lahoti and Kobayashi (1974) for Steckel rolling with $T_0 = 0$ are displayed in Fig. 3.30, where the lateral spread is plotted against the reduction in thickness for several values of w_0/h_0 and R/h_0, where R is the roll radius. The theoretical curves follow the same trends as those experimentally observed.

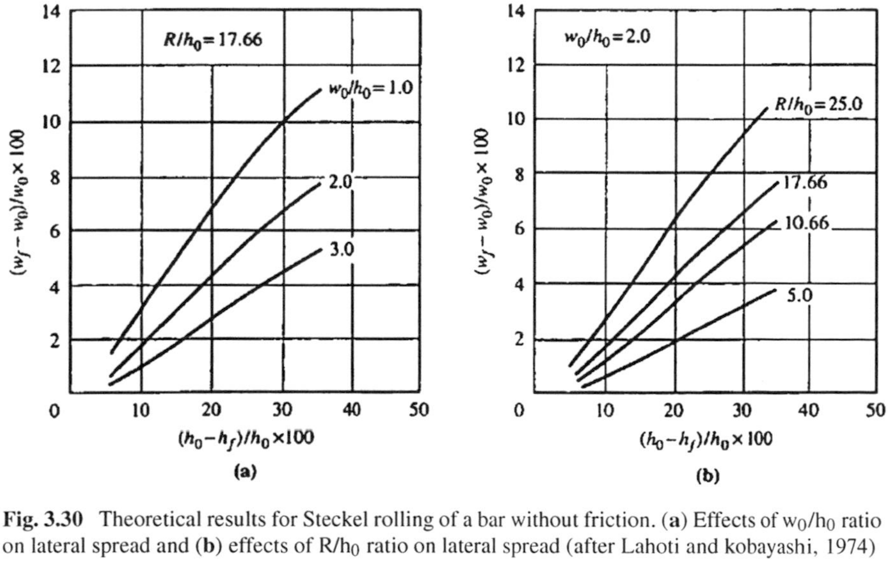

Fig. 3.30 Theoretical results for Steckel rolling of a bar without friction. (**a**) Effects of w_0/h_0 ratio on lateral spread and (**b**) effects of R/h_0 ratio on lateral spread (after Lahoti and kobayashi, 1974)

3.8.4 Compression of Noncircular Blocks

Consider the axial compression of a prismatic block of noncircular cross section by a pair of rough platens, involving a frictional stress of constant magnitude over the interfaces. The cross section is supposed to have two or more axes of symmetry, the shape of the boundary being otherwise arbitrary. An upper bound analysis will be presented here on the basis of a kinematically admissible velocity field that takes into account the incipient bulging of the material through the block thickness, which is denoted by $2h$. Choosing cylindrical coordinates (r, θ, z), with the origin at the geometrical center of the block, the simplest velocity field of this type may be written as

$$\left.\begin{aligned} u &= \frac{Ur}{2h}(1+\alpha n\cos 2n\theta)\,\phi'\left(\frac{z}{h}\right), \\ \text{v} &= \frac{Ur}{2h}(\alpha n\sin 2n\theta)\,\phi'\left(\frac{z}{h}\right), \\ w &= -U\phi\left(\frac{z}{h}\right), \end{aligned}\right\} \tag{3.129}$$

where U is the speed of each compression platen, α is a constant, $n \geq 1$ is an integer, and ϕ is a function z/h satisfying the conditions $\phi(0) = 0$ and $\phi(1) = -\phi(-1) = 1$. It may be noted that the circumferential velocity υ vanishes not only along $\theta = 0$ and $\theta = \pm\pi/2$ but also along $\theta = \pm\pi/2n$. The components of the strain rate associated with (3.129) are easily shown to be

$$\dot{\varepsilon}_r = \frac{U}{2h}(1+\alpha n\cos 2n\theta)\,\phi'\left(\frac{z}{h}\right), \quad \varepsilon_\theta = \frac{U}{2h}(1-\alpha n\cos 2n\theta)\,\phi'\left(\frac{z}{h}\right),$$

$$\dot{\varepsilon}_z = -\frac{U}{h}\phi'\left(\frac{z}{h}\right) = -(\dot{\varepsilon}_r + \dot{\varepsilon}_\theta), \quad \dot{\gamma}_{rz} = \frac{Ur}{4h^2}(1+\alpha n\cos 2n\theta)\,\phi''\left(\frac{z}{h}\right),$$

$$\dot{\gamma}_{r\theta} = -\frac{U}{2h}\left(\alpha n^2\sin 2n\theta\right)\phi'\left(\frac{z}{h}\right), \quad \dot{\gamma}_{\theta z} = -\frac{Ur}{4h^2}(\alpha\sin 2n\theta)\,\phi''\left(\frac{z}{h}\right).$$

The velocity field (3.129) is kinematically admissible since it satisfies the incompressibility condition and the velocity boundary conditions for the problem of block compression. It follows from above that the rates of extension are independent of r, according to the assumed deformation mode, and the rate of shear across the z-axis varies linearly with r. For a nonhardening material, the rate of dissipation of internal energy per unit volume is $Y\dot{\bar{\varepsilon}}$, where

$$\dot{\bar{\varepsilon}} = \frac{U}{h}\left\{f(\theta)\left[\phi'\left(\frac{z}{h}\right)\right]^2 + \frac{1}{3}\left(\frac{r}{2h}\right)^2 g(\theta)\left[\phi''\left(\frac{z}{h}\right)\right]^2\right\}^{1/2}, \tag{3.130}$$

denoting the effective strain rate, the functions f and g being defined as

$$\left.\begin{aligned} f(\theta) &= 1 + \frac{1}{3}\alpha^2 n^2 \left[1 + \left(n^2 - 1\right)\sin^2 2n\theta\right], \\ g(\theta) &= 1 + 2\alpha n \cos 2n\theta + \alpha^2 \left[1 + \left(n^2 - 1\right)\cos^2 2n\theta\right]. \end{aligned}\right\} \tag{3.131}$$

If the frictional stress along the die face is assumed to be of magnitude mY, where m is a constant, then the rate of energy dissipation due to friction per unit area of the interface is mu_0Y, where u_0 denotes the relative velocity of slip between the die and the block. It follows from (3.129) that

$$u_0 = \sqrt{u^2 + v^2} = \frac{Ur}{2h}\phi'(1)\sqrt{g(\theta)}. \tag{3.132}$$

An upper bound on the load P required for the incipient compression of the block is obtained by equating the rate of external work done to the rate of total internal energy dissipation. Thus

$$PU = Y\left\{\int \dot{\varepsilon}dV + m\int u_0 dA\right\}, \tag{3.133}$$

where the volume integral corresponds to the upper half of the block ($0 \le z \le h$), and the surface integral is taken over the upper face $z = h$. The integrals can be evaluated numerically using a suitable expression for ϕ (z/h) and an assumed value of α.

The velocity field (3.129) implies that plane sections normal to the z-axis remain plane during the compression. The variation of the radial velocity with z, represented by the function ϕ (z/h), indicates the expected barreling tendency. As a convenient expression for ϕ satisfying the required boundary conditions, we may take

$$\phi\left(\frac{z}{h}\right) = \frac{z}{h} + \frac{\beta}{\pi}\sin\frac{\pi z}{h}, \qquad \phi'\left(\frac{z}{h}\right) = 1 + \beta\cos\frac{\pi z}{h},$$

where β is another constant. It may be noted that the ratio of the radial velocities at $z = h$ and $z = 0$ have a constant value equal to $(1 - \beta)/(l + \beta)$ according to the above expression for ϕ (z/h). Alternatively, we may assume a parabolic mode of barreling expressed by

$$\phi\left(\frac{z}{h}\right) = \frac{z}{h}\left[1 + \beta\left(1 - \frac{z^2}{h^2}\right)\right], \qquad \phi'\left(\frac{z}{h}\right) = 1 + \beta\left(1 - \frac{3z^2}{h^2}\right).$$

The radial velocity at $z = h$, according to this choice, is $(1 - 2\beta)/(1 + \beta)$ times that at $z = 0$. Using the above representation of ϕ, and substituting (3.130) and (3.132) into (3.133), the upper bound load for compression can be expressed as

$$\frac{Ph}{Y} = \int\left\{\left[1 + \beta\left(1 - \frac{3z^2}{h^2}\right)\right]^2 f(\theta) + 3\beta^2\left(\frac{rz}{h^2}\right)^2 g(\theta)\right\}^{1/2} dV + m\left(\frac{1}{2} - \beta\right)\int r\sqrt{g(\theta)}dA \tag{3.134}$$

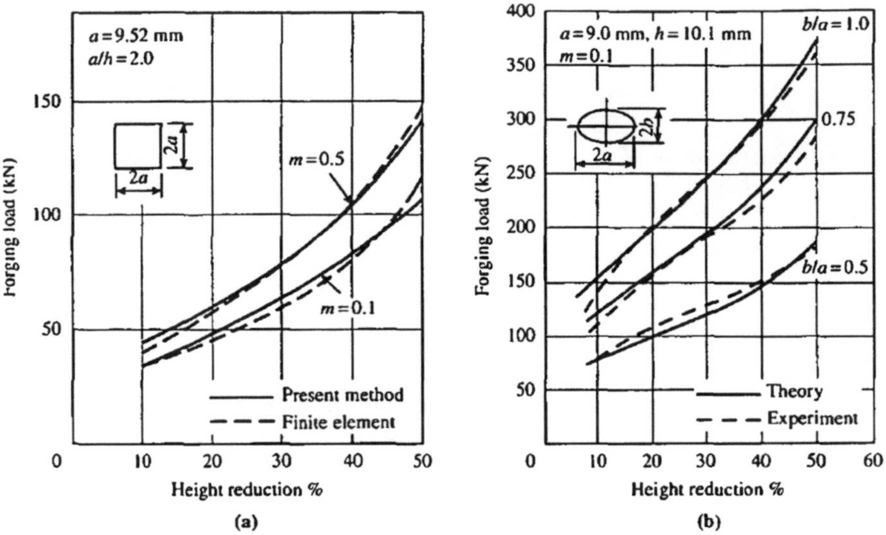

Fig. 3.31 Upper bounds of the forging load for prismatic blocks of annealed steel. (**a**) Square cross section and (**b**) elliptical cross section (after Yang and Kim, 1986)

where $f(\theta)$ and $g(\theta)$ are given by (3.131). The two unknown constants α and β must be determined by minimizing the upper bound with respect to these parameters. Numerical computations for a square cross section ($n = 2$) and an elliptic cross section ($n = 1$) have been carried out by Kim and Yang (1985a) and by Yang and Kim (1986a), respectively. In the latter case, it is convenient to replace α by $e\alpha$, where e denotes the eccentricity of the ellipse, in order to include the special case of a circular cross section ($e = 0$). These authors treated the problem of continued compression including work-hardening, their results being shown in Fig. 3.31.

Various approximate solutions for the compression of noncircular blocks, based on kinematically admissible velocity fields, have been presented by Nagpal (1977), Sagar and Juneja (1979), and Kim et al. (1987). An analysis for a three-dimensional extrusion process using conformal transformation has been presented by Yang and Lee (1978). Upper bound solutions for extrusion of polygonal sections have been discussed by Gunasekera and Hoshino (1982, 1985). The process of ring rolling, which is employed for reducing the thickness of a ring while increasing its diameter, has been investigated experimentally by Johnson and Needham (1968) and Mamalis et al. (1976) and theoretically by Hawkyard et al. (1973), Yang and Ryoo (1987), and Davey and Ward (2002). The rigid/plastic finite element method has been employed for solving the three-dimensional block compression problem by Park and Kobayashi (1984).

Problems

3.1 A rectangular block of thickness h, whose length $2b$ is large compared to its width, is compressed between a pair sof rigid platens. Considering the equilibrium of an element of the block under a vertical pressure p, longitudinal compressive stress q, and frictional stress μp, and using the von Mises yield criterion approximately in the form $2p - q = 2Y$, obtain an expression for the die pressure distribution. Show that the mean pressure on the block and its overall change in length are given by

$$\frac{\bar{p}_e}{Y} \approx 1 + \frac{\mu b}{2h}, \quad \frac{db}{b} = -\left(2 - \frac{3\bar{p}}{2Y}\right)\frac{dh}{h}$$

Verify that the longitudinal strain increment vanishes at the central section when $\mu b/h = 0.288$. Using the initial condition $b = b_0$ when $h = h_0$, integrate the above to obtain the expression

$$\frac{b}{b_0} = \left(\frac{1 - \mu b/2h}{1 - \mu b_0/2h_0}\right)\sqrt{\frac{h_0}{h}}, \quad \frac{\mu b}{h} \leq 0.288$$

3.2 The incipient necking of a cylindrical tensile bar made of a Tresca material may be represented by a deformation mode in which the plastic flow is confined in a region defined by $-\pi/4 \leq \psi \leq \pi/4$, where ψ is the angle made with the central section by a straight line drawn through the geometrical center of the bar. If the central section is assumed to move outward with a unit speed, while the end of the bar is held fixed, show that the components of velocity of a typical particle are given by

$$u = \frac{2}{\pi}\left(\frac{\sqrt{\cos 2\psi}}{\cos\psi}\right), \quad w = \frac{2}{\pi}\tan^{-1}\left(\frac{\sqrt{\cos 2\psi}}{\sin\psi}\right), \quad 0 \leq \psi \leq \frac{\pi}{4}$$

which are continuous across $\psi = \pi/4$. Verify that both u and w approach the quantity $(2/\pi)\sqrt{\pi - 4\psi}$ as ψ tends to $\pi/4$. Obtain a graphical plot for the variation of u and w with the angle ψ.

3.3 A solid cylindrical block of radius a and height $2h$ is axially compressed between rough parallel dies. An upper bound solution for the incipient compression may be obtained by using a deformation mode that involves a pair of central conical zones of rigid material bordering the dies, as shown in Fig. A. Assuming a unit speed of each compression die, show that the velocity field

$$u = \frac{a}{2h}, \quad w = -\frac{a}{2h}\left(\frac{z}{r}\right)$$

using the von Mises yield criteri, obtain the upper bound on the mean die pressure q in the form

$$\frac{q}{k} = \frac{1}{2}\sqrt{1 + \frac{h^2}{4a^2}} + \frac{a}{h}\ln\left(\frac{h}{2a} + \sqrt{1 + \frac{h^2}{4a^2}}\right) + \frac{1}{2}\left(\frac{a}{h} + \frac{h}{a}\right)$$

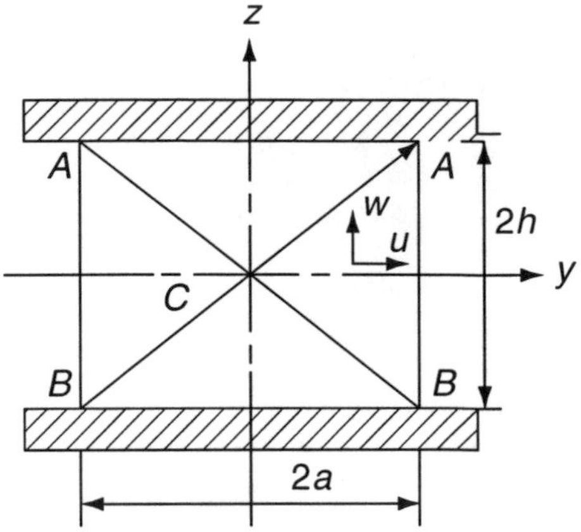

Fig. A

Show also that for h/a >.1.1 approximately, a better upper bound is given by the simple velocity field. $u = r/2a$, and $w = -z/h$. Obtain a graphical plot of q/k against h/a over the range $0.3 \leq h/a \leq 2.0$.

3.4 A circular disc of radius b and thickness zh is indented symmetrically by a pair of opposed flat dies of radius a, each die being assumed to move with a unit speed. When the ratio b/a is sufficiently small, the incipient deformation mode may be defined by the velocity field

$$u = \frac{r}{2h}, \quad w = -\frac{z}{h}, \quad (0 \leq r \leq a); \quad u = \frac{a^2}{2rh}, \quad w = 0, \quad (a \leq r \leq b)$$

If the dies are perfectly rough, and the material obeys the von Mises yield criterion, show that the associated upper bound on the mean die pressure q is given by

$$\frac{q}{k} = \sqrt{3} + \left(\frac{a}{3h} + \frac{h}{a}\right) + 2\ln\frac{b}{a}$$

For relatively large values of b/a, an improved upper bound involves a velocity discontinuity emanating from $r = a$ on $z = 0$ and inclined at 45° to the middle plane of the disc, the outer part of the disc being rigid. Using the velocity field $u = w = a/2rh$ within the outer plastic region prove that

$$\frac{q}{k} = 3 + \sqrt{3} + \left(\frac{a}{3h} + \frac{h}{a}\right) + \sqrt{5}\left\{\left(1 + \frac{a}{h}\right)\ln\left(1 + \frac{h}{a}\right) - 1\right\}$$

3.5 A hollow cylindrical block of height h, internal radius a, and external radius b, exactly fits into a rigid hollow container and is axially compressed by a pair of parallel dies. An upper bound solution for the compression of the block may be obtained by using a velocity field in which the radial and axial components are expressed as

$$u = -\frac{b^2 - r^2}{2hr}, \quad w = -\frac{z}{h}$$

The upper die is assumed to move downward with a unit speed, the lower die being stationary. If the container wall is smooth and the dies are perfectly rough, show that the mean die pressure q is given by

$$\left(1-\frac{a^2}{b^2}\right)\frac{q}{k}=2-\sqrt{1+\frac{3a^4}{b^4}}+\ln\left\{\frac{b^2}{3a^2}+\sqrt{\frac{1}{3}+\frac{b^4}{9a^4}}\right\}+\frac{2b}{3h}\left(1-\frac{a}{b}\right)\left(2-\frac{a}{b}-\frac{a^2}{b^2}\right)$$

Plot the variation of q/Y with h/b for $a/b = 0.3$ and 0.5 considered in the range $0.1 \leq h/b \leq 0.5$.

3.6 An approximate solution for the frictionless extrusion through a square-faced die can be obtained by a statistical analysis based on the stress discontinuity pattern of Fig. B, in which the material is allowed to be overstressed in certain parts of the triangular regions *ABC* and ABF in the axial plane. Using (3.82), show that the extrusion pressure and die pressure distribution are given by

$$\frac{p_e}{Y}=4\left(\frac{b^2-a^2}{b^2+a^2}\right),\quad \frac{q}{Y}=\frac{2b(a+b)}{b^2-a^2}\left(2-\frac{a}{r}\right)$$

Fig. B

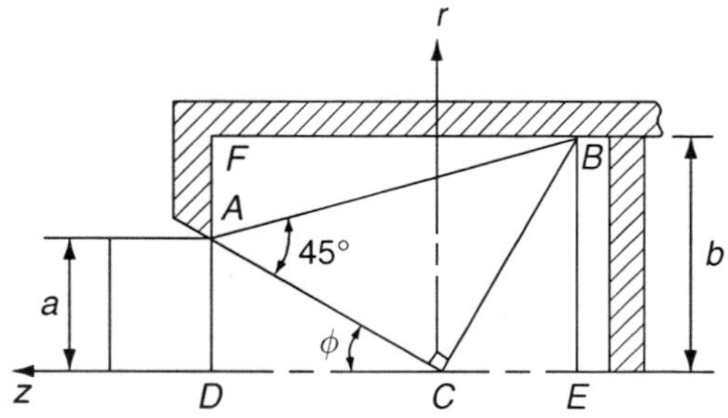

Verify that this formula gives higher values of the extrusion pressure than that given by (3.84), when the reduction in area is sufficiently large

3.7 For sufficiently large reductions in diameter, a reasonable upper bound solution for extrusion through a square-faced die is obtained by simplifying the velocity discontinuity pattern of Fig. 3.21a, in which the inclined discontinuity *AC* is replaced by a vertical discontinuity *AE,* and the dead metal boundary AB is assumed to extend from $z = 0$ to $z = c$. The material in the region *ADCE* deforms under the velocity field $u = vr/2a$ and $w = 1 + v\, z/a$, where $v = (b^2 - a^2)/ca$. Show that

$$\frac{p_e}{2k}=\ln\frac{b}{a}+\frac{1}{2}\left(1-\frac{a^2}{b^2}\right)\left(\frac{3b-a}{3c}+\frac{c}{a}+\sqrt{3}\frac{a^2}{b^2}-1\right)+\frac{2c}{3b^2}\left(\frac{b^3-a^3}{b^2-a^2}\right)$$

where *c/a* corresponds to the minimum extrusion pressure. Obtain a graphical plot of p_e/Y against the fractional reduction R over the range $0.2 \leq R \leq 0.9$.

3.8 An upper bound solution for extrusion through a conical die of semiangle α may be obtained by assuming a radially converging velocity field with discontinuities occurring across a pair of concentric spherical surfaces *AC* and *BD* as indicted in Fig. C.

Fig. C

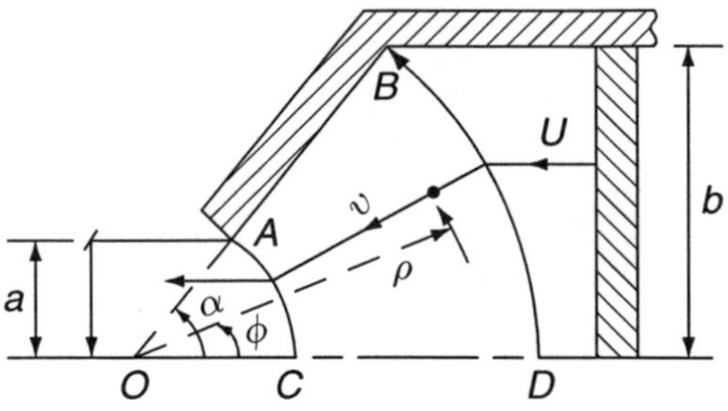

Show that the velocity field

$$\mathrm{v}_\rho = -U\left(\frac{b^2}{\rho^2}\right)\cos ec^2\alpha\cos\phi, \quad \mathrm{v}_\phi = 0$$

in spherical coordinates, (ρ,ϕ) is kinematically admissible, U being the speed of the undeformed billet. For a perfectly smooth die, show that the extrusion pressure is $p_e = 2Yf[(\alpha)\ln(b/a) + g(\alpha)]$, where

$$f(\alpha) = \left\{1 - \cos\alpha\sqrt{1 - c^2\sin^2\alpha} + \left(\frac{1-c^2}{c}\right)\ln\left(\frac{1+c}{c\cos\alpha + \sqrt{1 - c^2\sin^2\alpha}}\right)\right\}\cos ec^2\alpha$$

$$g(\alpha) = \frac{1}{\sqrt{3}}(\alpha\cos ec^2\alpha - \cot\alpha) \quad c = \sqrt{\frac{11}{12}}$$

Note that in the special case of a square-faced die ($\alpha = \pi/2$), the upper bound solution is independent of the frictional condition along the die face.

3.9 A kinematically admissible velocity field for extrusion through a conical die is one in which the velocity changes discontinuously across plane surfaces BC and AC and moves parallel to the die face within the deforming zone ABC, as shown in Fig. D. Using the geometry of the discontinuity pattern and of the associated hodograph, show that the resultant velocity of a typical particle is given by

Fig. D

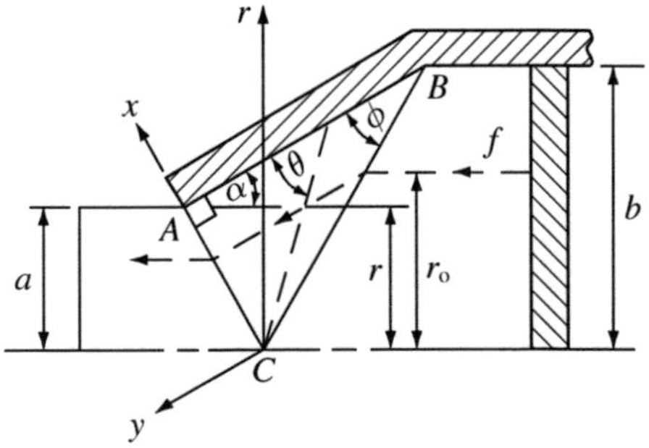

$$\mathrm{v} = \left(\frac{br_0}{ar}\right)\cos\alpha, \quad \frac{r_0}{r} = \frac{b}{a}\left(\frac{\cot\alpha}{\cot\alpha + \cot\theta}\right), \quad \frac{b}{a} = 1 + \frac{\cot\phi}{\cot\alpha}$$

where r is the current radius to a generic particle initially at a radius r_0, and θ the acute angle made with the die face by the line drawn from C to the given particle. Prove that the components of the strain increment, referred to the rectangular axes (x, y) as shown, are given by

$$d\varepsilon_x = 0, \quad d\varepsilon_y = -d\varepsilon_\theta, \quad 2d\gamma_{xy} = -(\cot\theta)\, d\varepsilon_\theta$$

3.10 Considering the velocity discontinuity pattern of the preceding problem, obtain an expression for the effective strain increment in the deforming region, and hence show that the internal energy dissipated per unit volume of the material passing through the die is equal to ke, where

$$e = \lambda\left\{2\ln\tfrac{b}{a} - \ln\left(\frac{4 - \cot\alpha\cot\phi + \lambda\sqrt{4+\cot^2\phi}}{2(2+\lambda)}\right)\right\} + \sqrt{4+\cot^2\phi} - \left[2 + \cot\alpha\sinh^{-1}\left(\tfrac{\cot\phi}{2}\right)\right]$$

and $\lambda = \sqrt{4+\cot^2\alpha}$. Assuming a frictional stress along the die face equal to mk $(0 \le m \le 1)$, show that the energy dissipated per unit volume at the discontinuities and the die face is equal to kf, where

$$f = \left(1 + \frac{a}{b}\right)\tan\alpha + \frac{b}{a}\left(1 - \frac{a}{b}\right)^2\cot\alpha + 2\left(\frac{b}{a} - 1\right)m\cot\alpha$$

Using the fact that the extrusion pressure is $p_e = k\,(e + f)$, obtain a graphical plot of $p_e/2\,k$ against the fractional reduction in area for $\alpha = 45°$, assuming $m = 0$ and $m = 1$.

Chapter 4
Plastic Bending of Plates

In this chapter, we shall be concerned with the yield point state of perfectly plastic plates whose thickness is small compared to the dimensions of its plane faces. The load acting on the plate is normal to its surface and is regarded as positive if it is pointing vertically downward. The vertical displacement of the middle surface is assumed to be generally small compared to the plate thickness, and plane vertical sections are assumed to remain plane during the bending. The deformation of the plate is therefore entirely defined by the vertical displacement of its middle surface, which remains effectively unstrained during the bending. A theory based on this model is found to be satisfactory not only in the elastic range but also in the plastic range of deflections. However, when the deflection of the plate exceeds the thickness, significant membrane forces are induced by the bending, the effect of which is to enhance the load carrying capacity of the plate.

4.1 Plastic Collapse of Circular Plates

4.1.1 The Basic Theory

Consider a circular plate of radius a and thickness h under rotationally symmetric conditions of loading and support. We choose cylindrical coordinates $(r, 0, z)$, where the z-axis is normal to the plane of the plate, and is directed vertically downward. The shear stresses $\tau_{r\theta}$ and $\tau_{\theta z}$ are identically zero because of the rotational symmetry, while σ_z *and* τ_{rz} *are* negligible for the assumed small values of h/a. Thus, σ_r and σ_θ are the only relevant principal stresses, acting in the radial and circumferential directions, respectively. The corresponding strain rates $\dot{\varepsilon}_r$ and $\dot{\varepsilon}_\theta$ vary linearly through the thickness and are given by

$$\dot{\varepsilon}_r = z\dot{\kappa}_r = -z\frac{d^2 w}{dr^2}, \quad \dot{\varepsilon}_\theta = z\dot{\kappa}_\theta = -\frac{z}{r}\frac{dw}{dr}, \tag{4.1}$$

where $\dot{k}_r$, $\dot{k}_\theta$ are the rates of curvature of the middle surface, and w is the downward velocity of the middle surface. The rate of energy dissipation across the thickness of the plate per unit area of the middle surface is

J. Chakrabarty, *Applied Plasticity, Second Edition*, Mechanical Engineering Series, DOI 10.1007/978-0-387-77674-3_4, © Springer Science+Business Media, LLC 2010

$$D = \int_{-h/2}^{h/2} (\sigma_r \dot{\varepsilon}_r + \sigma_\theta \dot{\varepsilon}_\theta) dz = M_r \dot{\kappa}_r + M_\theta \dot{\kappa}_\theta,$$

where M_r and M_θ are the bending moments per unit circumference at a generic section of the plate and are defined as

$$M_r = \int_{-h/2}^{h/2} z\sigma_r dz, \qquad M_\theta = \int_{-h/2}^{h/2} z\sigma_\theta dz. \tag{4.2}$$

In view of the above expression for D, the bending moments M_r and M_θ may be regarded as the generalized stresses for the problem, the corresponding generalized strain rates being the curvature rates $\dot{k}_r$ and $\dot{k}_\theta$. The generalized stress and strain variables are related to one another through the normality rule, based on the concept of generalized plastic potential introduced by Prager (1956a).

Since the strain rates $\dot{\varepsilon}_r$ and $\dot{\varepsilon}_\theta$ are proportional to z, the vector representing them has the same direction on one side of the middle surface, and an opposite direction on the other. For a convex yield surface, the state of stress is therefore constant on each side of the middle surface, across which σ_r and σ_θ discontinuously change their signs. In view of (4.2), the stresses in $z > 0$ may be written as

$$\frac{\sigma_r}{Y} = \frac{M_r}{M_0}, \qquad \frac{\sigma_\theta}{Y} = \frac{M_\theta}{M_0},$$

where M_0 is the fully plastic moment $Yh^2/4$ for a material having a uniaxial yield stress Y. The yield function expressed in terms of the dimensionless moments M_r/M_0 and M_θ/M_0 is therefore identical to that in terms of the dimensionless stresses σ_r/Y and σ_θ/Y. In particular, the von Mises yield criterion furnishes the yield condition

$$M_r^2 - M_r M_\theta + M_\theta^2 = M_0^2 \tag{4.3}$$

which defines an ellipse in the (M_r, M_θ)-plane. Since we are concerned here only with the yield point state, the material will be regarded as rigid/plastic. The associated flow rule then gives the curvature rates as

$$\kappa_r = \lambda(2M_r - M_\theta), \; \kappa_\theta = \lambda(2M_\theta - M_r). \tag{4.4}$$

where λ is a positive scalar factor of proportionality. The Tresca yield locus in the moment plane is a hexagon defined by the six equations

$$M_r = \pm M_0, \quad M_\theta = \pm M_0, \quad M_r - M_\theta = \pm M_0. \tag{4.5}$$

The hexagon is inscribed in the ellipse as shown in Fig. 4.1a. The flow rule associated with each side of the hexagon is given by the normality rule for the generalized strain rate vector. No plastic flow can occur, however, for any stress point lying on the sides AB and DE, since the condition $\dot{\kappa}_\theta = 0$ required by the flow rule implies $w =$ constant, which corresponds to rigid body motion of a part of the plate. At a

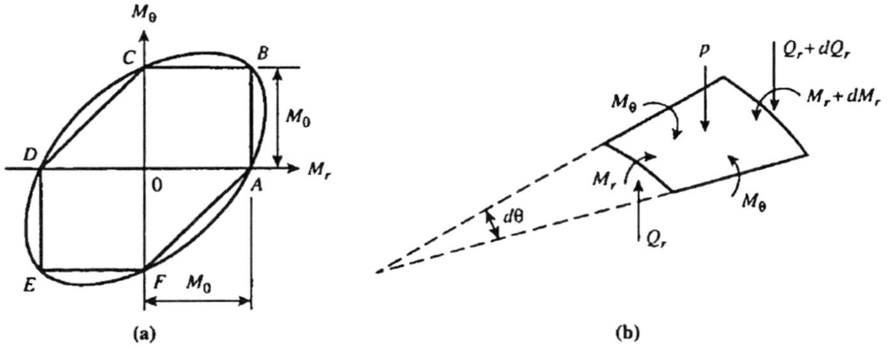

Fig. 4.1 Bending of circular plates (**a**) yield condition and (**b**) stress resultants acting on an element of the plate

corner such as B, the flow rule only requires the strain rate vector to be directed outward such that $\dot{\kappa}_r > 0$ and $\dot{\kappa}_\theta > 0$. The Tresca theory for the plastic collapse of circular plates is due to Hopkins and Prager (1953).

The yield condition and the flow rule must be supplemented by the equations of equilibrium, which involves not only the bending moments M_r and M_θ but also the shearing force Q transmitted across a vertical section. Considering an arbitrary plate element defined by adjacent radial lines and circular arcs, as shown in Fig. 4.1b, the conditions of force and moment equilibrium can be shown to be

$$\frac{d}{dr}(rQ) + rp = 0, \quad \frac{d}{dr}(rM_r) - M_\theta = rQ,$$

where $p = p(r)$ is the local intensity of the normal pressure. If there is no concentrated load acting at the center of the plate, the shearing force Q must vanish at $r = 0$. Integration of the first equilibrium equation and substitution of the result into the second furnish the equation

$$\frac{d}{dr}(rM_r) - M_\theta = \int_0^r rp\,dr, \tag{4.6}$$

which does not contain the shearing force Q. Equilibrium requires M_r and Q to be continuous, but M_θ may be discontinuous across a circle $r =$ constant. The velocity itself must be continuous, but the velocity slope dw/dr can be discontinuous across a circle, called a hinge circle, which is the limit of a narrow region of rapid but continuous change in slope. Since $\dot{\kappa}_r$ becomes infinitely large in the limit, the strain rate vector must be horizontal at a hinge circle. For the Tresca material, the radial moment M_r therefore has the absolute value M_0 at the position of the hinge circle.

4.1.2 Circular Plates Carrying Distributed Loads

Consider a solid circular plate of uniform thickness h which is built-in along its entire boundary $r = a$ and is loaded with a uniform normal pressure of intensity p over a concentric circle of radius b, as shown in Fig. 4.2. At the center of the plate, rotational symmetry requires $M_r = M_\theta = M_0$, and hence a central region of radius ρ should correspond to the plastic regime BC for which $M_\theta = M_0$. The equilibrium equation (4.6) under the assumed distribution of pressure becomes

$$\frac{d}{dr}(rM_r) = \begin{cases} M_\theta - \frac{1}{2}pr^2, & 0 \le r \le b, \\ M_\theta - \frac{1}{2}pb^2, & b \le r \le a. \end{cases} \tag{4.7}$$

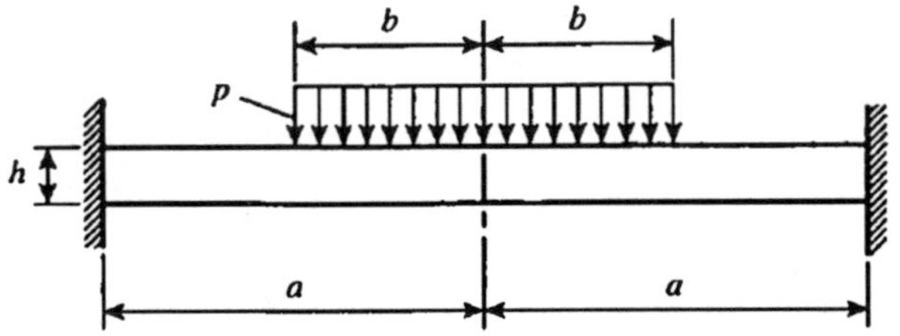

Fig. 4.2 Built-in circular plate carrying a uniformly distributed load over a concentric circular area

We begin with the situation where $b \le \rho \le a$. Introducing the dimensionless quantity $q = pb^2/2M_0$ and using the yield condition $M_\theta = M_0$, (4.7) can be integrated to give the solution

$$\begin{aligned} M_r &= M_0\left(1 - \frac{qr^2}{3b^2}\right), \quad 0 \le r \le b, \\ M_r &= M_0\left[1 - q\left(1 - \frac{2b}{3r}\right)\right], \quad b \le r \le \rho, \end{aligned} \tag{4.8}$$

in view of the condition of continuity of M_r across $r = b$. The radius ρ is obtained from the fact that it corresponds to the corner C of the yield hexagon. The condition $M_r = 0$ at $r = \rho$ therefore gives

$$q\left(1 - \frac{2b}{3\rho}\right) = 1 \text{ or } \rho = \frac{2}{3}\left(\frac{bq}{q-1}\right), \quad q \le 3. \tag{4.9}$$

Evidently, $q \le 3$ for $\rho \ge b$. Since the clamping moment must be negative, the outer region $\rho \le r \le a$ corresponds to the plastic regime CD. Substituting $M_\theta = M_r + M_0$ into the second equation of (4.7), and using the condition $M_r = 0$ at $r = \sigma$, we obtain the solution

$$M_r = -(q-1)M_0 \ln \frac{r}{\rho}, \quad \rho \le r \le a.$$

If there is no hinge circle at the clamped edge, the slope and velocity must both vanish at $r = a$. If turns out that these two conditions cannot be simultaneously satisfied for any nonzero velocity field. The clamped edge must therefore be a hinge circle, requiring $M_r = -M_0$ at $r = a$. This gives $(q-1)\ln(a/\rho) = 1$, which is equivalent to

$$\frac{3}{2}\left(1 - \frac{1}{q}\right)\exp\left(-\frac{1}{q-1}\right) = \frac{b}{a}, \quad q \le 3. \tag{4.10}$$

The greatest value of b/a for which this solution is valid is $\exp(-0.5) \approx 0.607$, corresponding to $q = 3$. Since q tends to unity when b tends to zero, a concentrated load P acting at the center is found to be equal to $2\pi M_0$ at the incipient collapse. It should be noted that the term concentrated load must be interpreted as a load that is uniformly distributed over a small circle whose radius is of order of thickness h.

When $q \ge 3$, or $b/a \ge 0.607$, the radial moment M_r vanishes at a radius $\rho \le b$. The regions $0 \le r \le \rho$ and $\rho \le r \le b$ of the loaded area then correspond to the plastic regimes BC and CD, respectively. The first equation of (4.8) therefore holds only over the region $0 \le r \le \rho$, and the condition $M_r = 0$ at $r = \rho$ therefore gives

$$q\rho^2/3b^2 = 1 \quad \text{or} \quad \rho = b\sqrt{3/q}, \quad q \le 3. \tag{4.11}$$

The solution for the outer part of the loaded area is obtained by setting $M_\theta = M_r + M_0$ in the first equation of (4.7) and integrating under the condition $M_r = 0$ at $r = \rho$, the result being

$$M_r = M_0\left[\ln\frac{r}{\rho} + \frac{1}{2}\left(3 - \frac{qr^2}{b^2}\right)\right], \quad \rho \le r \le b.$$

The entire unloaded region $b \le r \le a$ corresponds to the plastic regime CD, and the radial moment in this region is obtained by integrating the second equation of (4.7) with the substitution $M_\theta = M_r + M_0$. In view of the continuity of M_r across $r = b$, the solution becomes

$$M_r = M_0\left[1 + \frac{1}{2}\ln\frac{q}{3} - (q-1)\left(\frac{1}{2} + \ln\frac{r}{b}\right)\right], \quad b \le r \le a.$$

The existence of a hinge circle at the clamped edge requires the boundary condition $M_r = -M_0$ at $r = a$, giving the relationship

$$\exp\left\{\frac{1}{2} - \frac{4 + \ln(q/3)}{2(q-1)}\right\} = \frac{b}{a}, \quad q \ge 3. \tag{4.12}$$

When the load distribution extends over the entire plate ($b = a$), the expression in the curly brackets of (4.12) must vanish, and the collapse load is then given by

$$q - \ln(q/3) = 5 \qquad \text{or} \quad q = 5.63, \qquad \rho/a = 0.730.$$

The total load carried by the clamped plate is therefore equal to $11.267\pi\, M_0$ at the point of collapse. The collapse load steadily increases with increasing values $b/a \leq 1$, as indicated in Fig. 4.3. An analysis for the plastic collapse of a circular plate under combined concentrated and distributed loads has been given by Drucker and Hopkins (1954).

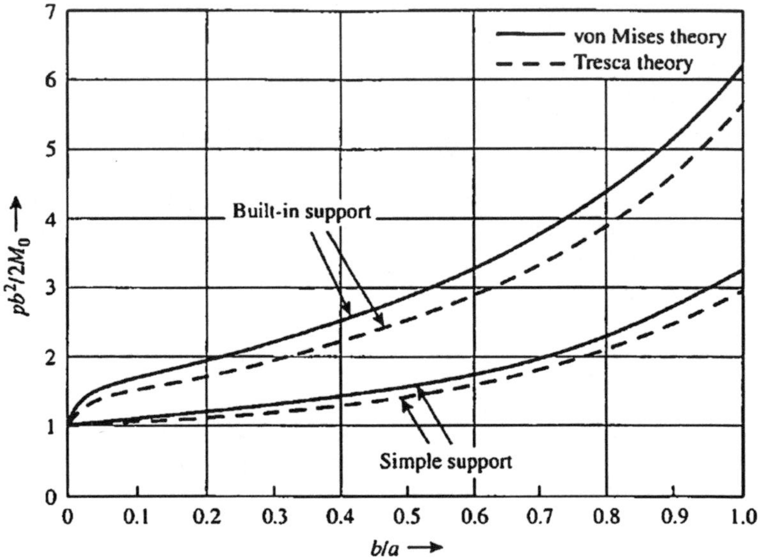

Fig. 4.3 Dimensionless collapse loads for simply supported and built-in circular plates uniformly loaded over a central circular area

The statically admissible stress field obtained in the preceding analysis will now be justified by the consideration of the velocity field. Since the flow rule associated with side BC of the yield hexagon gives $\kappa_r = 0$, $\kappa_\theta = 0$, while that associated with side CD is $\dot{\kappa}_r + \dot{\kappa}_\theta = 0$, the velocity equations are

$$\frac{d^2 w}{dr^2} = 0 \; (r \leq \rho), \quad \frac{d}{dr}\left(r \frac{dw}{dr} \right) = 0 \; (r \geq \rho).$$

Denoting by w_0 the vertically downward speed of the center of the plate at the incipient collapse, the velocity distribution is obtained by the integration of the above equations. Using the boundary condition $w = 0$ at $r = a$, and the conditions of continuity of w and dw/dr across $r = \rho$, the velocity field is found to be

$$\begin{aligned} w &= w_0(1 - \alpha r/\rho), \quad 0 \le r \le \rho, \\ w &= w_0 \alpha \ln(a/r), \quad \rho \le r \le a, \end{aligned} \tag{4.13}$$

where $\alpha = [1 + \ln(a/\rho)]^{-1}$. The velocity slope dw/dr at the clamped edge $r = a$ is seen to be negative, which justifies the assumption of a hinge circle at the boundary. Since the solution is both statically and kinematically admissible, it is therefore a complete solution giving the actual collapse load based on the Tresca theory.

If the plate is simply supported along its edge, the boundary condition is modified to $M_r = 0$ at $r = a$. The bending moment distribution is then given by (4.8) with $\rho = a$, the plastic regime BC being applicable to the entire plate. The collapse load and the incipient velocity field for the simply supported plate are therefore given by

$$q\left(1 - \frac{2b}{3a}\right)^{-1}, \quad w = w_0\left(1 - \frac{r}{a}\right). \tag{4.14}$$

A central concentrated load P causing plastic collapse is again found to be equal to $2\pi M_0$. By a simple extension of the stress field for a clamped circular plate in a statically admissible manner, it can be shown that the collapse load $P = 2tx\, M_0$ also applies to a built-in plate of arbitrary shape carrying a concentrated load (Haythornthwaite and Shield, 1958). The effect of kinematic hardening on the post-yield carrying capacity of circular plates has been examined by Prager (1956b) and Boyce (1959). Experimental evidences on the plastic bending theory have been provided by Lance and Onat (1962).

4.1.3 Other Types of Loading of Circular Plates

Consider a circular plate of radius a containing a central boss of radius b, which is used for loading the plate with a rigid cylindrical punch. The central boss moves down as a rigid body at the incipient collapse, which involves the formation of a hinge circle at $r = b$ corresponding to the corner B of the yield hexagon. If the plate is simply supported as shown in Fig. 4.4a, the entire stress profile lies along the side BC of the hexagon, giving $M_\theta = M_0$. Since the shearing force Q per unit circumference is equal to $-P/2\pi r$, where P is the applied punch load, the equilibrium equation becomes

$$\frac{d}{dr}(rM_r) = M_\theta + rQ = M_0 - \frac{P}{2\pi}.$$

Setting $q = P/2\pi M_0$, and using the boundary condition $M_r = 0$ at $r = a$, the solution for the radial bending moment is obtained on integration as

$$M_r = M_0(q - 1)\left(\frac{a}{r} - 1\right), \quad b \le r \le a.$$

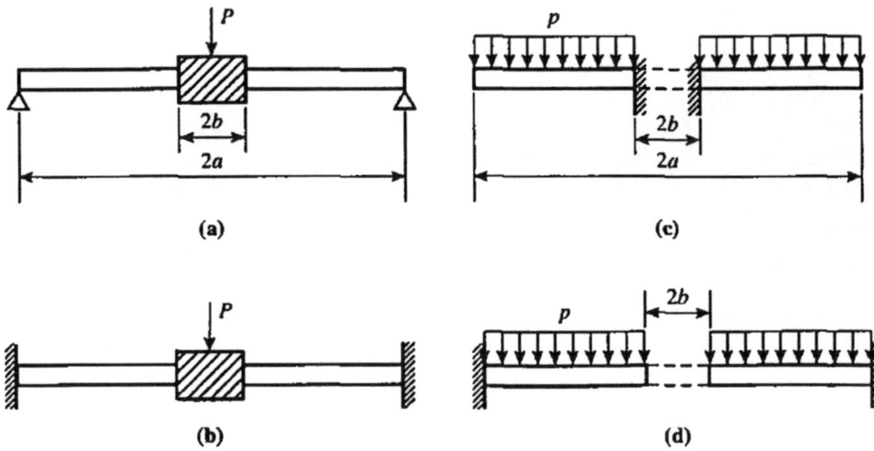

Fig. 4.4 Various loading and support condition for circular plates with and without central hole

The collapse load now follows from the condition $M_r = M_0$ at the hinge circle $r = b$. Since the yield condition $M_\theta = M_0$ implies that the plate deforms conically at the point of collapse, we have

$$q = \left(1 - \frac{a}{b}\right)^{-1}, \quad w = w_0\left(\frac{a-r}{a-b}\right), \tag{4.15}$$

where w_0 is the downward speed of the loading boss. A statically admissible stress field in the central boss can be constructed by assuming the punch load to be equivalent to a line load around its periphery $r = b$ and setting $M_r = M_\theta = M_0$ throughout the region $0 \leq r \leq b$ satisfying the equation of equilibrium. The collapse load given by (4.15) is therefore exact according to the Tresca theory.

When the plate is fully clamped, as shown in Fig. 4.4b, the plastic regime BC applies only to an inner region of the plate defined by $b \leq r \leq \rho$, where $r = \rho$ corresponds to $M_r = 0$. The bending moment distribution over this region is given by

$$M_r = M_0(q-1)\left(\frac{\rho}{r} - 1\right), \qquad M_\theta = M_0, \quad b \leq r \leq \rho.$$

In the outer region of the plate, defined by $\rho \leq r \leq a$, the plastic regime CD will hold, giving the yield condition $M_\theta - M_r = M_0$. The equilibrium equation then reduces to

$$r\frac{dM_r}{dr} = M_0 - \frac{P}{2\pi} = -(q-1)M_0 \tag{4.16}$$

Integrating (4.16), and using the continuity condition $M_r = 0$ at $r = \rho$, the solution is found as

$$M_r = M_0(q-1)\ln\left(\frac{r}{\rho}\right), \qquad M_\theta = M_r + M_0, \quad \rho \leq r \leq a.$$

The radius ρ to the regime interface and the collapse load parameter q can be determined by using the boundary conditions $M_r = M_0$ at $r = b$ and $M_r = -M_0$ at $r = a$. It follows from the preceding expressions for M_r that

$$(q-1)\left(\frac{\rho}{b}-1\right)=1, \qquad (q-1)\ln\left(\frac{a}{\rho}\right)=1,$$

and the elimination of ρ and q in turn between the preceding relations results in

$$\left(\frac{q-1}{q}\right)\exp\left(-\frac{1}{q-1}\right)=\frac{b}{a}, \quad \frac{\rho}{b}=\frac{q}{q-1}. \tag{4.17}$$

These equations define q and ρ/a as functions of b/a. The kinematically admissible velocity field associated with the stress field, which is statically admissible, is similar to that for the distributed loading. It is easily shown that

$$\begin{aligned} w &= w_0\left(\frac{\rho^2-br}{\rho^2-b^2}\right), \quad b \le r \le \rho, \\ w &= w_0\left(\frac{\rho b}{\rho^2-b^2}\right)\ln\left(\frac{a}{r}\right), \quad \rho \le r \le a, \end{aligned} \tag{4.18}$$

where w_0 is the downward speed of the central boss. When b tends to zero, ρ also tends to zero, and the collapse load tends to the limit $P = 2\pi M_0$ for both the edge conditions as expected.

As a next example, consider a uniformly loaded annular plate that is fully clamped along its inner edge $r = b$ and is free along its outer edge $r = a$, as shown in Fig. 4.4c. Since the shearing force vanishes at $r = a$, the equilibrium equation (4.6) is modified to

$$\frac{d}{dr}(rM_r) = M_\theta + \int_r^a pr\,dr,$$

where p is the applied normal pressure. It is natural to suppose that a hinge circle is formed at the clamped edge $r = b$ when the plate is loaded to the point of collapse. Since the radial bending moment varies from $M_r = 0$ at $r = a$ to $M_r = -M_0$ at $r = b$, the entire stress profile should lie on the side EF of the yield hexagon. Setting $M_\theta = -M_0$ in the above equation, and integrating, we get

$$M_r = -M_0\left(\frac{a}{r}-1\right)\left\{\frac{q}{3}\left(1-\frac{r}{a}\right)\left(2+\frac{r}{a}\right)-1\right\}, \quad b \le r \le a,$$

where $q = pa^2/2M_0$, the free-edge condition $M_r = 0$ at $r = a$ being incorporated in the solution. The radial bending moment M_r, which has a maximum value at a radius $r = a(2 - 3/q)^{1/3}$, steadily decreases inward from this radius toward the clamped edge. The boundary condition $M_r = -M_0$ at r $= b$ then furnishes

$$q = \frac{3}{(1-b/a)^2(2+b/a)}. \tag{4.19}$$

The parameter q increases as the ratio b/a is increased, the least value being $q = 1.5$ corresponding to $b = 0$. Since $\dot{k}_r = 0$ by the associated flow rule, the shape of the deformed plate is conical at the incipient collapse, and the solution is both statically and kinematically admissible.

As a final example, consider an annular plate which is clamped at its outer edge $r = a$ and free at its inner edge $r = b$, as shown in Fig. 4.4d. In this case, the shearing force vanishes at $r = b$, and the equilibrium equation takes the form

$$\frac{d}{dr}(rM_r) = M_\theta - \int_b^r pr\,dr.$$

The plate is expected to collapse with a hinge circle formed at $r = a$, requiring the stress profile to be along CD in an outer portion of the plate. In the remaining inner portion, defined by $b \leq r \leq \rho$, the stress profile should lie along the side CB, which corresponds to $M_\theta = M_0$. For a uniformly distributed pressure p, the above equation is readily integrated to give

$$M_r = M_0\left(1 - \frac{b}{r}\right)\left\{1 - \frac{q}{3}\left(\frac{r-b}{a}\right)\left(\frac{r+2b}{a}\right)\right\}, \quad b \leq r \leq \rho,$$

on using the boundary condition $M_r = 0$ at $r = b$ and setting $pa^2/2M_0 = q$. The radial bending moment M_r vanishes again at $r = \rho$ after attaining a maximum, the relationship between q and ρ being

$$\frac{3}{q} = \left(\frac{\rho - b}{a}\right)\left(\frac{\rho + 2b}{a}\right). \tag{4.20}$$

In the outer portion ($\rho \leq r \leq a$), the yield condition is $M_\theta - M_r = M_0$, and the equilibrium equation then becomes

$$r\frac{dM_r}{dr} = M_0 - \frac{p}{2}(r^2 - b^2).$$

Integrating, and using the continuity condition $M_r = 0$ at r $=\rho$, the bending moment distribution is obtained as

$$M_r = -M_0\left\{q\left(\frac{r^2 - \rho^2}{2a^2}\right) - \left(1 + \frac{qb^2}{a^2}\right)\ln\left(\frac{r}{\rho}\right)\right\}, \quad \rho \leq r \leq a.$$

The remaining boundary condition $M_r = -M_0$ at $r = a$ furnishes a second relationship between q and ρ as

$$1 - \frac{\rho^2}{a^2} - \frac{2b^2}{a^2}\ln\left(\frac{a}{\rho}\right) = \frac{2}{q}\left(1 + \ln\frac{a}{\rho}\right). \tag{4.21}$$

Equations (4.20) and (4.21) can be solved simultaneously to obtain ρ/a and q for any given b/a. The associated velocity field at the incipient collapse is given by (4.13), where

$$w_0 = w_b \left(1 - \alpha\frac{b}{\rho}\right)^{-1}, \quad \alpha \left(1 + \ln\frac{a}{\rho}\right)^{-1},$$

with w_b denoting the vertical speed of the free edge $r = b$. When b tends to zero, q tends to the limiting value 5.63, which is the dimensionless collapse pressure for a plate without a hole. Figure 4.5 shows the variation of q with b/a for all the four cases of loading considered above.

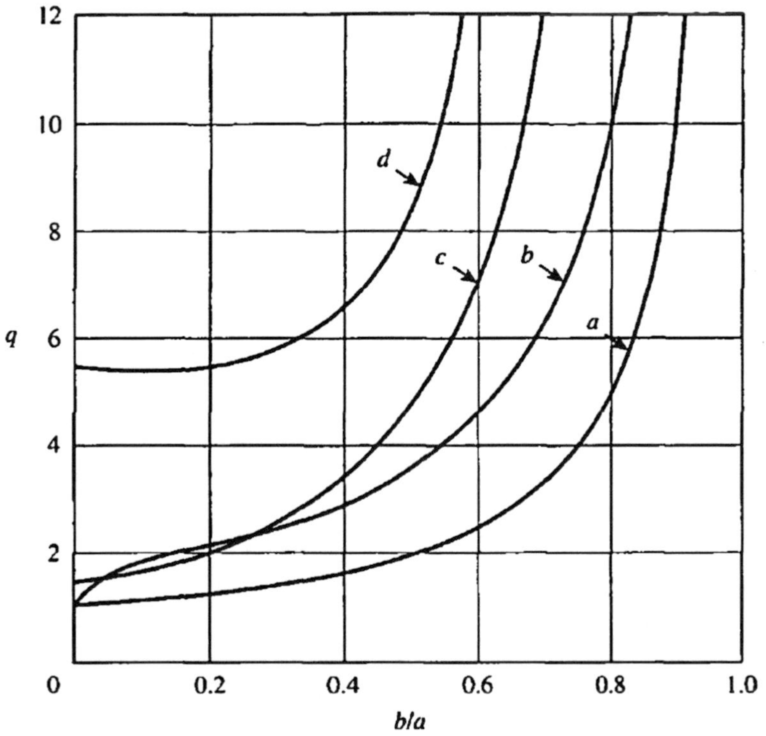

Fig. 4.5 Dimensionless collapse loads for circular and annular plates

The load-carrying capacity of an annular plate simply supported along both its edges has been examined by Hodge (1963). The limit analysis of reinforced circular plates has been discussed by Nemirovsky (1962), while the influence of shear on the limit load has been examined by Sawczuk and Duszek (1963). The plastic collapse of plates resting on elastic foundations has been investigated by Krajcinovic (1976). The plastic collapse of circular and annular plates under different loading conditions has been investigated by Cinquini and Zanon (1985). The solutions to a number of practical problems on the carrying capacity of plates have been presented by Sawczuk and Jaeger (1963) and by Sobotka (1989).

4.1.4 Solutions Based on the von Mises Criterion

Since the Tresca hexagon is inscribed within the von Mises ellipse, the collapse load based on the Tresca theory provides a lower bound for that based on the von Mises theory. By multiplying the Tresca collapse load by a factor of $2/\sqrt{3}$, we obtain an upper bound on the von Mises collapse load, since the corresponding yield hexagon circumscribes the ellipse. Bounds on the collapse load based on the von Mises criterion have been given by Pell and Prager (1951). To obtain the actual collapse load based on the von Mises theory, we write the yield criterion (4.3) in the form

$$2M_\theta = M_r \pm \sqrt{4M_0^2 - 3M_r^2},$$

where the positive or negative sign is to be taken according as the ratio M_θ/M_r is greater or less than $\frac{1}{2}$. The equilibrium equation (4.6) therefore becomes

$$2\frac{d}{dr}(rM_r) = M_r \pm \sqrt{4M_0^2 - 3M_r^2 -} - 2\int_0^r rp\,dr. \tag{4.22}$$

Consider a simply supported plate of radius a under a uniform normal pressure p over a circle of radius b. Introducing dimensionless variables

$$m = \frac{M_r}{M_0}, \quad \xi = r\sqrt{\frac{p}{2M_0}} = \frac{r}{b}\sqrt{q},$$

where $q = pb^2/2M_0$ is the dimensionless collapse load, (4.22) is reduced to

$$2\xi\frac{dm}{d\xi} = \begin{cases} -m + \sqrt{4 - 3m^2} - 2\xi^2, & 0 \le \xi \le \sqrt{q}, \\ -m + \sqrt{4 - 3m^2} - 2q, & \sqrt{q} \le \xi \le \xi_0, \end{cases} \tag{4.23}$$

where $\xi_0 = (a/b)\sqrt{q}$. This is a pair of first-order nonlinear differential equations which can be solved numerically under the boundary conditions $m = 1$ at $\xi = 0$ and $m = 0$ at $\xi = \xi_0$ using the condition of continuity of m at $\xi = \sqrt{q}$. A close approximation is achieved, however, by using the Tresca solution $m = 1 - \xi^2/3$ on the right-hand side of the first equation of (4.23), which becomes

$$2\frac{d}{d\xi}(\xi m) = 1 - \frac{7}{3}\xi^2 + \left[4 - 3\left(1 - \frac{\xi^2}{3}\right)\right]^{1/2}, \quad 0 \le \xi \le \sqrt{q}.$$

Integrating this equation under the boundary condition $m = 1$ at $\xi = 0$, the solution may be written as

$$m(\xi) = \frac{1}{2}\left(1 - \frac{7}{9}\xi^2\right) + \frac{1}{2}f(\xi), \quad 0 \leq \xi \leq \sqrt{q}. \tag{4.24}$$

The function $f(\xi)$ appearing in (4.24) can be expressed in terms of elliptic integrals, the result being

$$\begin{aligned}\sqrt{2}\xi f(\xi) = \left(\frac{2}{\sqrt{3}}\right)^{3/2} &\left\{\left(2 - \sqrt{3}\right)\left[F\left(\frac{\pi}{2}, k\right) - F(\phi, k)\right]\right. \\ &+ 2\sqrt{3}\left[E\left(\frac{\pi}{2}, k\right) - E(\phi, k)\right] \\ &\left. + \left(1 + \frac{\sqrt{3}}{2}\right)\sin 2\phi\sqrt{1 - k^2 \sin^2 \phi}\right\}\end{aligned} \tag{4.25}$$

where F and E denote elliptic integrals of the first and second kinds, respectively, the modulus k and variable ϕ being defined as

$$k = \frac{\sqrt{2+\sqrt{3}}}{2}, \quad \phi = \cos^{-1}\left\{\frac{\xi}{\sqrt{3 + 2\sqrt{3}}}\right\}.$$

In the unloaded region $\left(\sqrt{q} \leq \xi \leq \xi_0\right)$, the second equation of (4.23) can be written in the separable form

$$\frac{d\xi}{\xi} = \frac{2\,dm}{\sqrt{4 - 3m^2} - m - 2q}, \quad \sqrt{q} \leq \xi \leq \xi_0.$$

Using the boundary condition $m = 0$ *at* $\xi = \xi_0 = (a/b)\sqrt{q}$, it is straightforward to obtain the solution

$$\ln\left(\frac{a\sqrt{q}}{b\xi}\right) = \frac{1}{2}\ln\left\{\frac{\sqrt{3}q - 2\sin\alpha}{\sqrt{3}(q-1)}\right\} - \frac{\sqrt{3}}{2}\left(\frac{\pi}{3} - \alpha\right) + g(\alpha), \quad \sqrt{q} \leq \xi \leq \xi_0, \tag{4.26}$$

where $a = \pi/3 - \sin^{-1}\left(\sqrt{3}m/2\right)$, which takes the boundary value of $\pi/3$ at $r = a$, the function $g(\alpha)$ being defined as

$$
\left.
\begin{aligned}
g(\alpha) &= \frac{2\sqrt{3}q}{4-3q^2}\ln\left\{\left[\frac{(2-\eta)-\sqrt{3}q\tan(\alpha/2)}{(2+\eta)-\sqrt{3}q\tan(\alpha/2)}\right]\left(\frac{2+\eta-q}{2-\eta-q}\right)\right\}, \\
&\qquad\qquad\qquad\qquad\qquad\qquad\qquad\qquad\qquad q<\frac{2}{\sqrt{3}}, \\
g(\alpha) &= \frac{2\sqrt{3}q}{\sqrt{3q^2-4}}\left\{\tan^{-1}\left[\frac{2-\sqrt{3}q\tan(\alpha/2)}{\sqrt{3q^2-4}}\right]-\tan^{-1}\left(\frac{2-q}{\sqrt{3q^2-4}}\right)\right\}, \\
&\qquad\qquad\qquad\qquad\qquad\qquad\qquad\qquad\qquad q>\frac{2}{\sqrt{3}},
\end{aligned}
\right\}, \tag{4.27}
$$

with $\eta=\sqrt{4-3q^2}$. Equations (4.24) and (4.26) define the distribution of the radial moment over the loaded and unloaded regions, respectively, the parameter q being obtained from the condition of continuity of m at $\xi=\sqrt{q}$. For a given b/a ratio, the value of q can be computed by a trial and error procedure using the Tresca solution $q=(1-2b/3a)^{-1}$ as a first approximation. When $b/a=1$, equation (4.24) applies to the entire plate, and the boundary condition $m=0$ at $\xi=\sqrt{q}$ furnishes $q\approx 3.25$, which may be compared with the value $q=3.0$ given by the Tresca theory.

The analysis can be extended to deal with the built-in edge support, which requires the radial bending moment to have the numerically greatest value $-2M_0/\sqrt{3}$ at the boundary $r=a$. The first equation of (4.23) is solved as before using the Tresca solution as a first approximation, the solution to the second equation of (4.23) in this case being

$$
\ln\left(\frac{a\sqrt{q}}{b\xi}\right)=\frac{1}{2}\ln\left\{\frac{\sqrt{3}q-2\sin\alpha}{\sqrt{3}q+1}\right\}+\frac{\sqrt{3}}{2}\left(\alpha+\frac{\pi}{6}\right)+g(\alpha)-g\left(-\frac{\pi}{6}\right) \tag{4.28}
$$

in view of the modified boundary condition $\alpha=-\pi/6$ at $\xi=\xi_0$. In the special case $b/a=1$, the numerical solution furnishes $q\approx 6.25$. The variation of q with b/a, computed by Hopkins and Wang (1955) for both the simply supported and the clamped edge conditions, is included in Fig. 4.3 for comparison with that furnished by the Tresca theory. In the limiting case of a concentrated load P acting at the center, the condition $\alpha=0$ at $\xi=0$ gives $q=1$ or $P=2\pi M_0$ for a simply supported plate, and $q=2/\sqrt{3}$ or $p=4/\sqrt{3}\pi M_0$ for a built-in plate, according to the von Mises theory. An exact formulation of the problem of limit analysis of circular plates made of a von Mises material has been discussed by Eason (1958).

4.1.5 Combined Bending and Tension

Consider a circular plate that is simply supported along its outer edge $r=a$ and is subjected to a uniform normal pressure p as well as a radial tension t around its circumference Fig. 4.6a. The plate has a central circular cutout of radius b, the inner edge of the plate being either stress-free or required to support a rigid cutout plug that carries the same intensity of pressure. The radial and circumferential forces

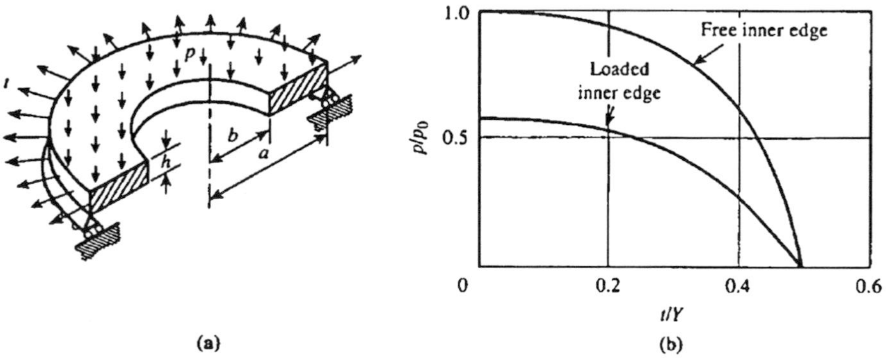

Fig. 4.6 Circular plate under combined bending and tension. (**a**) Loaded plate with a central cutout and (**b**) interaction curves for b/a = 0.5

per unit length, denoted by N_r and N_θ, respectively, must satisfy the equation of equilibrium

$$\frac{d}{dr}(rN_r) - N_\theta = 0. \tag{4.29}$$

The equilibrium equation satisfied by the bending moments M_r and M_θ is still given by (4.6). The generalized forces in this problem are N_r, N_θ, M_r, and M_θ, the corresponding generalized strain rates being the rates of extension $(\dot{\lambda}_r, \dot{\lambda}_\theta)$ and the rates of curvature $(\dot{k}_r, \dot{k}_\theta)$, which are given by

$$\dot{\lambda}_r = \frac{du}{dr}, \qquad \dot{\lambda}_\theta = \frac{u}{r}, \qquad \dot{\kappa}_r = -\frac{d^2w}{dr^2}, \qquad \dot{\kappa}_\theta = -\frac{1}{r}\frac{dw}{dr}, \tag{4.30}$$

where u denotes the radially outward velocity of the middle surface, and w the vertically downward velocity. The yield condition according to the Tresca theory is defined by several curved surfaces in a four-dimensional space (Section 5.3), the one relevant to the present problem being

$$\left(\frac{N_\theta}{N_0}\right)^2 + \frac{M_\theta}{M_0} = 1, \tag{4.31}$$

where $N_0 = hY$ denotes the fully plastic value of the direct force. The generalized strain rates associated with (4.31) can be written in terms of an arbitrary scalar factor using the normality rule of plastic flow.

We begin with the situation where the inner edge of the plate is totally unsupported. When $p = 0$, the bending moments are identically zero, and the yield condition reduces to $N_\theta = N_0$. The equilibrium equation (4.29) and the boundary conditions $N_r = ht$ at $r = a$ and $N_r = 0$ at $r = b$ then furnish the distribution of N_r and the collapse value of t, the results being

$$N_\theta = N_0, \qquad N_r = N_0\left(1 - \frac{b}{r}\right), \quad t = Y\left(1 - \frac{b}{a}\right). \tag{4.32}$$

Since $\dot{\lambda}_r = 0$ in view of the flow rule, the corresponding velocity is $u =$ constant. When $t = 0$, the direct forces are absent, and the yield condition is $M_\theta = M_0$. The equilibrium equation (4.6) for a constant p therefore becomes

$$\frac{d}{dr}(rM_r) = M_0 - \frac{1}{2}p(r^2 - b^2),$$

which is integrated under the boundary condition $M_r = 0$ at $r = a$, the remaining condition $M_r = 0$ at $r = b$ being used to determine p. The result may be expressed as

$$\left.\begin{aligned} M_\theta = M_0, \quad M_r &= M_0\left(1 - \frac{a}{r}\right)\left\{1 - q\left(1 + \frac{r}{a} + \frac{r^2 - 3b^2}{a^2}\right)\right\}, \\ q = \frac{pa^2}{2M_0} &= 3\left\{\left(1 - \frac{b}{a}\right)\left(1 + \frac{2b}{a}\right)\right\}^{-1}. \end{aligned}\right\} \tag{4.33}$$

The flow rule $\dot{\kappa}_r = 0$ indicates that dw/dr is a constant equal to $-B$ (say), giving $w = B(1 - r/a)$ in view of the boundary condition $w = 0$ at $r = a$.

When t and p are both nonzero, a statically admissible stress field will result if we combine the values of N_θ and N_r obtained by multiplying those in (4.32) by $(t/Y)/(1-b/a)$, with the values of M_θ, M_r, and p obtained by multiplying those in (4.33) by $1 - (t/Y)^2/(1 - b/a)^2$. The equilibrium equations, the boundary conditions, and the yield condition are then identically satisfied, and the relationship between t and p at the yield point then becomes

$$\left(1 - \frac{b}{a}\right)\left(1 + \frac{2b}{a}\right)\frac{p}{p_0} + \left(1 - \frac{b}{a}\right)^{-2}\left(\frac{t}{Y}\right)^2 = 1, \tag{4.34}$$

where $p_0 = 6M_0/a^2$. By the lower bound theorem of limit analysis, (4.34) provides a lower bound on the interaction curve. In order to see if (4.34) is also an upper bound on the interaction curve, it is only necessary to examine the velocity field associated with the stress field. The flow rule corresponding to (4.31) gives

$$\dot{\lambda}_r = 0, \qquad \dot{\kappa}_r = 0, \qquad \dot{\lambda}_\theta/\dot{\kappa}_\theta = hN_\theta/2N_0,$$

in view of the fact that $M_0/N_0 = h/4$. Using (4.30), and integrating, we obtain the velocity field satisfying the boundary conditions as

$$u = C, \quad w = B\left(1 - \frac{r}{a}\right), \quad C = B\left(\frac{h}{a - b}\right)\frac{t}{2Y},$$

where $B > 0$ for $\dot{k}_\theta$ to be positive. The existence of a kinematically admissible velocity field indicates that (4.34) is also an upper bound for the yield point state, and the actual interaction curve for a Tresca material has therefore been found (Haddow, 1969).

Suppose now that the inner edge of the plate is supported in such a way that it is free to bend and expand, while carrying a total shearing force equal to $\pi b^2 p$. The solution for $p = 0$ is evidently given by (4.32), while that for $t = 0$ is given by the integration of the differential equation

$$\frac{d}{dr}(rM_r) = M_0 - \frac{1}{2}pr^2.$$

Using the boundary conditions $M_r = 0$ at $r = a$ and at $r = b$, the bending moment distribution and the dimensionless load are easily shown to be

$$\left.\begin{aligned} M_\theta = M_0, \quad M_r = M_0\left(1 - \frac{a}{r}\right)\left\{1 - \frac{q}{3}\left[1 + \frac{r}{a} + \frac{r^2}{a^2}\right]\right\}, \\ q = \frac{pa^2}{2M_0} = 3\left(1 + \frac{b}{a} + \frac{b^2}{a^2}\right)^{-1}. \end{aligned}\right\} \quad (4.35)$$

The incipient velocity is $u = C$ when $p = 0$, and $w = B(1 - r/a)$ when $t = 0$, with C and B as arbitrary constants. A statically admissible stress field for the combined loading is again obtained by multiplying the values of N_θ and N_r in (4.32) by the factor $(t/Y)/(1 - b/a)$ and combining them with the values of M_θ and M_r in (4.35) multiplied by the factor $1 - (t/Y)^2/(1 - b/a)^2$. The pressure p is also multiplied by the same factor, and the interaction relation therefore becomes

$$\left(1 + \frac{b}{a} + \frac{b^2}{a^2}\right)\frac{p}{p_0} + \left(1 - \frac{b}{a}\right)^{-2}\left(\frac{t}{Y}\right)^2 = 1, \quad (4.36)$$

where $p_0 = 6M_0/a^2$. Equation (4.36) defines in fact the exact interaction curve according to the Tresca theory, since the velocity field obtained by combining those for the two special cases as before is kinematically admissible. The interaction curves defined by (4.34) and (4.36) are shown graphically in Fig. 4.6b. Upper and lower bounds for the yield point state for several other support conditions have been examined by Hodge and Sankaranarayanan (1960). The plastic interaction relationship for the combined bending and shear in plates has been discussed by Sawczuk and Duszek (1963).

4.2 Deflection of Circular Plates

4.2.1 Basic Equations

The estimation of the displacement that occurs in a loaded plate prior to the state of plastic collapse is of considerable interest in limit design. To simplify the elastic/plastic analysis, it is assumed that the plate is of ideal sandwich construction with a plastic moment M_0 (Haythornthwaite, 1954). Each cross-section of the plate is then either entirely elastic or entirely plastic during the loading. Over the elastic range, the relevant stress–strain relations are

$$\varepsilon_r = \frac{1}{E}(\sigma_r - \nu\sigma_\theta), \quad \varepsilon_\theta = \frac{1}{E}(\sigma_\theta - \nu\sigma_r),$$

when E denotes Young's modulus and ν Poisson's ratio. Multiplying both sides of the above equations by zdz and integrating between the limits $-h/2$ and $h/2$, we obtain the moment curvature relations

$$\left(\frac{Eh^3}{12}\right)\kappa_r = M_r - \nu M_\theta, \quad \left(\frac{Eh^3}{12}\right)\kappa_\theta = M_\theta - \nu M_r, \tag{4.37}$$

in view of the relations $\varepsilon_r = z\kappa_r$ and $\varepsilon_\theta = z\kappa_\theta$. In terms of the downward displacement, which is denoted here by w, the radial and circumferential curvatures are

$$\kappa_r = \frac{d^2w}{dr^2}, \quad \kappa_\theta = -\frac{1}{r}\frac{dw}{dr}.$$

Substituting from above into (4.37) and eliminating w between the two relations, we have

$$r\frac{d}{dr}(M_\theta - \nu M_r) = -(1+\nu)(M_\theta - M_r).$$

This is the compatibility equation expressed in terms of the bending moments. Using the equilibrium equation (4.6), the quantity $M_\theta - M_r$ can be eliminated from the right-hand side of the above equation, giving

$$r\frac{d}{dr}(M_r + M_\theta) = -(1+\nu)\int pr\,dr. \tag{4.38}$$

For any given pressure distribution $p(r)$, this equation can be integrated to give $M_r + M_\theta$ in terms of an arbitrary constant. The elimination of M_θ between the resulting equation and (4.6) then leads to a first-order differential equation for M_r. When p is a constant, (4.38) gives

$$M_r + M_\theta = 2A - \frac{1}{4}(1+\nu)pr^2,$$

where A is a constant of integration. Eliminating M_θ, the equilibrium equation (4.6) is then reduced to

$$\frac{r}{dr}(r^2 M_r) = 2Ar - \frac{1}{4}(3+\nu)pr^3$$

and a straightforward integration of this differential equation leads to the bending moment distribution

$$\left.\begin{aligned} M_r &= A - B\left(\frac{a}{r}\right)^2 - \tfrac{1}{16}(3+\nu)pr^2, \\ M_\theta &= A + B\left(\frac{a}{r}\right)^2 - \tfrac{1}{16}(1+3\nu)pr^2, \end{aligned}\right\} \tag{4.39}$$

where B is a second constant, and a denotes the plate radius. The substitution for M_r and M_θ in the second equation of (4.37) results in the equation for the displacement gradient dw/dr as

$$-\frac{Eh^3}{6a^2}\frac{dw}{dr} = r(1-\nu)\left[A - (1+\nu)\frac{pr^2}{16}\right] + (1+\nu)B\frac{a^2}{r}.$$

For the subsequent analysis, it is convenient to write the integral of the above equation in the form

$$\begin{aligned} \frac{Eh^3 w}{6a^2} &= (1-\nu)\left(1-\frac{r^2}{a^2}\right)\left\{A - (1+\nu)\frac{pa^2}{32}\left(1+\frac{r^2}{a^2}\right)\right\} \\ &\quad + 2(1+\nu)B\ln\left(\frac{a}{r}\right) + C, \end{aligned} \tag{4.40}$$

where C is another constant of integration. Equations (4.39) and (4.40) furnish the bending moments and the transverse displacement not only in an elastic plate but also in the elastic part of a partially plastic plate, the constants A, B, and C being obtained from the boundary and continuity conditions in each particular case.

In an elastic/plastic plate, the curvature rates in any plastic element are the sum of an elastic part, given by the differentiated form of (4.37), and a plastic part given by the normality rule of plastic flow. If the plate element has entered a certain plastic regime of the Tresca hexagon (Fig. 4.7) directly from an elastic state and has never left this regime during the subsequent deformation, the flow rule can be written in the integrated form, and the elastic/plastic analysis is considerably simplified. For example, when the stress point is on the side BC, the plastic part of the radial curvature vanishes, and the first equation of (4.37) continues to hold in the plastic range. If a finite region of the plate involves the plastic regime BC, the bending moments are found by the integration of (4.6) with $M_\theta = M_o$, and the vertical deflection is then obtained by the integration of the equation

$$\left(\frac{Eh^3}{12}\right)\frac{d^2w}{dr^2} = -(M_r - \nu M_0), \tag{4.41}$$

and the use of the boundary or continuity conditions across the appropriate radii. Similarly, when the stress point remains on the side CD after reaching it from the elastic state, the plastic part of the quantity k_r + k_θ is zero, and (4.37) then furnishes

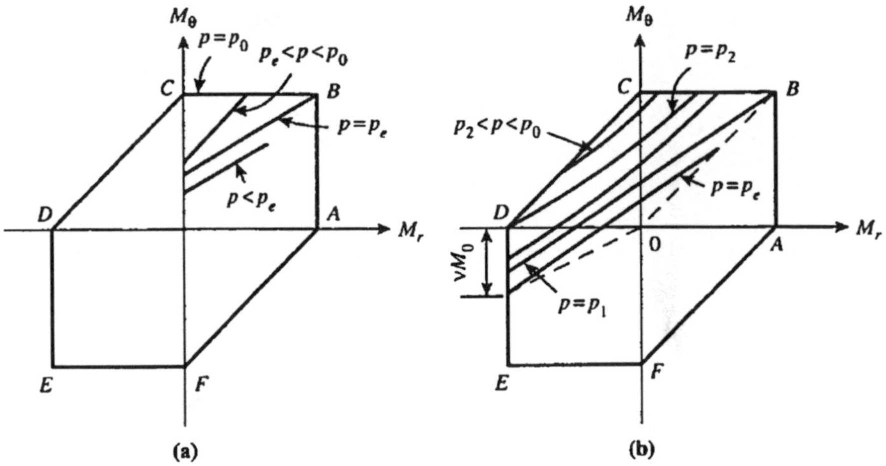

Fig. 4.7 Stress profiles for elastic/plastic circular plates. (**a**) Simple support and (**b**) built-in support

$$\left(\frac{Eh^3}{12}\right)\frac{d}{dr}\left(r\frac{dw}{dr}\right) = (1-\nu) - (2M_r - M_0)r \tag{4.42}$$

in view of the yield condition $M_\theta - M_r = M_0$. This equation is readily integrated, once M_r has been found from (4.6) together with the yield condition. If, on the other hand, a given plate element has a previous history involving different plastic regimes, the deflection analysis becomes much more complex.

4.2.2 Deflection of a Simply Supported Plate

Consider a simply supported circular plate of radius a under a uniformly distributed normal pressure of gradually increasing intensity p. For a sufficiently small value of p, the plate is entirely elastic, and the bending moment distribution is given by (4.39), where $B = 0$ for the bending moments to be finite at $r = 0$. The boundary condition $M_r = 0$ at $r = a$ gives $A = (3 + \nu)pa^2/16$, and the bending moments become

$$M_r = \frac{pa^2}{16}(3+\nu)\left(1-\frac{r^2}{a^2}\right), \quad M_\theta = \frac{pa^2}{16}\left[(3+\nu)-(1+3\nu)\frac{r^2}{a^2}\right]. \tag{4.43}$$

The displacement boundary condition $w = 0$ at $r = a$ gives $C = 0$ in (4.40). The substitution for A and B in (4.40) then furnishes the deflection w, the result being

$$\frac{Eh^3w}{a^2} = \frac{3}{16}(1-\nu)pa^2\left(1-\frac{r^2}{a^2}\right)\left\{(5+\nu)-(1+\nu)\frac{r^2}{a^2}\right\}. \tag{4.44}$$

Eliminating r^2 between the two equations of (4.43), we obtain the equation for the stress profile as

$$(3+\nu)M_\theta - (1+3\nu)M_r = (1-\nu)(3+\nu)(pa^2/8),$$

which defines a straight line extending from the point $M_r = 0$, $M_\theta = (1-\nu)pa^2/8$ to the point $M_r = M_\theta = (3+4\nu)pa/16$. The stress profiles for varying load intensities are parallel to one another lying within the yield locus, Fig. 4.7a. Plastic yielding begins at the center of the plate when $M_r = M_\theta = M_0$ at $r = 0$, the corresponding load intensity p_e and the central deflection δ_e being given by

$$\frac{p_e a^2}{M_0} = \frac{16}{3+\nu}, \quad \frac{Eh^3\delta_e}{a^2 M_0} = 3(1-\nu)\left(\frac{5+\nu}{3+\nu}\right)$$

in view of (4.43) and (4.44). The stress profile at this stage meets the yield locus at the corner B, specifying the elastic limit of the plate, as the applied pressure attains the value p_e.

For some value of p exceeding p_e, a part of the stress profile would coincide with side BC, the plate being rendered plastic within a radius c. The bending moments in the central plastic region are given by

$$M_\theta = M_0, \quad M_r = M_0 - \frac{1}{6}pr^2, \quad 0 \le r \le c,$$

in view of the equilibrium equation (4.6). In the elastic region ($c \le r \le a$), the bending moments are given by (4.39), the constants of integration A and B, and the relationship between p and c, being obtained from the boundary condition $M_r = 0$ at $r = a$, and the conditions of continuity of M_r and M_θ across $r = c$. It is easily shown that

$$\left.\begin{aligned} B &= \frac{c^2}{a^2}(1+3\nu)\frac{pc^2}{48}, \quad A = M_0 + (1+3\nu)\frac{pc^2}{24}, \\ \frac{pa^2}{6M_0} &= \left[1 + \frac{1+3\nu}{8}\left(1 - \frac{c^2}{a^2}\right)^2\right]^{-1}. \end{aligned}\right\} \tag{4.45}$$

The deflection in the elastic region is given by (4.40), where $C = 0$ in view of the boundary condition $w = 0$ at $r = a$, while A and B are those given for (4.45). In the plastic region of the elastic/plastic plate, the deflection w must be determined by the integration of (4.41), which becomes

$$\frac{Eh^3}{12}\frac{d^2w}{dr^2} = -(1-\nu)M_0 + \tfrac{1}{6}pr^2, \quad 0 \le r \le c.$$

The condition of continuity of the slope requires the value of $-(Eh^3/12)(dw/dr)$ at $r = c$ to be equal to $(1-\nu)cM_0 + pc^3/6$, in view of the second equation of (4.37), and the preceding expressions for M_r and M_θ. The integration of the above equation therefore gives

$$\left(\frac{Eh^3}{12}\right)\frac{dw}{dr} = -(1-\nu)M_0 r + \frac{pr^3}{18} - (1+3\nu)\frac{pc^3}{18}.$$

A straightforward integration of this differential equation results in the transverse deflection w, which can be expressed as

$$\left.\begin{aligned} \frac{Eh^3}{a^2}(\delta - w) &= \tfrac{2}{3}(1+3\nu)\frac{pc^3 r}{a^2} + \frac{r^2}{a^2}\left[6(1+\nu) - \frac{pr^2}{6}\right], 0 \leq r \leq c, \\ \frac{Eh^3\delta}{a^2} &= \left\{3(1-\nu)(5+\nu) + (1+\nu)(1+3\nu)\left(5 + 4\ln\frac{a}{c}\right)\frac{c^4}{a^4}\right\}\frac{pa^2}{16}, \end{aligned}\right\} \tag{4.46}$$

where δ is the central deflection obtained from the condition of continuity of w across $r = c$. Equations (4.40), (4.45), and (4.46) furnish the load–deflection relationship parametrically through the ratio c/a. The fully plastic state corresponds to $c/a = 1$, the load intensity and the central deflection at the incipient collapse being p_0 and δ_0, respectively, where

$$\frac{p_0 a^2}{M_0} = 6, \quad \frac{Eh^3\delta_0}{a^2 M_0} = \tfrac{3}{2}(5 + 2\nu + 3\nu^2).$$

The first result agrees with the rigid/plastic solution of the previous section. For a material with $\nu = \frac{1}{3}$, the load parameter pa^2/M_0 increases from 4.8 to 6.0, and the displacement parameter Eh^3w/M_0a^2 at the plate center increases from 3.2 to 9.0, over the elastic/plastic range. It may be noted that the collapse load in this case is attained for a finite deflection which is less than three times the deflection at the initial yielding. An analysis for the elastic/plastic deflection for a von Mises material has been given by Eason (1961). The transverse deflection of elastic/plastic solid circular plates has been investigated by Popov et al. (1967). Solutions based on the Hencky theory have been discussed by Sokolovsky (1969) and Mazumdar and Jain (1989).

4.2.3 Deflection of a Built-In Plate

An elastic/plastic analysis will now be carried out for a built-in circular plate uniformly loaded with a transverse pressure p which is gradually increased from zero. While the plate is completely elastic, the bending moments and the transverse deflection are obtained from (4.39) and (4.40), respectively, where B and C are zero. The boundary condition $dw/dr = 0$ at $r = 0$ requires $M_\theta = \nu M_r$ in view of the second equation of (4.37), giving $A = (1+\nu)pa^2/16$. The elastic solution therefore becomes

$$\left.\begin{aligned} M_r &= \frac{pa^2}{16}\left[(1+\nu)-(3+\nu)\frac{r^2}{a^2}\right],\ M_\theta = \frac{pa^2}{16}\left[(1+\nu)-(1+3\nu)\frac{r^2}{a^2}\right],\\ \frac{Eh^3w}{a^2} &= \frac{3}{16}(1-\nu^2)pa^2\left(1-\frac{r^2}{a^2}\right)^2. \end{aligned}\right\} \tag{4.47}$$

Both the moments change sign from positive to negative as the boundary is approached from the center of the plate. The stress profiles in the elastic range of bending are parallel straight lines given by the equation

$$(3+\nu)M_\theta - (1+3\nu)M_r = (1-\nu^2)(pa^2/8).$$

According to the Tresca criterion, plastic yielding begins at the built-in edge when the radial moment M attains the value $-M_0$. The load intensity p_e and the central deflection δ_e at this stage are given by

$$\frac{p_e a^2}{M_0} = 8, \quad \frac{Eh^3\delta_e}{a^2 M_0} = \frac{3}{2}\left(1-\nu^2\right).$$

The stress profile corresponding to the elastic limit meets the yield locus at a point on the side *DE*, as shown in Fig. 4.7b.

When p is increased beyond p_e, the plastic zone cannot immediately spread inward from the edge of the plate without violating the yield condition. The stress profile therefore involves a single point that moves on the yield locus along *ED* as the load increases. Since yielding is localized along the edge, the boundary condition $dw/dr = 0$ at $r = a$ must be replaced by the hinge condition $M_r = -M_0$ at $r = a$. The bending moment distribution in the entire plate therefore becomes

$$\left.\begin{aligned} M_r &= -M_0\frac{pa^2}{16}(3+\nu)\left(1-\frac{r^2}{a^2}\right),\\ M_\theta &= -M_0 + \frac{pa^2}{16}\left[(3+\nu)-(1+3\nu)\frac{r^2}{a^2}\right]. \end{aligned}\right\} \tag{4.48}$$

This is the same as that in a simply supported elastic plate with a superimposed all-round moment equal to $-M_0$. The expression for the deflection is easily shown to be

$$\frac{Eh^3w}{a^2} = (1-\nu)\left(1-\frac{r^2}{a^2}\right)\left\{-2M_0 + \frac{pa^2}{16}\left[(5+\nu+(1+\nu)\frac{r^2}{a^2}\right]\right\}. \tag{4.49}$$

The stress profile continues to be a straight line, the last profile of this type being the one that meets the hexagon at the corner *B*. The corresponding load intensity P_1, obtained by setting $M_r = M_\theta = M_0$ at $r = 0$ in (4.48), and the central deflection δ_1, obtained from (4.49) by setting $r = 0$, may be written as

$$\frac{p_1 a^2}{M_0} = \frac{32}{3+\nu}, \quad \frac{Eh^3\delta_1}{a^2 M_0} = 12\left(\frac{1-\nu}{3+\nu}\right).$$

If the load intensity is increased further, a part of the stress profile coincides with side BC of the yield hexagon. The plate then becomes plastic within some radius c, where the yield condition is $M_\theta = M_0$, giving the bending moment distribution

$$M_\theta = M_0, \quad M_r = M_0 - \frac{1}{6}pr^2, \quad 0 \le r \le c.$$

The bending moments in the elastic region ($c \le r \le a$) are given by (4.39), where A and B are now obtained from the conditions $M_\theta = M_0$ and $M_\theta - M_r = pc^2/6$ at $r = c$. The load intensity is subsequently obtained from the boundary condition $M_r = -M_0$ at $r = a$. The results may be collected as

$$\left.\begin{aligned} B &= \tfrac{1}{48}(1+3\nu)\frac{pc^4}{a^2}, \quad A = M_0 + (1+3\nu)\frac{pc^4}{24}, \\ \frac{pa^2}{6M_0} &= 2\left[1 + \frac{1+3\nu}{8}\left(1 - \frac{c^2}{a^2}\right)^2\right]^{-1}. \end{aligned}\right\} \tag{4.50}$$

The deflection in the elastic region ($c \le r \le a$) is obtained by setting $C = 0$ in (4.40) and using the values of A, B, and pa^2 from (4.50). In the plastic region ($0 \le r \le c$), the deflection is given by the same differential equation and boundary condition as those for a simply supported elastic/plastic plate. The expression for w in the plastic region of a clamped plate is therefore given by the first equation of (4.46), where the central deflection δ is now given by the modified expression

$$\frac{Eh^3\delta}{a^2} = \left\{(1-\nu)\left[3 + (1+3\nu)\frac{c^2}{a^2}\right] + (1+3\nu)\left[(4+\nu) + 2(1+\nu)\ln\frac{a}{c}\right]\frac{c^4}{a^4}\right\}\frac{pa^2}{8}, \tag{4.51}$$

in view of the continuity of w across $r = c$, and the fact that the value of w at $r = c$ is given by (4.40) with $C = 0$. The analysis remains valid until the stress point, which moves along the side ED, reaches the corner D of the yield hexagon, implying that $M_\theta = 0$ at $r = a$. By (4.39) and (4.50), the corresponding value of c/a is given by

$$\frac{c^4}{a^4} + \frac{2c^2}{3a^2} - \frac{5\nu - 1}{1+3\nu} = 0.$$

This is a quadratic in c^2/a^2, and the relevant root is $c^2/a^2 = \frac{1}{3}$ when $\nu = \frac{1}{3}$. The load intensity and central deflection at this stage become p_2 and δ_2, respectively, where

$$\frac{p_2 a^2}{M_0} = 10.8, \quad \frac{Eh^3\delta_2}{a^2 M_0} = 5.04 (\nu = \tfrac{1}{3}).$$

Since this load is only 4% lower than the collapse load, the continuation of the analysis is hardly necessary for practical purposes. However, the manner in which the elastic/plastic plate approaches the limiting state of collapse will be discussed below mainly because of its theoretical interest. Figure 4.8 shows the development of the deflected shape of the plate over the range $p \leq p_2$ of the loading process.

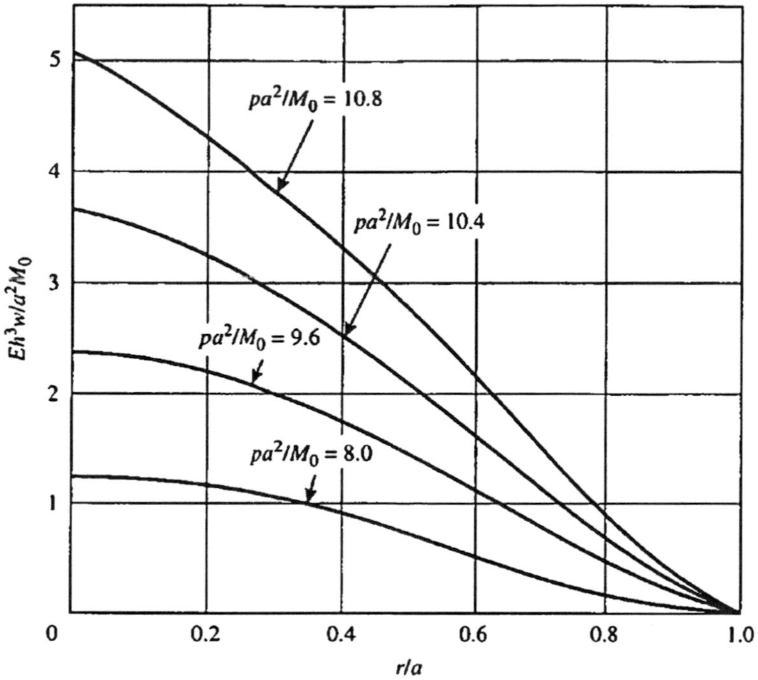

Fig. 4.8 Deformed shapes of a fully clamped circular plate uniformly loaded in elastic/plastic range ($\nu = 1/3$)

When p exceeds p_2, a second plastic zone spreads inward from the clamped edge, and a part of the stress profile then lies on the plastic regime *CD*. The clamped edge $r = a$ must remain at corner D in order that the discontinuity in slope is able to increase in the same sense. Since the yield condition in the outer plastic region is $M_\theta - M_r = M_0$, the equilibrium equation (4.6) and the boundary condition $M_r = -M_0$ at $r = a$ furnish the solution for the outer plastic region as

$$\left.\begin{aligned} M_r &= M_0\left(1 + \ln\left(\frac{a}{r}\right)\right) + \frac{pa^2}{4}\left(1 - \frac{r^2}{a^2}\right), \\ M_\theta &= M_0 \ln\frac{a}{r} + \frac{pa^2}{4}\left(1 - \frac{r^2}{a^2}\right), \end{aligned}\right\} \quad \rho \leq r \leq a, \tag{4.52}$$

where ρ denotes the inner radius of the plastic annulus. The bending moments in the elastic core ($c \leq r \leq \rho$) are still given by (4.39), where A and B are given by the first two equations of (4.50). The relationship between p, c, and ρ is obtained from the yield condition $M_\theta - M_r = M_0$ at $r = \rho$, and the condition of continuity of M_r across $r = \rho$. After some algebraic manipulation, the result can be put in the form

$$\left(1-\frac{c^2}{\rho^2}\right)\left(5+\frac{3c^2}{\rho^2}\right)+\left(1-\frac{c^4}{\rho^4}\right)\ln\left(\frac{a^2}{\rho^2}\right)=\frac{4}{1+3\nu}\left(5-\frac{3a^2}{\rho^2}+\ln\frac{a^2}{\rho^2}\right),$$
$$\frac{pa^2}{6M_0}=\left(4+\ln\frac{a^2}{\rho^2}\right)\left\{3-\frac{\rho^2}{a^2}\left[1-\frac{1+3\nu}{4}\left(1-\frac{c^2}{\rho^2}\right)^2\right]\right\}^{-1}. \tag{4.53}$$

For any given $\rho/a < 1$, the ratio c/ρ can be computed from the first equation of (4.53), and the ratio pa^2/M_0 would then follow the second equation. As the collapse load is approached, c/ρ tends to unity, and the limiting values of ρ/a and pa^2/M_0 are found to be 0.73 and 11.26, respectively, in agreement with the rigid/plastic analysis given in the preceding section.

The deflection of the plate for $p > p_2$ over the regions $c \le r \le \rho$ and $0 \le r \le c$ are given by (4.40) and the first equation of (4.46), respectively, where A and B are given by (4.50), while C is an arbitrary constant. In the plastic annulus $\rho \le r \le a$, the deflection is obtained by integrating (4.32), after substitution from (4.42), and using the boundary condition $w = 0$ at $r = a$ and the condition of the continuity

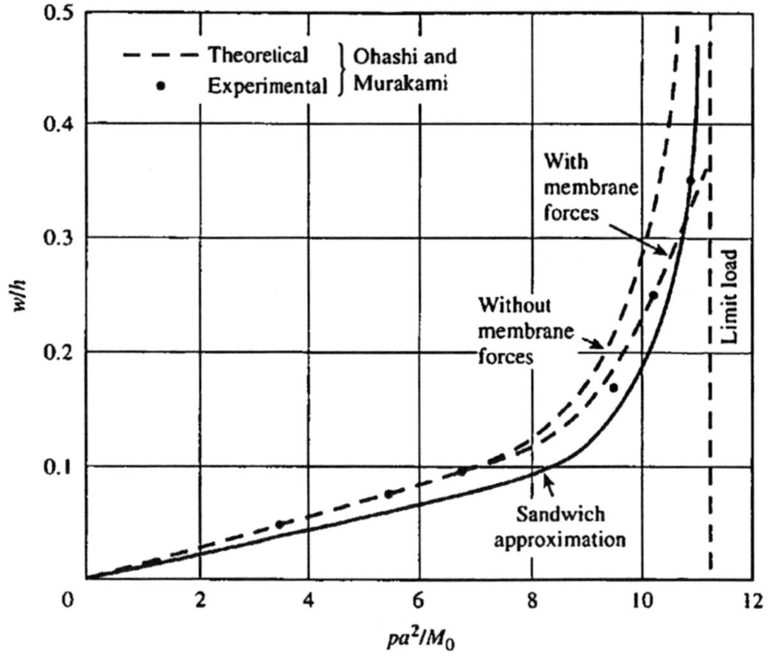

Fig. 4.9 Load–deflection curves for a fully clamped circular plate under uniformly distributed normal pressure (E/ν = 990, ν = 0.28, a/h = 12.5)

of dw/dr across $r = \rho$. The unknown constant C and the central deflection δ are subsequently determined from the condition of continuity of w across the interfaces $r = c$ and $r = \rho$. The deflected plate over the region $0 \leq r \leq \rho$ rapidly approaches the conical shape predicted by the rigid/plastic analysis.

As the deflection increases in the elastic/plastic range, the effect of membrane forces which are neglected in the preceding analysis become increasingly significant. An analysis for the elastic/plastic bending, including the membrane forces, has been presented by Ohashi and Murakami (1964) using Hencky's deformation theory of plasticity. The load–deflection curve obtained by these authors for the clamped plate is compared with those of the present analysis in Fig. 4.9, which includes the theoretical curve neglecting the membrane effect. The experimental points inserted in the figure indicate the importance of membrane forces as the deflection is increased in the plastic range.

4.3 Influence of Membrane Forces

In an elastic/plastic plate, as the applied load approaches the collapse value, the deflection increases fairly rapidly, and the membrane forces induced by the change in geometry soon become significant. The effect of the membrane forces is to enhance the load-carrying capacity of the plate, and the continued deflection of the plate requires the load to exceed the theoretical collapse value even when the rate of work-hardening is negligible. To establish the load–deflection relation in the presence of the membrane forces, the material will be assumed here as rigid/plastic with a constant uniaxial yield stress Y, obeying the Tresca criterion for yielding. The result may be subsequently modified in an approximate manner to include the elastic part of the deflection.

4.3.1 Simply Supported Circular Plates

We begin by considering the deflection of a uniformly loaded circular plate which is simply supported along its edge $r = a$, using rollers which can move radially during the bending. For sufficiently small deflections following the yield point, the plastic deformation of the plate involves a horizontal sheet of unstrained material situated at a height e above the original bottom face of the plate, Fig. 4.10a. On a diametral section of the plate, the hoop stress is a uniform compression Y above the unstrained plane CD and a uniform tension Y below this plane. The work done by the membrane forces will be zero if the associated bending moments are referred to the unstrained plane instead of the middle plane of the plate. At any radius r, where the vertical deflection is w, the circumferential bending moment per unit length about CD is easily shown to be

$$M'_\theta = 2M_0\left\{\left(1 - \frac{2e}{h}\right)\left(1 - \frac{2w}{h}\right) + 2\left(\frac{e^2 + w^2}{h^2}\right)\right\}, \tag{4.54}$$

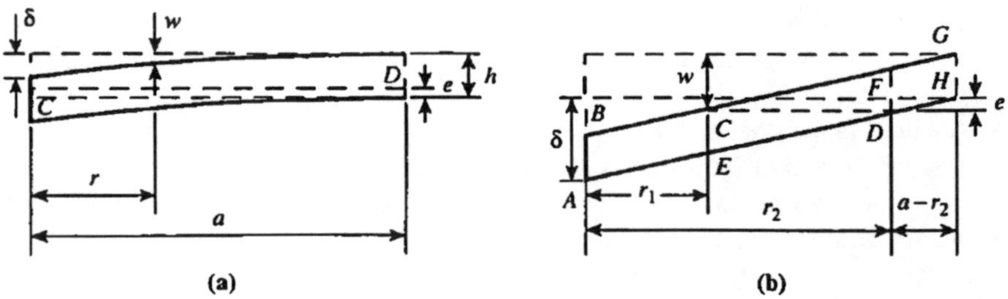

Fig. 4.10 Geometry of deformed circular plates including the influence of membrane force. (**a**) Smaller deflections and (**b**) larger deflections

where M_0 is the fully plastic moment equal to $Yh^2/4$. A prime is introduced to distinguish the considered bending moments from those referred to the middle plane. The analysis can be carried out with the help of the equilibrium equation

$$\frac{d}{dr}(rM'_r) = M'_\theta - \int_0^r pr\,dr = M'_\theta - \frac{1}{2}pr^2, \tag{4.55}$$

which can be integrated on substitution from (4.54), together with an assumed distribution of w. Using the boundary condition $M'_r = 0$ at $r = a$, where the radial membrane force also vanishes, the result may be written as

$$\frac{pa^2}{6M_0} = 2\left(1 - \frac{2e}{h}\right)\left(1 - \frac{\alpha\delta}{h}\right) + \left(\frac{4e^2}{h^2} + \frac{\beta\delta^2}{h^2}\right),$$

where δ is the deflection of the center of the plate, and α and β are dimensionless parameters defined as

$$\alpha = 2\int_0^1 \left(\frac{w}{\delta}\right) d\left(\frac{r}{a}\right), \quad \beta = 4\int_0^1 \left(\frac{w}{\delta}\right)^2 d\left(\frac{r}{a}\right). \tag{4.56}$$

The ratio e/h can be determined from the condition that the resultant membrane force across the diametral section must be zero for overall radial equilibrium. Thus

$$\int_0^a [h - 2(e + w)]dr = 0 \quad \text{or } e = (h - \alpha\delta)/2.$$

Substituting for e/h into the preceding expression for p, the load–deflection formula is obtained as

$$\frac{pa^2}{6M_0} = 3\left\{1 + (\beta - \alpha^2)\frac{\delta^2}{h^2}\right\}, \quad 0 \le \frac{\delta}{h} \le \frac{1}{2 - \alpha}. \tag{4.57}$$

The upper limit for δ/h in (4.57) follows from the condition $\delta \le h \le z$ required for the validity of (4.54). If the deflection surface is taken to be conical, as in the initial yield point state, we have $w/\delta = 1 - r/a$, which gives $\alpha = 1$ and $\beta = \frac{4}{3}$ in view of (4.56). The results in this particular case become

$$\frac{pa^2}{2M_0} = 1 + \frac{\delta^2}{3h^2}, \quad \frac{e}{h} = \frac{1}{2}\left(1 = \frac{\delta}{h}\right), \quad 0 \leq \frac{\delta}{h} \leq 1. \tag{4.58}$$

The method of analysis adopted here is essentially due to Calladine (1968). A different method of including the effect of membrane forces has been discussed earlier by Onat and Haythornthwaite (1956).

The analysis for higher deflections will now be carried out on the basis of the assumed conical deformation mode. Then for $\delta \geq h$, the unstrained plane CD is located at some distance e below the bottom face of the plate, as shown in Fig. 4.10b. The circumferential stress in the plate is purely tensile over the region $0 \leq r \leq r_1$ and purely compressive over the region $r_2 \leq r \leq a$. It follows from simple geometry that

$$\frac{r_1}{a} = 1 - \frac{e+h}{\delta}, \quad \frac{r_2}{a} = 1 - \frac{e}{\delta}.$$

Since the hoop stress is everywhere of magnitude Y, the resultant force across the diametral section will be zero if

$$r_1 = a - r_2 \text{ or } e = (\delta - h)/2.$$

The resultant moments about CD of the stresses acting over the areas $ABCE$ and $DFGH$ are equal to one another, the magnitude of these moments being readily obtained from geometrical considerations as

$$\int_0^{r_1} M'_\theta dr = Yhr_1\left(\frac{\delta - e}{2}\right) = \frac{1}{2}M_0 a\left(\frac{\delta}{h} - \frac{h}{\delta}\right) = \int_{r_2}^{a} M'_\theta dr.$$

The resultant moment about CD of the stresses acting over the remaining area $CEDF$ is similarly found to be

$$\int_{r_1}^{r_2} M'_\theta dr = \frac{1}{3}Yh^2(r_2 - r_1) = \frac{4}{3}M_0 a\left(\frac{h}{\delta}\right).$$

In view of the boundary condition $M'_r = 0$ at $r = a$, the integration of (4.55) between the limits $r = 0$ and $r = a$ furnishes the result

$$\frac{pa^2}{6M_0} = \int_1^1 \frac{M'_\theta}{M_0} d\left(\frac{r}{a}\right) = \frac{\delta}{h} + \frac{h}{3\delta}, \quad \frac{\delta}{h} \geq 1. \tag{4.59}$$

The ratio of the applied load to the yield point load for a simply supported plate based on the conical deformation mode coincides with that given by (4.58) and (4.59) even when the plate is only partially loaded (Kondo and Pian, 1981). In the special case of a concentrated load P acting at the center of the plate, the ratio

$P/2\pi M_0$ is given by the expressions for $pa^2/6M_0$ in (4.58) and (4.59). The theoretical load–deflection curve is compared with some experimental results on mild steel plates in Fig. 4.11. The initial part of the curve, based on the rigid/plastic model, is replaced by a pair of straight lines (shown broken), one of which is the elastic line and the other corresponds to $\delta = 2.5\delta_e$ at the yield point load $P = 2\pi M_0$.

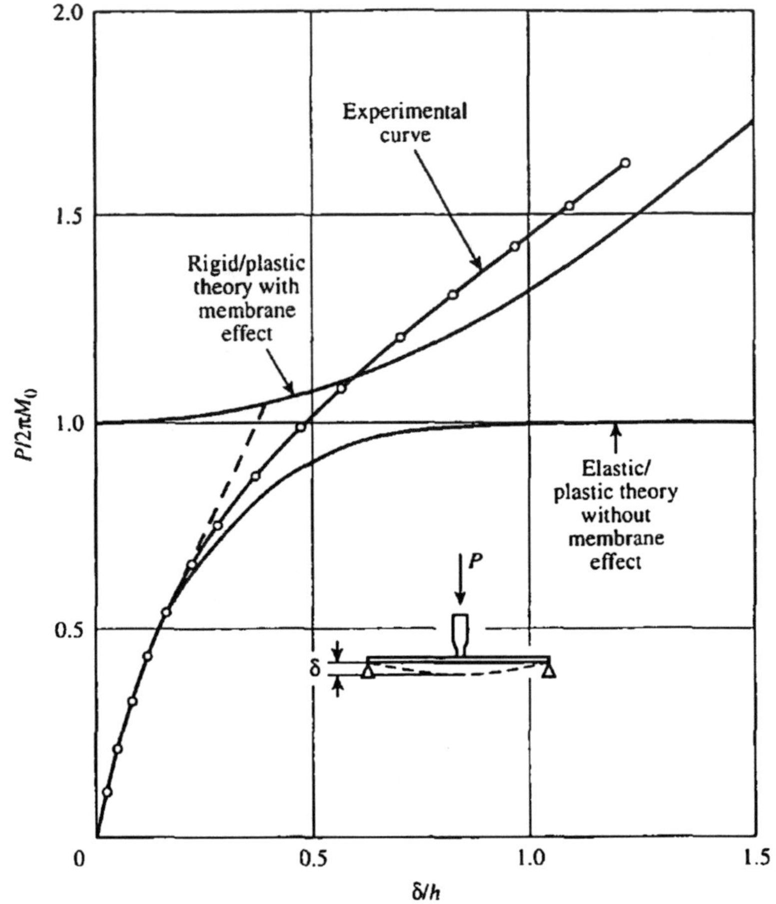

Fig. 4.11 Load–deflection curve for simply supported circular plate loaded at its center

4.3.2 Built-In Circular Plates

When the plate is clamped at its periphery $r = a$, there is a region near the edge where the magnitude of the hoop stress is less than Y, if the material yields according to the Tresca criterion. The analysis is considerably simplified, however, by adopting the square yield condition defined by $\sigma_r \pm Y$ and $\sigma_\theta = \pm Y$. The ratio of the current load to the initial collapse load, based on the square yield condition, may be regarded as a good approximation for both Tresca and von Mises materials.

The circumferential bending moment referred to the unstrained plane is then given by (4.54) throughout the plate even when the boundary is clamped. The integration of the equilibrium equation (4.55) therefore gives

$$\frac{pa^2}{6M_0} = 2\left(1 - \frac{2e}{h}\right)\left(1 - \frac{\alpha\delta}{h}\right) + \left(\frac{4e^2}{h^2} + \frac{\beta\delta^2}{h^2}\right) - \left(\frac{M_r'}{M_0}\right)_{r=a},$$

where α and β are given by (4.56). In view of the assumed yield condition, the existence of a hinge circle at the clamped edge requires $\sigma_r = -\sigma_\theta$ at $r = a$, which is equivalent to $M'_r = -M'_\theta$ at $r = a$. The preceding relation therefore becomes

$$\frac{pa^2}{6M_0} = 2 + 2\left(1 - \frac{2e}{h}\right)^2 - 2\left(1 - \frac{2e}{h}\right)\frac{\alpha\delta}{h} + \frac{\beta\delta^2}{2h^2}. \tag{4.60}$$

The position of the neutral plane, represented by the distance e, can be determined by minimizing q with respect to e for a given δ, the result being

$$2(h - 2e) = \alpha\delta \quad \text{or} \quad 2e = h - \alpha\delta/2.$$

Substituting in (4.60) and noting the fact that the initial value of p is $p_0 = 12M_0/a^2$ according to the square yield condition, the load–deflection formula for a clamped plate is obtained in the form

$$\frac{p}{p_0} = \left\{1 + (2\beta - \alpha^2)\frac{\delta^2}{4h^2}\right\}, \quad 0 \le \frac{\delta}{h} \le \frac{2}{4 - \alpha}. \tag{4.61}$$

The inequality arises from the fact that the configuration of Fig. 4.10a applies only over the range $\delta \le h - e$. If the conical deformation mode of the yield point state is assumed to persist during the continued loading, then $\alpha = 1$ and $\beta = \frac{4}{3}$. For a Tresca material, it is only necessary to set $p_0 = 11.26M_0/a^2$ in (4.61), and the corresponding results then become

$$\frac{pa^2}{2M_0} = 5.63\left(1 + \frac{5\delta^2}{12h^2}\right), \quad \frac{e}{h} = \frac{1}{2}\left(1 - \frac{\delta}{2h}\right), \quad 0 \le \frac{\delta}{h} \le \frac{2}{3}. \tag{4.62}$$

When the built-in plate is deformed by a concentrated load P acting at the center, the ratio $P/2\pi M_0$ is given by the expression in the first parenthesis of (4.62) over the specified range. A modified load–deflection formula for $\delta \ge h$, when the plate is fully clamped, has been given by Chakrabarty (1998).

The pressing of a simply supported circular plate into a hemispherical cavity using a matching punch has been investigated by Yu et al. (1985), who also discussed the spring back on unloading. As the deflection increases, the compressive hoop stresses near the periphery become large enough to cause wrinkling, and this has been examined by Yu and Stronge (1985) and Zhang and Yu (1991).

The elastic/plastic analysis for large deflections of circular plates based on the Hencky theory has been presented by Naghdi (1952), Ohashi and Murakami (1964, 1966), Ohashi and Kamiya (1967), Myszkowsky (1971), Sherbourne and Srivastava (1971), and Turvey (1979), among others. A numerical analysis for the elastic/plastic large deflection of plates, including the isotropic and kinematic hardening of the material, has been discussed by Tanaka (1972). The distribution of residual stresses on unloading the plate from a partially plastic state has been examined by Ohashi and Kawashima (1969). The elastic/plastic analysis for relatively thick plates has been considered by Oblak (1986).

4.4 Plastic Collapse of Noncircular Plates

4.4.1 General Considerations

The theory of plastic bending of noncircular plates can be developed on the same basic assumptions as those used for circular plates. We choose rectangular coordinates (x, y, z), where the x- and y-axes are taken in the middle plane of the plate, and the z-axis is normal to the plane in the downward sense. The nonzero components of the strain rate vary linearly with z according to the relations

$$\dot{\varepsilon}_x = z\dot{\kappa}_x, \quad \dot{\varepsilon}_y = z\dot{\kappa}_y, \quad \dot{\gamma}_{xy} = z\dot{\kappa}_{xy},$$

where $\dot{\kappa}_x$, $\dot{\kappa}_y$ are the rates of curvature, and $\dot{k}_{xy}$ is the rate of twist of the middle plane. In terms of the vertically downward velocity w of the middle plane, these quantities are given by

$$\dot{\kappa}_x = -\frac{\partial^2 w}{\partial w^2}, \quad \dot{\kappa}_y = -\frac{\partial^2 w}{\partial y^2}, \quad \dot{\kappa}_{xy} = -\frac{\partial^2 w}{\partial x \partial y}. \tag{4.63}$$

Let h denote the plate thickness which is assumed small compared to the other dimensions. The nonzero stress components are σ_x, σ_y, and τ_{xy} which give rise to the resultants

$$M_x = \int_{-h/2}^{h/2} \sigma_x z\, dz, \; M_y = \int_{-h/2}^{h/2} \sigma_y z\, dz, \; M_{xy} = \int_{-h/2}^{h/2} \tau_{xy} z\, dz.$$

Evidently, M_x, M_y are the bending moments per unit length across transverse sections perpendicular to the x and y-axes, respectively, while M_{xy} is the twisting moment per unit length acting on each of these sections. The moments are taken as positive when they act in the directions shown in Fig. 4.12a. The rate of dissipation of plastic energy through the plate thickness per unit area of the middle plane is

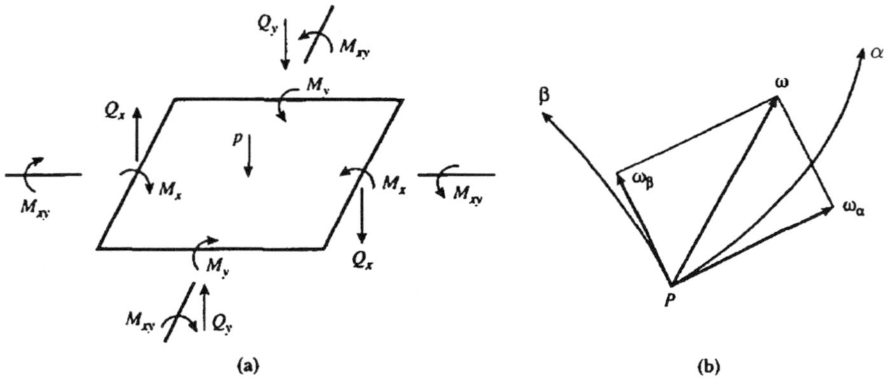

Fig. 4.12 Bending of noncircular plates. (**a**) Rectangular element with associated stress resultants and (**b**) local spin vector and its components

$$D = \int_{-h/2}^{h/2} (\sigma_x \dot{\varepsilon}_x + \sigma_y \dot{\varepsilon}_y + 2\tau_{xy}\dot{\gamma}_{xy})dz = M_x\dot{\kappa}_x + M_y\dot{\kappa}_y + 2M_{xy}\dot{\kappa}_{xy}. \qquad (4.64)$$

The generalized stresses and strain rates are evidently the bending and twisting moments and the corresponding rates of curvature and twist. The generalized strain rates can be expressed in terms of the components of spin (ω_x, ω_y), which are given by the gradients of the transverse velocity w. Thus

$$\begin{gathered} \omega_x = \frac{\partial w}{\partial y}, \quad \omega_y = \frac{\partial w}{\partial x}, \\ \dot{\kappa}_x = -\frac{\partial \omega_y}{\partial x}, \quad \dot{\kappa}_y = \frac{\partial \omega_x}{\partial y}, \quad \dot{\kappa}_{xy} = \frac{\partial \omega_x}{\partial x} = \frac{\partial \omega_y}{\partial y}. \end{gathered} \qquad (4.65)$$

Since the ratios of the components of the strain rate do not vary through the thickness, and the yield surface is convex, the state of stress must be constant on each side of the middle plane, across which the stresses become discontinuous. Consequently, the bending and twisting moments are directly proportional to the corresponding stress components. For $z > 0$, we have

$$M_x = \frac{1}{4}h^2\sigma_x, \quad M_x = \frac{1}{4}h^2\sigma_y, \quad M_{xy} = \frac{1}{4}h^2\tau_{xy}.$$

The yield condition satisfied by the dimensionless moments M_x/M_0, M_y/M_0, and M_{xy}/M_0, where $M_0 = Yh^2/4$, is exactly the same as that satisfied by the dimensionless stresses σ_x/Y, σ_y/Y, and τ_{xy}/Y. The von Mises yield criterion therefore gives

$$M_x^2 - M_xM_y + M_y^2 + 3M_{xy}^2 = M_0^2. \qquad (4.66)$$

The yield condition corresponding to the Tresca criterion depends on the sign of the quantity $M_xM_y - M_{xy}^2$ and may be written as

$$\left.\begin{aligned} &(M_x - M_y)^2 + 4M_{xy}^2 = M_0^2, \qquad 0 \le |\, M_x + M_y \,| \le M_0 \\ &M_{xy}^2 - M_xM_y + M_0 \,|\, M_x + M_y \,| = M_0^2, \qquad M_0 \le |\, M_x + M \,| \le 2M_0 \end{aligned}\right\}. \tag{4.67}$$

The flow rule associated with each of these yield conditions is obtained by considering the partial derivatives of the yield function with respect to the moments, replacing M_{xy}^2 by $\left(M_{xy}^2 + M_{yx}^2\right)/2$, and treating M_{xy} and M_{yx} as distinct components. Thus (4.66) gives

$$\dot{\kappa}_x = \lambda(2M_x - M_y),\ \dot{\kappa}_y = \lambda(2M_y - M_x),\ \dot{\kappa}_{xy} = 3\lambda M_{xy}, \tag{4.68}$$

where λ is a scalar factor which is necessarily positive for plastic flow. Similar expressions can be written for the flow rule associated with the Tresca yield functions (4.67), following the normality rule of plastic flow.

The equations of equilibrium can be derived by considering the forces and moments acting on a rectangular element having its sides parallel to the coordinate axes, as shown in Fig. 4.12a. The shearing forces Q_x and Q_y are developed by the application of transverse loads, and the conditions of equilibrium of the element are found to be

$$\left.\begin{aligned} &\frac{\partial M_x}{\partial x} + \frac{\partial M_{xy}}{\partial y} = Q_x, \qquad \frac{\partial M_{xy}}{\partial x} + \frac{\partial M_y}{\partial y} = Q_y, \\ &\frac{\partial Q_x}{\partial x} + \frac{\partial Q_y}{\partial y} = -p, \end{aligned}\right\} \tag{4.69}$$

where p is the local intensity of the transverse loading. The substitution for Q_x and Q_y from the first two equations of (4.69) into the third furnishes the differential equation

$$\frac{\partial^2 M_x}{\partial x^2} + 2\frac{\partial^2 M_{xy}}{\partial x \partial y} + \frac{\partial^2 M_y}{\partial y^2} = -p, \tag{4.70}$$

which involves only the bending and twisting moments. The bending of rigid/plastic plates based on Tresca's yield condition has been discussed in general terms by Hopkins (1957) and Schumann (1958), and the theory has been extended to an arbitrary yield condition by Hodge (1964a).

When the material yields according to the von Mises criterion, the dissipation function D is uniquely defined in terms of the curvature rates of the middle surface. Indeed, the substitution for the curvature rates from (4.68) into (4.64) gives

$$D = \lambda[M_x(2M_x - M_y) + M_y(2M_y - M_x) + 6M_{xy}^2] = 2\lambda M_0^2$$

in view of the yield criterion (4.66). On the other hand, the substitution for the bending moments in (4.64), after expressing them in terms of the curvature rates using (4.68), results in the alternative form

$$D = \frac{1}{3\lambda}[\dot{\kappa}_x(2\dot{\kappa}_x + \dot{\kappa}_y) + \dot{\kappa}_y(2\dot{\kappa}_y + \dot{\kappa}_x) + 2\dot{\kappa}_{xy}^2].$$

The elimination of λ between the two preceding expressions for D furnishes the dissipation function entirely in terms of the curvature rates, the result being

$$D = \frac{2M_0}{\sqrt{3}}\{\dot{\kappa}_x^2 + \dot{\kappa}_x\dot{\kappa}_y + \dot{\kappa}_y^2 + \dot{\kappa}_{xy}^2\}^{1/2} = M_0\dot{\kappa}_0,$$

where $\dot{\kappa}_o$ is an effective curvature rate at the yield point. Substituting from (4.63) into the above expression, we finally obtain the dissipation function in the form

$$D = \frac{2M_0}{\sqrt{3}}\left\{\left(\frac{\partial^2 w}{\partial x^2}\right)^2 + \frac{\partial^2 w}{\partial x^2}\frac{\partial^2 w}{\partial y^2} + \left(\frac{\partial^2 w}{\partial y^2}\right)^2 + \left(\frac{\partial^2 w}{\partial x \partial y}\right)^2\right\}^{1/2}. \qquad (4.71)$$

A curve across which the normal component of the velocity gradient is discontinuous is called a hinge line. The rate of energy dissipation per unit length of the hinge line for a von Mises material is equal to $2M_0\dot{\psi}/\sqrt{3}$, where

$$\dot{\psi} = \left\{\left(\frac{\partial w}{\partial x}\right)^2 + \left(\frac{\partial w}{\partial y}\right)^2\right\}^{1/2},$$

representing the magnitude of the normal discontinuity in the velocity slope. The existence of hinge lines plays an important part in the estimation of upper bounds on the collapse load in the limit analysis of plates.

4.4.2 Uniformly Loaded Rectangular Plates

A rectangular plate carrying a uniformly distributed load of intensity p is simply supported along its edges of lengths $2a$ and $2b$, where $a \leq b$. To obtain an upper bound on the collapse load, we assume a deformation mode involving five straight hinges as shown in Fig. 4.13a. The triangular and quadrilateral areas separated by the plastic hinges instantaneously rotate as rigid bodies with angular velocities ω_1 and ω_2 respectively, where $\omega_2 = \omega_1 \tan\phi$. Each of the hinges emanating from the corners is of length $a \sec\phi$ and undergoes a rotation at the rate $\sqrt{\omega_1^2 + \omega_2^2} = \omega_1 \sec\phi$. The central hinge, having a length $2(b - a \tan \phi)$, suffers a rate of rotation equal to $2\omega_2$. For a Tresca material, the bending moment at the yield hinge is M_0, giving the rate of internal energy dissipation as

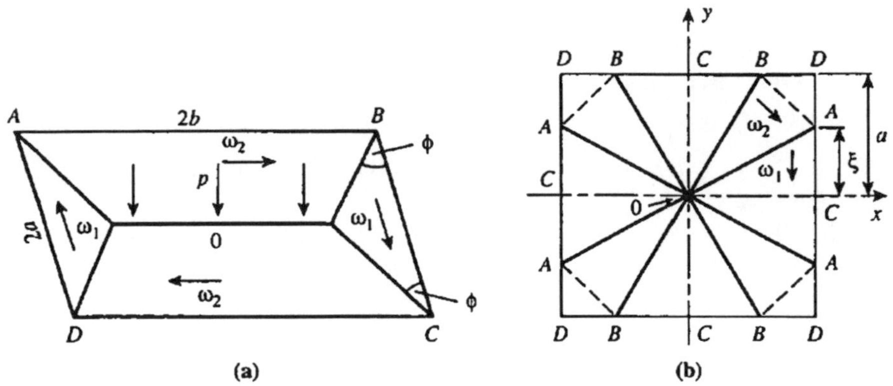

Fig. 4.13 Upper bound modes of collapse for simply supported plates. (**a**) Pyramidal modes for a rectangular plate and (**b**) lifting-up modes for a square plate

$$U = 4M_0(a \sec\phi)\omega_1 \sec\phi + 4M_0(b - a\tan\phi)\omega_1 \tan\phi.$$

The rate of external work done, denoted by W, is p times the volume enclosed by the deflected plate and its rectangular base when the depth of the central hinge is a ω_2. Thus

$$W = p\left(\tfrac{4}{3}a^2 \tan\phi\right) a\omega_2 + 2ap\,(b - a\tan\phi)\,a\omega_2.$$

Equating the rates of external and internal energy ($U = W$) at the incipient collapse, an upper bound on the collapse pressure is obtained as

$$\frac{pa^2}{6M_0} = \frac{1 + (a/b)\cot\phi}{3 - (a/b)\tan\phi}$$

giving p as a function of ϕ. The right-hand side of the above equation has a minimum value when

$$\tan\phi = \sqrt{3 + \frac{a^2}{b^2}} - \frac{a}{b}.$$

Using this value of tan ϕ in the preceding expression for $pa^2/6M_0$, the best upper bound for the assumed mode of deformation is obtained as

$$\frac{pa^2}{6M_0} = \left\{\sqrt{3 + \frac{a^2}{b^2}} - \frac{a}{b}\right\}^{-2}. \tag{4.72}$$

For a square plate ($b = a$), the right-hand side of the above equation reduces to unity. In the case of a von Mises material, it is only necessary to multiply the right-hand side of (4.72) by the factor $2/\sqrt{3}$ to obtain the upper bound.

A lower bound solution can be found by assuming a suitable distribution of bending and twisting moments throughout the plate. Considering a parabolic distribution of M_x and My with respect to x and y, respectively, the equilibrium equation (4.70) can be satisfied by taking

$$M_x = A\left(1 - \frac{x^2}{a^2}\right), \quad M_y = A\left(1 - \frac{y^2}{b^2}\right), \quad M_{xy} = -C\frac{xy}{ab},$$

where A and C are constants, the origin of coordinates being taken at the centroid O. The boundary conditions $M_x = 0$ along $x = \pm a$ and $M_y = 0$ along $y = \pm b$ are identically satisfied, and the relationship between A, C, and p is found to be

$$2A\left(1 + \frac{a^2}{b^2}\right) + 2C\frac{b}{a} = pa^2.$$

The assumed moment distribution will be statically admissible if the yield condition is nowhere violated. In view of the nature of this distribution, it is sufficient to consider the points (o, o), $(o, \pm b)$, $(\pm a, o)$ and $(\pm a, \pm b)$, which will simultaneously reach the yield limit according to the Tresca condition if $A = 2C = M_0$. The preceding equation then gives the lower bound solution as

$$\frac{pa^2}{2M_0} = 1 + \frac{a}{2b} + \frac{a^2}{b^2}. \tag{4.73}$$

For a square plate, the collapse load is therefore bounded by the inequalities $5Mq < pa^2 < 6Mq$. When the von Mises yield condition is employed, the second term on the right-hand side of (4.73) must be replaced by $a/\sqrt{3}b$ to get the modified lower bound. An improved lower bound for the Tresca material, based on the assumption $M_x = M_y$ everywhere in the plate, has been given by Shull and Hu (1963), and further discussed by Nielsen (1999).

If the simply supported plate is free to lift up at the incipient collapse, a better upper bound for a square plate can be found by using the deformation mode indicated in Fig. 4.13b. The regions AOA and $AOBD$ rotate with angular velocities ω_1 and ω_2 about the lines AA and AB, respectively, where $\sqrt{2}a\varpi_1 = (a + \xi)\,\varpi_2$ by the condition of continuity of the rate of deflection at O. Each of the eight plastic hinges is of length $\sqrt{a^2 + \xi^2}$ and involves a rate of change of angle, $\dot{\psi}$, which is the magnitude of the angular velocity of one side of the hinge line relative to the other. It follows from the vector triangle formed by the angular velocities that

$$\dot{\psi} = \left\{\left(\omega_1 - \frac{\omega_2}{\sqrt{2}}\right)^2 + \left(\frac{\omega_2}{\sqrt{2}}\right)^2\right\}^{1/2} = \frac{\omega_2}{\sqrt{2}}\left(1 + \frac{\xi^2}{a^2}\right)^{1/2}.$$

The rate of dissipation of internal energy along the plastic hinges is

$$D = 8M_0\dot{\psi}\sqrt{a^2+\xi^2} = 8M_0\left(\frac{a^2+\xi^2}{a+\xi}\right)\omega_1.$$

To obtain the rate of external work done, we write the velocity distributions in regions *AOA* and *AOBD* as

$$w_1 = (a-x)\omega_1, \quad w_2 = \left(1 - \frac{x+y}{a+\xi}\right)a\omega_1, \tag{4.74}$$

respectively. The above expression for w_2 follows from the fact that the perpendicular distance from *AB* of a typical point (x, y) is equal to $(a+\xi-x-y)/\sqrt{2}$ by simple geometry, the equation to the line AB being given by this expression set to zero. In view of the eight-fold symmetry, the rate of external work done is given by

$$W = 8p\int_0^a\left(\int_0^{y_0} w_1 dy + \int_{y_0}^{x} w_2 dy\right)dx$$

where $y_0 = \xi x/a$. Substituting from (4.74) and integrating, we obtain

$$W = \frac{4}{3}pa^2\xi\left(\frac{3a-\xi}{a+\xi}\right)\varpi_1$$

The upper bound on the collapse load, obtained by setting $D = W$, is therefore given by the expression (Hodge, 1981)

$$\frac{pa^2}{6M_0} = \frac{a^2+\xi^2}{\xi\,(3a-\xi)}. \tag{4.75}$$

The upper bound has a minimum value when $dp/d\xi = 0$, which corresponds to the best upper bound, and the result is the quadratic $3\xi^2 + 2a\xi - 3a^2 = 0$, having the solution

$$\frac{\xi}{a} = \frac{\sqrt{10}-1}{3} \approx 0.721, \quad \frac{pa^2}{6M_0} = \frac{2}{\sqrt{10}-1} \approx 0.925.$$

Consequently, the collapse load is now bounded by the inequalities $5M_0 < pa^2 < 5.55\ M_0$. Upper bound solutions for simply supported polygonal plates can be found in a similar manner.

In the approximate estimation of limit loads, it is often convenient to assume that the yield locus in the principal moment plane is a square whose sides are of length equal to M_0. For a simply supported rectangular plate, the upper bound solution holds equally well to the square yield condition, but the lower bound solution needs to be modified. It is easily shown that the second term on the right-hand side of (4.73) is multiplied by a factor of 2 when the square yield condition is used. The upper and lower bounds then become remarkably close to each other and are actually coincident when $a = b$. Bounds on the collapse load for a rectangular plate with different edge conditions have been given by Manolakos and Mamalis (1986). Upper bound solutions for several plate problems based on the square yield condition have been given by Sawczuk and Jaeger (1963), Johnson and Mellor (1983), and Skrzypek and Hetnarski (1993). These solutions are closely related to the yield line theory (Section 4.6), originally introduced by Johansen (1943) for the analysis of reinforced concrete slabs

4.4.3 Finite Element Analysis for Plate Bending

In the finite element method, the plate is considered as made up of a number of discrete elements of suitable size and shape. Within each element, either a velocity distribution or a bending moment distribution is assumed in terms of a finite number of parameters, which must satisfy the necessary continuity conditions. An upper or lower bound solution is then obtained in terms of the free parameters, which must be determined so as to minimize or maximize the collapse load. Consider first the kinematic approach, which is the most usual one with finite elements. The plate is subdivided into m triangular elements involving (n + 1) nodes, which are designated as the vertices and midpoints of sides of all the triangles. The normal velocity at any point (x, y) in a typical triangular element j is expressed in the polynomial form

$$w = A\frac{x^2}{a^2} + B\frac{xy}{ab} + C\frac{y^2}{b^2} + D\frac{x}{a} + e\frac{Y}{B} + f, \tag{4.76}$$

where a and b are suitable linear dimensions of the element. The six constants A and F are uniquely determined in terms of the six nodal velocities of the triangle. If, for example, the element j is a right-angled triangle as shown in Fig. 4.14a, and the associated rectangular axes are as indicated in the figure, then the constants in (4.76) are given by

$$\left.\begin{aligned} 2A &= w_1 - 2w_2 + w_3,\ B = w_1 - w_2 + w_4 - w_6, \\ 2C &= w_1 - 2w_6 + w_5,\ F = w_1, \\ 2D &= 4w_2 - 3w_1 + w_3,\ 2E = 4w_6 - 3w_1 - w_5. \end{aligned}\right\} \tag{4.77}$$

Since the velocity distribution is completely defined by the nodal velocities, w is necessarily continuous across each side of the element. However, the normal derivative $\partial w/\partial n$ can be discontinuous, and such a discontinuity is indeed permissible across the element edges which are potential hinge lines (Hodge, 1981).

Considering the hinge line represented by the side common to the two adjacent elements J and i in Fig. 4.14a, the magnitude of the velocity gradient discontinuity across this line may be written as

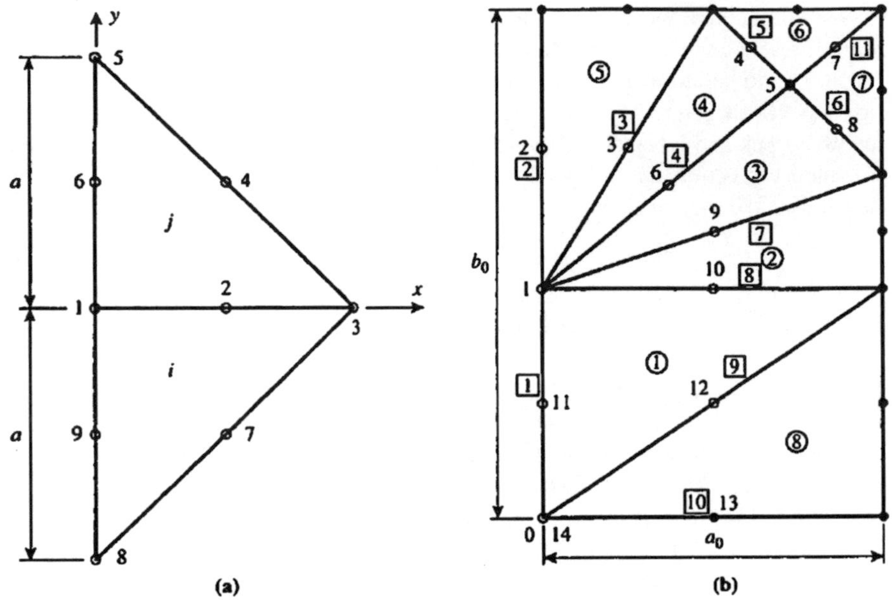

Fig. 4.14 Finite elements analysis of a rectangular plate. (**a**) Typical pair of elements having a common side and (**b**) subdivision of a plate quadrant

$$\dot{\psi} = \left|\Delta\left(\frac{\partial w}{\partial x}\right)\right| = \frac{1}{b}\left|\Delta\left(E + B\frac{x}{a}\right)\right| = \frac{2}{b}\left|\alpha + \beta\frac{x}{a}\right|,$$

where the prefix Δ denotes the jump in the quantity that follows. Substituting from (4.77) for triangle j and similar expressions for triangle i, it is easily shown that

$$\left.\begin{aligned} \alpha &= 2(w_6 + w_9) - 3w_1 - \frac{1}{2}(w_5 + w_8), \\ \beta &= 2(w_1 - w_2) + (w_4 - w_7) - (w_6 + w_9). \end{aligned}\right\} \tag{4.78}$$

For a von Mises material, the bending moment acting across the hinge line is of magnitude $2M_0/\sqrt{3}$, and consequently, the rate of energy dissipation along the hinge line is

$$H_{ij} = \frac{4M_0 a}{\sqrt{3}b}\int_0^{2a}\left|\alpha + \beta\frac{x}{a}\right| dx = \frac{4M_0 a}{\sqrt{3}b}|\alpha|\int_0^2\left|1 + \frac{\beta}{\alpha}\xi\right| d\xi,$$

where $\xi = x/a$. The quantity $1 + (\beta/\alpha)\xi$ is positive throughout the hinge line when $\beta/\alpha \geq -\frac{1}{2}$, but only over a part of this line when $\beta/\alpha \leq -\frac{1}{2}$. Using this fact, the energy rate for the considered hinge action is found as

$$H_{ij} = \frac{4M_0}{\sqrt{3}}\left(\frac{a}{b}\right)\begin{cases} |\alpha|(1 + \beta/\alpha), & \beta/\alpha \geq -\frac{1}{2}, \\ -|\alpha|[1 + (\beta/\alpha) + (\alpha/2\beta)] & \beta/\alpha \leq -\frac{1}{2}. \end{cases} \tag{4.79}$$

The rate of internal energy dissipation in each triangular element is a constant for the assumed velocity field, as may be seen from expression (4.71) for the dissipation function D. The total internal energy rate over triangle j is

$$E_j = \iint D\,dx\,dy = \left(\frac{8M_0}{\sqrt{3}}\right)\left\{\left(\frac{b}{a}\right)^2 A^2 + \left(\frac{a}{b}\right)^2 C^2 + AC + \frac{1}{4}B^2\right\}^{1/2}$$

in view of (4.71) and (4.76), the quantities A, B, and C being given by (4.77) in terms of the nodal velocities. It may be noted that the quadratic expression (4.76) represents the lowest-order velocity field that accounts for an internal energy dissipation other than that along the hinge lines.

Each element edge coinciding with the boundary of the plate, which is assumed to be polygonal, is considered as free, simply supported, or clamped along its entire length. The simply supported and clamped edge conditions require the associated nodal velocities to be zero. A clamped edge of the plate may be a hinge line with a nonzero normal derivative of w. If the vertical side of triangle j in Fig. 4.14a forms a part of the clamped edge, then the rate of energy dissipation along this hinge line is given by (4.79), except that $w_1 = w_5 = w_6 = 0$ in the corresponding expressions for α and β, which therefore become

$$\alpha = 2w_2 - \tfrac{1}{2}w_1, \quad \beta = w_4 - w_2.$$

The energy rate for the hinge action along the entire clamped edge of the plate is the sum of the contributions from all the associated triangular elements.

Let $p(x, y)$ denote the distribution of normal pressure acting on the plate, the intensity of the pressure at a selected point being p_0. Then the rate of external work done on the assumed velocity field (4.76) for triangle j is

$$W_j = p_0 ab \iint (A\xi^2 + B\xi\eta + C\eta^2 + D\xi + E\eta + F)(p/p_0)d\xi\, dn,$$

where $\xi = x/a$ and $n = y/b$, the double integral being extended over the whole area of the triangle. The upper bound on the pressure p_0 for plastic collapse in terms of the unknown nodal velocities is obtained by equating the external and internal energy rates for the entire plate. Thus

$$\sum W_j = \sum E_j + \sum H_{ij},$$

where the last term includes the contribution from the boundary hinges. Since the velocity field is indeterminate at collapse, the solution can be expressed in terms of normalized velocities, defined by selecting a suitable positive nodal velocity by which the remaining nodal velocities are divided. If there are $n + 1$ nonzero nodal velocities, the problem of finding the best upper bound becomes one of the unconstrained minimization in n variables. The minimization process is most conveniently carried out by the well-known simplex method which does not involve derivatives of the objective function. Since the simplex method decreases in effectiveness as the number of variables increases, relatively small elements should be used over the plastic region, but larger elements could be used where the material is expected to be non-plastic.

Figure 4.14b shows one quadrant of a simply supported rectangular plate which is loaded by a uniformly distributed normal pressure of intensity p_0, the sides of the rectangle being denoted by 2a$_0$ and 2a$_0$ *and* $2b_0$, with $a_0 \leq b_0$. The quadrant is divided into eight triangular elements involving 14 nodes where the transverse velocity is nonzero, and 11 sides representing possible hinge lines. The boundary nodes with prescribed zero velocity are not numbered, and the nonzero nodal velocities are normalized with respect to the central velocity w_{14}. The best upper bound, obtained by Hodge and Belytschko (1968) who also considered several other loading conditions, is shown as the upper solid curve in Fig. 4.15. The broken curves in the figure represent the upper and lower bound solutions expressed by (4.72) and (4.73), respectively.

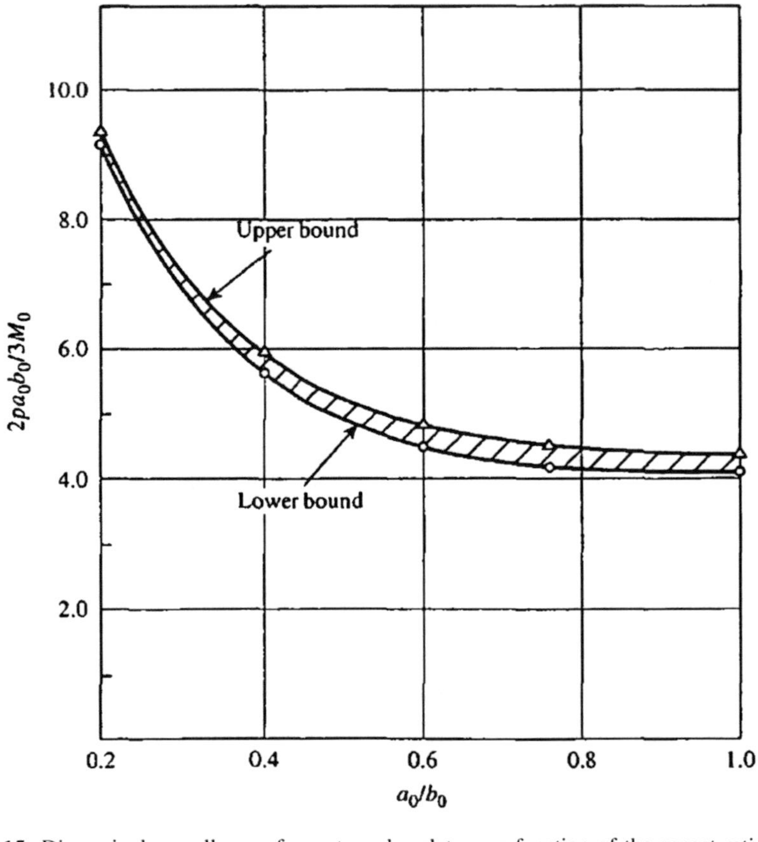

Fig. 4.15 Dimensionless collapse of a rectangular plate as a function of the aspect ratio (after P.G. Hodge 1981)

A lower bound solution based on the finite element procedure is much more difficult than the upper bound solution. In the static approach, the plate is again divided into a number of triangular regions, in each of which the components of the bending moment are assumed to be quadratic in x and y. Introducing dimensionless bending moments $m_x = M_x/M_0$, $m_y = M_y/M_0$, and $m_{xy} = M_{xy}/M_0$, we write

$$\left.\begin{aligned} m_x &= A_1\xi^2 + B_1\xi\eta + C_1\eta^2 + D_1\xi + E_1\eta + F_1, \\ m_y &= A_2\xi^2 + B_2\xi\eta + C_2\eta^2 + D_2\xi + E_2\eta + F_2, \\ m_{xy} &= A_3\xi^2 + B_2\xi\eta + C_3\eta^2 + D_3\xi + E_3\eta + F_3. \end{aligned}\right\} \tag{4.80}$$

The moment field (4.80) will be compatible with the equilibrium equation (4.70) if the transverse pressure p is constant over each element. Denoting by p_j the intensity of the constant pressure acting on a typical triangular element j, we get

$$p_j = -\frac{2M_0}{ab}\left(A_1\frac{b}{a} + C_2\frac{a}{b} + B_3\right) \tag{4.81}$$

If the plate is made up of m triangular elements, then (81) furnishes m equations, one of which may be taken as the definition of the collapse preserve, while treating the remaining equations as constraints. Hence there are $m - 1$ constraints arising from the equilibrium consideration of the individual elements.

Across an edge between two adjacent elements, the normal bending moment and the twisting moment must be continuous, while at each internal vertex, the resultant twisting moment must be zero. If there are e internal edges and q internal vertices, then there will be $5e + q$ additional constraints due to the necessary continuity restrictions (Hodge, 1981). Since the number of static boundary conditions is one for a simply supported edge, two for a free edge, and none for a clamped edge, there will be $3s + 5f$ constraints along the boundary of the plate involving s simply supported edges and f free edges of the triangular elements. Thus, the total number of constraints which apply to the $18m$ constants defining the moment field for the m triangular elements is

$$\mu = (m-1) + 5e + q + 3s + 5f.$$

The number of degrees of freedom therefore becomes $18m - \mu$. If these coefficients are taken as the components of a vector c, and the entire set of $18m$ coefficients is represented by a vector a, it is then possible to form a matrix L such that

$$c = La. \tag{4.82}$$

In the case of the quarter plate shown in Fig. 4.14b, where the boundary edges are simply supported, we have $m = 8$, $e = 11$, $q = 3$, $s = 5$, and $f = 0$, giving $\mu = 80$, which leaves us with $18(8) - 80 = 64$ degrees of freedom. These 64 coefficients of the moment field may be represented by the vector c, which is linearly dependent on the vector a defining all the 144 coefficients of the field.

The yield condition provides a number of inequality constraints necessary for establishing the lower bound. Since the yield condition is quadratic in the bending moments, which are linear in the coefficients of c and quadratic in ξ and η the yield inequality may be written as

$$\phi(c_i, \xi, \eta) \leq 0,$$

where the function ϕ is quadratic in the component c_i and quartic in the variables ξ and η. In the numerical analysis, the above inequality must be satisfied at a finite number of points without violating it at other points of the field. The analysis has been carried out by Hodge and Belytschko (1968), who found that it was necessary to satisfy the yield inequality not only at a number of fixed points but also at certain other points of relative maxima. The result of their analysis for a simply supported rectangular plate under a uniformly distributed load is shown as the lower solid curve in Fig. 4.15. The application of linear programming to the limit analysis of plates has been discussed by Koopman and Lance (1965) and Hodge (1981).

4.5 Plane Strain Analogy for Plate Bending

4.5.1 The Use of Square Yield Condition

At each point of the middle plane of the plate, there is a pair of orthogonal directions which are associated with the principal bending moments denoted by M_α and M_β. In terms of these principal moments, let the yield locus be represented by a square as shown in Fig. 4.16a, where M_0 denotes the fully plastic moment of the cross-section. The corners A and C of the yield locus, which correspond to $M_\alpha = M_\beta$ and is equivalent to $M_x = M_y$ and $M_{xy} = 0$, cannot represent a plastic state in any finite region of a transversely loaded plate. Consequently, only two types of plastic state, characterized by one of the remaining corners and one of the sides of the square, can arise in physical situations. The limit analysis of plates based on the square yield condition has been discussed by Mansfield (1957), Massonnet (1967), Sawczuk and Hodge (1968), Collins (1971, 1973), and Sawczuk (1989), among others.

There is an interesting analogy between the problem of plate bending and that of plane plastic strain when the state of stress in the plate corresponds to the corner B of the yield locus. The plane strain analogy was first recognized by Johnson (1969), and subsequently established by Collins (1971). Since $M_\alpha = -M_0$ and $M_\beta = M_0$ for the considered plastic regime, the rectangular components of the moment are

$$M_x = -M_0 \cos 2\phi, \ M_y = M_0 \cos 2\phi, \ M_{xy} = -M_0 \sin 2\phi, \qquad (4.83)$$

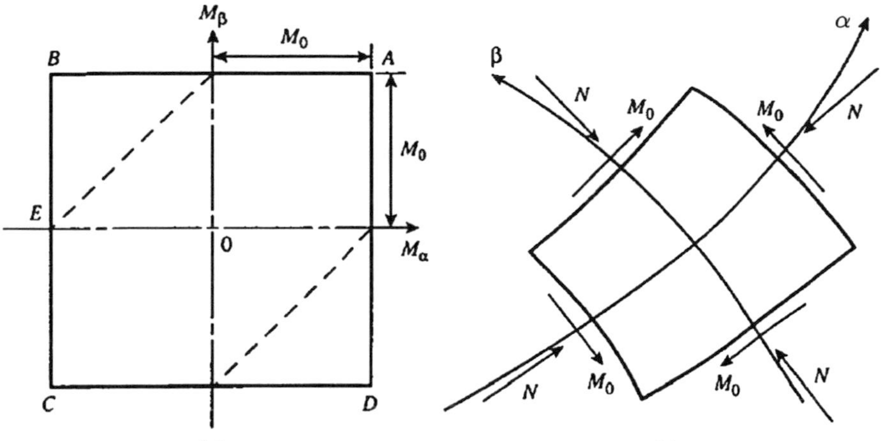

Fig. 4.16 The square yield condition for plate bending. (**a**) Simplified yield locus and (**b**) curvilinear element formed by neighboring characteristics

where ϕ denotes the counterclockwise angle made by the α-direction with the x-axis. In the special case of $p = 0$, the last equation of (4.69) is identically satisfied by introducing a stress function N, such that

$$Q_x = -\frac{\partial N}{\partial y}, \quad Q_y = \frac{\partial N}{\partial x}. \tag{4.84}$$

The change in the value of N around a closed curve Γ containing a concentrated force P is given by

$$[N] = \oint \frac{\partial N}{\partial s} ds = \oint Q_n ds = P, \tag{4.85}$$

where s is measured along Γ, and n refers to the direction of its inward normal. Evidently, N is not a single-valued function of x and y. In view of (4.83) and (4.84), the first two equations of (4.69) become

$$\frac{\partial N}{\partial x} + 2M_0 \cos 2\phi \frac{\partial \phi}{\partial x} + 2M_0 \sin 2\phi \frac{\partial \phi}{\partial y} = 0,$$
$$\frac{\partial N}{\partial y} + 2M_0 \sin 2\phi \frac{\partial \phi}{\partial x} - 2M_0 \cos 2\phi \frac{\partial \phi}{\partial y} = 0.$$

These equations are hyperbolic with the α- and β-curves as the characteristics. Setting $\phi = 0$ in the preceding equations, and integrating, we obtain the relations along the characteristics as

$$\left.\begin{aligned} N + 2M_0\phi &= \text{constant along an } \alpha - \text{line}, \\ N - 2M_0\phi &= \text{constant along an } \beta - \text{line}. \end{aligned}\right\} \tag{4.86}$$

These relations are analogous to the well-known Hencky relations in plane plastic strain, where N corresponds to the hydrostatic pressure in the element and M_0 to the shear yield stress of the material. The stress resultants acting across the faces of a curvilinear element bounded by the neighboring α and β lines shown in Fig. 4.16b consist of uniform bending and twisting moments of magnitudes M_0 and N, respectively.

In a plastically deforming element, let ω_α and ω_β denote the components of the spin vector with respect to the α and β directions, taken as a right-handed pair of curvilinear axes, as shown in Fig. 4.12b. The rectangular components of the spin are related to the (α, β)-components as

$$\begin{aligned} \omega_x &= \omega_\alpha \cos\phi - \omega_\beta \sin\phi \\ \omega_y &= \omega_\alpha \sin\phi + \omega_\beta \cos\phi. \end{aligned} \tag{4.87}$$

For an isotropic material, the rate of twist associated with the α-and β-axes must vanish, irrespective of the yield condition. If the x- and y-axes are taken along the tangents to the local α and β lines, respectively, then the condition $\dot{\kappa}_{xy} = 0$ gives $\partial\omega_x/\partial x = \partial\omega_y/\partial y = 0$ when $\phi = 0$. Substituting for ω_x and ω_y from (4.87), and setting $\phi = 0$ after taking the partial derivatives, we get

$$\frac{\partial\omega_\alpha}{\partial x} - \omega_\beta\frac{\partial\phi}{\partial x} = 0, \quad \frac{\partial\omega_\beta}{\partial y} + \omega_\alpha\frac{\partial\phi}{\partial y} = 0.$$

These equations are equivalent to a pair of differential relations holding along the characteristics, the result being

$$\left.\begin{aligned} d\omega_\alpha - \omega_\beta d\phi &= 0 \text{ along an } \alpha - \text{line}, \\ d\omega_\beta + \omega_\alpha d\phi &= 0 \text{ along an } \beta - \text{line}, \end{aligned}\right\} \tag{4.88}$$

which are analogous to the well-known Geiringer equations in plane plastic flow. The angular velocity ω at any point P in the physical plane may be represented by a point P' in the hodograph plane with position vector ω. As P traces out an α or a β-line in the physical plane, P' traces its image α' or β'-curve in the hodograph plane. The curves in the physical plane and corresponding curves in the hodograph plane are mutually orthogonal. Also, a discontinuity in angular velocity across a line, which coincides with an α or a β-line, remains constant along its length.

The principal curvature rates of a deforming element, denoted by $\dot{\kappa}_\alpha$ and $\dot{\kappa}_\beta$, can be directly obtained from the rectangular components $\dot{\kappa}_x$ and $\dot{\kappa}_y$. Inserting (4.87) into (4.65), and setting $\phi = 0$ after the differentiation, we obtain

$$\dot{\kappa}_\alpha = \frac{\partial\omega_\beta}{\partial s_\alpha} + \frac{\omega_\alpha}{\rho_\alpha}, \quad \dot{\kappa}_\beta = \frac{\partial\omega_\alpha}{\partial s_\beta} - \frac{\omega_\beta}{\rho_\beta}, \tag{4.89}$$

where ρ_α and ρ_β are the radii of curvature of the α-and β-lines, respectively, reckoned positive as shown in the figure. If ρ'_α and ρ'_β denote the radii of curvature of the α' and β'-curves, defined in the same way as those for the α-and β-curves, it can be shown that

$$\dot{\kappa}_\alpha = \frac{\rho'_\alpha}{\rho_\alpha} \leq 0, \quad \dot{\kappa}_\beta = \frac{\rho'_\beta}{\rho_\alpha} \geq 0.$$

The above inequalities, which follow from the flow rule associated with the corner regime *B*, must be satisfied in a complete solution for the plate problem.

The significance of the stress function *N* can be appreciated by considering the equilibrium equations in the form (4.69) together with $p = 0$. The substitution from (4.84) indicates that $N + M_{xy}$ and $N - M_{xy}$ represent the effective twisting moments along the edges directed as shown in Fig. 4.17a. The shearing forces Q_x and Q_y are therefore statically equivalent to an isotropic twisting moment *N* acting along the edges. When $M_x = M_y = 0$, the yield condition requires $M_{xy} = \pm M_0$ in a plastic element, and the difference between the two components of the effective twisting moment acting in the element becomes numerically equal to $2M_0$, as indicated in Fig. 4.17b

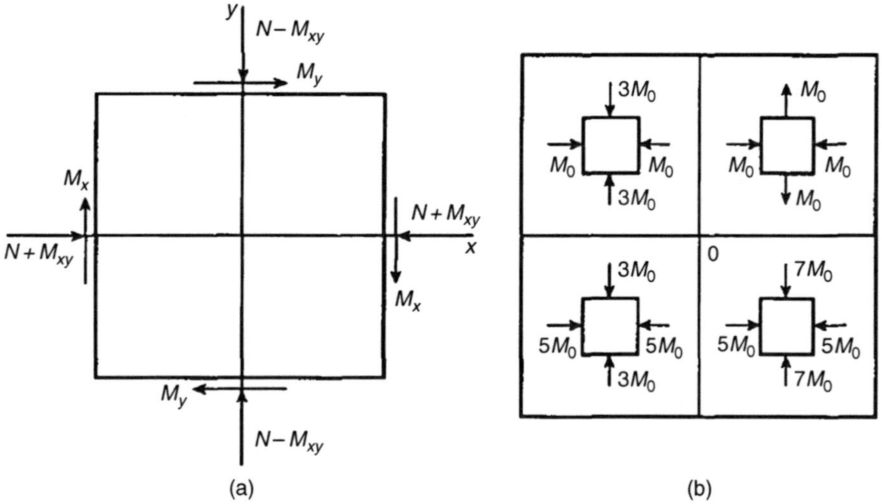

Fig. 4.17 (**a**) Bending and twisting moments acting on a plate element and (**b**) a discontinuous field involving jumps in the value of *N*

The isotropic twisting moment *N* may be discontinuous across a line whose inward normal and tangent are designated by *n* and *t*, respectively. Equilibrium requires the bending moment M_n and the effective shearing force $Q_n + \partial M_{nt}/\partial t$ to be continuous across the line. Since $Q_n = \partial N/\partial t$, any jump in the effective twisting moment $N + M_{nt}$ across such a line must be constant along its length. Since

N is undefined with respect to an arbitrary additive constant, the effective twisting moment may generally be taken as continuous across a line of discontinuity.

4.5.2 Application to Rectangular Plates

A simply supported rectangular plate, whose sides are of lengths $2a$ and $2b$ ($a \leq b$), carries a concentrated load P at its centroid O. When the aspect ratio b/a is sufficiently large, the mode of deformation at the incipient collapse may be assumed as that shown in Fig. 4.18. The triangles OAC and OBD undergo rigid body rotation about the respective boundaries with an angular velocity ω. The regions OAB and OCD, where the characteristics are radial lines and circular arcs, are deformed into conical surfaces. Plastic hinges are formed along the α-lines OA, OB, OC, OD, and along the β-lines AB, CD, the rate of change of slope across each hinge line being $\omega' = \omega/\sqrt{2}$. The difference between the isotropic twisting moments along the inner edges of each of the rigid triangles OAC and OBD is

$$N_2 - N_1 = N_4 - N_3 = 2M_0,$$

since the bending moment must vanish across the corresponding outer edges. The change in the value of N along each β-line within the 90° fans OAB and OCD is

$$N_3 - N_2 = N_1' - N_4 = \pi M_0$$

by the second equation of (4.86). Since the applied load P at O is equal to $N_1' - N_1$ in view of (4.85), the collapse load becomes

$$P = 4M_0 \left(1 + \frac{\pi}{2}\right). \tag{4.90}$$

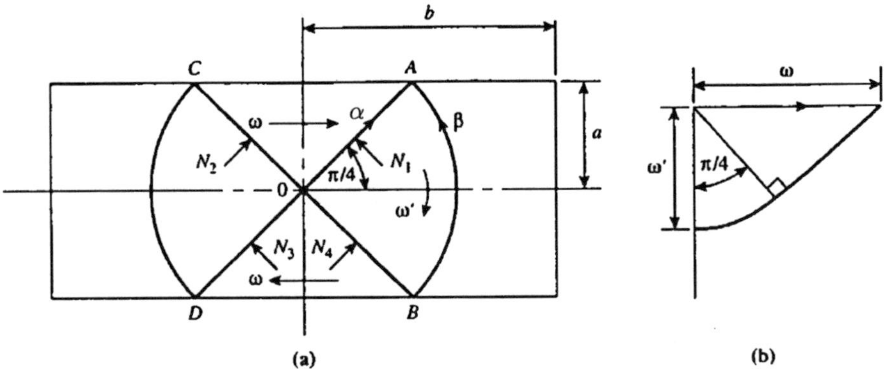

Fig. 4.18 Rectangular plate carrying a concentrated load at its center. (**a**) Discontinuous mode of collapse and (**b**) associated hodograph

This expression is also obtained by equating the rate of internal energy dissipation to the rate of external work done. Since $\dot{k}_\alpha = 0$ and $\dot{k}_\beta = \omega / r$ within the centered fans, where r is the radial distance from O, the rate of internal energy dissipation due to the incipient deformation is equal to $\pi M_0 a\omega$. The rate of energy dissipation along the hinge lines is $M_0\omega'$ times the total length of the discontinuities and is therefore equal to $(4 + \pi)M_0 a\omega$. Since the rate of work done by the external load is $Pa\omega$, we recover formula (4.90) for the collapse load.

Statically admissible states of stress in the triangular regions *OAC* and *OBD* are easily constructed by considering a vertical line of stress discontinuity through O and assuming plastic states in each separate region with $M_x = M_y = 0$. The stress field, which involves a jump in the value of N by an amount $2M_0$ across the discontinuity, can be continued into the rigid regions beyond the centered fans following the method used for plane strain extrusion (Chakrabarty, 2006). The construction involves the generation of trajectories of the maximum twisting moment through the corners A, B, C, and D. Since the α-lines in the plate problem must be continued as straight lines through O, these trajectories are logarithmic spirals intersecting at points E and F on the longitudinal axis of symmetry, where $OE = OF = \sqrt{2}a \exp(\pi/4)$. Assuming the trajectories to be lines of stress discontinuity, a statically admissible distribution of the effective twisting moment can be found in the rigid regions (Collins, 1971). The solution is therefore complete whenever $b/a \geq 3.10$.

A reasonably accurate upper bound solution for $1 \leq b/a \leq 2.21$ is furnished by the deformation mode shown in Fig. 4.19. The regions *OAC*, *OFH*, *OBD*, and *OKM* rotate rigidly about the respective boundaries with angular velocities ω_1 and ω_2. This is permitted by hinge lines forming along *OEF*, *OGH*, *OJK*, and *OLM*, the deformation in the remaining regions being continuous. The centered fan *OAE* is extended as the Hencky–Prandtl net *AEF*, which is defined by the circular arc *AE* and its reflection in *AF*. The fan angle Ψ is uniquely defined by the aspect ratio b/a and is given by the relationship

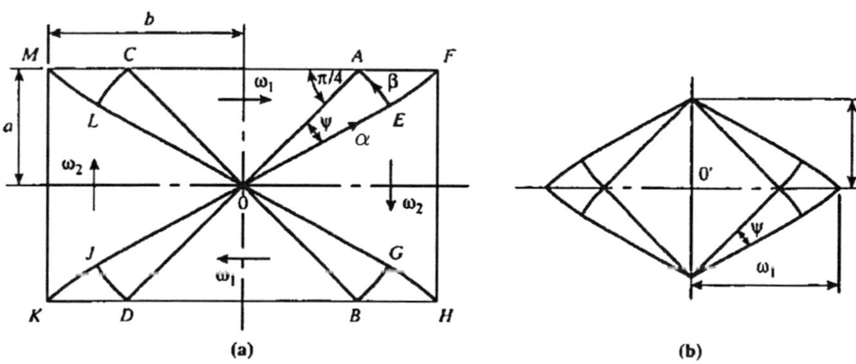

Fig. 4.19 Simply supported rectangular plate loaded at its center. (**a**) Characteristic field and (**b**) hodograph

$$\frac{b}{a} = I_0(2\psi) + \int_0^{2\psi} I_0(\alpha)d\alpha \tag{4.91}$$

according to a known result in the theory of slipline fields (Chakrabarty, 2006), where I_0 is the modified Bessel function of the first kind. A discontinuity of the constant amount $\sqrt{2}\omega_2$ in the angular velocity occurs along the hinge lines, and the hodograph indicates that ω_1/ω_2 is equal to the aspect ratio *b/a*.

To determine the variation of the isotropic twisting moment, we begin by setting $N = 0$ along *OA* and obtain $N = 2M_0$ along OC by the condition of overall moment equilibrium of the triangle *OAC*. The values of *N* along *OD*, *OB*, and *OA*, considered in a counterclockwise sequence, are denoted by N_1, N_2, and *P*, respectively, where *P* is the applied load at *O* and $N_2 - N_1 = 2M_0$. The load *P* is obtained from the conditions that the resolved component of the moments acting along *FAOBH* in the direction parallel to *FH*, and the resolved component of the moments acting along *KDOCM* in the direction parallel to *KM*, must individually vanish.

By (4.86), the isotropic twisting moments at generic points on *AF* and *BH* are $N = P - 4M_0\theta$ and $N' = N_2 + 4M_0\theta$, respectively, where (θ, θ) are the angular coordinates of these points with respect to *A* and *B*, respectively. The longitudinal distance *x* of the generic points from *O* is given by

$$\frac{x}{a} = I_0(2\theta) + \int_0^{2\theta} I_0(\alpha)d\alpha.$$

Considering the region *OAFHB* as a free body, the condition of the zero resultant moment about the boundary *FH* may be written as

$$(P - N_2)a - 2M_0 b + \int_a^b (N - N)dx = (P - N_2 - 2M_0)b - 8M_0 \int_a^b \theta dx = 0. \tag{4.92}$$

In view of the well-known relation $d[zI_1(z)]/dz = zI_0(z)$, it is readily shown

$$\frac{\theta}{a}\frac{dx}{d\theta} = 2\theta[I_0(2\theta) + I_1(2\theta)] = \frac{d}{d\theta}[\theta I_0(2\theta) + \theta I_1(2\theta)] - I_0(2\theta).$$

Using this expression to eliminate *dx* from the last integral of (4.92), and substituting from (4.91), we obtain

$$(P - N_1)\frac{b}{a} = 4M_0\{I_0(2\psi) + 2\psi[I_0(2\psi) + I_1(2\psi)]\},$$

where use has been made of the fact that $N_2 = N_1 + 2M_0$ Considering the region *OCMKD*, the condition of zero resultant moment about *KM* is similarly established, and $(b/a)N_1$ is then found to be given by the right-hand side of the above equation. The collapse load is therefore given by

$$\frac{P}{4M_0} = \frac{2a}{b}\{I_0(2\psi) + 2\psi[I_0(2\psi) + I_1(2\psi)]\}. \tag{4.93}$$

Equations (4.91) and (4.93) define the variation of collapse load with aspect ratio for sufficiently small values of *b/a*. The load steadily increases with the aspect ratio and attains the limiting value (4.90) when $b/a \approx 2.21$ ($\Psi \approx 28.5°$), as indicated by the graphical display in Fig. 4.20. The collapse load remains unchanged for all values of $b/a \geq 2.21$. In the special case of a square plate ($b = a$), a statically admissible stress field can be associated with the collapse load of $8Mo$ which is exact.

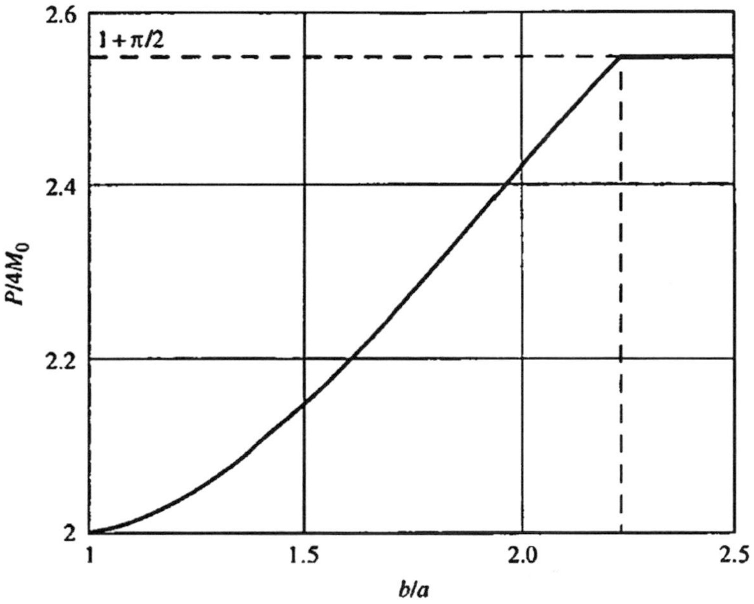

Fig. 4.20 Variation of the dimensionless collapse load with aspect ratio for a simply supported rectangular plate

Suppose that the point of application of the load lies on the axis of symmetry parallel to the shorter side and is situated at a distance $h \leq a$ from one of the longer sides. For sufficiently large values of *b/a*, the plate may collapse in one of the two modes shown in Fig. 4.21a,b, the former applying to relatively small *h/a* ratios and the latter to relatively large *h /a* ratios. The local mode of collapse indicated in Fig. 4.21a involves a rigid body rotation of the triangle *OAC* with an angular velocity ω, and a continuous deformation with hinge rotation in the fan *ABC*. The isotropic twist *N* increases by the amount $3\pi M_0$ along the circular arc *CBA*, and by the amount $2A/_0$ along the line *AC*, giving the dimensionless collapse load

$$\frac{P}{4M_0} = \frac{1}{2} + \frac{3\pi}{4} \approx 2.856.$$

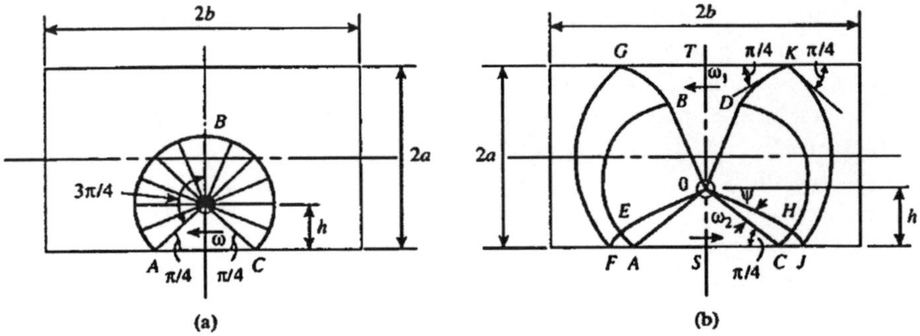

Fig. 4.21 Lines of principle bending moment in an asymmetrically loaded rectangular plate. (**a**) $0 \leq h/a \leq 0.51$ and (**b**) $0.51 < h/a \leq 1.0$

The same collapse load is obtained by the kinematical analysis, in which the rates of external work done, and internal energy dissipation, are found to be $Ph\omega$ and $(3\pi + 2)M_0 h\omega$, respectively.

When h/a exceeds a limiting value, the deformation spreads across the width of the plate at the incipient collapse, and the characteristic field of Fig. 4.21b becomes appropriate. The nature of the field is identical to the slipline field for an analogous problem of extrusion through a square-faced die with an extrusion ratio equal to $2a/h$ (Chakrabarty, 2006). The field on either side of the line of symmetry is generated from a centered fan with an angular span $\pi/2 + \psi$, where ψ depends on the ratio h/a. The regions OGK and OAC rotate as rigid bodies about their respective outer edges, while the longitudinal ends of the plate remain stationary. The only twisting moment transmitted across the line of symmetry is the isotropic twist N, whose value is set to zero along OT. Along the α-line OBG, N increases from a constant N_1 on OB to a value $N_1 - 2M_0\theta$ at a generic point of BG, where θ is the angular distance from B. Considering the region OGT as a free body, the condition of zero bending moment about GT may be written as

$$M_0 d - \int N\,dy = 0,$$

where d denotes the distance GT. Integrating by parts, and using the fact that $dN = 0$ along OB and $dN = -2M_0 d\theta$ along BG, we have

$$\left(\frac{2a}{h} - 1\right)\frac{N_1}{2M_0} = \left(\frac{2a}{h}\right)\psi + \frac{d}{2h} - \int_0^{\psi} d\theta \qquad (4.94)$$

Since N vanishes along OT, its value increases in the counterclockwise sense to become equal to $P/2$ along OS. Assuming $N = N_2$ along OA, the application of (4.86) along the β-line, BA, and the consideration of overall equilibrium of the triangular region OAS, lead to

$$N_2 - N_1 = M_0(\pi + 2\psi), \quad P/2 - N_2 = M_0.$$

The elimination of N_2 between these two equations gives the relationship between P and N, which can be combined with (4.94) to obtain the dimensionless collapse load as

$$\frac{P}{4M_0} = \frac{1+\pi}{2} + \left(\frac{4a-h}{2a-h}\right)\psi + \left(\frac{d}{2h} - \int_0^{\psi} \frac{y}{h} d\theta\right) \Big/ \left(\frac{2a}{h} - 1\right). \tag{4.95}$$

The comparison of (4.95) with a similar expression for the mean ram pressure in plane strain extrusion through a smooth container with a fractional reduction $r \geq 0.5$ indicates that the value of $P/4M_0$ in plate bending is equal to 1/r times the dimensionless extrusion pressure $p_e/2k$. The ratio $P/4M_0$ is therefore identical to the dimensionless punch pressure in the plane strain-piercing operation with a fractional reduction r. The analogy just established was noted by Johnson (1969), who employed a kinematical argument to derive the result. Since $P/4M_0 \approx 2.856$ when $h/a \approx 0.51$ according to (4.95), the collapse load must be the same for all $h/a \leq 0.51$.

4.5.3 Collapse Load for Triangular Plates

A uniform plate in the form of an equilateral triangle of height $3a$ is supported along its edges, and is loaded by a uniformly distributed transverse pressure of intensity p. The assumed mode of collapse, represented by Fig. 4.22a with a six-fold symmetry, consists of rigid-body rotation of triangles *OBC*, *ODE*, and *OFA* with an angular velocity ω, together with continuous deformation of the conical type in the sectors *OAB*, *OCD*, and *OEF*. Plastic hinge lines appear along the inner sides of

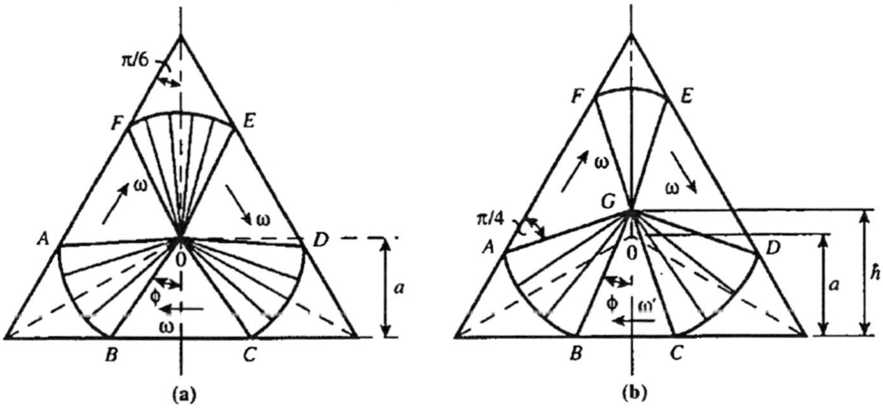

Fig. 4.22 Plastic collapse of an equilateral triangular plate under a concentrated load. (**a**) Symmetrically loaded and (**b**) asymmetrically loaded

the triangles, as well as along the circular arcs, the discontinuity in angular velocity being of amount ωsinϕ across the straight hinges and ωcosφ across the curved hinges. When the plate is clamped along its edges, plastic hinges are also formed along the lines *BC*, *DE*, and *AF.* Since the radial and circumferential curvature rates in the regions of deformation are zero and $(\omega/r)\cos\phi$, respectively, where r is the radial distance from 0, the rate of dissipation of total internal energy is found to be

$$U = 6M_0 a\omega\left[(1+\lambda)\tan\phi + 2\left(\frac{\pi}{3}-\phi\right)\right],$$

where $\lambda = 0$ for simply supported edges, and $\lambda = 1$ for perfectly clamped edges. The rate of external work done by the transverse pressure p is

$$W = p\omega a^3\left[\tan\phi + \left(\frac{\pi}{3}-\phi\right)\sec^2\phi\right].$$

An upper bound on the collapse pressure is obtained by equating the rates of internal and external works, resulting in

$$\frac{pa^2}{6M_0} = \frac{(1+\lambda)\sin\phi + 2(\pi/3-\phi)\cos\phi}{\sin\phi + (\pi/3-\phi)\sec\phi}. \tag{4.96}$$

The minimization of the upper bound pressure with respect to the angle ϕ furnishes the equation for ϕ as

$$\cos 2\phi + 2\left(\frac{\pi}{3}-\phi\right)(\operatorname{cos ec}2\phi - \lambda\cot 2\phi) + 4\left(\frac{\pi}{3}-\phi\right)^2 = \lambda.$$

For a simply supported plate ($\lambda = 0$), the solution to this equation is $\phi \approx 53.5°$, and the best upper bound is given by $pa^2/Mo \approx 5.66$. When the plate is perfectly clamped ($\lambda = 1$), p is a minimum when $\phi = 42.7°$, giving $pa^2/Mo \approx 9.92$ as the best upper bound on the collapse pressure.

The same mode of deformation may be assumed when a concentrated load P acts at the centroid of the plate. The rate of internal energy dissipation is unaffected, but the rate of external work is now modified to $Pa\omega$, giving the upper bound as

$$\frac{P}{6M_0} = (1+\lambda)\tan\phi + 2\left(\frac{\pi}{3}-\phi\right). \tag{4.97}$$

The collapse load has its minimum value when $\cos 2\phi = \lambda$. Thus $\phi \approx \pi/4$ for a simply supported plate, and the upper bound then becomes

$$\frac{P}{4M_0} = \frac{3}{2} + \frac{\pi}{4} \approx 2.285.$$

A clamped plate, on the other hand, requires $\phi = 0$, giving $P = 4\pi M_0$ with a purely conical mode of collapse. This is the collapse load for a clamped plate of arbitrary shape according to the square yield condition.

If a concentrated load P is applied on the axis of symmetry at a point G, which is away from the centroid, and the plate is simply supported along its edges, the deformation mode indicated in Fig. 4.22b may be used for the estimation of an upper bound. The rigid triangle on the base, having height h and included angle 2ϕ, rotates with an angular velocity ω', where $2h\omega' = (3a - h)\omega$ by the condition of continuity of the deflection at G. It follows from geometry that

$$\cos\phi = \frac{\sqrt{2}h}{3a-h}, \quad \frac{\omega}{\omega'} = \sqrt{2}\cos\phi.$$

The discontinuity in angular velocity is of amount $\omega' \sin\phi$ across the straight lines GB and GC, and $\omega' \cos\phi$ across the circular arcs BA and CD, each having an angular span equal to $5\pi/12 - \phi$. The jump in angular velocity across all other lines of discontinuity being of amount $\omega/\sqrt{2}$. The rate of dissipation of the internal energy is easily evaluated, the rate of external work being equal to $Ph\omega'$. The collapse load is finally obtained as

$$\frac{P}{4M_0} = 1 + \frac{1}{2}(\pi + \tan\phi) - \phi. \tag{4.97a}$$

The right-hand side of (4.97a) becomes equal to 2.285, as expected, when $\phi = \pi/4$ or $h/a = 1$. The collapse value of $P/4M_0$ is equal to 2.856 when $\phi \approx 72°$, giving $h/a \approx 0.538$, and the deformation mode of Fig. 4.21a becomes applicable for all smaller values of h/a. The angle ϕ vanishes when $h/a \approx 1.243$, and the triangle GBC disappears, the right-hand side of (97b) then being equal to $1 + \pi/2 \approx 2.571$. The deformation pattern and collapse load remain unchanged for higher values of h/a.

4.6 Yield Line Theory for Plates

The yield line theory, originally proposed by Johansen (1943) in the context of reinforced concrete plates, has been considered in details by Jones and Wood (1967), Save and Massonnet (1972), Sobotka (1989), and Nielsen (1999). The method of analysis, which tacitly assumes the maximum principal moment yield condition introduced in the preceding section, is essentially an upper bound approach for the estimation of the limit load. In this section, we shall begin by considering the complete yield line theory for the plastic collapse of plates, based on the square yield condition when the state of stress involves regular moment points, leading to solutions which are both statically and kinematically admissible.

4.6.1 Basic Yield Line Theory

Consider the plastic state associated with the regular regime *AB*, which corresponds to $M_\beta = M_0$ and $-M_0 \le M_\alpha \le M_0$, Fig. 4.23a. According to the normality rule of plastic flow, the principal curvature rates associated with this plastic regime are $\dot{k}_\alpha = 0$ and $\dot{k}_\beta > 0$. The plate can therefore bend by the rotation of each element about the β-direction with an angular velocity that remains constant in the α-direction. If ϕ denotes the counterclockwise angle made by the local α-direction with the *x*-axis, then by the transformation rule for tensors, we have

$$M_x = M_\alpha \cos^2\phi + M_0 \sin^2\phi, \quad M_y = M_\alpha \sin^2\phi + M_0 \cos^2\phi$$
$$M_{xy} = -(M_0 - M_\alpha)\sin\phi\cos\phi. \tag{4.98}$$

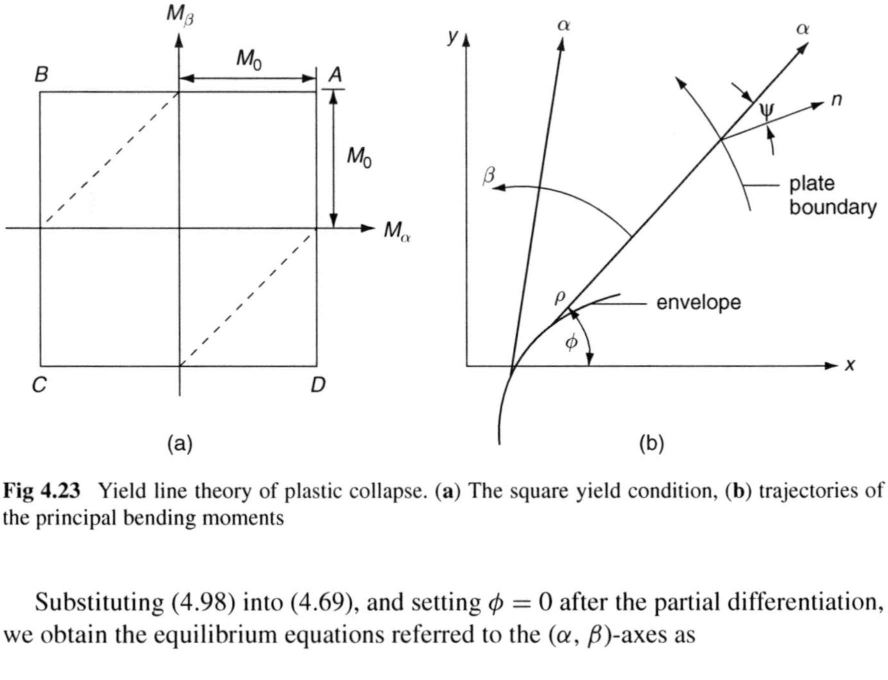

Fig 4.23 Yield line theory of plastic collapse. (**a**) The square yield condition, (**b**) trajectories of the principal bending moments

Substituting (4.98) into (4.69), and setting $\phi = 0$ after the partial differentiation, we obtain the equilibrium equations referred to the (α, β)-axes as

$$Q_\beta = 0, \quad Q_\alpha = \frac{\partial M_\alpha}{\partial s} + \frac{M_\alpha - M_0}{\rho}, \quad \frac{\partial Q_\alpha}{\partial s} + \frac{Q_\alpha}{\rho} = -p,$$

where *ds* is a line element of the straight α-lines and ρ is the radius of curvature of the β-lines. The last equation follows from the fact that $Q_x = Q_\alpha \cos\phi$ and $Q_y = Q_\alpha \sin\phi$, by the conditions of vertical equilibrium of the triangular elements formed by planes perpendicular to the principal axes, and those perpendicular to the x-and *y*-axes in turn. Since $ds = d\rho$ along an α-line, the last two equations can be solved in succession to give the stress resultants as

$$\begin{aligned} \rho Q_\alpha &= f(\phi) - \int p\rho d\rho \\ M_\alpha &= M_0 + \rho Q_\alpha + \frac{1}{\rho}\left[g(\phi) + \int p\rho^2 d\rho\right] \end{aligned} \tag{4.99}$$

The integrations can be carried out along the straight α-lines for any given distribution of transverse pressure p, the functions $f(\phi)$ and $g(\phi)$ being obtained from the boundary conditions, which involve the normal component of the bending moment given by

$$M_n = M\cos^2\psi + M_0\sin^2\psi$$

where ψ is the angle between the characteristic and the normal to the boundary. Since the yield condition requires $M \geq -M_0$, it follows that $\psi \leq \pi/4$ in a simply supported plate ($M_n = 0$) for the stress distribution to be statically admissible.

Since the directions of the principal bending moments and curvature rates coincide in an isotropic plate, the characteristics also define the directions of zero curvature rates. The plate therefore deforms into a developable surface with generators coinciding with the straight characteristics. In general, the characteristics run together to form an envelope, as shown in Fig. 4.23b, which is usually initiated at a point outside the area of the plate.

4.6.2 Elliptical Plate Loaded at the Center

Consider, as an example, an elliptical plate which is simply supported along its boundary and is subjected to a concentrated load P at its center O, as shown in Fig. 4.24a. The characteristics are radial lines emanating from O, giving a conical shape of the deflection surface. If ρ denotes the length of the radius vector to a typical point on the ellipse, and r the radius to any point on this line, then the deflection rate is given by

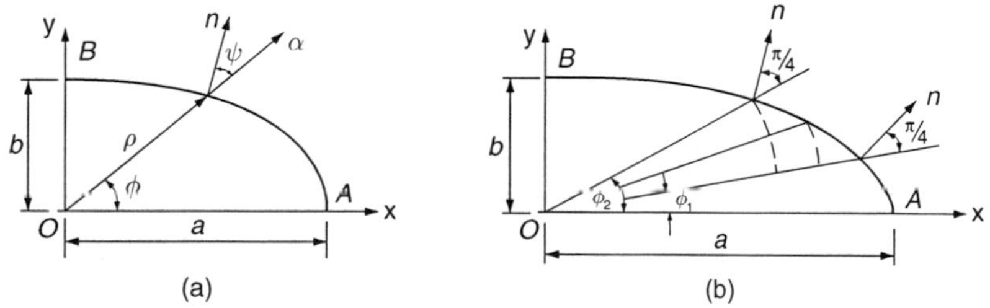

Fig. 4.24 Collapse mode for a simply supported elliptical plate under a concentrated load. (**a**) $a/b \leq \sqrt{2}+1$, (**b**) $a/b \geq \sqrt{2}+1$

$$w = w_0\left(1 - \frac{r}{\rho}\right), \quad \frac{a}{\rho} = \left(\cos^2\phi + \alpha^2\sin^2\phi\right)^{1/2}, \quad \alpha = \frac{a}{b}, \tag{4.100}$$

where w_0 is the velocity at the center of the plate, while $2a$ and $2b$ are the major and minor axes of the ellipse. The counterclockwise angle ψ, which the normal to a generic point of the boundary makes with the radius vector, is given by

$$\tan\psi = -\frac{1}{\rho}\frac{d\rho}{d\phi} = \frac{\left(\alpha^2 - 1\right)\tan\phi}{1 + \alpha^2\tan^2\phi}. \tag{4.101}$$

The condition $M_n = 0$ along a simply supported boundary gives $M_r = -M_0\tan^2\psi$ along the ellipse. Setting $\rho = r$ and $p = 0$ in (4.100), we therefore obtain

$$M_r = -M_0\tan^2\psi \quad rQ_r = -M_0\sec^2\psi.$$

throughout the plate. Integrating along any circle of radius r around O, the applied load at the incipient collapse may be written as

$$P = -4\int_0^{\pi/2} rQ_r d\phi = 4M_0\int_0^{\pi/2}\sec^2\psi\, d\phi = 4M_0\int_0^{\pi/2}\frac{(1 + \alpha^4\tan^2\phi)\sec^2\phi\, d\phi}{(1 + \alpha^2\tan^2\phi)^2}.$$

The integration is facilitated by the substitution $\tan\theta = \alpha\tan\phi$, giving $d\theta/d\phi = \alpha(\sec^2\phi/\sec^2\theta)$. The collapse load therefore becomes

$$P = \frac{4M_0}{\alpha}\int_0^{\pi/2}\left(\cos^2\theta + \alpha^2\sin^2\theta\right)d\theta = \pi M_0\left(\frac{\alpha^2 + 1}{\alpha}\right) = \pi M_0\left(\frac{a}{b} + \frac{b}{a}\right). \tag{4.102}$$

The validity of this solution requires $M_r \geq -M_0$, which is equivalent to $\psi \leq \pi/4$. Since the maximum value of $\tan\psi$ corresponds to $\cot\phi = \alpha$ in view of (4.101), we have the restriction

$$\alpha^2 - 1 \leq 2\alpha, \quad \text{or} \quad \alpha \leq \sqrt{2} + 1.$$

The limiting characteristic for which $\psi = \pi/4$ corresponds to $\phi = \pi/8$, and the collapse load associated with the limiting aspect ratio is $P = 2\sqrt{2}\pi M_0$. For the special case of a circular plate ($\alpha = 1$), we recover the well-known result $P = 2\pi M_0$.

For higher values of α, the preceding stress distribution continues to hold in a pair of regions defined by $0 \leq \phi \leq \phi_1$ and $\phi_2 \leq \phi \leq \pi/2$ over the first quadrant, as indicated in Fig. 4.24b. In the intervening region $\phi_1 \leq \phi \leq \phi_2$, the yield condition becomes $-M_r = M_\phi = M_0$, corresponding to a hyperbolic region defined by the corner B of the yield locus (Sawczuk and Hodge, 1968). Since $\psi = \pi/4$ along $\phi = \phi_1$ and $\phi = \phi_2$, these angles are given by the roots of the equation

$$\alpha^2 \tan^2 \phi - \left(\alpha^2 - 1\right) \tan \phi + 1 = 0$$

in view of (4.101). Thus, $\tan \phi_1$ and $\tan \phi_2$, which are the roots of the quadratic, satisfy the relations

$$\tan \phi_1 + \tan \phi_2 = \frac{\alpha^2 - 1}{\alpha^2}, \quad \tan \phi_1 \tan \phi_2 = \frac{1}{\alpha^2},$$
$$\tan \phi_2 - \tan \phi_1 = \frac{\sqrt{\alpha^4 - 6\alpha^2 + 1}}{\alpha^2}. \tag{4.103}$$

The first two relations of (4.103) indicate that $\tan(\phi_1 + \phi_2) = 1$. Since the last two relations of (4.103) furnish the expression fortan $(\phi_2 - \phi_1)$, we obtain the result

$$\phi_1 + \phi_2 = \frac{\pi}{4}, \quad \phi_2 - \phi_1 = \frac{\sqrt{\alpha^4 - 6\alpha^2 + 1}}{\alpha^2 + 1}. \tag{4.103a}$$

The shearing force in this case is obtained on setting $M_n = -M_0$, and $\rho = r$ in the second equation of (4.99), resulting in $r\,Q_r = -\,2M_0$. The load carried by the plate is given by

$$\frac{P}{4M_0} = -\int_0^{\pi/2} rQ_r d\phi = \int_0^{\phi_1} \sec^2\psi \; d\phi + 2\left(\phi_2 - \phi_1\right) + \int_{\phi_2}^{\pi/2_1} \sec^2\psi \; d\phi,$$

where ψ is given by (4.101). Using the substitution $\tan\theta = \alpha \tan\phi$ as before, we have

$$\int_{00}^{\phi} \sec^2\psi \quad d\phi = \frac{1}{\alpha}\int_0^{\theta} \left(\cos^2\theta + \alpha^2 \sin^2\theta\right) d\theta = \left(\frac{\alpha^2+1}{2\alpha}\right)\theta - \left(\frac{\alpha^2-1}{4\alpha}\right)\sin 2\theta.$$

The integrals in the preceding equation are therefore easily evaluated in terms of θ_1 and θ_2, which correspond to ϕ_1 and ϕ_2, respectively, and the result becomes

$$\frac{P}{4M_0} = 2\left(\phi_2 - \phi_1\right) + \left(\frac{\alpha^2+1}{2\alpha}\right)\left[\frac{\pi}{2} - \left(\theta_2 - \theta_1\right)\right] + \left(\frac{\alpha^2-1}{4\alpha}\right)\left(\sin 2\theta_2 - \sin 2\theta_1\right).$$

Since $\psi = \pi/4$ at both $\theta = \theta_1$ and $\theta = \theta_2$, we have $\sin 2\theta_1 = \sin 2\theta_2$ in view of (4.101) with $\alpha \tan\phi$ set equal to $\tan\theta$. Evaluating $\cot(\theta_2 - \theta_1)$ using (4.103), and substituting $\phi_2 - \phi_1$ from (4.103a), the collapse load is finally obtained as

$$\frac{P}{4M_0} = 2\tan^{-1}\left(\frac{\sqrt{\alpha^4 - 6\alpha^2 + 1}}{\alpha^2}\right) + \frac{\alpha^2+1}{2\alpha}\tan^{-1}\left[\frac{2\alpha}{\sqrt{\alpha^4 - 6\alpha^2 + 1}}\right], \alpha \geq \sqrt{2}+1. \tag{4.104}$$

When $\alpha = \sqrt{2} + 1$, the above expression reduces to $P = 2\sqrt{2}\pi M_0$. Also, by letting α tend to infinity, we recover the collapse load $P = 2(2+ \pi)\ M_0$, for plates with large aspect ratios. The variation of the collapse load parameter $P/2\pi M_0$ with the aspect ratio *a/b* over the range $1 \leq a/b \leq 5$ is displayed in Fig. 4.25. Several examples of complete solution to plate problems based on the yield line theory have been discussed by Save and Massonnet (1972).

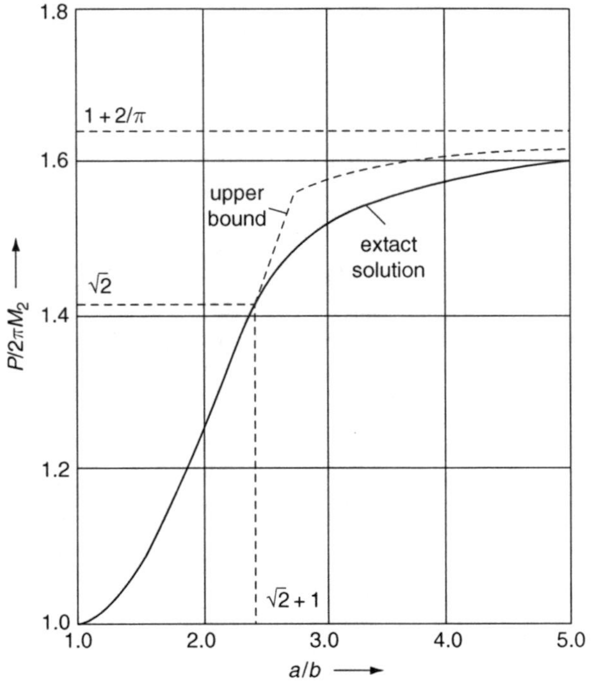

Fig. 4.25 Variation of the dimensionless collapse load for a simply supported elliptical plate with the aspect ratio

4.6.3 A Plate Under Distributed Loading

The analysis for the plastic collapse of plates under distributed loads is complicated due to the fact that the straight characteristics generally form an envelope, the shape of which is not known in advance. As an exceptional case, in which the characteristics meet in single points under a uniformly distributed load, consider a simply supported plate in which two opposite edges are formed by circular arcs of radius R the other two edges of the plate being assumed straight (Sawczuk, 1989), as shown in Fig. 4.26. Assuming the yield condition $M_\theta = M_0$ to hold for the entire plate, the expressions for Q_r and M_r are found from (4.99), where p is a constant. Consider first the fan of characteristics *BCA,* with r denoting the radial distance of a typical point from the center *C.* Using the boundary condition $Q_r = 0$ at $r = 0$, we obtain

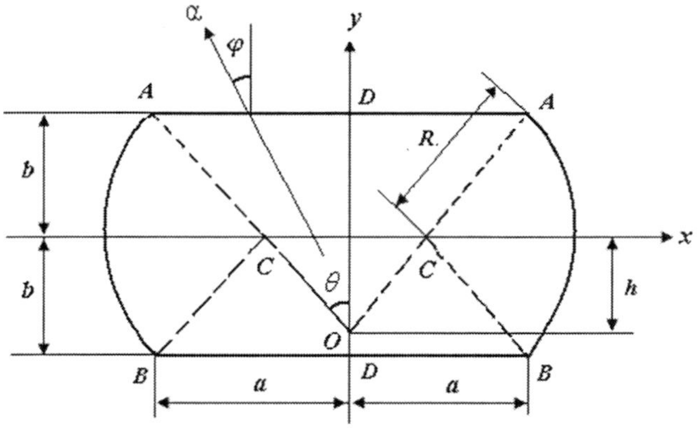

Fig. 4.26 Collapse mode for a simply supported plate under a uniformly distributed normal pressure

$$Q_r = -M_0 \frac{qr}{b^2}, \quad M_r = M_0 \left(1 - \frac{qr^2}{3b^2}\right), \quad q = \frac{pb^2}{2M_0}. \tag{4.105}$$

In the region *ACCA,* the straight characteristics meet in a point O when produced backward, where O is situated at a distance h from the central hinge line *CC.* Let ρ denote the radial distance of a any point in this region from O, and ϕ the angle made by it with the vertical through O. The functions $f(\phi)$ and $g(\phi)$ in (4.99) are obtained from the boundary conditions $Q_\rho = 0$ and $M_\rho = M_0$ along *CC,* where $\rho = h \sec \phi$, and the solution is easily shown to be

$$\rho Q_\rho = M_0 \left[\frac{h^2}{b^2} \sec^2 \phi - \frac{q\rho^2}{b^2}\right], \quad M_\rho = M_0 \left[1 + \frac{h^2}{b^2}\left(1 - \frac{2h}{3\rho} \sec \phi\right) \sec^2 \phi - \frac{q\rho^2}{3b^2}\right]. \tag{4.106}$$

Applying the remaining boundary conditions $M_r = 0$ along the circular arc AB, and $M_\rho = -M_0 \tan^2 \phi$ along the straight edge *AA,* where $\rho = (h + b) \sec \phi$, and using (4.105) and (4.106) for the respective regions of the plate, it is easily shown that

$$\frac{q}{3} = \frac{b^2}{R^2} = \frac{h + b}{3h + b},$$

giving

$$\frac{b}{h + b} = \frac{1}{2}\left(3 - \frac{R^2}{b^2}\right) = \frac{b}{a}\sqrt{\frac{R^2}{b^2} - 1},$$

where the last expression follows from the similarity of the triangles with apex at A. The relationship between the dimensionless collapse pressure p and the aspect ratio a/b is therefore given in the parametric form

$$\frac{pb^2}{6M_0} = \frac{b^2}{R^2}, \quad \frac{a}{b} = 2\left(\frac{R^2}{b^2} - 1\right)^{1/2}\left(3 - \frac{R^2}{b^2}\right)^{-1}. \tag{4.107}$$

It may be noted that the shearing force and the radial moment are discontinuous across the radial lines *CA* and *CB*, the amounts of the discontinuity being of variable intensities. The deformation mode at the incipient collapse consists of a pair of conical surfaces with vertices at points *C*, and a pair of flat faces on either side of *CC* undergoing rigid body rotation.

Since the stress field is statically admissible for $0 \leq \theta \leq \pi/4$, which is equivalent to $1 \leq R/b \leq \sqrt{2}$, the parabolic solution is valid only over the range $0 \leq a/b \leq 2$. When $a/b = 0$, the plate becomes circular with $R = b$, and the collapse load parameter $pb^2 2M_0$ reduces to the known value of 3.0. The collapse load steadily decreases with the aspect ratio *a/b*, and in the limiting case of $a/b = 2$, the collapse load is obtained as $pb^2/2M_0 = 1.5$. For higher values of *a/b*, Equation (4.107) provides only an upper bound on the collapse load.

4.6.4 Yield Line Upper Bounds

The yield line method of analysis is based on an assumed collapse mechanism in which finite portions of the plate rotate relative to one another as rigid bodies about straight plastic hinges called *yield lines*. The magnitude of the bending moment across each yield line is equal to the plastic moment M_0 in accordance with the square yield condition. The relative angles of rotation of the adjacent rigid regions at the incipient collapse form a vector polygon known as the *angular velocity hodograph* (Johnson, 1969). The application of the virtual work principle then furnishes an upper bound on the collapse load.

Consider a flat plate which is either simply supported or clamped along its boundary. When the boundary is rectilinear, the yield lines emanating from one or more points in the plate generally pass through the corners of the boundary. When the plate has a curvilinear boundary, we have to consider a continuous field of yield lines, in the form of a yield line fan, spreading out either from a single point in the plate or from an envelope. Referring to Fig. 4.27a, the boundary of the plate and the associated mode of collapse may be expressed by the equations

$$r = \rho(\phi), \quad w = w_0\left(1 - \frac{r}{\rho}\right),$$

where w_0 denotes the deflection rate at the center of the fan, which is taken as the origin of polar coordinates. For the yield condition, the associated curvature rates are

$$\dot{\kappa}_r = 0, \quad \dot{\kappa}_\phi = -\left(\frac{1}{r}\frac{\partial w}{\partial r} + \frac{1}{r^2}\frac{\partial^2 w}{\partial \phi^2}\right) = \frac{w_0}{r\rho}\left(1 + \frac{2\rho'^2}{\rho^2} - \frac{\rho''}{\rho}\right), \tag{4.108}$$

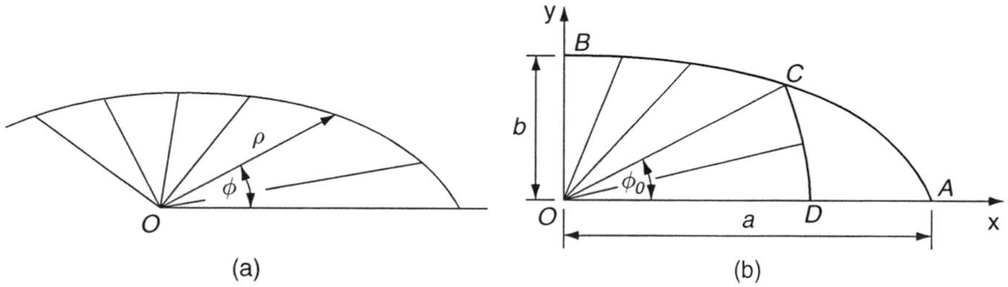

Fig. 4.27 Yield line modes of plastic collapse, (**a**) Arbitrary curvilinear boundary, (**b**) elliptical plate in partial collapse

where the prime denotes differentiation with respect to ϕ. If the plate is simply supported along its edge, there is no hinge action across the boundary, and the rate of plastic energy dissipation at the incipient collapse is given by (Sawczuk, 1989)

$$D = \iint M_\phi \dot{\kappa}_\phi r dr d\phi = M_0 w_0 \int \left(1 + \frac{2\rho'^2}{\rho^2} - \frac{\rho''}{\rho}\right) d\phi,$$

where the integral extends over the simply supported part of the boundary. In the case of complete collapse of a plate, which is simply supported along a closed contour, the rate of energy dissipation becomes

$$D = M_0 w_0 \int_0^{2\pi} \left[\left(1 + \frac{\rho'^2}{\rho^2}\right) - \frac{d}{d\phi}\left(\frac{\rho'}{\rho}\right)\right] d\phi = M_0 w_0 \int_0^{2\pi} \left(1 + \frac{\rho'^2}{\rho^2}\right) d\phi. \quad (4.109)$$

When the plate is fully clamped, the boundary of the plate coincides with a hinge line, and plastic energy is dissipated at the rate $M_0\omega$ per unit length of the boundary, where ω is the angular velocity of rotation about a typical line element ds of the boundary, we have

$$\omega \cos\psi = w_0/\rho, \qquad ds \cos\psi = \rho d\phi,$$

where ψ is the acute angle between the radius vector and the normal to the boundary, the additional amount of plastic energy dissipation due to this hinge rotation becomes

$$\Delta D = M_0 \int_0^{2\pi} \omega\, ds = M_0 w_0 \int_0^{2\pi} \sec^2\psi\, d\phi = M_0 w_0 \int_0^{2\pi} \left(1 + \frac{\rho'^2}{\rho^2}\right) d\phi.$$

The total rate of energy dissipation for a plate, which is clamped along a closed contour, is obtained by adding ΔD to the right-hand side of (4.109), the result being

$$D = M_0 w_0 \int_0^{2\pi} \left[2\left(1 + \frac{\rho'^2}{\rho^2}\right) - \frac{d}{d_\phi}\left(\frac{\rho'}{\rho}\right) \right] d\phi = 2M_0 w_0 \int_0^{2\pi} \left(1 + \frac{\rho'^2}{\rho^2}\right) d\phi. \tag{4.110}$$

It follows that for a plate undergoing complete collapse, the collapse load for the built-in plate under a given loading condition is exactly twice that for the simply supported plate. If, on the other hand, the collapse mode for a simply supported plate involves a state of partial collapse, then an integral of type (4.110) must be used over the part of the plate that has a modified boundary in the assumed collapse mechanism.

When the plate carries a concentrated load P, the center of the yield line fan coincides with the point of application of the load, and the rate of work done W by the applied load is then equal to Pw_0. For a distributed load acting on the plate, the center of the fan may be taken to correspond to the position of the center of gravity of the loading. The rate of external work done by a uniformly distributed load of intensity p is given by

$$W = \iint pwrdrd\phi = w_0 \iint p\left(1 - \frac{r}{\rho}\right) rdrd\phi = \frac{1}{6}pw_0 \int \rho^2 d\phi = \frac{1}{3}pw_0 A$$

where A is the area of the plate. The right-hand side of the above equation is evidently equal to $Pw_0/3$, where P denotes the total load carried by the plate. Similar results can be derived for any given non-uniform distribution of loading. The collapse load is finally obtained from the work equation $W = D$ in each particular case.

4.6.5 Examples of Upper Bounds

As a first example, consider a circular plate of radius a, which is loaded eccentrically by a concentrated load P at a distance c from the center of the plate, the associated yield line fan being that shown in Fig. 4.28a. By the well-known cosine law, the length of the radius vector from O to the boundary of the plate is given by.

$$\rho^2 - 2c\rho\cos\phi + c^2 = a^2, \text{ or } \rho = c\cos\phi + \sqrt{a^2 - c^2\sin^2\phi}.$$

The differentiation of this expression with respect to ϕ furnishes the result

$$\tan\psi = -\frac{1}{\rho}\frac{d\rho}{d\phi} = \frac{c\sin\phi}{\sqrt{a^2 - c^2\sin^2\phi}}. \tag{4.111}$$

When the plate is simply supported, substituting for ρ'/ρ into (4.109) and using the work equation $D = Pw_0$ furnishes the upper bound on the collapse load as

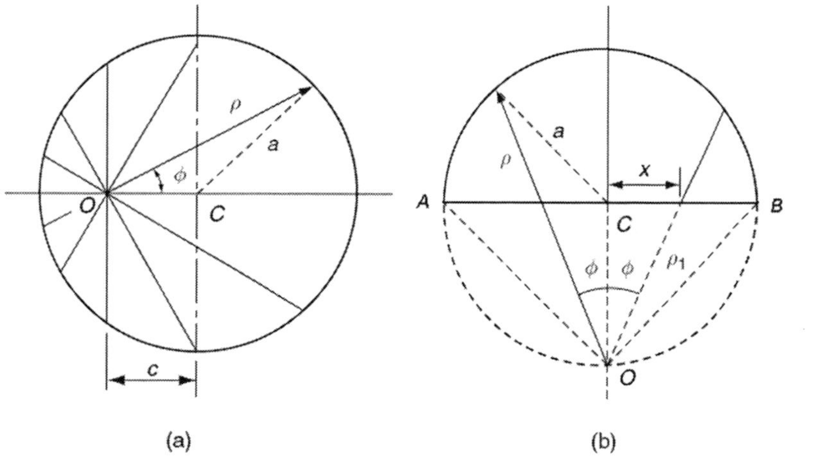

Fig. 4.28 Yield line patterns for plates, **(a)** Circular plate under eccentric loading, **(b)** semi-circular plate under line load

$$P = 2M_0 \int_0^{\pi} \left(1 + \frac{\rho'^2}{\rho^2}\right) d\phi = 2M_0 a^2 \int_0^{\pi} \frac{d\phi}{a^2 - c^2 \sin^2 \phi} = \frac{2\pi a M_0}{\sqrt{a^2 - c^2}}. \tag{4.112}$$

In the limiting case of c = 0, we recover the known result $P = 2\pi\ M_0$. It follows from (4.112) that tan ψ has a maximum value equal to $c/\sqrt{a^2 - c^2}$ then $\phi = \pi/2$. Since the solution is statically admissible when tan $\psi \leq 1$, the above solution provides the exact collapse load over the range $0 \leq c/a \leq 1/\sqrt{2}$, but only an upper bound for all higher values of *c/a*.

As a second example, consider the problem of a simply supported elliptical plate loaded by a concentrated load *P* at its center, an exact solution for which has been given in Section 4.6 (*ii*). The yield line method furnishes the collapse load

$$P = 4M_0 \int_0^{\pi/2} \left(1 + \frac{\rho'^2}{\rho^2}\right) d\phi = \frac{4M_0}{\alpha} \int_0^{\pi/2} \left(\cos^2\theta + \alpha^2 \sin^2\theta\right) d\theta = \left(\frac{\alpha^2 + 1}{\alpha}\right) \pi M_0, \tag{4.113}$$

which coincides with the exact solution over the range $1 \leq \alpha \leq \sqrt{2} + 1$. The collapse load given by (4.113) is definitely an upper bound for higher values of α. For sufficiently large values of α, an improved upper bound is obtained by assuming a partial collapse mode in which a hinge line appears in the plate in the form of a circular arc with center at the origin and meeting the boundary of the ellipse at a point having an angular position $\phi = \phi_0$ as shown in Fig. 4.27b. The contribution to the collapse load from the part of the plate suffering partial collapse is $8M_0\ \phi_0$ in view of (4.110), since $\rho' = 0$ over the circular arc, while that from the remainder of the plate is given by (4.109) with appropriate change of limits. The upper bound therefore becomes

$$\frac{P}{2M_0} = 4\phi_0 + \frac{2}{\alpha}\int_{\phi_0}^{\pi/2}\left(\cos^2\theta + \alpha^2\sin^2\theta\right)d\theta = 4\phi_0 + \frac{\alpha^2+1}{\alpha}\left(\frac{\pi}{2} - \theta_0\right) + \frac{\alpha^2-1}{2\alpha}\sin 2\theta_0. \tag{4.114}$$

where $\tan\theta_0 = \alpha\tan\phi_0$. The best upper bound corresponds to $dP/d\phi_0. = 0$. Using the fact that $d\theta_0/d\phi_0 = \alpha\sec^2\phi_0/\sec^2\theta_0$, the equation for $\tan\phi_0$is obtained as

$$\alpha^4\tan^4\phi_0 - \left[\alpha^2\left(\alpha^2-4\right)+1\right]\tan^2\phi_0 + 1 = 0.$$

Over the relevant range of values of α, it is sufficiently accurate to write down the solution approximately in the form

$$\tan\phi_0 = \frac{\beta}{\alpha}, \quad \beta \approx \left\{\frac{1}{2}\left(\alpha^2-4\right) + \frac{1}{2}\sqrt{\left(\alpha^2-2\right)\left(\alpha^2-6\right)}\right\}^{1/2}, \quad \alpha^2 \geq 6.$$

Substituting the above value of $\tan\phi_0$ into (4.114), the best upper bound is easily obtained in the dimensionless form

$$\frac{P}{2M_0} = 4\tan^{-1}\left(\frac{\beta}{\alpha}\right) + \frac{\alpha^2+1}{\alpha}\left(\cot^{-1}\beta\right) + \frac{\alpha^2-1}{\alpha}\left(\frac{\beta}{\beta^2+1}\right), \quad \alpha > 2.76. \tag{4.115}$$

For $1 \leq \alpha \leq 2.76$, Equation (4.113) gives a lower value of the collapse load for a simply supported plate. As α tend to infinity, the ratio β/α tends to unity, and the right-hand side of (4.115) tends to the asymptotic value of $2 + \pi$. A graphical plot of the complete upper bound solution, furnished by (4.113) and (4.115), is included as broken lines in Fig. 4.25 for a visual comparison with the exact solution.

The collapse load for a simply supported elliptical plate which carries a uniformly distributed load of intensity p may be obtained in a similar manner. In the case of complete collapse of the plate, the rate of internal energy dissipation is the same as that for the centrally loaded plate, but the rate of external work done is now becomes $(pw_0/3)(\pi\ a\ b)$. The work equation therefore furnishes the upper bound solution as

$$\frac{pb^2}{M_0} = 3\left(\frac{\alpha^2+1}{\alpha^2}\right) = 3\left(1 + \frac{b^2}{a^2}\right). \tag{4.116}$$

The upper bound solution defined by (4.116) does not differ significantly from the exact solution (Sawczuk, 1989), which is known only over the range $1 \leq \alpha \leq 3.52$. As indicated earlier, the upper bound load would be exactly doubled if the plate were fully clamped. The yield line upper bound for a uniformly loaded rectangular plate has been given in Section 4.5 (*ii*).

When a part of the boundary of the plate is unsupported, a realistic collapse mode requires the center of the yield line fan to be located outside the plate. It is necessary in this case to introduce the factor $(1 - \rho_1/\rho)$ in the last integral of (4.109), where ρ_I denotes the length of the radius vector to the point of intersection of a generic yield line with the free edge of the plate. Consider, for example, the plastic collapse of a semi-circular plate which is simply supported along the curved edge and is loaded by a uniform line load q per unit length along the straight edge AB, which is unsupported. Assuming the yield line pattern shown in Fig. 4.28b, in which the center O of the fan is taken at a distance a from the free edge, we have

$$D = M_0 w_0 \int_{-\pi/4}^{\pi/4} \left(1 - \frac{\rho_1}{\rho}\right)\left(1 + \frac{2\rho'^2}{\rho^2} - \frac{\rho''}{\rho}\right) d\phi = M_0 w_0 \int_{-\pi/4}^{\pi/4} \left(2 - \sec^2\phi\right)\sec^2\phi \, d\phi = \frac{4}{3} M_0 w_0. \tag{4.117}$$

in view of the relations $\rho_1 = a\sec\phi$, $\rho = 2a\cos\phi$. If the distance of a generic point of the free edge from its center is denoted by x, then the rate of external work done is

$$W = \int_{-a}^{a} qw dx = qw_0 \int_{-a}^{a} \left(1 - \frac{\rho_1}{\rho}\right) dx = qaw_0 \int_{-\pi/4}^{a} \left(1 - \frac{1}{2}\sec^2\phi\right)\sec^2\phi \, d\phi = \frac{2}{3} qaw_0.$$

The work equation $W = D$ finally gives the collapse load $q = 2M_0/a$, which is essentially due to Johansen (1943). Since the acute angle between the normal to the semi-circular edge and the radius vector nowhere exceeds $\pi/4$, the solution is also statically admissible, and the yield line load is therefore the actual collapse load for the considered plate.

4.7 Minimum Weight Design of Plates

4.7.1 Basic Principles

Consider the problem of design of a flat plate which is just at the point of collapse under given conditions of loading and support, the thickness of the plate being allowed to vary in such a way that the total volume of the plate is a minimum. The material is assumed to be homogeneous so that the design for minimum volume is identical to that for minimum weight. Let M_1, M_2 be the principal bending moments at any point in the plate that collapses under a distribution of transverse pressure p. If w denotes the rate of deflection of the plate whose middle surface is of area A, then

$$\int pw dA = \int (M_1\dot{\kappa}_1 + M_2\dot{\kappa}_2)\, dA = \int M_0\dot{\kappa}_0 dA,$$

where $\dot{\kappa}_1$,$\dot{\kappa}_2$ are the principal curvature rates of the middle surface, while $\dot{\kappa}_0$ is an effective curvature rate that depends on the yield function and is given by

$$\dot{\kappa}_0 = \left\{ \begin{array}{ll} \left(2/\sqrt{3}\right)\left[\dot{\kappa}_1^2 + \dot{\kappa}_1\dot{\kappa}_2 + \dot{\kappa}_2^2\right]^{1/2} & \textit{Mises} \\ (1/2)\left[|\dot{\kappa}_1| + |\dot{\kappa}_2| + |\dot{\kappa}_1 + \dot{\kappa}_2|\right], & \textit{Tresca} \end{array} \right\}. \tag{4.118}$$

The plastic moment M_0 depends on the local plate thickness, which is denoted by h in the optimum design. If M_1^*, M_2^* denote the principal moments in any other design which is capable of supporting the given distribution of pressure p, then by the principle of virtual work,

$$\int pwdA = \int (M_1^*\dot{\kappa}_1 + M_2^*\dot{\kappa}_2)dA \leq \int M_0^*\dot{\kappa}_0 dA,$$

where M_0^* is the plastic moment in the second design, characterized by a thickness distribution h^*. The inequality arises from the fact that $(\dot{\kappa}_1, \dot{\kappa}_2)$ need not be associated with $\left(M_1^*, M_2^*\right)$. It follows from the preceding relations that

$$\int \left(M_0 - M_0^*\right) \dot{\kappa}_0 dA \leq 0. \qquad (4.119)$$

Consider first a sandwich plate which has a light-weight core of constant thickness H between two identical face sheets of variable thickness h made of the given material $(h<<H)$. Since the core does not carry bending stresses, $M_0 = YHh$, $M_0^* = YHh^*$, where Y is the uniaxial yield stress of the material. If such a plate is designed to collapse in a mode involving a constant effective curvature rate $\dot{\kappa}_0$, then (4.119) reduces to

$$\int hdA \leq \int h^* dA,$$

indicating that the volume of the plate in the optimum design is an absolute minimum. When the weight of the plate is taken into account, the volume of the plate can be shown to be an absolute minimum if $\dot{\kappa}_0 - \beta^2\omega$ has a constant value at the incipient collapse, where $\beta^2 = 2\gamma/YH$, with γ denoting the specific weight of the material of the face sheets. The preceding theorem, which is essentially due to Drucker and Shield (1957), has been further examined by Mröz (1961).

For a solid plate, in which the plastic moment varies as the square of the plate thickness, we may consider any neighboring design with $h^* = h + \delta h$, where $\delta h << h$. If $h\dot{\kappa}_0$ is assigned a constant value at the point of collapse, then (4.119) furnishes

$$\int \left(h^{*2} - h^2\right) h^{-1} dA \approx 2 \int \delta h dA \geq 0. \qquad (4.120)$$

Thus, the volume of a solid plate, which collapses in such a way that the rate of effective curvature is inversely proportional to the local plate thickness, is a relative minimum. The continuously varying thickness predicted by the minimum volume criterion may not be a practical proposition, but it does provide a useful basis against which an actual design may be checked for material economy. The solid plate has been considered in detail by Kozlowski and Mröz (1969).

For simplicity, the material is assumed to obey Tresca′s yield criterion and its associated flow. The principal bending moments in any plastic element are then restricted to lie on the yield hexagon shown in Fig. 4.29a. Since the material is

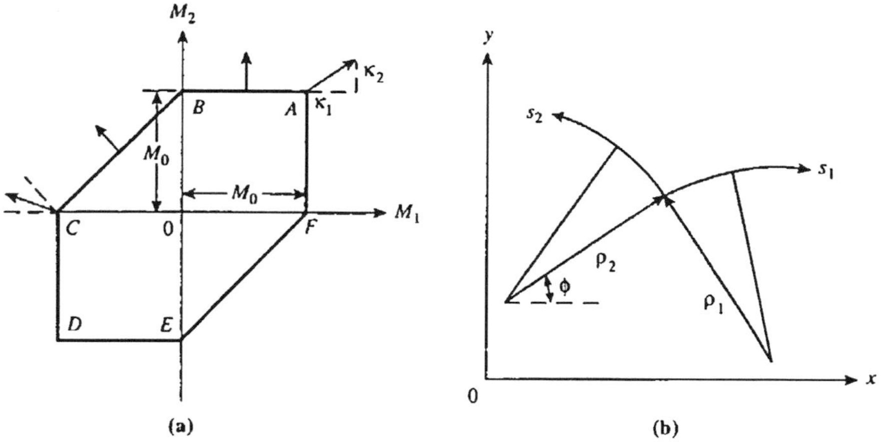

Fig. 4.29 Bending of rigid/plastic plates. **(a)** Tresca yield condition, **(b)** lines of principal stresses

isotropic, the lines of principal curvature rates $\dot{\kappa}_1$ and $\dot{\kappa}_2$ coincide with those of the principal bending moments M_l and M_2. The radii of curvature ρ_1 and ρ_2 of these lines being defined as

$$\frac{1}{\rho_1} = -\frac{\partial \phi}{\partial s_1}, \quad \frac{1}{\rho_2} = \frac{\partial \phi}{\partial s_2},$$

where ϕ is the counterclockwise angle made by the first principal moment direction with the x-axis, and s_1, s_2 are tne arc lengths along the principal lines, Fig. 4.29b. In terms of the deflection rate w of the middle surface, the curvature rates are

$$\dot{\kappa}_1 = -\left(\frac{\partial^2 w}{\partial s_1^2} + \frac{1}{\rho_1}\frac{\partial w}{\partial s_2}\right), \quad \dot{\kappa}_2 = -\left(\frac{\partial^2 w}{\partial s_2^2} + \frac{1}{\rho_2}\frac{\partial w}{\partial s_2}\right), \tag{4.121}$$

in view of (4.89) and the first two relations of (4.65). Since the rate of twist $\dot{\kappa}_{12}$ referred to the principal directions must be zero, we have the additional relations

$$\frac{\partial^2 w}{\partial s_1 \partial s_2} = \frac{1}{\rho_1}\frac{\partial w}{\partial s_1}, \quad \frac{\partial^2 w}{\partial s_2 \partial s_1} = \frac{1}{\rho_2}\frac{\partial w}{\partial s_2},$$

which follow from the equations preceding (4.88) on substitution for the components of the angular velocity. Combining the preceding relations with (4.121), it can be shown that

$$\frac{\partial \dot{\kappa}_1}{\partial s_2} = \frac{\dot{\kappa}_2 - \dot{\kappa}_1}{\rho_1}, \quad \frac{\partial \dot{\kappa}_2}{\partial s_1} = \frac{\dot{\kappa}_1 - \dot{\kappa}_2}{\rho_2}. \tag{4.122}$$

These relations represent the conditions of compatibility of the principal curvature rates at the incipient collapse of the plate.

Consider a sandwich plate, for which the effective curvature rate at the incipient collapse of the plate, having an absolute minimum volume, has a constant value equal to α. When the stress state is represented by the corner A of the yield locus, we have $M_1 = M_2 = M_0$, and the curvature rates are both positive. The second equation of (4.118) therefore gives $\dot{\kappa}_1 + \dot{\kappa}_2 = \alpha$ for the optimum design, the deflection rate w and the plastic moment M_0 being given by

$$\frac{\partial^2 w}{\partial x^2} + \frac{\partial^2 w}{\partial y^2} = -\alpha, \quad \frac{\partial^2 M_0}{\partial x^2} + \frac{\partial^2 M_0}{\partial y^2} = -p. \tag{4.123}$$

The second equation follows from (4.70), where $M_x = M_y = M_0$ and $M_{xy} = 0$ for the corner regime A. The partial differential equations (4.123) can be solved in each particular case using the appropriate boundary and continuity conditions.

When the stress state is represented by the corner C of the yield hexagon, we have $M_1 = -M_0$ and $M_2 = 0$, while the curvature rates satisfy the inequalities

$$\dot{\kappa}_1 \leq 0, \quad \dot{\kappa}_2 \geq 0, \quad \dot{\kappa}_1 + \dot{\kappa}_2 \leq 0.$$

Hence, the effective curvature rate is $-\dot{\kappa}_1 = \alpha$ in view of (4.118), and the first equation of (4.120) indicates that $\rho_1 = \infty$. The first principal lines are therefore straight, and their orthogonal trajectories form the second principal lines. Setting $\rho_i = \rho + s$, where $p(\phi)$ is the radius of curvature of the base curve $s = 0$, we have $ds_1 = ds$ and $ds_2 = \rho_2 \, d\phi$, giving

$$\frac{\partial^2 w}{\partial s^2} = \alpha, \quad \frac{\partial^2 w}{\partial \phi \partial s} = \frac{1}{\rho_2}\frac{\partial w}{\partial \phi},$$

in view of the first and second equations of (4.120) and (4.121), respectively. The solution for w may then be written in terms of an arbitrary function $f(\phi)$ as

$$w = \tfrac{1}{2}\alpha s^2 + sf(\phi) + \int \rho(\phi) f(\phi) \, d\phi. \tag{4.124}$$

The equilibrium equations are directly obtained from (4.82) on changing the subscripts α and β into 1 and 2, respectively. Setting $M_1 = -M_0$ and $M_2 = 0$, and eliminating Q_1 between the two equations, we have

$$\frac{\partial}{\partial s}\left(\rho_2^2 \frac{\partial M_0}{\partial s}\right) = p\rho_2^2,$$

in view of the relation $\partial \rho_2 / \partial s = 1$. The solution to this equation can be formally written down as

$$M_{(0)} = A(\phi) + \int p\rho_2 ds + \frac{1}{\rho_2}\left[B(\phi) - \int p\rho_2 ds\right], \tag{4.125}$$

where A and B are arbitrary functions of ϕ, to be determined from the boundary or continuity conditions. The results corresponding to the corner B of the yield hexagon would be similar to those for the corner C. The plastic regimes AB and BC' predict that both families of principal lines are straight and they are unlikely to play an important part in the solution of special problems on minimum weight design (Shield, 1960a).

4.7.2 Circular Sandwich Plates

In the case of rotationally symmetric circular plates, the principal bending moments M_1, M_2 may be taken as the radial and circumferential moments M_r, M_θ, respectively. It follows from the flow rule, and expressions (4.1) for the curvature rates, that the condition of constant effective curvature rate can be satisfied over a finite portion of the plate only if the moment state is represented by one of the corners A or D, and C or F of the yield hexagon. The problem has been discussed by Hopkins and Prager (1955), Freiberger and Tekinalp (1956), Onat et al. (1957), Eason (1960), Prager and Shield (1959), Marcal (1967), and Save and Massonnet (1972), among others.

We begin with a sandwich plate which is simply supported along its edge $r = a$ and loaded by a uniform transverse pressure of intensity p over a concentric circle of radius b. Since the curvature rates are likely to be both positive, the moment state for the entire plate may be represented by the corner A. Then, the condition for minimum volume becomes $\dot{\kappa}_r + \dot{\kappa}_\theta = \alpha$, which gives

$$\frac{d}{dr}\left(r\frac{dw}{dr}\right) = \alpha r \text{ or } w = \tfrac{1}{4}\alpha\left(a^2 - r^2\right), \tag{4.126}$$

in view of the conditions $w = 0$ at the edge $r = a$, and $dw/dr = 0$ at the center $r = 0$. The minimum volume plate therefore collapses in the form of a spherical surface with $\dot{\kappa}_r = \dot{\kappa}_\theta = \alpha/2$. To obtain the distribution of the plastic moment, we set $M_r = M_\theta = M_0$ in the equilibrium equation (4.7) to yield

$$\frac{dM_0}{dr} = -\frac{pr}{2}\ (0 \le r \le b), \quad \frac{dM_0}{dr} = -\frac{pb^2}{2r}\ (b \le r \le a).$$

Using the boundary condition $M_0 = 0$ at $r = a$, and the conditions of continuity of M_0 at $r = b$, the solution is obtained as

$$M_0 = YHh = \begin{cases} (p/2)\, b^2 \ln(a/b) + (p/4)\left(a^2 - r^2\right), & 0 \le r \le b, \\ (p/2)\, b^2 \ln(a/r), & b \le r \le a. \end{cases} \tag{4.127}$$

These relations define the variation of the thickness h of the face sheets with the radius r for the minimum volume design. The total volume of the face sheets is found to be

$$V = 4\pi \int_o^a hrdr = \left(1 - \frac{b^2}{2a^2}\right) \frac{\pi a^2 b^2 p}{2YH}. \tag{4.128}$$

For a given total load $\pi b^2 p$, the required volume decreases as the area of loading increases. In view of (4.14), a circular plate of constant thickness collapses under a uniformly distributed pressure p when the plastic moment is $M_0 = (p\, b^2/2)(1 - 2b/3a)$, the corresponding volume being

$$V = \left(1 - \frac{2b}{3a}\right) \frac{\pi p a^2 b^2}{YH}.$$

Comparison of this result with (4.128) shows that the saving of material involved in the minimum volume design over the constant thickness design is 25% when $b = a$, increasing to 50% as b tends to zero. In the case of a concentrated force P acting at the center of the plate, the results for the minimum volume design are obtained from (4.127) and (4.128) by letting $\pi b^2 p$ tend to P as b tends to zero.

For a circular plate with built-in support along its edge $r = a$, the plastic regime C will apply to an outer region $(c \le r \le a)$, where $M_r = -M_0$ and $M_\theta = 0$ by the yield condition, and $\dot{\kappa}_r = -\alpha$ by the minimum volume criterion, the deflection rate in this region being given by

$$\frac{d^2 w}{dr^2} = \alpha, \quad c \le r \le a.$$

The plastic regime A still applies to the inner region $(0 \le r \le c)$, where w is given by the differential equation in (4.126). Integrating and using the boundary conditions $dw/dr = 0$ at $r = 0$ and at $r = a$, and the condition of continuity of w at $r = c$, we find $c = 2a/3$, and the solution for w finally becomes

$$w = \begin{cases} (\alpha/6)\left(a^2 - 3r^2/2\right), & 0 \le r \le 2a/3, \\ (\alpha/2)(a - r)^2, & 2a/3 \le r \le a, \end{cases} \tag{4.129}$$

which satisfies the requirement $\dot{\kappa}_r + \dot{\kappa}_\theta < 0$ over the region $r \ge 2a/3$. The distribution of the plastic moment M_0 depends on whether the radius b of the loaded area is greater or less than $c = 2a/3$. For $b \le 2a/3$, the differential equation for M_0 over the loaded area $(0 \le r \le b)$ is the same as that for the simply supported plate, while the remainder of the plate involves the differential equations

$$\frac{dM_0}{dr} = \frac{pb^2}{2r} \quad (b \le r \le c), \qquad \frac{d}{dr}(rM_0) = \frac{pb^2}{2r} \qquad (c \le r \le a).$$

The last equation is obtained by setting $M_r = -M_0$ and $M_\theta = 0$ in the second equation of (4.7). The above equations can be integrated using the condition $M_0 = 0$ at $r = 2a/3$, since M_r must be continuous at this radius. The continuity of M_0 at $r = b$ then gives the solution for M_0 over the loaded region. The complete distribution of the plastic moment is therefore given by

$$M_0 = YHh = \begin{cases} (p/2)\, b^2 \ln(2a/3b) + (p/4)\left(b^2 - r^2\right), & 0 \le r \le b, \\ (p/2)\, b^2 \ln(2a/3r), & b \le r \le 2a/3, \\ (p/2)\, b^2 \ln(1 - 2a/3r), & 2a/3 \le r \le a. \end{cases} \quad (4.130)$$

For $b \geq 2a/3$, the differential equation of M_0 over the loaded region depends on whether the radius r to the element is greater or less than $c = 2a/3$. By the first equation of (4.7), we have

$$\frac{dM_0}{dr} = -\frac{pr}{2}\,(0 \le r \le c)\,, \quad \frac{d}{dr}\,(rM_0) = \frac{pr^2}{2}\,(c \le r \le b)\,.$$

These equations are easily integrated using the continuity condition $M_0 = 0$ at $r = 2a/3$. In the remaining part of the plate ($b \le r \le a$), the differential equation for M_0 is identical to that in the outermost part for $b \le 2a/3$. The complete variation of M_0 with r for $b \geq 2a/3$ is easily shown to be

$$M_0 = YHh = \begin{cases} (p/4)\left(4a^2/9 - r^2\right), & 0 \le r \le 2a/3, \\ (p/6)\left(r^2 - 8a^3/27r\right), & 2a/3 \le r \le b, \\ (p/6)\left(3b^2 - 2b^3/r - 8a^3/27r\right), & b \le r \le a, \end{cases} \quad (4.131)$$

in view of the condition of continuity of M_0 across $r = b$. The total volume of the face sheets in the optimum design is most conveniently obtained by equating the rate of internal energy dissipation to the rate of external work done. Consequently,

$$V = \frac{4\pi}{YH}\int_0^a M_0 r dr = \frac{4\pi}{YH\alpha}\int_0^b M_0 \dot{k}_0 r dr = \frac{4\pi p}{YH\alpha}\int_0^b w r dr,$$

where w is given by (4.129). The minimum volume depends on whether b is greater or less than 2a/3, the result being

$$V = \begin{cases} \left(1 - \dfrac{3b^2}{4a^2}\right)\dfrac{\pi p a^2 b^2}{3YH}, & 0 \le b \le \dfrac{2}{3}a, \\ \left(1 - \dfrac{4b}{3a} + \dfrac{b^2}{2a^2} - \dfrac{4a^2}{81b^2}\right)\dfrac{\pi p a^2 b^2}{YH}, & \dfrac{2}{3}a \le b \le a. \end{cases} \quad (4.132)$$

The minimum volume may be expressed as $V = \eta\pi p a^2 b^2/Y H$, where $\eta = \frac{1}{3}$ when $b = 0, \eta = \frac{2}{9}$ when $b\ 2a/3$, and $\eta = 0.117$ when $b = a$. A circular plate of constant thickness, on the other hand, requires $M_0 = 0.089pa^2$ for plastic collapse under uniform loading, the corresponding volume of the face sheets being $0.178\pi a^*/Y H$, indicating a saving of 34% in the minimum volume design. The nature of the minimum volume design of built-in circular plates under a uniformly distributed load covering the entire plate is indicated in Fig. 4.30. The influence of the weight of the plate itself on the optimum design has been considered by Onat et al. (1957).

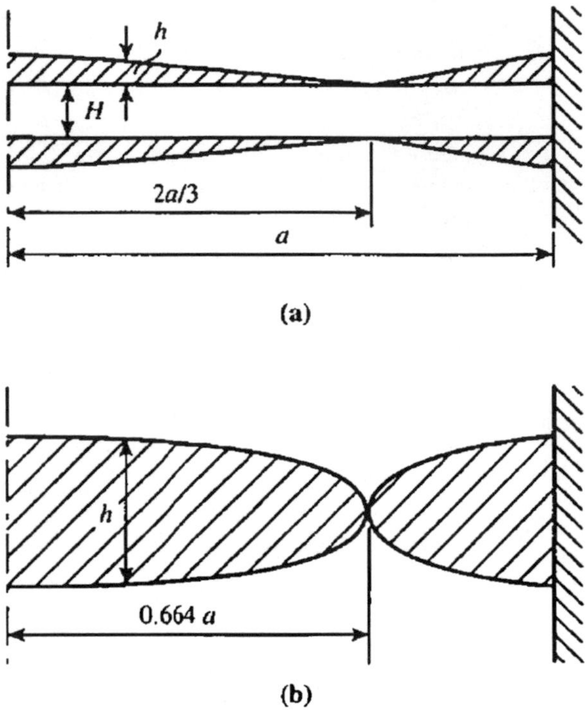

Fig. 4.30 Thickness profiles for uniformly loaded clamped circular plates designed for minimum weight. **(a)** Sandwich plate, **(b)** solid plate

4.7.3 Solid Circular Plates

In the case of a solid circular plate, the volume is a relative minimum when $h\dot{\kappa}_0$ has a constant value throughout the plate at the incipient collapse. Since the minimum criterion involves the design thickness h, the problem is evidently more complicated than that for the sandwich plate. For a simply supported plate under circular loading, if the plastic regime A is still assumed to hold, the distribution of the plastic moment M_0 is again given by (4.127), but the distribution of h is modified by the fact that $M_0 = Yh^2/4$ when the plate is solid. For a built-in plate, the adoption of plastic regimes A and C lead to a distribution of M_0 that is given by (4.130) provided the radius $2a/3$ to the regime interface is replaced by an unknown radius c. The assumed plastic regimes must of course be justified by the compatibility of the moment state with the associated flow rule.

Considering first the simply supported plate, the condition for the relative minimum volume may be written as $h\,(\dot{\kappa}_r + \dot{\kappa}_\theta) = \beta$, in view of the flow rule at the corner A of the yield hexagon, where B is a constant. The deflection rate at collapse therefore satisfies the equation

$$\frac{d}{dr}\left(r\frac{dw}{dr}\right) = -\frac{\beta r}{h}.$$

Since $dw/dr = 0$ at $r = 0$, and $w = 0$ at $r = 0$, the integration of this equation results in

$$\frac{dw}{dr} = -\frac{\beta}{r}\int_0^r \frac{\rho d\rho}{h(\rho)}, \quad w = \beta\int_r^a \frac{d\xi}{\xi}\int_0^r \frac{\rho d\rho}{h(p)}. \tag{4.133}$$

Since the curvature rates must be both nonnegative at the instant of collapse, the conditions $dw/dr \leq 0$ and $d^2w/dr^2 \leq 0$ must be satisfied. It is easily shown from (4.133) that

$$\frac{d^2w}{dr^2} = \frac{\beta}{r^2}\int_0^r \frac{\rho d\rho}{h(\rho)} - \frac{\beta}{h(r)} \leq \frac{\beta}{r^2 h(r)}\int_0^r \rho d\rho - \frac{\beta}{h(r)} = -\frac{\beta}{2h},$$

in view of the fact that $h(r)$ is a monotonically decreasing function of r. The distribution of M_0 given by (4.107) therefore provides a valid relative minimum design. In the particular case of a uniformly distributed pressure p acting over the entire plate, we have

$$h = a\left\{\frac{P}{Y}\left(1 - \frac{r^2}{a^2}\right)\right\}^{1/2}, \quad V = \frac{2}{3}\pi a^3\left(\frac{P}{Y}\right)^{1/2}. \tag{4.134}$$

A simply supported solid plate of constant thickness collapses under a uniform transverse pressure p when the plastic moment is $M_0 = pa^2/6 = Yh^2/4$, requiring a volume $V = \pi a^3\sqrt{2p/3Y}$. The saving of material effected by the minimum volume design over the constant thickness design is therefore about 18%.

When the plate is built-in at its edge $r = a$, regime A applies for $0 \leq r \leq c$ and regime C for $c \leq r \leq a$. The condition of continuity of the radial moment requires $M_0 = 0$ at the regime interface $r = c$. In the outer region $(c \leq r \leq a)$, the condition of relative minimum volume is $k\dot{\kappa}_r = -\beta$, which gives

$$\frac{d^2w}{dr^2} = \frac{\beta}{h} \quad \text{or} \quad \frac{dw}{dr} = -\beta\int_r^a \frac{dp}{h(p)}, \quad c \leq r \leq a,$$

since $dw/dr = 0$ at $r = a$. In the inner region $(0 \leq r \leq c)$, the velocity gradient is given by (4.133), and the condition of continuity of dw/dr at $r = c$ gives

$$\int_0^c \frac{\rho d\rho}{h(p)} = c\int_c^a \frac{dp}{h(p)}, \tag{4.135}$$

from which c can be determined for any given radius b of the circular loading. The value of c/a is found to decrease from 0.740 to 0.664 as b/a increases from zero to unity. Since $h(r)$ is monotonically increasing over the range $c \leq r \leq a$ we have

$$\frac{dw}{dr} \leq -\frac{\beta}{h(r)}\int_r^a dp = -\beta\left(\frac{a-r}{h}\right), \quad c \leq r \leq a.$$

This shows that $\dot{\kappa}_r + \dot{\kappa}_\theta \leq \beta\,(a - 2r)\,/rh$, which is negative for $c > a/2$, and the assumed moment state is therefore compatible with the flow rule. When the transverse pressure p is applied over the whole plate, the minimum volume design becomes

$$\begin{aligned} h &= a\left\{\frac{P}{Y}\left(\frac{c^2}{a^2} - \frac{r^2}{a^2}\right)\right\}^{1/2} \quad (0 \leq r \leq c), \\ h &= \left\{\frac{2P}{3Y}\left(\frac{r^2}{a^2} - \frac{c^3}{a^2 r}\right)\right\}^{1/2} \quad (c \leq r \leq a), \end{aligned} \tag{4.136}$$

where $c = 0.664$a. The volume of the plate can be computed by integration. The thickness variations for minimum volume design of uniformly loaded solid circular plates are indicated by the solid and broken lines in Fig. 4.31 for the simply supported and built-in edge conditions, respectively. The minimum volume design for a simply supported plate with a linearly varying load distribution has been considered by Sawczuk and Jaeger (1963). The plastic design of stepped plates with segment wise constant thickness has been studied by König and Rychlewsky (1966), and Sheu and Prager (1969). A variational method for the optimal plastic design has been proposed by Cinquini et al. (1977).

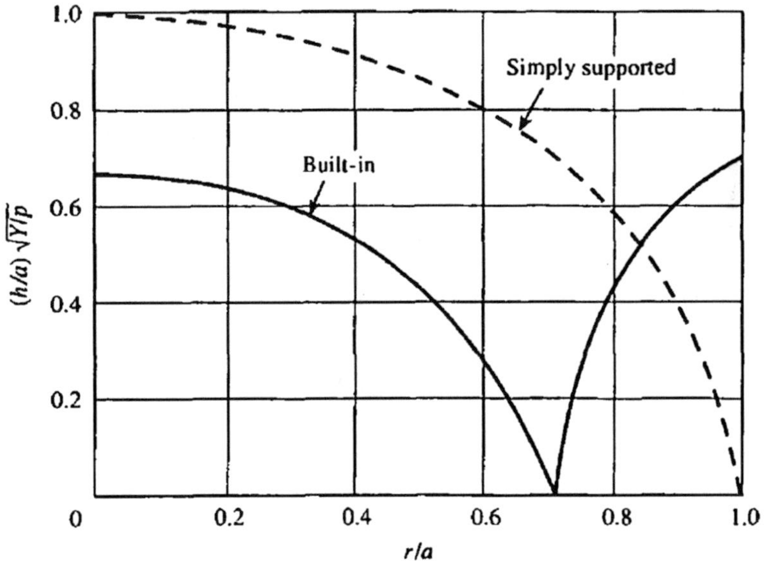

Fig. 4.31 Graphical plot of the thickness distribution in the optimum design of solid circular plates under uniform transverse loading

For a von Mises material with a given uniaxial yield stress *Y*, the volume of the plate in the optimum design must lie between those corresponding to the inscribed and circumscribed hexagons with respect to the von Mises ellipse. If *V(Y)* denotes the minimum volume corresponding to the Tresca theory with a uniaxial yield stress *Y*, then the minimum volume *V*(Y)* corresponding to the von Mises theory will lie between *V(Y)* and $V(2Y/\sqrt{3})$ For a sandwich plate, *V(Y)* is inversely proportional to *Y*, and *V*(Y)* can be estimated to within ±7.5% using the Tresca solution. For a solid plate, *V(Y)* is inversely proportional to the square root of *Y*, and a somewhat closer approximation to *V*(Y)* is achieved on the basis of the Tresca solution.

4.7.4 Elliptical Sandwich Plates

A plate of sandwich construction is bounded by an ellipse with semi-axes a and b, where $b \leq a$, and is loaded by a uniformly distributed transverse pressure p. If the plate is simply supported along its edge, the deflection rate w and the normal bending moment M_n must vanish at the boundary. Assuming the plastic regime A to apply over the entire plate, the minimum volume design is obtained by the solution of (4.103) under the boundary conditions $w = 0$ and $M_0 = 0$ along the boundary. Taking the x- and y-axes along the major and minor axes of the ellipse, respectively, the solutions for w and M_0 for the simply supported plate (Prager, 1955) are easily shown to be

$$w = w_0\left(1 - \frac{x^2}{a^2} - \frac{y^2}{b^2}\right), \quad M_0 = M_0'\left(1 - \frac{x^2}{a^2} - \frac{y^2}{b^2}\right), \tag{4.137}$$

where w_0 and M_0' denote the deflection rate and plastic moment, respectively, at the center of the plate. They are given by

$$w_0 = \frac{\alpha a^2 b^2}{2\left(a^2 + b^2\right)}, \quad M_0' = \frac{p a^2 b^2}{2\left(a^2 + b^2\right)}.$$

The thickness distribution in the minimum volume design is given by the relation $M_0 = YHh$, and the total volume of the face sheets is obtained from (4.137) as

$$V = 2\iint\left(\frac{M_o}{YH}\right)dxdy = \frac{\pi p a^3 b^3}{2YH(a^2 + b^2)}.$$

Since the principal curvature rates, which are equal to $-\partial^2 w/\partial x^2$ and $-\partial^2 w/\partial y^2$, are both nonnegative according to (4.137), the corresponding thickness distribution defines the absolute minimum volume for a Tresca material.

When the plate is clamped along its elliptical boundary, the deflection rate w and its normal derivative $\partial w/\partial n$ must vanish along the boundary, which is therefore a line of zero principal curvature rate. It is reasonable to assume that the plastic regime A applies to the region inside a concentric ellipse Γ, and the plastic regime C applies to the region between the edge of the plate and Γ. The lines of principal bending moment in the outer region are straight lines normal to the edge, and their orthogonal trajectories parallel to the edge, as shown in Fig. 4.32. If the base curve $s = 0$ is taken as the edge of the plate, and the direction of increasing s is taken along its outward normal, then $\partial w/\partial s = 0$ along $s = 0$. Consequently,

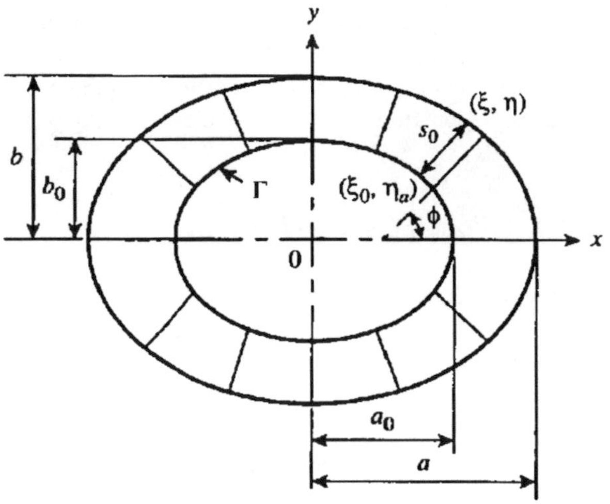

Fig. 4.32 Principal lines and the regime interface in a uniformly loaded clamped elliptical plate designed for minimum weight

$$w = \frac{1}{2}\alpha s^2, \quad \frac{\partial w}{\partial s} = \alpha s = \sqrt{2aw},$$

in the outer region of the plate, in view of (4.124). It follows that the deflection rate in this region is such that

$$\left(\frac{\partial w}{\partial x}\right)^2 + \left(\frac{\partial w}{\partial y}\right)^2 = 2\alpha w, \quad \tan\phi = \frac{\partial w/\partial y}{\partial w/\partial x}, \tag{4.138}$$

where ϕ is the angle of inclination of the straight principal lines with respect to the x-axis. If a typical principal line meets the external boundary and the regime interface at (ξ, η) and (ξ_0, η_0), respectively, then by simple geometry,

$$\left.\begin{aligned} \alpha(\xi - \xi_0) &= \alpha s_0 \cos\phi = -(\partial w/\partial x)\,\Gamma, \\ \alpha(\eta - \eta_0) &= \alpha s_0 \sin\phi = -(\partial w/\partial y)\,\Gamma, \end{aligned}\right\} \tag{4.139}$$

where s_0 denotes the value of $-s$ on Γ. Since the principal curvature rates are $\dot{\kappa}_1 = -\alpha$ and $\dot{\kappa}_2 = -\alpha s(\rho + s) \geq 0$, where ρ is the radius of curvature of the elliptical boundary, the plastic regime C is appropriate for the outer region. The deflection rate in the region inside Γ may be written in the form

$$w = \frac{\alpha c^2 d^2}{2(c^2 + d^2)}\left(1 - \frac{x^2}{c^2} - \frac{y^2}{d^2}\right), \tag{4.140}$$

where c and d are constants. The first equation of (4.123) is identically satisfied, and the principal curvature rates $-\partial^2 w/\partial x^2$ and $-\partial^2 w/\partial y^2$ are found to be both positive. Since w and its first derivatives must be continuous across Γ, the first equation of (4.139) must hold on this curve, and the substitution from (4.140) into this equation shows that Γ is indeed an ellipse with semi-axes a_0 and b_0, where

$$a_0^2 = c^2\left(\frac{c^2+d^2}{c^2+2d^2}\right), \quad b_0^2 = d^2\left(\frac{c^2+d^2}{2c^2+d^2}\right). \tag{4.141a}$$

Substituting from (4.140) into (4.139), and using the above relations, we get

$$\xi = \xi_0\left(c^2/a_0^2\right), \quad \eta = \eta_0\left(d^2/b_0^2\right).$$

Since (ξ, η) and (ξ_0, η_0) must satisfy the equations of ellipses with semi-axes (a, b) and (a_0, b_0), respectively, we have

$$a^2 = c^2\left(\frac{c^2+2d^2}{c^2+d^2}\right), \quad b^2 = d^2\left(\frac{2c^2+d^2}{c^2+d^2}\right). \tag{4.141b}$$

These equations can be solved for c and d to express them in terms of a and b, the result being (Shield, 1960)

$$3c^2 = 2a^2 - b^2 + \sqrt{a^4 - a^2b^2 + b^4},$$
$$3d^2 = 2b^2 - a^2 + \sqrt{a^4 - a^2b^2 + b^4}.$$

It is interesting to note that $c^2 - d^2 = a^2 - b^2$. Furthermore, it is easily shown from (4.141) that

$$aa_0 = c^2, \quad bb_0 = d^2, \quad (a - a_0)\,a = (b - b_0)\,b.$$

When the plate is circular $(b = a)$, we have $c = d = \sqrt{\frac{2}{3}}a$, and $a_0 = b_0 = 2a/3$, as found previously. As b/a tends to zero, the ratios c/a and a_0/a tend to unity, while d^2/b^2 and b_0/b tend to $\frac{1}{2}$ Fig. 4.33a shows the variation of the ratios a_0/a and b_0/b with the aspect ratio b/a of the elliptic boundary.

The distribution of the plastic moment in the region inside Γ is obtained from the solution of the second equation of (4.123), satisfying the condition $M_0 = 0$ on Γ. It is easily shown that

$$M_0 = YHh = \frac{pa_0^2b_0^2}{2\left(a_0^2 + b_0^2\right)}\left(1 - \frac{x^2}{a_0^2} - \frac{y^2}{b_0^2}\right) \tag{4.142}$$

inside Γ. In the annular region outside Γ, the plastic moment is given by (4.125), where $\rho_2 = \rho + s$. Using the condition $M_0 = 0$ at $s = -s_0$, the solution for a constant applied pressure p may be written as

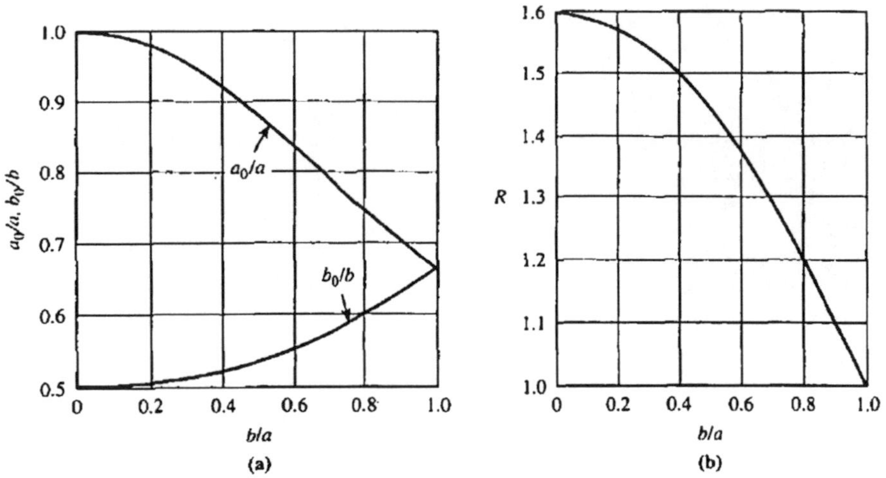

Fig. 4.33 Minimum weight design of a uniformly loaded elliptical plate. **(a)** semiaxes of the regime interface as function of the aspect ratio, **(b)** ratio of mean thickness of an elliptical plate to that if the circular plate.

$$M_0 = YHh = \frac{1}{3}p\left(s_0 + s\right)\left\{\rho - \frac{1}{2}\left(s_0 - s\right) - \frac{D\left(\phi\right)}{\left(\rho - s_0\right)\left(\rho + s\right)}\right\} \tag{4.143}$$

outside Γ. The function $D(\phi)$ must be determined from the condition that the shear force Q_n is continuous across Γ. It is convenient at this stage to introduce the eccentric angle ψ for the given elliptic boundary, so that

$$\xi = a\cos\psi, \quad \eta = b\sin\psi,$$
$$\xi_0 = a_0\cos\psi, \quad \eta_0 = b_0\sin\psi.$$

The angle of inclination ϕ of the normal to the boundary at any point (ξ, η), and the inclination θ of the normal to Γ at the corresponding point (ξ_0, η_0), are given by

$$\left.\begin{aligned} \tan\phi &= \frac{\eta - \eta_0}{\xi - \xi_0} = \frac{a}{b}\tan\psi, \\ \tan\theta &= \frac{a_0^2\eta_0}{b_0^2\xi_0} = \frac{a_0}{b_0}\tan\psi. \end{aligned}\right\} \tag{4.144}$$

The radius of curvature ρ of the elliptic boundary at (ξ, η) and the distance s_0 of this point from (ξ_0, η_0) are easily shown to be

$$\left.\begin{aligned} \rho &= a^2b^2\left(a^2\cos^2\phi + b^2\sin^2\phi\right)^{-3/2}, \\ s_0 &= a\left(a - a_0\right)\left(a^2\cos^2\phi + b^2\sin^2\phi\right)^{-1/2}. \end{aligned}\right\} \tag{4.145}$$

Let Q_n denote the shearing force per unit length acting on a transverse section tangential to Γ at any generic point (ξ_0, η_0). Since $M_x = M_y = M_0$ and $M_{xy} = 0$ for the region inside Γ, we have

$$Q_n = \frac{\partial M_0}{\partial n} = \frac{\partial M_0}{\partial x}\cos\theta + \frac{\partial M_0}{\partial y}\sin\theta,$$

where the partial derivatives are evaluated at (ξ_0,η_0). Substituting from (4.142), and using (4.144), we obtain

$$Q_n = -\frac{pa_0 b_0^2}{a_0^2 + b_0^2}\left(1 + \frac{b^2 a_0^2}{a^2 b_0^2}\tan^2\phi\right)\cos\theta\cos\psi.$$

In the region outside Γ the plastic regime C corresponds to $M_1 = -M_0$ and $M_2 = 0$, giving $Q_2 = 0$ and $Q_1 = -\partial M_0/\partial s$ in view of (4.82), and consequently,

$$Q_n = Q_1\cos(\theta - \phi) = -\frac{\partial M_0}{\delta s}\cos(\theta - \phi),$$

where M_0 is given by (4.143), and the derivative is considered at $s = -s_0$. In view of (4.144), the result becomes

$$Q_n = -\frac{p}{3}\left(\rho_0 - \frac{D}{\rho_0^2}\right)\left(1 + \frac{ba_0}{ab_0}\tan^2\phi\right)\cos\theta\cos\phi,$$

where $\rho_0 = \rho - s_0$, denoting the radius of curvature of the curve through (ξ_0, η_0) drawn parallel to the elliptic boundary. The condition of continuity of Qn across the interface Γ gives

$$\frac{D(\phi)}{\rho_0^2} = \rho_0 - \frac{3a_0 b_0}{a_0^2 + b_0^2}\left\{\frac{a^2 b_0^2\cos^2\phi + b^2 a_0^2\sin^2\phi}{(ab_0\cos^2\phi + ba_0\sin^2\phi)\sqrt{a^2\cos^2\phi + b^2\sin^2\phi}}\right\}. \tag{4.146}$$

Equations (4.143) and (4.146) furnish the required plastic moment at any point (s, ϕ) in the annular region between the clamped edge and the Γ. The thickness of the face sheets decreases from the center of the plate to zero on Γ, and thereafter increases toward the clamped edge.

The volume V of the face sheets in the optimum design can be obtained without having to determine the design thickness explicitly. Since the rate of dissipation of internal energy per unit area of the plate is $M_0\dot{\kappa}_0 = YH h\alpha$, we write

$$V = 2\int h dA = \frac{2}{YH\alpha}\int M_0\dot{k}_0 dA = \frac{2}{YH\alpha}\int w d\,\mathrm{A},$$

where the integral extends over the entire area of the plate. Using the expressions for w in the regions inside and outside the regime interface, we have

$$V = \frac{4pc^2d^2}{YH\left(c^2+d^2\right)}\iint\left(1-\frac{x^2}{c^2}-\frac{y^2}{d^2}\right)dxdy + \frac{4p}{YH}\iint(p+s)\,s^2 ds d\phi,$$

where ρ and s_0 are given by (4.143) as functions of ϕ. A straightforward evaluation of the integrals results in

$$V = \frac{\pi p a_0^2 b_0^2}{4YH}\left\{\frac{4ab - ab_0 - ba_0}{aa_0 + bb_0} + \frac{ba_0}{b_0^2}\left(\frac{a}{a_0}-1\right)^3\left(\frac{5b^2}{3a^2}+\frac{a_0}{a}+\frac{b_0}{b}-1\right)\right\}. \tag{4.147}$$

For a circular plate, $b = a$, and $b_0 = a_0 = 2a/3$, and the minimum volume reduces to that obtained previously. The ratio R of the average thickness $V/\pi ab$ for an elliptic plate of semi-axes a and b to that of a circular plate of radius b is represented by the curve in Fig. 4.33b, both plates being assumed to be of the same material and subjected to the same uniform pressure p. For plates of arbitrary shapes, a numerical method of optimization, based on the finite element technique, will be required for the optimal plastic design.

Problems

4.1 A solid circular plate of radius a and plastic moment M_0 carries a uniform normal pressure p_1 over its entire area, and an additional normal pressure p_2 over an inner circle of radius b. If the plate is simply supported along its boundary $r = a$, and is at the point of plastic collapse according to the Tresca theory, show that

$$\frac{p_1 a^2}{3M_0} + \left(1-\frac{2b}{3a}\right)\frac{p_2 b^2}{M_0} = 2$$

Assuming $p_1 = p_2 = p$, estimate the dimensionless collapse load $p(a^2 + b^2)/ 2M_0$, when the ratio b/a is equal to 0.5, 0.75, 0.9, and 1.0.
(*Answer:* 2. 5, 2. 542, 2. 754, 3. 0).

4.2 A solid circular plate of plastic moment M_0, which is clamped along its boundary $r = a$, carries a uniformly distributed normal pressure p over an annular area defined by $b \le r \le a$. Using the Tresca criterion and its associated flow rule, show that the collapse pressure p and the radius ρ where the radial bending moment is zero are given by

$$\frac{pa^2}{2M_0} = \frac{3a^2/\rho^2}{(1+2b/\rho)(1-b/\rho)^2} = \frac{2+\ln\left(a^2/\rho^2\right)}{\left(1-\rho^2/a^2\right)-\left(b^2/a^2\right)\ln\left(a^2/\rho^2\right)}.$$

Taking $b/a = 0.5$, compute the values of ρ/a and $p\,a^2/2M_0$ at the point of plastic collapse.

4.3 A solid circular plate of plastic moment M_0 is fully clamped along its edge $r = a$ and is loaded by normal pressures p_1 and p_2 in the same way as that in Problem 4.1 Show that the radius ρ to the circle where the radial bending moment vanishes corresponds to $r \geq b$ if $(p_1 + p_2)\,b^2 \leq 6M_0$. Setting $q_1 = p_1 a^2/2M_0$ and $q_2 = p_2 b^2/2M_0$, obtain the interaction relationship parametrically through ρ in the form

$$q_2\left(\frac{a^2}{\rho^2} - 1\right)\frac{b}{\rho} - (q_2 - 1)\left[\frac{3}{2}\left(\frac{a^2}{\rho^2} - 1\right) - \ln\frac{a}{\rho}\right] = 1,$$

$$q_1 = \frac{3a^2}{\rho^2}\left[1 - q_2\left(1 - \frac{2b}{3\rho}\right)\right].$$

Assuming $b/a = 0.5$, compute the values of ρ/a and q_1, when $q_2 = 3.0$, 1.0, and 0. (*Answer:* $\rho/a = 0.5, 0.657, 0.730$; $q_1 = 0, 3.526, 5.630$).

4.4 An annular plate of outer radius a and inner radius b is simply supported along both these edges, subjected to a uniform normal pressure p. Prove that the complete solution to the problem of plastic collapse involves the regime $M_\theta = M_0$ in an outer annulus $\rho \leq r \leq a$, the regime $M_r - M_\theta = M_0$ in an inner annulus $b \leq r \leq c$, and the regime $M_r = M_0$ in the intervening region. Using the equilibrium equation and the continuity conditions, and setting $q = pa^2/2M_0$, establish the relations

$$\frac{3}{q} = \left\{\left(1 + \frac{2\rho}{a}\right)\left(1 - \frac{\rho}{a}\right)^2\right\} = 3\left(\frac{\rho^2}{a^2} - \frac{c^2}{a^2}\right)$$

$$\frac{b^2}{c^2} + \ln\left(\frac{c^2}{b^2}\right) = 1. + \frac{2}{q}\left(\frac{a^2}{c^2}\right).$$

Note that c tends to zero as b tends to zero, and verify that the values of ρ/a and q in this limiting state are 0.442 and 5.12, respectively.

4.5 Consider the large deflection δ of a built-in circular plate of radius a over the range $\delta \geq h$, when subjected to a load P which is uniformly distributed over an inner circle of radius b. Assuming the special case of b tending to zero, and using the method of analysis presented in Section 4.2 (4.2.4), obtain the load–deflection formula

$$\frac{P}{4\pi M_0} = \frac{\delta}{h} + \frac{2h}{3\delta}, \quad \delta \geq h.$$

Indicate how this formula can be adopted to give an approximate load–deflection formula for nonzero values of b over the same range of values of δ.

4.6 A simply supported circular plate of radius a carries a uniformly distributed normal pressure p in the fully plastic range. The material yields according to the Tresca criterion and hardens kinematically with a constant slope c of the moment–curvature relationship. The solution involves a central portion of the plate of some radius ρ that corresponds to a corner of the yield hexagon. Using the equilibrium equation and the yield condition, show that the shape of the deformed plate is given by

$$\frac{w}{a} = \frac{pa^3}{8c}\left\{\left(1 - \frac{3\rho}{8a}\right)\frac{\rho^3}{a^3} - \frac{1}{4}\left(\frac{3\rho^2}{a^2} - \frac{r^2}{2a^2}\right)\frac{r^2}{a^2}\right\}, \quad 0 \le r \le \rho$$

$$\frac{w}{a} = \frac{p\rho^3}{8c}\left(1 - \frac{r}{a}\right), \quad \rho \le r \le a.$$

4.7 Referring to the simply supported strain-hardening circular plate considered in the preceding problem, show that the distribution of the radial bending moment in the inner region and the circumferential bending moment in the outer region are given by

$$M_r = M_0 + \frac{3p\rho^2}{16}\left(1 - \frac{r^2}{\rho^2}\right), \quad 0 \le r \le \rho.$$

$$M_\theta = M_0 + \frac{p\rho^2}{8r}, \quad \rho \le r \le a$$

Using the condition of continuity of M_r across $r = \rho$ shows that the ratio of the applied pressure p to that at the yield point state may expressed in the form

$$\frac{p}{p_0} = \left\{\left(1 - \frac{\rho^3}{a^3}\right) - \frac{3\rho^3}{4a^3}\ln\frac{a}{\rho}\right\}^{-1}.$$

4.8 A plate, having the shape of a regular polygon with n sides, is fully clamped along its boundary, and is subjected to a vertical load P. The plate may be assumed to collapse in the form of a pyramid with yield hinges formed along its edges Using the square yield condition for plate bending shows that the upper bound on the collapse load is $P = P_I$ when the load is concentrated at the centroid, and $P = P_2$ when the load is uniformly distributed over the entire area, where

$$P_1 = 4M_0 n \tan(\pi/n), \quad P_2 = 12M_0 n \tan(\pi/n).$$

4.9 A simply supported rectangular plate of sides $2a$ and $2b$ carries a concentrated load P at its centroid. Considering the collapse mechanism depicted in Fig. A, and using the square yield condition for plate bending, derive the upper bound formula.

$$\frac{P}{4M_0} = \sqrt{\frac{b^2}{a^2} - 1} + 2\sin^{-1}\left(\frac{a}{b}\right), \quad \frac{b}{a} \le \sqrt{2}.$$

Verify that the upper bound has a minimum value when $b/a = \sqrt{2}$, and indicate how the collapse mechanism should be modified to give the same collapse load for all higher values of b/a.

4.10 A clamped square plate of sides $2a$ carries a uniformly distributed normal pressure of intensity p. The assumed collapse mechanism consists of four identical yield line fans of angular span ϕ, as shown in Fig. B. Draw the associated angular velocity hodograph for the incipient collapse, and establish the work equation to obtain the upper bound on the collapse load in he form

$$\frac{pa^2}{12M_0} = \frac{\phi/2 + \tan(\pi/4 - \phi/2)}{(\phi/2)\sec^2(\pi/4 - \phi/2) + \tan(\pi/4 - \phi/2)}.$$

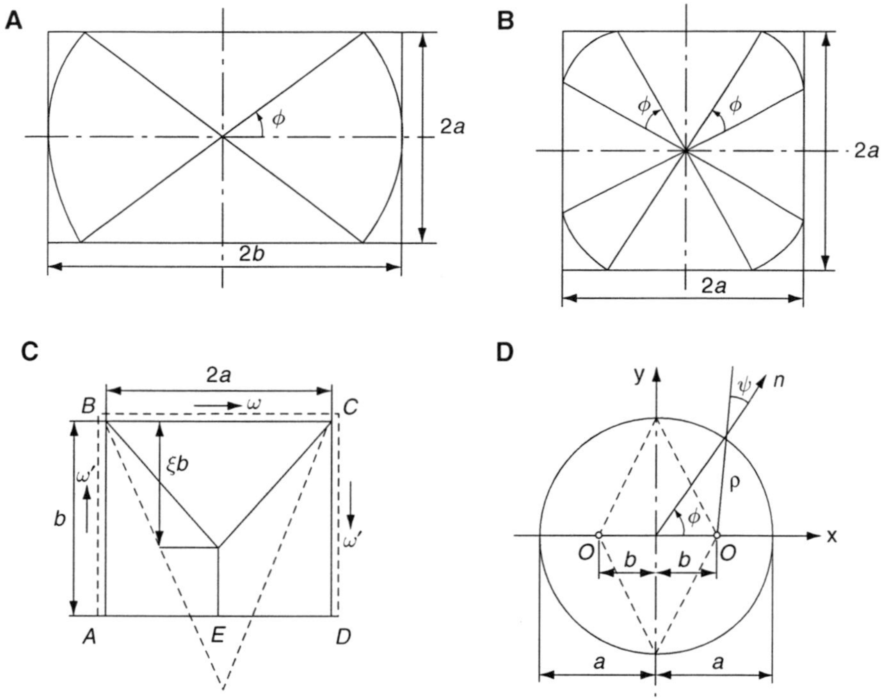

Fig. A

Verify that the minimum value of the load corresponds to $\phi = 29.9^\circ$ approximately, giving the best upper bound as $pa^2/M_0 = 10.87$.

4.11 A rectangular plate *ABCD*, which is free along the edge *AD* and simply supported along the other three edges, carries a uniform normal pressure of intensity *p*. For a range of values of the aspect ratio *b/2a*, the collapse mechanism may be taken as that indicated by solid lines in Fig. C. Using the yield line method, obtain the upper bound

$$\frac{pb^2}{6M_0} = \frac{1+\xi(b^2/a^2}{\xi(3-\xi)}, \quad \xi = \frac{a}{b}\left(\sqrt{3+\frac{a^2}{b^2}} - \frac{a}{b}\right), \quad \frac{b}{a} \geq 1.$$

For higher values of *b/a*, the yield lines intersect each other outside the plate, as indicated by the broken lines meeting at a vertical distance ξb from the side *BC*. Show that the upper bound is then modified to

$$\frac{pb^2}{6M_0} = \frac{1+(b^2/a^2)\xi^2}{3\xi - 1}, \quad \xi = \frac{1}{3}\left(1+\sqrt{1+\frac{9a^2}{b^2}}\right), \quad \frac{b}{a} \leq \sqrt{3}.$$

Verify that lower values of the collapse load are given by the first solution for $b/a \geq 1.56$, and by the second solution for $b/a \leq 1.56$.

4.12 A simply supported circular plate of radius a is loaded by two concentrated forces P, each acting on a diameter at a distance $b = \xi\, a$ from the center of the circle, as shown in Fig. D. For sufficiently small values of ξ, a parabolic state may be expected at the incipient collapse. Show that the equation to a typical characteristic and the angle made by it with the normal to the boundary, are given by

$$y = \frac{(x-b)\sin\theta}{a\cos\theta - b}, \qquad \tan\psi = \frac{\xi\sin\phi}{\sqrt{1-\xi^2\sin^2\phi}},$$

where $\phi = \theta + \psi$ is the characteristic angle. Prove that the collapse load is given by

$$\frac{P}{2M_0} = \frac{1}{\sqrt{1-\xi^2}}\left\{\pi - \tan^{-1}\left(\frac{\sqrt{1-\xi^2}}{\xi}\right)\right\}.$$

Establish the fact that this solution is statically admissible only over the range $0 \le \xi \le 1/\sqrt{2}$.

Chapter 5
Plastic Analysis of Shells

A shell is a thin-walled structure in which the material fills the space between two parallel or nearly parallel surfaces, the mean surface that halves the shell thickness being known as the middle surface of the shell. When the shell is loaded to the point of plastic collapse, the deformation proceeds in an unrestricted manner under constant load, provided geometry changes are disregarded and the material is considered as ideally plastic. As in the case of thin plates, the incipient deformation mode of the shell will be described in terms of that of the middle surface, and the associated stress field will be specified in terms of the stress resultants, acting across the shell thickness. The usual assumptions of rigid/plastic material will be made for the estimation of the collapse load.

5.1 Cylindrical Shells Without End Load

5.1.1 Basic Equations

Consider a circular cylindrical shell of uniform thickness h, subjected to an axially symmetrical loading without the presence of an axial load, Fig. 5.1a. We choose cylindrical coordinates (r, θ, x) in which the x-axis coincides with the longitudinal axis of the cylinder. Due to the rotational symmetry, the velocity at any point of the middle surface is specified by its components u and w in the axial and radially inward directions, respectively. Assuming that line elements normal to the middle surface remain normal during the deformation, the velocity field at any point in the shell may be written as

$$v_r = -w, \quad v_\theta = 0, \quad v_x = u - z\frac{dw}{dx},$$

where z is the radially inward distance from the middle surface. Denoting the radius to the middle surface by a, the nonzero components of the strain rate may be written as

$$\dot{\varepsilon}_\theta = -\frac{w}{a}, \quad \dot{\varepsilon}_x = \frac{du}{dx} - z\frac{d^2w}{dx^2}. \tag{5.1}$$

J. Chakrabarty, *Applied Plasticity, Second Edition*, Mechanical Engineering Series,
DOI 10.1007/978-0-387-77674-3_5, © Springer Science+Business Media, LLC 2010

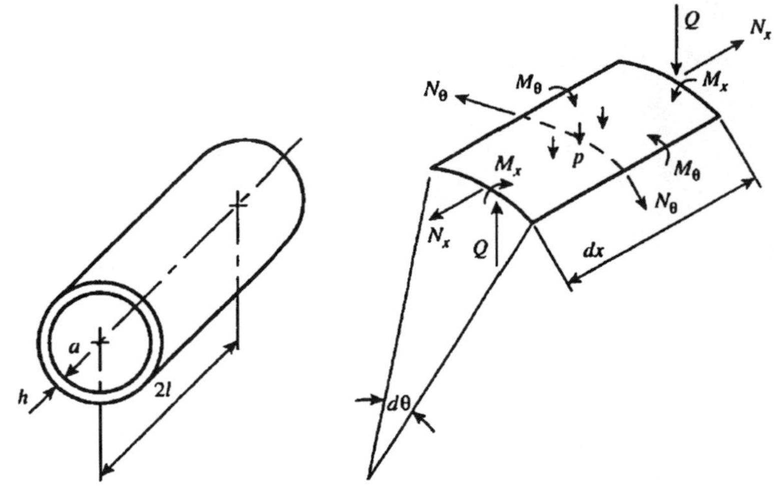

Fig. 5.1 Symmetrically loaded cylindrical shell. (**a**) Geometry of the shell and (**b**) equilibrium of shell element

Owing to symmetry, the velocity components u and w are functions of x only. In view of (5.1), the rate of dissipation of internal energy per unit area of the middle surface is

$$D = \int_{-h/2}^{h/s} (\sigma_x \dot{\varepsilon}_x + \sigma_\theta \dot{\varepsilon}_\theta) dz = N_x \dot{\lambda}_x + N_\theta \dot{\lambda}_\theta + M_x \dot{\kappa}_x,$$

where N_x, N_θ are the resultant membrane forces, and M_x the axial bending moment, defined as

$$N_x = \int_{-h/2}^{h/2} \sigma_x dz, \quad N_\theta = \int_{-h/2}^{h/2} \sigma_\theta dz \quad M_x = \int_{-h/2}^{h/2} \sigma_x z dz.$$

These are the generalized stresses for the problem, the corresponding generalized strain rates being

$$\dot{\lambda}_x = \frac{du}{dx}, \quad \dot{\lambda}_\theta = -\frac{w}{a}, \quad \dot{\kappa}_x = -\frac{d^2 w}{dx^2}, \tag{5.2}$$

which define the rates of extension and curvature of the middle surface. Since the circumferential curvature rate $\dot{\kappa}_\theta$ vanishes due to the assumed velocity field, the corresponding bending moment M_θ does not appear in the expression for D, and constitutes only a reaction for the problem.

Figure 5.1b shows the stress resultants acting in a typical shell element formed by adjacent cross sections and a pair of axial planes inclined at an angle d_θ with one another. Axial equilibrium requires N_x to be a constant, which must be zero when the end loads are absent. The conditions of radial and moment equilibrium are easily shown to be

$$\frac{dQ}{dx} + \frac{N_\theta}{a} = -p, \quad \frac{dM_x}{dx} - Q = 0, \tag{5.3}$$

where Q is the shearing force per unit length of the generator, and p is the radial pressure per unit area of the element. The elimination of Q between these two equations furnishes

$$\frac{d^2M_x}{dx^2} + \frac{N_\theta}{a} = -p.$$

This is the governing differential equation for equilibrium. Introducing the dimensionless variables

$$n_\theta = \frac{N_\theta}{N_0}, \quad m_x = \frac{M_x}{M_0}, \quad q = \frac{pa}{N_0}, \quad \xi = x\sqrt{\frac{2}{ah}}$$

where N_0 and M_0 are the fully plastic values of the stress resultants, equal to Yh and $Yh^2/4$, respectively, the equilibrium equation may be expressed as

$$\frac{d^2m_x}{d\xi^2} + 2n_\theta = -2q. \tag{5.4}$$

Across any section $x =$ constant, the bending moment M_x and the shearing force Q must be continuous, but N_θ and M_θ may be discontinuous without affecting equilibrium. It follows that the bending moment gradient dM_x/dx, which is equal to Q, must be a continuous function of x.

5.1.2 Yield Condition and Flow Rule

We begin with the situation where the material obeys the maximum shear criterion of Tresea (Drucker, 1953), the uniaxial yield stress of the material being denoted by Y. Possible distributions of stresses through the thickness of the shell, when M_x and N_θ are both positive, are shown in Fig. 5.2. The parameter η, defining the position of the neutral axis, must lie between 0 and 1 for the yield criterion not to be violated anywhere in the shell. When the cross section carries full moment as in Fig. 5.2a, the yield condition is $M_x = M_0$ with $0 \le N_\theta \le N_0/2$. For higher values of N_θ, considered in Fig. 5.2b, the stress resultants are

$$M_x = \left(1 - \eta^2\right) M_0, \quad N_\theta = \frac{1}{2}(1+\eta)\, N_0, \quad 0 \le \eta \le 1,$$

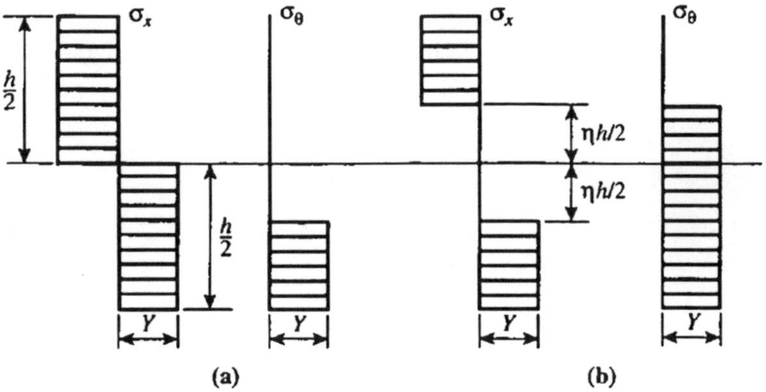

Fig. 5.2 Yield point state in a cylindrical shell without end load. (**a**) Section carrying the full moment and (**b**) section carrying less than full moment

which define the yield condition parametrically through η. In terms of the dimensionless variables m_x and n_θ, one quadrant of the interaction curve is therefore defined by the equations

$$m_x = 1 \quad \left(0 \le n_\theta \le \frac{1}{2}\right), \quad m_x = 4n_\theta\,(1 - n_\theta) \quad \left(\frac{1}{2} \le n_\theta < 1\right). \tag{5.5}$$

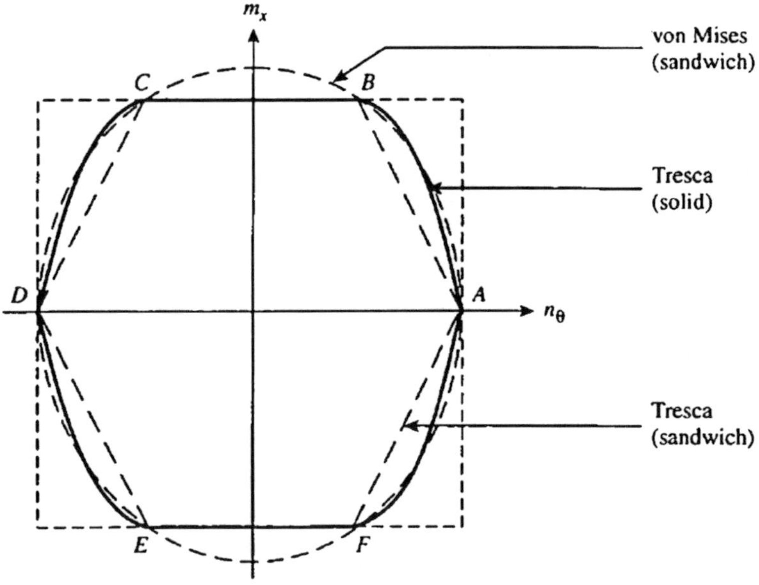

Fig. 5.3 Yield conditions for circular cylindrical shells without axial load

The other three quadrants of the interaction curve correspond to the stress distribution obtained by suitable reflections of those considered in Fig. 5.2. The complete interaction curve, which is symmetric with respect to m_x and n_θ, is shown by the solid lines in Fig. 5.3. Although it is based entirely on statically admissible stress states, the interaction curve is in fact actual since it represents the best lower bound.

It is instructive to rederive the interaction curve on the basis of a kinematically admissible velocity field. Consider the plane diagram in which the strain rates $\dot{\varepsilon}_x$ and $\dot{\varepsilon}_\theta$ are plotted as rectangular coordinates, Fig. 5.4. According to Tresea's associated flow rule, each of the six regions separated by the rectangular axes and the line $\dot{\varepsilon}_\theta = -\dot{\varepsilon}_x$ corresponds to a corner of the yield hexagon *ABCDEF.* The separating lines themselves correspond to the six sides of the hexagon. In view of the relations

$$\dot{\varepsilon}_x = \dot{\lambda}_x + z\dot{\kappa}_x, \qquad \dot{\varepsilon}_\theta = \dot{\lambda}_\theta,$$

which are obtained by combining (5.1) and (5.2), the variation of the strain rate across the wall thickness at any section is represented by a straight line parallel to the $\dot{\varepsilon}_x$-axis, the center R of the straight line corresponds to the middle surface. Since the axial force is zero, equal segments of this straight line must lie in regions B and D unless σ_x is identically zero over the whole cross section. It follows that R lies

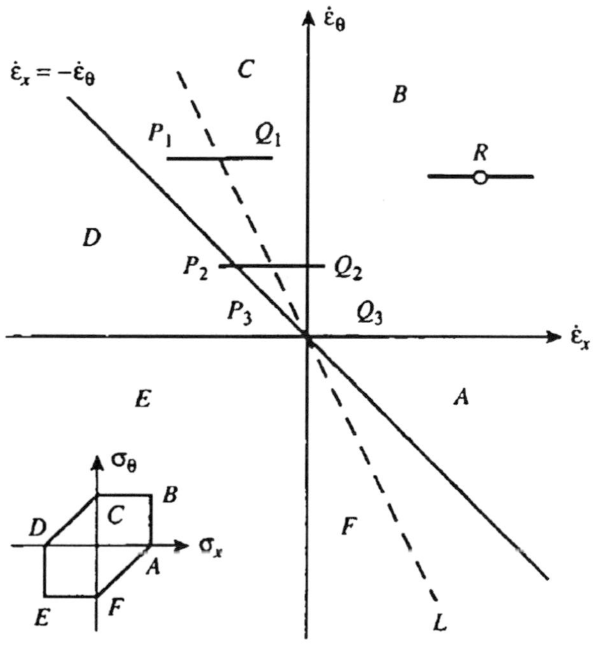

Fig. 5.4 Strain rate distribution through the wall thickness of a cylindrical shell according to the Tresca theory

on the broken straight line inclined at an angle $\tan^{-1}(\frac{1}{2})$ to the vertical. Considering the situation where $\dot{\lambda}_\theta$ and $\dot{\kappa}_x$ are both positive, we can envisage three different possibilities as indicated in the figure. The line $P_1 Q_1$ corresponds to $\sigma_\theta = Y$ and $\sigma_x = 0$ over the entire cross section, giving the single point $n_\theta = 1$, $m_x = 0$ on the interaction curve. The lines P_2Q_2 and P_3Q_3, on the other hand, correspond to the stress distributions of Fig. 5.2. It follows that the true interaction curve for a Tresca material is indeed that given by (5.5), obtained from statistical considerations.

In the solution of special problems, it is usually convenient to use simplified yield conditions based on the sandwich approximation of the uniform solid section. The sandwich section consists of two thin flanges each of thickness $t/2$ and yield stress σ_0, separated by a core of thickness h'. The core does not carry any membrane stresses, which are assumed to be uniformly distributed over each flange. If the sandwich shell is assumed to have the same yield force N_0 and yield moment M_0 as for the uniform shell, then

$$t\sigma_0 = hY, \qquad 2h' = h.$$

If the stresses in the interior and exterior flanges of the sandwich shell are denoted by σ^+ and σ^-, respectively, we have

$$N_x = \left(\sigma_x^+ + \sigma_x^-\right) t, \qquad M_x = \frac{1}{2}\left(\sigma_x^+ - \sigma_x^-\right) th',$$

$$N_\theta = \left(\sigma_\theta^+ + \sigma_\theta^-\right) t, \qquad M_\theta = \frac{1}{2}\left(\sigma_\theta^+ - \sigma_\theta^-\right) th'.$$

These equations may be solved for the stress components in terms of the stress resultants. Setting $N_x = 0$, the result may be expressed as

$$\sigma_x^\pm = \pm m_x \sigma_0, \quad \sigma_\theta^\pm = (n_\theta \pm m_\theta)\,\sigma_0. \tag{5.6}$$

When n_θ and m_x are both positive, the cross section will carry full moment according to the Tresca criterion if $\sigma_x^+ = \sigma_0, \sigma_x^- = -\sigma_0, \sigma_\theta^+ = \sigma_0, -\sigma_0 \leq \sigma_\theta^- \leq 0$. In view of (5.6), these conditions are equivalent to

$$m_x = 1, \quad n_\theta + m_\theta = 1, \quad -1 \leq n_\theta - m_\theta \leq 0.$$

For $0 \leq m_x \leq 1$, the values of σ_x^+ and σ_x^- will be numerically less than σ_0, and the yield criterion is then modified to $\sigma_\theta^+ = \sigma_0, \sigma_\theta^- - \sigma_x^- = \sigma_0, 0 \leq \sigma_x^+ \leq \sigma_0$, giving

$$n_\theta + m_\theta = 1, \quad n_\theta - m_\theta + m_x = 1, \quad 0 \leq m_x \leq 1.$$

The quantity m_θ, which does not represent a generalized stress, is easily eliminated from the last two sets of equations to obtain

$$m_x = 1 \quad \left(0 \le n_\theta \le \frac{1}{2}\right), \quad m_x + 2n_\theta = 2 \quad \left(\frac{1}{2} \le n_\theta \le 1\right). \tag{5.7}$$

The yield condition is therefore linearized by the sandwich approximation. Over the remaining quadrants, the yield equations are obtained from (5.7) by suitable changes in sign. The complete interaction curve for the sandwich shell is the inscribed hexagon shown in Fig. 5.3.

The flow rule associated with the yield condition requires the vector $(N_0\dot{\lambda}_\theta, M_0\dot{\kappa}_x)$ to be normal to the interaction curve defined by the dimensionless stress resultants (n_θ, m_x). For the hexagonal yield locus, the flow rule corresponding to sides AB and DE gives $N_0\dot{\lambda}_\theta/M_0\dot{\kappa}_x = 2$ and the substitution from (5.2) leads to the differential equation,

$$\frac{d^2w}{d\xi^2} - w = 0$$

in terms of the dimensionless variable ξ. When q is a constant, $q + n_\theta$ is seen to satisfy the same differential equation as that for w, for the same inclined side of the hexagon. The integration of the above equation gives

$$w = A\cosh\xi + B\sinh\xi,$$

where A and B are constants to be determined from the boundary conditions. The expression for w corresponding to sides CD and AF is similarly established. Since $\dot{\lambda}_\theta = 0$ corresponding to sides BC and EF, we have $w = 0$ along these sides indicating that no plastic flow can occur when the stress point lies on these sides. The yield condition, flow rule, and the velocity field associated with each of the six sides of the hexagon are listed in Table 5.1. Results for the nonlinear yield condition of type (5.5) are also included in the table for comparison.

Table 5.1 Yielding and flow in cylindrical shells without end load

	Linearized yield condition		Exact Tresca condition	
Line or Curve	w	m_x	m_x	$N_0\dot{\lambda}_\theta : M_0\dot{\kappa}_x$
AB	$A\cosh\xi + B\sinh\xi$	$2(1-n_\theta)$	$4n_\theta(1-n_\theta)$	$4(2n_\theta-1)$:1
BC	0	1	1	0:1
CD	$A\cos\xi + B\sin\xi$	$2(1+n_\theta)$	$-4n_\theta(1+n_\theta)$	$4(2n_\theta+1)$:1
DE	$A\cosh\xi + B\sinh\xi$	$-2(1+n_\theta)$	$4n_\theta(1+n_\theta)$	$4(2n_\theta+1)$: − 1
EF	0	−1	−1	0:−1
FA	$A\cos\xi + B\sin\xi$	$-2(1-n_\theta)$	$-4n_\theta(1-n_\theta)$	$4(2n_\theta-1)$: − 1

As in the case of plates, the velocity w must be everywhere continuous, but the velocity slope dw/dx may be discontinuous across a circular section called a hinge circle. The ratio of d^2w/dx^2 to w therefore tends to infinity along a hinge circle, indicating that a hinge circle must correspond to one of the end points of sides BC and EF. The sense of the discontinuity in slope at a hinge circle must be compatible with the sign of the local bending moment M_x in order to have a positive rate of plastic work.

When the material yields according to the von Mises criterion, the relationship between the generalized stresses at the yield point is easily obtained on the basis of the sandwich approximation. Since the yield criterion $\sigma_x^2 - \sigma_x\sigma_\theta + \sigma_\theta^2 = \sigma_0^2$ must be satisfied in both the inner and the outer flanges of the sandwich shell, we have

$$m_x^2 - m_x(n_\theta + m_\theta) + (n_\theta + m_\theta)^2 = 1,$$
$$m_x^2 + m_x(n_\theta - m_\theta) + (n_\theta - m_\theta)^2 = 1$$

in view of (5.6). These two equations are simultaneously satisfied only if $m_\theta = m_x/2$. The substitution for m_θ into either one of the above equations furnishes the yield condition

$$2f(n_\theta, m_x) = n_\theta^2 + \frac{3}{4}m_x^2 = 1. \tag{5.8}$$

The interaction curve defined by (5.8) is an ellipse passing through the corners of the hexagonal yield locus, as shown in Fig. 5.3. The flow rule associated with the yield condition (5.8) furnishes the generalized strain rates as

$$N_0\dot{\lambda}_\theta = \mu\frac{\partial f}{\partial n_\theta} = \mu n_\theta, \quad M_0\dot{\kappa}_x = \mu\frac{\partial f}{\partial m_x} = \frac{3}{4}\mu m_x, \tag{5.9}$$

where μ is necessarily positive for plastic flow and is equal to the dissipation rate D per unit area of the middle surface. Indeed, it follows from (5.8) and (5.9) that

$$D = N_\theta\dot{\lambda}_\theta + M_x\dot{\kappa}_x = \mu\left(n_\theta^2 + \frac{3}{4}m_x^2\right) = \mu.$$

The exact yield condition for a uniform shell made of the von Mises material has been derived by Hodge (1961). In terms of an auxiliary angle ψ, the exact yield condition is given by

$$\left.\begin{aligned} n_\theta &= \pm\cot\psi\tanh^{-1}(\sin\psi), \\ m_X &= \pm\frac{2}{\sqrt{3}}\left[\cot^2\psi\tanh^{-1}(\sin\psi) - \operatorname{cosec}\psi\right]. \end{aligned}\right\} \tag{5.10}$$

The corresponding interaction curve falls slightly outside the ellipse defined by (5.8), while passing through the corners A and D of the yield hexagon. The square yield condition indicated by broken lines in the figure is sometimes used as a convenient approximation.

5.1.3 Shell Under Uniform Radial Pressure

Consider the plastic collapse of a circular cylindrical shell of length 21, which is clamped at both ends and carries a uniform hydrostatic pressure p on the cylindrical surface. We choose one end of the shell as the section $x = 0$ and adopt the piecewise linear yield condition represented by the hexagon of Fig. 5.3. For a sufficiently short shell, we may expect it to collapse as a result of hinge circles formed at the ends and at the center (Hodge, 1954). Then the boundary conditions are $m_x = -1$ at $x = 0$ and $m_x = 1$ at $x = 1$, only one-half of the shell being considered. Since n_θ is everywhere compressive it is reasonable to suppose that the stress profile involves the sides ED and DC, the meeting point D corresponding to $x = x_0$ or $\xi = \xi_0$, which is to be determined from the stress field. Setting $m_x = -2(1 + n_\theta)$ in (5.4), the equilibrium equation for the outer part of the shell is reduced to

$$\frac{d^2 n_\theta}{d\xi^2} - n_\theta = q, \qquad 0 \leq \xi \leq \xi_0.$$

Integrating and using the boundary conditions $n_\theta = -\frac{1}{2}$ at $\xi = 0$ and $n_\theta = -1$ at $\xi = \xi_0$, we obtain the solution

$$n_\theta = -q + (q-1)\frac{\sinh \xi}{\sinh \xi_0} + \left(q - \frac{1}{2}\right)\frac{\sinh(\xi_0 - \xi)}{\sinh \xi_0}, \quad 0 \leq \xi \leq \xi_0. \tag{5.11a}$$

In the inner part of the shell, the yield condition is $m_x = 2(1 + n_\theta)$ and the equilibrium equation (5.4) becomes

$$\frac{d^2 n_\theta}{d\xi^2} + n_\theta = -q, \qquad \xi_0 \leq \xi \leq \omega,$$

where $\omega = l\sqrt{2/ah}$. Since $m_x = 1$ and $dm_x/d\xi = 0$ at the central section, the boundary conditions are $n_\theta = -\frac{1}{2}$ and $dn_\theta/d\xi = 0$ at $\xi = \omega$, and the solution is easily shown to be

$$n_\theta = -q + \left(q - \frac{1}{2}\right)\cos(\omega - \xi), \quad \xi_0 \leq \xi \leq \omega. \tag{5.11b}$$

The quantities q and ξ_0 can be determined from the conditions of continuity of m_x and $dm_x/d\xi$ across $\xi = \xi_0$. We therefore set $n_\theta = -1$ and $\xi = \xi_0$ in (5.11b) and equate the value of $-dn_\theta/d\xi$ at $\xi = \xi_0$ given by (5.11a) to that of $dn_\theta/d\xi$ at $\xi = \xi_0$ given by (5.11b), the result being

$$\frac{2q-1}{2q-2} = \frac{1}{\cos(\omega - \xi_0)} = \frac{\cosh \xi_0}{1 - \sinh \xi_0 \sin(\omega - \xi_0)}.$$

After some algebraic manipulation, the solution can be expressed in the more convenient form

$$\sinh \xi_0 = \frac{\sin(\omega - \xi_0)}{1 + \sqrt{2}\cos(\omega - \xi_0)}, \quad 2q = \frac{2 - \cos(\omega - \xi_0)}{1 - \cos(\omega - \xi_0)}. \tag{5.12}$$

Starting with an assumed value of $\omega - \xi_0$, a set of values of ω, ξ_0, and q satisfying (5.12) can be easily computed. It can be shown that the variation of m_x with ξ, given by (5.11) and the yield condition, is monotonic between $\xi = 0$ and $\xi = \omega$, justifying the initial assumption regarding the stress profile. The solution is therefore statically admissible and provides a lower bound on the collapse load for all values of ω.

The range of values of ω for which the solution is kinematically admissible can be determined by considering the velocity field. Denoting the velocity at $\xi = \xi_0$ by w_0, and using the boundary condition $w = 0$ at $\xi = 0$ and the condition of continuity of $dw/d\xi$ at $\xi = \xi_0$, the constants of integration for $w(\xi)$ given in Table 5.1 can be determined for each of the two plastic regimes *DE* and *CD*. The velocity distribution therefore becomes

$$\left.\begin{aligned} w &= w_0 \cos ech\, \xi_0 \sinh \xi, & 0 \le \xi \le \xi_0, \\ w &= w_0 \left[\cos(\xi - \xi_0) + \coth \xi_0 \sin(\xi - \xi_0)\right], & \xi_0 \le \xi \le \omega. \end{aligned}\right\} \tag{5.13}$$

The existence of the hinge circle at $\xi = \omega$ (where $m_x = 1$) requires the velocity slope $dw/d\xi$ to be positive at this section. By the second equation of (5.13), this condition is equivalent to

$$\coth \xi_0 \ge \tan(\omega - \xi_0) \quad \text{or} \quad \omega \le 1.65 \qquad (\xi_0 \le 0.5),$$

in view of the first equation of (5.12). The collapse load given by (5.12) is therefore the actual collapse load for $\omega \le 1.65$, but provides only a lower bound for $\omega > 1.65$.

When ω exceeds 1.65, the discontinuity in the velocity slope at $\xi = \omega$ disappears, and the statistical boundary condition $m_x = 1$ at $\xi = \omega$ should be replaced by the kinematical condition $d\omega/d\xi = 0$ at $\xi = \omega$. The stress profile begins at *E*, corresponding to $\xi = 0$, proceeds along *ED* to reach *D* for $\xi = \xi_0$, and then terminates at some point of *DC* for $\xi = \omega$. The velocity distribution is still given by (5.12), but the relationship between ω and ξ_0 is now modified to

$$\coth \xi = \tan(\omega - \xi_0), \quad \omega \ge 1.65, \tag{5.14}$$

which follows from the condition of zero velocity slope at $\xi = \omega$. The expression for n_θ in the region $0 \le \xi \le \xi_0$ is given by (5.11a), while that in the region $\xi \le \xi_0 \le \omega$ is obtained from the solution of the relevant differential equation using the boundary conditions $m_x = 0$ at $\xi = \xi_0$ and $dm_x/d\xi = 0$ at $\xi = \omega$. Since these conditions are equivalent to $n_\theta = -1$ at $\xi = \xi_0$ and $dn_\theta/d\xi = 0$ at $\xi = \omega$, we obtain the solution

$$n_\theta = -q + (q - 1)\sec(\omega - \xi_0)\cos(\omega - \xi), \quad \xi_0 \le \xi \le \omega. \tag{5.15}$$

The collapse load is furnished by the condition of continuity of $dm_x/d\xi$ across $\xi = \xi_0$, where $dn_\theta/d\xi$ changes sign without changing its magnitude. Using

(5.11a), (5.14), and (5.15), the relationship between q and ω can be expressed parametrically as

$$\left.\begin{aligned} 2q &= (4\cosh\xi_0 - 1)/(2\cosh\xi_0 - 1)\,, \\ \omega &= \xi_0 - \tan^{-1}(\coth\xi_0)\,, \end{aligned}\right\} \quad \omega \geq 1.65. \tag{5.16}$$

This is the actual collapse load for $\omega \geq 1.65$, since the solution is both statically and kinematically admissible. For $\pi/2 \geq \omega < 1.65$, the value of m_x, given by (5.15) and the yield condition is found to exceed unity near $\xi = \omega$. The solution may therefore be accepted as an upper bound in the range $\pi/2<\omega< 1.65$. The solution is not defined for $\omega < \pi/2$, since the value of ξo predicted by (5.14) then becomes negative. The variations of q and ξ_0/ω with ω are shown graphically in Fig. 5.5.

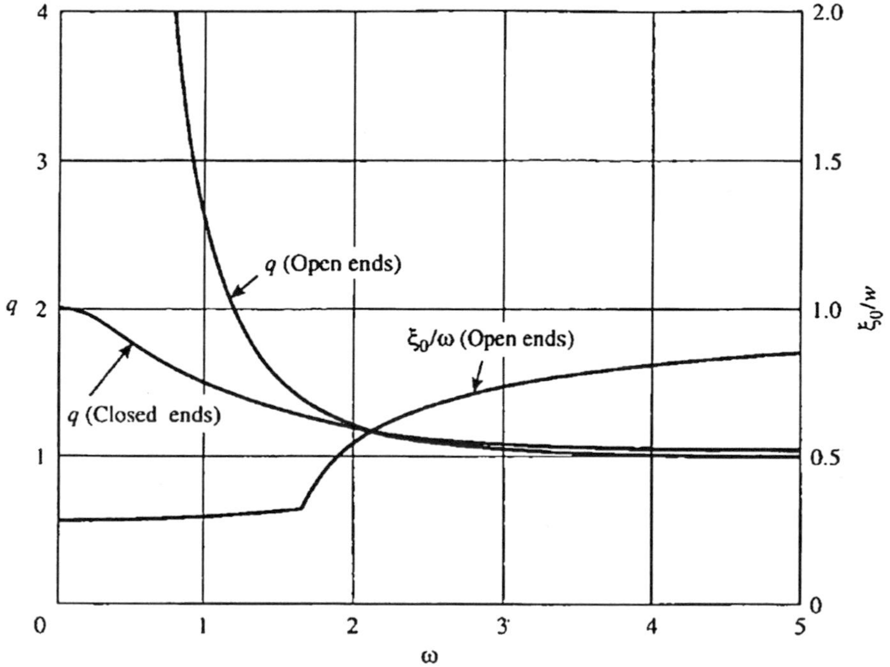

Fig. 5.5 Dimensionless collapse load for clamped cylindrical shells subjected to uniform external pressure

The plastic collapse of cylindrical shells under other types of boundary condition can be treated in a similar manner. The influence of reinforcing longitudinal ribs on the collapse pressure has been examined by Biron and Sawczuk (1967). The effect of geometry changes on the limit load has been examined by Coon and Gill (1968). Displacements in cylindrical shells, involving the complete elastic/plastic analysis, have been discussed by Hodge (1956a), Hodge and Nardo (1958), Montague and Horner (1968), Ohashi and Okouchi (1975), and Brooks (1987).

5.1.4 Shell Under a Band of Pressure

A cylindrical shell of length $2l$ is symmetrically loaded by a band of uniform external pressure p, the width of the band being denoted by $2b$. For simplicity, we shall analyze the problem on the basis of the square yield condition of Fig. 5.3, assuming that the ends of the cylinder are stress-free (Eason and Shield, 1955). Only the right-hand half of the cylinder will be considered, taking the origin of coordinate at the center of the band. A sufficiently short cylinder will collapse without a yield hinge, and consequently, the yield condition in this case is $n_\theta = -1$ over the entire length. The velocity ω has a constant value giving $\dot{\kappa}_x = 0$, which satisfies the flow rule associated with the yield condition. The equilibrium equation (5.4) becomes

$$\frac{d^2 m_x}{d\xi^2} = 2, \ \xi \geq \alpha; \ \frac{d^2 m_x}{d\xi^2} = 2\,(1-q)\,, \ \xi \leq \alpha, \tag{5.17}$$

where $\alpha = b\sqrt{2/ah}$. Using the free-end boundary conditions $m_x = dm_x/d\xi = 0$ at $\xi = \omega = l\sqrt{2/ah}$, the symmetry condition $dm_x/d\xi = 0$ at $\xi = 0$, and the condition of continuity of m_x across $\xi = \alpha$, we obtain the solution

$$m_x = (\omega - \xi)^2, \qquad \alpha \leq \xi \leq \omega,$$
$$m_x = (\omega - \alpha)^2 + (q-1)(\alpha^2 - \xi^2), \quad 0 \leq \xi \leq \alpha.$$

The collapse load is given by the condition of continuity of $dm_x/d\xi$ across $\xi = \alpha$, and the range of validity of the solution corresponds to $m_x \leq 1$ at $\xi = 0$. Thus

$$q\alpha = \omega, \ 2\omega \leq \alpha + \sqrt{4+\alpha^2}. \tag{5.18}$$

When the length of the shell is greater than the limit set by (5.18), the collapse mode would involve a hinge circle at the central section $\xi = 0$. We assume the stress profile to be on the side $n_\theta = -1$ over a central part of the cylinder, where $\dot{\lambda}_\theta \leq 0$ and $\dot{\kappa}_x = 0$ by the flow rule. The integration of the equation $d^2w/d\xi^2 = 0$ results in the velocity distribution

$$w = w_0\,(1 - \xi/\xi_0)\,. \qquad 0 \leq \xi \leq \omega,$$

where w_0 is the radial velocity at $\xi = 0$, the velocity being assumed to vanish at some section $\xi = \xi_0$. The velocity field implies that each half of the cylinder deforms into a conical surface. Since $\dot{\lambda}_\theta \geq 0$ for $\xi \geq \xi_0$ the circumferential force n_θ changes sign discontinuously across $\xi = \xi_0$. The yield condition therefore becomes $n_\theta = 1$ over the length $\xi_0 \leq \xi \leq \omega$, and the equilibrium equation (5.4) reduces to

$$\frac{d^2 m_x}{d\xi^2} = -2, \qquad \xi_0 \leq \xi \leq \omega.$$

In view of the boundary conditions $m_x = dm_x/d\xi = 0$ at the end $\xi = \omega$, the solution of this differential equation can be written as

$$m_x = -(\omega - \xi)^2, \qquad \xi_0 \le \xi \le \omega.$$

The equilibrium equations for $\alpha \le \xi \le \xi_0$ and $0 \le \xi \le \alpha$ are identical to the first and second equations, respectively, of (5.17). The first equation is integrated under the conditions of continuity of m_x and $dm_x/d\xi$ across $\xi = \xi_0$, while the second equation is integrated under the conditions $m_x = 1$ and $dm_x/d\xi = 0$ at $\xi = 0$, the result being

$$\left.\begin{aligned} m_x &= (\xi_0 - \xi)^2 - 2(\omega - \xi)(\xi_0 - \xi) - (\omega - \xi_0)^2, & \alpha \le \xi \le \xi_0, \\ m_x &= 1 - (q-1)\xi^2, & 0 \le \xi \le \alpha. \end{aligned}\right\} \qquad (5.19)$$

The quantities ξ_0 and q are now obtained from the conditions of continuity of m_x and $dm_x/d\xi$ across $\xi = \alpha$. It follows from (5.19) that

$$2\xi_0 = \alpha + \sqrt{2 + \omega^2 + (\omega - \alpha)^2}, \qquad q\alpha = 2\xi_0 - \omega. \qquad (5.20)$$

The bending moment m_x has a stationary minimum value of $-2(\omega - \xi_0)^2$, occurring at $\xi = 2\xi_0 - \omega$ in the region where external pressure is zero. The solution will therefore be statically admissible so long as $2(\omega - \xi_0)2 \le 1$. In view of (5.20), the range of validity of the preceding solution is found to be

$$\alpha + \sqrt{4 + \alpha^2} \le 2\omega \le 2\sqrt{2} + \alpha + \sqrt{8 + \alpha^2} \qquad (5.21)$$

As ω exceeds the critical value specified by (5.21), the plastic collapse of the shell would involve a rigid part on either side, the deformation mode being again conical in accordance with the velocity field

$$w = w_0(1 - \xi/\xi_1), \qquad 0 \le \xi \le \xi_1,$$

where ξ_1 represents the semilength of the incipient deformation. The velocity slope is discontinuous at $\xi = 0$ and at $\xi = \xi_1$, which specify the location of hinge circles requiring $m_x = 1$ and $\xi = 0$ and $m_x = -1$ at $\xi = \xi_1$. The yield condition compatible with the velocity field is $n_\theta = -1$, and the distribution of m_x over the loaded part of the shell is the same as that in (5.19). For the remainder of the deforming region, the bending moment is obtained from the solution of the first equation of (5.17), together with the conditions of continuity of m_x and $dm_x/d\xi$ across $\xi = \alpha$. Thus

$$\left.\begin{aligned} m_x &= 1 - (q-1)\xi^2, & 0 \le \xi \le \alpha, \\ m_x &= 1 - (q-1)(2\xi - \alpha)\alpha + (\xi - \alpha)^2, & \alpha \le \xi \le \xi_1. \end{aligned}\right\} \qquad (5.22)$$

The section $\xi = \xi_1$ corresponds to the position of minimum m_x given by the second expression of (5.22). Since the minimum value of m_x must be equal to –1 for the formation of the hinge circle, it is easy to show that

$$2\xi_1 = 2q\alpha = \alpha + \sqrt{8 + \alpha^2}. \tag{5.23}$$

The collapse load given by (5.23) is identical to that given by (5.20) when ω has the upper limit defined by (5.21). It follows from (5.23) that the collapse load is independent of ω in the supercritical range. Figure 5.6 indicates how the dimensionless collapse load varies with the parameters α and ω.

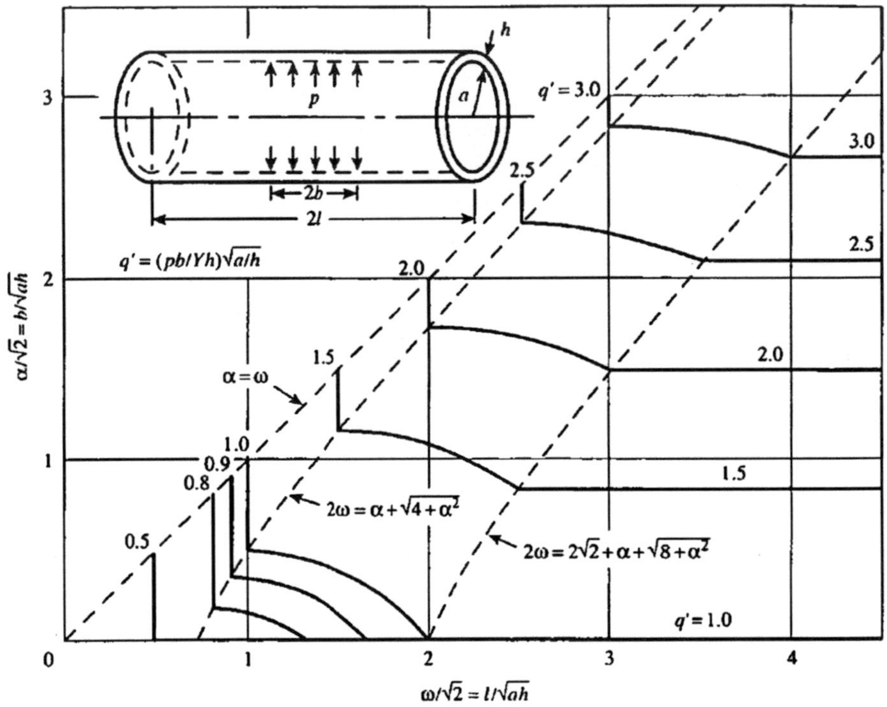

Fig. 5.6 Relationship between shell parameters at the incipient collapse of a cylindrical shell for different values of the collapse load parameter $q' = qa/\sqrt{2}$ (after Eason and Shield, 1955)

For a sufficiently long shell, the collapse load is actually independent of the end condition, which merely affects the minimum shell length for which the solution is applicable. When the shell is simply supported at both ends, a statically admissible stress field can be easily constructed in the rigid part of the shell, not only to complete the solution but also to establish the critical shell length (Hodge, 1981). The equilibrium equation (5.4), the boundary condition $m_x = 0$ at $\xi = \omega$, and the continuity conditions $m_x = -1$ and $dm_x/d\xi = 0$ at $\xi = \xi_1$ are simultaneously satisfied by taking

$$m_x = -1 + \left(\frac{\xi - \xi_1}{\omega - \xi_1}\right)^2, \quad n_\theta = -\frac{1}{(\omega - \xi_1)^2}, \quad \xi_1 \leq \xi \leq \omega. \tag{5.24}$$

The stress distribution will be statically admissible so long as n_θ is numerically less than unity. Using the expression for ξi from (5.23), this condition may be written as

$$\omega \geq 1 + \xi_1 = 1 + \frac{1}{2}\left(\alpha + \sqrt{8 + \alpha^2}\right).$$

For somewhat shorter shells with simple supports, the entire cylinder is plastic with $n_\theta = -1$. The distribution of m_x is again given by (5.22), which now holds for $0 \leq \xi \leq \omega$. The boundary condition $m_x = 0$ at $\xi = \omega$ furnishes the collapse load parameter q as

$$q\alpha = \left(1 + \omega^2\right) / (2\omega - \alpha), \quad \omega \leq 1 + \frac{1}{2}\left(\alpha + \sqrt{8 + \alpha^2}\right). \tag{5.25}$$

The value of $q\,\alpha$ corresponding to the critical value of ω is found to be identical to (5.23). It may be noted that $q\alpha$ is equal to $\omega - 1$ for a simply supported shell when the length is critical. The plastic analysis of cylindrical shells under a ring of force applied at some distance away from the central section has been discussed by Demir (1965).

When a cylindrical shell is subjected to a ring of force of intensity P per unit circumference at the central section $x = 0$, the collapse load is directly obtained from the solution for a band of pressure by letting qa tend to the limit $(P/N_0)\sqrt{a/2h}$, as α tends to zero. An elastic/plastic analysis for this particular case has been presented by Brooks (1988).

5.1.5 Solution for a von Mises Material

To estimate the degree of accuracy of the collapse load based on the assumed yield condition, it is instructive to compare the solution with that based on the von Mises yield condition. For simplicity, we consider the nonlinear yield condition (5.8) and apply it to the plastic collapse of a simply supported cylindrical shell centrally loaded with a ring of force of intensity P per unit circumference. For a relatively short shell, the collapse mode would involve a single hinge circle located at the central section $\xi = 0$ with $\dot{\lambda}_\theta < 0$ throughout the shell. Since $n_\theta < 0$ by the associated flow rule, the equilibrium equation (5.4) becomes,

$$\frac{d^2 m_x}{d\xi^2} = \sqrt{4 - 3m_x^2}, \qquad 0 \leq \xi \leq \omega.$$

The integration of this equation is facilitated by multiplying it with $dm_x/d\xi$ to obtain the differential equation

$$\frac{d}{d\xi}\left(\frac{dm_x}{d\xi}\right)^2 = 2\sqrt{4-3m_x{}^2}\left(\frac{dm_x}{d\xi}\right)$$

The existence of the hinge circle at the central section requires $n_\theta = 0$ or $m_x = 2/\sqrt{3}$ at $\xi = 0$, while the shearing force of magnitude $P/2$ acting at this section requires

$$\frac{dM_x}{dx} = -\frac{P}{2} \quad or \quad \frac{dm_x}{d\xi} = -\frac{P}{N_0}\sqrt{\frac{2\alpha}{h}} = -\sqrt{2}q$$

at $\xi = 0$. Using these boundary conditions, the integration of the differential equation results in

$$\left(\frac{dm_x}{d\xi}\right)^2 = 2\left(q^2 - \frac{\pi}{\sqrt{3}}\right) + m_x\sqrt{4-3m_x^2} + \frac{4}{\sqrt{3}}\sin^{-1}\left(\frac{\sqrt{3}}{2}m_x\right). \tag{5.26}$$

Since the right-hand side of (5.26) is always positive, and the support condition requires $m_x = 0$ at $\xi = \omega$, the solution is valid only for $q \geq \sqrt{\pi/\sqrt{3}} \approx 1.347$. Over this range, m_x is negative in a part of the shell and has a minimum value $m_x = -m_0$ at some section $\xi = \xi_0$, the bending moment gradient $dm_x/d\xi$ being negative for $\xi < \xi_0$. It follows from (5.26) that

$$\xi_0 = \int_{-m_0}^{2/\sqrt{3}} \{f(m_x)\}^{-1/2}\, dm_x, \quad \omega = \xi_0 + \int_{-m_0}^{0} \{f(m_x)\}^{-1/2}\, dm_x, \tag{5.27}$$

where $f(m_x)$ represents the expression on the right-hand side of (5.26). The integrals in (5.27) can be evaluated numerically for any assumed value of m_0, the corresponding q being given by the condition $f(-m_0) = 0$.

As ω increases from 1.096, q increases from the value 1.347 and attains a limiting value when $m_0 = 2/\sqrt{3}$. Setting $m_x = -2/\sqrt{3}$ and $dm_x/d\xi = 0$ in (5.26), we get $q = \sqrt{2\pi/\sqrt{3}}$ in the limit. The corresponding value of ξ_0 and ω is obtained from (5.27) by setting $m_0 = 2/\sqrt{3}$ and using (5.26) for $f(m_x)$, the final results being

$$q \approx 1.905, \quad \xi_0 \approx 2.076, \quad \omega \approx 3.662.$$

Longer shells will collapse by the formation of hinge circles at $\xi = 0$ and $\xi = 2.076$, the material in the region $2.076 \leq \xi \leq \omega$ remaining rigid at the incipient collapse. The collapse load for $\omega \geq 3.662$ is $q = 1.905$, which may be compared with the value $q = 2.0$ given by the square yield condition. The Tresca condition for a uniform shell and the von Mises condition for a uniform shell, on the other hand, are found to give $q = 1.826$ and $q = 1.949$, respectively (Sawczuk and Hodge, 1960). Since $\dot{\kappa}_x$ has the same sign as that of m_x according to the flow rule, a velocity field exists that is compatible with the collapse mode involving a hinge circle at the central section.

5.2 Cylindrical Shells with End Load

5.2.1 Yield Condition and Flow Rule

When a cylindrical shell is subjected to an axial force along with a radial loading, the problem is more complicated due to the presence of three nonzero stress resultants, namely N_x, N_θ, and M_x. Consider first the exact yield condition for a uniform shell when the material yields according to the Tresca criterion. We shall derive the yield condition following the static approach, using possible distributions of the stress components σ_x and σ_θ across the thickness (Hodge, 1954). When the cross section is fully plastic due to the hoop stress alone, a statically admissible stress distribution is that shown in Fig. 5.7a, where $-1 \leq \alpha \leq \beta \leq 1$. The resultant forces and moments in the dimensionless form are

$$n_\theta = 1, \quad n_x = 1 - \frac{1}{2}(\beta - \alpha), \quad m_x = \frac{1}{2}\left(\beta^2 - \alpha^2\right),$$

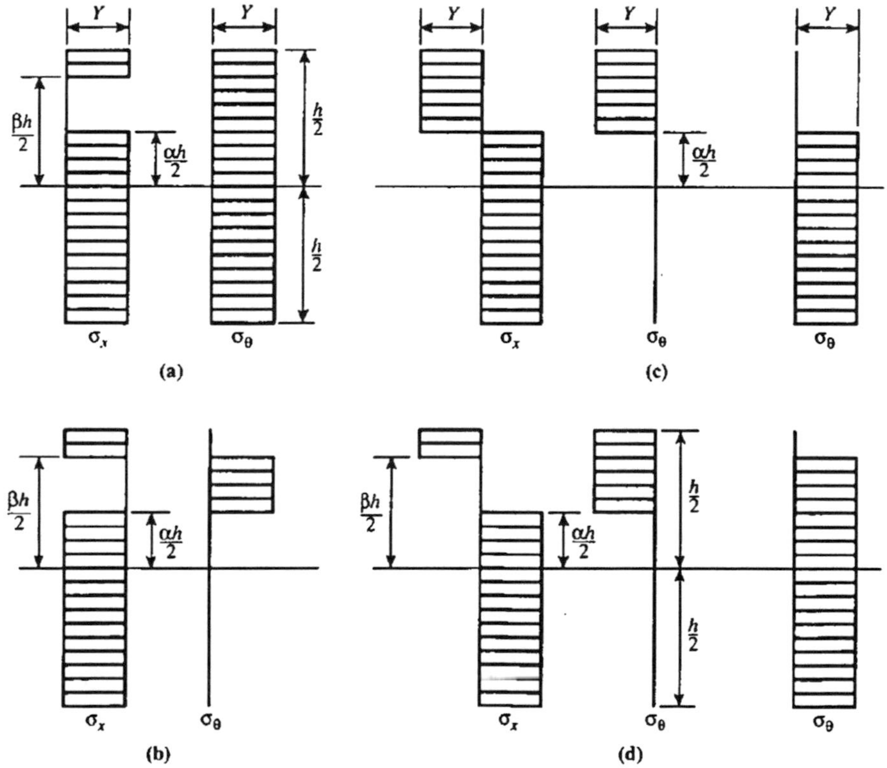

Fig. 5.7 Possible stress distributions through the thickness of an end-loaded circular cylindrical shell at the incipient collapse

where $n_\theta = N_\theta/N_0$, $n_x = N_x/N_0$, and $m_x = M_x/M_0$. In view of the assumed inequalities for α and B, the yield condition over this range may be written as

$$n_\theta = 1, \quad -2n_x(1-n_x) \le m_x \le 2n_x(1-n_x), \tag{5.28a}$$

where $n_x \ge 0$. Positive values of n_θ can be associated with negative values of n_x only for $n_\theta < 1$, the situation where the stresses do not change sign over the section being shown in Fig. 5.7b. Since $n_\theta = (\beta - \alpha)/2$, while $-n_x$ and m_x are unaffected, we have

$$n_\theta - n_x = 1, \quad 2n_x(1+n_x) \le m_x \le -2n_x(1+n_x), \tag{5.28b}$$

where $n_x \le 0$. Typical distributions of σ_x and σ_θ, when the cross section is fully plastic due to the axial stress alone, are shown in Fig. 5.7c, where $-1 < \alpha < 1$. The stress resultants are

$$n_x = \alpha, \quad m_x = 1 - \alpha^2, \quad n_\theta = -\frac{1}{2}(1-\alpha) \text{ and } \frac{1}{2}(1+\alpha).$$

Since n_θ can have any value between those for the two extreme cases shown, without violating the yield criterion, the elimination of α gives

$$m_x = 1 - n_x^2, \quad -\frac{1}{2}(1-n_x) \le n_\theta \le \frac{1}{2}(1+n_x), \tag{5.28c}$$

where $m_x \ge 0$. Figure 5.7c is a special case of the more general situation shown in Fig. 5.7d, where $-1 \le \alpha \le \beta \le 1$. Considering the σ_θ distribution corresponding to maximum n_θ, we obtain

$$n_x = \frac{1}{2}(\alpha+\beta), \; m_x = 1 - \frac{1}{2}\left(\alpha^2+\beta^2\right), \; n_\theta = \frac{1}{2}(1+\beta).$$

The elimination of α and β between these three relations furnishes the yield condition as

$$m_x = 1 - \frac{1}{2}\left[(2n_\theta - 1)^2 + (2n_\theta - 2n_x - 1)^2\right]. \tag{5.28d}$$

The inequalities $\alpha \ge -1$ and $\alpha \le \beta$ lead to the further restrictions $n_\theta \le 1$, $n_\theta - n_x \le 1$ and $2n_\theta - n_x \ge 1$ on the yield condition. Turning now to the σ_θ distribution for minimum n_θ and reversing the sign of the stresses so that n_θ becomes positive, we have

$$n_x = -\frac{1}{2}(\alpha+\beta), \quad m_x = -1 + \frac{1}{2}\left(\alpha^2+\beta^2\right), \quad n_\theta = \frac{1}{2}(1-\alpha).$$

Eliminating α and β as before, we obtain the yield condition in the form

$$m_x = -1 + \frac{1}{2}\left[(2n_\theta - 1)^2 + (2n_\theta - 2n_x - 1)^2\right]. \tag{5.28e}$$

The inequalities to be satisfied by n_θ and n_x for the validity of (5.28e) are identical to those for (5.28d). To complete the construction of the yield surface, a further set of five yield equations may be obtained by changing the sign of m_x in (5.28c), and the sign of n_θ and n_x in the remaining equations of the preceding set. The same yield conditions can be derived with a kinematic approach by considering the whole range of possible line segments in Fig. 5.4, rather than only those centered on the broken line (Onat, 1955). The yield surface obtained from the preceding analysis is therefore exact for a uniform shell according to Tresca's yield criterion and the associated flow rule.

The complete yield surface, which is symmetrical above the plane $m_x = 0$, is shown in Fig. 5.8, where the origin of coordinates is taken at the center of symmetry. The faces A and B of the yield surface are the planes $n_\theta = 1$ and $n_\theta - n_x = 1$, respectively, the bounding curves being the parabolic arcs $\pm 2n_x\ (1 - n_x)$ and $\pm\ 2n_x(l + n_x)$ for the faces A and B, respectively. Face C is the parabolic cylinder $m_x = 1 - n_x^2$, which is bounded by the planes $m_x = 0$ and $2n_\theta - n_x = \pm\ 1$. Faces D and E are paraboloids defined by (5.28d) and (5.28e), respectively, each being bounded by the planes A and B as well as the plane $2n_\theta - n_x = 1$. Face C' is the

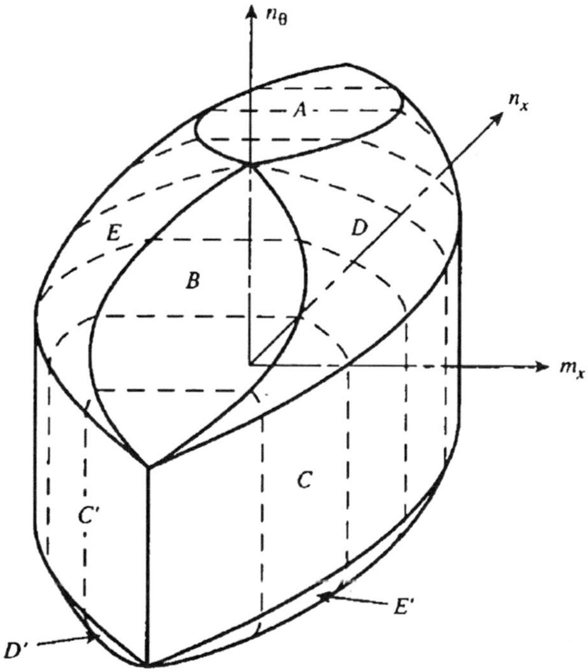

Fig. 5.8 Exact yield surface for an end-loaded cylindrical shell based on the Tresca criterion for yielding

reflection of C in the plane of symmetry, while A' and B' (not shown) are parallel to the planes A and B, respectively. The remaining faces D' and E' are convex areas on the lower side of the yield surface. The strain rate vector $\left(N_0\dot{\lambda}_x, N_0\dot{\lambda}_\theta, M_0\dot{\kappa}_x\right)$ associated with each face is defined by the usual normality rule, the normal being uniquely determined at all points except those along the boundaries of the plane faces A, B, A', and B'. A number of sections of the yield surface, made by planes parallel to $n_x = 0$, are indicated by broken lines in Fig. 5.8.

For practical applications, it is often convenient to employ a polygonal approximation of the yield surface, which is directly obtained by considering the idealized sandwich shell with a uniaxial yield stress σ_0. Since $n_x \neq 0$, (5.6) is modified to

$$\sigma_x^{\pm} = (n_x \pm m_x)\,\sigma_0,\ \ \sigma_\theta^{\pm} = (n_\theta \pm m_\theta)\,\sigma_0, \tag{5.29}$$

relating the stress components to the dimensionless stress resultants. We shall examine the conditions for which the inner and outer flanges of the sandwich shell simultaneously become plastic according to the Tresca criterion. Assuming n_θ to be positive, the following possibilities may be envisaged when the direct forces are predominant:

$$\begin{aligned} \sigma_\theta^+ = \sigma_\theta^- &= \sigma_0, & 0 \leq \sigma_\theta^+ \leq \sigma_0, \quad & 0 \leq \sigma_\theta^- \leq \sigma_0, \\ \sigma_\theta^+ - \sigma_x^+ = \sigma_\theta^- - \sigma_x^- &= \sigma_0, & \sigma_\theta^+ \leq 0 \leq \sigma_\theta^+, \quad & \sigma_x^- \leq 0 \leq \sigma_\theta^-. \end{aligned}$$

When n_θ is positive, but n_x takes both positive and negative values, we may consider the possibilities

$$\begin{aligned} \sigma_\theta^+ &= \sigma_0, & \sigma_\theta^- - \sigma_x^- &= \sigma_0, & 0 \leq \sigma_x^+ \leq \sigma_0 \quad & \sigma_x^- \leq 0 \leq \sigma_\theta^-, \\ \sigma_\theta^+ - \sigma_x^+ &= \sigma_0, & \sigma_x^- &= \sigma_0, & \sigma_x^- \leq 0 \leq \sigma_\theta^+, \quad & 0 \leq \sigma_x^- \leq \sigma_0. \end{aligned}$$

The remaining two possibilities for positive n_θ involve only positive values of n_x, the yield criterion being expressed as

$$\left.\begin{aligned} \sigma_\theta^+ &= \sigma_0, & \sigma_x^- - \sigma_\theta^- &= \pm\sigma_0, & 0 \leq \sigma_x^+ \leq \sigma_0, \\ \sigma_x^+ - \sigma_\theta^+ &= \pm\sigma_0, & \sigma_\theta^- &= \sigma_0, & 0 \leq \sigma_x^- \leq \sigma_0, \end{aligned}\right\} \ -\sigma_\theta^- \leq \sigma_\theta^+.$$

Substituting from (5.29), and eliminating m_θ, we obtain six equations defining six plane faces of the yield surface corresponding to $n_\theta \geq 0$, the result being

$$\left.\begin{aligned} n_\theta = 1\,(0 \leq n_x \leq 1)\,, \qquad & n_\theta - n_x = 1\ (n_x \leq 0 \leq n_\theta)\,, \\ \left.\begin{aligned} 2n\theta - n_x + m_x = 2\ (m_x \geq 0)\,, \\ 2n\theta - n_x - m_x = 2\ (m_x \leq 0)\,, \end{aligned}\right\} \qquad & -\tfrac{1}{2} \leq n_x \leq \tfrac{1}{2}, \\ n_x + m_x = 1\ (m_x \geq 0)\,, \qquad & n_x - m_x = 1\ (m_x \leq 0)\,. \end{aligned}\right\} \tag{5.30}$$

The last equation also holds for negative n_θ over the range $-1/2 \leq n_\beta \leq 0$. The yield equations for the remaining six faces of the yield surface, corresponding to $n_\theta \leq 0$, are obtained by simply changing the sign of all the generalized stress components in (5.30), as listed in Table 5.2.

Table 5.2 Linearized yield condition for cylindrical shells with end load

Face	Equation	$(N_0\dot{\lambda}_x, N_0\dot{\lambda}_\theta, M_0\dot{\kappa}_x)$	Face	Equation	$(N_0\dot{\lambda}_x, N_0\dot{\lambda}_\theta, M_0\dot{\kappa}_x)$
A	$n_\theta = 1$	$\mu(0,1,0)$	A'	$-n_\theta = 1$	$\mu(0,-1,0)$
B	$n_\theta - n_x = 1$	$\mu(-1,1,0)$	B'	$n_x - n_\theta = 1$	$\mu(1,-1,0)$
C	$-n_x + m_x = 1$	$\mu(-1,0,1)$	C'	$n_x - m_x = 1$	$\mu(1,0,-1)$
D	$2n_\theta - n_x + m_x = 2$	$\mu(-1,2,1)$	D'	$-2n_\theta + n_x - m_x = 2$	$\mu(1,-2,-1)$
E	$2n_\theta - n_x - m_x = 2$	$\mu(-1,2,-1)$	F'	$-2n_\theta + n_x + m_x = 2$	$\mu(1,-2,1)$
F	$n_x + m_x = 1$	$\mu(1,0,1)$	F'	$-n_x - m_x = 1$	$\mu(-1,0,-1)$

In a three-dimensional stress space, the complete yield surface is a convex polyhedron shown in Fig. 5.9, where some level curves of n_x are indicated by broken lines. The intersection of the polyhedron with the plane $n_x = 0$ is the hexagonal yield locus of Fig. 5.3. Each plane face of the polyhedron is designated by a letter, and the equation defining the plane can be found from Table 5.2. The strain rate vector for each face, defined by the normality rule, is also given in the table in terms of an arbitrary positive multiplier μ. At the intersection of two faces, the strain rate

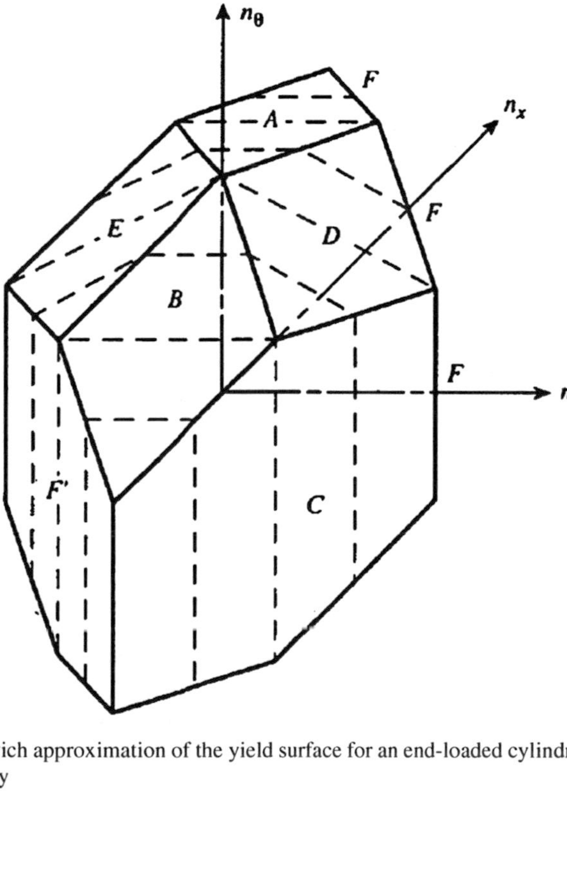

Fig. 5.9 Sandwich approximation of the yield surface for an end-loaded cylindrical shell based on the Tresca theory

vector is given by a linear combination of those for the individual faces forming the intersection.

The yield condition for the sandwich shell based on the von Mises criterion may be obtained by substituting from (5.29) into the yield criterion $\sigma_\theta^2 - \sigma_\theta\sigma_x + \sigma_x^2 = \sigma_0^2$, keeping in mind that both the inner and the outer flanges of the shell are simultaneously at the yield point during plastic collapse. The resulting pair of equations

$$(n_\theta - m_\theta)^2 - (n_\theta - m_\theta)(n_x - m_x) + (n_x - m_x)^2 = 1,$$
$$(n_\theta + m_\theta)^2 - (n_\theta + m_\theta)(n_x + m_x) + (n_x + m_x)^2 = 1,$$

can be solved for $n_\theta - m_\theta$ and $n_\theta + m_\theta$ in terms of $n_x - m_x$ and $n_x + m_x$, respectively, and the elimination of m_θ between the two solutions finally gives the yield condition

$$\pm 2(2n_\theta - n_x) = \sqrt{4 - 3(n_x + m_x)^2} + \sqrt{4 - 3(n_x - m_x)^2}, \qquad (5.31)$$

which represents a pair of surfaces circumscribing the Tresca polyhedron in a three-dimensional stress space (Hodge, 1961). The intersection of either of the two surfaces (5.31) with the plane $n_x = 0$ furnishes the ellipse represented by (5.8).

A hinge circle can occur at a section where the flow rule permits the velocity slope $d\omega/dx$ to be discontinuous. Since d^2w/dx^2 then becomes infinite for a finite ω, a hinge circle can only be associated with the faces C, F, C', and F' for the linearized yield surface, as may be seen from the components of the strain rate vector given in Table 5.2. The axial displacement u is also discontinuous at a hinge circle, since du/dx is infinitely large there. This situation is similar to the yield point state in a beam under combined bending moment and axial force.

5.2.2 Shell Under Radial Pressure and Axial Thrust

A cylindrical shell of length $2l$ is clamped at its ends $x = 0$ and $x = 2l$ and is subjected to a uniform radial pressure p as well as an axial compressive force of magnitude nN_0. Since axial equilibrium requires $n_x = -n$ throughout the shell, it is only necessary to consider a single level curve of n_x for the analysis of the problem. We shall use the piecewise linear approximation for the yield condition, the yield polyhedron for $n_\theta \leq 0$ being separately drawn for the purpose in Fig. 5.10. The qualitative nature of the solution will depend on whether n is greater or less than 1/2. These two cases will therefore be taken up separately to establish the relations between $q = pa/N_0$, $\omega = l\sqrt{2/ah}$, and n for plastic collapse of the cylinder (Hodge, 1954).

Consider first the situation where $-1/2 \leq n \leq 0$. It is natural in this case to expect the entire stress profile to lie on face A', where $n_\theta = -1, \dot{\lambda}_x = \dot{\kappa}_x = 0$, $\dot{\lambda}_\theta < 0$. The corresponding velocity field is

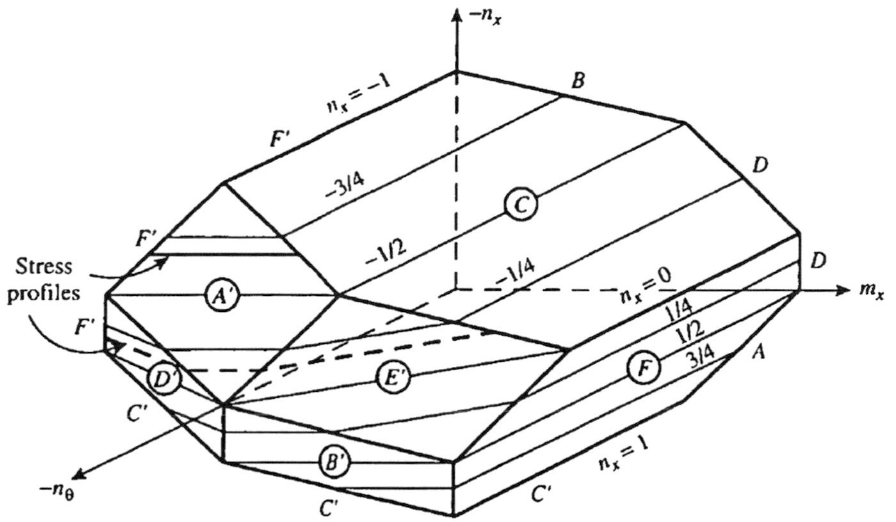

Fig. 5.10 Yield polyhedron for an end-loaded cylindrical shell involving compressive hoop force

$$u = -c, \quad w = w_0(x/l), \qquad 0 \leq x \leq l,$$

where c is a constant, and w_0 is the radial velocity at the central section $x = l$. The shell therefore deforms into a pair of conical parts, requiring hinge circles to be formed at the clamped edges and at the center. It follows that the section $x = 0$ and $x = l$ correspond to the intersection of face A' with C and F', respectively.

The equilibrium equation (5.4), which is unaffected by the axial force, becomes

$$\frac{d^2 m_x}{d\xi^2} = 2\,(1 - q)$$

after the substitution $n_\theta = -1$. The formation of hinge circles requires the boundary conditions

$$m_x = -1 + n \quad \text{at } \xi = 0, \quad m_x = 1 - n \quad \text{at } \xi = \omega,$$

where $\omega = l\sqrt{2/ah}$. Integrating and using these conditions, the solution for the bending moment distribution at the incipient collapse is obtained as

$$m_x = -\,(1 - n)\left(1 - \frac{2\xi}{\omega}\right) + (q - 1)\,\xi\,(\omega - \xi)\,, \quad 0 < \xi < \omega. \tag{5.32}$$

Each half of the shell moves axially outward with a speed c $= M_0 \omega_0 / N_0 l$ which follows from the integration of the flow rule $N_0 \dot{\lambda}_x = M_0 \dot{\kappa}_x$ that applies to the hinge circle at $x = 0$. The application of the symmetry condition $dm_x/d\xi = 0$ at $\xi = \omega$, using (5.32), furnishes the relationship between n, q, and ω as

$$(q-1)\,\omega^2 = 2\,(1-n)\,, \quad \frac{1}{2} \le n \le 1, \tag{5.33}$$

which is exact for $n \ge 1/2$, but provides only an upper bound for $n < 1/2$. In the special case of a closed-ended cylinder in a hydrostatic pressure field of intensity p, we have $n = pa/2N_0 = q/2$ and (5.33) gives

$$q = \left(2+\omega^2\right)\Big/\left(1+\omega^2\right) \tag{5.33a}$$

as the dimensionless collapse pressure. Since $q > 1$ (or $n > 1/2$) according to this relation, it furnishes the exact solution for all ω corresponding to the linearized yield condition. The variation of q with ω for the closed-ended shell is included in Fig. 5.5 for visual comparison. It may be noted that the collapse pressure is practically independent of the end condition for large values of ω.

For $0 \le n \le 1/2$, only a part of the shell will involve face A' of the yield polyhedron, since no hinge circle can be associated with its edges formed by the adjacent faces E' and D'. Consequently, the stress profile must extend into these faces to meet their boundaries as shown by the broken lines in Fig. 5.10. The sections $\xi = 0$ and $\xi = \omega$ correspond to the edges where faces E' and D' meet C and F', respectively. Let $\xi = \xi_1$ and $\xi = \xi_2$ denote the cross sections which correspond to the ends of the stress profile lying on face A'. For the part of the shell that corresponds to face E', the yield condition is $m_x = -2(1+n_\theta) - n$, and the equilibrium equation (5.4) becomes

$$\frac{d^2 n_\theta}{d\xi^2} - n_\theta = q, \qquad 0 \le \xi \le \xi_1.$$

The boundary condition $m_x = -(1-n)$ at $\xi = 0$ is equivalent to $n_\theta = -\left(n+\frac{1}{2}\right)$ at $\xi = 0$, the remaining boundary condition being $n_\theta = -1$ at $\xi = \xi_l$. The integration of the above equation therefore gives

$$n_\theta = -q + (q-1)\,\frac{\sinh\xi}{\sinh\xi_1} + \left(q - n - \frac{1}{2}\right)\frac{\sinh\left(\xi_1 - \xi\right)}{\sinh\xi_1} \qquad 0 \le \xi \le \xi_1. \tag{5.34}$$

The part of the shell that corresponds to face D' of the yield surface requires $m_x = 2(1+n_\theta) + n$ by the yield condition, and the equilibrium equation (5.4) then becomes

$$\frac{d^2 n_\theta}{d\xi^2} + n_\theta = -q, \quad \xi_2 \le \xi \le \omega.$$

The plastic hinge condition at the center requires $m_x = 1 - n$, or $n_\theta = -\left(n+\frac{1}{2}\right)$ at $\xi = \omega$, while the symmetry condition implies $dn_\theta/d\xi = 0$ at $\xi = \omega$. Using these boundary conditions, the solution to the above equation is obtained as

$$n_\theta = -q + \left(q - n - \frac{1}{2}\right)\cos\left(\omega - \xi\right), \qquad \xi_2 \le \xi \le \omega. \tag{5.35}$$

In the remaining portion of the shell, for which the stress profile lies on face *A'*, the yield condition $n_\theta = -1$ reduces the equilibrium equation to

$$\frac{d^2 m_x}{d\xi^2} = 2(1-q), \qquad \xi_1 \le \xi \le \xi_2.$$

The conditions of continuity of m_x and $dm_x/d\xi$ across $\xi = \xi_2$ imply $m_x = n$ at $\xi = \xi_2$ and $dm_x/d\xi$ is equal to the value of $2dn_\theta/d\xi$ given by (5.35) at $\xi = \xi_2$.

The solution is therefore found to be

$$m_x = n - 2\left(q - n - \frac{1}{2}\right)(\xi_2 - \xi)\sin(\omega - \xi_2) - (q-1)(\xi_2 - \xi), \qquad \xi_1 \le \xi \le \xi_2. \tag{5.36}$$

The hoop force n_θ has been automatically made continuous across $\xi = \xi_1$, while its continuity across $\xi = \xi_2$ is established by setting $n_\theta = -1$ in (5.35) when $\xi = \xi_2$, the result being

$$2(q-1)[\sec(\omega - \xi_2) - 1] = 1 - 2n, 0 \le n \le \tfrac{1}{2}. \tag{5.37}$$

The remaining continuity conditions are those of m_x and $dm_x/d\xi$ across $\xi = \xi_1$. The former implies $m_x = -n$ at $\xi = \xi_1$, while the latter is equivalent to $dm_x/d\xi$ given by (5.36) being equal to $2dn_\theta/d\xi$ given by (5.34) when $\xi = \xi_1$. The last condition furnishes

$$2(q-1)\left\{\frac{\cosh\xi_1 + (\xi_2 - \xi_1)\sinh\xi_1}{1 - \sin(\omega - \xi_2)\sinh\xi_1} - 1\right\} = 1 - 2n.$$

The elimination of $2(q-1)/(1-2n)$ between the last two equations gives us the relationship between ξ_1, ξ_2, and ω as

$$\cos(\omega - \xi_2)\cosh\xi_1 + [\sin(\omega - \xi_2) + (\xi_2 - \xi_1)\cos(\omega - \xi_2)]\sinh\xi_1 = 1. \tag{5.38}$$

Finally, setting $m_x = -n$ in (5.36) when $\xi = \xi_1$, and using (5.37) to eliminate $2q-2n-1$ and $q-1$, we obtain the relation

$$\frac{4n}{1-2n} = (\xi_2 - \xi_1)\left\{\frac{2\tan(\omega - \xi_2) + (\xi_2 - \xi_1)}{\sec(\omega - \xi_2) - 1}\right\}, \quad 0 \le n \le \tfrac{1}{2}. \tag{5.39}$$

For any given value of ω, (5.38) defines ξ_1 as a function of ξ_2. The load parameters n and q as functions of ξ_2 are then obtained from (5.39) and (5.37), respectively. This gives the relationship between n and q at the incipient plastic collapse corresponding to a given shell parameter ω. When $n = \frac{1}{2}$, we have $\xi_1 = 0$, $\xi_2 = \omega$, and (5.37) and (5.38) give $(q-1)\,\omega^2 = 1$, in agreement with that given by (5.33). When $n = 0$, we have $\xi_1 = \xi_2 = \xi_0$, and (5.37) and (5.38) then reduce to (5.12). Figure 5.11 shows the interaction curves involving n and q for several values of ω.

To obtain the distribution of the radial velocity ω associated with the stress distribution, let ω_1 denote the value of ω at $\xi = \xi_1$. Over the length $0 \le \xi \le \xi_1$ the

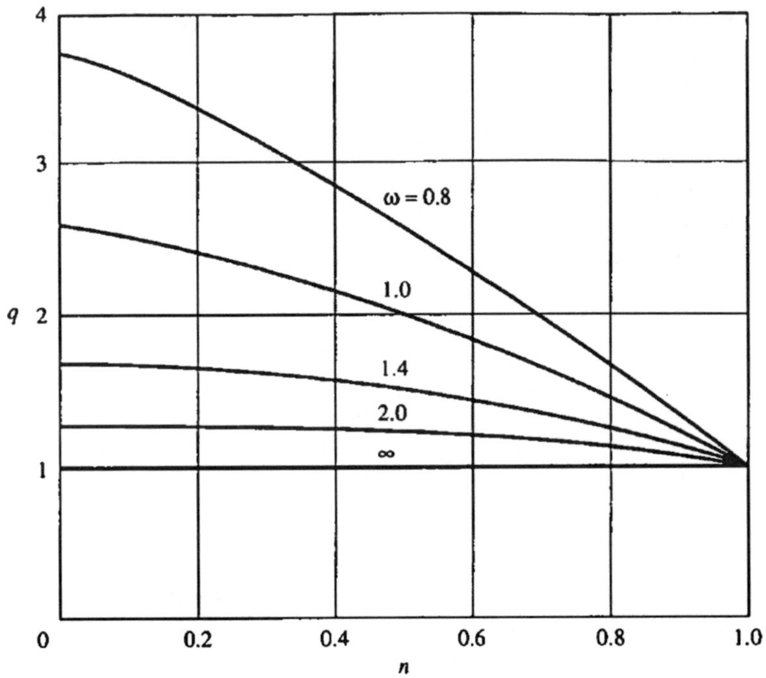

Fig. 5.11 Interaction curves for clamped cylindrical shells under combined external pressure and axial thrust

flow rule associated with the face *E'* gives $M_0\dot{\kappa}_x/N_0\dot{\lambda}_\theta = \frac{1}{2}$, and the substitution for the strain rates furnishes $d^2w/d\xi^2 = w$;. The solution to this differential equation, satisfying the boundary conditions $w = 0$ at $\xi = 0$ and $w = w_1$ at $\xi = \xi_1$, is found to be

$$w = w_1 (\sinh \xi / \sinh \xi_1), \quad 0 \leq \xi \leq \xi_1.$$

Since $\dot{\kappa}_x = 0$ over the length $\xi_1 \leq \xi \leq \xi 2$, which corresponds to the face *A'* of the yield surface, this part of the shell tends to be conical with the velocity distribution

$$w = w_1 [1 + (\xi - \xi_1) \coth \xi_1), \quad \xi_1 \leq \xi \leq \xi_2,$$

in view of the continuity of *w* and $dw/d\xi$; across $\xi = \xi_1$. The flow rule associated with the face *D'*, which applies over the length $\xi_2 \leq \xi \leq \omega$, gives the differential equation $d^2w/d\xi^2 = -w$, and the conditions of continuity across $\xi = \xi_2$ furnishes

$$w = w_1 \coth \xi_1 \sin (\xi - \xi_2) + w_2 \cos (\xi - \xi_2), \quad \xi_2 \leq \xi \leq \omega,$$

where w_2 is the value of w at $\xi = \xi_2$. The assumed hinge circle at $\xi = \xi_2$ requires $dw/d\xi \geq 0$ there, so that the discontinuity in slope is compatible with the positive bending moment. The range of values of ω for which the solution is complete is therefore given by the inequality

$$\cot(\omega - \xi_2) \geq (\xi_2 - \xi_1) + \tanh \xi_1, \quad 0 \leq n \leq \tfrac{1}{2}. \tag{5.40}$$

As n increases from zero, the minimum value of ω for the solution to be exact increases from 1.65 to infinity. When ω exceeds the range restricted by (5.40), the solution should be modified in the same way as that carried out in Section 5.1.3, but may be otherwise accepted as a lower bound on the collapse pressure. The problem of combined loading of cylindrical shells has also been discussed by Sankaranarayanan (1960) and by Hodge and Panarelli (1962). The plastic design of closed cylindrical shells under internal pressure has been discussed by Hodge (1964b) and Sayir (1966).

5.2.3 Influence of Elastic Deformation

The collapse load of cylindrical shells, based on the assumption of a rigid/plastic material, does not necessarily provide a good approximation for an elastic/plastic shell when an axial force is present. The discrepancy is due to the fact that the axial force produces a bending moment distribution throughout the shell while it is still elastic, as a result of the radial displacement caused by the radial pressure. The bending moments in the elastic/plastic shell are consequently higher than those in the corresponding rigid/plastic shell. This phenomenon, known as the beam-column effect, results in the plastic collapse at a load that is significantly lower than the rigid/plastic collapse load.

For the analysis of the elastic/plastic shell, we may assume the deformation to remain axially symmetrical and the displacement to be sufficiently small, so that the axial force is effectively constant along the length of the shell. The equation of radial equilibrium therefore remains unchanged, but that of moment equilibrium is modified by the presence of an additional bending moment produced by the axial force. The two equations of equilibrium may be written as

$$\frac{dQ}{dx} + \frac{N_\theta}{a} = -p\,\frac{dM_x}{dx} - N_x\frac{dw}{dx} - Q = 0,$$

where w denotes the radially inward displacement. Combining these equations to eliminate the shear force Q and introducing dimensionless quantities as before, we obtain the equation

$$\frac{d^2 m_x}{d\xi^2} + 2n_\theta + 4n_x\frac{d^2}{d\xi^2}\left(\frac{w}{h}\right) = -2q. \tag{5.41}$$

The rate of internal work per unit area must contain the additional term $(N_x/2)(dw/dx)^2$ due to the axial change in length of the surface element. Hence, the generalized strain rates in terms of the velocities $\dot{u}$ and $\dot{\omega}$ are

$$\dot{\lambda}_x = \frac{d\dot{u}}{dx} + \frac{1}{2}\left(\frac{d\dot{w}}{dx}\right)^2, \quad \dot{\lambda}_\theta = -\frac{\dot{w}}{a}, \quad \dot{\kappa}_x = -\frac{d^2\dot{w}}{dx^2}. \tag{5.42}$$

The ratios of the plastic components of the strain rates are given by the flow rule associated with the appropriate yield condition. The elastic components are given by

$$E\dot{\lambda}_x^e = \dot{N}_x - \nu\dot{N}_\theta, \quad E\dot{\lambda}_\theta^e = \dot{N}_\theta - \nu\dot{N}_x, \quad Eh^3\dot{\kappa}_x^e = 12\left(1-\nu^2\right)\dot{M}_x, \tag{5.43}$$

where E is Young's modulus, and ν is Poisson's ratio for the material of the shell. Starting with the elastic solution, and following the development of the plastic zone as the loading is continued, the state of plastic collapse can be reached by an analysis involving the above equations together with the appropriate yield condition and flow rule. However, the complete solution would generally involve several plastic regimes, and the requirement of continuity of the various physical quantities across their boundaries would involve considerable difficulties even for the idealized sandwich shell (Paul and Hodge, 1958; Paul, 1959).

Consider an approximate solution for the particular case of a simply supported cylindrical shell of length $2l$, which is closed at both ends and is subjected to a gradually increasing hydrostatic pressure p. At the point of collapse, the entire stress profile is assumed to be on face A', as in the corresponding rigid/plastic solution. The yield condition $n_\theta = -1$ corresponds to $\dot{\kappa}_x^p = 0$, which indicates that $\dot{\kappa}_x$ is entirely elastic, and the last equation of (5.43) gives

$$\frac{d^2\dot{w}}{dx^2} = -\left(1-\nu^2\right)\frac{12\dot{M}_x}{Eh^2} = -\left(1-\nu^2\right)\frac{3Y\dot{m}_x}{Eh}.$$

Since this equation is equally valid in the elastic range of deformation, while the initial condition is $m_x = w = 0$, the integrated form of the equation may be expressed as

$$\frac{d^2}{d\xi^2}\left(\frac{w}{h}\right) = \frac{a}{2}\frac{d^2w}{dx^2} = -\frac{3Ya}{2Eh}(1-\nu^2)m_x.$$

Inserting into the equilibrium equation (5.41) and setting $n_\theta = -1$ and $n = -q/2$, we have

$$\frac{d^2m_x}{d\xi^2} + \beta^2 m_x = 2(1-q), \tag{5.44}$$

$$\beta^2 = 3q\left(1-\nu^2\right)\frac{Ya}{Eh} = \frac{3}{2}q\omega^2\left(1-\nu^2\right)\frac{Ya^2}{El^2}, \tag{5.45}$$

where $\omega^2 = 2l^2/ah$. The shell collapses by the formation of a hinge circle at the central section, which is taken as $\xi = O$. Hence, the boundary conditions are

$$m'_x(0) = 0, \qquad m_x(\omega) = 0, \qquad m_x(0) = 1 - q/2.$$

Integrating (5.44) and using the first two boundary conditions, the solution for the bending moment is obtained as

$$m_x = \frac{2}{\beta^2}(q-1)\left(\frac{\cos\beta\xi}{\cos\beta\omega} - 1\right).$$

The remaining boundary condition at $\xi = 0$ now furnishes the dimensionless collapse pressure as

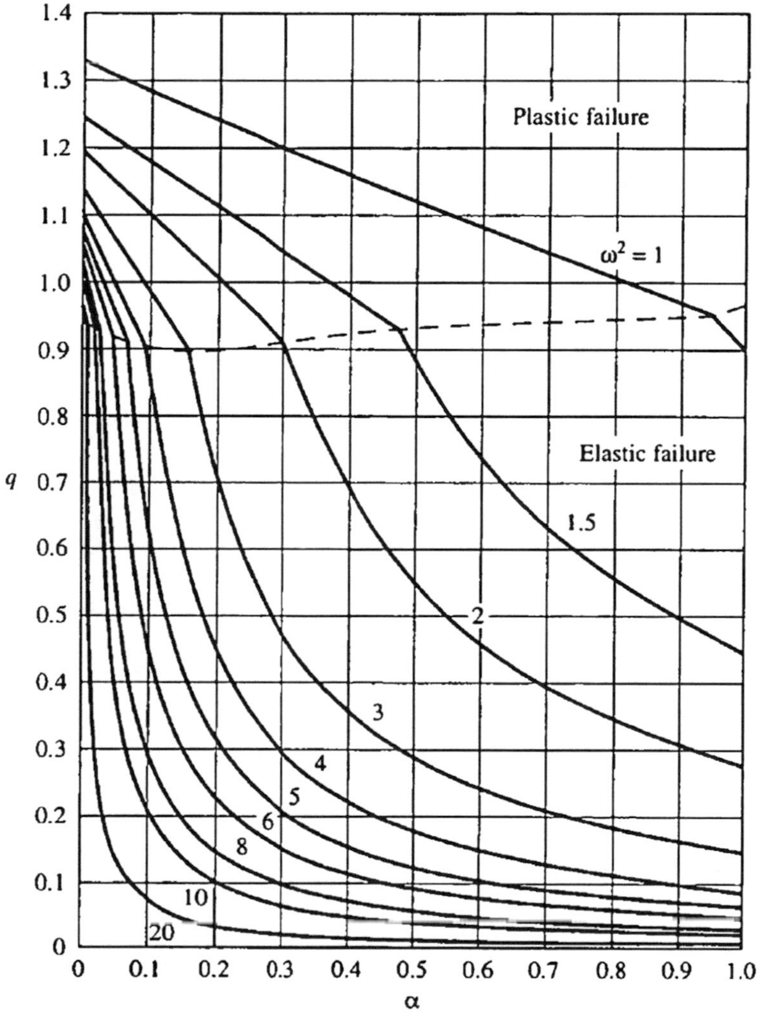

Fig. 5.12 Elastic and plastic ranges of failure of a cylindrical shell due to beam-column effect under radial external pressure (after Paul and Hodge, 1958)

$$q = 1 + \frac{\beta^2 \cos \beta\omega}{4\,(1 - \cos \beta\omega) + \beta^2 \cos \beta\omega}. \tag{5.46}$$

For a rigid/plastic shell, β tends to zero, and (5.46) reduces to $q = 2(1 + \omega^2)/(1 + 2\omega^2)$, which may be obtained by a direct rigid/plastic analysis.

Since β depends on q for an elastic/plastic shell, it is convenient to start with a selected value of β for a given ω, and then determine q from (5.46), the corresponding value of a/l for a given material being finally obtained from (5.45). In Fig. 5.12, the dimensionless collapse pressure q is plotted as a function of the shell parameter $\alpha = (1 - \nu^2)\,(Y/2E)\,(a/l)^2$ for a number of values of ω. For a given shell geometry, as the ratio Y/E increases from zero, q steadily decreases according to the elastic/plastic collapse formula (5.46), until a critical value is reached, below which the elastic buckling formula becomes applicable. Equation (5.46) appears to be a good approximation for relatively short shells.

5.3 Yield Point States in Shells of Revolution

5.3.1 Generalized Stresses and Strain Rates

Consider a thin shell of uniform thickness h whose middle surface forms a surface of revolution, characterized by its radial distance r from the axis of symmetry as a given function of the meridional angle ϕ. If the loading and support conditions are axially symmetrical, the state of stress in the shell is specified by the meridional and circumferential membrane forces N_ϕ, and N_θ, the meridional and circumferential bending moments M_ϕ and M_θ, and the shearing force Q. The applied load per unit area of the middle surface may be resolved into the components p_ϕ and p_r in the directions of the tangent and normal, respectively, to the meridian. Considering an element formed by neighboring meridian sections and parallel circles, Fig. 5.13, the equations of force equilibrium along the meridian and surface normal, and the equation of moment equilibrium, are easily shown to be d

$$\left.\begin{aligned}
&\frac{d}{d\phi}\left(rN_\phi\right) - r_1 N_\theta \cos\phi - rQ + r_1 r p_\phi = 0,\\
&\frac{d}{d\phi}(rQ) + r_1 N_\theta \sin\phi + rN_\phi + r_1 r p_r = 0,\\
&\frac{d}{d\phi}\left(rM_\phi\right) - r_1 M_\theta \cos\phi - r_1 r Q = 0,
\end{aligned}\right\} \tag{5.47}$$

where r_1 is the meridional radius of curvature of the middle surface. If the circumferential radius of curvature is denoted by r_2, then by simple geometry,

$$r = r_2 \sin\phi \quad \frac{dr}{d\phi} = r_1 \cos\phi.$$

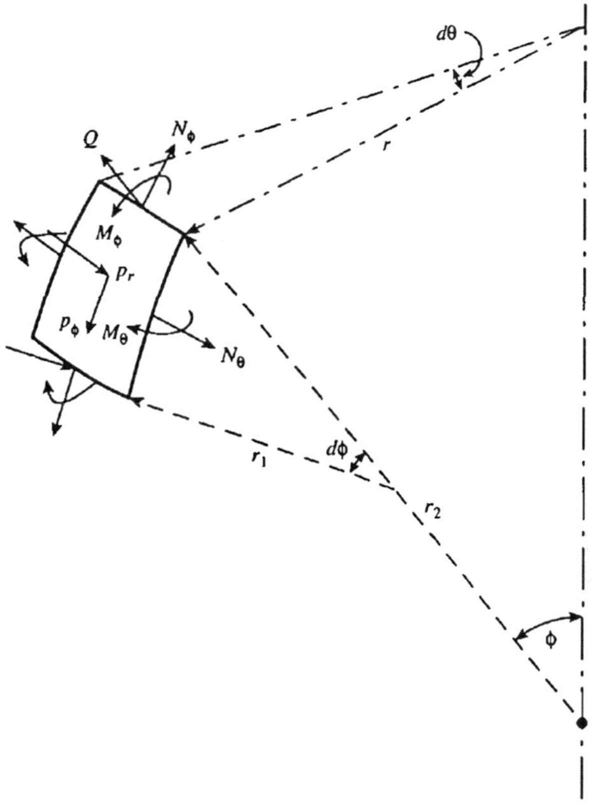

Fig. 5.13 Stress resultants acting on a typical element of a shell of revolution under axially symmetrical loading conditions

Multiplying the first two equations of (5.47) by sin ϕ and cos ϕ, respectively, and adding them together, we obtain a differential equation whose integral is

$$r\left(N_\phi \sin\phi + Q\cos\phi\right) = C - \int r_1\left(p_r\cos\phi + p_\phi\sin\phi\right)r\,d\phi, \qquad (5.48)$$

where C is a constant that depends on the magnitude of a concentrated or line load acting on the shell. Equation (5.48) may be directly obtained from the condition of vertical equilibrium of the portion of the shell lying above a generic circle of radius r.

The velocity of a typical particle on the middle surface has meridional and normal components u and ω, respectively, their positive directions being the same as those of p_ϕ and p_r shown in Fig. 5.13. In terms of u and w, the principal rates of extension and curvature of the middle surface (Fliigge, 1960) are found to be

$$\left.\begin{aligned} \dot{\lambda}_\theta &= \frac{u\cot\phi - w}{r_2}, & \dot{\lambda}_\theta &= \frac{1}{r_1}\left(\frac{du}{d\phi} - w\right), \\ \dot{\kappa}_\theta &= -\frac{1}{r_1}\left(u + \frac{dw}{d\phi}\right)\frac{\cot\phi}{r_2}, & \dot{\kappa}_\theta &= -\frac{1}{r_1}\frac{d}{d\phi}\left[\frac{1}{r_1}\left(u + \frac{dw}{d\phi}\right)\right]. \end{aligned}\right\} \tag{5.49}$$

Let z denote the normal distance of a generic point in the shell from the middle surface, reckoned positive in the inward direction. Then the meridional and circumferential strain rates are given by the relations

$$\dot{\varepsilon}_\theta = \dot{\lambda}_\theta + z\dot{\kappa}_\theta, \qquad \dot{\varepsilon}_\phi = \dot{\lambda}_\phi + z\dot{\kappa}_\phi,$$

which are associated with the stress components σ_θ and σ_ϕ, respectively. The rate of dissipation of energy during plastic flow per unit area of the middle surface is

$$D = \int_{-h/2}^{h/2} (\sigma_\theta\dot{\varepsilon}_\theta + \sigma_\phi\dot{\varepsilon}_\phi)\, dz = N_\theta\dot{\lambda}_\theta + N_\phi\dot{\lambda}_\phi + M_\theta\dot{\kappa}_\theta + M_\phi\dot{\kappa}_\phi \tag{5.50}$$

which is easily shown by substituting for $\dot{\varepsilon}_\theta$ and $\dot{\varepsilon}_\phi$, and using the fact that the stress resultants are given by

$$N_\alpha = \int_{-h/2}^{h/2} \sigma_\alpha dz, \qquad M_\alpha = \int_{-h/2}^{h/2} \sigma_\alpha z dz,$$

where α denotes either θ or ϕ. It follows that N_θ, N_ϕ, M_θ, and M_ϕ, are the four generalized stresses in the present problem, while $\dot{\lambda}_\theta$, $\dot{\lambda}_\phi$, $\dot{\kappa}_\theta$, and $\dot{\kappa}_\phi$ are the corresponding generalized strain rates. The shearing force Q, which does not occur in the expression for D, is not a generalized stress but has the nature of a reaction.

The normal velocity w must be everywhere continuous, but the velocity slope $dw/d\phi$ and/or the tangential velocity u may be discontinuous across a parallel circle known as a hinge circle. If such a circle is regarded as the limit of a narrow zone in which $dw/d\phi$ and/or u can change rapidly but continuously, then $d^2w/d\phi^2$ and/or $du/d\phi$ must be numerically large compared to their lower derivatives. It follows therefore from (5.49) that $\dot{\kappa}_\phi/\dot{\kappa}_\theta$ and/or $\dot{\lambda}_\phi/\dot{\lambda}_\theta$ must be infinitely large across a hinge circle, which often occurs at the point of plastic collapse.

5.3.2 Yield Condition for a Tresca Material

We begin our discussion by considering a uniform shell that yields according to the maximum shear stress criterion of Tresca. For an isotropic material, the yield condition expressed in terms of the stress resultants must be symmetric in the direct forces as well as in the bending moments irrespective of the yield criterion. It is therefore convenient to deal with the principal stresses and strain rates in a shell element without reference to the circumferential and meridional components (Onat and Prager, 1954). Denoting the principal rates of extension of the middle surface by $\dot{\lambda}_1$ and $\dot{\lambda}_2$, and the principal rates of curvature by $\dot{\kappa}_1$ and $\dot{\kappa}_2$, the principal strain rates at a generic point in the shell wall may be written as

$$\dot{\varepsilon}_1 = \dot{\lambda}_1 + z\dot{\kappa}_1, \ \dot{\varepsilon}_2 = \dot{\lambda}_2 + z\dot{\kappa}_2, \ \dot{\varepsilon}_3 = -(\dot{\varepsilon}_1 + \dot{\varepsilon}_2).$$

Typical distributions of the strain rates over the thickness of the shell are shown in Fig. 5.14a,b,c, the zeros of the strain rates $\dot{\varepsilon}_1$ $\dot{\varepsilon}_2$, and $\dot{\varepsilon}_3$ in the three cases being given by the points *P, R,* and *Q*, respectively, whose ordinates are

$$\alpha h = -\dot{\lambda}_1/\dot{\kappa}_1, \qquad \gamma h = -\dot{\lambda}_2/\dot{\kappa}_2, \qquad \beta h = -\left(\dot{\lambda}_1 + \dot{\lambda}_2\right)/\left(\dot{\kappa}_1 + \dot{\kappa}_2\right). \quad (5.51)$$

The three parameters α, β, and γ are adequate for the specification of the ratios of the principal strain rates throughout the shell. Since the state of stress at the yield point depends only on the strain rate ratios, the stress resultants can be expressed in terms of the above parameters throughout the shell.

According to Tresca's associated flow rule, the rate of plastic work per unit volume corresponding to any plastic state is *Y* times the magnitude of the numerically greatest principal strain rate e. The distribution of $|\dot{\varepsilon}|$ over the thickness of the shell, obtained by the superimposition of those of the three individual components, is shown by the heavy lines in Fig. 5.14d. The rate of energy dissipation *D* per unit

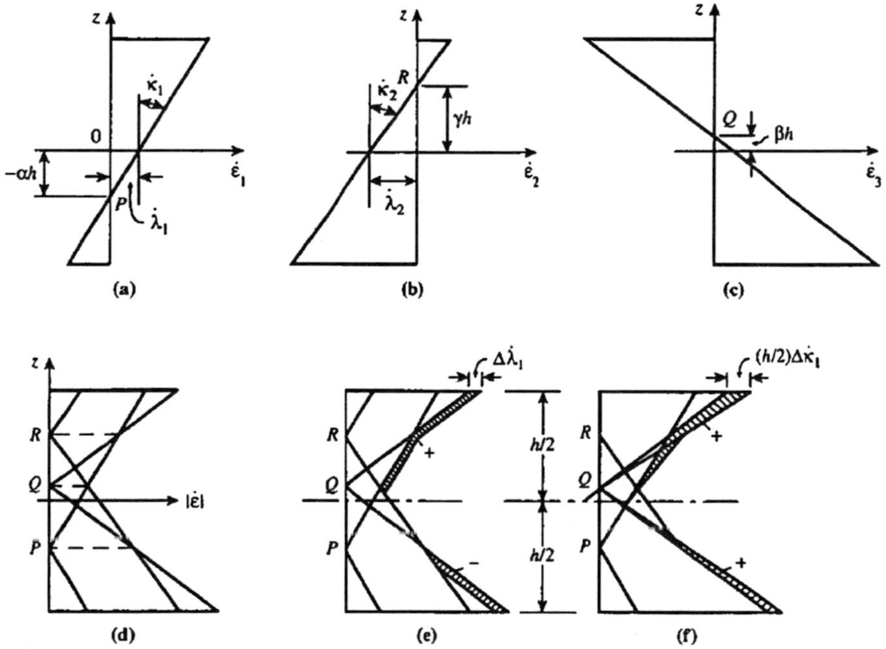

Fig. 5.14 Strain distribution through the thickness of a shell of revolution according to the Tresca theory in terms of parameters defining the yield condition

area of the middle surface is Y times the area bounded by the heavy lines. Since the material is inviscid, D must be a homogeneous function of degree one in the components of the generalized strain rates and by Euler's theorem on homogeneous functions,

$$D = \frac{\partial D}{\partial \dot{\lambda}_1}\dot{\lambda}_1 + \frac{\partial D}{\partial \dot{\lambda}_2}\dot{\lambda}_2 + \frac{\partial D}{\partial \dot{\kappa}_1}\dot{\kappa}_1 + \frac{\partial D}{\partial \dot{\kappa}_2}\dot{\kappa}_2.$$

Since D is also equal to the sum of the products of the generalized stresses and the corresponding generalized strain rates in view of (5.50), we have

$$N_1 = \frac{\partial D}{\partial_1 \dot{\lambda}_1}, \qquad N_2 = \frac{\partial D}{\partial \dot{\lambda}_2}, \qquad M_1 = \frac{\partial D}{\partial \dot{\kappa}_1}, \qquad M_2 = \frac{\partial D}{\partial \dot{\kappa}_2}.$$

These derivatives are most conveniently obtained by considering the change in area of the diagram for the distribution of $|\dot{\varepsilon}|$ due to a small change in one of the generalized strain rates. If, for instance, $\dot{\lambda}_1$ and $\dot{\kappa}_1$ are changed by the small amounts $\Delta\dot{\lambda}_1$ and $\Delta\dot{\kappa}_1$, respectively, the corresponding changes in D are Y times the shaded areas in Fig. 5.14e,f, giving

$$N_1 = Yh\left[\left(\frac{1}{2} - \beta\right) - \left(\frac{1}{2} - \alpha\right)\right] = -Yh\left(\alpha + \beta\right),$$

$$M_1 = \frac{1}{2}Yh^2\left[\left(\frac{1}{4} - \beta^2\right) + \left(\frac{1}{4} - \alpha^2\right)\right] = \frac{1}{4}Yh^2\left[1 - 2\left(\alpha^2 + \beta^2\right)\right].$$

Similarly, by considering small changes in the values of $\dot{\lambda}_2$ and $\dot{\kappa}_2$ and evaluating the corresponding changes in areas, it is found that

$$N_2 = -Yh\left(\beta + \gamma\right), \; M_2 = \frac{1}{4}Yh^2\left[1 - 2\left(\beta^2 + \gamma^2\right)\right].$$

Dividing the membrane forces by $N_0 = Yh$, and the moments by $M_0 = Yh^2/4$, and denoting the resulting dimensionless quantities by n and m with appropriate subscripts, the preceding relations may be written as

$$\left.\begin{aligned} n_1 &= -\left(\alpha + \beta\right), \quad m_1 = 1 - 2\left(\alpha^2 + \beta^2\right), \\ n_2 &= -\left(\beta + \gamma\right), \quad m_2 = 1 - 2\left(\beta^2 + \gamma^2\right). \end{aligned}\right\} \tag{5.52}$$

Equation (5.52) constitutes a parametric representation of a part of the yield surface in a four-dimensional stress space. If the strain rate distributions of Fig. 5.14a,b are assumed to represent the distribution of $\dot{\varepsilon}_2$ and $\dot{\varepsilon}_1$, respectively, the parameters α and γ are interchanged, and Fig. 5.14e,f would indicate the effects of small changes in $\dot{\lambda}_2$ and $\dot{\kappa}_2$ on the dissipation rate D. The resulting expressions for N_2 and M_2 would evidently be identical to the preceding ones for N_1 and M_1, respectively. It follows, therefore, that Equation (5.52) apply for the generalized stresses, so long as the points P, Q, R are distinct, and β lies between α and γ. Two further sets of relations for the stress resultants can be similarly obtained by taking P and R as the intermediate points. In all cases the ratios of the generalized strain rates are obtained from (5.51) as

$$N_0\dot{\lambda}_1 : N_0\dot{\lambda}_2 : M_0\dot{\kappa}_1 : M_0\dot{\kappa}_2 = -4\alpha\,(\beta-\gamma) : -4\gamma\,(\alpha-\beta) : (\beta-\gamma) : (\alpha-\beta)\,. \tag{5.53}$$

The results for the dimensionless stress resultants, when the parameters α, β, and γ are distinct, are collected in Table 5.3, which also includes those corresponding to the reversal of the sign of the stresses. These expressions presuppose that each of the points P, Q, R lies within the thickness of the shell. When one of the points, say P, falls outside the thickness, implying $\alpha < -\frac{1}{2}$, it is only necessary to replace α by $-\frac{1}{2}$, in the corresponding expressions for the stress resultants, although no such change should be made in the flow rule. In terms of the generalized stresses, the yield condition of Table 5.3 may be expressed as

$$\left.\begin{aligned} \pm(m_1 - m_2) + (n_1 - n_2)^2 + \left(\frac{m_2}{2n_2} \pm n_1\right)^2 &= 1, \\ \pm m_2 + n_2^2 + \left[\frac{1}{2}\left(\frac{m_1 - m_2}{n_1 - n_2}\right) \pm n_1\right]^2 &= 1, \\ \pm(m_2 - m_1) + (n_2 - n_1)^2 + \left(\frac{m_1}{2n_1} \pm n_2\right)^2 &= 1. \end{aligned}\right\} \tag{5.54}$$

The generalized strain rates associated with each of these yield functions may be directly obtained by considering the partial derivatives of (5.54) with respect to the generalized stresses. The range of applicability of each yield function is of course governed by a set of inequalities which ensure that the yield condition is nowhere violated (Lance and Onat, 1963).

Consider the situation where one of the parameters a, β, γ is indeterminate, making the other two parameters equal to one another in view of (5.51). Such a situation arises when the numerically greater principal stress has the magnitude Y at the yield point. Taking, for instance $\dot{\lambda}_1 = \dot{\kappa}_1 = 0$, we have $\beta = \gamma$, while α is indeterminate. Since in this case $\dot{\varepsilon}_1 = 0$ and $\dot{\varepsilon}_2 = -\dot{\varepsilon}_3$, the yield criterion is $\sigma_2 = Y$ when $\dot{\varepsilon}_2 > 0$ and $\sigma_2 = -Y$ when $\dot{\varepsilon}_2 < 0$. The last two rows of Table 5.3 indicate

Table 5.3 Tresca yield condition for shells of revolution (noncylindrical surfaces)

Intermediate parameter	$\pm n_1$	$\pm n_2$	$\pm m_1$	$\pm m_2$
α	$-(\alpha+\beta)$	$-(\beta-\gamma)$	$1-2(\alpha^2+\beta^2)$	$2(\gamma^2-\beta^2)$
β	$-(\alpha+\beta)$	$-(\beta+\gamma)$	$1-2(\alpha^2+\beta^2)$	$1-2(\beta^2+\gamma^2)$
γ	$-(\beta-\alpha)$	$-(\beta+\gamma)$	$2(\alpha^2-\beta^2)$	$1-2(\beta^2+\gamma^2)$

that the stress resultants n_1 and m_1 are not uniquely determined, while n_2 and m_2 depend on the single parameter $\beta = \gamma$. The elimination of this parameter between the expressions for m_1 and m_2 furnishes the yield condition

$$m_2 = \pm\left(1 - n_2^2\right).$$

Similar equations are obtained for the cases $\alpha = \beta$ and $\alpha = \gamma$, the latter requiring a combination of the first and third rows of Table 5.3. The results are summarized in Table 5.4, which includes the ratios of the generalized strain rates associated with each yield function according to the normality rule of plastic flow.

Table 5.4 Parabolic hypercylinders on the Tresca yield surface

Coincidence	Yield condition	$N_0\dot{\lambda}_1$	$N_0\dot{\lambda}_2$	$N_0\dot{\kappa}_1$	$N_0\dot{\kappa}_2$
$\alpha = \beta$	$\pm m_1 + n_1^2 = 1$	$2\mu n_1$	0	$\pm\mu$	0
$\alpha = \gamma$	$\pm(m_1 + m_2) + (n_1 + n_2)^2 = 1$	$2\mu(n_1 - n_2)$	$2\mu(n_2 - n_1)$	$\pm\mu$	$\mp\mu$
$\beta = \gamma$	$\pm m_2 = n_2^2 = 1$	0	$2\mu n_2$	0	$\pm\mu$

The yield functions given in Tables 5.3 and 5.4 define a closed convex hypersurface in a four-dimensional space where n_1, n_2, m_1, and m_2 are taken as rectangular coordinates. The 12 faces of the yield surface intersect along 25 edges, some of which are smooth, while others form ridges where the normal is not unique. The part of the yield surface covered by Table 5.4 consists of three pairs of parabolic cylindrical hypersurfaces and is found to occupy a large proportion of the total yield surface. It would therefore be a reasonable approximation in certain cases to assume the entire yield surface to be formed by these hypercylinders, extended to circumscribe the noncylindrical part of the actual yield surface. It may be shown that the approximate yield surface nowhere differs from the actual surface by more than 15% (Fliigge and Nakamura, 1965).

5.3.3 Approximations for a von Mises Material

The yield condition for a uniform shell, when the material obeys the von Mises criterion, becomes extremely complicated (Ilyushin, 1948; Hodge 1961). It is instructive, however, to consider the von Mises criterion in relation to the ideal sandwich shell, the principal stresses acting in the inner and outer flanges of the shell being

$$\sigma_1^{\pm} = \sigma_0\left(n_1 \pm m_1\right), \quad \sigma_2^{\pm} = \sigma_0\left(n_2 \pm m_2\right), \tag{5.55}$$

where σ_0 is the uniaxial yield stress for the sandwich shell. Since the yield criterion $\sigma_1^2 - \sigma_1\sigma_2 + \sigma_2^2 = \sigma_0^2$ cannot be violated in either flange, the yield condition is represented by the two nonlinear surfaces

$$\left(n_1 \pm m_1\right)^2 - \left(n_1 \pm m_1\right)\left(n_2 \pm m_2\right) + \left(n_2 \pm m_2\right)^2 = 1. \tag{5.56}$$

The upper or lower sign should be used according as m_1/m_2 is greater or less than $(n_1 - 2n_2)/(2n_1 - n_2)$. The generalized strain rates associated with (5.56) are

$$N_0\dot{\lambda}_1 = \pm M_0\dot{\kappa}_1 = \mu\left[2\left(n_1 \pm m_1\right) - \left(n_2 \pm m_2\right)\right],$$

$$N_0\dot{\lambda}_2 = \pm M_0\dot{\kappa}_2 = \mu\left[2\left(n_2 \pm m_2\right) - \left(n_1 \pm m_1\right)\right],$$

where μ is a nonnegative factor of proportionality, the upper and lower signs in the above expressions being in exact correspondence with those in the yield condition. A different approximation to the von Mises type of yield condition has been examined by Ivanov (1967) and further discussed by Robinson (1971).

From the practical point of view, a more convenient approximation to the yield condition for a von Mises material is obtained on the assumption that the vector representing the deviatoric component of the direct forces (n_1, n_2) in the four-dimensional space is orthogonal to the vector representing the bending moments (m_1, m_2), so that

$$m_1\left(2n_1 - n_2\right) + m_2\left(2n_2 - n_1\right) = 0.$$

The two surfaces defined by (5.56) then reduce to a single yield surface having the equation

$$n_1^2 - n_1 n_2 + n_2^2 + m_1^2 - m_1 m_2 + m_2^2 = 1, \tag{5.57}$$

which has been proposed by Shapiro (1961). In the case of a cylindrical shell without end load, (5.57) reduces to (5.8). The yield condition (5.57) is identically satisfied by expressing the generalized stresses in the form

$$\left.\begin{aligned} n_1 &= \frac{2}{\sqrt{3}}\bar{n}\cos\psi, & n_2 &= \frac{2}{\sqrt{3}}\bar{n}\sin\left(\frac{\pi}{6} + \psi\right), \\ m_1 &= \frac{2}{\sqrt{3}}\bar{m}\cos\eta, & m_2 &= \frac{2}{\sqrt{3}}\bar{m}\sin\left(\frac{\pi}{6} + \eta\right), \end{aligned}\right\} \tag{5.58}$$

where ψ and η are suitable auxiliary angles, while $\bar{n}$ and $\bar{m}$ are dimensionless equivalent forces and moments which can take both positive and negative values satisfying the equation

$$\bar{n}^2 + \bar{m}^2 = 1.$$

By the usual normality rule of plastic flow, the generalized strain rates associated with the yield condition (5.57) may be written as

$$N_0\dot{\lambda}_1 = \mu\left(2n_1 - n_2\right), \qquad N_0\dot{\lambda}_2 = \mu\left(2n_2 - n_1\right),$$

$$M_0\dot{\kappa}_1 = \mu\left(2m_1 - m_2\right), \qquad M_0\dot{\kappa}_2 = \mu\left(2m_2 - m_1\right),$$

with μ representing an arbitrary nonnegative multiplier. Inserting from (5.58), the preceding relations may be expressed as

$$\left.\begin{aligned} N_0\dot{\lambda}_1 &= 2\mu\bar{n}\cos\left(\frac{\pi}{6}+\psi\right), \quad N_0\dot{\lambda}_2 = 2\mu\bar{n}\sin\psi, \\ M_0\dot{\kappa}_1 &= 2\mu\bar{m}\cos\left(\frac{\pi}{6}+\eta\right), \quad M_0\dot{\kappa}_2 = 2\mu\bar{m}\sin\eta. \end{aligned}\right\} \tag{5.59}$$

The state of stress and strain rate in the shell is therefore completely determined by the four quantities ψ, η, μ, and $\bar{n}$ or $\bar{m}$. These variables together with the velocity components u and ω form a set of six unknowns which can be determined from the field equations and the prescribed boundary conditions.

5.3.4 Linearization and Limited Interaction

As in the case of cylindrical shells, the use of Tresca's yield criterion together with the sandwich approximation leads to a set of linear equations defining the yield surface. The yielding of either of the two flanges of the sandwich shell is governed by six equations corresponding to the six sides of the Tresca hexagon. Using (5.55), where subscripts 1 and 2 are now replaced by ϕ and θ, respectively, we obtain the relations

$$\left.\begin{aligned} n_\phi \pm m_\phi = 1, \quad & n_\theta \pm m_\theta = 1, \quad n_\phi - n_\theta \pm \left(m_\phi - m_\theta\right) = 1, \\ -n_\phi \mp m_\phi = 1, \quad & -n_\theta \mp m_\theta = 1, \quad n_\theta - n_\phi \pm \left(m_\theta - m_\phi\right) = 1. \end{aligned}\right\} \tag{5.60}$$

The yield surface defined by (5.60) consists of 12 hyperplanes in a four-dimensional stress space. The strain rate vector having components $N_0\dot{\lambda}_\phi$, $N_0\dot{\lambda}_\theta$, $M_0\dot{\kappa}_\phi$, and $N_0\dot{\kappa}_\theta$ is obtained from the fact that this vector is normal to the appropriate hyperplane. For example, the first pair of hyperplanes in (5.60) correspond to $\dot{\lambda}_\theta = \dot{\kappa}_\theta = 0, N_0\dot{\lambda}_\phi = \pm M_0\dot{\kappa}_\phi > 0$ and the first two of these relations give

$$u = -\frac{dw}{d\phi} = w\tan\phi$$

in view of (5.49). The velocity field therefore becomes $w = A\cos\phi$, $u = A\sin\phi$, representing a rigid-body translation. The remaining hyperplanes can be similarly shown to represent a mere rigid-body motion of the shell material. It follows that none of the 12 hyperplanes can be associated with a plastic flow field over a finite portion of the shell. The plastic state of stress at the incipient collapse will therefore be represented by the intersection of two or more hyperplanes in the four-dimensional space (Hodge, 1981).

The upper sign in (5.60) corresponds to yielding of the lower flange, and the lower sign to the yielding of the upper flange of the sandwich shell. Each of the six hyperplanes defined by the upper sign can therefore intersect each of those defined by the lower sign. These intersections define 36 fully plastic sides on the yield surface, since they correspond to both the flanges being at the yield point. In the plastic analysis of shells using the linearized yield condition (5.60), it is only necessary to consider these fully plastic sides and their intersections. Each fully plastic side corresponds to two relations connecting the generalized stresses, and two relations connecting the generalized strain rates. The vector representing the strain rate is formed by a combination of those for the intersecting hyperplanes with nonnegative scalar factors. A vertex of the yield surface is formed by the intersection of four hyperplanes and corresponds to a stress state in which both flanges are at appropriate corners of the Tresca hexagon.

In the solution of practical shell problems, it is usually convenient to employ a simplified linear approximation proposed by Hodge (1960a, 1963). In this approximation, all interactions between the direct forces and between the bending moments are retained, but all interactions between the forces and the moments are neglected. The resulting yield condition, known as the linearized limited interaction yield condition, is therefore given by

$$\left.\begin{aligned} n_\phi &= \pm 1, \quad n_\theta = \pm 1, \quad & n_\phi - n_\theta &= \pm 1, \\ m_\phi &= \pm 1, \quad m_\theta = \pm 1, \quad & m_\phi - m_\theta &= \pm 1. \end{aligned}\right\} \tag{5.61}$$

In a four-dimensional stress space, these relations define a region formed by the intersection of two hexagonal cylinders, the traces of which in the $(n_1\ n_2)$- and (m_1, m_2)-planes are the two hexagons shown in Fig. 5.15. The state of stress in the shell at the yield point would involve one of the sides of the force hexagon and one of the sides of the moment hexagon. The strain rate vector is a combination of those directed along the exterior normals to the two hexagons. A hinge circle must correspond to $\mathrm{m}_\phi = \pm 1$ and/or $\mathrm{n}_\phi = \pm 1$ in order to have the required ratios for the generalized strain rates. The proposed yield surface circumscribes the actual yield surface for a Tresca material and furnishes an upper bound on the collapse load. Reducing the size of the proposed surface by the factor $(\sqrt{5} - 1)/2 \approx 0.618$ produces an inscribed surface, providing a lower bound on the collapse load. A factor of 7/8 may be considered as a good approximation in most practical situations.

When the material yields according to the von Mises criterion, an approximation similar to the above is achieved by assuming the yield condition to be defined by a pair of ellipses in the force and moment planes. Each ellipse may be considered as either inscribed in the hexagon or circumscribing the hexagon. When both the ellipses circumscribe as shown in Fig. 5.15, the yield condition becomes

$$n_\phi^2 - n_\phi n_\theta + n_\theta^2 = 1, \qquad m_\phi^2 - m_\phi m_\theta + m_\theta^2 = 1. \tag{5.62}$$

Equation (5.62) may be regarded as defining the limited interaction yield condition for the von Mises material, and they are identically satisfied by introducing auxiliary angles ψ and η such that

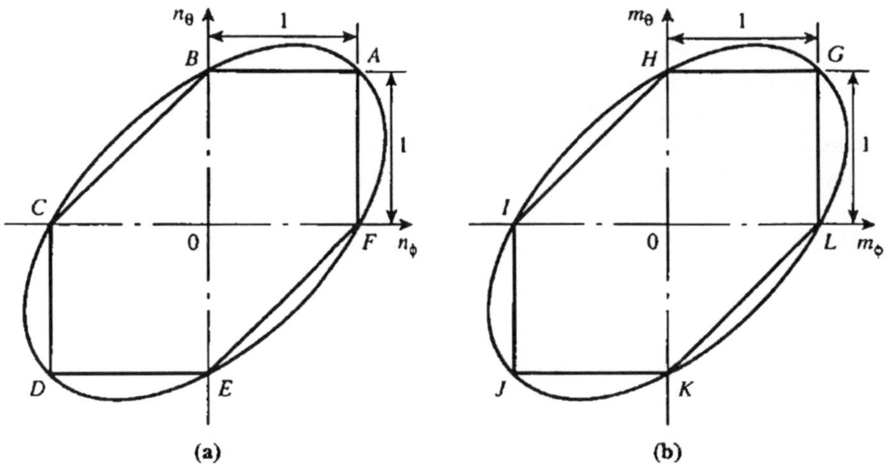

Fig. 5.15 Two-moment limited interaction yield conditions for shells of revolution corresponding to the Tresca and von Mises criteria for yielding

$$\left.\begin{aligned} n_\phi &= \pm\frac{2}{\sqrt{3}}\cos\psi, \quad n_\theta = \pm\frac{2}{\sqrt{3}}\sin\left(\frac{\pi}{6}+\psi\right), \\ m_\phi &= \pm\frac{2}{\sqrt{3}}\cos_\eta, \quad m_\theta = \pm\frac{2}{\sqrt{3}}\sin\left(\frac{\pi}{6}+\eta\right), \end{aligned}\right\} \tag{5.63}$$

where either all the upper signs or all the lower signs should be used for both the forces and the moments. The flow rule associated with (5.62) gives

$$\begin{aligned} \dot{\lambda}_\phi &= \mu\,(2n_1 - n_2), \quad \dot{\lambda}_\theta = \mu\,(2n_2 - n_1), \\ \dot{\kappa}_\phi &= v\,(2m_1 - m_2), \quad \dot{\kappa}_\theta = v\,(2m_2 - m_1), \end{aligned}$$

where μ and v are nonnegative quantities. The substitution from (5.63) furnishes the relations

$$\left.\begin{aligned} \dot{\lambda}_\phi &= \pm 2\mu\cos\left(\frac{\pi}{6}+\psi\right), \quad \dot{\lambda}_\theta = \pm 2\mu\sin\psi, \\ \dot{\kappa}_\phi &= \pm 2v\cos\left(\frac{\pi}{6}+\eta\right), \quad \dot{\kappa}_\theta = \pm 2v\sin\eta. \end{aligned}\right\} \tag{5.64}$$

For any given loading condition, (5.48) gives Q as a function of ψ and ϕ in view of the first equation of (5.63). The substitution for the forces and moments into the first and third equations of (5.47) then leads to a pair of differential equations for ψ and η which can be solved numerically using the boundary conditions. The velocity distribution can be determined from (5.49) on substitution for the strain rate ratios given by (5.64). It follows from (5.64) that a hinge circle would occur only at $\eta = 0$ and π and/or $\psi = 0$ and π, which correspond to $m_\phi = \pm 2/\sqrt{3}$ and/or $n_\phi = \pm 2/\sqrt{3}$, respectively. The nonlinear yield condition (5.62) is particularly useful in situations where the stress profile involves a large segment of the yield locus in the force and moment planes, requiring several plastic regimes for the linearized yield condition.

5.4 Limit Analysis of Spherical Shells

5.4.1 Basic Equations

An incomplete spherical shell of radius a is assumed to be loaded symmetrically by a prescribed distribution of radial pressure p. To carry out the statistical analysis, it is convenient to write the equations of equilibrium in terms of dimensionless forces and moments. Setting $p_r = p$, $p_\phi = 0$, $r_1 = a$, and $r_2 = a \sin\phi$ in (5.48) and introducing dimensionless variables s and q, the equation of overall vertical equilibrium can be written as

$$\left(n_\phi \sin\phi + s\cos\phi\right)\sin\phi = c - 2\int q \sin\phi\cos\phi d\phi, \tag{5.65}$$

where $s = Q/N_0$, $q = pa/2N_0$, and c is a dimensionless constant. The same substitution into the first and third equations of (5.47) reduces them to

$$\left.\begin{aligned} &\frac{d}{d\phi}\left(n_\phi \sin\phi\right) - n_\theta\cos\phi = s\sin\phi, \\ &\frac{d}{d\phi}\left(m_\phi \sin\phi\right) - m_\theta\cos\phi = \frac{s}{k}\sin\phi, \end{aligned}\right\} \tag{5.66}$$

where $k = M_0/aN_0 = h/4a$. Equation (5.65) and (5.66) form a set of three equations involving the generalized stresses and the shearing force. In terms of the velocities u and ω, the generalized strain rates are

$$\left.\begin{aligned} &\dot\lambda_\phi = \frac{1}{a}\left(\frac{du}{d\phi} - w\right), &&\dot\lambda_\theta = \frac{1}{a}\left(u\cot\phi - w\right), \\ &\dot\kappa_\phi = -\frac{1}{a^2}\frac{d}{d\phi}\left(u + \frac{dw}{d\phi}\right), &&\dot\kappa_\theta = -\frac{\cot\phi}{a^2}\left(u + \frac{dw}{d\phi}\right), \end{aligned}\right\} \tag{5.67}$$

in view of (5.49) with $r_1 = r_2 = a$. Eliminating the velocity components u and ω between the above relations, we obtain the equations of strain rate compatibility as

$$\frac{d}{d\phi}\left(\dot\lambda_\theta \sin\phi\right) = \left(\dot\lambda_\phi + a\dot\kappa_\theta\tan^2\phi\right)\cos\phi, \qquad \frac{d}{d\phi}\left(\dot\kappa_\theta\tan\phi\right) = \dot\kappa_\phi. \tag{5.68}$$

These equations must be supplemented by the yield condition and the flow rate. If the yield condition is of the limited interaction type, we have a pair of additional equations for the generalized stresses, and the integration of the stress equations can be carried out individually before considering the associated velocity field. The solution based on the linearized yield condition (5.60) can be similarly handled, although it is more involved owing to the difficulty in selecting the intersection of the appropriate hyperplanes.

The method of solution to the shell problem based on the nonlinear yield condition (5.57) of the von Mises type may be briefly outlined. In view of the parametric form (5.58) of the yield condition, with the subscripts 1 and 2 replaced by ϕ and θ, respectively, the differential equations (5.66) become

$$\begin{aligned}\frac{d}{d\phi}(\bar{n}\cos\psi)+\bar{n}\sin\left(\frac{\pi}{6}-\psi\right)\cot\phi&=\frac{\sqrt{3}s}{2},\\ \frac{d}{d\phi}(\bar{m}\cos\eta)+\bar{m}\sin\left(\frac{\pi}{6}-\eta\right)\cot\phi&=\frac{\sqrt{3}s}{2k}.\end{aligned} \tag{5.69}$$

Substituting for n_ϕ in (5.65), the dimensionless shear force s can be expressed in terms of the other variables as

$$\left(s+\frac{2}{\sqrt{3}}\bar{n}\cos\psi\tan\phi\right)\sin\phi\cos\phi=c-2\int q\sin\phi\cos\phi d\phi.$$

Inserting the expressions for the generalized strain rates from (5.59) into the compatibility equation (5.68) and eliminating the arbitrary multiplier μ between the resulting equations, we obtain an additional equation involving the stress variables:

$$\frac{d}{d\phi}\left\{\ln\left(\frac{\bar{m}\sin\eta}{\bar{n}\sin\psi}\right)\right\}+\frac{\tan\phi}{k}\left(\frac{\bar{m}\sin\eta}{\bar{n}\sin\psi}\right)=\frac{\sqrt{3}}{2}(\cot\eta-\cot\psi)\cot\phi-\tan\phi. \tag{5.70}$$

Equation (5.70) together with (5.69), from which s can be eliminated, constitute a set of three nonlinear ordinary differential equations in the three basic unknowns ψ, η, and $\bar{m}$, since $\bar{n}=\sqrt{1-\bar{m}^2}$. The integration can be carried out numerically using the boundary conditions $\psi=\eta=\pi/6$ at $\phi=0$, $\eta=\pi/2$ at a simply supported edge, and either $\eta=0$ or $\psi=0$ at a built-in edge that coincides with a hinge circle. Once the dependent variables have been found, the normal velocity w can be determined from the differential equation

$$\frac{d^2w}{d\phi^2}+w=\left(\frac{\dot{\lambda}_\phi+a\dot{\kappa}_\phi}{\dot{\lambda}_\theta+a\dot{\kappa}_\theta}\right)\left(\frac{dw}{d\phi}\cot\phi+w\right), \tag{5.71}$$

which follows from (5.67) and where $\dot{\kappa}_\phi$ and $\dot{\kappa}_\theta$ are given by (5.59). Using the appropriate boundary conditions, one of which is $w=0$ along the supporting edge, the solution can be carried out by a numerical procedure. The tangential velocity u may be subsequently obtained from the flow rule appropriate for the ratio $\dot{\lambda}_\phi/\dot{\lambda}_\theta$, and the condition $u=0$ along a circle of tangential constraint.

5.4.2 Plastic Collapse of a Spherical Cap

Consider, as an example, the plastic collapse of a spherical cap of angular span 2α under a uniform radial pressure p, the edge of the shell being either simply supported or clamped, Fig. 5.16a. From (5.65) with $c = 0$, the condition of overall vertical equilibrium is obtained as

$$s = -\left(q + n_\phi\right)\tan\phi. \tag{5.72}$$

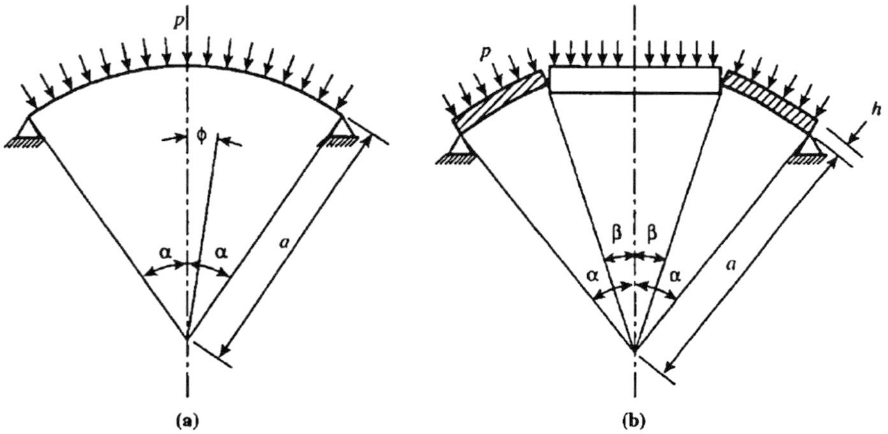

Fig. 5.16 Uniformly loaded spherical caps with and without a central cutout

This relation may be used to eliminate s from the remaining equilibrium equations (5.66). In view of (5.72), the first equation of (5.66) becomes

$$\frac{d}{d\phi}\left(n_\phi \tan\phi\right) - n_\theta = -q\tan^2\phi.$$

Following Hodge (1960a), we shall use the two-moment limited interaction yield condition to solve this problem. Since $n_\theta = n_\phi$ at $\phi = 0$ because of symmetry, it is natural to expect that the stress profile would be either on side DE or on side CD of the force hexagon, Fig. 5.15a. It turns out that side DC is the correct hypothesis, and the integration of the above equation with $n_\phi = -1$ gives

$$n_\theta = -\left[1 - (q-1)\tan^2\phi\right], \qquad n_\phi = -1. \tag{5.73}$$

The substitution from (5.72) into the second equation of (5.66), together with the yield condition $n_\phi = -1$, leads to the differential equation

$$\frac{d}{d\phi}\left(m_\phi \sin\phi\right) - m_\theta \cos\phi = -\left(\frac{q-1}{k}\right)\sin\phi\tan\phi.$$

For a simply supported shell, it is reasonable to suppose that the stress profile would be on side *HG* of the moment hexagon, Fig. 5.15b. Inserting $m_\theta = 1$ into the above equation and integrating, we obtain the solution

$$m_\phi = 1 - \frac{q-1}{k}\left[\text{cosec}\phi \ln\left(\sec\phi + \tan\phi\right) - 1\right] \tag{5.74}$$

in view of the symmetry condition $m_\phi = m_\theta$ at $\phi = 0$. The boundary condition $m_\phi = 0$ at $\phi = \alpha$ furnishes the dimensionless collapse pressure as

$$q = 1 + \frac{k\sin\alpha}{\ln\left(\sec\alpha + \tan\alpha\right) - \sin\alpha}. \tag{5.75}$$

Since $q > 1$ in view of (5.75), it follows from (5.73) and (5.74) that n_θ monotonically increases and m_ϕ, monotonically decreases as ϕ increases. It follows from the boundary conditions that the solution will be statically admissible only if $n_\theta \leq 0$ at $\phi = \alpha$. This condition is equivalent to $q \leq \text{cosec}^2\alpha$ and (5.75) gives

$$\frac{\ln\left(\sec\alpha + \tan\alpha\right)}{\sin\alpha} \geq 1 + k\tan^2\alpha. \tag{5.76}$$

Since k is usually small compared to unity, this condition will be satisfied in nearly all practical situations (Hodge, 1960a).

For a clamped shell, a hinge circle may be assumed to form at the edge $\phi = \alpha$, where m_φ is equal to –1. The moment distribution (5.74) therefore holds only over a central part of the shell defined by $0 \leq \phi \leq \beta$. The outer part of the shell corresponds to the side *IH* of the moment hexagon, which requires $m_\theta = 1 + m_\phi$. The differential equation for m_ϕ then becomes

$$\frac{dm_\phi}{d\phi} = \cot\phi - \left(\frac{q-1}{k}\right)\tan\phi, \qquad \beta \leq \phi \leq \alpha.$$

Integrating, and using the continuity condition $m_\phi = 0$ at $\phi = \beta$, we get

$$m_\phi = \ln\left(\frac{\sin\phi}{\sin\beta}\right) - \left(\frac{q-1}{k}\right)\ln\left(\frac{\cos\beta}{\cos\phi}\right), \quad \beta \leq \phi \leq \alpha. \tag{5.77}$$

The relationship between β, α, and q is obtained by using the boundary condition $m_\phi = -1$ at $\phi = \alpha$ in (5.77), and the continuity condition $m_\phi = 0$ at $\phi = \beta$ in (5.74), the result being

$$\left(\frac{q-1}{k}\right) = \frac{1 + \ln\left(\sin\alpha/\sin\beta\right)}{\ln\left(\cos\beta/\cos\alpha\right)} = \frac{\sin\beta}{\ln\left(\sec\beta + \tan\beta\right) - \sin\beta}. \tag{5.78}$$

Since m_ϕ monotonically decreases from 0 at $\phi = \beta$ to -1 at $\phi = \alpha$, the condition for the solution to be statically admissible is still $n_\theta < 1$ at $\phi = \alpha$. The required inequality is therefore (5.76) for both the simply supported and the built-in edge conditions.

Consider now the velocity field that can be associated with the stress field. Since the stress profile corresponding to the direct forces is everywhere on side DC, the flow rule gives $\dot{\lambda}_\theta = 0$ and $\dot{\lambda}_\phi < 0$. It follows then from the second equation of (5.67) that

$$u = w \tan \phi. \tag{5.79}$$

The resultant velocity is therefore everywhere parallel to the central axis for both the edge conditions. The simply supported shell, which involves the side HG of the moment hexagon, requires $\dot{\kappa}_\phi = 0$ and $\dot{\kappa}_\theta > 0$, and the last two equations of (5.67) give

$$\frac{dw}{d\phi} + w \tan \phi = -A$$

in view of (5.79), where A is a positive constant. If the radial velocity at the apex $\phi = 0$ is denoted by w_0, the integration of the above equation under the boundary condition $w = 0$ at $\phi = \alpha$ results in

$$\left.\begin{aligned} w &= [w_0 - A \ln(\sec\phi + \tan\phi)]\cos\phi, \\ A &= w_0 / \ln(\sec\alpha + \tan\alpha). \end{aligned}\right\} \tag{5.80}$$

For a clamped shell, the radial velocity in the inner plastic region ($0 \le \phi \le \beta$) is given by the first equation of (5.80), where A is still unknown. In the outer plastic region ($\beta \le \phi \le \alpha$), the flow rule associated with the yield condition is $m_\theta - m_\phi = 1$ given $\dot{\kappa}_\theta = -\dot{\kappa}_\phi > 0$, and it follows from (5.67) that

$$\frac{dw}{d\phi} + u = -B \operatorname{cosec}\phi, \quad \beta \le \phi \le \alpha,$$

where B is another positive constant. Substituting from (5.79), the above equation can be integrated using the boundary condition $w = 0$ at $\phi =$ a, resulting in

$$w = B \cos\phi \ln\left(\frac{\tan\alpha}{\tan\phi}\right), \quad \beta \le \phi \le \alpha. \tag{5.81}$$

The complete radial velocity distribution for a clamped shell is given by (5.80) and (5.81), the constants A and B being obtained from the conditions of continuity of w and $dw/d\phi$ across $\phi = \beta$. Thus

$$A = B = \operatorname{cosec}\beta = w_0 \left\{\ln(\sec\beta + \tan\beta) + \sin\beta \ln\left(\frac{\tan\alpha}{\tan\beta}\right)\right\}^{-1}.$$

Since A and B are both positive, the conditions $\dot{\kappa}_\phi < 0$ and $\dot{\kappa}_\theta > 0$ are satisfied throughout the shell. At $\phi = 0$ and $\phi = \alpha$, the velocity slope $dw/d\phi$ is negative, indicating that the discontinuities in slope are consistent with the nature of the bending moments existing at the hinge circles. Figure 5.17 shows the variation of the dimensionless collapse pressure with the cap semiangle. As α increases, each curve approaches the membrane solution $q = 1$ as expected. The plastic collapse of a spherical cap under a single axial load, which will be treated in Section 7.7, has been investigated by Palusamy (1971), Brooks and Leung (1989), and Yeom and Robinson (1996).

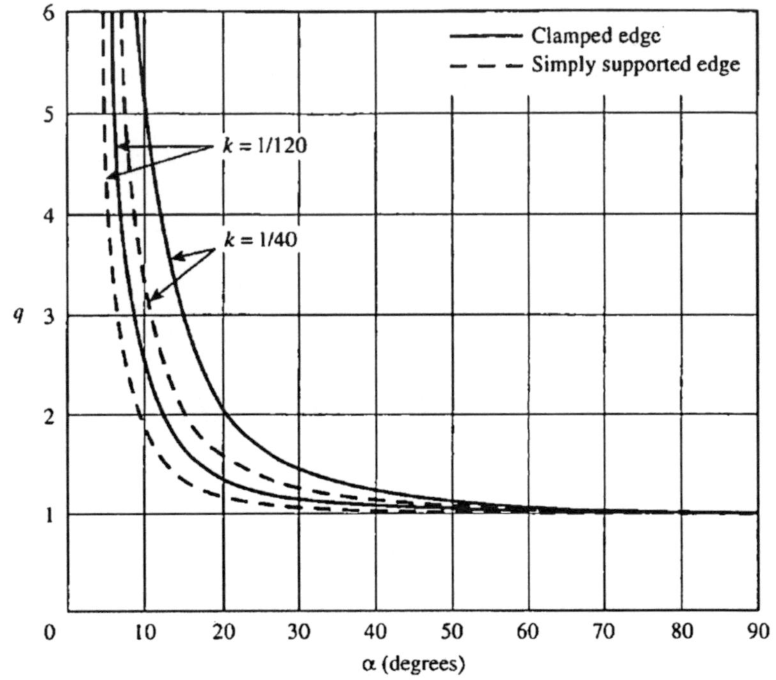

Fig. 5.17 Variation of dimensionless collapse pressure with the semiangle of an uniformly loaded spherical cap under simply supported and clamped edge conditions

5.4.3 Spherical Cap with a Covered Cutout

Consider a spherical shell having a central circular cutout which is covered by a rigid plate that is either hinged or rigidly connected to the shell (Hodge, 1964a). The shell is assumed simply supported along its outer edge $\phi = \alpha$ and is subjected to a uniform external pressure p which is extended to the rigid plate, Fig. 5.16b. The resultant vertical load acting on the rigid plate is therefore transmitted to the shell as a ring load at the cutout boundary $\phi = B$. The kinematic boundary conditions $u = w = 0$ at $\phi = \alpha$ are identically satisfied by taking the velocity field in the form

$$u = w \tan\phi, \qquad w = A\cos\phi \ln\left(\frac{\sec\alpha + \tan\alpha}{\sec\phi + \tan\phi}\right), \tag{5.82}$$

where A is a positive constant. The field is the same as that for a complete spherical cap given by (5.79) and (5.80). Since the resultant velocity is purely vertical according to (5.82), it is continuous across the cutout boundary, the downward speed of the rigid plate being

$$\delta = A\ln\left(\frac{\sec\alpha + \tan\alpha}{\sec\beta + \tan\beta}\right). \tag{5.83}$$

The cutout boundary is constrained to have zero horizontal velocity, being connected to the rigid plate. The components of the strain rate vector corresponding to (5.82) are

$$\dot{\lambda}_\phi = -\frac{A}{a}\tan\phi, \quad \dot{\lambda}_\theta = \dot{\kappa}_\phi = 0, \quad \dot{\kappa}_\theta = \frac{A}{a^2}\cot\phi.$$

If the two-moment limited interaction yield condition is adopted, the strain rate vector would be associated with the yield condition

$$n_\phi = -1, \quad m_\theta = 1, \quad \beta \leq \phi \leq \alpha.$$

The velocity slope $d\omega/d\phi$ changes discontinuously from a negative value to zero across the cutout boundary. The sign of the discontinuity is therefore positive when considered from outside to inside of the cutout area. The discontinuity has no effect on the hinged connection which simply requires $m_\phi = 0$ at $\phi = \beta$, but indicates the formation of a hinge circle along the inner edge for the rigid connection which requires $m_\phi = 1$ at $\phi = \beta$.

The condition of overall vertical equilibrium of the spherical cap between the cutout edge $\phi = \beta$ and any other parallel circle is given by (5.65), where $q = (pa/2N_0)$ and $c = q\sin^2\beta$. The result is easily shown to coincide with (5.72), derived for the complete spherical cap. Introducing the yield condition $n_\phi = -1$, the other two dimensionless forces may be written as

$$s = -1\,(q-1)\tan\phi, \qquad n_\theta = -\left[1 - (q-1)\tan^2\phi\right]. \tag{5.84}$$

In view of (5.84), and the yield condition $m_0 - 1$ for the bending moment, the differential equation for m_ϕ, becomes identical to that for the complete spherical cap, namely

$$\frac{d}{d\phi}\left(m_\phi \sin\phi\right) = \cos\phi - \left(\frac{q-1}{k}\right)\sin\phi\tan\phi.$$

Using the boundary condition $m_\phi = 0$ at the simply supported edge $\phi = \alpha$, the above equation is integrated to give

$$\left.\begin{aligned} m_\phi &= \left(1 - \frac{\sin\alpha}{\sin\phi}\right) + \frac{q-1}{k}\operatorname{cosec}\phi\left[f(\alpha) - f(\phi)\right], \\ \mathrm{f}(\phi) &= \ln(\sec\phi + \tan\phi) - \sin\phi. \end{aligned}\right\} \tag{5.85}$$

The dimensionless collapse pressure q is obtained from the remaining boundary condition, which is $m_\phi = 0$ at $\phi = \beta$ for the hinged plate and $m_\phi = 1$ at $\phi = \beta$ for the rigidly connected plate. The result may be expressed as

$$q = 1 + \frac{k(\sin\alpha - \varepsilon\sin\beta)}{f(\alpha) - f(\beta)}, \tag{5.86}$$

where $\varepsilon = 1$ for the hinged connection and $\varepsilon = 0$ for the rigid connection. For a complete spherical cap ($\beta = 0$), we have $f(\beta) = 0$, and (5.86) then coincides with (5.75). For a nonzero value of β, the right-hand side of (5.86) is greater than that of (5.75) when $\varepsilon = 0$ and less than that of (5.75) when $\varepsilon = 1$.

It remains to establish the range of validity of the solution. The assumed velocity field would correspond to a statically admissible distribution of stress only if

$$-1 \le n_\theta \le 0, \qquad 0 \le m_\phi \le 1 \qquad (\beta \le \phi \le \alpha).$$

We begin by noting the fact that $f(\phi)$ is a positive function that increases with increasing ϕ Indeed, by writing $f(\phi)$ in the alternative form

$$f(\phi) = \frac{1}{2}\ln\left(\frac{1+\sin\phi}{1-\sin\phi}\right) - \sin\phi,$$

and expanding it in ascending powers of $\sin\phi$ it is found that all the derivatives of $f(\phi)$ are positive. It follows therefore from (5.86) that $q > 1$, and (5.84) indicates that n_θ is always greater than -1, but $n_\theta \le 0$ requires $(q - 1)\tan^2\alpha \le 1$, which is equivalent to

$$\frac{f(\alpha) - f(\beta)}{\sin\alpha - \varepsilon\sin\beta} \ge k\tan^2\alpha \tag{5.87}$$

in view of (5.86). Since $f(\phi)/\sin\phi$ is a monotonic function, it is easily shown from (5.85) and (5.86) that monotonically decreases with increasing ϕ when the cutout is rigidly connected and has a maximum value of less than unity at a point in the

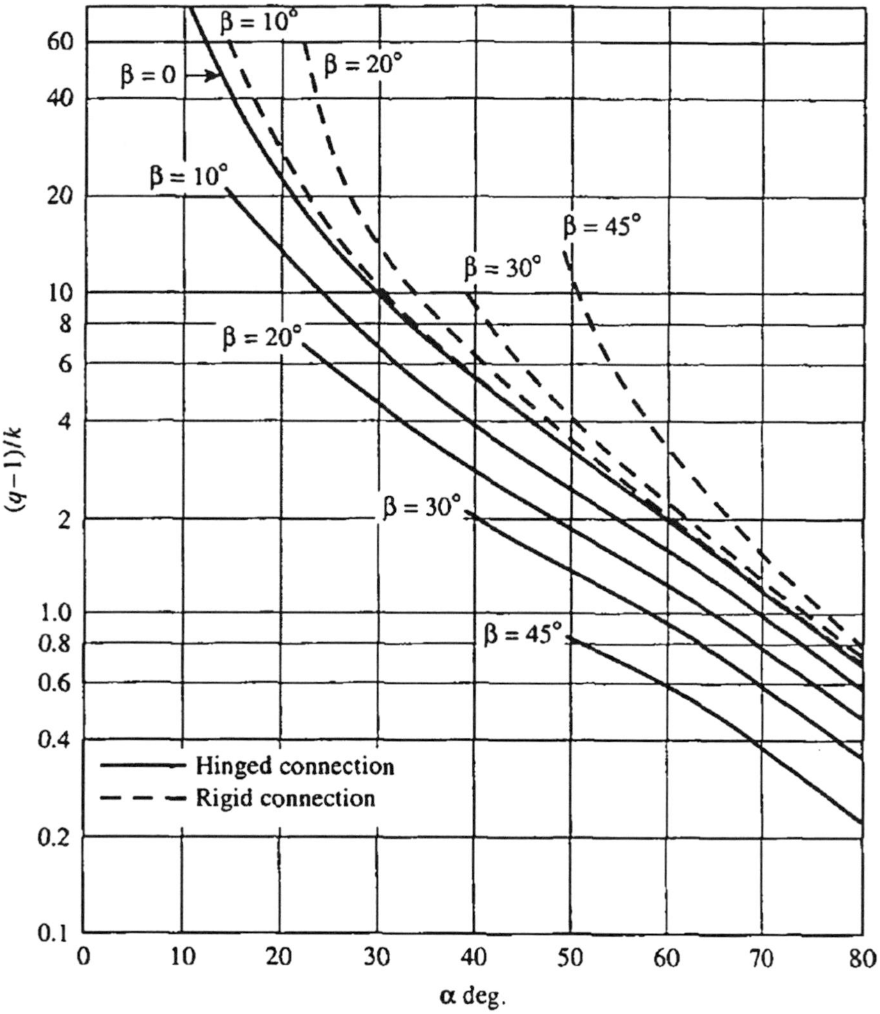

Fig. 5.18 Dimensionless collapse pressure for a spherical cap having a central cutout and fitted with a rigid cover plate (after Hodge, 1964a)

interval $\beta \leq \phi \leq \alpha$ when the plate is hinged. The condition $0 \leq m_\phi \leq 1$ is therefore satisfied in both cases. Unless α is very close to $\pi/2$, inequality (5.87) will be satisfied in most cases of practical interest. The parameter $(q-1)/k$, plotted against α in Fig. 5.18 for different values of β, indicates the influence of the mode of connection on the intensity of the collapse load.

The plastic collapse of a spherical cap having a central cutout without the reinforcement has been examined by Hodge and Lakshmikantham (1962).

5.4.4 Solution for a Tresca Sandwich Shell

Suppose that a uniformly loaded spherical cap of semi-angle α is built-in along its edge $\phi = \alpha$ and is composed of a material that obeys the Tresca criterion for yielding (Lee and Onat, 1968). It is assumed for simplicity that the shell is of sandwich construction so that the entire yield surface is defined by 12 hyperplanes given by (5.60). When the shell is sufficiently shallow, we may expect the membrane stresses to be small and the bending moment distribution similar to that in a clamped circular plate. Within a certain central portion of the shell, defined by $\phi \leq \phi_1$, the yield condition may therefore be assumed as

$$m_\theta = 1, \qquad n_\theta = 0 \qquad 0 \leq \phi \leq \phi_1.$$

In a four-dimensional space, this corresponds to the intersection of the hyperplanes $m_\theta \pm n_\theta = 1$, and the associated flow rule gives

$$\dot{\lambda}_\phi = 0, \qquad \dot{\kappa}_\phi = 0 \qquad N_0\dot{\lambda}_\theta = \mu_1 - \mu_2, \qquad M_0\dot{\kappa}_\theta = \mu_1 + \mu_2,$$

where μ_1 and μ_2 are nonnegative parameters. Substituting the yield condition into the equilibrium equations (5.66) and eliminating s by means of (5.72), the resulting equations are integrated under the boundary conditions $n_\phi = 0$ and $m_\phi = 1$ at $\phi = 0$ to give

$$n_\phi = -q\,(1 - \phi\,\cot\phi), \qquad m_\phi = 1 - \frac{q}{k}\,(1 - \phi\cot\phi) \qquad 0 \leq \phi \leq \phi_1. \quad (5.88)$$

The stress distribution (5.88) will be statically admissible so long as $m_\phi + n_\phi \geq 0$, which is easily shown to be the requirement for the yield condition (5.60) not being violated, considering the fact that $m_\phi > 0$ and $n_\phi < 0$. The angle $\phi = \phi_1$, where $m_\phi + n_\phi = 0$, is therefore given by

$$q\,(1 - \phi_1\cot\phi_1) = k/\,(1 + k)$$

Setting $\dot{\lambda}_\phi = 0$ and $\dot{\kappa}_\phi = 0$ in the velocity equations (5.67), and integrating, we obtain the velocity field

$$\left.\begin{aligned} u &= -A_1 + A_2\cos\phi + A_3\sin\phi, \\ w &= A_3\cos\phi - A_2\sin\phi, \end{aligned}\right\} 0 \leq \phi \leq \phi_1,$$

where A_1, A_2, and A_3 are constants. The components of the strain rate are now obtained from (5.67), the circumferential components being given by

$$a\dot{\lambda}_\theta = A_2\text{cosec}\phi - A_1\cot\phi, \qquad a^2\dot{\kappa}_\theta = A_1\cot\phi, \qquad 0 \leq \phi \leq \phi_1. \qquad (5.89)$$

The quantities μ_1 and μ_2 become infinite at $\phi = 0$, according to (5.89), but they remain nonnegative if A1 ≥ 0 and $A_2/A_1 \geq 1 - k$. Since u and $dw/d\phi$ do not vanish at $\phi = 0$, the center coincides with a plastic hinge which requires these quantities to be discontinuous.

Near the built-in edge $\phi = \alpha$, the state of stress over a certain portion of the shell, defined by $\phi_2 \le \phi \le \alpha$ may be assumed to be similar to that in a clamped circular plate. The yield condition in this region may therefore be written as

$$m_\theta - m_\phi = 1, \qquad n_\theta - n_\phi = 0, \qquad \phi_2 \le \phi \le \alpha.$$

so that the considered edge of the yield surface is the intersection of the hyperplanes $(m_\theta - m_\phi) \pm (n_\theta - n_\phi)$. The flow rule associated with the yield condition gives

$$\dot{\lambda}_\theta + \dot{\lambda}_\phi = 0, \qquad \dot{\kappa}_\theta + \dot{\kappa}_\phi = 0, \qquad M_0\dot{\kappa}_\theta \pm N_0\dot{\lambda}_\theta > 0.$$

In view of the yield condition, and the expression (5.72) for s, the equilibrium equations (5.66) can be easily integrated to obtain the stress distribution

$$\left.\begin{aligned} n_\theta &= n_\phi = -q + B_1 \cos\phi, \\ m_\phi &= m_\theta - 1 = \ln\sin\phi + \left(B_2 + \frac{B_1}{k}\cos\phi\right), \end{aligned}\right\} \phi_2 \le \phi \le \alpha, \tag{5.90}$$

where B_1 and B_2 are constants. The plastic collapse of the shell requires a hinge circle to be formed at $\phi = \alpha$, where u and $d\omega/d\phi$ are discontinuous. The built-in edge therefore corresponds to the meeting point of the considered edge of the yield surface with the hyperplane $n_\phi + m_\phi = -1$. The substitution from (5.67) into the flow relations, one of which can be immediately integrated, leads to the pair of equations

$$\frac{du}{d\phi} + u\cot\phi = 2w, \qquad \frac{dw}{d\phi} + u = -c\,\mathrm{cosec}\,\phi,$$

for the velocity components u and w, where c is a constant. The quantities $\dot{\lambda}_\theta$ and $\dot{\kappa}_\theta$ are not uniquely determined at the hinge circle $\phi = \alpha$, but the flow rule corresponding to the hinge circle at the boundary gives

$$N_0\dot{\lambda}_\phi = M\dot{\kappa}_\phi \qquad N_0\dot{\lambda}_\theta = M_0\dot{\kappa}_\theta \qquad \text{at } \phi = \alpha.$$

Inserting the expressions for $\dot{\lambda}_\theta$ and $\dot{\kappa}_\theta$ from (5.67), we obtain the velocity boundary conditions

$$u = ck\,\mathrm{cosec}\,\alpha \qquad w = 0 \qquad \text{at } \phi = \alpha.$$

In view of these conditions at the clamped edge $\phi = \alpha$, the integration of the preceding equations results in

$$\frac{u}{c} = (1+k)\left\{1+\cos\alpha \ln\left[\frac{\tan(\alpha/2)}{\tan(\phi/2)}\right]\right\}\sin\phi + (1+k)\cos\alpha\cot\phi - \mathrm{cosec}\phi,$$

$$\frac{\mathrm{w}}{\mathrm{c}} = (1+k)\left\{1+\cos\alpha \ln\left[\frac{\tan(\alpha/2)}{\tan(\phi/2)}\right]\right\}\cos\phi - (1+k)\cos\alpha.$$

The circumferential components of the rates of extension and curvature corresponding to this velocity field are given by

$$\left.\begin{aligned} a\dot{\lambda}_\theta &= c\,[(1+k)\cos\alpha - \cos\phi]\,\mathrm{cosec}^2\phi, \\ a^2\dot{\kappa}_\theta &= c\mathrm{cosec}\phi\cot\phi \end{aligned}\right\} \quad \phi_2 \le \phi \le \alpha. \tag{5.91}$$

The other two components of the generalized strain rate are simply the negative of the above quantities. The kinematic condition $M_0\dot{\kappa}_\theta \pm N_0\dot{\lambda}_\theta$ being nonnegative will be fulfilled when $c > 0$. We may take $c = 1$ without loss of generality.

In the region $\phi_1 \le\phi \le\phi_2$, the appropriate edge of the yield surface is formed by the intersection of the hyperplanes $m_\theta - m_\phi + n_\theta - n_\phi = 1$ and $m_\theta - n_\theta = 1$. The yield condition and the flow rule in this region may therefore be written as

$$m_\theta - n_\theta = 1, \qquad m_\phi - n_\phi = 2n_\phi, \qquad \phi_1 \le \phi \le \phi_2,$$
$$N_0\dot{\lambda}_\phi = M_0\dot{\kappa}_\phi = -\mu_1, \quad N_0\dot{\lambda}_\theta = \mu_1 - \mu_2, \qquad M_0\dot{\kappa}_\theta = \mu_1 + \mu_2,$$

where μ_1 and μ_2 are nonnegative. Eliminating m_θ and m_ϕ from (5.66) by means of the yield condition, and substituting from (5.72), the equilibrium equations are reduced to

$$\frac{dn_s}{d\phi} - n_\theta = q, \qquad \frac{dn_\theta}{d\phi} + \left(\frac{1+k}{2k}\right)n_s = \frac{1}{2}\cot\phi, \qquad \phi_1 \le \phi \le \phi_2, \tag{5.92}$$

where $n_s = -s = (q + n_\phi)\tan\phi$. Thus, we have a pair of simultaneous first-order ordinary differential equations (5.92), which can be solved numerically using the boundary conditions $n_\theta = 0$ and $n_s = q\phi_{1x}$ at $\phi = \phi_1$. The second condition being obtained from (5.88). The solution is based on an assumed value of q and is continued until n_θ and n_ϕ become equal to one another at an angular position $\phi = \phi_2$. The value of n_θ itself at this position furnishes the constant B_1 in (5.90), while the remaining constant B_2 follows from the condition $n_\theta = m_\phi$ at $\phi = \phi_2$. Thus

$$B_2 + B_1\left(\frac{1-k}{k}\right)\cos\phi_2 - \ln(\cos ec\phi_2) = -q.$$

Once the constants B_1 and B_2 are known, the cap semi-angle α corresponding to the assumed q is found from the relationship

$$B_2 + B_1\left(\frac{1+k}{k}\right)\cos\alpha - \ln(\cos ec\alpha) = q - 1$$

which is obtained by substituting from (5.90) into the yield condition $m_\phi + n_\phi = -1$ that applies to the clamped edge $\phi = a$. This completes the construction of a statically admissible stress field in the entire spherical cap.

To complete the solution to the plastic collapse problem, the deformation mode in the region $\phi_1 \leq \phi \leq \phi_2$ remains to be examined. This can be most conveniently achieved by considering the equations of compatibility (5.68) and substituting in them the expressions for the generalized strain rates in terms of μ_1 and μ_2. The result is the pair of differential equations

$$\left.\begin{aligned} \frac{d\mu_1}{d\phi} &= -\left[2\cot\phi - \left(\frac{1-k}{2k}\right)\tan\phi\right]\mu_1 + \left(\frac{1-k}{2k}\right)\mu_2\tan\phi, \\ \frac{d\mu_2}{d\phi} &= \left(\frac{1+k}{2k}\right)\mu_1\tan\phi - \left[\cot\phi + \left(\frac{1+k}{2k}\right)\tan\phi\right]\mu_2, \end{aligned}\right\} \qquad (5.93)$$

and the integration can proceed numerically from $\phi = \phi_2$, using boundary conditions that are based on the continuity of the velocity components (u, w) and the normal velocity gradient $dw/d\phi$. Since $\dot{\lambda}_\theta$ and $\dot{\kappa}_\theta$ must therefore be continuous in view of (5.67), the boundary conditions at $\phi = \phi_2$ are

$$\begin{aligned} \mu_1 &= \frac{1}{2}\left(N_0\dot{\lambda}_\theta + M_0\dot{\kappa}_\theta\right) = \frac{N_0}{2a}\left[(1+k)\cos\alpha - (1-k)\cos\phi_2\right]\operatorname{cosec}^2\phi_2, \\ \mu_2 &= \frac{1}{2}\left(M_0\dot{\kappa}_\theta + N_0\dot{\lambda}_\theta\right) = \frac{N_0}{2a}\left[(1+k)\cos\alpha - \cos\phi_2\right]\operatorname{cosec}^2\phi_2, \end{aligned}$$

in view of (5.91) with c=l. The values of μ_1 and μ_2 at $\phi = \phi_1$, obtained from the integration, furnish the corresponding strain rates by the relations $N_0\dot{\lambda}_\theta = \mu_1 - \mu_2$ and $M_0\dot{\kappa}_\theta = \mu_1 + \mu_2$, and the constants A_1 and A_2 then follow from (5.89) considered at $\phi = \phi_1$. When $k = 1/50$, these constants are found to satisfy the required inequalities for shells with $\alpha < 0.281$ radians, which is equivalent to $q > 1.39$. The bending moment and velocity distributions for the limiting case of $q = 1.39$ $(k = 1/50)$ are displayed in Fig. 5.19. The corresponding values of n_θ and n_ϕ are found to be small.

5.4.5 Extended Analysis for Deeper Shells

For sufficiently deep shells, we would expect the solution to require $n_\theta = n_\phi = -1$ and $m_\theta = m_\phi = 0$ at $\phi = 0$. Over a certain region near the center of the shell, which may be assumed to coincide with a plastic hinge, the yield condition would involve the intersection of the two hyperplanes

$$-m_\phi - n_\phi = 1, \quad m_\theta - n_\theta = 1, \quad 0 \leq \phi \leq \phi_1,$$

the entire shell being assumed to be plastic as before. The flow rule associated with the yield condition may be written as

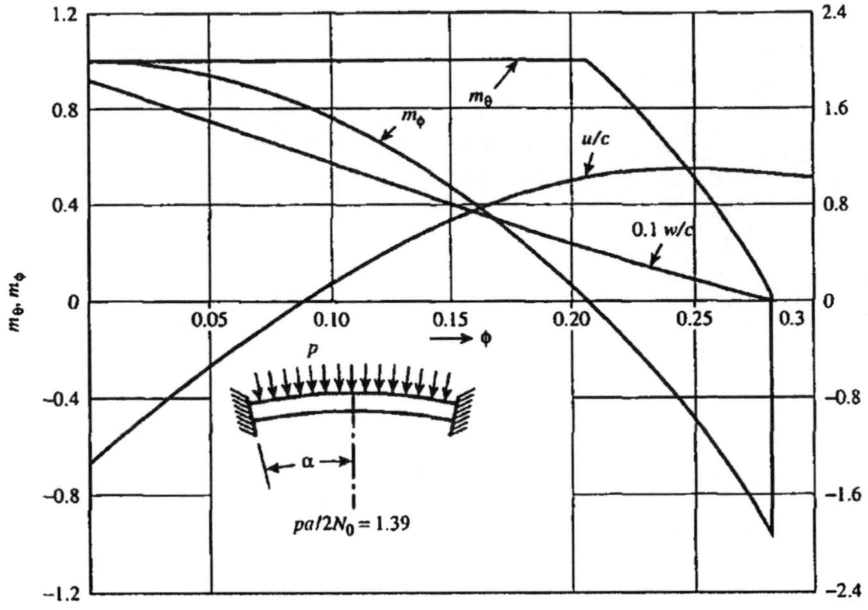

Fig. 5.19 Bending moment and velocity distributions at the incipient collapse of an uniformly loaded clamped spherical cap according to the exact yield condition for a Tresca material (after Lee and Onat, 1968)

$$N_0\dot{\lambda}_\phi = M_0\dot{\kappa}_\phi = -\mu_1, \quad N_0\dot{\lambda}_\theta = -M_0\dot{\kappa}_\theta = -\mu_2.$$

Adding the two equations of (5.66), substituting from (5.72), and using the yield condition, we obtain the relationship between n_θ and n_ϕ. The first equation of (5.66) then becomes

$$\frac{dn_s}{d\phi} - \left(\frac{1+k}{2k}\right) n_s \tan\phi = q - 1,$$

where $n_s = (q + n_\phi)\tan\phi$. This equation is readily integrated to give the distribution of the generalized stresses as

$$\left.\begin{aligned} -m_\phi &= 1 + n_\phi = (q-1)\left[f(\phi)\cot\phi - 1\right], \\ m_\theta &= 1 + n_\theta = (q-1)\,\gamma f(\phi)\tan\phi, \end{aligned}\right\} 0 \le \phi \le \phi_1, \qquad (5.94)$$

where

$$\gamma = (1+k)/2k, \quad f(\phi) = (\sec\phi)^\gamma \int_0^\phi (\cos\theta)^\gamma\, d\theta.$$

Since $f(\phi) = \cot\phi = 1$ at $\phi = 0$, the stress conditions at $\phi = 0$ are satisfied. Substituting for the generalized strain rates into the compatibility equations (5.68), we obtain a pair of equations in μ_1 and μ_2, and the solution is easily shown to be

$$\mu_2 \tan\phi = \left(\frac{2k}{1-k}\right)\mu_1 \cot\phi = D\,(\sec\phi)^{1-\gamma}, \quad 0 \le \phi \le \phi_1,$$

where D is a constant of integration to be determined from the condition of continuity of $\dot{\lambda}_\theta$ and $\dot{\kappa}_\theta$ across $\phi = \phi_1$. The considered plastic regime cannot be extended over the entire shell, since the flow condition $N_0\dot{\lambda}_\theta/M_0\dot{\kappa}_\theta = -1$ is incompatible with that required at the clamped edge, where a hinge circle is assumed to develop. Indeed, it follows from (5.67) that the condition to be satisfied at $\phi = \alpha$, where $w = 0$, is

$$\frac{N_0\dot{\lambda}_\theta}{M_0\dot{\kappa}_\theta} = \frac{u}{k\,(u + dw/s\phi)} = \frac{[u]}{k\left[u + dw/d\phi\right]} = \frac{N_0\dot{\lambda}_\phi}{M_0\dot{\kappa}_\phi} = 1,$$

where the square brackets denote the discontinuity, and the last step follows from the fact the hingle circle requires the yield condition $m_\phi + n_\phi = -1$ at $\phi = \alpha$. In the outer region $\phi_1 \le \phi \le \alpha$, it is therefore necessary to consider the intersection of the hyperplane $m_\theta - n_\theta = 1$ with the hyperplane $m_\theta - m_\phi + n_\theta - n_\phi = 1$, giving

$$m_\theta - n_\theta = 1, \quad m_\phi + n_\phi = 2n_\theta, \quad \phi_1 \le \phi \le \alpha,$$

as the appropriate yield condition. The normal component of the velocity in this region is given by the differential equation

$$-\frac{N_0}{a}\left(\frac{dw}{d\phi}\cot\phi + w\right) = \left(\frac{1+k}{k}\right)\mu_1 + \left(\frac{1-k}{k}\right)\mu_2,$$

and the boundary condition $\omega = 0$ at $\phi = \alpha$. The solution begins with the consideration of the stress equations (5.92), which are integrated under the conditions $n_\theta = -1/2$ and $n_s = (q-1)\,f(\phi_1)$ at $\phi = \phi_1$, a value of which is assumed for a given q, the associated value of α being found from the remaining boundary condition $n_\theta = -1/2$ at $\phi = \alpha$. The correct values of ϕ_1 and α are those which correspond to a maximum in the plot of ϕ_1, against α. The distribution of μ_1 and μ_2 is then obtained from the solution of (5.93) with the boundary conditions $\mu_1 = 0$ at $\phi = \phi_1$ and $\mu_2 = 0$ at $\phi = \alpha$. The solution obtained in this way involves discontinuities in n_θ and m_θ at $\phi = \phi_1$, and is statically admissible for $1.0 < q < 1.16$ when $k = \frac{1}{50}$, as shown by Lee and Onat (1968). The plastic collapse of shallow spherical shells has also been investigated by Haydi and Sherbourne (1974).

5.5 Limit Analysis of Conical Shells

5.5.1 Basic Equations

In the case of a conical shell, the meridional angle ϕ has a constant value α, the principal radii of curvature being $r_1 = \infty$ and $r_2 = \xi l \cot\alpha$, where ξl is the distance

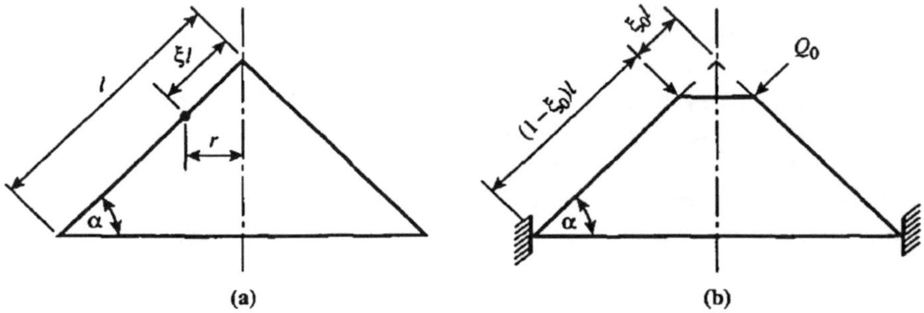

Fig. 5.20 Analysis of a conical shell: (**a**) geometry of the shell and (**b**) truncated shell under a ring load

of a generic point on the meridian measured from the apex, Fig. 5.20a. It is necessary to change the independent variable from ϕ to ξ by setting $r_1 d\phi = l\, d\xi$ in the general equations for shells of revolution. Using the relation $r = \xi l \cos\alpha$, and setting $p_r = p$ and $p_\phi = 0$, (5.48) may be written as

$$\xi\left(s + n_\phi \tan\alpha\right) = c\sec^2\alpha - 2\int q\xi d\xi, \qquad (5.95)$$

where $q = pl/2N_0$, $s = Q/N_0$, and c is a constant. Using (5.47), the differential equations of equilibrium for n_ϕ and m_ϕ are similarly reduced to

$$\frac{d}{d\xi}\left(\xi n_\phi\right) - n_\theta = 0, \quad \frac{d}{d\xi}\left(\xi m_\phi\right) - m_\theta = \frac{\xi s}{k}, \qquad (5.96)$$

where $k = M_0/N_0 l = h/4l$. Setting $r_1 d_\phi = ld\xi$ and $r_2 = \xi l\cot\alpha$ in (5.49), the expressions for the generalized strain rates in terms of the velocities u and w are obtained as

$$\left.\begin{aligned} l\dot{\lambda}_\phi &= \frac{du}{d\xi}, & l\dot{\lambda}_\theta &= \frac{1}{\xi}\left(u - w\tan\alpha\right), \\ l^2\dot{\kappa}_\phi &= -\frac{d^2w}{d\xi^2}, & l^2\dot{\kappa}_\theta &= -\frac{dw}{d\xi}, \end{aligned}\right\} \qquad (5.97)$$

The elimination of u and w between these relations furnishes the equation of strain rate compatibility as

$$\frac{d}{d\xi}\left(\xi\dot{\lambda}_\theta\right) = \dot{\lambda}_\phi + \xi l\dot{\kappa}_\theta\tan\alpha \quad \frac{d}{d\xi}\left(\xi\dot{\kappa}_\theta\right) = \dot{\kappa}_\phi. \qquad (5.98)$$

As in the case of spherical shells, the limited interaction yield condition and its associated flow rule are useful for the derivation of complete solutions. The stress and velocity equations then become uncoupled, and their integration becomes greatly simplified.

Consider now the approximate yield condition (5.57) that corresponds to the von Mises material, the subscripts 1 and 2 being identified as ϕ and θ, respectively. Using the parametric form (5.58), and inserting into the equilibrium equations (5.96), we have

$$\left.\begin{aligned} \xi \frac{d}{d\xi}(\bar{n}\cos\psi) + \bar{n}\sin\left(\frac{\pi}{6} - \psi\right) &= 0, \\ \xi \frac{d}{d\xi}(\bar{m}\cos\eta) + \bar{m}\sin\left(\frac{\pi}{6} - \eta\right) &= \frac{\sqrt{3}\xi s}{2k}. \end{aligned}\right\} \tag{5.99}$$

The quantity s can be eliminated from the second equation of (5.99) by means of the relation

$$\xi\left(s + \frac{2}{\sqrt{3}}\bar{n}\cos\psi\tan\alpha\right) = c\sec^2\alpha - 2\int q\xi\, d\xi,$$

which is obtained from (5.58) and (5.95). The substitution from (5.99) into the compatibility equation (5.98) results in

$$\begin{aligned} \frac{d}{d\xi}(\mu\bar{n}\sin\psi) - \frac{\sqrt{3}}{\xi}\mu\bar{n}\sin\left(\frac{\pi}{6} - \psi\right) &= \frac{\tan\alpha}{k}(\mu\bar{m}\sin\eta), \\ \frac{d}{d\xi}(\mu\bar{m}\sin\eta) - \frac{\sqrt{3}}{\xi}\mu\bar{m}\sin\left(\frac{\pi}{6} - \eta\right) &= 0. \end{aligned}$$

Eliminating the multiplier μ between these two equations, we obtain the required differential equation

$$\frac{d}{d\xi}\left\{\ln\left(\frac{\bar{m}\sin\eta}{\bar{n}\sin\psi}\right)\right\} + \frac{\tan\alpha}{\xi k}\left(\frac{\bar{m}\sin\eta}{\bar{n}\sin\psi}\right) = \frac{\sqrt{3}}{2\xi}(\cot\eta - \cot\psi). \tag{5.100}$$

In view of the preceding expression for s, and the relation $\bar{n} = \sqrt{1 - \bar{m}^2}$, (5.99) and (5.100) form a set of three nonlinear first-order ordinary differential equations for the three basic unknowns ψ, η, and $\bar{m}$. These equations must be integrated numerically using the boundary conditions which require $\eta = \pi/2$ at a simply supported edge, and either $\eta = 0$ or $\psi = 0$ at a clamped edge coinciding with a hinge circle. The velocity field then follows from the integration of equations obtained by suitable combinations of the velocity relations (5.97), as in the case of spherical shells.

5.5.2 Truncated Shallow Shell Under Line Load

A truncated conical shell of base angle α is clamped around its outer edge $\xi = 1$, and a line load of intensity $Q_0 = q_0 N_0$ per unit circumference is applied normal to the shell along its inner edge $\xi = \xi_0$, as shown in Fig. 5.20b. The shell is assumed to be sufficiently shallow so that the state of stress at the incipient collapse lies in the vicinity of that for a clamped circular plate under similar loading. If the entire yield surface is defined by the equations listed in Table 5.4 (Flügge and Nakamura, 1965), the appropriate yield condition for the analysis of the present problem is

$$(m_\theta - m_\phi) + (n_\theta - n_\phi)^2 = 1, \tag{5.101}$$

and the associated relations between the generalized strain rates are

$$\dot{\lambda}_\theta + \dot{\lambda}_\phi = 0 \quad \dot{\kappa}_\theta + \dot{\kappa}_\phi = 0.$$

The sign of $\dot{\kappa}_\theta$ is positive, while $\dot{\lambda}_\theta$ is of the same sign as that of $n_\theta - n_\phi$. In view of (5.97), the second strain rate relation $\dot{\kappa}_\theta + \dot{\kappa}_\phi = 0$ gives the differential equation for ω, having the solution

$$w = -A \ln \xi, \quad A > 0,$$

so that the boundary condition $\omega = 0$ at $\xi = 1$ is satisfied. Substituting from (5.97) into the first strain rate relation $\dot{\lambda}_\theta + \dot{\lambda}_\phi = 0$, we obtain the differential equation

$$\xi \frac{du}{d\xi} + u = -A \ln \xi \tan \alpha.$$

If a hinge circle is assumed to form at $\xi = 1$, involving discontinuities in both $dw/d\xi$ and u, the tangential velocity does not vanish at $\xi = 1$, and the solution to the above equation may be written as

$$u = [A(1 - \ln \xi) - B/\xi_j] \tan \alpha,$$

where B is another constant. The yield condition appropriate for the clamped edge $\xi = 1$ is

$$-m_\phi + n_\phi^2 = 1.$$

This relation does not represent plastic flow over any finite part of the shell, since the associated velocity field corresponds merely to a rigid-body translation. The above yield condition does, however, represent a hinge circle, since the associated strain rate ratios $\dot{\lambda}_\phi / \dot{\lambda}_\theta$ and $\dot{\kappa}_\phi / \dot{\kappa}_\theta$ are infinitely large.

The relationship between n_θ and n_ϕ at the yield point, necessary for the integration of the equation of force equilibrium, can be found from the flow rule associated with the yield condition (5.101), which holds throughout the shell. It follows from (5.97) and the normality rule that

$$n_\theta - n_\phi = \frac{N_0 \dot{\lambda}_\theta}{2M_0 \dot{\kappa}_\theta} = -\frac{u - w \tan\alpha}{2k\,(dw/d\xi)} = \left(\xi - \frac{B}{A}\right)\frac{\tan\alpha}{2k}, \tag{5.102}$$

in view of the preceding expressions for the velocity components. Since ω vanishes at the hinge circle $\xi = 1$, where u and $dw/d\xi$, are discontinuous, the associated flow rule indicates that the value of $-n_\phi$ at $\xi = 1$ is given by the right-hand side of (5.102). The boundary conditions may therefore be written as

$$n_\theta = m_\phi = 0 \text{ at } \xi = \xi_0, \quad n_\theta = m_\phi = 0 \text{ at } \xi = 1.$$

In view of (5.102), the first equation of (5.96) takes the form

$$z\frac{dn_\phi}{d\phi} = \left(1 - \frac{B}{A\xi}\right)\frac{\tan\alpha}{2K}, \quad \xi_0 \leq \xi \leq 1.$$

Integrating, and using the boundary conditions stated above, we get

$$B/A = (2 - \xi_0)\,/\,(1 - \ln\xi_0)\,,$$

and the distribution of the direct forces is then given by the solution tan α

$$\begin{aligned} n_\phi &= -\frac{\tan\alpha}{2k}, \left\{\frac{(2-\xi_0)\ln(\xi - \xi_0)}{1 + \ln(1/\xi_0)} - (\xi - \xi_0)\right\}, \\ n_\theta &= -\frac{\tan\alpha}{2k}, \left\{(2-\xi_0)\,\frac{1 + \ln(\xi/\xi_0)}{1 + \ln(1/\xi_0)} - (2\xi - \xi_0)\right\}, \end{aligned} \tag{5.103}$$

Setting $q = 0$ in (5.95), and using the boundary conditions $n_\phi = 0$ and $s = -q$ at $\xi = \xi_0$, we get

$$s = \frac{\tan^2\alpha}{2k}\left\{\frac{B}{A}\ln\left(\frac{\xi}{\xi_0}\right) - (\xi - \xi_0)\right\} - \frac{q_0\xi_0}{\xi}.$$

Substituting for s into the second equation of (5.96) and using (5.101) and (5.102), the differential equation for m_ϕ is obtained in the form

$$\frac{dm_\phi}{d\xi} = \frac{1}{\xi}\left(1 - \frac{q_0\xi_0}{k}\right) - \frac{\tan^2\alpha}{2k^2}\left\{\frac{1}{2\xi}\left(\xi - \frac{B}{A}\right)^2 - \frac{B}{A}\ln\left(\frac{\xi}{\xi_0}\right) + (\xi - \xi_0)\right\}.$$

Integrating, and using the boundary condition $m_\phi = 0$ at $\xi = \xi_0$, we obtain the distribution of m_ϕ as

$$m_\phi = \left(1 - \frac{q_0\xi_0}{k}\right)\ln\left(\frac{\xi}{\xi_0}\right) - \frac{\tan^2\alpha}{4k^2}\left\{\frac{B}{A}\left(\frac{B}{A} - 2\xi\right)\ln\left(\frac{\xi}{\xi_0}\right) + \frac{1}{2}(\xi - \xi_0)(3\xi - \xi_0)\right\}. \tag{5.104}$$

The circumferential bending moment follows from the yield condition (5.101), where m_ϕ, is given by (5.104) and $n_\phi - n_\theta$ is given by (5.102). The remaining boundary condition $m_\phi = -(1 - n_\phi^2)$ at the clamped edge $\xi = 1$ furnishes the collapse load

$$\left(\frac{q_0}{k}\right)\xi_0 \ln\left(\frac{1}{\xi_0}\right) = \left(1 + \ln\frac{1}{\xi_0}\right) + \frac{\tan^2\alpha}{4k^2}\left\{\frac{3 - \xi_0^2}{2} - \frac{(2 - \xi_0)^2}{1 - \ln\xi_0}\right\}. \tag{5.105}$$

If the dimensionless collapse load q_0 is plotted against the truncation ratio $1/\xi_0$ for a given value of $\tan\alpha$, the curve attains a minimum for a finite value of ξ_0. Since the collapse load cannot increase as ξ_0 is decreased, the part of the curve beyond the minimum cannot represent actual collapse loads. The minimum is found to occur at $\xi_0 \approx 0.6$ when $\tan\alpha = 0.1$, and at $\xi_0 \approx 0.4$ when $\tan\alpha = 0.04$. Within the range defined by the minimum, the remaining yield inequalities are found to be nowhere violated. A general theory for the plastic collapse of shallow shells has been developed by Lakshmikantham and Hodge (1963).

5.5.3 Shallow Shell Loaded Through Rigid Boss

A simply supported conical shell is subjected to a concentrated load P applied in the direction of the cone axis through a rigid circular boss which is built into the shell, Fig. 5.21a. In order to include the limiting case of an annular plate ($\alpha = 0$), the sign convention indicated in Fig. 5.21b will be adopted for positive stress resultants and velocity components. The differential equations (5.96) remain unchanged, but (5.95) for overall vertical equilibrium must be modified by changing the sign of s. Setting $q = 0$ and $c = P/2\pi N_0 l$ in (5.95), we get

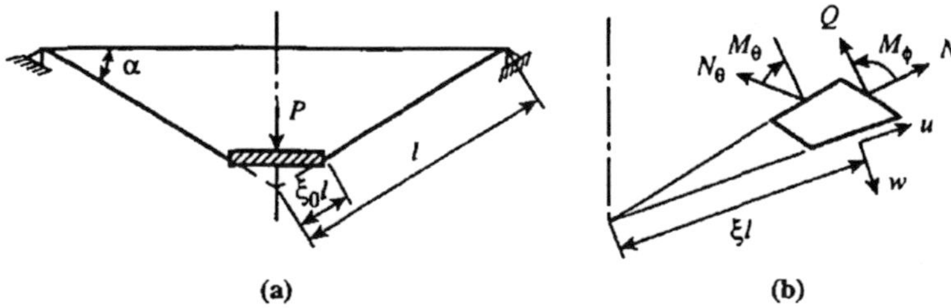

Fig. 5.21 Conical shell loaded through a rigid boss: (**a**) the loading condition and (**b**) adopted sign convention for the stress resultants

$$\xi \left(n_\phi \tan\alpha - s\right) = q_0\kappa, \tag{5.106}$$

where $k = M_0/N_0l$ and $c = P/2\pi M_0 cos^2\alpha$, denoting the dimensionless collapse load. Since the solution for the annular plate requires the yield condition $m_\theta = 1$ for the Tresca material, it is natural to expect that the yield condition and the flow rule in the shell problem for sufficiently small α would be

$$\dot{\lambda}_\phi = \dot{\kappa}_\phi = 0, \quad m_\theta + n_\theta^2 = 1, \quad N_0\dot{\lambda}_\theta/M_0\dot{\kappa}_\theta = 2n_\theta.$$

The boundary conditions along the outer edge of the shell are

$$n_\phi = 0, \quad m_\phi = 0, \quad w = 0 \text{ at } \xi = 1.$$

Using (5.97) for $\dot{\lambda}_\phi$ and $\dot{\kappa}_\phi$ which vanish throughout the shell, the velocity field is readily obtained by integration. Setting $\omega = \omega_0$ at $\xi = \xi_0$, the solution may be written as

$$u = u_0, \quad w = w_0\left(\frac{1-\xi}{1-\xi_0}\right). \tag{5.107}$$

The modified sign convention for the normal velocity requires a change in sign for ω in the expression for $\dot{\lambda}_\theta$. Using (5.97), and setting the ratio $N_0\dot{\lambda}_\theta/M_0\dot{\kappa}_\theta$ equal to $2n_\theta$ according to the flow rule, we get

$$2kn_\theta = (1-\xi_0)\frac{\mu_0}{w_0} + (1-\xi)\tan\alpha.$$

Inserting it into the first equation of (5.90), the resulting equation is readily integrated to give n_ϕ. Since $u = -\omega_0 \tan\alpha$ and $d\omega/d\xi = 0$ for the rigid toss at $\xi = \xi_0$, it follows from (5.107) that these quantities are discontinuous at the built-in edge, which therefore becomes a hinge circle that corresponds $m_\phi + n_\phi = 1$. This will be compatible with the yield condition in the shell if

$$n_\phi = n_\theta \quad \text{and} \quad m_\phi = m_\theta \text{ at } \xi = \xi_0.$$

The boundary conditions $n_\phi = n_\theta$ at $\xi = \xi_\theta$ and $n_\phi = 0$ at $\xi = 1$, applied to the differential equation for n_ϕ in (5.96), furnish

$$u_0/w_0 = -\frac{1}{2}(1+\xi_0)\tan\alpha.$$

The discontinuity in u at $\xi = \xi_0$ is consistent with the sign of n_ϕ. The expressions for the dimensionless membrane forces and circumferential bending moment then become

$$\left.\begin{aligned} n_\theta &= \beta\left(1+\xi_0^2-2\xi\right), \quad n_\phi = \beta\,(1-\xi)\left(1-\xi_0^2/\xi\right), \\ m_\theta &= 1-n_\theta^2 = 1-\beta^2\left(1+\xi_0^2-2\xi\right)^2, \end{aligned}\right\} \tag{5.108}$$

where $\beta = \tan\alpha/4k = (l/h)\tan\alpha$. As ξ increases from ξ_0, the forces n_θ and n_ϕ decrease from $\beta(l-\xi_0)^2$ to reach the values $-\beta\left(1-\xi_0^2\right)$ and zero, respectively, at $\xi = 1$, while m_θ has a maximum value of unity at $\xi = \left(1+\xi_0^2\right)/2$ where vanishes. It is easy to see that $\dot{\lambda}_\theta$ has the same sign as that of n_θ as required by the flow rule.

The distribution of m_ϕ must be found by solving the differential equation obtained by eliminating s from the second equation of (5.96) by means of (5.106), the result being

$$\frac{d}{d\xi}\left(\xi m_\phi\right) - (m_\theta) = 4\beta\xi n_\phi - q_0. \tag{5.109}$$

Substituting from (5.108), and using the boundary conditions $m_\phi = m_\theta = 1 - \beta^2(1-\xi_0)^4$ at $\xi = \xi_0$, the integration of the above equation results in

$$\begin{aligned} m_\phi = 1 - \left(1-\frac{\xi_0}{\xi}\right) q_0 - \beta^2 \left\{ \frac{\xi_0}{\xi}(1-\xi_0)^4 + \left(1-\frac{\xi_0}{\xi}\right)\left[\left(1+\xi_0^2\right)\right.\right. \\ \left.\left. -4\left(1+\xi_0^2\right)(\xi+\xi_0) + \frac{8}{3}\left(\xi^2+\xi\xi_0+\tfrac{5}{2}\xi_0^2\right)\right]\right\}. \end{aligned} \tag{5.110}$$

The remaining boundary condition $m_\phi = 0$ at the simply supported edge $\xi = 1$ furnishes the dimensionless collapse load as (Onat, 1960)

$$q_0 = (1-\xi_0)^{-1} + \frac{\beta^2}{3}(1+3\xi_0)(1-\xi_0)^2. \tag{5.111}$$

Since the incipient velocity field (5.107) maintains the conical shape, the preceding solution may be applied to the post-yield behavior of a circular plate of radius a, loaded through a central boss of radius b. The plate deforms into a cone, and the load corresponding to a small central deflection δ is given by (5.111), where

$$q_0 \approx \frac{P}{2\pi M_0}, \quad \beta = \frac{\delta}{h}, \quad \xi_0 = \frac{b}{a}.$$

It is assumed that a small overhang exists at the external support, preventing the plate from leaving the support as it continues to deform. The predicted load-deflection curve has the same trend as that experimentally observed, provided the elastic part of the deformation is disregarded.

The assumed yield condition holds only over a small range of values of β. Indeed, it can be shown that the preceding solution is strictly valid for $\beta < \beta_0$, where

$$\beta_0^2 = 3\xi_0\left\{(1-\xi_0)\left[2-\xi_0(4-\xi_0)\right]\right\}^{-1} \tag{5.112}$$

It is found that $\beta_0 \approx 0.455$ when $\xi_0 = 0.1$. Over the range $\beta_0 \leq \beta \leq \beta_1$, where $\beta_1 \approx 1.0$ when $\xi_0 = 0.1$, an outer part of the shell would require the yield condition $(m_\phi - m_\theta) + (n_\phi - n_\theta)^2 = 1$. The distribution of the generalized stresses for $\xi_0 = 0.1$ and $\beta = 1.0$, obtained by Lance and Onat (1963) using the exact yield surface for a Tresca material, is shown in Fig. 5.22. An extension of the solution for higher values of β has been discussed by Lance and Lee (1969). However, (5.111) gives a reasonable upper bound for all $\beta \geq \beta_0$.

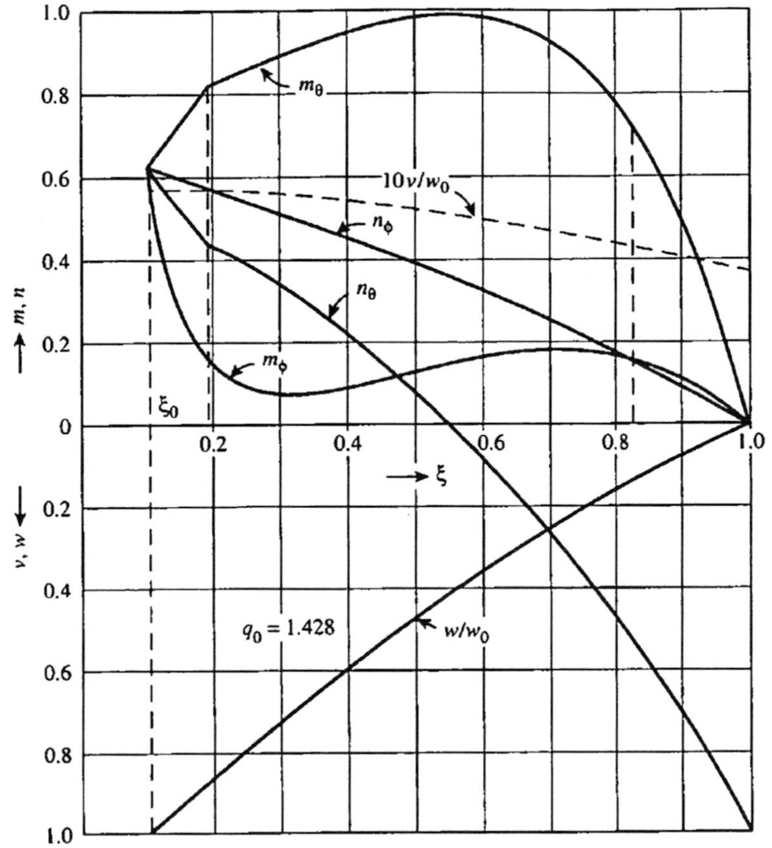

Fig. 5.22 Stress and velocity distributions in a shallow conical shell loaded through a rigid boss when $\xi_0 = 0.1$ and $\beta = 1.0$ (after Lance and Onat, 1963)

5.5.4 Centrally Loaded Shell of Finite Angle a

Suppose that the conical frustum of Fig. 5.21a is loaded by a central concentrated load P through small rigid boss which is hinged to the inner edge of the shell, the outer edge being simply supported against normal motion as shown. The boundary conditions along the outer edge are therefore unchanged but those along the inner edge are modified to

$$m_\phi = 0, \quad w = \delta \cos\alpha, \quad u = -\delta \sin\alpha \ \text{ at } \xi = \xi_0,$$

where δ is the downward speed of the rigid boss. We shall analyze the problem using the two-moment limited interaction yield condition, which involves two separate yield hexagons, one for the direct forces and the other for the bending moments (Hodge, 1960b).

Since n_ϕ vanishes at $\xi = 1$ and is positive at $\xi = \xi_0$, it is reasonable to suppose that it is positive throughout the shell. Since the outer part of the shell moves inward at the incipient collapse, n_θ is expected to be negative in this region. These two facts suggest that the stress profile will be on side *FA* of the force hexagon, Fig. 5.23a, at least over a part of the shell. In view of the yield condition $n_\phi - = 1$ and the boundary condition $n_\phi = 0$ at $\xi = 1$, the first equation of (5.96) is immediately integrated to give $n_\phi = \ln(1/\xi)$. Evidently, this solution holds only for $\xi \geq 1/e \approx 0.368$, so that the condition $n_\theta \leq 1$ is satisfied. For $\xi \leq 1/e$, the correct hypothesis is $n_\phi = 1$, giving $n_\theta = 1$, corresponding to the corner *B* of the hexagon. The solution for the direct forces therefore becomes

$$\left.\begin{aligned} &n_\phi = n_\theta = 1, \qquad \xi_0 \leq \xi \leq 1/e, \\ &n_\phi = \ln(1/\xi), \quad n_\theta = -1 + \ln(1/\xi), \quad 1/e \leq \xi \leq 1. \end{aligned}\right\} \tag{5.113}$$

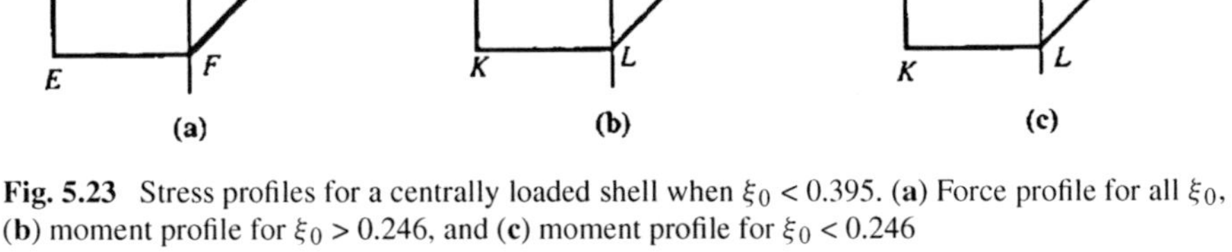

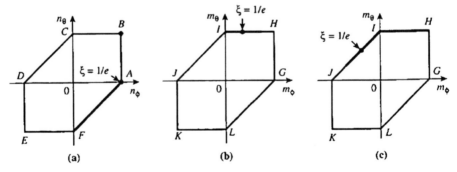

Fig. 5.23 Stress profiles for a centrally loaded shell when $\xi_0 < 0.395$. (**a**) Force profile for all ξ_0, (**b**) moment profile for $\xi_0 > 0.246$, and (**c**) moment profile for $\xi_0 < 0.246$

It may be noted that n_θ is discontinuous across $\xi = 1/e$, the discontinuity being of unit magnitude. The meridional force n_θ is, however, continuous as required.

To obtain the distribution of the bending moments, we observe that the corresponding solution for the flat plate requires $m_\theta = 1$, which suggests that the stress profile will be on side *HI* of the moment hexagon, Fig. 5.23b. Setting $m_\theta = 1$ in (5.109), inserting the expression for from (5.113), and using the boundary conditions $m_\phi = 0$ at $\xi = \xi_0$ and $\xi = 1$, we obtain the solution

$$\left.\begin{aligned} m_\phi &= -(q_0 - 1)(1 - \xi_0/\xi) + 2\beta\left(\xi - \xi_0^2/\xi\right), & \xi_0 \le \xi \le 1/e, \\ m_\phi &= -\left(q_\phi - 1\right)(1 - 1/\xi) + \beta\left(\xi - 1/\xi - 2\xi \ln \xi\right), & 1/e \le \xi \le 1, \end{aligned}\right\} \tag{5.114}$$

where $q_0 = P/2\pi M_0 \cos^2\alpha$ and $\beta = (1/h)\tan\alpha$. The condition of continuity of m_ϕ across $\xi = 1/e$ furnishes the collapse load as

$$q_0 = 1 + \beta\left(\frac{1 - 2\xi_0^2 - 1/e^2}{1 - \xi_0}\right). \tag{5.115}$$

The velocity field associated with the stress field can be determined from the plastic flow rule corresponding to the yield condition. Since the strain rates must satisfy the relations

$$\dot{\kappa}_\phi = 0 \ (\xi_0 \le \xi \le 1) \qquad \dot{\lambda}_\theta = 0 \ (\xi_0 \le \xi \le 1/e),$$
$$\dot{\lambda}_\theta + \dot{\lambda}_\phi = 0 \ (1/e \le \xi \le 1),$$

the substitution from (5.91), and integration under the boundary conditions $w = 0$ at $\xi = 1$, and $(\omega, u) = \delta(\cos\alpha, -\sin\alpha)$ at $\xi = \xi_0$, we have

$$\begin{aligned} \frac{w}{\delta} &= \left(\frac{1-\xi}{1-\xi_0}\right)\cos\alpha \ (\xi_0 \le \xi \le 1), \quad \frac{u}{\delta} = -\frac{w}{\delta}\tan\alpha\left(\xi_0 \le \xi \le \frac{1}{e}\right), \\ \frac{u}{\delta} &= -\frac{\sin\alpha}{1-\xi_0}\left[1 - \frac{1}{2}\left(\xi + \frac{1}{e^2\xi}\right)\right]\left(\frac{1}{e} \le \xi \le 1\right), \end{aligned} \tag{5.116}$$

in view of the continuity of u across $\xi = l/e$. It follows from (5.116) that $\dot{\lambda}_\phi \ge 0$ and $\dot{\kappa}_\phi \ge 0$ throughout, so that the velocity field is kinematically admissible. However, the stress field is not statically admissible when ξ_0 is sufficiently small, since m_ϕ then becomes negative near $\xi = \xi_0$. Indeed, it follows from (5.109) that the condition $dm_\phi/d\xi \ge 0$ at $\xi = \xi_0$ required for the validity of the solution will be satisfied if $q_0 \le 1 + 4\beta\xi_0$, which is equivalent to

$$e^2\left(1 - 4\xi_0 + 2\xi_0^2\right) \le 1, \quad \text{or} \quad \xi_0 \ge 0.246.$$

For $\xi_0 < 0.246$, the correct stress profile is that shown in Fig. 5.23c, which implies that it is negative in the region $\xi_0 \le \xi \le \xi_1$, vanishing at $\xi = \xi_0$ and ξ_1, as well as at $\xi = 1$. Since $dm_\phi/d\xi$ must be positive at $\xi = \xi_1$, it is reasonable to expect that $\xi_1 > 1/e$, for which $n_\phi < 1$. In (5.109), we therefore have to substitute $m_\theta = 1 + m_\phi$ and $n_\phi = 1$ for $\xi_0 \le \xi \le 1/e$, $m_\theta = 1 + m_\phi$ and $n_\phi = -\ln\xi$ for $1/e \le \xi \le \xi_1$, and $m_\phi = 1$ and $n_\phi = -\ln\xi$ for $\xi_1 \le \xi \le 1$. Integrating, and using the boundary conditions $m_\phi = 0$ at $\xi = \xi_0, \xi_1$ and 1, we obtain

$$\left.\begin{aligned} m_\phi &= -(q_0-1)\ln(\xi/\xi_0)+4\beta(\xi-\xi_0), && \xi_0\le\xi\le 1/e,\\ m_\phi &= -(q_0-1)\ln(\xi/\xi_1)+4\beta(\xi-\xi_1-\xi\ln\xi+\xi\ln\xi_1), && 1/e\le\xi\le\xi_1,\\ m_\phi &= -(q_0-1)(1-1/\xi)+\beta(\xi-1/\xi-2\xi\ln\xi), && \xi_1\le\xi\le 1. \end{aligned}\right\} \tag{5.117}$$

The parameters q_0 and ξ_1 are determined from the conditions of continuity of m_ϕ across $\xi = 1/e$ and $\xi = \xi_1$, the result being

$$\frac{q_0-1}{4\beta} = \frac{\xi_1(1-\ln\xi_1)-(\xi_0+1/e)}{\ln(\xi_1/\xi_0)} = \frac{1-\xi_1^2(1-2\ln\xi_1)}{4(1-\xi_1)}. \tag{5.118}$$

The velocity distribution for $\xi_0 \le 0.246$ is obtained from the plastic flow rule associated with the yield condition, which gives the relations

$$\dot{\kappa}_\theta+\dot{\kappa}_\phi = 0\,(\xi_0\le\xi\le\xi_1), \qquad \dot{\kappa}_\phi = 0\ (\xi_1\le\xi\le 1),$$
$$\dot{\lambda}_\theta = 0\ (\xi_0\le\xi\le 1/e), \quad \dot{\lambda}_\phi+\dot{\lambda}_\theta = 0\,(1/e\le\xi\le 1).$$

The normal velocity ω is directly obtained from these relations and (5.97), the constants of integration being determined from the boundary conditions $\omega = \delta\cos\alpha$ at $\xi = \xi_0$ and $\omega = 0$ at $\xi = 1$ and the conditions of continuity of ω and $d\omega/d\xi$ across $\xi = \xi_1$. Once the distribution of ω is known, u is obtained by using (5.97) and the continuity condition at $\xi = 1/e$, the final results being

$$\left.\begin{aligned} \frac{w}{\delta} &= \frac{1-f(\xi)}{1-f(\xi_0)}\cos\alpha\,(\xi_0\le\xi\le\xi_1),\\ \frac{w}{\delta} &= \frac{(1-\xi)\cos a}{1-f(\xi_0)}\ (\xi_0\le\xi\le\xi_1),\\ -\frac{u}{\delta} &= \frac{1-f(\xi)}{1-f(\xi_0)}\sin\alpha\,(\xi_0\le\xi\le 1/e),\\ -\frac{u}{\delta} &= \frac{\left[1-\xi_1/e\xi+g(\xi)\right]\sin\alpha}{1-f(\xi_0)}\ (1/e\le\xi\le 1), \end{aligned}\right\} \tag{5.119}$$

where $f(\xi) = \xi_1 - \xi_1\ \ln(\xi_1/\xi)$, while $g(\xi) = \xi_1\ln(\xi_1/\xi)$ for $1/e \le \xi \le \xi_1$ and $g(\xi) = \xi 1\ (1- \xi 1/\xi)/2$ for $\xi_1 \le \xi \le 1$. It is easy to verify that the nonzero strain rates are of the appropriate sign, indicating that the velocity field is kinematically admissible.

The range of β for which the stress field (5.117) is statically admissible will now be examined. The first derivative of the moment indicates that m_ϕ decreases from zero at $\xi = \xi_0$ to attain a relative minimum at $\xi = a = (q_0 - 1)/4\beta < 1/e$, and then increases through zero at $\xi = \xi_1$ to attain a relative maximum at a point $\xi = b$, such that

$$\beta[1-b^2(1+2\ln b)] = q_0 - 1 \tag{5.120}$$

before decreasing to zero at $\xi = 1$. Hence, the solution will be statically admissible if $m_\phi > -1$ at $\xi = a$ and $m_\phi < 1$ at $\xi = b$. From (5.117) and (5.120), these two conditions may be written as

$$4\beta\left[a - \xi_0 - a\ln(a/\xi_0)\right] \geq -1, \quad \beta\left[b^2 - 2b(2-b)\ln(b-1)\right] \leq 1.$$

For any given $\xi \leq 0.246$, these inequalities define the smallest value of β for the solution to be statically admissible. When ξ_0 is very small, we may use the approximation

$$a \approx (1 - 1/e)/(-\ln \xi_0), \quad b \approx 1 - a, \quad \xi_1 \leq 1 - 2a.$$

The second inequality then becomes trivial, while the first one reduces to $4\beta(1 - 1/e) \leq 1$, or $\beta \leq 0.395$. For higher values of β, the solution provides a reasonable upper bound for all $\xi_0 \leq l/e$.

By letting ξ_0 tend to zero in (5.118), we obtain $q_0 = 1$ as an upper bound on the collapse load for a complete conical shell of arbitrary angle α. To obtain a lower bound, we assume the stress field $n_\phi = n_\theta = m_\phi = 0$, $m_\theta = 1$, satisfying the equilibrium equations and the boundary conditions, giving $q_0 = 1$ in view of (5.109). Since $q_0 = 1$ is both an upper bound and a lower bound, and the assumed stress state does not violate the actual yield condition for a Tresca material, we have indeed found the actual concentrated collapse load. The modification of this formula due to the size of the loading boss has been investigated by DeRuntz and Hodge (1966). The plastic collapse of conical shells under uniform internal pressure has been studied by Kuech and Lee (1965), and a numerical method of solution has been discussed by Biron and Hodge (1967). Various numerical methods used in the literature for the limit analysis of shells have been reviewed by Xu et al. (1998).

5.6 Limit Analysis of Pressure Vessels

5.6.1 Plastic Collapse of a Toroidal Knuckle

Although pressure vessels behave primarily as membranes, their economical design requires the introduction of discontinuities in curvature where bending and shear are induced for the necessary continuity of stresses. Bending effects are also important in regions of relatively sharp curvature, such as a toroidal knuckle commonly employed between a cylindrical shell and a spherical cap, Fig. 5.24a. Since M_ϕ, M_θ and their algebraic difference cannot exceed M_0 in magnitude, and since the knuckle is sufficiently far away from the axis of symmetry, the quantity rQ will far exceed the magnitude of M_ϕ, or M_θ. The terms containing the bending moments themselves may therefore be omitted from the last equation of (5.47). Assuming a uniform internal pressure p and introducing dimensionless quantities

$$q = \frac{pa}{2N_0}, \quad k = \frac{h}{4r_1}, \quad \rho = \frac{r_1}{a},$$

the relevant equations of equilibrium for knuckle of radius r_1 and angle α, in terms of the dimensionless forces and moments, may be written from (5.47) and (5.48) as

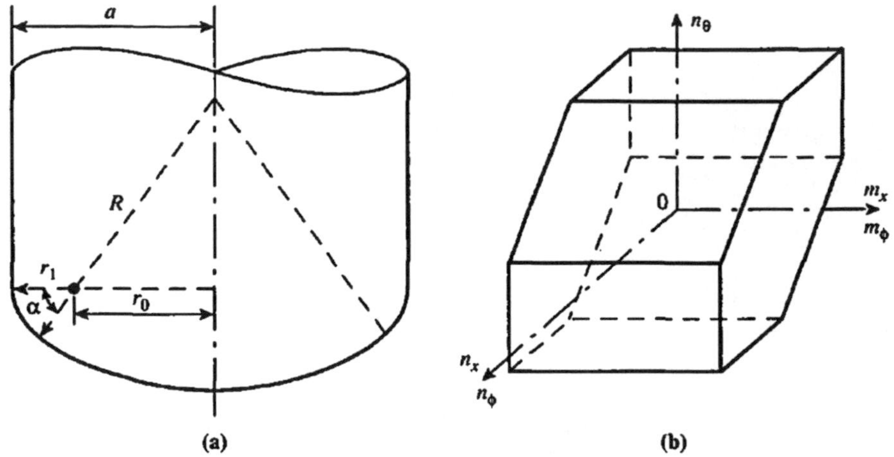

Fig. 5.24 Pressure vessel geometry and approximate yield surface for both the cylindrical and the spherical portions of the vessel

$$\frac{dn_\phi}{d\phi} + \frac{r_1}{r}\left(n_\phi - n_\theta\right)\cos\phi = s,\ \ k\,\frac{dm_\phi}{d\phi} = s, \tag{5.121}$$
$$n_\phi \sin\phi + s\cos\phi = q\,(1 + \rho\sin\phi)\,,$$

where $s = Q/N_0$. When the knuckle is considered as a membrane, $m_\phi = s = 0$, and the membrane forces are immediately obtained from (5.121) as

$$n_\phi = q\,(\operatorname{cosec}\phi + \rho)\,, \quad n_\theta = -n_\phi\left(\frac{\cot\phi}{\rho} - 1\right). \tag{5.122}$$

Thus n_ϕ is everywhere tensile, while n_θ is compressive for $\phi \leq \cot^{-1}\rho$. It may be seen that n_ϕ steadily decreases from the value $pR/2N_0$ at $\phi = \pi/2 - \alpha$ to $pa/2N_0$ at $\phi = \pi/2$, which indicates that n_ϕ is continuous and n_θ discontinuous across the interfaces with the cylinder and the sphere. For the usual situation $\pi/2 - \alpha \leq \cot^{-1}\rho$, the limit pressure according to the membrane theory corresponds to $n_\phi - n_\theta = 1$ at $\phi = \pi/2$, giving $pR/2N_0 = \rho\cot\alpha$. This pressure is considerably lower than that based on the theory which takes into account the bending resistance of the knuckle.

It may be shown that the rate of work done by the circumferential bending moment M_θ on the change in the corresponding curvature is small compared to that produced by the other generalized stresses, provided h/r is sufficiently small compared to unity. Since M_θ has been omitted from the equation of equilibrium, it is consistent therefore to omit M_θ from the yield condition as well. For practical purposes, the simplified yield surface shown in Fig. 5.24b would be more useful than the actual yield surface or its linearized version. The proposed yield surface is a hexagonal prism that circumscribes the actual one and is appropriate for both cylindrical shells and shells of revolution. Reducing the size of the yield surface by the factor $\left(\sqrt{5}-1\right)/2 \approx 0.618$ produces an inscribed surface, indicating that a factor of 7/8 may be used for a good overall approximation based on the proposed yield surface (Drucker and Shield, 1959).

For simplicity, we begin with the assumption that the knuckle is bounded by heavy members which do not deform at the incipient collapse. The deformation mode at collapse then involves hinge circles formed at $\phi = \pi t/2 - \alpha$ and $\phi = \pi/2$, as well as at an intermediate value of ϕ where m_ϕ is a relative maximum, Fig. 5.25a. Since $n\phi$ is tensile and n_θ is compressive throughout the knuckle, the stress point lies on the sloping plane of the yield surface, giving $n_\phi - n_\theta = 1$. Eliminating s between the first and last equations of (5.121), using the yield condition and rearranging, we have

$$\frac{d}{d\phi}\left(n_\phi \sec\phi\right) = q\sec^2\phi\,(1+\rho\sin\phi) - \rho\,(1+\rho\sin\phi)^{-1}.$$

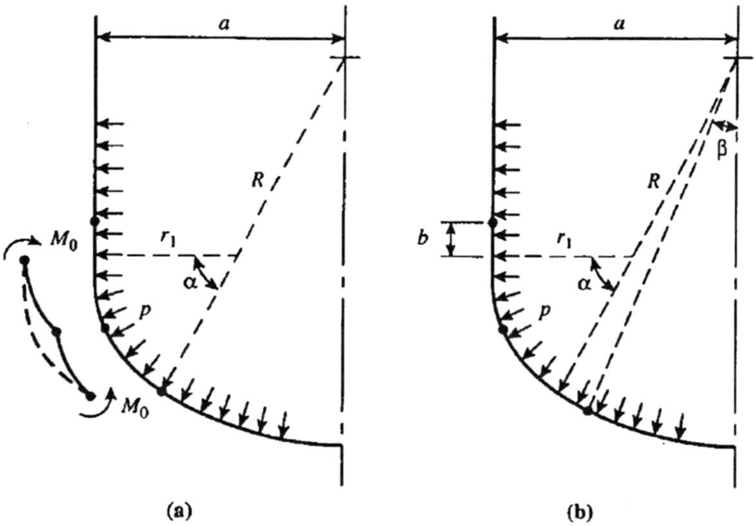

Fig. 5.25 Plastic collapse of a pressure vessel. (**a**) Simplified collapse mechanism and (**b**) the complete collapse problem

Since ρ sin ϕ is usually small, the last term of the above equation may be written as $\rho(1 - \rho \sin\phi)$ to a close approximation. The integration of the above equation then results in

$$\left.\begin{aligned} n_\phi &= A\cos\phi + q\,(\sin\phi + \rho) - \rho\cos\phi\,(\phi + \cos\phi)\,,\\ s &= -A\sin\phi + q\cos\phi - \rho\sin\phi\,(\phi + \cos\phi)\,, \end{aligned}\right\} \tag{5.123}$$

where A is a constant. Substituting for s into the second equation of (5.121), we obtain the differential equation for m_ϕ, which is readily integrated to give

$$km_\phi = B + A\cos\phi + (q+\rho)\sin\phi - \rho\cos\phi\left(\phi + \frac{1}{2}\rho\cos\phi\right), \qquad (5.124)$$

where B is another constant. The two constants of integration can be determined from the boundary conditions $m_\phi = -1$ at $\phi = \pi/2 - \alpha$ and $\phi = \pi/2$. The last equation therefore furnishes

$$A = (q+\rho)\tan\frac{\alpha}{2} + \rho\left(\frac{\pi}{2} - \alpha + \frac{\rho}{2}\sin\alpha\right), \quad B = -(q+\rho+k).$$

The collapse pressure is obtained from the fact that $m_\phi = 1$ at the position where $dm_\phi/d_\phi = 0$. Since the variation of m_ϕ is small in the region of maximum moment, it is a good approximation to set $= 1$ at $m_\phi = \pi/2 - \alpha/2$. Using (5.124), and substituting for A and B, we get

$$(q+\rho)\tan\frac{\alpha}{4} = 2k\cot\frac{\alpha}{2} + \frac{1}{2}\rho\cos\frac{\alpha}{2}\left[\alpha - \rho\left(\sin\alpha - \sin\frac{\alpha}{2}\right)\right]. \qquad (5.125)$$

Taking, for example, $\alpha = \pi/3$ and $\rho = 6/47$, giving $r_0/a = 47/53$, the collapse value of pa/N_0 is found to be $0.178 + 7.29\ h/r_1$, which is substantially higher than that predicted by the membrane theory.

The exterior normal to the considered inclined face of the yield surface indicates that $\dot\lambda_\theta$ is compressive and is equal in magnitude to $\dot\lambda_\phi$ which is tensile, while $\dot\kappa_\phi$ is identically zero. Using the relations $\dot\lambda_\phi + \dot\lambda_\theta = 0$ and $\dot\kappa_\phi = 0$, the boundary conditions $u = \omega = 0$ at $\phi = \pi/2$, and the condition of zero horizontal component of velocity at $\phi = \pi/2 - \alpha$, it is clearly possible to construct a velocity field that is kinematically admissible. Although it is somewhat difficult in this case to write the velocity solution explicitly, the mere existence of such a solution is sufficient to establish the validity of the equilibrium solution.

5.6.2 Collapse of a Complete Pressure Vessel

The solution for the plastic collapse of a pressure vessel involves consideration of the cylinder, the torus, and the sphere simultaneously. Relaxing the constraints at the boundaries of the toroidal knuckle will lower the magnitude of the bending moment acting across these interfaces. The forces and moments acting at these sections will cause the plastic flow field to extend over a length b in the cylindrical shell and spread into the spherical cap to reach a parallel circle $\phi = \beta$. The hinge circles will therefore move from the ends of the knuckle to the positions $x = b$ in the cylindrical shell and $\phi = \beta$ in the spherical cap, Fig. 5.25b, the net effect of which will be an appreciable reduction in the collapse pressure.

Consider first the cylindrical shell of radius a, where the axial force N_ϕ has the constant value $pa/2 = (qa/r_0)N_0$, which is identical to the meridional force in the torus at $\phi = \pi/2$, indicating that N_θ is automatically continuous across the junction of torus and cylinder. Since the circumferential force N_θ should be negative for continuity, the yield condition $n_\phi - n_\theta = 1$ should apply according to the hexagonal prism yield surface. Inserting the value $n_\theta = -(1 - qa/r_0)$ into the equilibrium equation (5.4), where $-q$ must be replaced by $2qa/r_0$ and m_ϕ written for m_x, we get

$$\frac{d^2 m_\phi}{d\xi^2} = 2\,(1 + qa/r_0)\,.$$

Since the bending moment has a relative minimum value of $-M_0$ at $x = b$, the boundary conditions are $m_\phi = -1$ and $dm_\phi/d\xi = 0$ at $\xi = \omega = b\sqrt{2/ah}$, and the solution is

$$\left.\begin{aligned} m_\phi &= -1 + \left(1 + \frac{qa}{r_0}\right)(\omega - \xi)^2\,, \\ s &= -\sqrt{\frac{h}{2a}}\left(1 + \frac{qa}{r_0}\right)(\omega - \xi)\,. \end{aligned}\right\} \tag{5.126}$$

The stresses in the knuckle are given by (5.123) and (5.124) with A and B yet unknown. Using (5.123), (5.124), and (5.126), and the conditions of continuity of m_ϕ and s across $\xi = 0$ or $\phi = \pi/2$, we have

$$\left.\begin{aligned} \omega &= \sqrt{\frac{2a}{h}}\left(A + \frac{\pi\rho}{2}\right) \Big/ \left(1 + \frac{qa}{r_0}\right), \\ B &= -\,(q + \rho + k) + k\omega^2\left(1 + \frac{qa}{r_0}\right). \end{aligned}\right\} \tag{5.127}$$

In the spherical cap of radius R, the relationship between η_ϕ and s is directly given by (5.72), where q must be replaced by $-qR/r_0$. Thus

$$n_\phi = \frac{qR}{r_0} - s\cot\phi.$$

Substituting into the first equation of (5.66), and using the yield condition $n_\phi - n_\theta = 1$, the differential equation for s is obtained as

$$\frac{d}{d\phi}(s\cot\phi) + s = \cot\phi.$$

In view of the boundary condition $s = 0$ at $\phi = \beta$ where m_ϕ is a minimum, the integration of the above equation results in

$$s = \left(\frac{qr}{r_0} - n_\phi\right)\tan\phi = \sin\phi\ln\left\{\frac{\tan(\phi/2)}{\tan(\beta/2)}\right\}. \tag{5.128}$$

The angle β defining the position of the hinge circle in the sphere is obtained from the condition of continuity of s across $\phi = \pi/2 - \alpha$. In view of (5.123) and (5.128), we have

$$\cot\frac{\beta}{2}=\cot\left(\frac{\pi}{4}-\frac{\alpha}{2}\right)\exp\left\{q\tan\alpha+\rho\,(\alpha\sin\alpha)-\left(A+\frac{\pi\rho}{2}\right)\right\}. \tag{5.129}$$

The distribution of bending moment in the sphere follows from the second equation of (5.121), where k must be replaced by kr_1/R. In view of (5.128), the differential equation from m_ϕ becomes

$$k\frac{dm_\phi}{d\phi}=\frac{R}{r_1}\sin\phi\ln\left\{\frac{\tan(\phi/2)}{\tan(\beta/2)}\right\}.$$

Integrating, and using the boundary condition $m_\phi = -1$ at $\phi = \beta$, the solution is obtained as

$$k\left(1+m_\phi\right)=\frac{R}{r_1}\left[\ln\left(\frac{\sin\phi}{\sin\beta}\right)-\cos\phi\ln\left\{\frac{\tan(\phi/2)}{\tan(\beta/2)}\right\}\right]. \tag{5.130}$$

The constant A is now obtained in terms of q and β from the condition of continuity of m_ϕ across $\phi = \pi/2 - \alpha$. Using (5.124), (5.127), and (5.130), and setting $\eta = (1 + qa/r_0)^{-1}$, we get

$$\frac{a\eta}{2r_1}\left(A+\frac{\pi\rho}{2}\right)^2\operatorname{cosec}\alpha+\left(\mathrm{A}+\frac{\pi\rho}{2}\right)-(q+\rho)\tan\frac{\alpha}{2}+\rho\left(\alpha-\frac{1}{2}\rho\sin\alpha\right)-\delta=0,$$

where,

$$\delta=\frac{R}{r_1}\left[\operatorname{cosec}\alpha\ln\left(\frac{\cos\alpha}{\sin\beta}\right)-\ln\left\{\frac{\tan(\pi/4-\alpha/2)}{\tan(\beta/2)}\right\}\right]. \tag{5.131}$$

It is not a good approximation in this case to assume the location of maximum bending moment in the knuckle to be at $\phi = \pi/2 - \alpha/2$. However, for practical purposes, it would be sufficiently accurate to use the approximate expression

$$(q+p)\tan\frac{\alpha}{4}=2k\cot\frac{\alpha}{2}+\frac{1}{2}\rho\cos\frac{\alpha}{2}\left[\alpha-\rho\left(\sin\alpha-\sin\frac{\alpha}{2}\right)\right]-\delta\cos\frac{\alpha}{2}, \tag{5.132}$$

specifying the condition $m_\phi = 1$ at the plastic hinge in the torus. Since δ is always positive, the collapse pressure predicted by (5.132) will be significantly lower than that given by (5.125). The computation may be carried out by assuming a value of β for a given a, and evaluating q from (5.132) and A from (5.131), the results being subsequently checked using (5.129). Setting the numerical values $\alpha = \pi/3$, $4k = 1/30$, and $\rho = 6/47$, giving $r_0/a = 47/53$, $r_1/a = 6/53$, and $r_1/R = 0.060$, the parameter pa/N_0 is found to be equal to 0.351 (corresponding to $\beta \approx 28.4°$), which may be compared with the value 0.421 obtained by considering the knuckle alone. A finite element method for the limit analysis of pressure vessels has been discussed by Liu et al. (1995).

5.6.3 Cylindrical Nozzle in a Spherical Vessel

A spherical pressure vessel of mean radius R and uniform thickness H is fitted with a flush cylindrical nozzle of mean radius a and thickness h. The nozzle is supposed to be long enough to permit a collapse mechanism involving a rigid outer part. If the nozzle is rigidly connected to the sphere, a possible mechanism of collapse under an internal pressure p would consist of a hinge circle at the nozzle/sphere intersection, a second hinge circle at a distance b from the plane of intersection, and a third hinge circle at an angle $\phi = \alpha$ in the spherical shell, Fig. 5.26. The problem has been discussed by Gill (1964, 1970), Leckie and Payne (1965), Dinno and Gill (1965), and Gill and Leckie (1968).

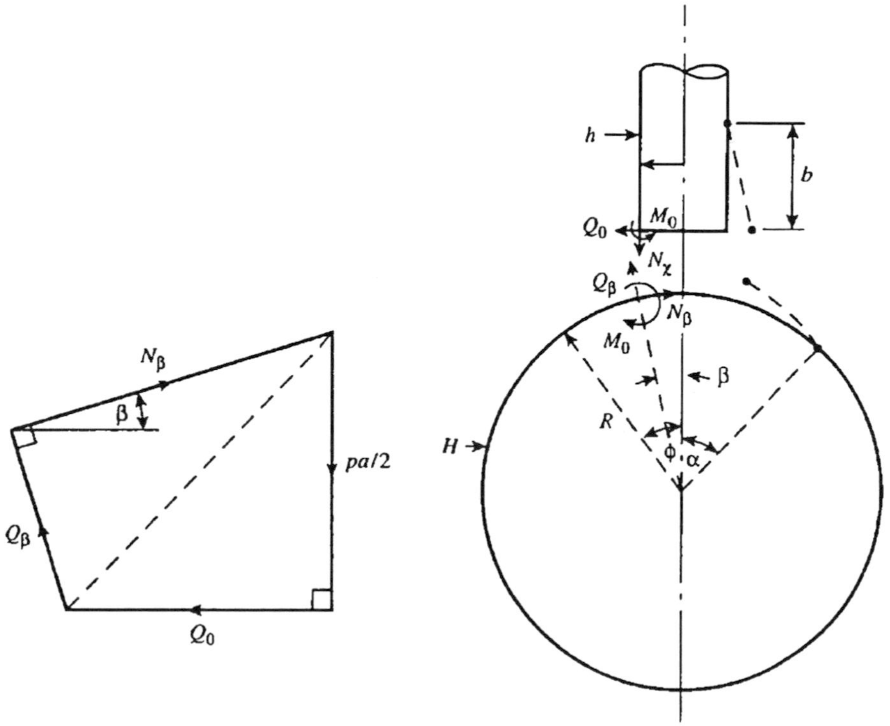

Fig. 5.26 Plastic collapse of a spherical pressure vessel fitted with a cylindrical flush nozzle

We shall use the two-moment limited interaction yield condition for the spherical shell and an analogous yield condition represented by the hexagonal prism of Fig. 5.24b for the cylindrical branch. The yield locus defined by the direct forces in the dimensionless form is therefore the same for both the cylinder and the sphere. Both of them are assumed to have the same uniaxial yield stress Y. It is convenient to introduce the dimensionless quantities

$$q = \frac{pR}{2YH}, \quad \eta = \frac{2aH}{hR}, \quad \omega = b\sqrt{\frac{2}{ah}}.$$

Evidently, q represents the dimensionless collapse pressure, whose value is unity for a complete spherical shell. Due to the outward motion of the material undergoing deformation at the incipient collapse, the circumferential forces would be everywhere positive. Over the practical range, it is reasonable to assume the yield condition to be $N_\theta = Yh$ in the cylinder and $N_\theta = YH$ in the sphere. Further, the nature of the discontinuity in the velocity slope indicates that M_x is equal to $-Yh^2/4$ at $x = 0$ and to $Yh^2/4$ at $x = b$ in the cylinder and is equal to $-YH^2/4$ *at* $\phi = \beta$ and to $YH^2/4$ at $\phi = \alpha$ in the spherical shell.

Consider first the plastic collapse of the cylindrical branch, in which the closed-end condition produces an axial force $N_x = pa/2 < N_\theta$. In the equilibrium equation (5.4), we set $n_\theta = 1$ and replace $-q$ by ηq, which represents the quantity pa/Yh, the result being

$$d^2 m_x / d\xi^2 = -2\,(1 - \eta q)\,.$$

This equation is readily integrated under the boundary conditions $m_x = -1$ *at* $\xi = 0$ and $m_x = 1$ at $\xi = \omega$. The generalized stresses in the cylinder therefore become

$$\left.\begin{aligned} n_\theta = 1, \qquad n_x = \eta q/2, \\ m_x = (1 - \eta q)\,\xi\,(\omega - \xi) - \left(1 - \frac{2\xi}{\omega}\right), \end{aligned}\right\} \quad 0 \le \xi \le \omega. \tag{5.133}$$

Since the shearing force Q is equal to dM_x/dx, the dimensionless quantity $s = Q/Yh$ is given by

$$2\sqrt{\frac{2a}{h}}s = \frac{dm_x}{d\xi} = \frac{2}{\omega} + (1 - \eta q)\,(\omega - 2\xi)\,.$$

The parameter ω is obtained from the condition that m_x is a maximum at $\xi = \omega$. Thus $s = 0$ at $\xi = \omega$, giving

$$\omega = \sqrt{\frac{2}{1 - \eta q}}, \quad s = \left(1 - \frac{\xi}{\omega}\right)\sqrt{(1 - \eta q)\,\frac{h}{a}}. \tag{5.134}$$

With this value of ω, the stress distribution (5.133) is statically admissible, provided $nq = pa/Yh$ is less than unity. The flow rule associated with the yield condition $n_\theta = 1$ gives $\dot{\lambda}_x = du/dx = 0$ and $\dot{\kappa}_x = -d^2 w/dx^2 = 0$. Hence

$$u = u_0, \quad w = -w_0\left(1 - \frac{\xi}{\omega}\right),$$

where ω_0 is the radially outward speed at $\xi = 0$, and u_0 represents a rigid-body motion of the nozzle. Since $\dot{\lambda}_\theta = -w/a > 0$, the velocity field is kinematically admissible.

For the analysis of the spherical vessel, it is instructive to consider the differential equation of equilibrium involving the shearing force Q. Setting $r_1 = R$, $r = R \sin \phi$ and $p_r = -p$ in the second equation of (5.47) and normalizing the forces by the yield force YH, we have

$$\frac{ds}{d\phi} + s \cot \phi + \left(n_\theta + n_\phi\right) = 2q.$$

Eliminating $s \cot \phi$ by means of (5.72), where $-q$ must be written for q, leads to the simple differential equation

$$\frac{ds}{d\phi} + n_\theta = q.$$

Introducing the yield condition $n_\theta = 1$, and using the boundary condition $s = 0$ at $\phi = \alpha$, where m_ϕ is a maximum, the above equation is readily integrated to give

$$\left.\begin{aligned} s = (1-q)(\alpha-\phi), \quad n_\theta = 1, \\ n_\phi = q - (1-q)(\alpha-\phi)\cot\phi, \end{aligned}\right\} \alpha \le \phi \le \beta. \tag{5.135}$$

For simplicity, the hexagonal yield condition for the bending moment will be replaced by the square yield condition defined by $m_\theta = \pm 1$ and $m_\phi = \pm 1$. Since $\dot{\kappa}_\theta$ is expected to be positive in view of the assumed mode of collapse, we set $m_\theta = 1$ in the second equation of (5.66) to obtain the differential equation

$$\frac{d}{d\phi}\left(m_\phi \sin\phi\right) = \cos\phi + \frac{s}{k}\sin\phi.$$

Substituting from (5.135), this equation is integrated under the boundary condition $m_\phi = 1$ at $\phi = \alpha$ to give

$$m_\phi = 1 - \frac{1-q}{k}\left\{1 - \frac{\sin\alpha}{\sin\phi} + (\alpha-\phi)\cot\phi\right\}, \quad \beta \le \phi \le \alpha. \tag{5.136}$$

The continuity condition $m_\phi = -h^2/H^2$ at junction $\phi = \beta$, where $\sin \beta = a/R$, furnishes the relationship between q and a as

$$(1-q)\left\{(\alpha-\beta) - \sqrt{1-\frac{a^2}{R^2}} - \sin\alpha + \frac{a}{R}\right\} = \frac{aH}{4R^2}\left(1+\frac{h^2}{H^2}\right). \tag{5.137}$$

Since $q < 1$ due to the weakening effect of the nozzle, the stress distribution predicted by (5.135) and (5.136) is statically admissible. The velocity distribution can be found from the associated flow rule, which gives $\dot{\lambda}_\phi = \dot{\kappa}_\phi = 0$, the corresponding velocity equations being

$$\frac{du}{d\phi} - w = 0, \quad \frac{dw}{d\phi} + u = c,$$

in view of (5.67), where c is a constant velocity. Using the boundary condition $u = w = 0$ at $\phi = \alpha$, the solution is obtained as

$$w = -c\sin(\alpha - \phi), \quad u = c\left[1 - \cos(\alpha - \phi)\right]. \tag{5.138}$$

It follows from (5.67) and (5.138) that $\dot{\lambda}_\theta > 0$ and $\dot{\kappa}_\theta > 0$ when c is positive, indicating that the velocity field is kinematically admissible. The continuity of the velocity vector across $\phi = \beta$ requires

$$w_0 = c(\cos\beta - \cos\alpha), \quad u_0 = c(\sin\alpha - \sin\beta).$$

The velocity distribution throughout the region of deformation is therefore completely determined in terms of a single constant c, which is of arbitrary magnitude at the incipient collapse.

The collapse pressure p must be determined from the condition that the resultant of the meridional and shearing forces is continuous across the interface $\phi = \beta$ (Fig. 5.26). Resolving horizontally and vertically, we get

$$N_\beta\cos\beta - Q_\beta\sin\beta = Q_0, \quad N_\beta\sin\beta + Q_\beta\cos\beta = pa/2,$$

where Q_0 is the value of Q in the cylinder at $x = 0$ and the subscript β refers to quantities at $\phi = \beta$. The second equation of the above pair is equivalent to (5.72), considered at $\phi = \beta$, with the necessary sign change of q. Since

$$N_\beta\cos\beta - Q_\beta\sin\beta = YH\left[q\cos\beta - (1-q)(\alpha - \beta)\operatorname{cosec}\beta\right]$$

in view of (5.135), while Q_0 is the value of Yhs at $\xi = 0$ given by (5.134), the collapse pressure is given by

$$\sqrt{\frac{a}{h}}\left\{q\sqrt{1 - \frac{a^2}{R^2}} - \frac{R}{a}(1-q)(\alpha - \beta)\right\} = \frac{h}{H}\sqrt{1 - \eta q}, \tag{5.139}$$

where $\beta = \sin^{-1}(a/R)$. For given values of the ratios h/a, H/R, and a/R, (5.137) and (5.139) can be solved simultaneously for q and a. In Fig. 5.27, the parameters q and $\alpha - \beta$ at the incipient collapse are plotted against a/R for different values of H/R in the special case of $\eta = 1$. The collapse pressure is actual for the assumed yield condition, but provides only an upper bound for a non-hardening Tresca material.

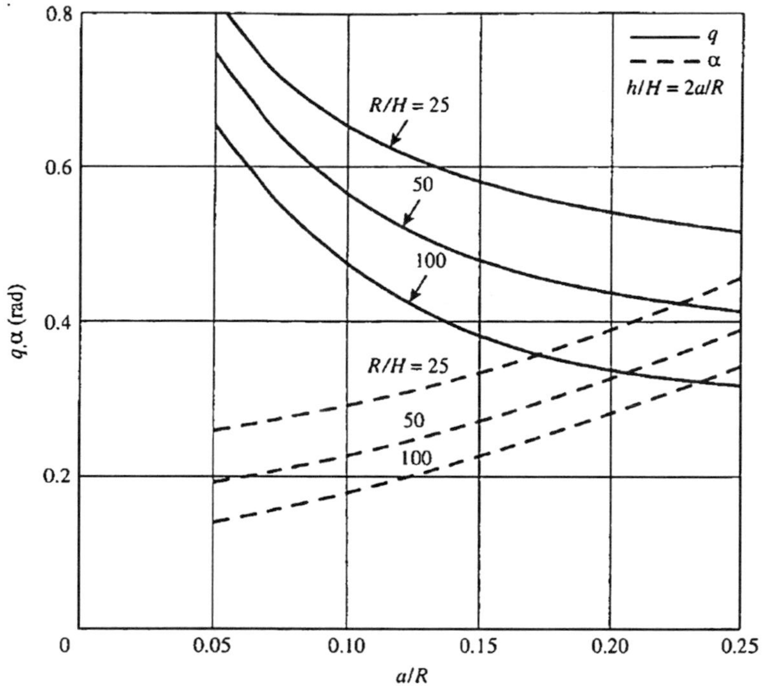

Fig. 5.27 Collapse pressure and the extent of the associated plastic region in a spherical pressure vessel with a projecting cylindrical nozzle

5.7 Minimum Weight Design of Shells

5.7.1 Principles for Optimum Design

The problem of optimum design considered here involves the requirement of minimizing the weight of the material of the shell that is capable of supporting given loads. For a homogeneous material, this is equivalent to finding the minimum volume of shell, which will provide us with a basis for comparison with any actual design. Since the shell carries both direct forces and bending moments, it is convenient to discuss the design criterion in terms of the rate of plastic energy dissipation D per unit area of the middle surface. The dissipation rate is a monotonically increasing function of the local shell thickness h and depends also on the velocity v_j of the middle surface. For a shell of variable thickness h which is just at the point of collapse under a boundary traction T_j and a velocity field v_j, we have

$$\int D\left(v_j\right) dA = \int T_j v_j dA \tag{5.140}$$

in the absence of body forces, the integrals being extended over the entire area of the middle surface. Consider now a neighboring thickness distribution $h^* = h + \delta h$ for the same geometry of the middle surface, where δh is an infinitesimally small variation in thickness. The dissipation rate for the shell of thickness h^* in the deformation mode v_j may be written as

$$D^*\left(v_j\right) = D\left(v_j\right) + \frac{\partial}{\partial h}\left[D\left(v_j\right)\right]\delta h$$

on neglecting higher-order terms, provided $D(v_j)$ is a continuously differentiable function of h. Since the corresponding distribution of the generalized stresses in the shell of thickness h^* is not necessarily in equilibrium, it follows from the upper bound theorem of limit analysis that

$$\int D^*\left(v_j\right) dA \geq \int T_j v_j dA.$$

Substituting for $D^*(v_j)$, and using (5.140), the preceding inequality is reduced to

$$\int \frac{\partial}{\partial h}\left[D\left(v_j\right)\right]\delta h dA \geq 0. \tag{5.141}$$

An immediate consequence of inequality (5.141) is that if $\delta D/\delta h$ is a positive constant for the shell of thickness distribution h, then

$$\int \delta h dA = \delta\left(\int h dA\right) \geq 0.$$

This inequality implies that the volume of the shell of thickness h, which is designed to collapse under the given loads, is a relative minimum when $\delta D/\delta h$ has a constant value over the middle surface. The minimum volume criterion established here, which is due to Shield (1960b), is an extension of that given earlier for the bending of plates (Section 4.5).

A stronger result can be obtained for the ideal sandwich shell which consists of a core of thickness H between two thin identical face sheets each of thickness $h/2$. The membrane forces and bending moments across any section are carried by the face sheets, while the core carries only the shearing force. The strain rate in each of the face sheets may be considered as constant, so that the dissipation rate D is proportional to h. The core thickness H is assumed to be prescribed, and the face sheet thickness is to be determined for minimum volume so that the shell is just at the point of collapse under the given loads T_j.

To obtain the condition for minimum volume of the sandwich shell, consider any sheet thickness distribution h^* for which the shell is at or below collapse under the loads T_j and in the velocity pattern v_j. Since the dissipation rate for this shell is $D^*(v_j) = (h^*/h)D(v_j)$, we have

$$\int \left(h^*/h\right) D\left(v_j\right) dA \geq \int T_j v_j dA$$

by the upper bound theorem of limit analysis. The elimination of the right-hand side of the above inequality by means of (5.140) results in

$$\int \left(h^* - h\right)\left[D\left(v_j\right)/h\right] dA \geq 0. \tag{5.142}$$

If the shell of thickness h is such that $D(v_j)/h$ is a positive constant, then (5.142) reduces to

$$\int h^* dA \geq \int h dA.$$

Thus, a sandwich shell designed to collapse in a mode that makes the ratio D/h a constant over the middle surface provides an absolute minimum for its volume under the prescribed loads (Shield, 1960). The condition D/h = constant for the optimum design of a sandwich shell is independent of the design thickness due to the linear dependence of D on h. For a solid shell, on the other hand, the condition $\partial D/\partial h$ = constant for the optimum design involves the thickness h, which renders the problem more complicated. The question of uniqueness of the optimum design has been examined by Hu and Shield (1961), and some criteria for minimum cost design have been discussed by Prager and Shield (1967).

5.7.2 Basic Theory for Cylindrical Shells

Consider the particular case of a circular cylindrical shell of sandwich construction under axially symmetrical loading. When the applied load is an internal pressure p, which may vary with the axial coordinate x, the axial force N_x is a constant, and equilibrium also requires

$$\frac{dM_x}{dx} = Q, \qquad \frac{dQ}{dx} = p - \frac{N_\theta}{a}.$$

These equations follow from (5.3) with a change in sign for p. The shear force Q is easily eliminated between the above equations to give

$$\frac{\partial^2 M_x}{\partial x^2} + \frac{N_\theta}{a} = p. \tag{5.143}$$

The circumferential bending moment M_θ is a passive or induced moment that arises from the fact that the corresponding curvature rate $\dot{\kappa}_\theta$ vanishes because of symmetry. The rate of dissipation of internal energy per unit area of the middle surface is

$$D = M_x \dot{\kappa}_x + N_\theta \dot{\lambda}_\theta + N_x \dot{\lambda}_x, \tag{5.144}$$

where $\dot{\kappa}_x$ is the axial curvature rate, and $\dot{\lambda}_\theta$,$\dot{\lambda}_x$ are the circumferential and axial extension rates of the middle surface. They are given by

$$\dot{\kappa}_x = \frac{d^2w}{dx^2}, \qquad \dot{\lambda}_\theta = \frac{w}{a}, \qquad \dot{\lambda}_x = \frac{du}{dx}, \tag{5.145}$$

where u and w are the axial and radially outward velocities of the middle surface.

Suppose that the material of the face sheets is ideally plastic obeying Tresca's yield criterion and the associated flow rule. The corresponding yield conditions in terms of the generalized stresses have been discussed in Sections 5.1 and 5.2. When the axial force is absent ($N_x = 0$), the condition $D/h =$ constant for minimum volume restricts the stress point to be at one of the corners of the yield hexagon shown in Fig. 5.28a. Indeed, D vanishes along the sides BC and EF of the hexagon and is a function of x along the remaining sides of the hexagon. For the stress point lying at the corners, the condition $D/h =$ constant may be written as

$$\left.\begin{array}{lll} \dot{\lambda} = \text{constant}, & \dot{\lambda} \geq H\dot{\kappa}, & \text{corners } A \text{ and } D, \\ \dot{\lambda} + H\dot{\kappa} = \text{constant}, & \dot{\lambda} \leq H\dot{\kappa}, & \text{points } B, C, E, F, \end{array}\right\} \tag{5.146}$$

where $\dot{\lambda}$ and $\dot{\kappa}$ denote the absolute values of $\dot{\lambda}_\theta$ and $\dot{\kappa}_x$, respectively. The above relations are directly obtained by setting the values N_θ and M_x in (5.144) for the considered stress point and using the fact that $N_0 = Yh$ and $M_0 = YHh/2$ for the sandwich shell having a uniaxial yield stress Y.

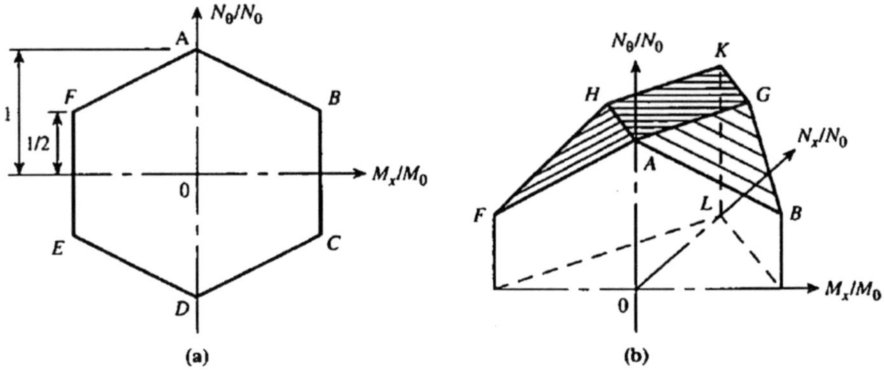

Fig. 5.28 Yield condition for a cylindrical sandwich shell. (**a**) Yield locus for no axial force and (**b**) a part of the yield surface for nonzero axial force

When the axial force is present ($N_x \neq 0$), the part of the yield surface for which the membrane forces N_0 and N_x are positive is shown in Fig. 5.28b. It can be shown that only the stress states represented by points on the edges AG, AH, BG, FH, GK, HK, and KL, and also those on the plane $AGKH$ can be associated with a rate of deformation for which the condition $D/h =$ constant is satisfied. Considering, for instance, stress points on the edges GK and HK, the condition $D/h =$ constant is found to imply

$$2\dot{\lambda} + H\dot{\kappa} = \text{constant}, \quad \dot{\lambda}_\theta \geq 0, \quad 2\dot{\lambda}_x = H\dot{\kappa}. \tag{5.147}$$

Similar relations can be established for the other plastic regimes relevant to the optimum design. The preceding results, as well as their applications to be discussed later in this section, are essentially due to Shield (1960).

When the material yields according to the von Mises criterion, the stress point has the freedom to move along the yield locus, which for zero axial load is an ellipse circumscribing the Tresca hexagon. By the flow rule (5.9) associated with the yield condition (5.8), the dimensionless stress resultants can be expressed in terms of the generalized strain rates. Thus

$$n_\theta = N_0\dot{\lambda}_\theta/D, \quad m_x = 4M_0\dot{\kappa}_x/3D, \quad n_\theta^2 + \frac{3}{4}m_x^2 = 1.$$

Since D/h is a constant for the optimum design, we write $D = chY$, where c is an arbitrary positive constant that can be associated with the mode of collapse. The above relations then give

$$n_\theta = \dot{\lambda}_\theta/c, \quad m_x = 2H\dot{\kappa}_x/3c, \quad \dot{\lambda}_\theta^2 + (H/3)\,\dot{\lambda}_x^2 = c^2. \tag{5.148}$$

In view of (5.145), the quantities n_θ and m_x are obtained from (5.148) as functions of the velocity w, the spatial distribution of which is determined by the numerical integration of the last equation of (5.148). The bending moment distribution can be found by the integration of (5.143) after eliminating N_θ by means of the relation

$$N_\theta/M_x = N_0 n_\theta/M_0 m_x = 3\dot{\lambda}_\theta/H^2\dot{\kappa}_x.$$

The design thickness h at each section is finally obtained from the ratio of M_x to m_x. Applications of the basic theory to the von Mises material have been discussed by Reiss and Megarefs (1969). The optimum design of cylindrical shells based on the calculus of variation and the Tresca theory has been considered by Freiberger (1956), following an earlier work by Onat and Prager (1955).

5.7.3 Simply Supported Shell Without End Load

As a first example illustrating the preceding theory, consider an open-ended cylindrical shell of length $2l$ loaded by a uniform internal pressure p. If the shell is simply supported at both ends, which correspond to $x = 0$ and $x = 2l$, the bending moment M_x and the radial velocity w must vanish at these sections. For relatively short shells, the mode of collapse should involve bending of the entire shell, and the stress point F in Fig. 5.28a will therefore apply throughout the shell. Using the second equation of (5.146), where $\dot{\lambda} = \dot{\lambda}_\theta$ and $\dot{\kappa} = \dot{\kappa}_x$, the differential equation for w in the optimum design is obtained as

$$aH\frac{d^2w}{dx^2} - w = -w_0 \tag{5.149}$$

in view of (5.145), the quantity w_0 being a positive constant velocity. Integrating, and using the boundary conditions $w = 0$ at $x = 0$ and $x = 2l$, the solution is found as

$$w = w_0\left\{1 - \frac{\cosh(\alpha - \xi)}{\cosh\alpha}\right\},$$

where $\xi = x/\sqrt{aH}$ and $\alpha = l/\sqrt{aH}$. The velocity field will be associated with the stress state at corner F if the first inequality of (5.146) is also satisfied. This gives

$$w \leq -d^2w/d\xi^2 \quad \text{or} \quad \cosh\alpha \leq 2\cosh(\alpha - \xi).$$

This condition will be satisfied throughout the shell if it holds at the central section $\xi = \alpha$. Hence

$$\alpha \leq \cosh^{-1} 2 \quad \text{or} \quad l \leq 1.317\sqrt{aH}.$$

The thickness distribution in the optimum design, when the shell is sufficiently short, is obtained by setting $M_x = -YHh/2$ and $N_\theta = Yh/2$ in the equilibrium equation (5.143), the resulting differential equation for h being

$$aH\frac{d^2h}{dx^2} = -\frac{2pa}{Y}. \tag{5.150}$$

Since the bending moment must vanish at the ends of the shell, the boundary conditions are $h = 0$ at $x = 0$ and $x = 2l$, and the solution is

$$\frac{hY}{pa} = 2\left\{1 - \frac{\cosh(\alpha - \xi)}{\cosh\alpha}\right\}, \quad \alpha \leq 1.317. \tag{151}$$

It is interesting to note that the variation of hY/pa with ξ in the case of short shells is identical to that of w/w_0 with ξ, except for a scale factor.

For longer shells ($\alpha \geq 1.317$), a central portion of the shell is stressed by N_θ alone, corresponding to corner A of the yield locus, Fig. 5.28a. The stress state in the remainder of the shell corresponds to point F and covers a length $d = \delta\sqrt{aH}$ on either side of the central portion, where δ is a constant. Over the central portion, the condition $\dot{\lambda}_\theta$ = constant for minimum volume requires w to be constant in the region. Considering the end portion $0 \leq x \leq d$, the velocity distribution is easily determined from the differential equation (5.149) and the boundary conditions $w = 0$ at $x = 0$ and $dw/dx = 0$ (and hence continuous) at $x = d$, the result being

$$w = w_0 \left\{ 1 - \frac{\cosh(\delta - \xi)}{\cosh \delta} \right\}, \quad 0 \leq \xi \leq \delta. \tag{5.152}$$

The constant δ is obtained from the requirement $\dot{\lambda} = H\dot{\kappa}$, or $w = -d^2w/d\xi^2$ at $\xi = \delta$, giving

$$\delta = \cosh^{-1} 2 = 1.317.$$

By the condition of continuity at $\xi = \delta$, the velocity in the central portion of the shell is $w = w_0/2$. The inequality $w \leq -d^2w/d\xi^2$ in the end portion and the inequality $w \geq -d^2w/d\xi^2$ in the central portion are identically satisfied.

The thickness distribution in each portion of the shell for minimum volume is determined from the equilibrium equation and the appropriate yield condition. In the end portion $0 \leq x \leq \delta$, the resulting differential equation is (5.150), which must be solved under the boundary conditions $h = 0$ at $x = 0$ and $x = d$, the bending moment at these sections being zero. The solution is easily shown to be

$$\frac{hY}{pa} = 2\left\{ 1 - \frac{1}{\sqrt{3}} [\sinh \xi + \sinh(\delta - \xi)] \right\}, \quad 0 \leq \xi \leq \delta. \tag{5.153a}$$

In the central portion of the shell, the stress point corresponds to $M_x = 0$ and $N_0 = Yh$, and (5.143) immediately furnishes

$$hY/pa = 1, \quad \delta \leq \xi \leq \alpha, \tag{5.153b}$$

only one-half of the shell being considered because of symmetry. The thickness changes discontinuously from 0 to pa/Y at $x = d$. Since the thickness gradient dh/dx is zero in the central portion, but is nonzero in the end portion, the shearing force $Q = dM_x/dx$ is also discontinuous at $x = d$. The discontinuity in Q is removed, however, by adding a flange of vanishingly small width but of finite area of cross section at $x = d$.

The total volume V of the face sheets in the optimum design is now determined by integration over the area of the middle surface, using the expression

$$V = 4\pi a \int_0^l h dx = 4\pi a \sqrt{aH} \int_0^a h d\xi. \tag{5.154}$$

Substituting from (5.151) and (5.153) for short and long shells, respectively, we obtain

$$V = \begin{cases} 8\pi a^2 l\,(p/Y)\,(1 - \tanh\alpha/\alpha) & \alpha \leq 1.317, \\ 8\pi a^2 l\,(p/Y)\left[1 - \left(\sqrt{3} - \delta\right)/\alpha\right], & \alpha \geq 1.317, \end{cases} \tag{5.155}$$

A term equal to $1/\sqrt{3}\alpha$ has been included in the square brackets of (5.155) to take account of the flanges so that volume is continuous at $\alpha = \delta = 1.317$.

It is instructive to compare the volume for the optimum design with that for the constant thickness design for the sandwich shell. To this end, we express the equilibrium equation (5.143) in the dimensionless form

$$\frac{d^2 m_x}{d\xi^2} + 2n_\theta = 2q,$$

where $q = pa/Yh_0$, with $h_0/2$ denoting the thickness of each face sheet. For a sufficiently short shell, the state of stress involves $m_x < 0$ and $n_\theta > 0$, so that side AF of the yield hexagon applies throughout the shell. Using the yield condition $2n_\theta - m_x = 2$ to eliminate n_θ from the equilibrium equation, we get

$$\frac{d^2 m_x}{d\xi^2} + m_x = 2\,(q-1)\,.$$

In view of the boundary conditions $m_x = 0$ at $\xi = a$ and $dm_x/d\xi = 0$ at $\xi = a$, the solution becomes

$$m_x = 2\,(q-1)\left\{1 - \frac{\cos(\alpha - \xi)}{\cos\alpha}\right\}, \quad \alpha \leq \frac{\pi}{2}. \tag{5.156}$$

The velocity equation associated with the yield condition gives $w = w_0 \sin \xi$ ($\xi \leq a$) in view of the boundary condition $w = 0$ at $\xi = 0$. The velocity slope $dw/d\xi$ is discontinuous at $\xi = \alpha < \pi/2$, giving rise to a hinge circle, where $m_x = -1$. By (5.156), this condition furnishes

$$q = \frac{2 - \cos\alpha}{2 - 2\cos\alpha}, \quad \alpha \leq \frac{\pi}{2}. \tag{5.157}$$

For longer shells ($\alpha \geq \pi/2$), the plastic state of stress is represented by corner A of the yield hexagon. Then $m_x = 0$ and $n_\theta = 1$, giving $q = 1$ for plastic collapse of the simply supported shell, the associated velocity field being

$$w = w_0 \sin\xi \;\; (\xi \leq \pi/2)\,, \quad w = w_0 \;\; (\xi \geq \pi/2)\,.$$

Since the total volume of the face sheet material in the constant thickness design is $V_0 = 4\pi a l h_0$, the ratio of the face sheet volumes for the optimum and constant thickness designs may be written from (5.157) as

$$\frac{V}{V_0} = \begin{cases} 2q\,(1 - \tanh\alpha/\alpha)\,, & \alpha \leq 1.317, \\ q\left[1 - \left(\sqrt{3} - \delta\right)/\alpha\right] & \alpha \geq 1.317. \end{cases} \tag{5.158}$$

The parameter q is given by (5.157) when $\alpha \leq \pi/2$ and is equal to unity when $\alpha \geq \pi/2$. The ratio V/V_0 is plotted as a function of α in Fig. 5.29 on the basis of (5.158) for the simply supported shell. The discontinuities in slope of the curves at $\alpha = 1.317$ and $\alpha = 1.571$ are due to the change in character of the solution for the

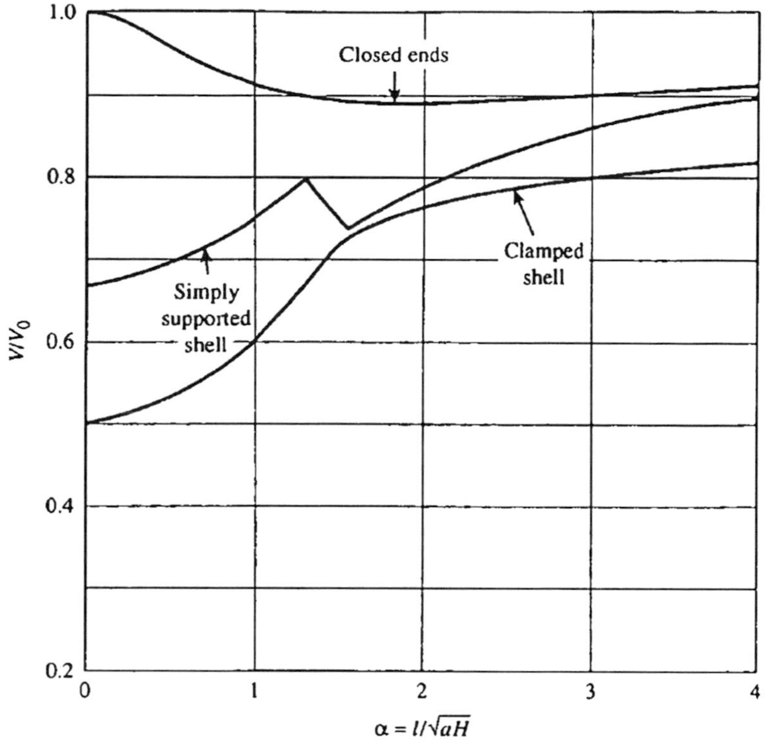

Fig. 5.29 Ratio of face sheet volume in the optimum design to that in the constant thickness design of a cylindrical shell under uniform internal pressure

optimum design and constant thickness design, respectively. The saving of material effected by the minimum volume design is quite appreciable for short shells.

5.7.4 Cylindrical Shell with Built-In Supports

A circular cylindrical shell of semilength $l = \alpha\sqrt{ah}$ is provided with built-in support at its open ends and is subjected to a uniform internal pressure p. The condition D/h = constant for the minimum volume design requires the velocity slope to vanish at the clamped edges when the shell is at the point of collapse. A portion of the shell of length $b = \beta\sqrt{ah}$ exists at each end where the state of stress is represented by point B in Fig. 5.28a. If the shell is sufficiently short, the remainder of the shell would correspond to point F of the yield hexagon. By the second equation of (5.146), the velocity equation for the optimum design is

$$aH\frac{d^2w}{dx^2} \pm w = \pm w_0,$$

where the upper sign applies to the end portions and the lower sign to the central portion of the shell. Considering only one-half of the shell and using the boundary conditions

$$w = dw/dx = 0 \text{ at } x = 0, \qquad dw/dx = 0 \text{ at } x = l,$$

and the condition of continuity of w at $x = b$, the solution for the velocity at the incipient collapse is obtained as

$$\left.\begin{aligned} &w = w_0\,(1 - \cos\xi)\,, \qquad 0 \leq \xi \leq \beta, \\ &w = w_0\left[1 - \cos\beta\cosh(\alpha-\xi)/\cosh(\alpha-\beta)\right], \quad \beta \leq \xi \leq \alpha. \end{aligned}\right\} \tag{5.159}$$

Since dw/dx must be continuous at $\xi = \beta$ for the minimum volume design, the relationship between β and α is

$$\tan\beta = \tanh(\alpha - \beta)\,. \tag{5.160}$$

The compatibility of the velocity field with the stress field associated with corners B and F requires

$$w \leq \frac{d^2w}{d\xi^2}\,(0 \leq \xi \leq \beta)\,, \qquad w \leq -\frac{d^2w}{d\xi^2}\,(\beta \leq \xi \leq \alpha)\,.$$

In view of (5.160), the above inequalities will be satisfied throughout the respective portions of the shell if

$$2\cos\beta \geq 1, \qquad \cosh(\alpha - \beta) \leq 2\cos\beta.$$

It turns out that the first inequality is satisfied when the second inequality is. Combining the second inequality with (5.160), it is easily shown that

$$\alpha - \beta \leq \frac{1}{2}\cosh^{-1}4, \qquad \beta \leq 0.659, \qquad \alpha \leq 1.691. \tag{5.161}$$

The thickness distribution in the shell for $\alpha \leq 1.691$ is given by the solution of the differential equation

$$aH\frac{d^2h}{dx^2} \pm h = \pm\frac{2pa}{Y}, \tag{5.162}$$

which is obtained by setting $m_x = \pm YHh/2$ and $N_\theta = Yh$ in (5.143). The upper sign applies to $0 \leq x \leq b$ and the lower sign to $b \leq x \leq a$. Since the bending moment must be continuous at $x = b$, it must vanish at this section. Further, symmetry requires the shearing force to vanish at $x = l$. Hence the boundary conditions are

$$h = 0 \text{ at } \xi = \beta, \qquad dh/d\xi = 0 \text{ at } \xi = \alpha.$$

These conditions, together with the fact that $dh/d\xi$ changes sign at $\xi = \beta$ while retaining its numerical values for the shearing force to be continuous, furnish the solution

$$\left.\begin{aligned} hY/pa &= 2\left[1 - (\cos\xi/\cos\beta) + 2\tan\beta\sin(\beta - \xi)\right], \quad 0 \le \xi \le \beta, \\ hY/pa &= 2\left[1 - \cosh(\alpha - \xi)/\cosh(\alpha - \beta)\right], \quad \beta \le \xi \le \alpha. \end{aligned}\right\} \qquad (5.163)$$

For longer shells with $\alpha \ge 1.691$, the minimum volume condition $D/h =$ constant can only be satisfied by introducing a central region of constant radial velocity associated with corner A in Fig. 5.28a. The stress points B and F of the yield hexagon then correspond to the regions $0 \le \xi \le \gamma$ and $\gamma \le \xi \le \delta$, respectively, where γ and δ are dimensionless constants. The velocity distribution in these two regions can be written as

$$\left.\begin{aligned} w &= w_0(1 - \cos\xi), \qquad 0 \le \xi \le \gamma, \\ w &= w_0\left[1 - \cos\gamma\cosh(\delta - \xi)/\cosh(\delta - \gamma)\right], \quad \gamma \le \xi \le \delta. \end{aligned}\right\} \qquad (5.164)$$

The velocity at $\xi = \gamma$ and the velocity slope at $\xi = \delta$ are automatically made continuous. The condition of continuity of $dw/d\xi$ at $\xi = \gamma$, and the requirement $w = -d^2w/d\xi^2$ at $\xi = \delta$ furnish

$$\tan\gamma = \tanh(\delta - \gamma), \quad 2\cos\gamma = \cosh(\delta - \gamma),$$

in view of (5.164). These two equations are easily solved for γ and δ to give

$$\delta - \gamma = \frac{1}{2}\cosh^{-1} 4, \quad \gamma = 0.659, \quad \delta = 1.691.$$

The velocity in the region $\delta \le \xi \le a$ is evidently $w = w_0/2$ by the condition of continuity. The thickness distribution over the region $0 \le \xi \le \delta$ is determined by solving (5.162) under the conditions $h = 0$ at $\xi = \gamma$ *and* $\xi = \delta$, so that the bending moment is continuous, and the fact that $dh/d\xi$ merely changes its sign at $\xi = \gamma$. The result is

$$\left.\begin{aligned} \frac{hY}{pa} &= 2\left[1 - 1.265\cos\xi + 1.249\sin(\gamma - \xi)\right], \qquad 0 \le \xi \le \gamma, \\ \frac{hY}{pa} &= 2\left[1 - \sqrt{\frac{2}{3}}\left\{\sinh(\xi - \gamma) + \sinh(\delta - \xi)\right\}\right], \quad \gamma \le \xi \le \delta, \end{aligned}\right\}. \qquad (5.165a)$$

Since $2\sin\gamma = \sinh(\delta - \gamma) = \sqrt{\frac{3}{2}}$. In the remainder of the considered half of the shell, the state of stress is given by $M_x = 0$, $N_\theta = Yh$ and (5.143) furnishes

$$hY/pa = 1, \quad \delta \le \xi \le \alpha. \tag{5.165b}$$

Since dh/dx is discontinuous at $\xi = \delta$, a flange of vanishingly small width but of finite area of cross section must be added at this section to maintain continuity of the shearing force Q.

To obtain the total volume V of the face sheets of the shell designed for minimum volume, it is only necessary to insert in (5.154) the expressions for h in (5.163) and (5.165) for short and long shells, respectively, and to evaluate the integral. Including the volume of the flanges appropriately, so that the results match when $\alpha = 1.691$, the minimum volume can be written in the dimensionless form

$$\frac{V}{V_0} = \begin{cases} 2q\,(1 - 2\sin\beta/\alpha), & \alpha \le 1.619, \\ q\left[1 - \left(\sqrt{6} - \delta\right)/\alpha, \right. & \alpha \ge 1.619, \end{cases} \tag{5.166}$$

where $V_0 = 4\pi a l h_0$ represents the volume of a shell of constant face sheet thickness h_0, and q denotes the quantity pa/h_0Y. The relationship between q and a for the sandwich shell is directly obtained from (5.12) and (5.16) with w replaced by α. When a is very small, $q \approx 2/\alpha^2$, while $\alpha \approx 2\tan\beta$ in view of (5.160), so that $V/V_0 = \frac{1}{2}$ in the limit $\alpha = 0$. The ratio V/V_0 is plotted as a function of a in Fig. 5.29, which indicates that the saving of material effected by the optimum design is higher for the clamped shell than for the simply supported shell.

5.7.5 Closed-Ended Shell Under Internal Pressure

Suppose now that a circular cylindrical shell of length $2l$ is closed at the ends by rigid plates and is subjected to a uniform internal pressure p. The rigid plates not only produce an axial force $N_x = pa/2$ per unit circumference but also give rise to clamped edge conditions at the ends of the shell. As in the case of a shell of constant thickness, the yield condition requires $N_\theta = N_0$ throughout the shell, only stress states represented by the edges GK and HK of Fig. 5.28b being involved in the minimum volume design. Hence the relationship between the axial force and the bending moment is

$$M_x = \pm M_0\left(1 - \frac{N_x}{N_0}\right) = \pm\frac{YH}{2}\left(h - \frac{pa}{2Y}\right), \tag{5.167}$$

where the upper sign holds for the line GK and the lower sign for the line HK. The former applies to two identical outer portions of the shell, each having a length $b = \beta\sqrt{aH}$, and the latter applies to a central portion of length $2\,(l - b) = 2\,(\alpha - \beta)\sqrt{aH}$. The condition D/h = constant, which is equivalent to (5.147) with $\dot{\lambda}$ and $\dot{\kappa}$ denoting $\dot{\lambda}_\theta$ and $\pm\dot{\kappa}_x$, respectively, leads to the differential equation

$$aH\frac{d^2w}{dx^2} \pm 2w = \pm 2w_0,$$

where w_0 is a positive constant. These two equations cover the entire shell for all values of l and are subject to the boundary conditions

$$w = dw/dx = 0 \text{ at } x = 0, \qquad dw/dx = 0 \text{ at } x = l.$$

Further, w and dw/dx must be continuous at the interface $x = b$. The solution for the velocity field is given by (5.159) with a factor of $\sqrt{2}$ for each of the quantities ξ, β, and α, the relationship between β and α being

$$\tan\left(\sqrt{2}\beta\right) = \tanh\left[\sqrt{2}\,(\alpha - \beta)\right]. \tag{5.168}$$

Inserting (5.167) into the equilibrium equation (5.143), and setting $N_\theta = Yh$, the differential equation for the thickness h is obtained as

$$aH\frac{d^2 h}{dx^2} \pm 2\,h = \pm\frac{2pa}{Y}.$$

Since M_x must vanish at $x = b$ by the condition of continuity, and the shearing force Q must vanish at $x = l$ because of symmetry, we have

$$h = \frac{pa}{2Y} \text{ at } \xi = \beta, \qquad \frac{dh}{d\xi} = 0 \text{ at } \xi = \alpha.$$

In addition, $dh/d\xi$ must change sign at $\xi = \beta$ without changing its absolute value for the shearing force to be continuous. The solution to the above equation therefore becomes

$$\left.\begin{aligned} hY/pa &= 1 - \tfrac{1}{2}\left(\cos\xi/\cos\beta\right) + \tan\left(\sqrt{2}\beta\right)\sin\left[\sqrt{2}\,(\beta - \xi)\right], \quad 0 \le \xi \le \beta,\\ hY/pa &= 1 - \tfrac{1}{2}\cosh\left[\sqrt{2}\,(\alpha - \xi)\right]/\cosh\left[\sqrt{2}\,(\alpha - \beta)\right], \quad \beta \le \xi \le \alpha. \end{aligned}\right\} \tag{5.169}$$

In Fig. 5.30, the thickness distribution for a closed-ended shell is compared with that for an open-ended shell when $\alpha = \sqrt{2}$. The ratio of the volume of the face sheets for the optimum design to that for the constant thickness design is obtained from (5.154) and (5.169) as

$$\frac{V}{V_0} = q\left\{1 - \frac{\sin\left(\sqrt{2}\beta\right)}{\sqrt{2}\alpha}\right\} = \left(\frac{2+\alpha^2}{1+\alpha^2}\right)\left\{1 - \frac{\sin\left(\sqrt{2}\beta\right)}{\sqrt{2}\alpha}\right\}, \tag{5.170}$$

where the last step follows from (5.33a) with ω replaced by α. When α tends to zero, β/a tends to $\frac{1}{2}$ and V/V_0 tends to unity. The variation of V/V_0 with α for the closed-ended shell, computed from (5.168) and (5.170), is included in Fig. 5.29.

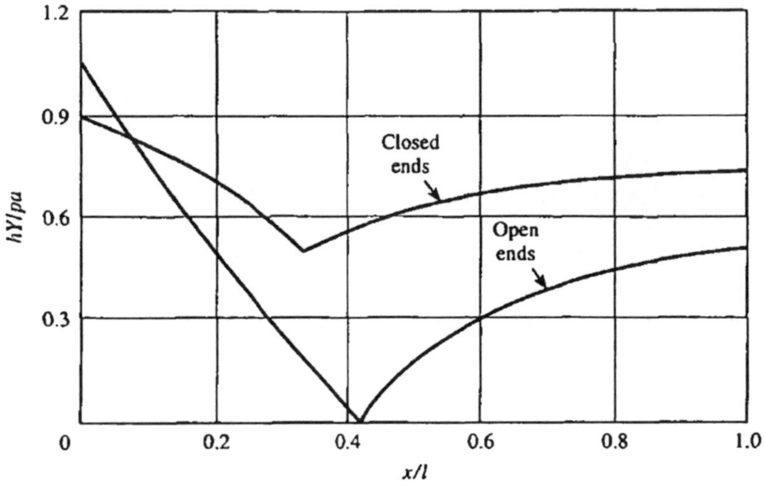

Fig. 5.30 Thickness distribution in a clamped cylindrical sandwich shell designed for minimum weight under uniform internal pressure

The presence of the axial force appreciably reduces the saving of material caused by the optimum design. The minimum weight design of cylindrical shells based on the von Mises yield criterion has been presented by Reiss and Magarefs (969).

The optimum design of closed pressure vessel heads has been discussed by Hoffman (1962) and by Save and Massonnet (1972). The minimum weight design of conical shells has been considered by Reiss (1974, 1979). The plastic design of shells of revolution for constant strength has been examined by Ziegler (1958), Issler (1964), and Dokmeci (1966). The minimum weight design of membrane shells of revolution subjected to uniform external pressure and vertical load has been investigated by Richmond and Azarkhin (2000).

Problems

5.1 A short cylindrical shell of mean radius a and wall thickness h is under a uniform internal pressure p, and is simply supported at both ends, the length of the shell being denoted by $2l$. Adopting the linearized yield condition for a Tresca sandwich shell, show that the dimensionless collapse pressure and the associated velocity field are given by.

$$\frac{pa}{Yh} = \frac{2 - \cos\omega}{2\,(1 - \cos\omega)}, \quad \frac{w}{w_0} = \frac{\sin(\omega - \xi)}{\sin\omega}$$

where $\xi = x\sqrt{2/ah}$, with x denoting the axial distance measured from the central section, ω is the value of ξ at $x = l$, and w_0 is the deflection rate at the center....

5.2 A cylindrical shell of length l, thickness h, and mean radius a is clamped at one end and is free at the other. The shell is subjected to a uniform radial pressure p to reach the point of plastic collapse. Using the yield condition for a Tresca sandwich shell, show that the intensity of the pressure and the velocity distribution are given by

$$\frac{pa}{N_0} = \frac{2\cosh\omega - 1}{2(\cosh\omega - 1)}, \quad \frac{w}{w_0} = \frac{\sinh(\omega - \xi)}{\sinh\omega}$$

where ξ and ω denote the same quantities as those in the preceding problem, and w_0 is the deflection rate at the free end of the shell.

5.3 A cylindrical shell of wall thickness h and mean radius a is free at both ends and is subjected to a radially outward ring load P per unit circumference applied at $x = 0$. Introducing the parameter $\xi = x\sqrt{2/ah}$, the ends of the shell are defined by $\xi = -\alpha$ and $\xi = \beta$. Assuming α and β to be sufficiently small, so that the shell can collapse without the formation of a hinge circle, and using the square yield condition defined by $m_x = \pm 1$ and $n_\theta = \pm 1$, show that the hoop force changes sign at $\xi = \lambda$, where $2\lambda^2 = \alpha^2 + \beta^2$, and that the collapse load is given by

$$\frac{P}{N_0} = \sqrt{\frac{h}{a}}\left\{\sqrt{\alpha^2 + \beta^2} - \frac{\beta - \alpha}{\sqrt{2}}\right\}, \qquad \alpha \leq 1, \quad \beta \leq \sqrt{2} + \sqrt{1 + \alpha^2}$$

5.4 For higher values of β in the preceding problem, show that a hinge circle must form at $\xi = \sqrt{1 + \alpha^2}$, the outer portion of the shell remaining rigid, the associated collapse load being given by the modified expression

$$\frac{P}{N_0} = \sqrt{\frac{h}{2a}}\left\{\alpha + \sqrt{1 + \alpha^2}\right\}, \qquad \alpha < 1, \quad \beta > \sqrt{2} + \sqrt{1 + \alpha^2}$$

Over the range $\alpha > 1$ and $\beta > 2\sqrt{2}$, prove that the collapse mode involves a hinge circle at $\xi = 0$, and the corresponding collapse load is given by

$$\frac{P}{N_0} = \sqrt{\frac{h}{a}}\left\{\sqrt{1 + \alpha^2} + \sqrt{1 + \beta^2} - \frac{\alpha + \beta}{\sqrt{2}}\right\}, \qquad 1 < \alpha < \beta \leq 2\sqrt{2}$$

5.5 A cylindrical shell of length $2l$, mean radius a, and wall thickness h is subjected to a uniform radial pressure p, and also a radially outward ring shearing force of intensity ql at either end. For sufficiently small values of q, the shell will collapse by uniform radial contraction. Using the square yield condition, obtain the interaction relation

$$(p - q)\,a = N_0, \quad 0 \leq qa/N_0 \leq 1/\omega^2$$

where $\omega = l\sqrt{2/ah}$,. For higher values of q, the collapse mode consists of a velocity field given by $w = w_0(1 - x/c)$, implying that the radial velocity changes sign across $x = c$. Prove that the interaction relation is then given parametrically by

$$\frac{pa}{N_0} = -1 + \frac{2c}{l}\left(2 - \frac{c}{l}\right) + \frac{1}{\omega^2}, \quad \frac{qa}{N_0} = \frac{2c}{l}\left(1 - \frac{c}{l}\right) + \frac{1}{\omega^2}, \quad \omega^2 \leq \frac{1}{2}$$

Show also that $qa/N_0 = 1/\omega$ when $p = 0$ for all ω. This result enables us to extend the interaction curve to $p = 0$ for $\omega^2 > 1/2$ approximately by a straight line.

5.6 A cylindrical shell of mean radius a, thickness h, and length $2\,l$ is built-in at both the ends, and is brought to the point of plastic collapse by the application of a uniform radial pressure p, and an independent axial load N which can take both positive and negative values. Adopting the yield condition for a Tresca sandwich shell, show that the interaction curve is closely approximated by

$$q = (1-2n)\,q_0 + n\left(1+\frac{2}{\omega^2}\right),\quad 0 \le n \le \frac{1}{2};\qquad q = (1-n)\left(1+\frac{2}{\omega^2}\right),\ \frac{1}{2} \le n \le 1$$

$$dq = (1+2n)\,q_0 - 2n\left(1+\frac{1}{\omega^2}\right),\ -\frac{1}{2} \le n \le 0;\qquad q = 1 + (1+n)\,\frac{2}{\omega^2},\ -1 \le n \le -\frac{1}{2}$$

where $q = pa/N_0$, $n = N/N_0$, and q_0 is the value of q given in (5.16), Chapter 5. Obtain a graphical plot of the complete interaction curve assuming $\omega^2 = 2$.

5.7 A spherical cap of angular span 2α contains a central circular cutout subtending an angle 2β, the radius of curvature and thickness of the cap being denoted by a and h, respectively. The cap is made of a Tresca material and carries a uniform normal pressure p distributed over its entire surface. If the outer edge of the cap is supported in such a way that it is free to rotate while being unable to translate, show that the collapse pressure based on the limited interaction yield condition is given by

$$\frac{pa}{2N_0} = \frac{(1+k)(\sin\alpha - \sin\beta) - (\alpha-\beta)\cos\alpha}{\sin(\alpha-\beta)\cos\beta - (\alpha-\beta)\cos\alpha}$$

where $k = h/4a$. Verify that this solution is statically and kinematically admissible over the relevant range of values of k when $\alpha < 35^0$

5.8 A lower bound solution for the collapse load of a simply supported spherical cap of angular span 2α under a uniform normal pressure p may be obtained by assuming $n_\theta = n_\phi$ throughout the cap. Show that the equilibrium equations (5.66) are satisfied by the stress distribution

$$n_\theta = n_\phi = kc_1\cos\phi - q,\quad m_\phi = c_1\cos\phi + c_2\sin\phi + c_3,\quad m_\theta = c_1\cos\phi + 2c_2\sin\phi + c_3$$

Applying the yield conditions $-m_\phi - n_\phi = 1$ and $m_\theta - m_\phi = 1$, at $\phi = \alpha$, and using the boundary condition $m_\phi = 0$ at $\phi = \alpha$, determine the constants c_1, c_2, and c_3. Assuming that the first yield condition also holds at $\phi = 0$, obtain the lower bound as

$$q = 1 + \left(\frac{k}{1+k}\right)\frac{\cos\alpha}{1-\cos\alpha}$$

Chapter 6
Plastic Anisotropy

During continued plastic deformation of a polycrystalline metal, certain crystallographic planes tend to rotate with respect to the direction of the greatest principal strain, and the texture of the metal assumes a fibrous appearance. The phenomenon is similar to that of a single crystal in which the slip planes tend to rotate in such a way as to become parallel to the direction of the maximum principal strain. Since the individual crystal grains in a polycrystalline aggregate cannot rotate freely due to their mutual constraints, the development of a preferred orientation in a polycrystalline metal is much more complex. As a result of progressive cold work, an initially isotropic metal therefore becomes anisotropic, and its mechanical properties vary with the direction. The yield strength of the metal in the direction of mechanical working may be greater or less than that in the transverse direction, depending on the type of preferred orientation that is produced by cold work. In a cold rolled sheet of brass, for instance, the tensile yield stress transverse to the direction of rolling can be considerably higher than that in the rolling direction. Although anisotropy and the Bauschinger effect always occur together, the latter can be largely removed by mild annealing, while the former can be altered only by carrying out the heat treatment above the recrystallization temperature.

6.1 Plastic Flow of Anisotropic Metals

6.1.1 The Yield Criterion

Consider the state of anisotropy that involves three mutually orthogonal planes of symmetry at each point, the lines of intersection of these planes being regarded as the principal axes of anisotropy. These axes may be fixed in direction throughout the specimen as in the case of a rolled sheet or vary continuously in direction as in a hollow tube expanded by internal pressure. For any given element of the orthotropic metal, we choose rectangular axes coinciding with the principal axes of anisotropy. The simplest regular yield criterion for the anisotropic material is that which reduces to the von Mises criterion when the state of anisotropy is vanishingly small. Assuming the yield function to be homogeneous of degree two in the stress components,

J. Chakrabarty, *Applied Plasticity, Second Edition*, Mechanical Engineering Series,
DOI 10.1007/978-0-387-77674-3_6, © Springer Science+Business Media, LLC 2010

the anisotropic yield criterion may be written in terms of the rectangular components of the stress as

$$F\left(\sigma_y-\sigma_z\right)^2+G\left(\sigma_z-\sigma_x\right)^2+H\left(\sigma_x-\sigma_y\right)^2+2L\tau_{yz}^2+2M\tau_{yz}^2+2N\tau_{xy}^2=1, \quad (6.1)$$

where F, G, H, L, M, and N are parameters defining the current state of anisotropy. The absence of linear terms in the yield function implies that there is no Bauschinger effect, while the presence of terms involving the normal stress differences ensures that yielding is unaffected by the hydrostatic stress.

Let X, Y, Z denote the uniaxial yield stresses in the principal directions of anisotropy and R, S, T the shear yield stresses with respect to these directions. Then it follows from (6.1) that

$$\left.\begin{aligned}&(G+H)\,X^2=(H+F)\,Y^2=(F+G)\,Z^2=1,\\&2LR^2=2MS^2=2NT^2=1.\end{aligned}\right\} \quad (6.2)$$

Thus, only one of the parameters F, G, H can be negative, while L, M, N are essentially positive. Moreover, $F > G$ when $X > Y$, together with two similar inequalities for the first three parameters. Evidently, (6.1) reduces to the von Mises criterion when $L = M = N = 3F = 3G = 3H = 3/2Y^2$, where Y is the uniaxial yield stress of the isotropic material. For an arbitrary orthotropic material, the first three relations of (6.2) give

$$2F=Y^{-2}+Z^{-2}-X^{-2},\quad 2G=Z^{-2}+X^{-2}-Y^{-2},$$
$$2H=X^{-2}+Y^{-2}-Z^{-2}.$$

The state of anisotropy in an element is, therefore, specified by the directions of the three principal axes of anisotropy, and the values of the six independent yield stresses X, Y, Z, R, S, and T, which depend on the degree of previous cold work as well as on the subsequent heat treatment (Hill, 1948).

The yield criterion is expressed in the form (6.1) only when the principal axes of anisotropy are taken as the axes of reference. For an arbitrary set of rectangular axes, the form of the yield criterion must be changed by the appropriate transformation of the stress components. When the state of anisotropy is rotationally symmetric about the z-axis, the form of the yield criterion must be independent of the choice of x-and y-axes. Since $X = Y$ and $R = S$ for such a symmetry, it is obviously necessary to set $F = G$ and $L = M$. To obtain additional conditions for the rotational symmetry, we rewrite (6.1) in the form

$$(F+H)\left(\sigma_x+\sigma_y\right)^2+2F\sigma_z\left(\sigma_z-\sigma_x-\sigma_y\right)-2\,(F+2H)\left(\sigma_x\sigma_y-\tau_{xy}^2\right)$$
$$+\,2L\left(\tau_{yz}^2+\tau_{zx}^2\right)+2\,(N-F-2H)\,\tau_{xy}^2=1.$$

Since the first four terms of the above expression are invariants for a fixed z-axis, the coefficient of the last term must be identically zero for the yield criterion to be unaffected by any rotation of the x- and y-axes. The necessary and sufficient conditions for the anisotropy to be rotationally symmetric about the z-axis therefore becomes

$$N = F + 2H = G + 2H,\ L = M.$$

The number of independent parameters defining a rotationally symmetrical state of anisotropy is therefore reduced to three, which may be taken as the uniaxial yield stresses along and perpendicular to the axis of symmetry and the shear yield stress with respect to these two directions.

6.1.2 Stress–Strain Relations

To derive the relations between the stress and the strain increments for an anisotropic material, we adopt the usual normality rule of plastic flow, assuming the plastic potential to be identical to the yield function. Referred to the principal axes of anisotropy, when the elastic strains are disregarded, the strain increment tensor for an orthotropic material may be written as

$$d\varepsilon_{ij} = \frac{\partial f}{\partial \sigma_{ij}} d\lambda,$$

where $2f(\sigma_{ij;})$ denotes the expression on the left-hand side of (6.1), and $d\lambda$ is necessarily positive for plastic flow in an element that is stressed to the yield point. Setting $2\tau^2{}_{yz} = \tau^2{}_{yz} + r^2{}_{zy}$, etc., in the field function (6.1), and treating all nine components of the stress tensor as independent, we obtain the strain increment relations

$$\left.\begin{aligned} d\varepsilon_x &= \left[H\left(\sigma_x - \sigma_y\right) + G\left(\sigma_x - \sigma_z\right)\right] d\lambda, & d\gamma_{xy} &= N\tau_{xy} d\lambda, \\ d\varepsilon_y &= \left[F\left(\sigma_y - \sigma_z\right) + H\left(\sigma_y - \sigma_x\right)\right] d\lambda, & d\gamma_{yz} &= L\tau_{yz} d\lambda, \\ d\varepsilon_z &= \left[G\left(\sigma_z - \sigma_x\right) + F\left(\sigma_z - \sigma_y\right)\right] d\lambda, & d\gamma_{zx} &= M\tau_{zx} d\lambda, \end{aligned}\right\}, \qquad (6.3)$$

which constitute a generalization of the Lévy–Mises flow rule. The sum of the three normal strain components is seen to be zero, satisfying the condition of plastic incompressibility. When the principal axes of stress coincide with those of anisotropy, the principal axes of the strain increment also occur in the same directions. In general, however, the principal axes of stress and strain increment do not coincide for an anisotropic material. The preceding results are due to Hill (1948), although similar relations have been given by Jackson et al. (1948) and Dorn (1949).

The ratios of the anisotropic parameters can be determined by carrying out tensile tests on specimens cut at suitable orientations with respect to the principal axes of anisotropy. It is, of course, necessary for this purpose that the anisotropy is uniformly distributed through a volume of sufficient extent in order to allow the preparation of the specimens. For a tensile specimen cut parallel to the x-axis of anisotropy, the ratios of the principal strain increments are

$$d\varepsilon_x{:}d\varepsilon_y{:}d\varepsilon_z = G + H{:} - H{:} - G. \tag{6.4}$$

A longitudinal extension is therefore accompanied by a contraction in each transverse direction, unless the yield stresses differ so much that one of the parameters G and H is negative. The magnitude of the incremental transverse strain is greater in the direction of the lesser yield stress. Tensile tests carried out on specimens cut parallel to the y- and z-axes of anisotropy similarly furnish the ratios F/H and G/F, respectively, providing an immediate test on the theory in view of the identity $(H/G)(G/F)(F/H) = 1$. Where the theory is applicable, the measurement of strain ratios in the appropriate tensile specimens provide an indirect method of finding the ratios of the yield stresses along the three principal axes of anisotropy.

6.1.3 Variation of Anisotropic Parameters

It is assumed at the outset that the material has a very pronounced state of anisotropy, and that further changes in anisotropy during cold work are negligible over the considered range of strains. The yield stresses of the material in the different directions then increase in strict proportion as the material deforms, the factor of proportionality being denoted by a parameter h which increases monotonically from unity to represent the amount of hardening. Thus $X = hX_0$, $Y = hY_0$, etc., where the subscript zero denotes the initial value, giving $F = F_0/h^2$, $G = G_0/h^2$, etc. The anisotropic parameters therefore decrease in strict proportion, and their ratios remain constant during the deformation. The scalar parameter h is a dimensionless form of the equivalent stress $\bar{\sigma}$ which may be defined as

$$\bar{\sigma} = h\sqrt{\frac{3}{2\left(F_0 + G_0 + H_0\right)}} = h\sqrt{\frac{3}{2c}}, \qquad c = F_0 + G_6 + H_0,$$

so that h is equal to $\bar{\sigma}/Y$ for an isotropic material with an initial yield stress Y. The substitution for $F = F_0/h^2$, etc., in (6.1) furnishes h^2 in terms of the initial values of the anisotropic parameters, and the expression for the equivalent stress becomes

$$\begin{aligned}\bar{\sigma} = \sqrt{\frac{3}{2c}}\Big\{&F_0\left(\sigma_y - \sigma_z\right)^2 + G_0\left(\sigma_z - \sigma_x\right)^2 + H_0\left(\sigma_x - \sigma_y\right)^2 \\ &+2L_0\tau_{yz}^2 + 2M_0\tau_{zx}^2 + 2N_0\tau_{xy}^2\Big\}^{1/2}.\end{aligned} \tag{6.5}$$

As in the case of isotropic solids, $\bar{\sigma}$ may be regarded as a function of an equivalent strain whose increment may be defined according to the hypothesis of strain equivalence as

$$d\bar{\varepsilon} = \sqrt{\tfrac{2}{3}}\left\{d\varepsilon_x^2 + d\varepsilon_y^2 + d\varepsilon_z^2 + 2d\gamma_{xy}^2 + 2d\gamma_{xz}^2\right\}^{1/2}. \tag{6.6}$$

When a uniaxial tension X is applied in the x-direction, the ratios of the nonzero components of the strain increment are given by (6.4), and the equivalent stress and strain increments become

$$\bar{\sigma} = \sqrt{\frac{3}{2}\left(\frac{G_0 + H_0}{F_0 + G_0 + H_0}\right)}X, \qquad \overline{d\varepsilon} = \frac{2\sqrt{G_0^2 + G_0H_0 + H_0^2}}{\sqrt{3}\,(G_0 + H_0)}d\varepsilon_x. \tag{6.7}$$

Similar expressions for $\bar{\sigma}$ and $\overline{d\varepsilon}$ are obtained for uniaxial tensions Y and Z applied in the y- and z-directions, respectively. A comparison of the stress–stress curves along the three principal axes of anisotropy provides a direct means of testing the hypothesis.

Consider the alternative hypothesis in which $\bar{\sigma}$ is assumed to be a function of the total plastic work per unit volume of the element. This has been proposed by Jackson et al. (1948) and was later followed by Hill (1950a). The increment of plastic work per unit volume is

$$dW = \sigma_{ij}d\varepsilon_{ij} = \sigma_{ij}\frac{\partial f}{\partial \sigma_{ij}}d\lambda = 2fd\lambda = d\lambda$$

by Euler's theorem of homogeneous functions and by the yield criterion expressed as $2f = 1$. From (6.3), we have

$$Gd\varepsilon_y - Hd\varepsilon_z = (FG + GH + HF)\left(\sigma_y - \sigma_z\right)d\lambda,$$

together with two similar relations obtained by cyclic permutation. The substitution for the normal stress differences into the yield criterion (6.1) then gives

$$\sum\left\{F\left(\frac{Gd\varepsilon_y - Hd\varepsilon_z}{FG + GH + HF}\right)^2 + \frac{2d\gamma_{yz}^2}{L}\right\} = (d\lambda)^2.$$

Since $dW = \bar{\sigma}d\bar{\varepsilon}$, where $d\bar{\varepsilon}$ is the equivalent strain increment according to the hypothesis of work equivalence, we have

$$d\bar{\varepsilon} = \frac{d\lambda}{\bar{\sigma}} = \left[\frac{2}{3}(F_0 + G_0 + H_0)\right]^{1/2}\frac{d\lambda}{h}.$$

Substituting for $d\lambda$, and using the relations $F = F_0/h^2$, etc., the equivalent strain increment according to the work-hardening hypothesis is finally obtained as

$$d\bar{\varepsilon} = \left[\frac{2}{3}(F_0 + G_0 + H_0)\right]^{1/2} \times\left\{F_0\left(\frac{G_0d\varepsilon_y + H_0d\varepsilon_z}{F_0G_0 + G_0H_0 + H_0F_0}\right)^2 \cdots + \frac{2\gamma_{yz}^2}{L_0} + \cdots\right\}^{1/2}. \tag{6.8}$$

For a uniaxial tension X parallel to the x-axis of anisotropy, the expressions for $\bar{\sigma}$ and $d\bar{\varepsilon}$ are

$$\bar{\sigma} = \sqrt{\frac{3}{2}\left(\frac{G_0 + H_0}{F_0 + G_0 + H_0}\right)}X, \qquad d\bar{\varepsilon} = \sqrt{\frac{2}{3}\left(\frac{F_0 + G_0 + H_0}{G_0 + H_0}\right)}d\varepsilon_x.$$

For an isotropic material, the equivalent strain increments defined by (6.6) and (6.8) are identical, and the two hypotheses for the hardening process are therefore equivalent. For an anisotropic material, the two hypothesis are distinct, and the predicted stress–strain curves along one axis of anisotropy derived from another will generally be different in the two cases. The choice of the appropriate expression for the equivalent strain increment for a particular material must be decided by experiment. The more general case of hardening of an orthotropic material, involving both expansion and translation of the yield surface in the stress space, has recently been discussed by Kojic et al. (1996).

6.2 Anisotropy of Rolled Sheets

6.2.1 Variation of Yield Stress and Strain Ratio

In a rolled sheet of metal, the principal axes of anisotropy are along the rolling, transverse, and through-thickness directions at each point of the sheet. Let the axes of reference be so chosen that the x-axis coincides with the direction of rolling, the y-axis with the transverse direction in the plane, and the z-axis with the normal to the plane. If the sheet is subjected to forces in its plane, the only nonzero stress components are σ_x, σ_y,and τ_{xy}, and the yield criterion (6.1) reduces to

$$(G + H)\sigma_x^2 - 2H\sigma_x\sigma_y + (H + F)\sigma_Y^2 + 2N\tau_{xy}^2 = 1. \tag{6.9}$$

Let σ denote the uniaxial yield stress of the sheet metal in a direction making a counterclockwise angle α with the rolling direction. The stress components corresponding to a uniaxial tension σ applied in the α-direction are

$$\sigma_x = \sigma\cos^2\alpha, \qquad \sigma_y = \sigma\sin^2\alpha, \qquad \tau_{xy} = \sigma\sin\alpha\cos\alpha. \tag{6.9a}$$

The substitution of (6.9a) into the yield criterion (6.9) furnishes σ as a function of α for any given state of anisotropy, the result being

$$\sigma = \left[F\sin^2\alpha + G\cos^2\alpha + H + (2N - F - G - 4H)\sin^2\alpha\cos^2\alpha\right]^{-1/2}. \tag{6.10}$$

The uniaxial yield stress σ can be shown to have maximum and minimum values along the anisotropic axes, and also in the directions $\alpha = \pm\alpha_0$, where

$$\alpha_0 = \tan^{-1}\sqrt{\frac{N-G-2H}{N-F-2H}}. \tag{6.11}$$

When N is greater than both $F + 2H$ and $G + 2H$, the yield stress has maximum unequal values in the x- and y-directions, and minimum equal values in the α_0-directions. If N is less than both $F + 2H$ and $G+2H$, the yield stress has minimum unequal values in the x- and y-directions, and maximum equal values in the α_0-directions. When N lies between $F + 2H$ and G + $2H$, there is no real α_0, and σ is a maximum in the x-direction and a minimum in the y-direction if $F > G$, and vice versa if $F < G$. These remarks are in broad agreement with the experiments.

Consider now the components of the strain increment corresponding to a uniaxial tension σ acting at an angle $\boldsymbol{a}$ to the rolling direction. Inserting the nonzero stress components (6.9a) into the stress–strain relations (6.3), the components of the strain increment in the plane of the sheet are obtained as

$$\begin{aligned} d\varepsilon_x &= \left[(G+H)\cos^2\alpha - H\sin^2\alpha\right]\sigma\, d\lambda, \\ d\varepsilon_y &= \left[(F+H)\sin^2\alpha - H\cos^2\alpha\right]\sigma\, d\lambda, \\ d\gamma_{xy} &= (N\sin\alpha\cos\alpha)\,\sigma\, d\lambda. \end{aligned} \tag{6.12}$$

By the condition of incompressibility, the thickness strain increment is $d\varepsilon_z = -(d\varepsilon_x + d\varepsilon_y)$. The principal axes of the strain increment in the plane of the sheet will coincide with the corresponding principal axes of the stress if

$$\left(d\varepsilon_x - d\varepsilon_y\right)/d\gamma_{xy} = \left(\sigma_x - \sigma_y\right)/\tau_{xy}.$$

Substituting for the stresses and strain increments for a uniaxial tension σ, the above condition is easily shown to be equivalent to

$$\left(\frac{G+2H}{N}\right)\cot\alpha - \left(\frac{F+2H}{N}\right)\tan\alpha = \cot\alpha - \tan\alpha.$$

This equation is satisfied when $\alpha = 0$, $\pi/2$, and α_0, where α_0 is given by (6.11). It follows that the principal axes of stress and strain increments coincide only when the applied tension is in one of the directions of the stationary yield stress.

If the ratio of the transverse strain increment to the thickness strain increment corresponding to a uniaxial tension in any given direction is denoted by R, it follows from (6.12) and the condition of incompressibility that

$$\begin{aligned} R &= -\frac{d\varepsilon_x\sin^2\alpha + d\varepsilon_y\cos^2\alpha - 2d\gamma_{xy}\sin\alpha\cos\alpha}{d\varepsilon_x + d\varepsilon_y} \\ &= \frac{H + (2N - F - G - 4H)\sin^2\alpha\cos^2\alpha}{F\sin^2\alpha + G\cos^2\alpha}. \end{aligned} \tag{6.13}$$

The ratios H/G, H/F, and N/G can be determined from the measurement of R-values of the sheet material along and perpendicular to the rolling direction, as well as in the directions making an angle of 45° with that of rolling. If these R-values are denoted by R_x, R_y, and R_s, respectively, then (6.13) furnishes

$$\frac{H}{G} = R_x, \quad \frac{H}{F} = R_y, \quad \frac{N}{G} = \left(\frac{1}{2} + R_s\right)\left(1 + \frac{R_x}{R_y}\right).$$

The R-value of a sheet metal can be determined from the measured longitudinal and transverse strains for a total elongation of more than 5%. The elastic strains are then negligible, and the thickness strain is obtained with sufficient accuracy from the condition of plastic incompressibility. The variation of the R-value with direction has been obtained experimentally for several materials by Klinger and Sachs (1948), and Bramley and Mellor (1966), who verified the assumed constancy of the R-value with the amount of plastic strain over the considered range. Over the elastic range of strains, the R-value varies rapidly with the magnitude of the strain, as has been shown by Taghavaipour et al. (1972). The situation where the R-value changes with continued plastic strain has been examined by Hill and Hutchinson (1992). The results based on the combined isotropic and kinematic hardening of the material have been discussed by Wu (2002).

6.2.2 *Localized and Diffuse Necking*

When a thin sheet of metal whose width is at least five times its thickness is pulled in tension, a localized neck inclined at an oblique angle to the direction of tension is often observed. As in the case of isotropic sheets, the neck coincides with a characteristic and the rate of extension vanishes along its length. The inception of the neck requires both the normal and the tangential components of the velocity to be discontinuous across it. If β denotes the angle of inclination of the neck with the direction of tension, measured in the counterclockwise sense as shown in Fig. 6.1, the condition of zero rate of extension along the neck may be written as

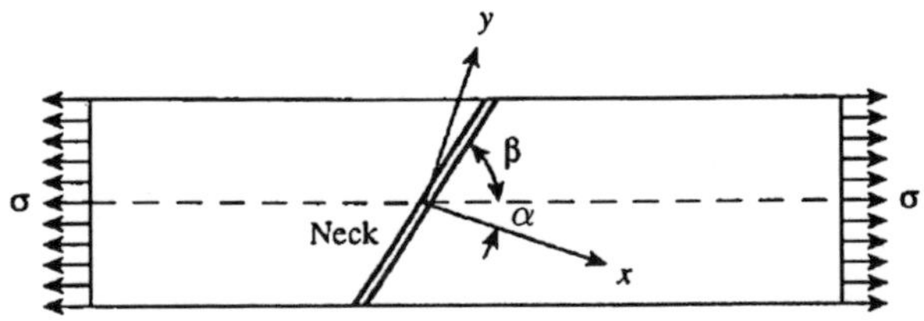

Fig. 6.1 Localized necking of a thin sheet under a uniaxial tension in the plane of the sheet

$$d\varepsilon_x \cos^2\theta + 2d\gamma_{xy}\sin\theta\cos\theta + d\varepsilon_y \sin^2\theta = 0,$$

where $\theta = \alpha + \beta$. Substituting from (6.12), and rearranging, this condition may be put in the form

$$\left(a\tan^2\beta + 2b\tan\beta\right)\sigma^2 = 1, \tag{6.14}$$

where σ is given by (6.10), while a and b depend on the parameters F, G, h, N as well as on the angle α according to the relations

$$a = H + (2N - F - G - 4H)\sin^2\alpha\cos^2\alpha,$$
$$b = \left[(N - F - 2H)\sin^2\alpha - (N - G - 2H)\cos^2\alpha\right]\sin\alpha\cos\alpha.$$

For an isotropic sheet, $F = G = h = N/3$, giving $a\sigma^2 = \frac{1}{2}$ and $b = 0$, so that $\beta = \pm\tan^{-1}\sqrt{2} \approx 54.7°$ as expected. When the material is anisotropic, there are still two equally possible directions of the neck corresponding to the two roots of the quadratic (6.14). The roots are numerically equal only when $b = 0$, which corresponds to $a = O$, $\pi/2$, and $\alpha_0 >$ where α_0 is given by (6.11) and the two possible necks are then equally inclined to the direction of the applied tension.

A localized neck is able to develop only if the rate of work-hardening of the material is lower than a certain critical value, for which it is exactly balanced by the rate of reduction of thickness in the neck. Since the normal stresses transmitted across the neck are proportional to the applied tension a, we have

$$\frac{d\sigma}{\sigma} = -d\varepsilon_z = \frac{d\varepsilon}{1+R}, \quad \text{or} \quad \frac{d\sigma}{d\varepsilon} = \frac{\sigma}{1+R}, \tag{6.15}$$

where $d\varepsilon$ is the longitudinal strain increment, and R is given by (6.13). The critical subtangent to the appropriate stress–strain curve for localized necking is therefore $(1+R)$ times that for diffuse necking. Hence, a localized neck can be expected to form on a superimposed diffuse neck as in the case of an isotropic sheet metal.

In the case of a plane sheet subjected to biaxial stresses $\sigma 1$ and σ_2 in the rolling and transverse directions, respectively, the condition of the zero rate of extension along the localized neck, which makes an angle β with the direction of $\sigma 1$, furnishes

$$\tan^2\beta = -\frac{d\varepsilon_1}{d\varepsilon_2} = \frac{(G+H)\sigma_1 - H\sigma_2}{H\sigma_1 - (F+H)\sigma_2},$$

in view of the flow rule (6.3). Assuming $\sigma_1 > \sigma_2$, the range of stress ratios for which the necking can occur is given by $\sigma_2/\sigma_2 \leq H/(F + H)$, which ensures $d\varepsilon_2 \leq 0$. An analysis for localized necking based on the total strain theory along with the assumption of a yield vertex has been presented by Storen and Rice (1975), the analysis being similar to that given in Section 2.1. The development of localized necks as a result of void growth has been considered by Needleman and Triantafyllidis (1978).

Consider now the initiation of diffuse necking in a thin sheet under biaxial tensile stresses σ_1 and σ_2 along the rolling and transverse directions, respectively. As in the case of isotropic sheets, the uniform deformation mode becomes unstable when the rate of hardening becomes critical. In terms of the initial values of the anisotropic parameters, the yield criterion (6.9) may be written as

$$(G_0 + H_0)\,\sigma_1^2 - 2H_0\sigma_1\sigma_2 + (F_0 + H_0)\,\sigma_2^2 = h^2. \tag{6.16}$$

It follows from the stress–strain relations and the differentiated form of (6.16) that

$$d\sigma_1 d\varepsilon_1 + d\sigma_2 d\varepsilon_2 = (dh/h)\,d\lambda = (d\bar{\sigma}/\bar{\sigma})\,d\lambda.$$

If the applied loads simultaneously attain their maximum at the onset of instability, then $d\sigma_1/\sigma_1 = d\varepsilon_1$ and $d\sigma_2/\sigma_2 = d\varepsilon_2$, and the preceding relation gives

$$\begin{aligned}\frac{1}{\bar{\sigma}}\frac{d\bar{\sigma}}{d\lambda} &= \sigma_1\left(\frac{d\varepsilon_1}{d\lambda}\right)^2 + \sigma_2\left(\frac{d\varepsilon_2}{d\lambda}\right)^2 \\ &= (G+H)^2\,\sigma_1^3 - H\,[(H+2G)\,\sigma_1 + (H+2F)\,\sigma_2]\,\sigma_1\sigma_2 + (F+G)^2\,\sigma_2^3.\end{aligned}$$

The hypothesis of work equivalence will be adopted here for simplicity. Using the relation $d\lambda = h\sqrt{3/2c}\,\overline{d\varepsilon}$ on the left-hand side, substituting for G, H, and F on the right-hand side, and denoting the stress ratio σ_2/σ_1 by ρ, the instability condition is finally obtained in the form

$$\frac{1}{\bar{\sigma}}\frac{d\bar{\sigma}}{\overline{d\varepsilon}} = \sqrt{\frac{3}{2}}\,\frac{(G_0+H_0)^2 - H_0\,(H_0+2G_0)\,\rho - H_0\,(H_0+2F_0)\,\rho^2 + (F_0+H_0)^2\,\rho^3}{(F_0+G_0+H_0)^{1/2}\left[(G_0+H_0) - 2H_0\rho + (F_0+H_0)\,\rho^2\right]^{3/2}} \tag{6.17}$$

in view of (6.16). The expression on the right-hand side, which is due to Moore and Wallace (1964), can be evaluated for any given p using the measured r-values in the rolling and transverse directions. An instability condition similar to (6.17) follows for the hypothesis of strain equivalence. The physical significance of the instability condition (6.17), which reduces to (2.38) when the material is isotropic, has been discussed by Dillamore et al. (1972).

6.2.3 Correlation of Stress–Strain Curves

Consider a uniaxial tension σ applied to a specimen cut of an angle a to the direction of rolling. When the hypothesis of strain equivalence is adopted, the equivalent strain increment is most conveniently obtained by using the property of its invariance. Thus, by taking the x-axis temporarily along the axis of the specimen, we get

$$d\varepsilon_x = d\varepsilon, \quad d\varepsilon_y = -\frac{R}{1+R}d\varepsilon, \quad d\varepsilon_z = -\frac{1}{1+R}d\varepsilon,$$

where r is given by (6.13). The substitution from above into (6.6) then gives the equivalent strain increment $d\bar{\varepsilon}$, while the equivalent stress $\bar{\sigma}$ is directly obtained by inserting (6.9a) in (6.5), the results being

$$\bar{\sigma} = \sqrt{\frac{3\,(1+R)\,F_0\xi}{2\,(F_0+G_0+H_0)}}\sigma, \qquad d\bar{\varepsilon} = \left\{\frac{2\sqrt{1+R+R^2}}{\sqrt{3}\,(1+R)}\right\}d\varepsilon,$$

where $\xi = \sin^2\alpha + (G_0/F_0)\cos^2\alpha$. Let the uniaxial stress–strain curve in the rolling direction be defined by the equation $\sigma = f(\varepsilon)$. In view of (6.7), the equivalent stress–strain curve is given by

$$\bar{\sigma} = \sqrt{\frac{3}{2}\left(\frac{G_0+H_0}{F_0+G_0+H_0}\right)}f\left\{\frac{\sqrt{3}\,(G_0+H_0)\,\bar{\varepsilon}}{2\sqrt{G_0^2+G_0H_0+H_0^2}}\right\}. \tag{6.18}$$

Substituting for $\bar{\sigma}$ and $\bar{\varepsilon}$ into (6.18), and introducing the R-values in the rolling and transverse directions, respectively, the equation for the stress–strain curve in the direction α is obtained as

$$\sigma = \sqrt{\frac{R_y}{\xi R_x}\left(\frac{1+R_x}{1+R}\right)}f\left\{\frac{1+R_x}{1+R}\sqrt{\frac{1+R+R^2}{1+R_x+R_x^2}}\varepsilon\right\}. \tag{6.19}$$

The stress–strain curve transverse to the rolling direction is obtained by setting $\xi = 1$ and $R = R_y$ in (6.19). When the hypothesis of work equivalence is adopted, the equivalent strain increment corresponding to a uniaxial tension σ is equal to $(\sigma/\bar{\sigma})d\varepsilon$, and (6.19) is then replaced by

$$\sigma = \sqrt{\frac{R_y}{\xi R_x}\left(\frac{1+R_x}{1+R}\right)}f\left\{\sqrt{\frac{R_y}{\xi R_x}\left(\frac{1+R_x}{1+R}\right)}\varepsilon\right\}.$$

It is evident that the stress–strain curve predicted by this relation will be generally different from that predicted by (6.19), except when the sheet is isotropic in its plane. The effective stress–strain behavior of anisotropic sheet metals has been investigated by Wagoner (1980) and Stout et al. (1983).

Consider now a state of balanced biaxial tension σ in the plane of the sheet, which is equivalent to a uniaxial compression σ normal to the plane. If the increment of the compressive thickness strain is denoted by $d\varepsilon$, then

$$d\varepsilon_x = \left(\frac{G_0}{F_0+G_0}\right)d\varepsilon, \qquad d\varepsilon_y = \left(\frac{F_0}{F_0+G_0}\right)d\varepsilon, \qquad d\varepsilon_z = -d\varepsilon,$$

on setting $\sigma_x = \sigma_y = \sigma$ in the stress–strain relations. The expressions for the equivalent stress $\bar{\sigma}$ and the equivalent strain increment $\overline{d\varepsilon}$ therefore become

$$\bar{\sigma} = \sqrt{\frac{3}{2}\left(\frac{F_0 + G_0}{F_0 + G_0 + H_0}\right)}\sigma, \qquad \overline{d\varepsilon} = \frac{G_0 + H_0}{F_0 + G_0}\sqrt{\frac{F_0^2 + F_0 G_0 + G_0^2}{G_0^2 + G_0 H_0 + H_0^2}}\,d\varepsilon,$$

in view of (6.5) and (6.6). The substitution into (6.18) shows that the stress–strain curve in the through-thickness direction according to the hypothesis of strain equivalence is given by

$$\sigma = \sqrt{R_y\left(\frac{1+R_x}{R_x+R_y}\right)} f\left\{\frac{1+R_x}{R_x+R_y}\sqrt{\frac{R_x^2 + R_x R_y + R_y^2}{1 + R_x + R_x^2}}\,\varepsilon\right\}. \tag{6.20}$$

The hypothesis of work equivalence, on the other hand, leads to the equation for the through-thickness stress–strain curve as

$$\sigma = \sqrt{R_y\left(\frac{1+R_x}{R_x+R_y}\right)} f\left\{\sqrt{R_y\left(\frac{1+R_x}{R_x+R_y}\right)}\,\varepsilon\right\}.$$

Experimentally, such a curve is most conveniently obtained by the bulge test, in which a thin circular blank of sheet metal is clamped round the periphery and deformed by uniform fluid pressure applied on one side. Due to the symmetry of the loading, a state of balanced biaxial tension exists at the pole of the bulge, where the compressive thickness strain e is equal to the sum of the two orthogonal surface strains.

In Fig. 6.2, the stress–strain curves in the thickness direction given by the above equations are compared with the bulge test curve obtained by Bramley and Mellor (1966). The derived curves are based on their measured R-values and the experimental stress–strain curve in the rolling direction. The hypothesis of strain equivalence is evidently in better agreement with experiment, at least for the materials used in this investigation, as has been shown by Chakrabarty (1970b). Further experimental results available in the literature tend to suggest that the strain-hardening hypothesis is preferable to the work-hardening hypothesis for most engineering materials. A crystallographic method of predicting the anisotropic behavior of sheet metals has been developed by Chan and Lee (1990).

6.2.4 Normal Anisotropy in Sheet Metal

In many applications, the anisotropy in the plane of the sheet is small and can be disregarded by considering a state of planar isotropy with a uniform mean R-value. This provides a radical simplification to the problem, since the yield criterion and the flow rule then become independent of the choice of coordinate axes in the plane of the sheet. Since $H/F = H/G = R$ and $N/F = 1+2R$, when the anisotropy is rotationally symmetric about the z-axis, the yield criterion and the flow rule become

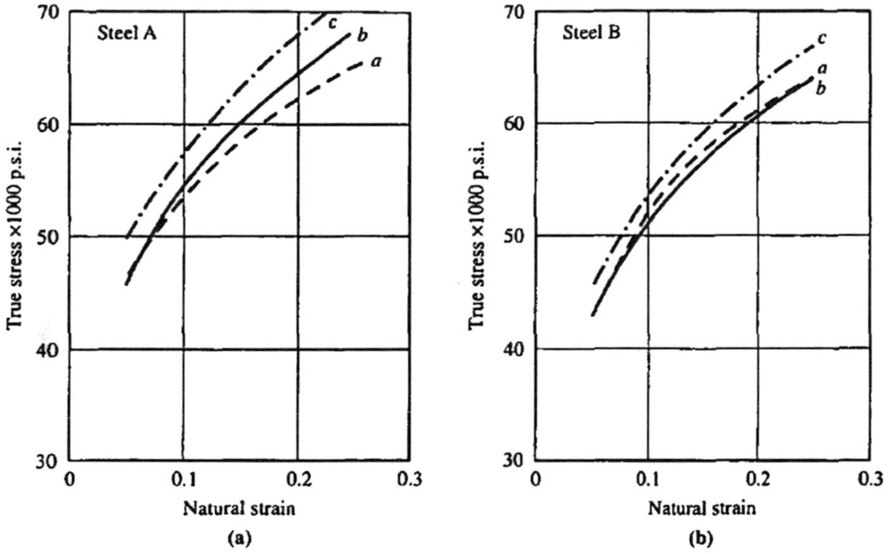

Fig. 6.2 Equibiaxial stress–strain curves for anisotropic sheets, (**a**) Experimental curve, (**b**) theoretical curve based on the uniaxial curve in the rolling direction and the hypothesis of strain equivalence, (**c**) theoretical curve derived from the uniaxial curve on the basis of work equivalence

$$\sigma_x^2 - \left(\frac{2R}{1+R}\right)\sigma_x\sigma_y + \sigma_y^2 + 2\left(\frac{1+2R}{1+R}\right)\tau_{xy}^2 = Y^2, \tag{6.21}$$

$$\frac{d\varepsilon_x}{(1+R)\,\sigma_x - R\sigma_y} = \frac{d\varepsilon_y}{(1+R)\,\sigma_y - R\sigma_x} = \frac{s d\gamma_{xy}}{(1+2R)\,\tau_{xy}} = \frac{d\lambda}{(1+R)\,Y}, \tag{6.22}$$

where Y is the uniaxial yield stress of the material in the plane of the sheet. For the present purpose, it is convenient to take the equivalent stress $\bar{\sigma}$ as identical to the current value of the planar yield stress and redefine the equivalent strain increment $d\bar{\varepsilon}$ according to the hypothesis of strain equivalence as

$$\overline{d\varepsilon} = \frac{1+R}{2}\left\{\frac{3\left(d\varepsilon_x + d\varepsilon_y\right)^2 + \left(d\varepsilon_x - d\varepsilon_y\right)^2 + 4d\gamma_{xy}^2}{1+R+R^2}\right\}^{1/2}, \tag{6.23a}$$

so that it reduces to the longitudinal strain increment for a uniaxial tension applied in the plane of the sheet. The corresponding expression for $\overline{d\varepsilon}$ according to the hypothesis of work equivalence takes the form

$$\overline{d\varepsilon} = \sqrt{\frac{1+R}{2}}\left\{(1+R)\left(d\varepsilon_x^2 + d\varepsilon_y^2\right) + 2R d\varepsilon_x d\varepsilon_y + 2d\gamma_{xy}^2\right\}^{1/2}. \tag{6.23b}$$

Expressed in terms of the principal stresses (σ_1, σ_2) in the plane of the sheet, (6.21) represents an ellipse whose major and minor axes coincide with those of the von Mises ellipse. For $R > 1$, the effect of anisotropy is to elongate the ellipse along the major axis, and slightly contract it along the minor axis for a given planar yield stress μY, Fig. 6.3 (a). The extremities of the major axis represent states of balanced biaxial stress of magnitude μY, where $\mu = \sqrt{(1+R)/2}$, as may be seen on setting $\sigma x = \sigma y = \sigma$ and $\tau_{xy} = 0$ in (6.21). A piecewise linear approximation to the nonlinear yield criterion is achieved by replacing the ellipse with a hexagon obtained by elongating the inclined sides of the Tresca hexagon (Chakrabarty, 1974). The new yield locus is defined by

$$\sigma_1 = \pm\mu Y, \ \sigma_2 = \pm\mu Y, \ \sigma_1 - \sigma_2 = \pm Y.$$

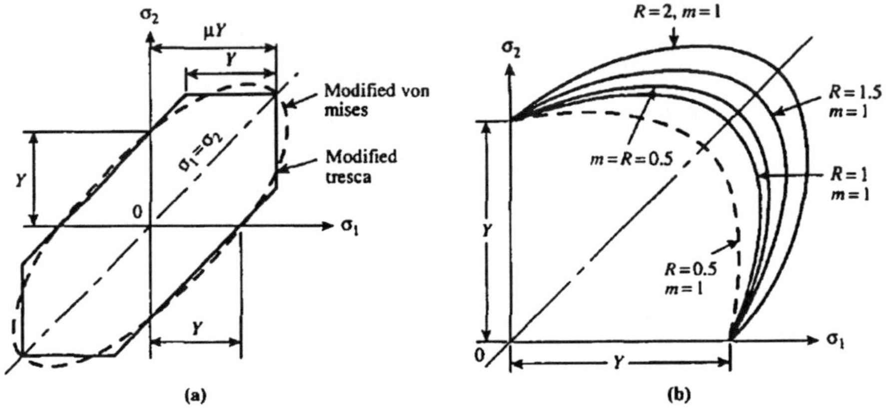

Fig. 6.3 Biaxial yield loci for sheet metals with normal anisotropy. (**a**) Quadratic yield function and its linearization for R 1, (**b**) quadratic and nonquadratic yield functions for different R-values

It may be noted that the hexagon meets the ellipse not only at points representing uniaxial and biaxial states of stress, but also at points where the principal stress ratio is $\mu^2(\mu^2 - 1)$, which is equal to $(R + l)/(R - l)$. The flow rule associated with the hexagonal yield locus requires the strain increment vector to be directed along the exterior normal when the stress point is on one of the sides of the hexagon. At a corner of the hexagon, the strain increment vector may assume any position between the normals to the two sides meeting in the corner. The increment of plastic work per unit volume is uniquely defined for all points of the hexagon.

The preceding theory, though generally adequate for $R > 1$, does not account for an anomalous behavior observed in materials with $R < 1$. Indeed, Woodthorpe and Pearce (1970) have found that the yield stress in balanced biaxial tension for rolled aluminum, having an R-value lying between 0.5 and 0.6, is significantly higher than the uniaxial yield stress in the plane of the sheet. A modification of the theory is therefore necessary for dealing with such materials. Following Hill (1979), the modified yield criterion in terms of the principal stresses σ_1 and σ_2 may be taken as

$$|\sigma_1 + \sigma_2|^{1+m} + (1 + 2R)\,|\sigma_1 - \sigma_2|^{1+m} = 2\,(1 + R)\,Y^{1+m}, \tag{6.24}$$

where m is an additional parameter that depends on the state of normal anisotropy. It is easy to show that (6.24) reduces to (6.21) when $m = 1$. The parameter R in (6.24) is indeed the customary strain ratio in simple tension, as may be seen from the plastic flow rule which may be expressed in the form

$$\frac{|d\varepsilon_1 + d\varepsilon_2|}{|\sigma_1 + \sigma_2|^m} = \frac{|d\varepsilon_1 - d\varepsilon_2|}{(1 + 2R)\,|\sigma_1 - \sigma_2|^m} = \frac{d\lambda}{(1 + R)\,Y^m}. \tag{6.25}$$

The signs of $d\varepsilon_1 + d\varepsilon_2$ *and* $d\varepsilon_1 - d\varepsilon_2$ are the same of those of $\sigma_1 + \sigma_2$ and $\sigma_1 - \sigma_2$, respectively. The parameter $d\lambda$ in (6.25) is a positive scalar equal to the equivalent strain increment $d\varepsilon$ defined by the hypothesis of work equivalence. Indeed, it follows from (6.24) that

$$2dW = (\sigma_1 + \sigma_2)\,(d\varepsilon_1 + d\varepsilon_2) + (\sigma_1 - \sigma_2)\,(d\varepsilon_1 - d\varepsilon_2) = 2Yd\lambda.$$

Eliminating $\sigma_1 + \sigma_2$ and $\sigma_1 - \sigma_2$ between (6.24) and (6.25), the corresponding expression for the equivalent strain increment is obtained as

$$\left(\overline{d\varepsilon}\right)^{1+n} = \frac{1}{2}\,(1 + R)^n \left\{|d\varepsilon_1 + d\varepsilon_2|^{1+n} + (1 + 2R)^{-n}\,|d\varepsilon_1 - d\varepsilon_2|^{1+n}\right\},$$

where $n = 1/m$. For $m = 1$, this expression reduces to (6.23b) when the principal components alone are considered. The equivalent strain increment defined by the hypothesis of strain equivalence is obtained directly from (6.23a). For a work-hardening material, it is only necessary to replace the planar yield stress Y by the equivalent stress $\bar{\sigma}$, which is a given function of the total equivalent strain.

In the case of an equal biaxial tension σ, the yield criterion (6.24) gives $(\sigma/Y)^{1+m} = (1 + R)/2^m$, which indicates that a can be greater than Y for $R < 1$ with suitable values of m. Parmar and Mellor (1978b) derived M-values of 0.5 for rim steel (R $\approx$ 0.44), 0.7 for soft aluminum (R $\approx$ 0.63), and 0.8 for software brass ($R \approx$ 0.86), on the basis of their experimental results using the bulge test and the hole expansion test. Similar experimental results for different materials have been examined by Kobayashi et al. (1985). Since the m-value depends on the R-value, these two parameters cannot be chosen independently for any given material. Other types of nonquadratic yield function for normal anisotropy of sheet metal have been considered by Logan and Hosford (1980), Vial et al. (1983), and Dodd and Caddell (1984).

6.2.5 A Generalized Theory for Planar Anisotropy

The inadequacy of the quadratic yield function in predicting the ratio of the equibiaxial and uniaxial yield stresses for materials with R-values less than unity has already been noted. In the case of planar anisotropy, similar discrepancies are expected when the average R-value of the sheet is less than unity. The difficulty can be overcome by considering a suitable extension of the nonquadratic yield criterion in (6.24) to include the effect of planar anisotropy. The simplest yield criterion of this type, when the x- and y-axes are taken along the rolling and transverse directions, respectively, may be taken in the form

$$\begin{aligned} &\left|\sigma_x+\sigma_y\right|^\nu + 2a\left(|\sigma_x|^\nu - \left|\sigma_y\right|^\nu\right) + b\left|2\tau_{xy}\right|^\nu \\ &\quad + c\left[\left(\sigma_x+\sigma_y\right)^2 + 4\tau_{xy}^2\right]^{\nu/2} = (2Z)^\nu, \end{aligned} \tag{6.26}$$

where a, b, c, and $\nu > 1$ are the dimensionless anisotropic parameters, while Z is the through-thickness yield stress of the material (Chakrabarty, 1993a). A state of planar isotropy requires $a = b = 0$, and (6.26) then reduces to (6.24) with $\nu = 1 + m$ and $c = 1 + 2R$. When $\nu = 2$, the above yield criterion becomes identical to (6.9) if we set

$$a = \frac{G-F}{G+F}, \quad b = 2\left(\frac{N-2H}{F+G}\right) - 1, \quad c = 1 + \frac{4H}{F+G},$$

and invoke the relation $f + g = 1/Z^2$. The anisotropic parameters appearing in (6.26) can be determined from the measurement of the R-values in the rolling, transverse, and 45° directions, as well as by the estimation of the ratio of the equibiaxial yield stress to the uniaxial yield stress in one of these directions. Using (6.26) the anisotropic coefficients can be expressed as

$$\left.\begin{aligned} 4a = \left(\frac{2Z}{X}\right)^\nu - \left(\frac{2Z}{Y}\right)^\nu, 2(1+c) = \left(\frac{2Z}{X}\right)^\nu + \left(\frac{2Z}{Y}\right)^\nu, \\ 2b = 2\left(\frac{2Z}{S}\right)^\nu - \left(\frac{2Z}{X}\right)^\nu - \left(\frac{2Z}{Y}\right)^\nu, \end{aligned}\right\} \tag{6.27}$$

where X, Y, and S denote the uniaxial yield stresses in the rolling, transverse, and 45° directions, respectively, in the plane of the sheet. The uniaxial yield stress σ in any direction making an angle α with the rolling direction is obtained by inserting (6.9a) into (6.26), the result being

$$\left(\frac{2Z}{\sigma}\right)^\nu = 1 + c + b(\sin 2\alpha)^\nu + 2a\left[\left(\cos^2\alpha\right)^\nu - \left(\sin^2\alpha\right)^\nu\right]. \tag{6.28}$$

Setting $d\sigma/d\alpha = 0$, the stationary values of the yield stress are found to occur along the anisotropic axes, as well as in the directions $\alpha = \pm\alpha_0$, where α_0 is given by the expression

$$\frac{(\tan\alpha_0)^{2-\nu} + (\tan\alpha_0)^{\nu}}{1 - \tan^2\alpha_0} = 2^{\nu-2}\left(\frac{b}{a}\right).$$

Thus, α_0 depends only on the parameters ν and a/b. When $a = 0$, we have $\alpha_0 = \pi/4$ irrespective of the value of ν, a result which is in accord with experiment. Other types of nonquadratic yield functions for planar anisotropy have been discussed by Barlat and Lian (1989), and Hill (1990).

Taking the plastic potential to be identical to the yield function (6.26), the components of the strain increment in the plane of the sheet can be written down from the normality rule. It is convenient to express the stress–strain relations as

$$\left.\begin{aligned} d\varepsilon_x + d\varepsilon_y &= \left\{a\left(\sigma_x\left|\sigma_x\right|^{\nu-2} - \sigma_y\left|\sigma_y\right|^{\nu-2}\right) + (\sigma_x + \sigma_y)\left|\sigma_x + \sigma_y\right|^{\nu-2}\right\} d\mu, \\ d\varepsilon_x + d\varepsilon_y &= \left\{a\left(\sigma_x\left|\sigma_x\right|^{\nu-2} + \sigma_y\left|\sigma_y\right|^{\nu-2}\right) + c(\sigma_x - \sigma_y)(2\tau)^{\nu-2}\right\} d\mu, \\ d\gamma_{xy} &= \left\{b\tau_{wy}\left|2\tau_{xy}\right|^{\nu-2} + c\tau_{xy}(2\tau)^{\nu-2}\right\} d\mu, \end{aligned}\right\} \tag{6.29}$$

where τ denotes the magnitude of the maximum shear stress in the plane of the sheet, and $d\mu$ denotes a positive scalar factor of proportionality. For a uniaxial tension σ acting at an angle α to the rolling direction, the transverse strain increment is given by the expression

$$\begin{aligned} d\varepsilon_w &= \frac{1}{2}\left(d\varepsilon_x + d\varepsilon_y\right) - \frac{1}{2}\left(d\varepsilon_x - d\varepsilon_y\right)\cos 2\alpha - d\gamma_{xy}\sin 2\alpha \\ &= -\frac{1}{2}\sigma^{\nu-1}\left\{c - 1 + b(\sin 2\alpha)^{\nu} - 2\alpha\left[\left(\cos^2\alpha\right)^{\nu-1}\sin^2\alpha\right.\right. \\ &\quad \left.\left. - \left(\sin^2\alpha\right)^{\nu-1}\cos^2\alpha\right]\right\} d\mu, \end{aligned}$$

which is obtained on substitution from (6.9a) and (6.29). The ratio of the transverse strain increment $d\varepsilon_\omega$ to the thickness strain increment $d\varepsilon_z = -(d\varepsilon_x + d\varepsilon_y)$ is the ft-value in the considered direction and is the R-value in the considered direction and is found as

$$R = \frac{(c-1) + b(\sin 2\alpha)^{\nu} - 2a\left[\left(\cos^2\alpha\right)^{\nu-1}\sin^2\alpha - \left(\sin^2\alpha\right)^{\nu-1}\cos^2\alpha\right]}{2 + 2a\left[\left(\cos^2\alpha\right)^{\nu-1} - \left(\sin^2\alpha\right)^{\nu-1}\right]}. \tag{6.30}$$

It is interesting to note that the R-value is independent of ν when $\alpha = 0$, $\pi/4$, and $\pi/2$. Denoting these three R-values by R_x, R_s, and R_y, respectively, and using (6.30), we obtain the relations

$$a = \frac{R_y - R_x}{R_y + R_x}, \quad b = 2\left(R_s - \frac{2R_xR_y}{R_x + R_y}\right), \quad c = 1 + \frac{4R_xR_y}{R_x + R_y}, \tag{6.31}$$

which are the same as those for the quadratic yield function. The remaining parameter v is given by the three independent relations

$$\left(\frac{2Z}{S}\right)^{\iota} = 2\left(1+R_s\right), \quad \left(\frac{2Z}{X}\right)^{\nu} = \frac{4R_y\left(+R_x\right)}{R_x+R_y}, \quad \left(\frac{2Z}{Y}\right)^{\iota} = \frac{4R_x\left(1+R_y\right)}{R_x+R_y}, \tag{6.31a}$$

which are easily obtained by suitable combinations of (6.27) and (6.31). The first expression in (6.31a) is generally the most appropriate one. It follows from (6.26) and (6.29) that the plastic work increment per unit volume is given by

$$2dW = \left(\sigma_x+\sigma_y\right)\left(d\varepsilon_x+d\varepsilon_y\right)+\left(\sigma_x-\sigma_y\right)\left(d\varepsilon_x-d\varepsilon_y\right)+4\tau_{xy}d\gamma_{xy} = (2Z)^{\nu}\,d\mu,$$

which immediately indicates that $d\mu$ is a measure of the equivalent strain increment based on the hypothesis of work equivalence.

If the equivalent stress $\bar{\sigma}$ is defined in such a way that it reduces to the applied stress when a uniaxial tension acts in the rolling direction, then in view of (6.28) and (6.31), we can express it in the form

$$(2Z/\bar{\sigma})^{\nu} = 1+2a+c = 4R_y\left(1+R_x\right)/\left(R_x+R_y\right).$$

For a uniaxial tension σ applied in any direction α in the plane of the sheet, the ratio $\bar{\sigma}/\sigma$ is obtained by combining (6.33) with (6.28). The equivalent strain increment $\overline{d\varepsilon}$, based on the hypothesis of strain equivalence, is given by (6.23a) with R replaced by R_x. The results for the rolling, transverse, and 45° directions are

$$\left.\begin{aligned} \left(\frac{\bar{\sigma}}{\sigma}\right)^{\nu} &= \frac{1+R}{1+R_x}\left\{\left(\frac{R_x+R_y}{2R_y}\right)+\eta\left(\frac{R_y-R_x}{2R_y}\right)\right\}, \\ \frac{\overline{d\varepsilon}}{d\varepsilon} &= \frac{1+R_x}{1+R}\sqrt{\frac{1+R+R^2}{1+R_x+R_x^2}}, \end{aligned}\right\} \tag{6.32}$$

where R is the strain ratio associated with the given direction; $d\varepsilon$ is the longitudinal strain increment; and η is a parameter whose value is 1 for $\alpha = 0$, is –1 for $\alpha = \pi/2$, and is 0 for $\alpha = -\pi/4$. In the case of an equibiaxial tension σ, the equivalent stress directly follows from (6.33) with $Z = \sigma$. The equivalent strain increment is obtained from (6.23a), using the fact that $d\varepsilon_x + d\varepsilon_y = d\varepsilon$, $d\varepsilon_x - d\varepsilon_y = \beta\, a\, d\varepsilon$ and $d\gamma_{xy} = 0$, where $d\varepsilon$ is the compressive thickness strain increment at any stage of the loading, and $\beta = 2^{\nu-2}$.

In Figs. 6.4 and 6.5, some comparison has been made of the theoretical predictions with the experimental data reported by Naruse et al. (1993) and Lin and Ding (1995). An interesting explanation of the observed anomaly in biaxial tension has been advanced by Wu et al. (1997) with the help of kinematic hardening along with the quadratic yield function.

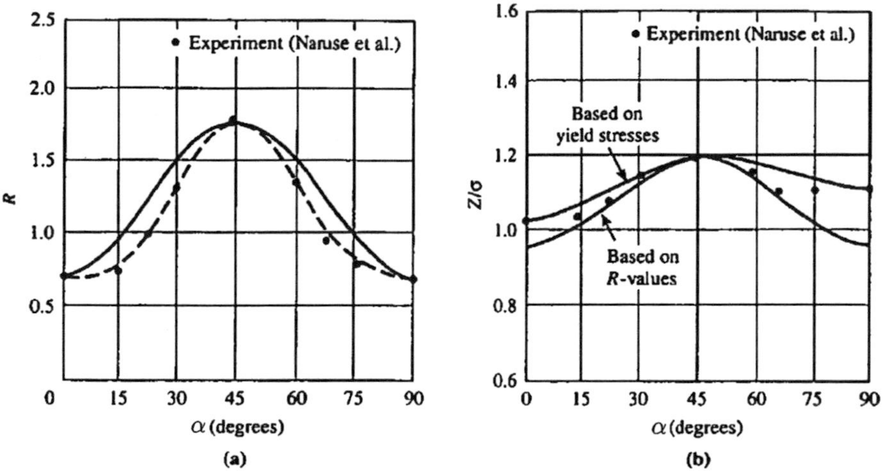

Fig. 6.4 Theoretical and experimental variations of the R-value and yield stress ratio in the plane of the sheet for annealed aluminum

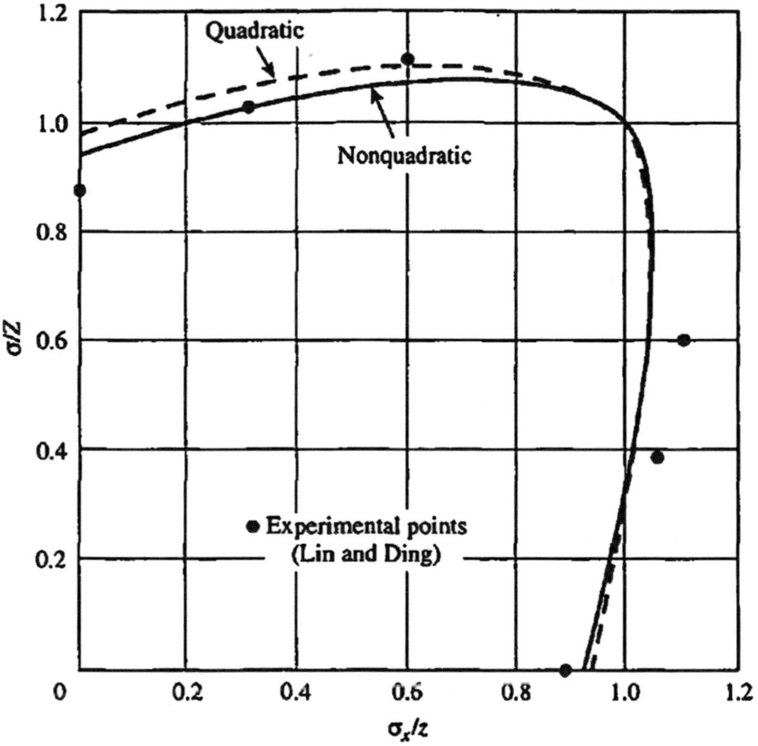

Fig. 6.5 Theoretical and experimental yield loci for cold rolled aluminum sheet exhibiting planar anisotropy

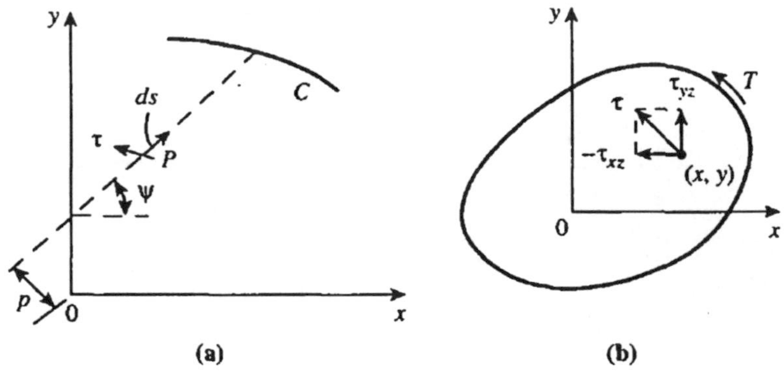

Fig. 6.6 Plastic torsion of an anisotropic bar. (**a**) Direction of shear stress in relation to that of the characteristic, components of shear stress on a transverse section

Better agreement with experimental data can be achieved by considering a higher order yield function, such as the biquadratic yield function proposed by Gotoh (1977). Taking the axes of reference along the principal axes of anisotropy, the yield. criterion may be expressed as

$$A\sigma_x^4+B\sigma_x^3\sigma_y+C\sigma_x^2\sigma_y^2+D\sigma_x\sigma_y^3+E\sigma_y^4+(F\sigma_x^2+G\sigma_x\sigma_y+H\sigma_y^2)\tau_{xy}^2+K\tau_{xy}^4=1. \quad (6.33)$$

The nine anisotropic coefficients appearing in (6.33) can be uniquely determined from the measurement of the uniaxial yield stresses and R-values in directions making angles of 0°, 22.5° (or 67.5°), 45°, and 90° with the direction of rolling, in addition to that of the through-thickness yield stress. The R-value of the sheet metal corresponding to a uniaxial tension in any given direction follows from the flow rule associated with the yield criterion. Evidently, the distribution of the yield stresses and R-values, according to (6.33), is defined by curves which pass through all the experimental points included in Figs. 6.6 and 6.7. An isotropic material corresponds to $B=D=-2A$, $C=3A$, $E=A$, $F=-G=H=6A$, $K=9A$, and the yield function (6.33) then becomes a perfect square of the von Mises yield function.

6.3 Torsion of Anisotropic Bars

6.3.1 Bars of Arbitrary Cross Section

A prismatic bar of arbitrary cross section is twisted by terminal couples about an axis parallel to the generator. The material is considered as rigid/plastic, and the applied torque is assumed sufficient to render the bar fully plastic. The state of anisotropy is orthotropic with the z-axis of anisotropy parallel to the generator. Choosing the x- and y-axes in an end section of the bar, the components of the velocity may be taken as

$$u = -yz, \quad v = xz, \quad w = w(x, y),$$

the rate of twist per unit length being taken as unity. The nonzero components of the strain rate corresponding to the above velocity field are given by

$$2\dot{\gamma}_{xy} = \frac{\partial w}{\partial x} - y, \quad 2\dot{\gamma}_{yz} = \frac{\partial w}{\partial y} + x.$$

The nonzero stress components are τ_{xz} and τ_{yx} which must satisfy the equilibrium equation and the yield criterion, these equations being

$$\frac{\partial \tau_{xz}}{\partial x} + \frac{\partial \tau_{yz}}{\partial y} = 0, \quad \frac{\tau_{xz}^2}{k_1^2} + \frac{\tau_{yz}^2}{k_2^2} = 1, \tag{6.34}$$

where $k_1 = (2M)^{-/1/2}$ and $k_2 = (2L)^{-/1/2}$ are the yield stresses in simple shear in the x- and y-directions, respectively. Differentiating the second equation of (6.34) partially with respect to x and y, and using the first equation, we get

$$\frac{\tau_{yz}}{k_2^2}\frac{\partial \tau_{xz}}{\partial x} - \frac{\tau_{xz}}{k_1^2}\frac{\partial \tau_{xz}}{\partial y} = 0, \quad \frac{\tau_{yz}}{k_2^2}\frac{\partial \tau_{yz}}{\partial x} - \frac{\tau_{xz}}{k_1^2}\frac{\partial \tau_{yz}}{\partial y} = 0.$$

These equations are seen to be hyperbolic, and the characteristic through a generic point P in the (x, y)-plane is in the direction

$$\frac{dy}{dx} = \tan\psi = -\left(\frac{k_2}{k_1}\right)^2 \frac{\tau_{xz}}{\tau_{yz}}, \tag{6.35}$$

where ψ is the counterclockwise angle made by the tangent to the characteristic with the positive x-axis. Using the preceding relations, it is easy to show that

$$d\tau_{xz} = d\tau_{yz} = 0$$

along a characteristic (Hill, 1954). It follows that the characteristics are straight, and the resultant shear stress r along each one of them is constant in direction and magnitude. Since the lateral surface of the bar is stress free, the resultant shear stress must be directed along the tangent to the boundary C at the point where it is cut by the characteristic. The magnitude of the shear stress along any characteristic can be calculated from (6.35) and the second equation of (6.34). The characteristics meet the boundary at varying angles when the bar is anisotropic.

The resultant shear strain rate at a point of the cross section is orthogonal to the characteristic, as may be seen from the flow rule associated with the yield function.

Indeed, the components of the rate of shear are in the ratio

$$\frac{\gamma_{yz}}{\gamma_{yz}} = \left(\frac{k_1}{k_2}\right)^2 \frac{\tau_{yz}}{\tau_{xz}} = -\cot\psi$$

in view of (6.35), establishing the condition of orthogonality of the strain rate. The component of the rate of shear in the characteristic direction is therefore zero. The substitution for $\dot{\gamma}_{xz}$ and $\dot{\gamma}_{yz}$ above relation gives

$$\left(\frac{\partial w}{\partial x} - y\right)\cos\psi + \left(\frac{\partial w}{\partial x} - x\right)\sin\psi = 0.$$

This equation is hyperbolic with characteristics identical to (6.35). The variation of the rate of warping along a characteristic is given by

$$\frac{\partial w}{\partial s} = \frac{\partial w}{\partial x}\cos\psi + \frac{\partial w}{\partial y}\sin\psi = y\cos\psi - x\sin\psi,$$

where ds is a line element along the characteristic. If w_0 denotes the value of w at any given point on the characteristic, the integration of the above equation furnishes

$$\omega = \omega_0 + \int (ydx - xdy) = \omega_0 + \int pds, \tag{6.36}$$

where p is the perpendicular distance from the origin to the considered element, reckoned positive when the vector ds has a clockwise moment about the z-axis, Fig. 6.6(a). The integration begins from the point of intersection of the characteristic with a stress discontinuity Γ, formed by the intersection of characteristics. Since the shear strain rate vanishes along Γ, (6.36) also holds along Γ. Any projecting corner of C is the source of one of the branches of Γ, while a reentrant corner is a point of singularity that generates a fan of characteristics.

For the evaluation of the torque T in the fully plastic state it is convenient to introduce a stress function ϕ such that the shear stress components are given by

$$\tau_{xz} = \frac{\partial \phi}{\partial y}, \quad \tau_{yz} = -\frac{\partial \phi}{\partial x}, \tag{6.37}$$

satisfying the equilibrium equation identically. The lines of constant ϕ are the shearing stress trajectories in the (x, y)-plane. Theexternal boundary C is also a contour line of ϕ, and we may choose $\phi = 0$ along this boundary. Referring to Fig. 6.6(b), the applied torque is found to be

$$T = \iint \left(x\tau_{yz} - y\tau_{xz} \right) dxdy = 2 \iint \phi dxdy. \tag{6.38}$$

The second expression of (6.38) follows on substituting from (6.37), integrating by parts, and using the boundary condition $\phi = 0$. If we now introduce the transformation

$$k_2 x = k\xi, \quad k_1 y = k\eta,$$

where k is an arbitrary constant having the dimension of stress, the yield criterion expressed by the second equation of (6.34) becomes

$$\left(\frac{\partial \phi}{\partial \xi} \right)^2 + \left(\frac{\partial \phi}{\partial \eta} \right)^2 = k^2.$$

In the (ξ, η)-plane, the stress function therefore satisfies the same equation as that for an isotropic bar with a shear yield stress k. The characteristics in the (ξ, η)-plane are normal to the transformed contours of ϕ, and correspond to the characteristics in the (x, y)-plane. Equation (6.38) now becomes

$$T = \frac{2k^2}{k_1 k_2} \iint \phi d\xi d\eta = \left(\frac{k^2}{k_1 k_2} \right) T^*, \tag{6.39}$$

where T^* denotes the fully plastic torque for an isotropic bar having the transformed cross section. The stress function for the anisotropic bar is obtained directly from the fact that $\partial\phi/\partial s$ along a characteristic is of magnitude $\left(k_1^2 \sin^2 \psi + k_2^2 \cos^2 \psi \right)^{1/2}$ which is easily obtained from (6.35), (6.37), and (6.34).

6.3.2 Some Particular Cases

As a simple application of the theory, let us consider a bar of elliptical cross section whose semiaxes are in the ratio $k_1{:}k_2$, the equation of the ellipse being

$$\frac{x^2}{a^2} + \frac{y^2}{b^2} = 1, \quad \frac{a}{b} = \frac{k_1}{k_2}.$$

The corresponding contour in the (ξ, η)-plane becomes the circle $\xi^2 + \eta^2 = c^2$, where $kc = k_1 b = k_2 a$. Since $T^* = \frac{2}{3}\pi kc^3$ for the isotropic bar, (6.39) furnishes the fully plastic torque for the anisotropic bar as

$$T = \frac{2}{3}\pi \left(\frac{k^3 c^3}{k_1 k_2} \right) = \frac{2}{3}\pi k_1 ab^2 = \frac{2}{3}\pi k_2 a^2 b.$$

Since the lines of shearing stress in the (ξ, η)-plane are concentric circles, the stress trajectories in the *(x, y)*-plane are concentric ellipses. The characteristics are radial lines in both planes, and the warping is absent.

When the cross section of the anisotropic bar is a circle of radius *c*, the transformed contour is an ellipse with semiaxes $a = ck_2/k$ and $b = ck_1/k$. When $k_1 > k_2$, a stress discontinuity extends along the η-axis over a length $2b(1-a^2/b^2)$, which is the distance between the centers of curvatures at the extremities of the major axis. The corresponding discontinuity along the *y*-axis covers a length $2c(1 - k_2^2/k_1^2)$. If the degree of anisotropy is sufficiently small, it is a good approximation to take the yield point torque for the transformed section as the mean of those for circular sections of radii *a* and *b*. Thus

$$T \approx \frac{\pi}{3}\left(\frac{k^3}{k_1 k_2}\right)\left(a^3 + b^3\right) = \frac{\pi}{3}c^3\left(\frac{k_1^3 + k_2^3}{k_1 k_2}\right).$$

It follows from (6.36), and the nature of the characteristic field, that the warping is positive in the first and third quadrants, and negative in the others.

As a final example, consider a bar whose cross section is a rectangle having sides of lengths *a* and *b* parallel to the *x*- and *y*-axes, respectively. The transformed contour is also a rectangle with sides $\alpha = ak_2/k$ and $\beta = bk_1/k$, parallel to the ξ and η-axes, respectively. Since $T^* = k\beta^2(3\alpha-\beta)/6$ for $\alpha \geq \beta$, the actual yield point torque is obtained from (6.39) as

$$T = \frac{1}{6}\left(\frac{k^3\beta^2}{k_1 k_2}\right)(3\alpha - \beta) = \frac{1}{6}k_1 b^3\left(\frac{3a}{b} - \frac{k_1}{k_2}\right), \quad \frac{a}{b} \geq \frac{k_1}{k_2} \tag{6.40}$$

The solution for $a/b \leq k_1/k_2$ is obtainedby merely interchanging *a, b* and k_1, k_2 in (6.40). The stress discontinuities emanating from the corners of the rectangle meet on the *x*- or *y*-axes accordingly as *a/b* is greater or less than k_1/k_2, the axial discontinuity being of length $|ak_2 - bk_1|/k$. The warping of the cross section can be calculated from (6.36) in the same way as that employed for an isotropic bar (Chakrabarty, 2006).

6.3.3 Length Changes in Twisted Tubes

A thin-walled cylindrical tube is twisted in the plastic range, the ends of the tube being supported in such a way that it is free to extend or contract in the axial direction. We begin with the situation where the tube may be assumed to remain isotropic during the deformation. Consider a small element of the tube formed by cross sections at a unit distance apart, so that the displacement of a particle on one edge due to its rotation relative to the other is equal to the engineering shear strain γ. Thus, a typical material line element *OP*, inclined at an angle ψ to the direction of

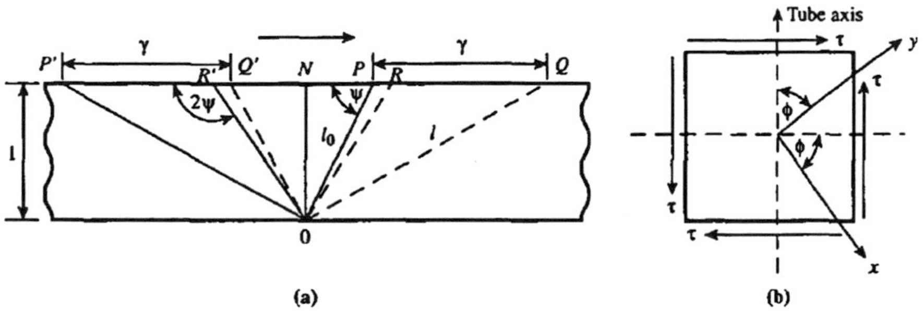

Fig. 6.7 Torsion of an anisotropic tube. (**a**) Geometry of finite shear deformation, (**b**) stress acting on an element of the tube wall

shear, rotates to a new position OQ such that $PQ = \gamma$. If the lengths OP and OQ are denoted by l_0 and l, respectively, Fig. 6.7(a), then from simple geometry,

$$l^2 = l_0^2 + 2l_0\gamma \cos\psi + \gamma^2 = l_0^2\left(1 + 2\gamma \sin\psi \cos\psi + \gamma^2 \sin^2\psi\right)$$

in view of the relation $l_0 \sin\psi = 1$. Differentiating this expression with respect to ψ, it is found that the ratio l/l_0 has a maximum value when

$$-2\cot 2\psi = \gamma, \quad l = l_0 \tan\psi, \quad l\cos\psi = 1.$$

It follows that the angles of inclination of OP and OQ to the direction of shear are complementary when *i/iq* is a maximum. The same conclusion holds for the line elements OP' and OQ' in the initial and final states, respectively, for which the length ratio is a minimum. The directions OR and OR', which correspond to zero resultant change in length, must be each inclined at an angle 2ψ to the direction of shear, since $R'N = RN = \gamma/2$, where ON is parallel to the tube axis.

Suppose, now, that the tube becomes progressively anisotropic during the torsion, and that the anisotropic axes coincide at each stage with the directions of the greatest relative extension and contraction. Since the axial strain ε is small compared to the shear strain *y*, these directions may be assumed identical to those for an isotropic tube. Let the x- and y-axes of anisotropy, which correspond to OQ' and OQ, respectively, at any given stage, make an angle ϕ with the direction of twist and the axis of the tube, respectively, Fig. 6.7(b). The nonzero components of the stress are

$$\sigma_x = -\sigma_y = -\tau \sin 2\phi, \quad \tau_{xy} = \tau \cos 2\phi,$$

where τ denotes the applied shear stress. The substitution into the yield criterion (6.9) then furnishes

$$\tau^2 = \left[2N + (F + G + 4H - 2N)\sin^2 2\phi\right]^{-1}.$$

The ratios of the components of the strain increment associated with the yield criterion are obtained from (6.3) as

$$\frac{d\varepsilon_x}{G + 2H} = \frac{d\varepsilon_y}{F + 2H} = \frac{d\varepsilon_z}{G - F} = \frac{d\gamma_{xy}}{N}\tan 2\phi.$$

The axial strain increment $d\varepsilon$ and the engineering shear strain increment $d\gamma$ produced by the torsion are

$$d\varepsilon = d\varepsilon_x \sin^2\phi + d\varepsilon_y \cos^2\phi - 2d\gamma_{xy}\sin\phi\cos\phi,$$
$$d\gamma = \left(d\varepsilon_y - d\varepsilon_x\right)\sin 2\phi + 2d\gamma_{xy}\cos 2\phi.$$

Substituting for the strain increment components appearing on the right-hand side of these relations, we obtain the ratio of $d\varepsilon$ to $d\gamma$ in the form

$$\frac{d\varepsilon}{d\gamma} = \frac{\left[(N - G - 2H)\sin^2\phi - (N - F - 2H)\cos^2\phi\right]\sin 2\phi}{2N + (F + G + 4H - 2N)\sin^2\phi}. \qquad (6.41)$$

The value of $d\varepsilon/d\gamma$ is initially zero, when the tube is isotropic ($\phi = \pi/4$). For small angles of twist, ϕ is slightly greater than $\pi/4$, and $d\varepsilon$ has the same sign as that of $F - G$, or equivalently of X–Y. With increasing angle of twist, ϕ eventually approaches $\pi/2$, and $d\varepsilon$ is finally positive if $N > G + 2H$. For many metals, the anisotropic parameters vary in such a way that the tube lengthens continuously during the torsion (Swift, 1947; Bailey et al. 1972). In exceptional cases, however, the length of the tube may progressively decrease with increasing angle of twist (Toth et al., 1992).

As the tube becomes increasingly anisotropic during the torsion, the ratios F/H, G/H, and $N/3H$ vary continuously from their common initial value of unity to approach some limiting values in an asymptotic manner. These limiting ratios can be found by direct measurements after subjecting the tube to a sufficiently large angle of twist. Each ratio of the anisotropic parameters at a generic stage of the torsion depends on its asymptotic value and the magnitude of the shear strain, according to a mathematical function which may be assumed to be the same for all three ratios. A suitable choice of such a function enables us to determine the length change at any stage by the integration of (6.41). Conversely, an actual measurement of $d\varepsilon$ and $d\gamma$ for each increment of torque provides a means for determining the variation of the ratios of the anisotropic parameters under increasing strain.

6.3.4 Torsion of a Free-Ended Tube

The change in length that takes place during the free-end twisting of a thin-walled tube should be taken into consideration for the derivation of the shear stress–strain curve from the torsion test. Let a_0 and t_0 denote the initial mean radius and wall thickness, respectively, of a thin-walled tube which is twisted in the plastic range with a torque T and an angle of twist θ per unit length of the tube. If the current internal and external radii of the tube are denoted by a_1 and a_2, respectively, then the applied torque is

$$T = 2\pi \int_{a_1}^{a_2} \tau r^2 dr = \frac{2\pi}{\theta^3} \int_{a_1\theta}^{a_2\theta} \tau \gamma^2 d\gamma.$$

Multiplying both sides of this equation by θ^3 differentiating with respect to θ, and considering a mean shear stress $\bar{\tau}$ through the thickness of the tube, we get

$$3T + \theta \frac{dT}{d\theta} = 2\pi \bar{\tau} \left\{ a_2^2 \frac{d}{d\theta}(a_2\theta) - a_1^2 \frac{d}{d\theta}(a_1\theta) \right\}.$$

Since the wall thickness is small compared to the mean radius a and remains practically unchanged during the torsion, it is reasonable to introduce the approximations

$$da_1/a_1 \approx da_2/a_2 \approx -d\varepsilon, \qquad a_2^3 - a_1^3 \approx 3t_0a^2 \approx 3t_0a_0^2e^{-2\varepsilon}$$
$$\tau_2 - \tau_1 \approx \frac{t_0}{a}\left(\theta \frac{d\bar{\tau}}{d\theta}\right) \approx \frac{1}{2\pi a^3}\left(\theta \frac{dT}{d\theta}\right), \quad \tau_1 + \tau_2 \approx 2\bar{\tau}$$

where ε denotes the axial strain, which is approximately equal to the magnitude of the hoop strain, the elastic deformation being disregarded. The expression for the mean shear stress is therefore closely approximated by the formula

$$\bar{\tau} \approx \frac{T}{2\pi t_0 a_0^2}\left(1 + 2\varepsilon + \theta \frac{d\varepsilon}{d\theta}\right) \tag{6.42}$$

to a sufficient accuracy. The evaluation of the mean shear stress $\bar{\tau}$ requires the measurement of T, θ, and ε simultaneously at each stage of the loading. The corresponding mean shear strain $\bar{\gamma}$ is approximately equal to $\alpha_0\theta$, the change in radius being small. A more elaborate expression for the shear stress in the free-end torsion of a thin-walled tube has been given by Wu et al. (1997).

The shear stress–strain curve obtained from the torque-twist curve in the free-end torsion of a thin-walled tube, using (6.42), is found to coincide with that derived from the fixed-end torsion of the tube. The fixed-end torsion naturally gives rise to an axial compressive stress that suppresses the axial strain, while producing small amounts of circumferential and thickness strains in the twisted tube. Figure 6.8 shows the observed variation of the axial and hoop strains with the shear strain during the free-end torsion of an extruded aluminum tube, obtained experimentally by Wu (1996). The accumulation of axial strain in the finite torsion of tubular specimens with both free and fixed ends has been investigated by Wu et al. (1998).

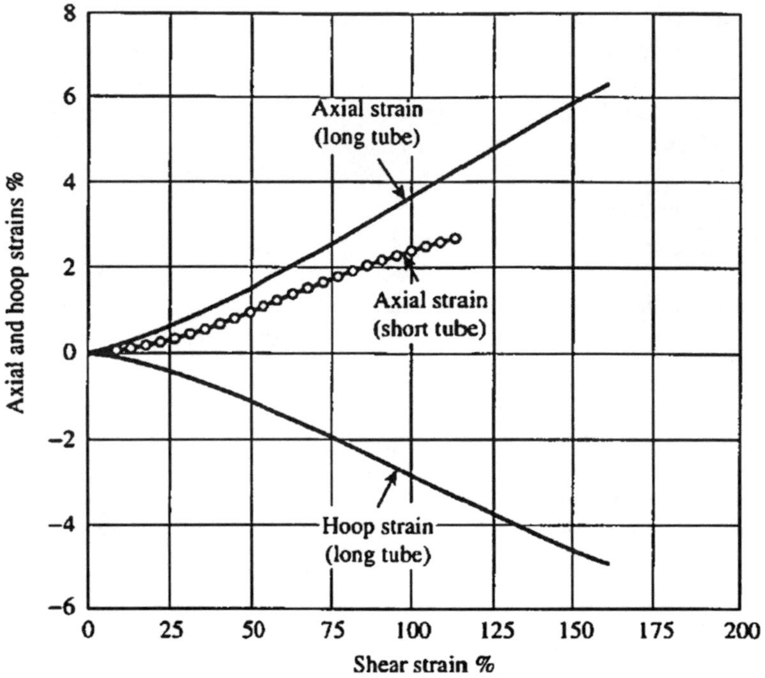

Fig. 6.8 Development of axial and hoop strains in finitely twisted thin-walled tubes in the plastic range (after H.C. Wu, 1996)

6.4 Plane Strain in Anisotropic Metals

6.4.1 Basic Equations in Plane Strain

Consider the class of problems in which the plastic flow is restricted in the plane perpendicular to the z-axis of anisotropy, which coincides with a principal axis of the strain rate. The condition $\dot{\gamma}_{xz} = \dot{\gamma}_{yz} = 0$ requires $\tau_{xz} = \tau_{yz} = 0$ by the flow rule, indicating that σ_z is a principal stress. Using the plane strain condition $d\varepsilon_z = 0$ in the third equation of (6.3), we get

$$\sigma_z = \frac{G\sigma_x + F\sigma_y}{G + F}.$$

In the special case when $\sigma_y = 0$, the above relation indicates that σ_z is greater or smaller than $\sigma_x/2$ according as G is greater or smaller than F. Substituting for σ_z in (6.1), and setting $\tau_{xz} = \tau_{yz} = 0$, the yield criterion is reduced to

$$\left(H + \frac{FG}{F + G}\right)(\sigma_x - \sigma_y)^2 + 2N\tau_{xy}^2 = 1.$$

It is convenient at this stage to introduce a dimensionless parameter c defined as

$$c = 1 - \frac{N(F+G)}{2(FG+GH+HF)} \quad (-\infty < c < 1). \tag{6.43}$$

Thus, c is positive when N is less than both $F + 2H$ and $G + 2H$, and negative when N is greater than both $F + 2H$ and $G + 2H$. If the anisotropy is rotationally symmetrical about the z-axis, $c = 0$. In view of (6.43), the yield criterion for anisotropic materials is expressed as

$$\frac{\left(\sigma_x - \sigma_y\right)^2}{4(1-c)} + \tau_{xy}^2 = k^2, \tag{6.44}$$

where $k = (2N)^{-1/2}$ is the yield stress in pure shear associated with the x- and y-axes. The yield stress σ in plane strain tension in the direction making an angle α with the x-axis is obtained by inserting (6.9a) into (6.44), the result being

$$\sigma = 2k\left\{\frac{1-c}{1-c\sin^2 2\alpha}\right\}^{1/2}.$$

Since σ has equal values in the directions ±α and $\pm(\pi/2-\alpha)$, the angular variation of σ is symmetrical about the x- and y-axes, as well as about their bisectors. When c is positive, σ has a minimum value of $2k\sqrt{1-c}$ along the anisotropic axes, and a maximum value of $2k$ in the 45° directions. When c is negative, σ has a maximum value of $2k\sqrt{1-c}$ along the anisotropic axes, and a minimum value of $2k$ in the 45°. A more general theory for plane strain has been discussed by Rice (1973).

Let v_x and v_y denote the components of velocity of a typical particle with respect to the anisotropic axes. The components of the strain rate in the plane of plastic flow referred to the anisotropic axes are

$$\dot{\varepsilon}_x = \frac{\partial v_x}{\partial x}, \quad \dot{\varepsilon}_y = \frac{\partial v_y}{\partial y}, \quad \dot{\gamma}_{xy} = \frac{1}{2}\left(\frac{\partial v_x}{\partial y} + \frac{\partial v_y}{\partial x}\right).$$

It follows from the flow rule associated with the yield criterion (6.44) that

$$\frac{2\dot{\gamma}_{xy}}{\dot{\varepsilon}_x - \dot{\varepsilon}_y} = (1-c)\left(\frac{2\tau_{xy}}{\sigma_x - \sigma_y}\right). \tag{6.45}$$

If ψ and ψ' denote the angles of inclination of a principal stress direction and the corresponding principal strain rate direction, respectively, with respect to the x-axis, then $2\psi' = (1-c)\tan 2\psi$ in view of (6.45). When $c \neq 0$, it follows that $\psi = \psi'$ only for $\psi = 0$, π/4, and π/2, as expected. In terms of the velocity gradients, (6.45) can be expressed as

$$\left(\sigma_x - \sigma_y\right)\left(\frac{\partial v_x}{\partial y} + \frac{\partial v_y}{\partial x}\right) = 2(1-c)\tau_{xy}\left(\frac{\partial v_x}{\partial x} - \frac{\partial v_y}{\partial y}\right).$$

The last equation, together with the incompressibility condition, the equilibrium equations, and the yield criterion, constitutes a set of five equations for the three stress components and the two velocity components. The parameters k and c in the theoretical framework can be experimentally determined by carrying out the plane strain compression test at 0° and 45° to one of the axes of anisotropy. As in the case of isotropic solids, the governing equations are hyperbolic with characteristics in the directions of maximum rate of shear at each point of the deforming zone. The characteristics for the stress are identical to those for the velocity, but these curves do not generally coincide with the trajectories of the maximum shear stress in the plane of plastic flow. Due to the incompressibility of the material, the rate of extension vanishes along the characteristics, which are known as the sliplines.

6.4.2 Relations Along the Sliplines

The two orthogonal families of sliplines will be designated by α and β following the convention that the acute angle made by the algebraically greater principal stress in the considered plane is measured counterclockwise with respect to the α-direction. If ϕ denotes the counterclockwise angle made by a typical α-line with the x-axis, as shown in Fig. 6.9(a), then $dy/dx = \tan \phi$ along an α-line and $dy/dx = -\cot \phi$ along a β-line. Since the left-hand side of (6.45) is equal to $-\cot \mathbf{2\phi}$, we have

$$\left(\sigma_x - \sigma_y\right) \cos 2\phi + 2\,(1 - c)\,\tau_{xy} \sin 2\phi = 0. \tag{6.46}$$

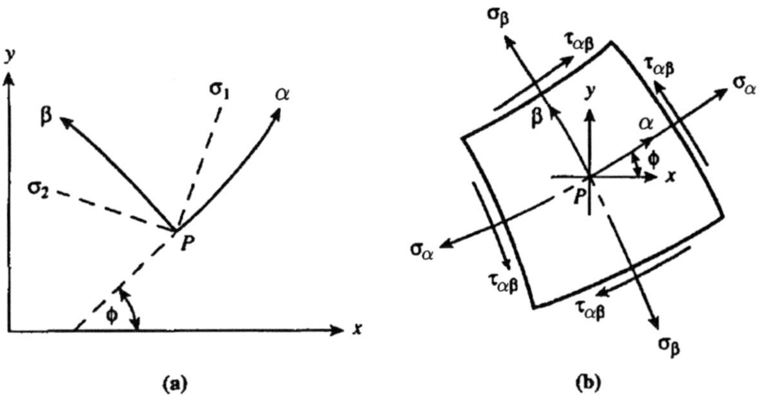

Fig. 6.9 Orientation of sliplines, and the stresses acting across them in a curvilinear element

It follows from (6.44) and (6.46) that the outward drawn normal to the yield locus, obtained by plotting τ_{xy} against $(\sigma_x - \sigma_y)/2$, makes a counterclockwise angle of 2ϕ with respect to the τ_{xy}-axis. This result actually holds for any convex yield function, under condition of plane strain, as has been shown by Rice (1973).

Equation (6.46) can be solved simultaneously with the yield criterion (6.44) to give

$$\frac{\tau_{xy}}{k} = \frac{\cos 2\phi}{\sqrt{1 - c\sin^2 2\phi}}, \quad \frac{\sigma_x - \sigma_y}{k} = -\frac{2(1-c)\sin 2\phi}{\sqrt{1 - c\sin^2 2\phi}}. \tag{6.47}$$

Let σ_α, σ_β, and $\tau_{\alpha\beta}$ denote the stress components referred to the local sliplines taken as a pair of curvilinear axes in the plane, as shown in Fig. 6.9(b). Then

$$\sigma_\alpha - \sigma_\beta = \left(\sigma_x - \sigma_y\right)\cos 2\phi + 2\tau_{xy}\sin 2\phi,$$
$$\tau_{\alpha\beta} = \left(\sigma_x - \sigma_y\right)\sin 2\phi - 2\tau_{xy}\cos 2\phi.$$

Substituting from (6.47), and setting $p = -(\sigma_\alpha + \sigma_\beta)/2$, which is the mean compressive stress in the plane of plastic flow, the (α, β) components of the stress can be expressed as

$$\left.\begin{aligned} \sigma_\alpha = -p - k\frac{ds}{d\phi}, \quad \sigma_\beta = -p + k\frac{ds}{d\phi}, \; \tau_{\alpha\beta} = 2ks, \\ 2s = \sqrt{1 - c\sin^2 2\phi}. \end{aligned}\right\} \tag{6.48}$$

If the radii of curvature of the α- and β-lines at a generic point are denoted by R_α and R_β, respectively, the equilibrium equations in the curvilinear coordinates (α, β) may be written as

$$\frac{\partial \sigma_\alpha}{\partial s_\alpha} + \frac{\sigma_\alpha - \sigma_\beta}{R_\beta} + \frac{\partial \tau_{\alpha\beta}}{\partial s_\beta} - \frac{2\tau_{\alpha\beta}}{R_\alpha} = 0,$$
$$\frac{\partial \sigma_\beta}{\partial s_\beta} + \frac{\sigma_\alpha - \sigma_\beta}{R_\alpha} + \frac{\partial \tau_{\alpha\beta}}{\partial s_\alpha} + \frac{2\tau_{\alpha\beta}}{R_\beta} = 0,$$

where s_α and s_β are the arc lengths along the respective sliplines. Since the curvatures $1/R_\alpha$ and $1/R_\beta$ are equal to $\partial\phi/\partial S_\alpha$ and $\partial\phi/\partial S_\beta$, respectively, while the normal stress difference $\sigma_\alpha - \sigma_\beta$ is equal to $= -d\tau_{\alpha\beta}/d\phi$ in view of (6.48), the second and third terms of each of the above equations cancel one another, while the remaining terms give

$$\frac{\partial}{\partial s_\alpha}\left(p + k\frac{ds}{d\phi}\right) + 4ks\frac{\partial \phi}{\partial s_\alpha} = 0,$$
$$\frac{\partial}{\partial s_\beta}\left(p - k\frac{ds}{d\phi}\right) + 4ks\frac{\partial \phi}{\partial s_\beta} = 0,$$

in view of (6.48). These equations can be expressed in the integrated form (Hill, 1949) to obtain the characteristic relations

$$\left.\begin{aligned} p + 2\,k\omega &= \text{constant along an}\,\alpha\;\;\text{line,} \\ p - 2\,k\omega &= \text{constant along a}\,\beta\;\;\text{line,} \end{aligned}\right\} \tag{6.49}$$

The parameter ω depends on the angle ϕ and the anisotropic parameter c according to the relation

$$2\omega = \frac{ds}{d\phi} + 4\int_0^{\phi} s d\phi = -\frac{c \sin 2\phi \cos 2\phi}{\sqrt{1 - c \sin^2 2\phi}} + E\left(2\phi, \sqrt{c}\right), \tag{6.50}$$

where E denotes the standard elliptic function of the second kind and is defined as

$$E\left(\alpha, m\right) = \int_0^{\alpha} \sqrt{1 - m^2 \sin^2\theta}\, d\theta.$$

For an isotropic material, $\omega = \phi$, and (6.49) reduces to the well-known Hencky equations for plane strain.

If the components of the velocity along the α- and β-lines are denoted by u and v, respectively, the condition of zero rate of extension along the sliplines can be easily reduced to

$$\left.\begin{aligned} du - v d\phi &= 0\,\text{along an}\,\alpha\text{–line,} \\ dv + u d\phi &= 0\,\text{along a}\,\beta\text{–line,} \end{aligned}\right\} \tag{6.51}$$

These relations are the same as the Geiringer equations for isotropic solids. In analogy to Hencky's first theorem, it is easily shown that the difference in the values of ω or p between a pair of points, where two given sliplines of one family are intersected by a slipline of the other family, remains constant along their lengths. It follows that if one segment of a slipline is straight, the corresponding segments of sliplines of the same family are also straight, and the straight segments are consequently of equal lengths.

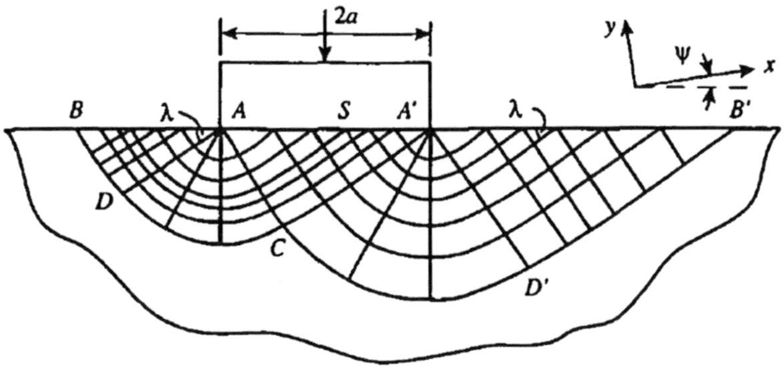

Fig. 6.10 Indentation of the plane surface of a semi-infinite anisotropic medium by a rigid flat punch

6.4.3 Indentation by a Flat Punch

Consider, as an example, the indentation of the plane surface of a large block of metal by a rigid flat punch. At the incipient plastic flow, the sliplines emanating from the corners A and A' of the punch reach the free surface of the block. The slipline field shown in Fig. 6.10 is an extension of the Prandtl field for isotropic solids. In the triangular regions ABD and $A'B'D'$, the state of stress is a uniform compression $2p_0$ parallel to the surface, where p_0 is the value of p in these regions. If ψ denotes the angle of inclination of the plane surface with the x-axis of anisotropy, it follows from (6.44) that

$$p_0 = k\left(\frac{1-c}{1-c\sin^2 2\psi}\right)^{1/2}. \tag{6.52}$$

If λ denotes the acute angle made by the α-lines with the plane surface, then $\phi = \lambda - \psi$ along AB and $A'B'$. Since the direction of the algebraically greater principal stress is normal to the plane, making an angle $\pi/2 - \psi$ with the x-axis, (6.46) gives

$$\cot 2(\lambda - \psi) + (1-c)\tan(\pi - 2\psi) = 0,$$

or (6.53)

$$\lambda = \psi + \tfrac{1}{2}\cot^{-1}[(1-c)\tan 2\psi].$$

It is easily shown that λ lies between $\cot^{-1}\sqrt{1-c}$ and $\pi/2 \cot^{-1}\sqrt{1-c}$ whatever the value of ψ. Since the state of anisotropy cannot change appreciably during the deformation preceding the yield point, we are justified in treating c as a constant throughout the plastic region.

It is assumed that the material beneath the punch is uniformly stressed with the α sliplines inclined at an angle λ to the punch face, regardless of the frictional condition. This is compatible with a rigid-body motion of the plastic triangle ACA' which is attached to the punch. The slipline field is completed by introducing the 90° centered fans ACD and $A'CD'$, where the sliplines are radial lines and circular arcs. From (6.49), the value of p in ACD increases from p_0 on AD to $p_0 + 2kE$ on AC, where

$$E = E\left(\frac{\pi}{2}, \sqrt{c}\right) = \int_0^{\pi/2} \sqrt{1 - c\sin^2 2\theta}\, d\theta.$$

The principal compressive stresses in the region ACA', directed normal and parallel to the punch face, are $2(p_0 + kE)$ and $2kE$, respectively. The normal pressure q on the punch face is therefore given by

$$\frac{q}{2k} = \frac{p_0}{k} + E = \left(\frac{1-c}{1-c\sin^2 2\psi}\right)^{1/2} + E\left(\frac{\pi}{2}, \sqrt{c}\right) \tag{6.54}$$

in view of (6.52). When the degree of anisotropy is small, it is convenient to expand P_0/k and E in ascending powers of c, and neglect all terms of order c^2 and above. Then the result becomes

$$\frac{q}{2k} \approx \left(1 + \frac{\pi}{2}\right) - \frac{c}{2}\left(\frac{\pi}{4} + \cos^2 2\psi\right).$$

For an isotropic solid ($c = 0$), the punch pressure reduces to the Prandtl value $2k(1+\pi/2)$. The punch pressure for an anisotropic solid is less than this when c is positive and greater than this when c is negative. It may be noted that the punch pressure is the same for orientations ψ and $\pi/2 - \psi$ of the axes of anisotropy, which is a consequence of the symmetry of the anisotropy about the directions equally inclined to the anisotropic axes.

The incipient deformation mode consists of plastic flow parallel to the sliplines CDB and $CD'B'$, the material below these boundaries being held rigid. The velocity is of magnitude $U \sin \lambda$ parallel to CDB, and of magnitude $U \cos \lambda$ parallel to $CD'B'$, where U is the downward speed of the punch. These are also the magnitudes of the velocity discontinuities across $A'C$ and AC, respectively. In an alternative solution, valid only for a smooth punch (Hill, 1950a), the bounding sliplines meet at a point S on the punch face. The magnitudes of the velocity discontinuity initiated at this point are then modified to U cosec λ and $U \sec \lambda$, while the punch pressure is still given by (6.54).

6.4.4 Indentation of a Finite Medium

Consider the indentation of the plane surface of a medium of finite depth h resting on a rigid smooth foundation. When the ratio h/a is sufficiently small, where a is the semiwidth of the punch, the plastic zone spreads downward from the punch face to reach the foundation at the yield point state. The slipline field shown in Fig. 6.11 (a) is developed from the triangle ABC, which is not an isosceles triangle unless AB coincides with an anisotropic axis. For an arbitrary angle ψ made by the x-axis of anisotropy with the free surface, the centered fans ACE and BCD are of unequal radii, their angular spans being θ and δ, respectively. The angle λ which the β-line makes with the punch face is given by (6.53). Since the foundation is smooth, the bounding β-line BDF meets the foundation at the same angle λ. The existence of an incipient velocity field at the yield point can be easily established following the method used for isotropic materials. By the analogue of

Hencky's first theorem, which is obtainable from (6.49), we have

$$\omega_D = \omega_F + \omega_C - \omega_E,$$

where the subscripts refer to the points in the slipline field. The right-hand side of the above equation can be evaluated for any assumed angle θ, using (6.50) and the relations

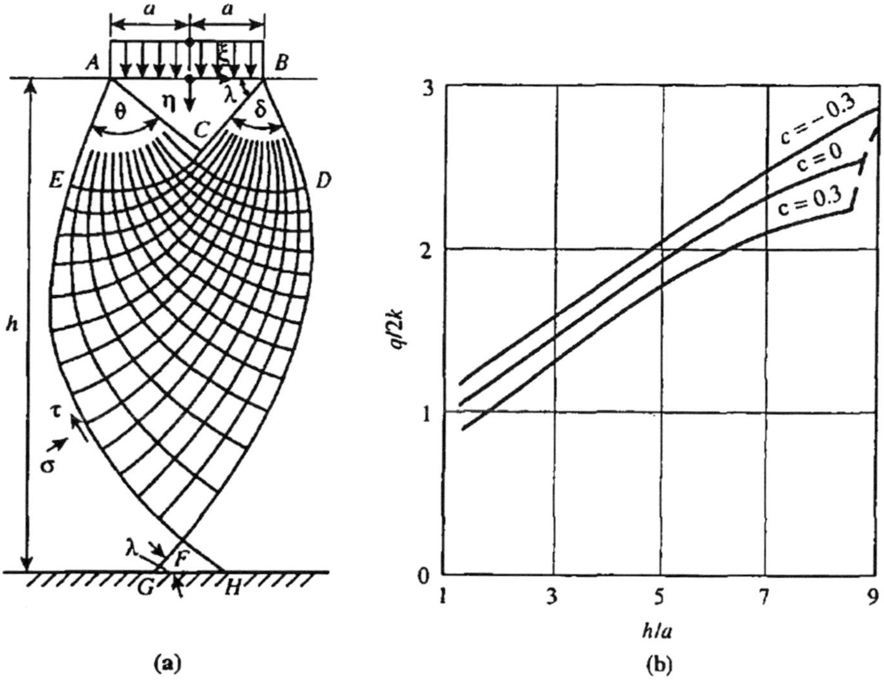

Fig. 6.11 Indentation of a block of anisotropic material of finite depth. (**a**) Slipline field, (**b**) variation of punch pressure with block thickness

$$\phi_F = \phi_C = -\left(\frac{\pi}{2} - \lambda + \psi\right), \quad \phi_C = \phi_E = \theta.$$

The other fan angle δ is then computed from (6.50) by a trial-and-error procedure, using the fact that ϕ_D–ϕ_D–$\phi_C = \delta$. The slipline field is generated from the given intersecting sliplines *CD* and *CE* by the usual small arc process of approximation. The construction is similar to that for isotropic solids (Chakrabarty, 1987), except that the value of ϕ at each unknown nodal point must be determined from the computed slipline variable ω.

The mean compressive stress p_F at F must be determined from the condition of zero horizontal force transmitted across the bounding sliplines *AF* and *BF*. Considering the α-line *AEF*, and a pair of rectangular axes (ξ, η) through the midpoint of *AB*, the resultant horizontal thrust exerted on *AEF* by the rigid material on its left may be written as

$$F = -\int \tau d\xi + \int \sigma d\eta = 2k\int s d\xi + \int p d\eta - k\int \left(\frac{ds}{d\phi}\right) d\eta,$$

where $\tau = \tau_{\alpha\beta}$ and $\sigma = -\sigma_\beta$, denoting the tangential stress and normal pressure on a slipline element, and the integrals are taken along the entire slipline *AEF*. Integrating the second term on the right-hand side by parts, and noting that $dp = -2k\,d\omega$ along an α-line, the boundary condition $F = 0$, which holds when point F lies on the foundation, is reduced to

$$h\left(\frac{p_F}{2k}\right) = \int_A^F s d\xi + \frac{1}{2}\int_A^F \left(\frac{ds}{d\phi}\right) d\eta - \int_E^F \eta d\omega \tag{6.55}$$

where $s(\phi)$ is given by (6.48). The integrals can be evaluated numerically for a selected value of θ since the values of ϕ, ω, ξ, and η are known along the slipline, to obtain a value of p_F using (6.55).

As h increases to a critical value h^*, the pressure p_F decreases to the value $-p_0$, where p_0 is given by (6.52). The normal pressure on the foundation then vanishes at F, an element at F being under a uniaxial tension $2p_0$ parallel to the foundation. For $h > h^*$, it is necessary to introduce a triangle FGH of height t with the inclined sides tangential to the sliplines at F. The triangle contributes a horizontal tensile force $2p_0t$, and the condition of zero resultant force across the boundary $AEFH$ again leads to (6.55), which furnishes h (and hence t) for a given θ. The indentation pressure q, uniformly distributed over the punch face, is given by

$$\frac{q}{2k} = \frac{p_C + p_0}{2k} = \frac{p_F + p_0}{2k} + (\omega_F + \omega_C - 2\omega_E)\,.$$

The first term on the right-hand side is positive for $h < h^*$ and zero for $h \geq h^*$. The solution holds only for those values of h/a for which $q/2k$, given by the above expression, is less than or equal to that given by (6.54). For higher h/a ratios, the slipline field of Fig. 6.10 applies and the indentation pressure then has the constant value (6.54). The variation of $q/2k$ with h/a for $\psi = 0$, obtained by Venter et al. (1971), is shown in Fig. 6.11(b) for three different values of c. The effect of a nonzero value of ψ is significant for the shape of the slipline field but has negligible effects on the indentation pressure for a given state of anisotropy. The theory is found to be in broad agreement with experiment. An analysis for the incipient extrusion through a wedge-shaped die has been given by Johnson et al. (1973).

6.4.5 Compression Between Parallel Platens

A block of anisotropic metal is compressed between a pair of perfectly rough platens whose width 2ω is large compared to the thickness $2h$. The face of each platen is assumed to be parallel to the xz-plane of anisotropy, and the compressed material between the platens is assumed to be completely plastic at the yield point state. In view of the large ω/h ratio, it is reasonable to look for a solution in which the slipline directions are independent of x. It follows from (6.45) that $\sigma_x - \sigma_y$ and x_{xy} are also independent of x, and we may write

$$\tau_{xy} = kf(y)\,, \quad \sigma_x - \sigma_y = 2k\sqrt{(1-c)\left(1-f^2\right)},$$

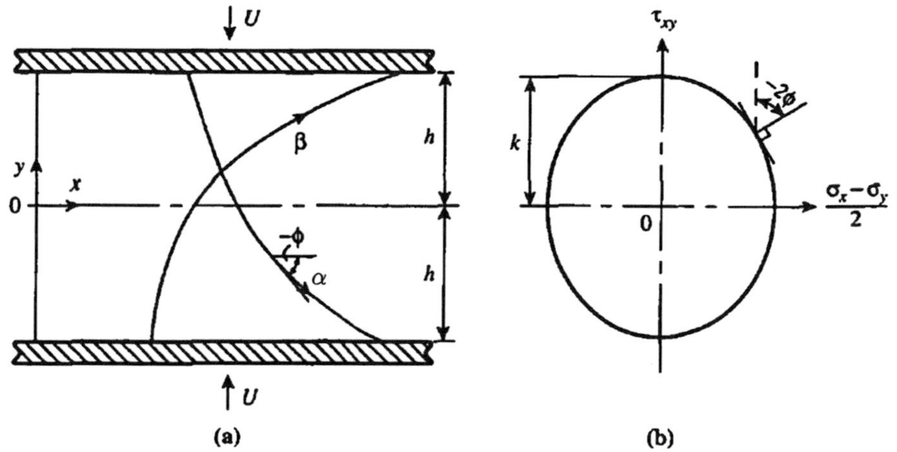

Fig. 6.12 Compression of a wide block between rough parallel platens. (**a**) A pair of intersecting sliplines, (**b**) plane strain yield locus

so that the yield criterion (6.44) is identically satisfied, the origin of coordinates being taken at the left-hand edge of the block, as shown in Fig. 6.12 (a). The conditions of equilibrium of a typical element along the coordinate axes then furnish

$$\frac{\partial \sigma_y}{\partial x} + kf'(y) = 0, \quad \frac{\partial \sigma_y}{\partial y} = 0.$$

These equations indicate that the function (y) is a linear function of y. In view of the boundary conditions $\tau_{xy} = 0$ along $y = 0$ and $\tau_{xy} = k$ along $y = h$, we have $f(y) = y/h$, and the stress distribution in the block becomes

$$\frac{\sigma_x}{k} = -\left(A + \frac{x}{h}\right) + 2\sqrt{(1-c)\left(1 - \frac{y^2}{h^2}\right)},$$
$$\frac{\sigma_y}{k} = -\left(A + \frac{x}{h}\right), \quad \frac{\tau_{xy}}{k} = \frac{y}{h}, \tag{6.56}$$

where A is a constant whose value is determined from the condition that the resultant horizontal thrust acting on any vertical section is balanced by the frictional resistance exerted by the platens. Thus

$$\int_{-h}^{h} \sigma_x dy = -kx \quad \text{or} \quad A = \frac{\pi}{2}\sqrt{1-c}.$$

It follows that the normal pressure on the platen is distributed linearly from $x = 0$ to $x = w$, the intensity of the pressure at the edge $x = 0$ being equal to $(\pi k/2)\sqrt{1-c}$ The mean pressure is

$$\bar{q} = -\frac{1}{w}\int_0^w \sigma_y dx = \frac{k}{2}\left(\pi\sqrt{1-c} + \frac{w}{h}\right).$$

The stress distribution given by (6.56) provides a good approximation for a block of finite width, except near the edge of the block where the solution is significantly modified by the presence of the stress-free boundary.

The counterclockwise angle ϕ, which the α-line at a generic point makes with the x-axis, can be determined from the fact that each ratio in (6.45) is equal to — cot 2ϕ, a geometrical representation of this result being indicated in Fig. 6.12(b). Substituting from (6.56) into (6.45), we have

$$\cot 2\phi = -y\sqrt{\frac{1-c}{h^2 - y^2}} = -\sqrt{1-c}\tan\theta,$$

where the last step from the substitution $y = h\ sin\theta$. The slope of the α-lines is therefore given by

$$\frac{dy}{dx} = \tan\phi = \sqrt{1-c}\ \tan\theta - \sqrt{\sec^2\theta - c\tan^2\theta}.$$

Let d denote the distance from the origin of the point of intersection of the considered slipline with the x-axis. Then the solution to the above equation can be written as

$$x = d - h\int_0^\theta \frac{\cos^2\theta\, d\theta}{\sqrt{1-c\sin^2\theta} - \sqrt{1-c}\sin\theta}, \tag{6.57a}$$

and the integration can be carried out numerically for any given c. The slope of the β-lines, on the other hand, is given by the expression

$$\frac{dy}{dx} = -\cot\phi = \frac{\cos\theta}{\sqrt{1-c\sin^2\theta} - \sqrt{1-c}\sin\theta}$$

On changing the dependent variable from y to $\theta = \sin^{-1}(y/h)$, this equation is readily integrated to give

$$x = d + h\left[E\left(\theta, \sqrt{c}\right) - \sqrt{1-c}\,(1-\cos\theta)\right], \tag{6.57b}$$

where $E\left(\theta,\sqrt{c}\right)$ represents the elliptic integral of the second kind. Equations (6.57) define the sliplines passing through any point $x = d$ on the axis of symmetry. The sliplines meet the platens normally and tangentially as in the case of compression of an isotropic block.

Let the speed of the platens moving toward each other be denoted by U. It is natural to look for a velocity solution in which the vertical component v_y is a linear function of y. In view of the incompressibility condition and the boundary condition, the velocity field may therefore be written as

$$v_x = U\left[\frac{x}{h} + g(y)\right], \quad v_y = -U\frac{y}{h}.$$

The function *g(y)* can be determined by substituting from above in (6.45) and integrating, the result being easily shown to be

$$g(y) = B - \sqrt{(1-c)\left(1 - \frac{y^2}{h^2}\right)},$$

where B is a constant. The velocity distribution at the incipient compression of the block therefore becomes

$$\frac{v_x}{U} = B + \frac{x}{h} - 2\sqrt{(1-c)\left(1 - \frac{y^2}{h^2}\right)}, \quad \frac{v_y}{U} = -\frac{y}{h}, \tag{6.58}$$

The constant B must be determined from the condition that the rate of horizontal flow of material across a vertical section is equal to the rate of displacement of material by the platens between this section and the central sections $x = w$. Thus

$$-\int_0^h v_x dy = (w - x)\,U, \quad \text{or} \quad B = -\frac{w}{h} + \frac{\pi}{2} + \sqrt{1-c}.$$

The velocity field (6.58) does not hold near the central vertical section, where a pair of wedge-shaped zones of rigid material must exist as part of the solution. A slipline field solution for the compression problem has been presented by Chitkara and Collins (1974), using a graphical technique.

The preceding solution also applies to the continued compression of a non-hardening block, provided *2w* and *2 h* are taken as the instantaneous width and thickness, respectively. Since the vertical velocity of a typical particle is proportional to *y/h,* the particle remains at the same relative distance from the platen and the axis of symmetry. At a generic stage of the compression process, let *f* denote the horizontal displacement of this particle relative to the one on the platen, originally situated in the same vertical section. Since the rate of change of *f* is equal to the difference between the horizontal velocities of the two considered particles, it follows from (6.58) that

$$\frac{1}{U}\frac{d\xi}{dl} = -\frac{d\xi}{dh} = \frac{\xi}{h} + 2\sqrt{(1-c)\left(1 - \frac{y^2}{h^2}\right)},$$

where *y/h* is a constant during the motion of the particle. Using the initial condition £ $= 0$ when $h = h_0$, which is the initial semithickness of the block, the above equation is integrated to give

$$\left\{\frac{\xi h}{\left(h_0^2 - h^2\right)\sqrt{1-c}}\right\}^2 + \frac{y^2}{h^2} = 1.$$

This is the equation of an ellipse into which an original vertical line is distorted during the compression. The ellipse is tangential to the platens at the extremities of its major axis as in the case of compression of an isotropic block. A solution to the problem of continued compression of a work-hardening anisotropic block has been given by Collins and Meguid (1977). The influence of rotation of the anisotropic axes during compression has been examined by Markin and Yakovlev (1996). The plane strain bending of anisotropic sheets for finite strains has been investigated by Wang et al. (1993), Tan et al. (1995). The elastic/plastic bending of anisotropic sheets under conditions of plane strain has been treated by Chakrabarty et al. (2000, 2001).

6.5 Anisotropy in Stretch Forming

6.5.1 Basic Equations for Biaxial Stretching

In the stretch forming of sheet metals, using either a fluid pressure or a rigid punch, the principal surface strains are both positive with their ratio lying between zero and unity over the whole surface. The analysis for the forming process when the sheet has normal anisotropy is similar to that for isotropic sheets (Section 2.7), the only difference being due to the inclusion of the R-value in the stress and velocity equations. The quadratic yield criterion (6.21) and the associated flow rule (6.22) for a work-hardening material can be identically satisfied by writing the principal stresses and strain rates as

$$\left.\begin{aligned} \sigma_1 &= \bar{\sigma}\left(\frac{1+R}{\sqrt{1+2R}}\cos\psi\right), \sigma_2 = \bar{\sigma}\left(\frac{R}{\sqrt{1+2R}}\cos\psi + \sin\psi\right), \\ \dot{\varepsilon}_1 &= \dot{\lambda}\left(\frac{\sqrt{1+2R}}{1+R}\cos\psi - \frac{R}{1+R}\sin\psi\right), \quad \dot{\varepsilon}_2 = \dot{\lambda}\sin\psi, \end{aligned}\right\} \tag{6.59}$$

where ψ is an auxiliary angle varying between 0 and $\cot^{-1}\sqrt{1+2R}$ when $\sigma_1 \geq \sigma_2$, and $\dot{\lambda}$ is a positive scalar equal to the equivalent strain rate $\dot{\bar{\varepsilon}}$ according to the hypothesis of work equivalence. If, on the other hand, the hypothesis of strain equivalence is adopted, $\dot{\lambda}$ is related to $\dot{\bar{\varepsilon}}$ according to the equation

$$\dot{\bar{\varepsilon}} = \dot{\lambda}\left\{\frac{(1+R)(2+R) - (R-1)\left(R\cos 2\psi + \sqrt{1+2R}\sin 2\psi\right)}{2\left(1+R+R^2\right)}\right\}, \tag{6.60a}$$

which is obtained by substituting from (6.59) into (6.23a), expressed in terms of the principal components. For $R \leq 1$, the modified yield criterion (6.24) and the associated flow rule (6.25) can be identically satisfied by taking

$$\left.\begin{aligned}\frac{\sigma_1+\sigma_2}{\bar\sigma}&=\left\{(1+R)\left(\frac{\dot\varepsilon_1+\dot\varepsilon_2}{\dot\lambda}\right)\right\}^{1/m}=\left(\sqrt{1+2R}\cos\psi+\sin\psi\right)^{1/\mu},\\ \frac{\sigma_1-\sigma_2}{\bar\sigma}&=\left\{\left(\frac{1+R}{1+2R}\right)\left(\frac{\dot\varepsilon_1-\dot\varepsilon_2}{\dot\lambda}\right)\right\}^{1/m}=\left(\frac{\cos\psi}{\sqrt{1+2R}}-\sin\psi\right)^{1/\mu},\end{aligned}\right\} \tag{6.61}$$

where $\mu = (1 + m)/2$. When the hypothesis of work equivalence is employed, $\dot\lambda$ can be replaced by the equivalent strain rate $\dot{\bar\varepsilon}$. For the strain-hardening hypothesis, on the other hand, the equivalent strain rate is given by (6.23), (6.25), and (6.61) as

$$\dot{\bar\varepsilon}=\dot\lambda\left\{\frac{3\,(c\cos\psi+\sin\psi)^{2m/\mu}+(\cos\psi/C-\sin\psi)^{2m/\mu}}{4\left(1+R+R^2\right)}\right\}, \tag{6.60b}$$

where $c = \sqrt{1+2R}$. It is convenient in this case to treat $\dot\lambda$ as a basic unknown of the problem. Since the state of stress can vary from balanced biaxial tension to that corresponding to plane strain, the relevant range of values of Ψ is

$$\tan^{-1}\left(\sqrt{1+2R}\right)^{1/m}-\tan^{-1}\sqrt{1+2R}\le\psi\le\cot^{-1}\sqrt{1+2R}.$$

The problem of expanding a central hole in a circular blank using a rigid flat punch has been investigated by Yamada and Koide (1968) and Parmar and Mellor (1978b). The related problem of radially symmetric loading of an infinite flat plate containing a circular hole has been treated by Durban (1986).

The theoretical analysis for stretch forming over a hemispherical punch, using the quadratic yield function and numerical techniques, has been considered by Kaftanoglu and Alexander (1970), and Wang (1970), using the quadratic yield criterion, and by Wang (1984) using the nonquadratic yield criterion. For the hydrostatic bulging process, the theory indicates that the polar height and the polar thickness depend only marginally on the R-value of the material. Numerical solutions to this forming process have been presented by Wang and Shammamy (1969) using the quadratic yield function, and by Ilahi et al. (1981) and Wang (1984) using the nonquadratic yield function. The initiation and growth of wrinkling in the general problem of sheet metal forming has been investigated by Hassani and Neale (1991) and Kim et al. (2000).

6.5.2 Plastic Instability in Tension

Consider the plastic instability of a plane sheet under biaxial tension based on the nonquadratic yield criterion (6.24), which may be used for all values of R with the appropriate choice of m. The stress ratio σ_1/σ_2, which may vary during the loading, can have any value between zero and unity. During an incremental deformation, the stress and strain increments are such that

$$d\sigma_1 d\varepsilon_1 + d\sigma_2 d\varepsilon_2 = d\sigma\,(d\varepsilon_1 + d\varepsilon_2) + d\tau\,(d\varepsilon_1 - d\varepsilon_2) = d\bar{\sigma}\,d\lambda,$$

where σ and τ are the mean normal stress and maximum shear stress in the plane, equal to $(\sigma_1+\sigma_2)/2$ and $(\sigma_1-\sigma_2)/2$, respectively. The last step follows from the flow rule (6.25), and the differentiated form of the yield criterion (6.24). If the applied loads simultaneously attain their maximum values at the onset of instability, $d\sigma_1/\sigma_1 = d\varepsilon_1$ and $d\sigma_2/\sigma_2 = d\varepsilon_2$ at this stage, and the preceding relation furnishes

$$\frac{d\bar{\sigma}}{d\lambda} = (\sigma + \tau)\left(\frac{d\varepsilon_1}{d\lambda}\right)^2 + (\sigma - \tau)\left(\frac{d\varepsilon_2}{d\lambda}\right)^2.$$

Inserting the expression for $d\varepsilon_1/d\lambda$ and $d\varepsilon_2/d\lambda$ in terms of σ and τ using the flow rule (6.25), we get

$$4\,(1+R)^2\,\frac{d\bar{\sigma}}{d\lambda} = \bar{\sigma}\left(\frac{2\sigma}{\bar{\sigma}}\right)^{1+2m}\left\{1 + 2\,(1+2R)\left(\frac{\tau}{\sigma}\right)^{1+m} + (1+2R)^2\left(\frac{\tau}{\sigma}\right)^{2m}\right\}.$$

Adopting the hypothesis of strain equivalence, and substituting from (6.25) into (6.23a), it is easily shown that (Chakrabarty, 1997)

$$2\frac{d\bar{\varepsilon}}{d\lambda} = \left(\frac{2\sigma}{\bar{\sigma}}\right)^m\left\{\frac{3 + (1+2R)^2\,(\tau/\sigma)^{2m}}{1+R+R^2}\right\}. \tag{6.62}$$

Combining the last two equations, and eliminating $2\sigma/\bar{\sigma}$ by means of the yield criterion (6.24), the condition for plastic instability is obtained as

$$\frac{1}{\bar{\sigma}}\frac{d\bar{\sigma}}{d\bar{\varepsilon}} = \frac{\sqrt{1+R+R^2}\left[1 + 2\,(1+2R)\,\beta^{1+m} + (1+2R)^2\,\beta^{2m}\right]}{(1+R)\left[1 + (1+2R)\,\beta^{1+m}\right]\sqrt{3 + (1+2R)^2\,\beta^{2m}}}, \tag{6.63}$$

where $m = 1$ for $R \geq 1$ and $m = R$ for $R \leq 1$, while β is a dimensionless quantity defined as

$$\beta = \tau/\sigma = (\sigma_1 - \sigma_2)\,/\,(\sigma_1 + \sigma_2)\,.$$

When the hypothesis of work equivalence is adopted, an analysis similar to above gives

$$\frac{1}{\bar{\sigma}}\frac{d\bar{\sigma}}{d\bar{\varepsilon}} = \frac{1 + 2(1+2R)\beta^{1+m} + (1+2R)^2\beta^{2m}}{[2\,(1+R)]^{1/(1+m)}\left[1 + (1+2R)\beta^{1+m}\right]^{(2+m)/(1+m)}} \tag{6.63a}$$

For an isotropic sheet metal, (6.63) and (6.63a) reduce to (2.38). If the stress–strain curve in the plane of the sheet is represented by the power law $\sigma = C\varepsilon^n$, the left-hand side of (6.63) and (6.63a) is equal to $n/\bar{\varepsilon}$. When the stress ratio is maintained constant throughout the loading, the ratios $\varepsilon_1/\bar{\varepsilon}$ and $\varepsilon_2/\bar{\varepsilon}$ associated with (6.63) follow from (6.25) and (6.62), and the instability strains become

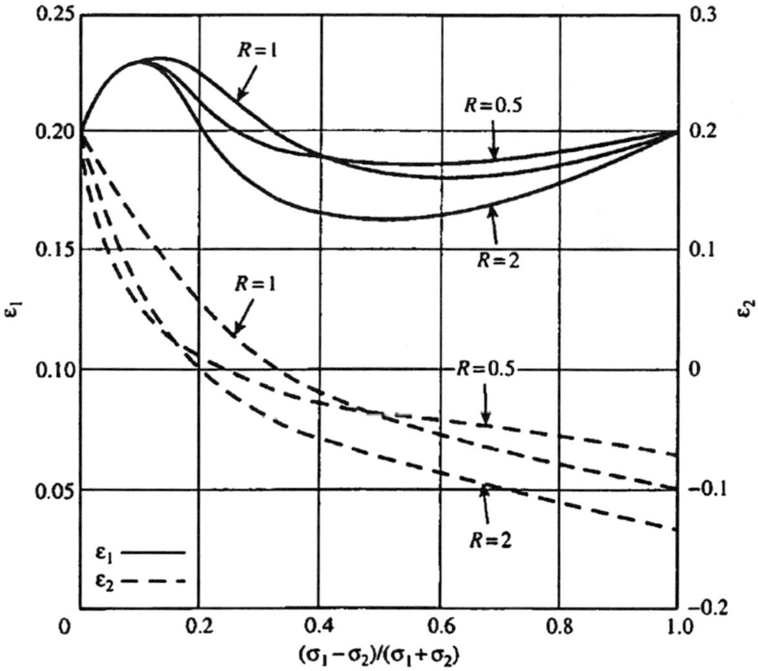

Fig. 6.13 Principal surface strains at the onset of plastic instability of a plane sheet under biaxial tension

$$\varepsilon_1, \varepsilon_2 = \frac{n\left[1 + (1 + 2R)\,\beta^{l+m}\right]\left[1 \pm (1 + 2R)\,\beta^m\right]}{1 + 2\,(1 + 2R)\,\beta^{1+m} + (1 + 2R)^2\,\beta^{2m}}.$$

In Fig. 6.13, the principal strains ε_l and ε_2 at instability for $n = 0.2$ are plotted against β over the range $0 \leq B \leq 1$, which covers stress states varying between uniaxial tension and balanced biaxial tension. It is interesting to note that ε_1 is equal to n when $\beta = 0$, 1, and $(1{+}2R)^{-1/m}$, representing equibiaxial tension, uniaxial tension, and plane strain, respectively. Plastic instability of pressurized tubes has been examined by Yamada and Aoki (1966), and Stout and Hecker (1983).

6.5.3 Forming Limit Diagram

In the plastic forming of sheet metals over rigid punch heads, the strain to failure cannot generally be predicted by the instability condition for a plane sheet. Experiments have revealed that the formability of the sheet metal depends largely on the stress ratio existing in the neighborhood of the failure site (Keeler, 1965). The presence of friction between the material and the punch produces a large thickness gradient in a critically stressed element where the deformation becomes localized as the stretching continues. A graphical plot of one of the principal surface strains against the other at the instant of failure represents the forming limit diagram, which defines the boundary between the safe and the unsafe domains. The construction of the forming limit diagram involves the measurement of major and minor axes of the appropriate ellipses generated from circles marked on the undeformed sheet. Various experimental techniques for the derivation of the forming limit diagram have been described by Azrin and Backofen (1970), Hecker (1975), Tadros and Mellor (1978), and Raghavan (1995).

For the theoretical prediction of the limit strain, which depends on a number of factors including the frictional condition and the strain history for the critical element, it is convenient to consider an equivalent problem of in-plane stretching of a sheet containing an initial imperfection. Following Marciniak and Kuczyniski (1967), the material imperfection in the idealized problem is assumed to be in the form of a narrow groove where the thickness is slightly smaller than that of the parent sheet. The simplified model also assumes the groove to run perpendicular to the direction of the major principal stress $\sigma_\backslash$. The groove and the adjacent material deform simultaneously, the eventual failure being caused by a process of strain localization in the groove. Adopting the modified yield criterion (6.24) and the associated flow rule (6.25), we have

$$(1+R)\left(\frac{d\varepsilon_1 + d\varepsilon_2}{d\lambda}\right) = \left(\frac{\sigma_1 + \sigma_1}{\bar{\sigma}}\right)^m \quad \text{or} \quad (1+R)\frac{d\varepsilon}{d\lambda} = \left(\frac{2\sigma}{\bar{\sigma}}\right)^m,$$

where $d\varepsilon$ is the compressive thickness strain increment, equal to $d\varepsilon_I + d\varepsilon_2$, and σ is the mean tensile stress in the plane, equal to $(\sigma_I + \sigma_2)/2$. Combining the above relation with (6.62) gives

$$\frac{\overline{d\varepsilon}}{d\varepsilon} = \frac{1+R}{2}\left\{\frac{3+(1+2R)^2(\tau/\sigma)^{2m}}{1+R+R^2}\right\}^{1/2}, \tag{6.64}$$

where $r = (\sigma_I—\sigma_2)/2$ is the maximum shear stress in the plane. Since m should approach unity as R tends to 1, we shall take $m = 1$ for $R \geq 1$ and $m = R$ for $R \leq 1$ in the numerical evaluation of the limit strains.

The stresses and strains in the groove and the adjacent material, represented by regions A' and A in Fig. 6.14, will be denoted by primed and unprimed quantities, respectively. If the strain increment parallel to the groove is assumed to be continuous, then $d\varepsilon_2' = d\varepsilon_2$, which leads to the relation

$$d\varepsilon - d\varepsilon' = (d\varepsilon_1 - d\varepsilon_2) - \left(d\varepsilon_1' - d\varepsilon_2'\right).$$

In view of the flow rule (6.25) applied to the material in the regions outside and inside the groove, we have

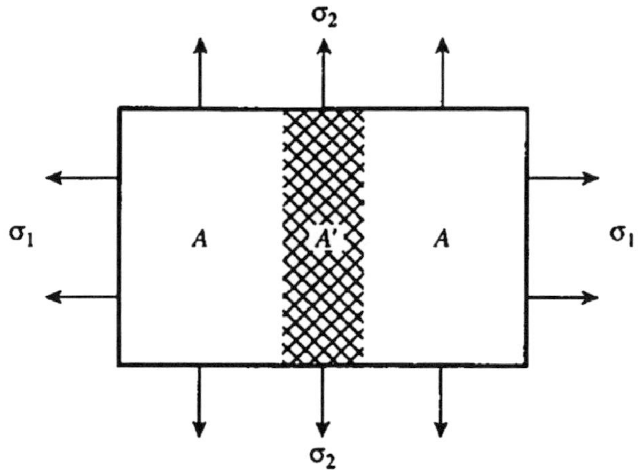

Fig. 6.14 Biaxial stretching of a sheet element having an initial inhomogeneity in thickness

$$d\varepsilon_1 - d\varepsilon_2 = (1+2R)\left(\frac{\sigma_1 - \sigma_2}{\sigma_1 + \sigma_2}\right)^m d\varepsilon, \quad d\varepsilon_1' - d\varepsilon_2' = (1+2R)\left(\frac{\sigma_1' - \sigma_2'}{\sigma_1' - \sigma_2'}\right)^m d\varepsilon'.$$

Inserting these expressions into the preceding one, the ratio of the two thickness strain increments is obtained as

$$\frac{d\varepsilon}{d\varepsilon'} = \frac{1-(1+2R)\left(\tau'/\sigma'\right)^m}{1-(1+2R)\left(\tau/\sigma\right)^m} = \frac{1-(1+2R)\,\beta^m}{1-(1+2R)\,\alpha^m}, \tag{6.65}$$

where

$$\alpha = \frac{\tau}{\sigma} = \frac{\sigma_1 - \sigma_2}{\sigma_1 + \sigma_2}, \quad \beta = \frac{\tau'}{\sigma'} = \frac{\sigma_1' - \sigma_2'}{\sigma_1' + \sigma_2'}.$$

Let the initial thickness of the sheet and the groove be denoted by t_0 and ηt_0, respectively, where η is an empirically assigned factor less than unity. If t and t' denote their current thicknesses, outside and inside the groove, then

$$t = t_0 \exp\left(-\varepsilon\right), \quad t' = \eta t_0 \exp\left(-\varepsilon'\right).$$

The condition of force equilibrium in the direction normal to the groove requires $t\sigma_1 = t'\sigma'$. Since $\sigma_1 = \sigma + \tau$ and $\sigma_1' = \sigma' + \tau'$, the result is

$$\frac{t'}{t} = \frac{\sigma + \tau}{\sigma' + \tau'} = \left(\frac{1+\alpha}{1+\beta}\right)\frac{\sigma}{\sigma'}.$$

The yield criterion (6.24) applied to the material inside and outside the groove furnishes the relation

$$\left(\frac{\sigma}{\sigma'}\right)^{1+m}\left\{\frac{1+(1+2R)\,\alpha^{1+m}}{1+(1+2R)\,\beta^{1+m}}\right\}=\left(\frac{\bar{\sigma}}{\bar{\sigma}'}\right)^{1+m}.$$

Combining the last three equations, and using the power law $\bar{\sigma} = C\bar{\varepsilon}^n$ for the strain hardening, it is easily shown that

$$\frac{1+(1+2R)\,\beta^{1+m}}{1+(1+2R)\,\alpha^{1+m}}=\left\{\eta\left(\frac{1+\beta}{1+\alpha}\right)\left(\frac{\bar{\varepsilon}'}{\bar{\varepsilon}}\right)^n\exp\left(\varepsilon-\varepsilon'\right)\right\}^{1+m}. \tag{6.66}$$

If α is maintained constant throughout the process, then the logarithmic differentiation of (6.66) with respect to ε' furnishes

$$\begin{aligned}\frac{d\beta}{d\varepsilon'}&=(1+\beta)\left[1+(1+2R)\,\beta^{1+m}\right]\\&\quad\times\left\{\frac{1-\left(n/\bar{\varepsilon}'\right)\left(d\bar{\varepsilon}'/d\varepsilon'\right)}{1-(1+2R)\,\beta^m}-\frac{1-(n/\varepsilon)}{1-(1+2R)\,\alpha^m}\right\},\end{aligned} \tag{6.66a}$$

in view of (6.65) and the fact that $d\bar{\varepsilon}/d\varepsilon = \bar{\varepsilon}/\varepsilon$ for a constant α. Equations (6.65), (6.66a), and the primed counterpart of (6.64) form a set of three first-order differential equations for finding β, ε, and $\bar{\varepsilon}'$ as functions of ε'. The initial conditions are $\varepsilon = \bar{\varepsilon}' = 0$ and $\beta = \beta_0$ when $\varepsilon' = 0$, where β_0 follows from (6.66) with $\bar{\varepsilon}' / \bar{\varepsilon} = 1$. The principal surface strains outside the groove are finally obtained from the relations

$$\varepsilon_1, \varepsilon_2 = \left[1 \pm (1+2R)\,\alpha^m\right](\varepsilon/2).$$

The thickness strain ε attains a limiting value when $d\varepsilon/d\varepsilon'$ tends to zero. The computation is therefore terminated when β is sufficiently close to the plane strain value $\beta^* = (1+2R)^{-1/m}$ wich corresponds to $d\varepsilon_2' = 0$. When the entire deformation occurs under plane strain, so that $\alpha = \beta = \beta^*$ throughout the straining, the limit strain corresponds to plastic instability of the material in the groove and is given by

$$(\varepsilon/n)^n \exp(n-\varepsilon) = \eta \quad \text{for all } R.$$

A similar analysis has been presented by Parmar and Mellor (1978a). Figure 6.15 shows the forming limit diagram obtained by plotting the limiting principal strain $\varepsilon_1$and ε_2 against one another. The solid curves are derived from the above analysis using $n = 0.2$ and $\eta = 0.98$, and assuming $m = 1$. The lower broken curve for $R = 0.5$ is based on $m = 0.5$. The upper broken curve for $R = 0.5$, which is based on $m = 1$, indicates the limitation of the quadratic law for low R-value materials. The marked dependence of the limit strain on the R-value, which is not supported

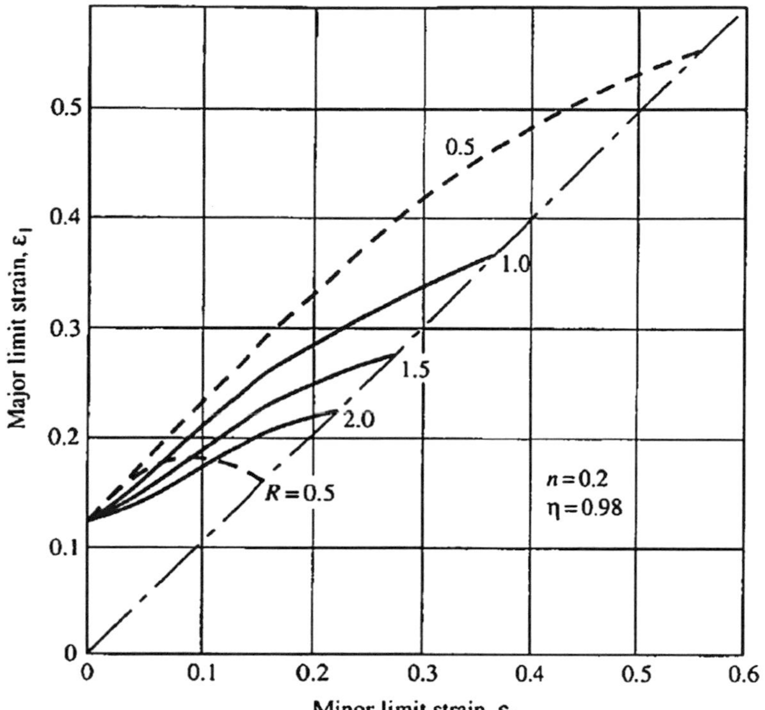

Fig. 6.15 Theoretical forming limit diagram showing the influence of normal anisotropy on the magnitude of the limit strain (after Sowerby and Duncan, 1971)

by experiments, can be reduced by assuming an appropriate variation of η with α and R for any given n-value of the material.

Theoretical results for the limit strain based on the quadratic yield function have been examined by Sowerby and Duncan (1971), Marciniak et al. (1973), Ghosh (1977), Lee and Zaverl (1982), Marciniak and Duncan (1992), among others. Solutions based on the Hencky stress–strain relations along with the assumption of the development of a yield vertex have been given by Hutchinson and Neale (1978) and Neale and Chater (1980). Other types of nonquadratic yield functions have been considered by Chan (1985), Lian et al. (1989), and Graf and Hosford (1990). The effect of void growth on the limit strain has been discussed by Kim and Kim (1983). The influence of strain path on the forming limit has been examined by Chu (1984), Graf and Hosford (1993), and Zhao et al. (1996). The effect of grain size on the limit strain has been examined in detail by Yamaguchi and Mellor (1976), and Wilson et al. (1981), while a crystallographic method of finding the limit strain has been discussed by Lee and Chan (1991). A stress-based forming limit diagram, as opposed to the strain-based ones, has been proposed by Stoughton (2000).

6.6 Anisotropy in Deep Drawing

6.6.1 The Radial Drawing Process

The theoretical investigation of the deep drawing of anisotropic circular blanks is similar to that for isotropic blanks (Section 2.8). A great deal of useful information can be derived by considering the radial drawing of the flange in the presence of normal anisotropy. Since the state of stress in the flange varies between uniaxial compression and uniaxial tension, the quadratic yield function may be employed not only for $R \geq 1$ but also for $R \leq 1$ as a reasonable approximation. The yield criterion and the flow rule are then identically satisfied by writing the radial and circumferential components of the stress and strain rate as

$$\left.\begin{array}{ll} \sigma_r = \sigma\left(\frac{1+R}{\sqrt{1+2R}}\sin\psi\right) & \sigma_\theta = -\sigma\left(\cos\psi - \frac{R}{\sqrt{1+2R}}\sin\psi\right) \\ \dot{\varepsilon}_r = \frac{\dot{\lambda}}{1+R}\left(R\cos\psi + \sqrt{1+2R}\sin\psi\right) & \dot{\varepsilon}_\theta = -\dot{\lambda}\cos\psi \end{array}\right\}, \tag{6.67}$$

where σ is the current uniaxial yield stress in the plane of the sheet, $\dot{\lambda}$ is a positive scalar, and Ψ is an auxiliary angle lying between 0 and $\cot^{-1}\left[R/\sqrt{1+2R}\right]$. The equations of equilibrium and strain rate compatibility are

$$\frac{\partial}{\partial r}(h\sigma_r) = \frac{h}{r}(\sigma_\theta - \sigma_r), \quad \frac{\partial \dot{\varepsilon}_\theta}{\partial r} = \frac{1}{r}(\dot{\varepsilon}_r - \dot{\varepsilon}_\theta), \tag{6.68}$$

where σ is the current radius to an element, and h its current thickness. The initial radius and thickness of the flange are denoted by b_0 and h_0, respectively, and the rim displacement $b_0 - b$ is taken as the time scale, where b is the current radius of the flange. The stress boundary condition requires $\sigma_r = 0$ at $r = b$.

Let r_0 denote the initial radius to a particle which is currently at a radius r. The components of the total true strain in a typical element, so long as it remains in the flange, are given by the expressions

$$\varepsilon_r = \ln\left(\frac{\partial r}{\partial r_0}\right), \quad \varepsilon_0 = \ln\left(\frac{r}{r_0}\right), \quad \varepsilon_z = \ln\left(\frac{h}{h_0}\right) = -(\varepsilon_r + \varepsilon_\theta).$$

Substituting from (6.67) into (6.68), and using the above relations, we obtain the first-order differential equations

$$\begin{aligned} \xi\frac{\partial}{\partial\xi}\left[s\sin\psi\exp(\varepsilon_z)\right] &= -\frac{s}{1+R}(\sin\psi + c\cos\psi)\exp(-2\varepsilon_\theta), \\ \xi\frac{\partial}{\partial\xi}\left(\dot{\lambda}\cos\psi\right) &= -\frac{c\dot{\lambda}}{1+R}(\sin\psi + c\cos\psi)\exp(\varepsilon_r - \varepsilon_\theta), \end{aligned} \tag{6.69}$$

where c denotes the constant $\sqrt{1+2R}$ while $\xi$and s are dimensionless variables defined as

$$\xi = r_0/b_0, \quad s = \sigma/\sigma_0 = \varepsilon^n,$$

with σ_0 and n denoting empirical constants in the assumed power law of hardening. The effective strain ε is such that $\dot{\varepsilon} = \dot{\lambda}$ when the work-hardening hypothesis is adopted. The strain-hardening hypothesis, on the other hand, requires

$$\dot{\varepsilon} = \dot{\lambda}\left\{1 + \frac{(R-1)\sin\psi\,(c\cos\psi - R\sin\psi)}{1+R+R^2}\right\}^{1/2} \tag{6.70}$$

in view of (6.23) and (6.67). Equations (6.69) must be solved numerically under the boundary conditions $\Psi = 0$, $\dot{\lambda} = \dot{\varepsilon} = 1/b$, and $\varepsilon = \ln(b_0/b)$ at $\xi = 1$. The solution at each stage is relevant over the range of $a \le r \le b$, where a is the radius of the die throat where the drawn material leaves the flange to enter into the die. Useful experimental results on deep drawing have been reported by Yoshida and Miyauchi (1978).

As in the case of drawing of isotropic blanks (Section 2.8), the solution for sufficiently small strains can be obtained analytically by setting $\varepsilon_r = \varepsilon_\theta - \varepsilon_z = 0$ in (6.69). Assuming $\dot{\lambda} = \dot{\varepsilon}$ for the initial problem, and using the total strain theory that allows $\dot{\varepsilon}$ to be replaced by ε, equations (6.69) are reduced to

$$\frac{n}{\varepsilon}\frac{\partial\varepsilon}{\partial\xi} + \cot\psi\frac{\partial\psi}{\partial\xi} = -\frac{1+c\cot\psi}{(1+R)\,\xi},$$
$$\frac{1}{\varepsilon}\frac{\partial\varepsilon}{\partial\xi} - \tan\psi\frac{\partial\psi}{\partial\xi} = -\frac{c\,(c+\tan\psi)}{(1+R)\,\xi},$$

in view of the hardening law $s = \varepsilon^n$ Eliminating $\partial\varepsilon/\partial\xi$ and $\partial\psi/\partial\xi$ between these equations in turn, we get

$$\frac{\partial\psi}{\partial\xi} = -\frac{(1-nc\tan\psi)\,(c+\tan\psi)}{\xi\,(1+R)\left(1+n\tan^2\psi\right)}, \quad \frac{\partial\varepsilon}{\partial\xi} = -\frac{\varepsilon\,(c+\tan\psi)^2}{\xi\,(1+R)\left(1+n\tan^2\psi\right)}. \tag{6.71}$$

The first equation of (6.71) can be integrated using the boundary condition $\psi = 0$ at $\xi = 1$, the result being

$$\xi^2 = \frac{\exp\left[-c\left(1-n^2\right)\psi/\left(1+c^2n^2\right)\right]}{\cos\psi + c^{-1}\sin\psi}\,(\cos\psi - nc\sin\psi)^{2n(1+R)/\left(1+c^2n^2\right)}. \tag{6.72}$$

This equation gives f as a function of 0. The elimination of f between the two equations of (6.71) leads to

$$\frac{d\varepsilon}{d\psi} = \left[\frac{c\cos\psi + \sin\psi}{\cos\psi - nc\sin\psi}\right]\varepsilon,$$

and the integration of this equation with the boundary condition $\varepsilon = ln\ (b_0/b)$ at $\xi = 1$ furnishes

$$\varepsilon = \ln\left(\frac{b_0}{b}\right)\left\{\frac{\exp\left[c\left(1-n^2\right)\psi/\left(1+c^2n\right)\right]}{\cos\psi - nc\sin\psi}\right\}^{\left(1+c^2n\right)/\left(1+c^2n^2\right)}. \qquad (6.73)$$

The distributions of stresses and strains for an arbitrary small value of $ln(b_0/b)$ are given by (6.67), (6.72), and (6.73). The solution may be continued for large strains by the numerical integration of (6.69) using a step-by-step procedure. For $n > 0$, the drawing load per unit circumference, which is the value of $h\sigma_r$ at $r = a$, attains a maximum at some stage of the drawing. The maximum load, computed by Budiansky and Wang (1966), is plotted in Fig. 6.16 as a function of the inverse drawing ratio for various values of the strain-hardening exponent n.

For an ideally plastic material, the stress and velocity distributions are obtained in closed form if the thickness variation in the equilibrium equation is disregarded.

The substitution from (6.67) into the first equation of (6.68) and the elimination of $\dot{\lambda}$ between the strain rate relations furnish

$$r\frac{\partial\psi}{\partial r} = -\frac{c+\tan\psi}{1+R}, \quad r\frac{\partial v}{\partial r} = -\left(\frac{R+c\tan\psi}{1+R}\right)v, \qquad (6.74)$$

where v is the radial velocity of a typical particle. The first equation of (6.74) is readily integrated under the boundary condition $\Psi = 0$ at $r = b$, giving

$$\frac{r}{b} = \left(\cos\psi + \frac{1}{c}\sin\psi\right)^{-1/2}\exp\left(-\frac{c\psi}{2}\right). \qquad (6.75)$$

Eliminating r between the two equations of (6.74), we obtain $dv/d\psi$ in terms of v and ψ. The integration of this equation, using the boundary condition $v = -1$ at $\psi = 0$, furnishes

$$v = -\left(\cos\psi + \frac{1}{c}\sin\psi\right)^{-1/2}\exp\left(\frac{c\psi}{2}\right). \qquad (6.76)$$

The thickness strain rate $\dot{h}/h$ equal to $-(\dot{\varepsilon}_r + \dot{\varepsilon}_\theta)$ by the condition of incompressibility. The substitution from (6.67) therefore gives

$$\frac{dh}{h} = -\left(1+\frac{\dot{\varepsilon}_r}{\dot{\varepsilon}_\theta}\right)\frac{dr}{r} = \left\{-2+\frac{c\left(c+\tan\psi\right)}{1+R}\right\}\frac{dr}{r}.$$

Taking the derivative of (6.75) following the particle, and using the fact that $dr/db = (r/b)\exp\left(\Psi\right)$ in view of (6.75) and (6.76), we get

$$\frac{dr}{r} = -\frac{(1+R)\,d\psi}{(c+\tan\psi)\left[1-\exp\left(-c\psi\right)\right]} = \exp\left(c\psi\right)\frac{db}{b}. \qquad (6.77)$$

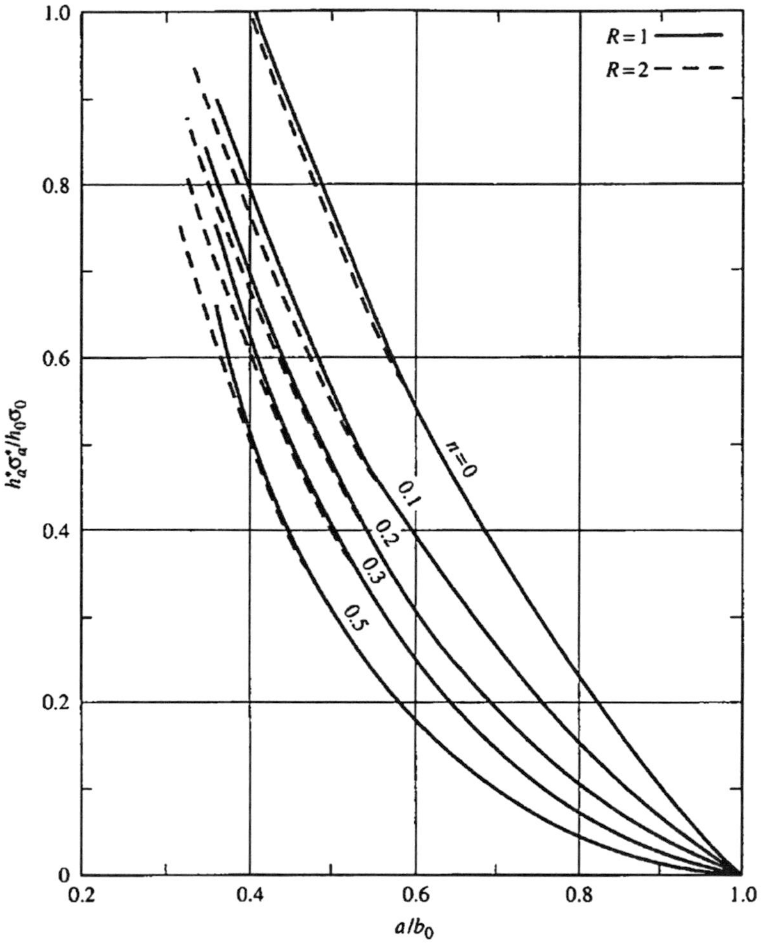

Fig. 6.16 Dimensionless maximum drawing load as a function of inverse of the drawing ratio (after Budiansky and Wang, 1966)

This is the relationship between *dr*, *db*, and dΨ in a typical element following its motion. The preceding equation for the thickness change may therefore be expressed as

$$\frac{dh}{h} = -2\frac{db}{b} + \frac{2(1+R)\,d\psi}{c+\tan\psi} - \frac{c d\psi}{1-\exp(-c\psi)}.$$

Using the initial conditions $h = h_0$ and $\Psi = \Psi_0$ when $b = b_0$, this equation can be integrated to give

$$\frac{h}{h_0} = \left(\frac{b_0}{b}\right)^2 \left(\frac{\sin\psi + c\cos\psi}{\sin\psi_0 + c\cos\psi_0}\right) \left\{\frac{1-\exp(-c\psi_0)}{1-\exp(-c\psi)}\right\}. \tag{6.78}$$

The relationship between Ψ_0 and Ψ for any given value of b/b_0 can be found by the numerical integration of (6.77). The initial radius r_0 to any given element, specified by the angle ψ_0, can be found from (6.75) considered in the initial state $b = b_0$. Detailed numerical investigations on the deep drawing process, including anisotropy and work-hardening, have been reported by Saran et al. (1990) and Keck et al. (1990).

6.6.2 Use of the Linearized Yield Condition

Over the range of stress ratios that are relevant to the radial drawing process, the von Mises-type quadratic yield criterion for an anisotropic sheet metal with $R > 1$ can be closely approximated by the Tresca-type yield criterion, resulting in a considerable simplification of the analysis. We shall use the Tresca criterion in the modified form

$$\sigma_r - \sigma_\theta = mY, \quad 1 < m < (1 + R)/\sqrt{+2R}$$

the material being assumed as nonhardening to begin with. If the thickness variation is disregarded in the equation of equilibrium, we immediately obtain the stress distribution

$$\sigma_r = mY \ln\left(\frac{b}{r}\right), \quad \sigma_0 = -mY\left[1 - \ln\left(\frac{b}{r}\right)\right], \tag{6.79}$$

as in the case of isotropic materials. The associated velocity distribution for the anisotropic sheet metal is obtained by integrating the differential equation

$$\frac{\partial v}{\partial r} = \left\{\frac{(1+R)\,\sigma_r - R\sigma_\theta}{(1+R)\,\sigma_\theta - R\sigma_r}\right\}\frac{v}{r} = -\left\{\frac{R + \ln(b/r)}{(1+R) - \ln(b/r)}\right\}\frac{v}{r},$$

which follows from the flow rule corresponding to the stress distribution (6.79). In view of the boundary condition $v = -1$ at $r = b$, the solution is easily shown to be

$$v = -\frac{r}{b}\left\{1 - \frac{1}{1+R}\ln\left(\frac{b}{r}\right)\right\}^{-(1+2R)}. \tag{6.80}$$

To obtain the paths of the particles during the radial drawing, consider the variation of the parameter $\rho = ln(b/r)$ following the motion of the particle. Since $v = -dr/db$, we have

$$d\rho = \frac{db}{b} - \frac{dr}{r} = -\left(1 + \frac{r}{vb}\right)\frac{dr}{r}.$$

Substituting from (6.80), the preceding relation can be expressed with close approximation as

$$\frac{dr}{r} \approx \frac{(1+R)\,d\rho}{\gamma\rho\,[(1+R)-R\rho]}, \quad \gamma = \frac{1+2R}{1+R}.$$

Integrating, and using the initial condition r = r_0 when $\rho = \rho_0 = \ln(b_0/r_0)$, we obtain the solution

$$\frac{r_0}{r} = \left\{ \frac{\ln(b/r)\left[(1+R) - R\ln(b_0/r_0)\right]}{\ln(b_0/r_0)\left[(1+R) - R\ln(b/r)\right]} \right\}^{1/\gamma}. \tag{6.81}$$

The computation of the ratio b_0/r_0 for any given b/r and b/b_0 is greatly simplified by substituting an approximate value of b_0/r_0 on the right-hand side of (6.81), based on the assumption of a constant flange thickness equal to that at $r = b$. Then, by the condition of plastic incompressibility,

$$b_0^2 - r_0^2 = \left(b^2 - r^2\right)\frac{h}{h_0} \approx \left(b^2 - r^2\right)\left(\frac{b_0}{b}\right)^{1/(1+R)},$$

where the last step follows from the fact that the outer radius is always in a state of uniaxial compression. This equation can be rearranged to give

$$\left(\frac{r_0}{r}\right)^2 = \left(\frac{b_0}{b}\right)^{2-\gamma}\left\{1 + \left[\left(\frac{b_0}{b}\right)^{\gamma} - 1\right]\frac{b^2}{r^2}\right\}, \tag{6.82}$$

which is a direct relationship between b/r, b_0/r_0, and b_0/b. From (6.82), the ratio b_0/b can be expressed in terms of b/r and b_0/r_0 as

$$\frac{b_0}{b} = \left\{ \frac{\left(b_0^2/r_0^2\right)\left(b^2/r^2 - 1\right)}{\left(b^2/r^2\right)\left(b_0^2/r_0^2 - 1\right)} \right\}^{1/\gamma}.$$

This relation provides a very useful approximation for the paths of the particles in the radial drawing process. It can be seen that the ratio b/r in a given element progressively increases as the drawing continues.

The change in thickness that occurs in any element while it is drawn toward the die throat can be determined by using (6.79) in the appropriate stress–strain relation, the resulting differential equation being

$$\begin{aligned}\frac{dh}{h} &= -\left\{\frac{1 - 2\ln(b/r)}{(1+R) - \ln(b/r)}\right\}\frac{dr}{r}\\ &= -\frac{1}{1+R}\frac{dr}{r} + \frac{(1+R)\,d\rho}{[(1+R)-\rho]\,[(1+R)-R\rho]}.\end{aligned}$$

Integrating, and using the initial condition $h = h_0$ when p–p_0 = \n(b_0/r_0), the flange thickness at any radius is obtained as

$$\frac{h}{h_0} = \left(\frac{r_0}{r}\right)^{1/(1+R)} \left\{ \frac{\left[(1+R) - R\ln(b/r)\right]\left[(1+R) - R\ln(b_0/r_0)\right]}{\left[(1+R) - R\ln(b_0/r_0)\right]\left[(1+R) - R\ln(b/r)\right]} \right\}^{1/(R-1)}. \tag{6.83}$$

For an isotropic material (R = 1), this expression becomes indeterminate, but considering the limit as R tends to unity, (6.83) is easily shown to coincide with equation (2.124). The thickness variations predicted by (6.83) are compared with those given by (6.78) in Fig. 6.17, which indicates the usefulness of the linearization of the yield criterion.

The flange element that is currently at the die throat $r = a$ enters the die to form an elemental ring of the same thickness h_a, and having a length $dz = v_a\, db$, where v_a is the particle velocity at $r = a$. Substituting from (6.80) and integrating, the amount of punch travel z when the flange radius has decreased from $\mathbf{b_0}$ **to** b is obtained as

$$\frac{z}{a} = \frac{1+R}{2R}\left\{\left[1 - \frac{\ln(b_0/a)}{1+R}\right]^{-2R} - \left[1 - \frac{\ln(b/a)}{1+R}\right]^{-2R}\right\}. \tag{6.84}$$

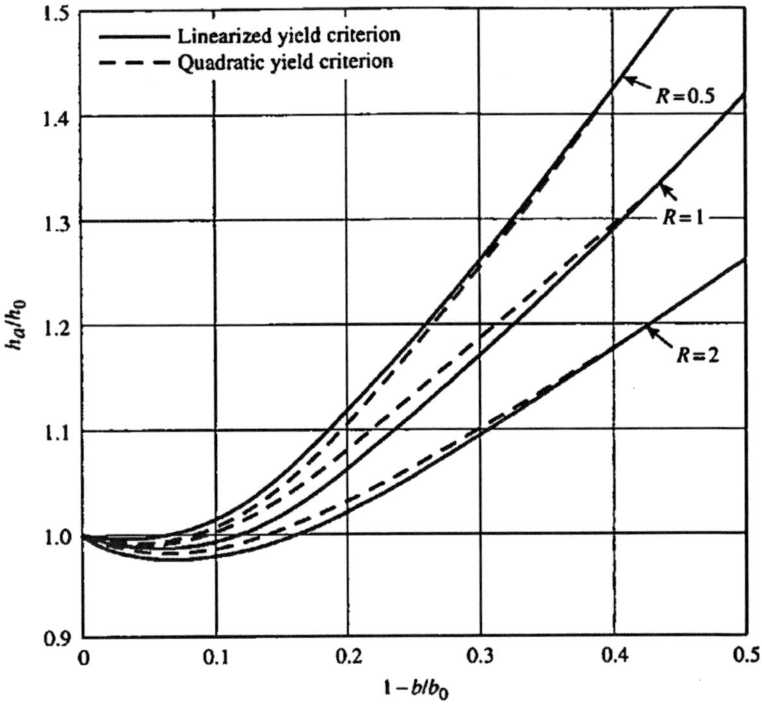

Fig. 6.17 Variation of flange thickness at the die throat with continued radial drawing of an anisotropic circular blank ($n = 0$)

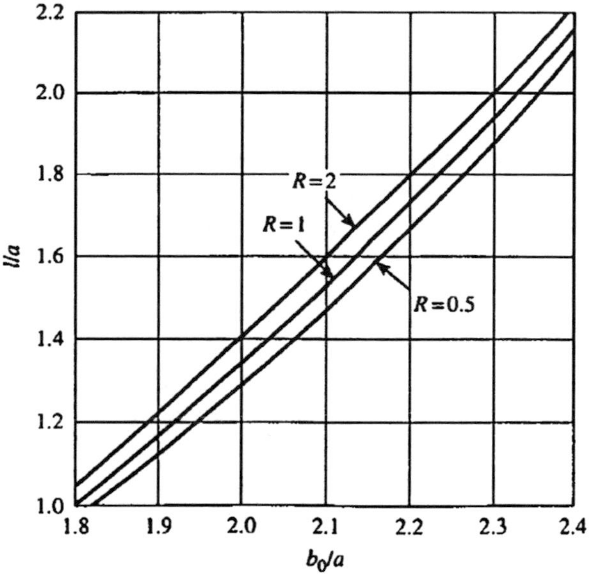

Fig. 6.18 Length of the deep-drawn cup against the drawing ratio for different R-values of the sheet metal ($n = 0$)

For a given drawing ratio, the punch travel increases as R increases. When $b_0/a = 2$, the length l of the drawn cup, which is the value of z when $b = a$, increases from 1.342a to 1.395a as R is increased from 1 to 2. Figure 6.18 indicates how the ratio l/a varies with the drawing ratio for materials with different values of R. Neglecting the effects of bending and friction, the punch load at any stage may be written as

$$P = 2\pi a h_a Y \ln(b/a).$$

As in the drawing of isotropic blanks, the total equivalent strain in a given element may be assumed to be independent of the rate of work-hardening. When the radius to the element changes by dr, the effective strain increment based on the concept of work equivalence becomes

$$d\varepsilon = -\frac{(1+R)\,Y d\varepsilon_\theta}{(1+R) - \ln(b/r)} = -\frac{dr}{r} + \frac{(1+R)\,d\rho}{\gamma\left[(1+R) - R_\rho\right]\left[(1+R) - \rho\right]}.$$

Since $\varepsilon = 0$ when $r = r_0$ and $\rho = \rho_0$, the integration of the above equation results in

$$\varepsilon = \ln\left(\frac{r_0}{r}\right) + \frac{1}{\gamma(R-1)} \ln\left\{\frac{[(1+R) - R\rho_0][(1+R) - \rho]}{[(1+R) - R\rho][(1+R) - \rho_0]}\right\} \qquad (6.85)$$

where r_0/r is given by (6.82). It follows from (6.83) and (6.85) that

$$\varepsilon = \frac{1}{\lambda}\left\{2\ln\left(\frac{r_0}{r}\right) - \ln\left(\frac{h}{h_0}\right)\right\}$$

which relates the equivalent strain to the hoop strain and the thickness strain in a particularly simple way. Evidently, the equivalent strain exceeds the magnitude of the hoop strain by an amount that steadily increases from $r = b$ to $r = a$.

When the material work-hardens, the linearized yield criterion for the radial drawing of an anisotropic blank with $R > 1$ is approximately the same as that for an isotropic material. It is therefore sufficiently accurate to write

$$\sigma_r - \sigma_\theta = m\sigma = mF(\varepsilon),$$

where σ is the current uniaxial yield stress of the material in the plane of the sheet, the associated stress–strain curve being defined by the function $F(\varepsilon)$. The substitution in the equilibrium equation furnishes the radial stress as

$$\sigma_r = m\int_r^b \sigma\frac{dr}{r} = m\int_0^\rho F(\varepsilon)d\rho, \tag{6.86}$$

where the integral is evaluated numerically for any given b/a. The velocity distribution and the thickness change are obtained by the integration of the equations

$$\frac{\partial v}{\partial r} = -\left\{\frac{mR\sigma + \sigma_r}{m(1+R)\sigma - \sigma_r}\right\}\frac{v}{r}, \quad \frac{dh}{h} = -\left\{\frac{m\sigma - 2\sigma_r}{m(1+R)\sigma - \sigma_r}\right\}\frac{dr}{r}, \tag{6.87}$$

using the boundary condition $v = -1$ at $r = b$, and the initial condition $h = h_0$ when $r = r_0$. The blank holding force necessary to suppress wrinkling of the flange is found to decrease with increasing normal anisotropy, and to increase with increasing planar anisotropy of the blank (Naziri and Pearce, 1968). A detailed analysis for the wrinkling process has been presented by Carreia et al. (2002).

6.6.3 The Limiting Drawing Ratio

The deep drawing of a circular blank by a flat-headed punch is commonly employed as a method for determining the drawability of sheet metals. The limiting drawing ratio, which is the greatest ratio of the blank diameter to the punch diameter that can be successfully drawn through the die, provides a measure of the drawability. The experimental investigations by Whiteley (1960) and Wilson (1966) have revealed that the limiting drawing ratio depends strongly on the normal anisotropy of the sheet, but very little on the state of work-hardening of the material. A drastic simplification to the problem is therefore achieved by considering a nonhardening material, for which the radial drawing load has its greatest value at the beginning of the process. If the effect of bending is neglected, failure would occur in uniaxial tension at the die throat when the drawing stress becomes equal to Y.

The effect of the blank holding friction is most conveniently taken into account by assuming a radial tension ηY existing at the rim, where η is an empirical friction factor that depends on the blank holding force and the friction coefficient. The first equation of (6.74) must therefore be solved under the modified boundary condition $\Psi = \alpha$ at $r = b_0$, where

$$\alpha = \sin^{-1}\left[\eta\sqrt{1+2R}/\,(1+R)\right] \tag{6.88}$$

in view of the first equation of (6.67) with $\sigma = Y$ and $\sigma_r = \eta Y$ at $\Psi = \alpha$. The distribution of Ψ in the initial state, when friction is included in the analysis, is given by the appropriate modification of (6.75), the result being

$$\frac{b_0}{r} = \left(\frac{\sin\psi + c\cos\psi}{\sin\alpha + c\cos\alpha}\right)^{1/2} \exp\left[\frac{c}{2}(\psi - \alpha)\right].$$

For the limiting blank size, the condition to be satisfied at $r = a$ is $\sin\Psi = c/(1+R)$ or $\cos\Psi = R/(\backslash + R)$. The preceding relation for the limiting drawing ratio then becomes

$$\frac{b_0}{a} = \left(\cos\alpha + \frac{\eta}{1+R}\right)^{-1/2} \exp\left\{\frac{\sqrt{1+2R}}{2}\left[\cos^{-1}\left(\frac{R}{1+R}\right) - \alpha\right]\right\}. \tag{6.89}$$

When the material work-hardens, failure would occur in the most critical element after some drawing has taken place, but (6.89) may still be used to predict the limiting drawing ratio, which is significantly affected by the blank holding friction. In an isotropic sheet, for instance, the limiting drawing ratio is decreased from 2.48 to 2.27 as the value of η is increased from 0 to 0.1.

For sufficiently large values of n, the failure would occur by plastic instability in the cup wall under conditions of plane strain. The stress ratio in the critical element varies continuously from an initial value, which lies between those corresponding to equibiaxial tension and plane strain. The variation of the stress ratio may be simulated by assuming that the ratio of the loads acting in the principal directions remains constant during the drawing (Chakrabarty and Chen, 2005). In terms of a dimensionless parameter β, the constancy of the load ratio may be written as

$$d\varepsilon_1 - d\varepsilon_2 = \frac{d\sigma_1}{\sigma_1} - \frac{d\sigma_2}{\sigma_2} = \frac{2d\beta}{1-\beta^2}, \quad \beta = \frac{\sigma_1 - \sigma_2}{\sigma_1 + \sigma_2}$$

Combining this expression with that for $d\varepsilon_2/d\varepsilon_1$ given by the flow rule, we obtain

$$(1+2R)\frac{d\varepsilon_1}{d\beta} = \frac{1+(1+2R)\beta}{\beta\left(1-\beta^2\right)}, \quad (1+2R)\frac{d\varepsilon_2}{d\beta} = \frac{1-(1+2R)\beta}{\beta\left(1-\beta^2\right)}.$$

The differential equation for the thickness change in the element is given by $dh/h = -(d\varepsilon_1 + d\varepsilon_2)$. Substituting from above and integrating, and using the initial condition $h = h_0$ when $\beta = \beta_0$, we have

$$\frac{h}{h_0} = \left\{ \frac{\beta_0^2 \left(1 - \beta^2\right)}{\beta^2 \left(1 - \beta_0^2\right)} \right\}^{1/(1+2R)} = \left\{ \frac{4R\,(1+R)\,\beta_0^2}{1 - \beta_0^2} \right\}^{1/(1+2R)} \tag{6.90}$$

in view of the plane strain condition $\beta = 1/(1 + 2R)$ at the onset of instability. If we adopt the hypothesis of work equivalence, the effective instability strain is found as

$$\varepsilon = \frac{n\sqrt{2\,(1+R)}\left[1 + (1+2R)\,\beta^2\right]^{3/2}}{1 + (1+2R)\,(3+2R)\,\beta^2} \tag{6.90a}$$

in view of (6.63a). Thus, $\varepsilon = n(1 + R)/\sqrt{1+2R}$ when $\beta = 1/(1 + 2R)$. It follows from (6.21) and (6.22) in relation to a work-hardening material that

$$\frac{d\varepsilon}{d\varepsilon_1} = \frac{\sigma}{\sigma_1}\left\{1 - \frac{R}{1+R}\left(\frac{\sigma_2}{\sigma_1}\right)\right\}^{-1} = \frac{\sqrt{2\,(1+R)\left[1 + (1+2R)\,\beta^2\right]}}{1 + (1+2R)\,\beta}$$

Consequently, the effective strain ε changes with β according to the differential equation

$$\frac{d\varepsilon}{d\beta} = \frac{d\varepsilon}{d\varepsilon_1}\frac{d\varepsilon_1}{d\beta} = \frac{\sqrt{2\,(1+R)}}{1+2R}\left\{\frac{\sqrt{1 + (1+2R)\,\beta^2}}{\beta\left(1 - \beta^2\right)}\right\}$$

using the values of ε and β at the point of instability, and setting $\alpha_0 = \sqrt{1 + (1+2R)\,\beta_0^2}$ and $c = \sqrt{2(1+R)}$ the integration of the above equation results in

$$\left(\frac{\alpha_0 - 1}{\alpha_0 + 1}\right)\left(\frac{c + \alpha_0}{c - \alpha_0}\right)^c = \frac{\left(1 + R + \sqrt{1+2R}\right)^c}{R^c\left(c + \sqrt{1+2R}\right)^2}\exp\left(-cn\sqrt{1+2R}\right) \tag{6.91}$$

The punch load attained at the onset of plastic instability is $P = 2\pi\, ah\, \sigma_1$, where $\sigma_1 = \sigma\,(1+R)\,/\sqrt{1+2R}$, and h is given by (6.90). Adopting the power law of hardening given by, $\sigma = \sigma_0\varepsilon^n$, the maximum punch load may be expressed as

$$\frac{P}{2\pi a h_0 \sigma_0} = \frac{1+R}{\sqrt{1+2R}}\left(\frac{h}{h_0}\right)\varepsilon^n = n^n\left(\frac{1+R}{\sqrt{1+2R}}\right)^{1+n}\left\{\frac{4R\,(1+R)\,\beta_0^2}{1 - \beta_0^2}\right\}^{1/(1+2R)} \tag{6.92}$$

For any given values of R and n, the dimensionless maximum punch load at the onset of instability can be computed from (6.91) and (6.92). Introducing a die friction factor λ, the instability load given by (6.92) must be equated to the maximum radial drawing load, which is equal to $(1 + \lambda)$ times that for the frictionless drawing. Assuming $m = (2/3)(2+3R)/(1+2R)$, and using a modification of equation (2.149), the limiting drawing ratio may be written approximately as

$$\left(\ln \frac{b_0}{a}\right)^{1+\sqrt{n}} \approx \frac{3}{2}(3)^{n/2}\left(\frac{1+2R}{2+3R}\right)\left(\frac{1+n}{1+\lambda}\right)\left(\frac{P}{2\pi a h_0 \sigma_0}\right) \tag{6.93}$$

where P is expressed by (6.92). For relatively small values of n, the failure would occur by necking in uniaxial tension at the die throat when the drawing load attains a maximum. The limiting drawing ratio is still given by (6.93), but the load P is now given by the condition $\sigma_r = \sigma_0 n^n$ at $r = a$.

In Fig. 6.19, the computed limiting drawing ratio b_0/a is plotted against the strain-hardening exponent n for various values of R, assuming frictionless conditions ($\lambda = \eta = 0$). It can be seen that the limiting drawing ratio is practically independent of the n-value except for low work-hardening materials. The predicted values of the limiting drawing ratio are found to be appreciably lower than those based on the assumption of a constant stress ratio, as reported by Chiang and Kobayashi (1966), and also by El-Sebaie and Mellor (1972). Higher values of the limiting drawing ratio can be achieved in the pressure assisted deep drawing process, for which the failure would always occur at the die throat, as has been shown by El-Sebaie and Mellor (1973).

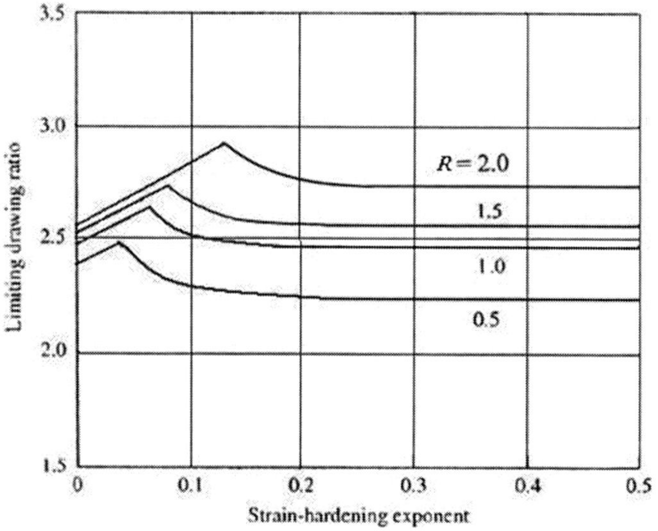

Fig. 6.19 Influence of n-values and R-values on the limiting drawing ratio in the deep drawing of flat-based cylindrical cups

6.6.4 Earing of Deep-Drawn Cups

When a circular blank of sheet metal having some degree of planar anisotropy is drawn through a die to form a cylindrical cup, the height of the rim above the base is found to be nonuniform due to the formation of ears and hollows, which are symmetrically situated with respect to the direction of rolling in the original sheet. From the theoretical point of view, the ears and hollows begin to form at those points on the rim where the radial direction is a principal direction of the strain increment. Since the state of stress at the rim is a uniaxial compression in the circumferential direction, the location of ears and hollows correspond to the points where the tangent to the rim is in the direction of stationary yield stress. It follows that, according to the quadratic yield function (6.9), the ears and hollows can only form at angular positions 0, $\pi/2$, and $\pi/2 \pm\alpha_0$ with respect to the direction of rolling, where α_0 is given by (6.11). Since α_0 may or may not be real, the quadratic yield function predicts four ears at most.

The yield function must be generalized to deal with certain metals which are known to produce more than four ears. For instance, in rolled sheets of cartridge brass, as many as six ears have been observed at angles of 0° and 60° to the rolling direction. Whatever the yield criterion, the ears and hollows must develop from points on the rim where the tangents are in the directions of stationary yield stress. This may be shown by writing the yield criterion as

$$f\left(\sigma_x,\sigma_y,\tau_{xy}\right) = 1,$$

where the x-axis is taken in the direction of rolling and the y-axis in the transverse direction. For a uniaxial tension σ applied at an angle α to the x-axis,

$$\sigma_x = \sigma\cos^2\alpha, \quad \sigma_y = \sigma\sin^2\alpha, \quad \tau_{xy} = \sigma\sin\alpha\cos\alpha.$$

The directions of stationary yield stress are given by $d\sigma/d\alpha = 0$, which is equivalent to $\partial f/\partial\alpha = 0$ in view of the relation

$$df = \frac{\partial f}{\partial\alpha}d\alpha + \frac{\partial f}{\partial\sigma}d\sigma = 0.$$

Substituting for the partial derivatives of σ_x, σ_y, and τ_{xy} with respect to a, it is easily shown that

$$\frac{\partial f}{\partial\alpha} = \sigma\left\{\left(\frac{\partial f}{\partial\sigma_y} - \frac{\partial f}{\partial\sigma_x}\right)\sin 2\alpha + 2\frac{\partial f}{\partial\tau_{xy}}\cos 2\alpha\right\} = 0.$$

If the plastic potential is taken as identical to the yield function, the last equation furnishes the relation

$$\tan 2\alpha = \frac{2\partial f/\partial\tau_{xy}}{\partial f/\partial\sigma_x - \partial f/\partial\sigma_y} = \frac{2d\gamma_{xy}}{d\varepsilon_x - d\varepsilon_y},$$

which is the condition for the principal axes of the strain increment to coincide with those of the stress. The corresponding angles αe therefore define points on the rim where the yield stress has maximum and minimum values.

Suppose now that the yield function and the plastic potential can be represented by a polynomial of degree n in the deviatoric stress components. In the special case of plane stress ($\sigma_z = 0$), the polynomial can be expressed in terms of the actual stress components in the $(x\ y)$-plane, and the yield function assumes the form

$$f = \sum C_{klm} \sigma_x^k \sigma_y^l \tau_{xy}^m, \tag{6.94}$$

where the powers k,l,m are positive integers or zero, with $k + I + m = n$ for a homogeneous yield function of degree n. If the (x, y)-directions coincide with the principal axes of anisotropy, m must be an even integer. The components of the strain increment associated with the yield function are

$$\begin{aligned} d\varepsilon_x &= \frac{\partial f}{\partial \sigma_x} d\lambda = d\lambda \sum k C_{klm} \sigma_x^{k-1} \sigma_y^l \tau_{xy}^m, \\ d\varepsilon_y &= \frac{\partial f}{\partial \sigma_y} d\lambda = d\lambda \sum l C_{klm} \sigma_x^k \sigma_y^{l-1} \tau_{xy}^m, \\ 2d\gamma_{xy} &= \frac{\partial f}{\partial \tau_{xy}} d\lambda = d\lambda \sum m C_{klm} \sigma_x^k \sigma_y^l \tau_{xy}^{m-1}, \end{aligned}$$

where λ is a positive scalar, and the factor of 2 in the last equation arises from the fact that $r_{xy} = r_{yx}$. For a uniaxial tension σ acting at an angle α to the praxis, (6.91) provides the directions for stationary values of the yield stress. Substituting for the stress components in terms of σ and α, and rearranging, we obtain the equation

$$\sum C_{klm} (\cos\alpha)^{2k+m-1} (\sin\alpha)^{2l+m-1} \left[(2l+m)\cos^2\alpha - (2k+m)\sin^2\alpha\right] = 0. \tag{6.95}$$

Since m is even, each term of (6.95) has a factor $\sin\alpha \cos\alpha$, so that $\alpha = 0$ and $\alpha = \pi/2$ are two possible roots of the equation. There are altogether 4n roots in the range $0 < \alpha < 2\pi$, giving rise to $2n$ ears when all the roots are real, these ears being symmetrically disposed with respect to the anisotropic axes. The position of an ear or a hollow is inclined at $\pi/2 - \alpha_0$ to the x-axis, where αo is a typical root of (6.92). For certain values of the anisotropic parameters C_{klm}, some of the roots will be imaginary, and the number of ears will then be less than $2n$. The special case of $n = 3$, giving 6 ears at most, has been discussed by Bourne and Hill (1950).

Based on the fourth order yield function (6.33), which is a special case of (6.91), the conditions under which 4, 6, and 8 ears will form in a deep drawn cup have been examined by Gotoh (1977), Using a finite element analysis for the flange, the ear configuration and the ear height in a deep drawn cylindrical cup has also been evaluated by Gotoh (1980), neglecting friction over the blank holder. For a given initial diameter of the blank, the ear is prone to grow in the direction of the larger

R-value of the sheet metal, and the ear height decrease with increasing n-value of the material.

6.7 Anisotropy in Plates and Shells

The effect of normal anisotropy in which the R-value is greater than unity is significant in improving the load-carrying capacity of plates and shells. As in the theory of isotropic solids, a piecewise linear approximation of Hill's quadratic yield criterion would be useful for the estimation of limit loads. The yield stress of the material in the thickness direction is assumed to be μ times that in any direction in the plane of the applied stresses, where $\mu \geq 1$. The modified Tresca criterion of Fig. 6.3(a) and its associated flow rule will be taken as the basis for the estimation of collapse loads in the presence of normal anisotropy (Chakrabarty, 1974). A different type of linearization of the yield function in relation to anisotropic plates has been considered by Sawczuk et al. (1975).

6.7.1 Bending of Circular Plates

The load-carrying capacity of isotropic circular plates under rotationally symmetric transverse loading has been discussed in Chapter 4. The deformation of the plate is also rotationally symmetric, even when there is normal anisotropy, and is defined by the normal velocity w of its middle surface. For a rigid/plastic plate, the radial and circumferential strain rates in a typical element are $\dot{\varepsilon}_r = z\dot{\kappa}_r$, and $\dot{\varepsilon}_\theta = z\dot{\kappa}_\theta$ where z is the normal distance from the middle surface measured positive in the direction of w, while $\dot{\kappa}_r$ and $\dot{\kappa}_\theta$ are the curvature rates of the middle surface, which are given by

$$\dot{\kappa}_r = -\frac{d^2w}{dr^2}, \quad \dot{\kappa}_\theta = -\frac{1}{r}\frac{dw}{dr}.$$

The nonzero stress components σ_r and σ_θ have constant values on each side of the middle surface, and the bending moments M_r and M_θ are equal to $\sigma_r h^2/4$ and $\sigma_\theta h^2/4$, respectively, where h is the uniform thickness of the plate. The linearized yield condition in terms of the bending moments therefore becomes

$$M_r = \pm\mu M_0, \quad M_\theta = \pm\mu M_0, \quad M_r - M_\theta = \pm M_0, \tag{6.96}$$

where M_0 is the fully plastic moment equal to $Yh^2/4$, and $\mu = \sqrt{(1+R)/2}$, the yield locus defined by (6.96) being that shown in Fig. 6.20(a). The velocity slope dw/dr may be discontinuous across a circle, called a hinge circle, which must be regarded as the limit of a narrow annulus of rapid change in slope. It follows that κ_r is infinitely large and M_r is of magnitude μmq at a hinge circle.

Consider, as an example, a solid circular plate of radius a which is supported round the periphery and loaded by a uniform normal pressure p over a concentric

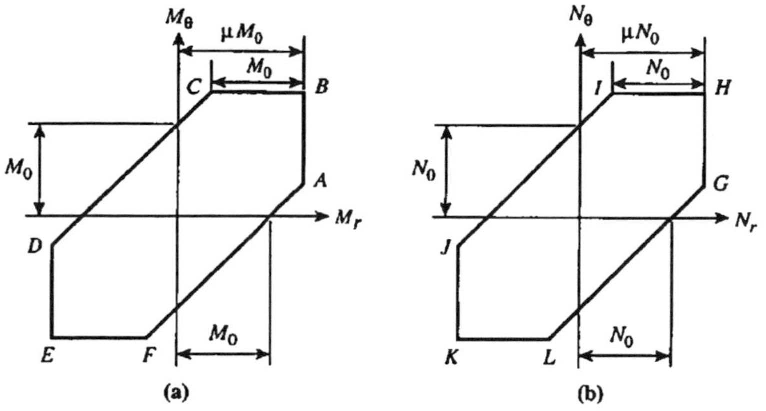

Fig. 6.20 Linearized yield conditions for transversely isotropic plates and shells. (**a**) Yield locus in the moment plane, (**b**) yield locus in the force plane

circle of radius b, where $b \leq a$. Since $M_r = M_\theta$ at the center of the plate, it is natural to expect that an inner region of the plate would correspond to the plastic regime BC. If the plate is either simply supported or clamped, the outer part of the plate should correspond to the plastic regime CD in order to satisfy the boundary condition. The equilibrium equation becomes

$$\frac{d}{dr}(rM_r) = \begin{cases} M_\theta - pr^2/2, & 0 \leq r \leq b, \\ M_\theta - pb^2/2, & r \leq b \leq a. \end{cases} \tag{6.97}$$

Depending on the values of b/a and μ, the circle $r = \rho$ separating the two plastic regimes may fall either outside or inside the loaded area. In the former situation, the entire loaded region corresponds to the plastic regime BC, for which $M_{\theta e} = \mu M_0$. Introducing the dimensionless parameter $q = pb^2/2\ M_0$, equation (6.97) is readily integrated to give

$$\left.\begin{aligned} M_r &= M_0\left(\mu - \frac{qr^2}{3b^2}\right), && 0 \leq r \leq b, \\ M_r &= M_0\left\{\mu - q\left(1 - \frac{2b}{3r}\right)\right\}, && b \leq r \leq \rho, \end{aligned}\right\} \tag{6.98}$$

in view of the continuity of M_r across $r = b$, and the substitution $q = pb^2/2\ M_0$-For $r \geq \rho$, the yield criterion is modified to $M_g - M_r - M_0$, and the second equation of (6.97) furnishes the solution

$$M_r = M_0\left[(\mu - 1) - (q - 1)\ln\left(\frac{r}{\rho}\right)\right], \quad \rho \leq r \leq a, \tag{6.99}$$

in view of the condition $M_r = (\mu-1)m\text{Q}$ at the radius $r = \rho$, which corresponds to the corner C of the yield hexagon. The continuity of M_r across $r = \rho$ requires

$$q\left(1 - \frac{2b}{3\rho}\right) = 1 \quad \text{or} \quad \rho = \frac{2}{3}\left(\frac{bq}{q-1}\right).$$

The assumption $\rho \geq b$ therefore implies $q \leq 3$. The relationship between b/a and q is obtained from the boundary condition $M_r = 0$ at $r = a$ for a simply supported plate, and $M_r = -\mu M_0$ at $r = a$ for a clamped plate when there is a hinge circle formed at the boundary. It follows from (6.99) that

$$\frac{b}{a} = \frac{3}{2}\left(1 - \frac{1}{q}\right)\exp\left(-\frac{\varepsilon\mu - 1}{q-1}\right), \quad q \leq 3, \tag{6.100}$$

where $\varepsilon = 1$ for a simply supported plate, and $\varepsilon = 2$ for a clamped plate. The greatest value of b/a for which the restriction $\rho \geq b$ holds is seen to be equal to exp $[(1—\varepsilon\mu p)/2]$, which corresponds to $q = 3$.

For $\rho \leq b$, an outer annulus of the loaded area corresponds to the plastic regime *CD*. The solution to the radial bending moment for the inner region $0 \leq r \leq p$ is given by the first equation of (6.98), and the condition $M_r = (\mu-\backslash)M_0$ at $r = \rho$ furnishes

$$q\rho^2 / 3b^2 = 1 \quad \text{or} \quad \rho = b\sqrt{3/q}$$

The solution for the outer region $\rho \leq r \leq b$ of the loaded area is obtained by integrating the first equation of (6.97) with $M_\theta = M_r + M_0$, and using the condition of continuity of M_r across $r = \rho$, the result being

$$M_r = M_0\left\{\mu + \ln\left(\frac{r}{\rho}\right) - \frac{1}{2}\left(\frac{qr^2}{b^2} - 1\right)\right\}, \quad \rho \leq r \leq b.$$

The solution for the outer annulus $b \leq r \leq a$ is obtained by setting $M_\theta = M_r + M_0$ in the second equation of (6.97), and integrating under the condition of continuity of M_r across $r = b$. Using the relationship between ρ/b and q, we have

$$M_r = M_0\left\{\mu + \frac{1}{2}\ln\left(\frac{q}{3}\right) - (q-1)\left(\frac{1}{2} + \ln\left(\frac{r}{b}\right)\right)\right\}, \quad b \leq r \leq a.$$

In view of the boundary conditions $M_r = 0$ at $r = a$ for a simply supported plate and $M_r = -\mu M_0$ for a clamped plate, the relationship between b/a and q is obtained in the form

$$\frac{b}{a} = \exp\left\{\frac{1}{2} - \frac{2\varepsilon\mu + \ln(q/3)}{2(q-1)}\right\}, \quad q \geq 3, \tag{6.101}$$

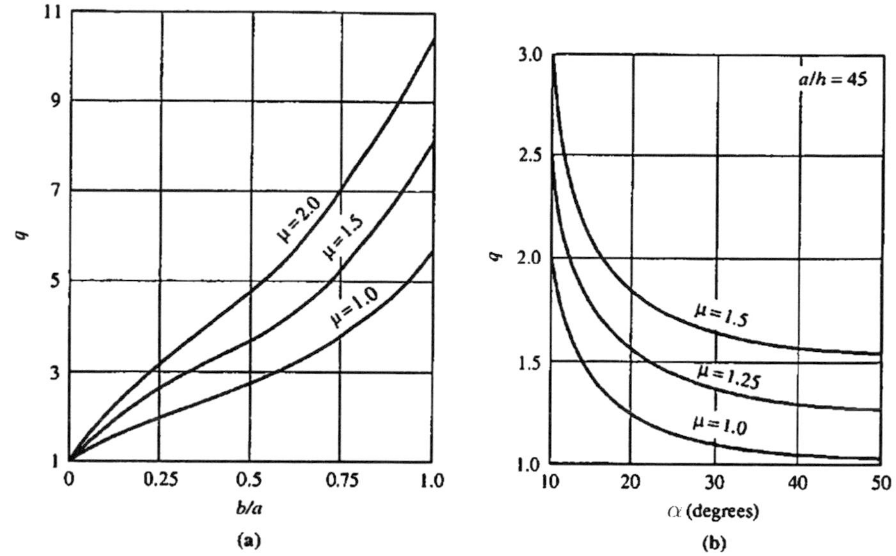

Fig. 6.21 Dimensionless collapse pressure for anisotropic plates and shells. (**a**) Clamped circular plate, (**b**) clamped spherical cap

where ε equals 1 for the simple support and 2 for the built-in support. Equations (6.100) and (6.101) become identical when $q = 3$. When the load is distributed over the entire plate $(b = a)$, the collapse load is given by the transcendental equation

$$q - \ln(q/3) = 1 + 2\varepsilon\mu.$$

In the case of an isotropic plate ($\mu = 1$), this equation furnishes $q = 3.0$ for a simply supported plate and $q = 5.63$ for a clamped plate. When b tends to zero and p tends to infinity so that the total load has a finite value p, the collapse load is given by $q = 1$ or $P = 2\pi M_0$, irrespective of the edge condition. Figure 6.21(a) shows the variation of the dimensionless collapse load with the b/a ratio for clamped plates with different values of μ. For a given b/a ratio, the velocity field in an anisotropic plate is identical to that in an isotropic plate clamped at the edge.

6.7.2 Plastic Collapse of a Spherical Cap

In the case of rotationally symmetric shells of revolution, the state of stress is com pletely specified by the meridional and circumferential forces N_r, N_θ, the corresponding bending moments M_r, M_θ, and the transverse shearing force Q. Owing to the customary thin-shell approximation, the direct forces and the bending moments constitute the only generalized stresses in the problem, the corresponding generalized strain rates being the unit extension rates $\dot{\lambda}_r$, $\dot{\lambda}_\theta$and the curvature rates $\dot{\kappa}_r$, $\dot{\kappa}_\theta$of the middle surface of the shell. For simplicity, we shall adopt the limited interaction yield condition, in which all interactions between force and moment are disregarded for the yielding of an element. The linearized yield condition for the shell is then represented by two similar hexagons, one of which is defined by (6.94) and the other by

$$N_r = \pm\mu N_0, \qquad N_\theta = \pm\mu N_0, \qquad N_r - N_\theta = \pm N_0, \tag{6.102}$$

where N_0 is the yield force *hY* for a shell of thickness *h*, the new yield hexagon being shown in Fig. 6.20(b). The incipient plastic flow of any element involves one of the sides of the force hexagon and one of the sides of the moment hexagon, the associated strain rate vector being a combination of the vectors directed along the outward normal to the two separate hexagons. In terms of the tangential velocity *u* and the inward normal velocity *w*, the generalized strain rates in a spherical shell of radius *a* are

$$\left.\begin{aligned} \dot{\lambda}_\theta &= \frac{1}{a}\left(u\cot\phi - w\right), & \dot{\lambda}_r &= \frac{1}{a}\left(\frac{du}{d\phi} - w\right), \\ \dot{\kappa}_\theta &= -\frac{\cot\phi}{a^2}\left(u + \frac{dw}{d\phi}\right), & \dot{\kappa}_r &= -\frac{1}{a^2}\frac{d}{d\phi}\left(u + \frac{dw}{d\phi}\right), \end{aligned}\right\} \tag{6.103}$$

where ϕ denotes the angle of inclination of the surface normal with the vertical axis of symmetry. The normal velocity *w* must be continuous, but its derivative $dw/d\phi$ and the tangential velocity *u* can be discontinuous across a hinge circle, where M_r has the magnitude μm_0 and/or N_r has the magnitude μN_0.

Consider a uniform spherical cap of radius of curvature *a* and angular span 2α, subjected to a uniform normal pressure of intensity *p* per unit area, and supported along the horizontal base defined by $\phi = a$. The generalized stresses at the incipient collapse must satisfy the yield conditions (6.96) and (6.102), and the equilibrium equations

$$\left.\begin{aligned} \frac{d}{d\phi}\left(N_r\tan\phi\right) - N_\theta &= -\frac{pa}{2}\tan^2\phi, \\ \sec\phi\frac{d}{d\phi}\left(M_r\sin\phi\right) - M_\theta &= -a\left(N_r + \frac{pa}{2}\right)\tan^2\phi. \end{aligned}\right\} \tag{6.104}$$

For an incomplete shell, it is natural to expect the entire shell to correspond to the side *JK* of the force hexagon. Setting $N_r = -\mu N_0$, and integrating, we get

$$N_\theta = -N_0\left[\mu - (q-\mu)\tan^2\phi\right], \quad N_r = -\mu N_0, \tag{6.105}$$

where $q = pa/2\,N_0$. The symmetry condition $N_r = N_\theta$ at $\phi = 0$ is satisfied. Since $M_r = M_\theta$ at $\phi = 0$, it is apparent that a central part of the cap of angular span 2ψ would correspond to the side *BC* of the moment hexagon. Setting $M_\theta = \mu M_0$ and $n_r = -\mu N_0$ in (6.104), and integrating, we obtain the solution

$$M_r = \mu M_0 - a(q-\mu)N_0[\text{cosec}\phi \ln(\sec\phi + \tan\phi) - 1], \quad 0 \le \phi \le \psi.$$

The element at $\phi = \psi$ corresponds to the corner *C*, where $M_r = (\mu{-}1)M_0$ Since $M_0/N_0 = h/4$, the above equation gives

$$(q-\mu)[\ln(\sec\psi + \tan\psi) - \sin\psi] = \frac{h}{4a}\sin\psi. \tag{6.106}$$

The stress profile for the outer part of the cap must involve the side *CD* of the moment hexagon. The second equation of (6.104) is therefore to be solved with the substitution $M_\theta = M_r + M_0$ and $N_r = -\mu N_0$. In view of the continuity of M_r across $\phi = \psi$, the integration results in

$$M_r = M_0\left\{(\mu - 1) + \ln\left(\frac{\sin\phi}{\sin\psi}\right)\right\} - a(q-u)N_0 \ln\left(\frac{\cos\psi}{\cos\phi}\right), \qquad \psi \le \phi \le \alpha.$$

If the edge of the cap is simply supported, we have $M_r = 0$ at $\phi = \alpha$. For a clamped edge condition, on the other hand, $M_r = -\mu\ m_0$ at $\phi = \alpha$, provided a hinge circle is formed at the boundary. Inserting these boundary conditions into the above equation, the dimensionless collapse pressure is obtained as

$$q = \mu + \frac{h}{4a}\left\{\frac{\varepsilon\mu - 1 + \ln(\sin\alpha/\sin\psi)}{\ln(\cos\psi/\cos\alpha)}\right\}, \tag{6.107}$$

where $\varepsilon = 1$ for simple support and $\varepsilon = 2$ for built-in support. Since q is always greater than μ, equation (6.105) indicates that N_θ steadily increases from $\phi = 0$ to $\phi = \alpha$. The solution will therefore be statically admissible if $N_e \le -(\mu{-}1)\,N_0$, which is satisfied if $q \le \mu + \cot^2\alpha$. This condition will be satisfied in most cases of practical interest. The variation of the collapse pressure with the cap semiangle for a clamped spherical cap, computed from (6.106) and (6.107), is displayed in Fig. 6.21(b), which indicates the beneficial effect of higher value of μ.

Since the stress profile corresponding to the direct forces is everywhere on the plastic regime *JK*, the associated flow rule gives $\dot\lambda_\theta = 0$ and $\dot\lambda_r = 0$ and it follows from the first relation of (6.103) that

$$u = w\tan\phi, \quad 0 \le \phi \le \alpha,$$

throughout the cap for both the edge conditions. In the inner and outer plastic regions, which correspond to the regimes *BC* and *CD*, respectively, on the moment hexagon, we have $\dot\kappa_r = 0\ \dot\kappa_\theta > 0$for $0 \le \phi \le \psi$, and $kg = \dot\kappa_\theta = -\dot\kappa_r > 0$for $\psi < \phi < \alpha$. The last two relations of (6.103) therefore give the differential equations

$$\frac{dw}{d\phi} + u = -A \qquad (0 \le \phi \le \psi), \qquad \frac{dw}{d\phi} + u = -B\text{cosec}\phi \qquad (\psi \le \phi \le \alpha),$$

where *A* and *B* are constants of integration. These equations hold for both the simply supported and the clamped shells, and the boundary condition $w = 0$ at $\phi = \alpha$ apply in both cases. The incipient velocity field in the anisotropic shell is therefore the same for both the edge conditions and is identical to that in the isotropic clamped shell obeying the limited interaction yield condition (Section 5.4).

6.7.3 Reinforced Circular Plates

The load-carrying capacity of circular plates can also be enhanced by introducing a circumferential reinforcement that constitutes an axially symmetrical planar anisotropy (Markowitz and Hu, 1964), the radial and circumferential yield moments being denoted by M_0 and kM_0, respectively, where $k > 1$. The analysis can be greatly simplified by employing a modified form of the Tresca criterion, for which the yield locus in the moment plane is that shown in Fig. 6.22 (a) and is defined by the equations

$$M_r = \pm M_0, \qquad M_\theta = \pm kM_0, \qquad M_\theta - kM_r = \pm kM_0. \tag{6.108}$$

The flow rule associated with each side of the hexagon can be immediately written following the normality rule. As in the case of isotropic plates, no plastic flow can be associated with the sides AB and DE of the hexagon over any finite portion of the plate.

We begin with the situation where a central load of magnitude P is applied to a circular plate of radius a through a rigid circular boss of radius b. As in the case of isotropic plates (Section 4.1), the deformation mode at the incipient collapse may

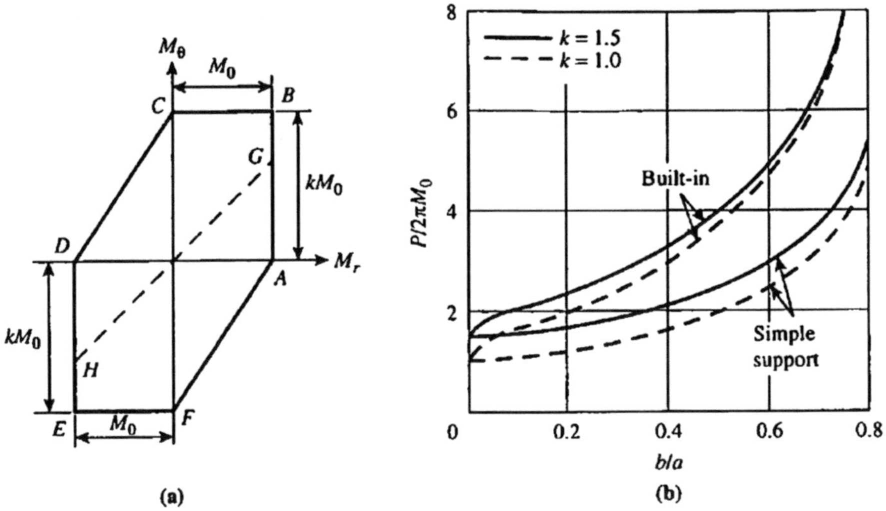

Fig. 6.22 Plastic collapse of rib-reinforced circular plates centrally loaded by a rigid punch. (**a**) Linearized yield locus, (**b**) variation of collapse load with punch radius

be assumed to involve a hinge circle coinciding with radius $r = b$. The stress point for the inner boundary $r = b$ is therefore at the corner B of the yield hexagon. If the plate is simply supported along its edge $r = a$, the radial bending moment vanishes along this boundary, indicating that the stress profile CB applies to the entire plate at the yield point. Setting $M_\theta = kM_0$ in the equilibrium equation

$$\frac{d}{dr}(rM_r) = M_\theta - qM_\theta, \tag{6.109}$$

where $q = P/2\pi\ M_0$ denotes the dimensionless collapse load, and using the boundary condition $M_r = 0$ at $r = a$, the above equation is integrated to give

$$M_r = M_0\,(q-k)\left(\frac{a}{r}-1\right), \quad b \le r \le a. \tag{6.110}$$

The remaining boundary condition $M_r = kM_0$ at $r = b$ then furnishes the collapse load as

$$q = k\left(\frac{a}{a-b}\right) - (k-1)\left(\frac{b}{a-b}\right). \tag{6.111}$$

Since the flow rule associated with the yield condition gives $\dot{\kappa}_r = 0$, the plate collapses in the form of a frustum of a cone, exactly like an isotropic circular plate. In the case of a central concentrated load ($b = 0$), the collapse load for the anisotropic plate is seen to be k times that for the isotropic plate.

When the plate is fully clamped around its edge $r = a$, the radial bending moment over an inner region is given by (6.110) with the plate radius a replaced by some radius $\rho < a$ that corresponds to $M_r = 0$. In the outer region defined by $p \le r \le a$, the plastic regime CD must hold in order to satisfy the boundary condition at $r = a$. Since the yield condition corresponding to the side CD is $M_\theta = k(M_r + M_0)$, the equilibrium equation (6.109) becomes

$$r^k \frac{d}{dr}\left(r^{1-k} M_r\right) = -\,(q-k)\,M_0.$$

The integration of this equation under the continuity condition $M_r = 0$ at r = ρ results in

$$M_r = -M_0\left(\frac{q-k}{q-1}\right)\left\{\left(\frac{r}{\rho}\right)^{k-1} - 1\right\}, \qquad \rho \le r \le a. \tag{6.112}$$

Introducing the usual assumption of a hinge circle formed at the clamped edge, we have the boundary conditions $M_r = -M_0$ at $r = a$ and $M_r = M_0$ at $r = b$. Equation (6.112) and the modified form of (6.110) then provide a pair of relations which are combined to give

$$\frac{a}{b} = \left(1 + \frac{1}{q-k}\right)\left(\frac{q-1}{q-k}\right)^{1/(k-1)}, \quad \frac{\rho}{b} = 1 + \frac{1}{q-k}. \tag{6.113}$$

According to the associated flow rule, $\dot{\kappa}_r = 0$ for the inner plastic region, and $\dot{\kappa}_\theta \,/\, \dot{\kappa}_r = -k$ for the outer plastic region. Consequently, by the condition of continuity of the velocity slope across $r = \rho$, we have

$$\frac{dw}{dr} = -c \quad (b \leq r \leq \rho), \quad \frac{dw}{dr} = -c\left(\frac{\rho}{r}\right)^{1/k} \quad (\rho \leq r \leq a),$$

where c is the absolute value of the slope for the inner plastic region. Integrating the above equations, and using the boundary condition $w = 0$ at $r = a$, the velocity field is obtained as

$$w = w_0 - c\,(r-b)\,, b \leq r \leq \rho,$$
$$w = \frac{ck\rho}{k-1}\left\{\left(\frac{a}{\rho}\right)^{(k-1)/k} - \left(\frac{r}{\rho}\right)^{(k-1)/k}\right\}, \quad \rho \leq r \leq a, \tag{6.114}$$

where w_0 is the velocity of the rigid boss. The relationship between w_0 and c follows from the condition of continuity of w across $r = \rho$. The variation of the dimensionless collapse load for the simply supported and clamped plates is displayed in Fig. 6.22(b), which indicates the influence of rib-reinforcement on the carrying capacity of the plate.

As a final example, consider a solid circular plate that is uniformly loaded with a normal pressure of intensity p. The radial and circumferential bending moments in each element must satisfy the equilibrium equation

$$\frac{d}{dr}\,(rM_r) = M_\theta - qM_0\left(\frac{r^2}{a^2}\right), \tag{6.115}$$

where $q = pa^2/2Mo$ The center of the plate must correspond to point G of the yield hexagon in order to satisfy the condition $M_r = Me$ at r— 0. For a simply supported plate, the boundary condition requires $M_r = 0$ at r—a, and the plastic collapse of the plate involves the sides GB and BC, which correspond to

$$M_r = M_0 \quad (0 \leq r \leq \rho)\,, \quad M_\theta = kM_0 \quad (\rho \leq r \leq a)\,,$$

where ρ is the radius to the interface between the two plastic regions. Substituting into the equilibrium equation (6.115), and integrating it under the boundary conditions stated above, we obtain the solution

$$M_\theta = M_0\left(1+\frac{qr^2}{a^2}\right), \quad 0 \le r \le \rho,$$

$$M_r = \frac{1}{3}M_0\left\{q\left(\frac{a}{r}-\frac{r^2}{a^2}\right)-3k\left(\frac{a}{r}-1\right)\right\}, \quad \rho \le r \le a.$$

The dimensionless collapse load q and the interface radius ρ are determined from the condition of continuity of M_r and M_θ across $r = \rho$, the result being

$$q = (k-1)\frac{a^2}{\rho^2}, \quad 1-\left(\frac{3k}{k-1}\right)\frac{\rho^2}{a^2}+\frac{2\rho^3}{a^3} = 0. \tag{6.116}$$

Equations (6.116) indicate that the collapse load is increased by increasing the value of k. When k tends to unity, ρ tends to zero, and q tends to the limiting value 3.0. When $k = 2$, q is found to increase to 5.12. In the case of a clamped circular plate, there is a third plastic region that corresponds to the regime *CD*, and the analysis is required to be modified accordingly. Further results on the plastic collapse of anisotropic plates have been given by Save and Massonnet (1972).

Problems

6.1 A block of anisotropic material is uniformly compressed in the x-direction under conditions of plane strain in the z-direction, the stress in the y-direction being zero. The material is assumed to be elastically isotropic with Young's modulus E and Poisson's ratio ν. Adopting the yield criterion in the form $\sigma_x = -X$, show that the strain in the x-direction at any stage of the deformation is give by

$$-\frac{E}{X}\varepsilon_x = \left(1-\nu^2\right)+\left(\frac{G}{F+G}-\nu\right)(s-\nu)+\frac{FG+GH+HF}{(F+G)^2}\ln\left\{\frac{G-(F+G)\nu}{G-(F+G)s}\right\}$$

where $s = -\sigma_z/X$. Note that s rapidly approaches the asymptotic value of $G/(F+G)$ as the deformation continues in the elastic/plastic range.

6.2 For a state of plane stress in a sheet metal exhibiting planar anisotropy, show that the quadratic yield criterion and its associated flow rule are identically satisfied by writing the principal stresses and strain increments in the plane in terms of an angle ψ as

$$\sigma_x = X\sqrt{1+\alpha^2}\cos\psi, \qquad \sigma_z = Y(\alpha\cos\psi - \sin\psi)$$

$$d\varepsilon_x = \left(\frac{\cos\psi+\alpha\sin\psi}{\sqrt{1+\alpha^2}}\right)d\lambda, \quad d\varepsilon_z = -\left(\frac{X}{Y}\sin\psi\right)d\lambda$$

where X and Y are the uniaxial yield stresses in the principal directions, and α denotes the parameter $H/\sqrt{FG+GH+HF}$. Note that in the case of normal anisotropy, we have $\alpha = R/\sqrt{1+2R}$, and $\sqrt{1+\alpha^2} = (1+R)/\sqrt{1+2R}$, where R is the constant strain ratio.

6.3 In the plane strain compression considered in Prob. 6.1, suppose that the material exhibits normal anisotropy with a uniform R-value. Using the expressions given in the preceding problem with a reversal of the sign shows that yielding begins in the block when $\sigma_x = -m\,Y$ and $\psi = \psi_0$, where

$$m = \left\{1 - \left(\frac{2R}{1+R}\right)\nu + \nu^2\right\}^{-1/2}, \quad \psi_0 = \tan^{-1}\left\{\frac{R - \nu\,(1+R)}{\sqrt{1+2R}}\right\}.$$

If the compression is continued in the elastic/plastic range, prove the longitudinal strain in the direction of compression at a generic stage is given by

$$-\frac{E}{Y}\varepsilon_x = \frac{1}{m} + 2\left(\frac{R}{1+R} - \nu\right)\left(\frac{R}{\sqrt{1+2R}}\cos\psi - \sin\psi\right) + \frac{\sqrt{1+2R}}{R}\ln\left\{\frac{\tan(\psi_0/2)}{\tan(\psi/2)}\right\}.$$

6.4 An infinite plate containing a circular hole of radius a is subjected a balanced biaxial tension σ in its plane. The material is incompressible and exhibits normal anisotropy with a planar initial yield stress Y. Adopting the deformation theory of plasticity, assuming the ratios of the total strains to be approximately the same as those of the plastic strains, and using the approximate yield criterion $\sigma_\theta = \bar{\sigma}$, where $\bar{\sigma}$ varies with $\bar{\varepsilon}$ according to the strain-hardening law$\bar{\sigma} = Y\,(E\bar{\varepsilon}/Y)^n$, show that $\sigma_\theta = \lambda\sigma$ at $r = a$, where

$$\lambda = \frac{Y}{\sigma}\left\{\frac{2\sigma}{Y} + (1-n)\left(\frac{1+2R}{1+R}\right)\left(1 - \frac{2\sigma}{Y}\right)\right\}^{-n(1+R)/[R-n(1+2R)]}.$$

In the special case of $n = R/(1+2R)$, for which the right-hand side of the above equation becomes indeterminate, obtain the expression for λ by an independent analysis.

6.5 A rectangular sheet of metal, exhibiting normal anisotropy, is subjected to biaxial loading parallel to the opposite sides, while the load ratio is held constant during the stretching. Assuming the hypothesis of work equivalence, show that the parameter $\beta = (\sigma_1 - \sigma_2)\,/\,(\sigma_1 + \sigma_2)$ increases from its initial value β_0, while the in-plane principal strains at any stage are

$$\varepsilon_1 = \frac{1}{1+2R}\left\{\ln\frac{\beta}{\beta_0} + R\ln\left(\frac{1+\beta}{1+\beta_0}\right) + (1+R)\ln\left(\frac{1-\beta_0}{1-\beta}\right)\right\}$$

$$\varepsilon_2 = \frac{1}{1+2R}\left\{\ln\frac{\beta}{\beta_0} - R\ln\left(\frac{1-\beta_0}{1-\beta}\right) - (1+R)\ln\left(\frac{1+\beta}{1+\beta_0}\right)\right\}.$$

6.6 Suppose that the biaxial loading of the preceding problem is continued up to the point of plastic instability, when the loads simultaneously attain their maximum. To derive the relationship between the initial and the final values of β, it is convenient to use the substitution $\lambda = \left[1 + (1+2R)\,\beta^2\right]^{1/2}$. Establish the appropriate differential equation for$\bar{\varepsilon}$ in terms of λ and integrate it to obtain the result

$$c\ln\left\{\frac{(c+\lambda)\,(c-\lambda_0)}{(c-\lambda)\,(c+\lambda_0)}\right\} + \ln\left\{\frac{(\lambda-1)\,(\lambda_0+1)}{(\lambda+1)\,(\lambda_0-1)}\right\} = \frac{2n\,(1+2R)\,\lambda^3}{(3+2R)\,\lambda^2 - c^2},$$

where $c = \sqrt{2\,(1+R)}$, and n is the strain-hardening exponent, while λ_0 is the initial value of λ. Note that each side of the above equation is equal to $(2/c)$ $(1 + 2R)\bar{\varepsilon}$

6.7 A thin-walled cylindrical tube is closed at both ends and is subjected to a combined internal pressure p and independent axial loads N, the material of the tube being in a state of normal anisotropy with a uniform R-value. Setting $\alpha = N/\pi\, a^2\, p$, where a denotes the current mean radius of the tube, show that the condition for plastic instability according to the hypothesis of work equivalence is given by

$$\frac{1}{\bar{\sigma}}\frac{d\bar{\sigma}}{d\bar{\varepsilon}} = \frac{(5+R)^2 - (R-1)\,\alpha\,[(11+R)+(3+R)\,\alpha] + (1+R)^2\,\alpha^3}{\sqrt{1+R}\left[(5+R) - 2\,(R-1)\,\alpha + (1+R)\,\alpha^2\right]^{3/32}}.$$

Assume that the onset of plastic instability corresponds to the internal pressure p and additional axial load N simultaneously attaining their maximum...

6.8 In a flat sheet of metal subjected to biaxial tension in its plane, a localized neck is initiated in the direction coinciding with that of the minimum principal strain $\varepsilon_2 > 0$. If the Hencky type of stress–strain relation holds in its incremental form at the point of localized necking, and the stress ratio is maintained constant through the biaxial stretching, show that the principal strains at the incipient necking are

$$\varepsilon_1 = \frac{n\,(1+2R)\,(1+\alpha)^2 + [1-(1+2R)\,\alpha]^2}{(1+2R)\,(1+\alpha)\left[1+(1+2R)\,\alpha^2\right]}, \quad \varepsilon_2 = \left\{\frac{1-(1+2R)\,\alpha}{1+(1+2R)\,\alpha}\right\}\varepsilon_1$$

according to the hypothesis of work equivalence, where α denotes the ratio $(\sigma_1 - \sigma_2)\,/\,(\sigma_1 + \sigma_2)$, and R is the usual plastic strain ratio for the material exhibiting normal anisotropy.

6.9 Using the linearized yield criterion for a sheet metal with normal anisotropy, show that the rates of change of the thickness h and effective strain ε at the initial stage of radial drawing of a circular blank of radius b_0 are given by

$$\dot{h} = \frac{h_0}{1+R}\left(1 - 2\ln\frac{b_0}{r}\right)\dot{\varepsilon}, \qquad \dot{\varepsilon} = \frac{1}{b_0}\left(1 - \frac{1}{1+R}\ln\frac{b_0}{r}\right)^{-2(1+R)}.$$

When the change in radius b_0–b is taken as the timescale. If the stress–strain curve is expressed by $\sigma = \mathrm{F}\,(\varepsilon)$, show that the radial drawing load will initially tend to increase with the amount of drawing if

$$\frac{F'(0)}{F(0)}. > \frac{1+2R}{1+R}\left\{\frac{R-(1-\alpha)\left[(1+R) - 2R\,(1-\alpha)^{2R}\right]}{R\left[1-(1-\alpha)^{1+2R}\right]}\right\}, \qquad \alpha = \frac{1}{1+R}\ln\frac{b_0}{a}.$$

6.10 A solid circular plate of radius a, whose through-thickness yield stress is μ times the planar yield stress, is loaded by a punch through a concentric rigid boss of radius b. Using the linearized yield condition of the Tresca type, show that the dimensionless collapse load $q = P/2\pi\, M_0$ is given implicitly by the relation

$$\left(\frac{q-1}{q}\right)\exp\left(-\frac{\alpha\mu - 1}{q-1}\right) = \frac{b}{a},$$

where $\alpha = 1$ when the plate is simply supported, and $\alpha = 2$ when it is fully clamped. Obtain a graphical comparison of the variation of the dimensionless collapse load for the simply supported and clamped plates when. $\mu = 1$ and $\mu = 2$.

Chapter 7
Plastic Buckling

In a typical boundary value problem involving prescribed nominal traction rates on a part S_F of the boundary surface, and prescribed velocities on the remainder S_v, more than one mode of deformation may be possible when the applied load reaches a critical value. The lack of uniqueness of the deformation mode under given boundary conditions is commonly referred to as bifurcation, the current shape and mechanical state of the body being supposed to be given or previously determined. For a linear solid, in which the strain rate is a unique linear function of the stress rate during both loading and unloading, a bifurcation mode corresponds to an eigensolution of the field equations and represents a mode quasi-statically possible under constant loads on S_F and rigid constraints on S_v. In dealing with the conventional elastic/plastic solid, which is bilinear in the sense that the strain rate is related to the stress rate by separate linear functions for loading and unloading, it is convenient to introduce a linear comparison solid with identical boundary conditions (Section 1.5). While bifurcation in the linearized solid can occur under any given traction rates on S_F and velocities on S_v when the load becomes critical, bifurcation in the actual elastic/plastic solid would occur only under those traction rates for which there is no instantaneous unloading of the material that is currently plastic. The incremental theory of plasticity will be almost exclusively used in this chapter for the estimation of the critical load.

7.1 Buckling of Axially Loaded Columns

A uniform column of length l and cross-sectional area A is assumed to be currently straight, with no restriction placed on the magnitude of its prestrain. The column is built-in at its lower end, and a compressive load P acts at its upper end along the centroidal axis. The slenderness ratio r/l is assumed to be small compared to unity, where Ar^2 denotes the second moment of area of the cross section A about its weaker principal axis. The origin of rectangular coordinates (x, y, z) is taken at the centroid of the fixed end, with the x-axis along the line of centroids and z-axis along the weaker axis of the section. We suppose that the previous straining has resulted in

J. Chakrabarty, *Applied Plasticity, Second Edition*, Mechanical Engineering Series,
DOI 10.1007/978-0-387-77674-3_7, © Springer Science+Business Media, LLC 2010

a uniform orthotropic anisotropy in the column with identical mechanical properties in the y- and z-directions (Hill and Sewell, 1960).

7.1.1 Analysis for Bifurcation

For the estimation of the critical load for bifurcation, we consider the linearized elastic/plastic solid having identical loading and unloading responses. Since the current state of stress in the column is a uniaxial compression of magnitude $\sigma = P/A$ in the axial direction, it would be sufficient to consider the deformation mode for which the components of the strain rate satisfy

$$\dot{\varepsilon}_{yy} = \dot{\varepsilon}_{zz} = -\eta \dot{\varepsilon}_{xx}, \qquad \dot{\varepsilon}_{yz} = 0, \tag{7.1}$$

where η denotes the contraction ratio corresponding to a uniaxial tension applied in the x-direction. The true strain rate is $\dot{\varepsilon}_{ij}$ related to the Jaumann stress rate, denoted here by $\dot{\tau}_{ij}$ for the linearized elastic/plastic solid, through the constitutive law. For the assumed isotropy in the yz-plane, (7.1) implies $\dot{\tau}_{yy} = \dot{\tau}_{zz}$ and $\dot{\tau}_{yz} = 0$ throughout the column, and the relevant constitutive relations may be written as

$$T\dot{\varepsilon}_{xx} = \dot{\tau}_{xx} - \eta\left(\dot{\tau}_{yy} + \dot{\tau}_{zz}\right) = \dot{\tau}_{xx} - 2\eta\dot{\tau}_{yy},$$
$$2G\dot{\varepsilon}_{xy} = \dot{\tau}_{xy}, \qquad 2G\dot{\varepsilon}_{xz} = \dot{\tau}_{xz},$$

where G is the shear modulus in the yz-plane, and T is the tangent modulus equal to the slope of the uniaxial stress–strain curve in the x-direction at the current stress σ. The preceding relations together with (7.1) furnish

$$\dot{\tau}_{ij}\dot{\varepsilon}_{ij} = T\dot{\varepsilon}_{xx}^2 + 4G\left(\dot{\varepsilon}_{xy}^2 + \dot{\varepsilon}_{xz}^2\right). \tag{7.2}$$

A sufficiently wide class of admissible velocity fields corresponding to (7.1), representing an instantaneous deflection of the centroidal axis in the xy-plane due to bending about the weaker principal axis, may be taken as

$$\left.\begin{aligned} \upsilon_x &= -\frac{y}{l}f'(\xi), \quad \upsilon_z = \frac{\eta yz}{l^2}f''(\xi), \\ \upsilon_y &= f(\xi) + \eta\left(\frac{y^2 - z^2}{2l^2}\right)f''(\xi), \end{aligned}\right\} \tag{7.3}$$

where $\xi = x/l$, and the primes denote derivatives with respect to ξ (Hill and Sewell, 1960). At the built-in end $x = 0$, the boundary condition $\upsilon_x = 0$ is satisfied by taking $f'(0) = 0$. The remaining boundary conditions $\upsilon_y = \upsilon_z = \partial\upsilon_y/\partial x = 0$ at $x = 0$ are exactly satisfied at the origin and approximately elsewhere if we also take $f(0) = 0$. The velocity field (7.3) furnishes

$$\left.\begin{aligned} \frac{\partial \upsilon_x}{\partial x} &= \dot{\varepsilon}_{xx} = -\frac{y}{l^2} f''(\xi), & \frac{\partial \upsilon_z}{\partial x} &= 2\dot{\varepsilon}_{xz} = \frac{\eta y z}{l^3} f'''(\xi), \\ \frac{\partial \upsilon_y}{\partial x} &= \frac{1}{l} f'(\xi) + \eta \left(\frac{y^2 - z^2}{2 l^3} \right) f'''(\xi), & 2\dot{\varepsilon}_{xy} &= \eta \left(\frac{y^2 - z^2}{2 l^3} \right) f''(\xi). \end{aligned}\right\} \quad (7.4)$$

Evidently, the bending is accompanied by nonuniform shearing unless $f'''(\xi)$ is identically zero. For a slender column, the shear strain rates are always small compared to the rates of extension, the three derivatives of $f(\xi)$ being supposed to be comparable with one another.

For a wide class of admissible velocity fields υ_j, a sufficient condition for uniqueness of the deformation mode for the linearized solid is given by inequality (1.79), where $\omega_{jk} = \dot{\varepsilon}_{jk} - \partial \upsilon_k / \partial x_j$ giving the alternative form

$$\int \left\{ \dot{\tau}_{ij} \dot{\varepsilon}_{ij} - \sigma_{ij} \left(2\dot{\varepsilon}_{ik} \dot{\varepsilon}_{jk} - \frac{\partial \upsilon_k}{\partial x_i} \frac{\partial \upsilon_k}{\partial x_j} \right) \right\} dV > 0 \quad (7.5)$$

for any admissible field υ_j. Using (7.2), and the fact that the only nonzero stress component is $\sigma_{xx} = -\sigma$, the inequality is reduced to

$$\int \left\{ T\dot{\varepsilon}_{xx}^2 + 4G\left(\dot{\varepsilon}_{xy}^2 + \dot{\varepsilon}_{xz}^2\right) - \sigma \left[\left(\frac{\partial \upsilon_y}{\partial x} \right)^2 + \left(\frac{\partial \upsilon_z}{\partial x} \right)^2 - \left(\frac{\partial \upsilon_x}{\partial x} \right)^2 \right] \right\} dV > 0,$$

with a minor approximation since σ is small compared to $2G$. Substituting from (7.4), and integrating through the cross section of the column, the uniqueness condition is reduced approximately to

$$\int_0^1 \left\{ \left(\frac{r}{l} \right)^2 (f'')^2 + \frac{\eta^2 G}{4T} \left(\frac{c}{l} \right)^4 (f''')^2 - \frac{\sigma}{T} (f')^2 \right\} d\xi > 0, \quad (7.6)$$

where r and c have the dimensions of length and are given by the relations

$$Ar^2 = \iint y^2 \, dydz, \qquad Ac^4 = \iint (Y^2 + z^2)^2 dydz.$$

Thus, Ar^2 is the second moment of area about the weaker axis, and Ac^4 is the fourth moment of area about the line of centroids. The terms neglected in the derivation of (7.6) are at most of order $(r/l)^2$ in comparison with the terms retained. The minimum value of the functional on the left-hand side of (7.6) must vanish at the point of bifurcation for the linearized material.

When the tangent modulus T is of the order G, the second term in the integral of (7.6), representing the contribution of shear, can be disregarded. In that case, the minimum value of the integral is given by the Euler–Lagrange differential equation

$$\frac{d^2}{d\xi^2} \left(\frac{d^2 f}{d\xi^2} + k^2 f \right) = 0, \qquad k^2 = \frac{\sigma}{T} \left(\frac{l}{r} \right)^2.$$

The solution to this differential equation, satisfying the boundary conditions $f(0) = f'(0) = f''(0) = 0$ is

$$f(\xi) = \upsilon_0 (1 - \cos k\xi),$$

where υ_0 is a constant. The remaining boundary condition $f'(1) = 0$, which ensures zero bending moment at the free end of the column to the present approximation, furnishes $k = \pi/2$, and the critical stress for bifurcation becomes

$$\sigma = \frac{\pi^2 T}{4}\left(\frac{r}{l}\right)^2.$$

This is the familiar tangent modulus formula for the buckling of a concentrically loaded straight column (Shanley, 1947). The shear contribution becomes increasingly significant as the ratio G/T increases. In the limit when G is infinitely large, with T remaining finite, the minimum value of the functional is obtained by taking $f(\xi) = \upsilon_0 \xi^2$. The substitution in (7.6) then gives $\sigma = 3T(r/l)^2$ as the bifurcation stress for a linearized rigid/plastic solid in which T is identical to the plastic modulus.

In an elastic/plastic column, as the ratio G/T increases, the tangent modulus formula increasingly underestimates the bifurcation load by a factor that has the limiting value $12/\pi^2$. The discrepancy is therefore quite significant in a structural metal when the rate of hardening is small compared to the shear modulus. Over the relevant range of values of G/T, it would be a close approximation to take

$$f(\xi) = \upsilon_0 \left[1 - \cos\left(\frac{1}{2}\pi\xi\right) + \frac{3}{4}\pi\alpha\xi^2\right], \tag{7.7}$$

where α is a disposable constant whose value must be such that the left-hand side of (7.6) is a minimum. It is easily shown that the critical stress and the best value of α are given by

$$\frac{\sigma}{T} = \pi^2\left(\frac{1 + 3\alpha}{4 + \pi^2\alpha}\right)\left(\frac{r}{l}\right)^2, \qquad \left(\frac{12}{\pi^2} - 1\right)\alpha = \frac{\eta^2 G}{4T}\left(\frac{c^4}{r^2 l^2}\right). \tag{7.8}$$

The error involved in the approximation for α never exceeds 0.1%. At one extreme, when G/T is of order unity, α is of order $(r/l)^2$, and the expression for the critical stress in (7.8) practically coincides with the tangent modulus formula. At the other extreme, when G/T is of order $(l/r)^2$, the parameter α is of order unity and the critical stress given by (7.8) is approximately the same as that for the rigid/plastic column. For any given material, the critical stress in the plastic range can be calculated by using the Ramberg–Osgood relations in the form

$$\varepsilon = \frac{\sigma}{E}\left\{1 + \frac{1}{4}\left(\frac{\sigma}{\sigma_0}\right)^{m-1}\right\}, \qquad \frac{E}{T} = 1 + \frac{m}{4}\left(\frac{\sigma}{\sigma_0}\right)^{m-1}, \tag{7.9}$$

for the uniaxial stress–strain curve, where E is Young's modulus, σ_0 is a nominal yield stress, and m is a constant. In Fig. 7.1, the critical stress given by (7.8) is plotted as a function of r/l for a square cross section ($c^4 = 1.8r^4$), assuming $G = E/3$, $\eta = \frac{1}{2}$, $E/\sigma_0 = 750$, and using three different values of m. The broken curve represents the Euler formula for elastic buckling. For the type of material chosen, the results do not differ significantly from those given by the tangent modulus theory.

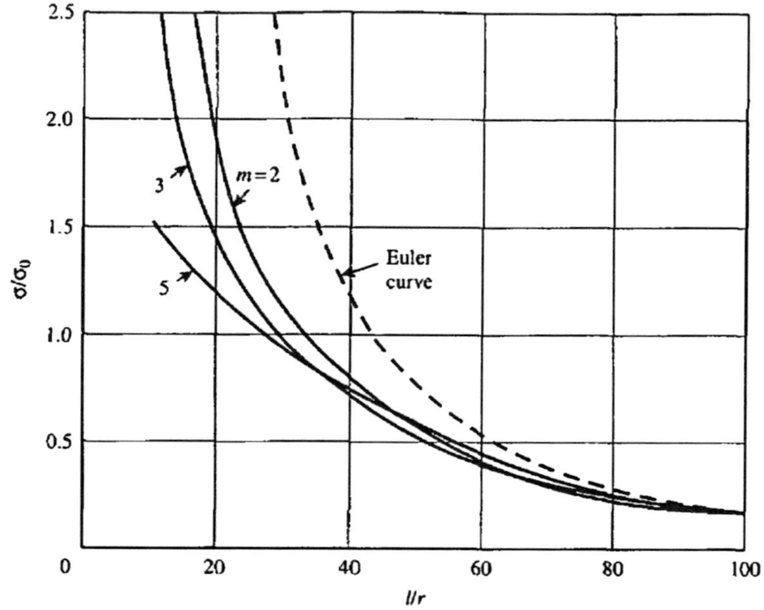

Fig. 7.1 Critical stress for bifurcation of an axially loaded column built-in at its base ($E/\sigma_0 = 750$)

The analysis for bifurcation in the linearized solid is also relevant to the bilinear elastic/plastic solid in the sense that the critical stress is certainly not lower than that given by (7.8). When this critical value is reached, the elastic/plastic column continues to deform under an increasing axial load, such that the material nowhere unloads. Since a uniform straining associated with a uniaxial compression is always a possible solution, the axial velocity of an elastic/plastic column at bifurcation may be written as

$$v_x = -\left\{\frac{v_0 x}{l} + \frac{\rho y}{l} f'(\xi)\right\},$$

where ρ is an arbitrary constant, the axial speed at the centroid of the free end being set equal to v_0. The loading condition $\dot{\varepsilon}_{xx} < 0$ therefore gives

$$1 + \frac{\rho y}{v_0 l} f''(\xi) > 0, \qquad -a \le y \le b,$$

where (a, b) are the greatest distances of the boundary of the section from the weaker axis on opposite sides. Since $f'(\xi)$ is positive throughout the column in view of (7.7), its greatest value being at $\xi = 0$, we have

$$-\frac{l}{b} \leq \frac{\pi^2}{4}\left(1 + \frac{6\alpha}{\pi}\right)\rho \leq \frac{l}{a}. \tag{7.10}$$

The column therefore begins to deflect in a range of possible ways with ρ lying in the interval (7.10), the axial stress σ given by (7.8) being critical in this sense for the actual elastic/plastic column. A variational method of treating the plastic buckling problem has been discussed by Hill and Sewell (1962). The influence of initial imperfections on the critical stress has been examined in details by Hutchinson (1972).

7.1.2 Analysis for Instability

A position of equilibrium is stable if during any geometrically possible small additional displacement the energy absorbed in the distortion exceeds the work done by the external loads. A sufficient condition for stability of an elastic/plastic solid established in Section 1.5 indicates that the associated functional is analogous to that for uniqueness, except that the stress rate appearing in the stability problem corresponds to the actual constitutive law and not its linearized version. Owing to the difference in material response for loading and unloading, the instantaneous neutral axis of bending of the column will be at some distance d from the centroidal axis measured on the convex side. The admissible velocity field may therefore be taken as

$$\left.\begin{aligned} \upsilon_x &= -\frac{y+d}{l}g'(\xi), \qquad \upsilon_z = \frac{\eta z}{l^2}(y+d)\,g''(\xi), \\ \upsilon_y &= g(\xi) + \frac{\eta}{2l^2}\left(y^2 - z^2 + 2yd\right)g''(\xi), \end{aligned}\right\} \tag{7.11}$$

where $\xi = x/l$ and $g(0) = g'(0) = 0$. The strain rate ratios given by (7.1) are maintained throughout the column, and the fixed-end conditions at $\xi = 0$ are satisfied to the present approximation. The surface of momentarily unstretched fibers, according to (7.11), is a plane parallel to the axis of the column, and we get

$$\left.\begin{aligned} \frac{\partial \upsilon_x}{\partial x} &= \dot{\varepsilon}_{xx} = -\frac{y+d}{l}g''(\xi), \qquad \frac{\partial \upsilon_z}{\partial x} = \frac{\eta z}{l^3}(y+d)\,g'''(\xi) \\ \frac{\partial \upsilon_y}{\partial x} &= \frac{1}{l}g'(\xi) + \frac{\eta}{2l^3}\left(y^2 - z^2 + 2yd\right)g'''(\xi), \\ 2\dot{\varepsilon}_{xy} &= \frac{\eta}{2l^3}\left(y^2 - z^2 + 2yd\right)g'''(\xi), \qquad 2\dot{\varepsilon}_{xz} = \frac{\eta z}{l^3}(y+d)\,g'''(\xi). \end{aligned}\right\} \tag{7.12}$$

The relationship between the components of the true strain rate $\dot{\varepsilon}_{ij}$ and the Jaumann stress rate $\dot{\sigma}_{ij}$ is identical to that for the linearized solid in the region of loading ($y \geq -d$) and is given by the generalized Hooke's law in the region of unloading ($y \leq -d$). In order to make the subsequent analysis tractable, Poisson's ratio is

$$\dot{\sigma}_{xx} - 2\eta\dot{\sigma}_{yy} = T\dot{\varepsilon}_{xx} \quad (y \geq -d)\,, \qquad \dot{\sigma}_{xx} - 2\eta\dot{\sigma}_{yy} = E\dot{\varepsilon}_{xx} \quad (y \leq -d)\,,$$
$$\dot{\sigma}_{xy} = 2G\dot{\varepsilon}_{xy}, \qquad \dot{\sigma}_{xz} = 2G\dot{\varepsilon}_{xz}, \qquad \dot{\sigma}_{yz} = 0 \text{ all y},$$

where E is Young's modulus and T is the tangent modulus for the stress–strain curve in the x-direction for the orthotropic column. Inserting from the above relations, and using (7.1), it is easily shown that

$$\dot{\sigma}_{ij}\dot{\varepsilon}_{ij} = \begin{cases} T\dot{\varepsilon}_{xx}^2 + 4G\left(\dot{\varepsilon}_{xy}^2 + \dot{\varepsilon}_{xz}^2\right), & y \geq -d, \\ E\dot{\varepsilon}_{xx}^2 + 4G\left(\dot{\varepsilon}_{xy}^2 + \dot{\varepsilon}_{xz}^2\right), & y \leq -d. \end{cases} \tag{7.13}$$

The actual surface separating the regions of loading and unloading in the column is unlikely to be a plane, but the optimum value of d obtained from the analysis should provide a good approximation to the mean distance of the unstretched fibers from the line of centroids.

For a sufficiently wide class of continuous velocity fields vanishing on the constraints, the stability of an elastic/plastic solid is guaranteed by inequality (7.5), with $\dot{\sigma}_{ij}$ written for $\dot{\tau}_{ij}$ Since the existing stress in the column is the uniaxial compression $\sigma_{xx} = -\sigma$, the inequality becomes

$$\int \left\{\dot{\sigma}_{ij}\dot{\varepsilon}_{ij} - \sigma\left[\left(\frac{\partial v_y}{\partial x}\right)^2 + \left(\frac{\partial v_z}{\partial x}\right)^2 - \left(\frac{\partial v_x}{\partial x}\right)^2\right]\right\} dV > 0.$$

The leading term in the above integral is given by (7.13). Substituting from (7.12), and carrying out the integration over the cross section of the column, the stability condition is reduced to

$$\int_0^1 \left\{\left(\frac{r}{l}\right)^2 (g'')^2 + \frac{\eta^2 G}{4\bar{E}}\left(\frac{\bar{c}}{l}\right)^4 (g''')^2 - \frac{\sigma}{E}(g')^2\right\} d\xi > 0, \tag{7.14}$$

where Ar^2 is the second moment of area about the weaker axis, $\bar{E}$ is a weighted modulus, and $\bar{c}$ is a parameter having the dimension of length. The last two quantities are defined by the relations

$$Ar^2\bar{E} = T\iint_{A_1} (y+d)^2\,dy\,dz + E\iint_{A_2} (y+d)^2 dy\,dz,$$
$$A\bar{c}^4 = \iint_A \left(y^2 + z^2\right)\left[(y+2d)^2 + z^2\right] dy\,dz, \tag{7.15}$$

where A_l and A_2 are parts of the area of cross section corresponding to $y \geq d$ and $y \leq d$, respectively. The critical stress for instability is that for which the minimum value of the integral in (7.14) is zero. For any given d, the extremum problem for $g(\xi)$ in (7.14) is exactly the same as that for $f(\xi)$ in (7.6). The correspondence between the conditions for bifurcation and instability may therefore be stated in terms of the parameters

$$\gamma = \frac{\eta^2 G}{4T}\left(\frac{c^4}{r^2 l^2}\right), \qquad \bar{\gamma} = \frac{\eta^2 G}{4\bar{E}}\left(\frac{\bar{c}^4}{r^2 l^2}\right).$$

Thus, if the quantity $(\sigma/T)(l/r)^2$ for bifurcation is denoted by $\phi(\gamma)$, then the quantity $(\sigma/E)(l/r)^2$ for instability is given by $\phi(\bar{\gamma})$, the distanced d being obtained by minimizing the instability stress (Hill and Sewell, 1960). Using the approximate formula (7.8) for bifurcation, the critical stress for instability may be written as

$$\frac{\sigma}{E} = \pi^2\left(\frac{1+3\bar{\alpha}}{4+\pi^2\bar{\alpha}}\right)\left(\frac{r}{l}\right)^2, \quad \left(\frac{12}{\pi^2}-1\right)\bar{\alpha} = \frac{\pi^2 G}{4\bar{E}}\left(\frac{\bar{c}^4}{r^2 l^2}\right), \tag{7.16}$$

where $\bar{c}$ and $\bar{E}$ are functions of d, to be determined from the condition that the optimum value of d makes the critical stress a minimum.

For a fairly wide range of values of G/T, the minimum instability stress is obtained to a close approximation by assuming that the x-component of the rate of change of load across any section vanishes at the onset of instability. Since $\dot{\sigma}_{yy} = \dot{\sigma}_{zz} = 0$ by the condition of zero normal component of the traction rate on the lateral surface, we have

$$\iint_A \dot{\sigma}_{xx} dy\, dz = T\iint_{A_l} \dot{\varepsilon}_{xx} dy\, dz + E\iint_{A2} \dot{\varepsilon}_{xx} dy\, dz = 0.$$

Substituting from (7.12) and dropping the factor $g''(\xi)$, which is constant for any given cross-section, the equation for d is obtained as

$$T\iint_{A_1} (y+d) dy\, dz + E\iint_{A_2} (y+d) dy\, dz = 0. \tag{7.17}$$

The modulus $\bar{E}$ corresponding to this particular value of d is known as the reduced modulus, which is actually the absolute minimum value of the weighted modulus. For a rectangular cross-section, $2\sqrt{E/\bar{E}} = 1 + \sqrt{E/T}$, defining the reduced modulus for the plastic buckling of ideal columns.

When E/T is of order unity, the shear effect is negligible, and the critical stress formula (7.16) coincides with the well-known formula $\sigma/\bar{E} = \left(\pi^2/4\right)(r/l)^2$, due to Engesser (1889) and von Karman (1910), for the eigenstate buckling with $\bar{E}$ denoting the reduced modulus. In the limit when E/T tends to infinity, the region A_2 disappears, and (7.16) furnishes the rigid/plastic formula $\sigma/\bar{E} = 3(r/l)^2$, with $\bar{E} = T/(1 + d^2/r^2)$ in view of (7.15), where $y = -d$ is a tangent to the section parallel to the weaker axis. For a critical review of the plastic buckling of columns, the reader is referred to Hutchinson (1974).

7.2 Behavior of Eccentrically Loaded Columns

In a real structure, a compression member is scarcely ever subjected to pure axial load acting along the line of centroids. The eccentricity of the loading, that is unavoidable in practice, causes the member to deflect as the load is increased from zero. Due to the combined effect of bending and axial compression, yielding occurs on the concave side of the member under a relatively small load, and the customary elastic analysis ceases to hold beyond this point. In a structural member made of ductile material, the deformation can continue well into the elastic/plastic range before the load attains a critical value to produce plastic collapse. The problem has been investigated by von Karman (1910), Chawalla (1937), Baker et al. (1949), Home (1956), and Chen and Astuta (1976), among others.

7.2.1 Moment-Curvature Relations

In the initial stages of deformation of an elastic/plastic column, a plastic zone in simple compression exists on the concave side, until the bending stress becomes large enough to cause yielding in tension on the convex side. The plastic zones formed on the concave and convex sides of the column will be referred to as primary and secondary plastic zones, respectively, which coexist in the later stages of the elastic/plastic deformation. The analysis for the part of the bent column in which only one side is plastic differs from that for the part in which both sides are plastic. The elastic analysis which holds in the nonplastic portion of the column, must of course be invoked to complete the elastic/plastic solution.

Consider a column whose cross-section is a rectangle of width b and depth $2\,h$ across the weaker axis, the material being ideally plastic with a uniaxial yield stress Y. A typical section of the column is under a normal compressive force P acting at the centroid, together with a bending moment M, the local curvature of the longitudinal axis of the column being denoted by κ. It is convenient to introduce the dimensionless variables

$$n = \frac{P}{P_e}, \quad m = \frac{M}{M_e}, \quad \psi = \frac{\kappa}{\kappa_e},$$

where P_e is the yield force in pure compression, equal to $2bhY$, while M_e and κ_e are the values of M and κ at the initial yielding under pure bending, and are equal to $\frac{2}{3}bh^2Y$ and Y/Eh, respectively. Evidently, $m = \psi$ for a purely elastic cross-section, irrespective of the value of $n < 1$. The greatest compressive stress due to the combined bending and axial load is of magnitude $(m + n)Y$, occurring along the edge of the rectangle on the concave side. Yielding begins therefore along this edge when $m = \psi = 1 - n$ as the column is increasingly loaded.

For $m > 1 - n$, the primary plastic zone formed on the concave side gradually spreads inward, the depth of the elastic material at any stage being denoted by $2\xi h$. The stress distribution shown in Fig. 7.2(a) indicates that the greatest tensile stress σ, which acts in the bottommost fiber, is equal to $Y(2\xi - \rho)/\rho$. Since the resultant force produced by the distribution of stress must be of magnitude $P = 2bhYn$, we have

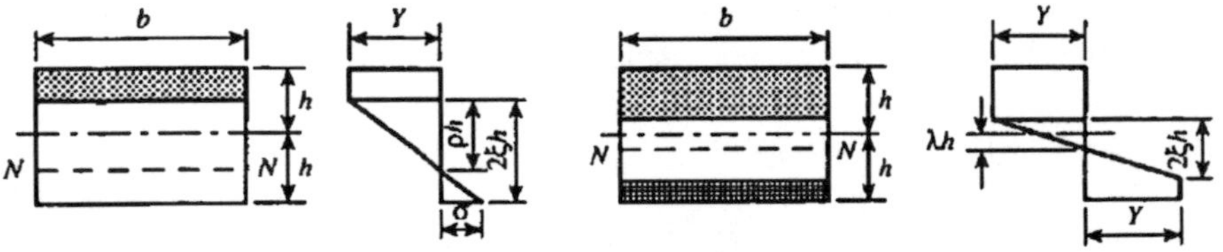

Fig. 7.2 Elastic/plastic stress distribution due to combined bending and axial thrust. (**a**) Primary plastic zone, (**b**) primary and secondary plastic zones

$$\rho^2 (2\xi - \rho)^2 + 4(1 - \xi)\rho = 4n\rho \qquad or \quad \rho = \xi^2 / (1 - n)\,.$$

Since the longitudinal strain at the elastic/plastic boundary is $-\kappa\rho h$, the corresponding stress is $-E\kappa\rho h$, which must be equal to $-Y$ for the boundary to be at the point of yielding. This gives $\psi\rho = 1$ during the primary plastic phase. Since the moment of the stress distribution about the centroidal axis must be equal to the bending moment M, the dimensionless moment and curvature become

$$m = (1 - n)(3 - 2\xi)\,, \qquad \psi = (1 - n)/\xi^2. \tag{7.18}$$

The tension side begins to yield when $\sigma = Y$, which corresponds to $\rho = \xi = 1-n$. Equations (7.18) therefore hold in the range $1 \geq \xi \geq 1 - n$, the corresponding range of values of m and ψ being

$$1 - n \leq m \leq (1 - n)(1 - 2n)\,, \qquad 1 - n \leq \psi \leq (1 - n)^{-1}\,.$$

For $\xi \leq 1 - n$, the secondary plastic zone appears on the convex side where the material yields in tension. The stress distribution over the cross-section is then modified to that shown in Fig. 7.2(b). The total depth of the elastic core is $2\xi h$, the neutral axis being midway between the two plastic boundaries. If the distance between the neutral axis and the centroidal axis is denoted by λh, then the condition of the resultant force across the section being equal to $2bhYn$ furnishes $\lambda = n$.

The curvature of the longitudinal axis is such that $E\kappa\xi h = Y$, while the bending moment is given by the resultant moment about the centroidal axis. Thus

$$m = \frac{3}{2}(1-n) - \frac{1}{2}\xi^2, \quad \psi = \xi^{-1}, \quad \xi \leq 1-n. \tag{7.19}$$

As the cross-section tends to be fully plastic, ψ tends to infinity, while m approaches the finite value $\frac{3}{2}\left(1-n^2\right)$, which is in agreement with the result for a rigid/plastic bar stressed to the yield point (Chakrabarty, 2006). The moment–curvature relation obtained by eliminating ξ from each of the two pairs of (7.18) and (7.19), with appropriate adjustment of coefficients, serves as a good approximation for a wide range of cross sections.

7.2.2 Analysis for a Pin-Ended Column

An initially straight column of length $2l$ is pin-supported at its ends and is subjected to equal and opposite forces P acting at a constant eccentricity e as shown in Fig. 7.3(a). The column has a uniform rectangular cross section whose width b is perpendicular to the line of eccentricity. Since the transverse shear forces are always small compared with longitudinal forces, the transverse reactions at the supports may be neglected. It is convenient to take the origin of coordinates at the upper end of the column, with the x-axis along the line of action of the forces, and the y-axis along the line of eccentricity. Over the range of deflections considered here, it is sufficiently accurate to write $k = -d^2v/dx^2$, where $y = v$ corresponds to the deformed centre line of the column. Hence

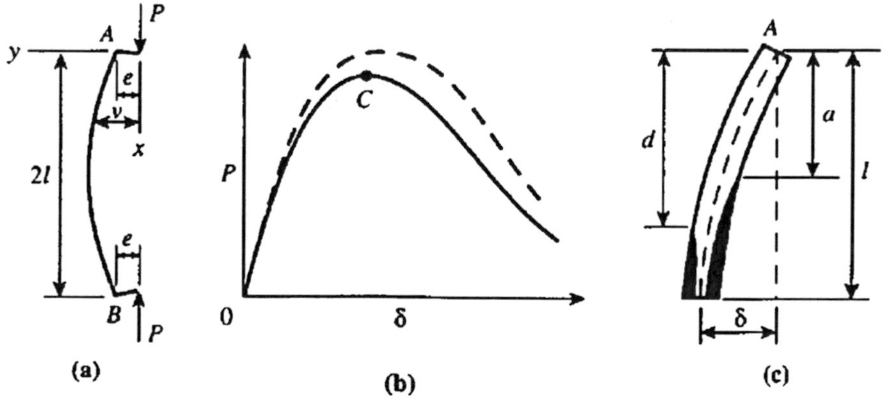

Fig. 7.3 Bent configuration, load–deflection curve, and growth of plastic zones during eccentric loading of an initially straight column

$$\psi = -h\sqrt{\frac{E}{Y}}\left(\frac{d\phi}{dx}\right), \quad \phi = \sqrt{\frac{E}{Y}}\left(\frac{dv}{dx}\right).$$

Evidently, ϕ is a measure of the local slope of the deflection curve. Combining the above two relations furnishes the expression for $d\phi/d\upsilon$. Since the bending moment at any section is $M = P\upsilon$, we have

$$\phi\frac{d\phi}{d\upsilon} + \frac{\psi}{h} = 0, \quad \frac{\upsilon}{h} = \frac{M}{Ph} = \frac{m}{3n}. \tag{7.20}$$

Equations (7.20) hold throughout the column in both the elastic and the plastic ranges of loading. Over the elastic length of the column, $\psi = m = 3nv/h$, and the differential equation for ϕ becomes

$$\phi\frac{d\phi}{d\upsilon} + \frac{3n\upsilon}{h^2} = 0.$$

This equation is immediately integrated, and the substitution for ϕ then gives the differential equation

$$\frac{d\upsilon}{dx} = k\sqrt{\alpha^2 - \frac{\upsilon^2}{h^2}}, \qquad k = \sqrt{\frac{3Yn}{E}}, \tag{7.21}$$

where α is a constant that depends on n. Since an elastic length would always exist near each end of the column, the boundary condition $\upsilon = e$ at $x = 0$ would apply to (7.21) in all stages. The integration of the above equation therefore results in

$$\frac{kx}{h} = \cos^{-1}\left(\frac{e}{\alpha h}\right) - \cos^{-1}\left(\frac{v}{\alpha h}\right). \tag{7.22}$$

So long as the column is entirely elastic, the boundary conditions $d\upsilon/dx = 0$ and $\upsilon = \delta + e$ *at* $x = l$ may be used in (7.21) and (7.22), which give $a = (\delta + e)/h$, and consequently,

$$\delta = e\left(\sec\frac{kl}{h} - 1\right) \quad \frac{kl}{h} < \frac{\pi}{2}.$$

This is the well-known secant formula for the deflection of an eccentrically loaded elastic column. Yielding commences at $x = l$ when $m = 1 - n$ at this section. The elastic solution therefore holds over the range $(\delta + e)/h \leq (1-n)/3n$, in view of (7.20), and this condition is equivalent to

$$\frac{3n}{1-n}\sec\left(\frac{kl}{h}\right) \leq \frac{h}{e}. \tag{7.23}$$

The deflection of the elastic column tends to infinity when kl/h approaches the value $\pi/2$, which corresponds to the Euler load for elastic buckling. However, for the usual values of l/h and e/h ratios, the elastic range of loading is restricted by (7.23).

When the load is increased beyond the elastic limit, the part of the column that is entirely elastic progressively decreases, the situation where the primary and secondary plastic zones coexist being shown in Fig. 7.3(c). Since the cross section $x = a$ is just at the point of yielding, $m = 1 - n$ and $\upsilon/h = (1-n)/3n$ at $x = a$ in view of (7.20). Equation (7.22) therefore gives

$$\frac{ka}{h} = \cos^{-1}\left(\frac{e}{\alpha h}\right) - \cos^{-1}\left(\frac{1-n}{3\alpha n}\right), \tag{7.24}$$

throughout the elastic/plastic loading, where a is a constant to be determined. The deflection of the center line of the column in the primary plastic zone of the elastic/plastic column is given by

$$\frac{\upsilon}{h} = \frac{1-n}{n}\left(1 - \frac{2}{3}\xi\right), \quad 1 \geq \xi \geq 1-n, \tag{7.25}$$

in view of (7.20) and (7.18). Changing the independent variable in (7.20) from v to ξ using (7.25), we obtain

$$\phi\frac{d\phi}{d\xi} - \frac{c^2}{2\xi^2} = 0, \qquad c = \frac{2(1-n)}{\sqrt{3n}},$$

in view of (7.18). This equation is readily integrated, and the result is put in the form

$$\xi = \frac{c^2}{\beta^2 - \phi^2}, \quad \psi = \frac{\sqrt{3n}}{2c^3}\left(\beta^2 - \phi^2\right)^2, \quad \xi \geq 1-n, \tag{7.26}$$

where β is a constant of integration. Since $\psi = -h\sqrt{E/Y}\,(d\phi/dx)$, the second relation of (7.26) furnishes the differential equation

$$2c^3\frac{d\phi}{dx} + \frac{k}{h}\left(\beta^2 - \phi^2\right)^2 = 0.$$

If ϕ_1 denotes the value of ϕ at $x = \alpha$, where $\xi = 1$, then the solution to the above equation may be written as

$$k\left(\frac{x-a}{h}\right) = \frac{c}{\beta}\left\{\left(\frac{\phi_1 - \xi\phi}{\beta}\right) + \frac{c^2}{\beta^2}\ln\left[\left(\frac{\beta+\phi_1}{\beta+\phi}\right)\xi^{-1/2}\right]\right\}, \quad \xi \geq 1-n, \tag{7.27}$$

on substitution from (7.26), which also gives $\phi_1 = \sqrt{\beta^2 - c^2}$, the quantity ka/h being given by (7.24). Equations (7.25), (7.26), and (7.27) define the shape of the deflection curve in the primary plastic region parametrically through ξ.

Let ξ_0 denote the value of ξ at $x = l$. For $\xi_0 \geq 1$ the secondary plastic region is absent, and the solution is then completed by finding a and β as functions of ξ_0. This is easily done by using the condition $\phi = 0$ at $\xi = \xi_0$. and the condition of continuity of ϕ across $x = a$. The first condition gives $\xi_0 = c^2/\beta^2$, while the second condition furnishes the relationship between a and β. Thus

$$\beta^2 - 3n\alpha^2 = (1-n)^2/n, \ \xi_0 = c^2/\beta^2,$$

obtained by setting $v/h = (1-n)/3n$ and $\xi = 1$ in (7.21) and (7.26), respectively. The result therefore becomes

$$\alpha = \frac{1-n}{3n}\sqrt{\frac{4}{\xi_0} - 3}, \ \beta = \frac{2(1-n)}{\sqrt{3n\xi_0}}, \ \xi_0 \geq 1-n. \tag{7.28}$$

The relationship between $(l-a)/h$ and ξ_0 at any stage is obtained by setting $x = l$, $\phi = 0$ and $\xi = \xi_0$ in (7.27). Using the fact that $c/\beta = \sqrt{\xi_0}$ and $\phi_1/\beta = \sqrt{1-\xi_0}$, we get

$$k\left(\frac{l-a}{h}\right) = \sqrt{\xi_0}\left[\sqrt{1-\xi_0} + \xi_0 \cosh^{-1}\left(\xi_0^{-1/2}\right)\right], \ \xi_0 \geq 1-n \tag{7.29}$$

It is interesting to note that the length of the plastic zone during the primary plastic phase depends only on ξ_0. When ξ_0 has been found from (7.24), (7.28), and (7.29), the corresponding central deflection δ follows from (7.25) with $\xi = \xi_0$. The value of n that marks the end of the primary plastic phase is obtained from (7.24), after ka/h has been computed from (7.29) with $\xi_0 = 1-n$.

For $\xi < 1-n$, the secondary plastic zone is formed over a length $d \leq x \leq l$, while the elastic region $0 \leq x \leq a$ is further reduced. The length of the primary plastic zone in each half of the column is given by

$$k\left(\frac{d-a}{h}\right) = \frac{c}{\beta}\left\{\frac{\phi_1 - (1-n)\phi_2}{\beta} + \frac{c^2}{\beta^2}\ln\left[\left(\frac{\beta+\phi_1}{\beta+\phi_2}\right)(1-n)^{-1/2}\right]\right\}, \tag{7.30}$$

in view of (7.27), the quantities ϕ_1 and ϕ_2 being the values of ϕ given by (7.26) at $\xi = 1$ and $\xi = (1-n)$, respectively. The lateral deflection in the primary plastic part ($a \leq x \leq d$) is still given by (7.25), while that in the secondary plastic part ($d \leq x \leq l$) is given by

$$\frac{v}{h} = \frac{1}{2}\left(\frac{1-n^2}{n} - \frac{\xi^2}{3n}\right), \quad 1-n \geq \xi \geq \xi_0, \tag{7.31}$$

in view of (7.19) and (7.20). Using (7.19) and (7.31), the differential equation in (7.20) is transformed into $3n\phi(d\phi/d\xi) = 1$, and in view of the boundary condition $\phi = 0$ at $\xi = \xi_0$, the solution is obtained as

$$\sqrt{\frac{E}{Y}}\frac{dv}{dx} = \phi = \sqrt{\frac{2}{3}\left(\frac{\xi-\xi_0}{n}\right)}, \quad 1-n \geq \xi \geq \xi_0. \tag{7.32}$$

Combining this equation with the expression for $d\upsilon/d\xi$ given by (7.31), we obtain the differential equation

$$\xi \frac{d\xi}{dx} + \frac{k}{h}\sqrt{2\,(\xi - \xi_0)} = 0.$$

Integrating, and using the boundary condition $\xi = \xi_0$ at $x = l$, the variation of ξ with x is obtained implicitly in the form

$$k\left(\frac{l-x}{h}\right) = \frac{\sqrt{2}}{3}\,(\xi + 2\xi_0)\,\sqrt{\xi - \xi_0}, \quad 1 - n \geq \xi \geq \xi_0.$$

The boundary condition $x = d$ at $\xi = 1 - n$ finally gives the semilength of the secondary plastic zone as

$$k\left(\frac{l-d}{h}\right) = \frac{\sqrt{2}}{3}\,(1 - n + 2\xi_0)\,\sqrt{1 - n - \xi_0}. \tag{7.33}$$

To complete the solution, the expressions for α and β during the secondary plastic phase must be found. The condition of continuity of ϕ across $x = d$ furnishes β in view of (7.26) and (7.32), while that across $x = a$ gives the same relationship between α and β as before. The result can be expressed as

$$\alpha = \frac{\sqrt{3\left(1 - n^2\right) - 2\xi_0}}{3n}, \quad \beta = \sqrt{\frac{2}{n}\left(1 - n - \frac{\xi_0}{3}\right)}, \quad \xi_0 \leq 1 - n. \tag{7.34}$$

Equations (7.28) and (7.34) coincide when $\xi_0 = 1 - n$ as expected. The total semilength of the plastic region at any stage is given by (7.30), (7.33), and (7.34). The value of ξ_0 for a given n is found from (7.24) and (7.28), the corresponding central deflection δ being obtained from (7.31) with $\xi = \xi_0$.

If the load acting at a constant eccentricity is increased from zero, the graph of load against central deflection shows a maximum at some point C as shown in Fig. 7.3(b). The load corresponding to this point represents the collapse load for the elastic/plastic column and is given by the condition $dn/d\xi_0 = 0$. For the graphical presentation of results, it is convenient to derive l/h for selected values of ξ_0, using given values of e/h and n. The relationship between the dimensionless terminal moment $2ne/h$ and the dimensionless length parameter $kl/h\sqrt{n}$ at the point of collapse is shown in Fig. 7.4 for constant values of n and e/h. At zero eccentricity, the critical kl/h ratio reduces to the Euler value $\pi/2$ for all values of $n \leq 1$. The effect of work-hardening of the material is to increase the critical stress for a given eccentricity, the corresponding load–deflection behavior being that indicated diagrammatically in Fig. 7.3(b) by the broken curve.

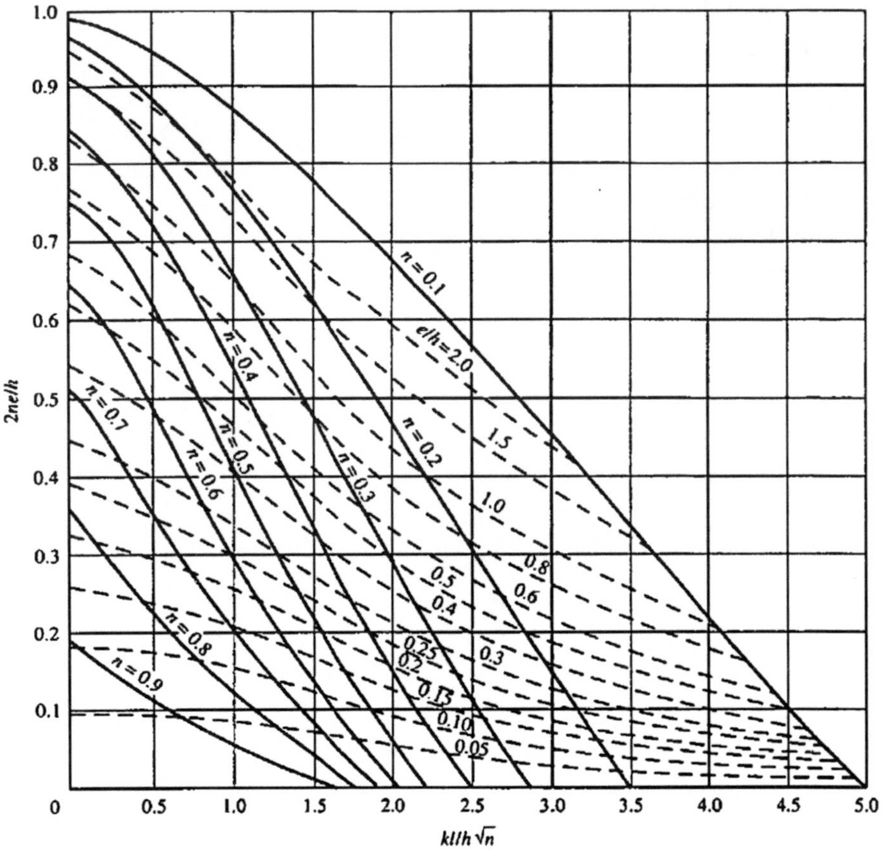

Fig. 7.4 Collapse load curves for eccentrically loaded columns made of ideally plastic material (after M.R. Horne, 1956)

7.2.3 Solution for an Inelastic Beam Column

Suppose that a lateral load $2Q$ acts at the midspan of an initially straight column of length $2l$, which is loaded eccentrically by equal and opposite axial forces P, as shown in Fig. 7.5. The axial load is assumed to be applied first, and maintained constant as the lateral load is gradually increased from zero. This assumption ensures that the material stressed into the plastic range does unload at any stage. Because of symmetry, only one-half of the beam-column will be considered, and the origin of the x-axis taken at the left-hand support. If the bending moment and curvature at any section of the beam-column are denoted by M and κ respectively, then the equation of transverse equilibirum is easily shown to be

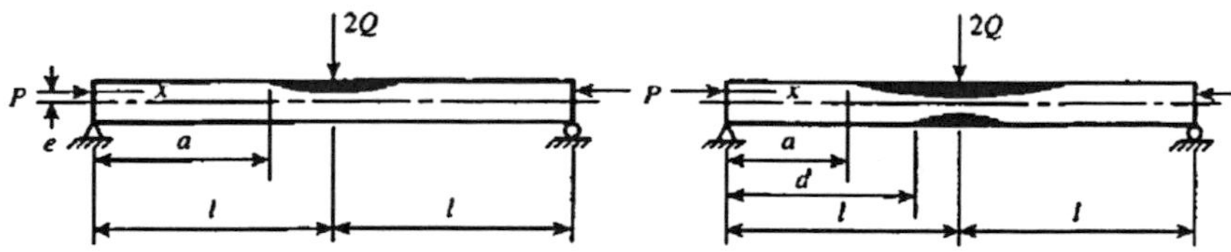

Fig. 7.5 Spread of plastic zones in an elastic/plastic beam-column, (**a**) Primary plastic phase, (**b**) secondary plastic phase

$$\frac{d^2 M}{dx^2} + P\kappa = 0.$$

In terms of the dimensionless variables m, n, and ψ already defined, the equilibrium equation becomes

$$\frac{d^2 m}{dx^2} + \frac{k^2}{h^2}\psi = 0, \quad k = \sqrt{\frac{3nY}{E}}. \tag{7.35}$$

For the elastic part of the beam-column, the substitution $\psi = m$ results in a second-order linear differential equation for m. If the end of the beam-column remains elastic, the boundary condition $M = Pe$ *or* $m = 3ne/h$ *at* $x = 0$ applies to the elastic solution, for which the bending moment distribution is given by

$$m = \frac{3ne}{h}\cos\left(\frac{kx}{h}\right) + A\sin\left(\frac{kx}{h}\right), \tag{7.36}$$

where A is a constant. When the beam-column is completely elastic, the condition $dM/dx = Q$ or $dm/dx = q/l$ at $x = l$ may be applied to (7.36), where q denotes the quantity Ql/M_e, which is the ratio of the applied lateral load to that at the initial yielding under zero axial thrust. This condition gives the constant A in (7.36), which therefore becomes

$$m = \left\{\frac{3ne}{h}\cos\left(\frac{k(l-x)}{h}\right) + \frac{qh}{kl}\sin\left(\frac{kx}{h}\right)\right\}\sec\left(\frac{kl}{h}\right) \tag{7.37}$$

for the fully elastic beam-column. The central deflection δ in the elastic range is most conveniently obtained from the relation

$$\delta = \frac{Y}{Eh}\int_0^1 \psi x dx = \frac{Y}{Eh}\int_0^1 m x dx,$$

which may be verified by using the relation $Y\psi/Eh = -d^2 \upsilon/dx^2$, and the boundary condition $d\upsilon/dx = 0$ at $x = l$. Substituting from (7.37), and integrating, we have

$$\frac{\delta}{h} = \frac{e}{h}\left[\sec\left(\frac{kl}{h}\right) - 1\right] + \frac{q}{3n}\left[\frac{h}{kl}\tan\left(\frac{kl}{h}\right) - 1\right]. \tag{7.38}$$

Since the bending moment has its greatest value at the central section, yielding begins when $m = 1 - n$ *at* $x = l$, giving

$$\frac{qh}{kl} = \left[(1-n)\cos\left(\frac{kl}{h}\right) - \frac{3ne}{h}\right]\operatorname{cosec}\left(\frac{kl}{h}\right) \tag{7.39}$$

at the initial yielding in view of (7.37). The corresponding central deflection follows from (7.38) on substitution from (7.39). It may be noted that the deflection in the elastic range tends to infinity when kl/h approaches the Euler value $\pi/2$. Solutions for other types of loading on the elastic beam-column have been given by Timoshenko and Gere (1961). The elastic/plastic beam-column has been treated by Ketter (1961) and Chen (1970).

When q exceeds the value specified by (7.39), a primary plastic zone extends from $x = l$ to $x = a$ at a generic stage, the elastic part of the beam-column being of length a, as shown in Fig. 7.5(a). Equation (7.36) holds for the elastic part with the constant of integration obtained from the condition $m =$ 1–n at $x = a$. Thus

$$m = \left\{\frac{3ne}{h}\sin\left(\frac{k(a-x)}{h}\right) + (1-n)\sin\left(\frac{kx}{h}\right)\right\}\operatorname{cos\,ec}\left(\frac{ka}{h}\right), \tag{7.40}$$

which is valid for $0 \leq x \leq a$. For any given values of l/h and n, the relationship between a/h and q can be found from the solution for the plastic part. Substituting from (7.18) into (7.35), and multiplying it by $d\xi/dx$, the equilibrium equation may be expressed as

$$\left\{\frac{d}{d\xi}\left(\frac{d\xi}{dx}\right)^2 - \frac{k^2}{h^2\xi^2}\right\}\frac{d\xi}{dx} = 0.$$

Since $d\xi/dx \neq 0$, the expression in curly brackets must vanish for equilibrium. Integrating and taking the square root, we get

$$\frac{d\xi}{dx} = -\frac{k}{h}\sqrt{\frac{1}{\alpha} - \frac{1}{\xi}}, \quad 1 \geq \xi \geq 1 - n, \tag{7.41}$$

where α is a constant of integration. In view of the boundary condition $\xi = 1$ at $x = a$, the integration of the above equation results in

$$\begin{aligned} k\left(\frac{x-a}{h}\right) = \sqrt{\alpha}\Bigg\{&\sqrt{1-\alpha} - \xi\sqrt{1 - \frac{\alpha}{\xi}} \\ &+ \alpha\tanh^{-1}\sqrt{1-\alpha} - \alpha\tanh^{-1}\sqrt{1-\frac{\alpha}{\xi}}\Bigg\}, \end{aligned} \tag{7.42}$$

which holds for $1 \geq \xi \geq 1 - n$. The constant a follows from the condition of continuity of dm/dx across $x = a$. Since $dm/dx = -2(1 - n)\, d\xi/dx$ in the primary plastic region in view of (7.18), (7.40) and (7.41) furnish

$$\sqrt{\frac{1-\alpha}{\alpha}} = \frac{1}{2}\left\{\cos\left(\frac{ka}{h}\right) - \left(\frac{3n}{1-n}\right)\frac{e}{h}\right\}\operatorname{cosec}\left(\frac{ka}{h}\right), \qquad (7.43)$$

At any given stage, ξ steadily decreases from unity at $x = a$ to some value ξ_0 at $x = l$. When a has been found from (7.43) for a selected value of $a/h, \xi_0$ can be determined from (7.42), considered at $x = l$. The corresponding value of q is finally obtained from the condition $dm/dx = q/l$ at $x = l$. It follows from (7.18) and (7.41) that

$$\frac{qh}{kl} = 2(1-n)\sqrt{\frac{1}{\alpha} - \frac{1}{\xi_0}}, \quad 1 \geq \xi_0 \geq 1 - n. \qquad (7.44)$$

The central deflection of the beam-column during the primary plastic phase can be determined from the relation

$$\frac{\delta}{h} = \frac{Y}{E}\int_0^a \left(\frac{mx}{h}\right)\frac{dx}{h} + \frac{1-n}{3n}\int_{\xi_0}^1 \left(\frac{kx}{h}\right)\sqrt{\frac{\alpha\xi}{\xi - \alpha}}\frac{d\xi}{\xi^2}.$$

The second integral follows on substitution for ψ and $dx/d\xi$, using (7.18) and (7.41), respectively. The integration is facilitated by a change of variable to $t = \sqrt{1 - \alpha/\xi}$. Substituting from (7.40) and (7.42), and integrating, we obtain the result

$$\begin{aligned}\frac{\delta}{h} = {} & \frac{e}{h}\left[\frac{ka}{h}\operatorname{cosec}\left(\frac{ka}{h}\right) - 1\right] + \frac{1-n}{3n}\left[1 - \frac{ka}{h}\cot\left(\frac{ka}{h}\right)\right] \\ & + 2\left(\frac{1-n}{3n}\right)\left[(1-\xi_0) + \frac{ka}{h}\sqrt{\frac{1-\alpha}{\alpha}} - \frac{kl}{h}\sqrt{\frac{1}{\alpha} - \frac{1}{\xi_0}}\right],\end{aligned} \qquad (7.45)$$

which holds for $1 \geq \boldsymbol{\xi_0} \geq 1 - \text{n}$, the quantity kl/h being that given by (7.42) with $x = l$ and $\xi = \boldsymbol{\xi}_0$. When $\boldsymbol{\xi}_0 = 1$, the expression in the last pair of square brackets vanishes, and (7.45) reduces to (7.38) with $l = \alpha$ in view of (7.39).

When q exceeds the value given by (7.44) with $\xi_0 = 1 - \text{n}$, the secondary plastic zone forms in a central part extending from $x = d$ to $x = l$, as shown in Fig. 7.5(b). An elastic part still exists over the length $0 \leq x \leq l$, provided n is sufficiently small. By (7.42), the length of the primary plastic zone at any stage is given by

$$\begin{aligned}k\left(\frac{d-a}{h}\right) = \sqrt{\alpha}\Bigg\{ & \sqrt{1-\alpha} - (1-n)\sqrt{1 - \frac{\alpha}{1-n}} + \alpha\tanh^{-1}\sqrt{1-\alpha} \\ & -\alpha\tanh^{-1}\sqrt{1 - \frac{\alpha}{1-n}}\Bigg\},\end{aligned} \qquad (7.46)$$

in view of (7.42). The bending moment distribution in the primary plastic region ($a \leq x \leq d$) is given by (7.18) and (7.42), where α is given by (7.43). To obtain the distribution of m in the secondary plastic region ($d \leq x \leq l$), we insert (7.19) into the equilibrium equation (7.35), multiply it by $2\xi(d\xi/dx)$ and rearrange it as

$$\left\{\frac{d}{d\xi}\left(\xi\frac{d\xi}{dx}\right)^2 - \frac{2k^2}{h^2}\right\}\frac{d\xi}{dx} = 0.$$

Setting the expression in curly brackets to zero, integrating this equation, and taking the square root, we obtain the differential equation

$$\xi\frac{d\xi}{dx} = -\sqrt{2}\frac{k}{h}\sqrt{\beta+\xi}, \quad \xi \leq 1-n, \tag{7.47}$$

where β is a constant of integration. Integrating again, and using the boundary condition $\xi = 1-n$ at $x = d$, the solution is found to be

$$k\left(\frac{x-d}{h}\right) = \frac{\sqrt{2}}{3}\left[(2\beta-\xi)\sqrt{\beta+\xi} - (2\beta-1+n)\sqrt{\beta+1-n}\right], \tag{7.48}$$

which applies in the range $1-n \geq \xi \geq \xi_0$. The constant β may now be determined from the condition that dm/dx is continuous across $x = d$. Equating the right-hand side of (7.47) to $2(1-n)$ times the right-hand side of (7.41), and setting $\xi = 1-n$, we obtain the relationship between α and β as

$$\beta = (1-n)\left[2(1-n)\left(\frac{1-\alpha}{\alpha}\right) - (1+2n)\right]. \tag{7.49}$$

The value of q at any stage of the secondary plastic phase is finally obtained from the relation $q/l = dm/dx$ at $x = l$. Using (7.19) and (7.47), it is easily shown that

$$\frac{qh}{kl} = \sqrt{2(\beta+\xi_0)}, \quad \xi_0 \leq 1-n. \tag{7.50}$$

For given values of l/h, e/h, and n, the relationship a/h and ξ_0 is given by (7.46), and the equation

$$k\left(\frac{1-d}{h}\right) = \frac{\sqrt{2}}{3}\left[(2\beta-\xi_0)\sqrt{\beta+\xi_0} - (2\beta-1+n)\sqrt{\beta+1-n}\right]$$

obtained from (7.48), the constants α and β being given by (7.43) and (7.49), respectively. The central deflection of the beam-column during the secondary plastic phase is obtained from the relation

$$\frac{\delta}{h} = \int_0^a \left(\frac{mx}{h}\right)\frac{dx}{h} + \frac{1-n}{3n}\int_{1-n}^{1}\left(\frac{kx}{h}\right)\sqrt{\frac{\alpha\xi}{\xi-\alpha}\frac{d\xi}{\xi^2}}$$
$$+\frac{\sqrt{2}}{3}\int_{\xi_0}^{1-n}\left(\frac{kx}{2h}\right)\frac{d\xi}{\sqrt{\beta+\xi}},$$

which follows from (7.18), (7.19), (7.41), and (7.47). The first integral is directly obtained on substitution from (7.40), while the other two integrals are evaluated by integration by parts using (7.41) and (7.47). The result may be expressed as

$$\frac{\delta}{h} = \frac{e}{h}\left[\frac{ka}{h}\operatorname{cos\,ec}\left(\frac{ka}{h}\right)-1\right] + \frac{1-n}{n}\left[\frac{1+n}{2} - \frac{ka}{3h}\cot\left(\frac{ka}{h}\right)\right] - \frac{\xi_0^2}{6n}$$
$$+2\left(\frac{1-n}{3n}\right)\left[\frac{ka}{h}\sqrt{\frac{1-\alpha}{\alpha}} - \frac{kd}{h}\sqrt{\frac{1}{\alpha}-\frac{1}{1-n}}\right] \qquad (7.51)$$
$$+\frac{\sqrt{2}}{3n}\left[\frac{kd}{h}\sqrt{\beta+1-n} - \frac{kl}{h}\sqrt{\beta+\xi_0}\right],$$

which holds for $\xi_0 \leq l\text{–}n$. Since n is held constant during the loading, the parameter kl/h has a constant value in any given beam-column problem. As ξ_0 tends to zero, the lengths of the elastic and plastic parts of the beam-column approach their limiting values. When the end loading is sufficiently large, the beam-column becomes plastic on one side before the lateral load is applied, a slight modification of the solution being necessary to deal with such a situation (Chen, 1970).

The variations of lateral load, central deflection, and elastic length with central curvature in dimensionless form, based on the preceding analysis, are shown by solid curves in Fig. 7.6, assuming $n = 0.2$, $\sqrt{3}e/h = 1$, $l/h = 23.1$, and $\sqrt{E/Y} = 29.7$. The broken curve corresponds to $n = 0.4$, which renders the beam-column partially plastic before the lateral load is applied, the expression in the square brackets of (7.39) being negative in this case. The failure of the beam-column in each case corresponds to a maximum value of q, which occurs before the central cross-section becomes fully plastic. A numerical solution for the beam-column made of a strain-hardening material, has been discussed by Lu and Kamalvand (1968), and also by Chen and Astuta (1976).

7.3 Lateral Buckling of Beams

When a beam is bent in the plane of greatest flexural resistance, it may buckle lat erally under a certain critical value of the applied load. Unless appropriate lateral supports are provided, the possibility of this kind of buckling must be taken into account in the design of beams. Over the practical range of beam dimensions, buckling frequently occurs after the beam has been rendered partially plastic in bending. We shall begin by considering the lateral buckling of beams of narrow rectangular

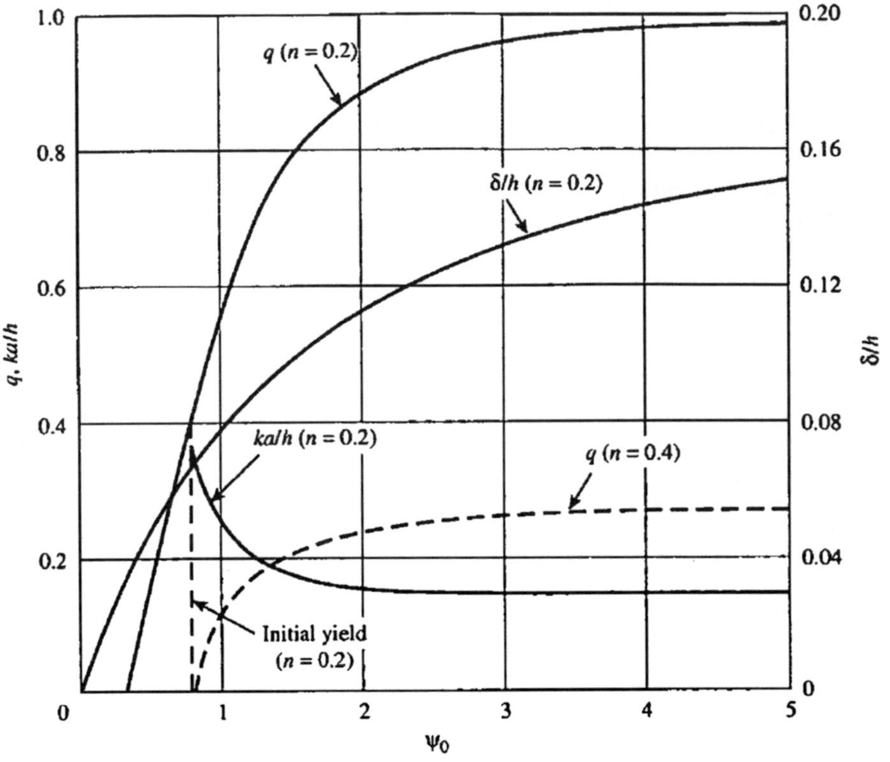

Fig. 7.6 Elastic/plastic response of simply supported beam-columns under an eccentric axial load of constant magnitude

cross section subjected to pure bending. The extension of the theory to the lateral buckling of beams under transverse loading will be subsequently discussed.

7.3.1 Pure Bending of Narrow Beams

Consider a uniform beam of rectangular cross section having its width $2b$ small compared to its depth $2h$. We choose rectangular coordinates (x, y, z) in which the z-axis coincides with the center line of the beam, while the x- and y-axes are parallel to the shorter and longer sides, respectively, of the cross section, Fig. 7.7. Let $2c$ denote the depth of the elastic core of the cross section at the incipient buckling under the action of terminal couples M. The mode of buckling consists of instantaneous bending in both the horizontal and the vertical planes, as well as a nonuniform twisting caused by rotation of the cross sections about the centroidal axis.

Each element of the beam is currently under a uniaxial stress σ in the z-direction with its magnitude varying with the distance of the element from the xz-plane. For

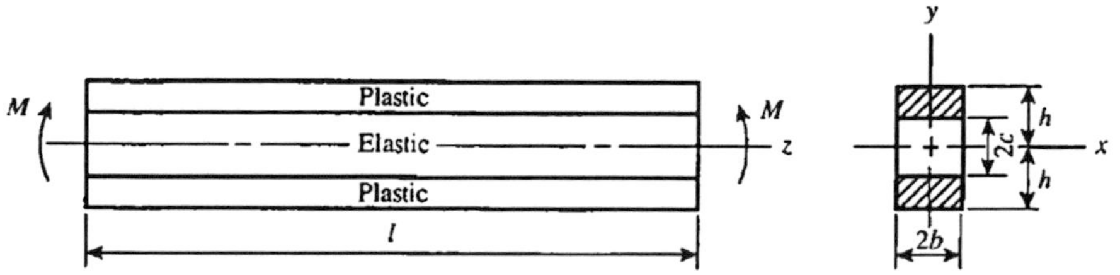

Fig. 7.7 Lateral buckling of an elastic/plastic beam of narrow rectangular cross section subjected to pure bending

a regular isotropic yield surface, it is natural to consider a deformation mode which corresponds to

$$\dot{\varepsilon}_{xx} = \dot{\varepsilon}_{yy} = -\eta \dot{\varepsilon}_{zz}, \ \dot{\varepsilon}_{xy} = 0, \tag{7.52a}$$

where η is the contraction ratio which is equal to the Poisson ratio ν in the elastic region, and which depends on the magnitude of σ. The above relations imply $\dot{\tau}_{xx} = \dot{\tau}_{yy}$ and $\dot{\tau}_{xy} = 0$, the remaining constitutive relations for the linearized elastic/plastic material being

$$T\dot{\varepsilon}_{zz} = \dot{\tau}_{zz} - 2\eta \dot{\tau}_{xx}, \quad 2G\dot{\varepsilon}_{xz} = \dot{\tau}_{xz}, \quad 2G\dot{\varepsilon}_{yz} = \dot{\tau}_{yz}, \tag{7.52b}$$

where T is the usual tangent modulus, which is equal to E over the elastic range. It follows from (7.52), together with the relation $\dot{\tau}_{xx} = \dot{\tau}_{yy}$, that

$$\dot{\tau}_{ij}\dot{\varepsilon}_{ij} = E\dot{\varepsilon}_{zz}^2 + 4G\left(\dot{\varepsilon}_{xz}^2 + \dot{\varepsilon}_{yz}^2\right).$$

A sufficient condition for uniqueness of the deformation mode is given by inequality (7.5), and the substitution $\sigma_{zz} = \sigma$ as the only nonzero stress component reduces this inequality to

$$\int \left\{ (T - \sigma)\,\dot{\varepsilon}_{zz}^2 + 4G\left(\dot{\varepsilon}_{xz}^2 + \dot{\varepsilon}_{yz}^2\right) + \sigma \left[\left(\frac{\partial v_x}{\partial z}\right)^2 + \left(\frac{\partial v_y}{\partial z}\right)^2 \right] \right\} dV > 0 \tag{7.53}$$

to a close approximation, the integral being extended over the entire volume of the material of the beam. The uniqueness is guaranteed when (7.53) is satisfied for all possible velocity fields consistent with (7.52).

A sufficiently wide class of admissible velocity fields, representing simultaneous bending in two orthogonal planes together with nonuniform twisting and warping of the cross sections, would be too complicated for practical purposes. On the other hand, the velocity gradients appearing in (7.53) can be determined with sufficient accuracy from the relatively simple field

$$\upsilon_x = u - \dot{\phi}y, \ \upsilon_y = \upsilon - \dot{\phi}x, \qquad \upsilon_z = \left(x\frac{du}{dz} + y\frac{d\upsilon}{dz}\right), \tag{7.54}$$

where (u, υ) denotes the velocity of the axis of the beam, and $\dot{\phi}$ denotes the rate of twist about this axis, all these quantities being functions of z only. It follows from (7.54) that

$$\left.\begin{aligned} \frac{\partial \upsilon_z}{\partial z} &= -\left(x\frac{d^2u}{dz^2} + y\frac{d^2\upsilon}{dz^2}\right), \\ \frac{\partial \upsilon_x}{\partial z} &= \frac{du}{dz} - y\frac{d\phi}{dz}, \ \frac{\partial \upsilon_y}{\partial z} = \frac{d\upsilon}{dz} + x\frac{d\dot{\phi}}{dx}. \end{aligned}\right\} \tag{7.55}$$

The assumed velocity field (7.54) is consistent with the fact that $\dot{\varepsilon}_{xy} = 0$ throughout the beam, but it does not give a nonzero value of $\dot{\varepsilon}_{xx} = \dot{\varepsilon}_{yy.}$ The field is also unsuitable for calculating the shear rates $\dot{\varepsilon}_{xz}$ and $\dot{\varepsilon}_{yz}$, since the rate of warping of the cross section has been disregarded. It is reasonable, however, to assume the relations

$$\dot{\varepsilon}_{xz} = 0, \quad \dot{\varepsilon}_{yz} = x\frac{d\dot{\phi}}{dz}, \tag{7.56}$$

which apply to the elastic torsion of bars of narrow rectangular cross section. The usefulness of such simplified velocity fields with appropriate adjustments has been noted by Pearson (1956) in the context of elastic buckling. The substitution from (7.55) and (7.56) into (7.53) gives

$$\iiint (T-\sigma)\left(x\frac{d^2u}{dz^2} + y\frac{d^2\upsilon}{dz}\right)^2 + 4G\left(x\frac{d\dot{\phi}}{dz}\right)^2 + \left\{\sigma\left[\left(\frac{du}{dz} - y\frac{d\dot{\phi}}{dz}\right)^2 + \left(\frac{d\upsilon}{dz} + x\frac{d\dot{\phi}}{dz}\right)^2\right]\right\} dx\,dy\,dz > 0,$$

where x varies between $-b$ and b, y varies between $-h$ and h, and z varies between 0 and l. The evaluation of the above integral is greatly simplified by the fact that $\sigma(-y) = -\sigma(y)$ and $T(-y) = T(y)$, giving

$$\int_{-h}^{h} \sigma\,dy = \int_{-h}^{h} \sigma y^2 dy = \int_{-h}^{h} Ty dy = 0.$$

Since σ and T are independent of x, and the stress distribution across each vertical section is statically equivalent to a bending moment equal to M, the condition for uniqueness becomes

$$\int_0^l \left\{ \alpha EI_y \left(\frac{du}{dz}\right)^2 + \beta EI_x \left(\frac{d\upsilon}{dz}\right)^2 + GJ\left(\frac{d\dot{\phi}}{dz}\right)^2 + 2M\left(\frac{du}{dz}\frac{d\dot{\phi}}{dz}\right)\right\} dz > 0, \tag{7.57}$$

where I_x and I_y are the principal moments of inertia of the cross section about the x- and y-axes, respectively, and GJ is the torsional rigidity with $J = 4I_y$, the constants α and β being defined as

$$\alpha Eh = \frac{1}{2}\int_{-h}^{h} T dy = R\sigma_0, \ \ \beta Eh^3 = \frac{3}{2}\int_{-h}^{h} Ty^2 dy,$$

where R is the radius of curvature of the bent axis at the incipient buckling, and σ_0 the magnitude of the numerically largest stress that occurs at $y = \pm h$. The deformation prior to buckling is assumed to be small, so that the longitudinal strain is equal to $-y/R$ throughout the elastic/plastic bending.

The minimum value of the functional in (7.57) with respect to the variations of u, υ, and $\dot{\phi}$ must vanish at the point of bifurcation. The Euler–Lagrange differential equations characterizing this variational problem are easily established as

$$\left.\begin{aligned} &\alpha EI_y \frac{d^4u}{dz^4} - M\frac{d^2\dot{\phi}}{dz^2} = 0, \quad \frac{d^4\upsilon}{dz^4} = 0, \\ &GJ\frac{d^2\dot{\phi}}{dz^2} + M\frac{d^2u}{dz^2} = 0. \end{aligned}\right\} \tag{7.58}$$

The ends of the beam are assumed to be supported in such a way that they cannot rotate about the z-axis. Then the curvature rate in the xz-plane also vanishes at these sections, and the boundary conditions become

$$\dot{\phi} = \frac{d^2u}{dz^2} = 0 \ \text{at}\, z = 0 \ \text{and} \ z = l.$$

Due to the symmetry of the loading, the curvature rate in the yz-plane must have identical values at the sends of the beam. The first two equations of (7.58) may therefore be integrated twice to give

$$\alpha EI_y = \frac{d^2u}{dz^2} - M\dot{\phi} = 0, \quad \frac{d^2\upsilon}{dz^2} = \text{constant}. \tag{7.59}$$

The second equation of (7.59) merely states that the axis of the beam continues to bend into a circular arc at the point of bifurcation. The elimination of d^2u/dz^2 between the first equation of (7.59) and the last equation of (7.58) results in

$$\frac{d^2\dot{\phi}}{d\xi^2} + k^2\dot{\phi} = 0, \ k = \frac{Ml}{EI_y}\sqrt{\frac{1+\nu}{2\alpha}}, \tag{7.60}$$

where $\xi = z/l$ and $I_y = 4bh^3/3$. The solution to this differential equation, subject to the boundary conditions $\phi = 0$ at $\xi = 0$ and $\xi = 1$, is readily obtained as

$$\dot{\phi} = A \sin \pi\xi, \quad k = \pi,$$

where A is a constant. Over the elastic range, we have $\sigma_0 = Eh/R$ giving $\alpha = 1$, and the critical moment is given by $Ml/EI_y = \pi\sqrt{2/(1+\upsilon)}$, which is in agreement with the well-known result for elastic buckling of narrow rectangular beams.

When the buckling occurs in the plastic range, α depends on the magnitude of the critical moment through the strain-hardening characteristic of the material. Assuming the power law $\sigma/Y = (E\varepsilon/Y)^n$ for the uniaxial stress–strain curve in the plastic range, the stress distribution for $y \geq 0$ may be written as

$$\frac{\sigma}{Y} = -\frac{y}{c}\ (0 \leq y \leq c), \quad \frac{\sigma}{Y} = -\left(\frac{y}{c}\right)^n (c \leq y \leq h).$$

The radius of curvature of the bent axis, when the elastic/plastic boundary is at $y = c$, is $R = Ec/Y$. The bending moment across any section is

$$M = -4b \int_0^h \sigma y dy = \frac{4}{3} bh^2 Y \left\{ \frac{3}{2+n} \left(\frac{h}{c}\right)^n - \frac{1-n}{2+n} \left(\frac{c}{h}\right)^2 \right\}.$$

Introducing the initial yield moment $M_e = 4bh^2Y/3$, and using the fact that h/c is equal to the curvature ratio k/k_e, the moment–curvature relationship may be expressed in the dimensionless form

$$\frac{M}{M_e} = \frac{3}{2+n} \left(\frac{\kappa}{\kappa_e}\right)^n - \frac{1-n}{2+n} \left(\frac{\kappa_e}{\kappa}\right)^2. \tag{7.61}$$

Since the greatest bending stress at any stage is $\sigma_0 = Y(h/c)^n$ during the elastic/plastic bending, we have

$$\alpha = \frac{R\sigma_0}{Eh} = (c/h)^{1-n} = (\kappa_e/\kappa)^{1-n}.$$

Substituting for α and M in (7.60) and setting $k = \pi$, the equation for the critical curvature ratio for lateral buckling is obtained as

$$\frac{3}{2+n} \left(\frac{\kappa}{\kappa_e}\right)^{(1+n)/2} - \frac{1-n}{2+n} \left(\frac{\kappa_e}{\kappa}\right)^{(3+n)/2} = \frac{\pi Eh}{Yl} \sqrt{\frac{2}{1+\nu}} = \lambda \text{ (say)}. \tag{7.62}$$

When κ/κ_e is computed from (7.62) for any given values of n and Eh/Yl, the critical bending moment follows from (7.61). This solution is valid for $\kappa/\kappa_e \geq 1$, which is equivalent to the condition $\lambda \geq 1$. In the case of $\lambda \leq 1$, the buckling will occur in the elastic range when $M/M_e = \lambda$, obtained by setting $n = 1$ in (7.62). Figure 7.8 shows the variation of the dimensionless critical bending moment with the parameter μ for several values of n.

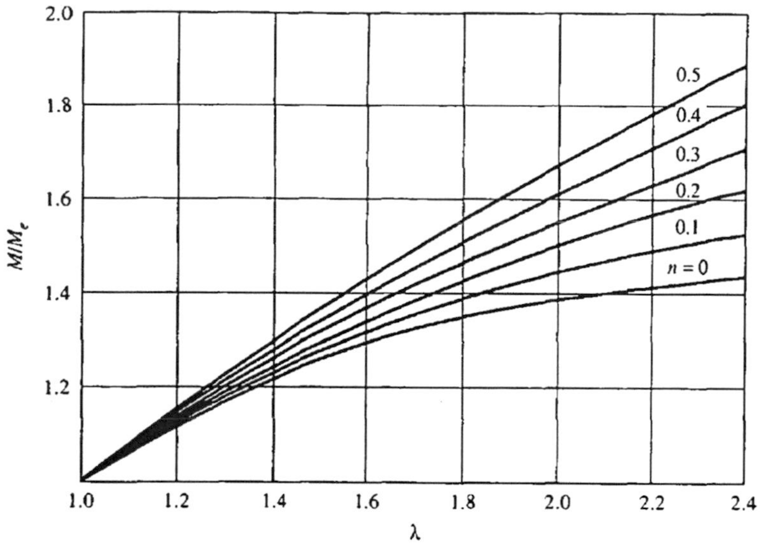

Fig. 7.8 Dimensionless critical bending couple for the lateral buckling in the plastic range as a function of a parameter $\lambda\,(Eh/Yl)\,\sqrt{2/(1+\nu)}$

Equations (7.59) are equally applicable to the lateral buckling of beams of arbitrary doubly symmetric cross sections $\left(J \neq 4I_y\right)$ subjected to terminal couples in the *yz*-plane, which is considered as the plane of maximum flexural rigidity. The last equation of (7.58) must be modified, however, by the inclusion of the term $-EC_w\left(d^4\dot{\phi}/dz^4\right)$ on the left-hand side to take account of the warping rigidity, the warping constant C_w being identical to that for elastic buckling (Timoshenko and Gere, 1961). The governing differential equation (7.60) is then replaced by a fourth-order equation for $\dot{\phi}$, which can be solved in terms of trigonometric and hyperbolic functions.

7.3.2 Buckling of Transversely Loaded Beams

We begin with a cantilever beam of narrow rectangular cross section carrying a terminal load P applied in the *yz*-plane, as shown in Fig. 7.9. As the load is increased to its critical value, the beam can buckle laterally in the *xz*-plane, and this is accompanied by the rotation of cross sections by varying degrees along the beam. Following the customary treatment for the elastic buckling, the basic equations for the plastic analysis under transverse loads are assumed to be the same as those in pure bending, provided M is regarded as the local bending moment equal to $-P(l-z)$. The basic differential equations involving u and $\dot{\phi}$ therefore become

$$\alpha EI_y\frac{d^2u}{dz^2} + P\,(l-z)\,\dot{\phi} = 0, \qquad GJ\frac{d^2\dot{\phi}}{dz^2} - P\,(l-z)\,\frac{d^2u}{dz^2} = 0.$$

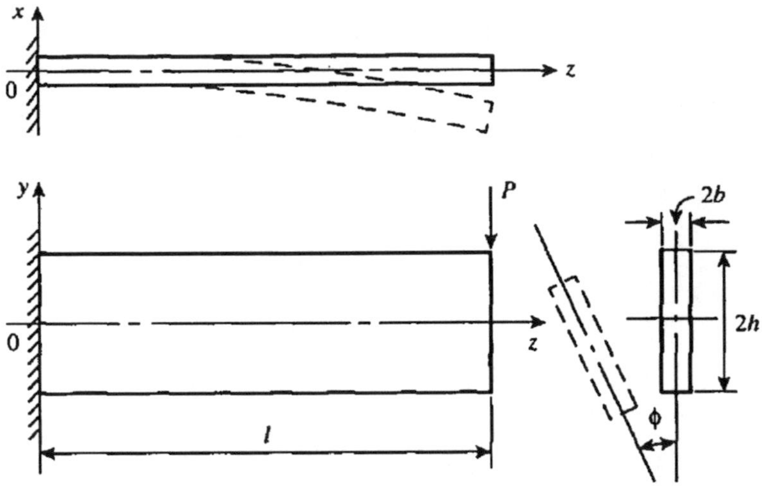

Fig. 7.9 Lateral plastic bucking of an end-loaded cantilever of narrow rectangular cross section

Eliminating d^2u/dz^2 between these two relations and setting $z = l(1-\xi)$, we obtain the differential equation for $\dot{\phi}$ as

$$\alpha \frac{d^2\dot{\phi}}{d\xi^2} + \beta^2\xi^2\dot{\phi} = 0, \; \beta = \frac{Pl^2}{EI_y}\sqrt{\frac{1+\nu}{2}}, \tag{7.63}$$

where α is a function of ξ in the elastic/plastic portion of the beam. Evidently, $\alpha = 1$ over the part of the beam that is entirely elastic. If the length of the elastic portion of the beam is denoted by a, then

$$\alpha = 1 \quad (0 \leq \xi \leq a/l)\,, \; \alpha = \rho^{-(1-n)} \qquad (a/l \leq \xi \leq 1)\,,$$

where ρ denotes the absolute value of the curvature ratio κ/κ_e, the variation of ρ with ξ in the elastic/plastic portion of the beam being given by (7.61) with the substitution $Pl\xi$ for M. Since $Pa = M_e$, the result is

$$s = \frac{l}{a}\xi = \left(\frac{3}{2+n}\right)\rho^n - \left(\frac{1-n}{2+n}\right)\rho^{-2}. \tag{7.64}$$

It is convenient at this stage to change the independent variable from ξ to s. Since $M_e = 4bh^2Y/3$ and $I_y = 4hb^3/3$, the differential equation (7.63) is transformed into

$$\frac{d^2\dot{\phi}}{ds^2} + k^2\lambda^2\dot{\phi}, \qquad k = \frac{Yah}{Eb^2}\sqrt{\frac{1+\nu}{2}}, \tag{7.65}$$

$$\lambda = \left(\frac{3}{2+n}\right)\rho^{(1+n)/2} - \left(\frac{1-n}{2+n}\right)\rho^{-(3+n)/2}. \tag{7.66}$$

Since elastic bending corresponds to $n = 1$, it is only necessary to set $\lambda = s$ in the above differential equation over the elastic portion $(0 \leq s \leq 1)$.

At the built-in end of the beam, the angle of twist vanishes, requiring $\dot{\phi} = 0$, while at the free end of the beam, the torque is zero requiring $d\dot{\phi}/ds = 0$. The boundary conditions may therefore be written as

$$\frac{d\dot{\phi}}{ds} = 0 \quad \text{at } s = 0, \ \dot{\phi} = 0 \ \text{at } s = \frac{l}{a}.$$

Over the elastic portion of the beam, where $\lambda = s$, the solution to the differential equation (7.65) can be expressed in terms of Bessel functions. In view of the first boundary condition $d\dot{\phi}/ds = 0$at $s = 0$, the solution is easily shown to be

$$\dot{\phi} = \sqrt{s}J_{-1/4}\left(\frac{1}{2}ks^2\right), \qquad 0 \leq s \leq 1. \tag{7.67a}$$

The constant of integration has been arbitrarily set to unity for convenience. By a well-known formula for the first derivative of the Bessel functions, we have

$$\frac{d\dot{\phi}}{ds} = -ks^{3/2}J_{3/4}\left(\frac{1}{2}ks^2\right), \qquad 0 \leq s \leq 1. \tag{7.67b}$$

Over the elastic/plastic portion of the beam $(1 \leq s \leq l/a)$ the relationship between λ and s is given by (7.64) and (7.66) parametrically through ρ. For a given b/h and a selected value of a/h, the differential equation (7.65) can be integrated numerically starting from $s = 1$, the values of $\dot{\phi}$ and $d\dot{\phi}/ds$ at this section being those given by (7.67). The remaining boundary condition $\dot{\phi} = 0$ at $s = l/a$ finally gives the value l/a, furnishing the solution for a definite value of l/h. The corresponding value of the critical load is obtained from the relation

$$P = \frac{M_e}{a} = \frac{4}{3}bhY\left(\frac{h}{a}\right) = Y\left(\frac{hI_y}{ab^2}\right). \tag{7.68}$$

The greatest value of hl/b^2 for which the elastic/plastic analysis is applicable corresponds to $a = l$, for which $\dot{\phi} = 0$ at $s = 1$ giving $J_{-1/4}(k/2) = 0$ or $k = 4.013$. Buckling will therefore occur in the plastic range if

$$\frac{hl}{b^2} \leq 4.013\frac{E}{Y}\sqrt{\frac{2}{1+\nu}}.$$

For higher values of hl/b^2, buckling will occur in the elastic range, and the critical load is then obtained by setting $a = l$ in the last expression of (7.68) and using the greatest value of hl/b^2 defined by the above inequality.

In the case of a simply supported beam of length 2 l bent in the yz-plane by a concentrated load $2P$ acting at the midspan, the governing equations for lateral buckling are still (7.64), (7.65), and (7.66), provided z is measured from the center of the beam and a denotes the length of the elastic part on each side of the central section. Assuming that the ends of the beam are prevented from rotation about the z-axis by appropriate constraints, the boundary conditions may be written as

$$\dot{\phi} = 0 \ \text{at}\, s = 0, \ \frac{d\dot{\phi}}{ds} = 0 \ \text{at} \ s = \frac{l}{a}.$$

The solution for the elastic portion on each side is obtained by setting $\lambda = s$ in the differential equation (7.65) and using the boundary condition $\dot{\phi} = 0$ at $s = 0$. The rate of twist and its derivative over the elastic region ($0 \leq s \leq 1$) therefore become

$$\dot{\phi} = \sqrt{s} J_{1/4}\left(\frac{1}{2}ks^2\right), \qquad \frac{d\dot{\phi}}{ds} = ks^{3/2} J_{-3/4}\left(\frac{1}{2}ks^2\right). \tag{7.69}$$

The integration of the differential equation (7.65) over the elastic/plastic lengths ($1 \leq s \leq l/a$) can be carried out in the same way as that indicated in the case of the cantilever, using the continuity conditions

$$\dot{\phi} = J_{1/4}\left(\frac{k}{2}\right), \qquad \frac{d\dot{\phi}}{ds} = kJ_{-3/4}\left(\frac{k}{2}\right) \qquad \text{at}\, s = 1,$$

in view of (7.69). The symmetry condition $d\dot{\phi}/ds = 0$ at $s = -l/a$ finally gives the ratio l/h for any assumed value of a/h, while the critical value of P follows from (7.68). When $a = l$, the plastic zones disappear, and (7.69) holds for the entire beam. Then $d\dot{\phi}/ds$ must vanish at $s = 1$, requiring $J_{-3/4}(k/2) = 0$, which gives $k = 2.117$. It follows that the elastic/plastic analysis applies only over the range

$$\frac{hl}{b^2} \leq 2.117\frac{E}{Y}\sqrt{\frac{2}{1+\nu}}.$$

Outside this range, buckling would occur while the beam is still elastic, and the corresponding value of P is obtained by inserting in the last expression of (7.68) the value of h/ab^2 that corresponds to $a = l$. The solution to the plastic buckling problem for other kinds of loading may be obtained in a similar manner. The analysis can be extended to the buckling of beams of arbitrary doubly symmetric cross sections in the manner indicated for pure bending.

7.4 Buckling of Plates Under Edge Thrust

7.4.1 Basic Equations for Thin Plates

Consider a thin plate of arbitrary shape in which the material is bounded between the planes $z = \pm h/2$, and the middle surface $z = 0$ is bounded by a closed curve that defines the edge of the plate. The bounding planes are unstressed, while uniform compressive stresses of magnitudes σ_1 and σ_2 act in the x-and y-directions, respectively, to represent the plastic state. If the transverse shear rates on the incipient deformation mode at bifurcation are disregarded, the admissible velocity field may be written as

$$v_x = u - z\frac{\partial w}{\partial x}, \qquad v_y = v - z\frac{\partial w}{\partial y}, \qquad v_z = w, \tag{7.70}$$

where u, v, and w are functions of x and y, representing the components of the velocity of the middle surface. The velocity field (7.70) is adequate for expressing all the strain rate components except the through-thickness one, which follows from the relation

$$\dot{\varepsilon}_{zz} = -\eta\left(\dot{\varepsilon}_{xx} + \dot{\varepsilon}_{yy}\right),$$

where η is the contraction ratio in the current state. If the yield surface is assumed to be that of von Mises, the relevant components of the associated unit normal are $n_{xx} = -\left(2\sigma_1 - \sigma_2\right)/\sqrt{6}\bar{\sigma}$ and $n_{yy} = -\left(2\sigma_1 - \sigma_2\right)/\sqrt{6}\bar{\sigma}$, where $\bar{\sigma}$ is the equivalent stress defined as

$$\bar{\sigma}^2 = \sigma_1^2 - \sigma_1\sigma_2 + \sigma_2^2.$$

Using the rate form of the constitutive law defined by (1.37), and considering the linearized solid for which the Jaumann stress rate is denoted by $\dot{\tau}_{ij}$ we obtain the relations

$$\begin{aligned} T\dot{\varepsilon}_{xx} &= \left(1 - \frac{3\sigma_2^2 T}{4\bar{\sigma}^2 H}\right)\dot{\tau}_{xx} - \left(\eta - \frac{3\sigma_1\sigma_2 T}{4\bar{\sigma}^2 H}\right)\dot{\tau}_{yy}, \\ T\dot{\varepsilon}_{yy} &= \left(1 - \frac{3\sigma_1^2 T}{4\bar{\sigma}^2 H}\right)\dot{\tau}_{yy} - \left(\eta - \frac{3\sigma_1\sigma_2 T}{4\bar{\sigma}^2 H}\right)\dot{\tau}_{xx}, \end{aligned} \tag{7.71}$$

where T is the tangent modulus related to the plastic modulus H by $T = EH/(E + H)$, while the contraction ratio η satisfies the relation $1 - 2\eta = (1 - 2v)T/E$. The constitutive relations (7.71) are readily inverted to express the stress rates in terms of the strain rates, the result being

$$\dot{\tau}_{xx} = E\left(\alpha\dot{\varepsilon}_{xx} + \beta\dot{\varepsilon}_{yy}\right), \qquad \dot{\tau}_{yy} = E\left(\beta\dot{\varepsilon}_{xx} + \gamma\dot{\varepsilon}_{yy}\right), \qquad \dot{\tau}_{xy} = 2G\dot{\varepsilon}_{xy},$$

where the last relation follows from the fact that the shear strain rate is purely elastic, the parameters α, β, and γ being given by

$$\left.\begin{aligned}
\alpha &= \rho^{-1}\left[4 - 3\,(1 - T/E)\,\sigma_1^2/\bar{\sigma}^2\right],\\
\beta &= \rho^{-1}\left[2 - 2\,(1 - 2\nu)\,T/E - 3\,(1 - T/E)\,\sigma_1\sigma_2/\bar{\sigma}^2\right],\\
\gamma &= \rho^{-1}\left[4 - 3\,(T/E)\,\sigma_2^2/\bar{\sigma}^2\right],\\
\rho &= (5 - 4\nu) - (1 - 2\nu)^2\,T/E - 3\,(1 - 2\nu)\,(1 - T/E)\,\sigma_1\sigma_2/\bar{\sigma}^2.
\end{aligned}\right\} \tag{7.72}$$

These quantities evidently depend on the current state of stress. It follows from the preceding expressions for the stress rates that

$$\dot{\tau}_{ij}\dot{\varepsilon}_{ij} = E\left(\alpha\dot{\varepsilon}_{xx}^2 + 2\beta\dot{\varepsilon}_{xx}\dot{\varepsilon}_{yy} + \gamma\dot{\varepsilon}_{yy}^2\right) + 4G\dot{\varepsilon}_{xy}^2. \tag{7.73}$$

To obtain the condition for bifurcation of the plate in the elastic/plastic range, consider the uniqueness criterion in the form (7.5). Since the only nonzero components of the stress tensor are $\sigma_{xx} = -\sigma_1$ and $\sigma_{yy} = -\sigma_2$, which are small compared to the modulus of elasticity E, the condition for uniqueness becomes

$$\int\left\{\dot{\tau}_{ij}\dot{\varepsilon}_{ij} - \sigma_1\left[\left(\frac{\partial v_y}{\partial x}\right)^2 + \left(\frac{\partial v_z}{\partial x}\right)^2\right] - \sigma_2\left[\left(\frac{\partial v_x}{\partial y}\right)^2 + \left(\frac{\partial v_z}{\partial y}\right)^2\right]\right\}dV > 0.$$

Inserting (7.73) into the above inequality and substituting for the strain rates and the velocity gradients which are given by

$$\dot{\varepsilon}_{xx} = \frac{\partial u}{\partial x} - z\frac{\partial^2 w}{\partial x^2}, \quad \dot{\varepsilon}_{yy} = \frac{\partial v}{\partial y} - z\frac{\partial^2 w}{\partial y^2}, \qquad 2\dot{\varepsilon}_{xy} = \frac{\partial u}{\partial y} + \frac{\partial v}{\partial x} - 2z\frac{\partial^2 w}{\partial x\partial y},$$

$$\frac{\partial v_y}{\partial x} = \frac{\partial v}{\partial x} - z\frac{\partial^2 w}{\partial x\partial y}, \qquad \frac{\partial v_x}{\partial y} = \frac{\partial u}{\partial y} - z\frac{\partial^2 w}{\partial x\partial y},$$

in view of (7.70), and integrating through the thickness of the plate, the condition for uniqueness is reduced to

$$\begin{aligned}
&\iint\left\{\alpha\left(\frac{\partial u}{\partial x}\right)^2 + 2\beta\frac{\partial^2 u}{\partial x^2}\frac{\partial^2 v}{\partial y^2} + \gamma\left(\frac{\partial v}{\partial y}\right)^2 + \frac{G}{E}\left(\frac{\partial v}{\partial y} + \frac{\partial v}{\partial x}\right)^2\right\}dx\,dy\\
&+\frac{h^2}{12}\iint\left\{\alpha\left(\frac{\partial^2 w}{\partial x^2}\right)^2 + 2\beta\frac{\partial^2 w}{\partial x^2}\frac{\partial^2 w}{\partial y^2} + y\left(\frac{\partial^2 w}{\partial y^2}\right)^2 + \frac{4G}{E}\left(\frac{\partial^2 w}{\partial x\partial y}\right)^2\right\}dx\,dy\\
&-\iint\left\{\frac{\sigma_1}{E}\left[\left(\frac{\partial v}{\partial x}\right)^2 + \left(\frac{\partial w}{\partial x}\right)^2\right]\frac{\sigma_2}{E}\left[\left(\frac{\partial u}{\partial y}\right)^2 + \left(\frac{\partial w}{\partial y}\right)^2\right]\right\}dx\,dy > 0,
\end{aligned}$$

with a minor approximation that is perfectly justified for thin plates. All the integrals appearing in the above expression extend over the middle surface of the plate. The left-hand side of the above inequality is seen to have a minimum value when

$u = v = 0$, and the Euler–Lagrange differential equation, associated with the minimization with respect to arbitrary variations of w, is easily shown to be

$$\alpha \frac{\partial^4 w}{\partial x^4} + 2(\beta + \mu) \frac{\partial^4 w}{\partial x^2 \partial y^2} + \gamma \frac{\partial^4 w}{\partial y^4} = -\frac{12}{h^2} \left(\frac{\sigma_1}{E} \frac{\partial^2 w}{\partial x^2} + \frac{\sigma_2}{E} \frac{\partial^2 w}{\partial y^2} \right), \tag{7.74}$$

where $\mu = 1/(1+v)$. The solution to the bifurcation problem is therefore reduced to the solution of the differential equation (7.74) under appropriate boundary conditions. When the bifurcation occurs in the elastic range ($T = E$), we have $\alpha = \beta = \mu = \gamma = 1/(1-v^2)$ and (7.74) reduces to the well-known governing equation for elastic buckling.

In order to establish the static boundary conditions in terms of w, it is convenient to take the components of the nominal stress rate $\dot{s}_{ij}$ as approximately equal to those of $\dot{\tau}_{ij}$. This is justified by the fact that the stresses at bifurcation will be small compared to the elastic and plastic moduli. Then the rates of change of the resultant bending and twisting moments per unit length are given by

$$\left.\begin{aligned}
\dot{M}_x &= \int_{-h/2}^{h/2} \dot{\tau}_{xx} z dz = -\frac{Eh^2}{12} \left(\alpha \frac{\partial^2 w}{\partial x^2} + \beta \frac{\partial^2 w}{\partial y^2} \right), \\
\dot{M}_y &= \int_{-h/2}^{h/2} \dot{\tau}_{yy} z dz = -\frac{Eh^3}{12} \left(\beta \frac{\partial^2 w}{\partial x^2} + \gamma \frac{\partial^2 w}{\partial y^2} \right), \\
\dot{M}_{xy} &= \int_{-h/2}^{h/2} \dot{\tau}_{xy} z dz = -\frac{Eh^2}{12(1+v)} \frac{\partial^2 w}{\partial x \partial y},
\end{aligned}\right\} \tag{7.75}$$

in view of the relations (7.72) and (7.70) with $u = v = 0$. These expressions furnish the bending moment rate $\dot{M}_n$ and the twisting moment rate $\dot{M}_{nt}$ across a typical boundary element by the rule of tensor transformation. While $\dot{M}_n$ must vanish along a simply supported edge, the quantity $\dot{Q}_n + \partial \dot{M}_{nt}/\partial S$ is required to vanish along a free edge, where $\dot{Q}_n$ is the transverse shear force rate per unit length and ds is an arc element of the boundary.

7.4.2 Buckling of Rectangular Plates

To illustrate the preceding theory, consider a rectangular plate whose sides are of lengths a and b, the origin of coordinates being taken at one of the corners of the rectangle as shown in Fig. 7.10(a). The plate is assumed to be simply supported along two opposite sides $x = 0$ and $x = a$, while different edge conditions may apply to the remaining two sides $y = 0$ and $y = b$. The plate is compressed by equal and opposite forces in the direction perpendicular to the simply supported sides. Since the deflection rate w and the bending moment rate $\dot{M}_x$ must vanish along the simply supported edges, the boundary conditions are

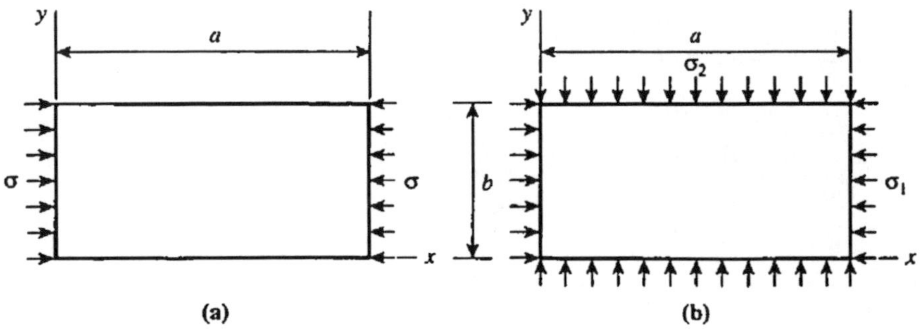

Fig. 7.10 Buckling of uniformly compressed rectangular flat plates. (**a**) Uniaxial edge thrust, (**b**) biaxial edge thrust

$$w = 0, \frac{\partial^2 w}{\partial x^2} = 0 \qquad \text{on } x = 0 \text{ and } x = a.$$

It may be noted that $w = 0$ implies $\partial^2 w/\partial y^2 = 0$ along these sides. Assuming that the plate buckles in m sinusoidal half-waves, the deflection rate is taken in the form

$$w = f(y) \sin\left(\frac{m\pi x}{a}\right),$$

where $f(y)$ is an unknown function of y. The boundary conditions along $x = 0$ and $x = a$ are identically satisfied, and the substitution for w in the differential equation (7.74) with $\sigma_1 = \sigma$ and $\sigma_2 = 0$ results in the ordinary differential equation

$$\frac{d^2 f}{dy^4} - \frac{2\phi}{b^2}\frac{d^2 f}{dy^2} - \frac{\psi f}{b^4} = 0, \tag{7.76}$$

where ϕ and ψ are dimensionless parameters expressed by the relations

$$\left.\begin{aligned} \phi &= \frac{1}{4}\left(\frac{m\pi b}{a}\right)^2 \left[\left(\frac{7-2\nu}{1+\nu}\right) - 3\left(\frac{1-2\nu}{1+\nu}\right)\frac{T}{E}\right], \\ \psi &= \frac{1}{4}\left(\frac{m\pi b}{a}\right)^2 \left[\frac{12\rho\sigma}{E}\left(\frac{b}{h}\right)^2 - \left(1+\frac{3T}{E}\right)\left(\frac{m\pi b}{a}\right)^2\right], \end{aligned}\right\} \tag{7.77}$$

in view of (7.72). The physical constraints along the edges $y = 0$ and $y = b$ are usually such that $\psi > 0$ at bifurcation. The solution to (7.76) may then be expressed in terms of two parameters k_1 and k_2 which are defined as

$$k_1 = \sqrt{-\phi + \sqrt{\phi^2 + \psi}}, \qquad k_2 = \sqrt{\phi + \sqrt{\phi^2 + \psi}}. \tag{7.78}$$

The general solution to the above fourth-order differential equation can be written in the form

$$f(y) = A\cos\left(\frac{k_1 y}{b}\right) + B\sin\left(\frac{k_1 y}{b}\right) + C\cosh\left(\frac{k_2 y}{b}\right) + D\sinh\left(\frac{k_2 y}{b}\right), \quad (7.79)$$

where A, B, C, and D are constants of integration, the ratios of which can be determined from the boundary conditions on $y = 0$ and $y = b$, which also furnish the critical stress σ for bifurcation. An incremental theory for the plastic buckling of plates has been discussed earlier by Pearson (1950). The analysis given here is essentially due to Sewell (1963, 1964).

As a first example, consider the situation where the sides $y = 0$ and $y = b$ are also simply supported, so that all four sides of the plate have identical supports. Then the additional boundary conditions are

$$w = 0, \qquad \partial^2 w/\partial y^2 = 0 \;\; on \;\; y = 0 \;\text{ and }\; y = b.$$

In view of the assumed expression for w, these conditions are equivalent to $f = d^2 f/dy^2 = 0$ along $y = 0$ and $y = b$. They are satisfied by taking $A = C = D = 0$ in the general solution (7.79) and by setting $k_l = n\pi$ where n is an integer. The first relation of (7.78) therefore gives

$$\psi - 2\pi^2 n^2 \phi - \pi^4 n^4 = 0,$$

and the substitution from (7.77) then furnishes the critical compressive stress for bifurcation as

$$\frac{\sigma}{E} = \frac{\pi^2 h^2}{3\rho b^2}\left\{2 + \left(1 + \frac{3T}{E}\right)\left(\frac{mb}{2a}\right)^2 + \frac{3}{2}\left(\frac{1-2\nu}{1+\nu}\right)\left(1 - \frac{T}{E}\right) + \left(\frac{a}{mb}\right)^2\right\}, \quad (7.80)$$

$$\rho = 3 + (1-2\nu)\left[2 - (1-2\nu)\,T/E\right]$$

where we have set $n = 1$ to minimize σ. It remains to choose the value of m for given a/b, h/b, and T/E ratios, so that the right-hand side of (7.80) is a minimum. When the ratio a/b is not too large, it is natural to expect the bifurcation mode to involve a single half-wave in the direction of compression ($m = 1$), which requires $2a^2/b^2 \le \sqrt{1 + 3T/E}$. For sufficiently large values of a/b, the critical stress is closely approximated by

$$\frac{\sigma}{E} = \frac{\pi^2 h^2}{3\rho b^2}\left\{2 + \sqrt{1 + \frac{3T}{E}} + \frac{3}{2}\left(\frac{1-2\nu}{1+\nu}\right)\left(1 - \frac{T}{E}\right)\right\}, \quad (7.81)$$

which is obtained by setting $2a^2/m^2 b^2 = \sqrt{1 + 3T/E}$ in (7.80), although the corresponding value of m is not generally an integer. The graphical plot in Fig. 7.11

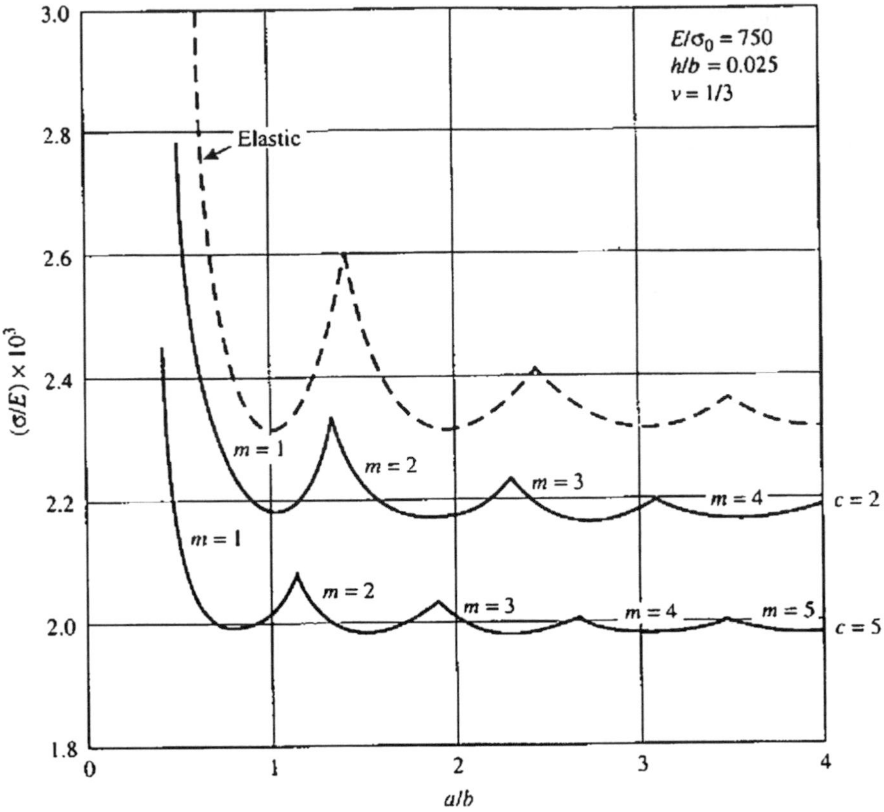

Fig. 7.11 Dimensionless critical stress for buckling of rectangular plates under unidirectional compression. The parameter c denotes the exponent of the Ramberg–Osgood stress–strain law

shows how the critical stress varies with the ratio a/b, when the stress-strain curve is represented by (7.9) with m being replaced by c. The critical stress is considerably lowered by using the rate form of the Hencky stress-strain relation, as has been shown by Shrivastava (1979), Durban and Zuckerman (1999), and Wang et al. (2001).

As a second example, suppose that the sides $y = 0$ and $y = b$ are rigidly clamped so that these edges are prevented from rotation. The boundary conditions along these edges therefore take the form

$$w = 0, \quad \partial w/\partial y = 0 \quad on\ y = 0 \text{ and } y = b.$$

These conditions are evidently equivalent to $f = df/dy = 0$ along $y = 0$ and $y = b$ where $f(y)$ is given by (7.79). The consideration of the side $y = 0$ indicates $C = -A$ and $D = -(k_1 - k_2)\, B$, and the function f then becomes

$$f(y) = A\left[\cos\left(\frac{k_1 y}{b}\right) - \cosh\left(\frac{k_2 y}{b}\right)\right] + B\left[\sin\left(\frac{k_1 y}{b}\right) - \frac{k_1}{k_2}\sinh\left(\frac{k_2 y}{b}\right)\right].$$

The application of the boundary conditions to the remaining side $y = b$ leads to the pair of equations

$$A(\cos k_1 - \cosh k_2) + B\left(\sin k_1 - \frac{k_1}{k_2}\sinh k_2\right) = 0,$$
$$-A(k_1 \sin k_1 + k_2 \sinh k_2) + Bk_1(\cos k_1 - \cosh k_2) = 0,$$

which can be satisfied by nonzero values of A and B if the determinant of their coefficients is zero. The relationship between k_l and k_2 at the point of bifurcation therefore becomes

$$2(1 - \cos k_1 \cosh k_2) = \left(\frac{k_1}{k_2} - \frac{k_2}{k_1}\right)\sin k_1 \sinh k_2. \tag{7.82}$$

For a given stress–strain curve and the ratio *a/b*, we may assume a value of σ/E, guided by the corresponding elastic solution (Timoshenko and Gere, 1961). The parameters k_1 and k_2 are then computed from (7.77) and (7.78) with an appropriate value of m. If the computed values do not satisfy (7.82), the initial assumption must be altered and the procedure repeated until a consistent value is obtained for the critical stress.

As a final example, we consider the buckling of a rectangular plate in which the sides $x = 0$, $x = a$, and $y = 0$ are simply supported, while the remaining side $y = b$ is free. The boundary conditions along the edges $y = 0$ and $y = b$ may be written as

$$w = 0, \qquad \partial^2 w/\partial y^2 = 0 \text{ along } y = 0$$
$$\dot{M}_y = 0, \quad \dot{Q}_y + \partial \dot{M}_{xy}/\partial x = 0 \text{ along } y = b$$

The condition of moment equilibrium of a typical element of the plate, the shear force rate $\dot{Q}_y$, is given by the differential equation

$$\dot{Q}_y = \frac{\partial \dot{M}_{xy}}{\partial x} + \frac{\partial \dot{M}_y}{\partial y} = -\frac{Eh^3}{12}\frac{\partial}{\partial y}\left[(\beta + \mu)\frac{\partial^2 w}{\partial x^2} + \gamma\frac{\partial^2 w}{\partial y^2}\right],$$

in view of (7.75). Using (7.72), the free-edge boundary conditions may be expressed in terms of the deflection rate w in the form

$$\frac{\partial^2 w}{\partial y^2} + \eta\frac{\partial^2 w}{\partial x^2} = 0, \quad \frac{\partial}{\partial y}\left(\frac{\partial^2 w}{\partial y^2} + \xi\frac{\partial^2 w}{\partial x^2}\right) = 0 \quad \text{on } y = b, \tag{7.83}$$

where η is the contraction ratio, and $\xi = (1 + \eta)(2 - \nu)/(1 + \nu)$. Due to the simply supported edge conditions along $y = 0$, it is necessary to set $A = C = 0$ in (7.79), and the function $f(y)$ then becomes

$$f(y) = B\sin\left(\frac{k_1 y}{b}\right) + D\sinh\left(\frac{k_2 y}{b}\right).$$

The remaining boundary conditions (7.83), corresponding to the free edge $y = b$, furnish the relations

$$-B\left(k_1^2 + \eta\frac{m^2\pi^2 b^2}{a^2}\right)\sin k_1 + D\left(k_2^2 - \eta\frac{m^2\pi^2 b^2}{a^2}\right)\sin k_2 = 0,$$

$$-Bk_1\left(k_1^2 + \xi\frac{m^2\pi^2 b^2}{a^2}\right)\cos k_1 + Dk_2\left(k_2^2 - \xi\frac{m^2\pi^2 b^2}{a^2}\right)\cos h\, k_2 = 0.$$

Setting the determinant of the coefficients of A and B in the above equations to zero, which is required by nonzero values of these constants, the relationship between k_1 and k_2 is obtained as

$$\begin{aligned} &k_2\left(k_1^2 + \eta\frac{m^2\pi^2 b^2}{a^2}\right)\left(k_2^2 - \xi\frac{m^2\pi^2 b^2}{a^2}\right)\tan k_1 \\ &= k_2\left(k_2^2 - \eta\frac{m^2\pi^2 b^2}{a^2}\right)\left(k_1^2 + \xi\frac{m^2\pi^2 b^2}{a^2}\right)\tanh k_2. \end{aligned} \tag{7.84}$$

For specified values of the ratios *a/b* and *h/b*, and a given stress–strain curve, the least value of the critical compressive stress that satisfies (7.84) can be determined by a trial-and-error procedure, following the same method as explained before. The direction of the yield surface normal has a marked influence on the critical stress, as has been shown by Sewell (1973).

7.4.3 Rectangular Plates Under Biaxial Thrust

A rectangular plate, which is simply supported along all its edges, is subjected to compressive stresses σ_1 and σ_2 uniformly distributed along the sides perpendicular to the x- and y-axes, respectively, Fig. 7.10(b). All the boundary conditions are identically satisfied by taking the deflection rate in the form

$$w = w_0 \sin\left(\frac{m\pi x}{a}\right)\sin\left(\frac{n\pi y}{b}\right),$$

where w_0 is a constant. The substitution for w into the differential equation (7.74) furnishes

$$m^2\frac{\sigma_1}{E} + n^2\left(\frac{a}{b}\right)^2\frac{\sigma_2}{E} = \frac{\pi^2 h^2}{12b^2}\left\{\alpha m^4\left(\frac{b}{a}\right)^2 + 2(\beta+\mu)m^2 n^2 + \gamma n^4\left(\frac{a}{b}\right)^2\right\}. \tag{7.85}$$

This is the required relationship between the applied stresses σ_1 and σ_2 at the point of bifurcation. The integers m and n should be such that for a given value of one of these stresses, the other one is a minimum. When the tangent modulus is independent of the stress, as in the case of an elastic material, the critical combination of stresses defined by (7.85) lie on a concave polygon in the (σ_1, σ_2)-plane, as has been shown by Timoshenko and Gere (1961).

Consider any sequence of states satisfying (7.85) and lying in the neighborhood of the state $\sigma_2 = 0$, so that the mode of bifurcation corresponds to $n = 1$. Then for a given value of σ_2, the critical value of σ_1 is established by the equation

$$\frac{\sigma_1}{E} = \frac{\pi^2 h^2}{12b^2} \left\{ \alpha \left(\frac{mb}{a} \right)^2 + 2(\beta + \mu) + \left(\gamma - \frac{12b^2 \sigma_2}{\pi^2 h^2 E} \right) \left(\frac{a}{mb} \right)^2 \right\} \tag{7.86}$$

for an appropriate value of m that minimizes the right-hand side of this equation. When $\sigma_2/E \leq \gamma(\pi^2 h^2/12b^2)$, the range of values of a/b, for which $m = 1$ is applicable, is given by

$$\frac{a^2}{b^2} \leq \sqrt{\frac{\alpha}{\lambda}}, \qquad \lambda = \gamma - \frac{12b^2 \sigma_2}{\pi^2 h^2 E} \geq 0.$$

For higher values of the ratio a/b, the critical state corresponds to $m = 2$. When a/b is sufficiently large, the minimization is very closely achieved by setting $(a/mb)^2 = \sqrt{\alpha/\lambda}$. The critical state is then independent of a/b and is given by

$$\frac{\sigma_1}{E} = \frac{\pi h}{3b} \left\{ \sqrt{\alpha \left(\frac{\gamma \pi^2 h^2}{4b^2} - \frac{3\sigma_2}{E} \right)} + (\beta + \mu) \frac{\pi h}{2b} \right\}. \tag{7.87}$$

This expression is an immediate generalization of (7.81) for bifurcation under biaxial compression satisfying the condition $\lambda \geq 0$. For $\lambda < 0$, the bifurcation state will correspond to $n = 1$ only for a certain range of values of the aspect ratio a/b.

For arbitrary combinations of σ_1 and σ_2, the conditions under which the critical state corresponds to $m = n = 1$, so that the bifurcation mode involves only one half-wave in both directions of compression, can be established by using (7.85). The bifurcation state in this case is given by

$$\frac{\sigma_1}{E} + \left(\frac{a}{b} \right)^2 \frac{\sigma_2}{E} = \frac{\pi^2 h^2}{12b^2} \left\{ \alpha \left(\frac{b}{a} \right)^2 + 2(\beta + \mu) + \gamma \left(\frac{a}{b} \right)^2 \right\}. \tag{7.88}$$

The validity of (7.88) requires that for $m = 2$, $n = 1$ and for $m = 1$, $n = 2$ the equality sign in (7.85) must be replaced by an inequality in which the left-hand side is less than the right-hand side. Thus,

$$4\frac{\sigma_1}{E} + \left(\frac{a}{b}\right)^2 \frac{\sigma_2}{E} \leq \frac{\pi^2 h^2}{12b^2}\left\{16\alpha\left(\frac{b}{a}\right)^2 + 8(\beta+\mu) + \gamma\left(\frac{a}{b}\right)^2\right\},$$

$$\frac{\sigma_1}{E} + 4\left(\frac{a}{b}\right)^2 \frac{\sigma_2}{E} \leq \frac{\pi^2 h^2}{12b^2}\left\{\alpha\left(\frac{b}{a}\right)^2 + 8(\beta+\mu) + 16\gamma\left(\frac{a}{b}\right)^2\right\}.$$

These inequalities together with (7.88) lead to the necessary restrictions on σ_1 and σ_2 for which the bifurcation state is defined by (7.88), the continued inequalities to be satisfied by the stress σ_1 being

$$\alpha - 4y\left(\frac{a}{b}\right)^4 \leq \frac{12a^2\sigma_1}{\pi^2 h^2 E} \leq 5\alpha + 2(\beta+\mu)\left(\frac{a}{b}\right)^2.$$

The value of σ_1 defined by the lower limit and the value of σ_2 furnished by the upper limit in the above inequalities are negative when $2(a/b)^2 > \sqrt{\alpha/\gamma}$ and $(a/b)^2 < 2\sqrt{\alpha/\gamma}$, respectively. When the state of stress is an equibiaxial compression defined by $\sigma_1 = \sigma_2 = \sigma$, the parameters $\rho\alpha$, $\rho(\beta+\mu)$, and $\rho\gamma$ are each equal to $1+3T/E$ in view of (7.72) with $\bar{\sigma} = \sigma$, and the expression for the critical stress then becomes

$$\frac{\sigma}{E} = \frac{\pi^2 h^2}{12a^2}\left\{\frac{(1+3T/E)\left(1+a^2/b^2\right)}{2(1+\nu)\left[1+(1-2\nu)T/E\right]}\right\}. \tag{7.89}$$

For a given aspect ratio *a/b*, the bifurcation stress in equibiaxial compression is found to be considerably lower than that in unidirectional compression. Over the elastic range of buckling ($T = E$), the equibiaxial critical stress for a square plate ($a = b$) is exactly one-half of the unidirectional critical stress. The bifurcation stress predicted by (7.89) is plotted in Fig. 7.12 as a function of the aspect ratio, assuming the stress–strain curve to be given by (7.9) with different values of the exponent *m*, which is here replaced by *c*. The influence of edge restraints produced by friction on the critical stress has been investigated by Gjelsvik and Lin (1985) and Tugcu (1991).

Solutions for the critical stress based on the Hencky stress–strain relations have been presented by Illyushin (1947), Bijlaard (1949, 1956), Gerard (1957), and El-Ghazaly and Sherbourne (1986). The buckling loads predicted by the total strain theory, despite its physical shortcomings, are found to be in better agreement with experiments, whereas those based on the incremental theory are found to be significantly higher than the experimental ones. This apparent paradox is partly due to the presence of geometrical and other imperfections which are not considered in the theory. It is also possible for some kind of non-associated flow rule, resembling the Hencky relations, to apply at the point of bifurcation. The plastic buckling of plates based on the slip theory of plasticity has been discussed by Batdorf (1949) and by Inoue and Kato (1993). A detailed investigation of the plastic buckling of relatively thick plates has been carried out by Wang et al. (2001).

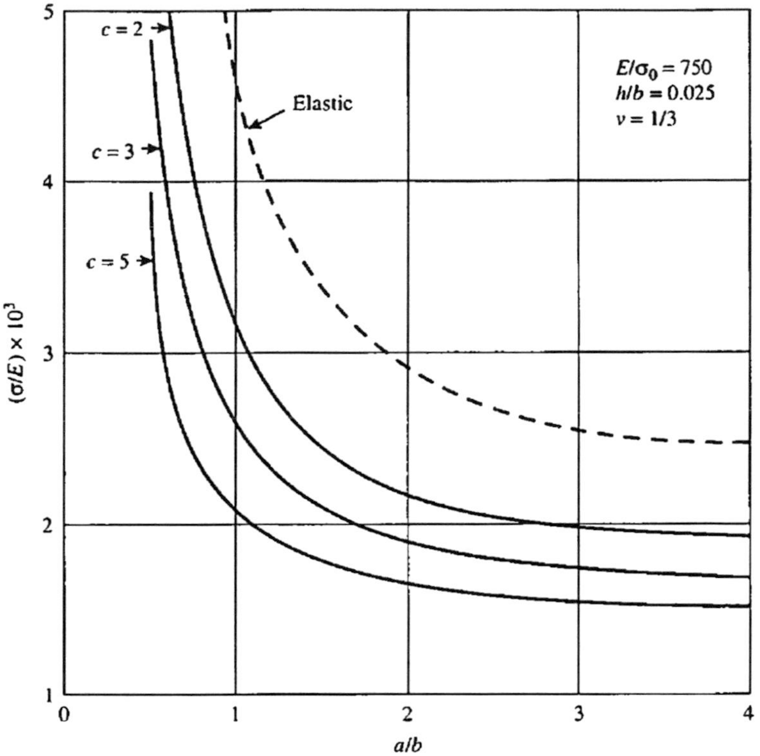

Fig. 7.12 Variation of critical stress with aspect ratio for the buckling of rectangular plates under equibiaxial compression

7.4.4 Buckling of Circular Plates

A circular plate of thickness h and radius a is submitted to a radial compressive stress σ uniformly distributed around the periphery. The incipient deformation mode at the point of bifurcation is assumed to be such that the deformed middle surface is a surface of revolution in which the deflection rate w is a function of the radius r only. Since the radial velocity of particles on the middle surface may be set to zero in the investigation for bifurcation, the velocity field may be written as

$$v_r = -z\frac{d^2w}{dr^2}, \qquad v_\theta = 0, \qquad v_z = w, \tag{7.90a}$$

in cylindrical coordinates (r, θ, z). The shear strain rates are identically zero, and the relevant components of the velocity gradient are

$$\frac{\partial v_r}{\partial r} = -x\frac{d^2w}{dr^2}, \qquad \frac{\partial v_r}{\partial z} = -\frac{dv_z}{dr} = -\frac{dw}{dr}. \tag{7.90b}$$

The radial and circumferential components of the linearized constitutive relation, which are similar to (7.71), can be written as

$$\dot{\tau}_{rr} = E\left(\alpha\dot{\varepsilon}_{rr} + \beta\dot{\varepsilon}_{\theta\theta}\right), \qquad \dot{\tau}_{\theta\theta} = E\left(\beta\dot{\varepsilon}_{rr} + \gamma\dot{\varepsilon}_{\theta\theta}\right),$$

where α, β, and γ are given by (7.72) with $\sigma_1 = \sigma_2 = \sigma$, the result being

$$\left.\begin{aligned} \alpha = \frac{1}{\rho}\left(1 + \frac{3T}{E}\right), \qquad & \beta = \alpha - \frac{1}{1+\nu}, \quad \gamma = \alpha, \\ \rho = 4\,(1+\nu)\,(1-\eta)\,, \quad & 2\eta = 1 + (1-2\nu)\,\frac{T}{E}. \end{aligned}\right\} \tag{7.91}$$

Since the radial and circumferential strain rates are $\dot{\varepsilon}_{rr} = \partial v_r/\partial r$ and $\dot{\varepsilon}_{\theta\theta} = v_r/r$, the criterion for uniqueness in this case takes the form

$$\int \left\{ E\left[\alpha\left(\frac{\partial v_r}{\partial r}\right)^2 + 2\beta\frac{v_r}{r}\frac{\partial v_r}{\partial r} + \gamma\left(\frac{v_r}{r}\right)^2\right] - \sigma\left(\frac{\partial v_z}{\partial r}\right)^2 \right\} dV > 0,$$

where the integral extends over the entire volume of the plate. Substituting from (7.90) and integrating through the thickness, we get

$$\int_0^a \left\{ \alpha\left(\frac{d^2w}{dr^2}\right)^2 + 2\beta\left(\frac{1}{r}\frac{dw}{dr}\right)\left(\frac{d^2w}{dr^2}\right) + \gamma\left(\frac{1}{r}\frac{dw}{dr}\right)^2 - \frac{12\sigma}{h^2E}\left(\frac{dw}{dr}\right)^2 \right\} r\,dr > 0.$$

The deflection rate w should be such that the left-hand side of the above inequality is a minimum. The associated variational problem is characterized by the Euler–Lagrange differential equation

$$\alpha\frac{d^2}{dr^2}\left(r\frac{d^2w}{dr^2}\right) - \gamma\left(\frac{1}{r}\frac{dw}{dr}\right)^2 + \frac{12\sigma}{h^2E}\frac{d}{dr}\left(r\frac{dw}{dr}\right) = 0.$$

Since the coefficients of this equation have constant values throughout the plate at the point of bifurcation, the first integral of the above equation can be immediately written down, and the constant of integration can be set to zero. Introducing the notation

$$\phi = \frac{dw}{dr}, \qquad \xi = \frac{2r}{h}\sqrt{\frac{3\sigma}{\alpha E}},$$

the governing differential equation for the occurrence of bifurcation is easily shown to be

$$\xi^2\frac{d^2\phi}{d\xi^2} + \xi\frac{d\phi}{d\xi} + \left(\xi^2 - 1\right)\phi = 0, \tag{7.92}$$

which is recognized as Bessel's differential equation having the general solution

$$\phi = AJ_1(\xi) + BY_1(\xi), \tag{7.93}$$

where $J_1(\xi)$ and $Y_1(\xi)$ are Bessel functions of the first order, and of the first and second kinds, respectively. Since ϕ must be zero at the center of the plate $\xi = 0$, we must set $B = 0$ in (7.93). The critical stress for bifurcation evidently depends on the condition of support of the circular edge $r = a$.

As a first application of the preceding theory, consider the plastic buckling of a circular plate which is fully clamped around its edge. Then the boundary condition is $dw/dr = 0$ along $r = a$, which is equivalent to $\phi = 0$ along $\xi = k$, and the critical stress is then given by

$$\frac{\sigma}{E} = \frac{\alpha k^2 h^2}{12a^2} = \frac{k^2 h^2}{12a^2}\left\{\frac{1+3T/E}{2(1+\nu)\left[1+(1-2\nu)T/E\right]}\right\}, \tag{7.94}$$

in view of (7.91). It follows from the boundary condition applied to (7.93), where B is identically zero, that k is the smallest root of the equation $J_1(k) = 0$. Since T/E is a function of σ/E for any given stress–strain curve of the material, the solution must be found by a trail-and-error procedure. In the case of elastic buckling, we have $T = E$ and $\alpha = 1/(1-\nu^2)$, giving $k \approx 3.832$ and $(1-\nu^2)\sigma/E \approx 1.224h^2/a^2$ at the point of bifurcation.

As a second example, let the circular plate be simply supported along its edge, so that the rate of change of the radial bending moment vanishes along $r = a$. Since the bending moment rate is very closely given by the expression

$$\dot{M}_r = \int_{-h/2}^{h/2} \dot{\tau}_{rr} z\,dz = -\frac{Eh^3}{12}\left(\alpha\frac{d^2w}{dr^2} + \frac{\beta}{r}\frac{dw}{dr}\right),$$

the simply supported edge condition $\dot{M}_r = 0$ can be written down as

$$\alpha\frac{d\phi}{dr} + \beta\frac{\phi}{r} = 0 \quad \text{or} \quad \frac{d\phi}{d\xi} + \frac{\beta\phi}{\alpha\xi} = 0$$

at $r = a$ or $\xi = k$, where ϕ is given by (7.93) with $B = 0$. Using the well-known derivative formula

$$\xi\left[J_0(\xi) - J_1'(\xi)\right] = J_1(\xi)$$

and applying the boundary condition to the above expression for ϕ, the bifurcation state is given by

$$\frac{kJ_0(k)}{J_1(k)} = 1 - \frac{\beta}{\alpha} = \frac{4(1-\eta)}{1+3T/E}, \quad k = \frac{2a}{h}\sqrt{\frac{3\sigma}{\alpha E}}, \tag{7.95}$$

in view of (7.91). Since the right-hand sides of these equations are functions of σ/E, the critical stress must be computed by trial and error, using a table of Bessel functions. When the bifurcation occurs in the elastic range, $T = E$, and

$k^2 = 12(1 - \nu^2)a^2\sigma/h^2E$, and the critical stress is then given by $(1 - \nu^2)\,\sigma/E \approx 0.35h^2/a^2$ when $\nu = 0.3$. The bifurcation stress is therefore strongly dependent on the edge condition. A detailed investigation of the plastic buckling of circular plates has been carried out by Hamada (1985). The influence of the transverse shear on the critical stress has been examined by Wang et al. (2001).

If a radially compressed circular plate has a concentric circular hole, which is assumed as stress free, a uniform radial compressive stress applied around the outer edge produces a nonuniform distribution of stress within the plate. The differential equation (7.92) is therefore modified, and the critical stress for bifurcation then depends on the ratio b/a, where b denotes the radius of the hole. The plastic buckling of annular plates under pure shear has been discussed by Ore and Durban (1989). An analysis for the plastic buckling of relatively thick annular plates under uniform compression has been presented by Aung et al. (2005).

7.5 Buckling of Cylindrical Shells

7.5.1 Formulation of the Rate Problem

Consider a circular cylindrical shell of uniform thickness h and mean radius a, subjected to the combined action of an axial compressive stress σ and a uniform lateral pressure p. The cylinder is supported in such a way that it is free to expand or contract radially during the loading. For certain critical values of the applied loads, a point of bifurcation is reached, and the shell no longer retains its cylindrical form. Let (x, θ, r) be a right-handed system of cylindrical coordinates, in which the x-axis is taken along the axis of the shell, the origin of coordinates being taken at one end of the shell. At a generic point in the material of the shell, situated at a radially outward distance z from the middle surface, the components of the velocity vector may be written as

$$\upsilon_x = u + z\omega_\theta, \qquad \upsilon_\theta = \upsilon - z\omega_x, \qquad \upsilon_x = w, \tag{7.96}$$

where (u, v, w) are the velocities at the middle surface, and ω_x, ω_θ are the rates of rotation of the normal to the middle surface about the positive x- and θ-axes, respectively. Within the framework of thin-shell theory, the latter quantities are directly obtained from the fact that the through-thickness shear rates $\dot{\varepsilon}_{rx}$ and $\dot{\varepsilon}_{r\theta}$ are identically zero. Denoting the remaining component of the spin vector by ω_r, we have

$$\omega_x = \frac{1}{a}\left(\frac{\partial w}{\partial \theta} - \upsilon\right), \qquad \omega_\theta = -\frac{\partial w}{\partial x}, \qquad \omega_r = \frac{1}{2}\left(\frac{\partial \upsilon}{\partial x} - \frac{1}{a}\frac{\partial u}{\partial \theta}\right). \tag{7.97}$$

The nonzero components of the anti-symmetric spin tensor ω_{ij} are related to the components of the spin vector as

$$-\omega_{x\theta} = \omega_{\theta x} = \omega_r, \qquad -\omega_{rx} = \omega_{xr} = \omega_\theta, \qquad -\omega_{\theta r} = \omega_{r\theta} = \omega_x.$$

The nonzero components of the strain rate, except the through-thickness one, that are associated with the velocity field (7.96), may be written as

$$\dot{\varepsilon}_{xx} = \dot{\lambda}_x - z\dot{\kappa}_x, \quad \dot{\varepsilon}_{\theta\theta} = \dot{\lambda}_\theta - z\dot{\kappa}_\theta, \quad \dot{\varepsilon}_{x\theta} = \dot{\lambda}_{x\theta} - z\dot{\kappa}_{x\theta}, \tag{7.98}$$

where $\dot{\lambda}_x$, $\dot{\lambda}_\theta$, and $\dot{\lambda}_{x\theta}$ are the rates of extension and shear of the middle surface, while $\dot{\kappa}_x$, $\dot{\kappa}_\theta$, and $\dot{\kappa}_{x\theta}$ are the rates of change of curvature and twist of the middle surface. It is easily shown that

$$\left.\begin{aligned} \dot{\lambda}_x &= \frac{\partial u}{\partial x}, \quad \dot{\lambda}_\theta = \frac{1}{a}\left(\frac{\partial v}{\partial \theta} + w\right), \quad \dot{\lambda}_{x\theta} = \frac{1}{2}\left(\frac{\partial v}{\partial x} + \frac{1}{a}\frac{\partial u}{\partial \theta}\right), \\ \dot{\kappa}_x &= \frac{\partial^2 w}{\partial x^2}, \quad \dot{\kappa}_\theta = \frac{1}{a^2}\frac{\partial}{\partial \theta}\left(\frac{\partial w}{\partial \theta} - v\right), \quad \dot{\kappa}_{x\theta} = \frac{1}{a}\frac{\partial}{\partial x}\left(\frac{\partial w}{\partial \theta} - v\right). \end{aligned}\right\} \tag{7.99}$$

The strain rates given by (7.98) and (7.99) are consistent with the customary thin-shell approximation and are adequate for the investigation of bifurcation.

The material is assumed to obey the von Mises yield criterion and the associated Prandtl–Reuss flow rule. Since the nonzero components of the current stress tensor σ_{ij} are $\sigma_{xx} = -\sigma$ and $\sigma_{\theta\theta} = -pa/h$, the outward drawn unit normal n_{ij} to the yield surface has the nonzero components

$$n_{xx} = -\frac{2\sigma - pa/h}{\sqrt{6}\bar{\sigma}}, \quad n_{\theta\theta} = -\frac{2pa/h - \sigma}{\sqrt{6}\bar{\sigma}}, \quad n_{rr} = \frac{\sigma + pa/h}{\sqrt{6}\bar{\sigma}},$$

where $\bar{\sigma}$ is the equivalent stress, which is equal to the current yield stress in simple compression, and is given by

$$\bar{\sigma}^2 = \sigma^2 - \sigma\,(pa/h) + (pa/h)^2\,.$$

Introducing the linear comparison solid, for which the Jaumann stress rate is denoted by $\dot{\tau}_{ij}$, the constitutive equation may be written as

$$\dot{\varepsilon}_{ij} = \frac{1}{E}\left[(1+v)\,\dot{\tau}_{ij} - v\dot{\tau}_{kk}\delta_{ij}\right] + \frac{3}{2}\left(\frac{1}{T} - \frac{1}{E}\right)\dot{\tau}_{kl}n_{kl}n_{ij}, \tag{7.100}$$

where T is the tangent modulus in the current state of hardening, E is Young's modulus, and v is Poisson's ratio for the material. Since $\dot{\tau}_{rr}$ is identically zero, the axial and circumferential components of the rate of extension are given by

$$T\dot{\varepsilon}_{xx} = \left[1 - \frac{3}{4}\left(1 - \frac{T}{E}\right)\left(\frac{pa}{\bar{\sigma}h}\right)^2\right]\dot{\tau}_{xx} - \left[\frac{vT}{E} + \frac{1}{2}\left(1 - \frac{T}{E}\right)\left(1 - \frac{3\sigma pa}{2\bar{\sigma}^2 h}\right)\right]\dot{\tau}_{\theta\theta},$$

$$T\dot{\varepsilon}_{\theta\theta} = \left[\frac{vT}{E} + \frac{1}{2}\left(1 - \frac{T}{E}\right)\left(1 - \frac{3\sigma pa}{2\bar{\sigma}^2 h}\right)\right]\dot{\tau}_{xx} + \left[1 - \frac{3}{4}\left(1 - \frac{T}{E}\right)\left(\frac{\sigma}{\bar{\sigma}}\right)^2\right]\dot{\tau}_{\theta\theta}.$$

The nonzero shear strain rate $\dot{\varepsilon}_{x\theta}$ is purely elastic and is equal to $(1+\nu)\,\dot{\tau}_{x\theta}/E$ by Hooke's law. The preceding pair of equations can be solved for $\dot{\tau}_{xx}$ and $\dot{\tau}_{\theta\theta}$ to give

$$\dot{\tau}_{xx} = \frac{E}{1+\nu}\left(\alpha\dot{\varepsilon}_{xx} + \beta\dot{\varepsilon}_{\theta\theta}\right), \quad \dot{\tau}_{\theta\theta} = \frac{E}{1+\nu}\left(\beta\dot{\varepsilon}_{xx} + \gamma\dot{\varepsilon}_{\theta\theta}\right), \quad \dot{\tau}_{x\theta} = \frac{E\dot{\varepsilon}_{x\theta}}{1+\nu}, \tag{7.101}$$

where

$$\left.\begin{aligned}
\alpha &= \rho^{-1}(1+\nu)\left[4 - 3(1-T/E)\left(\sigma/\bar{\sigma}^2\right)\right], \\
\beta &= \rho^{-1}(1+\nu)\left[2 - 2(1-2\nu)\,T/E - 3(1-T/E)\left(\sigma pa/\bar{\sigma}^2 h\right)\right], \\
\gamma &= \rho^{-1}(1+\nu)\left[4 - 3(1-T/E)\left(pa/\bar{\sigma}^2 h\right)^2\right], \\
\rho &= (5-4\nu) - (1-2\nu)^2\,T/E - 3(1-2\nu)(1-T/E)\left(\sigma pa/\bar{\sigma}^2 h\right),
\end{aligned}\right\} \tag{7.102}$$

The parameters α, β, γ, and ρ are easily calculated for any given state of stress and rate of hardening. Using (7.101) for the nonzero stress rates, we get

$$\dot{\tau}_{ij}\dot{\varepsilon}_{ij} = \frac{E}{1+\nu}\left\{\alpha\dot{\varepsilon}_{ij}^2 + 2\beta\dot{\varepsilon}_{xx}\dot{\varepsilon}_{\theta\theta} + \gamma\dot{\varepsilon}_{\theta\theta}^2 + 2\dot{\varepsilon}_{x\theta}^2\right\}. \tag{7.103}$$

The complete rate problem also involves the rate equations of equilibrium in which geometry changes are duly allowed for. A set of equilibrium equations in terms of the rate of change of the stress resultants have been developed by Batterman (1964). Such equation are, however, not required in the present analysis, which involves a variational principle based on the appropriate criterion for uniqueness.

7.5.2 Bifurcation Under Combined Loading

A sufficient condition for uniqueness of the deformation in an elastic/plastic body under the combined action of incremental dead loading and uniform fluid pressure is given by inequality (1.82). Since the plastic modulus H would be large compared to the applied stresses at the onset of buckling, the quantity $\sigma_{ij}\dot{\varepsilon}_{jk}\dot{\varepsilon}_{ik}$ may be omitted in the uniqueness criterion, which therefore becomes

$$\int\left[\dot{\tau}_{ij}\dot{\varepsilon}_{ij} + \sigma_{ij}\left(\omega_{ik}\omega_{jk} - 2\dot{\varepsilon}_{ik}\omega_{jk}\right)\right]dV - p\int l_k\left(\dot{\varepsilon}_{kj} + \omega_{kj}\right)\upsilon_j dS_f > 0, \tag{7.104}$$

for all continuous velocity fields vanishing at the constraint. The volume integral extends throughout the material of the body, while the surface integral extends over the boundary that is submitted to fluid pressure. Using (7.103), and remembering that the only nonzero stress components are $\boldsymbol{\sigma_{xx}} = -\boldsymbol{\sigma}$ and $\boldsymbol{\sigma_{\theta\theta}} = -pa/h$, the condition for uniqueness is reduced to

$$\int\int\int \left(\alpha\dot{\varepsilon}_{xx}^2 + 2\beta\dot{\varepsilon}_{xx}\dot{\varepsilon}_{\theta\theta} + \gamma\dot{\varepsilon}_{\theta\theta}^2 + 2\dot{\varepsilon}_{x\theta}^2\right)dx\, d\theta dz$$
$$- (1+\nu)\frac{\sigma h}{E}\int\int \left(\omega_r^2 + \omega_\theta^2 + 2\omega_r\dot{\lambda}_{x\theta}\right)dx\, d\theta$$
$$- (1+\nu)\frac{pa}{E}\int\int \left(\omega_x^2 + \omega_\theta^2 - 2\omega_\theta\dot{\lambda}_{x\theta}\right)dx\, d\theta$$
$$+ (1+\nu)\frac{p}{E}\int\int \left[\left(\dot{\lambda}_x + \dot{\lambda}_\theta\right)w + u\omega_\theta - \nu\omega_x\right]dx\, d\theta > 0,$$

with sufficient accuracy, since σ/E and p/E are small compared to unity. Substituting from (7.97), (7.98), and (7.99) and introducing the dimensionless parameters

$$s = (1+\nu)\frac{\sigma}{E}, \quad q = (1+\nu)\frac{pa}{Eh}, \quad k = \frac{h^2}{12a^2}, \quad \xi = \frac{x}{a},$$

and integrating through the thickness of the shell, we obtain

$$\iiint \left\{\alpha\left(\frac{\partial u}{\partial \xi}\right) + 2\beta\frac{\partial u}{\partial \xi}\left(\frac{\partial \upsilon}{\partial \theta} + w\right) + \gamma\left(\frac{\partial \upsilon}{\partial \theta} + w\right)^2 + \frac{1}{2}\left(\frac{\partial \upsilon}{\partial \xi} + \frac{\partial u}{\partial \theta}\right)^2\right\} d\xi d\theta$$
$$+ k\iint \left\{\alpha\left(\frac{\partial^2 w}{\partial \xi^2}\right)^2 + 2\beta\frac{\partial^2 w}{\partial \xi^2}\left(\frac{\partial^2 w}{\partial \theta^2} - \frac{\partial \upsilon}{\partial \theta}\right) + \gamma\left(\frac{\partial^2 w}{\partial \theta^2} - \frac{\partial \upsilon}{\partial \theta}\right)^2\right.$$
$$\left. + 2\left(\frac{\partial^2 w}{\partial \xi \partial \theta} - \frac{\partial \upsilon}{\partial \xi}\right)^2\right\} d\xi d\theta - s\iint \left\{\left(\frac{\partial \upsilon}{\partial \xi}\right)^2 + \left(\frac{\partial w}{\partial \xi}\right)^2\right\} d\xi d\theta$$
$$- q\iint \left\{\left(\frac{\partial u}{\partial \theta}\right)^2 + \left(\frac{\partial w}{\partial \theta}\right)^2 - w\left(\frac{\partial u}{\partial \xi} + w\right) + u\frac{\partial w}{\partial \xi}\right\} d\xi d\theta > 0,$$

where use has been made of the fact that the velocities are single-valued functions of θ. The terms containing the quantities $s\dot{\lambda}_{x\theta}^2$ and $q\dot{\lambda}_{x\theta}^2$ have been neglected in the last two integrals to be consistent with the basic approximation. The occurrence of bifurcation is marked by the vanishing of the above functional, which must be minimized with respect to the admissible velocities u, υ, and w. The Euler–Lagrange differential equations associated with this variational problem are easily found to be

$$\alpha\frac{\partial^2 u}{\partial \xi^2} + \frac{1}{2}\frac{\partial^2 u}{\partial \theta^2} + \left(\beta + \frac{1}{2}\right)\frac{\partial^2 \upsilon}{\partial \xi \partial \theta} + \beta\frac{\partial w}{\partial \xi} + q\left(\frac{\partial w}{\partial \xi} - \frac{\partial^2 u}{\partial \theta^2}\right) = 0,$$

$$\left(\beta + \frac{1}{2}\right)\frac{\partial^2 u}{\partial \xi^2 \partial \theta} + \frac{1}{2}\frac{\partial^2 \upsilon}{\partial \xi^2} + \gamma\left(\frac{\partial^2 \upsilon}{\partial \theta^2} + \frac{\partial w}{\partial \theta}\right) - s\frac{\partial^2 \upsilon}{\partial \xi^2}$$
$$+ k\left\{2\frac{\partial^2 \upsilon}{\partial \xi^2} + \gamma\left(\frac{\partial^2 \upsilon}{\partial \theta^2} - \frac{\partial^3 w}{\partial \theta^3}\right) - (\beta + 2)\frac{\partial^3 w}{\partial \xi^2 \partial \theta}\right\} = 0,$$

$$\beta \frac{\partial u}{\partial \xi} + \gamma \left(\frac{\partial \upsilon}{\partial \theta} + w \right) + s \frac{\partial^2 \upsilon}{\partial \xi^2} + q \left(\frac{\partial u}{\partial \xi} + w + \frac{\partial^2 w}{\partial \theta^2} \right)$$
$$+ k \left\{ - (\beta + 2) \frac{\partial^3 \upsilon}{\partial \xi^2 \partial \theta} - \gamma \frac{\partial^4 \upsilon}{\partial \theta^4} + 2 (\beta + 1) \frac{\partial^4 w}{\partial \xi^2 \partial \theta^2} + \gamma \frac{\partial^4 w}{\partial \theta^4} \right\} = 0. \tag{7.105}$$

In the case of elastic buckling, we have $\boldsymbol{\alpha} = \boldsymbol{\gamma} = 1/(1 - \nu)$ and $\beta = \nu/(l - \nu)$, and (7.105) reduce to those given by Timoshenko and Gere (1961), except for the coefficients of certain small-order terms, the effects of which are insignificant in the final result. The above equations provide a systematic generalization of the eigenvalue problem when buckling occurs in the plastic range (Chakrabarty, 1973).

The class of admissible velocity fields for the investigation of bifurcation, which is characterized by a nonuniform mode of deformation, may be considered as that in which the radial velocity vanishes at the ends of the shell. Denoting the length of the shell by l, the solution of (7.105) may therefore be sought in the form

$$\left. \begin{array}{c} u = U \cos \lambda \xi \cos m\theta, \qquad \upsilon = V \sin \lambda \xi \sin m\theta, \\ w = W \sin \lambda \xi \cos m\theta, \end{array} \right\} \tag{7.106}$$

where $\lambda = r\pi a/l$, and m and r are integers, while U, V, and W are arbitrary constant velocities. The virtual velocity field (7.106) evidently satisfies the boundary conditions

$$w = \partial^2 w / \partial \xi^2 = 0 \qquad \text{at} \qquad \xi = 0 \text{ and } \xi = 1,$$

which correspond to a shell with simply supported edges. For sufficiently long shells, the results based on (7.106) can be applied to other types of edge condition without appreciable error. The velocity field (7.106) implies that the generator of the shell is subdivided into r half-waves and the circumference into $2m$ half-waves at the onset of bifurcation. Substituting (7.106) into (7.105), these equations are found to be satisfied everywhere in the shell if

$$\left. \begin{array}{l} \left[\alpha\lambda^2 + \left(\frac{1}{2} - q \right) m^2 \right] U - \left(\frac{1}{2} + \beta \right) \lambda m V - (\beta + q) \lambda W = 0, \\ - \left(\frac{1}{2} + \beta \right) \lambda m U + \left[\left(\frac{1}{2}\lambda^2 + \gamma m^2 - \lambda^2 s \right) + k \left(2\lambda^2 + \gamma m^2 \right) \right] V \\ + \left[\gamma m + km \left\{ (2 + \beta) \lambda^2 + \gamma m^2 \right\} \right] W = 0, \\ - (\beta + q) \lambda U + \left[\gamma m + km \left\{ (2 + \beta) \lambda^2 + \gamma m^2 \right\} \right] V \\ + \left[\gamma - \lambda^2 s - \left(m^2 - 1 \right) q + k \left\{ \alpha\lambda^4 + 2 (1 + \beta) \lambda^2 m^2 + \gamma m^4 \right\} \right] W = 0. \end{array} \right\} \tag{7.107}$$

This is a system of three linear homogeneous equations for the unknown velocities U, V, and W. For nontrivial solutions to these quantities, the determinant of their coefficients must vanish. It is interesting to note that the matrix of this determinant is symmetric. Expanding the determinant, and neglecting the small-order terms involving the squares and products of s, q, and k, the result may be expressed in the form

$$A + Bk = Cs + Dq, \tag{7.108}$$

where

$$\left.\begin{aligned}
A &= \delta\lambda^4, \qquad \delta = \alpha\gamma - \beta^4 = (4/\rho)(1+\nu)^2(T/E),\\
B &= \left[\alpha\lambda^4 + 2(\delta-\beta)\lambda^2 m^2 + \gamma m^4\right]\left[\alpha\lambda^4 + 2(1+\beta)\lambda^2 m^2 + \gamma m^4\right]\\
&\quad - 2m^2\left[(2+\beta)\lambda^2 + \gamma m^2\right]\left[(2\delta-\beta)\lambda^2 + \gamma m^2\right],\\
C &= \lambda^2\left[\alpha\lambda^4 + 2(\delta-\beta)\lambda^2 m^2 + \gamma m^2\right] + \lambda^2\left(2\delta\lambda^2 + \gamma m^2\right),\\
D &= \left(m^2 - 1\right)\left[\alpha\lambda^4 + 2(\delta-\beta)\lambda^2 m^2 + \gamma m^4\right] + \lambda^2\left(2\beta\lambda^2 - \gamma m^2\right).
\end{aligned}\right\} \tag{7.109}$$

For a given material and shell geometry, and with selected values of r and m, (7.108) defines the relationship between the critical values of s and q for bifurcation. Using a constant value of r and successive values of m, a series of curved segments may be obtained in a graphical plot of s against q. The value of m appropriating over a certain range is that which provides the smallest ordinate for a given abscissa. A set of curves may be constructed in this way for various values of $\lambda = r\pi a/l$, similar to those presented by Flügge (1932) for the buckling of an elastic shell.

7.5.3 Buckling Under Axial Compression

Consider the special case where the lateral pressure is absent and the axial compressive stress σ is increased to a critical value to cause buckling in the plastic range. The critical stress is then obtained by setting $q = 0$ in (7.108) and using the values of λ and m which correspond to a minimum value of s. When the cylindrical shell is relatively short, we may expect it to buckle into short longitudinal waves, so that λ^2 is sufficiently large. Retaining only the first terms in the expressions for B and C in (7.109), the critical stress may be written in the simplified form

$$S = \frac{\delta\lambda^2}{\alpha\lambda^4 + 2(\delta-\beta)\lambda^2 m^2 + \gamma m^4} + \frac{k\left[\alpha\lambda^4 + 2(1+\beta)\lambda^2 m^2 + \gamma m^4\right]}{\lambda^2}, \tag{7.110}$$

where $\delta = 4(1 + \nu)^2(T/\rho E)$, the coefficients α, β, γ, and ρ being given by (7.102) with $\bar{\sigma} = \sigma$ and $\rho = 0$. The value of λ^2 which makes the right-hand side of (7.110) a minimum is found to be given by

$$\alpha\lambda^4 - \left[\sqrt{\delta/k} - 2(\delta - \beta)m^2\right]\lambda^2 + \gamma m^4 = 0,$$

and the corresponding expression for the critical stress for buckling then becomes

$$s = 2\left[\sqrt{kg} + (1 - \delta + 2\beta)m^2\right].$$

Since $\delta - 2\beta \leq 1$ for $0 < T/E < 1$, the smallest value of s for plastic buckling corresponds to $m = 0$, which represents a symmetrical mode of buckling. The preceding relations therefore reduce to

$$\alpha\lambda^2 = \sqrt{\delta/k}, \qquad s = 2\sqrt{\delta k}.$$

Substituting for k, α, and δ, the critical values of σ/E and λ^2 for plastic buckling are finally obtained as

$$\frac{\sigma}{E} = \frac{2h}{a}\sqrt{\frac{T}{3\rho E}}, \qquad \lambda^2 = \frac{4a}{h}\left(1 + \frac{3T}{E}\right)^{-1}\sqrt{\frac{3\rho T}{E}} \tag{7.111}$$

This is the true tangent modulus formula for the plastic buckling of relatively short cylindrical shells under axial compression. The same formula has been obtained by Batterman (1965) using a different method. Setting $T/E = 1$ and $\rho = 4(1 - \nu^2)$ reduces (7.111) to the well-known formulas for elastic buckling in which case m need not be zero. The influence of a singular yield criterion in lowering the critical stress has been examined by Ariaratnam and Dubey (1969). Plastic buckling formulas based on the total strain theory together with some experimental results have been given by Bijlaard (1949) and also by Jones (2009). The effect of an initial imperfection on the critical stress based on the total strain theory has been investigated by Lee (1962) and by Bardi et al. (2006).

Since the eigenmode is symmetrical, the second equation of (7.107) is identically satisfied, and either of the two remaining equations furnishes $W/U = \alpha\lambda/\beta$. The eigenfield may therefore be written in terms of an arbitrary constant velocity u_0 as

$$u = u_0\beta\cos\lambda\xi, \qquad v = 0, \qquad w = u_0\alpha\lambda\sin\lambda\xi.$$

Since a uniform axial compression is always a possible mode, the nonzero components of the actual velocity at bifurcation may be expressed as

$$u = -u_0\left[\frac{x}{l} + c\beta\left(1 - \cos\frac{\lambda x}{l}\right)\right], \qquad w = u_0\left(\frac{\eta a}{l} + c\alpha\lambda\sin\frac{\lambda x}{l}\right), \tag{7.112}$$

where c is a constant and η is the contraction ratio, the axial velocity being assumed to vanish at the end $x = 0$. The condition of no instantaneous unloading of the plastically compressed cylinder may be written as $\dot{\varepsilon}_{xx} < 0$, and it follows from (7.98) and (7.112) that

$$1 + \frac{c\lambda l}{a}\left(\beta - \frac{z}{a}\sqrt{\frac{\delta}{k}}\right)\sin\lambda\xi > 0$$

in view of the result $\alpha\lambda^2 = \sqrt{\delta/k}$. This inequality will be satisfied throughout the shell if

$$-\frac{a}{l} < \left\{\beta + 2(1+\nu)\sqrt{\frac{3T}{\rho E}}\right\}\lambda c < \frac{a}{l},$$

where β and ρ are given by (7.102). The buckling may therefore occur in a range of possible modes, when the axial stress σ attains the value given by (7.111), with the load increasing as the shell continues to deform in the postbuckling range.

In the case of long cylindrical shells, the generators are expected to buckle into relatively long waves in the longitudinal direction, so that λ *is* sufficiently small. We may then neglect in the expressions for B and C all powers of λ higher than the second to obtained the result

$$A = \delta\lambda^4, \qquad C = \gamma\lambda^2 m^2\left(m^2+1\right),$$

$$B = \gamma m^2\left(m^2-1\right)^2\left[\gamma m^2 + 2(1+\delta)\lambda^2\right].$$

Substituting in (7.108), where q is set to zero, the critical stress for buckling may be expressed as

$$s = \frac{\delta\lambda^2}{\gamma m^2\left(m^2+1\right)} + \frac{k\left(m^2-1\right)^2}{m^2+1}\left\{\frac{\gamma m^2}{\lambda^2} + 2(1+\delta)\right\}. \tag{7.113}$$

Minimizing this expression with respect to λ, the associated values of λ^2 and s are found to be

$$\lambda^2 = \gamma m^2\left(m^2-1\right)\sqrt{k/\delta},$$

$$s = 2\left(\frac{m^2-1}{m^2+1}\right)\left[\sqrt{k\delta} + k(1+\delta)\left(m^2-1\right)\right].$$

Evidently, the smallest value of the critical stress corresponds to $m = 2$, giving the final results

$$\frac{\sigma}{E} = \frac{6h}{5a}\left\{\sqrt{\frac{T}{3\rho E}} + \frac{h}{4a}\left(\frac{1+\delta}{1+\nu}\right)\right\}, \quad \lambda^2 = \frac{4h}{a}\sqrt{\frac{3E}{\rho T}}, \tag{7.114}$$

where δ is given by the first equation of (7.109), and ρ by the last equation of (7.102) with the last term set to zero. For an elastic material, $T = E, \delta = (1+\nu)/(1-\nu)$, and

$\rho = 4(1 - \nu^2)$, reducing the above expressions to those given by Timoshenko and Gere (1961). A limit of applicability of (7.114) is marked by the buckling of the shell as a column, the critical stress then being given by the formula $\sigma = \pi^2 a^2 T/2l$ The plastic buckling of axially compressed cylindrical shells has also been considered by Gellin (1979), Shrivastava (1979), and Reddy (1980).

The preceding discussion is based on the assumption that the shell is thick enough for the critical stress σ to exceed the yield stress of the material. The results for the whole range of values of *h/a* may, however, be presented on the basis of the plastic buckling formula if we adopt the relations

$$\varepsilon = \frac{\sigma}{E}\left\{1 + \frac{3}{7}\left(\frac{\sigma}{\sigma_0}\right)^{n-1}\right\}, \qquad \frac{E}{T} = 1 + \frac{3n}{7}\left(\frac{\sigma}{\sigma_0}\right)^{n-1}, \tag{7.115}$$

for the uniaxial stress–strain curve, where σ_0 and n are empirical constants. The assumed curve has an initial slope E, but the tangent modulus T steadily decreases as the stress increases from zero. Inserting (7.115) into the critical stress formula (7.111) or (7.114), a relationship between σ/E and h/a is obtained for any given values of σ_0/E and n. The relationship between λ and h/a then follows from the corresponding formula for λ^2. The results are displayed in Fig. 7.13 for several values of n, assuming $\nu = 0.3$ and $\sigma_0/E = 0.002$. Due to geometrical imperfections and other uncertainties, the experimental buckling loads are found to be appreciably lower than the theoretical ones. A useful review of the subject has been reported by Babcock (1983). The buckling problem for a square-section tube, based on the total strain theory, has been investigated by Li and Reid (1992).

7.5.4 Influence of Frictional Restraints

When a cylindrical shell is axially compressed between a pair of rigid platens, lateral displacement of the elements in contact with the platens is prevented by friction, and the shell is therefore subjected to simultaneous bending and compression right from the beginning of the loading. The bending moment at the crests of the longitudinal waves nearest to the platens rapidly increases with increasing load until the yield point state is reached. The first convolution that appears at one end of the cylinder, where the frictional restraint is more predominant, then collapses under decreasing load. When the first convolution nearly flattens out, a second convolution begins to form on top of the other one, and the load starts to increase again with further compression. If the shell is relatively thick, the result is a concertina-type of deformed shape, for which an average buckling load has been given by Alexander (1960b). For thinner shells, there is generally a polygonal-type of final configuration treated by Pugsley and Macualay (1960), and Pugsley (1979). An experimental investigation on the energy absorption of tubular structures during buckling has been made by Balen and Abdul-Latif (2007).

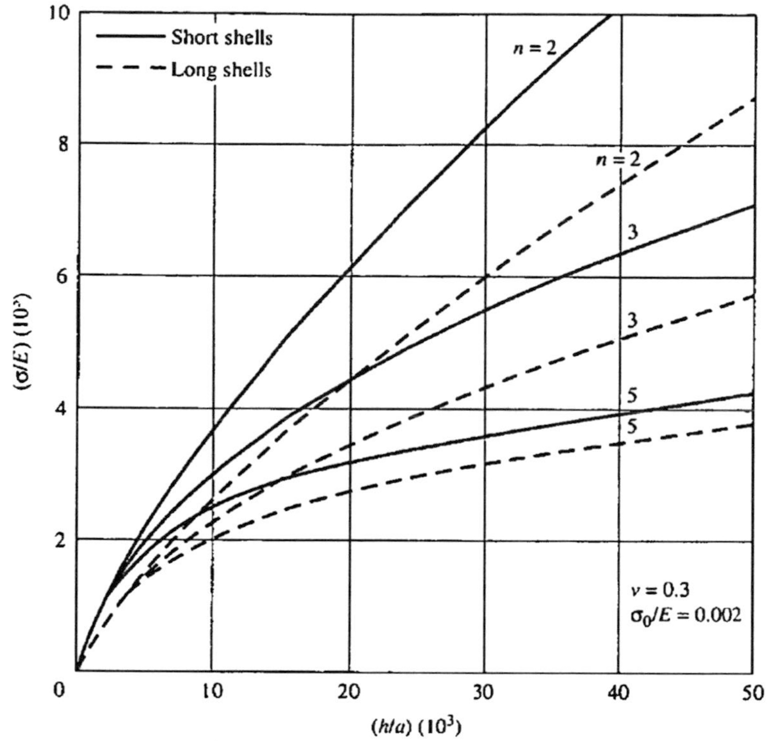

Fig. 7.13 Critical stress for plastic bucking of an axially compressed cylindrical shell as a function of the wall thickness

Consider the concertina-type of collapse in which each convolution is approximated by two identical conical surfaces as shown in Fig. 7.14. In this simplified model, the convolutions are taken to be purely external to the cylinder, though in actual practice they are formed partly outward and partly inward. The formation of each convolution is associated with three circular hinges which allow rotation of the conical bands, each having a constant slant height b. For simplicity, the small change in thickness that occurs in the formation of convolution will be disregarded. Neglecting elastic strains and work-hardening, and denoting the current semivertical angle of the conical bands by ψ the increment of plastic energy dissipation in the hinge circles during a small change in angle $d\psi$ is found as

$$dW_1 = 4\pi b M_0 \left(2a + b\sin\psi\right)\ d\psi = 2\pi a h^2 Y\left(1 + \frac{b}{2a}\sin\psi\right) d\psi,$$

the influence of the meridional force on the fully plastic moment being disregarded. Since the mean circumferential strain in the material between the hinge circles during the incremental change in angle $d\psi$ is $d\varepsilon_\theta = b\cos\psi d\psi(2a + bsin\psi)$, the increment of plastic work done during the circumferential stretching of the material is

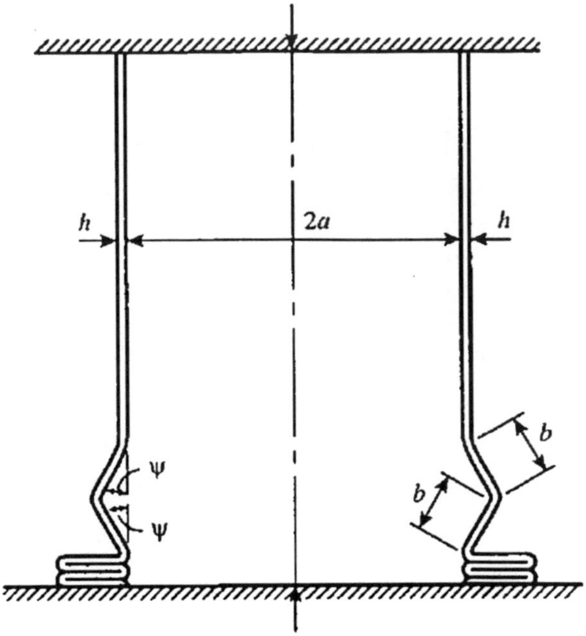

Fig. 7.14 Concertina-type of plastic collapse of a circular cylindrical shell under axial compression with frictional restraint

$$dW_2 = 2\pi b N_0 (2a + b\sin\psi)\ \ d\varepsilon_\theta = 2\pi h b^2 Y \cos\psi d\psi.$$

This expression involves the approximate yield condition $N_\theta \approx N_0$ and the fact that the meridional strain increment is assumed to be zero. The total plastic work done in collapsing a typical convolution, as ψ increases from 0 to $\pi/2$, is

$$W = \int (dW_1 + dW_2) = \pi h^2 (\pi a + b) Y + 2\pi b^2 h Y. \tag{7.116}$$

The applied compressive stress decreases from an initial value σ_0 during the collapse of a convolution, the stress at a generic stage being denoted by σ. Available theoretical and experimental evidences (Chakrabarty and Wasti, 1971) tend to suggest that the variation of the stress may be written as

$$\sigma = \sigma_0 \left(\cos^2\psi + \sqrt{\frac{h}{a}} \sin^2\psi \right).$$

The total work done by the external load must be equal to W. Since the amount of axial compression corresponding to a change $d\psi$ in the semi-vertical angle is equal to $2b\ \sin\psi,\ d\psi$, we have

$$W = 4\pi a h b \sigma_0 \int_0^{\pi/2} \left(\cos^2\psi + \sqrt{\frac{h}{a}} \sin^2\psi \right) \sin^2\psi d\psi.$$

Integrating, and inserting the expression (7.116) for *W*, the critical stress is found to be given by

$$\frac{\sigma_0}{Y}\left(1+2\sqrt{\frac{h}{a}}\right)=\frac{3h}{4a}\left(\frac{\pi a}{b}+1+\frac{2b}{h}\right).$$

The quantity b is still an unknown but can be determined by minimizing σ_0 the result being $b=\sqrt{\pi ah/2}$. The critical stress formula therefore becomes

$$\frac{\sigma_0}{Y}=3\frac{3h}{4a}\left(\frac{\pi a}{b}+1+\frac{2b}{h}\right)\Big/\left(1+2\sqrt{\frac{h}{a}}\right). \tag{7.117}$$

The critical stress for the initiation of the first convolution would be somewhat smaller than (7.117) since the hinge circle at the base of the deformation zone is absent during its formation. The analysis has practical importance in the design of buffers bringing moving bodies to a stop without appreciable damage. Further results on the axial crushing of cylindrical shells have been given by Mamalis and Johnson (1983). The influence of the meridional stress on the magnitude of the buckling load has been examined by Wierzbicki and Abramowicz (1983).

7.5.5 Buckling Under External Fluid Pressure

We begin with the situation where a circular cylindrical shell is submitted to a uniform external pressure acting on the lateral surface only. The critical hoop stress at the incipient buckling is then directly given by (7.108), where s must be set to zero. The parameters α, β, γ, and δ appearing in (7.109) are obtained from (7.102) on setting $\sigma=0$ and $pa/h=\bar{\sigma}$. If the length of the cylinder is greater than about twice its diameter, the ratio λ^2/m^2 will be a small fraction, and we may omit all terms containing λ^2 and λ^4 in the expressions for B and D in (7.109). Then the result for the critical pressure is easily shown to be

$$q\approx\frac{\delta\lambda^4}{\gamma m^4\left(m^2-1\right)}+k\gamma\left(m^2-1\right). \tag{7.118}$$

Since q increases with λ, the least value of the critical hoop stress corresponds to $r=1$, giving $\lambda=\pi a/l$. Substituting for γ, δ, λ, and k into the above expression, we finally obtain the buckling formula

$$\frac{pa}{Eh}=\frac{(4T/E)\,(\pi a/l)^4}{(1+3T/E)\,m^4\left(m^2-1\right)}+\frac{\left(h^2/12a^2\right)(1+3T/E)\left(m^2-1\right)}{(5-4\nu)-(1-2\nu)^2\,T/E}. \tag{7.119}$$

In the case of elastic buckling ($T=E$), this formula reduces to that originally given by Southwell (1913). Equation (7.119), which is due to Chakrabarty (1973), represents a rigorous extension of the result when buckling occurs in the plastic range. For a very long tube, the first term on the right-hand side of (7.119) may be disregarded, and the least value of the critical stress then corresponds to $m = 2$. In a wide range of values of l/a and h/a would generally require $m = 3$ for a lower value of the critical stress, and the condition for this to happen is

$$\frac{\pi^4 a^4}{l^4} > \frac{27\,h^2}{5a^2}\,\frac{(E/T)\,(1+3T/E)^2}{(5-4\nu)-(1-2\nu)^2\,T/E}.$$

For shorter tubes, the critical stress corresponding to $m = 4$ would be lower than that given by $m = 3$. For exceptionally short tubes, however, the approximation leading to (7.118) is no longer valid.

For the numerical evaluation of the critical stress, T/E may be eliminated from (7.119) by using (7.115), to obtain the relationship between the critical hoop stress pa/h and the shell parameters. The resulting equation can be most conveniently solved by selecting a value of pa/Eh for a given value of l/a, and calculating the corresponding value of h^2/a^2 required for bifurcation. The computation may be carried out with a suitable value of m, changing it if necessary to obtain the least value of the critical stress. Figure 7.15 shows the results of the computation based on $\nu = 0.3$, $n = 3.0$, and $\sigma_0/E = 0.001$. The curves for plastic buckling are represented by the solid lines, while those for elastic buckling are indicated by the broken lines.

Only two of the equations in (7.107) are independent when U, V, and W correspond to the eigenfield. Setting $s = 0$ in the last two equations of (7.107), and using (7.118), it is easily shown that

$$U = \lambda w_0, \qquad V = m w_0, \qquad W = -m^2 w_0,$$

to a close approximation, where w_0 is an arbitrary constant velocity. Introducing a constant parameter c, the components υ and w of the actual velocity at bifurcation may be written as

$$\left.\begin{aligned} \upsilon &= m w_0 c \sin\left(\frac{\pi x}{l}\right)\sin m\theta, \\ w &= -m^2 w_0\left[1 + c\sin\left(\frac{\pi x}{l}\right)\cos m\theta\right], \end{aligned}\right\} \tag{7.120}$$

the radial velocity at each end of the shell being taken as equal to $-m^2 wo$. The substitution from (7.120) into the second equation of (7.98) furnishes the hoop strain rate

$$\dot{\varepsilon}_{\theta\theta} = -\frac{m^2 w_0}{a}\left\{1 + \frac{cz}{a}\left(m^2-1\right)\sin\left(\frac{\pi x}{l}\right)\cos m\theta\right\},$$

which must be negative for no incipient unloading at the point of bifurcation. Since the value of the second term in the curly brackets of the above expression varies between $-(m^2 - 1)(ch/2a)$, and $(m^2 - 1)(ch/2a)$, the loading condition $\dot{\varepsilon}_{\theta\theta} < 0$ gives

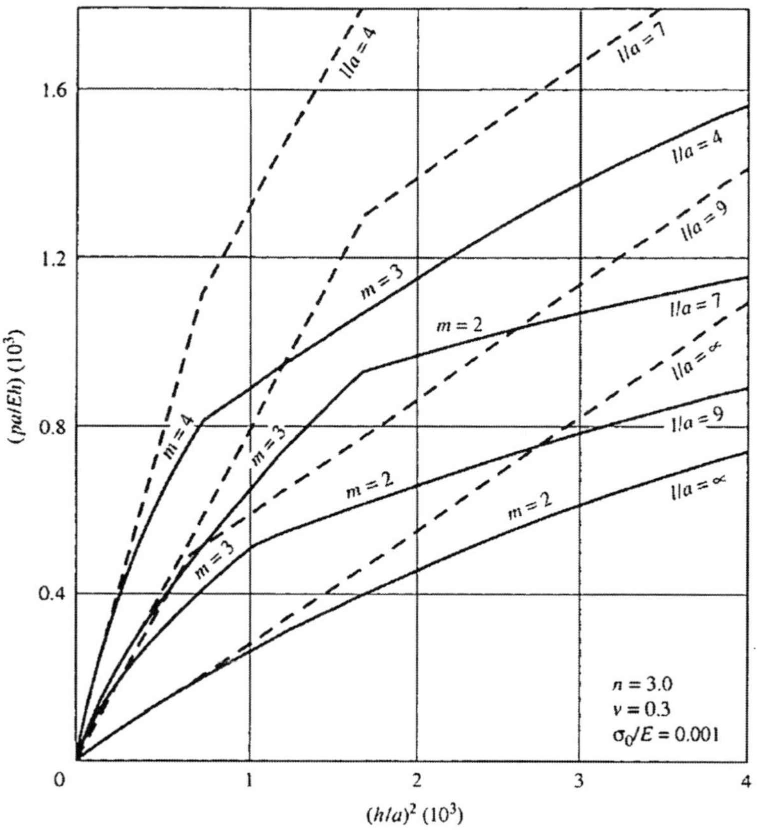

Fig. 7.15 Critical stress for plastic buckling of a circular cylindrical shell under uniform external pressure on the lateral surface

$$-\frac{2a}{h} < \left(m^2 - 1\right) c < \frac{2a}{h}.$$

The nonlinear elastic/plastic solid may therefore buckle in a range of possible ways, and the external pressure must continue to increase from its critical value (7.118) as the deformation continues in the post-buckling range.

Consider now the situation where a circular cylindrical shell is closed at both ends by rigid plates, and is submitted to an all-round external pressure of intensity p. The uniform deformation that precedes buckling involves $\sigma = pa/2h = \bar{\sigma}/\sqrt{3}$, and (7.102) become

$$\left.\begin{aligned} \alpha &= (1+\nu)(3+T/E)\,\rho^{-1}, \qquad & \beta &= \nu\,(1+\nu)(4T/\rho E), \\ \gamma &= (1+\nu)(4T/\rho E) = \delta/(1+\nu), \qquad & \rho &= 3 + \left(1 - 4\nu^2\right) T/E. \end{aligned}\right\} \tag{7.121}$$

The critical pressure for buckling is obtained by setting $s = q/2$ in (7.108), the coefficient of q in the resulting expression being $D + C/2$. Neglecting small-order terms, and using the fact that $\delta = (1 + \nu)\gamma$ and $\delta - \beta = \gamma$ in view of (7.121), we obtain the simplified formula

$$\frac{pa}{Eh} = \frac{\lambda^4}{m^2\left(m^2 - 1\right)\left(m^2 + 2.5\lambda^2\right)} + \frac{k\left(m^2 - 1\right)\left[\gamma m^2 + 2\left(1 + \delta\right)\lambda^2\right]}{\left(1 + \nu\right)\left(m^2 + 2.5\lambda^2\right)} \tag{7.122}$$

to a close approximation, where $\lambda = \pi a/l$. For given l/a and h/a ratios, the value m that makes the right-hand side of (7.122) a minimum should be used for calculating the critical pressure. In Fig. 7.16, the results for a closed-ended cylinder under an all-round hydrostatic pressure are compared with those for an open-ended cylinder under a radial pressure alone, using (7.115) with $n = 3$ and $\sigma_0/E = 0.002$. For relatively thick tubes, the critical values of the equivalent stress in the two cases are found to be approximately the same. The plastic buckling of an initially imperfect cylindrical shell under internal pressure and axial compression, based on the total strain theory, has been considered by Paquette and Kyriakides (2006).

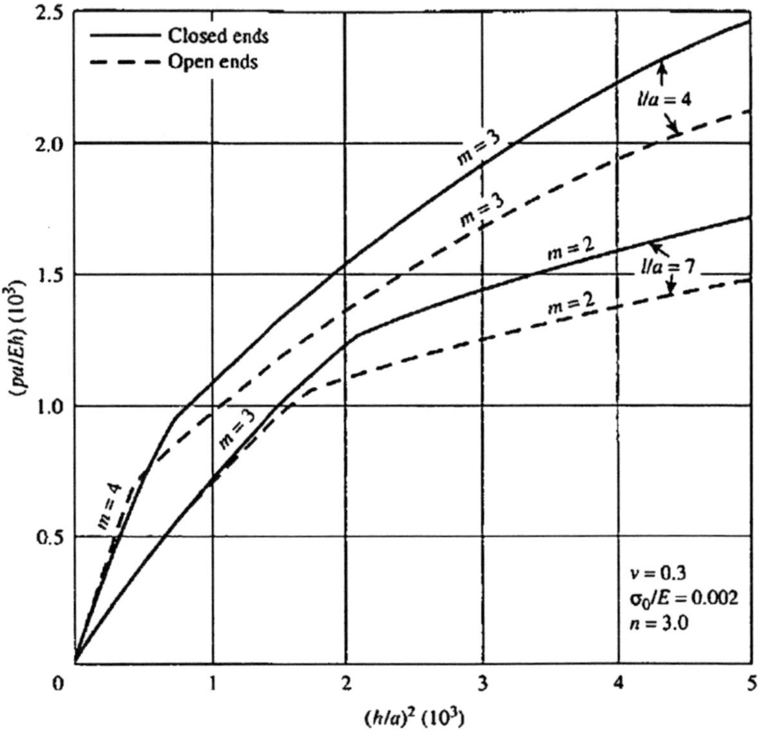

Fig. 7.16 Comparison of critical external pressure for the plastic buckling of cylindrical shell with open and closed end conditions

7.6 Torsional and Flexural Buckling of Tubes

7.6.1 Bifurcation Under Pure Torsion

In the case of a thin cylindrical shell subjected to equal and opposite twisting moments at its ends, the deformation mode ceases to be uniform when the applied shear stress τ attains a certain critical value. The thickness h and length l in relation to the mean radius a are assumed to be such that the bifurcation occurs beyond the elastic limit. Since the only nonzero stress at the onset of buckling is $\sigma_{x\theta} = \tau$, the rates of extension in the longitudinal and circumferential directions are purely elastic, and the corresponding normal components of the stress rate are

$$\dot{t}_{xx} = \frac{E}{1-\nu^2}\left(\dot{\varepsilon}_{xx} + \nu\dot{\varepsilon}_{\theta\theta}\right), \qquad \dot{t}_{\theta\theta} = \frac{E}{1-\nu^2}\left(\dot{\varepsilon}_{\theta\theta} + \nu\dot{\varepsilon}_{\theta\theta}\right).$$

The only nonzero rate of shear is $\dot{\varepsilon}_{x\theta}$ whose plastic component is $(3/2H)\,\dot{t}_{x\theta}$ and consequently,

$$\dot{\varepsilon}_{x\theta} = \left(\frac{1}{2G} + \frac{3}{2H}\right)\dot{t}_{x\theta} = \left(\frac{3}{2T} - \frac{1-2\nu}{2E}\right)\dot{t}_{x\theta},$$

where T is the tangent modulus at the current state of hardening of the material. The shear stress rate may therefore be written as

$$\dot{t}_{x\theta} = \frac{2E\alpha}{1-\nu^2}\dot{\varepsilon}_{x\theta}, \qquad \alpha = \frac{\left(1-\nu^2\right)T/E}{3-(1-2\nu)\,T/E}. \tag{7.123}$$

During the uniform twisting, α progressively decreases from the elastic value $(1-\nu)/2$ with increasing plastic strain. It follows from the above expressions that

$$\dot{t}_{ij}\dot{\varepsilon}_{ij} = \frac{E}{1-\nu^2}\left(\dot{\varepsilon}_{xx}^2 + 2\nu\dot{\varepsilon}_{xx}\dot{\varepsilon}_{\theta\theta} + \dot{\varepsilon}_{\theta\theta}^2 + 4\alpha\dot{\varepsilon}_{x\theta}^2\right).$$

The condition for uniqueness of the deformation mode is given by inequality (7.104) with $\rho = 0$. Since $\sigma_{x\theta} = \tau$, while all other components are identically zero, the inequality becomes

$$\iiint \left\{\left(\dot{\varepsilon}_{xx}^2 + 2\nu\dot{\varepsilon}_{xx}\dot{\varepsilon}_{\theta\theta} + \dot{\varepsilon}_{\theta\theta}^2 + 4\alpha\dot{\varepsilon}_{x\theta}^2\right) + 2\left(1-\nu^2\right)\frac{\tau}{E}\left(\omega_{xr}\omega_{\theta r} + \dot{\varepsilon}_{\theta\theta}\omega_{\theta x}\right)\right\} dx\,d\theta\,dz > 0 \tag{7.124}$$

to a close approximation, the possibility of sideways buckling of the tube being excluded. In terms of the velocities of the middle surface, the components of the strain rate are given by (7.98) and (7.99), while those of the rate of spin are given by (7.97), in view of the relations $\omega_{xr} = \omega_\theta$, $\omega_{\theta r} = -\omega_x$, and $\omega_{\theta x} = \omega_r$. Substituting into (7.124), and integrating through the shell thickness, we obtain

$$\iint \left\{ \left(\frac{\partial u}{\partial \xi}\right)^2 + 2\nu \frac{\partial u}{\partial \xi}\left(\frac{\partial \upsilon}{\partial \theta} + w\right) + \left(\frac{\partial \upsilon}{\partial \theta} + w\right)^2 + \alpha\left(\frac{\partial \upsilon}{\partial \xi} + \frac{\partial u}{\partial \theta}\right)^2 \right.$$
$$+ k\left[\left(\frac{\partial^2 w}{\partial \xi^2}\right)^2 + 2\nu \frac{\partial^2 w}{\partial \xi^2}\left(\frac{\partial^2 w}{\partial \theta^2} - \frac{\partial \upsilon}{\partial \theta}\right) + \left(\frac{\partial^2 w}{\partial \theta^2} - \frac{\partial \upsilon}{\partial \theta}\right)^2\right.$$
$$\left. + 4\alpha\left(\frac{\partial^2 w}{\partial \xi \partial \theta} - \frac{\partial \upsilon}{\partial \xi}\right)^2\right] + 2\phi\left[\frac{\partial w}{\partial \xi}\left(\frac{\partial w}{\partial \theta} - \upsilon\right)\right.$$
$$\left.\left. + \frac{\partial \upsilon}{\partial \xi}\left(\frac{\partial \upsilon}{\partial \theta} + w\right)\right]\right\} d\xi d\theta > 0,$$

on neglecting certain small-order terms and on introducing the dimensionless quantities

$$\phi = \left(1 - \nu^2\right)\frac{\tau}{E}, \quad k = \frac{h^2}{12a^2}, \quad \xi = \frac{x}{a}.$$

The bifurcation would occur when the functional on the left-hand side of the above inequality vanishes for some distribution of the velocities u, υ, and w that makes the functional a minimum. The Euler–Lagrange differential equations associated with this variational problem are easily shown to be

$$\begin{aligned}
&\frac{\partial^2 u}{\partial \xi^2} + \alpha \frac{\partial^2 u}{\partial \theta^2} + (\alpha + \nu)\frac{\partial^2 \upsilon}{\partial \xi \partial \theta} + \upsilon \frac{\partial w}{\partial \xi} = 0,\\
&\frac{\partial^2 \upsilon}{\partial \theta^2} + \alpha \frac{\partial^2 \upsilon}{\partial \xi^2} + (\alpha + \nu)\frac{\partial^2 \upsilon}{\partial \xi \partial \theta} + \frac{\partial w}{\partial \theta} + 2\phi\left(\frac{\partial^2 \upsilon}{\partial \xi \partial \theta} + \frac{\partial w}{\partial \xi}\right)\\
&\quad + k\left[\frac{\partial^2 \upsilon}{\partial \theta^2} + 4\alpha \frac{\partial^2 \upsilon}{\partial \xi^2} - (4\alpha + \nu)\frac{\partial^3 w}{\partial \xi^2 \partial \theta} - \frac{\partial^3 w}{\partial \theta^3}\right] = 0,\\
&\frac{\partial \upsilon}{\partial \theta} + \upsilon \frac{\partial u}{\partial \xi} + w + 2\phi\left(\frac{\partial \upsilon}{\partial \xi} - \frac{\partial^2 \upsilon}{\partial \xi \partial \theta}\right)\\
&\quad + k\left[\frac{\partial^4 w}{\partial \xi^2} + 2\upsilon \frac{\partial^4 w}{\partial \xi^2 \partial \theta^2} + \frac{\partial^4 w}{\partial \theta^2} - (4\alpha + \nu)\frac{\partial^3 w}{\partial \xi^2 \partial \theta} - \frac{\partial^3 w}{\partial \theta^3}\right] = 0.
\end{aligned} \tag{7.125}$$

In the case of an elastic material, (7.125) reduce to those given by Timoshenko and Gere (1961), except for certain small-order terms which do not affect the final results significantly. The problem of the buckling of a cylindrical shell under pure torsion is reduced to the integration of (7.125), using the boundary conditions.

In each of the three equations of (7.125), there are both odd- and even-order derivatives of a given velocity component with respect to the same independent variable. These equations cannot therefore be satisfied by assuming solutions in the form of products of sines and cosines of angles involving ξ and θ. When the cylinder is sufficiently large, so that the critical stress is practically independent of the edge constraints, we may assume the simple velocity field

$$\left.\begin{aligned} u = A\cos(\lambda\xi - m\theta), \qquad v = B\cos(\lambda\xi - m\theta), \\ W = C\sin(\lambda\xi - m\theta), \end{aligned}\right\} \tag{7.126}$$

where $\lambda = r\pi a/l$, with r denoting the number of longitudinal waves. The mode of buckling involves m circumferential waves which run helically along the cylinder. The substitution of (7.126) into (7.125) results in

$$\begin{aligned} &-\left(\lambda^2 + \alpha m^2\right)A + (\alpha + \nu)\lambda m B + \nu\lambda C = 0, \\ &(\alpha + \nu)\lambda m A - \left[(1+k)m^2 + (1+4k)\alpha\lambda^2 - 2\lambda m\phi\right]B \\ &\qquad - \left[m + km^3 + k(4\alpha + \nu)\lambda^2 m - 2\lambda\phi\right]C = 0, \\ &\nu\lambda A - \left[m + km^3 + k(4\alpha + \nu)\lambda^2 m - 2\lambda\phi\right]B \\ &\qquad - \left[1 + k\left(\lambda^4 + 2\nu\lambda^2 m^2 + m^4\right) - 2\lambda m\phi\right]C = 0. \end{aligned}$$

This system of linear homogeneous equations can have nontrivial solutions for A, B, and C only if the determinant of their coefficients vanishes. Equating the determinant to zero, and neglecting terms containing k^2, $k\phi$, ϕ^2, etc., the result may be expressed as

$$T\phi = R + kQ, \tag{7.127}$$

where R, Q, and T are dimensionless parameters given by the expressions

$$\begin{aligned} R &= \left(1 - \nu^2\right)\alpha\lambda^4, \\ Q &= m^2\left(m^2 - 1\right)^2\left[\alpha m^2 + \left(1 + 4\alpha^2 - \nu^2\right)\lambda^2\right] \\ &\quad + 2\lambda^2 m^4\left[\left(1 - \nu^2\right)(\alpha + \nu)\lambda^2 - 2\alpha^2 m^2\right], \\ T &= 2\lambda m\left\{\left(m^2 - 1\right)\left[\alpha m^2 + \left(1 - 2\alpha\nu - \nu^2\right)\lambda^2\right] + \alpha\lambda^4\right\}. \end{aligned}$$

The terms involving higher powers of λ have been omitted in the above expression for Q. Substituting for R, Q, and T into (7.127), the expression for the critical shear stress may be written with sufficient accuracy as

$$\frac{\tau}{E} = \frac{\lambda^3}{2m\left(m^2 - 1\right)\left(m^2 + \beta\lambda^2\right)} + \frac{km\left(m^2 - 1\right)}{2\lambda\left(1 - \nu^2\right)}, \tag{7.128}$$

where $\beta = (1 - \nu^2)/\alpha - 2\nu$. The value of λ that minimizes the critical stress may be approximately taken as that for elastic buckling, in which case $\beta\lambda^2$ is negligible in comparison with m^2. The optimum value of λ is then given by

$$\lambda^2 = \frac{m^2\left(m^2-1\right)}{6\sqrt{1-\nu^2}}\frac{h}{a}. \tag{7.129}$$

The critical stress formula for plastic buckling, obtained by omitting the quantity $\beta\lambda^2$ in the first term of (7.128) and substituting from (7.129), may be written as

$$\frac{\tau}{E} = \sqrt{\frac{3}{2}}\left(1-\nu^2\right)^{-3/4}\left(\frac{h}{3a}\right)^{3/2}$$

where we have set $m = 2$ to minimize τ. The critical stress for plastic buckling can be computed from (7.128) and (7.129) by setting $m = 2$ and using the relation

$$\alpha = \left(1-\nu^2\right)/\left\{2(1+\nu) + \frac{9n}{7}\left(\frac{\sqrt{3}\tau}{\sigma_0}\right)^{n-1}\right\}, \tag{7.130}$$

which is obtained from (7.123) and (7.115) with the substitution $\sigma = \sqrt{3}\tau$. The results for plastic buckling are displayed as solid curves in Fig. 7.17, using $\nu = 0.3$,

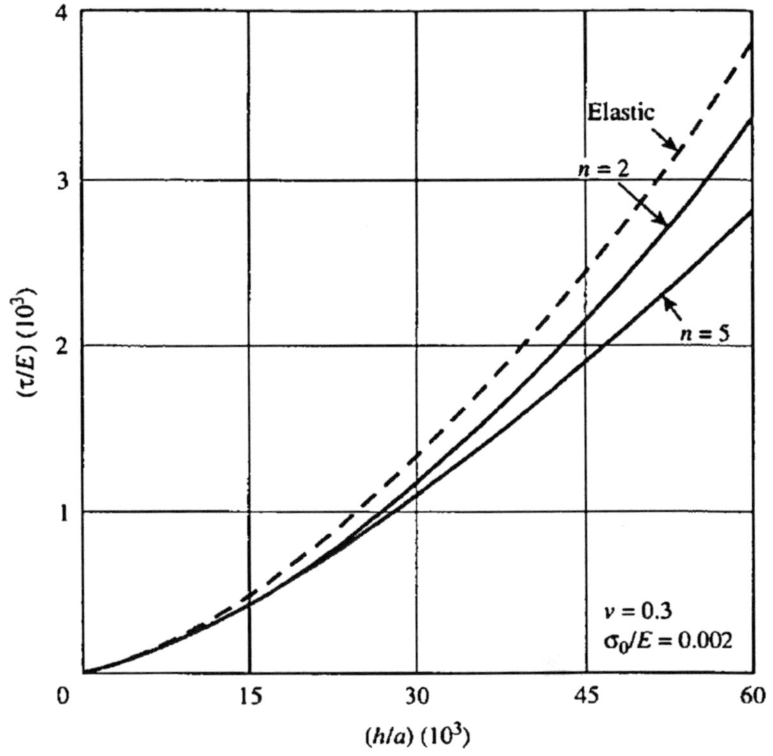

Fig. 7.17 Critical shear stress for plastic buckling of a long cylindrical tube under pure torsion

$\sigma_0/E = 0.002$, and two different values of n. The broken curve represents the elastic buckling formula which is valid only for very small thicknesses.

In the case of relatively short cylindrical shells, the conditions of support at the ends of the cylinder must be taken into account for the estimation of the critical stress. For the elastic range of buckling, an approximate analysis of the problem has been discussed by Donnell (1933), who introduced several simplifications into the basic differential equations (7.125). By applying appropriate boundary conditions on v and w, while permitting some axial motion of the ends of the cylinder, Donnell obtained buckling formulas that were found to be in good agreement with experiment. An empirical extension of the elastic buckling formula to cover the plastic range of loading has been suggested by Gerard (1962). Assuming the cylindrical shell to be clamped at both ends, the critical stress given by Gerard may be written as

$$\frac{\tau}{S} = 0.82\left(1 - \eta^2\right)^{-5/8}\left(\frac{h}{a}\right)^{5/4}\left(\frac{a}{l}\right)^{1/2}, \tag{7.131}$$

where S is the secant modulus at the onset of buckling and η is the contraction ratio given by the relation $1 - 2\eta = (1 - 2v)S/E$. For a given h/a ratio, (7.131) should be used for the range of values of l/a, which corresponds to a lower value of the critical stress than that given by (7.128). Adopting the Ramberg–Osgood equation for the stress–strain curve, the values S and η are easily computed by using (7.115).

A modified buckling formula that includes the minor effect of preventing the axial strain in an approximate manner has been proposed by Rees (1982), who also produced some experimental evidence in support of the theory.

7.6.2 Buckling Under Pure Bending

When an initially straight cylindrical tube is subjected to a gradually increasing bending moment, the circular cross section becomes increasingly oval until the applied moment starts to decrease after attaining a maximum, which constitutes a state of collapse due to buckling of the tube. This phenomenon is particularly significant for sufficiently long tubes, and the maximum compressive stress at the point of buckling is found to be considerably lower than the critical stress in an axially compressed cylindrical shell. The problem has been investigated by Brazier (1926) for the elastic range of buckling, and by Ades (1957), Gellin (1980), and Zhang and Yu (1987) for the plastic range of buckling. The effect of an internal pressure on the plastic collapse of a bent tube has been examined by Corona and Kyriakides (1988).

Figure 7.18 shows the deformed configuration of the cross section of the middle surface, in which the position of a typical particle is specified by the angular coordinate θ measured from the crown of the circle. The circumferential and radial components of the displacement of any particle will be denoted by v and w, respectively, and the curvature of the neutral surface of the bent tube will be denoted by

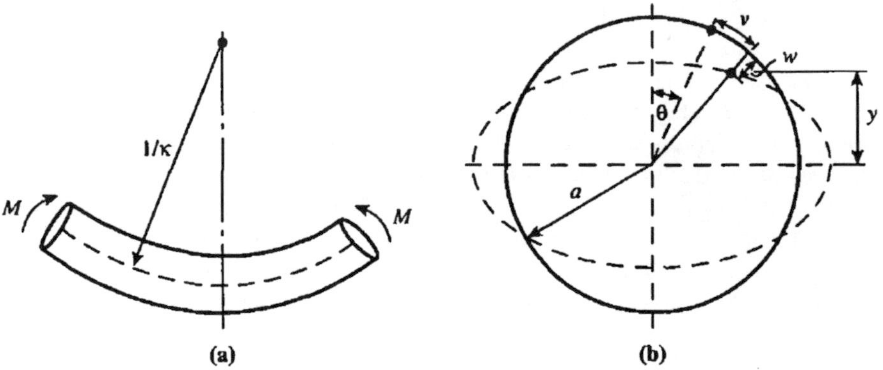

Fig. 7.18 Buckling of a cylindrical tube under pure bending. (**a**) Overall shape of the bent tube, (**b**) cross section before and after buckling

v, which is measured positive as indicated in the figure. By simple geometry, the distance of any point on the deformed middle surface from the neutral plane is

$$y = (a + w)\cos\theta - \upsilon\sin\theta, \tag{7.132}$$

where α is the mean radius of the cylindrical tube. If the deformation of the middle surface is assumed to be inextensional in the circumferential direction, and the deformation and rotation are everywhere small, the longitudinal and circumferential components of the strain are

$$\left.\begin{aligned} \varepsilon_x &= -y\kappa,\ \varepsilon_\theta = -z\kappa_\theta, \\ \kappa_\theta &= -\frac{1}{a^2}\frac{d}{d\theta}\left(\upsilon - \frac{dw}{d\theta}\right), \end{aligned}\right\} \tag{7.133}$$

where κ_θ is the circumferential curvature of the middle surface, and z is the radially outward distance from the middle surface. The velocity components υ and w are related to one another through the equation $d\upsilon/d\theta + w = 0$, which is the condition of zero circumferential strain at $z = 0$.

Since the variation of the stress ratio during the prebuckling deformation is expected to be sufficiently small, the total strain theory of Hencky may be used without introducing significant error. Denoting the axial and circumferential stresses by σ_x and σ_θ, respectively, the stress–strain relations may be written as

$$S\varepsilon_x = \sigma_x - \eta\sigma_\theta, \quad S\varepsilon_\theta = \sigma_\theta - \eta\sigma_x,$$

where S is the secant modulus of the effective stress–strain curve corresponding to the current effective stress $\bar{\sigma}$, and

$$\eta = \frac{1}{2}\left[1 - (1-2\nu)\frac{S}{E}\right], \qquad \frac{E}{S} = 1 + \left(\frac{3}{7}\right)\left(\frac{\bar{\sigma}}{\sigma_0}\right)^{m-1}, \tag{7.134}$$

when the stress–strain curve is represented by the Ramberg–Osgood equation (7.115). The stress–strain relations are readily inverted to express the stresses as

$$\sigma_x = \frac{S}{1-\eta^2}\left(\varepsilon_x + \eta\varepsilon_\theta\right), \quad \sigma_\theta = \frac{S}{1-\eta^2}\left(\varepsilon_\theta + \eta\varepsilon_x\right), \tag{7.135}$$

The significant components of the resultant force and moment are the axial force N_x and the circumferential bending moment M_θ acting per unit perimeter. They are defined as

$$N_x = \int_{-h/2}^{h/2} \sigma_x \, dz, \quad M_\theta = -\int_{-h/2}^{h/2} \sigma_\theta z \, dz, \tag{7.136}$$

where h is the uniform wall thickness of the bent tube. The variation of the internal virtual work per unit length of the shell may be written to a close approximation in the form

$$\delta U \approx a \int_0^{2\pi} \left(N_x \, \delta\varepsilon_x + M_\theta \, \delta\kappa_\theta\right) d\theta, \tag{7.137}$$

where $\delta\varepsilon_x$ and $\delta\kappa_\theta$ are the variations of ε_x and κ_θ, respectively. The contributions to the work from the remaining stress resultants N_θ and M_x are neglected.

For a prescribed curvature κ of the neutral surface, the displacements v and w should be such that $\delta U = 0$, which is a variational statement of equilibrium of the bent tube. While evaluating N_x, and M_θ, it would be a good approximation to assume a constant value of S equal to that on the middle surface. Then, in view of (7.135) and (7.136), and the fact that $\varepsilon_\theta = -zk_\theta$, we have

$$N_x = \left(\frac{Sh}{1-\eta^2}\right)\varepsilon_x, \qquad M_\theta = \left(\frac{Sh}{1-\eta^2}\right)\frac{h^2}{12}\kappa_\theta.$$

Inserting in (7.137), and setting $\delta U = 0$, the variational relation is reduced to the compact form

$$\int_0^{2\pi} \frac{S}{1-\eta^2}\left\{\kappa^2 y \, \delta y + \frac{h^2}{12}\kappa_\theta \, \delta\kappa_\theta\right\} d\theta = 0. \tag{7.138}$$

To evaluate the integral, the secant modulus S must be determined as a function of θ. Since $\sigma_x = S\varepsilon_x/(1-\eta^2)$ and $\sigma_\theta = \eta\sigma_x$ at $z = 0$, we have

$$\bar{\sigma} = \sqrt{\sigma_x^2 - \sigma_x\sigma_\theta + \sigma_\theta^2} = \frac{S\sqrt{1-\eta+\eta^2}}{1-\eta^2}\,|\varepsilon_x|$$

at $z = 0$. The substitution for $\bar{\sigma}$ and η in terms of S/E into the above relation, using (7.134), leads to the equation

$$\left(\frac{3E}{4S} + \mu - \frac{\mu^2 S}{E}\right)\left(\frac{3}{4} + \frac{\mu^2 S^2}{E^2}\right)^{-1/2}\left\{\frac{7}{3}\left(\frac{E}{S} - 1\right)\right\}^{1/(m-1)} = \frac{Ea\kappa}{\sigma_0}\left|\frac{y}{a}\right|, \quad (7.139)$$

where $\mu = \frac{1}{2} - \nu$, and y/a is a known function for any given displacement field. The applied bending moment M increases with the curvature κ in the prebuckling stages, the magnitude of the bending couple being given by

$$M = -a\int_0^{2\pi} N_x y\, d\theta = a^3 hk \int_0^{2\pi} \frac{S}{1-\eta^2}\left(\frac{y}{a}\right) d\theta, \quad (7.140)$$

where $S/(1-\eta^2)$ corresponds to $z = 0$ as before. The critical moment M_c is the maximum value of M for varying κ during the bending.

The components of the displacement of the middle surface, satisfying the inextensibility condition $d\upsilon/d\theta = -w$ and the fourfold symmetry condition, may be expressed in the general form

$$w = -\sum_{m=1}^{N} w_m \cos 2m\theta, \qquad \upsilon = \sum_{m=1}^{N} (w_m/2m)\sin 2m\theta. \quad (7.141)$$

The analysis for the buckling problem will be given here for the special case of $N = 2$, which is a good approximation for practical purposes. We therefore write

$$w = -\left(w_1 \cos 2\theta + w_2 \cos 4\theta\right), \qquad \upsilon = \frac{1}{2}\left(w_1 \sin 2\theta + \frac{1}{2} w_2 \sin 4\theta\right),$$

where w_1 and w_2 are constants. Considering only the first quadrant, υ is seen to vanish at $\theta = 0$ and $\theta = \pi/2$, the value of w at these two points being $-(w_1 + w_2)$ and $(w_1 - w_2)$, respectively. The substitution for υ and w into (7.132) and (7.133) gives

$$\left.\begin{aligned} y &= \cos\theta\left\{a - \left[w_1 - (2 - 3\cos 2\theta)\, w_2\right]\cos^2\theta\right\}, \\ \kappa_\theta &= \frac{3}{a^2}\left(w_1 \cos 2\theta + 5 w_2 \cos 4\theta\right). \end{aligned}\right\} \quad (7.142)$$

The constant displacements w_1 and w_2 can be determined by inserting (7.142) into the variational equation (7.138). It is convenient at this stage to set $\lambda = (S/E)/(1-\eta^2)$ and denote the various integrals as

$$\left.\begin{aligned}
A_1 &= \int_0^{\pi/2} \lambda \cos^4 \theta d\theta, & A_2 &= \int_0^{\pi/2} \lambda (2 - 3\cos 2\theta) \cos^6 \theta d\theta, \\
B_1 &= \int_0^{\pi/2} \lambda \cos^6 \theta d\theta, & B_2 &= \int_0^{\pi/2} \lambda (2 - 3\cos 2\theta) \cos^6 \theta d\theta, \\
C_1 &= \frac{3}{4}\int_0^{\pi/2} \lambda \cos^2 2\theta d\theta, & C_2 &= \frac{15}{4}\int_0^{\pi/2} \lambda \cos 2\theta \cos 4\theta d\theta, \\
C_3 &= \frac{75}{4}\int_0^{\pi/2} \lambda \cos^2 4\theta d\theta, & B_3 &= \int_0^{\pi/2} \lambda (2 - 3\cos 2\theta)^2 \cos^6 \theta d\theta,
\end{aligned}\right\} \tag{7.143}$$

where λ is a known function of θ in view of (7.139) and (7.142). Using (7.138) and (7.142), and the fact that δw_1 and δw_2 are arbitrary variations of w_1 and w_2, respectively, we thus obtain the pair of equations

$$\left(C_1 + B_1 \frac{\kappa^2 a^4}{h^2}\right)\frac{w_1}{a} + \left(C_2 - B_2 \frac{\kappa^2 a^4}{h^2}\right)\frac{w_2}{a} = A_1,$$

$$\left(C_2 - B_2 \frac{\kappa^2 a^4}{h^2}\right)\frac{w_1}{a} + \left(C_3 + B_3 \frac{\kappa^2 a^4}{h^2}\right)\frac{w_2}{a} = -A_2,$$

for the two dimensionless constants w_1/a and w_2/a. Setting $\rho = \kappa a^2/h$, the solution can be written in the form

$$\left.\begin{aligned}
\frac{w_1}{a} &= \frac{A_1\left(C_3 + B_3\rho^2\right) + A_2\left(C_2 - B_2\rho^2\right)}{\left(C_1 + B_1\rho^2\right)\left(C_3 + B_3\rho^2\right) - \left(C_2 - B_2\rho^2\right)^2}, \\
\frac{w_2}{a} &= -\frac{A_1\left(C_2 - B_2\rho^2\right) + A_2\left(C_1 + B_1\rho^2\right)}{\left(C_1 + B_1\rho^2\right)\left(C_3 + B_3\rho^2\right) - \left(C_2 - B_2\rho^2\right)^2}.
\end{aligned}\right\} \tag{7.144}$$

When the ratios w_1/a and w_2/a have been computed from (7.143) and (7.144) for a selected value of ρ, the corresponding bending moment can be determined from (7.140). In view of the relations (7.142) and (7.143), the bending moment can be expressed in the dimensionless form

$$\begin{aligned}
\frac{M}{M^*} = \frac{4hE\rho}{\pi a\sigma^*}\Bigg\{\int_0^{\pi/2} \lambda \cos^2 \theta d\theta - 2A_1\left(\frac{w_1}{a}\right) + 2A_2\left(\frac{w_2}{a}\right) + B_1\left(\frac{w_1}{a}\right)^2 \\
-2B_2\left(\frac{w_1 w_2}{a^2}\right) + B_3\left(\frac{w_2}{a}\right)^2\Bigg\},
\end{aligned} \tag{7.145}$$

where $M^* = \pi a^2 h\sigma^*$, with σ^* denoting the critical stress $E\,(h/a)\,/\sqrt{3\left(1-\nu^2\right)}$ for the elastic buckling of the shell under uniform axial compression (Section 7.5). The critical bending moment M_c under pure bending is the maximum value of M predicted by (7.145) for varying values of ρ. Over the elastic range, $S = E$ and

$\eta = \nu$, giving the critical moment $M_c \approx 9.545M^*$ as the maximum bending moment that corresponds to $p \approx 0.495$ when $\nu = 0.3$.

The solution is marginally improved if the nonlinearity of the strain–displacement relations is taken into consideration. This has been demonstrated by Gellin (1980), who computed the bending moment and curvature of the tube at the onset of buckling for different values of *n*, using a series of values of the parameter

$$\frac{\sigma_0}{\sigma^*} = \sqrt{3\left(1-\nu^2\right)}\frac{a\sigma_0}{hE}$$

and assuming $\nu = 0.3$. His results are displayed in Fig. 7.19, which indicates that the critical curvature is relatively insensitive to the variation of σ_0/σ^* for usual values of *n*. The theory has been found to be in reasonable agreement with available experimental data.

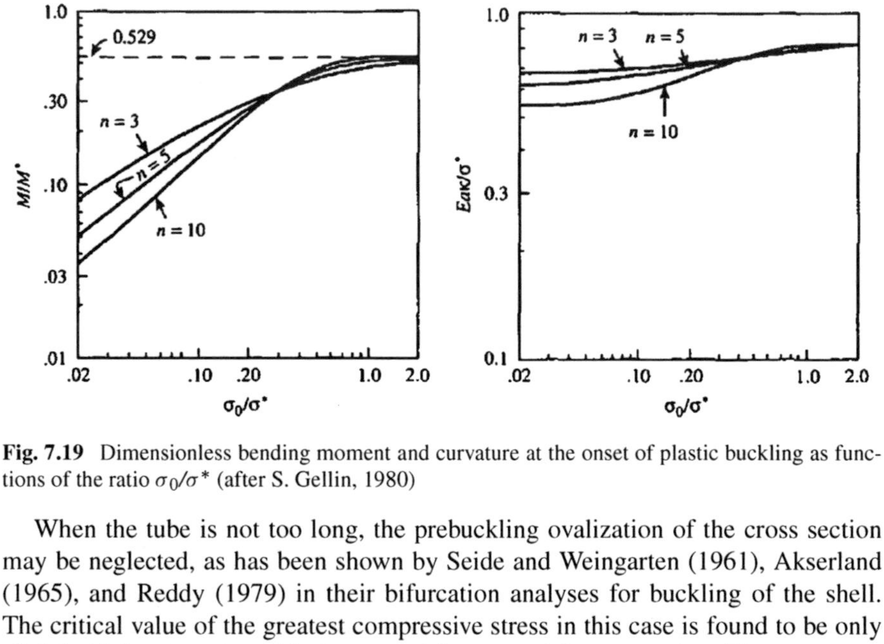

Fig. 7.19 Dimensionless bending moment and curvature at the onset of plastic buckling as functions of the ratio σ_0/σ^* (after S. Gellin, 1980)

When the tube is not too long, the prebuckling ovalization of the cross section may be neglected, as has been shown by Seide and Weingarten (1961), Akserland (1965), and Reddy (1979) in their bifurcation analyses for buckling of the shell. The critical value of the greatest compressive stress in this case is found to be only slightly higher than the critical stress σ^* in pure axial compression. The same conclusion may be assumed to hold for the flexural buckling in the plastic range when the tube is relatively short.

7.7 Buckling of Spherical Shells

7.7.1 Analysis for a Complete Spherical Shell

A complete spherical shell of uniform small thickness *h* is subjected to a uniform external pressure of intensity *p* per unit area of the middle surface. When the pressure is increased to a certain critical value, the spherical form of equilibrium is no longer guaranteed, and a nonuniform mode of deformation is possible as a result of buckling. It is assumed at the outset that the buckled shape of the middle surface is symmetrical with respect to a diameter of the sphere. The velocity field at the incipient buckling, referred to the spherical coordinates (r, ϕ, θ), where ϕ is measured from one of the poles, may be written as

$$v_r = w, \qquad v_\phi = u + z\omega, \qquad v_\theta = 0,$$

where z is the radially outward distance from the middle surface, and ω is the rate of spin about the θ-axis. The thin shell approximation $\dot{\varepsilon}_{r\phi} = 0$ gives

$$\omega = \frac{1}{a}\left(u - \frac{dw}{d\phi}\right),$$

where a denotes the radius of curvature of the middle surface. The nonzero components of the strain rate at a generic point of the shell material are

$$\dot{\varepsilon}_{\theta\theta} = \dot{\lambda}_\theta - z\dot{\kappa}_\theta, \qquad \dot{\varepsilon}_{\phi\phi} = \dot{\lambda}_\phi - z\dot{\kappa}_\phi. \tag{7.146}$$

The quantities $\dot{\lambda}_\theta$,$\dot{\lambda}_\phi$ are the rates of extension of the middle surface, and $\dot{\kappa}_\theta$,$\dot{\kappa}_\phi$ are the rates of curvature of the middle surface. By (5.67), with the necessary change in sign for w, we have

$$\left.\begin{aligned} \dot{\lambda}_\theta &= \frac{1}{a}\left(u \cot\phi + w\right), & \dot{\lambda}_\phi &= \frac{1}{\alpha}\left(\frac{du}{d\phi} + w\right), \\ \dot{\kappa}_\theta &= \frac{1}{a^2}\left(\frac{dw}{d\phi} - u\right)\cot\phi, & \dot{\kappa}_\phi &= \frac{1}{a^2}\frac{d}{d\phi}\left(\frac{dw}{d\phi} - u\right). \end{aligned}\right\} \tag{7.147}$$

The current state of stress prior to buckling is a balanced biaxial compression represented by $\sigma_{\theta\theta} = \sigma_{\phi\phi} = -pa/2h$ the remaining stress components being identically zero. The rate of change of the stress at the incipient buckling varies, however, along the meridian due to the nonuniformity of the mode of deformation.

We consider a regular isotropic yield surface for the material which is stressed in the plastic range with a current uniaxial yield stress σ and tangent modulus T. The unit normal to the yield surface, considered in a nine-dimensional space, has the nonzero components

$$n_{\theta\theta} = n_{\phi\phi} = -\frac{1}{\sqrt{6}}, \qquad n_{rr} = \sqrt{\frac{2}{3}}.$$

Considering a linearized elastic/plastic solid having the constitutive equation (7.100), the expressions for the circumferential and meridional rates of extension can be written as

$$4T\dot{\varepsilon}_{\theta\theta} = \left(1 + \frac{3T}{E}\right)\dot{\tau}_{\theta\theta} + \left[1 - (1+4\nu)\frac{T}{E}\right]\dot{\tau}_{\phi\phi},$$

$$4T\dot{\varepsilon}_{\phi\phi} = \left(1 + (1+4\nu)\frac{T}{E}\right)\dot{\tau}_{\theta\theta} + \left[1 + \frac{3T}{E}\right]\dot{\tau}_{\phi\phi}.$$

Setting $T = E$ in these equations, we recover the rate form of stress–strain relations for an isotropic elastic material. The above equations are easily solved for the stress rates to give

$$\dot{\tau}_{\theta\theta} = \frac{E}{1+\nu}\left(\alpha\dot{\varepsilon}_{\theta\theta} + \beta\dot{\varepsilon}_{\phi\phi}\right), \qquad \dot{\tau}_{\phi\theta} = \frac{E}{1+\nu}\left(\beta\dot{\varepsilon}_{\theta\theta} + \alpha\dot{\varepsilon}_{\phi\phi}\right), \tag{7.148}$$

where

$$\alpha = \frac{1+3T/E}{2\left[1+(1-2\nu)\,T/E\right]}, \qquad \beta = \frac{-1+(1+4\nu)\,T/E}{2\left[1+(1-2\nu)\,T/E\right]}. \tag{7.149}$$

The remaining components of the stress rate are identically zero. It follows from (7.148) that

$$\dot{\tau}_{ij}\dot{\varepsilon}_{ij} = \frac{E}{1+\nu}\left(\alpha\dot{\varepsilon}_{\theta\theta}^2 + 2\beta\dot{\varepsilon}_{\theta\theta}\dot{\varepsilon}_{\phi\phi} + \alpha\dot{\varepsilon}_{\phi\phi}^2\right),$$

which gives the leading term in the uniqueness functional (7.104), the strain rates appearing in this expression being given by (7.146) and (7.147).

The condition for uniqueness of the deformation mode is easily established by using the facts that the only nonzero components of the stress tensor are $\sigma_{\theta\theta}$ and $\sigma_{\phi\phi}$, each being equal to $-pa/2h$, and that the only nonzero components of the spin tensor are $\omega_{r\phi} = -\omega$ and $\omega_{\phi r} = \omega$. Since the shear components of the strain rate are identically zero, the uniqueness criterion (7.104) furnishes

$$\begin{aligned}\int_{-h/2}^{h/2}\int_0^{\pi}&\left(\alpha\dot{\varepsilon}_{\theta\theta}^2 + 2\beta\dot{\varepsilon}_{\theta\theta}\dot{\varepsilon}_{\phi\phi} + \alpha\dot{\varepsilon}_{\phi\phi}^2\right)\left(1 + \frac{2z}{a}\right)\sin\phi d\phi dz \\ &- (1+\nu)\frac{pa}{E}\int_0^{\pi}\omega^2 \sin\phi d\phi \\ &+ (1+\nu)\frac{p}{E}\int_0^{\pi}\left[\left(\dot{\lambda}_\theta + \dot{\lambda}_\phi\right)w + uw\right]\sin\phi d\phi > 0\end{aligned}$$

to a close approximation. Substituting for the strain rates and the rate of spin, integrating through the thickness of the shell, and omitting the quantity $d(uw \sin\phi)$ which does not contribute to the integral, the inequality may be expressed as

$$\int_0^\pi \left[\alpha \left(u\cot\phi + w\right)^2 + 2\beta\left(u\cot\phi + w\right)\left(\frac{du}{d\phi} + w\right) + \alpha\left(\frac{du}{d\phi} + w\right)^2 \right] \sin\phi d\phi$$
$$+ \frac{h^2}{12a^2}\int_0^\pi \left[\alpha\left(\frac{dw}{d\phi} - u\right)^2 \cot^2\phi + 2\beta\left(\frac{dw}{d\phi} - u\right)\left(\frac{d^2w}{d\phi^2} - \frac{du}{d\phi}\right)\cot\phi + \alpha\left(\frac{d^2w}{d\phi} - \frac{du}{d\phi}\right)^2 \right]\sin\phi d\phi$$
$$- (1+\nu)\frac{pa}{2Eh}\int_0^\pi \left[\left(\frac{dw}{d\phi}\right)^2 - 2w^2\right]\sin\phi d\phi > 0$$

on neglecting terms of order $(h/a)^2$ times the square of the rate of extension of the middle surface. The bifurcation would occur when the left-hand side of the above inequality is zero, the corresponding velocity field being that which minimizes the functional. Introducing the dimensionless quantities

$$q = (1+\nu)\frac{pa}{2Eh}, \quad k = \frac{h^2}{12a^2},$$

the Euler–Lagrange differential equations satisfied by the two velocity components u and w can be easily written down following the standard technique of the calculus of variations. Taking due account of the factor $\sin\phi$ that appears outside the square brackets in each integral, we obtain

$$\begin{aligned}(1+k)&\left[\alpha\left(\frac{d^2u}{d\phi^2} + \cot\phi\frac{du}{d\phi}\right) - \left(\beta + \alpha\cot^2\phi\right)u\right] + (\alpha+\beta)\frac{dw}{d\phi}\\ &- k\left[\alpha\left(\frac{d^3w}{d\phi^3} + \cot\phi\frac{d^2w}{d\phi^2}\right) - \left(\beta + \alpha\cot^2\phi\frac{dw}{d\phi}\right)\right] = 0,\end{aligned} \tag{7.150}$$

$$\begin{aligned}&(\alpha+\beta)\left(\frac{du}{d\phi} + u\cot\phi + 2w\right) + k\left[-\alpha\left(\frac{d^3u}{d\phi^3} + 2\cot\phi\frac{d^2u}{d\phi^2}\right)\right.\\ &+ \left(\beta + \alpha\mathrm{cosec}^2\phi\right)\frac{du}{d\phi} - \cot\phi\left(\alpha - \beta + \alpha\mathrm{cosec}^2\phi\right)u\\ &+ \alpha\left(\frac{d^4w}{d\phi^4} + 2\cot\phi\frac{d^3w}{d\phi^3}\right) - \left(\beta + \alpha\mathrm{cosec}^2\phi\right)\frac{d^2w}{d\phi^2}\\ &\left. + \cot\phi\left(\alpha - \beta + \alpha\mathrm{cosec}^2\phi\right)\frac{dw}{d\phi}\right] + q\left(\frac{d^2w}{d\phi^2} + \cot\phi\frac{dw}{d\phi} + 2w\right) = 0.\end{aligned} \tag{7.151}$$

An immediate simplification of (7.150) is achieved by omitting k in the first factor, since it is small compared to unity. The above equations can be expressed in a more convenient form by setting $u = dv/d\phi$ where v is a new variable, and by introducing an operator H defined as

$$H = \frac{d^2}{d\phi^2} + \cot\phi \frac{d}{d\phi} + 2.$$

In terms of the dependent variables v and w, the differential equation (7.150) then becomes

$$\frac{d}{d\phi}\left[\alpha H(v) + (\alpha+\beta)(w-v) - k\alpha H(w) + k(\alpha+\beta)w\right] = 0$$

The last term in the square brackets of this equation is seen to be negligible compared to the second. Integrating this equation, and setting the constant of integration to zero, we have

$$\alpha H(v) + (\alpha+\beta)(w-v) - k\alpha H(w) = 0. \tag{7.152}$$

Equation (7.151) can be similarly expressed in terms of the operator H and the new variable v. The analysis is considerably simplified by noting the fact that the expressions in v and w appearing in the square brackets of (7.151) are identical to one another except for the sign. The final result is easily shown to be

$$(\alpha+\beta)[H(v) + 2(w-v)] + qH(w) + k\alpha HH(w-v) - (\alpha+\beta)H(w-v)] = 0. \tag{7.153}$$

on neglecting a term of order k compared to unity. The analysis for the bifurcation problem is therefore reduced to the solution of the pair of differential equations (7.152) and (7.153), the only restriction on the admissible velocity field being $dv/d\phi = dw/d\phi = 0$ at $\phi = \pi$ and $0 = \pi$ in view of the symmetry of the field about the diameter passing through these points.

7.7.2 Solution for the Critical Pressure

The nature of the differential equations (7.152) and (7.153) indicates that the solution may be expressed in terms of Legendre functions P_m of integer orders m. These functions are defined as

$$\begin{aligned} P_m(\cos\phi) = \frac{(2m)!}{2^m (m!)^2}\Bigg\{(\cos\phi)^m - \frac{m(m-1)}{2.(2m-1)}(\cos\phi)^{m-2} \\ + \frac{m(m-1)(m-2)(m-3)}{2.4.(2m-1)(2m-3)}(\cos\phi)^{m-4} - \cdots\Bigg\}. \end{aligned} \tag{7.154}$$

The expression on the right-hand side of (7.154) is a polynomial of degree m in the variable $\cos\phi$. The Legendre functions of the first few orders are

$$P_0(\cos\phi) = 1, \qquad P_1(\cos\phi) = \cos\phi, \qquad P_2(\cos\phi) = \frac{1}{2}\left(3\cos^2\phi - 1\right),$$
$$P_3(\cos\phi) = \frac{1}{2}\cos\phi\left(5\cos^2\phi - 3\right), \qquad P_4(\cos\phi) = \frac{5}{8}\left(7\cos^4\phi - 6\cos^2\phi + \frac{3}{5}\right).$$

All these functions are found to satisfy a second-order linear differential equation, known as Legendre's equation, which is

$$\frac{d^2P_m}{d\phi^2} + \cot\phi\frac{dP_m}{d\phi} + m(1+m)P_m = 0. \tag{7.155}$$

Applying the operator H on the Legendre function P_m, and using (7.155), it is readily shown that

$$H(P_m) = -\lambda_m P_m, \quad HH(P_m) = \lambda_m^2 P_m, \quad \lambda_m = m(1+m) - 2. \tag{7.156}$$

The usefulness of the operator H when dealing with Legendre functions now becomes obvious.

To obtain a general solution to the eigenvalue problem, it is convenient to express the quantities v and w in the form of the infinite series

$$\upsilon = \sum_{m=0}^{\infty} A_m P_m(\cos\phi), \qquad w = \sum_{m=0}^{\infty} B_m P_m(\cos\phi),$$

where A_m and B_m are arbitrary constants. Substitute these expressions into the differential equations (7.152) and (7.153), and using (7.156), we obtain the relations

$$\sum_{m=0}^{\infty}\{[\alpha+\beta+\alpha\lambda_m]A_m - [\alpha+\beta+k\alpha\lambda_m]B_m\}P_m = 0,$$
$$\sum_{m=0}^{\infty}\{[(\alpha+\beta)(2+\lambda_m) + k\lambda_m(\alpha+\beta+\alpha\lambda_m)]A_m$$
$$-\left[2(\alpha+\beta) - q\lambda_m + k\lambda_m(\alpha+\beta+\alpha\lambda_m)\right]B_m\}P_m = 0.$$

These equations will be satisfied only if the expressions appearing on the left-hand side for each value of m individually vanish. Hence, omitting the summation sign, we obtain the pair of equations

$$\left.\begin{aligned} (\alpha+\beta+\alpha\lambda_m)A_m - (\alpha+\beta+k\alpha\lambda_m)B_m = 0,\\ [(\alpha+\beta)(2+\lambda_m) + k\lambda_m(\alpha+\beta+\alpha\lambda_m)]A_m\\ -\left[2(\alpha+\beta) - q\lambda_m + k\lambda_m(\alpha+\beta+\alpha\lambda_m)\right]B_m = 0, \end{aligned}\right\} \tag{7.157}$$

for the two typical constants A_m and B_m. For the bifurcation to occur, (7.157) must admit nonzero values of these constants for some value of m, in which case the determinant of the coefficients of A_m and B_m must vanish. Neglecting the small terms involving q^k and k^2, the result may be expressed as

$$q\lambda_m\left(\alpha+\beta+\alpha\lambda_m\right)-\lambda_m\left(\alpha^2-\beta^2\right)-k\alpha\lambda_m\left[\alpha\lambda_m^2-2\left(\alpha+\beta\right)\right]=0$$

Since $\lambda_m = 0$ implies $m = 1$, representing a rigid-body motion, only $\lambda_m \neq 0$ will be relevant for bifurcation. Using the fact that $\alpha—\beta = 1$, the dimensionless critical pressure is obtained as

$$q=\frac{\left(\alpha+\beta\right)+k\alpha\left[\alpha\lambda_m^2-2\left(\alpha+\beta\right)\right]}{\left(\alpha+\beta\right)+\alpha\lambda_m}. \tag{7.158}$$

To obtain the smallest value of q for which bifurcation may occur, it is convenient to regard the right-hand side of (7.158) as a continuous function of λ_m. Then setting $dq/d\lambda_m = 0$ to minimize q, the result is easily shown to be

$$\left.\begin{aligned}\alpha\lambda_m&=-\left(\alpha+\beta\right)+\sqrt{\frac{\alpha+\beta}{k}},\\ q&=2\sqrt{k\left(\alpha+\beta\right)}-2k\left(\alpha+\beta\right),\end{aligned}\right\} \tag{7.159}$$

to a close approximation. Since λ_m is a large number, the critical pressure given by (7.159) cannot differ significantly from the smallest pressure based on successive integer values of m. For practical purposes, the second term on the right-hand side for q in (7.159) may be disregarded without any significant error. Using (7.149) for α and β we obtain the critical compressive stress for bifurcation in the form

$$\frac{\sigma}{E}=\frac{pa}{2Eh}=\frac{\sqrt{2}\left(h/a\right)}{\sqrt{3\left(1+\nu\right)\left(1-2\nu+E/T\right)}}, \tag{7.160}$$

which is the true tangent modulus formula for the plastic buckling of a spherical shell under external pressure. The same formula has been obtained earlier by Batterman (1969) using the rate equations of equilibrium instead of a variational principle. For any given material, the critical stress can be easily computed from (7.160) and (7.115) as a function of the ratio h/a, the results of the computation being presented graphically in Fig. 7.20.

The deformation mode at the incipient buckling consists of a uniform radial contraction superposed on the eigenmode with m representing the nearest integer corresponding to the value of λ_m predicted by (7.159). Since

$$A_m/B_m\approx\sqrt{k\left(\alpha+\beta\right)}$$

in view of (7.157) and (7.159), the actual velocity field at the point of bifurcation may be written as

$$w=-w_0\left(1+cP_m\right),\qquad u=-cw_0\sqrt{k\left(\alpha+\beta\right)}P'_m, \tag{7.161}$$

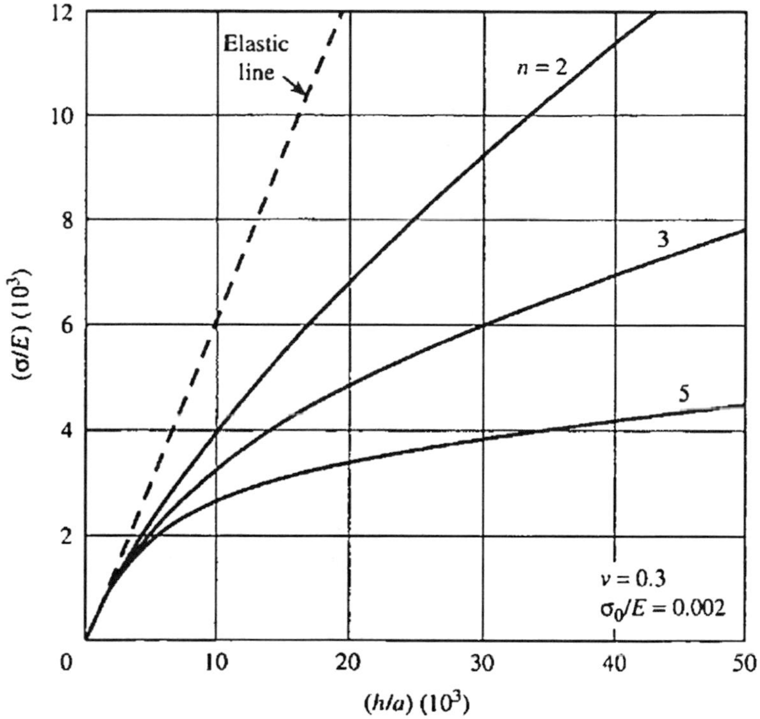

Fig. 7.20 Variation of critical stress with wall thickness for the plastic buckling of a complete spherical shell under uniform external pressure

where c is a constant and the prime denotes differentiation with respect to ϕ the quantity $(1 + c)w_0$ being the radially inward velocity at $\phi = 0$. The condition of continued loading of the radially compressed shell may be expressed as $\dot{\varepsilon}_{\theta\theta} + \dot{\varepsilon}_{\theta\theta} < 0$, which is equivalent to the restriction

$$\left(\dot{\lambda}_\theta + \dot{\lambda}_\phi\right) - z\left(\dot{\kappa}_\theta + \dot{\kappa}_\phi\right) < 0$$

for all values of ϕ and z. Substituting from (7.147) and (7.161), and using the Legendre equation (7.155), this inequality is reduced to

$$cm\,(1+m)\left[\mu + \frac{h}{a}\,(1-\mu)\right] P_m < 2\,(1 + cP_m)\,,$$

where $\mu = \sqrt{k\,(\alpha + \beta)}$. This inequality can be satisfied for a range of values of c, giving the possible modes of deformation at bifurcation which occurs under increasing external pressure.

It is interesting to note that the critical stress for buckling of an externally pressurized spherical shell is identical to that of an axially compressed short cylindrical shell, not only for an elastic material but also for an incompressible elastic/plastic material ($\nu = 0.5$). As in the case of cylindrical shells, slight geometrical imperfections in a spherical shell have a significant effect of lowering the critical pressure from the theoretical value (Hutchinson, 1972). An analysis for the plastic buckling of spherical shells based on the total strain theory has been discussed by Gerard (1962). The plastic buckling of shells of revolution has been investigated by Bushnell and Galletly (1974) and Bushnell (1982).

7.7.3 Snap-Through Buckling of Spherical Caps

Consider a deep spherical cap of thickness h and radius a, which is subjected to an inward load P at the apex applied through a rigid circular boss of radius b, as shown in Fig. 7.21(a). As the load is gradually increased from zero, the material response is initially elastic, and the deflection at the apex increases monotonically with the intensity of loading. Over the practical range of values of h/a, plastic deformation inevitably begins before the failure occurs due to buckling. As the loading is continued in the plastic range, the load–deflection curve continues to rise with decreasing slope until the load attains a maximum, the corresponding deflection being about half the shell thickness. Subsequently, the load falls rapidly with increasing deflection and reaches a minimum when the deflection is about twice the shell thickness. Thereafter, the load increases again with deflection due to the strengthening effect of the membrane forces. If the shell is subjected to an incremental dead loading, a snap-through type of buckling would occur when the maximum value of the load is reached.

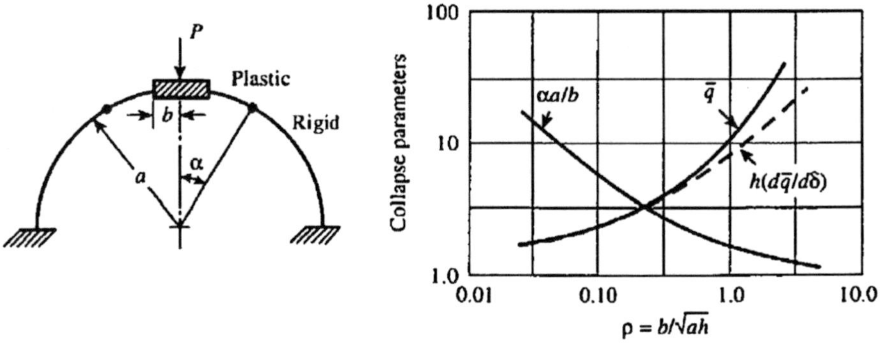

(b) (b)

Fig. 7.21 Plastic collapse of a spherical cap centrally loaded through a rigid boss. (**a**) Geometry and loading, (**b**) conditions at incipient collapse

A complete elastic/plastic analysis of the problem, taking due account of the geometry changes, is actually required to obtain the load–deflection behavior of the shell, in order to predict the initiation of the snap-through action. For practical purposes, however, the maximum load can be approximately estimated by finding the point of intersection of the load–deflection curves obtained on the basis of purely elastic and rigid/plastic behaviors of the shell. The rigid/plastic curve begins with the theoretical collapse load, which is most conveniently determined by using the limited interaction yield condition. The analysis is similar to that used for a spherical cap under external pressure (Section 5.4) and involves the yield condition in the dimensionless form

$$n_\phi = -1, \; -1 < n_\theta < 0 \;\; (\phi_0 \le \phi \le \alpha),$$
$$m_\theta = 1 \;(\phi_0 \le \phi \le \beta), \; m_\theta - m_\phi = 1 \;\; (\beta \le \phi \le \alpha),$$

where $\phi_0 = \sin^{-1}(b/a)$. The angles α and β represent the extent of the deformation zone and the interface between the two plastic regimes, respectively. The shearing force s and the circumferential force n_θ in the dimensionless form are easily shown to be

$$\left.\begin{aligned} s &= \left(1 - k\bar{q}\,\text{cosec}^2\phi\right)\tan\phi, \\ n_\theta &= -\left(1 - k\bar{q}\right)\sec^2\phi, \end{aligned}\right\} \qquad \phi_0 \le \phi \le \alpha,$$

where $\bar{q} = P/2\pi M_0$ and $k = h/4a$. These relations follow from the equations of force equilibrium under zero surface loading. In view of the preceding relations, the equation of moment equilibrium becomes

$$\left.\begin{aligned} \frac{d}{d\phi}\left(m_\phi \sin\phi\right) &= \left(\frac{1}{k} - \bar{q}\right)\sec\phi - \left(\frac{1}{k} - 1\right)\cos\phi, \qquad \phi_0 \le \phi \le \beta, \\ \frac{dm_\phi}{d\phi} &= \left(\frac{1}{k} - \bar{q}\right)\tan\phi - (\bar{q} - 1)\cot\phi, \qquad \beta \le \phi \le \alpha. \end{aligned}\right\} \tag{7.162}$$

The deformation mode at the incipient collapse involves hinge circles at $\phi = \phi_0$ and $\phi = \alpha$, requiring $m\phi = 1$ and $m\phi, = -1$ at these two sections, respectively. Using these as boundary conditions, (7.162) can be readily integrated to given the distribution of $m\phi$. Since $m\phi$ must vanish at $\phi = \beta$, we obtain

$$\left.\begin{aligned} (1 - k)\sin\beta - \sin\phi_0 &= (1 - k\bar{q})\ln\left(\frac{\sec\beta + \tan\beta}{\sec\phi_0 + \tan\phi_0}\right), \\ k\left\{(\bar{q} - 1)\ln\left(\frac{\sin\alpha}{\sin\beta}\right) - 1\right\} &= (1 - k\bar{q})\ln\left(\frac{\cos\beta}{\cos\alpha}\right). \end{aligned}\right\} \tag{7.163}$$

The angle α, defining the position of the plastic boundary, must be such that the bending moment $m\phi$ is a minimum at $\phi = \alpha$. The second equation of (7.162) therefore gives the relation

$$\bar{q} = \cos^2\alpha + \frac{1}{k}\sin^2\alpha, \tag{7.164}$$

from which the collapse load parameter $\bar{q}$ can be determined for any assumed α. The corresponding values of $\boldsymbol{\phi_0}$ and β then follow from (7.163). For a given value of k, the collapse load increases as the boss size is increased, approaching a limiting value equal to $1/k$ as α tends to $\pi/2$. Since $n_\theta < 1$ for $\alpha < \pi/2$ the preceding solution is statically admissible. The results of the calculation are shown in Fig. 7.21(b), which indicates that the region of plastic deformation is increasingly confined in the neighborhood of the loading boss as the radius of the boss is increased.

The solution is also kinematically admissible, the associated velocity field being similar to that for a clamped spherical cap under uniform external pressure. Denoting the axial speed of the boss by $\dot{\delta}$, the normal component of velocity of the material around it can be written as

$$w = \dot{\delta}\left\{1 - \eta \ln\left(\frac{\sec\phi + \tan\phi}{\sec\phi_0 + \tan\phi_0}\right)\right\}\frac{\cos\phi}{\cos\phi_0}, \qquad \phi_0 \le \phi \le \beta, \tag{7.165}$$

where

$$\eta = \left\{\ln\left(\frac{\sec\beta + \tan\beta}{\sec\phi_0 + \tan\phi_0}\right) + \sin\beta \ln\left(\frac{\tan\alpha}{\tan\beta}\right)\right\}^{-1}.$$

In the outer plastic region ($\beta \le \phi \le \alpha$), the normal component of velocity is obtained by replacing the expression in the curly brackets of (7.165) with η sin β ln(tan α/tan ϕ). The velocity field implies the formation of hinge circles at $\phi = \phi_0$ and $\phi = \alpha$ with appropriate discontinuities in the velocity gradient. The slope of the load–deflection curve at the incipient collapse has been determined by Leckie on the basis of a method due to Batterman (1964). A graphical plot of the large initial slope, included in Fig. 7.21(b), indicates that the carrying capacity would be considerably lowered due to the effect of the elastic deformation of the shell.

The load–deflection curves shown in Fig. 7.22 have been obtained by Leckie and Penny (1968) using a series of carefully controlled tests. While these curves have the same general trend, the variation of load with deflection is seen to be more pronounced for higher values of the parameter $\rho = b/\sqrt{ah}$. The theoretical collapse loads and the associated initial slopes of the load–deflection curves, predicted by the rigid/plastic analysis, are indicated by broken lines in the figure. We may consider a typical elastic load–deflection curve in which the plastic yielding is disregarded, the derivation of such a curve being that discussed by Ashwell (1959). The point of intersection of this curve with the corresponding rigid/plastic line is seen to provide a reasonable approximation for the maximum load at which the snap-through action is initiated.

The snap-through buckling of a spherical cap results in an inversion of a central portion of the cap, having a radius of order $\sqrt{hw_0}$, where w_0 is the central deflection of the cap, and h the shell thickness The inverted cap has the same geometrical configuration as that obtained by removing a central portion of the cap, turning it over, and reuniting it with the parent cap around its edge. In actual practice, the transition between the inverted cap and the parent cap occurs through a narrow region in the form of a toroidal knuckle the width of which is of the order $\sqrt{ah}$, where a is the mean radius of the shell. Over the initial stages of the post-buckling behavior, the variation of the load P with the central deflection w_0 may be expressed by

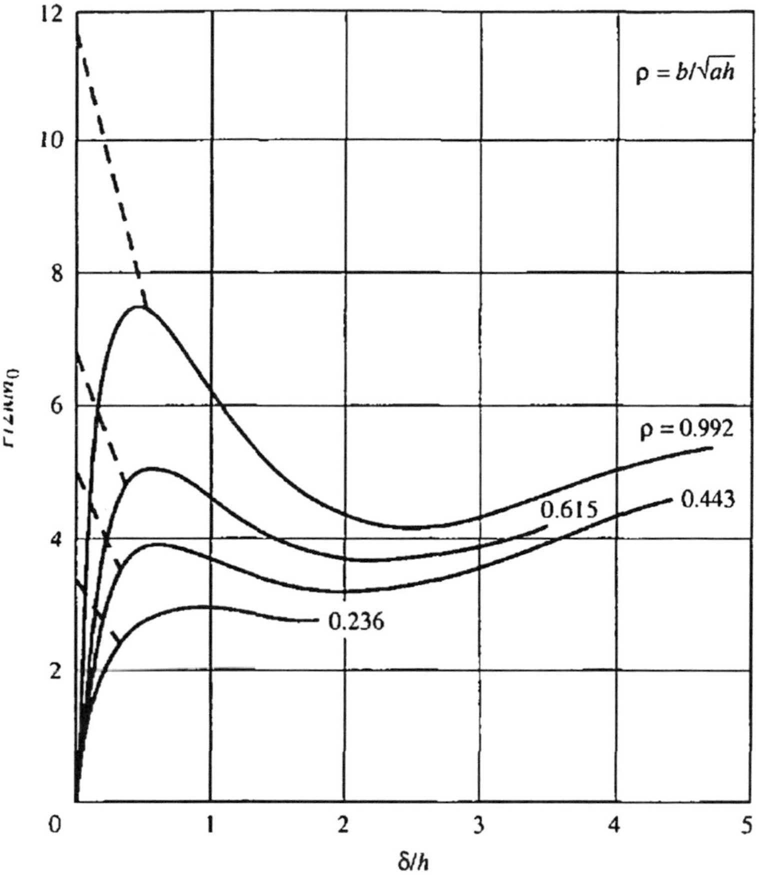

Fig. 7.22 Experimental load–deflection curves for a spherical cap loaded inwardly through a rigid boss (after Leckie and Penny, 1968)

the formula $Pa = 1.7\lambda E w_0^{0.5} h^{2.5}$, where E denotes Young's modulus, while λ is a dimensionless parameter whose value is unity for purely elastic buckling (Calladine, 2001) Over the plastic range of buckling,, λ should depend on the ratio T/E, where T is the tangent modulus at the incipient buckling of the shell.

Problems

7.1 A uniform straight column of slenderness ratio k is pin-supported at both ends, and is subjected to axial compression by equal and opposite forces P applied along the centroidal axis, The stress–strain curve of the material in the plastic range may be expressed by the equation

$$\varepsilon = \frac{\sigma}{E} + \frac{\alpha Y}{E}\left(\frac{\sigma}{Y} - 1\right)^m, \qquad \sigma \geq Y,$$

where α and m are dimensionless constants, the slope of the stress–strain curve being equal to E at the yield point. If the cross section of the column is rectangular, and the critical stress is denoted by σ according the tangent modulus theory, and by $\sigma*$ according to the reduced modulus theory, show that

$$\frac{\sigma}{Y}\left\{1 + m\alpha\left(\frac{\sigma}{Y} - 1\right)^{m-1}\right\} = \frac{\pi^2}{k^2}\left(\frac{E}{Y}\right), \quad \sqrt{\frac{\sigma*}{Y}}\left\{1 + \sqrt{1 + m\alpha\left(\frac{\sigma*}{Y} - 1\right)^{m-1}}\right\} = \frac{2\pi}{k}\sqrt{\frac{E}{Y}}$$

7.2 A work-hardening elastic/plastic material yields according to the von Mises yield criterion, but obeys the non-associated flow rule furnished by the rate form of the Hencky stress–strain relation. Show that the constitutive equations for a balanced biaxial state of stress may be written as

$$\dot{\sigma}_{xx} = S\left(\alpha\dot{\varepsilon}_{xx} + \beta\dot{\varepsilon}_{yy}\right), \qquad \dot{\sigma}_{yy} = S\left(\beta\dot{\varepsilon}_{xx} + \alpha\dot{\varepsilon}_{yy}\right), \qquad \dot{\tau}_{xy} = S\dot{\varepsilon}_{xy}/\left(1 + \eta'\right),$$

where S is the current secant modulus of the uniaxial stress–strain curve of the material, η' is the modified contraction ratio obtained by replacing T with S in the expression for η, the parameters α and β being given by the expressions

$$\alpha = \frac{1}{\rho}\left(1 + \frac{3T}{S}\right), \qquad \beta = \alpha - \frac{1}{1 + \eta'}, \qquad \rho = 4\left(1 - \eta\right)\left(1 + \eta'\right)$$

7.3 Considering a state of plane stress, in which the current state is defined by $\sigma_x = -\sigma_1$, $\sigma_y = -\sigma_2$, and $\tau_{xy} = 0$, and assuming the rate form the Hencky stress–strain relation, derive the constitutive equations for an isotropic elastic/plastic material in the form

$$\dot{\sigma}_{xx} = S\left(\alpha\dot{\varepsilon}_{xx} + \beta\dot{\varepsilon}_{yy}\right), \qquad \dot{\sigma}_{yy} = S\left(\beta\dot{\varepsilon}_{xx} + \gamma\dot{\varepsilon}_{yy}\right), \qquad \dot{\tau}_{xy} = S\dot{\varepsilon}_{xy}/\left(1 + \eta'\right)$$

where S is the secant modulus of the uniaxial stress–strain curve, while α, β, and γ are dimensionless parameters expressed in terms of the ratios of the applied stresses to the effective stress $\bar{\sigma}$ as

$$\rho\alpha = 4 - 3\left(1 - \frac{T}{S}\right)\frac{\sigma_1^2}{\bar{\sigma}^2}, \quad \rho\beta = 4\eta - 3\left(1 - \frac{T}{S}\right)\frac{\sigma_1\sigma_2}{\bar{\sigma}^2}, \quad \rho\gamma = 4 - 3\left(1 - \frac{T}{S}\right)\frac{\sigma_2^2}{\bar{\sigma}^2}$$

$$\rho = 3 + (1 - 2\nu)\frac{S}{E}\left[(1 + 2\eta) - 3\left(1 - \frac{T}{S}\right)\frac{\sigma_1\sigma_2}{\bar{\sigma}^2}\right], \quad 2\eta' = 1 - (1 - 2\nu)\frac{S}{E}$$

7.4 A rectangular plate, whose sides are of lengths a and b, respectively, where $b \leq a$, is subjected to normal compressive stresses of magnitude σ along the shorter sides of the rectangle. Using the non-associated constitutive equations given in the preceding problem, show that the critical compressive stress for buckling is given by

$$\frac{\sigma}{S} = \frac{\pi^2 h^2}{3\rho b^2}\left\{2 + \left(1 + \frac{3T}{S}\right)\left(\frac{mb}{2a}\right)^2 + \frac{3}{2}\left(\frac{1 - 2\nu}{1 + \eta'}\right)\left(\frac{S}{E} - \frac{T}{E}\right) + \left(\frac{a}{mb}\right)^2\right\},$$

where m is an integer that minimizes the right-hand side of the above equation. Obtain the corresponding that holds for sufficiently large values of the aspect ratio *a/b*.

7.5 A rectangular plate of sides a and b is subjected to equibiaxial compressive stresses of magnitude σ normal to the edges of the plate. Using the non-associated constitutive equations of Prob.7.2, show that the critical stress for buckling is given by the expression

$$\frac{\sigma}{S} = \frac{\pi^2 h^2}{3a^2} \left\{ \frac{(1 + 3T/S)\left(1 + a^2/b^2\right)}{\left[3 - (1 - 2\nu) S/E\right]\left[1 + (1 - 2\nu) T/E\right]} \right\}$$

Using the Ramberg–Osgood equation with $\alpha = 3/7$, $E/\sigma_0 = 750$ and $c = 3$, draws a graph of $\sigma b^2/Eh^2$ against a/b when $\nu = 0.3$, and compare it with that given by the Prandtl–Reuss theory.

7.6 A solid circular plate of radius a is subjected to a uniform radial compressive stress σ along its boundary. Using the constitutive equations given in Prob.7.2, obtain he critical stress for buckling of a clamped plate in the form

$$\frac{\sigma}{S} = \frac{k^2 h^2}{12a^2} \left\{ \frac{1 + 3T/S}{4(1 - \eta)\left(1 + \eta'\right)} \right\}, \qquad 2\eta' = 1 - (1 - 2\nu)\frac{S}{E}$$

where k is the smallest root of the equation $J_I(k) = 0$. Adopting the Ramberg–Osgood stress–strain curve of the preceding problem, and using $\nu = 0.3$, obtain the variation of σ/E with h/a, and compare it with that given by the Prandtl–Reuss theory.

7.7 If the circular plate of the preceding problem is simply supported along its boundary $.r = a$, show that the critical stress for buckling under a uniform radial compression σ is given by

$$\frac{kJ_1(k)}{J_0(k)} = \frac{4(1 - \eta)}{1 + 3T/S}, \qquad k = \frac{4a}{h} \left\{ \frac{3\sigma}{S} \left(\frac{(1 - \eta)\left(1 + \eta'\right)}{1 + 3T/S} \right) \right\}^{1/2}$$

in view of the results given in Prob. 7.1. Assuming the Ramberg–Osgood equation of the preceding problem, compute the critical value of σ/E when $\nu = 0.3$ and $h/a = 0.05$, and compare the result with that given by the Prandtl–Reuss theory.

Chapter 8
Dynamic Plasticity

In this chapter, we shall be concerned with the class of problems in which the plastic deformation is so rapid that the inertia effects cannot be disregarded. Problems of dynamic plasticity arise in the high-velocity forming of metals, penetration of high-speed projectiles into fixed targets, enlargement of cavities by underground explosion, and design of crash barriers related to collisions, to name only a few. The rate of loading and the size of the components are usually such that the deformation process can be described in terms of the propagation of elastic/plastic waves. However, simplified theories which disregard the wave propagation phenomenon are generally capable of providing useful information for practical purposes. In the case of structural members subjected to impact loading, the mode of plastic deformation can be most conveniently represented by the existence of discrete yield hinges that rapidly move away from the point of loading. The concept of moving yield hinges is a useful device for the dynamic analysis of structures.

8.1 Longitudinal Stress Waves in Bars

8.1.1 Wave Propagation Without Rate Effects

The propagation of longitudinal elastic/plastic waves in thin rods or wires was first discussed independently by von Karman (1942), Taylor (1942), and Rakhmatulin (1945), although a theoretical treatment in a restricted sense was given earlier by Donnel (1930). Following von Karman and Duwez (1950), the theory will be developed here in terms of the nominal stress and strain, denoted here by σ and ε, respectively, and the initial coordinate x measured along the axis of the bar, which is assumed to have a uniform cross section. If the longitudinal velocity of the particle at any instant is denoted by v, and transverse plane sections are assumed to remain plane, the equation of motion is

$$\frac{\partial \sigma}{\partial x} = \rho \frac{\partial \upsilon}{\partial t},$$

where ρ is the initial density of the material, and σ is a known function of ε, tensile stress being taken as positive. Since $\upsilon = \partial u/\partial t$ and $\varepsilon = \partial u/\partial x$, where u is the

DOI 10.1007/978-0-387-77674-3_8, © Springer Science+Business Media, LLC 2010

longitudinal displacement, the differential equation for a rate insensitive material becomes

$$\frac{\partial^2 u}{\partial t^2} = c^2 \frac{\partial^2 u}{\partial x^2}, \quad c^2 = \frac{1}{\rho}\frac{d\sigma}{d\varepsilon}, \tag{8.1}$$

which is a quasi-linear wave equation governing the motion of the particle, with $c = c(\varepsilon)$ denoting the speed of propagation of the wave. Equation (8.1) applies only during the loading process in which the stress continuously increases with time, and for common engineering materials, the wave speed decreases with increasing strain. If the loading is tensile, a critical stage would be reached when the total strain is equal to that at the onset of necking, and the plastic wave speed is then reduced to zero.

The solution of the differential equation (8.1) is simplified by the fact that this equation is actually hyperbolic. Indeed, along any curve considered in the (x, t)-plane, the variation of the velocity v is given by

$$dv = \frac{\partial v}{\partial x}dx + \frac{\partial v}{\partial t}dt = \frac{\partial \varepsilon}{\partial t}dx + c^2\frac{\partial \varepsilon}{\partial x}dt, \tag{8.2}$$

in view of the identity $\partial v/\partial x = \partial \varepsilon/\partial t$, and the relation $\partial v/\partial t = c^2\,(\partial \varepsilon/\partial x)$ which follows from (8.1). The variation of the strain ε along this curve is

$$d\varepsilon = \frac{\partial \varepsilon}{\partial t}dt + \frac{\partial \varepsilon}{\partial x}dx.$$

For given values of dv and $d\varepsilon$ along the curve, the derivatives $\partial\varepsilon/\partial t$ and $\partial \varepsilon/\partial x$ can be uniquely determined from the last two equations unless the determinant of their coefficients vanishes. The considered curve will therefore be a characteristic if $(dx)^2 = c^2(dt)^2$. Equation (8.1) is therefore hyperbolic, the characteristic directions and the differential relations holding along them being

$$\frac{dx}{dt} = \pm c, \qquad dv = \pm c\,d\varepsilon. \tag{8.3}$$

where the second result follows on substitution for dx and dt from the first into the second expression of (8.2). Since c is a function of ε, the characteristic lines generally consist of two families of curves corresponding to the upper and lower signs in (8.3). When the characteristic is a straight line, c is a constant, which means that the stress, strain, and velocity remain constant along its length.

Unloading begins in a given cross section as soon as the stress begins to decrease after reaching a certain maximum value σ^*, corresponding to a strain ε^*. The stress–strain relation for the unloading process may be written as

$$\sigma - \sigma^* = E\left(\varepsilon - \varepsilon^*\right),$$

where E is Young's modulus for the material. Substituting in the equation of motion, and using the fact that $\upsilon = \partial u/\partial t$, we obtain the differential equation

$$\frac{\partial^2 u}{\partial t^2} = c_0^2 \frac{\partial^2 u}{\partial x^2} + \frac{d}{dx}\left(\frac{\sigma^*}{\rho} - c_0^2 \varepsilon^*\right), \tag{8.4}$$

where $c_0 = \sqrt{E/\rho}$ representing the speed of propagation of the elastic wave. Equation (8.4) is also hyperbolic with two families of straight characteristics in the $(x,\, t)$-plane, the characteristic relations being obtained in the same way as before with the result

$$\frac{dx}{dt} = \pm c_0, \qquad \rho c_0 d\upsilon = \pm d\sigma. \tag{8.5}$$

Since σ^* and ε^* are functions of x, and are not known in advance, the shape of the loading/unloading boundary must be determined as a part of the solution. For a bar of finite length, waves are reflected from the ends of the bar, and the solution is strongly dependent on the nature of the boundary conditions.

If a bar is subjected to impact loading at one end $x = -l$, the first wave to reach the other end $x = 0$ is always an elastic wave. The displacement of the particles during the propagation of the direct elastic wave is governed by (8.1) with c replaced by the elastic wave speed c_0. A general solution of this wave equation is

$$u_1 = f_1\,(c_0 t - x)\,,$$

where f_1 is an arbitrary function that can be determined from the prescribed initial conditions. For the reflected wave, which propagates in the opposite direction, the displacement is

$$u_2 = f_2\,(c_0 t + x)\,.$$

The total displacement of a particle traversed by the incident and reflected waves is $u = u_1 + u_2$, and the corresponding stress is

$$\sigma = E\frac{\partial}{\partial x}\,(u_1 + u_2) = E\left\{f_2'\,(c_0 t + x) - f_1'\,(c_0 t - x)\right\},$$

where the prime denotes differentiation with respect to the argument $c_0 t - x$. When the end $x = 0$ is free, the stress vanishes there for all t, and consequently $f_1'\,(c_0 t) = f_2'\,(c_0 t)$, which gives $f_1 = f_2 = f$. The displacement and velocity at the free end therefore become

$$u = 2f\,(c_0 t)\,, \qquad \upsilon = 2c_0 f'\,(c_0 t)\,.$$

Thus, the displacement and velocity at the free end are doubled due to the reflection, a compression wave being reflected as a tension wave and vice versa. If the end $x = 0$ is fixed, the total displacement u_1+u_2 vanishes there for all t, and we have $f_2 = -f_1 = f$ and the stress at the fixed end becomes $\sigma = 2Ef'(c_0 t)$. The stress is therefore doubled in magnitude due to the reflection, and an elastic wave may therefore be reflected as a plastic wave.

Across a wave front, the first derivatives of v and ε are necessarily discontinuous. In the case of *weak waves*, υ and ε are themselves continuous, and the wave front then coincides with a characteristic. In problems of plastic wave propagation, one frequently encounters *shock waves* in which the wave fronts are surfaces of discontinuity even for υ and ε. Shock waves are usually generated by a sudden change in velocity imposed at one end of the bar. If a shock wave front moves through a distance dx during a time interval dt, the displacement on either side of the wave front changes by the amount $du = \upsilon\, dt + \varepsilon\, dx$. The condition of continuity of the displacement across the wave front therefore gives

$$[\upsilon] + c_s\,[\varepsilon] = 0, \tag{8.6a}$$

where c_s is the speed of propagation of the shock wave, and the square brackets represent the jump in the enclosed quantity when the front has passed a given cross section. The momentum equation for the element of length dx traversed by the wave front during the time dt yields

$$\rho c_s\,[\upsilon] + [\sigma] = 0. \tag{8.6b}$$

The two jump conditions established in (8.6), together with the stress–strain equation, are sufficient to study the propagation of shock waves, provided the change in temperature due to impact is negligible. The elimination of $[\upsilon]$ between the two relations of (8.6) furnishes $\rho c_s^2 = [\sigma]/[\varepsilon]$. The speed of propagation of the shock wave coincides with that of the weak wave only when the stress–strain curve is linear.

8.1.2 Simple Wave Solution with Application

Consider the situation where ε and v are functions of the ratio x/t, but not of x and t individually. Then all the characteristics of positive slope are straight lines passing through the origin, the result being a centered fan of linear characteristics, having the equation $x/t = c$, known as simple waves. Since v and ε are functions of c only, we have

$$\frac{d\upsilon}{dc} = t\frac{\partial \upsilon}{\partial x} = t\frac{\partial \varepsilon}{\partial t},$$

the right-hand side of this expression being $-c$ times the derivative of ε with respect to c. Thus

$$\frac{dv}{dc} = -c\frac{d\varepsilon}{dc} = -\frac{1}{\rho c}\frac{d\sigma}{dc}.$$

Using the boundary condition $v = v_0$ along the characteristic that corresponds to $\sigma = Y$, the above equation is immediately integrated to give

$$\rho c_0 (v - v_0) = -\int_Y^\sigma (c_0/c)\, d\sigma. \tag{8.7}$$

Simple waves generally appear under instantaneous loading by a uniform stress or velocity at one end of the bar and exist in a region adjacent to one of constant stress or strain, which evidently satisfies (8.1) and the boundary condition. The initial loading of the bar is generally followed by an unloading process in which the stress progressively decreases in magnitude.

To illustrate the basic principles, consider the normal impact of a cylindrical bar of length l, which is moving parallel to its longitudinal axis with a velocity $-U$ against a stationary rigid target $x = 0$ (Lee, 1953). For mathematical convenience, we superimpose a constant velocity U to the whole system in the opposite direction and write the initial and boundary conditions as

$$v = 0, \quad \sigma = 0 \ at\ t = 0, \qquad 0 < x \leq t,$$
$$v = U, at\ x = 0, \qquad \sigma = 0\ at\ x = l, \qquad t > 0.$$

The stress and strain will be considered positive in compression. The characteristic field is shown in Fig. 8.1(a), where *OA* is the elastic wave front across which the stress rises instantaneously from zero to the yield stress *Y*. The fan *OAD* consists of plastic waves with *OD* representing the characteristic for $v = U$. Within the triangle *ODE*, the material has a constant velocity *U* and is subjected to a constant stress that corresponds to the characteristic *OD*.

At the free end $x = l$, the elastic loading wave is reflected as an unloading shock wave which propagates with a speed c_0 and has the equation $x = 2l - c_0 t$. At *A*, the value of $\rho c_0 v$ is equal to *Y* before the reflection, in view of (8.6b) with $c_s = co$ and a change in sign for σ. The physical quantities along *ABC* after the reflection will be denoted by a prime. Since the velocity at the free end is doubled by the reflected wave, $\rho c_0 v'$ is equal to $2Y$ at *A*, and the second relation of (8.5) applied to the boundary wave *AC* gives

$$\rho c_0 v' - 2Y = \sigma'$$

as the stress vanishes at the free end after the reflection. At a generic point of *AC*, we may apply the jump condition (8.6b) with $c_s = -c_0$ and $-\sigma$ written for σ, the result being

$$-\rho c_0 \left(v - v'\right) = \sigma - \sigma'$$

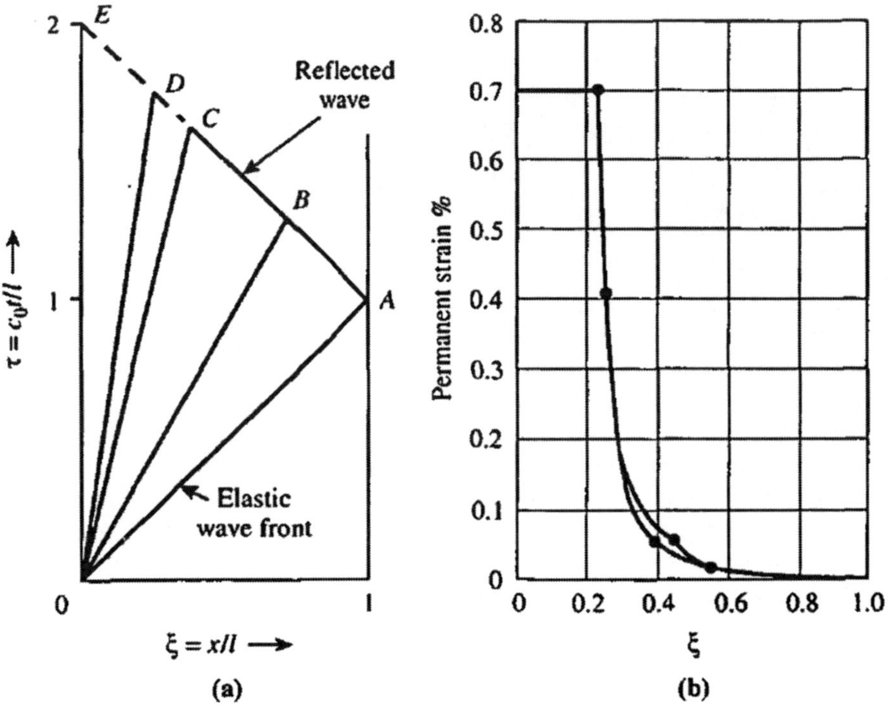

Fig. 8.1 Longitudinal impact on a bar of finite length. (**a**) Characteristics in the plastic region and (**b**) distribution of plastic strain along its length

Eliminating υ' between the above relations, the amount of stress discontinuity across the unloading wave may be expressed in terms of the stress and velocity before the reflection as

$$\sigma - \sigma' = Y - \frac{1}{2}\left(\rho c_0 \upsilon - \sigma\right). \tag{8.8a}$$

The relationship between υ and σ is obtained from (8.7) by setting $\rho c_0 \upsilon_0 = Y$ and changing the sign of σ. Thus

$$\rho c_0 \upsilon = Y + \int_Y^\sigma (c_0 - c)\, d\sigma, \tag{8.8b}$$

where $\rho c^2 = d\sigma/d\varepsilon$, given by the stress–strain curve. Since $c \leq c_0$, the integral in (8.8b) is increasingly greater than $\sigma - Y$, and the expression in the parenthesis of (8.8a) steadily increases along AC from its minimum value zero at A. Thus the unloading shock wave is progressively absorbed during its propagation. If the impact velocity U is sufficiently high, the unloading wave will be completely absorbed at a point C, where the stress discontinuity is reduced to zero. When the impact velocity is below a certain critical value, the unloading wave can continue through to the impact face at E, and the material is unloaded throughout the length of the bar.

In the case of a supercritical impact velocity that terminates the unloading wave at C, the plastic region spreads from this point into an area above AE. The solution to this part of the impact problem involves discontinuities in stress and velocity derivatives, which must be admitted for the continuation of the loading–unloading boundary. For a subcritical impact velocity, which ensures that the unloading wave traverses the entire length of the bar, a new plastic region is initiated at a point

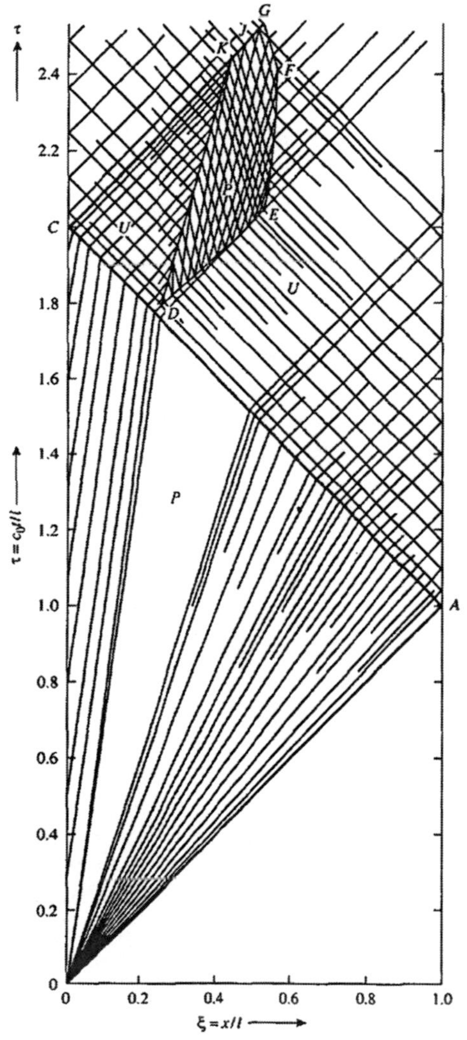

Fig. 8.2 Characteristic field in the propagation of longitudinal plastic waves involving regions of loading and unloading (after E. H. Lee, 1953)

where the unloaded bar has a stress equal to the maximum stress previously attained in the same section during loading. The complete characteristic field for this particular case obtained by Lee is shown in Fig. 8.2, where D is the point of initiation of the new plastic zone $DEFGK$. At point K, the loading–unloading boundary is intersected by the unloading wave reflected from the impact face. The permanent strain produced by the secondary plastic region is found to be small compared to that due to the primary plastic loading wave, as may be seen from Fig. 8.1(b), which displays the permanent strain distribution in the bar.

8.1.3 Solution for Linear Strain Hardening

The problem of plastic wave propagation is greatly simplified by the assumption of a linear strain-hardening law, which makes the stress–strain curve consist of a pair of straight lines of slopes E and T in the elastic and plastic ranges. The elastic and plastic waves then propagate with constant speeds c_0 and c_p, respectively, where $c_0 = \sqrt{E/\rho}$ and $c_\rho = \sqrt{T/\rho}$. To illustrate the simplicity of this approach, we consider the same problem as that discussed before, a uniform bar of length l being assumed to strike a rigid wall $x = 0$ with a velocity U from right to left at time $t = 0$. For sufficiently large values of U, the elastic and plastic waves are simultaneously generated at $x = 0$ and are propagated along the length of the bar, the positions of the wave fronts at subsequent times being shown in Fig. 8.3. The solution, which is due to Lensky (1949), has been discussed by Cristescu (1967).

During the time interval $0 \leq t \ll l/c_0$, there are three distinct regions in the bar as indicated in Fig. 8.3(a). In the outer region, the material is undisturbed with $\upsilon_1 = -U$ and $\sigma_1 = \varepsilon_1 = 0$, while in the central region that is traversed by the elastic waves only, we have $\sigma_2 = Y$, $\varepsilon_2 = Y/E = \varepsilon_0$(say), and $\upsilon_2 = -U + c_0\varepsilon_0$, in view of the jump conditions (8.6). In the region adjacent to the wall, where the material is brought to rest ($\upsilon_3 = 0$) by the plastic shock wave, the compressive stress and strain are similarly obtained as

$$\sigma_3 = Y + \rho c_p \left(U - c_0\varepsilon_0\right), \qquad \varepsilon_3 = \varepsilon_0 + (U - c_0\varepsilon_0)/c_\rho.$$

It follows that the bar will be rendered plastic only if the velocity of impact satisfies the condition $U > c_0\varepsilon_0$.

For $t > l/c_0$, the elastic wave front moves backward after reflection from the free end of the bar, while the plastic wave front advances further to the right as indicated in Fig. 8.3(b). The reflected wave front completely unloads the region traversed by it, giving $\sigma_4 = \varepsilon_4 = 0$, and its velocity is reduced in magnitude to $\upsilon_4 = 2c_0\varepsilon_0 - U$. At time $t = t_s$ the reflected elastic wave front meets the advancing plastic wave front at some section S, which is at a distance x_s from the wall. Then $t_s = x_s/c_\rho = 2(2\,l - x_s)/c_0$, giving the relations

$$x_s = 2lc_p/\left(c_0 + c_p\right), \qquad t_s = 2\,l/\left(c_0 + c_p\right).$$

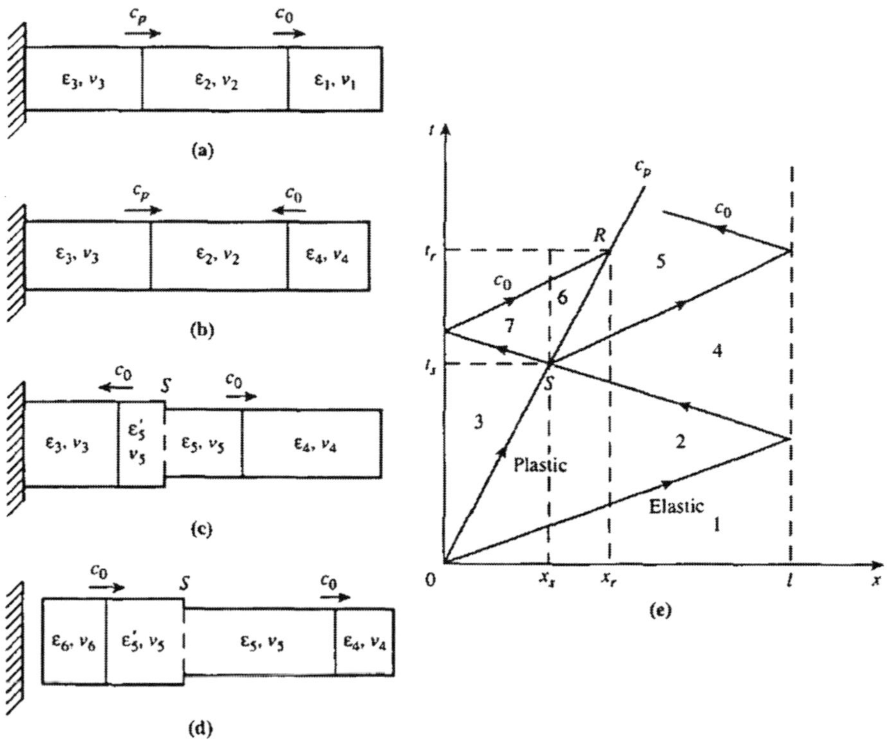

Fig. 8.3 Propagation of shock waves in an elastic/plastic bar of finite length striking a rigid target and the associated reflection and interaction of elastic and plastic waves

If the impact velocity U is not too large, the plastic wave will not propagate any further for $t > t_s$, and reflected elastic waves will spread in both directions from S, as shown in Fig. 8.3(c). In the region between the two reflected waves, the particle velocities on both sides of section S must be the same, but the strains in the two portions will be different. By (6), the associated stresses and strains are given by

$$\sigma_5 = \rho c_0 (\upsilon_5 - \upsilon_4), \qquad \sigma_5' = \sigma_3 - \rho c_0 \upsilon_s,$$
$$\varepsilon = (\upsilon_5 - \upsilon_4)/c_0, \qquad \varepsilon_5' = \varepsilon_3 - ((\upsilon_5 - \upsilon_3)/c_0).$$

Substituting for σ_3, ε_3, υ_3, and υ_4 and noting the fact that $\sigma_5 = \sigma'_5$ to a close approximation for longitudinal equilibrium, we obtain

$$\upsilon_5 = U - \frac{1}{2}(U - c_0\varepsilon_0)\left(3 - \frac{c_p}{c_0}\right),$$

$$\varepsilon_5 = (U - c_0\varepsilon_0)\left(\frac{c_0 + c_p}{2c_0^2}\right), \qquad \varepsilon_5' = (U - c_0\varepsilon_0)\left(\frac{(c_0 + c_p)(2c_0 - c_p)}{2c_0^2 c_p}\right). \tag{8.9}$$

If the plastic strain stops at $x = x_s$, the region to the right of section S must remain elastic, which requires $\varepsilon_5 < \varepsilon_0$. The preceding solution is therefore valid for

$$1 < \frac{U}{c_0\varepsilon_0} < 1 + \frac{2c_0}{c_0 + c_p}.$$

These inequalities ensure that elastic waves will propagate in both directions from section S. Since $v_5 > 0$ over this range, the bar will be pulled away from the wall as soon as the elastic wave that propagates to the left reaches the plane of contact. After the separation takes place, both the elastic wave fronts move outward along the bar, Fig. 8.3(d), and the inner end of the bar becomes stress free ($\sigma_6 = 0$). The strain and velocity in this region are

$$\varepsilon_6 = \varepsilon_5' - \frac{\sigma_5}{E} = (U - c_0\varepsilon_0)\frac{c_0^2 - c_p^2}{c_0^2 c_p}, \qquad v_6 = v_4 = 2c_0\varepsilon_0 - U. \tag{8.10}$$

The separation condition may therefore be written as $U < 2c_o\varepsilon_o$. It may be noted that ε_6 is actually the residual plastic strain in the region to the left of section S and is therefore identical to the quantity $\varepsilon_3 - \sigma_3/E$. Since the plastic strain vanishes outside this region, the bar will be left with a sudden change in cross section as shown in the figure. The time during which the bar remains in contact with the wall is evidently equal to $2l/c_o$.

Consider now the situation where the impact velocity is so large that the plastic wave continues to propagate beyond section S to its right. Depending on the ratio c_o/c_ρ, when the impact velocity satisfies the condition $U > (3c_0 + c_p)(c_0 + c_p)$, the advancing plastic wave will be intercepted either by the elastic wave which propagates to the left and is subsequently reflected at $x = 0$ or by the elastic wave which propagates to the right and is subsequently reflected at $x = l$. The sequence of events leading to the first of these two possibilities is represented in the characteristic plane shown in Fig. 8.3(e). The reflected wave front propagating from $x = 0$ meets the plastic wave front at a section R when $t = t_r$, the distance of the section from the rigid wall being $x = x_r$. It is easily shown that

$$x_r = 2lc_p/\left(c_0 - c_p\right), \qquad t_e = 2l/\left(c_0 - c_p\right).$$

In the case of the second possibility, the total distance traveled by the elastic wave front before it meets the plastic wave front is $2(2l - x_s) - x_r$, which leads to the relations

$$x_r = 4lc_0c_p/\left(c_0 + c_p\right)^2, \qquad t_r = 4lc_0/\left(c_0 + c_p\right)^2.$$

In the critical situation, when both the elastic waves meet the plastic waves simultaneously, we have

$$c_0^2 - 4c_0c_p - c_p^2 = 0 \quad or \quad c_0/c_p = 2 + \sqrt{5} \approx 4.24.$$

Thus, the leftward moving elastic wave from S meets the plastic wave on reflection when $c_0/c_p > 4.24$, and the rightward moving elastic wave from S meets the plastic wave on reflection when $c_0/c_p > 4.24$.

During the time interval $t_s < t < 2l/c_0$, the stress, strain, and velocity in regions 1–4 are identical to those previously given, while in region 5, these quantities are $\sigma_5 = Y$, $\varepsilon_5 = \varepsilon_0$, and $v_5 = 3c_0\varepsilon_0 - U$. In the remaining regions 6 and 7, the stresses and strains are given in terms of the velocities as

$$\sigma_6 - Y = \rho c_p (v_6 - v_5), \quad \sigma_7 - \sigma_3 = -\rho c_0 v_7,$$

$$\varepsilon_6 - \varepsilon_0 = (v_6 - v_5)/c_p, \quad \varepsilon_7 - \varepsilon_3 = -v_7/c_0.$$

Substituting for v_5, σ_3, and ε_3 and using the equilibrium and continuity conditions $\sigma_6 \approx \sigma_7$ and $v_6 = v_7$, respectively, we obtain the relations

$$v_6 = v_7 = \frac{2c_0c_p\varepsilon_0}{c_0 + c_p}, \quad \varepsilon_7 = \frac{U}{c_p} - \frac{\varepsilon_0 c_0^2 + c_p^2}{c_p\left(c_0 + c_p\right)}, \quad \varepsilon_6 = \varepsilon_7 - 2\left(\frac{c_0}{c_p} - 1\right)\varepsilon_0. \quad (8.11)$$

The solution can be continued in a similar manner for any given value of c_0/c_p. The final configuration of the bar will contain two stationary discontinuities occurring at $x = x_s$ and $x = x_r$, where there are abrupt changes in the cross section.

Other examples of longitudinal wave propagation in bars have been discussed by White and Griffis (1947), Rakhmatulin and Shapiro (1948), De Juhasz (1949), Lebedev (1954), Ripperger (1960), Clifton and Bodner (1966), and Cristescu (1970), among others. The problem of combined longitudinal and torsional plastic waves in thin-walled tubes has been discussed by Clifton (1966), Goel and Malvern (1970), Ting (1972), and Wu and Lin (1974). The dynamic plastic behavior of extensible strings has been examined by Craggs (1954) and Cristescu (1964), and discussed at great length by Cristescu (1967).

8.1.4 Influence of Strain-Rate Sensitivity

It is well known that the yield stress of engineering materials can be considerably higher under dynamic loading than under quasi-static loading. For example, the dynamic yield stress of annealed mild steel has been found to be more than double the quasi-static yield stress. The relevant experimental evidence has been provided by Duwez and Clark (1947), Campbell (1954), Goldsmith (1960), and Davies and Hunter (1963), along with a number of other investigators. The dynamic elastic modulus is essentially the same as the quasi-static elastic modulus, but the overall dynamic stress–strain curve is generally much higher than the quasi-static curve as shown in Fig. 8.4, which is based on the experimental results of Kolsky and Douch (1962). The magnitude of the strain rate at which a given material begins to be rate sensitive varies from material to material, but rate dependence of the stress–strain curve is an important factor which must be included in the theoretical framework for a realistic prediction of the dynamic behavior (Campbell, 1972).

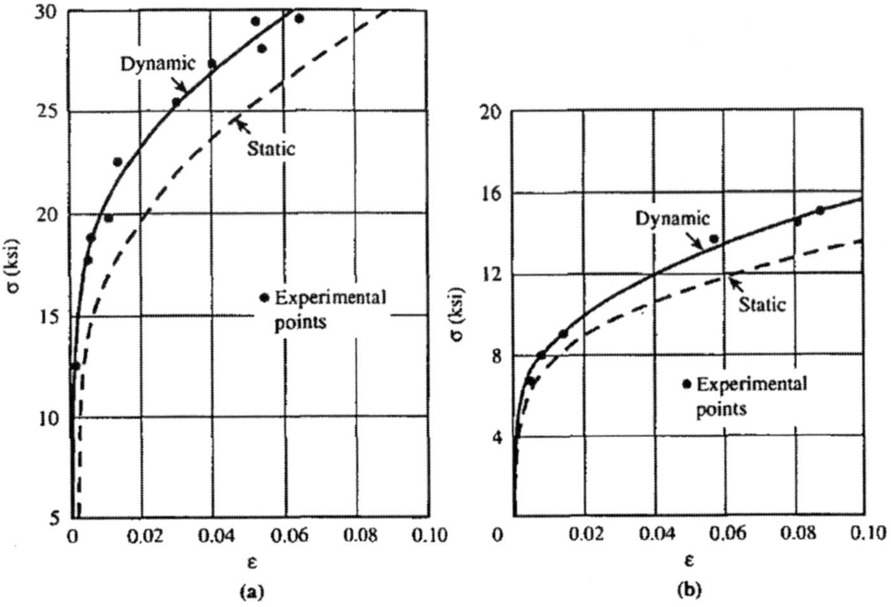

Fig. 8.4 Comparison of static and dynamic stress–strain curves for (**a**) copper and (**b**) aluminum (due to Kolsky and Douch, 1962)

For a work-hardening material having its uniaxial stress–strain curve given by $\sigma = f(\varepsilon)$ under quasi-static conditions, the simplest constitutive equation relating the stress and strain to the rate of straining may be obtained on the assumption that the plastic part of the strain rate is a function of the overstress $\sigma - f(\varepsilon)$, which is the difference between the dynamic and quasi-static yield stresses corresponding to a given strain. Since the elastic part of the strain rate is related to the stress rate by Hooke's law, the constitutive equation becomes

$$E\dot{\varepsilon} = \dot{\sigma} + F\left[\sigma - f(\varepsilon)\right]. \tag{8.12}$$

The function $F(z)$ must be positive for $z > 0$ and zero for $z \leq 0$. Equation (8.12), which has been proposed by Malvern (1951), is a generalization of one due to Sokolovsky (1948b), who assumed the material to be ideally plastic in the quasi-static state. The expression $F(z) = kz^n$, where κ and n are positive constants, has been found to fit experimental data reasonably well by Kukudjanov (1967). Lindholm (1964) has experimentally verified the constitutive relation

$$\sigma = \sigma_0 (\varepsilon) + \sigma_1 (\varepsilon) \ln\dot{\varepsilon},$$

where $\sigma_0(\varepsilon)$ defines the stress–strain curve under a unit strain rate. A different type of constitutive equation based on microstructural considerations has been examined by Steinberg and Lund (1989).

In the high-velocity impact or explosive loading, of components, most of the heat generated by the plastic deformation remains in the specimen, causing a thermal softening of the material. Using a Hopkinson pressure bar recovery technique, the isothermal stress–strain curves for materials at high strain rates have been obtained by Nemat-Nasser et al. (1991), who also estimated the effect of the adiabatic rise in temperature on the dynamic response. Their experimental data over the plastic range of strains can be fitted by the constitutive equation

$$\sigma = \sigma_0 \varepsilon^n \left\{ 1 + \left(\frac{\dot{\varepsilon}}{\dot{\varepsilon}_0} \right)^m \right\} \exp \left\{ -\lambda \left(\frac{T}{T_0} - 1 \right) \right\} \tag{8.13}$$

where σ_0, ε_0, and T_0 are reference values of the stress, strain rate, and temperature, respectively, while m, n, and λ are material parameters. The above equation is very similar to that suggested by Nemat-Nasser et al. (1994) and is found to be sufficiently accurate over the practical range.

In general, both the elastic and plastic strain increments would be involved in a prescribed stress increment, whether or not the material is rate sensitive. Considering this fact, a constitutive equation has been proposed by Cristescu (1963) and Lubiner (1965), in which the instantaneous strain rate is expressed in terms of the stress rate in the generalized form

$$E \frac{\partial \varepsilon}{\partial t} = \phi (\sigma,\varepsilon) \frac{\partial \sigma}{\partial t} + \psi (\sigma,\varepsilon) . \tag{8.14}$$

When the material is not rate sensitive, $\psi = 0$ and ϕ is independent of t, while the derivatives in (8.14) may be considered with respect to any monotonically increasing parameter. For a highly rate-sensitive material, on the other hand, it is a good approximation to set $\phi = 1$, and (8.14) then reduces to one of type (8.12). The problem of longitudinal wave propagation in an elastic/plastic bar with an arbitrary material response is therefore governed by (8.14), as well as the relations

$$\frac{\partial \sigma}{\partial x} = \rho \frac{\partial v}{\partial t}, \qquad \frac{\partial v}{\partial x} = \frac{\partial \varepsilon}{\partial t},$$

for the three unknowns σ, ε, and v. The first equation above is the equation of motion, while the second is the equation of compatibility. This system of equations is hyperbolic, as may be shown by considering the variation of σ, ε, and v along any curve in the (x, t)-plane. The conditions under which the first derivatives of these quantities may be discontinuous across the curve furnishes three families of real characteristics (Hopkins, 1968), which are given by

$$dx = 0, \qquad dx/dt = \pm c, \qquad c = \sqrt{E/\rho\phi}, \tag{8.15}$$

where c denotes the wave speed which is generally much higher than that given by (8.1). The differential relations holding along the characteristic curves $dx/dt = \pm c$ are easily shown to be

$$d\sigma = \pm\rho c d\upsilon - (\psi/\phi)\,dt, \tag{8.16a}$$

where the upper and lower signs correspond with those in (8.15). The relationship that holds along the remaining family of characteristics $dx = 0$ is found to be

$$E d\varepsilon = \phi d\sigma + \psi dt. \tag{8.16b}$$

Across a stationary wave front $dx = 0$, only $\partial\varepsilon/\partial x$ may be discontinuous, while all other first-order derivatives of σ, ε, and υ must be continuous. This is easily established from the condition of continuity of these quantities across the wave front, and the fact that $d\varepsilon = (\partial\varepsilon/\partial t)\,dt$ and so on along this line. The continuity of the time derivatives $\partial\varepsilon/\partial t$ and $\partial\upsilon/\partial t$ implies the continuity of $\partial\upsilon/\partial x$ and $\partial\sigma/\partial x$, but no information is available for $\partial\varepsilon/\partial x$ which may therefore be discontinuous. Across a moving wave front defined by the second relation of (8.15), all the first-order derivatives of σ, ε, and υ may become discontinuous. The method of numerical integration of (8.16a) and (8.16b) along the characteristics has been expounded by Cristescu (1967).

In the special case when $\psi = 0$ and ϕ is independent of t, the results are similar to those for the rate-independent material, the only difference being the existence of a third family of characteristics arising from the differential form of the constitutive relation. In the other extreme case of $\phi = 1$, the plastic wave speed becomes identical to the elastic wave speed $c_0 = \sqrt{E/\rho}$, while the characteristic lines (which are straight) and the differential relations along them are still given by (8.15) and (8.16) with $c = c_0$. It may be noted that the first term on the right-hand side of (8.13), when $\phi = 1$, represents the elastic strain increment, and the second term represents the plastic strain increment. The stress and elastic strain increase as a result of the wave propagation, causing the plastic strain to increase along the stationary characteristics.

More refined constitutive equations for the propagation of one-dimensional plastic waves with strain rate effects have been examined by Cristescu (1974). Generalized constitutive equations for viscoplastic solids, from the standpoint of continuum mechanics, have been presented by Perzyna (1963) and Mandel (1972). Other types of constitutive equations for the dynamic behavior of materials, based on the dislocation motion, have been examined by Nemat-Nasser and Guo (2003).

8.1.5 Illustrative Examples and Experimental Evidence

Some quantitative results for the plastic wave propagation in a rate-sensitive material have been given by Malvern (1951), who discussed an example in which the end $x = 0$ of a semi-infinite bar lying along the positive x-axis is instantaneously pulled with a longitudinal velocity $v = -U$ at $t = 0$. The constitutive law adopted by Malvern corresponds to (8.14) with

$$\phi = 1, \qquad \psi = k\left[\sigma - f(\varepsilon)\right], \qquad f(\varepsilon) = Y(2 - Y/E\varepsilon) \tag{8.17}$$

where k is a material constant. The numerical computation is based on the values $k = 10^6$/s, $Y/E = 10^{-3}$, $E = 70$ GPa, $c = c_0 = 5$ km/s, and $U = 15$ m/s. The predicted dynamic yield stress for this material is about 10% higher than the quasi-static value for a constant strain rate of 200/s.

Along the leading wave front $x = c_0 t$, the stress and strain are not constant in the solution with strain rate effect, which allows the stress to rise above Y without exceeding the elastic limit. The stress attained in an element just after the passage of the leading wave front is given by

$$d\sigma = -\frac{k}{2\sigma}(\sigma - Y)^2\, dt$$

in view of (8.16a) and (8.17), together with the relations $E\varepsilon = \sigma$ and $\rho c_0 v = -\sigma$ which hold along this shock wave. Integrating, and using the initial condition $\sigma = \rho c_0 U$ at $t = 0$, which follows from (8.6b), we obtain the solution

$$t = \frac{2}{k}\left\{\frac{(\rho c_0 U - \sigma)\, Y}{(\sigma - Y)(\rho c_0 U - Y)} + \ln\left(\frac{(\rho c_0 U - Y)}{(\sigma - Y)}\right)\right\}. \tag{8.18}$$

It may be noted that $\rho c_0 U = 3Y$ according to the assumed numerical data. Following the impulsive loading, the stress jumps instantaneously from 0 to $3Y$, but rapidly falls to a value very nearly equal to Y as the wave propagates along the bar. If, on the other hand, the strain rate effect is disregarded, the stress would have a constant value Y all along the leading wave front.

The solid curves in Fig. 8.5(a) represent the lines of constant strain in the (x, t)-plane according to rate-dependent solution, while the broken straight lines passing through the origin correspond to the solution that neglects strain rate effects. The region of constant strain, that exists in the rate-independent solution above the wave front of the maximum strain $\varepsilon = 0.0074$, disappears in the rate-dependent solution, which predicts a gradual transition from plastic to elastic response across the chain-dotted line. Along this unloading boundary, the quasi-static stress–strain relation $\sigma = f(\varepsilon)$ holds, where v and ε slowly decrease with increasing x and t.

The stress–strain relations at various sections of the bar are displayed in Fig. 8.5(b), where the broken curve represents the quasi-static stress–strain relation. It is seen that the plastic deformation occurs with decreasing stress in sections closer to the impact end $x = 0$, and with increasing stress in sections farther from $x = 0$. This is a

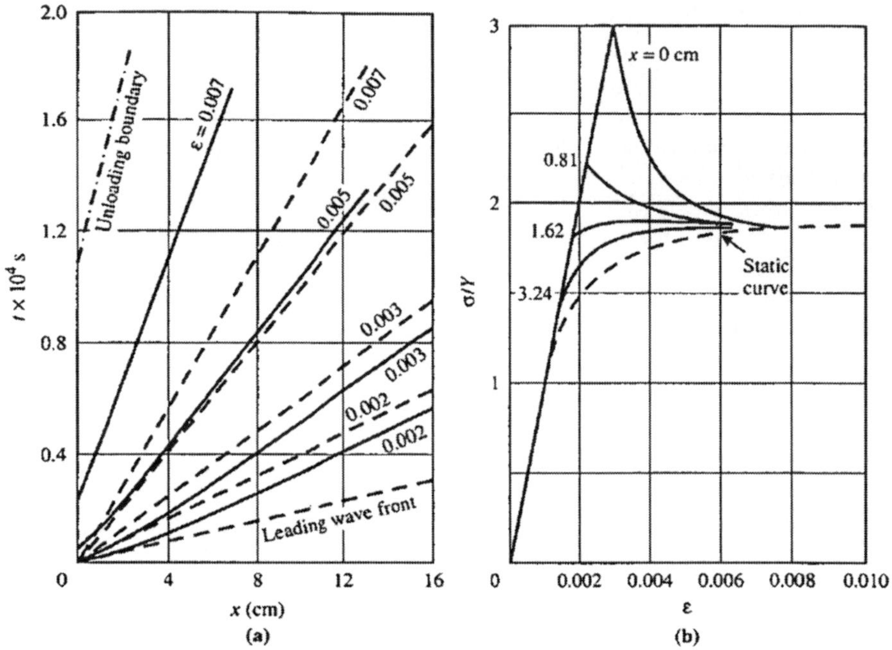

Fig. 8.5 Calculated results for the plastic wave propagation along a long bar made of rate-sensitive material. (**a**) Lines of constant strain and (**b**) stress–strain relations at different sections of the bar (after L.E. Malvern, 1951)

consequence of the high initial stress instantaneously attained at $x = 0$, and the fact that the strain rate at any section rapidly decreases with time. At sections sufficiently far away from the end $x = 0$, the stress–strain curves resemble the quasi-static curve. The strain distribution shown in Fig. 8.6 (a) indicates that the plateau of constant strain near the impact end disappears, which is contrary to that experimentally observed (Bell, 1968). The existence of the strain plateau can be theoretically predicted, however, by using a suitable modification of the constitutive equation.

The plastic wave propagation in a bar of finite length, initially prestressed to the yield point with a longitudinal stress $\sigma = Y$, has been treated by Cristescu (1965), using the relations (8.17) with the same material constants as those employed in the preceding example. The bar is assumed to have an initial length $l = 7.5$ cm with one end fixed and the other end prescribed to move with a velocity

$$\upsilon = -Ut/t_0\,(0 < t \leq t_0)\,, \quad \upsilon = -U\;\;(t \geq t_0)\,,$$

where $U = 30$ m/s and $t_0 = 12$ μs. The solution has been obtained numerically by using the finite difference form of the differential relations (8.16) along the characteristics. Some of the computed results are displayed in Fig. 8.7(a) and (b), which exhibit the same general trends as those in the previous example. There is, of course, no strain plateau near the impact end of the bar.

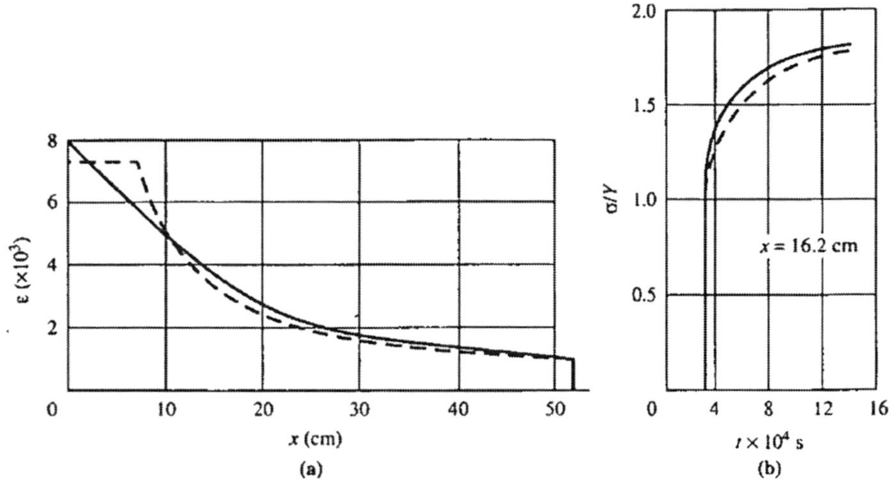

Fig. 8.6 Strain rate effects on the longitudinal plastic wave propagation. (**a**) Strain distribution at $t = 102.4$ μs and (**b**) stress–time relation at a distance $x = 16.2$ cm from impact end

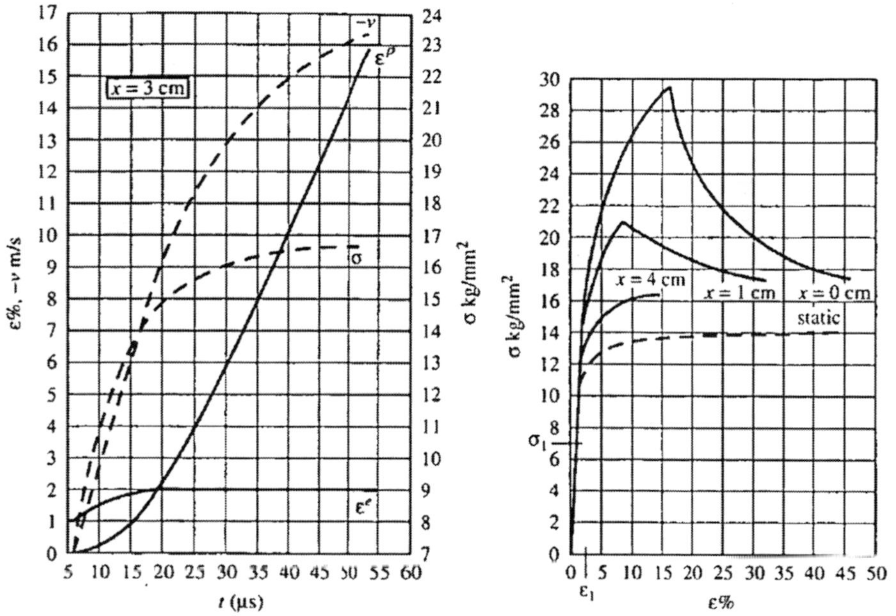

Fig. 8.7 Results for a bar of finite length subjected to a prescribed velocity at $x = 0$. (**a**) Time dependence of physical quantities at $x = 3$ cm and (**b**) dynamic stress–strain relations (after N. Cristescu, 1967)

Several other problems involving the strain rate effect have been discussed by Cristescu (1972), Banerjee and Malvern (1975), and Cristescu and Suliciu (1982). The problem of plastic wave propagation due to the longitudinal impact of two identical bars, one of which is at rest and the other one fired from an air gun, has been considered by Cristescu and Bell (1970). The effects of strain rate history on the dynamic response have been examined by Klepaczko (1968) and Nicholas (1971). The application of the endochronic theory to the propagation of one-dimensional plastic waves has been discussed by Lin and Wu (1983).

One of the earliest experimental investigations on the propagation of plastic waves, reported by von Karman and Duwez (1950), consisted in measuring the longitudinal strain in annealed copper wires subjected to impact loading by means of a drop hammer. Although strain rate effects were disregarded in their theoretical calculations, the overall agreement between theory and experiment was found to be reasonably good. Sternglass and Stuart (1953) applied small amplitude strain pulses to prestressed copper wires to study the propagation of plastic waves. The speed of propagation of the wave front was found to be that of elastic waves, and the wave speed of any part of the pulse was found to be much higher than the value given by the rate-independent theory. Further experimental supports for the rate-dependent theory have been provided by Alter and Curtis (1956), Malyshev (1961), Bianchi (1964), Dillons (1968), and others, and the results obtained by them were consistent with the differential constitutive law of type (8.12). Critical reviews of the published experimental work on the plastic wave propagation have been reported by Bell (1973), Clifton (1973), and Nicholas (1982).

The development of plastic strains with time at various distances from the impact end of bars, made of mild steel and aluminum, has been experimentally determined by Bell (1960) and Bell and Stein (1962), using a diffraction grating method. The split Hopkinson pressure bar has been used by Lindholm (1964), following a method proposed by Kolsky (1949), to establish the dynamic properties and stress–strain curves for lead, aluminum, and copper under different values of the rate of strain. An electromagnetic method of measurement of the particle velocity at any section of the bar during the propagation of plastic waves along its length has been discussed by Malvern (1965) and Efron and Malvern (1969). Both these methods are quite accurate and can be used for obtaining the overall dynamic response from the moment of impact to the moment when the unloading begins. Further references to the theoretical and experimental investigations on the plastic wave propagation have been given by Nicholas (1982).

8.2 Plastic Waves in Continuous Media

8.2.1 Plastic Wave Speeds and Their Properties

Consider an isolated wave front, not necessarily plane, which is advancing through an elastic/plastic material that work-hardens isotropically without exhibiting strain rate effects. We shall be concerned here with the propagation of weak waves across which discontinuities in the derivatives of stress, strain, and particle velocity may exist. Since the governing equations are quasi-linear, the weak waves are necessarily characteristics of the hyperbolic system. Any physical quantity in the immediate neighborhood of the wave front may be regarded as a function of the surface coordinates, the surface normal, and time t. Since the surface derivatives of a typical variable f must be continuous across the wave front, which has a velocity c in the direction of the normal n, the jump in the space and material derivatives of f across the wave front may be written as

$$\left[\frac{\partial f}{\partial x_i}\right] = \left[\frac{\partial f}{\partial n}\right] n_i \qquad [\dot{f}] = -\left[\frac{\partial f}{\partial n}\right] c, \tag{8.19}$$

where n_i denotes the unit vector in the direction of n. The second expression follows from the fact that the rate of change of f relative to the motion of the wave front in the x_i space is continuous, since f itself must be continuous. Each pair of square brackets in (8.19) will be taken to represent the value of the enclosed quantity behind the wave front in excess of that ahead of the front.

In the analysis of the problem of plastic wave propagation, all rotational effects will be disregarded. Consequently, the stress rate entering into the constitutive equation will be taken as the material derivative of the true stress σ_{ij}. If the material is prestressed in the plastic range, and obeys the von Mises yield criterion and the associated Prandtl–Reuss flow rule, the constitutive equation may be written in the form of equation (1.38), where $n_{ij} = \sqrt{\frac{3}{2}} s_{ij}/\bar{\sigma}$ The stress rate is therefore given by

$$\dot{\sigma}_{ij} = 2G\left\{\dot{\varepsilon}_{ij} + \frac{\nu}{1-2\nu}\dot{\varepsilon}_{kk}\delta_{ij} - \frac{3\alpha}{2\bar{\sigma}^2}\dot{\varepsilon}_{kl}s_{kl}s_{ij}\right\}, \quad \dot{\varepsilon}_{kl}s_{kl} \geq 0, \tag{8.20}$$

where G is the shear modulus, ν is Poisson's ratio, $\bar{\sigma}$ is the equivalent stress, s_{ij} is the deviatoric stress tensor, and $\alpha = 3G/(H + 3G)$ with H denoting the current plastic modulus. In the case of unloading, indicated by $\dot{\varepsilon}_{kl}s_{kl} < 0$, it is only necessary to set $\alpha = 0$ in (8.20). Introducing the expression

$$\dot{\varepsilon}_{ij} = \frac{1}{2}\left(\frac{\partial v_i}{\partial x_j} + \frac{\partial v_j}{\partial x_i}\right)$$

for the true strain rate, where v_i denotes the particle velocity, (8.20) may be expressed as

$$\dot{\sigma}_{ij} = G\left\{\left(\frac{\partial v_i}{\partial x_j} + \frac{\partial v_i}{\partial x_i}\right) + \frac{2\nu}{1-2\nu}\frac{\partial v_k}{\partial x_k}\partial_{ij} - \frac{3\alpha}{\bar{\sigma}^2}\frac{\partial v_k}{\partial x_l}s_{kl}s_{ij}\right\}.$$

Admitting possible discontinuities in the stress rate and the velocity gradient across the wave front, and using (8.19), we obtain the discontinuity relation

$$-c\left[\frac{\partial \sigma_{ij}}{\partial n}\right] = G\left\{\lambda_i n_j + \lambda_j n_i + \frac{2\nu}{1-2\nu}\lambda_k n_k \delta_{ij} - \frac{3\alpha}{\bar{\sigma}^2}\lambda_k n_l s_{kl}s_{ij}\right\},$$

where λ_i, denotes the quantity $[\partial v_i/\partial n]$. Multiplying the preceding equation by n_j following the summation convention, and using the relations

$$\sigma_{ij}n_j = \sigma_i, \qquad s_{ij}n_j = s_i, \qquad n_j n_j = 1,$$

where s_i and σ_i are the deviatone and actual stress vectors acting across the wave front, we get

$$-c\left[\frac{\partial \sigma_i}{\partial n}\right] = G\left\{\lambda_i + (1-2\nu)^{-1}\lambda_j n_j n_i - \frac{3\alpha}{\bar{\sigma}^2}\lambda_k s_k s_i\right\}. \tag{8.21}$$

The scalar parameter $\lambda_k s_k$, which represents the discontinuity in the deviatoric work rate, must be positive for continued loading, in view of the sign convention for λ_i. When $\lambda_k s_k$ is negative, the work rate in the element is decreased by the passage of the wave front, and unloading would occur as a result.

In order to complete the analysis, we must consider the equation of motion of the material element. Denoting the density of the material by ρ, which may be assumed constant since only elastic changes in volume are involved, the equation of motion may be written as

$$\frac{\partial \sigma_{ij}}{\partial x_j} - \rho\dot{v}_i = 0. \tag{8.22}$$

Using (8.19), the discontinuity relation corresponding to (8.22) is readily obtained, and the substitution $\sigma_{ij}n_j = \sigma_i$ then results in

$$\left[\frac{\partial \sigma_i}{\partial n}\right] + \rho c \lambda_i = 0.$$

The elimination of the stress gradient discontinuity $[\partial \sigma_i/\partial n]$ between this equation and (8.21) furnishes the connection equation (Jansen et al., 1972) as

$$\left(c^2 - c_2^2\right)\lambda_i - \left(c_1^2 - c_2^2\right)\lambda_j n_j n_i + c_2^2\left(\alpha/k^2\right)\lambda_k s_k s_i = 0, \tag{8.23}$$

where $k = \bar{\sigma}/\sqrt{3}$ is the current yield stress in shear, and c_1, c_2 are the speeds of propagation of elastic dilatational and shear waves, given by

$$c_1^2 = \left(\frac{1-\nu}{1-2\nu}\right)\frac{2G}{\rho}, \quad c_2^2 = \frac{G}{\rho}$$

Equation (8.23) constitutes a set of three linear homogeneous equations in the unknown component of the discontinuity vector λ_i. The existence of nontrivial solutions for these unknowns requires the determinant of their coefficients in these equations to vanish. The result is a cubic equation for c^2 having three distinct roots (Craggs, 1961; Mandel, 1962). One of these roots is obtained by a mere inspection of (8.23), the corresponding solution being

$$c^2 = c_2^2, \quad \lambda_j n_j = 0, \quad \lambda_k s_k = 0.$$

The wave front therefore advances at the speed of the elastic shear wave. The associated discontinuity in the velocity gradient vector is tangential to the wave front and is perpendicular to the plane containing the deviatoric stress vector and the normal to the wave front.

The other two roots of c^2 are most conveniently obtained by forming the scalar products of (8.23) with the vectors n_i and s_i, in turn. Setting

$$\lambda_j n_j = \lambda_n, \quad s_i n_i = s_n, \quad \lambda_k s_k = \omega,$$

and denoting the magnitude of the deviatoric stress vector by s, the two scalar equations involving the unknown quantities λ_n and ω are found as

$$\left.\begin{aligned} \left(c^2 - c_1^2\right)\lambda_n + c_2^2\left(\alpha/k^2\right)s_n\omega = 0, \\ -\left(c_1^2 - c_2^2\right)s_n\lambda_n + \left\{c^2 - c_2^2\left(1 - \alpha s^2/k^2\right)\right\}\omega = 0. \end{aligned}\right\} \tag{8.24}$$

If τ denotes the magnitude of the shear stress transmitted across the wave front, then $s^2 = s_n^2 + \tau^2$, and the two equations in (8.24) will be simultaneously satisfied for nonzero values of λ_n and ω if

$$\left(c^2 - c_2^2\right)\left(c^2 - c_1^2\right) + c_2^2\left\{\left(c^2 - c_2^2\right)\left(\alpha s_n^2/k^2\right) + \left(c^2 - c_1^2\right)\left(\alpha\tau^2/k^2\right)\right\} = 0 \tag{8.25}$$

This equation indicates that there are two real roots for c^2, one of which lies between 0 and c_2^2, and the other between c_2^2 and c_1^2. The former corresponds to slow waves and the latter to fast waves, their speeds of propagation being denoted by c_s and cf, respectively. The roots of the above quadratic may be expressed as

$$\left.\begin{aligned} c_f^2, c_s^2 = \frac{1}{2}\left[\left(c_1^2 + c_2^2\right) - c_2^2\left(\alpha s^2/k^2\right) \pm \sqrt{n}\right], \\ \eta = \left[\left(c_1^2 - c_2^2\right) - c_2^2\left(\alpha s^2/k^2\right)\right] + 4c_2^2\left(c_1^2 - c_2^2\right)\left(\alpha\tau^2/k^2\right). \end{aligned}\right\} \tag{8.25a}$$

In the special case of an elastic material ($\alpha = 0$), the wave speeds c_f and c_s reduce to c_l and c_2, respectively, the discontinuities in the velocity gradient in the two cases being normal and tangential, respectively, to the wave front. The related problem of acceleration waves in solids has been studied by Thomas (1961) and Hill (1962).

8.2.2 A Geometrical Representation

For an elastic/plastic material, a useful geometrical interpretation of the wave speed equation (8.25) is obtained by writing it in the alternative form

$$\left(\frac{c_1^2}{c_1^2 - c^2}\right)\frac{\alpha s_n^2}{k^2} + \left(\frac{c_2^2}{c_2^2 - c^2}\right)\frac{\alpha \tau^2}{k^2} = 1 \tag{8.26}$$

This relation indicates that if $\sqrt{\alpha} s_n$ and $\sqrt{\alpha}\tau$ are plotted as rectangular coordinates, the locus of constant wave speed is an ellipse when $c = c_s$ and is a hyperbola when $c = c_f$. The semifocal distance in each case is equal to $k\sqrt{c_1^2/c_2^2 - 1}$, and consequently, the family of ellipses is orthogonal to the family of hyperboles, as shown in Fig. 8.8 Moreover, it follows from (8.25a) that the sum of the squares of c_s and c_f is constant along any circle $\alpha\left(s_n^2 + \tau^2\right) = \text{constant}$, having its center at the origin of the stress plane (Ting, 1977).

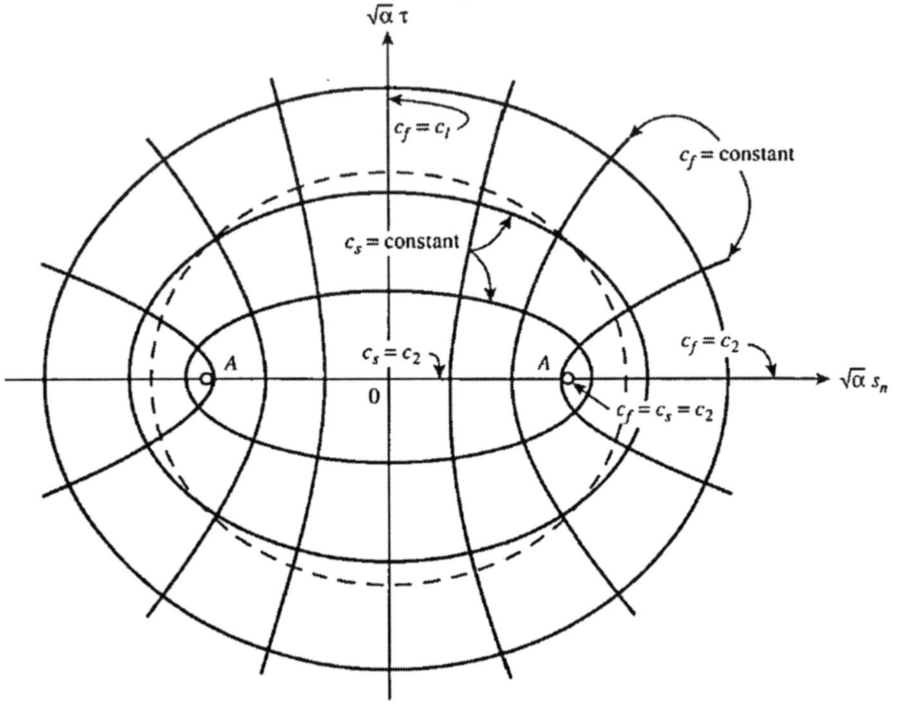

Fig. 8.8 Elastic/plastic wave speeds in a continuous medium and their dependence on the deviatoric normal and shear stresses acting across the wave front

For a given deviatoric stress tensor s_{ij} and the direction of propagation n_i, the normal and tangential components of the deviatoric stress vector acting across the wave front can be found from the relations

$$s_n = s_{ij}n_in_j, \quad s^2 = s_{ij}s_{jk}n_in_k, \quad \tau^2 = s^2 - s_n^2$$

The plastic wave speeds then follow from (8.25a), and the associated directions of the velocity gradient discontinuity are given by (8.23). When the deviatoric stress vector is tangential to the wave front ($s_n = 0$, $s = \tau$), (8.25a) furnishes

$$c_f^2 = c_1^2, \qquad c_s^2 = c_2^2\left(1 - \alpha\tau^2/k^2\right).$$

Since ω vanishes for the fast wave and λ_n vanishes for the flow wave in view of (8.24), the discontinuities in the velocity gradient in the two cases are directed along the normal and tangent to the wave front, respectively. When the deviatoric stress vector is normal to the wave front ($\tau = 0$), the wave propagates along a principal axis of the stress, and we have

$$c_s^2 = c_2^2, \qquad c_f^2 = c_1^2 - c_2^2\left(\alpha s_n^2/k^2\right) \qquad \alpha s_n^2/k^2 \leq c_1^2/c_2^2 - 1.$$

It follows from (8.23) and the first equation of (8.24) that the directions of the velocity gradient discontinuity again coincide with the normal and tangent to the wave front for the fast and slow waves, respectively. When the above inequality is reversed, the values of c_f and c_s and the associated discontinuity directions are simply interchanged.

When the rectangular axes of reference are taken along the principal axes of the stress, the deviatoric state of stress at any point can be represented by the appropriate Mohr circle. If the deviatoric principal stresses are denoted by s_1, s_2, and s_3, with s_2 assumed to lie between s_1 and s_3, we have the identity

$$s_1 + s_2 + s_3 = 0, \qquad s_1 > s_2 > s_3.$$

Evidently, $s_1 > 0$ and $s_3 < 0$, while s_2 can be either positive or negative. For given values of s_1, s_2, and s_3, the stress point (s_n, τ) lies within a region bounded by three circles, the largest of which has the equation

$$\left(s_n - \frac{s_1 + s_2}{2}\right)^2 + \tau^2 = \left(\frac{s_1 - s_2}{2}\right)^2. \tag{8.27}$$

The two principal components s_1 and s_3 appearing in this equation cannot be chosen arbitrarily, since they are required to satisfy the yield criterion which may be expressed as

$$3\,(s_1 + s_3)^2 + (s_1 - s_3)^2 = 4\,k^2$$

This restriction makes the family of Mohr's circles (8.27) bounded by an envelope, which is easily shown to be an ellipse (Ting, 1977) having the equation

$$\frac{3}{4}\left(\frac{s_n}{k}\right)^2 + \left(\frac{\tau}{k}\right)^2 = 1. \tag{8.28}$$

The broken curve in Fig. 8.8 represents this ellipse for a typical value of α. The focal points A and B of the ellipse defined by (8.26) with $c = c_s$ lie inside or outside the ellipse defined by (8.28) depending on whether $4\alpha/3$ is greater or smaller than $\left(c_1^2/c_2^2 - 1\right)$. The situation $c_f = c_s = c_2$ can arise only when the former condition is satisfied.

The smallest value of c_f^2, which corresponds to $\tau = 0$, is equal to c_2^2 when the focal points are inside the envelope, and to $c_1^2 - \frac{4}{3}\alpha_2^2$ when they are outside the envelope. The smallest value of c_s^2, on the other hand, corresponds to the point of tangency of the largest possible Mohr circle with the appropriate ellipse of constant c_s. It can be obtained as a function of s_2 by solving (8.26) and (8.27) simultaneously for s_n and τ, after setting $c = c_s, s_1 + s_3 = -s_2, (s_1 - s_3)^2 = 4k^2 - 3s_2^2$, and establishing the condition for the quadratics to have equal roots. When $s_2 = 0$, the smallest value of c_s^2 is equal to $(1 - \alpha)\, c_2^2$, corresponding to $s_n = 0$.

8.2.3 Plane Waves in Elastic/Plastic Solids

The preceding results are directly applicable to the propagation of plane waves in which the wave front is a plane surface advancing in a uniformly prestressed elastic/plastic medium. When the state of stress is a pure shear in which the normal stress vanishes across the plane ($s_n = 0, \tau = k$), the normal and tangential components of the velocity gradient discontinuity propagate with speeds c_l and $c_2\sqrt{1 - \alpha}$ respectively, in the direction of the normal to the plane. The strength of the former remains unchanged, while that of the latter steadily decreases as the plastic strain increases. In the case of a uniaxial prestress normal to the plane ($s_n = 2k/\sqrt{3}, \tau = 0$), the discontinuities in the tangential and normal velocity gradients propagate with speeds c_2 and $\sqrt{c_1^2 - 4\alpha c_2^2/3}$, respectively. The latter wave speed depends on the value of α, which increases with increasing plastic strain.

Consider now the rectilinear propagation of a plane wave in an isotropic elastic/plastic body, which is in the form of a thick plate whose lateral dimensions are infinitely large. The wave travels through the thickness of the plate along the x-axis, the kinematical restrictions being

$$\dot{\varepsilon}_y = \dot{\varepsilon}_z = 0, \qquad \dot{\varepsilon}_y{}^e = -\dot{\varepsilon}_y{}^p = \frac{1}{2}\dot{\varepsilon}_x{}^p$$

which follow from the relation $\dot{\sigma}_y = \dot{\sigma}_z$ holding throughout the body. The elastic stress–strain relations therefore give the elastic and plastic parts of $\dot{\varepsilon}_x$

$$E\dot{\varepsilon}_x{}^e = \dot{\sigma}_x - 2\nu\dot{\sigma}_y, \qquad E\dot{\varepsilon}_x{}^p = 2\left[(1 - \nu)\,\dot{\sigma}_y - \nu\dot{\sigma}_x\right]$$

Combining these two relations, the total rate of extension in the x-direction is expressed as

$$E\dot{\varepsilon}_x = (1 - 2\nu)\left(\dot{\sigma}_x + 2\dot{\sigma}_y\right) \tag{8.29}$$

The material is assumed to be rate sensitive, with the plastic part of the strain rate satisfying the quasi-linear constitutive equation

$$E\dot{\varepsilon}_x^p = \phi_1\dot{\sigma}_x + \phi_2\dot{\sigma}_y + \psi,$$

where ϕ_1 and ϕ_2 are functions of stress and strain, and ψ is a function of the dynamic overstress. In view of the expression for $\dot{\varepsilon}_x^\rho$ given above, the preceding relation becomes

$$(2\nu + \phi_1)\,\dot{\sigma}_x - \left[2\,(1-\nu) - \phi_{2_x}\right]\dot{\sigma}_y + \psi = 0$$

The elimination of $\dot{\sigma}_y$ between (8.29) and the preceding equation leads to the relationship between $\dot{\sigma}_x$ and $\dot{\varepsilon}_x$ in the form

$$\left[2\,(1-\nu) - \phi_{2_x}\right]E\dot{\varepsilon}_y = (1-2\nu)\,\{[2\,(1+\nu) + 2\phi_1 - \phi_2]\,\dot{\sigma}_x + 2\psi\} \tag{8.30}$$

The characteristics of this hyperbolic system are given by $dx/dt = \pm c$, where c is the speed of propagation of the wave. Along the characteristics, we have

$$\dot{\sigma}_x = \frac{\partial\sigma_x}{\partial t} = \pm c\frac{\partial\sigma_x}{\partial x} = \pm\rho c\frac{\partial \upsilon_x}{\partial t} = \pm\rho c^2\frac{\partial \upsilon_x}{\partial x} = \rho c^2\dot{\varepsilon}_x.$$

It should be noted that $\partial\sigma_x/\partial t = \dot{\sigma}_x$ when geometry changes are disregarded. Since $\dot{\varepsilon}_x$ and $\dot{\sigma}_x$ are not uniquely determined along the characteristics, the preceding two relations involving $\dot{\varepsilon}_x$ and $\dot{\sigma}_x$ furnish the result

$$c^2 = \frac{E}{\rho\,(1-2\nu)\,(1+2\lambda)}, \quad \lambda = \frac{2\nu + \phi_1}{2\,(1-\nu) - \phi_2} \tag{8.31}$$

The range of values of ϕ_1 and ϕ_2 for which the wave speed is real must satisfy the conditions

$$\phi_2 = 2\,(1-\nu)\,, \quad \phi_2 - 2\phi_1 < 2\,(1+\nu)$$

as well as those obtained by reversing these inequalities. It is easy to see that

$$\nu/(1-\nu) < \lambda < 1, \quad \sqrt{K/\rho} < c < c_1$$

where K is the bulk modulus for the material. When $\phi_1 = \phi_2 = 0$, there is no instantaneous plastic strain in the material response, and (8.31) gives $c = c_{1-}$. When $\psi = 0$, implying the absence of strain rate effects, the constitutive law gives $\phi_1 = -\phi_2 = E/H$, and the plastic wave speed given by (8.31) coincides with the value of c_f given by (8.25a) with $\tau = 0$ and $s = 2k/\sqrt{3}$, where k is the yield stress in shear.

In the (ϕ_1, ϕ_2)-plane, λ (and hence c) remains constant along straight lines passing through the point $[-2\nu, 2\,(1-\nu)]$, where the wave speed is indeterminate. It may be noted that the relation $c = c_1$ holds all along the straight line $\nu\phi_2 + (1-\nu)\,\phi_1 = 0$, not merely at $\phi_1 = \phi_2 = 0$. In the $(\dot{\sigma}_x, \dot{\sigma}_y)$-plane, (8.30) represents a straight line for given values of ϕ_1, ϕ_2, and ψ. The slope of this straight line is equal to λ, and it intersects the axis $\dot{\sigma}_y = 0$ at a point that depends on ϕ_1. No plastic flow is possible along the straight line $(1-\nu)\,\dot{\sigma}_y - \nu\dot{\sigma}_x = 0$, which therefore represents elastic unloading.

The propagation of plane waves excluding the strain rate effects, but including the variation of the bulk modulus with the hydrostatic pressure, has been considered by Morland (1959), who also studied the interaction of loading and unloading waves when a pressure pulse is applied on the free surface. The propagation of cylindrical waves in an infinite medium has been discussed by Cristescu (1967) under torsional loading and by Jansen et al. (1972) under radially symmetric loading. The propagation of spherically symmetric waves in an unbounded medium has been investigated by Hunter (1957) and Hopkins (1960).

8.3 Crumpling of Flat-Ended Projectiles

One of the simplest methods of studying the effect of high strain rates on the dynamic yield strength of metals consists in firing flat-ended cylindrical projectiles against rigid targets. The axial stress at the impact end immediately attains the yield limit, and a plastic wave moves away from the target plate rendering the projectile partially plastic. An elastic wave front initiated at the same time moves ahead of the plastic wave front, the region between the two wave fronts being stressed to the yield point. Due to the reflections of the elastic wave front from the free end of the bar and the advancing plastic wave front, the rear part of the projectile rapidly decelerates and comes to rest within a distance equal to the difference between the initial and final lengths of the cylinder. The extent of the deformed and undeformed portions of the projectile after impact depends on its kinetic energy before impact, as well as on the dynamic yield strength of the material.

8.3.1 Taylor's Theoretical Model

Let U denote the velocity of normal impact of a cylindrical projectile having an initial length L and an initial cross-sectional area A_0. At a generic stage of the dynamic process, the overall length of the projectile is reduced to l due to the piling up of material over a length h, leaving a nonplastic rear part of length x as shown in Fig. 8.9(b). For simplicity, the material is assumed to have a constant dynamic yield stress Y, and the radial inertia is disregarded so that the stress distribution may be considered as uniform over any given cross section (Taylor, 1948b).

Within a small time interval dt, a nonplastic element of length $-dx$ and area A_0 passes through the advancing plastic boundary to come to rest as a plastic element of length dh and area A. Neglecting elastic strains, the continuity equation may be written as $A\,dh = -A_0 dx$, which is equivalent to

$$A\upsilon = A_0\,(u+\upsilon)\,, \tag{8.32}$$

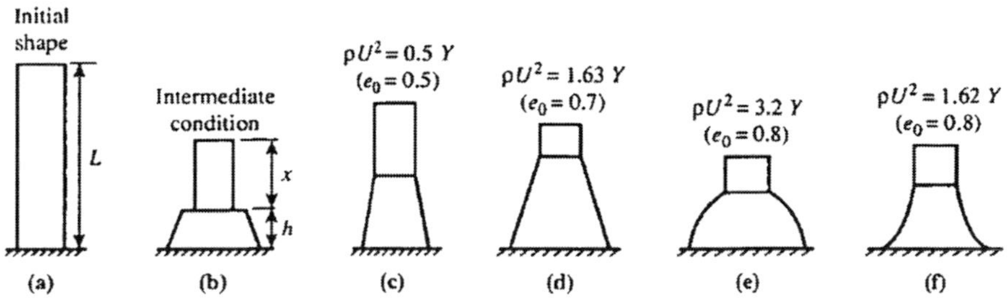

Fig. 8.9 Deformation of a flat-ended projectile fired at a speed U against a flat rigid target

where u is the current velocity of the rear part of the projectile, and υ is the velocity of the plastic boundary, the velocity of the nonplastic part relative to the plastic boundary being $u + \upsilon$, Thus

$$\upsilon = \frac{dh}{dt}, \qquad u + \upsilon = -\frac{dx}{dt}. \tag{8.33}$$

The momentum of the elemental volume $-A_0dx$ changes from $-\rho A_0 u dx$ to zero during the time dt, where ρ is the density of the material. Since the net force acting on the element is of magnitude $(A–A_0)Y$, the equation of motion becomes

$$\rho A_0(u + \upsilon)u = (A - A_0)\, Y. \tag{8.34}$$

Equations (8.32) and (8.34) are sufficient to express $\rho u^2/Y$ and υ/u in terms of a variable $e = 1 - (A_0/A)$, the result being

$$\frac{\rho u^2}{Y} = \frac{e^2}{1-e}, \quad \frac{\upsilon}{u} = \frac{1-e}{e}. \tag{8.35}$$

To obtain the expressions for x and h at any instant, we consider the equation of motion of the rear portion of the projectile, which moves as a rigid body with a velocity u under an opposing force equal to A_0Y. Since the change in momentum of this portion is $\rho A_0 x du$ during the time interval dt, the equation of motion is

$$\rho x \frac{du}{dt} = -Y. \tag{8.36}$$

Combining this equation with the second equation of (8.33), and using (8.35), we obtain the differential equation

$$\frac{dx}{du} = \frac{\rho u x}{Ye} \quad or \quad \frac{dx}{de} = \frac{x\,(2-e)}{2\,(1-e)^2}.$$

Let e_0 be the value of e at the moment of impact when $x = L$. Using this initial condition, the last equation is readily integrated to give

$$\left(\frac{x}{L}\right)^2 = \left(\frac{1-e_0}{1-e}\right) \exp\left\{\frac{e-e_0}{(1-e_0)(1-e)}\right\}. \tag{8.37}$$

The quantity e_0 depends on the impact velocity U according to the first equation of (8.35) with $e = e_0$ and $u = U$. When the projectile is brought to rest, $e = 0$ and $x = x^*$ (say), the relationship between x^*/L and $\rho U^2/Y$ being obtained as

$$\left(\frac{x*}{L}\right) = (1-e_0) \exp\left(-\frac{e_o}{1-e_0}\right), \qquad \frac{\rho U^2}{Y} = \frac{e_0^2}{1-e_o}, \tag{8.38}$$

in view of (8.37) and (8.35). It may be noted that x^*/L decreases and e_0 increases as the parameter $\rho U^2/Y$ is increased. The shape of the deformed part of the projectile at any instant can be determined by the integration of the equation

$$\frac{dh}{dx} = -\frac{\upsilon}{u+\upsilon} = -(1-e),$$

which is obtained from (8.33) and (8.35). In view of the initial conditions $h = 0$ and $e = e_0$ at $x = L$, the solution may be written as

$$\frac{h}{L} = (1-e_0) - (1-e)\frac{x}{L} + \int_e^{e_o} \left(\frac{x}{L}\right) de. \tag{8.39}$$

The integral is evaluated numerically for any given value of e_0 and selected values of e, using (8.37) for x/L. Each value of e furnishes a radius $a = a_0/\sqrt{1-e}$ of the deformed part corresponding to a distance h from the target plate, where a_0 is the initial radius of the cylinder. The shapes of the projectile after impact, predicted by the present theory for $e_0 = 0.5$, 0.7 and $e = 0.8$, are shown in Fig. 8.9(c)–(e) for the case where the diameter was initially 0.3 of the height. The calculated value of h^*/L and l^*/L, where h^* is the final length of the deformed part and l^* is the final overall length, is plotted against $\rho U^2/Y$ in Fig. 8.10, which includes some experimental points obtained by Whiffen (1948).

To determine how the various physical quantities vary with time t, measured from the beginning of the impact, it is necessary to integrate the differential equation for de/dt, which is most conveniently obtained from (8.36) and the first equation of (8.35). Indeed, the time derivative of the latter equation gives

$$\sqrt{\frac{\rho}{Y}}\frac{du}{dt} = \frac{2-e}{2(1-e)^{3/2}}\frac{de}{dt}.$$

Eliminating du/dt by means of (8.36), and substituting for $\sqrt{Y/\rho}$ obtained by setting $u = U$ and $e = e_0$ in the first equation of (8.35), we obtain the differential equation

$$\frac{de}{dt} = -\frac{2U}{x}\frac{\sqrt{1-e_0}}{e_o}\frac{(1-e)^{3/2}}{2-e}.$$

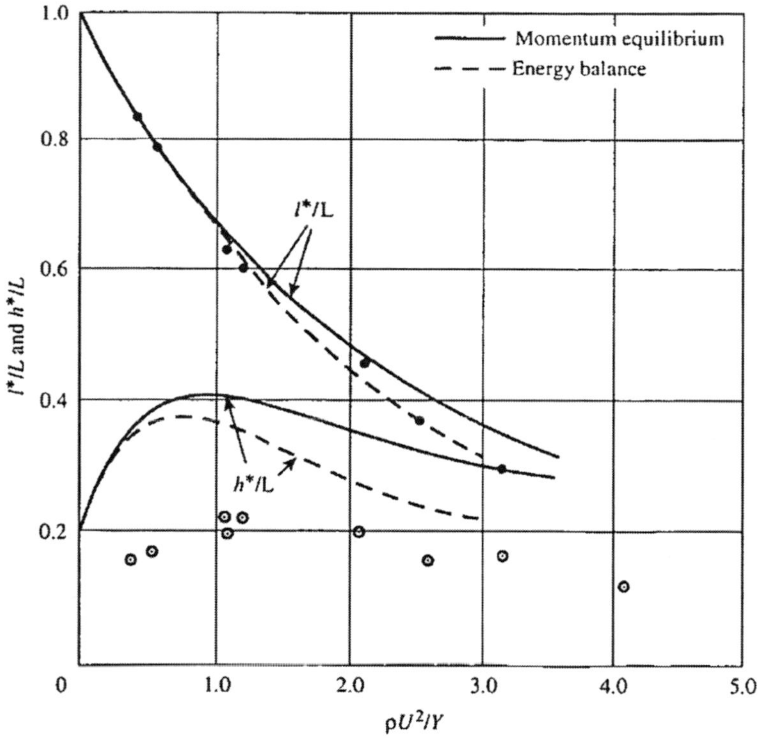

Fig. 8.10 Results of calculation for the longitudinal impact of a flat-ended projectile. The measures values of l^*/L and h^*/L are indicated by • and ⊙. respectively

Since $e = e_0$ at the moment of impact $t = 0$, the time interval t can be found numerically from the relation

$$\frac{Ut}{L} = \frac{e_o}{\sqrt{(1-e_o)}} \int_e^{e_o} \frac{x\,(2-e)\,de}{2L\,(1-e)^{3/2}}, \tag{8.40}$$

where x/L is given by (8.37) as a function of e and e_0. The parameter Ut/L, computed by numerical integration for the cases $e_0 = 0.5$ and 0.7, is plotted against h/L in Fig. 8.11, which indicates how the plastic boundary moves away from the target plate with the time interval. The same problem has been analyzed by Lee and Tupper (1954) on the basis of the elastic and plastic waves in the projectile, taking into account the strain hardening of the material. The Taylor anvil test has been used to establish the constitutive modeling of materials by Johnson and Holmquist (1988) and Nemat-Nasser et al. (1994).

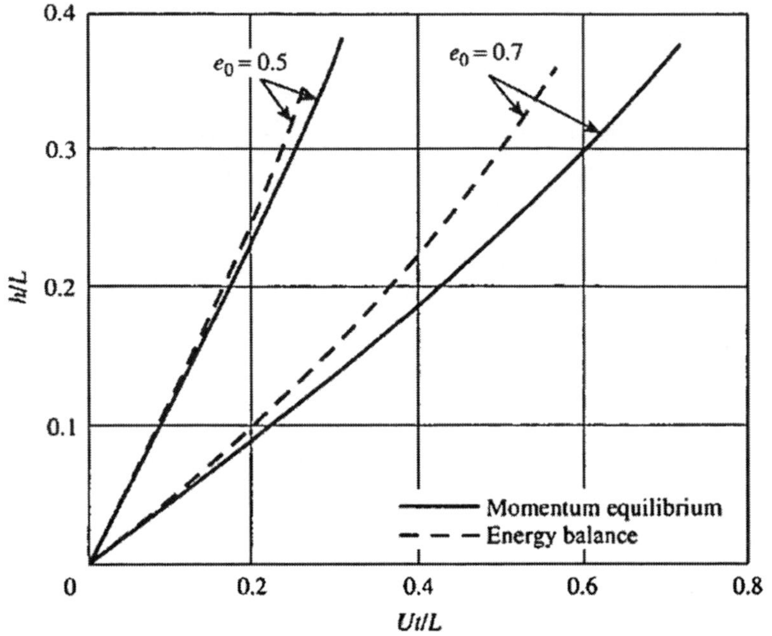

Fig. 8.11 Advancement of the plastic boundary during the longitudinal impact of a flat-ended projectile

8.3.2 An Alternative Analysis

The preceding theory, developed by Taylor (1948b), has been found to be in reasonable agreement with experiment for relatively low-impact velocities $\left(\rho U^2/Y \leq 0.5\right)$. The shapes of the slugs fired at greater speeds generally have a concave profile over the plastically strained part, instead of a convex profile predicted by Taylor's theory. To explain this mushrooming effect associated with the high-speed impact of projectiles, Hawkyard (1969) proposed an alternative method in which the rate of plastic work done on the projectile is equated to the total external energy supplied to it. The resultant equations defining the final geometry of the projectile are fairly simple, and the predicted profiles are found to be in closer agreement with experiment.

The amount of plastic work dW, which is done on an element of length $-dx$ in changing its cross-sectional area from A_0 to A before it comes to rest, is equal to $-A_0 Y \ln (A/A_0) \, dx$. Since the time taken for this change is dt, the rate of plastic work done is

$$\dot{W} = A_0 Y (u + \upsilon) \, \ln (A/A_0)$$

in view of (8.33). During this time interval, the kinetic energy of the element is reduced from $\frac{1}{2}\rho u^2 (-A_0 dx)$ to zero, while the work done by the external force becomes equal to $-A_0 Y (dx + dh)$. Hence, the rate at which the total energy is supplied to the system is

$$\dot{E} = A_0 Y \left[u + (u + \upsilon) \frac{\rho u^2}{2Y} \right].$$

Neglecting the loss of kinetic energy due to impact, we may equate the rate of plastic work $\dot{W}$ to the rate of supply of energy $\dot{E}$ to give

$$(u + \upsilon) \left(1n \frac{A}{A_0} - \frac{\rho u^2}{2Y} \right) = u. \tag{8.41}$$

This equation replaces (8.34) obtained on the basis of momentum equilibrium. Solving (8.32) and (8.41), and setting $e = 1 - (A_0/A)$ as before, we obtain the relations

$$\frac{\rho u^2}{2Y} = 1n \left(\frac{1}{1-e} \right) - e, \qquad \frac{\upsilon}{u} = \frac{1-e}{e}. \tag{8.42}$$

The equation of motion (8.36) for the undeformed part of the projectile is now combined with the time derivative of the first equation of (8.42) to give

$$\frac{de}{dt} = \frac{\rho u}{Y} \left(\frac{1-e}{e} \right) \frac{du}{dt} = - \left(\frac{1-e}{e} \right) \frac{u}{x}. \tag{8.43}$$

In view of the second equation of (8.42), (8.43) can be combined with (8.33) to give

$$\frac{dx}{de} = \frac{x}{1-e}, \qquad \frac{dh}{de} = -x.$$

Integrating these two equations in succession, and using the initial conditions $x = L$ and $h = 0$ when $e = e_0$, we obtain the solution

$$\frac{x}{L} = \frac{1-e_0}{1-e}, \quad \frac{h}{L} = (1 - e_0)\, 1n \left(\frac{1-e}{1-e_0} \right). \tag{8.44}$$

The values of x and h at the end of the impact, denoted by x^* and h^*, respectively, are given by (8.44) on setting $e = 0$. The impact velocity U is related to e_0 by the equation

$$\frac{\rho U^2}{2Y} = 1n \left(\frac{1}{1-e_0} \right) - e_o,$$

which follows from (8.42). The calculated shape of the projectile after impact, corresponding to $e_0 = 0.8$ $(\rho U^2/Y = 1.62)$ according to (8.44), is shown in Fig. 8.9(f). The profile is seen to be concave in form, resembling what is experimentally observed. The calculated values of h^*/L and l^*/L predicted by the above analysis are compared with those given by Taylor's theory in Fig. 8.10. The influence of strain hardening of the material has also been examined by Hawkyard (1969).

The variation of the strain parameter e with time can be determined by the integration of (8.43) after substitution for u/U given by (8.42), the final differential equation being

$$\frac{de}{dt} = -\frac{U}{L}\frac{(1-e)^2}{e\,(1-e_0)}\sqrt{\frac{e + \ln(1-e)}{e_o + \ln(1-e_0)}}.$$

In view of the initial condition $e = e_0$ at $t = 0$, the solution may be written as

$$\frac{Ut}{L} = (1-e)\int_e^{e_o}\sqrt{\frac{e_o + \ln(1-e_o)}{e + \ln(1-e)}}\,\frac{e\,de}{(1-e)^2}. \tag{8.45}$$

The integral can be evaluated numerically for any given value of e_0. The results corresponding to $e_0 = 0.5$ and $e_0 = 0.7$ are compared in Fig. 8.11 with those predicted by Taylor's theory. The two solutions do not seem to differ substantially from one another, though the shape of the projectile is predicted more realistically by the analysis based on the energy principle, particularly for $\rho U^2/Y \geq 0.5$.

8.3.3 Estimation of the Dynamic Yield Stress

The measurement of the final overall length and the position of the plastic boundary, after the impact, provides a convenient means of estimating the dynamic yield stress of the material. For practical purposes, a simple formula for the yield point Y can be developed on the assumption that the velocity of the undeformed part of the projectile relative to the advancing plastic boundary has a constant magnitude c. Then $dx/dt = -c$ in view of (8.33), and the equation of motion (8.36) gives

$$\frac{du}{dx} = \frac{Y}{\rho c x} \quad or \quad u = U - \frac{Y}{c\rho}\ln\left(\frac{L}{x}\right), \tag{8.46}$$

in view of the condition $u = U$ when $x = L$. If the distance traveled by the rear of the projectile at any instant is denoted by $du/dt = u\,(du/ds)$, then (8.36) gives

$$u\frac{du}{ds} = -\frac{Y}{\rho x} = -\frac{Y}{\rho L}\exp\left(-\frac{c\rho}{Y}(u-U)\right)$$

in view of (8.46). Integrating, and using the conditions $u = U$ when $s = 0$, and $u = 0$ when $s = L - l^*$, we obtain the solution

$$\frac{\rho c^2}{Y}\left(1 - \frac{l^*}{L}\right) = \left(\frac{c\rho U}{Y} - 1\right) + \exp\left[-\frac{c\rho U}{Y}\right].$$

The parameter $c\rho U/Y$ is equal to $\ln(L/x^*)$ in view of (8.46) with $u = 0$ and $x = x^*$. The elimination of c from the above relation therefore gives

$$\frac{Y}{\rho U^2} = \frac{\ln(L/x^*) - (1 - x^*/L)}{(1 - l^*/L)\left[\ln(L/x^*)\right]^2}. \tag{8.47}$$

If the deceleration of the rear of the projectile is assumed uniform, the formula for $Y/\rho U^2$ would become that given by Taylor. Equation (8.47) should be sufficiently accurate for the estimation of the dynamic yield stress Y for a given impact velocity U, using the measured values of x^*/L and l^*/L. The experimental results of Whiffen (1948) and Hawkyard (1969) indicate that the values of Y computed from (8.47) for different sets of values of U, x^*/L and l^*/L are approximately the same for a given material.

It is possible to define a mean strain rate as the ratio of the overall longitudinal plastic strain to the duration of the impact. If the rear portion of the projectile is assumed to move with a constant deceleration, the duration of impact is equal to $2(L - l^*)/U$ by the simple kinematics of the rigid-body motion. Since the initial and final lengths of the plastically deformed part of the projectile are $L - x^*$ and $l^* - x^*$, respectively, the mean strain rate $\dot{\lambda}$ may be written as

$$\dot{\lambda} = \frac{U}{2(L - l^*)} \ln\left(\frac{L - x^*}{l^* - x^*}\right). \tag{8.48}$$

The value of $\dot{\lambda}$ in each particular case may therefore be obtained from the direct measurement of the final undeformed length and the final overall length of the projectile. There seems to be very little variation of the duration of impact with the impact velocity, although there are fairly large variations in the other physical quantities. The buckling that would occur in a sufficiently long projectile due to the impact has been examined by Abrahamson and Goodier (1966) and Jones (1989).

The related problem of dynamic plastic buckling of columns has been treated by Lee (1981).

8.4 Dynamic Expansion of Spherical Cavities

The formation of spherically symmetric cavities in an infinitely extended medium under quasi-static conditions, originally discussed by Bishop et al. (1945), has been presented elsewhere (Chakrabarty, 2006). In this section, the problem of spherical cavity formation under dynamic conditions will be discussed for a material which obeys an arbitrary regular yield condition. The pressure applied at the cavity surface is supposed to be a given function of the current cavity radius. In the practical situation of cavity formation caused by high explosives, some kind of return motion following the expansion phase would be expected. Since the expansion process involves large plastic strains, it would be reasonable to begin by neglecting the elastic compressibility not only in the plastic region but also in the elastic region

(Hopkins, 1960). The compressibility of the material will be allowed for in a subsequent treatment of the cavity expansion process.

8.4.1 Purely Elastic Deformation

A spherical cavity of initial radius a_0 is expanded into an infinitely extended medium which is assumed to be completely incompressible. The internal pressure $p = p(t)$ is supposed prescribed at each instant of the expansion. Since the density ρ of the material remains constant by hypothesis, the equation for the conservation of mass may be written as

$$\frac{\partial}{\partial r}\left(r^2 \upsilon\right) = 0,$$

where ν is the radial velocity of a typical particle currently situated at a radius r. This equation is immediately integrated to give

$$\upsilon = \left(a^2/r^2\right)\dot{a}, \quad \dot{a} = da/dt. \tag{8.49}$$

The particle velocity is therefore everywhere determined in terms of the velocity of the cavity surface. The associated components of the strain rate are

$$\dot{\varepsilon}_r = \frac{\partial \upsilon}{\partial r} = -\left(\frac{2a^2}{r^3}\right)\dot{a}, \qquad \dot{\varepsilon}_\theta = \dot{\varepsilon}_\phi = \frac{\upsilon}{r} = \left(\frac{a^2}{r^2}\right)\dot{a}. \tag{8.50}$$

For an isotropic material, the result $\dot{\varepsilon}_\theta = \dot{\varepsilon}_\phi$ implies $\sigma_\theta = \sigma_\phi$ throughout the medium, and the equation of motion in terms of the stresses and velocity becomes

$$\frac{\partial \sigma_r}{\partial r} + \frac{2}{r}(\sigma_r - \sigma_\theta) = \rho\left(\frac{\partial \upsilon}{\partial t} + \upsilon\frac{\partial \upsilon}{\partial r}\right). \tag{8.51}$$

The convective term represented by the second term in parenthesis will be retained even when the deformation is small, as it is not necessarily negligible under conditions of high-speed cavity formation.

In the case of purely elastic deformation of the entire medium, the stress difference $\sigma_\theta - \sigma_r$ in (8.51) must be expressed as a function of r using the stress–strain relations before the integration can be carried out. Since the displacement is small in the elastic range, we may write $\upsilon \approx \partial u/\partial t$, where u is the radial displacement, and (8.49) then integrates to

$$u = \left(a^3 - a_0^3\right)/3r^2 \approx \left(a^2/r^2\right)u_a \tag{8.52}$$

in view of the initial condition $u = 0$ when $a = a_0$, the quantity u_a being the displacement a–a_0 of the cavity surface. The strain components corresponding to (8.52) are

$$\varepsilon_r = -\left(\frac{2a^2}{r^3}\right)u_a, \quad \varepsilon_\theta = \varepsilon_\phi = \left(\frac{a^2}{r^3}\right)u_a.$$

These results may be obtained by the direct integration of (8.50) using the same order of approximation. By the elastic stress–strain relations for an incompressible material ($\nu = 0.5$), we have

$$\sigma_\theta - \sigma_r = \frac{2}{3}E\left(\varepsilon_\theta - \varepsilon_r\right) = 2E\left(\frac{a^2}{r^3}\right)u_a, \tag{8.53}$$

where E is Young's modulus of the material. Substituting this into (8.51), and using (8.49), the equation of motion is reduced to

$$\frac{\partial \sigma_r}{\partial r} = 4E\left(\frac{a^2}{r^4}\right)u_a + \rho\left\{\left(\frac{a^2}{r^2}\right)\ddot{a} + \frac{2a}{r^2}\left(1 - \frac{a^3}{r^3}\right)\dot{a}^2\right\}.$$

Integrating, and using the condition that the stress vanishes at infinity, the solution for the radial stress is obtained in the form

$$\sigma_r = -\frac{4}{3}E\left(\frac{a^2}{r^3}\right)u_a - \frac{\rho a}{r}\left\{a\ddot{a} + \left(2 - \frac{a^3}{2r^3}\right)\dot{a}^2\right\}, \tag{8.54}$$

which involves the velocity $\dot{a}$ and acceleration $\ddot{a}$ of the cavity surface. The applied pressure p at the cavity surface is given by the boundary condition $\sigma_r = -p$ at $r = a$, the result being

$$p = \frac{4}{3}E\left(1 - \frac{a_0}{a}\right) + \rho\left(a\ddot{a} + \frac{3}{2}\dot{a}^2\right). \tag{8.55}$$

If the initial pressure is denoted by p_0, the initial acceleration is equal to $p_0/\rho a_0$, the initial velocity of the cavity being assumed to vanish. Multiplying both sides of (8.55) by $a^2\,da$, and using the identity $\left(2a\ddot{a} + 3\dot{a}^2\right)a^2 da = d\left(a^3\dot{a}^2\right)$, the resulting expressions can be integrated between the limits a_0 and a to obtain

$$\int_{a_o}^{a} pa^2 da = \left\{\frac{2}{3}E\left(1 - \frac{a_0}{a}\right)^2 + \frac{1}{2}\rho\dot{a}^2\right\}a^3. \tag{8.56}$$

with a minor approximation. The evaluation of the integral is straightforward when $p(a)$ is prescribed. A further quadrature of (8.56) after multiplying it by da furnishes the relationship between $\dot{a}$ and a, permitting the cavity radius $a(t)$ to be determined by integration.

Plastic yielding begins when the yield condition $\sigma_\theta - \sigma_r = Y$ is first satisfied during the elastic expansion. Since $\sigma_\theta - \sigma_r$ has its greatest value at $r = a$ in view of (8.53), yielding first occurs at the cavity surface when $a = a_1$ and $\dot{a} = \dot{a}_1$, such that

$$a_1 \approx a_o \left(1 + \frac{Y}{2E}\right), \qquad \rho \dot{a}_1^2 \approx \frac{Y}{E}\left(p_o - \frac{Y}{3}\right).$$

The last result follows from (8.56) with the approximation $p \approx p_0$ during the elastic loading. It may be noted that the radial displacement at the cavity surface at the onset of yielding is quite independent of inertial effects. Although the displacement at this stage is quite small, the corresponding expansion velocity may be high.

8.4.2 Large Elastic/Plastic Expansion

Subsequent to the commencement of yielding, an elastic/plastic boundary defined by $r = c$ spreads outward from the cavity surface. The material outside this radius is elastic, while that within this radius is rendered plastic with the stresses satisfying the yield criterion

$$\sigma_\theta - \sigma_r = Y, \qquad a \leq r \leq c.$$

The material is assumed to be nonhardening, so that the dynamic yield stress Y has a constant value throughout the plastic region. The velocity is still given by (8.49) everywhere in the medium, but the displacement in the elastic region is now given by

$$\frac{\partial u}{\partial t} = \left(\frac{c^2}{r^2}\right) v_c \qquad \text{or} \qquad u = \left(\frac{c^2}{r^2}\right) u_c, \qquad r \geq c,$$

where u_c is the displacement of the particles that are currently at the elastic/plastic boundary. The stress difference in the elastic region is

$$\sigma_\theta - \sigma_r = 2E\left(\frac{c^2}{r^3}\right) u_c = Y\left(\frac{c^3}{r^3}\right), \qquad r \geq c, \tag{8.57}$$

where the last expression follows from the fact that material at the elastic/plastic boundary is just at the point of yielding. Denoting by c_0 the initial radius to the particle which is currently at radius c, the condition of incompressibility may be written as

$$a^3 - a_0^3 = c^3 - c_0^3 \approx 3c^2 u_c = (3Y/2E)\, c^3. \tag{8.58}$$

This equation relates the current cavity radius to the radius of the elastic/plastic boundary. The strains in the plastic region are given by

$$\varepsilon_r = -\ln\left(\frac{\partial r_0}{\partial r}\right), \qquad \varepsilon_\theta = \varepsilon_\phi = \ln\left(\frac{r}{r_0}\right),$$

where r_0 is the initial radius to a typical particle, the relationship between r and r_0 being obtained by integrating the equation $dr/dt = v$ along the path of the particle.

The stress distribution in the elastic and plastic regions must be determined by integrating the equation of motion (8.51). Using (8.49) to eliminate v, this equation is reduced to

$$\frac{\partial \sigma_r}{\partial r} + \frac{2}{r}(\sigma_r - \sigma_\theta) = \rho\left\{\left(\frac{a^2}{r^2}\right)\ddot{a} + \frac{2a}{r}\left(1 - \frac{a^3}{r^3}\right)\dot{a}^2\right\}, \qquad (8.59)$$

where $\sigma_\theta - \sigma_r$ is given by (8.53) in the elastic region ($r \geq c$), and by (8.57) in the plastic region ($r \leq c$). The integration in the elastic region is based on the condition that σ_r is finite at $r = \infty$, while that in the plastic region involves the boundary condition $\sigma_r = -p$ at $r = a$. The result is easily shown to be

$$\left.\begin{aligned} \sigma_r &= -\frac{2}{3}Y\left(\frac{c^3}{r^3}\right) - \rho\frac{a}{r}\left\{a\ddot{a} + \left(2 - \frac{a^3}{2r^3}\right)\dot{a}^2\right\}, && r \geq c, \\ \sigma_r &= -p + 2Y\ln\left(\frac{r}{a}\right) + \rho\left\{\left(1 - \frac{a}{r}\right)a\ddot{a} - \frac{a}{r}\left(2 - \frac{a^3}{2r^3}\right)\dot{a}^2 + \frac{3}{2}\dot{a}^2\right\}, && r \leq c. \end{aligned}\right\} \qquad (8.60)$$

The fact that the stress in the plastic region contains a logarithmically divergent term indicates that the plastic region is always finite in extent. The condition of continuity of σ_r across $r = c$ furnishes

$$p = \frac{2}{3}Y + 2Y\ln\left(\frac{c}{a}\right) + \rho\left(a\ddot{a} + \frac{3}{2}\dot{a}^2\right)$$

in view of (8.60). Eliminating c/a from the above equation by means of (8.58), the applied pressure can be finally expressed as

$$p = \frac{2}{3}Y\left(1 + \ln\frac{2E}{3Y}\right) + \frac{2}{3}Y\ln\left(1 - \frac{a_0^3}{a^3}\right) + \rho\left(a\ddot{a} + \frac{3}{2}\dot{a}^2\right). \qquad (8.61)$$

The first two terms on the right-hand side of (8.61) correspond to the quasi-static result for an ideally plastic incompressible material. Multiplying (8.61) by $a^2\,da$, integrating between the limits a_1 and a, and using (8.56) corresponding to $a = a_1$, we get

$$\begin{aligned} \int_{a_0}^{a} pa^2\,da = \frac{2}{9}Y&\left\{\frac{3Y}{4E}a_0^3 + \left(a^3 - a_0^3\right)\left(1 + \ln\frac{2E}{3Y}\right)\right\} \\ &+ a^3\ln\left(1 - \frac{a_0^3}{a^3}\right) - a_0^3\ln\left(\frac{a^3}{a_0^3} - 1\right)\Bigg\} + \frac{1}{2}\rho a^3\dot{a}^2. \end{aligned} \qquad (8.62)$$

This equation expresses the energy balance from the commencement of cavity expansion. The process comes to a stop when $a = a_2$ and $\dot{a} = 0$, the maximum cavity radius a_2 being given by

$$\int_{a_0}^{a} pa^2 da = \frac{2}{9} Y \left\{ \frac{3Y}{4E} a_0^3 + \left(a^3 - a_0^3\right)\left(1 + \ln\frac{2E}{3Y}\right) + a_2^3 \ln\left(1 - \frac{a_0^3}{a^3}\right) - a_0^3 \ln\left(\frac{a_2^3}{a_0^3} - 1\right)\right\}$$

(8.63)

Since $\dot{\varepsilon}_r^p = -2\dot{\varepsilon}_\theta^p$, and $\dot{\varepsilon}_\theta^e - \dot{\varepsilon}_r^e = 0$ in the plastic region, the rate of plastic work done per unit volume is equal to $(2Y/3)\,(\dot{\varepsilon}_\theta - \dot{\varepsilon}_r)$,which is nonnegative for $\dot{a} \geq 0$ in view of (8.50). The preceding analysis is therefore valid at all stages of the expansion. When the expansion is so large that a_0/a is small compared to unity, the ratio c^3/a^3 is approximately equal to $2E/3Y$, which is independent of the cavity radius. In this case, (8.63) reduces approximately to

$$U = \frac{8}{9}\pi Y a_2^3 \left(1 + \ln\frac{2E}{3Y}\right), \tag{8.64}$$

where U denotes the total internal energy of the explosion products utilized in the cavity formation. Equation (8.64) is a particularly simple result that furnishes the maximum cavity radius a_2 from known values of U, Y, and E, when the cavity expansion is sufficiently large. The problem of fragmentation of a spherical shell due to an internal explosion has been discussed by Al-Hassani and Johnson (1969).

It may be noted that (8.63) involves the total work done by the applied pressure and is quite independent of the rate of doing this work. Let the law of expansion of the volume of the explosion products be assumed isentropic, according to the relation

$$p = p_0\,(a_0/a)^{3\gamma},$$

where p_0 is the initial pressure and $\gamma > 1$ is the index whose value is assumed constant over the considered range of expansion. The substitution of the above expression into the left-hand side of (8.63) results in

$$\int_{a_0}^{a_2} pa^2 da = \frac{p_0 a_0^3}{3(\gamma - 1)} \left\{1 - \left(\frac{a_0}{a_2}\right)^{3(\gamma-1)}\right\}. \tag{8.65}$$

Using (8.63) and (8.65), the ratio a_2/a_0 can be calculated for any given values of p_0/Y, E/Y and γ. In actual practice, γ generally decreases during the expansion, and consequently a suitable mean value of γ should be used in (8.65). In the case of TNT, it is a good approximation to assume $\gamma \approx 3$ for moderate expansions (Johnson, 1972). The variation of a_2/a_0 with p_0/Y is shown graphically in Fig. 8.12 for $E/Y =$ 450 and different values of γ. It is important to note that the final radius of the cavity achieved under dynamic conditions is the same as that under quasi-static conditions

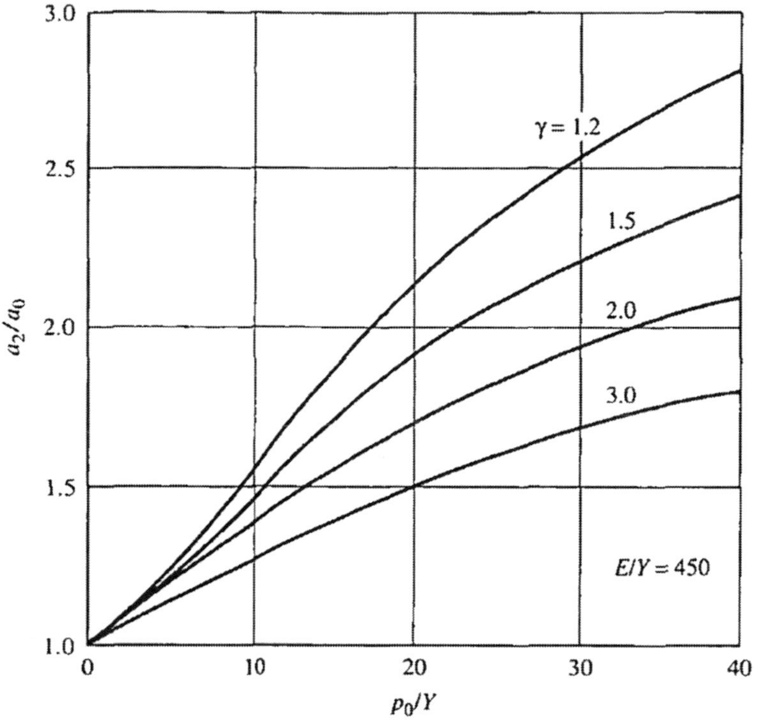

Fig. 8.12 Dependence of the maximum cavity radius on the intensity of initial pressure generated by an explosion

for the same expenditure of energy. An analysis for the cavity formation by deep underground explosion has been presented by Chadwick et al. (1964).

Consider now the nature of the return motion following the end of the first expansion phase. Due to the isochronous nature of the process, the entire plastic region instantaneously unloads, and consequently, the initial part of the first contraction phase is purely elastic. This part of the return motion is terminated when plastic yielding again occurs at the cavity surface $r = a$, the yield condition then being $\sigma_r - \sigma_\theta = Y$ in the absence of the Bauschinger effect. Since $\sigma_\theta - \sigma_r = Y$ in the plastic region at the instant of unloading, and $\sigma_\theta - \sigma_r = 2Eu_0$, at $r = a$ due to the superposed elastic stresses during the unloading, plastic flow at the cavity surface requires $u_a = -Y/E$, giving a cavity radius

$$a_3 = (1 - Y/E)\, a_2$$

at the end of the elastic unloading. In the subsequent part of the first contraction phase, a new elastic/plastic boundary spreads outward creating a plastic zone where the yield condition is $\sigma_r - \sigma_\theta = Y$. Since the return motion is very limited in extent, a detailed analysis of the contraction phase is of minor practical interest.

The internal energy of the explosion products is mostly dissipated into plastic work, and the available energy at the end of the first expansion phase is relatively small. The details of the analysis for the subsequent motion, consisting of alternate contraction and expansion phases, becomes progressively more complicated. Eventually, a final shakedown state of purely elastic oscillating motion results. The initial motion is so highly damped by the plastic deformation produced in the first expansion phase that the subsequent motion is reduced to one of small oscillations with minor significance. It should be noted that in the hypothetical case of cavity formation from zero radius, only the first expansion phase is geometrically similar, while the subsequent phases are not.

8.4.3 Influence of Elastic Compressibility

To simplify the analysis for the dynamic cavity expansion, when the compressibility of the elastic/plastic material is duly taken into account, we assume a constant speed U of the cavity surface which is expanded from zero radius. The plastic region at any instant is defined by $a \leq r \leq$ c, where b $= Ut$, and the elastic region is defined by $c \leq r \leq b$, where $b = Vt$, with V denoting the speed of the elastic dilatational wave in the material. In terms of the radial displacement u, the stress–strain relations in the elastic region may be written as

$$E\frac{\partial u}{\partial r} = \sigma_r - 2\nu\sigma_\theta, \qquad E\frac{u}{r} = (1-\nu)\,\sigma_\theta - \nu\sigma_r,$$

in view of the symmetry condition $\sigma_\theta = \sigma_\phi$, holding everywhere in the material. From the above relations, the elastic stress components are expressed as

$$\sigma_r = \frac{3K}{1+\nu}\left[(1-\nu)\frac{\partial u}{\partial r} + 2\nu\frac{u}{r}\right], \qquad \sigma_\theta = \frac{3K}{1+\nu}\left(\nu\frac{\partial u}{\partial r} + \frac{u}{r}\right), \tag{8.66}$$

where K denotes the elastic bulk modulus, equal to $E/[3\,(1-2\nu)]$. Substituting from (8.66) into the equation of motion (8.51), where υ is replaced by $\partial u/\partial t$, and neglecting the convective term $\upsilon\,(\partial u/\partial r)$, we obtain the wave equation

$$\frac{\partial^2 u}{\partial r^2} + \frac{2}{r}\frac{\partial u}{\partial r} - \frac{2u}{r^2} = \frac{1}{V^2}\frac{\partial^2 u}{\partial t^2},$$

due to Forrestal and Luk (1988). The quantity V denotes the speed of propagation of the elastic wave front of radius b and is given by

$$V^2 = \left(\frac{1-\nu}{1-2\nu}\right)\frac{2G}{\rho} = \left(\frac{1-\nu}{1+\nu}\right)\frac{3K}{\rho}$$

where ρ denotes the initial density of the material. Since the configuration maintains geometrical similarity during the expansion process, the dimensionless stresses σ_r/Y and σ_θ/Y, and the dimensionless displacement, u/c must be functions of the single variable r/c. Setting

$$\xi = \frac{r}{c}, \qquad \bar{u} = \frac{\dot{c}}{V} = c\sqrt{\left(\frac{1+\nu}{1-\nu}\right)\frac{\rho}{3K}},$$

and using the fact that the first derivatives of the displacement are

$$\frac{\partial u}{\partial r} = \frac{d\bar{u}}{d\xi}, \qquad \frac{\partial u}{\partial t} = \dot{c}\left(\bar{u} - \xi\frac{d\bar{u}}{d\xi}\right), \tag{8.67}$$

where $\dot{c}$ is a constant, the wave equation is transformed into the ordinary differential equation

$$\left(1-\alpha^2\xi^2\right)\frac{d^2\bar{u}}{d\xi^2} + \frac{2}{\xi}\frac{d\bar{u}}{d\xi} - \frac{2\bar{u}}{\xi^2} = 0. \tag{8.68}$$

The integration of (8.68) is facilitated by the substitution $\bar{u} = \xi\psi$, where ψ is a function of ξ, resulting in

$$\xi\left(1-\alpha^2\xi^2\right)\frac{d^2\psi}{d\xi^2} + 2\left(2-\alpha^2\xi^2\right)\frac{d\psi}{d\xi} = 0.$$

This is a first-order differential equation in $d\psi/d\xi$, which is easily integrated under the boundary condition $d\psi/d\xi = 0$ at the wave front $r = b$ or $\xi = 1/\alpha$, since $\bar{u}$ and $d\bar{u}/d\xi$ both vanish at this boundary. A second integration then furnishes the quantity ψ in view of the boundary condition $\psi = 0$ at $\xi = 1/\alpha$. Consequently,

$$\frac{d\psi}{d\xi} = -\frac{D\left(1-\alpha^2\xi^2\right)}{\xi^4}, \qquad \psi = \frac{D\left(1-\alpha\xi\right)^2\left(1+2\alpha\xi\right)}{3\xi^3},$$

where D is a constant of integration. These expressions yield the function $\bar{u}$ and its derivative as

$$\bar{u} = \frac{D\left(1-\alpha\xi\right)^2\left(1+2\alpha\xi\right)}{3\xi^2}, \qquad \frac{d\bar{u}}{d\xi} = -\frac{2D\left(1-\alpha^3\xi^3\right)}{3\xi^3}. \tag{8.69}$$

The constant D is now obtained from the condition that the material at the elastic/plastic boundary $r = c$ is at the point of yielding. Substituting from (8.69) into (8.66) with $u = c\bar{u}$, and setting $\sigma_\theta - \sigma_r = Y$ at $\xi = 1$, we get

$$D = (Y/E)(1+\nu)\,/\left(1-\alpha^2\right)$$

The stress distribution in the elastic region ($1 \leq \xi \leq 1/\alpha$) therefore becomes

$$\left.\begin{aligned}
\frac{\sigma_r}{Y} &= -\frac{2(1-\alpha\xi)\left[(1-2\nu)(1+\alpha\xi)+(1+\nu)\alpha^2\xi^2\right]}{3(1-2\nu)(1+\alpha^2)\xi^3}, \\
\frac{\sigma_\theta}{Y} &= \frac{(1-\alpha\xi)\left[(1-2\nu)(1+\alpha\xi)-2(1+\nu)\alpha^2\xi^2\right]}{3(1-2\nu)(1+\alpha^2)\xi^3},
\end{aligned}\right\} \tag{8.70}$$

It may be noted that both the stress components vanish at $\xi = 1/\alpha$ as expected. The velocity distribution in the elastic region ($1 \leq \xi \leq 1/\alpha$), by the second equation of (8.67), is given

$$\frac{\upsilon}{c} = -\xi^2 \frac{d\psi}{d\xi} = \frac{Y(1+\nu)\left(1-\alpha^2\xi^2\right)}{E\left(1-\alpha^2\right)\xi^2} \tag{8.71}$$

The elastic/plastic interface velocity $\dot{c}$, which is c/a times the cavity velocity U, must be determined by considering the plastic region. In an incompressible material, the elastic wave front does not exist ($\alpha = 0$), and the elastic region is then defined by $\xi \geq 1$. The effect of compressibility on the dynamic expansion was first considered by Hunter and Crozier (1968). The dynamic expansion of an elastic/plastic spherical shell under blast loading has been examined by Baker (1960).

Since the change in density is only of the elastic order, a constant value of ρ will be assumed in the equation of motion (8.51). The other equation necessary for the analysis in the plastic region is the compressibility equation

$$\frac{\partial \upsilon}{\partial c} + \frac{2\upsilon}{r} = \left(\frac{\partial}{\partial t} + \upsilon \frac{\partial}{\partial r}\right)\left(\frac{\sigma_r + 2\sigma_\theta}{3K}\right). \tag{8.72}$$

The left-hand side of this equation is equal to $-\dot{\rho}/\rho$, where the dot denotes the rate of change following the particle. In view of the yield criterion $\sigma_\theta - \sigma_r = Y$, the two governing equations in the plastic region become

$$\left.\begin{aligned}
\frac{\partial \sigma_r}{\partial r} - \frac{2Y}{r} &= \rho\left(\frac{\partial \upsilon}{\partial t} + \upsilon \frac{\partial \upsilon}{\partial r}\right), \\
\frac{\partial \upsilon}{\partial R} + \frac{2\upsilon}{r} &= \frac{1}{K}\left(\frac{\partial \sigma_r}{\partial t} + \upsilon \frac{\partial \sigma_r}{\partial r}\right).
\end{aligned}\right\} \quad a \leq r \leq c$$

Using the similarity transformation as before with the help of the dimensionless quantities

$$\xi = \frac{r}{c}, \quad s = \frac{\sigma_r}{Y}, \quad w = \frac{\upsilon}{c}, \quad \beta = c\sqrt{\frac{\rho}{K}},$$

and noting the fact that $c(\partial \upsilon/\partial t) = -c^2\xi\,(dw/d\xi)$, the preceding equations can be expressed as

$$\left.\begin{aligned} \frac{ds}{d\xi} - \frac{2}{\xi} &= \frac{K}{Y}\beta^2\left(\beta^2 - \xi\right)\frac{dw}{d\xi}, \\ \frac{dw}{d\xi} + \frac{2w}{\xi} &= \frac{Y}{K}(w - \xi)\frac{ds}{d\xi}, \end{aligned}\right\} \tag{8.73}$$

It has been shown by Forrestal and Luk (1988) that the velocity distribution can be estimated with insignificant error by omitting the convective terms on the right-hand side of (8.73). Introducing this approximation, and eliminating s between the two equations of (8.73), we obtain the differential equation

$$\left(1 - \beta^2\xi^2\right)\frac{dw}{d\xi} + \frac{2w}{\xi} = -\frac{2Y}{K}. \tag{8.74}$$

Integrating, and using the boundary condition $w = (1+\nu)\,Y/E$ at $\xi = 1$, in view of (8.71), the solution for the dimensionless velocity is obtained as

$$w = \frac{Y}{K\beta^2}\left\{-\frac{1}{\xi} + \frac{1-\beta^2\xi^2}{\xi^2}\left[\frac{1+\mu\beta^2}{1+\beta^2} + \frac{1}{2\beta}\ln\left(\frac{(1+\beta\xi)(1-\beta)}{(1-\beta\xi)(1+\beta)}\right)\right]\right\}, \tag{8.75}$$

where $\mu = (1+\nu)/[3(1-2\nu)]$. The elimination of $\xi\,(dw/d\xi)$ from the first equation of (8.73) with the help of (8.74) leads to the differential equation

$$\frac{ds}{d\xi} = \frac{K\beta^2}{Y}\left(w\frac{dw}{d\xi} + \frac{2w}{1-\beta^2\xi^2}\right) + \frac{2}{\xi\left(1-\beta^2\xi^2\right)}.$$

On substituting for w, in the second term of the parenthesis, and integrating under the boundary condition $s = s_0$ and $w = w_0$ at $\xi = 1$, the solution is found as

$$\begin{aligned} s = s_0 &+ \frac{K\beta^2}{2Y}\left(w^2 - w_0^2\right), \\ &- \left\{\frac{2\left(1+\mu\beta^2\right)(1-\xi)}{\left(1-\beta^2\right)\xi} + \ln\left(\frac{\left(1-\beta^2\xi^2\right)}{\left(1-\beta^2\right)\xi^2}\right) + \frac{1}{\beta\xi}\ln\left(\frac{(1+\beta\xi)(1-\beta)}{(1-\beta\xi)(1+\beta)}\right)\right\}. \end{aligned} \tag{8.76}$$

The quantities s_0 and w_0, obtained by setting $\xi = 1$ in (8.70) and (8.76), respectively, may be written as

$$s_0 = -\frac{2}{3}\left\{1 + \frac{(1+\nu)\alpha^2}{(1-2\nu)(1+\alpha)}\right\}, \qquad w_0 = \frac{Y(1+\nu)}{3K(1-2\nu)} = \frac{\mu Y}{K}.$$

Since the boundary condition $v = U$ at the cavity surface $r = a$ is equivalent to $w = a/c$ at $\xi = a/c$ in view of the relation $\dot{c} = (c/a)\,U$, the ratio c/a is obtained as the solution of the transcendental equation

$$\frac{c}{a}\left(\frac{a^2 + \mu\kappa^2c^2}{a^2 - \lambda^2c^2/a^2}\right) + \frac{1}{2\lambda}\ln\left\{\frac{(a-\lambda c)(1+\lambda)}{(a+\lambda c)(1-\lambda)}\right\} = \frac{1+\rho U^2/Y}{1-\lambda^2}. \tag{8.77}$$

where $\lambda = U\sqrt{\rho/K}$. This equation can be solved for $\lambda c/a$ with any given value of λ, ν and Y/K, and the ratio c/a then follows as the end result. The cavity pressure p is finally obtained by setting $s = -p/Y$ and $\xi = a/c$ in (8.76), the result being

$$\frac{p}{Y} = \frac{2}{3}\left(1 + \frac{(1+\nu)\,\mu\lambda^2 c^2}{(1-\nu)\,(1+\alpha)\,\alpha^2}\right) - \frac{1}{2}\left\{\frac{\rho U^2}{Y} - (1+\nu)\,\frac{\lambda^2 c^2}{3a^2}\right\}$$
$$+\,2\left(\frac{c}{a} - 1\right)\left(\frac{a^2 + \mu\lambda^2 c^2}{a^2 - \lambda^2 c^2}\right) + \ln\left\{\frac{(1-\lambda^2)\,c^2}{a^2 - \lambda^2 c^2}\right\} + \ln\left\{\frac{(a-\lambda c)\,(1+\lambda)}{(a+\lambda c)\,(1-\lambda)}\right\}. \tag{8.78}$$

Although this expression is more complicated than the corresponding formula for an incompressible material, the evaluation of the cavity pressure from (8.78) is straightforward, once c/a has been computed from (8.77) by a trial-and-error procedure. The dynamic expansion of a cylindrical cavity in a compressible elastic/plastic medium has been treated by Luk and Amos (1991).

The results for the cavity expansion from zero radius with a constant velocity U in an incompressible elastic/plastic material may be directly obtained by setting $a_0 = 0$, $\dot{a} = U$, and $\ddot{a} = 0$ in the more general solution given earlier, the cavity pressure in this case being expressible in the form

$$\frac{p}{Y} = \frac{2}{3}\left[1 + \ln\left(\frac{c^3}{a^3}\right)\right] + \frac{9\rho U^2}{4E}\left(\frac{c^3}{a^3}\right), \qquad \frac{c^3}{a^3} = \frac{2E}{3Y}. \tag{8.79}$$

In view of (8.61) and (8.58), the relationship between the interface velocity $\dot{c} = (c/a)\,U$ and the cavity velocity U for a compressible elastic/plastic material is compared with that for the incompressible material in Fig. 8.13(a), assuming $\nu = \frac{1}{3}$ and $Y/K = 0.00435$. The cavity pressures in the two cases are compared with one another in Fig. 8.13(b) for identical material properties. The cavity pressure obtained by replacing in (8.79) the ratio c^3/a^3 by its quasi-static value $E/[3\,(1-\nu)\,Y]$ for a compressible material, almost coincides with the exact numerical solution represented by the broken curve. A numerical solution for a compressible work-hardening material has been given by Luk et al. (1991) assuming a power law of hardening.

8.5 Mechanics of Projectile Penetration

The problem considered here is that of penetration into sufficiently thick targets caused by the motion of rigid projectiles striking at normal incidence with a velocity v_0. The depth of penetration that occurs before the projectile comes to rest depends on the mechanical properties of the target material as well as on the geometry of the projectile. The projectile produces a cylindrical tunnel in the target, about the size of the shank diameter, and a thin layer of material between the target and the nose of the projectile is melted during the penetration. Consequently, the resistance to penetration due to friction is usually small compared to that due the normal pressure

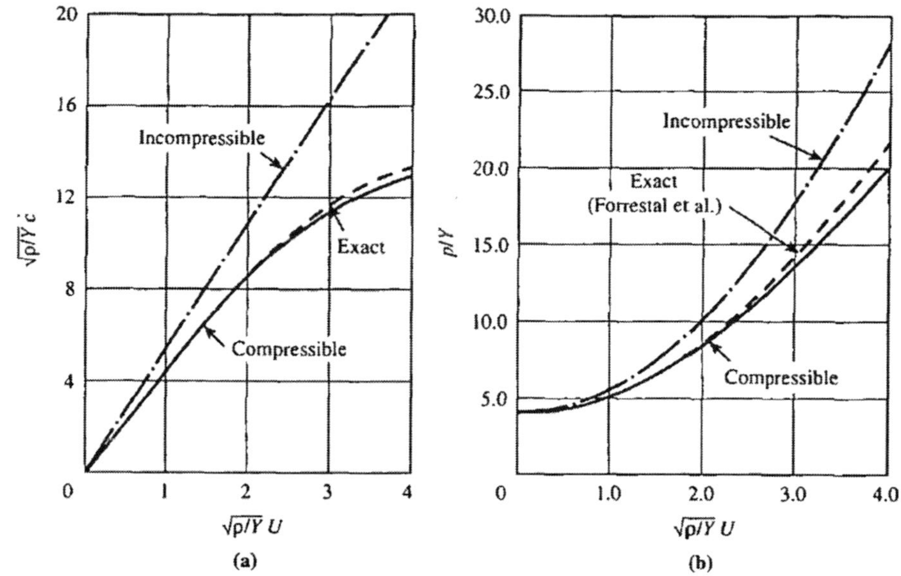

Fig. 8.13 Expansion of a spherical cavity in an elastic/plastic medium. (**a**) Interface velocity against cavity velocity and (**b**) cavity pressure against cavity velocity

exerted on the surface of the projectile. The problem has been treated by Backman and Goldsmith (1978) for an incompressible target material using the results for the expansion of spherical cavities in an elastic/plastic medium. The elastic compressibility of the material has been taken into account by Forrestal et al. (1988), who developed empirical formulas for the penetration mechanics based on spherical and cylindrical cavity expansion processes, and performed terminal-ballistic experiments to support their theory.

8.5.1 A Simple Theoretical Model

We begin by considering an ideally plastic target material, and an ogival shape of the projectile nose defined by a pair of circular arcs meeting the shank tangentially as shown in Fig. 8.14(b). The radius of each circular arc is denoted by na, where a is the radius of the shank and n is a constant factor. A generic point on the nose surface is subjected to a normal pressure p and a tangential traction μp, where μ is the coefficient of friction between the nose and the surrounding material. The resultant axial thrust acting on the ogival nose in the positive z-direction is given by the expression

$$P = 2\pi an \int_{\theta_0}^{\pi/2} (\cos\theta + \mu \sin\theta)\, prd\theta,$$

where θ denotes the angle of inclination of the normal to the surface with the axis of symmetry, and $\theta = \theta_0$ corresponds to the apex O. From simple geometry, the coordinates of a generic point on the nose are

$$r = a\,[n \sin\theta - (n-1)]\,, \qquad z = 1 - na\cos\theta.$$

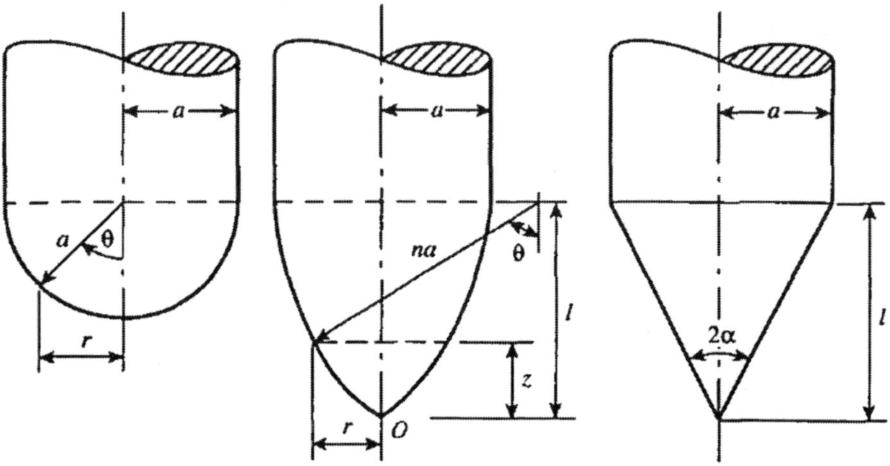

Fig. 8.14 Various shapes of projectile. (**a**) Spherical nose projectile, (**b**) ogival nose projectile, and (**c**) conical nose projectile

The first of these relations immediately gives $\sin\theta_0 = (n-1)/n$, and the preceding expression for the axial force becomes

$$P = 2\pi a^2 n^2 \int_{\theta_0}^{\pi/2} (\cos\theta + \sin\theta)\,(\sin\theta - \sin\theta_0)\,p d\theta, \tag{8.80}$$

where p is an unknown function of θ having its greatest value at the apex $\theta = \theta_0$. The height of the ogival nose is

$$l = na\cos\theta_0 = a\sqrt{2n-1}.$$

The hemispherical nose shown in Fig. 8.14(a) is the limiting case of an ogival nose, and corresponds to $n = 1$ giving $\theta_0 = 0$. The conical nose shown in Fig. 8.14(c) cannot be obtained by a limiting process, since the nose in this case meets the shank with a discontinuous slope.

To obtain an approximate solution to the penetration problem, it is assumed that the normal pressure p acting on the nose is equal to the internal pressure necessary to expand a spherical cavity from zero radius in the same medium with a radial velocity equal to the local particle velocity in the direction normal to the nose surface. The normal pressure distribution on the nose penetrating into a compressible elastic/plastic target is therefore obtained from (8.79) on setting $U = \upsilon \cos\theta$, where ν is the current projectile velocity, and replacing the quantity c^3/a^3 by its quasi-static value, the result being expressed in the dimensionless form

$$\frac{p}{Y} = \frac{2}{3}\left\{1 + \ln\left[\frac{E}{3(1-\upsilon)Y}\right]\right\} + \frac{3\rho\upsilon^2\cos^2\theta}{4(1-\nu)Y}, \tag{8.81}$$

where ρ is the density of the target material. Equation (8.81) implies that the pressure acting at the base $\theta = \pi/2$ is equal to the quasi-static expansion pressure. Substituting (8.81) into (8.80) and integrating, the penetration force is obtained as

$$P = \pi a^2 Y\left\{A + B\left(\frac{\rho\upsilon^2}{Y}\right)\right\}, \tag{8.82}$$

where A and B are dimensionless constants, given in terms of the mechanical properties of the target material, the geometry of the projectile nose and the friction coefficient as

$$\left.\begin{aligned} A &= \frac{2}{3}\left\{1 + \ln\left[\frac{E}{3(1-\nu)Y}\right]\right\}\left\{1 + \mu\left[n^2\left(\frac{\pi}{2} - \theta_0\right) - (n-1)\frac{l}{a}\right]\right\}, \\ B &= \frac{q}{1-\nu}\left\{\frac{4n-1}{8n^2} + \frac{\mu}{8}\left[\frac{3}{2}n^2\left(\frac{\pi}{2} - \theta_0\right) - (n-l)\frac{l}{a}\left(\frac{3}{2} + \frac{l^2}{n^2a^2}\right)\right]\right\} \end{aligned}\right\}. \tag{8.82a}$$

The parameters θ_0 and l/a in (8.82a) are defined by the value of n. When the frictional effect is disregarded ($\mu = 0$), the constant A reduces to the quasi-static value of p/Y for the spherical cavity expansion process.

The variation of the projectile velocity ν with time t and the depth of penetration h can be determined by considering the equation of rigid-body motion of the projectile. Since the rate of change of momentum is $m(d\upsilon/dt)$, where m is the mass of the projectile, the equation of motion is

$$P = -m\frac{d\upsilon}{dt} = -m\upsilon\frac{d\upsilon}{dh}.$$

Substituting P from (8.82), and integrating under the initial condition $\upsilon = \upsilon_0$ when $t = 0$ and $h = 0$, we obtain

$$\left.\begin{aligned} \left(\frac{\pi a^2 Y}{m}\right)t &= \sqrt{\frac{Y}{B\rho}}\left\{\tan^{-1}\left[\upsilon_0\sqrt{\frac{B\rho}{AY}}\right] - \tan^{-1}\left[\upsilon\sqrt{\frac{B\rho}{aY}}\right]\right\}, \\ \left(\frac{\pi\rho a^2}{m}\right)h &= \frac{1}{2B}\ln\left\{\frac{A + B\left(\rho\upsilon_0^2/Y\right)}{A + B\left(\rho\upsilon^2/Y\right)}\right\} = \frac{1}{2B}\ln\left(\frac{P_0}{P}\right), \end{aligned}\right\} \tag{8.83}$$

where P_0 denotes the initial value of the force. It follows that the force decreases exponentially with the increase in depth of penetration. The projectile comes to rest when $\upsilon = 0$, the corresponding depth of penetration h^* being given by

$$\left(\frac{\pi \rho a^2}{m}\right) h^* = \frac{1}{2B} \ln\left(1 + \frac{B\rho \upsilon_0^2}{AY}\right), \tag{8.84}$$

where A and B are obtained from (8.82a) for any given n. When the nose is hemispherical ($n = 1$), these expressions are simplified to

$$A = \frac{2}{3}\left(1 + \frac{\mu\pi}{2}\right)\left\{1 + \ln\left[\frac{E}{3(1-\nu)Y}\right]\right\}, \qquad B = \frac{3(1 + \mu\pi/4)}{8(1-\nu)}.$$

The total time of penetration t^* is obtained from the first relation of (8.83) by setting $\upsilon = 0$, and using the appropriate value of n to compute the constants A and B.

When the nose of the projectile is a circular cone of vertex angle 2α, the normal component of velocity at any instant has a constant value equal to $\upsilon \sin\alpha$. The normal pressure p, which is also a constant, is given by (8.81) with $\theta = \pi/2 - \alpha$, and the axial force acting on the conical surface is

$$P = \pi a^2 p (1 + \mu \cot\alpha)$$

for a coefficient of friction μ. The substitution of (8.81) with $\theta = \pi/2 - \alpha$ into the above relation results in (8.82), where the constants are now given by

$$\left.\begin{aligned} A &= \frac{2}{3}(1 + \mu\cot\alpha)\left\{1 + \ln\left[\frac{E}{3(1-\nu)Y}\right]\right\}, \\ B &= \frac{3}{4}(1 + \mu\cot\alpha)\ln\left[\frac{E}{3(1-\nu)Y}\right]\tan^2\alpha. \end{aligned}\right\}. \tag{8.85}$$

The variations of the projectile velocity and depth of penetration with time are defined by (8.83) and (8.85). The final penetration is still given by (8.84) as a function of the striking velocity v_0, provided A and B are determined from (8.85).

In Fig. 8.15, the theoretical predictions for the ogival nose ($n = 6$) and conical nose ($\alpha = 18.4°$) projectiles are compared with some available experimental results (Forrestal et al., 1988). The mechanical properties of the target material, which is prestrained aluminum, consist of $\nu = \frac{1}{3}, Y = 400\,\text{Mpa}, E = 68.9\,\text{Gpa}$, and $\rho = 2.71\left(10^3\right)\,\text{kg/m}^3$, the projectiles being made of managing steel with $a = 15.2\,mm, l = 3a$, and $m = 0.024\,\text{kg}$. The theoretical calculations are based on the assumption that the coefficient of friction μ would lie between 0.02 and 0.1 in the dynamic penetration. The agreement between theory and experiment is seen to be reasonably good, considering the uncertainty that exists in selecting the appropriate value of μ.

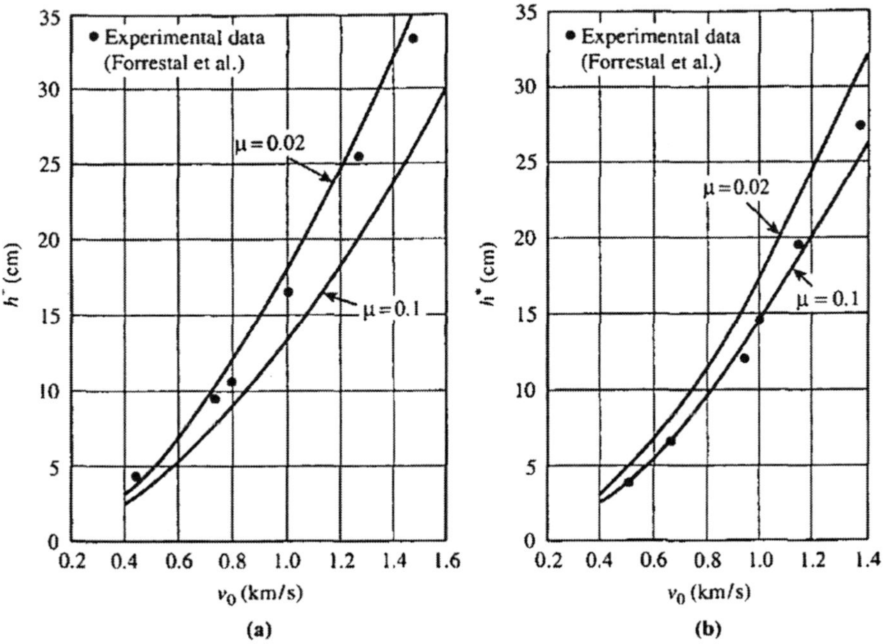

Fig. 8.15 Depth of penetration of rigid projectiles into thick metallic targets as a function of the striking velocity. (**a**) Ogival nose projectile ($n = 6$) and (**b**) conical nose projectile ($\alpha = 18.4°$)

The simplest way of taking into account the work-hardening property of the target material is to assume that the internal pressure in the dynamic cavity expansion is increased by the same amount as that for the quasi-static expansion in an incompressible elastic/plastic medium. If the stress–strain curve in the plastic range is approximated by a straight line of slope T, then (8.81), for the normal pressure on the projectile nose, is modified to

$$\frac{p}{Y} = \frac{2}{3}\left\{1 + \ln\left[\frac{E}{3(1-\nu)Y}\right] + \frac{\pi^2 T}{9Y}\right\} + \frac{3\rho v^2 \cos^2\theta}{4(1-\nu)Y} \qquad (8.86)$$

in view of a known result for the quasi-static expansion of spherical cavities (Chakrabarty, 2006). A similar expression can be written for a nonlinear hardening. Empirical formulas based on a numerical solution to the spherical cavity expansion process, using a power law of hardening, have been given by Forrestal et al. (1991), who obtained good agreement of their theoretical predictions with experimental results. The high-speed impact and penetration of long rods has been investigated by Tate (1969) and Rosenberg and Dekel (1994), while the associated ricochet problem has been examined by Johnson et al. (1982). Explicit formulas for the penetration dynamics of rigid projectiles into thick plates have been presented by Chen and Li (2003).

8.5.2 The Influence of Cavitation

At sufficiently high velocities of penetration, the target material sometimes flows away due to its own inertia to produce a hole which is larger in diameter than that of the projectile, Fig. 8.16. This phenomenon, known as cavitation, has the effect of enhancing the resistance to penetration, since a part of the kinetic energy is absorbed by the plastic deformation involved in the enlargement of the hole. The problem of cavitation in the penetration process has been discussed by Hill (1980), who assumed the normal pressure distribution over the projectile nose, defined by a convex function $r = r(z)$, in the form

$$p = q + k\rho v^2 \frac{d}{dz}\left(r\frac{dr}{dz}\right), \tag{8.87}$$

where q is the pressure corresponding to the quasi-static process, and k is a dimensionless constant that depends on the shape of the nose. Cavitation would occur over the region where the value of p given by (8.87) is found to be negative. The resistive force acting on the projectile in the absence of cavitation is

$$P = 2\pi \int_0^a prdr = 2\pi \int_0^l p\left(r\frac{dr}{dz}\right) dz = \pi q a^2.$$

provided $r\,dr/dz$ vanishes at both $z = 0$ and $z = l$. The resistance to penetration in this case is therefore independent of the dynamic factor. Since for a conventional nose, p decreases monotonically from $z = 0$ to $z = l$, cavitation would begin at the base of the nose when the velocity attains a critical value v_c. If the radius of curvature of the profile at this point is denoted by na, then $d^2r/dz^2 = -1/na$ at $z = 1$, where $p = 0$ at the incipient cavitation. When $d^2r/dz^2 = 0$ at $z = 1$, the critical velocity is

$$v_c = \sqrt{nq/kp}.$$

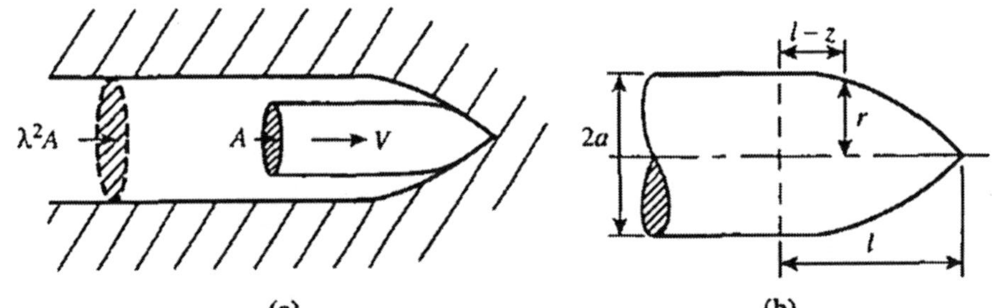

Fig. 8.16 Projectile penetration with cavitation. (**a**) Steady-state cavity formation and (**b**) geometry of conventional projectiles

For $\upsilon > \upsilon_c$ the target loses contact with the nose at a point which progressively moves toward the tip, and the resistive force becomes

$$P = \pi r_0^2 \left\{ q + k\rho\upsilon^2 \left(\frac{dr}{dz}\right)^2 \right\}_{r=r_0}, \qquad \upsilon \geq \upsilon_c, \tag{8.88a}$$

where $r = r_0$ corresponds to the point on the nose at which $p = 0$ according to (8.87). Consequently,

$$\left\{ q + k\rho\upsilon^2 \frac{d}{dz}\left(r\frac{dr}{dz}\right) \right\}_{r=r_0} = 0, \qquad \upsilon \geq \upsilon_c, \tag{8.88b}$$

Let the local radius of the hole due to cavitation be denoted by λa. Since the enlarged cavity may be imagined to have been produced by a projectile of radius λa without cavitation, we may write

$$p = \pi\lambda^2 a^2 q.$$

Inserting this expression in (8.88a), and using the relation $nq = kp\upsilon_c$, we get

$$n\left(\frac{dr}{dz}\right)^2 = \left(\frac{\lambda^2 a^2}{r_o^2} - 1\right)\left(\frac{\upsilon_c}{\upsilon}\right)^2, \qquad n\frac{d^2 r}{dz^2} = \frac{\lambda^2 a^2}{r_o^3}\left(\frac{\upsilon_c}{\upsilon}\right)^2, \tag{8.89}$$

at the point where the material breaks contact with the nose. The second equation of (8.89) follows from the first on combining it with (8.88b). Equations (8.89) are sufficient to determine r_0/a and λ in terms of the velocity ratio v/v_c for a given shape of the nose.

In the case of an ogival nose, the profile of the nose has a constant radius of curvature na, the equation of the circular profile being

$$[r + (n-1)\,a]^2 + (z - l)^2 = n^2 a^2,$$

where l is the height of the nose, equal to $a\sqrt{2n-1}$. In view of the above relation, equation (8.89) yields

$$\left.\begin{aligned} \left(\frac{\lambda\upsilon_c}{\upsilon}\right)^{2/3}\left(\frac{a}{r_0}\right) &= \left[n - \left(\frac{\lambda\upsilon_c}{\upsilon}\right)^{2/3}\right]\left(\frac{1}{n-1}\right) \\ \left(\frac{\lambda\upsilon_c}{\upsilon}\right)^{4/3}\left(\frac{a}{r_0}\right)^2 &= \left[n - \left(\frac{\lambda\upsilon_c}{\upsilon}\right)^2\right]\Big/\left(n - \left(\frac{\lambda\upsilon_c}{\upsilon}\right)^{2/3}\right) \end{aligned}\right\} \tag{8.90}$$

The elimination of a/r_0 between these two equations furnishes the relationship between λ and υ/υ_c as

$$\left(\frac{\lambda v_c}{v}\right)^{2/3} = n - (n-1)^{2/3}\left(n - \left(\frac{v_c^2}{v^2}\right)^{1/3}\right), \quad v \geq v_c \tag{8.91}$$

As v tends to infinity, the parameter $\lambda v_c/v$ tends to a limiting value that depends on n, the limiting position $r = r_0^*$ of the point where cavitation is initiated being given by

$$\frac{r_0^*}{na} = \left(\frac{n-1}{n}\right)^{1/3}\left\{1 - \left(\frac{n-1}{n}\right)^{2/3}\right\}$$

in view of (8.90). Consequently, the contact is always maintained over a finite part of the nose. The pressure distribution on the nose over the region of contact according to (8.87) is given by

$$\frac{p-q}{k\rho v^2} = \frac{(n-1)\,n^2a^3}{[r+(n-1)\,a]^3} - 1, \qquad 0 \leq r \leq r_0. \tag{8.92}$$

The pressure decreases monotonically from its greatest value at the tip, vanishing at $r = r_0 \leq a$, when $v \geq v_c$. The cavitation velocity v_c can be determined by an optimal fit of (8.92) with the experimentally measured value λ for a given material and any particular value of n defining the ogival nose.

When $n = 1$, the nose is hemispherical, and the projectile could be a spherical ball. Considering this as a limiting case of an ogival nose, the pressure distribution is uniform and can be written as

$$p = q - k\rho v^2, \quad 0 \leq r \leq a,$$

so long as there is no cavitation ($v < \underline{v_c}$). When v exceeds $v_c = \sqrt{q/k\rho}$ the pressure is zero everywhere except at the pole, where there is a concentrated force $P = \pi\lambda^2 a^2 q$ with $\lambda = v/v_c$. The nose therefore makes contact with the target only at the pole over the cavitation range ($v > v_c$).

The results for a conical nose cannot be obtained from those for the ogival nose by a limiting process, because of the sharp corner existing at the base $r = a$. The pressure distribution is, however, uniform according to (8.87) and is given by

$$p = q + k\rho v^2 \tan\alpha,$$

where α is the semiangle at the vertex. The cavitation in this case is only possible at the base, where there is a singularity in pressure and may occur at any velocity of the projectile. Since the resistive force is $pA = q\lambda^2 A$ where $A = \pi a^2$ is the area of the base, we have

$$\lambda^2 = \frac{p}{q} = 1 + \left(\frac{kp}{q}\right) v^2 \tan^2\alpha.$$

Assuming q/Y to be given by the first term on the right-hand side of (8.81), the constant k can be determined from the experimentally measured values of the cavity radius λa.

It remains to establish how the velocity υ of the projectile varies with the depth of penetration h, the striking velocity υ_0 being given. For a mass m of the projectile, the equation of motion is

$$m\upsilon \frac{d\upsilon}{dh} = -\lambda^2 qA.$$

For a conventional nose, $\lambda = 1$ when $\upsilon \leq \upsilon_c$ but is a function of υ when $\upsilon > \upsilon_c$. A direct integration of the above equation therefore results in

$$h^* - h = \frac{m}{qA}\int_0^{\upsilon} \upsilon d\upsilon = \frac{m\upsilon_c^2}{2qA}\left\{1 + \int_1^{\xi} \frac{d\xi}{\lambda^2}\right\}, \quad \upsilon \geq \upsilon_c, \tag{8.93}$$

where $\xi = (\upsilon_0/\upsilon_c)^2$ and h^* are the final penetration when the projectile comes to rest. Since $\upsilon = \upsilon_0$ when $h = 0$, we have

$$h^* = \frac{mn}{2\,k\rho A}\left\{1 + \int_1^{\beta} \frac{d\xi}{\lambda^2}\right\}, \quad \upsilon_0 \geq \upsilon_c,$$

where $\beta = (\upsilon_0/\upsilon_c)^2$. When the incidence velocity υ_0 is less than υ_c, the quantity in the curly brackets must be replaced by β. Indeed, in the absence of friction and cavitation, $qah^* = m\upsilon_0^2/2$ for a conventional nose. In the particular case of an ogival nose, the integral in (8.93) is evaluated by using (8.91) for $\lambda.(\xi)$, and the results are displayed graphically in Fig. 8.17. For a hemispherical nose, on the other hand, we have $\lambda^2 = \xi$, and the integral is then equal to In β. For a conical nose, an analysis similar to above gives

$$h^* = \frac{mn}{2\,k\rho A}\ln\left(1 + \frac{c\rho}{q}\upsilon_0^2\right),$$

where $c = k\tan^2\alpha$. The results for a composite nose, in which an ogival frustum is surmounted by a circular cone, can be similarly obtained. Some experimental evidence in support of the theoretical prediction involving cavitation has been reported by Hill (1980), who used copper targets and steel bullets with various head shapes.

In the case of hypervelocity impact, for which the speed of the projectile exceeds the elastic wave speed in the target material, the pressure generated on the cavity surface is large compared to the yield stress of the material. Since the rise in temperature caused by the impact is extremely high, the material can be treated as a fluid for the analysis of the penetration problem. When a meteor traveling in a highly eccentric orbit strikes the surface of a planet, there is complete volatilization of some material during the formation of the crater, and the situation is effectively similar to that encountered in a subsurface explosion (Johnson, 1972).

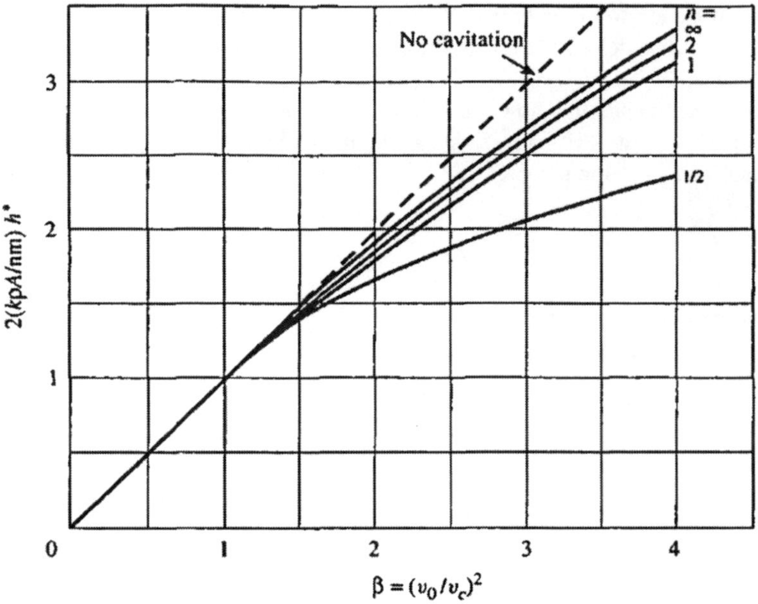

Fig. 8.17 Relationship between depth of penetration and squared striking velocity for projectiles with ogival heads

8.5.3 Perforation of a Thin Plate

A projectile having a conical nose or a sharp ogival nose strikes a target in the form of a thin plate of uniform thickness h_0 with a sufficiently high velocity v_0, the axis of the projectile being normal to the plane of the plate, Fig. 8.18(a). It is assumed that the projectile penetrates the plate without shattering it and produces a lip of height b as it leaves behind a circular hole of radius equal to the shank radius a. In the simplified model, each element of the lip is assumed to reach its final position through rotation under a uniaxial tension in the circumferential direction. If the initial radius to a typical element is denoted by s, and the final distance of the element from the outer edge of the lip is denoted by x, then the condition of plastic incompressibility requires

$$h_0 s\, ds = ha\, dx \quad \text{or} \quad dx/ds = h_0 s/ha,$$

where h is the local thickness of the lip. In view of the assumed uniaxial state of stress, we have the additional relation

$$\ln\left(\frac{h}{h_0}\right) = -\frac{1}{2}\ln\left(\frac{a}{s}\right) \quad \text{or} \quad \frac{h}{h_0} = \sqrt{\frac{s}{a}}.$$

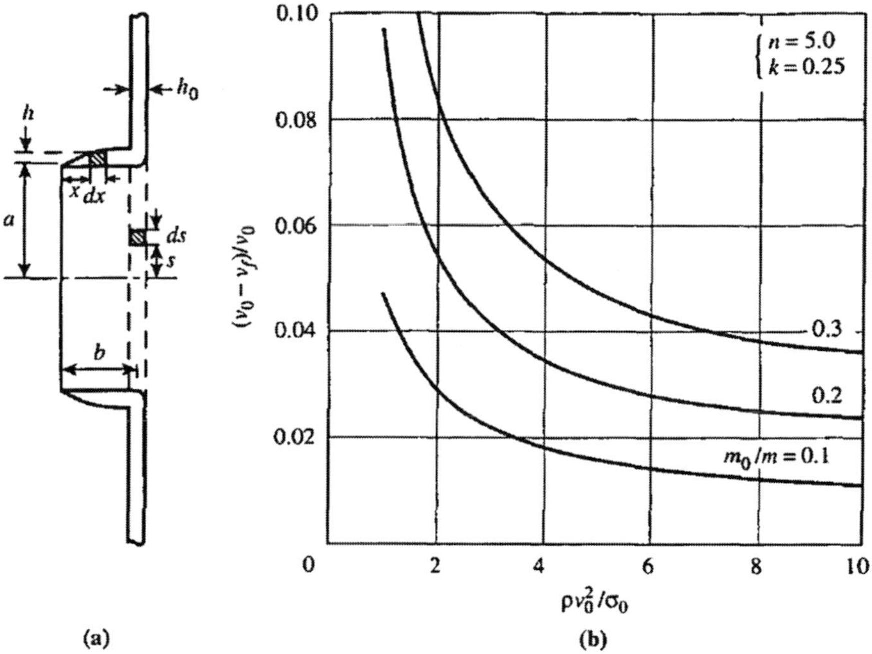

Fig. 8.18 Perforation of a flat plate by a projectile with pointed nose. (**a**) Geometry of lip formation and (**b**) fractional change in velocity against square of the striking velocity

Eliminating h/h_0 between the preceding relations, the resulting differential equation for x can be integrated to give

$$\frac{x}{a} = \frac{2}{3}\left(\frac{s}{a}\right)^{3/2}, \quad \frac{h}{h_0} = \left(\frac{3x}{2a}\right)^{1/3}, \tag{8.94}$$

in view of the boundary condition $x = 0$ when $s = 0$. The remaining boundary condition $x = b$ at $s = a$ indicates that the total height of the lip is $b = 2a/3$. The radial pressure exerted by the projectile when the lip is fully formed is given by

$$p = \frac{h}{a}Y = \frac{h_0}{a}Y\left(\frac{x}{b}\right)^{1/3},$$

when the material is nonhardening with a uniaxial yield stress equal to Y. The pressure varies, therefore, from zero at $x = 0$ to h_0Y/a at $x = b$. It may be noted that the strain at $x = 0$ is infinitely large.

Suppose that the material work-hardens according to the power law $\sigma = \sigma_0\varepsilon^k$, where σ_0 and k are constants. Then the plastic work done per unit volume in a typical element is $\sigma\varepsilon/(1 + k)$, where σ is the hoop stress corresponding to the final hoop strain $\varepsilon = \ln(a/s)$. Hence, the total plastic work expended during the formation of the lip is

$$W_1 = \frac{2\pi h_0\sigma_0}{1+k}\int_0^1 \varepsilon^{1+k} s ds = \frac{2\pi a^2 h_0\sigma_0}{1+k}\int_0^\infty e^{-2e}\varepsilon^{1+k} d\varepsilon.$$

The integral on the right-hand side is equal to Γ(2 + k)/2^{2+k}, where Γ (ξ) is the well-known gamma function of a positive variable |. The expression for the plastic work therefore becomes

$$W_1 = \pi a^2 h_0\sigma_0 \left\{\frac{\Gamma(1+k)}{2^{1+k}}\right\}. \tag{8.95a}$$

To obtain the work done by the inertia forces, let *r* denote the radius of the hole at any instant of time *t*. Since the mass of material displaced at time *t* is equal to $\pi\rho h_0 r^2$, where ρ is the density of the material, the work done by the distribution of the radial accelerating force *F* is

$$W_2 = \int_z^a F dr = \pi\rho h_0 \int_0^a \frac{d}{dt}\left(r^2\frac{dr}{dt}\right) dr.$$

The work expended in overcoming the frictional resistance will be disregarded, and W_2 will be evaluated on the basis of a constant projectile velocity equal to υ_0 (Thomson, 1955). Considering an ogival nose defined by

$$r = na(\sin\theta - \sin\theta_0), \quad \sin\theta_0 = (n-1)/n, \quad n > 1,$$

and setting dr/dt = $\upsilon_0 \cot\theta$ and $d\theta/dt = (\upsilon_0/na)$ cosec θ, we obtain the expression

$$W_2 = \pi\rho h_o \upsilon_0^2 \int_\theta^{\pi/2} \left(2nar\cot\theta\cos\theta - r^2\cos ec^2\theta\right)\cot\theta d\theta.$$

Inserting the expression for r(θ), and carrying out the integration, the result is found to be

$$W_2 = \pi\rho h_0 \upsilon_0^2 \left\{n^2 \ln\left(\frac{n}{n-1}\right) - \frac{1}{2}(1+2n)\right\}. \tag{8.95b}$$

Let the speed of the projectile decrease from υ_0 to υ_f during the perforation. Equating the loss of kinetic energy of the projectile to the total work done on the material, we have

$$m\left(\upsilon_0^2 - \upsilon_f^2\right) = 2(W_1 + W_2),$$

where m_0 is the mass of the projectile. The substitution of W_1 and W_2 from (8.95) into the preceding relation furnishes the square of the velocity ratio as

$$\left(\frac{\upsilon_f}{\upsilon_0}\right)^2 = 1 - \frac{m_0}{m}\left\{\frac{\Gamma(1+k)}{2^k\eta} + 2n^2\ln\left(\frac{n}{n-1}\right) - (1+2n)\right\}, \tag{8.96}$$

where $\eta = \rho v_0^2/\sigma_0$ and $m_0 = \pi \rho h_0 a^2$ is the mass of the displaced material. When the projectile has a conical nose, $dr/dt = v_0 a/l$, and an independent analysis leads to the formula

$$\left(\frac{v_f}{v_0}\right)^2 = 1 - \frac{m_0}{m}\left\{\frac{\Gamma\,(1+k)}{2^k \eta} + \frac{2a^2}{l^2}\right\}.$$

The residual velocity v_f for a standard ogival nose is somewhat higher than that for a conical nose having the same *a/l* ratio. Figure 8.18(b) shows how the ratio $(v_0 - v_f)/v_0$ varies with the parameter η in the case of an ogival nose projectile for different values of m_0/m. Due mainly to the neglect of plastic bending of the plate beyond the radius $r = a$, the theoretical prediction is found to underestimate the velocity change (Goldsmith and Finnegan, 1971) An experimental investigation on the perforation of target plates by the normal and oblique impact of projectiles has been reported by Piekutowski et al (1996).

In the case of perforation of a thin plate by the normal impact of a flat-ended cylinder, a plate plug is generally formed by shearing and is ejected from the target as the projectile passes through the plate. The diameter of the plug is approximately equal to that of the projectile, and the velocity of its ejection differs only marginally from the residual velocity of the projectile (Recht and Ipson, 1963). The analysis of the perforation problem involving truncated projectiles has been discussed by Zaid and Paul (1958). Useful experimental results for impact on finite plates have been reported by Calder and Goldsmith (1971). For very high velocities of impact, the effect of the rate of straining on the resistance to shear becomes significant, as has been shown by Chou (1961). The problem of ricochet of the deforming projectile after impact with plates has been investigated by Zukas and Gaskill (1996).

8.6 Impact Loading of Prismatic Beams

8.6.1 Cantilever Beam Struck at Its Tip

Consider a uniform cantilever of length l which is struck transversely at the end by an object of mass m_0 moving with velocity U, the mass being assumed to be attached to the beam during the plastic deformation that follows. The material is assumed to be rigid/plastic with constant uniaxial yield stress Y. The kinetic energy of the moving object is absorbed by a plastic hinge which is initially formed at the tip of the cantilever. As the inertia effect progressively decreases, the plastic hinge moves along the beam toward the built-in end, causing a permanent change in curvature over the distance traversed by the hinge. This problem has been investigated theoretically and experimentally by Parkes (1955) based on the rigid/plastic model and by Symonds and Fleming (1984) on an elastic/plastic model.

At any instant of time t measured from the moment of impact, let the plastic hinge H be situated at a distance ξl from the loaded end of the cantilever, Fig. 8.19(a). The bending moment at the hinge is equal to $-M_0$, where M_0 is the yield moment that

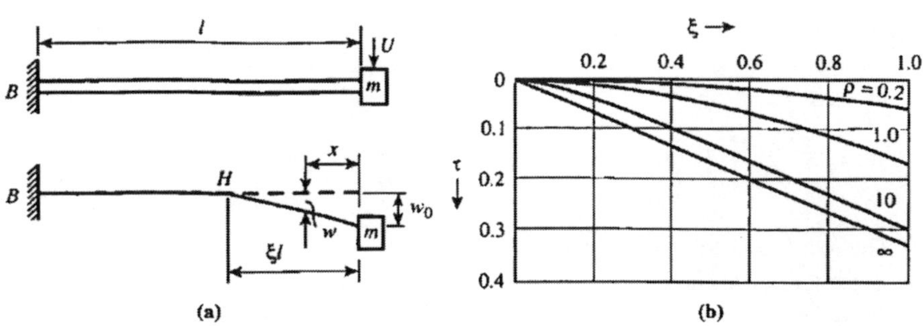

Fig. 8.19 Uniform cantilever struck at its tip. (**a**) Deformed configuration at any instant and (**b**) position of plastic hinge as a function of time

depends on the yield stress, the area of cross section, and the shape factor. Since the shearing force is zero at the hinge, where the bending moment is a relative maximum, we may analyze the portion of the beam between the hinge and the tip as a rigid body. Since the inertia force per unit length acting at a generic point of the beam is $(m/l)(\partial^2 w/\partial t^2)$ acting in the upward sense, where m is the total mass of the beam and w the downward deflection, the application of D'Alembert's principle for the dynamic equilibrium of forces and moments gives

$$\left.\begin{aligned} m_0\frac{d^2w_0}{dt^2} + \frac{m}{l}\int_0^{\xi t}\frac{d^2w}{dt^2}dx = 0,\\ M_0 + m_0\xi l\frac{d^2w_0}{dt^2} + m\int_0^{\xi t}\left(\xi - \frac{x}{l}\right)\frac{d^2w}{dt^2}dx = 0,\end{aligned}\right\} \tag{8.97}$$

where $w0(t)$ denotes the deflection at $x = 0$ and is assumed to be sufficiently small. Since the deformed part of the beam between the tip and the hinge at any instant rotates about $x = \xi l$ as a rigid body, the velocity of a generic point is given by

$$\frac{\partial w}{\partial t} = \left(1 - \frac{x}{\xi l}\right)\frac{dw_0}{dt}. \tag{8.98}$$

Substituting in (8.97), and introducing a constant deflection δ together with a set of dimensionless quantities ρ, z, and τ, which are defined as

$$\delta = \frac{m_0 l U^2}{2M_0}, \quad \rho = \frac{m}{2m_0}, \quad z = \frac{x}{l}, \quad \tau = \frac{tU}{2\delta}$$

we obtain a pair of ordinary differential equations for w_0 in a mathematically convenient form. The result may be expressed as

$$\left.\begin{aligned} \ddot{w}_0 + \rho\left(\xi\ddot{w}_0 + \dot{\xi}\dot{w}_0\right) &= 0, \\ 2\delta + \xi\ddot{w}_0 + \frac{\rho}{3}\left(2\xi^2\ddot{w}_0 + \xi\dot{\xi}\dot{w}_0\right) &= 0, \end{aligned}\right\} \tag{8.99}$$

where the dot denotes differentiation with respect to the dimensionless time τ. Since the expression in parenthesis of the first equation of (8.99) forms an exact differential, this equation is immediately integrated once to give

$$(1 + \rho\xi)\,\dot{w}_0 = 2\delta$$

in view of the initial conditions $\xi = 0$ and $dw_0/dt = U$ (or $\dot{w}_0 = 2\delta$) when $\tau = 0$. A suitable combination of the two equations in (8.99) gives

$$\rho\left(\xi^2\ddot{w}_0 + 2\xi\dot{\xi}\dot{w}_0\right) = 6\delta$$

The expression in parenthesis of the above equation is an exact differential, and integration gives

$$\rho\xi^2\dot{w}_0 = 6\delta\tau,$$

on using the initial condition $\xi = 0$ when $\tau = 0$. The relationship between the velocity of the tip of the cantilever and the time interval over the range $0 \leq \xi \leq 1$ is therefore given parametrically as

$$\begin{aligned} &\dot{w}_0 = 2\delta/\left(1 + \rho\xi\right), \\ &3\tau = \rho\xi^2/\left(1 + \rho\xi\right), \qquad 0 \leq \tau \leq \tau_0 = \rho/\left(3 + 3\rho\right), \end{aligned} \tag{8.100}$$

where τ_0 is the dimensionless time when the plastic hinge reaches the built-in end of the cantilever. Figure 8.19(b) indicates how the hinge position varies with time during the initial phase of bending.

For $\tau > \tau_0$, the plastic hinge remains fixed at the built-in end, and the hinge moment is no longer a relative maximum. The nonzero shear force that exists at the hinge can be determined from the equation of vertical motion. Considering the angular motion, we set $\xi = 1$ and $\dot{\xi} = 0$ in the second equation of (8.99) to have

$$\left(1 + \frac{2}{3}\rho\right)\ddot{w}_0 + 2\delta = 0.$$

Integrating, and using the condition that w_0 is continuous at $\tau = \tau_0$, we obtain the solution

$$\dot{w}_0 = \frac{6\delta\left(1 - \tau\right)}{3 + 2\rho}, \qquad \tau_0 \leq \tau \leq 1. \tag{8.101}$$

Since $\dot{w}_0 = 0$ when $\tau = 1$, the motion stops at this value of τ. Hence the duration of the motion is $t^* = 2\delta/U = m_0 lU/M_0$, which is independent of the mass of the beam.

The shape of the deformed beam at any instant can be determined by solving the appropriate differential equation for w. It follows from (8.98), (8.100), and (8.101) that

$$\left.\begin{aligned} \frac{\partial w}{\partial \tau} &= \frac{2\delta\,(\xi - z)}{\xi\,(1+\rho\xi)}, \quad 0 \le \tau \le \tau_0, \\ \frac{\partial w}{\partial \tau} &= \frac{6\delta\,(1-\tau)\,(1-z)}{3+2\rho\xi}, \quad \tau_0 \le \tau \le 1, \end{aligned}\right\} \tag{8.102}$$

the relationship between τ and ξ in the interval $0 \le \tau \le \tau 0$ being given by the second equation of (8.100). The differentiation of this equation with respect to ξ gives

$$\frac{d\tau}{d\xi} = \frac{\rho\xi\,(2+\rho\xi)}{3\,(1+\rho\xi)^2}, \quad 0 \le \tau \le \tau_0.$$

The above relation can be used to change the independent variable from τ to ξ in the first equation of (8.102), the result being

$$\frac{\partial w}{\partial \xi} = \frac{2\delta\rho\,(\xi - z)\,(2+\rho\xi)}{3\,(1+\rho\xi)^3}, \quad 0 \le z \le \xi.$$

Integrating, and using the boundary condition $w = 0$ at the plastic hinge $z = \xi$, we obtain the solution

$$\frac{w}{\delta} = \frac{2}{3\rho}\ln\left(\frac{1+\rho\xi}{1+\rho z}\right) - \frac{1}{3}\left(\frac{\xi - z}{1+\rho\xi}\right)\left[2 - \frac{\rho\,(\xi - z)}{(1+\rho\xi)\,(1+\rho z)}\right], \quad 0 \le \tau \le \tau_0, \tag{8.103}$$

which holds over the length $0 \le z \le \xi$. The remainder of the beam is undeformed, giving $w = 0$ for $\xi \le z \le 1$. When $\tau > \tau_0$, the expression for w is readily obtained by integrating the second equation of (8.102), and using the condition of continuity of w at $\tau = \tau_0$ $(\xi = 1)$. It is easily shown that

$$\begin{aligned} \frac{w}{\delta} &= 3\tau\,(2-\tau)\left(\frac{1-z}{3+2\rho}\right) + \frac{2}{3\rho}\ln\left(\frac{1+\rho}{1+\rho z}\right) \\ &\quad - \frac{1}{3}\left(\frac{1-z}{1+\rho}\right)\left[2 + \frac{\rho z}{(1+\rho z)} + \frac{3\rho}{(3+2\rho)}\right], \quad \tau_0 \le \tau \le 1, \end{aligned} \tag{8.104}$$

Equations (8.103) and (8.104) furnish the vertical displacement of the bent beam throughout the motion following the impact. The slope of the deflection curve at $z = 1$ is nonzero for $\tau_0 \le \tau \le 1$, implying a discontinuity which occurs in the same sense as that permitted by the plastic hinge.

The final shape of the deformed cantilever and the deflection of the tip of the beam as a function of time are of special practical interest. Setting $\tau = 1$ in (8.104), and denoting the limiting deflection by w^*, we obtain the relation

$$\frac{w^*}{\delta} = \frac{2}{3\rho} \ln\left(\frac{1+\rho}{1+\rho z}\right) + \frac{1-z}{3\,(1+\rho)\,(1+\rho z)} \tag{8.105}$$

giving the shape of the cantilever when it has come to rest. The deflection at the tip of the cantilever during its motion is obtained by setting $z = 0$ in (8.103) and (8.104), the result being

$$\left.\begin{aligned} \frac{w_0}{\delta} &= \frac{3}{2\rho} \ln\,(1+\rho\xi) - \frac{\xi\,(2+\rho\xi)}{3\,(1+\rho\xi)^2}, \quad 0 \le \tau \le \tau_0, \\ \frac{w_0}{\delta} &= \frac{3\tau\,(2-\tau)}{3+2\rho} + \frac{2}{3\rho} \ln\,(1+\rho) - \frac{2+7\rho/3}{(1+\rho)\,(3+2\rho)}, \quad \tau_0 \le \tau \le 1 \end{aligned}\right\} \tag{8.106}$$

where ξ is given by (8.100) as a function of $\tau < \tau_0$. If the mass of the beam is small compared to that of the striking object, so that ρ tends to zero, the plastic hinge moves almost instantaneously to the built-in end, and (8.104) reduces to

$$w/\delta = \tau\,(2-\tau)\,(1-z)\,, \quad 0 \le \tau \le 1.$$

It follows that δ is identical to the final deflection of the tip of the cantilever when the mass of the beam is negligible in comparison with the striking mass. The beam rotates in this case as a rigid body with the plastic hinge at the built-in end, the time taken by the striking object to come to rest being twice the time required by it to travel the same distance with a constant velocity U. The influence of the value of ρ on the final shape of the beam and the time dependence of the tip deflection are shown in Figs. 8.20 and 8.21, respectively. The problem of central impact of a simply supported beam can be similarly treated (Ezra, 1958). The transverse impact of a beam built-in at both ends has been considered by Parkes (1958) and Jones (1989). The influence of elastic deformation of the beam has been investigated by Symonds and Fleming (1984).

8.6.2 Rate Sensitivity and Simplified Model

When the ratio of the kinetic energy input to the greatest possible elastic energy in the beam is sufficiently large, the strain-rate dependence of the yield stress must be included in the analysis for a realistic prediction of the dynamic behavior. Since the strain-rate influence generally changes the mode of deformation, a simple correction factor for the yield stress could not be applied to the rate-independent theory without discrimination. To illustrate the procedure, we consider the cantilever beam of Fig. 8.19(a) and assume the mass m_0 to be attached to the tip instead of being dropped on the beam. Following Bodner and Symonds (1962), the relationship between the uniaxial stress σ and the plastic strain rate $\dot{\varepsilon}$ will be taken in the form of (8.13), specialized by setting $n = \lambda = 0$, and by replacing m with *1/n*. The power law,

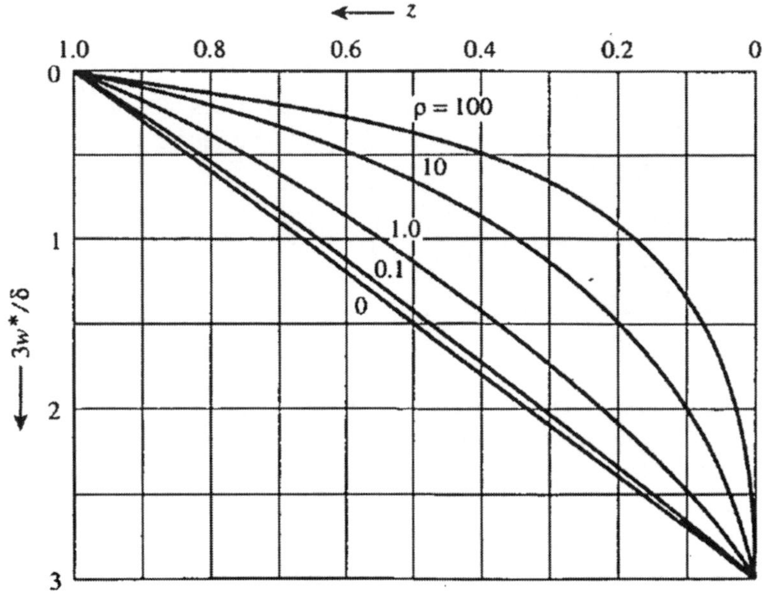

Fig. 8.20 Final shape of the cantilever beam due to impact loading with different ratios of the attached mass to the mass of the beam

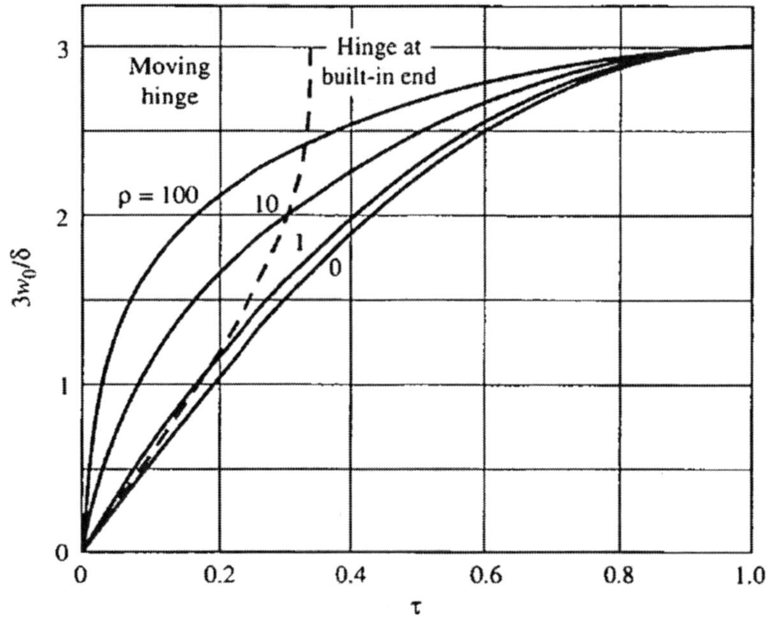

Fig. 8.21 Deflection of the cantilever tip as a function of time for various mass ratios

$$\frac{\sigma}{Y} = 1 + \left(\frac{\dot{\varepsilon}}{\alpha}\right)^{1/n}, \tag{8.107}$$

where α and n are empirical constants, has been found to fit with Manjoine's experimental data (1944) reasonably well by taking $a \approx 400/s$, $n \approx 5$ for mild steel, and $a \approx 6500/s$, $n \approx 4$ for aluminum. In the case of high strain rates with mild steel specimens, different values of these constants have been suggested by Hashmi (1980). In view of (8.107), and the fact that $\dot{\varepsilon}$ varies linearly with the vertical distance y measured from the axis of the beam, the bending moment for a beam of rectangular section of width b and depth h is given by

$$M = 2b\int_0^{h/2} \sigma\, dy = M_0 \left\{1 + \frac{2}{\dot{\varepsilon}_0^2}\int_0^{\varepsilon_0} \left(\frac{\dot{\varepsilon}}{\alpha}\right)^{1/n} \dot{\varepsilon}\, d\dot{\varepsilon}\right\},$$

where M_0 is the quasi-static value of the fully plastic moment, equal to $bh^2Y/4$, and $\dot{\varepsilon}_0$ is the maximum strain rate in the cross section, equal to $\dot{\kappa}h/2$, with $\dot{\kappa}$ denoting the curvature rate of the bent axis. The relationship between the bending moment M and the curvature rate $\dot{\kappa}$ may therefore be written as

$$\frac{M}{M_0} = 1 + \left(\frac{\dot{\kappa}}{\beta}\right)^{1/n}, \qquad \beta = \frac{2\alpha}{h}\left(1 + \frac{1}{2n}\right)^n. \tag{8.108}$$

The nature of the $(M, \dot{\kappa})$ curve is therefore identical to that of the $(\sigma, \dot{\varepsilon})$ curve and is completely defined by the empirical constants α and n for a given depth of the beam. Due to the effect of strain hardening of the material, the actual bending moment will depend not only on the curvature rate but also on the curvature of the beam, as has been shown experimentally by Apsden and Campbell (1966).

For the cantilever beam subjected to a tip impulse, as we have seen, a plastic hinge starts at the tip and moves toward the built-in end, when the material is ideally plastic. A rate-sensitive material, on the other hand, requires the plastic region to initially extend over the whole length of the beam without the formation of a localized plastic hinge. This is a consequence of the fact that the bending moment is a continuous function of the curvature rate, which vanishes at the tip of the cantilever and has its greatest value at the built-in end. The plastic zone continually shrinks during the motion and becomes zero when the beam comes to rest. The shape of the rate-sensitive cantilever at an intermediate stage of motion is compared with that of the ideally plastic cantilever during its first phase of motion in Fig. 8.22. For simplicity, the analysis will be carried out on the assumption that the outer portion of the cantilever is one of constant slope ψ that varies with time t as the motion continues.

It would be instructive at first to derive the rate-independent solution corresponding to the above simplified model (Mentel, 1958). Since the total momentum of the beam together with the attached mass has a constant value equal to the applied tip impulse I, while the resultant angular momentum about the built-in end is equal to the angular impulse $Il - M_0t$, we have

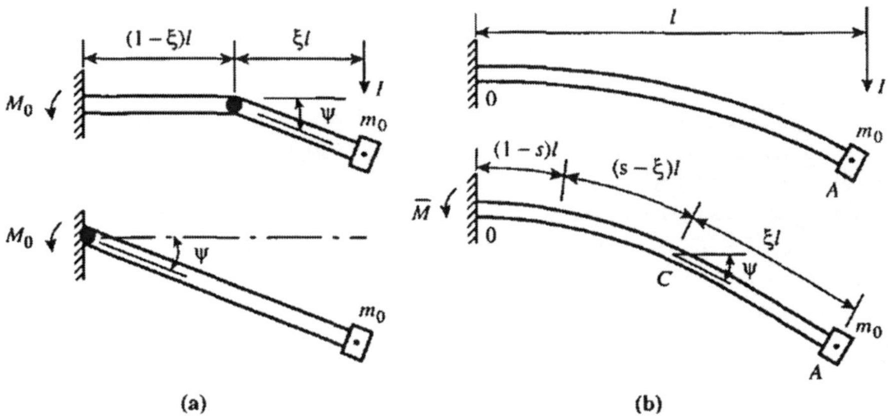

Fig. 8.22 Simplified deformation mode for a cantilever subjected to tip impulse. (**a**) Perfectly plastic material and (**b**) rate-sensitive material

$$\left.\begin{aligned} &\frac{1}{2}m\dot{\psi}\xi^2\, l + m_0\dot{\psi}\xi l = I, \\ &\frac{1}{2}m\dot{\psi}\xi^2\left(l - \frac{\xi}{3}\right)l^2 + m_0\dot{\psi}\xi l^2 = Il - M_0 t, \end{aligned}\right\} \tag{8.109}$$

where m is the mass of the beam, m_0 is the attached mass, and the dot denotes the time derivative. These equations are immediately solved for $\dot{\psi}$ and t as functions of ξ, the result being

$$\dot{\psi} = \frac{1}{m_0\xi\,(1+\rho\xi)\,l}, \qquad t = \frac{Il}{3M_0}\left(\frac{\rho\xi^2}{1+\rho\xi}\right),$$

where ρ denotes the ratio $m/2m_0$. The time taken by the hinge to reach the built-in end ($\xi = 1$) and the corresponding angle of rotation are found to be

$$t_0 = \left(\frac{\rho}{1+\rho}\right)\frac{Il}{3M_0}, \quad \psi_0 = \frac{\rho\,(4+3\rho)\,I^2}{6\,(1+\rho)^2\,M_0 m_0}. \tag{8.110}$$

It is interesting to note that the duration of the first phase is identical to that given by (8.100), if we set $I = m_0 U$. Since the second equation of (8.109) continues to hold (with $\xi = 1$) in the second phase, during which the whole beam rotates about the fixed end, the angular velocity is

$$\dot{\psi} = \frac{3\,(Il - M_{0t})}{(3+2\rho)\,m_0 l^2}, \quad t \geq t_0.$$

In view of the initial condition $\psi = \psi_0$ when $t = t_0$, this equation is readily integrated to give

$$\psi - \psi_0 = \frac{3\,(t - t_0)}{(3 + 2\rho)\,m_0 l}\left\{1 - \frac{M_0\,(t + t_0)}{2\,l}\right\}, \quad t \geq t_0 \tag{8.111}$$

The beam comes to rest when $\dot{\psi} = 0$, the total time of impact t^* and the final angle of rotation ψ^* being found as

$$t^* = \frac{Il}{M_0}, \quad \psi^* = \frac{I^2}{2M_0 m_0}.$$

Surprisingly, both t^* and ψ^* are independent of the mass ratio ρ according to this simplified analysis, the result for t^* being in agreement with that obtained in the previous solution. The plastic work done during the second phase is $M_0(\psi^* - \psi_0)>$ which is used up in absorbing a part of the kinetic energy input equal to $I_2/2m_0$. For sufficiently small values of ρ, most of the kinetic energy is absorbed in the second phase.

8.6.3 Solution for a Rate-Sensitive Cantilever

The inclusion of rate dependence of the yield moment completely changes the kinematics of the dynamic response of the beam, Fig. 8.22(b). The outer portion *CA* of constant slope ψ, at any instant, has unloaded from the plastic state, while the inner segment *OC* has its bending moment increasing from M_0 at *C* to a magnitude M_0' at 0. Let *M* denote the magnitude of the bending moment at a typical section in the plastic region at a distance *sl* from the free end. If the inertia effects in the plastic region are disregarded, the shearing force has a constant value *R*, and the bending moment distribution may be written as

$$M = M_0 + R\,(s - \xi)\,l = M_0' - R\,(1 - s)\,l.$$

Eliminating *R* and *M* between these relations in turn, and using (8.108) to express the ratios M/M_0 and M_0'/M_0 in terms of the corresponding curvature rates, we have

$$\dot{\kappa} = \dot{\kappa}_0 \left(\frac{s - \xi}{1 - \xi}\right)^n, \quad \frac{Rl}{M_0} = \frac{(\dot{\kappa}_0/\beta)^{1/n}}{1 - \xi}. \tag{8.112}$$

The curvature rate increases from zero at $s = \xi$ to attain its greatest value κ_0 at the fixed end $s = 1$. Since $\partial\theta/\partial s = -l\kappa$, where θ is the local slope of the plastic segment, the local angular velocity is

$$\dot{\theta} = l \int_s^1 \dot{\kappa}\,ds = l\dot{\kappa}_0 \left(\frac{1 - \xi}{1 + n}\right)\left\{1 - \left(\frac{s - \xi}{1 - \xi}\right)^{1+n}\right\} \tag{8.113}$$

in view of the boundary condition $\dot{\theta} = 0$ at $s = 1$. Setting $s = \xi$ in (8.113) furnishes the angular velocity $\dot{\psi}$ of the rigid outer segment as

$$\dot{\psi} = l\dot{\kappa}_0 \left(\frac{1-\xi}{1+n}\right).$$

Similarly, integrating the equation $\partial\dot{w}/\partial s = -l\dot{\theta}$, where $\dot{w}$ is the particle velocity, and using the boundary condition $\dot{w} = 0$ at $s = 1$, we obtain the velocity distribution in the plastic region as

$$\dot{w} = l^2\dot{\kappa}_0 \left(\frac{1-\xi}{1+n}\right)\left\{(1-s) - \left(\frac{1-\xi}{2+n}\right)\left[1 - \left(\frac{s-\xi}{1-\xi}\right)^{2+n}\right]\right\}. \qquad (8.114)$$

In particular, the velocity $\dot{w}_c$ at the rigid/plastic interface $s = \xi$ and the free-end velocity $\dot{w}_a$ which exceeds $\dot{w}_c$ by the amount $\xi\dot{\psi}$ are given by

$$\dot{w}_c = l^2\dot{\kappa}_0 \frac{(1-\xi)^2}{2+n}, \qquad \dot{w}_a = l^2\dot{\kappa}_0 \left(\frac{1-\xi}{2+n}\right)\left(1 + \frac{\xi}{1+n}\right). \qquad (8.115)$$

In all the preceding relations, ξ is a function of t to be determined. Once this is known, along with $\dot{\kappa}_0$, the physical quantities κ, θ, and w can be found by time integration, the initial conditions being $\kappa = \theta = w = 0$ at $t = 0$.

The two other equations necessary for the mathematical formulation of the problem are furnished by the principle of impulse and momentum. The equation of linear momentum in the vertical direction and the equation of angular momentum about the built-in end are easily shown to be

$$\left.\begin{aligned} I - \int_0^t R dt &= \left[m_0\dot{w}_a + \frac{1}{2}m\xi\,(\dot{w}_c + \dot{w}_a)\right]\cos\psi + G, \\ Il - \int_0^t R dt &= \frac{1}{2}ml\xi\left[(\dot{w}_c + \dot{w}_a) - \frac{\xi}{3}(\dot{w}_a + 2\dot{w}_c)\right] + m_0\dot{w}_a l + H \end{aligned}\right\}, \qquad (8.116)$$

where G and H represent the linear and angular momentum, respectively, of the plastic region OC and are given by

$$G \approx m\int_\xi^1 \dot{w}\,ds, \qquad H \approx ml\int_\xi^1 \dot{w}\,(1-s)\,ds.$$

It turns out that G and H make only minor contributions to the total momentum during the motion. For practical purposes, it is therefore a good approximation to take

$$G \approx \frac{ml^2\dot{\kappa}_0\,(1-\xi)^3}{3\,(2+n)}, \quad H \approx \frac{ml^3\dot{\kappa}_0\,(1-\xi)^4}{3\,(2+n)}. \qquad (8.117)$$

The first expression of (8.117) is obtained by assuming $\dot{w}$ to be given by the right-hand side of the first equation of (8.115) with j written for ξ, so that the fixed-end conditions $\dot{w} = \partial\dot{w}/\partial s = 0$ at $s = 1$ are identically satisfied. The expression for II is also obtained in the same way, but the numerical factor in the denominator is adjusted in such a way that both equations in (8.116) furnish identical initial values of $\dot{\kappa}_0$. Indeed, setting $t = \xi = 0$ in (8.116) results in

$$(\dot{\kappa}_0)_{t=0} = \frac{3\,(2+n)\,l}{(3+2)\,m_0 l^2} = cn\beta \quad (say)\,, \tag{8.118}$$

in view of (8.115) and (8.117). The integrals appearing on the left-hand side of (8.116) may be evaluated in an approximate manner, without introducing significant errors, by using the initial value of $\dot{\kappa}_0$ for expressing the integrands, which are then obtained from (8.112) as

$$\frac{Rl}{M_0} \approx \frac{c}{1-\xi}, \qquad \frac{M}{M_0} \approx 1 + c.$$

Introducing this approximation, and using (8.115) for the velocities, the momentum equations (8.116) are easily reduced to

$$\lambda - c\int_0^\tau \frac{d\tau}{1-\xi} = \phi\left(\frac{1-\xi}{2+n}\right)\left\{1 + \frac{2\rho}{3} + \left(\frac{2\rho}{3} + \frac{1}{1+n}\right)\xi - \frac{\rho}{3}\left(\frac{n-1}{n+1}\right)\xi^2\cos\psi\right\}$$

$$\lambda - (1+c)\,\tau = \phi\left(\frac{1-\xi}{2+n}\right)\left\{1 + \frac{2\rho}{3} + \frac{\xi}{1+n} + \frac{\rho}{3}\left(\frac{3-\xi}{1+n}\right)\xi^2\right\}, \tag{8.119}$$

the nondimensional quantities λ, τ, and ϕ introduced here being defined as

$$\lambda = \frac{I}{\sqrt{M_0 m}}, \qquad \tau = \frac{t}{l}\sqrt{\frac{M_0}{m}}, \qquad \phi = l^2\dot{\kappa}_0\sqrt{\frac{m}{M_0}}.$$

By eliminating the common factor $\phi(1 - \xi)/(2 + n)$ between the two equations (8.119), and solving the resulting equation numerically under the initial condition $\xi = 0$ when $\tau = 0$, we obtain τ as a function of ξ. The second equation of (8.119) then gives ϕ as a function of ξ, and the shape of the deformed cantilever is finally obtained from the distribution of θ determined by the integration of (8.113). The free-end deflection of the cantilever is given by

$$\frac{w_a}{l} = \int_0^t \left(\frac{\dot{w}_a}{l}\right) dt = \int_0^t \phi\left(\frac{1-\xi}{2+n}\right)\left(1 + \frac{\xi}{1+n}\right) d\tau, \tag{8.120}$$

which follows from (8.115). The slope $\dot{\psi}$ of the rigid portion is similarly computed as a function of time. The total time t^* required by the beam to come to rest corresponds to $\phi = 0$, or $\tau = \lambda/(1 + c)$ in view of the second equation of (8.119). Hence

$$t^* = \frac{Il}{M_0}\left\{1 + \frac{2n}{1+2n}\left[\frac{3\,(2+n)\,lh}{2\alpha\,(3+2\rho)\,m_0 l^2}\right]^{1/n}\right\}^{-1}. \tag{8.121}$$

Some of the computed results, obtained by Ting (1964) using $\lambda = 293$, $\rho = 0.305$, $n = 5.0$, $\alpha = 1036/\text{s}$, and $\eta = l\sqrt{m/M_0} = 0.039\text{s}$, are displayed graphically in Fig. 8.23, the duration of impact in this case being $t^* = 0.064$ s ($\psi^* = 1.033$ radians). The broken curves are based on the approximation $\cos\psi \approx 1$, which neglects the change in geometry. The theory has been found to be in good overall agreement with experiment by Bodner and Symonds (1962). The agreement is not so good, however, when the rate-independent theory is used.

The preceding theory indicates that the angular velocity $\dot{\psi}$ of the rigid segment decreases almost linearly with time. Using the linear relationship between $\dot{\psi}$ and t as an approximation (Ting, 1964), and assuming an initial value of $\dot{\psi}$ equal to $3I/(3 + 2\rho)m_0 l$, it is easily shown that

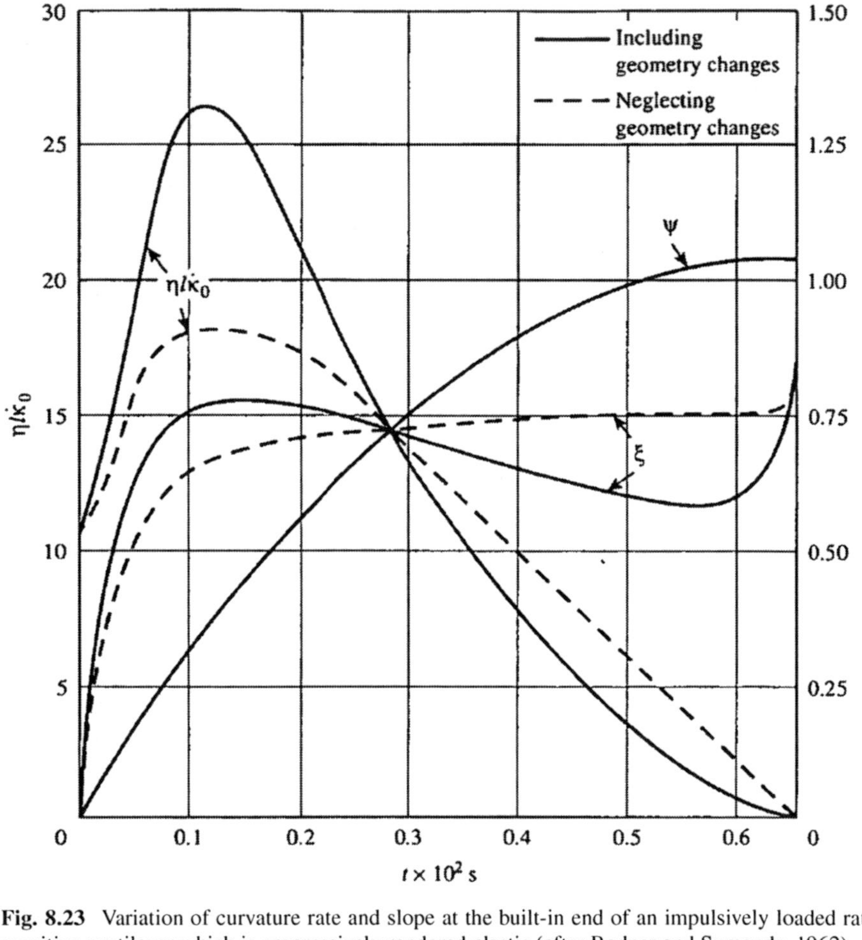

Fig. 8.23 Variation of curvature rate and slope at the built-in end of an impulsively loaded rate-sensitive cantilever which is progressively rendered plastic (after Bodner and Symonds, 1962)

$$\psi \approx \frac{3It}{2(3+2\rho)\,m_0 l}\left(2-\frac{t}{t^*}\right), \quad \psi^* = \frac{3I^2}{3(3+2\rho)(1+c)\,M_0 m_0}, \tag{8.122}$$

in view of the terminal condition $\dot{\psi} = 0$ at $t = t^*$ and the initial condition $\psi = 0$ at $t = 0$. The terminal slope ψ^* predicted by (8.122) is somewhat smaller than that given by (8.111). Further solutions on the dynamics of beams including strain-rate effects, using simplified models, have been discussed by Perrone (1965) and Lee and Martin (1970), among others. The influence of strain hardening of the material has been considered by Jones (1967) and Perrone (1970).

As a consequence of the elastic response of the material, an elastic flexural wave develops at the point of impact and propagates along the length of the cantilever before it is reflected from the fixed end. The reflected wave moves back toward the tip of the cantilever and meets the traveling plastic hinge midway along the beam. The interaction between the primary bending wave and the reflected wave significantly modifies the deformation mode from that predicted by the rigid/plastic theory. The subsequent deformation of the cantilever depends to a large extent on the ratio of the tip mass to the mass of the beam material. At the base of the cantilever, the bending moment oscillates in magnitude and sense, producing some reversed bending in this region. The final deflection of the tip of the cantilever is found to be about the same as that predicted by the rigid/plastic theory (Stronge and Yu, 1993).

8.6.4 Transverse Impact of a Free-Ended Beam

A uniform beam of mass $2m$ and length $2l$, initially at rest, is subjected to a concentrated impact load at its midpoint, such that the central section instantaneously attains a velocity U which is subsequently maintained constant. For simplicity, we propose to analyze the equivalent problem of a beam moving with a uniform normal velocity U, the central section of the beam being suddenly brought to rest by mean of a rigid stop, Fig. 8.24(a). Over a sufficiently small time interval after the beam strikes the stop, plastic hinges occur not only at the central section but also at two other sections each at a distance ξl from the center. The problem has been treated by Symonds and Leth (1954) for a beam of finite length, the corresponding problem for infinitely extended beams having been discussed by Lee and Symonds (1952), Conroy (1952, 1956), and Shapiro (1959).

Because of symmetry, it is only necessary to consider one-half of the beam with segments OH and HA subjected to end moments of magnitude M_0, which is assumed to have a constant value. The shear force vanishes at the moving hinge H, where the bending moment has a relative maximum, but a shear force of magnitude $R/2$ exists at O, where R denotes the reaction exerted by the stop at any instant t. The segment OH rotates about O with an angular velocity $d\phi/dt$ while HA rotates with an angular velocity $d\psi/dt$ and translates with zero acceleration of its mass center. OH acquires a permanent deformation from the traveling plastic hinge H, but the material to the right of this hinge at any instant is undeformed. Since the moments of inertia of

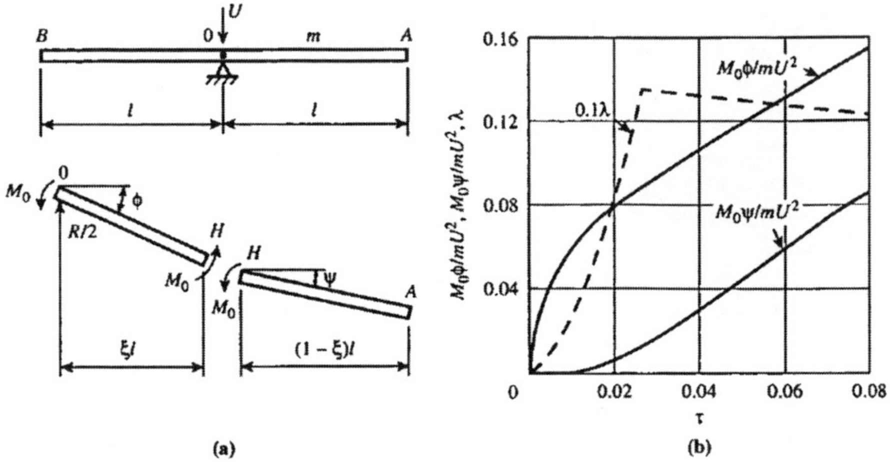

Fig. 8.24 Beam of infinite length subjected to transverse impact. (**a**) Deformation mode following impact and (**b**) graphical presentation of results

OH and HA about the respective mass centers are $m\xi^3 l^2/12$ and $m(1-\xi)^3 l^2/12$, respectively, the equations of angular rigid-body motion of these two segments may be written as

$$\frac{1}{3}m\xi^2 l^2 \frac{d^2\phi}{dt^2} = -2M_0, \qquad \frac{1}{12}m(1-\xi)^3 l^2 \frac{d^2\psi}{dt^2} = M_0. \tag{8.123}$$

Since the deformed segment OH is actually curved, ϕ must be interpreted as the angle made by the chord joining O and H. The velocity of the mid-section of the outer segment is always a constant, and its value must be equal to the impact velocity U. It follows from the kinematics of the motion that

$$\xi \frac{d\phi}{dt} + \frac{1}{2}(1-\xi)\frac{d\psi}{dt} = \frac{U}{l}. \tag{8.124}$$

It may be noted that the linear acceleration is discontinuous across the moving hinge H. Equations (8.123) and (8.124) form the basis for finding the three unknown quantities ϕ, ψ, and ξ. Introducing the dimensionless quantities

$$\omega = \frac{l}{U}\frac{d\phi}{dt}, \quad \lambda = \frac{l}{U}\frac{d\psi}{dt}, \quad \tau = \frac{M_0 t}{mlU},$$

and using a dot to denote differentiation with respect to τ, the preceding equations (8.123) and (8.124) can be expressed in the more convenient form

$$\left.\begin{aligned} \dot{\omega} = -6/\xi^3, \qquad \dot{\lambda} = 12/(1-\xi)^3, \\ \xi\omega + \frac{1}{2}(1-\xi)\lambda = 1 \end{aligned}\right\} \tag{8.125}$$

At the moment of impact, H coincides with O, and the angular velocity of HA vanishes. Hence the initial conditions are

$$\xi = 0, \qquad \lambda = 0, \qquad \phi = \psi = 0 \qquad \textit{when } \tau = 0.$$

The last equation of (8.125) therefore indicates that ω tends to infinity as ξ tends to zero, such that $\xi\omega = 1$ in the initial state, implying a singularity in the angular velocity at the point of impact.

To obtain the solution of (8.125), we begin by differentiating the last of these equations with respect to τ, and using the other two equations to eliminate $\dot{\omega}$ and $\dot{\lambda}$, the results being easily shown to be

$$\left(\omega - \frac{1}{2}\lambda\right)\dot{\xi} = \frac{6\,(1 - 2\xi)}{\xi^2\,(1 - \xi)^2}. \tag{8.126}$$

Differentiating this equation with respect to τ, and substituting again for $\dot{\omega}$ and $\dot{\lambda}$ using (8.125), we obtain the differential equation for ξ as

$$\xi\,(1 - \xi)\,(1 - 2\xi)\,\ddot{\xi} + \left(1 - 3\xi + 3\xi^2\right)\dot{\xi}^2 = 0.$$

Since the independent variable τ does not appear explicitly, a first integration can be easily carried out by setting $\dot{\xi} = \eta$, so that $\ddot{\xi} = \eta(d\eta/d\xi)$. The preceding equation then becomes

$$\frac{dn}{d\xi} + \frac{\left(1 - 3\xi + 3\xi^2\right)\eta}{\xi\,(1 - \xi)\,(1 - 2\xi)} = 0.$$

Since we are concerned only with the situation $\xi < \frac{1}{2}$, the integration of the above equation results in

$$\eta = \frac{d\xi}{d\tau} = \frac{\sqrt{1 - 2\xi}}{C\xi\,(1 - \xi)}, \tag{8.127}$$

where C is a constant to be determined later. Integrating again, and using the initial condition $\xi = 0$ when $\tau = 0$, we obtain the solution

$$\tau = \frac{C}{5}\left[1 - \left(1 + \xi - \xi^2\right)\sqrt{1 - 2\xi}\right]. \tag{8.128}$$

Substituting (8.127) for $\dot{\xi}$ into (8.126), and combining the resulting expression with the last equation of (8.125), the dimensionless angular velocities are found as

$$\omega = 1 + \frac{\sqrt{1 - 2\xi}}{\xi}, \qquad \lambda = 2\left[1 - \frac{\sqrt{1 - 2\xi}}{1 - \xi}\right], \tag{8.129}$$

on setting $C = \frac{1}{6}$ to satisfy the condition $\lambda = 0$ when $\tau = 0$. The angles ϕ and ψ at any instant during the motion of the outer hinge are

$$\left.\begin{aligned} \phi &= \frac{U}{l}\int_0^t \omega dt = \frac{mU^2}{M_0}\int_0^\tau \omega d\tau = \frac{mU^2}{M_0}\left[\tau + \frac{\xi\,(2-\xi)}{12}\right], \\ \psi &= \frac{U}{l}\int_0^t \lambda dt = \frac{mU^2}{M_0}\int_0^\tau \lambda d\tau = \frac{mU^2}{M_0}\left[2\tau - \frac{\xi^2}{16}\right], \end{aligned}\right\} \tag{8.130}$$

in view of (8.127) and (8.129), the dimensionless time τ being given by (8.128) with $C = \frac{1}{6}$. The reaction R at the stop is most conveniently obtained from the condition of moment equilibrium of the segment OH about its mass center. Thus

$$\frac{1}{3}m\xi^3\, l^2\frac{d^2\phi}{dt^2} = R\xi l - 8M_0.$$

The elimination of $d^2\phi/dt^2$ between this equation and the first equation of (8.123) immediately furnishes

$$R = 6M_0/\xi l.$$

For sufficiently small values of ξ, the beam behaves as being infinitely long. In this case, (8.127) and (8.128) give $\xi \approx 6/\xi$ and $\tau \approx \xi^2/12$, so that the parameter Rl/M_0 becomes equal to $\sqrt{3/\tau}$ approximately. An elastic/plastic analysis for an infinitely long beam with an arbitrary moment–curvature relation has been given by Duwez et al. (1950).

The preceding analysis remains valid until the angular velocities of the inner and outer segments of the beam become equal to one another. Setting $\omega = \lambda$, and using (8.129) and (8.128), the corresponding values of ξ and τ are found to be

$$\xi_0 = \left(\sqrt{5} - 2\right)^{1/2} \approx 0.486, \quad \tau_0 \approx 0.0265.$$

During the subsequent motion ($\tau > \tau_0$), there is a single plastic hinge occurring at the central section of the beam, and the two halves rotate as rigid bodies with an angular velocity $d\phi/dt$, the relevant equations in dimensionless form being

$$\dot{\omega} = -3, \qquad \dot{\phi} = \left(mU^2/M_0\right)\omega, \qquad \tau \geq \tau_0.$$

The first equation defines the angular motion, while the second equation follows from the definition of ω. The integration of the above equations gives

$$\omega = A - 3\tau, \qquad \phi = \frac{mU^2}{M_0}\left(B + A\tau - \frac{3}{2}\tau^2\right), \qquad \tau \geq \tau_0, \tag{8.131}$$

where A and B are constants. Since $\omega = \omega_0 \approx 1.3456$ and $\phi = \phi_0 \approx 0.0877mU^2/M_0$ when $\tau = \tau_0$, in view of (8.129) and (8.130), we have A $\approx$ 1.425 and B $\approx$ 0.051.

The motion stops when $\omega = 0$, and this corresponds to $\tau \approx$ s 0.495, giving $\phi \approx 0.389mU^2/M_0$ as the limiting angle of rotation. The reaction R at the support for $\tau > \tau_0$ discontinuously changes to a constant value equal to $3M_0/l$. Figure 8.24(b) shows the variation of ϕ, ψ, and R with time over the range $0 \leq \tau \leq 0.05$.

The neglect of geometry changes, which is implicit in the analysis, would be justified if the value of ϕ_0 is sufficiently small, preferably less than about 0.15 radians (say). The assumption of a rigid/plastic material, on the other hand, requires that the kinetic energy absorbed in the plastic deformation greatly exceeds the elastic energy stored in the beam. The range of validity of the solution may therefore be approximately defined as

$$\frac{kM_0 l}{EI} < \frac{mU^2}{M_0} < \frac{5}{3},$$

where EI is the flexural rigidity of the beam, and k is a numerical factor (presumably of the order of 10) that may be found from experiments. Outside this range, the elastic deformation of the beam must be considered for a realistic prediction of the dynamic behavior.

A variety of related problems on the dynamic plastic behavior of beams have been considered by Lee and Symonds (1952), Symonds (1953), and Seiler and Symonds (1954) using force pulses; by Symonds (1954), Johnson (1972), and Nonaka (1977) for blast loading; by Cotter and Symonds (1955), Martin and Symonds (1966), and Yu and Jones (1989) for impulsive loading. The dynamics of elastic/plastic beams has been treated by Bleich and Salvadori (1953), Seiler et al. (1956), Martin and Lee (1968), Symonds and Fleming (1984), and Reid and Gui (1987). The influence of axial restraints has been examined by Symonds and Mentel (1958), and that of shear has been considered by Karunes and Onat (1960).

The dynamic load characteristics of curved bars have been discussed by Owens and Symonds (1955), Perrone (1970), and Stronge et al. (1990). The dynamic plastic response of frames has been considered by Rawlings (1964), Symonds (1980), and Raphanel and Symonds (1984). The dynamic plastic behavior of a circular beam subjected to impact loading has been discussed by Yu et al. (1985), and that of a right-angled bent cantilever loaded at its tip has been examined by Reid et al. (1995). Lower and upper bound principles in the dynamic loading of structures have been discussed by several authors including Kalisky (1970) and Stronge and Yu (1993). A great deal of published work on the dynamic failure of structural members caused by severe plastic deformations has been reported by Wierzbicki and Jones (1989).

8.7 Dynamic Loading of Circular Plates

8.7.1 Formulation of the Problem

The theory of plastic bending of circular plates, developed in Chapter 4, can be extended to include inertia effects which arise under dynamic loading conditions. In accordance with the usual assumption for thin plates, shearing stresses normal to the plate surface are neglected in comparison with the bending stresses parallel to the surface. The resultant shearing force must be included, however, in the equation of dynamic equilibrium. For a rotationally symmetric state of stress, the nonzero stress resultants M_r, M_θ, and Q, denoting the radial and circumferential bending moments and the transverse shearing force, respectively, are functions of the radial coordinate r and time t. If the downward deflection of the plate is denoted by w, which is also a function of r and t, the inertia force per unit area acting in the upward sense is $-\mu(\partial^2 w/\partial t^2)$, where μ is the surface density of the plate. Then the equations of dynamic equilibrium may be written as

$$\frac{\partial}{\partial r}(rM_r) - M_\theta - rQ = 0, \qquad \frac{\partial}{\partial r}(rQ) + rp = \mu r \frac{\partial^2 w}{\partial t^2}, \tag{8.132}$$

where ρ denotes the local intensity of the normal pressure acting on the plate. Eliminating Q from the first equation of (8.132) by means of the integrated form of the second equation, we obtain the governing differential equation

$$\frac{\partial}{\partial r}(rM_r) - M_\theta = -\int_0^r \left(p - \mu\frac{\partial^2 w}{\partial t^2}\right) r\,dr, \tag{8.133}$$

which is independent of the mechanical properties of the plate material. For a plastically deforming plate made of an ideally plastic isotropic material, these properties are specified by the yield condition involving M_r, M_θ, and a fully plastic moment M_0, and the associated flow rule defining the ratio of the rates of change of the principal curvatures κ_r and κ_θ, which are given by

$$\kappa_r = -\frac{\partial^2 w}{\partial r^2}, \qquad \kappa_\theta = -\frac{1}{r}\frac{\partial w}{\partial r}. \tag{8.134}$$

The rate of change of all physical quantities is specified by the partial derivative with respect to t, the effect of geometry changes being disregarded. If the material yields according to the maximum shear stress criterion of Tresca, the yield locus in the moment plane is the hexagon $ABCDEF$ shown in Fig. 8.25(a). The generalized strain rate having components $\dot{\kappa}_r = \partial\dot{\kappa}_r/\partial t$ and $\dot{\kappa}_\theta = \partial\dot{\kappa}_\theta/\partial t$, corresponding to each side of the hexagon, is represented by a vector directed along the exterior normal to the side. If the stress state is represented by a vertex of the hexagon, the flow vector may have any direction between those defined by the two limiting normals.

In most physical problems, the plate may be divided into a central circular region together with surrounding annular regions, each of which corresponds to a different plastic regime. Across a circle Γ separating two plastic regimes, the radial moment M_r and the shear force Q must be continuous but the circumferential moment Mg may be discontinuous. The deflection w and the velocity $\partial w/\partial t$ must also be continuous across Γ, but all the second derivatives of w may become discontinuous as will be shown later. When the velocity slope $\partial^2 w/\partial r\,\partial t$ and hence the circumferential curvature rate $\dot{\kappa}_\theta$ are discontinuous across Γ, it is called a hinge circle, which may be either stationary or moving with respect to the plate. Since the discontinuity

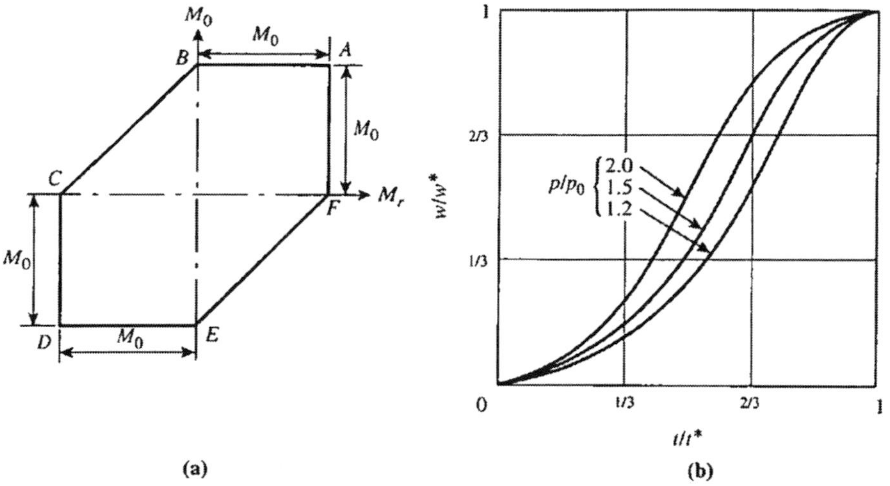

Fig. 8.25 Dynamic of pulse-loaded circular plates. (**a**) Yield condition and (**b**) growth of central deflection with time ($p \le 2\,p_0$)

must be considered as the limit of a narrow annulus of rapid change in slope, the ratio $\dot{\kappa}_r/\dot{\kappa}_\theta$ becomes infinite at the hinge circle. It follows that a hinge circle can be associated only with the side *CD* and *FA*, as well as the corners *A*, *C*, *D*, and *F* of the yield hexagon (Hopkins and Prager, 1954).

To establish the relations between the possible discontinuities across a hinge circle Γ, we represent this circle by the equation $r = \rho(t)$. Since w and $\partial w/\partial t$ are continuous across Γ, the derivative of these quantities along this curve is also continuous. Consequently,

$$\left[\frac{\partial w}{\partial t}\right] + \frac{dp}{dt}\left[\frac{\partial w}{\partial t}\right] = 0, \quad \left[\frac{\partial^2 w}{\partial t^2}\right] + \frac{dp}{dt}\left[\frac{\partial^2 w}{\partial r \partial t}\right] = 0, \tag{8.135}$$

where the square brackets denote the jump in the enclosed quantities. Since $\partial w/\partial t$ is continuous, the first equation of (8.135) indicates that the slope $\partial w/\partial r$ can be discontinuous only across a stationary hinge circle ($d\rho/dt = 0$). The second equation of (8.135), on the other hand, shows that the acceleration $\partial^2 w/\partial t^2$ must be continuous across a stationary hinge circle, but discontinuous across a moving hinge circle. Since $\partial w/\partial t$ is continuous across a moving hinge circle, its tangential derivative gives

$$\left[\frac{\partial^2 w}{\partial r \partial t}\right] + \frac{dp}{dt}\left[\frac{\partial^2 w}{\partial^2 r}\right] - 0, \qquad \frac{dp}{dt} \neq 0.$$

The radial curvature $\partial^2 w/\partial t^2$ is therefore discontinuous across a moving hinge circle. Due to the continuity of the radial bending moment, the space and time derivatives of M_r are both discontinuous across a moving hinge circle.

8.7.2 Simply Supported Plate Under Pressure Pulse

A circular plate which is simply supported round its edge $r = a$ is subjected to a uniformly distributed normal pressure of intensity ρ that is brought on suddenly at $t = 0$. The pressure is maintained constant during a small time interval $0 < t < t_0$, and then suddenly removed at $t = t_0$. The dynamic plastic action will occur only if ρ exceeds the value ρ_0 necessary for quasi-static plastic collapse (Section 4.1). We are therefore concerned here with

$$p > p_0 = 6M_0/a^2.$$

For $\rho = \rho_0$, the inertia forces do not arise because the deformation occurs indefinitely slowly. The dynamic behavior of the plate for $\rho > \rho_0$ is found to depend on whether or not ρ is greater or less than $2\rho_0$. Following Hopkins and Prager (1954), the former will be referred to as high load and the latter as medium load.

We begin by considering the medium load, characterized by $\rho_0 \leq \rho \leq 2\rho_0$ The first phase of the dynamic behavior of the plate corresponds to the time interval $0 \leq t \leq t_0$, during which the applied pressure is held constant throughout the plate. The equilibrium equation (8.133) therefore simplifies to

$$\frac{\partial}{\partial r}(rM_r) = M_\theta - M_0\left(\frac{3pr^2}{p_0a^2} - \lambda\right), \tag{8.136}$$

where

$$\lambda(r,t) = \frac{\mu}{M_0}\int_0^r\left(\frac{\partial^2 w}{\partial t^2}\right)rdr \tag{8.137}$$

Here, use has been made of the relation $p_0a^2 = 6M_0$. Equations (8.136) and (8.137) hold for all values of the ratio p/p_0. Under medium loads, the middle surface of the entire plate may be assumed to deform into a conical surface, as in the case of plastic collapse under the pressure $p = p_0$. We therefore take

$$w(r,t) = \frac{p_0}{\mu}\left(1 - \frac{r}{a}\right)f(t), \tag{8.138}$$

where $f(t)$ is a function of t to be determined. This expression for w satisfies the boundary condition $w = 0$ at $r = a$ for all $t \geq 0$. The rates of curvature associated with (8.138) are

$$\dot{\kappa}_r = -\frac{\partial^3 w}{\partial r^2 \partial t} = 0, \qquad \dot{\kappa}_\theta = -\frac{1}{r}\frac{\partial^2 w}{\partial r \partial t} = \frac{p_0}{\mu a r}f'(t).$$

For $f'(t) > 0$, these relations correspond to the flow rule associated with side AB of the yield hexagon, requiring $M_\theta = M_0$ and $0 \leq M_r \leq M_0$. It follows from (8.137) and (8.138) that

$$\lambda(r,t) = \frac{r^2}{a^2}\left(3 - \frac{2r}{a}\right) f''(t).$$

Substituting this into (8.133), and using the yield condition $M_\theta = M_0$, the resulting equation can be readily integrated. Since $M_r = M_0$ at $r = 0$, we get

$$\frac{M_r}{M_0} = 1 - \frac{pr^2}{p_0 a^2} + \frac{r^2}{a^2}\left(1 - \frac{r}{2a}\right) f''(t).$$

The boundary condition $M_r = 0$ at $r = a$ will be satisfied for all t if $f''(t)$ is a constant equal to $2(p/p_0 - 1)$. Since $w = \partial w/\partial t = 0$ at $t = 0$, this gives

$$f(t) = \left(\frac{p}{p_0} - 1\right) t^2. \tag{8.139}$$

The deflection of the plate as a function of r and t is completely defined by (8.138) and (8.139). The bending moment distribution is, however, independent of t, the radial moment being finally given by

$$\frac{M_r}{M_0} = 1 - \left[\frac{r^2}{a^2}\left(2 - \frac{r}{a}\right) - \frac{p}{p_0}\left(1 - \frac{r}{a}\right)\right]. \tag{8.140}$$

The bending moment M_r according to (8.140) monotonically decreases from M_0 at $r = 0$ when $p_0 < p < 2p_0$, and the yield condition is therefore nowhere violated. This completes the solution for the first plastic phase of the motion.

At $t = t_0$, the load is suddenly removed, but the motion continues until the kinetic energy acquired during the application of the load is dissipated by plastic work. The second plastic phase therefore corresponds to $p = 0$ and $t > t_0$. All conditions of the problem in the second phase can be satisfied by assuming w to be given by (8.138) together with $f''(t) = -2$. In view of the yield condition $M_\theta = M_0$, and the conditions of continuity of w and $\partial w/\partial t$ at $t = t_0$, we have

$$\left.\begin{aligned} f(t) &= \frac{pt_0}{p_0}(2t - t_0) - t^2, \\ \frac{M_r}{M_0} &= 1 - \frac{r^2}{a^2}\left(2 - \frac{r}{a}\right). \end{aligned}\right\} t \geq t_0 \tag{8.141}$$

Since M_r lies between 0 and M_0, the yield condition is nowhere violated, the only restriction on the solution being

$$f'(t) = 2\left(\frac{pt_0}{p_0} - t\right) \geq 0, \quad or \quad t \leq t^* = \frac{pt_0}{p_0}.$$

At the instant $t = t^*$, the entire plate comes to rest, since $f'(t) = 0$ implies that the velocity w is identically zero. The final deflection of the plate is found from (8.138) and (8.141) as

$$w^* = \frac{p}{\mu}\left(\frac{p}{p_0} - 1\right) t_0^2 \left(1 - \frac{r}{a}\right), \qquad (8.142)$$

which indicates that the deflection of the plate in the final stage at any radius r is p/p_0 times that at $t = t_0$. Figure 8.25(b) shows the manner in which the ratio w/w^* varies with t/t^* for given values of p/p_0 in the medium load range. The dynamic bending problem for circular and annular plates under a linearly distributed pressure pulse has been considered by Jones (1968a,b).

8.7.3 Dynamic Behavior Under High Loads

The bending moment distribution (8.140) is not admissible for $p > 2p_0$, since the predicted value of M_r then exceeds M_0 in the neighborhood of the plate center. Indeed, $\partial^2 M_r/\partial r^2$ is positiveat $r = 0$ according to (8.140) when $p > 2p_0$, indicating that M_r is a relative minimum at $r = 0$, when $M_r = M_0$. We may therefore expect $M_r = M_\theta = M_0$ in a central region of the plate within some radius p_0, which corresponds to the corner A of the yield locus. It follows from (8.136) and (8.137) that this region moves down with a constant acceleration equal to p/μ. Thus

$$\partial^2 w/\partial t^2 = p/\mu, \qquad 0 \le r \le \rho_0.$$

In the remainder of the plate, the plastic regime AB should apply, and a conical mode of deflection may be assumed as before. In view of the initial conditions $w = \partial w/\partial t = 0$ at $t = 0$, and the condition of continuity of w across $r = \rho_0$, the expression for the deflection during the initial phase of the motion may be written as

$$w\,(r,t) = \begin{cases} \dfrac{pt^2}{2\mu}, & 0 \le r \le \rho_0, \\ \dfrac{pt^2}{2\mu}\left(\frac{a-r}{a-\rho_0}\right), & \rho_0 \le r \le a, \end{cases} \qquad (8.143)$$

where ρ_0 is independent of t but depends on the applied pressure p. The boundary condition $w = 0$ at $r = a$ is identically satisfied, and the curvature rates associated with the velocity field (8.143) are

$$\dot{\kappa}_r = \dot{\kappa}_\theta = 0, \qquad 0 \le r \le \rho_0,$$
$$\dot{\kappa}_r = 0, \qquad \dot{\kappa}_\theta = \frac{pt}{\mu r\,(a - \rho_0)}, \qquad \rho_0 \le r \le a.$$

Since κ_θ is discontinuous across $r = \rho_0$, this radius coincides with a stationary hinge circle. The above relations are seen to be compatible with the assumed plastic regimes inside and outside the circle of radius ρ_0. It may be noted that the slope $\partial w/\partial r$ is discontinuous across the hinge circle.

To obtain the bending moment distribution in the region $\rho_0 \le r \le a$, we observe that the acceleration $\partial^2 w/\partial t^2$ is equal to p/μ inside the circle $r = \rho_0$, and equal to

$(p/\mu)(a-r)/(a-\rho_0)$ outside the circle $r=\rho_0$. The substitution in (8.137) therefore results in

$$\lambda\,(r,t)=\frac{p}{p_0}\left\{\frac{3r^2-(r-\rho_0)^2\,(2r+\rho_0)}{a^2\,(r-\rho_0)}\right\},\qquad \rho_0\le r\le a.$$

Equation (8.136) is now readily integrated to obtain an expression for M_r outside the circle $r=\rho_0$. Since $M_r=M_0$ at $r=\rho_0$ in view of the continuity of the bending moment, and $M_r=0$ at $r=a$ in view of the boundary condition, we get

$$\frac{p}{p_0}=\frac{2a^3}{(a-\rho_0)^2\,(a+\rho_0)},\qquad 0\le t\le t_0, \tag{8.144}$$

$$\frac{M_r}{M_0}=1-\frac{a\,(r-\rho_0)^3\,(r+\rho_0)}{r\,(a-\rho_0)^3\,(a+\rho_0)},\qquad \rho_0\le r\le a. \tag{8.145}$$

The last equation indicates that M_r decreases monotonically from M_0 at $r=p_0$ to 0 at $r=a$, and the yield condition is nowhere violated. The central deflection of the plate at $t=t_0$ is $\delta_0=pt_0^2/2\mu$. It may be noted that ρ_0/a tends to unity as p/p_0 tends to infinity.

A second plastic phase begins with the sudden removal of the load at $t=t_0$. During this phase, the radius of the hinge circle decreases from p_0 to some radius $p=p(t)$ at any instant, so that the kinetic energy of the plate decreases due to dissipation by plastic work. The velocity field in the second phase may therefore be taken in the form

$$\frac{\partial w}{\partial t}=\begin{cases}\dfrac{pt_0}{\mu}, & 0\le r\le\rho,\\[2mm] \dfrac{pt_0}{\mu}\left(\dfrac{a-r}{a-\rho}\right), & \rho\le r\le a.\end{cases} \tag{8.146}$$

Since $\rho(t_0)=\rho_0$, the velocity is automatically continuous at $t=t_0$. The acceleration $\partial^2w/\partial t^2$ vanishes inside the circle $r=\rho$ and assumes the value $(pt_0/\mu)\dot\rho$ $(a-r)/(a-p)^2$ outside this circle. It follows from (8.137) that

$$\lambda=\frac{p}{p_0}t_0\dot\rho\left\{\frac{3a\left(r^2-\rho^2\right)-2\left(r^3-\rho^3\right)}{a^2\,(a-\rho)^2}\right\},\qquad \rho\le r\le a.$$

Setting $M_\theta=M_0$ and $p=0$ in (8.136), substituting the above expression for λ, and using the boundary conditions $M_r=M_0$ and $\partial M_r/\partial r=0$ at $r=\rho$ and $M_r=0$ at $r=a$, the integration of the resulting equation furnishes

$$(a-\rho)\,(a+3\rho)\,\frac{d\rho}{dt}=-\frac{2p_0a^3}{pt_0},\qquad t_0\le t\le t_1, \tag{8.147}$$

where $t=t_1$ represents the instant when the hinge circle reaches the center of the plate. The function $\rho(t)$ is obtained by integrating (8.147) under the initial condition $\rho=\rho_0$ at $t=t_0$, the result being

$$\frac{M}{M_0} = 1 - \frac{a}{r}\left(\frac{r-\rho}{a-\rho}\right)^2 \left\{\frac{\rho\,(4a-3\rho)+2\,(a-\rho)\,r-r^2}{(a-\rho)\,(a+3\rho)}\right\}, \rho \leq r \leq a, \quad (8.148)$$

in view of (8.144). Evidently, $t_1 = pt_0/2p_0$, corresponding to $\rho = 0$. It may be noted that (8.149) establishes the same relationship between ρ/a and t/t_1 as (8.144) does between ρ_0/a and $p/2p_0$. Since $0 < M_r < M_0$ according to (8.148), the yield condition is nowhere violated.

Consider now the deflection of the plate during the time interval $t_0 < t < t_1$. Since the region inside the circle $r = \rho$ has been associated with the single plastic regime A, the deflection in this region is found by a straightforward integration of the first equation of (8.146), using the initial condition $w = pt_0^2/2\mu$ at $t = t_0$. Thus

$$\left(1-\frac{\rho}{a}\right)\left(1-\frac{\rho^2}{a^2}\right) = \frac{2p_0 t}{pt_0} = \frac{t}{t_1}, \qquad t_0 \leq t \leq t_1, \quad (8.149)$$

in view of (8.149). For the region outside the circle of radius ρ, it is convenient to change the independent variable from t to ρ in the second equation of (8.146) using

$$w = \frac{pt_0}{2\mu}\,(2t-t_0) = \frac{pt_0^2}{2\mu}\left\{\frac{p}{p_0}\left(1-\frac{\rho}{a}\right)\left(1-\frac{\rho^2}{a^2}\right)-1\right\}, \qquad 0 \leq r \leq \rho, \quad (8.150)$$

(8.147), the result being

$$a\frac{\partial w}{\partial \rho} = -\frac{p^2 t_0^2}{2\mu p_0}\left(1-\frac{r}{a}\right)\left(1+\frac{3\rho}{a}\right), \qquad \rho \leq r \leq a.$$

For the annular region $\rho_0 \leq r \leq a$, which has been associated only with the plastic regime AB, the preceding equation is readily integrated under the initial condition $w = (pt_0^2/2\mu)(a-r)/(a-\rho_0)$ at $\rho = \rho_0$ to obtain the solution

$$w = \frac{pt_0^2}{2\mu}\left(1-\frac{r}{a}\right)\left\{\frac{a}{a-\rho_0}+\frac{p}{p_0}\left(\frac{\rho_0-\rho}{a}\right)\left[1+\frac{3}{2}\left(\frac{\rho_0+\rho}{a}\right)\right]\right\}, \qquad \rho_0 \leq r \leq a. \quad (8.151)$$

Let τ denote the time when the contracting hinge circle coincides with a given radius r. Setting $t = \tau$ and $\rho = r$ in (8.149), we get

$$\tau = \frac{pt_0}{2p_0}\left(1-\frac{r}{a}\right)\left(1-\frac{r^2}{a^2}\right).$$

The material at radius r belongs to plastic regime A for $t \leq \tau$ and to plastic regime AB for $t > \tau$, the deflection at $t = \tau$ being

$$w\,(r,\tau) = \frac{pt_0}{2\mu}\left\{\frac{p}{p_0}\left(1-\frac{r}{a}\right)\left(1-\frac{r^2}{a^2}\right)-1\right\},$$

which is obtained by setting $\rho = r$ in (8.150). The integration of the deflection equation under the condition $w = w(r, \tau)$ when $\rho = r$ then furnishes

$$w = \frac{pt_0^2}{2\mu}\left\{\frac{p}{p_0}\left(1-\frac{r}{a}\right)\left[\left(1-\frac{\rho^2}{a^2}\right)+\frac{r-\rho}{a}\left(1+\frac{r+\rho}{2a}\right)\right]-1\right\}, \qquad \rho \leq r \leq \rho_0. \tag{8.152}$$

Equations (8.150), (8.151), and (8.152) provide the complete solution for the deflection of the plate during the second plastic phase. The deflection at the end of this phase within the circle of radius ρ_0 is

$$w\,(r,t_1) = \frac{pt_0}{2\mu}\left\{\frac{p}{p_0}\left(1-\frac{r}{a}\right)\left(1+\frac{r}{a}+\frac{r^2}{2a^2}\right)-1\right\}, \qquad 0 \leq r \leq \rho_0. \tag{8.153}$$

The slope $\partial w/\partial r$ is continuous across $r = \rho$ but discontinuous across $r = \rho_0$, the discontinuity being of amount $pt_0^2/2\mu(a - \rho_0)$. The jump condition (8.135) is found to be satisfied across $r = \rho$ at each stage of the interval $t_0 \leq t \leq t_1$. The central deflection of the plate is $\delta_1 = (pt_0^2/2\mu)(p/p_0 - 1)$ when $t = t_1$, obtained by setting $r = 0$ in the preceding equation. The shape of the deflected circular plate at the instant $t = t_1$ for different values of p/p_0 is displayed in Fig. 8.26.

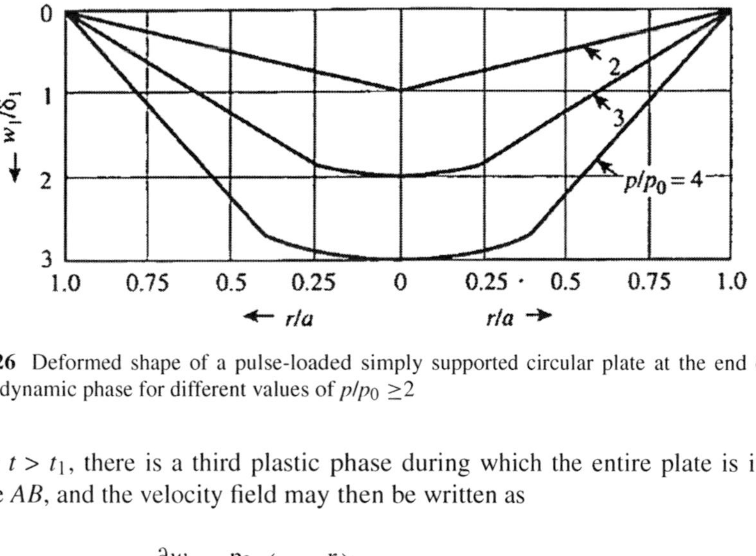

Fig. 8.26 Deformed shape of a pulse-loaded simply supported circular plate at the end of the second dynamic phase for different values of $p/p_0 \geq 2$

For $t > t_1$, there is a third plastic phase during which the entire plate is in the regime AB, and the velocity field may then be written as

$$\frac{\partial w}{\partial t} = \frac{p_0}{\mu}\left(1-\frac{r}{a}\right)f\,(t), \qquad t_1 \leq t \leq t^*, \tag{8.154}$$

which is identical in form to that given by (8.146), the instant when the plate comes to rest being denoted by t^*. The integration of the above equation results in

$$w\,(r,t) = w\,(r,t_1) + \frac{p_0}{\mu}\left(1-\frac{r}{a}\right)f\,(t),$$

where $f(t)$ is defined in such a way that $f(t_1) = 0$. The equation of dynamic equilibrium (8.136), where $p = 0$ and $M_\theta = M_0$, can be satisfied along with the boundary conditions by taking $f''(t) = -2$. Integrating, and using the initial conditions $f(t_1) = pt_0/p_0 = 2t_1$ and $f(t_1)=0$, which ensure the continuity of the velocity and deflection at $t = t_1$, we get

$$f(t) = (t - t_1)(3t_1 - t), \qquad f'(t) = 2(2t_1 - t), \qquad t^* = 2t_1$$

The bending moment distribution during this phase is the same as that given by (8.148) with $\rho = 0$, while the deflection of the plate is expressed as

$$w(r,t) = w(r,t_1) + \frac{p_0}{\mu}\left(1 - \frac{r}{a}\right)(t - t_1)(3t_1 - t), \qquad t_1 \le t \le t^*. \tag{8.155}$$

The final deflection w^* (which corresponds to $t = t^*$), considered over the region inside the circle $r = \rho_0$, is obtained from (8.153) and (8.155) as

$$w^* = \frac{pt_0^2}{2\mu}\left\{\frac{p}{2p_0}\left(1 - \frac{r}{a}\right)\left(3 + \frac{2r}{a} + \frac{r^2}{a^2}\right) - 1\right\} \qquad 0 \le r \le \rho_0. \tag{8.156}$$

Outside this circle, the shape of the deformed plate is conical, and the deflection at any radius r is $(a - r)/(a - \rho_0)$ times that at $r = \rho_0$, which is directly obtained from (8.156). The central deflection of the plate in the final stage is $\delta^* = (pt_0^2/2\mu)(3p/2p_0 - 1)$. Figure 8.27 shows the variation of w^*/δ^* with r/a for several values of the ratio $2p_0/p$ and indicates how the central curved part of the plate increases with increasing pressure. The corresponding problem for a built-in circular plate has been analyzed by Perzyna (1958) and Florence (1966). The dynamic behavior of circular plates under a central circular loading has been investigated by Conroy (1969) and Liu and Strange (1996).

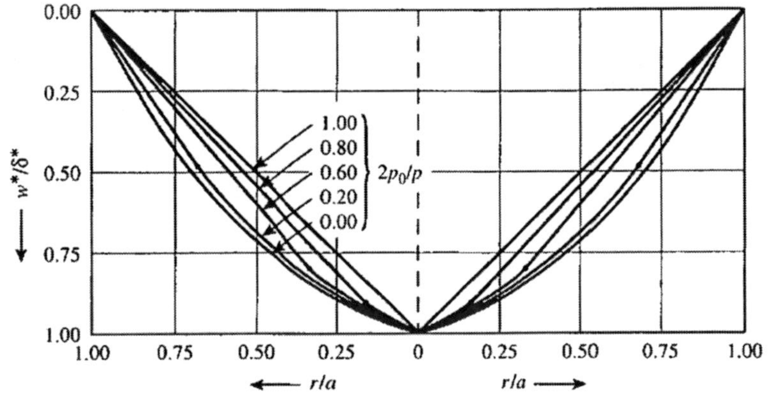

Fig. 8.27 Final shape of the deformed plate in relation to the central deflection for $p/p_0 \ge 2$. *The solid circles* indicate the positions of hinge circles

8.7.4 Solution for Impulsive Loading

A simply supported circular plate is subjected to a blast-type loading which instantaneously imparts a uniform transverse velocity U to the entire plate except at $r = a$. The impulsive action is immediately withdrawn so that the plate is free from transverse loads thereafter. During the first phase of the dynamic plastic deformation, a central part of the plate of steadily decreasing radius ρ continues to move with velocity U, while the surrounding annulus involves a conical flow field with $w = U$ at $r = \rho$. The velocity field may therefore be written as

$$\frac{\partial w}{\partial t} = \begin{cases} U, & 0 \le r \le \rho \\ U\,(a-r)\,/\,(a-\rho)\,, & \rho \le r \le a. \end{cases} \tag{8.157}$$

There is a plastic moving hinge at $r = \rho$ across which the velocity slope is discontinuous. Both the principal curvature rates vanish in the circular region $0 \le r \le \rho$, which corresponds to the plastic regime A, giving $M_r = M_\theta = M_0$. Since $k_r = 0$ and $\dot{\kappa}_\theta > 0$ in the annular region $\rho \le r \le a$, it corresponds to the plastic regime AB for which $M_\theta = M_0$, and the differential equation (8.133) for the bending moment becomes

$$\frac{\partial}{\partial r}\,(rM_r) = M_0 + \frac{\mu\rho U}{(a-\rho)^2}\int_\rho^r (a-r)\,rdr, \quad \rho \le r \le a.$$

The bending moment must satisfy the boundary conditions $M_r = 0$ at $r = a$ and $M_r = M_0$ at $r = p$, while $\partial M_r/\partial r$ must vanish at $r = \rho$ for the shearing force to be continuous. The integration of the above equation under these conditions furnishes M_r, which is the same as (8.148), while the differential equation for ρ is found to be

$$(a-\rho)\,(a+3\rho)\,\frac{dp}{dt} = -\frac{2M_0 a}{\mu U}. \tag{8.158}$$

Integrating, and using the initial condition $\rho = a$ when $t = 0$, the solution is obtained as

$$\left(1-\frac{\rho}{a}\right)\left(1-\frac{\rho^2}{a^2}\right) = \frac{12M_0 t}{\mu a^2 U} = \frac{t}{t_1}, \tag{8.159}$$

where t_1 denotes the time corresponding to $\rho = 0$. The defection of the plate during the time interval $0 \le t \le t_1$ is obtained by the integration of (8.157). Using (8.158), to change the independent variable from t to ρ, the second equation of (8.157) may be rewritten as

$$\frac{\partial w}{\partial \rho} = -\frac{Ut_1}{a}\left(1-\frac{r}{a}\right)\left(1+\frac{3\rho}{a}\right), \quad \rho \le r \le a.$$

If $t = \tau$ denotes the time at which the moving plastic hinge coincides with a given circle of radius r, then

$$\tau = t_1 \left(1 - \frac{r}{a}\right)\left(1 + \frac{3\rho}{a}\right)$$

in view of (8.159). Since $w = U\tau$ when $t = \tau$ or $\rho = r$, the integration of the above differential equation for w furnishes

$$w = Ut_1 \left(1 - \frac{r}{a}\right)\left\{\left(1 - \frac{\rho^2}{a^2}\right) + \frac{r-\rho}{a}\left(\frac{r+\rho}{2a}\right)\right\}, \quad \rho \leq r \leq a. \qquad (8.160)$$

Within the circle $0 \leq r \leq \rho$, the deflection at any instant has a constant value equal to Ut. The slope $\partial w/\partial r$ is continuous across $r = p$, although the velocity gradient is not. The deflection at the end of this phase is obtained by setting $p = 0$ in (8.160).

The rest of the analysis for impulsive loading is essentially the same as that for the high-pressure pulse considered before. The deflection during the second phase may therefore be written as

$$w = Ut_1 \left(1 - \frac{r}{a}\right)\left\{\left(1 + \frac{r}{a} + \frac{r^2}{2a^2}\right) + \frac{1}{2}\left(\frac{t}{t_1} - 1\right)\left(3 - \frac{t}{t_1}\right)\right\}, \quad t_1 \leq t \leq t^* \qquad (8.161)$$

where t_1 is given by (8.159). The acceleration at $r = 0$ has a constant value equal to $-U/t_1$, and the bending moment distribution satisfying the differential equation and the boundary conditions is given by (8.148) with $\rho = 0$. The motion stops when $t = t^*$, where

$$t^* = 2t_1 = \frac{\mu a^2 U}{6M_0}.$$

Setting $t = 2t_1$ in (8.161), the shape of the deformed plate after it has finally come to rest is obtained as (Wang, 1955)

$$w^* = \frac{1}{2}Ut_1 \left(1 - \frac{r}{a}\right)\left(3 + \frac{2r}{a} + \frac{r^2}{a^2}\right).\beta \qquad (8.162)$$

It may be noted that the slope of the deflected plate is discontinuous at $r = 0$ over the range $t_1 \leq t \leq t^*$. The central deflection of the plate finally attains the value $\delta^* = 3Ut_1/2$. The ratio w/δ^* is plotted as a function of r/a in Fig. 8.28 for $t/t_1 =$ 0.5, 1.0, and 2.0, the last two values defining the ends of the two plastic phases. The corresponding solution for a clamped circular plate has been given by Wang and Hopkins (1954). The solution for impact loading of an annular plate clamped at the inner radius has been discussed by Shapiro (1959), Florence (1965), and Johnson (1972). An analysis for a clamped circular plate impulsively loaded over a central circular area has been presented by Weirzbicki and Nurick (1996).

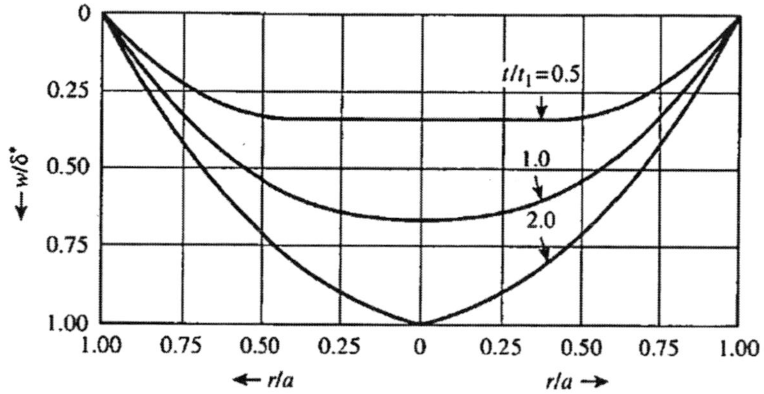

Fig. 8.28 Deformed shape of a simply supported circular plate under blast loading at different instants of time ($t_1 = \mu a^2 U/12M_0$)

The influence of the shape of the pressure pulse on the dynamic behavior has been examined by Youngdahl (1971) and Krajcinovic (1972). The dynamic behavior of circular plates under a central pulse loading has been investigated by Florence (1977). The effect of transverse shear on the dynamic plastic response has been studied by Kumar and Reddy (1986). An analysis for the dynamic bending problem of square plates has been presented by Cox and Morland (1959), and that of rectangular plates by Jones (1970). The use of mode approximation in predicting the dynamic plastic response of plates has been discussed by Chon and Symonds (1977). The influence of rate sensitivity on the dynamic behavior has been considered by Perrone (1967) and Perrone and Bhadra (1984). The effect of membrane forces on shape changes in dynamically loaded plates has been investigated by Jones (1971) and Symonds and Wierzbicki (1979). The dynamic buckling of rectangular plates in the plastic range has been treated by Goodier (1968). A variety of other problems on the dynamic plastic behavior of plates have been considered by Nurick et al. (1987), Jones (1989), Yu and Chen (1992), and Zhu (1996).

8.8 Dynamic Loading of Cylindrical Shells

8.8.1 Defining Equations and Yield Condition

Consider a circular cylindrical shell of radius a and thickness h, subjected to a uniform radial pressure p, whose initial value is greater than the quasi-static collapse pressure under identical boundary conditions. Such a loading will produce accelerated plastic flow that requires the inclusion of inertia effects in the theoretical framework. The state of stress in the shell is characterized by the axial bending moment M_x and the circumferential force N_θ acting per unit length of the circumference. The state of strain rate, on the other hand, is defined by the radially inward velocity w. As in the case of static analysis (Section 5.1), M_x will be taken as positive if it corresponds to tensile stresses on the inner surface, while N_θ will be reckoned positive when it is tensile in nature. The duration of the applied pressure is assumed small enough to justify the neglect of geometry changes in the analysis.

If the surface density of the material of the shell is denoted by μ, the inertia force per unit area of the middle surface is $\mu(\partial^2 w/\partial t^2)$ acting in the radially outward sense, and the equations of dynamic equilibrium are

$$\frac{\partial M_x}{\partial x} - Q = 0, \qquad \frac{\partial Q}{\partial x} + \frac{N_\theta}{a} + p = \mu \frac{\partial^2 w}{\partial t^2},$$

where Q is the shearing force per unit circumference and x is the distance measured along the length of the shell. The elimination of Q between the above equations gives the differential equation

$$\frac{\partial M_x}{\partial x} + \frac{N_\theta}{a} + p = \mu \frac{\partial^2 w}{\partial t^2} \tag{8.163}$$

for simplicity, the material is assumed to have a constant uniaxial yield stress Y, the fully plastic values of the resultant force and moment being $N_0 = Yh$ and $M_0 = Yh^2/4$, respectively. It is convenient at this stage to introduce the dimensionless qualities

$$m_x = \frac{M_x}{M_0}, \qquad n_\theta = \frac{N_\theta}{N_0}, \qquad q = \frac{pa}{Yh}, \qquad \xi = x\sqrt{\frac{2}{ah}}, \qquad \tau = \frac{t}{t_0}.$$

Denoting the differentiation with respect to τ by a superimposed dot, (8.163) can be expressed in the dimensionless form

$$\frac{\partial^2 m_x}{\partial \xi^2} + 2\,(n_\theta + q) = \frac{\ddot{w}}{\delta}, \tag{8.164}$$

where $\delta = Yht_0^2/2\mu a$ denotes a representative constant deflection of the middle surface. This equation can be integrated with the help of an appropriate yield condition and a suitable choice of the deflection function $w(\xi, \tau)$.

The analysis of the dynamic problem is greatly simplified by the use of the square yield condition shown in Fig. 8.29(a). It is an approximation not only to the Tresca yield condition (shown broken) but also to the von Mises yield condition (not shown). Referred to the dimensionless variables ξ and τ, the generalized strain rates may be defined as

$$\dot{\lambda}_\theta = -\frac{\dot{w}}{a}, \qquad \dot{\kappa}_x = -\frac{2}{ah}\frac{\partial^2 \dot{w}}{\partial \xi^2}. \tag{8.165}$$

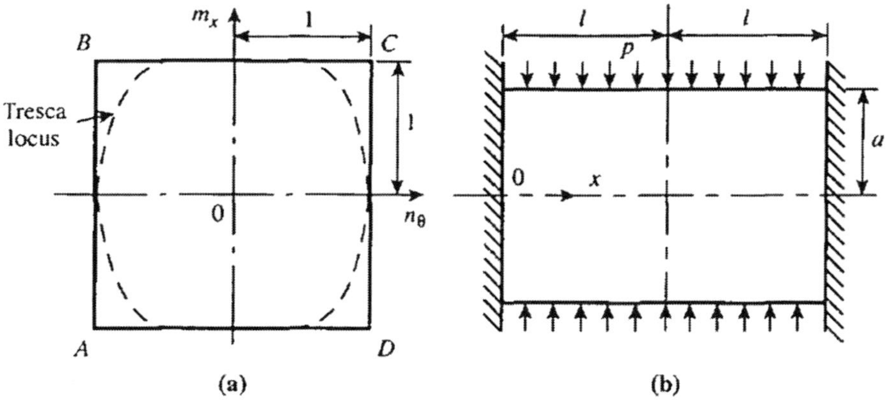

Fig. 8.29 Dynamic loading of a cylindrical shell. (**a**) Yield condition and (**b**) clamped shell with a pressure pulse

When the stress point lies on one of the sides of the yield locus, the vector $(N_0\dot{\lambda}_\theta, M_0\dot{\kappa}_x)$ is directed along the outward normal to this side. At a corner of the yield locus, the vector must lie between the two extreme normals defined there. In any particular problem, the stress profile generally includes two or more plastic regimes, and the flow rule in each case can be easily established (Hodge, 1955).

Considerations of equilibrium require the bending moment and the shearing force to be continuous, while cohesion of the material demands that the deflection and the velocity are continuous. Thus, m_x,$\partial m_x/\partial\xi$,$\dot{w}$, and w are all continuous across any boundary, moving or stationary. The velocity slope $\partial\dot{w}/\partial\xi$ may, however, be discontinuous across a hinge circle, which corresponds to a finite value of $\dot{w}$ and an infinitely large value of $\partial^2\dot{w}/\partial\xi^2$. Such conditions can only hold at the corners A and B of the field locus when n_θ is compressive.

8.8.2 Clamped Shell Loaded by a Pressure Pulse

A cylindrical shell of length $2\ l$ is rigidly clamped at both ends, Fig. 8.29(b), and is instantaneously loaded at $t = 0$ by a uniform radial pressure p which is held constant for a sufficiently small time interval t_0. The static collapse pressure p_0, which must be exceeded for dynamic actions, can be determined in the same way as that using the hexagonal yield condition (Section 5.1). Considering one-half of the shell defined by $0 \leq x = l$, the incipient velocity at collapse is taken as $\dot{w} = \xi$, which gives $\dot{\kappa}_x = 0$ and $\dot{\lambda} < 0$. The stress profile is therefore entirely on side AB, with $x = 0$ corresponding to point A and $x = l$ to point B. Setting $n_\theta = -1$, $q = q_0$, and $w = 0$ in (8.164), and integrating it under the boundary conditions $m_x = 1$ at $\xi = 0$, and $m_x = 1$ at $\xi = \omega$, we obtain the solution

$$m_x = (q_0 - 1)\,\xi\,(\omega - \xi) + 2\,(\xi/\omega) - 1.$$

Since the bending moment is a relative maximum at $x = l$, the derivative $\partial m_x/\partial\xi$ must vanish at $\xi = \omega$, giving the dimensionless collapse pressure

$$q_0 = 1 + \frac{2}{\omega^2}, \qquad \omega = l\sqrt{\frac{2}{ah}}.$$

For a range of values of $q > q_0$, it is reasonable to suppose that the same plastic regime AB is applicable for dynamic loading. Since the velocity at each instant is then proportional to ξ, the expressions for $\dot{w}$ and w may be written as

$$\dot{w} = \dot{w}_0\,(\xi/\omega)\,, \qquad w = w_0\,(\xi/\omega)\,, \tag{8.166}$$

where $w_0(t)$ is the deflection at the central section $\xi = \omega$. Substituting into the equilibrium equation (8.164), and setting $n_\theta = -1$, we obtain the differential equation for m_x as

$$\frac{d^2 m_x}{d\xi^2} = -2\,(q - 1) + \frac{\ddot{w}\xi}{\delta\omega}. \tag{8.167}$$

The assumed mode of deformation implies the formation of hinge circles at $\xi = 0$ and $\xi = \omega$. The preceding equation may therefore be integrated under the boundary conditions $\partial m_x/\partial\xi = 0$ at $\xi = \omega$ and $m_x = -1$ at $\xi = 0$, resulting in

$$m_x = -1 + (q - 1)\,\xi\,(2\omega - \xi) - \frac{\ddot{w}_0}{2\delta}\xi\left(\omega - \frac{\xi^2}{3\omega}\right).$$

The remaining boundary condition $m_x = 1$ at $\xi = \omega$ furnishes the central acceleration

$$\ddot{w}_0 = 3\,(q - q_0)\,\delta, \qquad 0 \le \tau \le 1, \tag{8.168}$$

in view of the expression for q_0, and the bending moment distribution in the first plastic phase becomes

$$m_x = -1 + (q - 1)\,\xi\,(2\omega - \xi) - \frac{1}{2}\,(q - q_0)\,\xi\left(3\omega - \frac{\xi^2}{\omega}\right), \quad 0 \le \tau \le 1 \tag{8.169}$$

Evidently, $m_x = m_x\,(\xi)$ in this phase, being independent of time. Integrating (8.168), and using the initial conditions $\dot{w}_0 = w_0 = 0$ at $\tau = 0$, we get

$$\frac{\dot{w}}{\delta} = 3\,(q - q_0)\,\tau, \qquad \frac{w_0}{\delta} = \frac{3}{2}\,(q - q_0)\,\tau^2, \qquad 0 \le \tau \le 1. \tag{8.170}$$

The preceding solution will be acceptable if the maximum bending moment does occur at $\xi = \omega$. Since $m_x''(\omega)$has the value $q - (3q_0 - 2)$, which is negative for $q < 3q_0 - 2$, the applied pressure must satisfy the inequalities

$$q_0 = 1 + \frac{2}{\omega^2} < q < 1 + \frac{6}{\omega^2} = q_c \quad (say). \tag{8.171}$$

When q exceeds the upper limit q_c over this range, considered as the medium load range, m_x is a relative minimum at $\xi = \omega$, and its value exceeds unity in the neighborhood of this section, thereby violating the yield condition.

Suppose that the applied pressure is instantaneously removed at $t = t_0$, which marks the beginning of a second plastic phase. For a sufficiently short shell subjected to medium load, the stress profile should continue to be on side AB. The central acceleration and the bending moment distribution are therefore given by (8.168) and (8.169), respectively, with $q = 0$, the latter quantity being

$$m_x = -\left(1 + 2\omega\xi - \xi^2\right) + \frac{1}{2}q_0\xi\left(3\omega - \frac{\xi^2}{\omega}\right). \tag{8.172}$$

The yield condition will not be violated in the neighborhood of $\xi = 0$ so long as $m'_x \leq 0$. By (8.172), this condition is equivalent to

$$q_0 \geq \frac{4}{3} \qquad or \qquad \omega \leq \sqrt{6}.$$

Such shells will be regarded as short shells, as opposed to long shells for which the above inequalities are reversed.

Since $m'_x(\omega) < 0$ during the second plastic phase, the bending moment distribution (8.172) is acceptable for short shells. Integrating (8.168) after setting $q = 0$, and using the conditions of continuity of $\dot{w}_0$ and w_0 at $\tau = 1$, we get

$$\left.\begin{aligned} \dot{w}_0 &= 3\left(q - q_0\tau\right)\delta, \\ \dot{w}_0 &= \frac{3}{2}\left[q\left(2\tau - 1\right) - q_0\tau^2\right]\delta, \quad 1 \leq \tau \leq q/q_0, \end{aligned}\right\} \tag{8.173}$$

in view of (8.170). The shell comes to rest ($\dot{w} = 0$) when $\tau = \tau^* = q/q_0$. The final deflection of the shell is $w^* = (\xi/\omega)w_0^*$, and it follows from (8.173) that

$$\frac{w^*}{\delta} = \frac{3q}{2}\left(\frac{q}{q_0} - 1\right)\frac{\xi}{\omega}, \quad q_0 < q < q_c, \quad \omega^2 \leq 6. \tag{8.174}$$

The final value of the central deflection is therefore equal to 1.5δ when q and ω have their limiting values of 2 and $\sqrt{6}$, respectively.

For long shells ($\omega^2 > 6$) under medium loads ($q < q_c$), the preceding solution still holds for $\tau < 1$, but that for $\tau > 1$ the solution is modified since the plastic regime does not apply throughout the shells. It is natural to expect that a region $0 < \xi < \rho$ near the built-in end would correspond to the plastic regime AD, for which $m_x = -1$ and $\dot{w} = 0$. This portion of the shell therefore becomes rigid after being previously deformed. The section $\xi = \rho$ defines the instantaneous position of the hinge circle which requires $m_x = -1$ there. The remaining portion $\rho \leq \xi \leq \omega$ of the half-shell corresponds to the plastic regime AB and involves continuation of the plastic deformation. The velocity in this region may be written as

$$\dot{w} = \dot{w}_0 \left(\frac{\xi - \rho}{\omega - \rho} \right), \quad \rho \leq \xi \leq \omega, \quad \tau \geq 1, \tag{8.175}$$

so that $\dot{\kappa}_x = 0$, and $\dot{w}$ is automatically made continuous across $\xi = \rho$. The distribution of m_x in the deforming region can be determined by integrating (8.167) with $q = 0$, $n_\theta = -1$, and the expression

$$\ddot{w} = \ddot{w}_0 \left(\frac{\xi - \rho}{\omega - \rho} \right), - \frac{\dot{\rho}\ddot{w}_0}{\omega - \rho} \left(\frac{\omega - \xi}{\omega - \rho} \right), \quad \rho \leq \xi \leq \omega.$$

Using the boundary conditions $\partial m_x/\partial \xi = 0$ and $\xi = \rho$ at the central section £ = co, the solution to the differential equation (8.167) in the region $\rho < \xi < \omega$ is obtained in the form

$$m_x = 1 + (\omega - \xi)^2 \left\{ \left(1 + \frac{\ddot{w}_0}{2\delta} \right) - \frac{1}{6\delta} \left(\ddot{w}_0 + \frac{\dot{\rho}_0 \dot{w}_0}{\omega - \rho} \right) \frac{\omega - \xi}{\omega - \rho} \right\}. \tag{8.176}$$

Since the bending moment and its derivative must be continuous across the hinge circle, the conditions $m_x = -1$ and $\partial m_x/\partial \xi = 0$ at $\xi = \rho$ must also be satisfied. Hence

$$\frac{\dot{\rho}\dot{w}_0}{\omega - \rho} - \ddot{w}_0 = 4\delta, \quad \frac{\dot{\rho}\dot{w}_0}{\omega - \rho} - 2\ddot{w}_0 = 6\delta \left[1 + \frac{2}{(\omega - \rho)^2} \right].$$

These two relations may be combined together to express $\ddot{w}_0$ and $\dot{\rho}\dot{w}_0$ in terms of ρ, the result being

$$\ddot{w}_0 = -2\delta \left[1 + \frac{6}{(\omega - \rho)^2} \right], \quad \frac{\dot{\rho}\dot{w}_0}{\omega - \rho} = 2\delta \left[1 - \frac{6}{(\omega - \rho)^2} \right]. \tag{8.177}$$

The substitution from (8.177) into (8.176) finally gives the bending moment distribution in the form

$$m_x = 1 - 2 \left[3 - 2 \left(\frac{\omega - \xi}{\omega - \rho} \right) \right] \left(\frac{\omega - \xi}{\omega - \rho} \right)^2, \quad \rho \leq \xi \leq \omega. \tag{8.178}$$

It follows from the boundary conditions on m_x that $-1 \leq m_x \leq 1$ for $\rho < \xi < \omega$. Since $\ddot{w}_0 = \dot{\rho}(d\dot{w}/d\rho)$, the elimination of $\dot{\rho}$ between the two relations of (8.177) leads to the differential equation

$$\frac{d\,\dot{w}_0}{d\,\rho} + \left\{ \frac{(\omega - \rho)^2 + 6}{(\omega - \rho)^2 - 6} \right\} \frac{\dot{w}_0}{\omega - \rho} = 0.$$

Since $\rho = 0$ at $\tau = 1$, when $\dot{w}_0 = 3(q-q_0)\delta$, the integration of the above equation results in

$$\frac{\dot{w}_0}{\delta} = \frac{3\omega\,(q-q_0)\left[(\omega-\rho)^2-6\right]}{\left(\omega^2-6\right)(\omega-\rho)}, \quad 0 \le \rho \le \omega - \sqrt{6}. \tag{8.179}$$

To determine the variation of ρ with time, we substitute (8.179) into the second equation of (8.177) and obtain the differential equation

$$\frac{d\rho}{d\tau} = \frac{2}{3\omega}\left(\frac{\omega^2-6}{q-q_0}\right) = \frac{2\omega}{3}\left(\frac{4-3q_0}{q-q_0}\right).$$

Thus $\dot{\rho}$ is a constant, which means that ρ varies linearly with the time, the result of integration of the above equation being

$$\frac{\rho}{\omega} = \frac{2}{3}\left(\frac{4-3q_0}{q-q_0}\right)(\tau-1), \quad 1 \le \tau \le \tau^*. \tag{8.180}$$

The velocity everywhere vanishes when $\rho = \omega - \sqrt{6}$, and the motion is terminated, the duration of the motion τ^* being obtainable from (8.180). The central deflection of the shell for $\tau \ge 1$ can be found by the integration of (8.179), using the fact that $\dot{w} = \dot{\rho}(dw_0/d\rho)$, and substituting for ρ. The result is easily shown to be

$$\frac{w_0}{\delta} = \frac{3}{2}(q-q_0)\left\{1 + \frac{3\,(q-q_0)}{4-3q_0}\left[\frac{\rho\,(2\omega-\rho)}{2\left(\omega^2-6\right)} + \frac{6}{\omega^2-6}\ln\left(1-\frac{\rho}{\omega}\right)\right]\right\}, \quad \omega^2 > 6 \tag{8.181}$$

The deflection at a generic section can be determined by integrating a similar equation obtained from (8.175) and (8.179), together with the change of variable to ρ. The final shape of the shell obviously corresponds to $\rho = \omega - \sqrt{6}$. The ratio w_0/δ is plotted against t/t^* in Fig. 8.30 for $\omega^2 = 3$ and 6, and for three different values of P/P_0 The derivation of the uppermost broken curve is based on the analysis for high loads which is given below.

8.8.3 Dynamic Analysis for High Loads

Consider the range of loads for which $q > q_c$, applied to sufficiently short shells characterized by $\omega^2 \le 6$. In this case, a central portion of the shell is in regime B, while the remainder of the shell is in regime AB, the two portions being separated by a hinge circle. During the first plastic phase ($0 \le \tau \le 1$), the hinge circle is fixed at a section $\xi = \alpha_0$, the velocity and deflection of the outer portion of the shell being given by

$$\dot{w} = \dot{w}_0\xi/\alpha_0, \quad w = w_0\xi/\alpha_0, \quad 0 \le \xi \le \alpha_0, \tag{8.182}$$

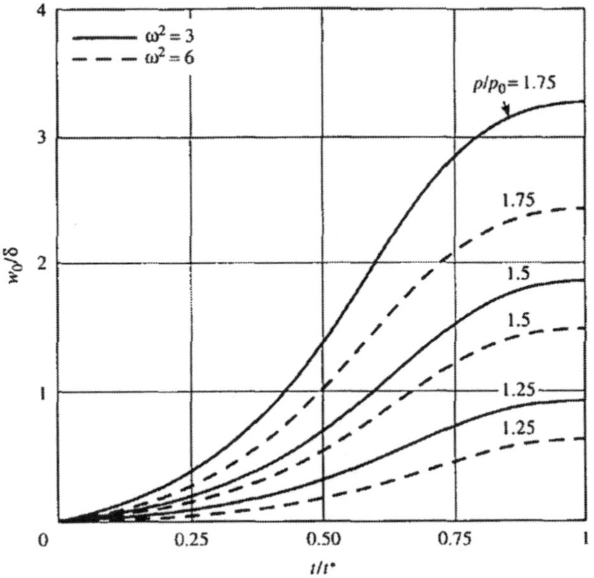

Fig. 8.30 Central deflection of a clamped cylindrical shell as a function of time under a uniform pressure pulse

where w_0 represents a uniform deflection of the central portion $\alpha_0 \leq \xi \leq \omega$. Since $m_x = 1$ and $n_\theta = -1$ over the length $\alpha_0 \leq \xi \leq \omega$, it follows from (8.164) that $\ddot{w}_0 = 2(q-1)\delta$, which gives on integration

$$\frac{\dot{w}_0}{w} = 2(q-1)\tau, \quad \frac{w_0}{\delta} = (q-1)\tau^2, \quad 0 \leq \tau \leq 1. \tag{8.183}$$

The bending moment distribution in the region $0 \leq \xi \leq \alpha_0$ and the quantity α_0 are determined by the integration of (8.167), where α_0 is written for ω, using the boundary conditions $m_x = -1$ at $\xi = 0$ and $m_x = 1$, $\partial m_x/\partial\xi = 0$ at $\xi = \alpha_0$. Thus

$$m_x = 1 - 2(1 - \xi/\alpha_0)^3, \quad n_\theta = -1, \quad 0 \leq \xi \leq \alpha_0,$$

$$m_x = 1, \quad n_\theta = -1, \quad \alpha_0 \leq \xi \leq \omega, \tag{8.184}$$

$$(q-1)\alpha_0^2 = 6, \quad or \quad \alpha_0 = \sqrt{6/(q-1)}.$$

For $\tau > 1$, the load is absent, and the hinge circle separating the two regions progressively moves toward the central section, its position at any instant being denoted by $\xi = \alpha$. The velocity (but not the deflection) is still given by (8.182) with a written for α_0, the central acceleration being given by $w_0 = -2\delta$. Integrating, and using the initial conditions $w_0 = 2(q-1)\delta$ at $\tau = 1$, we get

$$\dot{w}_0 = 2(q-\tau)\delta, \quad 1 \le \tau \le \tau_1,$$

where τ_1 is the value of r when the hinge circle reaches $\xi = \omega$. The velocity and displacement in the central region during the second plastic phase are given by

$$\frac{\dot{w}}{\delta} = 2(q-\tau), \quad \frac{w}{\delta} = (2\tau - 1)\, q - \tau^2, \quad \alpha \le \xi \le w, \tag{8.185}$$

The generalized stresses in this region are $m_x, = 1$ and $n_\theta = -1$. In the region $0 \le \xi \le \alpha$, the bending moment distribution can be directly written from the conditions $m_x(0) = -1, m_x(\alpha) = 1, m_x'(\alpha) = 0$, and $m_x''(0) = 2$, the result being

$$m_x = -1 + \left(3 - \frac{\alpha^2}{2}\right)\frac{\xi}{\alpha} + \xi^2 - \left(1 + \frac{\alpha^2}{2}\right)\frac{\xi^3}{\alpha^3}, \quad 0 \le \xi \le \alpha. \tag{8.186}$$

Inserting this expression into the differential equation (8.164), setting $n_\theta = -1$ and $q = 0$, and using the fact that

$$\frac{\dot{w}}{\delta} = 2(q-\tau)\frac{\xi}{\alpha}, \quad \frac{\dot{w}}{\delta} = -2\left[1 + (q-\tau)\frac{\dot{\alpha}}{\alpha}\right]\frac{\xi}{\alpha}, \quad 0 \le \xi \le \alpha,$$

(8.164) is found to be satisfied if α is given by the differential equation

$$(q-\tau)\frac{d\alpha}{d\tau} = \frac{6+\alpha^2}{2\alpha},$$

which is readily integrated under the initial condition $\alpha = \alpha_0$ when $\tau = 1$ to give

$$\frac{6+\alpha^2}{6+\alpha_0^2} = \frac{q-1}{q-\tau} \quad or \quad \alpha = \sqrt{\frac{6\tau}{q-\tau}}, \tag{8.187}$$

in view of the last equation of (8.184). Since $\alpha = \omega$ when $\tau = \tau_1$, (8.187) furnishes

$$\tau_1 = q\omega^2 / \left(6 + \omega^2\right) = q/q_c$$

in view of (8.171). The instant $\tau = \tau_1$ marks the end of the second plastic phase, since the hinge circle reaches the central section and can go no further.

In order to complete the solution for the second plastic phase ($1 \le \tau < \tau_1$), it is necessary to find the deflection in the region $0 \le \xi \le \alpha$ by the integration of the differential equation

$$\frac{\partial w}{\partial \tau} = 2\delta(q-\tau)\frac{\xi}{\alpha} = \delta\sqrt{\frac{2}{3}}\xi\,(q-\tau)\sqrt{\frac{q}{\tau} - 1}, \quad 0 \le \xi \le \alpha.$$

The solution is straightforward for the region $0 \leq \xi \leq \alpha_0$, which has always been in the plastic regime AB, the initial condition for this region being $w_0/\delta = (q-1)\xi/\alpha$ at t = 1, in view of (8.182) and (8.183). The integration of the above equation therefore gives

$$\frac{w}{\delta} = \frac{\xi}{\sqrt{6}}\left\{\frac{3}{2}q^2\left(\sin^{-1}\sqrt{\frac{\tau}{q}} - \sin^{-1}\frac{1}{\sqrt{q}}\right)\right.$$
$$\left.+\left(\frac{5}{2}q-\tau\right)\sqrt{\tau(q-\tau)}-\frac{3}{2}q\sqrt{q-1}\right\}, \quad 0 \leq \xi \leq \alpha_0. \tag{8.188}$$

The elements in the region $\alpha_0 \leq \xi \leq \alpha$ have passed from regime B to regime AB at different instants as they have been traversed by the moving hinge circle. Let τ' be the value of τ for which a typical section of the shell coincides with the hinge circle. Setting $\alpha = \xi$ and $\tau = \tau'$ in (8.187), we have

$$\tau' = q\xi^2/\left(6+\xi^2\right).$$

The deflection of the element at this instant is given by (8.185) with $\tau = \tau'$. Using this as the initial condition, the solution for the deflection in the region $\alpha_0 \leq \xi \leq \alpha$ is easily shown to be

$$\frac{w}{\delta} = \frac{\xi}{\sqrt{6}}\left\{\frac{3}{2}q^2\left(\sin^{-1}\sqrt{\frac{\tau}{q}} - \sin^{-1}\frac{\xi}{\sqrt{6}}\right) + \left(\frac{5}{2}q-\tau\right)\sqrt{\tau(q-\tau)}\right\}$$
$$-q\left(1+\frac{0.5\xi^2}{6+\xi^2}\right), \quad \alpha_0 \leq \xi \leq \alpha. \tag{8.189}$$

It is readily verified that the deflection is continuous at $\xi = \alpha_0$. The continuity of the deflection at $\xi = \alpha$ is also ensured by the fact that the right-hand side of (8.189) at $\xi =$ a coincides with that given by (8.185).

For $\tau > \tau_1$, the entire stress profile is in plastic regime AB, and the velocity distribution throughout the shell is given by the first equation of (8.182) with w to written for α_0. The bending moment distribution over the entire shell becomes

$$m_x = -1 + 3\left(3-\frac{\omega^2}{2}\right)\frac{\xi}{\omega} + \xi^2 - \left(1+\frac{\omega^2}{2}\right)\frac{\xi^2}{\omega^3}, \quad 0 \leq \xi \leq \omega, \tag{8.190}$$

obtained by simply replacing α by w in (8.186). The substitution in the differential equation (8.167) with $q = 0$ then gives the central acceleration as

$$\frac{\ddot{w}}{\delta} = -\left(1+\frac{2}{\omega^2}\right) = -3q_0.$$

In view of the continuity of the velocity $\dot{w}$ at $\tau = \tau_1$, the integration of the above equation results in the velocity field

$$\frac{\dot{w}}{\delta} = 3\,(q - q_0\tau)\,\frac{\xi}{\omega}, \quad \tau_1 \leq \tau \leq \tau^*, \quad 0 \leq \xi \leq \omega,$$

where $\tau^* = q/q_0$, representing the instant when the motion is terminated. A straightforward integration furnishes the deflection at any point as

$$\frac{w}{\delta} = \frac{3}{2}\tau\,(2q - q_0\tau)\,\frac{\xi}{\omega} + \lambda\,(\xi)\,, \quad \frac{q}{q_c} \leq \tau \leq \frac{q}{q_0}, \tag{8.191}$$

where $\lambda(\xi)$ must be determined from the condition of continuity of w at $\tau = \tau = q/q_c$. Using (8.188) and (8.189), it is easily shown that

$$\left.\begin{aligned} \lambda\,(\xi) &= \frac{q\xi}{2}\sqrt{\frac{3}{2}}\left\{q\left(\sin^{-1}\frac{1}{q_c} - \sin^{-1}\frac{1}{\sqrt{q}}\right) - \sqrt{q-1}\right\}, \quad 0 \leq \xi \leq \alpha_0, \\ \lambda\,(\xi) &= \frac{q\xi}{2}\left\{\sqrt{\frac{3}{2}}q\left(\tan^{-1}\frac{\omega}{\sqrt{6}} - \tan^{-1}\frac{\xi}{\sqrt{6}}\right) - \frac{q\xi}{6+\xi^2}\right\} - q, \quad \alpha_0 \leq \xi \leq \omega. \end{aligned}\right\} \tag{8.192}$$

Setting $\tau = q/q_0$ in (8.191), the final shape of the deformed middle surface of the shell is obtained as

$$\frac{w^*}{\delta} = \frac{3q^2\xi}{2q_0\omega} + \lambda\,(\xi)\,, \quad q \geq q_c, \quad \omega^2 \leq 6. \tag{8.193}$$

When $q{=}q_c$, we have $\alpha_0{-}w$ and $\lambda(\xi) = -3q\xi/2\omega$, which reduces (8.193) to (8.174) as expected. Setting $\xi = \omega$ in (8.193) and using (8.192), the central deflection in the final phase is found to be given by

$$\frac{w_0}{\delta} = q\left(3\tau - \frac{q}{2q_c} - 1\right) - \frac{3}{2}q_0\tau^2, \quad \frac{q}{q_c} \leq \tau \leq \frac{q}{q_0}, \tag{8.194}$$

where q_0 and q_c depend only on ω and are given by (8.171). Figure 8.31 displays the final deformation pattern of the clamped shell when $\omega^2 = 3$, each curve being based on a definite value q. For longer shells ($\omega^2 > 6$), the stress and velocity distributions are identical to those for short shells during the period of application of the load, but the subsequent part of the solution, following the load removal, is modified, due to the presence of a second hinge circle which begins at $\xi - 0$ and moves along the length of the shell. The effects of blast loading under different end conditions on the dynamic plastic behavior have been discussed by Hodge (1956b, 1959).

The dynamic plastic response of cylindrical shells under a band of pressure has been discussed by Eason and Shield (1956), Kuzin and Shapiro (1966), Youngdahl (1972), and Li and Jones (2005). The influence of membrane forces has been examined by Jones (1970) and Galiev and Nechitailo (1985). The dynamic plastic response of spherical caps under pulse and impact, loadings has been investigated by Sankaranarayanan (1963, 1966). The dynamics of an impulsively loaded

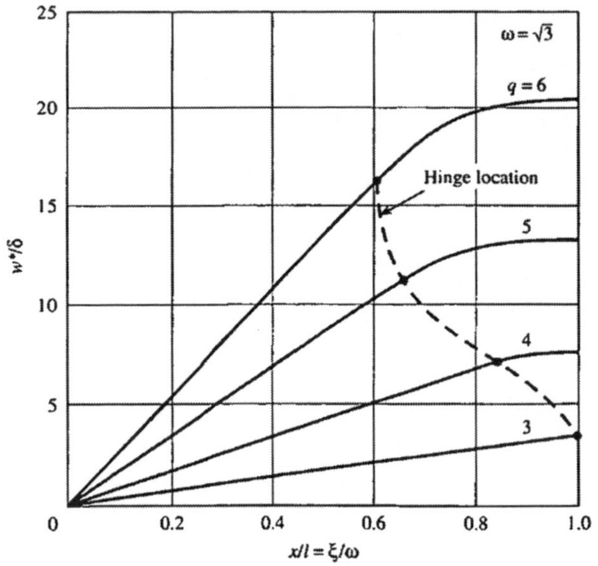

Fig. 8.31 Final shape of the deformed meridian of a clamped cylindrical shell subjected to a high-pressure pulse

cylindrical shell based on a generalized yield condition has been investigated by Lellep and Torn (2004).

The progressive crumpling of cylindrical tubes under axial compression has been discussed by Abramowicz and Jones (1984) and Jones (1989). The dynamic plastic buckling of cylindrical shells has been considered by Vaughan and Florence (1970) and Jones and Okawa (1976) and that of a complete spherical shell by Jones and Ahn (1974). A simplified method of analysis for the plastic buckling, based on energy considerations, has been discussed by Gu et al. (1996).

8.9 Dynamic Forming of Metals

8.9.1 High-Speed Compression of a Disc

Consider the rapid compression of a short circular cylinder between a pair of parallel platens, or dies, the speed of compression being such that the inertia effects are significant. The lower die $z = 0$ is stationary, while the upper die $z = h$ is assumed to move down with a constant speed U, the coefficient of friction μ between the dies and the cylindrical block being assumed to be constant (Haddow, 1965). The elastic and plastic stress waves initiated at the upper die travel up and down the block several times during the compression process. The load acting on the upper platen, which is significantly in excess of the quasi-static value in the early part of the process, appreciably decreases during the final stages. Only the incipient compression of a thin block of uniform diameter will be considered in what follows, ignoring the effect of wave propagation on the dynamic process.

Since the distribution of stresses and strains is symmetrical about the axis of the block, which is assumed to coincide with the z-axis as shown in Fig. 8.32(a), the equation of radial motion of a typical element in cylindrical coordinates (r, θ, z) may be written as

$$\frac{\partial \sigma_r}{\partial r} + \frac{\sigma_r - \sigma_\theta}{r} + \frac{\partial \tau_{rz}}{\partial z} = \rho \left(\frac{\partial u}{\partial t} + u\frac{\partial u}{\partial r} + \omega\frac{\partial u}{\partial z} \right),$$

where u and w are the radial and axial components of the velocity, and ρ is the density of the material. If the influence of barreling is disregarded, σ_r, σ_θ, σ_z, and u are independent of z, but the presence of die friction requires τ_{rz} to vary with z so that $\tau_{rz} = \mu p$ at $z = 0$ and $\tau_{rz} = -\mu p$ at $z = h$, where p is the die pressure. The multiplication of the preceding equation by dz and integration between the limits 0 and h therefore result in

$$\frac{\partial \sigma_r}{\partial r} + \frac{\sigma_r - \sigma_\theta}{r} - \frac{2\mu p}{h} = \rho \left(\frac{\partial u}{\partial t} + u\frac{\partial u}{\partial r} \right). \tag{8.195}$$

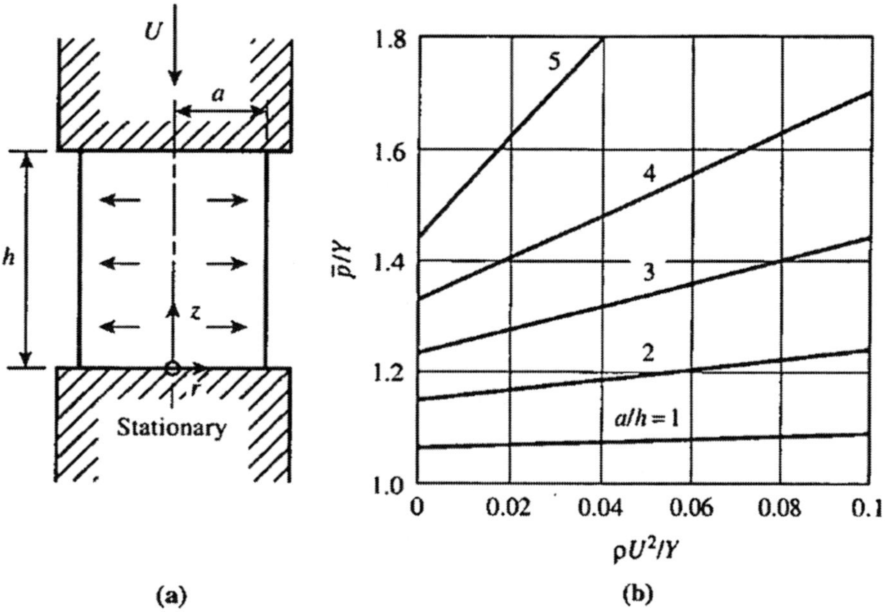

Fig. 8.32 High-speed compression of short cylinders. (**a**) Condition of loading and (**b**) mean die pressure against kinetic energy of impact

In the absence of barreling, the velocity field corresponds to a uniform compression of the block and is consequently given by

$$u = \frac{Ur}{2h}, \quad w = -\frac{Uz}{h}, \tag{8.196}$$

satisfying the condition of plastic incompressibility. Since $\dot{\varepsilon}_r = \dot{\varepsilon}_\theta$ according to (8.196), we have $\sigma_r = \sigma_\theta$ for an isotropic material, and the yield criteria of both Tresca and von Mises reduce to

$$\sigma_r - \sigma_z = \sigma_\theta - \sigma_z = Y,$$

where $\sigma_z = -p$, and the material is considered as ideally plastic with a uniaxial yield stress Y. Substituting from (8.196) into (8.195), setting $\sigma_e = \sigma_r$, and using the yield criterion, we obtain the differential equation

$$\frac{dp}{dr} + \frac{2\mu p}{h} = -\frac{3\rho U^2 r}{4h^2}.$$

Since σ_r must vanish along the cylindrical surface, the boundary condition is $p = Y$ at $r = a$. The integration of the above equation therefore furnishes

$$\frac{p}{Y} = \left\{1 - \frac{3\rho U^2}{8Y\mu}\left(\frac{1}{2\mu} - \frac{a}{h}\right)\right\} \exp\left[\frac{2\mu(a-r)}{h}\right] + \frac{3\rho U^2}{8Y\mu}\left(\frac{1}{2\mu} - \frac{r}{h}\right). \tag{8.197}$$

This equation predicts the die pressure distribution as a function of r. The inertia effect represented by the parameter $\rho U^2/Y$ is therefore to increase the die pressure at any given radius. The mean die pressure $\bar{p}$ corresponding to (8.197) is easily shown to be given by

$$\frac{\bar{p}}{Y} = \frac{h}{q\mu}\left\{1 - \frac{3\rho U^2}{8Y\mu}\left(\frac{1}{2\mu} - \frac{a}{h}\right)\right\}\left\{\frac{h}{2\mu a}\exp\left[\frac{2\mu a}{h}\right] - \left(1 + \frac{h}{2\mu a}\right)\right\} + \frac{\rho U^2}{4Y\mu}\left(\frac{3}{4\mu} - \frac{a}{h}\right). \tag{8.198}$$

When the ratio $\rho U^2/Y$ is vanishingly small, this formula reduces to (3.57), obtained for the quasi-static compression. Expanding $\exp(2\mu a/h)$ in ascending powers of $\mu a/h$, and neglecting terms containing powers of μ higher than the fourth, (8.198) can be reduced to

$$\frac{\bar{p}}{Y} = 1 + \frac{2\mu a}{3h} + \frac{3\rho U^2}{16Y}\left(\frac{a}{h}\right)^2,$$

which is sufficiently accurate for small values of μ. Figure 8.32(b) shows the variation of the mean die pressure with $\rho U^2/Y$ for different values of a/h in the special case of $\mu = 0.1$. In the case of steel, for instance, the inertia effect is significant when the speed of compression exceeds about 50 m s^{-1}. The situation where the speed of compression varies with time has been considered by Lippmann (1966) and Dean (1970).

The inclusion of the strain rate sensitivity of the material, based on the homogeneous deformation mode (8.196), is quite straightforward. Since the effective strain rate $\dot{\varepsilon}$ has a constant value equal to *U/h*, it is only necessary to replace Y in the preceding analysis by the modified yield stress

$$Y\left\{1+\left(\frac{\dot{\varepsilon}}{\alpha}\right)^n\right\}=Y\left\{1+\left(\frac{U}{\alpha h}\right)^n\right\},$$

where α and η are material constants to be determined by experiment. In particular, the simplified formula for small μ is easily shown to be modified to

$$\frac{\bar{p}}{Y}=\left(1+\frac{2\mu a}{3h}\right)\left\{1+\left(\frac{U}{\alpha h}\right)^n\right\}+\frac{3\rho U^2}{16Y}\left(\frac{a}{h}\right)^2. \tag{8.199}$$

This is a complete generalization of the well-known Siebel formula for the plastic compression of short cylinders. It involves the estimation of the coefficient of die friction, the speed of compression, and the empirical constants characterizing the dynamic response of the material. A upper bound analysis for the dynamics of a closed die forging process has been discussed by Scrutton and Marasco (1995).

8.9.2 Dynamic Response of a Thin Diaphragm

A thin circular diaphragm of initial thickness h_0 is rigidly held along its periphery $r = a$ and is subjected to a uniform velocity U normal to its plane (Hudson, 1951). An elastic wave front immediately sweeps inward from the edge, producing a radially outward motion of the material particles. At any later instant, a plastic bending wave generated at the edge travels some distance toward the center, producing a bulged shape of the diaphragm as shown in Fig. 8.33(a). The annular region swept over by the bending wave forms a surface of revolution, which is assumed to have come to rest, while the flat central region yet unaffected by the wave retains its normal velocity U. Since we are dealing with large plastic strains, all elastic effects other than those of the initial stress wave will be neglected, the material being effectively considered as rigid/plastic in the dynamic analysis of the process.

At any time t after the beginning of the process, let b denote the radius of the central flat portion of the diaphragm that has been uniformly deformed to a thickness h under the action of a constant normal velocity U and a variable radial velocity υ induced at $r = b$. During a time internal dt, an elemental ring of width ds just ahead of the bending wave impulsively rotates to form a part of the bulge after being swept over by the wave. Since the radial velocity of the ring relative to that of the wave is equal to $\upsilon - \dot{b}$, we have $ds = (\upsilon - \dot{b})\, dt$. The radial and transverse components of the displacement of the inner edge of the ring are $-\dot{b}\, dt$ and $U\, dt$, respectively, giving $ds = \sqrt{U^2 + \dot{b}^2}dt$. Consequently,

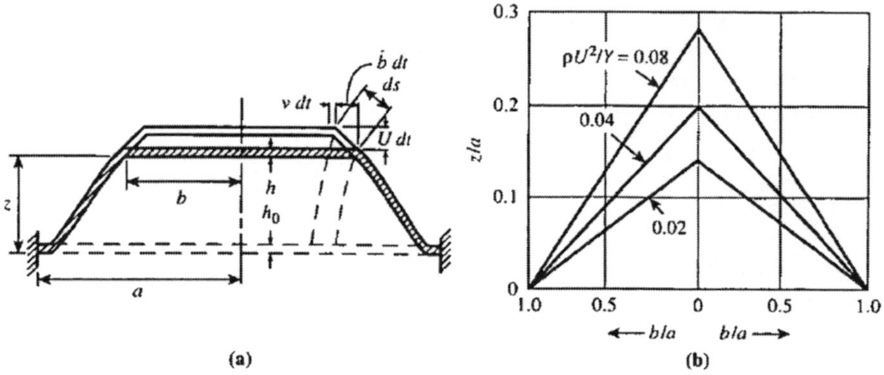

Fig. 8.33 Impact loading of a circular diaphragm. (**a**) Geometry of deformation and (**b**) deformed shape for different initial velocities

$$\frac{ds}{dt} = \upsilon - \dot{b} = \sqrt{U^2 + \dot{b}^2} \quad or \quad \dot{b} = \frac{\upsilon^2 - U^2}{2\upsilon}. \tag{8.200}$$

Since $\dot{b}$ is negative, it follows from (8.200) that $v < U$ throughout the deformation. The assumed uniformity of thickness in the flat central region requires a state of balanced biaxial tension $\sigma_r = \sigma_\theta$ to exist in this region. The radial and circumferential strain rates are therefore equal to one another, and the rate equation of incompressibility is

$$\frac{\dot{h}}{h} + 2\dot{\varepsilon}_\theta = 0 \quad or \quad \frac{\dot{h}}{h} + \frac{2\upsilon}{b} = 0. \tag{8.201}$$

when considered at the interface $r = b$. If r_0 denotes the initial radius to a particle that is currently at a radius r in the interior of the central region, then the integrated form of the flow rule gives $\varepsilon_\theta = \varepsilon_\mathrm{r} = \frac{1}{2} \ln (h_0/h)$. Hence

$$\frac{r}{r_0} = \frac{\partial r}{\partial r_0} = \sqrt{\frac{h_0}{h}}.$$

The assumed uniformity of the stress and strain in the central region is incompatible with the existence of inertia forces, the effect of which is disregarded in the yielding and flow of the material. Consequently, σ_r is equal to the current yield stress σ, which is a function of the total compressive thickness strain equal to $\ln(h_0/h)$.

Let q denote the resultant radial stress, including the inertia effect, in the material just ahead of the bending wave at $r = b$. As the wave sweeps inward over an annular element of width ds, the work done by the stress during the time interval dt is equal to $-qhbd\theta(\upsilon\, dt)$, where $d\theta$ is the angle subtended by the element at the center of the disc. Since the corresponding change in kinetic energy of the element is $-(\rho/2)h\, ds\, d\theta(\upsilon^2 + U^2)$, we have

$$q = p\left(\frac{\upsilon^2 + U^2}{2\upsilon}\right)\frac{ds}{dt} = \rho\left(U^2 + \dot{b}^2\right) \tag{8.202}$$

in view of (8.200). If the radially outward accelerating force acting on a typical element of mass *phr dr dθ* in the central region at any instant is denoted by $dF\, d\theta = (\partial F/\partial r)$, $dr\, d\theta$, then the equation of motion for this element may be written as

$$\frac{\partial F}{\partial r} = \rho h r \frac{\partial^2 r}{\partial t^2} = \rho h r^2 \sqrt{\frac{h}{h_0}}\frac{d^2}{dt^2}\left(\sqrt{\frac{h_0}{h}}\right).$$

It is reasonable to suppose that the rate of work done by the distribution of this force over the central region is equal to that produced by the stress difference $q - \sigma_r$ occurring at $r = b$. Then

$$(q - \sigma_r)\, hb\upsilon = \int \frac{\partial r}{\partial t} dF = \frac{\upsilon}{b}\int_0^b r\frac{\partial F}{\partial r} dr.$$

Substituting for $\partial F/\partial r$ and integrating, we obtain the result

$$q = \sigma + \frac{1}{4}\rho b^2 \sqrt{\frac{h}{h_0}}\frac{d^2}{dt^2}\left(\sqrt{\frac{h_0}{h}}\right). \tag{8.203}$$

The right-hand side of this equation depends only on the thickness ratio h/h_0. The solution to the system of equations (8.201) to (8.203) requires the specification of an initial value of v at the bending wave $r = b$. Assuming a linear distribution of velocity in the central region, Hudson (1951) obtained the initial condition

$$\upsilon = 2Y\sqrt{\frac{1-\nu}{\rho E}} \qquad at\, t = 0,$$

where E is Young's modulus, and ν is Poisson's ratio for the material of the diaphragm. The motion begins at $t = 0$ when the stress at the clamped edge rises suddenly from zero to the initial yield stress Y.

We begin with the situation where the material is nonhardening, so that $\sigma - Y$ throughout the motion. It is convenient to introduce the dimensionless parameters

$$\xi = \frac{Ut}{a} = \frac{z}{a}, \quad \alpha = \frac{b}{a}, \quad \beta = \frac{\upsilon}{U}, \quad \eta = \sqrt{\frac{h_0}{h}}, \quad \lambda = \frac{4Y}{\rho U^2},$$

where z is the current height of the central flat part above the initial plane of the diaphragm. Equations (8.200) and (8.201) immediately become

$$2\beta\frac{d\alpha}{d\xi} = -\left(1 - \beta^2\right), \quad \alpha\frac{dn}{d\xi} = \xi\eta, \tag{8.204}$$

while the elimination of q between (8.202) and (8.203) leads to the differential equation

$$4\left(\frac{d\alpha}{d\xi}\right)^2 = (\lambda - 4) + \frac{\alpha^2}{\eta}\frac{d^2\eta}{d\xi},$$

which can be combined with the second equation of (8.204) to eliminate η. After some algebraic manipulation using the first equation of (8.204), the resulting equation is reduced to

$$\alpha\frac{d\beta}{d\xi} = -\lambda + \frac{\left(1+\beta^2\right)\left(2+\beta^2\right)}{2\beta^2}. \qquad (8.205)$$

Equations (8.204) and (8.205) form a set of three basic differential equations for the three unknowns α, η, and β as functions of ξ, the initial conditions being

$$\alpha = \eta = 1, \quad \beta = \beta_0 = \sqrt{\lambda\,(1-\nu)\,Y/E} \quad when\ \xi = 0.$$

To carry out the integration, we combine (8.205) with the two equations of (8.204) in turn to obtain the results

$$\frac{d\alpha}{\alpha} = -\frac{\left(1-\beta^2 d\beta\right)}{2-(2\lambda-3)\,\beta^2+\beta^4}, \quad \frac{d\eta}{\eta} = \frac{2\beta^3 d\beta}{2-(2\lambda-3)\,\beta^2+\beta^4}.$$

Although these equations can be integrated exactly, it is more convenient for practical purposes to introduce a minor approximation (since β^4 is usually a small fraction) to express the solution in the form

$$\left.\begin{aligned} \alpha &= \left(\frac{m\beta^2-1}{m\beta_0^2-1}\right)^{(m-1)/4m^2} \exp\left(\frac{\beta_0^2-\beta^2}{4m}\right), \\ \eta &= \left(\frac{m\beta_0^2-1}{m\beta^2-1}\right)^{1/2m^2} \exp\left(\frac{\beta^2-\beta_0^2}{2m}\right), \end{aligned}\right\} \qquad (8.206)$$

where $m = \lambda - \frac{3}{2}$. These relations furnish η, as a function of α parametrically through β. Finally, ξ can be found as a function of β by the numerical integration of (8.205). When λ is sufficiently large, so that $m \approx \lambda$, the thickness h and the height z of the central flat part are closely approximated by the relations

$$\frac{h}{h_0} \approx \left(\frac{b}{a}\right)^{4/\lambda}, \quad \frac{z}{a} \approx U\sqrt{\frac{\rho}{Y}}\left(1-\frac{b}{a}\right). \qquad (8.207)$$

The diaphragm is therefore deformed into a conical shape, the central deflection being proportional to the initial velocity U. The radial velocity v of the flat part rapidly increases to its terminal value $U/\sqrt{\lambda}$. When the diaphragm is completely formed

($\alpha = 0$), thickness vanishes at the center, no doubt as a result of the mathematical idealization. The total time for the deflection, which is known as the swing time, is approximately equal to $a\sqrt{\rho/Y}$. Figure 8.33(b) shows the shape of the profiles of the completely deformed diaphragm for $\beta_0 = 0.04\sqrt{\lambda}$ with different values of λ.

The simplest way of taking into account the effect of work-hardening of the material is to consider a mean yield stress based on the given stress–strain curve. It is therefore only necessary to replace Y in the expression for λ by a quantity that depends on $\varepsilon = \ln(h_0/h)$. The general effect of work-hardening is to decrease the central deflection for a given initial velocity U. The predicted shape of the profile is found to be in complete qualitative agreement with observation. Figure 8.34, which has been obtained experimentally by Keil (1960), indicates the difference between the static and dynamic behaviors of a clamped circular diaphragm. An energy method of analysis for a circular membrane subjected to impact loading has been discussed by Boyd (1966). The propagation of plastic waves in an impulsively loaded circular membrane has been discussed by Munday and Newitt (1963) and Cristescu (1967).

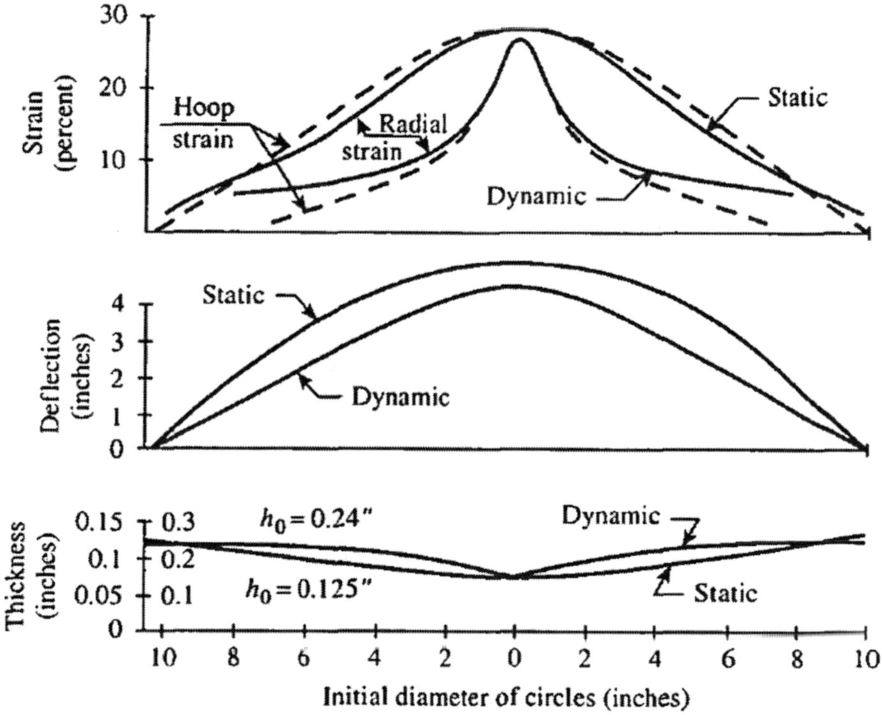

Fig. 8.34 Comparison of static and dynamic responses of clamped circular diaphragms (after A.H. Keil, 1960)

8.9.3 High-Speed Forming of Sheet Metal

A variety of techniques have been developed in recent years for the forming of sheet metal using loading rates that are large enough to have a dominating effect on the deformation process. Chemical explosives are commonly used to generate the energy which is transmitted to the workpiece through an intervening medium such as water. The forming die, together with the workpiece, is immersed in water contained in a forming tank, and the air in the die cavity is often evacuated with the help of a vacuum pump. The explosive charge is located at a suitable stand-off distance from the workpiece, and the kinetic energy released by the detonation of the charge is utilized in forming the component. Detailed accounts of the high-rate forming of metals have been presented by Rinehart and Pearson (1965).

One of the common methods of explosive forming of sheet metals is indicated in Fig. 8.35(a). An air-backed circular blank is clamped around the periphery and is subjected to a shock wave generated by the detonation of an explosive at a stand-off distance approximately equal to the blank radius. The shock wave provides the necessary kinetic energy which is dissipated in doing the plastic work. The mean initial velocity necessary for the material to attain a given polar strain in the deformed state can be approximately estimated from an assumed distribution of the thickness strain in the bulge (Johnson, 1972). Experiments tend to indicate that the final hoop strain is approximately equal to the radial strain, which means that the equivalent strain is approximately equal to the compressive thickness strain whose greatest value occurs at the pole.

In the explosive-forming process, the final shape of the blank depends on such factors as the hydrostatic head, the stand-off distance, the type of explosive, and the

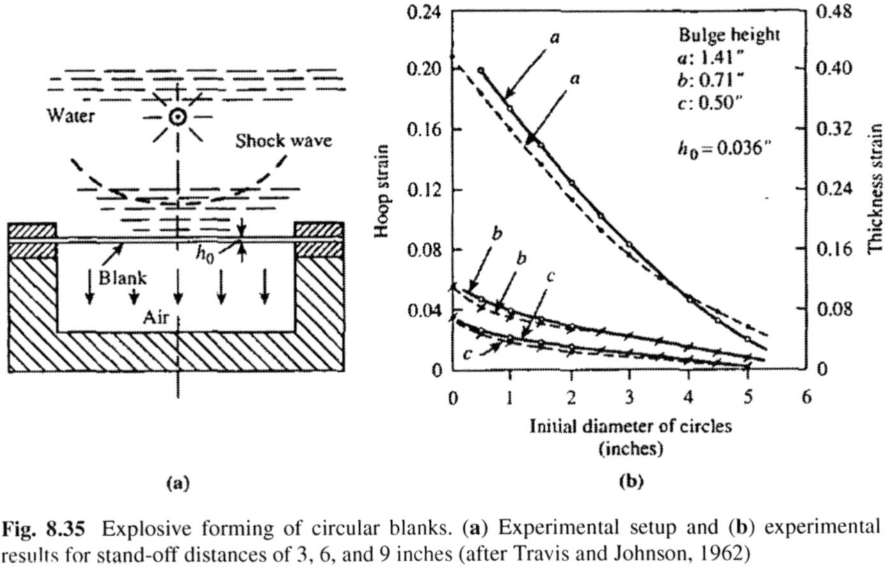

Fig. 8.35 Explosive forming of circular blanks. (**a**) Experimental setup and (**b**) experimental results for stand-off distances of 3, 6, and 9 inches (after Travis and Johnson, 1962)

weight of the charge. Figure 8.35(b), which is due to Travis and Johnson (1962), indicates how the hoop strain and thickness strain distributions in the deformed blank are affected by the ratio of the stand-off distance to the blank radius. When the size of the blank is relatively small, a closed system in which the water is contained in the die that is closed by a plate is sometimes used for maintaining the pressure for a longer interval of time. The related problem of deep drawing of cylindrical cups using an explosive device has been investigated by Johnson et al. (1965).

An interesting situation arises in the high-speed blanking of sheet metal, using a typical setup of die and punch. The width of the zone of shearing, which is confined near the area of clearance between the die and punch, decreases as the speed of the operation increases. The energy required for the separation of the blank from the stock generally increases with an increase in the speed of blanking, and this phenomenon may be attributed to the strain rate sensitivity of the material. The nonuniformity of the sheared edge, which is an essential feature of the process, constitutes a major challenge for its improvement. Useful experimental results on high-speed blanking have been reported by Johnson and Travis (1966), Balendra and Travis (1970), and Dowling et al. (1970). An electromagnetic method of high-speed bulging of tubes has been investigated by Aizawa et al. (1990).

Problems

8.1 A vertical wire of length l and cross-sectional area A is made of a rigid-perfectly plastic material with a rate-sensitive yield stress. σ. The wire is suspended from a ceiling and has an attached mass m at the lower end which is given an initial axial velocity u_0. The axial displacement and velocity of the mass at any instant t are denoted by x and u, respectively. Neglecting the mass of the wire, form the equation of motion, and using (8.107) show that the final displacement δ when the motion comes to a stop is given by

$$\frac{\delta}{l} = \mu \int_0^{\xi_0} \frac{\xi \, d\xi}{1 + \xi^{1/n}}, \quad \xi = \frac{u}{\alpha l}, \quad \mu = \frac{ml\alpha^2}{YA}$$

where ξ_0 denotes the initial value of ξ. Assuming $n = 5$ and $\xi_0 = 32$, determine the numerical value of δ/l in terms of μ Obtain also the result based on the approximation $\xi \approx \xi_0$ in the denominator of the integrand and verify that it differs only marginally from the exact value.

8.2 A thick-walled spherical shell made of a nonhardening rigid/plastic material with a uniaxial yield stress Y, and having an internal radius a_0 and external radius b_0, is subjected to an explosive internal pressure that produces large plastic expansion of the shell. Let p denote the magnitude of the pressure at any instant when the radii of the shell have increased to a and b respectively. Using the equation of motion and the yield criterion, together with the incompressibility condition, show that

$$\frac{d\lambda}{d\xi} + \left\{4 - \left(1 + \frac{a}{b}\right)\left(1 + \frac{a^2}{b^2}\right)\right\}\lambda = \frac{(p/Y) - 2\ln(b/a)}{1 - a/b}, \quad \lambda = \frac{\rho v_a{}^2}{2Y}, \quad \xi = \ln\frac{a}{a_0}$$

where ρ is the density of the material and $v_a = da/dt$ is the radial velocity at $r = a$. Using the relation $p = p_e (a_0/a)^{3\gamma}$, this equation can be integrated numerically with the initial condition $\lambda = 0$ when $\xi = 0$ for any given values of b_0/a_0, p_0/Y, and γ. Verify that the hoop stress is tensile throughout the shell, vanishing at $r = a$ when $p = Y$, which may be regarded as the criterion for the dynamic rupture.

8.3 A thin spherical shell having an initial thickness h_0, mean radius a_0, and density ρ is subjected to an internal explosion by detonating a concentric spherical charge. The explosion pressure is large enough to be approximated by the relation $p = \rho h \dot{v}$, where v is the radial velocity and h the instantaneous thickness when the shell radius is a. Using the relation $p = p_0 (a_0/a)^{3\gamma}$, where p_0 is the effective detonation pressure and γ is a constant isentropic expansion index, show that

$$\left(\frac{v}{U}\right)^2 = \frac{2p_0}{3(\gamma-1)Y}\left\{1 - \left(\frac{a_0}{a}\right)^{3(\gamma-1)}\right\}, \quad U = \sqrt{\frac{Ya_0}{\rho h_0}}$$

Note that the velocity v rapidly approaches a constant value obtained by omitting the second term in the curly brackets. Based on this constant velocity, show the time taken by the shell to rupture (defined by $p = Y$) and the corresponding shell radius are given by

$$t_f = \frac{U}{a_0}\sqrt{(\gamma-1)\frac{3Y}{2p_0}}\left(\frac{a_f}{a_0} - 1\right), \quad \frac{p_0}{Y} = \left(\frac{a_f}{a_0}\right)^{3\gamma}$$

8.4 A lateral load P is suddenly applied at the tip of a cantilever of length l and mass m at $t = 0$, and held constant for a short period. For moderate values of P exceeding the static collapse load $P_0 = M_0/l$, the beam rotates as a rigid body about a plastic hinge formed at the built-in end. Considering the equations of motion of a segment of length x measured from the tip, and using th boundary conditions, obtain the distribution of shearing force Q and bending moment M in terms of $p = P/P_0$ as

$$\frac{Q}{P_0} = -p + (p-1)\frac{3x}{2l}\left(2 - \frac{x}{l}\right), \quad \frac{M}{M_0} = -\frac{x}{l}\left\{p - (p-1)\frac{x}{2l}\left(3 - \frac{x}{l}\right)\right\}, \quad 1 \le p \le 3$$

Verify that the reaction at the built-in end vanishes when $p = 3$, and must continue to do so when $p > 3$, for which the plastic hinge is located at a distance $b_0 < l$ from the tip of the cantilever. Considering the equations of motion of the segment of length b_0 - x, and using the boundary conditions, show that

$$\frac{Q}{P_0} = -p\left(1 - \frac{x}{b_0}\right)^2, \quad \frac{M}{M_0} = -\frac{x}{l}\left(3 - \frac{3x}{b_0} + \frac{x^2}{b_0^2}\right), \quad \frac{b_0}{l} = \frac{3}{p}, \quad p \ge 3$$

over the length $0 \le x \le b_0$, the remainder of the cantilever being at rest. Verify that the tip acceleration is constant during this phase, giving a tip deflection of $w_0 = p^2 t^2 (P_0/3m)$ when $p \ge 3$.

8.5 Considering the range of loading $p \ge 3$, applied over a short period $0 \le t \le t_0$, suppose that the load is suddenly removed. The plastic hinge then moves away from $x = b_0$ toward the built-in end to assume a position $x = b$ at any instant t, the part of the cantilever traversed by the plastic hinge being curved. Form the equations of linear and angular momentum to show that $b = 3t/pt_0$ during this phase. Setting $\delta = P_0 t_0^2/m$, prove that the tip deflection and the shape of the curved portion are given by

$$\frac{w_0}{\delta} = \frac{p^2}{3}\left(1 + 2\ln\frac{b}{b_0}\right), \quad \frac{w}{\delta} = \frac{2p^2}{3}\left(\ln\frac{b}{x} + \frac{x}{b} - 1\right), \quad (b_0 \leq x \leq b)$$

On reaching the built-in end $x = l$, the plastic hinge becomes anchored at the root, while the cantilever rotates as a rigid whole with a constant angular acceleration until the motion ceases. Show that the tip velocity during this phase and the associated tip deflection are given by

$$\frac{\dot{w}_0}{\delta} = \frac{2}{t_0}\left(2p - \frac{3t}{t_0}\right), \quad \frac{w_0}{\delta} = \left(\frac{3t}{t_0} - p\right)\left(p - \frac{t}{t_0}\right) + \frac{p^2}{3}\left(1 + 2\ln\frac{p}{3}\right), \quad \frac{p}{3} \leq \frac{t}{t_0} \leq \frac{2p}{3}$$

8.6 A free-ended beam of length $2l$, made of a rigid/plastic material with a fully plastic moment M_0, is subjected to a central impact load P which increases with time. Show that a plastic hinge begins to form at the middle of the beam when $q = Pl/4\,M_0$ is equal to unity. For $q > 1$, each half of the beam rotates as a rigid body with an angular acceleration α, while the central hinge moves on with a linear acceleration a. Considering the equations of motion of one-half of the beam, show that

$$\frac{\mu l^2 a}{M_0} = 2\,(4q - 3)\,, \quad \frac{\mu l^3 \alpha}{M_0} = 12\,(q - 1)$$

where μ is the mass per unit length of the beam. Derive an expression of the bending moment at a distance x from the middle of the beam, and show that the maximum value of the moment has the magnitude M_0 when $3x/l = q/(q-1)$, the value of q at this stage being 5.725 given by the cubic

$$4\left(\frac{q}{3}\right)^3 - 10.5\left(\frac{q}{3}\right)^2 + 6\left(\frac{q}{3}\right) - 1 = 0$$

8.7 An annular plate, made of a rigid/plastic Tresca material, is clamped along its inner edge $r = a$ and is given a constant normal velocity U along its free outer edge $r = b$. A plastic hinge initiated along the outer edge at $t = 0$ travels inward to some radius $\rho = \lambda b$ at a generic instant. Assuming a velocity field $\dot{w} = U(r - \rho)/(b - \rho)$ for $\rho \leq r \leq b$, using the equations of dynamic equilibrium with $M_\theta = -M_0$, and setting $t_0 = \mu\, b^2 U/12\, M_0$, where μ denotes the surface density of the material, show that

$$t_0\dot{\lambda} = -\frac{1}{(1-\lambda)(1+3\lambda)}, \quad q = -\frac{2\,(1+2\lambda)}{(1-\lambda)(1+3\lambda)}$$

where q is the value of Q/bM_0 at $r = b$. Show also that the time interval t and the deflection w at any radius r can be expressed in terms of the parameter λ as

$$t = t_0\,(1-\lambda)\left(1-\lambda^2\right), \quad w = \frac{1}{2}Ut_0\,(\xi - \lambda)^2\,(1 + \xi + 2\lambda)\,, \quad \lambda \leq \xi \leq 1$$

where $\xi = r/b$. Derive also the radial bending moment distribution in this part of the plate in the form

$$\frac{M_r}{M_0} = -1 + \left(\frac{\xi - \lambda}{1 - \lambda}\right)\left\{\frac{(2 - \lambda - \xi)\left(\lambda^2 + \xi^2\right) + 2\lambda\xi - 2\,(3 - 2\lambda)\,\lambda^2}{(1-\lambda)^3\,(1+3\lambda)},\right\} \quad \lambda \leq \xi \leq 1$$

8.8 In the preceding problem, let the imposed velocity U be suddenly removed at the instant $t = t_1$, when the plastic hinge reaches the clamped edge $r = a$. The plate then rotates about this edge with a deceleration, the deflection of the plate at any subsequent instant being expressed in the form

$$w(\xi, t) = w_1 + U\left(\frac{\xi - \alpha}{1 - \alpha}\right) f(t), \quad t \geq t_1 = t_0 (1 - \alpha)^2 (1 + \alpha),$$

where $\alpha = a/b$, and w_1 is given by the preceding expression of w considered at $\lambda = \alpha$. The function $f(t)$ must satisfy the initial conditions $f'(t_1) = 1$ and $f(t_1) = 0$. Show that the equation of dynamic equilibrium for the bending moment, along with the boundary conditions, can be satisfied if

$$f(t) = (t - t_1)\left\{1 - \frac{1}{2}\left(\frac{t - t_1}{t_{22} - t_1}\right)\right\}, \quad t_2 - t_1 = t_0 (1 - \alpha)^2 (3 + \alpha)$$

where t_2 denotes the instant when the motion stops. Verify that the final deflection of the plate exceeds w_1 by the amount $(Ut_0/2)(\xi - \alpha)(1 - \alpha)(3 + \alpha)$, and hence obtain the final shape of the plate.

8.9 A rigid/plastic circular cylindrical shell of length $2l$, mean radius a, and thickness h, is clamped at both ends and subjected to an explosive blast loading, in which the external pressure p rises instantaneously to a peak value and then decays exponentially with time according to the law

$$q = \frac{pa}{Yh} = \lambda q_0 e^{-\tau}, \quad q_0 = 1 + \frac{2}{\omega^2}, \quad \omega = l\sqrt{\frac{2}{ah}}, \quad \tau = \frac{t}{t_0}$$

where $\lambda > 1$ is a constant, and t_0 is a constant time. Let $x = \xi l$ denote the axial distance measured from the left-hand edge of the shell. Using the square yield condition, show that the axial bending moment, radial velocity, and deflection of the shell for moderate values of λ are given by

$$m_x = -1 + \xi\left(3 - \xi^2\right) + \left(\frac{\lambda q_0 e^{-\tau} - 1}{q_0 - 1}\right)\xi(1 - \xi)^2, \quad 1 \leq \lambda \leq 3 - \frac{2}{q_0}$$

$$\frac{\dot{w}}{\delta} = \frac{3}{2} q_0 \left[\lambda\left(1 - e^{-\tau}\right) - \tau\right]\xi, \quad \frac{w}{\delta} = \frac{3}{2} q_0 \left[\tau\left(\lambda - \frac{\tau}{2}\right) - \lambda\left(1 - e^{-\tau}\right)\right]\xi$$

where $\delta = Yht_0^2/2a\mu$ with μ denoting the surface density, while the dot denotes the derivative with respect to τ. Determine the central deflection of he shell when the motion comes to a stop.

8.10 Consider the range $\lambda \geq 3 - 2/q_0$, for which there is a moving hinge circle located at $\xi = \rho$. The stress point remains on the same side of the locus over the region $0 \leq \xi \leq \rho$, while the remainder of the shell corresponds to the corner of the yield locus. Integrating the equation of motion, obtain the radial acceleration for the two portions of the shell, and hence show that

$$\frac{\dot{w}}{\delta} = \frac{3\xi}{2}\int_0^{\tau}\left[\left(\lambda q_0 e^{-\tau} - 1\right) - \frac{q_0 - 1}{\rho^2}\right]\frac{d\tau}{\rho} \qquad (0 \leq \xi \leq \rho),$$

$$\frac{\dot{w}}{\delta} = \lambda q_0\left(1 - e^{-\tau}\right) - \tau \qquad (\rho \leq \xi \leq 1)$$

Invoking the continuity of the radial velocity at $\xi = \rho$, dividing the resulting equation by ρ, and then differentiating with respect to τ, show that the parameter ρ, the time τ_1 for which $\rho = 1$, and the final deflection of the central section are given by

$$\rho^2 = \frac{3\,(q_0 - 1)\,\tau}{\lambda q_0\,(1 - e^{-\tau}) - \tau} \quad (0 \leq \tau \leq \tau_1), \quad \lambda\left(\frac{1 - e^{-\tau_1}}{\tau_1}\right) = 3 - \frac{2}{q_0}$$

$$\frac{w_0}{\delta} = \frac{3}{2}q_0\left\{\lambda\left(\tau_2 - \frac{\tau_1}{3}\right) - \tau_2\left(1 + \frac{\tau_2}{2}\right)\right\} + \left(\frac{3}{2}q_0 - 1\right)\tau_1\left(1 + \frac{\tau_1}{2}\right)$$

where $\tau_2 > \tau_1$ denotes the instant when the motion stops. Assuming $q_0 = 1.4$, obtain a graphical plot for the variation of w_0/δ with λ, covering the range $1 \leq \lambda \leq 10$.

Chapter 9
The Finite Element Method

In the numerical solution of engineering problems, it is often convenient to assume the physical domain to consist of an assemblage of a finite number of subdomains, called finite elements, which are connected with one another along their interfaces. The distribution of a governing physical parameter within each element is approximated by a suitable continuous function, which is uniquely defined in terms of its values at a specified number of nodal points that are usually located along the boundary of the element. The solution to the original boundary value problem is often reduced to that of a variational problem involving the nodal point values of the unknown parameter. In this chapter, we shall be concerned with a rigid/plastic formulation of the finite element method, a complete elastic/plastic formulation of the problem being available elsewhere (Chakrabarty, 2006).

9.1 Fundamental Principles

The technological forming of metals generally involves plastic strains which dominate over the elastic strains, and the rigid/plastic approximation of material response is therefore appropriate in most cases. At any stage of the deformation process, the current shape of the workpiece and the associated strain distribution are supposed to be known from the previous computation. It is further assumed that the influence of geometry changes during an incremental loading of the workpiece may be disregarded in a variational formulation for the estimation of the strain increment suffered by each material element. The geometry of the workpiece can be subsequently updated on the basis of the computed strain increment, and the solution can be continued in a stepwise manner. The simplest rigid/plastic formulation of the finite element method, widely used for the analysis of metal-forming processes (Kobayashi et al., 989), will be described in what follows.

9.1.1 The Variational Formulation

In a typical boundary value problem, the surface traction F_j is prescribed over a part of the boundary, and the velocity v_j is prescribed over the remainder. Let $\dot{\varepsilon}_{ij}$ denote

J. Chakrabarty, *Applied Plasticity, Second Edition*, Mechanical Engineering Series,
DOI 10.1007/978-0-387-77674-3_9, © Springer Science+Business Media, LLC 2010

the true strain rate corresponding to any kinematically admissible velocity field, and let σ_{ij} denote the associated true stress that is not necessarily in equilibrium. Among a sufficiently wide class of admissible velocity fields, the actual field corresponds to a stationary value of the functional

$$U = \int \sigma_{ij}\dot{\varepsilon}_{ij}\, dV - \int F_j v_j\, dS.$$

The variational principle is analogous to the conventional upper bound technique for the estimation of the yield point load. The finite element approach, because of its discretization, allows the consideration of a much wider class of velocity fields than that possible in the usual upper bound analysis. The restriction imposed on the admissible velocity field by the incompressibility of the material can be removed by introducing a large positive constant Γ, known as the penalty constant, which allows the variational functional to be written in the modified form

$$U = \int \sigma_{ij}\dot{\varepsilon}_{ij}\, dV + \frac{1}{2}\Gamma \int \dot{\varepsilon}_{kk}^2\, dV - \int F_j v_j\, dS. \tag{9.1}$$

Denoting the volumetric strain rate $\dot{\varepsilon}_{kk}$ by $\dot{\lambda}$, and setting the first variation δU to zero, we get

$$\int \bar{\sigma}\delta\dot{\bar{\varepsilon}}\, dV + \Gamma \int \dot{\lambda}\delta\dot{\lambda}\, dV - \int F_j \delta v_j\, dS = 0, \tag{9.2}$$

where $\bar{\sigma}$ and $\dot{\bar{\varepsilon}}$ are the effective stress and strain rate, respectively. The parameter Γ, which is similar to the elastic bulk modulus, must be carefully chosen, since too large a value of Γ would cause difficulties in the convergence, while too small a value of Γ would result in unusually large changes in volume. An appropriate choice of Γ seems to be that for which $\dot{\lambda}$ is restricted to the order of 10^{-3} times the mean effective strain rate in the material (Kobayashi et al., 1989).

Rigid zones, which generally coexist with plastically deforming zones in most metal-forming processes, may be identified by the occurrence of effective strain rates that are smaller than a certain limiting value $\dot{\varepsilon}_0$. These regions can be approximately included in the analysis by setting $\bar{\sigma} = h\dot{\bar{\varepsilon}}$ in the first integral of (9.2), where h is a constant. Such a modification of the variational equation is necessary only over those regions which are considered as nearly rigid. Realistic estimates of the extent of the deforming zone can be achieved, without adversely affecting the convergence of the numerical analysis, by taking $\bar{\varepsilon}$ to have an assigned limiting value of 10^{-2} approximately.

9.1.2 Velocity and Strain Rate Vectors

The finite element analysis of the boundary value problem begins with the specification of a velocity distribution within each element. The velocity must have continuous first derivatives within the element and must satisfy the condition of continuity across its interfaces with the adjacent elements. The components of the velocity at any point within an element are completely defined by those at a sufficient number of nodal points, which are generally located along the boundary of the element. Considering a three-dimensional velocity field with rectangular components *u, v,* and *w,* it is possible to express them in the form

$$u = \sum N_\alpha u_\alpha, \quad \mathrm{v} = \sum N_\alpha \mathrm{v}_\alpha, \quad w = \sum N_\alpha w_\alpha \tag{9.3}$$

where u_α, v_α, and w_α are the velocity components at the αth node, N_α is the associated *shape function,* and the summation extends over all the nodal points of the element. The explicit forms of the shape function in specific cases will be discussed later. Equation (9.3) can be conveniently written in the matrix form

$$\mathbf{v} = \mathbf{N}\mathbf{q} \tag{9.4}$$

where **N** is the shape function matrix, **v** is the velocity vector for a generic particle, and **q** is the nodal velocity vector. These vectors are defined as

$$\mathbf{v}^T = \{u, \mathrm{v}, w\}, \qquad \mathbf{q}^T = \{u_1, \mathrm{v}_1, w_1, u_2, \mathrm{v}_2, w_2, \ldots\},$$

where the superscript T denotes the transpose. The shape function matrix assumes the form

$$\mathbf{N} = \begin{bmatrix} N_1 & 0 & 0 & N_2 & 0 & 0 & N_3 & 0 & 0 & \ldots \\ 0 & N_1 & 0 & 0 & N_2 & 0 & 0 & N_3 & 0 & \ldots \\ 0 & 0 & N_1 & 0 & 0 & N_3 & 0 & 0 & N_3 & \ldots \end{bmatrix} \tag{9.5}$$

The total number of columns in the N-matrix is equal to the nodal degrees of freedom, defined by the number of nodal velocity components.

The components of the true strain rate within the element can be expressed in terms of the nodal velocities and the derivatives of the shape functions. If a typical component of the velocity at the αth node is denoted by $q_i(\alpha)$ then

$$v_i = N_\alpha q_i(\alpha)$$

in view of (9.4), and the expression for the true strain rate tensor becomes

$$\dot{\varepsilon}_{ij} = \frac{1}{2}\frac{\partial \mathrm{v}_i}{\partial x_j} + \frac{\partial \mathrm{v}_j}{\partial x_i} = \frac{1}{2}\frac{\partial N_\alpha}{\partial x_i} q_i(\alpha) + \frac{\partial N_\alpha}{\partial x_i} q_j(\alpha),$$

Denoting the rates of extension in the coordinate directions by $\dot{\varepsilon}_x$, $\dot{\varepsilon}_y$, and $\dot{\varepsilon}_z$, and the associated rates of engineering shear by $\dot{\gamma}_{xy}$, $\dot{\gamma}_{yz}$, and $\dot{\gamma}_{yz}$, the strain rate vector $\dot{\varepsilon}$ may be defined by its transpose

$$\dot{\boldsymbol{\varepsilon}}^{\mathrm{T}} = \left\{\dot{\varepsilon}_x, \dot{\varepsilon}_y, \dot{\varepsilon}_z, \dot{\varepsilon}_{xy}, \dot{\varepsilon}_{yz}, \dot{\varepsilon}_{zx}\right\}.$$

In view of the preceding expression for the strain rate, the components of the vector are given by

$$\begin{aligned}\dot{\varepsilon}_x &= \sum P_\alpha u_\alpha, \quad \dot{\varepsilon}_y = \sum Q_\alpha u_\alpha, \quad \dot{\varepsilon}_z = \sum R_\alpha w_\alpha, \\ \dot{\gamma}_{xy} &= \sum \left(Q_\alpha u_\alpha + P_\alpha v_\alpha\right), \quad \dot{\gamma}_{yz} = \sum \left(R_\alpha v_\alpha + Q_\alpha w_\alpha\right),\end{aligned} \tag{9.6}$$

together with a similar expression for $\dot{\gamma}_{zx}$, where we have introduced the notation

$$P_\alpha = \frac{\partial N_\alpha}{\partial x}, \quad Q_\alpha = \frac{\partial N_\alpha}{\partial y}, \quad R_\alpha = \frac{\partial N_\alpha}{\partial z}. \tag{9.7}$$

It may be noted that the parameters P_α, Q_α, and R_α are generally functions of the space variables. Equation (9.6) may be written in the matrix form

$$\dot{\boldsymbol{\varepsilon}} = \mathbf{Bq} \tag{9.8}$$

where $\mathbf{B}$ denotes the strain rate matrix, which can be written explicitly as

$$\mathbf{B} = \begin{bmatrix} p_1 & 0 & 0 & P_2 & 0 & 0 & P_3 & 0 & 0 & \ldots \\ 0 & Q_1 & 0 & 0 & Q_2 & 0 & 0 & Q_3 & 0 & \ldots \\ 0 & 0 & R_1 & 0 & 0 & R_2 & 0 & 0 & R_3 & \ldots \\ Q_1 & P_1 & 0 & Q_2 & P_2 & 0 & Q_3 & P_3 & 0 & \ldots \\ 0 & R_1 & Q_1 & 0 & R_2 & Q_2 & 0 & R_3 & Q_3 & \ldots \\ R_1 & 0 & P_1 & R_2 & 0 & P_2 & R_3 & 0 & P_3 & \ldots \end{bmatrix}.$$

The number of columns in the **B**-matrix is evidently identical to the nodal degrees of freedom of the considered element.

The equivalent or effective strain rate $\dot{\bar{\varepsilon}}$ and the volumetric strain rate λ, which occur in the finite element analysis, need to be expressed in matrix forms involving the nodal velocity vector $\mathbf{q}$. Since

$$\dot{\bar{\varepsilon}} = \left\{\frac{2}{3}\left(\dot{\varepsilon}_{ij}\dot{\varepsilon}_{ij}\right)\right\}^{1/2} \tag{9.9}$$

when the material is isotropic, it follows from (9.8) that

$$\dot{\bar{\varepsilon}}^{\mathbf{2}} = \dot{\boldsymbol{\varepsilon}}^T \mathbf{D} \dot{\boldsymbol{\varepsilon}} = \mathbf{q}^T \mathbf{S} \mathbf{q}., \quad \mathbf{S} = \mathbf{B}^\mathbf{T} \mathbf{D} \mathbf{B} \tag{9.10}$$

For a three-dimensional deformation mode, $\mathbf{D}$ is a diagonal matrix whose first three diagonal elements are equal to $\frac{2}{3}$ and the last three diagonal elements are equal to $\frac{1}{3}$. In the special case of plane stress, as we shall see later, the matrix $\mathbf{D}$ is not diagonal. In view of the first three relations of (9.6), the volumetric strain rate is

$$\dot{\lambda} = \mathbf{C}^T \mathbf{q}$$

where $\mathbf{C}^T$ is a row vector obtained by adding the vectors represented by the first three rows of **B**. Thus

$$\mathbf{C}^T = \{P_1, Q_1, R_1, P_2, Q_2, R_2, \ldots\}. \tag{9.11}$$

It is interesting to note that this vector is also obtained by premultiplying the matrix B by a row vector whose first three elements are unity and the remaining three elements are zero.

9.1.3 Elemental Stiffness Equations

The global integral appearing in the variational equation (9.2) is actually an assembly of integrals taken over the individual elements in the deforming body. The derivation of the stiffness equation in the matrix form at the elemental level is therefore essential for the establishment of the global stiffness equation. Denoting typical elements of the matrices *N, S,* and *C* by the quantities N_{ij}, S_{ij}, and C_j, respectively, and using (9.4), (9.10), and (9.11), we have

$$\bar{\sigma}\,\delta\dot{\bar{\varepsilon}} = \frac{\bar{\sigma}}{\dot{\bar{\varepsilon}}}\delta\left(\frac{1}{2}\dot{\bar{\varepsilon}}^2\right) = \frac{\bar{\sigma}}{\dot{\bar{\varepsilon}}}\delta\left(\frac{1}{2}S_{ij}q_iq_j\right) = \frac{\bar{\sigma}}{\dot{\bar{\varepsilon}}}\left(\frac{1}{2}S_{ij}q_i\right)\delta q_j,$$

$$\dot{\lambda}\,\delta\dot{\lambda} = C_jq_j\,\delta\left(C_iq_i\right) = C_iC_jq_j\,\delta q_i,$$

$$F_j\,\delta v_j = F_j\delta\left(N_{ji}q_i\right) = N_{ji}F_j\,\delta q_i.$$

The expressions on the left-hand side of the above relations are actually the successive integrands in the variational equation (9.2) in relation to a typical element. Since δq_j is arbitrary, the result becomes

$$\partial U/\partial q_i = \int \left(\bar{\sigma}\,/\dot{\bar{\varepsilon}}\right) S_{ij}q_j\,dV + \Gamma\int C_iC_jq_j\,dV - \int N_{ji}F_j\,dS = 0,$$

In matrix notation, the equation for the minimization of the functional U therefore takes the form

$$\int (\bar{\sigma}\,/\dot{\bar{\varepsilon}})\mathbf{S}\mathbf{q}\,dV + \Gamma\int \mathbf{C}\mathbf{C}^T\mathbf{q}\,dV - \int \mathbf{N}^T\mathbf{F}\,dS = 0 \tag{9.12}$$

where **F** is a column vector representing the applied force *(F_x, F_y, F_z)* on the considered element, the last integral being the equivalent nodal point force. The preceding result represents a set of nonlinear simultaneous equations for the nodal point velocities. By assembling equations of type (9.12) over all the elements in the deforming body, we obtain the global equation for the boundary value problem.

The solution to the stiffness equation (9.12) is generally obtained by an iterative procedure based on a linearization with the help of Taylor's expansion of the functional U in the neighborhood of an assumed solution point $\boldsymbol{q} = \boldsymbol{q}_0$. The condition $\delta U = 0$ may therefore be written as

$$\frac{\partial U}{\partial q_i} + \frac{\partial^2 U}{\partial q_i \partial q_j}\Delta q_j = -f_i + k_{ij}\Delta q_j = 0$$

where the derivatives are considered at $\mathbf{q} = \mathbf{q}_0$, while $\Delta\mathbf{q}$ denotes the first-order correction to the nodal velocity $\boldsymbol{q}_0$. The matrix form of the preceding equation is

$$\mathbf{k}\Delta\mathbf{q} = \mathbf{f} \tag{9.13}$$

where $\mathbf{k}$ denotes the elemental stiffness matrix, and $\mathbf{f}$ is the residual of the nodal point force vector obtained by setting $\mathbf{q} = \mathbf{q}_0$ on the left-hand side of (9.12). Since the first derivative $\partial U/\partial q_i$ is represented by (9.12), it is easily shown that

$$\frac{\partial^2 U}{\partial q_i \partial q_j} = \int \frac{\bar{\sigma}}{\dot{\bar{\varepsilon}}} S_{ij}\, dV + \int \frac{1}{\dot{\bar{\varepsilon}}} \frac{\partial}{\partial \dot{\bar{\varepsilon}}} \left(\frac{\bar{\sigma}}{\dot{\bar{\varepsilon}}}\right) S_{ik} q_k S_{jm} q_m\, dV + \Gamma \int c_i c_j\, dV$$

in view of (9.10). Setting $\partial\left(\bar{\sigma}/\dot{\bar{\varepsilon}}\right)/\dot{\bar{\varepsilon}} = \eta$, the elemental stiffness matrix can be expressed as

$$\mathbf{k} = \int (\bar{\sigma}/\dot{\bar{\varepsilon}})\mathbf{S}\, dV + \int (\eta/\dot{\bar{\varepsilon}})\mathbf{S}\mathbf{q}\mathbf{q}^T\mathbf{S}^T dV + \Gamma \int \mathbf{C}\mathbf{C}^T dV \tag{9.14}$$

The stiffness equations are most conveniently solved by an iterative method in which $\bar{\sigma}/\dot{\bar{\varepsilon}}$ is assumed constant ($\eta = 0$) during each iteration (Oh, 1982). The computation may begin by assuming that that the effective strain rate $\dot{\bar{\varepsilon}}$ in each element is the same as that in the previous step. Since $\bar{\sigma}$ then follows from the computed value of $\bar{\varepsilon}$, the ratio $\bar{\sigma}/\dot{\bar{\varepsilon}}$ is easily evaluated in each element, leading to the stiffness matrix $\mathbf{k}$. The elemental stiffness equations are then assembled to form the global stiffness equations (Section 9.4), which are solved under the prescribed boundary conditions. The solution for the velocity correction furnishes an updated nodal velocity field, and a modified strain rate in each element, which can be used to test the convergence of the solution.

When the effective strain rate $\dot{\bar{\varepsilon}}$ in a given element is found to be less than a preassigned value, it may be considered as nonplastic with an effective stress proportional to $\dot{\bar{\varepsilon}}$. The stiffness equation for such an element should be modified by replacing $\bar{\sigma}/\dot{\bar{\varepsilon}}$ in the leading integral of (9.12) with a constant h. The stiffness matrix $\mathbf{k}$ is similarly obtained by setting $\bar{\sigma}/\dot{\bar{\varepsilon}} = h$ and $\eta = 0$ in (9.14). The penalty function therefore enables us to separate the deforming region from the undeforming one.

9.2 Element Geometry and Shape Function

9.2.1 Triangular Element

It is evident from the preceding discussion that the shape function N_α for any given geometry of the element is a fundamental quantity in the finite element analysis. Equation (9.3) indicates that if $\left(x_\beta, y_\beta\right)$ denote the rectangular coordinates of the βth node, then

$$N_\alpha\left(x_\beta, y_\beta\right) = \delta_{\alpha\beta},$$

where $\delta_{\alpha\beta}$ is the familiar Kronecker delta. It should be noted that any scalar function $f(x, y)$ can be expressed in the same way as the velocity components are, using the same shape functions.

In the case of two-dimensional problems, the simplest finite element is a triangle whose vertices are defined by the coordinates (x_1, y_1), (x_2, y_2), and (x_3, y_3). The coordinates of any point P within the triangle can be expressed in terms of those of its vertices using the transformation

$$\begin{aligned} x &= L_1 x_1 + L_2 x_2 + L_3 x_3, \\ y &= L_1 y_1 + L_2 y_2 + L_3 y_3, \end{aligned} \tag{9.15}$$

where L_1, L_2, and L_3 are the ratios of the areas of the three triangles, formed by joining the generic point P to the vertices 1, 2, and 3, to the total area A of the triangular element shown in Fig. 9.1. It follows from this definition that $L_1 = 0$ along the side 2–3, $L_2 = 0$ along the side 3–1, and $L_3 = 0$ along the side 1–2. In view of the identity $L_1 + L_2 + L_3 = 1$, (9.15) can be solved for the area coordinates L_1, L_2 and L_3 to give

$$\begin{aligned} L_\alpha &= \left(a_\alpha + b_\alpha x + c_\alpha y\right)/2A, \\ 2A &= (x_1 - x_2)(y_2 - y_3) - (x_2 - x_3)(y_1 - y_2), \end{aligned} \tag{9.16}$$

where a_α, b_α, and c_α depend on the coordinates of the vertices of the triangle 1–2–3 and are given by

$$\left.\begin{aligned} a_1 &= x_2 y_3 - x_3 y_2,\ b_1 = y_2 - y_3,\ c_1 = x_3 - x_2, \\ a_2 &= x_3 y_1 - x_1 y_3,\ b_2 = y_3 - y_1,\ c_2 = x_1 - x_3, \\ a_3 &= x_1 y_2 - x_2 y_1,\ b_3 = y_1 - y_2,\ c_3 = x_2 - x_1. \end{aligned}\right\} \tag{9.17}$$

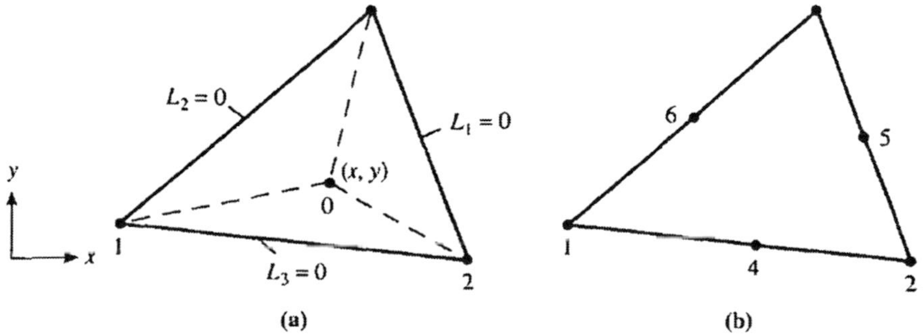

Fig. 9.1 Triangular elements. (**a**) Linear element and (**b**) quadratic element

It is important to note that $L_\alpha\left(x_\beta,y_\beta\right) = \delta_{\alpha\beta}$, which represents a fundamental property of the area coordinates, similar to that for the shape functions.

A linear triangular element consists of three nodes located at its vertices, as shown in Fig. 9.1 (a), the velocity components at any point within the triangle being assumed to vary linearly with x and y. The continuity of the velocity at the nodal points therefore ensures its continuity along the sides of the triangle. The velocity distribution within the triangle may be written as

$$u = N_1 u_1 + N_2 u_2 + N_3 u_3,$$
$$v = N_1 v_1 + N_2 v_2 + N_3 v_3,$$

where the shape functions N_1, N_2, and N_3 are linear functions of x and y. It follows from (9.15) and the linearity of the area coordinates that these functions are identical to L_1, L_2, and L_3 respectively. Thus

$$N_1 = (a_1 + b_1 x + c_1 y)/2A, \quad N_2 = (a_2 + b_2 x + c_2 y)/2A, \quad N_3 = (a_3 + b_3 x + c_3 y)/2A. \tag{9.18}$$

Elements which involve shape functions that are identical to the functions defining the coordinate transformation of type (9.15) are known as *isoparametric elements.* The linear triangular element is therefore isoparametric. The relevant components of the strain rate matrix B for the linear triangular element are

$$\begin{aligned} P_1 &= \frac{y_2 - y_3}{2A}, \quad P_2 = \frac{y_3 - y_1}{2A}, \quad P_3 = -\left(P_1 + P_2\right), \\ Q_1 &= \frac{x_3 - x_2}{2A}, \quad Q_2 = \frac{x_1 - x_3}{2A}, \quad Q_3 = -\left(Q_1 + Q_2\right), \end{aligned} \tag{9.19}$$

in view of (9.7) and (9.18). It may be noted that the components of the strain rate corresponding to the linear triangular element are constant over each element. A useful integral involving exponents of the area coordinates taken over the area of the triangle is

$$\iint L_1^m L_2^n L_3^p \, dxdy = \frac{m!\, n!\, p!(2A)}{(m+n+p+2)!} \tag{9.20}$$

where m, n, and p are integers. The result follows from the fact that the Jacobian of the transformation of coordinates is

$$J = \frac{\partial x}{\partial L_1}\frac{\partial y}{\partial L_2} - \frac{\partial x}{\partial L_2}\frac{\partial y}{\partial L_1} = 2A$$

The linear triangular element can be used for treating finite deformation problems by using a linear interpolation of the Lagrangian strains from the midpoints of the sides of the triangle, as has been shown by Flores (2006).

In a quadratic triangular element, there are three primary nodes located at the vertices of the triangle, and three secondary nodes at the midpoints of the sides of the triangle, as shown in Fig. 9.1(b). The velocity components u and v, which are assumed to be quadratic functions of x and y, are again continuous across the sides of the triangle and are expressed in terms of the nodal values as

$$u = N_1 u_1 + N_2 u_2 + N_3 u_3 + N_4 u_4 + N_5 u_5 + N_6 u_6$$
$$v = N_1 v_1 + N_2 v_2 + N_3 v_3 + N_4 v_4 + N_5 v_5 + N_6 v_6$$

where (u_α, v_α) denote the velocity vector at a typical nodal point α. The shape functions associated with the six nodal velocities can be expressed in terms of the area coordinates L_1, L_2, and L_3, the result being easily shown to be

$$\begin{aligned} N_1 &= L_1(2L_1 - 1), \quad N_2 = L_2(2L_2 - 1), \quad N_3 = L_3(2L_3 - 1), \\ N_4 &= 4L_1L_2, \qquad N_5 = 4L_2L_3, \qquad N_6 = 4L_3L_1. \end{aligned} \tag{9.20}$$

The elements of the strain rate matrix can be determined in the same way as that for the linear triangle. It can be seen that the quadratic element with straight sides is not isoparametric. It is possible, however, to construct curvilinear triangles in this case to form isoparametric elements (Zienkiewicz, 1977).

9.2.2 Quadrilateral Element

In the solution of special problems, it is often convenient to use quadrilateral elements with nodal points located at the corners. The element is generally defined parametrically in terms of auxiliary coordinates (ξ, η), known as natural coordinates, so that the quadrilateral is transformed into a square defined by $\xi \pm 1$ and $\eta = \pm 1$, as shown in Fig. 9.2. The shape functions are bilinear in ξ and η according to the relations

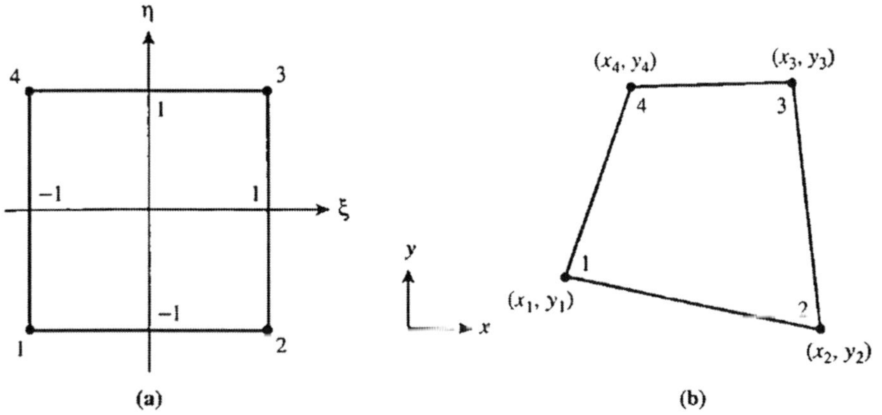

Fig. 9.2 Linear quadratic element. (**a**) Natural coordinates and (**b**) physical coordinates

$$\begin{aligned} N_1 &= \tfrac{1}{4}(1-\xi)(1-\eta), \quad N_2 = \tfrac{1}{4}(1+\xi)(1-\eta), \\ N_3 &= \tfrac{1}{4}(1+\xi)(1+\eta), \quad N_4 = \tfrac{1}{4}(1-\xi)(1+\eta). \end{aligned} \tag{9.21}$$

It is easy to see that $N_\alpha\,(\xi_\alpha,\eta_\alpha) = \delta_{\alpha\beta}$, where (ξ_α,η_α) are the natural coordinates of a typical node. The transformation of the natural coordinates (ξ,η) into the physical coordinates (x, y) is given by

$$\begin{aligned} x &= N_1x_1 + N_2x_2 + N_3x_3 + N_4x_4, \\ y &= N_1y_1 + N_2y_2 + N_3y_3 + N_4y_4. \end{aligned} \tag{9.22}$$

The condition of constancy of the slope dy/dx of each side of the quadrilateral is therefore identically satisfied. Since the coordinate transformation involves the same shape functions as those in the velocity relations

$$u = \sum N_\alpha u_\alpha, \quad v = \sum N_\alpha v_\alpha,$$

where α varies from 1 to 4, the linear quadrilateral element is isoparametric. If we consider an 8-node isoparametric element in which the secondary nodes are located at the midpoints of the sides of the square in the (ξ, η)-plane, the shape of the element in the physical plane is a curvilinear quadrilateral defined by shape functions that are obtained on multiplying the right-hand side of (9.23) by suitable linear functions of ξ and η.

The elements of the strain rate matrix B involve the derivatives of the shape functions with respect to x and y. Since the corresponding derivatives with respect to ξ and η are given by

$$\begin{Bmatrix} \partial N_\alpha/\partial \xi \\ \partial N_\alpha/\partial \eta \end{Bmatrix} = \begin{bmatrix} \partial x/\partial \xi & \partial y/\partial \xi \\ \partial x/\partial \eta & \partial y/\partial \eta \end{bmatrix} \begin{Bmatrix} \partial N_\alpha/\partial \xi \\ \partial N_\alpha/\partial y \end{Bmatrix},$$

where the square matrix is the well-known Jacobian matrix having a determinant J, which is the Jacobian of the transformation, and is given by

$$J = \frac{\partial x}{\partial \xi}\frac{\partial y}{\partial \eta} - \frac{\partial x}{\partial \eta}\frac{\partial y}{\partial \xi} \tag{9.23}$$

A straightforward process of inversion of the preceding matrix equation furnishes

$$\begin{Bmatrix} \partial N_\alpha/\partial x \\ \partial N_\alpha/\partial y \end{Bmatrix} = \frac{1}{J}\begin{bmatrix} \partial y/\partial \eta & -\partial y/\partial \xi \\ -\partial x/\partial \eta & \partial x/\partial \xi \end{bmatrix} \begin{Bmatrix} \partial N_\alpha/\partial \xi \\ \partial N_\alpha/\partial \eta \end{Bmatrix}.$$

In the case of a liner quadrilateral element, the partial derivatives appearing in the Jacobian matrix and its inverse are readily obtained from (9.22) and (9.23), the result being

$$8J = (x_{13}\,y_{24} - x_{24}\,y_{13}) + (x_{34}\,y_{12} - x_{12}\,y_{34})\,\xi + (x_{23}\,y_{14} - x_{14}\,y_{23})\,\eta, \tag{9.24}$$

where $x_{ij} = x_i - x_j$ and $y_{ij} = y_i - y_j$. Since $\partial N_\alpha/\partial x$ and $\partial N_\alpha/\partial y$ are equal to P_α and Q_α, respectively, according to (9.7), we have

$$\left\{\begin{matrix} P_1 \\ P_2 \\ P_3 \\ P_4 \end{matrix}\right\} = \frac{1}{8J}\left\{\begin{matrix} y_{24} & -y_{34}\xi & -y_{23}\eta \\ -y_{13} & +y_{34}\xi & +y_{14}\eta \\ -y_{24} & +y_{12}\xi & -y_{14}\eta \\ y_{13} & -y_{12}\xi & +y_{23}\eta \end{matrix}\right\},$$
$$\left\{\begin{matrix} Q_1 \\ Q_2 \\ Q_3 \\ Q_4 \end{matrix}\right\} = \frac{1}{8J}\left\{\begin{matrix} -x_{24} & +x_{34}\xi & +x_{23}\eta \\ x_{13} & -x_{34}\xi & -x_{14}\eta \\ x_{24} & -x_{12}\xi & +x_{14}\eta \\ -x_{13} & +x_{12}\xi & -x_{23}\eta \end{matrix}\right\}. \tag{9.25}$$

The results for the 8-node isoparametric quadrilateral element are evidently more complex. It is more convenient in this case to evaluate P_α and Q_α numerically for selected values of ξ and η, using the method employed for the linear element.

9.2.3 Hexahedral Brick Element

For treating three-dimensional problems, the quadrilateral element must be replaced by a brick element with eight corners. The simplest isoparametric element involves a node at each of the eight corners of the element, which assumes the form of a cube defined by $\xi = \pm\ 1$, $\eta = \pm\ 1$, and $\zeta = \pm\ 1$, in the associated natural coordinate system (ξ, η, ζ). The shape functions are defined as

$$N_\alpha = \frac{1}{8}\left(1 + \xi_\alpha\xi\right)\left(1 + \eta_\alpha\eta\right)\left(1 + \zeta_\alpha\zeta\right), \tag{9.26}$$

where (ξ_α, η_α, ζ_α) are the natural coordinates of a typical node α. Since $\xi_\alpha^2 = \eta_\alpha^2 = \zeta_\alpha^2 - 1$, the above expression indicates that $N_\alpha = 1$ at the αth node, while $N_\alpha = 0$ at all other nodes due to the vanishing of at least one of its factors. The velocity distribution is given by (9.3), while the coordinate transformation is

$$x = \sum N_\alpha x_\alpha, \quad y = \sum N_\alpha y_\alpha, \quad z = \sum N_\alpha z_\alpha, \tag{9.27}$$

where (x_α, y_α, z_α) are the rectangular coordinates of the αth node. Figure 9.3 shows the brick element defined in both the natural coordinates and the physical coordinates. The velocity field for the brick element can be expressed as

$$u = \sum N_\alpha u_\alpha, \quad \mathrm{v} = \sum N_\alpha \mathrm{v}_\alpha, \quad w = \sum N_\alpha w_\alpha$$

where (u_α, v_α, w_α) are the rectangular components of the velocity vector at the αth node. The 8-node brick element is therefore isoparametric. The Jacobian matrix

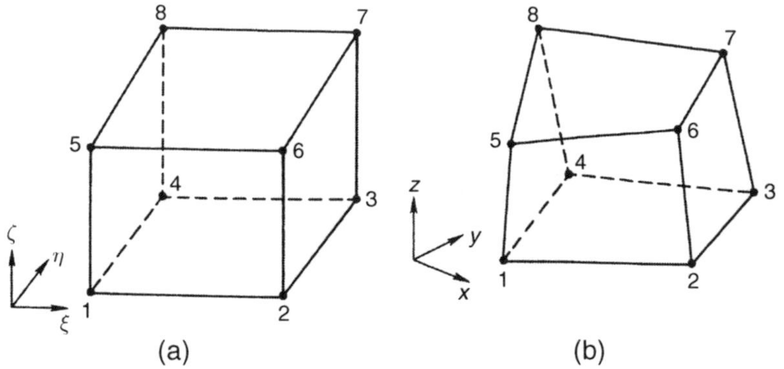

Fig. 9.3 Three–dimensional brick element depicted in (**a**) natural coordinate system, and (**b**) rectangular coordinate system

$$\mathbf{J} = \begin{bmatrix} \partial x/\partial\xi & \partial y/\partial\xi & \partial z/\partial\xi \\ \partial x/\partial\eta & \partial y/\partial\eta & \partial z/\partial\eta \\ \partial x/\partial\zeta & \partial y/\partial\zeta & \partial z/\partial\zeta \end{bmatrix} \tag{9.28}$$

for the transformation of coordinates is easily formed for any selected values of $(\xi,\ \eta,\zeta)$, using (9.26) and (9.27). This matrix and can be inverted numerically to determine the quantities P_α, Q_α, and R_α, using the expression

$$\begin{Bmatrix} P_\alpha \\ Q_\alpha \\ R_\alpha \end{Bmatrix} = \begin{Bmatrix} \partial N_\alpha/\partial x \\ \partial N_\alpha/\partial y \\ \partial N_\alpha/\partial z \end{Bmatrix} = \mathbf{j}^{-1} \begin{Bmatrix} \partial N_\alpha/\partial\xi \\ \partial N_\alpha/\partial\eta \\ \partial N_\alpha/\partial\zeta \end{Bmatrix} \tag{9.29}$$

Since the column vector on the right-hand side of (9.29) is easily obtained using (9.26), the column vector on the left-hand side can be evaluated for any selected node α. The determinant of the Jacobian matrix (9.28) furnishes the value of J in each particular case.

9.3 Matrix Forms in Special Cases

9.3.1 Plane Strain Problems

We begin with the situation where the resultant velocity of each particle is parallel to a given plane, which is considered as the $(x,$ y)-plane, the rectangular components of the velocity of a typical particle being denoted by $(u,\ v)$. Adopting the 4-node isoparametric quadrilateral element shown in Fig. 9.2, the nodal velocity vector q and the shape function matrix N may be written as

$$\mathbf{q}^T = \{u_1\ v_1\ u_2\ v_2\ u_3\ v_3\ u_4\ v_4\},$$
$$\mathbf{N} = \begin{bmatrix} N_1 & 0 & N_2 & 0 & N_3 & 0 & N_4 & 0 \\ 0 & N_1 & 0 & N_2 & 0 & N_3 & 0 & N_4 \end{bmatrix}, \tag{9.30}$$

where N_1, N_2, N_3, and N_4 are given by (9.21) in terms of the natural coordinates (ξ, η). The vectors representing the particle velocity and the strain rate are

$$\mathbf{v} = \begin{Bmatrix} u \\ v \end{Bmatrix} = \mathbf{Nq}, \qquad \dot{\boldsymbol{\varepsilon}} = \begin{Bmatrix} \dot{\varepsilon}_x \\ \dot{\varepsilon}_y \\ \dot{\gamma}_{xy} \end{Bmatrix} = \mathbf{Bq},$$

where $\mathbf{B}$ is the strain rate matrix having the form

$$\mathbf{B} = \begin{bmatrix} P_1 & 0 & P_2 & 0 & P_3 & 0 & P_4 & 0 \\ 0 & Q_1 & 0 & Q_2 & 0 & Q_3 & 0 & Q_4 \\ Q_1 & P_1 & Q_2 & P_2 & Q_3 & P_3 & Q_4 & P_4 \end{bmatrix}, \tag{9.31}$$

where P_ε and Q_4 are given by (9.25) as functions of (ξ, η) in terms of the global coordinates of the nodal points. The equivalent strain rate $\dot{\bar{\varepsilon}}$ and the volumetric strain rate $\dot{\lambda}$ can be evaluated from (9.10) and (9.11), respectively, the relevant matrix and vector being of the form

$$\mathbf{D} = \frac{2}{3}\begin{bmatrix} 1 & 0 & 0 \\ 0 & 1 & 0 \\ 0 & 0 & \frac{1}{2} \end{bmatrix}, \qquad \mathbf{C} = B^T \begin{Bmatrix} 1 \\ 1 \\ 0 \end{Bmatrix}. \tag{9.32}$$

The set of nonlinear equations for the unknown nodal velocities are finally obtained by considering (9.12) for each individual element. The solution is most conveniently obtained by using linearized stiffness equations of type (9.13) as explained before in general terms.

9.3.2 Axially Symmetrical Problems

In problems of axial symmetry, the finite element is taken in the form of a ring whose cross section is identical to the two-dimensional element. For an isoparametric ring element, the global coordinates (r, z) of a generic point inside the element are related to the natural coordinates (ξ, n) by the transformation

$$r = \sum N_\alpha r_\alpha, \quad z = \sum N_\alpha z_\alpha, \tag{9.33}$$

where (r_α, z_α) are the coordinates of a typical node, and N_α is the corresponding shape function that depends on (ξ, η). For the 4-node quadrilateral element shown in Fig. 9.4, the associated shape functions are given by (9.30). The radial and circumferential components of the velocity of a typical particle are given by

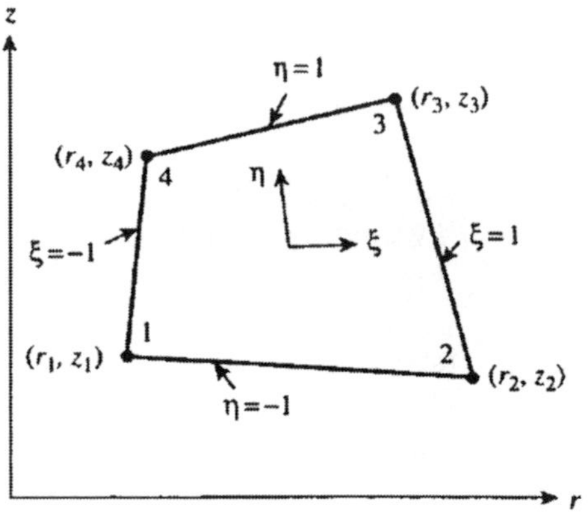

Fig. 9.4 Axisymmetric ring element having the cross section of a 4-node quadrilateral in the meridian plane

$$u = \sum N_\alpha u_\alpha, \quad w = \sum N_\alpha w_\alpha,$$

where $(u_\alpha\ w_\alpha)$ are the nodal velocity components, the shape function matrix N being identical to that in (9.27). The vector representing the strain rate then becomes

$$\dot{\boldsymbol{\varepsilon}} = \begin{Bmatrix} \dot{\varepsilon}_r \\ \dot{\varepsilon}_z \\ \dot{\varepsilon}_\theta \\ \dot{\gamma}''_{rz} \end{Bmatrix} = \begin{bmatrix} \partial/\partial r & 0 \\ 0 & \partial/\partial z \\ 1/r & 0 \\ \partial/\partial z & \partial/\partial r \end{bmatrix} \begin{Bmatrix} u \\ w \end{Bmatrix} = \mathbf{Bq}. \tag{9.34}$$

The nodal velocity vector $\mathbf{q}$ is given by (9.30) with v_α replaced by w_α, and the strain rate matrix $\mathbf{B}$ has the modified form

$$\mathbf{B} = \begin{bmatrix} P_1 & 0 & P_2 & 0 & P_3 & 0 & P_4 & 0 \\ 0 & Q_1 & 0 & Q_2 & 0 & Q_3 & 0 & Q_4 \\ B_1 & 0 & B_2 & 0 & B_3 & 0 & B_4 & 0 \\ Q_1 & P_1 & Q_2 & P_2 & Q_3 & P_3 & Q_4 & P_4 \end{bmatrix}, \tag{9.35}$$

where P_α and Q_α are given by (9.26) with x_{ij} and y_{ij} replaced r_{ij} by and z_{ij}, respectively, the expression for J in (9.24) being similarly modified, while the third row of (9.35) is given by

$$B_\alpha = N_\alpha/r = N_\alpha/(N_1 r_1 + N_2 r_2 + N_3 r_3 + N_4 r_4)\,.$$

The evaluation of $\dot{\bar{\varepsilon}}$ and $\dot{\lambda}$ is identical to that for the plane strain case, the matrix $\mathbf{D}$ and the vector $\mathbf{C}$ being similar to those given by (9.32). The final stiffness equations can be handled in the same way as those for plane strain.

9.3.3 Three-Dimensional Problems

In the three-dimensional finite element analysis, it is generally convenient to use the 8-node isoparametric brick element shown in Fig. 9.3. In the natural coordinate system (ξ, η, ζ), this element is a cube, and the deformation mode is specified by the nodal velocity vector $\mathbf{q}$, where

$$\mathbf{q}^{\mathbf{T}} = [u_1 \quad v_1 \quad w_1 \quad u_2 \quad v_2 \quad \ldots \quad u_8 \quad v_8 \quad w_8]$$

The transformation between the natural and global coordinates defined by (9.27) in terms of the shape functions are given by (9.26), the shape function matrix being

$$\mathbf{N} = \begin{bmatrix} N_1 & 0 & 0 & N_2 & 0 & 0 & \ldots. & N_8 & 0 & 0 \\ 0 & N_1 & 0 & 0 & N_2 & 0 & \ldots. & 0 & N_8 & 0 \\ 0 & 0 & N_1 & 0 & 0 & N_2 & \ldots. & 0 & 0 & N_8.. \end{bmatrix} \tag{9.36}$$

The velocity field within the element and the associated strain rate vector, having the six rectangular components given by (9.6), may be written as

$$\mathbf{u} = \begin{Bmatrix} u \\ v \\ w \end{Bmatrix} = \mathbf{Nq}, \quad \dot{\boldsymbol{\varepsilon}} = \mathbf{Bq}$$

where $\boldsymbol{B}$ is the strain rate matrix for the 8-node brick element, in which the components of train rate are given by (9.26). It is easily shown that

$$\mathbf{B} = \begin{bmatrix} P_1 & 0 & 0 & P_2 & 0 & 0 & . & P_8 & 0 & 0 \\ 0 & Q_1 & 0 & 0 & Q_2 & 0 & .. & 0 & Q_8 & 0 \\ 0 & 0 & R_1 & 0 & 0 & R_2 & .. & 0 & 0 & R_8 \\ Q_1 & P_1 & 0 & Q_2 & P_2 & 0 & .. & Q_8 & P_8 & 0 \\ 0 & R_1 & Q_1 & 0 & R_2 & Q_2 & .. & 0 & R_8 & Q_8 \\ R_1 & 0 & P_1 & R_2 & 0 & P_2 & .. & R_8 & 0 & P_8 \end{bmatrix} \tag{9.37}$$

The Jacobian of the transformation of coordinates, given by the determinant of the matrix (9.28), is easily evaluated numerically in each particular case. The square matrix **S**, which defines the stiffness matrix, is given by (9.10), where **D** is a diagonal matrix in which each of the six diagonal element is 2/3.

9.4 Sheet Metal Forming

9.4.1 Basic Equations for Sheet Metals

In the plastic forming of sheet metal, the stress component in the thickness direction is generally disregarded. A state of plane stress therefore exists in each element of the sheet, which is assumed to be orthotropic with the anisotropic axes coinciding with the rolling, transverse, and thickness directions. Considering a biaxial loading of the sheet, we choose a set of rectangular axes in which the x- and y-axes are directed along the rolling and transverse directions, respectively, the z-axis being taken along the normal to the sheet. The effective stress in a material element is given by the yield criterion, and may be defined in such a way that it reduces to the current uniaxial yield stress in the rolling direction. Using the quadratic yield criterion for simplicity (Section 6.2), we write

$$\bar{\sigma} = \left\{ {\sigma_x}^2 - \left(\frac{2H}{G+H} \right) \sigma_x \sigma_y + \left(\frac{F+H}{G+H} \right) \sigma_y^2 + \left(\frac{2N}{G+H} \right) {\tau_{xy}}^2 \right\}^{1/2} \tag{9.38}$$

where F, G, H, and N are parameters defining the state of planar anisotropy of the sheet metal. When the hypothesis of strain equivalence is adopted for the hardening of the material, the effective strain rate in a deforming element may be written as

$$\dot{\bar{\varepsilon}} = (G+H) \left\{ \frac{\dot{\varepsilon}_x^2 + \dot{\varepsilon}_x \dot{\varepsilon}_y + \dot{\varepsilon}_y^2 + \dot{\gamma}_{xy}^2/4}{G^2 + GH + H^2} \right\}^{1/2},$$

so that $\dot{\bar{\varepsilon}}$ becomes identical to the longitudinal strain rate $\dot{\varepsilon}_x$ in the case of a simple tension applied in the rolling direction. This is easily verified by setting $\sigma = \sigma$ as the only nonzero stress in the associated flow rule, which may be written as

$$\frac{\dot{\varepsilon}_x}{(G+H)\,\sigma_x - H\sigma_y} = \frac{\dot{\varepsilon}_y}{(F+H)\,\sigma_y - H\sigma_x} = \frac{\dot{\gamma}_{xy}}{N} = \dot{\lambda} \tag{9.39}$$

where $\dot{\lambda}$ is a positive scalar. The thickness strain rate follows from the incompressibility condition $\dot{\varepsilon}_z = -(\dot{\varepsilon}_x + \dot{\varepsilon}_y)$ Since the rate of plastic work per unit volume is equal to $\bar{\sigma}\dot{\lambda}$ in view of (9.39) and (9.38), the effective strain rate according to the hypothesis of work equivalence is equal to $\dot{\lambda}$, giving

$$\dot{\bar{\varepsilon}} = \sqrt{G+H} \left\{ \frac{(F+H)\,\dot{\varepsilon}_x^2 + 2H\dot{\varepsilon}_x\dot{\varepsilon}_y + (G+H)\,\dot{\varepsilon}_y^2}{FG + GH + HF} + \frac{\dot{\gamma}_{xy}^2}{2N} \right\}^{1/2}, \tag{9.40}$$

which is obtained by expressing the stresses in terms of the strain rates using the flow rule (9.39), and substituting them into the field criterion (9.38).

The ratios of the anisotropic parameters F, G, H, and N can be determined from the measured R-values of the sheet in the rolling, transverse, and at 45° to the rolling direction. The strain rate vector in the plane stress formulation is identical to those for plane strain. It is convenient to express the effective strain rate in the matrix form

$$\dot{\bar{\varepsilon}} = \sqrt{\dot{\boldsymbol{\varepsilon}}^{\mathbf{T}} \mathbf{D} \dot{\boldsymbol{\varepsilon}}}$$

as before, but the forms of the square matrix **D** for the anisotropic material depends on the hardening hypothesis and is expressed by

$$\mathbf{D} = \frac{(G+H)^2}{2S}\begin{bmatrix} 2 & 1 & 0 \\ 1 & 2 & 0 \\ 0 & 0 & \frac{1}{2} \end{bmatrix}, \quad \mathbf{D} = \frac{G+H}{T}\begin{bmatrix} F+H & H & 0 \\ H & G+H & 0 \\ 0 & 0 & T/2N \end{bmatrix}, \tag{9.41}$$

where $S = G^2+GH+H^2$ and $T = FG+GH+HF$. The first expression in (9.41) corresponds to the hypothesis of strain equivalence and the second expression to that of work equivalence. When the material exhibits normal anisotropy with a uniform R-value, it is only necessary to set $F = G$, $H = RG$, $N = (1+2R)G$, and $T = NG$ in the preceding relations. The shape function matrix and the strain rate matrix for a given shape of the element are identical to those for plane strain. The application of the preceding theory to the flange drawing and bore-expanding processes has been reported by Lee and Kobayashi (1975). A finite element formulation of the problem based on the biquadratic yield criterion (Section 6.2) has been discussed by Gotoh (1978, 1980).

9.4.2 Axisymmetric Sheet Forming

In the case of out-of-plane deformations of the sheet metal, such as in the hydraulic bulging and punch stretching, the deformed sheet at each stage may be regarded as a membrane with a state of plane stress existing in each element. Additional equations are obviously necessary to determine the deformed shape of the sheet metal, and the distribution of stress and strain in the workpiece.

When the deformed sheet forms a surface of revolution, it may be approximated by a succession of conical frustums, each frustum being treated as a finite element, Fig. 9.5. Consider a line element along the meridian, extending between the nodal points 1 and 2 with coordinates (r_1, z_1) and (r_2, z_2) respectively. If the radial and axial coordinates of a generic particle are denoted by r and z, and the corresponding components of the velocity are denoted by u and w, respectively, then

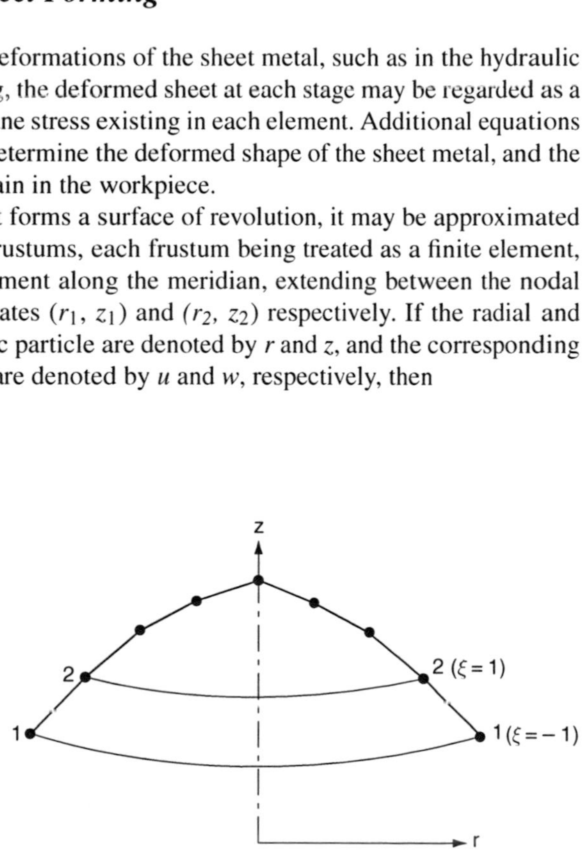

Fig. 9.5 Finite element approximation of a surface of revolution developed in the axisymmetric forming of a sheet metal

$$u = \frac{1}{2}(u_1 + u_2) + \frac{1}{2}(u_2 - u_1)\xi, \quad r = \frac{1}{2}(r_1 + r_2) + \frac{1}{2}(r_2 - r_1)\xi, \quad -1 \le \xi \le 1$$

where u_l and u_2 are the radial velocities at the nodal points 1 and 2, respectively. Similar relations may be written down for the components w and z. If ϕ is the angle made by the surface normal with the axis of symmetry, which coincides with the z-axis, then the circumferential and meridional components of the strain rate are given by

$$\dot{\varepsilon} = \frac{u}{r} = \frac{(u_2 + u_1) + (u_2 - u_1)\xi}{(r_2 + r_1) + (r_2 - r_1)\xi}$$

$$\dot{\varepsilon}_\phi = \frac{\partial u}{\partial r} + \dot{\phi}\tan\phi = \left(\frac{u_2 - u_1}{r_2 - r_1}\right) + \dot{\alpha}, \quad \alpha = \left\{1 + \left(\frac{z_2 - z_1}{r_2 - r_1}\right)^2\right\}^{1/2} \qquad (9.42)$$

The effective stress and strain rate for a uniform R-value material with normal anisotropy, according to the hypothesis of work equivalence, may be written as

$$\bar{\sigma} = \left\{\sigma_\theta^2 - \frac{2R}{1+R}\sigma_\theta\sigma_\phi + \sigma_\phi^2\right\}^{1/2}$$

$$\dot{\bar{\varepsilon}} = \frac{1+R}{\sqrt{1+2R}}\left\{\dot{\varepsilon}_\theta^2 + \frac{2R}{1+R}\dot{\varepsilon}_\theta\dot{\varepsilon}_\phi + \dot{\varepsilon}_\phi^2\right\}^{1/2}$$

in view of (9.38) and (9.40). The effective strain can be expressed in a matrix form as before in terms of a square **D**, the result being

$$\bar{\varepsilon} = \sqrt{\dot{\boldsymbol{\varepsilon}}^{\mathbf{T}}\mathbf{D}\dot{\boldsymbol{\varepsilon}}}, \quad \dot{\boldsymbol{\varepsilon}} = \begin{Bmatrix} \dot{\varepsilon}_\theta \\ \dot{\varepsilon}_\phi \end{Bmatrix}, \quad \mathbf{D} = \frac{l+R}{l+2R}\begin{bmatrix} l+R & R \\ R & l+R \end{bmatrix} \qquad (9.43)$$

The finite element procedure is based on a variational principle similar to (9.1). Since the condition of incompressibility need not be dealt with, the penalty function may be omitted for the variational formulation. It is convenient in this case to multiply the integrands in ( ) by a small time increment Δt and write the functional in the modified form

$$U = \int \vec{\sigma}(\Delta\vec{\varepsilon})\, dV - \int F_j\,(\Delta u_j)\, dS,$$

where $\Delta\bar{\varepsilon} = \dot{\bar{\varepsilon}}\Delta t$, and $\Delta u_j = v_j\Delta t$, not to be confused with the radial velocity. Considering the variation of U, and using the fact that $\delta\bar{\sigma} = H\delta\,(\Delta\bar{\varepsilon},)$, where H is the plastic modulus, we have

$$\delta U = \int h\left(\frac{\vec{\sigma}}{\Delta\vec{\varepsilon}} + H\right)(\Delta\bar{\varepsilon})\,\delta\,(\Delta\bar{\varepsilon})\, dA - \int F_j\, N_{ij}\,\delta(\Delta q_j)\, dS \qquad (9.44)$$

where h is the local sheet thickness, A is the surface area of the element, N_{ji} is a typical element of the shape function transpose matrix of which the nonzero components are $N_{11} = N_{22} = (1 - \xi)/2$ and $N_{31} = N_{42} = (1 + \xi)/2$, and Δq_j is a typical component of the nodal displacement increment vector, similar to the nodal velocity vector.

The effective strain increment $\Delta\bar{\varepsilon}$ can be expressed in terms of the components of the strain increment by replacing the strain rates appearing on the right-hand side of (9.39) by the corresponding strain increments. In matrix notation, the effective strain increment becomes

$$\Delta\bar{\varepsilon} = \left\{(\Delta\boldsymbol{\varepsilon}^{\mathrm{T}})\mathbf{D}(\Delta\boldsymbol{\varepsilon})\right\}^{1/2}, \quad \Delta\boldsymbol{\varepsilon}^{\mathrm{T}} = \left[\Delta\varepsilon_\theta \ \Delta\varepsilon_\phi\right],$$

in view of (9.43). The strain increments are evidently given by the strain rates in (9.43) multiplied by Δt. It follows from (9.43) that (Toh and Kobayashi, 1985)

$$\frac{\partial U}{\partial q_{i,}} = \int h\left(\frac{\bar{\sigma}}{\Delta\bar{\varepsilon}} + H\right) a_i \, dA - \int F_j N_{ji} \, dS, \quad a_i = \left\{\frac{\partial}{\partial q_i}\left(\Delta\boldsymbol{\varepsilon}^{\mathrm{T}}\right)\right\}\mathbf{D}\left(\Delta\boldsymbol{\varepsilon}\right) \quad (9.45)$$

The second derivative of the functional U with respect to a nodal velocity component follows from (9.45) and is easily shown to be

$$\frac{\partial^2 U}{\partial q_{i,}\partial q_j} = \int h\left(\frac{\bar{\sigma}}{\Delta\bar{\varepsilon}} + H\right)\left(b_{ij} + c_{ij}\right) dA - \int \frac{h\bar{\sigma}}{(\Delta\bar{\varepsilon})^2}\left(a_i a_j\right) dA,$$

$$b_{ij} = \left\{\frac{\partial^2}{\partial q_i \partial q_j}\left(\Delta\boldsymbol{\varepsilon}^{\mathrm{T}}\right)\right\}\mathbf{D}(\Delta\boldsymbol{\varepsilon}), \quad c_{ij} = \left\{\frac{\partial}{\partial q_i}\left(\Delta\boldsymbol{\varepsilon}^{\mathrm{T}}\right)\right\}\mathbf{D}\left\{\frac{\partial}{\partial q_i}(\Delta\boldsymbol{\varepsilon})\right\} \quad (9.46)$$

The element stiffness equations for the sheet metal-forming process now given by

$$k_{ij}\Delta q_j = f_i$$

where k_{ij} and $-f_i$ are given by the first equations of (9.46) and (9.45), respectively, while Δq_j represents the displacement correction vector. The formation of the global stiffness equations and their solution can be carried out in the same way as described before. A finite element formulation of the problem based on a nonlinear membrane shell theory has been considered by Wang and Budiansky (1978). An implementation of the nonquadratic yield criterion (Section 6.2) into the finite element formulation has been presented by Wang (1984).

9.4.3 Sheet Forming of Arbitrary Shapes

In general, the out-of-plane deformation of a sheet metal is complicated by the fact that the principal axes of the stress and strain increments are not known in advance. It is therefore necessary to extend the analysis given above to deal with the general sheet forming process. Assuming a state of normal anisotropy of the sheet metal as before, the effective stress and strain increments are obtained from (9.38) and (9.40) in the form

$$\bar{\sigma} = \left\{ \sigma_x^2 - \left(\frac{2R}{1+R} \right) \sigma_x \sigma_y + \sigma_y^2 + 2 \left(\frac{1+2R}{1+R} \right) \tau_{xy}^2 \right\}^{1/2}$$

$$\Delta\bar{\varepsilon} = \sqrt{\frac{1+R}{1+2R}} \left\{ (1+R) \left[(\Delta\varepsilon_x)^2 + (\Delta\varepsilon_y)^2 \right] + 2R (\Delta\varepsilon_x)(\Delta\varepsilon_y) + (\Delta\varepsilon_y)^2 + \frac{1}{2} (\Delta\gamma_{xy})^2 \right\}^{1/2} \tag{9.47}$$

according to the hypothesis of work equivalence. Introducing a square matrix D, the effective strain increment can be expressed in the matrix form

$$\Delta\bar{\boldsymbol{\varepsilon}} = \left\{ \left(\Delta\boldsymbol{\varepsilon}^T \right) \mathbf{D} (\Delta\boldsymbol{\varepsilon}) \right\}^{1/2}, \quad \mathbf{D} = \frac{l+R}{l+2R} \begin{bmatrix} l+R & R & 0 \\ R & l+R & 0 \\ 0 & 0 & 1/2 \end{bmatrix} \tag{9.48}$$

It is convenient to discretize the sheet metal into an assemblage of linear triangular elements and consider a set of rectangular axes in which the x- and y-axes are taken along the plane of the sheet, and the z-axis along the normal. The components of the increment of displacement at any point of the element during a time increment Δt may be expressed in terms of the nodal values in the matrix form

$$\Delta\mathbf{u} = \begin{Bmatrix} \Delta u \\ \Delta v \\ \Delta w \end{Bmatrix} = \mathbf{N}\Delta\mathbf{q}, \quad \mathbf{N} = \begin{bmatrix} N_1 & 0 & 0 & N_2 & 0 & 0 & N_3 & 0 & 0 \\ 0 & N_1 & 0 & 0 & N_2 & 0 & 0 & N_3 & 0 \\ 0 & 0 & N_1 & 0 & 0 & N_2 & 0 & 0 & N_3 \end{bmatrix} \tag{9.49}$$

where N_1, N_2, and N_3 are the shape functions, which are identical to the area coordinates and are given by (9.19), while $\Delta\mathbf{q}$ is the nodal displacement increment vector given by

$$\Delta\mathbf{q}^T = [\Delta u_1 \quad \Delta v_1 \quad \Delta w_1 \quad \Delta u_2 \quad \Delta v_2 \quad \Delta w_2 \quad \Delta u_3 \quad \Delta v_3 \quad \Delta w_3]$$

Each normal component of the true strain in the coordinate directions is the logarithm of the ratio of the final and initial material line elements originally coinciding with each coordinate axis. These are the logarithmic normal components of the Lagrangian strain tensor, and their sufficiently small increments in the surface, if the deforming sheet may be written with sufficient accuracy as

$$\Delta\varepsilon_x = \frac{\partial}{\partial x}(\Delta u) + \frac{1}{2} \left[\frac{\partial}{\partial x}(\Delta w) \right]^2, \quad \Delta\varepsilon_y = \frac{\partial}{\partial y}(\Delta v) + \frac{1}{2} \left[\frac{\partial}{\partial y}(\Delta w) \right]^2 \tag{9.50a}$$

where $\{u, v, w\}$ are the components of the displacement of a generic particle. The increment of the surface shear strain, to the same order of approximation, may be written as

$$\Delta\gamma_{xy} = \frac{\partial}{\partial x}(\Delta v) + \frac{\partial}{\partial y}(\Delta u) + \frac{\partial}{\partial x}(\Delta w) \frac{\partial}{\partial y}(\Delta w) \tag{9.50b}$$

In view of (9.49), the various derivatives appearing in the above equations may be expressed as

$$\frac{\partial}{\partial x}(\Delta u) = \sum P_\alpha \Delta u_\alpha, \quad \frac{\partial}{\partial x}(\Delta v) = \sum P_\alpha \Delta v_\alpha, \quad \frac{\partial}{\partial x}(\Delta w) = \sum P_\alpha \Delta w_\alpha$$

$$\frac{\partial}{\partial y}(\Delta u) = \sum Q_\alpha \Delta u_\alpha, \quad \frac{\partial}{\partial y}(\Delta v) = \sum Q_\alpha \Delta v_\alpha, \quad \frac{\partial}{\partial y}(\Delta w) = \sum Q_\alpha \Delta w_\alpha$$

where P_α and Q_α are given by (9.19), where α varies from 1 to 3. Substituting from the above into equation (9.49), the vector representing the strain increment may be written in the matrix form

$$\Delta \boldsymbol{\varepsilon} = \begin{Bmatrix} \Delta \varepsilon_\mathbf{x} \\ \Delta \varepsilon_\mathbf{y} \\ \Delta \gamma_\mathbf{xy} \end{Bmatrix} = \mathbf{B}\Delta\mathbf{q} + \begin{Bmatrix} \Delta \varepsilon'_\mathbf{x} \\ \Delta \varepsilon'_\mathbf{y} \\ \Delta \gamma'_\mathbf{xy} \end{Bmatrix}, \quad \mathbf{B} = \begin{bmatrix} P_1 & 0 & 0 & P_2 & 0 & 0 & P_3 & 0 & 0 \\ 0 & Q_1 & 0 & 0 & Q_2 & 0 & 0 & Q_3 & 0 \\ Q_1 & P_1 & 0 & Q_2 & P_2 & 0 & Q_3 & P_3 & 0 \end{bmatrix} \tag{9.51}$$

where **B** is the strain increment matrix. The second column vector in (9.50) arises from the change of the deforming sheet metal and is given by

$$\Delta\varepsilon'_x = \frac{1}{2}\left(\sum P_\alpha w_\alpha\right)^2, \quad \Delta\varepsilon'_y = \frac{1}{2}\left(\sum Q_\alpha w_\alpha\right)^2, \quad \Delta\gamma'_{xy} = \left(\sum P_\alpha w_\alpha\right)\left(\sum Q_\alpha w_\alpha\right)$$

Using the same variational principle as that for the axisymmetric forming process, we arrive at the element stiffness equation, which is still governed by (9.45) and (9.46), but the strain increment vector $\Delta\boldsymbol{\varepsilon}$ and the associated matrix appearing in these equations now correspond to (9.51) and (9.48), respectively. It may be noted that the first derivatives appearing in (9.44) can be expressed in terms of P_α, Q_α, and the associated components of the nodal displacement increment.

A finite element formulation for sheet metal forming, including planar anisotropy of the sheet, has been presented by Yang and Kim (1987). A simplified method of finite element analysis based on the total strain theory of plasticity has been discussed by Majlessi and Lee (1988). The influence of bending of the sheet, which is locally important in a variety of sheet-forming processes, has been incorporated in the finite element formulation by Huh et al. (1994).

9.5 Numerical Implementation

9.5.1 Numerical Integration

The elemental stiffness equation involves volume and surface integrals which generally require some kind of numerical integration in which the integrand is evaluated at a finite number of points, called *integration points*, within the limits of integration. We begin with the one-dimensional situation in which a scalar function $f(x)$ is to be integrated over the range $a \le x \le b$. Introducing the natural coordinate ξ, such that $2x = (b + a) + (b - a)\xi$, the formula for the numerical integration can generally be expressed as

$$\int_a^b f(x)dx = \frac{1}{2}(b-a)\int_{-1}^{1} F(\xi)d\xi = \frac{1}{2}(b-a)\sum_{i=1}^{n} w_i F(\xi_i) \tag{9.52}$$

where w_i is a *weight factor* associated with the integration point $\xi = \xi_i$ and n is the number of integration points. Simpson's one-third rule of integration is a special case of (9.52), where $n = 3$, and $w_1 = w_3 = 1/3$, $w_2 = 4/3$, the integration points being $\xi_1 = -1$, $\xi_2 = 0$, $\xi_3 = 1$.

In the finite element analysis, it is customary to employ the *Gaussian quadrature*, as it requires the minimum number of integration points for the same degree of accuracy, The Gaussian quadrature formula for n integration points gives the exact result when $F(\xi)$ is a polynomial of degree less than or equal to $2n - 1$. Setting $F(\xi) = \xi^s$ in (9.52), and integrating, we have

$$\sum_{i=1}^{n} w_i \xi_i^s = 0 \quad (s = 1, 3, 5, ..., 2n-1), \quad \sum_{i=1}^{n} w_i \xi_i^s = \frac{2}{s+1} \quad (s = 0, 2, 4, ..., 2n-2) \tag{9.53}$$

These relations enable us to determine the integration points and weight factors for any selected value of n. Considering $n = 2$, and setting $s = 0, \ldots, 3$, we have $w_1 + w_2 = 2$, $w_1\xi_1 + w_2\xi_2 = w_1\xi_1^3 + w_2\xi_2^3 = 0$, $w_1\xi_1^2 + w_2\xi_2^2 = 2/3$, giving the solution

$$w_1 = w_2 = 1, \quad -\xi_1 = \xi_2 = \frac{1}{\sqrt{3}} \ (n = 2)$$

Similarly, considering $n = 3$, and setting $s = 0, \ldots, 5$ in (9.53), we obtain a set of six equations which are easily solved to give

$$w_1 = w_3 = \frac{5}{9}, \quad w_2 = \frac{8}{9}, \quad -\xi_1 = \xi_3 = \sqrt{\frac{3}{5}}, \quad \xi_2 = 0$$

From the geometrical point of view, the Gaussian integration formulas corresponding to $n = 2$ and $n = 3$ are equivalent to linear and quadratic approximations, respectively, of the given function $F(\xi)$. The integration points in the Gaussian quadrature for any given value of n are in fact the roots of the equation $P_n(\xi) = 0$, where $P_n(\xi)$ denotes the Legendre polynomial of degree n.

Consider now a scalar function $f(x, y)$, which is defined over a two-dimensional isoparametric element with natural coordinates (ξ, η). If the number of integration points in the ζ and η directions be taken as m and n respectively, the integral of $f(x, y)$ over the area of the element may be written as

$$I = \iint f(x,y)dx\,dy = \int_{-1}^{1}\int_{-1}^{1} F(\xi,\eta)J(\xi,\eta)\,d\xi d\eta = \sum_{i=1}^{m}\sum_{j=1}^{n} w_i\,w_j\,F(\xi_i,\eta_j)\,J(\xi_i,\eta_j) \tag{9.54}$$

where $J(\xi,\eta)$ is the Jacobian of the transformation given by (9.23), while w_i and w_j are the weight factors corresponding to the integration points ξ_i and η_j, respectively.

In the case of axial symmetry, involving the volume integration of a function $f(r, z)$ defined over a ring element shown n Fig. 9.4, the integration formula becomes

$$I = 2\pi \iint f(r,z)\,r\,dr\,dz = 2\pi \sum_{i=l}^{m}\sum_{j=l}^{n} w_i\,w_j\,F(\xi_i,\eta_j)\,r(\xi,\eta)\,J(\xi_i,\eta) \tag{9.55}$$

where $J(\xi,\eta)$ is given by (9.23) with x and y replaced by r and z, respectively, while $r = N_1 r_1 + N_2 r_2 + N_3 r_3 + N_4 r_4$. The numerical integration formula for the general three-dimensional situation can be written down as a straightforward extension of (9.55).

In the Gaussian quadrature for two- or three-dimensional cases, the integration points and the weight factors in each coordinate direction for a given number of integration points are the same as those in the one-dimensional case. Setting

$$\phi(\xi,\eta) = F(\xi,\eta)J(\xi,\eta),$$

in the double integral (9.55), and assuming $m = n = 2$, we have four Gaussian integration points, each having a weight factor of unity, and the integration formula then becomes

$$I = \phi(-\alpha,-\alpha) + \phi(-\alpha,\alpha) + \phi(\alpha,-\alpha) + \phi(\alpha,\alpha), \quad \alpha = 1/\sqrt{3} \tag{9.56a}$$

The assumption $m = n = 3$, on the other hand, gives us nine Gauss points, one of which is located at the center of the element with a weight factor of 64/81, four of which are located along the diagonals $\eta = \pm\xi$ with a weight factor of 25/81 for each one, the remaining four being along the axes $\xi = 0$ and $\eta = 0$ each with a weight factor of 40/81. The Gaussian integration formula then becomes

$$\begin{aligned} I = {} & \frac{64}{81}\phi(0,0) + \frac{40}{81}\left[\phi(-\alpha,0) + \phi(0,-\alpha) + \phi(\alpha,0) + \phi(0.\alpha)\right] \\ & + \frac{25}{81}\left[\phi(-\alpha,-\alpha) + \phi(-\alpha,\alpha) + \phi(\alpha,-\alpha) + \phi(\alpha,\alpha)\right], \quad \alpha = \sqrt{0.6} \end{aligned} \tag{9.56b}$$

In problems of axial symmetry using quadrilateral ring elements, the Gaussian integration formula for the volume integral can be expressed exactly in the same forms as (9.56), provided we set

$$\phi(\xi,\eta) = 2\pi F(\xi,\eta)\,r(\xi,\eta)\,J(\xi,\eta),$$

In the case of a linear triangular element, for which $J = 2A$, it is customary in the finite element analysis to consider a single integration point located at the centroid of the triangle, and the integral becomes $I = 2A\ F_0$, where F_0 is the value of F at the centroid, which corresponds to $L_1 = L_2 = L_3 = 1/3$. When the same element is used as an axisymmetric ring element, a similar approximation to the volume integral gives $I = 4\pi A r_0\ F_0$, where r_0 is the radius to the centroid of the triangle

In a linear quadrilateral element, the condition of constancy of volume cannot be satisfied at all points except for a uniform mode of deformation. In the finite element formulation, this difficulty is usually overcome by using a single-point integration scheme for dealing with the volumetric strain rate term. If, on the other hand, four linear triangular elements are arranged to form a quadrilateral, then the plastic incompressibility condition can be satisfied over the entire quadrilateral (Nagtegaal et al., 1974). In the solution of metal-forming problems, the reduced integration scheme is frequently used, particularly for the evaluation of the stresses

9.5.2 Global Stiffness Equations

The finite element analysis of a physical problem is based on dividing the body into a large number of finite elements which are joined together at their nodal points. It is customary to assign the global node numbers and the element numbers sequentially as shown in Fig. 9.6(a). The physical constraints require the velocity vector at any nodal point to be identical to that of the individual elements sharing the same nodal point. The force vector at a given nodal point, on the other hand, is the sum of the forces associated with the elements having this nodal point in common (Desai and Abel, 1972).

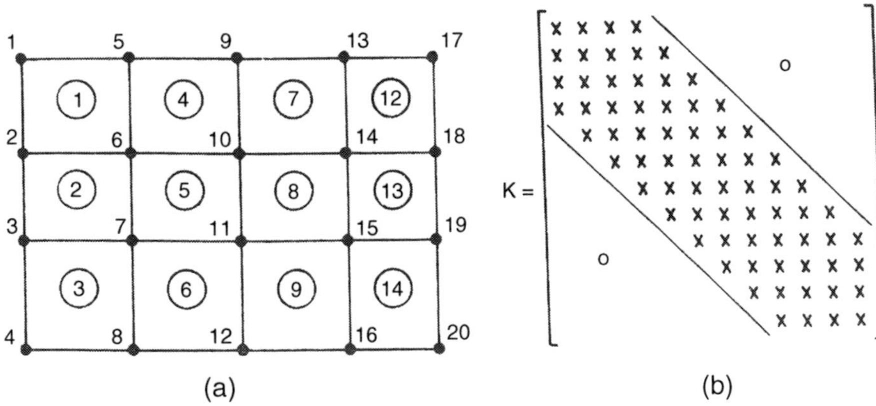

Fig. 9.6 Finite element mesh and associated global stiffness matrix. (**a**) Element node numbering, (**b**) typical banded matrix

It is customary to have the elemental nodes numbered in the same sequence for each element in the assemblage. Adopting the elemental node numbering to be indicated in Fig. 9.2, the elemental stiffness equation (9.13) may be expressed in terms of 2 × 2 submatrices in the form

$$\begin{bmatrix} [k_{11}{}^e] & [k_{12}{}^e] & [k_{13}{}^e] & [k_{14}{}^e] \\ [k_{21}{}^e] & [k_{22}{}^e] & [k_{23}{}^e] & [k_{24}{}^e] \\ [k_{31}{}^e] & [k_{32}{}^e] & [k_{33}{}^e] & [k_{34}{}^e] \\ [k_{41}{}^e] & [k_{42}{}^e] & [k_{43}{}^e] & [k_{44}{}^e] \end{bmatrix} \begin{Bmatrix} \{\Delta q_1{}^e\} \\ \{\Delta q_2{}^e\} \\ \{\Delta q_3{}^e\} \\ \{\Delta q_4{}^e\} \end{Bmatrix} = \begin{Bmatrix} \{f_1{}^e\} \\ \{f_2{}^e\} \\ \{f_3{}^e\} \\ \{f_4{}^e\} \end{Bmatrix}, \quad [k_{ji}{}^e] = [k_{ij}{}^e]^T \tag{9.57}$$

where the last relation follows from the symmetry of the stiffness matrix. The submatrices introduced in (9.57) are defined as

$$[k_{ij}{}^e] = \begin{bmatrix} k^e{}_{2i-1,\,2j-1} & k^e{}_{2i-1,\,2j} \\ k^e{}_{2i,\,2j-1} & k^e{}_{2i,\,2j} \end{bmatrix}, \quad \{q_j{}^e\} = \begin{Bmatrix} u_j{}^e \\ v_j{}^e \end{Bmatrix}, \quad \{f_j{}^e\} = \begin{Bmatrix} f_j{}^e \\ g_j{}^e \end{Bmatrix} \tag{9.58}$$

The global stiffness equation is formed by a suitable combination of the elemental stiffness equations, taking into account the connectivity of the elements, and may be written as

$$\mathbf{K}\Delta\mathbf{V} = \mathbf{T} \tag{9.59}$$

where $\mathbf{K}$ is the global stiffness matrix, $\mathbf{T}$ is the global load vector, and $\Delta\mathbf{V}$ the global velocity change vector, which differs from the elemental vector $\Delta\mathbf{q}$ only in the node numbering. Referring to Fig. 9.6(a), and considering, for example, element 2, whose connectivity with the neighboring elements is defined by the nodal points (2, 3, 7, 6), the submatrices of $\mathbf{K}$ associated with these nodes are found as

$$\begin{aligned} &[K_{22}] = [k_{11}{}^1] + [k_{44}{}^2], \quad &&[K_{23}] = [k_{23}{}^2], \quad &&[K_{26}] = [k_{12}{}^1] + [k_{43}{}^2] \\ &[K_{33}] = [k_{11}{}^2] + [k_{44}{}^3], \quad &&[K_{36}] = [k_{36}{}^2], \quad &&[K_{37}] = [k_{12}{}^2] + [k_{43}{}^3] \\ &[K_{66}] = [k_{22}{}^1] + [k_{33}{}^2] + [k_{11}{}^4] + [k_{44}{}^5], && &&[K_{67}] = [k_{32}{}^2] + [k_{41}{}^5] \\ &[K_{77}] = [k_{22}{}^2] + [k_{33}{}^3] + [k_{11}{}^5] + [k_{44}{}^6], && &&[K_{27}] = [k_{27}{}^2] \end{aligned}$$

The remaining submatrices of the global stiffness matrix can be similarly established by considering the connectivity of the other elements. When any two nodal points do not belong to the same element, the corresponding submatrix becomes a null matrix.

Due to the limited influence of the element connectivity, the global stiffness matrix is a sparse matrix, which can be arranged in a banded form, as indicated in Fig. 9.6(b). With the help of an appropriate node numbering, the band width can be kept down to a minimum. The global stiffness equations are most conveniently solved by the Gaussian elimination technique using a linear equation solver. In a skyline solver, the matrix coefficients are stored column-wise, starting from the first diagonal element and ending with the last nonzero element. The computational time required to solve the matrix equation is found to be proportional to the square of the semi-bandwidth of the matrix. It is therefore necessary to number the nodes in such a way that the band width is a minimum.

9.5.3 Boundary Conditions

The solution of the global stiffness equations requires due consideration of the boundary conditions. In general, the boundary surface of the workpiece consists of a part on which the traction is prescribed, a part S_v on which the velocity is prescribed, and a part S_T which is the tool–workpiece interface. The imposition of the traction boundary condition on S_F in the form of nodal point forces is straightforward. Consider, for example, the three-dimensional brick element shown in Fig. 9.3, and suppose that the lower surface 1–2–3–4 is subjected to normal and tangential tractions specified by **F**. Since $\zeta = 1$ over this surface, the nonzero shape functions are given by (9.21), and the shape function matrix reduces to

$$\mathbf{N} = \begin{bmatrix} N_1 & 0 & N_2 & 0 & N_3 & 0 & N_4 & 0 \\ 0 & N_1 & 0 & N_2 & 0 & N_3 & 0 & N_4 \end{bmatrix}$$

The associated nodal point force vector is easily determined from the expression

$$\mathbf{f}_0 = \iint (\mathbf{N}^T\mathbf{F})\, dx\, dy = \iint (\mathbf{N}^T\mathbf{F})\, J\, d\xi\, d\eta \tag{9.60}$$

where J is the Jacobian of the transformation, given by (9.28), and the integral extends over the area of the entire surface S_F

For a nodal point on S_v over which the velocity is prescribed, the velocity correction is zero, and the corresponding stiffness equation needs to be omitted. In the finite element solution, the simplest way to impose the velocity boundary condition $\Delta V_m = 0$ at a nodal point m is to set the diagonal element of the mth row of the stiffness matrix to unity and replace the remaining elements in the corresponding row and column by zeros, as indicated below.

$$\begin{bmatrix} K_{11} & K_{12} & . & . & 0 & . & . & K_{1n} \\ K_{21} & K_{22} & . & . & 0 & . & . & K_{2n} \\ . & . & . & . & . & . & . & . \\ 0 & 0 & ; & . & 1 & . & . & 0 \\ . & . & . & . & . & . & . & . \\ K_{n1} & K_{n2} & . & . & 0 & , & . & K_{nn} \end{bmatrix} \begin{Bmatrix} \Delta V_1 \\ \Delta V_2 \\ . \\ \Delta V_m \\ . \\ \Delta V_n \end{Bmatrix} = \begin{Bmatrix} T_1 \\ T_2 \\ . \\ 0 \\ . \\ T_n \end{Bmatrix}$$

On the remainder of the surface, S_T, representing the interface between the die and the workpiece, the boundary condition is of the mixed type, as the velocity is prescribed along the normal to the interface of contact, while the tangential traction is prescribed in the direction of relative sliding between the die and the workpiece. The tangential stress is usually specified in terms of a constant coefficient of friction, or as a constant frictional stress mk, where $0 \leq m \leq 1$. When the element surface does not conform with the die surface, an additional approximation is necessary to obtain the associated distribution of nodal forces.

Once the velocity solution is obtained for the entire workpiece, the geometry of the workpiece must be updated by changing the coordinates of the various nodes.

For a two-dimensional problem, the rectangular coordinates of a typical nodal point (x_j, y_j) are changed by the amounts

$$\Delta x_j = u_j \Delta t, \quad \Delta y_j = v_j \Delta t$$

where Δt denotes the increment of time scale. The nodal point strains are similarly updated from the available values of the strain rate. In metal-forming analysis, the time increment may be taken as that for which the next free node of the workpiece comes in contact with the die surface.

Since the deformation that occurs in metal-forming processes is generally large, the size and shape of the element soon become unacceptable as the deformation continues, and the imposition of the boundary conditions also becomes increasingly difficult. In order to overcome theses difficulties, it is necessary to modify the mesh system periodically, so that the mesh size remains sufficiently small, and also to transfer the information from the old mesh system to the new one through interpolation. The difficulty can be largely overcome by using a spatially fixed meshing scheme, as has been discussed by Derbalian et al. 1978) and by Mori et al. (1983). An area-weighted averaging method of evaluating a parameter at a node internal to an original linear quadrilateral element, which is found to be sufficiently accurate in metal-forming analysis, has been discussed by Kobayashi et al. (1989). The complete elastic/plastic formulation for large strain finite element analysis has been discussed by McMeeking and Rice (1975) and Nagtegaal and DeJong (1981). A useful discussion of the elastic/plastic formulations in relation to metal-forming problems has been made by Rebelo and Wertheimer (1986).

9.6 Illustrative Examples

Numerous solutions to the metal-forming problems, based on the various types of finite element formulation presented in Section 9.1, have been given in detail by Kobayashi et al. (1989). A number of these solutions are based on the variational method, and a few of these will be briefly discussed in what follows in order to illustrate the application of the preceding theory.

9.6.1 Compression of a Cylindrical Block

In the axial compression of a cylindrical block between a pair of flat dies, the plastic deformation is inhomogeneous due to the presence of friction at the interfaces, and the mean compressive stress exceeds the uniaxial yield stress of the material (Section 3.4). The deformation of the block is characterized by a barreling of the free surface, a part of which comes in contact with the die during the compression. When the ratio of the height of the block to its diameter is sufficiently small, the barreling consists of a single bulge in which the maximum diameter occurs at the central cross section of the block. For fairly large values of the height/diameter ratio, a double bulge is sometimes observed. The analysis may be based on a constant frictional stress equal to *mk* along the interface between the die and the workpiece. The finite element analysis carried out by Lee and Kobayashi (1971) for a cylindrical block with an initial height/diameter ratio of 2.5 reveals the formation of a double bulge, as depicted in Fig. 9.7. The double bulge gives way to a single bulge as the height/diameter ratio progressively decreases to sufficiently small values during the continued compression.

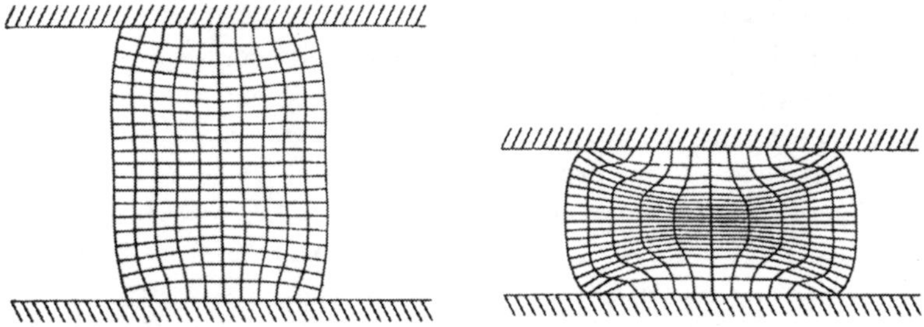

33 percent reduction in height

66 percent reduction in height

Fig. 9.7 Grid distortion patterns in the axial compression of a cylinder with an initial hight/diameter ratio of 2.5 (after Lee and Kobayashi, 1971)

The analysis may also be used to predict the limit of workability of ductile materials due to the formation of surface cracks during the compression. This has been investigated experimentally by Kudo and Aoi (1967), who measured the equatorial surface stains in upsetting solid cylindrical specimens under various frictional conditions until the surface cracks were observed. The computed strain paths of the critical element, obtained from the finite element analysis for a block of a unit initial height/diameter ratio under various frictional conditions, are plotted in Fig. 9.8, which also includes the experimental results referred to above. The limit set by the occurrence of surface cracks, based on the experimental data, may be approximated by the criterion $2\varepsilon_\theta + \varepsilon_z = 0.8$ to a close approximation. Finite element solutions to the axial compression of hollow cylinders have been discussed by Chen and Kobayashi (1978) and also by Hartley et al. (1979). A finite. element analysis of the upsetting process based on the total strain theory of plasticity has been reported by Vertin and Majlessi (1993).

9.6.2 Bar Extrusion Through a Conical Die

Consider the axisymmetric extrusion of a cylindrical billet through a conical die, along which the frictional stress has a constant value equal to *mk*, the container wall being assumed to be perfectly smooth. The material in the container approaches the die with a uniform unit speed, and it leaves the die with a uniform speed equal to

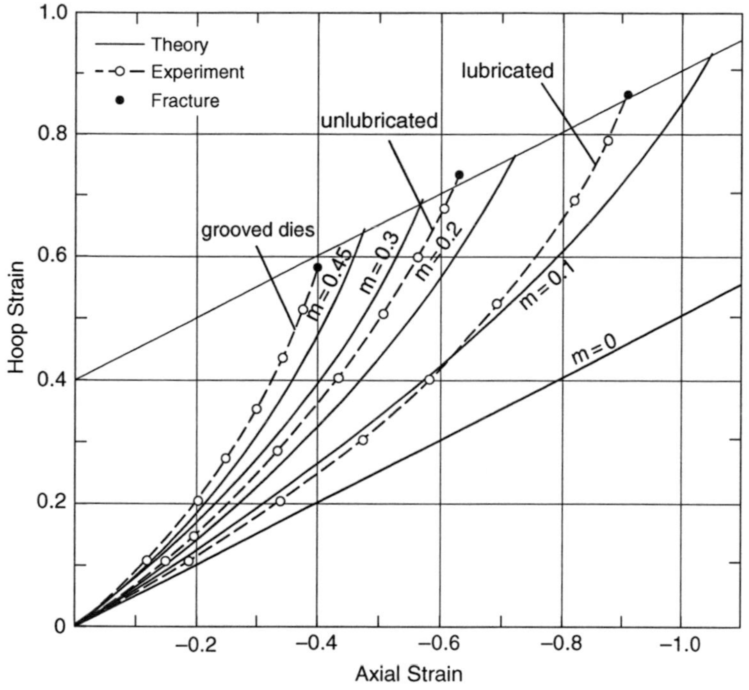

Fig. 9.8 Strain paths of an equatorial element during axial compression of a cylinder with an initial height/diameter ration of unity

b^2/a^2, where a and b denote the final and initial radii of he billet. The assumed finite element mesh for the extrusion problem is shown in Fig. 9.9, where the corners of the die have been slightly modified by straight lines joining he nodal points closest to the corners, in order to avoid singularities of velocity components near the edge of the die. The origin of the coordinate (r, z) is taken on the axis of symmetry at O with the r-axis coinciding with the exit plane of the die, as shown in the figure. Along the die face AB, the boundary conditions are

$$\tau = mk, \quad w = u \cot\alpha, \quad \text{along} \quad z = (r-a)\cot\alpha, \quad a \le r \le b$$

where α denotes the semiangle of the die. The extruded part of the billet, which moves as a rigid body, is entirely free of surface tractions. The remaining boundary conditions may be written as

$$u = 0, \quad F_z = 0, \quad \text{along} \quad r = 0 \quad \text{and} \quad r = b$$
$$w = -1, \quad F_r = 0 \quad \text{on} \quad z = c; \quad w = -(b^2/a^2) \quad \text{on} \quad z = 0$$

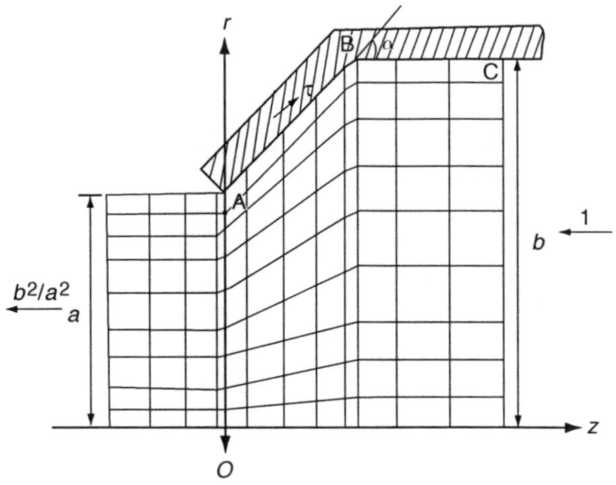

Fig. 9.9 Geometry and finite element grid pattern for the axisymmetric extrusion of a cylindrical billet through a conical die

The problem can be treated as one of steady state in which the geometrical configuration does not change with time. A complete elastic/plastic analysis for extrusion through a sigmoidal die until the attainment of the steady state has been discussed by Lee et al. (977). The results presented here have been obtained by the rigid/plastic method by Chen and Kobayashi (1978) and Chen et al. (1979).

In the finite element analysis for the extrusion of a work-hardening material, the components of the strain rate at the center of each element are initially assumed to be same as those in a nonhardening material. Starting from a selected point on the plane of entry, where the effective strain is zero, the rate of change of the effective strain is determined from the known values at the surrounding element centers. Since the velocity of the selected point is found from the element interpolation formula, the effective strain and the new position of the particle are then easily obtained from the increment of time. This procedure is sequentially repeated, following the path of the particle, until the exit plane is reached. The flow lines emanating from different points on the entry plane determined this way furnish the shape of the distorted grid, and also the distribution of the effective strain throughout the deforming region. Since the distribution of the effective stress follows from the given stress–strain curve of the material, the new distribution of nodal point velocities can be computed in order to carry out the next iteration. When the velocity solution converges after a few iterations, the mean extrusion pressure and the distribution of the die pressure can be determined for the given frictional condition, die angle, and fractional reduction in area.

The steady-state grid distortion pattern for frictionless extrusion through a 90° conical die, obtained by the finite element solution, is displayed in Fig. 9.10, where the upper half holds for a nonhardening material and the lower half for a work-hardening material (SAE 1112 steel). The difference between the two patterns is due to the restriction of metal flow that occurs in a work-hardening material. The distribution of radial and axial velocities within the die for $\alpha = 45^\circ$ and $b/a = 2$,

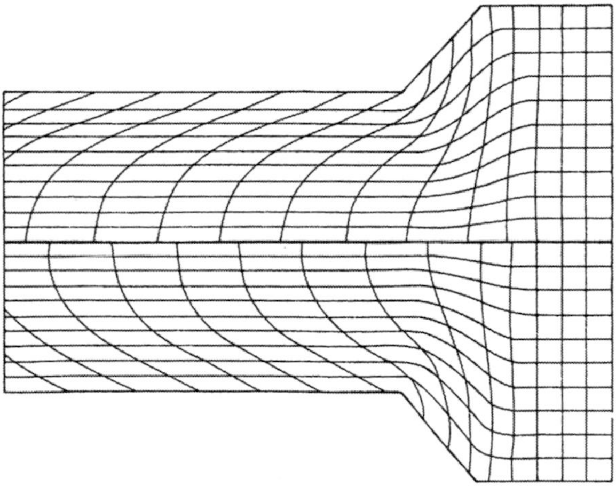

Fig. 9.10 The distortion of an initial square grid in a cylindrical billet extruded through a 90° conical die (after Chen et al., 1978)

furnished by the finite element solution, is shown in Fig. 9.11. The material adopted in this solution is the same as that used in an experimental investigation by Shabaik and Thomsen (1968), who obtained remarkably similar results for the velocity distribution. The computation also reveals that the hydrostatic part of the stress in a region near the center of the deforming region becomes tensile at sufficiently large reductions, leading to the possibility of a central crack which is frequently observed.

9.6.3 Analysis of Spread in Sheet Rolling

In the rolling of sheets and slabs, in which the width of the workpiece is less than about five times the length of the arc of contact, the usual assumption of plane strain is not justifiable. The amount of lateral spread that occurs in such cases is quite appreciable, and must be taken into consideration in the analysis of the rolling process. A finite element analysis of the problem using the three-dimensional brick element has been carried out by Li and Kobayashi (1982), who adopted the nonsteady-state approach for the solution. Figure 9.12 shows a narrow strip of the workpiece with the arrangement of an element in the upper half of the material within the arc of contact. The deformation of the material entering the roll gap with a bite is considered in a step-by-step manner based on a constant frictional stress, while updating the material properties and the coordinates of the nodal points at the end of each step. A steady state is assumed to be reached when the associated roll torque has attained a steady value, and the spread contour has become stationary.

The computation has been carried out with a friction stress equal to 0.5 k, and using $R/h_0 = 160$, two different values of w_0/h_0, and several values of the reduction in thickness, where R denotes the roll radius, 2 h_0 is the initial slab thickness, and $2w_0$ is the initial slab width, the material used in the analysis being annealed

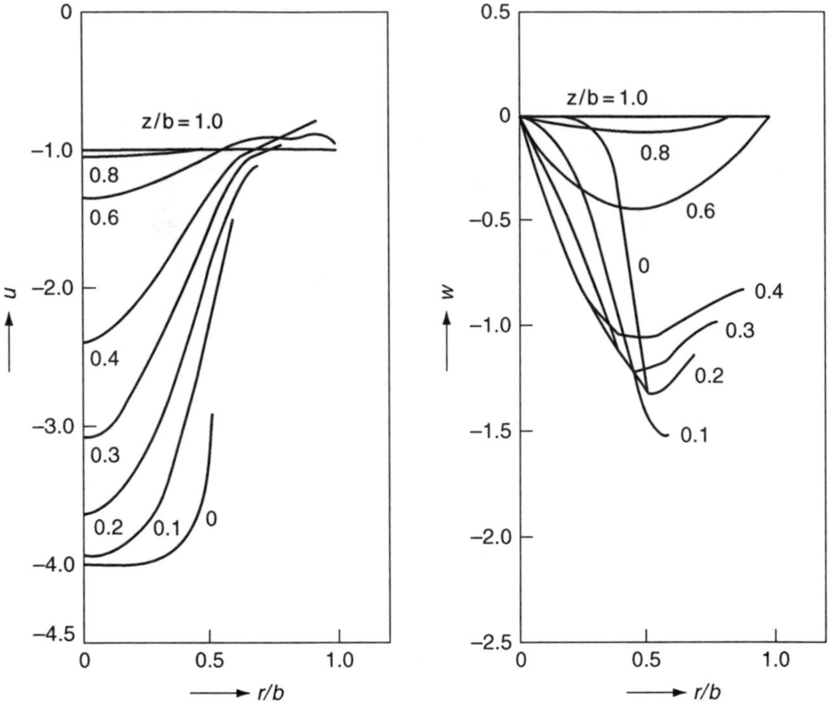

Fig. 9.11 Distribution of radial and axial velocities of particles moving through a 90° conical die with $b / a = 2$ (after Chen and Kobayashi, 1982)

AISI 1018 steel. The final values of the thickness and width of the rolled stock are denote by $2h_f$ and $2w_f$, respectively. The computed value of the mean lateral spread is plotted against the final reduction in height in Fig. 9.13a, which shows excellent agreement with some experimental results reported by Kobayashi et al. (1989). During the rolling process, not only the thickness but also the cross-sectional area of the rolled stock progressively decreases due to the effect of the lateral spread. The solid curves in Fig. 9.13(b) show the variation of the reduction in cross section of the rolled stock with the reduction in height for $w_0/h_0 = 1$ and 3, while the broken straight line indicates the plane strain situation in which the reduction in cross-sectional area is equal to the reduction in height. The spread in rolling has also been investigated approximately by Lahoti and Kobayashi (1974), and, by a finite element analysis, by Kanazawa and Marcal (1982).

The finite element solution for the compression of a rectangular block has been discussed by Park and Kobayashi (1984), and that of a ring of square cross section has been considered by Park and Oh (1987). The shape rolling of bars of various cross sections, using the finite element method, has been investigated by Park and Oh (1990).

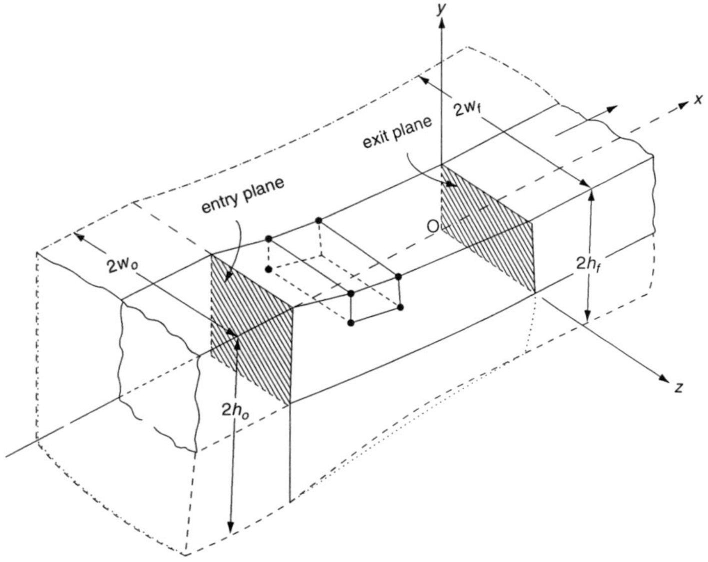

Fig. 9.12 A schematic view of sheet rolling with lateral spread indicating the location of a typical finite brick element

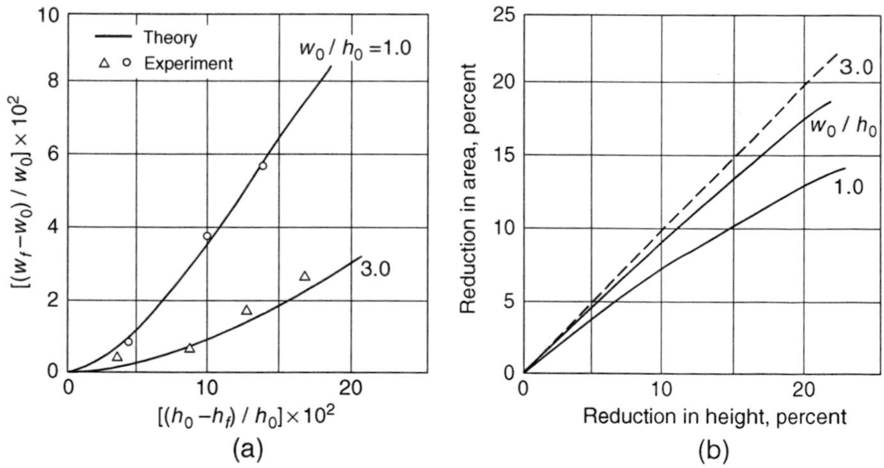

Fig. 9.13 Results for sheet rolling with lateral spread. (**a**) Variation of overall spread with reduction in height, (**b**) variation of change in cross section with change in height

9.6.4 Deep Drawing of Square Cups

As a final example, consider the deep drawing of a square cup using a flat punch, the base of the punch being a square of sides $2a$ The schematic view of the process is similar to that shown in Fig. 2.28(a). The cup is drawn from a square blank whose sides have an initial length equal to $2b_0$, the initial blank thickness being denoted by h_0. In the finite element formulation, the continuous blank holding force is replaced by a set of concentrated forces acting at the nodal points along the periphery of the blank. The frictional condition at the interfaces between the tools and the sheet metal is assumed to be governed by Coulomb's law with a constant coefficient of friction. Denoting the die and punch profile radii by r_d and r_p respectively, and the punch corner radius by r_c, the geometry of the process is defined as

$$\frac{b_0}{a} = 2.75, \quad \frac{h_0}{a} = 0.043, \quad \frac{r_d}{a} = \frac{r_p}{a} = 0.25, \quad \frac{r_c}{a} = 0.16.$$

The material is aluminum killed steel having a uniform R-value equal to 1.6, the planar stress–strain curve being given by the power law $\sigma = C\varepsilon^n$, where $n = 0.228$ and $C = 739$ MPa. The blank-holding force is taken as 4.9 kN, the friction coefficient being 0.2 over the punch and 0.04 over the die.

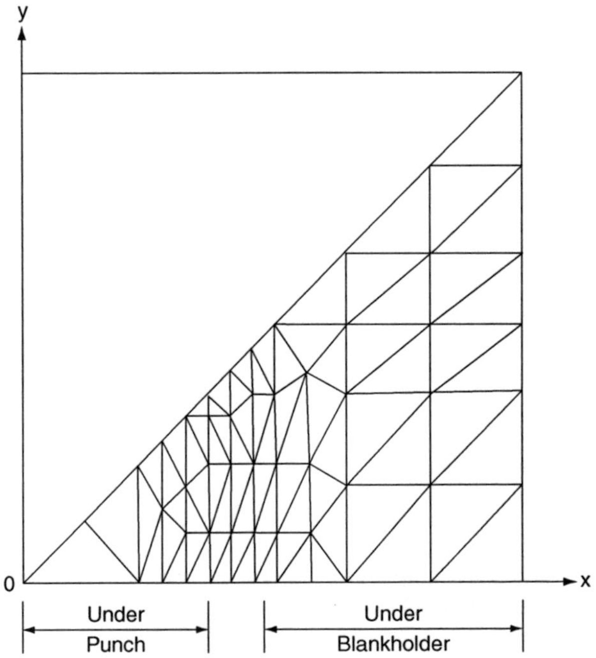

Fig. 9.14 Finite element mesh in a square blank to be drawn into a cup with square base

The finite element mesh used in the analysis of the square cup drawing is shown in Fig. 9.14, which indicates a choice of finer mesh over the region where the thickness is expected to vary rapidly. The thinning of the sheet is found to have maximum values over the punch and die profile radii, particularly along the diagonal of the square. In Fig. 9.15, the computed distribution of the thickness strain across the diagonal of the formed cup, based on $a = 20$ mm, is compared with that obtained experimentally by Thomson (1975) for a given punch load. The predicted strain distribution has the same trend as that of the experimental one, though there is an appreciable difference in the magnitude of the strain. The discrepancy is due to the fact that the observed punch penetration is 1.50 a, which is significantly higher than the value 1.01 a predicted by the finite element solution and may be attributed to the difference in the frictional conditions existing in the experiment. Similar results based on a simplified finite element analysis have been reported by Majlessi and Lee (1993).

Rigid/plastic finite element solutions for the hydrostatic bulging of circular diaphragms have been given by Lee and Kobayashi 1975) and Kim and Yang (1985a), and of elliptical diaphragms by Chung et al. (1988). The bulging of rectangular diaphragms has been investigated experimentally by Duncan and Johnson (1968) and numerically by Yang and Kim (1987). The large strain elastic/plastic

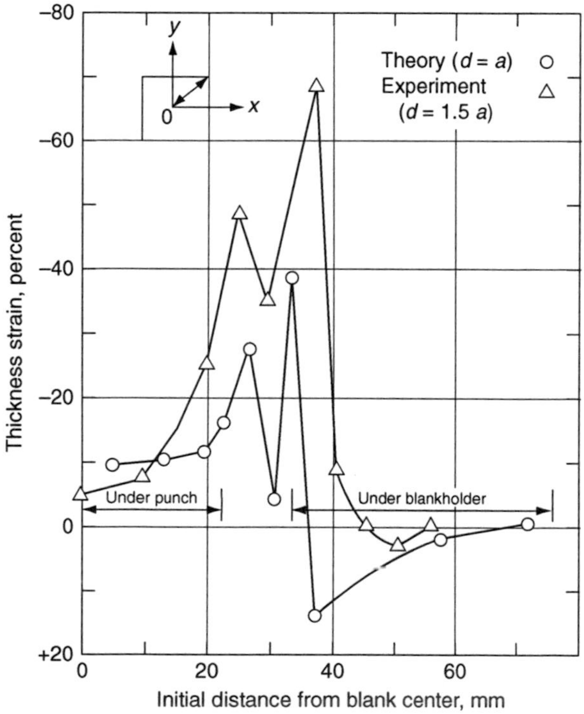

Fig. 9.15 Comparison of theoretical and experimental distributions of thickness strain along the diagonal of a square blank (after Toh and Kobayashi, 1985)

finite element formulation has been applied to the axisymmetric punch stretching problem by Kim et al. (1978) and to the plane strain bending of sheets by Oh and Kobayashi (1980). A rigid/plastic finite element solution to the punch stretching problem, based on the nonquadratic yield criterion, has been presented by Wang (1984). An elastic/plastic finite element analysis for the deep drawing of anisotropic cups has bee reported by Saran et al. (1990), and the associated problem of flange wrinkling has been investigated by Kim et al. (2000) and Correia et al. (2003).

Appendix: Orthogonal Curvilinear Coordinates

Cylindrical Coordinates

The position of a typical particle is defined by the coordinates (r, θ, z) taken in the radial, circumferential, and axial directions, respectively. If the associated components of the velocity are denoted by (u, v, w), respectively, then the components of the true strain rate are

$$\dot{\varepsilon}_r = \frac{\partial u}{\partial r} \qquad \dot{\gamma} r\theta = \frac{1}{2}\left(\frac{\partial \mathrm{v}}{\partial r} - \frac{\mathrm{v}}{r} + \frac{1}{r}\frac{\partial u}{\partial \theta}\right),$$

$$\dot{\varepsilon}_\theta = \frac{1}{r}\left(u + \frac{\partial \mathrm{v}}{\partial \theta}\right), \quad \dot{\gamma}\theta_z = \frac{1}{2}\left(\frac{\partial \mathrm{v}}{\partial z} + \frac{1}{r}\frac{\partial w}{\partial \theta}\right),$$

$$\dot{\varepsilon}_z = \frac{\partial w}{\partial z}, \qquad \dot{\gamma} rz = \frac{1}{2}\left(\frac{\partial u}{\partial z} + \frac{\partial w}{\partial r}\right).$$

If σ_r, σ_θ, and σ_z denote the normal stresses and $\tau_{r\theta}$, $\tau_{\theta z}$, and τ_{rz} the shear stresses, then the equations of equilibrium in the absence of body forces are

$$\frac{\partial \sigma_r}{\partial r} + \frac{1}{r}\frac{\partial \tau_{r\theta}}{\partial \theta} + \frac{\partial \tau_{rz}}{\partial z} + \frac{\sigma_r - \sigma_\theta}{r} = 0,$$

$$\frac{\partial \tau_{r\theta}}{\partial r} + \frac{1}{r}\frac{\partial \sigma_\theta}{\partial \theta} + \frac{\partial \tau_{\theta z}}{\partial z} + \frac{2\tau_{r\theta}}{r} = 0,$$

$$\frac{\partial \tau_{rz}}{\partial z} + \frac{1}{r}\frac{\partial \tau_{\theta z}}{\partial \theta} + \frac{\partial \sigma_z}{\partial z} + \frac{\tau_{rz}}{r} = 0.$$

Spherical Coordinates

The coordinate system is defined by (r, ϕ, θ), where r is the length of the radius vector, ϕ is the angle made by the radius vector with a fixed axis, and θ is the angle measured round this axis. If the velocity components in the coordinate directions are denoted by (u, v, w), then the components of the true strain rate are

J. Chakrabarty, *Applied Plasticity, Second Edition*, Mechanical Engineering Series,
DOI 10.1007/978-0-387-77674-3,

$$\dot{\varepsilon}_r = \frac{\partial u}{\partial r}, \qquad \dot{\gamma}_{r\phi} = \frac{1}{2}\left(\frac{\partial v}{\partial r} - \frac{v}{r} + \frac{1}{r}\frac{\partial u}{\partial \phi}\right),$$

$$\dot{\varepsilon}_\phi = \frac{1}{r}\left(u + \frac{\partial v}{\partial \phi}\right), \qquad \dot{\gamma}_{r\phi} = \frac{1}{2}\left(\frac{1}{r}\frac{\partial w}{\partial \phi} - \frac{w}{r}\cot\phi + \frac{1}{r\sin\phi}\frac{\partial v}{\partial \theta}\right),$$

$$\dot{\varepsilon}_\theta = \frac{1}{r}\left(u + v\cot\phi + \operatorname{cosec}\phi\frac{\partial w}{\partial \theta}\right), \qquad \dot{\gamma}_{r\theta} = \frac{1}{2}\left(\frac{\partial w}{\partial r} - \frac{w}{r} + \frac{1}{r\sin\phi}\frac{\partial u}{\partial \theta}\right).$$

Denoting the normal stresses by σ_r, σ_ϕ, and σ_θ and the shear stresses by $\tau_{r\phi}$, $\tau_{\phi\theta}$, and $\tau_{r\theta}$, the equations of equilibrium in the absence of body forces can be written as

$$\frac{\partial \sigma_r}{\partial r} + \frac{1}{r}\frac{\partial \tau_{r\phi}}{\partial \phi} \frac{1}{r\sin\phi}\frac{\partial \tau_{r\theta}}{\partial \theta} + \frac{1}{r}\left(2\sigma_r - \sigma_\phi - \sigma_\theta + \tau_{r\phi}\cot\phi\right) = 0,$$

$$\frac{\partial \tau_{r\phi}}{\partial r} + \frac{1}{r}\frac{\partial \sigma_\phi}{\partial \phi} \frac{1}{r\sin\phi}\frac{\partial \tau_{\phi\theta}}{\partial \theta} + \frac{1}{r}\left\{(\sigma_\phi - \sigma_\theta)\cot\phi + 3\tau_{r\phi}\right\} = 0,$$

$$\frac{\partial \tau_{r\theta}}{\partial r} + \frac{1}{r}\frac{\partial \tau_{\phi\theta}}{\partial \phi} + \frac{1}{r\sin\phi}\frac{\partial \sigma_\theta}{\partial \theta} + \frac{1}{r}\left(3\tau_{r\theta} + 2\tau_{\phi\theta}\cot\phi\right) = 0.$$

When the deformation is infinitesimal, the preceding expressions for the components of the strain rate may be regarded as those for the strain itself, provided the components of the velocity are interpreted as those of the displacement.

References

Abrahamson, G.R. and Goodier, J.N. (1966), Dynamic Flexural Buckling of Thin Rods Within an Axial Plastic Compressive Wave, *J. Appl. Mech.*, **33**, 241.

Abramowicz, W. and Jones, N. (1984), Dynamic Axial Crushing of Circular Tubes, *Int. J. Impact Engng*, **2**, 263.

Ades, C.S. (1957), Bending Strength of Tubes in the Plastic Range, *J. Aeronaut. Sci.*, **24**, 505.

Adie, J.F. and Alexander, J.M. (1967), A Graphical Method of Obtaining Hodographs to Axisymmetric Problems, *Int. J. Mech. Sci.*, **9**, 349.

Aizawa, T., Murata, M. and Suzuki, H. (1990), Electromagnetic Tube Bulging by the Direct Electrode Contact Method and the Solenoidal Coil Method, *Proc. 3rd Int. Conf. Technol. Plasticity*, Kyoto, Japan, p. 1593.

Akserland, E.L. (1965), Refinement of Upper Critical Loading of Pipe Bending Taking Account of Geometric Nonlinearity (in Russian), Mekhanika: Machinostroenic, *Izv. AkadNauk, SSSR*, **4**, 123.

Al-Hassani, S.T.S. and Johnson, W. (1969), The Dynamics of the Fragmentation Process for Spherical Bombs, *Int. J. Mech.*, **11**, 811.

Alexander, J.M. (1960a), An Appraisal of the Theory of Deep Drawing, *Metall. Rev.*, **5**, 349.

Alexander, J.M. (1960b), An Approximate Analysis of the Collapse of Thin Cylindrical Shells Under Axial Loading, *Quart. J. Mech. Appl. Math.*, **13**, 10.

Alexander, J.M., Brewer, R.C. and Rowe, G.W. (1987), *Manufacturing Technology*, Vols. 1 and 2, Ellis Horwood, Chichester.

Alexander, J.M. and Ford, H. (1954), On Expanding a Hole from Zero Radius in a Thin Infinite Plate, *Proc. Roy. Soc. London Ser. A*, **226**, 543.

Alexander, J.M and Lengyel, B. (1971), *Hydrostatic Extrusion*, Mills and Boon, London.

Alter, B.E.K. and Curtis, C.W. (1956), The Effect of Strain Rate on the Propagation of a Plastic Strain Pulse Along a Lead Bar, *J. Appl. Mech.*, **27**, 1079.

Anderson C.A. and Shield, R.T. (1966), On the Validity of the Plastic Theory of Structures for Collapse Under Highly Localized Loading, *J. Appl. Mech.*, **33**, 629.

Apsden, R.J. and Campbell, J.D. (1966), The Effect of Loading Rate on the Elastic-Plastic Flexure of Steel Beams, *Proc. Roy. Soc. London Ser. A*, **290**, 266.

Ariaratnam, S.T. and Dubey, R.N. (1969), Instability in an Elastic-Plastic Cylindrical Shell Under Axial Compression, *J. Appl. Mech.*, **36**, 47.

Asaro, R.J. (1983), Micromechanics of Crystals and Polycrystals, *Adv. in Appl. Mech.*, **23**, 1.

Ashwell, D.G. (1959), On the Large Deflection of a Spherical Shell with an Inward Point Load, *Proc. IUTAM on Theory of Thin Elastic Shells, Delft.*

Atkins, A.G. and Caddell, R.M. (1968), The Incorporation of Work-Hardening and Redundant Work in Rod Drawing Analysis, *Int. J. Mech. Sci.*, **10**, 15.

Aung, M., Wang, C.M. and Chakrabarty, J. (2005), Plastic Buckling of Moderately Thick Annular Plates, *Int. J. Struct. Stab. Dynamics*, **5**, 337.

Avitzur, B. (1965), Analysis of Metal Extrusion, *J. Engng. Indust., Trans. ASME*, **86**, 305.

Avitzur, B. (1968), Analysis of Central Bursting Defects in Drawing and Extrusion, *J. Engng. Indust., Trans. ASME,* **90**, 79.

Avitzur, B. (1980), *Metal Forming, the Application of Limit Analysis*, Marcfel Dekker, New York.

Avitzur, B. (1983), *Handbook of Metal Forming Processes*, Wiley, New York.

Azrin, M. and Backofen, W.A. (1970), The Deformation and Failure of Biaxially Stretched Sheets, *Metall. Trans.*, **1A**, 2857.

Babcock, CD. (1983), Shell Stability, *J. Appl. Mech.*, **50,** 935.

Backman, ME. and Goldsmith, W. (1978), The Mechanics of Penetration of Projectiles into Targets, *Int. J. Engng. Sci.*, **16**, 1.

Backof–en, W.A., Turner, I.R., and Avery, D.H. (1964), Superplasticity in an Al-Zn Alloy, *Trans. Amen Soc. Metals*, **57**, 980.

Bailey, J.A., Naos, S.L., and Nawab, K.C. (1972), Anisotropy in Plastic Torsion, *J. Basic Engng., Trans. ASME*, **94**, 231.

Baker, J.F., Hörne, M.R., and Roderick, J.W. (1949), The Behaviour of Continuous Stanchions, *Proc. Roy. Soc. London Ser. A.*, **198,** 493.

Baker, W.E. (1960), The Elastic-Plastic Response of Thin Spherical Shells to Internal Blast Loading, *J. Appl. Mech.*, **27**, 139.

Balen, R. and Abdul-Latif, A. (2007), Quasi-Static Plastic Buckling of Tubular Structures Used as an Energy Absorber, *J. Appl. Mech.*, **74**, 628.

Balendra, R. and Travis, F.W. (1970), Static and Dynamic Blanking of Varying Hardness, *Int. J. Mack Tool. Des. Res.*, **10**, 249.

Baltov, A. and Sawczuk, A. (1965), A Rule of Anisotropic Hardening, *Acta Mechanica*, **1**, 81.

Banerjee, A.K. and Malvern, L.E. (1975), Computation of Incremental Torsional Plastic Waves with Rate-Dependent Models, *Int. J. Solids Struct.*, **11**, 347.

Bardi, F.C. and Kyriakides, S. (2006), Plastic Buckling of Circular Tubes Under Axial Compression, Part 1, Experiments, *Int. J. Mech. Sci.*, **48**, 830.

Bardi, F.C. Kyriakides, S. and Yun, H.D. (2006), Plastic Buckling of Circular Tubes Under Axial Compression, Part 2: Analysis, *Int. J. Mech. Sci.*, **48**, 842.

Barlat, F. and Lian, J. (1989), Plastic Behavior and Strechability of Sheet Metals, Part 1, A Yield Function for Orthotropic Sheets Under Plane Stress Conditions, *Int. J. Plasticity*, **5**, 51.

Batdorf, S.R. (1949), Theories of Elastic-Plastic Buckling, *J. Aeronaut. Sci.*, **16**, 405.

Batterman, S. (1964), Load-Deflection Behavior of Shells of Revolution, *J. Engng. Mech. Div., Proc. ASCE*, **90**, EM6, 4167.

Batterman, S. (1965), Plastic Buckling of Axially Compressed Cylindrical Shells, *AIAA J.*, **3**, 316.

Batterman, S. (1968), Free Edge Buckling of Axially Compressed Cylindrical Shells, *J. Appl. Mech.*, **35**, 73.

Batterman, S. (1969), Plastic Buckling of an Externally Pressurized Complete Spherical Shell, *J. Engng. Mech. Div., Proc. ASCE*, **95**, 433

Bell, J.F. (1960), The Propagation of Large Amplitude Waves in Annealed Aluminum, *J. Appl. Phys.*, **31**, 277.

Bell, J.F. (1968), *The Physics of Large Deformation of Crystalline Solids*, Springer Tracts in Natural Philosophy, Vol. 14, Springer-Verlag, Berlin.

Bell, J.F. (1973), *The Experimental Foundations of Solid Mechanics*, Handbuch derPhysik, Vol. VI, Springer-Verlag, Berlin.

Bell, J.F. and Stein, A. (1962), The Incremental Loading Wave in the Pre-stressed Plastic Field, *J. Mecan.*, **1**, 395.

Berveiller, M. and Zaoui, A. (1979), An Extension of the Self-Consistent Scheme to Plastically Flowing Polycrystals, *J. Mech. Phys. Solids*, **26**, 325.

Bianchi, G. (1964), Some Experimental and Theoretical Studies on the Propagation of Longitudinal Plastic Waves in a Strain-Rate Dependent Material, in *IUTAM Symp. Stress Waves in Anelastic Solids* (eds., H. Kolsky and W. Prager), Springer-Verlag, Berlin, p. 101.

Biron, A. and Hodge, P.G. (1967), Limit Analysis of Rotationally Symmetric Shells Under Central Boss Loading by a Numerical Method, *J. Appl. Mech.*, **34**, 644.

Biron, A. and Sawczuk, A. (1967), Plastic Analysis of Rib-Reinforced Cylindrical Shells, *J. Appl. Mech.*, **34**, 37.

Bishop, J.F.W. (1953), On the Complete Solution to Problems of Deformation of a Plastic-Rigid Material, *J. Mech. Phys. Solids*, **2**, 43.

Bishop, J.F.W. (1958), On the Effect of Friction on Compression and Indentation Between Flat Dies, *J. Mech. Phys. Solids*, **6**, 132.

Bishop, J.F.W. and Hill, R. (1951), A Theory of the Plastic Distortion of a Polycrystalline Aggregate Under Combined Stresses, *Phil. Mag. Ser.* 7, **42**, 414.

Bishop, J.F.W., Hill, R. and Mott, N.F. (1945), The Theory of Indentation and Hardness Tests, *Proc. Roy. Phys. Soc.*, **57**, 147.

Bland, D.R. (1957), The Associated Flow Rule of Plasticity, *Mech. Phys. Solids*, **6**, 71.

Bleich, H.H. and Salvadori, N.G. (1953), Impulsive Motion of Elastic-Plastic Beams, *Proc. ASCE*, **79**, 287.

Bijlaard, P.P. (1949), Theory and Tests on the Plastic Stability of Plates and Shells, *J. Aeronaut. Sci.*, **16**, 529.

Bijlaard, P.P. (1956), Theory of Plastic Buckling of Plates and Application to Simply Supported Plates Subjected to Bending or Eccentric Compression in Their Plane, *Appl. Mech.*, **23**, 27.

Blazynski, T.Z. (1971), Optimization of Die Design in the Extrusion of Rod, *Int. J. Mech. Sci.*, **13**, 113.

Blazynski, T.Z. (1976), *Metal Forming: Tool Profiles and Flow*, Macmillan, London.

Blazynski, T.Z. and Cole, I.M. (1960), An Investigation of the Plug Drawing Process, *Proc. Inst. Mech. Engrs.*, **174**, 797.

Bodner, S.R. and Symonds, PS. (1962), Experimental and Theoretical Investigation of the Plastic Deformation of Cantilever Beams Subjected to Impulsive Loading, *J. Appl. Mech.*, **29**, 719.

Bourne, L. and Hill, R. (1950), On the Correlation of the Directional Properties of Rolled Sheet in Tension and Cupping Tests, *Phil. Mag. Ser. 7*, **41**, 671.

Boyce, W.E. (1959), A Note on Strain-Hardening Circular Plates, *J. Mech. Phys. Solids*, **7**, 114.

Boyd, D.E. (1966), Dynamic Deformation of Circular Membranes, *J. Engng. Mech. Div., Trans. ASCE*, **92**, 1.

Bramley, A.N. and Mellor, P.B. (1966), Plastic Flow in Stabilized Sheet Steel, *Int. J. Mech. Sci.*, 8 101.

Brazier, L.G. (1926), On the Flexure of Thin Cylindrical Shells and Other Sections, *Proc. Roy. Soc. London Ser. A*, **116**, 104.

Bridgman, P.W. (1944), Stress Distribution in the Neck of a Tension Specimen, *Trans. Amer. Soc. Metals*, **32**, 553.

Bridgman, P.W. (1945), Effects of High Hydrostatic Pressure on the Plastic Properties of Metals, *Rev. Mod. Phys.*, **17**, 3.

Bridgman, P.W. (1952), *Studies in Large Plastic Flow and Fracture*, McGraw-Hill, New York.

Brooks, G.M. (1987), Elastic-Plastic Ring Loaded Cylindrical Shells, *J. Appl. Mech.*, **54**, 597.

Brooks, G.M. (1988), Elastic-Plastic Ring-Loaded Cylindrical Shells, *J. Appl. Mech.* **55**, 761.

Brooks, G.M. and Leung, C.P. (1989), Elastic-Plastic Analysis of a Radially Loaded Spherical Shell, *J. Press. Vess. Technol., Trans. ASME*, **111**, 39.

Budiansky, B. (1959), A Reassessment of Deformation Theories of Plasticity, *J. Appl. Mech.*, **26**, 259.

Budiansky, B. and Wang, N.M. (1966), On the Swift Cup Test, *J. Mech. Phys. Solids*, **14**, 357.

Budiansky, B. and Wu, T.Y. (1962), Theoretical Prediction of Plastic Strains in Polycrystals, *Proc. 4th US Nat. Congr. Appl. Mech.*, p. 1175.

Bushnell, D. (1982), Plastic Buckling of Various Shells, *J. Pressure Vessels Technol, Trans. ASME*, **104**, 51.

Bushnell, D. and Galletly, G.D. (1974), Comparison of Test and Theory for Nonsymmetric Elastic-Plastic Buckling of Shells of Revolution, *Int. J. Solids Struct.*, **10**, 1271.

Caddell, R.M. and Atkins, A.G. (1968), Optimum Die Angles and Maximum Attainable Reductions in Rod Drawing, *J Engng. Indust., Trans. ASME*, **91**, 664.

Calder, C.A. and Goldsmith, W. (1971), Plastic Deformation and Perforation of Thin Plates Resulting from Projectile Impact, *Int. J. Solids Struct.*, **7**, 863.

Calentano, D.J., Cabezas, E.E. and Garcia, C.M. (2005), Analysis of the Bridgman Procedure to Characterize Mechanical Behavior of Materials in the Tensile Test, *J. Appl. Mech.*, **72**, 149.

Calladine, C.R. (1968), Simple Ideas in the Large Deflection Plastic Theory of Plates and Slabs, *Engineering Plasticity* (eds., J. Heyman and F. Leckie), p. 93, Cambridge University Press, UK.

Calladine, C.R. (1985), *Plasticity for Engineers*, Ellis Horwood, Chichester.

Calladine, C.R. (2000), A Shell Buckling Paradox, *Advances in the Mechanics of Plates and Shells* (ed., D. Durban and J.G. Symmonds), p. 119, Kulwer Academic Publishers, Netherland.

Campbell, J.D. (1954), The Yield of Mild Steel Under Impact Loading, *J. Mech. Phys. Solids*, **3**, 54.

Campbell, J.D. (1972), *Dynamic Plasticity of Metals*, Springer-Verlag, New York.

Campbell, J.D. (1973), Dynamic Plasticity: Macroscopic and Microscopic Aspects, *Mater. Sci. Engng.*, **12**, 3.

Carter, W.T. and Lee, D. (1986), Further Analysis of Axisymmetric Upsetting, *J. Manuf. Sci. Eng., Trans. ASME*, **108**, 165.

Casey, J. and Naghdi, P.M. (1981), On the Characterization of Strain Hardening in Plasticity, *J. Appl. Mech.*, **48**, 285.

Celentano, J., Cabezas, E.E., and Garcia, G.M. (2005), Analysis of the Bridgeman Procedure to Characterize Mechanical Behavior of Materials in the Tensile Test: Experiments and Simulation, *J. Appl. Mech.*, **72**, 149.

Chadwick, P., Cox, A.D., and Hopkins, H.G. (1964), Mechanics of Deep Underground Explosions, *Phil. Trans. Roy. Soc*, **256**, 235.

Chaboche, J.L. (1986), Time Independent Constitutive Theories for Cyclic Plasticity, *Int. J. Plasticity*, **2**, 149.

Chakrabarty, J. (1969a), On Uniqueness and Stability in Rigid/Plastic Solids, *Int. J. Mech. Sci*, **11**, 723.

Chakrabarty, J. (1969b), On the Problem of Uniqueness Under Pressure Loading, *Zeit. angew. Math. Phys*, **20**, 696.

Chakrabarty, J. (1970a), A Theory of Stretch Forming Over Hemispherical Punch Heads, *Int. J. Mech. Sei.*, **12**, 315.

Chakrabarty, J. (1970b), A Hypothesis of Strain-Hardening in Anisotropic Plasticity, *Int. J. Mech. Sei.*, **12**, 169.

Chakrabarty, J. (1971), Elastic-Plastic Expansion of a Circular Hole in an Infinite Plate of Variable Thickness, *Int. J. Mech. Sei.*, **13**, 439.

Chakrabarty, J. (1972), The Necking of Cylindrical Bars Under Lateral Fluid Pressure, *Proc. 13th MTDR Conference, Birmingham UK*, p. 565.

Chakrabarty, J. (1973), Plastic Buckling of Cylindrical Shells Subjected to External Fluid Pressure, *Z. angew. Math. Phys.*, **24**, 270.

Chakrabarty, J. (1974), Influence of Transverse Anisotropy on the Plastic Collapse of Plates and Sheels, *J. Eng. Indust., Trans. ASME*, **97**, 125.

Chakrabarty, J. (1993a), A New Yield Function for Planar Anisotropy in Sheet Metals, *Proc. Int. Conf. Adv. Mater. Process. Technol.*, Dublin City University Press, Dublin, p. 1181.

Chakrabarty, J. (1993b), An Analysis of Axisymmetric Extrusion Including Redundant Shear and Strain-Hardening, *Advances in Engineering Plasticity and its Applications* (ed., W.B.Lee), Elsevier, Amsterdam, p. 901.

Chakrabarty, J. (1997), Plastic Instability and Forming Limits in Anisotropic Sheet Metal, Proc. 32nd Int. MATADOR Conf, Macmillan Press, p. 373.

Chakrabarty, J. (1998), Large Deflections of a Clamped Circular Plate Pressed by a Hemispherical-Headed Punch, *Metals and Materials*, **4**, 680.

Chakrabarty, J. (2002), Influence of Anisotropy on the Plastic Buckling of Rectangular Plates, *Proc. 2nd int. conf. on Structural Stability and Dynamics*, World Scientific, p. 448.

Chakrabarty, J. (2006), *Theory of Plasticity*, third edition, Elsevier, Oxford.

Chakrabarty, J. and Alexander, J.M. (1970), Hydrostatic Bulging of Circular Diaphragms, *J. Strain Anal.*, **5**, 155.

Chakrabarty, J. and Chen F.K. (2005), Influence of Loading Path on the Plastic Instability Strain in Anisotropic Plane Sheets, *J. Mater. Process. Technol.*, **166**, 218.

Chakrabarty, J., Lee, W.B. and Chan, K.C. (2000), An Analysis for the Plane Strain Bending of an Anisotropic Sheet in the Elastic/Plastic Range, *J. Mater. Process. Technol.*, **104**, 48.

Chakrabarty, J., Lee, W.B. and Chan, K.C. (2001), An Exact Solution for the Elastic/Plastic Bending of Anisotropic Sheet Metal Under Conditions of Plane Strasin, *Int. J. Mech. Sci.*, **43**, 21871.

Chakrabarty, J. and Mellor, P.B. (1968) A New Approach for Predicting the Limiting Drawing Ratio, *La Metullugia, Italiana*, **60**, 791.

Chakrabarty, J. and Wasti, S.T. (1971), On the Plastic Collapse of Cylindrical Shells Under Axial Compression, *METU J. Pure Appl. Sci.*, **4**, 1.

Chan, K.C. and Lee, W.B. (1990), A Theoretical Prediction of the Strain Path of Anisotropic Sheet Metal Deformed Under Uniaxial and Biaxial Stress State, *Int. J. Mech. Sci.*, **32**, 497.

Chan, K.S. (1985), Effects of Plastic Anisotropy and Yield Surface Shape on Sheet Metal Stretchability, Metall. Trans., **16A**, 629.

Chawalla, E. (1937), Aussermittig Gedruckte Baustahlstabe mit Elastisch Eingespannten Enden und Verschieden Grossen Angriffschebeln, *Der Stahlbau*, **15**, 49.

Chen, C.C. and Kobayashi, S. (1978), Deformation Analysis of Multi-Pass Bar Drawing and Extrusion, *Annals. CIRP*, **27**, 151.

Chen, C.T. and Ling, F.F. (1968), Upper Bound Solutions to Axisymmetric Extrusion Problems, *Int. J. Mech. Sci.*, **10**, 863.

Chen, C.C., Oh, S.I. and Kobayashi, S. (1979), Ductile Fracture in Axisymmetric Extrusion and Drawing, Part 1, Deformation Mechanics of Extrusion and Drawing, *J. Engng. Indust. Trans. ASME*, **101**, 23.

Chen, F.K. and Liu, J.H. (1997), Analysis of an Equivalent Model for the Finite Element Simulations of a Stamping Process, *Int. J. Mach. Tools Manuf*, **37**, 409.

Chen, W.F. and Han, D.J (1988), *Plasticity for Structural Engineers*, Springer-Verlag, New York.

Chen, W.F. (1970), General Solution of Inelastic Beam-Column Problems, *Engng. J. Mech. Div., Proc. ASCE*, **96**, EM4, 421.

Chen, W.F. and Astuta, T. (1976), *Theory of Beam Columns*, Vols. 1 and 2, McGraw-Hill, New York.

Chen, X.W. and Li, Q.M. (2003), Perforation of a Thick Plate by Rigid Projectiles, *Int. J. Impact Engng.* **28**, 743.

Cheng, S.Y., Ariaratnam, S.T., and Dubey, R.N. (1971), Axisymmetric Bifurcation in an Elastic/Plastic Cylinder Under Axial Load and Lateral Hydrostatic Pressure, *Quart. Appl. Math.*, **29**, 41.

Chenot, L. and Wagoner, R.H. (2001), *Metal Forming Analysis*, Cambridge University Press, Cambridge.

Chenot, J.L. Feigeres, L., Lavarenne, B., and Salecon, J. (1978), A Numerical Application of the Slipline Field Method to Extrusion Through Conical Dies, *Int. J. Engng. Sci.*, **16**, 263.

Chern, J. and Nemat-Nasser, S. (1969), The Enlargement of a Hole in a Rigid-Workhardening Disk of Non-Uniform Initial Thickness, *J. Mech. Phys. Solids*, **17**, 271.

Chiang, D.C. and Kobayashi, S. (1966), The Effect of Anisotropy and Work-Hardening Characteristics on the Stress and Strain Distribution in Deep Drawing, J. Engng. Indust., **88**, 443.

Chitkara, N.R. and Bhutta, M.A. (1996), Near Net Shape Forging of Spur Gear Forms. An Analysis and Some Experiments, *Int. J. Mech. Sci.*, **38**, 891.

Chitkara, N.R. and Collins, I.F. (1974), A Graphical Technique for Constructing Anisotropic Slipline Fields, *Int. J. Mech. Sci.*, **16**, 241.

Chon, C.T. and Symonds, P.S. (1977), Large Dynamic Plastic Deflection of Plates by Mode Method, *J. Engng. Mech. Div., Trans. ASCE*, **103**, 169.

Chou, P.C. (1961), Perforation of Plates by High Speed Projectiles, in *Developments in Mechanics* (ed. J.E. Lay and L.E. Malvern), North Holland, 286.

Chu, C.C. (1984)), An Investigation of Strain Path Dependence of the Forming Limit Curve, Int. J. Solids Struct., **18**, 205.

Chu, E, and Xu, Y. (2001), An Elastoplastic analysis of Flange Wrinkling in Deep Drawing Process, *Int. J. Mech. Sci.*, **43**, 1421.

Chu, F.P.E. (1995), Springback in Stretch/Draw Sheet Forming, *Int. J. Mech. Sci.*, **37**, 327.

Chung, S.Y. and Swift, H.W. (1951), Cup Drawing from a Flat Blank, *Proc. Instn. Mech. Engrs.*, **165**, 199 (J.E. Lay and L.E. Malvern), North-Holland, Amsterdam, p. 286.

Chung, S.W. and Swift, H.W. (1952), A Theory of Tube Sinking, *J. Iron Steel Inst.*, **170**, 29.

Chung, W.J., Kim, Y.J. and Yang, D.Y. (1988), A Rigid/Plastic Finite Element Analysis of Hydrostatic Bulging of Elliptical Diaphragms Using Hill's New Yield Criterion, *Int. J. Mech. Sci.*, **31**, 391.

Cinquini, C, Lamblin, D., and Guerlement, G. (1977), Variational Formulation of the Optimal Plastic Design of Circular Plates, *Computer Methods Appl. Mech. Engng.*, **11**, 19.

Cinquini, C. and Zanon, P. (1985), Limit Analysis of Circular and Annular Plates, *Ingenieur-Archiv*, **55**, 157.

Clifton, R.J. (1966), An Analysis of Combined Longitudinal and Torsional Plastic Waves in a Thin-Walled Tube, *Proc. 5th U.S. Nat. Congr. Appl. Mech.*, 456.

Clifton, R.J. (1973), Plastic Waves: Theory and Experiment, in *Mechanics Today* (ed., Nemal-Nasser), Pergamon Press, New York, p. 102.

Clifton, R.J. and Bodner, S.R. (1966), An Analysis of Longitudinal Elastic-Plastic Pulse Propagation, *J. Appl. Mech.*, **33**, 248.

Cline, C.B. and Jahsman, W.E. (1967), Response of a Rigid-Plastic Ring to Impulsive Loading, *J. Appl. Mech.*, **34**, 329.

Collins, I.F. (1968), The Upper Bound Theorem for Rigid/Plastic Solids Generalized to Include Coulomb Friction, *J. Mech. Phys. Solids*, **17**, 323.

Collins, I.F. (1971), On the Analogy Between Plane Strain and Plate Bending Solutions in Rigid/perfectly Plasticity Theory, *Int. J. Solids Struct.*, **7**, 1037.

Collins, I.F. (1973), On the Theory of Rigid/Perfectly Plastic Plates Under Uniformly Distributed Loads *Acta Mech.*, **18**, 233.

Collins, I.F. and Meguid, S.A. (1977), On the Influence of Hardening and Anisotropy on the Plane Strain Compression of Thin Metal Strips, J. Appl. Mech., **52**, 271.

Collins, I.F. and Williams, B.K. (1985), Slipline Fields for Axisymmetric Tube Drawing, *J. Mech. Eng. Sci.*, **27**, 225.

Conroy, M.F. (1952), Plastic-Rigid Analysis of Long Beams Under Transverse Impact Loading, *J. Appl. Mech.*, **19**, 465.

Conroy, M.E (1956), Plastic Deformation of Semi-Infinite Beams Subject to Transverse Impact Loading, *J. Appl. Mech.*, **23**, 239.

Conroy, M.F. (1969), Rigid-Plastic Analysis of a Simply Supported Circular Plate due to Dynamic Circular Loading, *J. Franklin Inst.*, **288**, 121.

Coon, M.D. and Gill, S.S. (1968), Effect of Change of Geometry on the Rigid/Plastic Limit Load of Cylinders, *Int. J. Mech. Sei.*, **10**, 355.

Corona, E. and Kyriakides, S. (1988), On the Collapse of Inelastic Tubes Under Combined Bending and Pressure, *Int. J. Solids Struct.*, **24**, 505.

Correia, J.P.D., Ferron, G. and Moreira, L.P. (2003), Analytical and Experimental Investigation of Wrinkling for Deep Drawn Anisotropic Metal Sheets, *Int. J. Mech. Sci.*, **45**, 1167.

Cotter, B.A. and Symonds, P.S. (1955), Plastic Deformation of a Beam Under Impulsive Loading, *Proc. ASCE*, **81**, 675.

Cox, A.D. and Morland, L.W. (1959), Dynamic Plastic Deformations of Simply Supported Square Plates, *J. Mech. Phys. Solids*, **7**, 229.

Coulomb, C.A. (1773), Test on the Applications of the Rules of Maxima and Minima to Some Problems of the Statics Related to Architecture, *Mem. Math. Phys.*, **7**, 343.

Craggs, J.W. (1954), Wave Motion in Plastic-Elastic Strings, *J. Mech. Phys. Solids*, **2**, 286.

Craggs, J.W. (1957), The Propagation of Infinitesimal Plane Waves in Elastic-Plastic Materials, *J. Mech. Phys. Solids*, **5**, 115.

Craggs, J.W. (1961), Plastic Waves, in *Progress in Solid Mechanics* (eds., I.N. Sneddon and R. Hill), Vol. 2, Chap. 4, North-Holland, Amsterdam.

Cristescu, N. (1960), Some Observations on the Propagation of Plastic Waves in Plates, in *Plasticity* (eds., E.H. Lee and P.S. Symonds), Pergamon Press, New York, p. 501.

Cristescu, N. (1963), On the Propagation of Elastic-Plastic Waves in Metallic Rods, *Bull. Acad. Polon. Sci.*, **11**, 129.

Cristescu, N. (1964), Some Problems on the Mechanics of Extensible Strings, in *Stress Waves in Anelastic Solids* (eds., H. Kolsky and W. Prager), Springer-Verlag, Berlin, p. 118.

Cristescu, N. (1965), Loading/Unloading Criteria for Rate Sensitive Materials, *Arch. Mech. Stos.*, **11**, 291.

Cristescu, N. (1967), *Dynamic Plasticity*, North-Holland, Amsterdam.

Cristescu, N. (1970), The Unloading in Symmetric Longitudinal Impact of Two Elastic-Plastic Bars, *Int. J. Mech. Sci.*, **12**, 723.

Cristescu, N. (1972), A Procedure for Determining the Constitutive Equations for Materials Exhibiting Both Time-Dependent and Time-Independent Plasticity, *Int. J. Solids Struct.*, **8**, 511.

Cristescu, N. (1974), Rate-Type Constitutive Equations in Dynamic Plasticity, in *Problems of Plasticity* (ed., A. Sawczuk), Noordhoff International, Groningen, p. 287.

Cristescu, N. and Bell, J.F. (1970), On Unloading in the Symmetrical Impact of Two Aluminium Bars, in *Inelastic Behaviour of Solids* (eds., M.F. Kannienen et al.), McGraw-Hill, New York, p. 397.

Cristescu, N. and Suliciu, I. (1982), *Viscoplasticity*, Martinus Nijhoff, The Hague.

Crossland, B. (1954), The Effect of Fluid Pressure on Shear Properties of Metals, *Pwc. Instn. Mech. Engrs.*, **169**, 935.

Derbalian, K.A., Lee, E.H., Mallett, R.I. and McMeeking, R.M. (1978), Finite Element Metal Forming Analysis with Spatially Fixed Mesh, Applications of Numerical Methods to Forming Processes, *ASME, AMD*, **28**, 39.

Dafalias, Y.F. (1993), Planar Double-Slip Micromechanical Model for Polycrystal Plasticity, *Engng. Mech., Trans. ASCE*, **119**, 1260.

Dafalias, Y.F. and Popov, E.P. (1975), A Model for Nonlinearly Hardening Materials for Complex Loading, *Acta Mech.*, **21**, 173.

Danyluk, H.T. and Haddow, J.B. (1965), Elastic/Plastic Flow Through a Converging Conical Channel, *Acta Mech.*, **7**, 35.

Davey, K. and Ward, M.J. (2002), A Practical Method for Finite Element Ring Rolling Simulation Using the ALE Flow Formulation, *Int. J. Mech. Sci.*, **44**, 165.

Davidenkov, N.N. and Spiridonova, N.I. (1946), Analysis of Tensile Stress in the Neck of an Elongated Test Specimen, *Proc. ASTM.*, **46**, 1147.

Davies, R.M. and Hunter, S.C. (1963), The Dynamic Compression Testing of Solids by the Method of the Split Hopkinson Pressure Bar, *J. Mech. Phys. Solids*, **11**, 155.

Dean, T.H. (1970), Influence of Billet Inertia and Die Friction in Forging Processes, *Proc. 11th Int. M.T.D.R. Conf.*, Pergamon Press, Oxford.

De Juhasz, K.J. (1949), Graphical Analysis of Impact of Bars Stressed above the Elastic Range, *J. Franklin Inst.*, **248**, 15. and 113.

Demir, H.H. (1965), Cylindrical Shells Under Ring Loads, *Proc. ASCE, Struct. Engng. Div.*, **91**, 71.

Denton, A.A. and Alexander, J.M. (1963), On the Determination of Residual Stresses in Tubes, *J. Mech. Engng. Sci.*, **5**, 75.

Derbalian, K.A., Lee, E.H., Mallett, R.l. and McMeeking, R.M. (1978), Finite Element Metal Forming Analysis with Spatially Fixed Mesh, Applications of Numerical Methods to Forming Processes, *ASME, AMD*, **28**, 39.

DeRuntz, J.A. and Hodge, P.G. (1966), Significance of the Concentrated Load on the Limit Analysis of Conical Shells, *J. Appl. Mech.*, **33**, 93.

Desai, C.S. and Abel, J.F. (1972), *Introduction to the Finite Element Method*, Van Nostrand Reinhold, New York.

Desdo, D. (1971), Principal and Slipline Methods of Numerical Analysis in Plane and Axially Symmetric Deformations of Rigid/Plastic Media, *J. Mech. Phys. Solids*, **19**, 313.

Dillamore, I.L., Mella, P. and Hazel, R.I. (1972), Preferred Orientation and the Plastic Behavior of Sheet Metal, *Inst. Metal.*, **100**, 50.

Dillons, O.W. (1968), Experimental Data on Small Plastic Deformation Waves in Annealed Aluminium, *Int. J. Solids Struct.*, **4**, 197.

Dinno, K.S. and Gill, S.S. (1965), The Limit Analysis of a Pressure Vessel Consisting of the Junction of a Cylindrical and a Spherical Shell, *Int. J. Mech. Sei.*, **7**, 21.

Dokmeci, M.C. (1966), A Shell of Constant Strength, Z. *Angew Math. Phys.*, **17**, 545.

Dodd, B. and Caddell, R.M. (1984), On the Anomalous Behaviour of Anisotropic Sheet Metals, *Int. J. Mech. Sci.*, **26**, 113.

Donnell, L.H. (1933), Stability of Thin Walled Tubes Under Torsion, NACA Report 479, pp. 1–24.

Dorn, J.F. (1949), Stress-Strain Relations for Anisotropic Plastic Flow, *J. Appl. Phys.*, **20**, 15.

Donnel, L.H. (1930), Longitudinal Wave Transmission and Impact, *Trans. ASME*, **52**, 153.

Dowling, A.R., Harding, J. and Campbell, J.D. (1970), The Dynamic Punching of Metals, *J. Inst. Metals*, **98**, 215.

Drucker, D.C. (1951), A More Fundamental Approach to Plastic Stress-Strain Relations, *Proc. 1st. U.S. Nat. Congr. Appl. Mech.*, **1**, 487.

Drucker, D.C. and Palgen, L. (1981), On the Stress-Strain Relations Suitable for Cyclic and Other Loading, *J. Appl. Mech.*, **48**, 479.

Drucker, D.C., Prager, W. and Greenberg, H.J. (1952), Extended Limit Design Theorems for Continuous Media, *Quart. Appl. Math.*, **9**, 381.

Drucker, D.C. (1953), Limit Analysis of Cylindrical Shells Under Axially-Symmetric Loading, *Proc. 1st Midwestern Conf. Solid Mech., Urbana,* p. 158.

Drucker, D.C. and Hopkins, H.G. (1954),Combined Concentrated and Distributed Load on Ideally Plastic Circular Plates, *Proc. 2nd US Nat. Congr. Appl. Mech. (Ann Arbor)*, p. 517.

Drucker, D.C. and Shield, R.T. (1957), Bounds on Minimum Weight Design, *Quart. Appl. Math.*, **15**, 269.

Drucker, D.C. and Shield, R.T. (1959), Limit Analysis of Symmetrically Loaded Thin Shells of Revolution, *J. Appl. Mech.*, **26**, 61.

Drucker, D.C. and Shield, R.T. (1961), Design of Torispherical and Toriconical Pressure Vessel Heads, *J. Appl. Mech.*, **28**, 292.

Duffill, A.W. and Mellor P.B. (1969), A Comparison Between Conventional and Hydrostatic Extrusion Methods of Cold Extrusion Through Conical Dies, *Ann. of CIRP*, **17**, 97.

Dugdale, D.S. (1954), Cone Indentation Experiments, *J. Mech. Phys. Solids*, **2**, 265.

Duncan, J.L. and Johnson, W. (1968), The Ultimate Strength of Rectangular Anisotropic Diaphragms, *Int. J. Mech. Sci.*, **10**, 143.

Durban, D. (1979), Axially Symmetric Radial Flow of Rigid-Linear Hardening Material, *J. Appl. Mech.,* **46**, 322.

Durban, D. (1986), Radial Stressing of Thin Sheets with Plastic Anisotropy, *Int. J. Mech. Sci.*, **28**, 801.

Durban, D., Davidi, G. and Lior, D. (2001), Plastic Forming Processes Through Rotating Conical Dies, *J. Appl. Mech.*, **68**, 894.

Durban, D. and Fleck, N.A. (1992), Singular Plastic Fields in Steady Penetration of a Rigid Cone, *J. Appl. Mech.*, **59**, 706.

Durban, D. and Zuckerman, Z. (1999), Elastoplastic Buckling of Rectangular Plates in Biaxial Compression/Tension, *Int. J. Mech. Sci.*, **41**, 751.

Duwez, P.E. and Clark, D.S. (1947), An Experimental Study of the Propagation of Plastic Deformation Under Conditions of Longitudinal Impact, *Proc. ASTM*, **47**, 502.

Duwez, P.E., Clark, D.S., and Bohenblust, H.F. (1950), the Behaviour of Long Beams Under Impact Loading, *J. Appl. Mech.*, **7**, 27.

Eason, G. (1958), Velocity Fields for Circular Plates with the von Mises Yield Condition, *J. Mech. Phys. Solids*, **6**, 231.

Eason, G. (1959), The Load Carrying Capacity of Cylindrical Shells Subjected to a Ring of Force, *J. Mech. Phys. Solids*, **1**, 169.

Eason, G. (1960), The Minimum Weight Design of Circular Sandwich Plates, Z. *Angew. Math. Phys.*, **11**, 368.

Eason, G. (1961), The Elastic-Plastic Bending of a Simply Supported Plate, *J. Appl. Mech.*, **28**, 395.

Eason, G. and Shield, R.T. (1955), The Influence of Free Ends on the Load Carrying Capacities of Cylindrical Shells, *J. Mech. Phys. Solids*, **4**, 17.

Eason, G. and Shield, R.T. (1956), Dynamic Loading of Rigid-Plastic Cylindrical Shells, *J. Appl. Mech.*, **4**, 53.

Eason, G. and Shield, R.T. (1960), The Plastic Indentation of a Semi-Infinite Solid by a Rough Circular Punch, Z. *Angew. Math. Phys.*, **11**, 33.

Efron, L. and Malvern, L.E. (1969), Electromagnetic Velocity Transducer Studies of Plastic Waves in Aluminum Bars, *Experimental Mech.*, **9**, 255.

Eisenberg, M.A. (1976), A Generalization of the Plastic Flow Theory with Application of Cyclic Hardening and Softening Phenomena, *J. Engng. Mater. Technol, Trans. ASME*, **98**, 221.

El-Ghazaly, H.A. and Sherbourne, A.N. (1986), Deformation Theory for Elastic-Plastic Buckling Analysis of Plates Under Nonproportional Planar Loading, *Comput. Struct.*, **22**, 131.

El-Sebaie, M.G. and Mellor, P.B. (1972), Plastic Instability Conditions in the Deep-Drawing of a Circular Blank of Sheet Metal, *Int. J. Mech. Sci.*, **14**, 535.

El-Sebaie, M.G. and Mellor, P.B. (1973), Plastic Instability When Deep Drawing Into a High Pressure Medium, *Int. J. Mech. Sei.*, **15**, 485.

Engesser, F. (1889), Zur Knikung verwundener Stabe bei zusamengesetzter Belastung, *Z. Archit, Ingenieur,* **35**, 455.

Ewing, D.J.F. and Richards, C.E. (1973), The Yield Point Loads of Singly Notched Pin-Loaded Tensile Strips, *J. Mech. Phys. Solids*, **21**, 27.

Ewing, D.J.F. and Spurr, C.E. (1974), The Yield Point Loads of Symmetrically Notched Metal Strips, *J. Mech. Phys. Solids*, **22**, 37.

Ezra, A.A. (1958), The Plastic Response of a Simply Supported Beam Under Impact Loading, *Proc. 3rd U.S. Nat. Congr. Appl. Mech.*, 513.

Fliigge, W. (1960), Stresses in Shells, Springer-Verlag, Berlin.

Florence, A.L. (1965), Annular Plate Under a Transverse Line Impulse, *AIAA J.*, **3**, 1726.

Florence, A.L. (1966), Clamped Circular Rigid-Plastic Plates Under Blast Loading, *J. Appl. Mech.*, **33**, 256.

Florence, A.L. (1977), Response of Circular Plates to Central Pulse Loading, *Int. J. Solids Struct.*, **13**, 1091.

Flores, G. (2006), A Two-Dimensional Linear Assumed Strain Triangular Element for Finite Element Analysis, *J. Appl. Mech.*, **73**, 970.

Fogg, B. (1968), Theoretical Analysis for the Redrawing of Cylindrical Cups Through Conical Dies, *J. Mech. Engng. Sei.*, **10**, 141.

Flügge, W. (1932), Die Stabilität der Kreiszylinderschale, *Ingenieur-Archiv*, **3**, 24.

Fliigge, W. and Nakamura, T. (1965), Plastic Analysis of Shells of Revolution Under Axisymmetric Loads, *Ingenieur-Archiv*, **34**, 238.

Forrestal, M.J., Barr, N.S., and Luk, V.K. (1991), Penetrating of Strain-Hardening Targets with Rigid Spherical Nose Rods, *J. Appl. Mech.*, **58**, 7.

Forrestal, M.J. and Luk, V.K. (1988), Dynamic Spherical Cavity Expansion in a Compressible Elastic- Plastic Solid, *J. Appl. Mech.*, **55**, 275.

Forrestal, M.J., Okajima, K., and Luk, V.K. (1988), Penetration of 6061-T651 Aluminum Targets with Rigid Long Rods, *J. Appl. Mech.*, **55**, 755.

Freiberger, W. (1952), A Problem of Dynamic Plasticity. The Enlargement of a Circular Hole in a Flat Sheet, *Proc. Cambridge Philos. Soc.*, **48**, 135.

Freiberger, W. (1956), Minimum Weight Design of Cylindrical Shells, *J. Appl. Mech.*, **23**, 576.

Freiberger, W. and Tekinalp, B. (1956), Minimum Weight Design of Circular Plates, *J. Mech. Phys. Solids*, **4**, 294.

Fukui, S. and Hansson, A. (1970), Analytical Study of Wall Ironing Considering Work-Hardening, *Ann. Of CIRP*, **18**, 593.

Gaydon, F.A. (1952), An Analysis of the Plastic Bending of a Thin Strip in its Plane, *J. Mech. Phys. Solids*, **1**, 103.

Gaydon, F.A. (1954), On the Yield Point Loading of a Square Plate with Concentric Circular Hole, *J. Mech. Phys. Solids*, **2**, 170.

Gaydon, F.A. and McCrum, A.W. (1954), A Theoretical Investigation of the Yield Point Loading of a Square Plate with a Central Circular Hole., *J. Mech. Phys. Solids*, **2**, 156.

Galiev, S.U. and Nechitailo, N.V. (1985), The Dynamics of Shape Changes, in Thin Plates and Shells of Revolution, *Acad. Sci. Ukrainian SSR*, Kiev.

Geckeier, J.W. (1928), Plastic Folding of the Walls of Hollow Cylinders and Some Other Folding Phenomena in Bowls and Sheets, *Zeit angew. Math. Mech.*, **8**, 341.

Gellin, S. (1979), Effect of an Axisymmetric Imperfection on the Plastic Buckling of an Axially Compressed Cylindrical Shell, *J. Appl. Mech.*, **46**, 125.

Gellin, S. (1980), The Plastic Buckling of Long Cylindrical Shells Under Pure Bending, *Int. J. Solids Struct.*, **16**, 397.

Gerard, G. (1957), Handbook of Structural Stability, Part I, Buckling of Flat Plates, NACA TN3781.

Gerard, G. (1962), *Introduction to Structural Stability Theory*, McGraw-Hill, New York.

Gill, S.S. (1964), The Limit Pressure for a Flush Cylindrical Nozzle in a Spherical Pressure Vessel, *Int. J. Mech. Sei.*, **6**, 105.

Gill, S.S. (1970), *The Stress Analysis of Pressure Vessels and Pressure Vessel Components*, Chapter 3, Pergamon Press, Oxford, U.K.

Gill, S.S. and Leckie, F.A. (1968), The Effect of Geometry Changes on the Application of Limit Analysis to the Design of Pressure Vessel Nozzles, *Int. J. Mech. Sei.*, **10**, 989.

Ghosh, A.K. (1977), The Influence of Strain Hardening and Strains Rate Sensitivity on Sheet Metal Forming, *J. Engng. Mat. Tech., Trans. ASME*, **99**, 264.

Ghosh, A.K. and Backofen, W.A. (1973), Strain Hardening and Instability in Biaxially Stretched Sheets, *Metall. Trans.*, **4A**, 1113.

Gjelsvik, A. and Lin, G.S. (1985), Plastic Buckling of Plates with Edge Frictional Shear Effects, *J. Engng. Mech. Div., Trans. ASCE*, **113**, 953.

Goel, R.P. and Malvern, L.E. (1970), Biaxial Plastic Simple Waves with Combined Isotropic and Kinematic Work-Hardening, *J. Appl. Mech.*, **37**, 1100.

Gordon, W.A., Van Tyne, D.J. and Sriram, S. (2000), Extrusion Through Spherical Dies- An Upper Bound Analysis, *J. Manuf. Engng., Trans. ASME*, **124**, 92.

Gotoh, M. (1977), A Theory of Plastic Anisotropy Based on a Yield Function of Fourth Order, *Int. J. Mech. Sci.*, **19**, 505.

Gotoh, M. (1978), A Finite Element Analysis of the Rigid/Plastic Deformation of the Anisotropic Deep Drawing Process Based on a Fourth Degree Yield Function – I, *Int. J. Mech. Sci.*, **20**, 423.

Gotoh, M. (1980), A Finite Element Analysis of the Rigid/Plastic Deformation of the Anisotropic Deep Drawing Process Based on a Fourth Degree Yield Function – II, *Int. J. Mech. Sci.*, **22**, 367.

Graf, A. and Hosford, W.F. (1990), Calculations of Forming Limit Diagrams, *Metall. Trans.*, **21A**, 87.

Graf, A. and Hosford, W.F. (1993), Calculations of Forming Limits for Changing Strain Paths, *Metall. Trans.*, **24A**, 2497.

Green, A.P. (1954a), A Theory of Plastic Yielding due to the Bending of Cantilevers and Fixed-Ended Beams—Part I, *J. Mech. Phys. Solids*, **3**, 1.

Green, A.P. (1954b), A Theory of Plastic Yielding due to Bending of Cantilevers and Fixed-Ended Beams—Part II, *J. Mech. Phys. Solids*, **3**, 143.

Goldsmith, W. (1960), *Impact*, Edward Arnold, London.

Goldsmith, W. and Finnegan, S.A. (1971), Penetration and Perforation Processes in Metal Targets at and above Ballistic Velocities, *Int. J. Mech. Sci.*, **13**, 843.

Goodier, J.N. (1968), Dynamic Buckling of Rectangular Plates in Sustained Plastic Compressive Flow, in *Engineering Plasticity* (eds., J. Heyman and FA. Leckie), Cambridge University Press, Cambridge, UK, p. 183.

Gu, W., Tang, W. and Liu, T. (1996), Dynamic Pulse Buckling of Cylindrical Shells Subjected to External Impulsive Loading, *J. Pressure Vessel Technoi*, **118**, 33.

Gurson, A.L. (1977), Continuum Theory of Ductile Rupture by Void Nucleation and Growth, *J. Engng. Mater. Technol.*, **99**, 2.

Gunasekera, J.S. and Hoshino, S. (1982), Analysis of Extrusion or Drawing of Polygonal Sections Through Straightly Converging Dies , *J. Engng. Indust., Trans. ASME*, **104**, 38.

Gunasekera, J.S. and Hoshino, S., (1985), Analysis of Extrusion of Polygonal Sections Through Streamlined Dies, *J. Engng. Indust., Trans. ASME*, **107**, 229.

Gunasekera, J.S. 1989, *Cad/Cam of Dies*, Ellis Horwood, Chicheste

Haar, A. and von Karman, Th. (1909), *Nachr. Ges. Wiss. Göttingen, Math. Phys. Kl.*, 204.

Haddow, J.B. (1962), Ideal Die Pressures for Axisymmetric Tube Extrusion, *Int. J. Mech. Sci.*, **4**, 447.

Haddow, J.B. (1965), On the Compression of a Thin Disc, *Int. J. Mech. Sci.*, **7**, 657.

Haddow, J.B. (1969), Yield Point Loading Curves for Circular Plates, *Int. J. Mech.* 5ci., **11**, 455.

Haddow, J.B. and Danyluk, H.T. (1964), Indentation of a Rigid-Plastic Semi-Infinite Medium by a Smooth Rigid Cone Fitted in a Prepared Cavity, *J. Mech. Sci.*, **5**, 1.

Haddow, J.B. and Johnson, W. (1962), Bounds for the Load to Compress Plastically a Square Disc Between Rough Dies, *Appl. Sci. Res.*, **A10**, 476.

Hailing, J. and Mitchell, L.A. (1965), An Upper Bound for Axisymmetric Extrusion, *Int. J. Mech. Sci.*, **7**, 277.

Hamada, H. (1985), In-Plane Buckling of Circular Plates, *Proc. J. Soc. Mech. Engrs.*, **51**, 1928.

Hardy, C, Baronnet, C.N. and Tordion, G.V. (1971), The Elasto-Plastic Indentation of a Half Space by a Rigid Sphere, *Int. J. Numer. Methods Engng.*, **3**, 451.

Hart, E.N. (1967), Theory of the Tensile Test, *Acta Metall.*, **15**, 351.

Hartley, P., Sturgess, C.E.N. and Rowe, G.W. (1979), Friction in Finite Element Analysis of Metal Forming Processes, *Int. J. Mech. Sci.* **21**, 301.

Hashmi, M.S.J. (1980), Strain Rate Sensitivity of Mild Steel at Room Temperature and Strain Rate of up to 10^5 s^{-1}, Analysis, **15**, 201.

Hassani, H.A. and Neale, K.W. (1991), On the Analysis of Sheet Metal Wrinkling, *Int. J. Mech. Sci.*, **33**, 13.

Haydi, H.A. and Sherbourne, A.N. (1974), Plastic Analysis of Shallow Spherical Shells, *J. Appl. Mech.*, **41**, 593.

Hawkyard, J.B. (1969), A Theory for the Mushrooming of Flat Ended Projectiles Impinging on a Flat Rigid Anvil Using Energy Considerations, *Int. J. Mech. Sci.*, **11**, 313.

Hawkyard, J.B. and Johnson, W. (1967), An Analysis of the Changes in Geometry of a Short Hollow Cylinder During Axial Compression, *Int. J. Mech. Sci.*, **9**, 163.

Hawkyard, J.B., Johnson, W. and Kirkland, J. (1973), Analysis for Roll Force and Torque in Ring Rolling with Some Supporting Experiments, *Int. J. Mech. Sci.*, **15**, 473.

Haythornthwaite, R.M. (1954), The Deflection of Plates in the Elastic-Plastic Range, *Proc. 2nd US Nat. Congr. Appl. Mech. (Ann Arbor)*, 521.

Haythornthwaite, R.M. and Shield, R.T. (1958), A Note on the Deformable Region in a Rigid-Plastic Structure, *J. Mech. Phys. Solids*, **6**, 127.

Hecker, S.S. (1975), A Simple Technique for Determining Forming Limit Curves, *Sheet Metal Indust.*, **52**, 671.

Hencky, H. (1924), Zur Theorie Plastischer Deformationen and der Hierdurch im Material Herforgerufenen, *Z. Angew. Math. Mech.*, **4**, 323.

Hill, R. (1948), A Theory of Yielding and Plastic Row in Anisotropic Metals, *Proc. Roy. Soc. London Sen. A*, **193**, 281.

Hill, R. (1949), The Theory of Plane Plastic Strain for Anisotropic Metals, *Proc. Roy. Soc. London Ser. A*, **194**, 428.

Hill, R. (1949), Plastic Distortion of Non-Uniform Sheets, *Phil. Mag.* (Sen 7), **40**, 971.

Hill, R. (1950a), *The Mathematical Theory of Plasticity*, Clarendon Press, Oxford, UK.

Hill, R (1950b), A Theory of Plastic Bulging of a Metal Diaphragm by Lateral Pressure, *Phil. Mag.* (Ser. 7), **41**, 1133.

Hill, R. (1950c), On the Inhomogeneous Deformation of a Plastic Lamina in a Compression Test, *Phil. Mag., Ser.*, 7, **41**, 733.

Hill, R. (1951), On the State of Stress in a Plastic/Rigid Body at the Yield Point, *Phil. Mag.*, **42**, 868.

Hill, R. (1952), On Discontinuous Stress States with Special References to Localized Necking in Thin Sheets , *J. Mech. Phys. Solids*, **1**, 19.

Hill, R. (1953), A New Method for Determining the Yield Criterion and Plastic Potential of Ductile Metals, *J. Mech. Phys. Solids*, **1**, 271.

Hill, R. (1954), The Plastic Torsion of Anisotropic Bars, *J. Mech. Phys. Solids*, **3**, 87.

Hill, R. (1956), On the Problem of Uniqueness in the Theory of a Rigid/Plastic Solid, Parts 1 and 2, *J. Mech. Phys. Solids*, **4**, 247 and 5, 1.

Hill, R. (1957), On the Problem of Uniqueness in the Theory of a Rigid/Plastic Solid, Parts 3 and 4, *J. Mech. Phys. Solids*, **5**, 153 and 163.

Hill, R. (1958), A General Theory of Uniqueness and Stability of Elastic/Plastic Solids, *J. Mech. Phys. Solids*, **6**, 236.

Hill, R. (1959), Some Basic Problems in the Mechanics of Solids Without a Natural Time, *J. Mech. Phys. Solids*, **7**, 209.

Hill, R. (1962), Acceleration Waves in Solids, *J. Mech. Phys. Solids*, **10**, 1.

Hill, R. (1963), A General Method of Analysis for Metal Working Processes, *J. Mech. Phys. Solids*, **11**, 305.

Hill, R. (1965), Continuum Micromechanics of Elastoplastic Polycrystals, *J. Mech. Phys. Solids*, **13**, 89 and 213.

Hill, R. (1966), Generalized Constitutive Relations for Incremental Deformation of Metal Crystals by Multislip, *J. Mech. Phys. Solids*, **14**, 95.

Hill, R. (1967), Ideal Forming for Perfectly Plastic Solids, *J. Mech. Phys. Solids*, **15**, 223.

Hill, R. (1976), Plastic Analysis of Pressurized Cylinders Under Axial Load, *Int. J. Mech. Sci.*, **18**, 145.

Hill, R. (1979), Theoretical Plasticity of Textured Aggregates, *Math. Proc. Cambridge Philos. Soc.*, **85**, 179.

Hill, R. (1980), Cavitation and the Influence of Headshape in Attack of Thick Targets by Nondeforming Projectiles, *J. Mech. Phys. Solids*, **28**, 249.

Hill, R. (1990), Constitutive Modelling of Orthotropic Plasticity in Sheet Metals, *J. Mech. Phys. Solids*, **38**, 405.

Hill, R. and Hutchinson, J.W. (1992), Differential Hardening in Sheet Metal Under Biaxial Loading, *J. Appl. Mech.*, **59**, SI.

Hill, R. and Rice, J.R. (1972), Constitutive Analysis of Elastic/Plastic Crystals at Arbitrary Strain, *J. Mech. Phys. Solids*, **20**, 401.

Hill, R. and Storakers, B. (1980), Plasticity and Creep of Pressurized Membranes, *J. Mech. Phys. Solids*, **28**, 27.

Hill, R., Storakers, B., and Zdunek, A.B. (1989), A Theoretical Study of the Brinell Hardness Test, *Proc. Roy. Soc. London Ser. A*, **423**, 301.

Hill, R. and Sewell, M.J. (1960), A General Theory of Inelastic Column Failure—I and II, *J. Mech. Phys. Solids*, **8**, 105 and 112.

Hill, R. and Sewell, M.J. (1962), A General Theory of Inelastic Column Failure—III, *J. Mech. Phys. Solids*, **10**, 185.

Hillier, M.J. (1966), Instability Strains in Plane Sheet Under Biaxial Stresses, *J. Appl. Meek*, **23**, 256.

Hodge, P.G. (1954), The Rigid-Plastic Analysis of Symmetrically Loaded Cylindrical Shells, *Appl. Mech.*, **21**, 336.

Hodge, P.G. (1955), Impact Pressure Loading of Rigid-Plastic Cylindrical Shells, *J. Mech. Phys. Solids*, **3**, 176.

Hodge, P.G. (1956a), Displacements in an Elastic-Plastic Cylindrical Shell, *J. Appl. Mech.*, **23**, 73.

Hodge, P.G. (1956b), The Influence of Blast Characteristics on the Final Deformation of Circular Cylindrical Shells, *Appl. Mech.*, **23**, 617.

Hodge, P.G. (1957), A General Theory of Piecewise Linear Plasticity Based on Maximum Shear, *J. Mech. Phys. Solids*, **5**, 242.

Hodge, P.G. (1959), The Effect of End Conditions on the Dynamic Loading of Plastic Shells, *J. Mech. Phys. Solids*, **7**, 258.

Hodge, P.G. (1960a), Yield Conditions for Rotationally Symmetric Shells Under Axisymmetric Loading, *J. Appl. Mech.*, **27**, 323.

Hodge, P.G. (1960b), Plastic Analysis of Circular Conical Shells, *J. Appl. Mech.*, **27**, 696.

Hodge, P.G. (1961), The Mises Yield Condition for Rotationally Symmetric Shells, *Quart. Appl. Math.*, **18**, 305.

Hodge, P.G. (1963), *Limit Analysis of Rotationally Symmetric Plates and Shells*, Chap. 7, Prentice Hall, Englewood Cliffs, NJ.

Hodge, P.G. (1964a), Rigid-Plastic Analysis of Spherical Caps with Cutouts, *Int. J. Mech. Sci.*, **6**, 177.

Hodge, P.G. (1964b), Plastic Design of a Closed Cylindrical Structure, *J. Mech. Phys. Solids*, **12**, 1.

Hodge, P.G. (1964c), Full Strength Reinforcement of a Cutout in a Cylindrical Shell, *J. Appl. Mech.,* **31**, 667.

Hodge, P.G. (1980), A Piecewise Linear Theory of Plasticity for an Initially Isotropic Material in Plane Stress, *Int. J. Mech. Sci.*, **22**, 21.

Hodge, P.G. (1981), *Plastic Analysis of Structures*, Krieger, New York.

Hodge, P.G., Bathe, K.J., and Dvorkin, E.N. (1986), Causes and Consequences of Nonuniqueness in Elastic/Perfectly Plastic Truss, *J. Appl. Mech.*, **53**, 235.

Hodge, P.G. and Belytschko, T. (1968), Numerical Methods for the Limit Analysis of Plates, *J. Appl. Mech.*, **35**, 796.

Hodge, P.G. and Lakshmikantham, C. (1962), Yield Loads of Spherical Caps with Cutouts, *Proc. 4th US Nat. Congr. Appl. Mech. Berkeley*, **2**, 951.

Hodge, P.G. and Lakshmikantham, C. (1963), Limit Analysis of Shallow Shells of Revolution, *J. Appl. Mech.*, **30**, 215.

Hodge, P.G. and Nardo, S.V.N. (1958), Carrying Capacity of an Elastic-Plastic Cylindrical Shell with Linear Strain-Hardening, *J. Appl. Mech.*, **25**, 79.

Hodge, P.G. and Panarelli, J. (1962), Interaction Curve for Circular Cylindrical Shells According to the Mises or Tresca Yield Criteria, *J. Appl. Mech.*, **29**, 375.

Hodge, P.G. and Perrone, N. (1957), Yield Loads of Slabs with Reinforced Cutouts, *J. Appl. Meek*, **24**, 85.

Hodge, P.G. and Sankaranarayanan, S. (1958), On Finite Expansion of a Hole in a Thin Infinite Plate, *Quart. Appl. Math*, **16**, 73.

Hodge, P.G. and Sankaranarayanan, S. (1960), Plastic Interaction Curves of Annular Plates in Tension and Bending, *J. Mech. Phys. Solids*, **8**, 153.

Hoffman, G.A. (1962), Minimum weight Proportions of Pressure Vessel Heads, *J. Appl. Mech.*, **29**, 662.

Hong, H.K. and Liou, J.K. (1993a), Research Report, Department of Civil Engineering, National Taiwan University, Taipei.

Hopkins, H.G. (1957), On the Plastic Theory of Plates, *Proc. Roy. Soc. London Ser. A*, **241**, 153.

Hopkins, H.G. (1960), Dynamic Expansion of Spherical Cavities, in *Metals, Progress in Solid Mechanics* (eds., I.N. Sneddon and R. Hill), Vol.1, Chap. 3, North Holland Publishing Company, Amsterdam.

Hopkins, H.G. (1968), The Method of Characteristics and its Application to the Theory of Stress Waves in Solids, *Engineering Plasticity* (eds., J. Heyman and F.A. Leckie), Cambridge University Press, Cambridge, UK, p. 277.

Hopkins, H.G. and Prager, W. (1953), The Load Carrying Capacity of Circular Plates, *J. Mech. Phys. Solids*, **2**, 1.

Hopkins, H.G. and Prager, W. (1954), On the Dynamics of Plastic Circular Plates, Z. *Angew. Math. Mech.*, **5**, 317.

Hopkins, H.G. and Prager, W. (1955), Limits of Economy of Material in Plates, *J. Appl. Mech., Trans. ASME*, **22**, 372.

Hopkins, H.G. and Wang, A.J. (1955), Load Carrying Capacities of Circular Plates of Perfectly-Plastic Material with Arbitrary Yield Condition, *J. Mech. Phys. Solids*, **3**, 117.

Horne, M.R. (1956), The Elastic-Plastic Theory of Compression Members, *J. Mech. Phys. Solids*, **4**, 104.

Horne, M.R. and Merchant, W. (1965), *The Stability of Frames,* Pergamon Press, Oxford, UK.

Hosford, W.F. and Caddell, R.M. (1993), *Metal Forming, Mechanics and Metallurgy*, second edition, Prentice Hall, Englewood Cliffs, NJ.

Huh, H., Han, S.S., and Yang, D.Y. (1994), Modified Membrane Finite Element Formulation Considering Bending Effects in Sheet Metal Analysis, *Int. J. Mech. Sci.*, **36**, 659.

Hu, T.C. and Shield, R.T. (1961), Uniqueness in the Optimum Design of Structures, *J. Appl. Mech.*, **28**, 284.

Hudson, G.E. (1951), A Theory of Dynamic Plastic Deformation of a Thin Diaphragm, *J. Appl. Phys.,* **22**, 1.

Humphreys, J.S. (1965), Plastic Deformation of Impulsively Loaded Straight Clamped Beams, *J. Appl. Mech.*, **32**, 7.

Hundy, B.B. and Green, A.P. (1954), A Determination of Plastic Stress-Strain Relations, *J. Mech. Phys. Solids*, **3**, 16.

Hunter, S.C. (1957), The Propagation of Spherically Symmetric Disturbances in Ideally Plastic Materials, *Proc. Conf. Prop. Mat. High Rates Strain, Instn. Mech. Engrs. (London)*, 147.

Hunter, S.C. and Crozier, R.J.M. (1968), Similarity Solution for the Rapid Uniform Expansion of a Spherical Cavity in a Compressible Elastic-Plastic Solid, *Quart. J. Mech. Appl. Math.*, **21**, 467.

Hutchinson, J.W. (1970), Elasto-Plastic Behaviour of Polycrystalline Metals and Composites, *Proc. Roy. Soc. London Ser. A*, **319**, 247.

Hutchinson, J.W. (1972), On the Postbuckling Behavior of Imperfection Sensitive Structures in Plastic Range, *J. Appl. Mech.*, **39**, 155.

Hutchinson, J.W. (1974), Plastic Buckling, *Adv. in Appl. Mech.*, **14**, 67.

Hutchinson, J.W. and Miles, J.P. (1974), Bifurcation Analysis of the Onset of Necking in an Elastic/Plastic Cylinder Under Uniaxial Tension, *J. Mech. Phys.* Solids, **22**, 61.

Hutchinson, J.W. and Neale, K.W. (1978), in Mechanics of Sheet Metal Forming (eds., D.P. Koistinen and N.M. Wang), Plenum Press, New York, p. 172.

Ilahi, M.F., Parmar, A., and Mellor, P.B. (1981), Hydrostatic Bulging of a Circular Aluminium Diaphragm, *Int. J. Mech. Sci.*, **23**, 221.

Illyushin, A.A. (1946), The Theory of Small Elastic-Plastic Deformations, *Prikl. Math. Mekh.*, **10**, 347.

Illyushin, A.A. (1947), The Elastic-Plastic Stability of Plates, Technical Note, NACA-1188.
Ilyushin, A.A. (1948), Plasticity (in Russian), Gesteckhizdat, Moscow.
Inoue, T. and Kato, B. (1993), Analysis of Plastic Buckling of Steel Plates, *Int. J. Solids Struct.*, **15**, 567.
Ishlinsky, A. (1944), *Prikl. Mat. Mekh. Leningrad*, **8**, 201.
Ishlinsky, A., (1954), A General Theory of Plasticity with Linear Strain-Hardening (in Russian) *Ukrain. Mat. Zh.*, **6**, 314.
Issler, W. (1964), Membranschalen Gleicher Festigkeit, *Ingenieur-Archiv*, **33**, 330.
Ivanov, G.V. (1967), *Inzh. Zh. Mekh. Tuerdogo, Tele*, **6**, 74.
Jackson, L.R., Smith, K.F. and Lankford, W.T. (1948), Plastic Flow in Anisotropic Sheet Metal, Metals Technology, Tech. Pub. 2440.
Jahsman, WE. (1974), Reflection and Refraction of Weak Elastic-Plastic Waves, *J. Appl. Mech.*, **41**, 117.
Jansen, D.M. Datta, S.K., and Jahsman, W.E. (1972), Propagation of Weak Waves in Elastic-Plastic Solids, *J. Mech. Phys. Solids*, **20**, 1.
Jaumann, G. (1911), Geschlossenes System physikalischer und chemischer differentialge-setze, *sitzber. kais. akad. wiss., wien*, Abt. II, **120**, 385.
Jiang, W. (1993), The Elastic/Plastic Response of Thin Walled Tubes Under Combined Axial and Torsional Loads, *J. Press. Vess. Technol., Trans. ASME*, **115**, 283, 291.
Jiang, W. (1994), Study of Two-Surface Plasticity Theory, *J. Engng. Mech., Trans. ASCE*, **120**, 2179.
Jiang. W. (1995), A New Constitutive Model in the Theory of Plasticity, *J. Pressure Vessel Technol.*, **117**, 365, 371.
Johansen, K.W. (1943), Brudlinieteorier, Gjellerup, Copenhagen.
Johnson, G.R. and Holmquist, T.J. (1988), Evaluation of Cylinder Impact Test Data for Constitutive Model Constants, *J. Appl. Phys.*, **64**, 3901.
Johnson, K.L. (2001), *Contact Mechanics*, Cambridge University Press, Cambridge.
Johnson, R.W. and Rowe, G.W. (1968), Redundant Work in Drawing Cylindrical Stock, *J. Inst. Metals*, **96**, 97.
Johnson, W. (1957), The Pressure for the Cold Extrusion of Lubricated Rod Through Square Dies of Moderate Reduction at Slow Speeds, *J. Inst, of Metals*, **85**, 403.
Johnson, W. (1969), Upper Bounds to the Load for the Transverse Bending of Flat Rigid Perfectly Plastic Plates, *Int. J. Mech. Sci.*, **11**, 913.
Johnson, W. (1972), *Impact Strength of Materials*, Edward Arnold, London.
Johnson, W., Chitkara, N.R., and Ranshi, A.S. (1974), Plane Stress Yielding of Cantilevers in Bending due to Combined Shear and Axial Load, *J. Strain Anal.*, **9**, 67.
Johnson, W., de Malherbe, M.C., and Venter R. (1973), Some Slipline Field Results for the Plane Strain Extrusion of Anisotropic Material Through Frictionless Wedge-Shaped Dies, *Int. J. Mech. Sei.*, **15**, 109.
Johnson, W., Kormi, K. and Travis, F.W. (1965), The Explosive Drawing of Square and Flat-Bottomed Circular Cups and Double Pulsation Phenomenon, *Proc. 5th Int. M.T.D.R. Conf*, Pergamon Press, Oxford, p. 293.
Johnson, W. and Kudo, H. (1962), *Mechanics of Metal Extrusion*, Manchester University Press, Manchester.
Johnson, W. and Mellor, P.B. (1983), *Engineering Plasticity*, Ellis Horwood, Chichester.
Johnson, W and Needham, G. (1968), Experiments on Ring Rolling, *Int. J. Mech. Sci.*, **10**, 95.
Johnson, W., Sengupta, A.K. and Ghosh, S.K. (1982), Plasticine Modelled High Velocity Impact and Ricochet of Long Rods, *Int. J. Mech. Sci.*, **24**, 437.
Johnson, W., Slater, R.A.C., and Yu, A.S. (1966), The Quasi-Static Compression of Non-Circular Prismatic Blocks Between Very Rough Platens, *Int. J. Mech. Sci.*, **8**, 731.
Johnson, W. and Travis, F.W. (1966), High Speed Blanking of Copper, *Proc. Inst. Mech. Engrs.* **180**, 16.

Jones, L.L. and Wood, R.H. (1967), *Yield Line Analysis of Slabs*, Thames and Hudson, London.

Jones, N. (1967), Influence of Strain-Hardening and Rate-Sensitivity on the Permanent Deformation of Impulsively Loaded Rigid Plastic Beams, *Int. J. Mech. Sci.*, **9**, 111.

Jones, N. (1968a), Impulsive Loading of a Simply Supported Circular Rigid-Plastic Plate, *J. Appl. Mech.*, **35**, 59.

Jones, N. (1968b), Finite Deflections of a Simply Supported Rigid Plastic Annular Plate Loaded Dynamically, *Int. J. Solids Struct.*, **4**, 593.

Jones, N. (1970), The Influence of Large Deflections on the Behaviour of Rigid/Plastic Cylindrical Shells Loaded Impulsively, *J. Appl. Mech.*, **37**, 416.

Jones, N. (1971), A Theoretical Study of the Dynamic Plastic Behaviour of Beams and Plates with Finite Deflections, *Int. J. Solids Struct.*, **7**, 1007.

Jones, N. (1989), *Structural Impact*, Cambridge University Press, Cambridge, UK.

Jones, N. and Ahn, C.S. (1974), Dynamic Elastic and Plastic Buckling of Complete Spherical Shells, *Int. J. Solids Struct.*, **10**, 1357.

Jones, N. and Okawa, D.M. (1976), Dynamic Plastic Buckling of Rings and Cylindrical Shells, *Nuclear Engng. Design*, **37**, 125.

Jones, N, Uran, T and Tekin, S.A. (1970), The Dynamic Plastic Behavior of Fully Clamped Rectangular Plates, *Int. J. Solids Struct.*, **6**, 1499.

Jones, R.M. (1967), Plastic Buckling of Eccentrically Stiffened Circular Cylindrical Shells, *AIAA Journal*, **5**, 1147.

Jones, R.M. (2009), *Deformation Theory of Plasticity*, Bull Ridge Publishing, Blacksburg, VA.

Kachanov, A. (1971), *Foundations of the Theory of plasticity*, North-Holland, Amsterdam.

Kadashevitch, I. and Novozhilov, V.V. (1959), The Theory of Plasticity that Takes into Account Residual Microstresses (in Russian), *Prikl. Mat. Mekh.*, **22**, 104.

Kaftanoglu, B. and Alexander J.M. (1970), On Quasi-Static Axisymmetric Stretch Forming, *Int. J. Mech. Sci.*, **12**, 1065.

Kalisky, S. (1970), Approximate Solutions for Dynamically Loaded Inelastic Structures and Continua, *Int. J. Nonlinear Mech.*, **5**, 143.

Kalisky, S. (1989), *Plasticity Theory and Engineering Applications*, Elsevier, Amsterdam.

Kanazawa, R. and Marcal, P.V. (1982), Finite Element Analysis of Plane Strain Rolling, ****J. Engng. Indust., Trans. ASME*, **104**, 95.

Karunes, B. and Onat, E.T. (1960), On the Effect of Shear on Plastic Deformation of Beams Under Transverse Impact Loading, *J. Appl. Mech.*, **27**, 10.

Kasuga, Y. and Tsutsumi, S. (1965), Pressure Lubricated Deep Drawing, *Bull. Japan Soc. Mech. Engrs.*, **8**, 120.

Keck, P., Wilhelm, M., and Lange, K. (1990), Applications of the Finite Element Method to the Simulation of Sheet Metal Processes: Comparison of Calculation and Experiment, *Int. J. Numer. Methods. Engng.*, 1419.

Keeler, S.P. (1965), Determination of Forming Limits in Automotive Stamping, *Sheet Metal Ind.*, **42**, 683.

Keil, A.H. (1960), Problems of Plasticity in Naval Structures, Explosive and Impact Loading, *Plasticity* (eds., E.H. Lee and PS. Symonds), Pergamon Press, Oxford, UK, p. 22.

Kelly, J.M. and Wilshaw, T.R. (1968), A Theoretical and Experimental Study of Projectile Impact on Clamped Circular Plates, *Proc. Roy. Soc. London Ser. A*, **306**, 435.

Khan, A.S. and Huang, S. (1995), *Continuum Theory of Plasticity*, Wiley, New York.

Keeler, S.P. and Backofen, W.A. (1963), Plastic Instability and Fracture in Sheet Stretched Over Rigid Punches, *Trans. Amer. Soc. Metals*, **56**, 25.

Ketter, R.L. (1961), Further Studies of the Strength of Beam Columns, *J. Struct. Div., Proc. ASCE*, **87**, ST6, 135.

Kim, K.K. and Kim, D.W. (1983), The Effect of Void Growth on the Limit Strains of Steel Sheets, *Int. J. Mech. Sci.*, **25**, 293.

Kim, J.H., Oh, S.I. and Kobayashi, S. (1978), Analysis of Stretching of Sheet Metals with Hemi Spherical Punch, *Int. J. Mech. Tool Des. Res.*, **19**, 209.

Kim, J.H. and Kobayashi, S. (1978), Deformation Analysis of Axisymmetric Sheet Metal Forming Processes by Rigid-Plastic Finite Element Methods, in Mechanics of Sheet Metal Forming (eds., D.P. Koistinen and N.M. Wang), Plenum Press, New York, p. 341.

Kim, J.H., Oh, S.I. and Kobayashi, S. (1978), Analysis of Stretching of Sheet Metals with Hemispherical Punch, *Int. J. Mach. Tool Des. Res.*, **18**, 209.

Kim, J.H. and Yang, D.Y. (1985a), A Rigid/Plastic Finite Element Formulation Considering Effect of Geometry Change and its Application to Hydrostatic Bulging, *Int. J. Mech. Sci.*, **27**, 453.

Kim, J.H. and Yang, D.Y. (1985b), An Analysis of Upset Forging of Square Blocks Considering the Three-Dimensional Bulging of Sides, *Int. J. Mech. Tool Des. Res.*, **25**, 327.

Kim, J.H., Yang, D.Y and Kim, M.U. (1987), Analysis of Three-Dimensional Upset Forging of Arbitrary Shaped Rectangular Blocks, *Int. J. Mack Tool Manufact.*, **27**, 311.

Kim, J.B., Yoon, J.W. and Yang, D.Y. (2000), Wrinkling Initiation and growth in Modified Yoshida Test-:Finite Element Analysis and Experimental Comparison, *Int. J. Mech. Sci.*, **42**, 1683.

Klepaczko, J. (1968), Strain Rate History Effects for Polycrystalline Aluminium and Theory of Intersection, *J. Mech. Phys. Solids*, **16**, 255.

Klinger, L.G. and Sachs, G. (1948), Dependence of the Stress-Strain Curves of Cold-Worked Metals Upon the Testing Direction, *J. Aeronaut. Sci.*, **15**, 599.

Kliushnikov, V.D. (1959), New Concepts in Plasticity and Deformation Theory (in Russian), *Prikl. Mat. Mekh.*, **23**, 405.

Kobayashi, S. (1964), Upper Bound Solution to Axisymmetric Forming Problems, *J. Engng. Indust.,* Trans. ASME, **83**, 326.

Kobayashi, S., Caddell, R.M., and Hosford, W.F. (1985), Examination of Hills Latest Yield Criterion Using Experimental Data for Various Anisotropic Sheet Metals, *Int. J. Mech. Sci.*, **27**, 509.

Kobayashi, S. Oh, S.I., and Altan, T. (1989), *Metal Forming and the Finite Element Method*, Oxford University Press, New York.

Kobayashi, S. and Thomsen, E.G. (1965), Upper and Lower Bound Solutions to Axisym-metric Compression and Extrusion Problems, *Int. J. Mech. Sci.*, **7**, 127.

Kolsky, H. (1949), An Investigation of the Mechanical Properties of Materials at Very High Rates of Loading, *Proc. Phys. Soc*, **B62**, 676.

Koiter, W.T. (1953), Stress-Strain Relations, Uniqueness and Variational Theorems for Elastic-Plastic Materials with a Singular Yield Surface, *Quart. Appl. Math.*, **11**, 350.

Koiter, W.T. (1960), General Theorems, *progress in solid mechanics* (eds., I. Sneddon and R. Hill), Vol. I, North-Holland, Amsterdam.

Kojic, M., Grijovic, N., Slavkovic, R. and Zivkovic, M. (1996), A General von Mises Plasticity Material Model with Mixed Hardening; Model Definition and Implicit Stress Integration Procedure, *J. Appl. Mech.*, **63**, 376.

Kolsky, H. (1953), *Stress Waves in Solids*, Clarendon Press, Oxford, UK.

Kolsky, H. and Douch, L.S. (1962), Experimental Studies in Plastic Wave Propagation, *J. Mech. Phys. Solids*, **10**, 195.

Kondo, K. and Pian, T.H.H. (1981), Large Deformation of Rigid/Plastic Circular Plates, *Int. J. Solids Struct.*, **17**, 1043.

König, J.A. and Rychlewsky, R. (1966), Limit Analysis of Circular Plates with Jump Nonhomogeneity, *Int. J. Solids Struct.*, **2**, 493.

Koopman, D.C.A. and Lance, R.H. (1965), On Linear Programming and Plastic Limit Analysis, *J. Mech. Phys. Solids*, **13**, 77.

Kosel, F. and Bremec, B. (2004), Elastoplastic Buckling of Circular Annular Plates Under Uniform In- Plane Loading, *Thin-Walled Struct*, **42**, 101.

Kozlowski, W. and Mröz, Z. (1969), Optimal Design of Solid Plates, *Int. J. Solids Struct.*, **5**, 781.

Krajcinovic, D. (1972), Dynamic Analysis of Clamped Plastic Circular Plates, *Int. J. Mech. Sci.*, **14**, 225.

Krajcinovic, D. (1976), Rigid-Plastic Circular Plates on Elastic Foundation, *J. Engng. Mech. Div., Trans. ASCE*, **102**, 213.

Kröner, E. (1961), Zur Plastischen Verformung des Vielkristalls, *Acta Metall*, **9**, 153.

Kudo, H. (1960), Some Analytical and Experimental Studies of Axisymmetric Cold Forging and Extrusion I, *Int. J. Mech. Sci.*, **2**, 102.

Kudo, H. (1961), Some Analytical and Experimental Studies of Axisymmetric Cold Forging and Extrusion II, *Int. J. Mech. Sci.,* **3**, 91.

Kudo, H. and Aoi, K. (1967), Effect of Compression Test Condition upon Fracturing of Medium Carbon Steel; Study on Cold Forgeability Test, *Japan Soc. Tech. Plast.*, **8**, 17.

Kuech, R.W. and Lee, S.L. (1965), Limit Analysis of Simply Supported Conical Shells Subjected to Uniform Internal Pressure, *J. Franklin Inst.*, **280**, 71.

Kukudjanov, V.N. (1967), Propagation of Elastic-Plastic Waves in Rods Taking into Account the Rate Influence (in Russian), *Computing Centr. Acad. Sci., USSR*, Moscow.

Kumar, A. and Reddy, V.V.K. (1986), Dynamic Plastic Response of Circular Plates with Transverse Shear, *J. Appl. Mech.*, **53**, 952.

Kummerling, R. and Lippmann, H. (1975), On Spread in Rolling, *Mech. Res. Commun.*, **2**, 113.

Kunogi, M. (1954), On Plastic Deformation of Hollow Cylinders Under Axial Compressive Loading, *J. Sci. Res. Inst., Tokyo*, p. 2.

Kuzin, P.A. and Shapiro, G.S. (1966), On Dynamic Behaviour of Plastic Structures, *Proc. 11th Int. Congr. Appl. Mech. (Munich, 1964)*, Springer-Verlag, New York, p. 629.

Kwasczynska, K. and Mröz, Z. (1967), A Theoretical Analysis of Plastic Compression of Short Circular Cylinders, *Arch. Mech. Stos.*, **19**, 787.

Lahoti, G.D. and Kobayashi, S. (1974), On Hill's General Method of Analysis for Metal Working Processes, *Int. J. Mech. Sei.*, **16**, 521.

Lakshmikantham, C. and Hodge, P.G. (1963), Limit Analysis of Shallow Shells of revolution, *J. Appl. Mech.*, **30**, 215.

Lamba, H.S. and Sidebottom, O.M. (1978), Cyclic Plasticity for Nonproportional Paths: Comparison with Predictions of Three Incremental Plasticity Models, *J. Engng. Mat. Technol.*, **100**, 104.

Lambert, E.R. and Kobayashi, S. (1969), An Approximate Solution for the Mechanics of Axisymmetric Extrusion, *Proc. 9th M.T.D.R. Conference,* Pergamon Press, Oxford.

Lance, R.H. and Lee, C.H. (1969), The Yield Point Load of a Conical Shell, *Int. J. Mech. Sci.*, **11**, 129.

Lance, R.H. and Onat, E.T. (1962), A Comparison of Experiments and Theory in the Plastic Bending of Plates, *J. Mech. Phys. Solids*, **10**, 301.

Lance, R.H. and Onat, E.T. (1963), Analysis of Plastic Shallow Conical Shells, *J. Appl. Mech.*, **30**, 199.

Landgraf, R.W. (1970), The Resistance of Metals to Cyclic Deformation, *Advancement of High Fatigue Resistance in Metals and Alloys,* STP-467, ASTM, Philadelphia.

Lange, K. (1985), *Handbook of Metal Forming*, McGraw-Hill, New York.

Lebedev, N.F. (1954), *Prikt Mat. Mekh.*, **18**, 167.

Leckie, F.A. and Payne, D.J. (1965), Some Observation on the Design of Spherical Pressure Vessels with Flush Cylindrical Nozzles, *Proc. Instn. Mech. Engrs.*, **180**, 497.

Leckie, F.A. and Penny, R.K. (1968), Plastic Instability of Spherical Shells, in *Engineering Plasticity* (eds., J. Heyman and F.A. Leckie), Cambridge University Press, UK, p. 401.

Lee, C.H. and Kobayashi, S. (1971), Analysis of Axisymmetric Upsetting and Plane Strain Side Pressing of Solid Cylinders by the Finite Element Method, *J. Engng. Indust., Trans. ASME*, **93**, 445.

Lee, D. and Zaverl, R. (1982), Neck Growth and Forming Limits in Sheet Metals, *Int. J. Mech. Sci.*, **24**, 157.

Lee, E.H. (1953), A Boundary Value Problem in the Theory of Plastic Wave Propagation, *Quart. Appl. Math.*, **10**, 335.

Lee, E.H. (1960), The Theory of Wave Propagation in Anelastic Materials, in *Stress Wave Propagation in Materials* (ed. N. Davids), Interscience, New York, p. 199.

Lee, E.H. (1969), Large Elasti-Plastic Deformation at Finite Strains, *J. Aappl. Mech.*, **36**, 1.

Lee, E.H., Mallett, R.L. and Wertheimer, T.B. (1983), Stress Analysis for Anisotropie Hardening in Finite Deformation Plasticity, *J. Appl. Mech.*, **50**, 554.

Lee, E.H. and Symonds, P.S. (1952), Large Plastic Deformation of Beams Under Transverse Impact Loading. *J. Appl. Mech.*, **19**, 308.

Lee, E.H. and Tupper, S.J. (1954), Analysis of Plastic Deformation in a Steel Cylinder Striking a Rigid Target, *J. Appl. Mech.*, **21**, 63.

Lee, L.C. and Onat, E.T. (1968), Analysis of Plastic Spherical Shells, *Engineering Plasticity* (eds., J. Heyman and F.A. Leckie), Cambridge University Press, U.K., p. 413.

Lee, L.H.N. (1962), Inelastic Buckling of Initially Imperfect Cylindrical Shells Subject to Axial Compression, *J. Aeronaut. Sci.*, **29**, 87.

Lee, L.H.N. (1977), Quasi-Bifurcation in Dynamics of Elastic-Plastic Continua, *J. Appl. Mech.*, **44**, 413.

Lee, L.H.N. (1981), Dynamic Buckling of an Inelastic Column, *Int. J. Solids Struct.*, **17**, 271.

Lee, L.S.S. and Martin, J.B. (1970), Approximate Solutions of Impulsively Loaded Structures of a Rate Sensitive Material, *Z. Angew. Math. Phys*, **21**, 1011.

Lee, S.H. and Kobayashi, S. (1975), Rigid-Plastic Analysis of Bore Expanding Flange Drawing with Anisotropic Sheet Metals by Matrix Methods, Proc. 15th Int. MTDR Conf., p. **561**.

Lee, S.L. and Thorn, B.J. (1964), Deformation of Axisymmetrically Loaded Cylindrical Shells at Incipient Plastic Collapse, *Int. J. Mech. Sei.*, **6**, 247.

Lee, W.B. and Chan, K.C. (1991), Effect of Local Texture on the Development of M.K-Groove in Biaxial Deformation, *Textures and Microstructures*, **14**, 1221.

Lee, W.B. (1993), Methodology and Applications of Mesoplasticity in Manufacturing Science, *Int. J. Mech. Sei*, **35**, 147.

Lellep, J. and Torn, K. (2004), Plastic Response of a Circular Cylindrical Shell in Dynamic Loadings, *Int. J. Impact Engng.*, **30**, 555.

Lemaitre, J. and Chaboche, J.L. (1989), *Mechanics of Solid Structures*, Cambridge University Press.

Lensky, V.S. (1949), *Prikt. Mat. Mekh.*, **13**, 165.

Lévy, M. (1870), Mémoire surles équations générales des mouvements intérieurs des corps solides ductiles ou delà des limites ou l'élasticité pourrait les ramener à leur premier état, *C. R. Acad. Sei.* (Paris) **70**, 1323.

Li, G.J. and Kobayashi, S. (1982), Spread Analysis in Rolling by the Rigid/Plastic Finite Element Method, Numerical Method in Industrial Forming Processes, Pineridge Press, Swansea, p. 777.

Li, Q.M. and Jones, N. (2005), Foundation of Correlation Parameters for Eliminating Pulse Shape effects on Dynamic Plastic Response of Structures, *J. Appl. Mech.*, **72**, 172.

Li, S. and Reid, S.R. (1992), The Plastic Buckling of Axial Compressed Square Tubes, *Appl. Mech.*, **59**, 276.

Lian, J., Barlat, F., and Baudelet, B. (1989), Plastic Behavior and Stretchability of Sheet Metals, Part II: Effect of Yield Surface Shape on Sheet Forming Limit, *Int. J. Plasticity*, **5**, 131.

Lianis, G. and Ford, H. (1957), An Experimental Investigation of the Yield Criterion and the Stress-Strain Law, *J. Mech. Phys. Solids*, **5**, 215.

Limb, M.E., Chakrabarty, J., Garber, S. and Roberts, W.T. (1973), Hydraulic Forming of Tubes, *Sheet Metal Industries*, **53**, 418.

Lin, H.C. and Wu, H.C. (1983), On the Rate Dependent Endochronic Theory of Viscoplasticity, and its Application to Plastic Wave Propagation, *J. Appl. Mech.*, **43**, 92.

Lin, S.B. and Ding, J.L. (1995), Experimental Study of the Plastic Yielding of Rolled Sheet Metals with the Cruciform Plate Specimen, Int. J. Plasticity, **11**, 583.

Lin, T.H. (1957), Analysis of Elastic and Plastic Strains of a Face Centered Cubic Crystal, *J. Mech. Phys Solids*, **5**, 143.

Lin, T.H. (1971), Physical Theory of Plasticity, *Adv. in Appl. Mech.*, **11**, 255.

Lin, T.H. (1984), A Physical Theory of Plasticity and Creep, *J. Engng. Mater. Technol., Trans. ASME*, **106**, 290.

Lindholm, U.S. (1964), Some Experiments with the Split Hopkinson Pressure Bar, *J. Mech. Phys. Solids*, **12**, 317.
Lippmann, H. (1962), Principal Line Theory of Axially Symmetrical Plastic Deformation, *J. Mech. Phys. Solids*, **10**, 111.
Lippmann, H. (1965), Statics and Dynamics of Axially Symmetric Plastic Row, *J. Mech. Phys. Solids*, **13**, 29.
Lippmann, H. (1966), On the Dynamics of Forging, *Proc. 7th Int. Conf. MTDR*, Pergamon Press, Oxford, UK.
Lippmann, H. (1981), *Mechanik des Plastischen Fliessens*, Springer-Verlag, Berlin.
Lippmann, H. and Mahrenholtz, O, (1967), *The Mechanics of Plastic Forming of Metals* (in German), Springer-Verlag, Berlin.
Liu, Y.H., Cen, Z.Z., and Xu, B.Y. (1995), Numerical Method for Plastic Limit Analysis of 3-D Structures, *Int. J. Solids Struct.*, **32**, 1645.
Liu, D. and Strange, W.J. (1996), Deformation of Simply Supported Circular Plate by Central Pressure Pulse, *Int. J. Solids Struct.*, **33**, 283.
Lo, S.W., Hsu, I.C, and Wilson, W.R.D. (1993), An Analysis of the Hemispherical Punch Hydroforming Processes, *J. Mater. Process. Technol.*, **37**, 225.
Lockett, F.J. (1963), Indentation of a Rigid/Plastic Material by a Conical Indenter, *J. Mech. Phys. Solids*, **11**, 345.
Lode, W. (1926), Versuche über den Einfluss der mittleren Hauptspannung auf das Fliessen der Metalle, Eisen, Kupfer und Nickel, *Z. Phys.*, **36**, 913.
Logan, R.W. and Hosford, W.F. (1980), Upper Bound Anisotropic Yield Locus Calculations Assuming (111) Pencil Glide, *Int. J. Mech. Sci.*, **22**, 419.
Lu, L.W. and Kamalvand, H. (1968), Ultimate Strength of Laterally Loaded Columns, *J. Struct. Div., Proc. ASCEZ*, **94**, 1505.
Lubiner, J. (1965), The Strain Rate Effect in Plastic Wave Propagation, *J. Mecan.*, **4**, 111.
Ludwik, P. (1909), *Elemente der Technologischen Mechanik*, Springer Verlag, Berlin.
Luk, V.K. and Arnos, D.E. (1991), Dynamic Cylindrical Cavity Expansion of Compressible Strain-Hardening Material, *J. Appl. Mech.*, **58**, 334.
Luk, V.K. Forrestal, M.J., and Amos, D.E. (1991), Dynamic Spherical Cavity Expansion of Strain-Hardening Material, *J. Appl. Mech.*, **58**, 1.
MacDonald, A., Kobayashi, S. and Thomsen, E.G. (1960), Some problems of Press Forging Lead and Aluminum, *J. Engng. Indust., Trans. ASME*, **82**, 246.
Maclellan, G.D.S. (1952), Some Friction Factors in Wire Drawing, *J. Inst. Metals*, **81**, 1.
Majlessi, S.A. and Lee, D. (1988), Further Development of Sheet Metal Forming Analysis Method, *J. Engng. Indust., Trans. ASME,* **109**, 330.
Majlessi, S.A. and Lee, D. (1993), Deep Drawing of Square Shaped Sheet Metal Parts, Part I, Finite Element Analysis, *J. Engng. Indust., Trans. ASME*, **115**, 102.
Male, A.T. and Cockroft, M.G. (1964), The Ring Test, *J. Inst. Metals*, **93**, 38.
Male, A.T. and DePierre, V. (1970), The Validity of Mathematical Solutions for Determining Friction from the Ring Compression Test, *J. Lubrication Technol., Trans. ASME*, **92**, 389.
Malvern, L.E. (1951), The Propagation of Longitudinal Waves of Plastic Deformation in a Bar of Material Exhibiting Strain-Rate Effect, *J. Appl. Mech.*, **18**, 203.
Malvern, L.E. (1965), Experimental Studies of Strain Rate Effects and Plastic Wave Propagation in Annealed Aluminum, in *Behavior of Materials Under Dynamic Loading,* ASME. p. 81.
Malyshev, V.M. (1961), *J. Prikl. Mekh. Tech. Fiz.*, **2**, 104.
Mamalis, A.G. and Johnson, W. (1983), Quasi-Static Crumpling of Thin-Walled Circular Cylinders and Frusta Under Axial Compression, *Int. J. Mech. Sci.*, **25**, 713.
Mamalis, A.G., Johnson, W. and Hawkyard, J.B. (1976), Pressure Distribution, Roll Force and Torque In Cold Ring Rolling, *J. Mech. Engng. Sci.*, **18**, 196.
Ma, X., Barnett, M.R. and Kim, Y.H. (2004), Forward Extrusion Through Steadily Rotating Coonical Dies, Parts 1 and 2, *Int. J. Mech. Sci.*, **46**, 449. 465.

Mandel, J. (1962), Ondes Plastiques dans Milieu Indéfini a Trois Dimensions, in *Extrait du Seminaire de Plasticité*, Publ. Sei. Techn. Min. Air. NT 116, Paris, p. 151.

Mandel, J. (1972), *Plasticité Classique et Viscoplasticite,* Springer-Verlag, Wien.

Manjoine, M.J. (1944), The Influence of Rate of Strain and Temperature on Yield Stresses of Mild Steel, *J. Appl. Mech.*, **11**, 211.

Mansfield, E.H. (1957), Studies in Collapse Analysis of Rigid-Plastic Plates with a Square Yield Diagram, *Proc. Roy. Soc. London Ser. A*, **241**, 311.

Marcal, P.V. (1967), Optimal Plastic Design of Circular Plates, *Int. J. Solids Struct.*, **3**, 427.

Marciniak, and Duncan, J.L. (1992). The Mechanics of Sheet Metal Forming, Edward Arnold, London.

Marciniak, Z. and Kuczynski, K. (1967), Limit Strains in the Process of Stretch Forming Sheet Metals, *Int. J. Mech. Sci.*, **9**, 609.

Marciniak, Z., Kuczynski, K. and Pakora, T. (1973), Influence of Plastic Properties of a Material on the Forming Limit Diagram for Sheet Metal in Tension, *Int. J. Mech. Sci.*, **15**, 789.

Markin, A.A. and Yakovlev, S.S. (1996), Influence of Rotation of the Principal Orthotropy on the Deformation of Anisotropic Ideally Plastic Materials, Mechanics of Solids, **31**, 58.

Markowitz, J. and Hu, L.W. (1964), Plastic Analysis of Orthotropic Circular Plates, J. Engng. Mech., Trans. ASCE, **90**, 251.

Martin, J.B. and Lee, L.S.S. (1968), Approximate Solutions for Impulsively Loaded Elastic-Plastic Beams, *J. Appl. Mech.*, **35**, 803.

Martin, J.B. and Symonds, P.S. (1966), Mode Approximations for Impulsively Loaded Rigid-Plastic Structures, *J. Engng. Mech. Div., Proc. ASCE*, **92**, 43.

Martolakos, D.E. and Mamalis, A.G. (1986), Upper and Lower Bounds for Rectangular Plates Transversely Loaded, *Int. J. Mech. Sci.*, **12**, 815.

Marshall, E.R. and Shaw, M.C. (1952), The Determination of Flow Stress from a Tensile Specimen, *Trans. Amer. Soc. Metals*, **44**, 705.

Massonnet, C.E. (1967), Complete Solutions Describing the Limit State in Reinforced Concrete Slabs, *Mag. Conc. Res.*, **19**, 58.

Mazumdar, J. and Jain, R.K. (1989), Elastic-Plastic Bending of Plates of Arbitrary Shape – A New Approach, *Int. J. Plasticity*, **5**, 463.

McDowell, D.L. (1985), A Two-Surface Model for Transient Nonproportional Cyclic Plasticity, *J. Appl. Mech.*, **52**, 298. 303.

McMeeking, R.M. and Rice, J.R. (1975), Finite Element Formulation for Problems of Large Elastic/Plastic Deformation, *Int. J. Solids Struct.*, **11**, 601.

Mellor, P.B. (1954), Stretch Forming Under Fluid Pressure, *J. Mech. Phys., Solids*, **5**, 41.

Mellor, P.B. (1981), Sheet Metal Forming, International Metals Reviews, **26**, 1.

Mellor, P.B. (1982), Experimental Studies of Plastic Anisotropy in Sheet Metal, in *Mechanics of Solids* (eds., H.G. Hopkins and M.J. Sewell), Pineridge Press, p. **383**.

Mendelson, A. (1968), *Plasticity: Theory and Applications*, Macmillan, New York.

Mentel, T.J. (1958), The Plastic Deformation due to Impact of a Cantilever Beam with an Attached Tip Mass, *J. Appl. Mech.*, **25**, 515.

Meyer, O.E. (1908), Untersuchungen über Härteprüfung und Harte, Z *ver. deutsche Ing.*, **52**, 645.

Miles, J.P. (1969), Bifurcation in Rigid/Plastic Materials with Spherically Symmetric Loading Conditions, *J. Mech. Phys. Solids*, **17**, 363.

Mises, R. von (1913), Mechanik der Fasten Korper im plastisch deformablen Zustand, *Göttingen Nachrichten, Math. Phys. KL*, **582**.

Mises, R. von (1928), Mechanik der plastischen Formänderung von Kristallen, Z. *angew. Math. Mech.*, **8**, 161.

Montague, P. and Horner, M.R. (1968), Elastic-Plastic Axisymmetric Analysis of a Thin-Walled Cylindrical Shell Subjected to Radial Pressure Differences, *J. Appl. Mech.*, **35**, 787.

Moore, G.G. and Wallace, J.F. (1964), The Effect of Anisotropy on Instability in Sheet Metal Forming, *J. Inst. Metals*, **93**, 33.

Moore, G.G. and Wallace, J.F. (1967), Theories and Experiments on Hollow Sinking Through Conical Dies, *Proc. Inst. Mech. Engrs.*, **182**, 19.

Mori, K., Osakada, K. and Fukuda, M. (1983), Simulation of Severe Plastic Deformation by Finite Element Method with Spatially Fixed Elements, *Int. J. Mech. Sci.*, **25**, 775.

Morland, L.W. (1959), The Propagation of Plane Irrotational Waves Through an Elasto-plastic Medium, *Philos. Trans. Roy. Soc. London Ser. A*, **251**, 341.

Mröz, Z. (1961), On a Problem of Minimum Weight Design, *Quart. Appl. Math.*, **19**, 3.

Mröz, Z. (1967a), On the Description of Anisotropic Hardening, *J. Mech. Phys. Solids*, **15**, 163.

Mröz, Z. (1967b), Graphical Solution of Axially Symmetric Problems of Plastic Flow, *Z. angew. Math. Mech.*, **18**, 219.

Mröz, Z., Shrivastava, H.P., and Dubey, R.N. (1976), A Nonlinear Hardening Model and its Application to Cyclic Loading, *Acta Mech.*, **25**, 51.

Mukherjee, S. and Lieu, J.K. Isotropic Work-Hardening Rate-Independent Plasticity, *J. Appl. Mech.* **70**, 644.

Munday, G. and Newitt, D.M. (1963), The Deformation of Transversely Loaded Discs Under Dynamic Loads, *Philos. Trans. Roy. Soc. London Ser. A*, **256**, 1.

Myszkowsky, J. (1971), Endliche Durchbiegungen Beliebig Eigenspannter Dünner Kreis, Ingenieur-Archiv, **40**, 1.

Nadai, A. (1950), *Theory of Flow and Fracture of Solids*, McGraw-Hill, New York.

Nagamatsu, A., Murota, T. and Jimma, T. (1970), On the Non-Uniform Deformation of Block in Plane Strain Compression Caused by Friction, *Bull. Japan Soc. Mech. Engrs.* **13**, 1389.

Naghdi, P.M. (1952), Bending of Elastoplastic Circular Plate with Large Deflection, *J. Appl. Mech.*, **19**, 293.

Naghdi, P.M. (1960), Stress-Strain Relations in Plasticity and Thermoplasticity, in *Plasticity* (eds., E.H. Lee and P.S. Symonds), Pergamon Press, New York, p. 121.

Naghdi, P.M. (1990), A Critical Review of the State of Finite Plasticity, Z. *angew. Math. Phys.*, **41**, 315.

Naghdi, P.M., Essenberg, F. and Koff, W. (1958), An Experimental Study of Subsequent Yield Surfaces in Plasticity, *J. Appl. Mech.*, **25**, 101.

Nagpal, V. (1977), On the Solution of Three-Dimensional Metal Forming Processes, *J. Engng. Indust., Trans. ASME*, **99**, 624.

Nagtegaal, J.C. and DeJong, J.E. (1981), Some Computational Aspects of Elastic/Plastic Large Strain Analysis, *Int. J. Num. Methods Engr.,* **17**, 15.

Nagtegaal, J.C., Parks, D.M., and Rice, J.C. (1974), On Numerically Accurate Finite Element Solutions in the Fully Plastic Range, *Comput. Meth. Appl. Mech. Engg.*, **4**, 153.

Naruse, K., Dodd, B. and Motoki, Y. (1993), Evaluation of Yield Criteria for Planar Anisotropy in Sheet Metal Using Experimental Results, Advances in Engineering Plasticity and its Applications (ed. W.B. Lee), Elsevier, Amsterdam, p. 235.

Naziri, H. and Pearce, R. (1968), The Effect of Plastic Anisotropy on Flange Wrinkling Behaviour During Sheet Metal Forming, *Int. J. Mech. Sci.*, **10**, 681.

Neal, B.G. (1961), Effect of Shear Force on the Fully Plastic Moment of an I-Beam, *J. Mech. Engng. Sci.*, **3**, 258.

Neale, K.W. and Chater, E. (1980), Limit Strain Predictions for Rate Sensitive Anisotropic Sheets, *Int. J. Mech. Sci.*, **22**, 563.

Needleman, A. (1972), A Numerical Study of Necking in Circular Cylindrical Bars, *J. Mech., Phys. Solids*, **20**, 111.

Needleman, A. (1975), Post-Bifurcation Behavior and Imperfection Sensitivity of Elastic/Plastic Circular Plates, *Int. J. Mech. Sci.*, **17**, 1.

Needleman, A. and Triantafyllidis, N. (1978), Void Growth and Local Necking in Biaxially Stretched Sheets, *J. Engng. Mat. Tech.*, **100**, 164.

Nemat-Nasser, S. (1968), Finite Expansion of a Hole in a Rigid/Work-Hardening Disk of Initially Nonuniform Thickness, *J. Mech. Phys. Solids*, **16**, 195.

Nemat-Nasser, S. and Guo, W. (2003), Thermomechanical Response of DH-36 Structural Steel Over a Wide Range of Strain Rates and Temperatures, *Mech. Mater.*, **35**, 1023.

Nemat-Nasser, S., Isaacs, J.B. and Starrett, J.E. (1991), Hopkinson Techniques for Dynamic Recovery Experiments, *Proc. R. Soc. London*, **A20**, 371.

Nemat-Nasser, S., Li, Y.F. and Isaacs, J.B. (1994), Experimental/computational Evaluation of Flow Stress at High Strain Rates with Application to Adiabatic Shear Banding, *Mech. Mater.*, **17**, 111.

Nemirovsky, U.V. (1962), Carrying Capacity of Rib-Reinforced Circular Plates (in Russian), Izv. Nauk. USSR, *Mekh. Mack*, **2**, 163.

Ng, P, Chakrabarty, J. and Mellor, P.B. (1976), A Reappraisal of the Wedge-Drawing Test, *Int. J. Mech. Sci.*, **18**, 249.

Nicholas, T. (1971), Strain Rate History Effects in Several Metals in Torsion, *Experimental Mech.*, **11**, 153.

Nicholas, T. (1982), Elastic-Plastic Stress Waves, in *Impact Dynamics* (eds., A. Zukas et al.), Wiley, New York, pp. 95–153.

Nieh, T.G., Wadsworth, J. and Sherby, O.D. (1997), *Superplasticity in Metals and Ceramics*, Cambridge University Press, Cambridge, UK.

Nine, H.D. (1978), Drawbead Forces in Sheet Metal Forming, *Mechanics of Sheet Metal Forming* (eds., D.P. Koistinen and N.M. Wang), Plenum Press, New York, p. 179.

Nonaka, T. (1977), Shear and Bending Response of a Rigid Plastic Beam in Blast Type Loading, *Ingenieur-Archiv*, **46**, 35.

Norbury, A.L. and Samuel, T. (1928), The Recovery and Sinking in or Piling up of Material in the Brinell Test, and the Effects of these Factors on the Correlation of the Brinell Test with Certain Other Hardness Tests, *J. Iron. Steel Inst.*, **117**, 673.

Nordgren, R.P. and Naghdi, P.M. (1963), Finite Twisting and Expansion of a Hole in a Rigid/Plastic Plate, *J. Appl. Mech.*, **30**, 605.

Nurick, G.N., Pearce, H.T. and Martin, J.B. (1987), Prediction of Transverse Deflection and In-Plane Strain in Impulsively Loaded Thin Plates, *Int. J. Mech. Sei.*, **29**, 435.

Oblak, M. (1986), Elastoplastic Bending Analysis for Thick Plate, Z. *Angew. Math. Mech.*, **66**, 320.

Oh, S.I. (1982), Finite Element Analysis of Metal Forming Problems with Arbitrarily Shaped Dies, *Int. J. Mech. Sci.*, **24**, 479.

Oh, S.I. (1990), Application of Three Dimensional Finite Element Analysis to Shape Forming Processes, *J. Eng. Indust., Trans. ASME*, **112**, 36.

Oh, S.I. and Kobayashi, S. (1980), Finite Element Analysis of Plane Strain Sheet Bending, *Int. J. Mech. Sci.*, **22**, 583.

Oh, S.I., Chen, C.C. and Kobayashi, S. (1979), Ductile Fracture in Axisymmetric Extrusion and Drawing, Part 2, Workability in Extrusion and Drawing, *J. Engng. Indust. Trans. ASME*, **101**, 36.

Oh, S.I. and Kobayashi, S. (1975), An Approximate Method for a Three-Dimensional Analysis of Rolling, *Int. J. Mech. Sei.*, **17**, 293.

Ohashi, Y, and Kamiya, N. (1967), Bending of Thin Plates of Material with a Nonlinear Stress-Strain Relation, *Int. J. Mech. Sci.*, **9**, 183.

Ohashi, Y. and Kawashima, I. (1969), On The Residual Deformation of Elastoplastically Bent Circular Plate After Perfect Unloading, Z. *Angew. Math. Mech*, **49**, 275.

Ohashi, Y. and Murakami, S. (1964), The Elasto-Plastic Bending of a Clamped Thin Circular Plate, *Proc. 11th Int. Conf. Appl. Mech*, 212.

Ohashi, Y. and Murakami, S. (1966), Large Deflection of Elastoplastic Bending of a Simply Supported Circular Plate Under a Uniform Load, *J. Appl. Mech., Trans. ASME*, **33**, 866.

Ohashi, Y. and Okouchi, T. (1975), The Elastic-Plastic Deformation of a Supported Short Cylindrical Shell of Mild Steel Under Internal Pressure, *Int. J. Mech. Sci.*, **17**, 267.

Ohno, N. and Kachi, Y. (1986), A Constitutive Model of Cyclic Plasticity for Nonlinear Hardening Materials, *J. Appl. Mech.*, **53**, 395.

Olszak, W. and Sawczuk, A. (1967), *Inelastic Behaviour in Shells*, Noordhoff, Groningen.

Onat, E.T. (1955), Plastic Collapse of Cylindrical Shells Under Axially Symmetrical Loading, *Quart. Appl. Math.*, **13**, 63.

Onat, E.T. (1960), Plastic Analysis of Shallow Conical Shells, *Proc. ASCE*, **86**, EM6-1.

Onat, E.T. and Prager, W. (1954), Limit Analysis of Shells of Revolution, *Proc. Roy. Netherlands Acad. Sei., B*, **57**, 534.

Onat, E.T. and Prager, W. (1955), Limits of Economy of Material in Shells, *De Ingenieur*, **67**, 46.

Onat, E.T. and Haythornthwaite, R.M. (1956), The Load Carrying Capacity of Circular Plates at Large Deflections, *J. Appl. Mech.*, **23**, 49.

Onat, E.T., Schumann, W. and Shield, R.T. (1957), Design of Circular Plates for Minimum Weight, *Z. angew. Math. Phys.*, **8**, 485.

Ore, E. and Durban, D. (1989), Elastoplastic Buckling of Annular Plates in Pure Shear, *J. Appl. Mech.*, **56**, 644.

Osakada, K. and Nimi, Y. (1975), A Study of Radial Flow Fields for Extrusion Through Conical Dies, *Int. J. Mech. Sei.*, **17**, 241.

Owens, R.H. and Symonds, P.S. (1955), Plastic Deformation of a Free Ring Under Concentrated Dynamic Loading, *J. Appl. Mech.*, **22**, 523.

Padmanabhan, K.A. and Davies, C.J. (1980), *Superplasticity-Mechanical and Structural Aspects, Environmental Effects, Fundamentals and Applications*, Springer-Verlag, Berlin.

Palusamy, S. (1971), Limit Analysis of Spherical Shells Subjected to External Axial Force, *Nuclear Engng. Design*, **16**, 13.

Palusamy, S. and Luid, N.C. (1972), Limit Analysis of Non-Axisymmetrically Loaded Spherical Shells, *J. Appl. Mech.*, **39**, 422.

Paquette, J.A. and Kyriakides, S. (2006), Plastic Buckling of Tubes Under Axial Compression and Internal Pressure, *Int. J. Mech. Sci.*, **48**, 855.

Park, J.J. and Kobayashi, S. (1984), Three-Dimensional Finite Element Analysis of Block Compression, *Int. J. Mech., Sci.*, **26**, 165.

Park, J.J. and Oh, S.I. (1990), Application of Three-Dimensional Finite Element Analysis to Shape Rolling Processes, *J. Engng. Indust. Trans. ASME*, **112**, 36.

Parkes, E.W. (1955), The Permanent Deformation of a Cantilever Struck Transversely at its Tip, *Proc. Roy. Soc. London, Ser. A*, **228**, 462.

Parkes, E.W. (1958), The Permanent Deformation of an Encastre Beam Struck Transversely at any Point in its Span, *Proc. Inst. Civil Engrs.*, **10**, 277.

Parmar, A. and Mellor, P.B. (1978a), Prediction of Limit Strains in Sheet Metal Using a More General Yield Criterion, *Int. J. Mech. Sci.*, **20**, 385.

Parmar, A. and Mellor, P.B. (1978b), Plastic Expansion of a Circular Hole in Sheet Metal Subjected to Biaxial Tensile Stress, *Int. J. Mech. Sci.*, **20**, 707.

Paul, B. (1959), Carrying Capacity of Elastic-Plastic Shells with Various End Conditions Under Hydrostatic Compression, *J. Appl. Mech.*, **26**, 553.

Paul, B. and Hodge, P.G. (1958), Carrying Capacity of Elastic-Plastic Shells Under Hydrostatic Pressure, *Proc. 3rd US Nat. Congr. Appl. Mech, Providence*, p. 631.

Pearson, C.E. (1950), Bifurcation Criteria and Plastic Buckling of Plates and Columns, *J. Aeronaut. Sci.*, **7**, 417.

Pearson, C.E. (1956), A General Theory of Elastic Stability, *Quart. Appl. Math.*, **14**, 133.

Pell, W.H. and Prager, W. (1951), Limit Design of Plates, *Proc. 1st US Nat. Congr. Appl. Mech. (Chicago)*, 547.

Perrone, N. (1965), On a Simplified Method of Solving Impulsively Loaded Structures of Rate-Sensitive Materials, *J. Appl. Mech.*, **32**, 489.

Perrone, N. (1967), Impulsively Loaded Strain Rate Sensitive Plates, *J. Appl. Mech.*, **34**, 380.

Perrone, N. (1970), Impulsively Loaded Strain-Hardened Rate Sensitive Rings and Tubes, *Int. J. Solids Struct.*, **6**, 119.

Perrone, N. and Bhadra, P. (1984), Simplified Large Deflection Mode Solutions for Impulsively Loaded Viscoplastic Circular Membranes, *J. Appl. Mech.*, **51**, 505.

Petryk, H. (1983), A Stability Postulate for Quasi-Static Processes of Plastic Deformation, *Arch. Mech.*, **35**, 753.

Perzyna, P. (1958), Dynamic Load Carrying Capacity of a Circular Plate, *Arch. Mech. Stos.*, **10**, 635.

Perzyna, P. (1963), The Constitutive Equations for Rate Sensitive Plastic Solids, *Quart. Appl. Math.*, **2d**, 321.

Phillips, A. and Lee, C.W (1979), Yield Surface and Loading Surface, Experiments and Recommendations, *Int. J. Solids Struct.*, **15**, 715.

Phillips, A. and Weng. G.J. (1975), An Analytical Study of an Experimentally Verified Hardening Law, *J. Appl. Mech.*, **42**, 375.

Piekutowski, A.J., Forrestal, M.J., Poormon, K.L. and Warren, T.L. (1996), Perforation of Aluminum Plates with give-Nose Steel Rods at Normal and Oblique Impacts, *Int. J. Impact Engng.*, **18**, 877.

Popov, E.P., Khojestch-Bakht, M. and Yaghmai, S. (1967), Analysis of Elastic-Plastic Circular Plates, *J. Engng. Mech. Div., Trans. ASCE*, **93**, 49.

Prager, W. (1953), On the Use of Singular Yield Conditions and Associated Flow Rules, *J. Appl. Mech.*, **20**, 317.

Prager, W. (1955), Minimum Weight Design of Plates, *De. Ingenieur (Amsterdam)*, **67**, 141.

Prager, W. (1956a), The General Theory of Limit Design, *Proc. 8th Int. Congr. Appl. Mech. (Istanbul, 1952)*, **2**, 65.

Prager, W. (1956b), A New Method of Analyzing Stress and Strain in Work-Hardening Plastic Solids, *J. Appl. Mech.*, **23**, 493.

Prager, W. (1959), *An Introduction to Plasticity*, Chap. 3, Addison-Wesley, Reading, MA.

Prager, W. (1961a), An Elementary Discussion of the Definition of Stress Rate, *Quart. Appl. Math.*, **18**, 403.

Prager, W. (1961b), *Introduction to Mechanics of Continua*, Ginn, Boston.

Prager, W. and Hodge, P.G. (1951), *Theory of Perfectly Plastic Solids*, Wiley, New York.

Prager, W. and Shield, R.T. (1959), Minimum Weight Design of Circular Plates Under Arbitrary Loading, Z. *Angew. Math. Phys.*, **10**, 421.

Prager, W. and Shield, R.T. (1967), A General Theory of Optimal Plastic Design, *J. Appl. Mech.*, **34**, 184.

Prandtl, L. (1924), Spanmungsverteilung in Plastischen Körpern, *Proc. 1st Int. Conf. Appl. Mech., Delft*, p. **43**.

Prandtl, L. (1928), Ein Gedankemodell Zur Kinetischen Theorie der fester körper, Z. *Angew. Math. Mech.*, **8**, 85.

Presnyakov, A.A. (1976), *Superplasticity of Metals and Alloys*, The British Library, Wetherby, UK.

Pugh, H. and Li. D. (1970), *Mechanical Behavior of Metals under Pressure*, Elsevier, Amsterdam.

Pugsley, A. (1979), On the Crumpling of Thin Cylindrical Tubes, *Quart. J. Mech. Appl. Math.*, **32**, 1.

Pugsley, A. and Macaulay, M. (1960), Large Scale Crumpling of Thin Cylindrical Columns, *Quart. J. Mech. Appl. Math.*, **13**, 1.

Raghavan, K.S. (1995), A Simple Technique to Generate In-Plane Forming Limit Curves and Selected Applications, *Metall. Trans.*, **26A**, 2075.

Rakhmatulin, H.A. (1945), On the Propagation of Waves of Unloading (in Russian), *Prikl. Mat. Mekh.*, **9**, 91.

Rakhmatulin, H.A. and Shapiro, G.S. (1948), *Prikl. Mat. Mekh.*, **12**, 369.

Ramberg, W. and Osgood, W. (1943), Description of stress-strain curves by three parameters, NACA Tech. Note No. **902**.

Ranshi, A.S., Chitkara, N.R. and Johnson, W. (1974), Plane Stress Plastic Collapse Loads for Tapered Cantilevers and Haunched Beams, *Int. J. Mech. Sci.*, **16**, 867.

Ranshi, A.S., Chitkara, N.R. and Johnson, W. (1976), Plastic Yielding of I-Beams under Shear, and Shear and Axial Load, *Int. J. Mech. Sci.*, **18**, 375.

Raphanel, J.L. and Symonds, PS. (1984), The Estimation of Large Deflections of a Portal Frame Under Asymmetric Pulse Loading, *J. Appl. Mech.*, **51**, 494.

Rawlings, B. (1964), Mode Changes in Frames Deforming Under Impulsive Loads, *J. Mech. Engng. Sci.*, **6**, 327.

Rebelo, N., and Wertheimer, T.B. (1986), General Purpose Procedures for Elastic/Plastic Analysis of Metal Forming Processes, *Proc 14 th NAMRC*, Minneapolis, MN, p. 414.

Recht, R.F. and Ipson, T.W (1963), Ballistic Performance Dynamics, *J. Appl. Mech.*, **30**, 384.

Reid, S.R. and Gui, X.G. (1987), On the Elastic-Plastic Deformation of Cantilever Beams Subjected to Tip Impact, *Int. J. Impact Engng.*, **6**, 109.

Reddy, B.D. (1979), Plastic Buckling of a cylindrical Shell in Pure Bending, *Int. J. Mech. Sci.*, **21**, 671.

Reddy, B.D. (1980), Buckling of Elastic-Plastic Discretely Stiffened Cylinders in Axial Compression, *Int. J. Solids Struct.*, **16**, 313.

Reddy, N.V., Dixit, P.M. and Lal, G.K. (1996), Central Burst and Optimal Die Profile for Axisymmetric Extrusion, *J. Manuf. Sci. Eng., Trans. ASME*, **118**, 579.

Rees, D.W.A. (1981) Anisotropic Hardening Theory and the Bauschinger Effect, *J. Strain Anal.*, **16**, 85.

Rees, D.W.A. (1982), Plastic Torsional Buckling of Thin-Walled Cylinders, *Appl. Mech.*, **49**, 663.

Reid, S.R., Wang, B. and Yu, T.X. (1995), Yield Mechanism for a Bent Cantilever Beam Subjected to a Suddenly Applied Constant Out-of-Plane Tip Force, *Int. J. Impact Engng.*, **16**, 89.

Reiss, R. (1974), Minimum Weight Design of Conical Shells, *J. Appl. Mech.*, **41**, 599.

Reiss, R. (1979), Minimum Weight Design of Shallow Conical Shells, *J. Appl. Mech.*, **46**, 599.

Reiss, R. and Megarefs, G.J. (1969), Minimal Design of Axisymmetric Cylindrical Shells Obeying Mises Criterion, *Acta Mech.*, **7**, 72.

Reuss, E. (1930), Berücksichtigung der elastishen formänderungen in der Plastizitätstheorie, Z. *Angew. Math. Mech.*, **10**, 266.

Rice, J.R. (1973), Plane Strain Slipline Field Theory for Anisotropic Rigid/Plastic Materials, *Mech. Phys. Solids*, **21**, 63.

Richmond, O. (1965), Theory of Streamlined Dies for Drawing and Extrusion, *J. Mech. Phys. Solids*, **13**, 154.

Richmond, O. and Azarkhin, A. (2000), Minimum Weight Axisymmetric Shell Structures, *Int. J. Mech. Sci.*, **42**, 2439.

Richmond, O. and Morrison, H.L. (1967), Streamlined Wire Drawing Dies of Minimum Length, *J. Mech. Phys. Solids*, **15**, 195.

Rinehart, J.S. and Pearson, J. (1965), *Behavior of Metals Under Impulsive Loads*, Dover, New York.

Ripperger, E.A. (1960), Experimental Studies of Plastic Wave Propagation in Bar, in *Plasticity* (eds., E.H. Lee and PS. Symonds), Pergamon Press, New York, p. 475.

Robinson, M. (1971), A Comparison of Yield Surfaces for Thin Shells, *Int. J. Mech. Sei*, **13**, 345.

Rogers, T.G. (1967), Finite Expansion and Subsequent Unloading of Hole in Elastic-Plastic Plate of Initially Varying Thickness, *Quart. J. Mech. Appl. Math.*, **20**, 137.

Ronay, M. (1968), Second Order Elongation of Metal Tubes in Cyclic Torsion, *Int. J. Solids Struct.*, **4**, 509.

Rosenberg, Z. and Dekel, E. (1994), The Relation Between the Penetration Capability of Long Rods and Their Length to Diameter Ratios, *Int. J. Impact Engng.*, **15**, 125.

Ross, E.W. and Prager, W. (1954), On the Theory of the Bulge Test, *Quart. Appl. Math.*, **12**, 86.

Rowe, G.W. (1977), *Introduction to the Principles of Industrial Metalworking*, Arnold, London.

Ruiz, C. and Chukwujckwu, S.E. (1967), Limit Analysis and Design of Ring-Reinforced Radial Branches in Cylindrical and Spherical Vessels, *Int. J. Mech. Sei.*, **9**, 11.

Sachs, G. and Baldwin, W.M. (1946), Stress Analysis of Tube Sinking, *Trans. Amer. Soc. Mech. Engrs.*, **68**, 655.

Sagar, R. and Juneja, B.L. (1979), An Upper Bound Solution for Rat Tool Forging Taking into Account the Bulging of Sides, *Int. J. Mech. Tool Des. Res.*, **19**, 253.

Samanta, S.K. (1968), The Application of the Upper Bound Theorem to the Prediction of Indenting and Compressing Leads, *Acta Polytech. Scand.*, Mc-**38**.

Samanta, S.K. (1971), A New Die Profile with Higher Process Efficiencies, *Appl. Sei. Res.*, **25**, 54.

Sanchez, L.R. and Weinmann, K.J. (1996), An Analytical and Experimental Study of the Flow of Sheet Metal Between Circular Drawbeads, *J. Engng. Indust., Trans. ASME*, **118**, 45.

Sankaranarayanan, R. (1960), Plastic Interaction Curves for Circular Cylindrical Shells Under Combined Lateral and Axial Pressures, *J. Franklin Inst.*, **270**, 5.

Sankaranarayanan, R. (1963), On the Dynamics of Plastic Spherical Shells, *J. Appl. Mech.*, **30**, 87.

Sankaranarayanan, R. (1966), On the Impact Pressure Loading of a Plastic Spherical Cap., *J. Appl. Mech.*, **33**, 704.

Saran, M.J., Schedin, E., and Samuelsson, A. (1990), Numerical and Experimental Investigations of Deep Drawing of Metal Sheets, *J. Eng. Indust., Trans. ASME*, **112**, 272.

Save, M. (1961), On Yield Conditions in Generalized Stresses, *Quart. Appl. Math*, **19**, 3.

Save, M.A. and Massonnet, C.E. (1972), *Plastic Analysis and Design of Plates, Shells and Disks*, North-Holland, Amsterdam.

Sawczuk, A. (1982), On Plastic Shell Theories at Large Strains and Displacements, *Int. J. Mech. Sei.*, **24**, 231.

Sawczuk, A. (1989), *Mechanics and Plasticity of Structures*, Ellis Horwood, Chichester.

Sawczuk, A. and Duszek, M. (1963), A Note on the Interaction of Shear and Bending in Plastic Plates, *Arch. Mech. Stos.*, **15**, 411.

Sawczuk, A. and Hodge, P.G. (1960), Comparison of Yield Conditions for Circular Cylindrical Shells, *J. Franklin Ints.*, **269**, 362.

Sawczuk, A. and Hodge, P.G. (1968), Limit Analysis and Yield Line Theory, *J. Appl. Mech., Trans. ASME*, **35**, 357.

Sawczuk, A. and Jaeger, T. (1963), *Crenztragfähigkeits Theorie der Platten*, Springer-Verlag, Berlin.

Sawczuk, A., O'Donnell, W.J., and Porowsky, J. (1975), Plastic Analysis of Perforated Plates for Orthotropic Yield Criteria, *Int. J. Mech. Sci.*, **17**, 411.

Sayir, M. (1966), Kollapsbelastung von Rotationsymmetrischen Zylinderschalen, *Z. Angew. Math. Phys.*, **17**, 353.

Scrutton, R.F. and Marasco, A. (1995), The Dynamics of Closed Die Forging with Mechanical Presses, *J. Engng. Indust.*, **117**, 357.

Seide, P. and Weingarten, V.I. (1961), On the Buckling of Circular Cylindrical Shells Under Pure Bending, *J. Appl. Mech.*, **28**, 112.

Sewell, M.J. (1963), A General Theory of Elastic and Inelastic Plate Failure, Part I, *J. Mech. Phys. Solids*, **11**, 377.

Sewell, M.J. (1964), A General Theory of Elastic and Inelastic Plate Failure, Part II, *J. Mech. Phys. Solids*, **12**, 279.

Sewell, M.J. (1973), A Yield Surface Corner Lowers the Buckling Stress of an Elastic/Plastic Plate,*/Mech. Phys. Solids*, **21**, 19.

Seweryn, A. (1992), Analysis of Axisymmetric Steady State Extrusion Through Dies of Large Cone Angle by the Slipline Field Method, *Int. J. Mech. Sci.*, **33**, 997.

Schumann, W. (1958), On Limit Analysis of Plates, *Quart. Appl. Math.*, **16**, 61.

Seiler, J.A., Cotter, B.A. and Symonds, RS. (1956), Impulsive Loading of Elastic-Plastic Beams, *J. Appl. Mech.*, **23**, 516.

Seiler, J.A. and Symonds, PS. (1954), Plastic Deformation of Beams Under Distributed Dynamic Loads, *J. Appl. Phys.*, **25**, 556.

Senior, B.W. (1956), Flange Wrinkling in Deep Drawing Operations, *J. Mech. Phys. Solids*, **4**, 235.

Shah, S.N. and Kobayashi, S. (1977), A Theory on Metal Row in Axisymmetric Piercing and Extrusion, *J. Prod. Engng.*, **1**, 73.

Shabaik, A.H. and Thomsen, E.G. (1968), Investigation of the Application of the Visioplasticity Methods of Analysis to Metal Deformation Processes, Part II, US. Department of the Navy.

Shanley, F.R. (1947), Inelastic Column Theory, *J. Aeronaut. Sei.*, **14**, 251.

Shapiro, G.S. (1959), On a Rigid-Plastic Annular Plate Under Impulsive Load (translation from Russian), *J. Appl. Math. Mech.*, **23**, 234.

Shapiro, G.S. (1961), On Yield Surfaces for Ideally Plastic Shells, *Problems of Continuum Mechanics*, SIAM, Philadelphia, PA.

Shepherd, W.M. and Gaydon, F.A. (1957), Plastic Bending of a Ring Sector by End Couples, *Mech. Phys. Solids*, **5**, 296.

Sherbourne, A.N. and Srivastava, A. (1971), Elastic-Plastic Bending of Restrained Pin-Ended Circular Plates, *Int. J. Mech. Sei.*, **13**, 231.

Sheu, C.Y. and Prager, W. (1969), Optimal Plastic Design of Circular and Annular Plates with Piecewise Constant Cross Section, *J. Mech. Phys. Solids*, **17**, 11.

Shield, R.T. (1955a), Plastic Flow in a Converging Conical Channel, *J. Mech. Phys. Solids*, **3**, 246.

Shield, R.T. (1955b), On the Plastic Flow of Metals Under Conditions of Axial Symmetry, *Proc. Roy. Soc. London Sec. A*, **233**, 267.

Shield, R.T. (1955c), The Plastic Indentation of a Layer by a Rat Punch, *Quart. Appl. Math.*, **13**, 27.

Shield, R.T. (1960a), Plate Design for Minimum Weight, *Quart. Appl. Math.*, **18**, 131.

Shield, R.T. (1960b), Optimum Design of Shells, *Appl. Mech.*, **27**, 316.

Shield, R.T. (1963), Optimum Design Methods for Multiple Loading, *Z. Angew. Math. Phys.*, **14**, 38.

Shield, R.T. and Drucker, D.C. (1953), The Application of Limit Analysis to Punch Indentation Problems, *J. Appl. Mech., Trans. ASME*, **20**, 453.

Shield, R.T. and Ziegler, H. (1958), On Prager's Hardening Rule, *Z. Angew. Math. Phys.*, **9**, 260.

Shrivastava, H.R (1979), Inelastic Buckling of Axially Compressed Cylindrical Shells, *Int. J. Solids Struct.*, **15**, 567.

Shull, H.E. and Hu, L.W. (1963), Load Carrying Capacity of Simply Supported Rectangular Plates, *J. Appl. Mech.*, **30**, 617.

Siebel, E. (1923), *Stahl und Eisen*, Düsseldorf, **43**, 1295.

Siebel, E. (1947), Der derzeitige stand der Erkenntnisse über die mechanischen Vorgange beim Drahtziehen, *Stahl und Eisen*, **66**, 171.

Sinclair, G.B., Follansbee, P.S. and Johnson, K.L. (1985), Quasi-Static Normal Indentation of an Elastic-Plastic Half Space by a Rigid Sphere, *Int. J. Solids Struct.*, **21**, 865.

Skrzypek, J.J. and Hetnarski, R.B. (1993), *Plasticity and Creep, Theory, Examples and Problems,* CRC Press, Boca Raton, Florida.

Slater, R.A.C. (1979), *Engineering Plasticity. Theory and Application to Metal Forming,* Macmillan, London.

Skrzypek, J.J. and Hetnarski, R.B. (1993), *Plasticity and Creep: Theory, Examples and Problems*, CRC Press, Boca Raton, Florida.

Sobotka, Z. (1989), *Theory of Plasticity and Limit Design of Plates*, Academia, Prague.

Sokolovsky, W.W. (1948a), Elastic-Plastic Bending of Circular and Annular Plates (in Russian), *Prikl. Mat. Mekh.*, **12**, 141.

Sokolovsky, W.W. (1948b), The Propagation of Elastic Viscoplastic Waves in Bars (in Russian), *Prikl. Math. Mekh.*, **12**, 261.

Sokolovsky, V.V. (1969), *Theory of Plasticity* (in Russian), 3rd Edition, Moscow.

Sortais, H.C. and Kobayashi, S. (1968), An Optimum Die Profile for Axisymmetric Extrusion, *Int. J. Mach. Tool Des. Res.*, **8**, 61.

Southwell, R.V. (1913), On the Collapse of Tubes by External Pressure, *Phil. Mag.*, **25**, 687.

Sowerby, R. and Duncan, J.L. (1971), Failure of Sheet Metal in Biaxial Tension, *Int. J. Mech. Sci.*, **13**, 217.

Spencer, A.J.M. (1964), The Approximate Solution of Certain Problems of Axially Symmetric Plastic Row, *J. Mech. Phys. Solids*, **12**, 231.

Steinberg, D. and Lund, O. (1989), A Constitutive Model for Strain Rates from 10^{-4} to 10^{6} s^{-1}, *J. Appl. Phys.*, **65**, 1528.

Sternglass, E.J. and Stuart, D.A. (1953), An Experimental Study of Propagation of Transient Longitudinal Deformation in Elastic-Plastic Media, *J. Appl. Mech.*, **20**, 427.

Stören, S. and Rice, J. (1975), Localized Necking in Thin Sheets, *J. Mech. Phys. Solids*, **23**, 421.

Storakers, B. (1966), Finite Plastic Deformation of a Circular Membrane Under Hydrostatic Pressure, *Int. J. Mech. Sei.*, **8**, 619.

Storakers, B. and Larson, P. (1994), On Brinell and Boussinesq Indentation of Creeping Solids., *J. Mech. Phys. Solids*, **42**, 307.

Stoughton, T.B. (2000), General Forming Limit Criterion for Sheet Metal Forming, *Int. J. Mech. Sci.*, **42**, 1.

Stout, M.G. and Hecker, S.S. (1983), Role of Geometry in Plastic Instability and Fracture of Tubes and Sheets, *Mech. Mat.*, **2**, 23.

Stout, M.G., Hecker, S.S., and Bourcier, R. (1983), An Evaluation of Anisotropic Effective Stress-Strain Criteria for the Biaxial Yield and Flow of 202 Aluminium Tubes, *J. Engng. Mat. Technol*, **105**, 242.

Stronge, W.J. (2006), *Impact Mechanics*, Cambridge University Press, Cambridge.

Stronge, W.J., Shu, D. and Shim, V.P.W. (1990), Dynamics Modes of Plastic Deformation for Suddenly Loaded Curved Beams, *Int. J. Impact Engng.*, **9**, 1.

Stronge, W.J. and Yu, T.X. (1993), *Dynamic Models for Structural Plasticity*, Springer-Verlag, London.

Swift, H. (1947), Length Changes in Metals Under Torsional Overstrain, Engineering, **163**, 253.

Swift, H.W. (1949), Stresses and Strains in Tube Drawing, *Phil. Mag., Ser.*, **11**, 883.

Swift, H.W. (1952), Plastic Instability Under Plane Stress, *J. Mech. Phys. Solids*, **1**, 1.

Symonds, P.S. (1953), Dynamic Load Characteristics in Plastic Bending of Beams, *J. Appl. Mech.*, **20**, 475.

Symonds, PS. (1954), Large Plastic Deformation of Beams Under Blast Type Loading, *Proc. 2nd U.S. Nat. Congr. Appl. Mech., Ann Arbor*, p. 506.

Symonds, PS. (1980), Finite Elastic and Plastic Deformations of Pulse Loaded Structures by an Extended Mode Technique, *Int. J. Mech. Sci.*, **22**, 597.

Symonds, P.S. and Fleming, W.T. (1984), Parkes Revisited: On Rigid-Plastic and Elastic-Plastic Dynamic Structural Analysis, *Int. J. Impact Engng.*, **2**, 1.

Symonds, P.S. and Leth, C.F. (1954), Impact of Finite Beams of Ductile Metal, *J. Mech. Phys. Solids*, **2**, 92.

Symonds, P.S. and Mentel, T.J. (1958), Inpulsive Loading of Plastic Beams with Axial Restraint, *J. Mech. Phys. Solids*, **6**, 183.

Symonds, P.S. and Wierzbicki, T. (1979), Membrane Mode Solutions for Impulsively Loaded Circular Plates, *J. Appl. Mech.*, **46**, 58.

Szczepinski, W. (1962), The Method of Successive Approximation of Some Strain-Hardening Solutions, *Proc. 4th U.S. Nat. Congr. Appl. Mech.*, p. 131.

Szczepinski, W., Dietrich, L., Drescher, E., and Miastkowski, J. (1966), Plastic Flow of Axially Symmetric Notched Bars Pulled in Tension, *Int. J. Solid Struct.*, **2**, 543.

Tabor, D. (1951), *The Hardness of Metals*, Clarendon Press, Oxford, UK.

Tadros, A.K. and Mellor, P.B. (1978), An Experimental Study of the In-Plane Stretching of Sheet Metal, *Int. J. Mech. Sci.*, **20**, 121.

Taghvaipour, M. and Mellor, P.B. (1970), Plane Strain Compression of Anisotropic Sheet Metal, *Proc. Instn. Mech. Engrs.*, **185**, 593.

Taghvaipour, M., Chakrabarty, J. and Mellor, P.B. (1972), The Variation of the R-value of Titanium with Increasing Strain, *Int. J. Mech. Sci.*, **14**, 117.

Tan, Z., Persson, B., and Magnusson, C. (1995), Plastic Bending of Anisotropic Sheet Metals, *Int. J. Mech. Sci.*, **37**, 405.

Tanaka, M. (1972), Large Deflection Analysis of Elastic-Plastic Circular Plates with Combined Isotropic and Kinematic Hardening, *Ingenieur-Archiv*, **41**, 342.

Tang, S.C. (1981), Large Elastic/Plastic Analysis of a Flanged Hole Forming, *Comput. Struct.*, **13**, 363.

Tate, A. (1969), Further Results on Long Rod Penetration, *J. Mech. Phys. Solids*, **17**, 141.

Taylor, G.I. (1938), Plastic Strain in Metals, *J. Inst. Metals*, **62**, 307.

Taylor, G.I. (1942), The Plastic Wave in a Wire Extended by an Impact Load, British Official Report, RC 329.

Taylor, G.I. (1948a), The Formation and Enlargement of a Circular Hole in a Thin Plastic Sheet, *Quart. J. Mech. Appl. Math.*, **1**, 101.

Taylor, G.I. (1948b), The Use of Flat-Ended Projectiles for Determining Dynamic Yield Stress, I, *Proc. Roy. Soc. London Ser. A*, **194**, 289.

Taylor, G.I. and Quinney, H. (1931), The Plastic Distortion of Metals, *Phil. Trans. Roy. Soc. London Ser. A*, **230**, 323.

Tekinalp, B. (1957), Elastic-Plastic Bending of a Built-in Circular Plate Under Uniformly Distributed Load, *J. Mech. Phys. Solids*, **5**, 135.

Thomas, H.K. 1961), *Large Plastic Flow and Fracture in Solids*, Thames Hudson, London.

Thomson, T.R. (1975), Influence of Material Properties in the Forming of Square Shells, *J. Australian Inst. Metals*, **20**, 106.

Thomson, W.T. (1955), An Approximate Theory of Armour Penetration, *J. Appl. Phys.*, **26**, 80.

Thomsen, E.G., Yang, C.T., and Kobayashi, S. (1965), *Mechanics of Plastic Deformation in Metal Processing*, Macmillan, New York.

Timoshenko, S.P. and Gere, J.M. (1961), *Theory of Elastic Stability,* 2nd ed., McGraw-Hill, New York.

Ting, T.C.T. (1964), The Plastic Deformation of a Cantilever Beam with Strain Rate Sensitivity Under Inpulsive Loading, *J. Appl. Mech.*, **11**, 38.

Ting, T.C.T. (1972), The Initiation of Combined Stress Waves in a Thin-Walled Tube due to Impact Loading. *Int. J. Solids Struct.*, **8**, 269.

Ting, T.C.T. (1977), Plastic Wave Speeds in Isotropically Work-Hardening Material, *J. Appl. Mech.*, **44**, 68.

Tirosh, J. (1971), Dead Zone Formation in Plastic Axially Symmetric Converging Row, J *Mech. Phys. Solids*, **19**, 237.

Tirosh, J., Tylis, A. and Davidi, G. (2008), Upper and Lower Bounds to Indentation of Rigid Ball into Semi Infinite Solid, *Int. J. Mech. Sci.*, **50**, 328.

Toh, C.H. and Kobayashi, S. (1985), Deformation Analysis and Blank Design in Square Cup Drawing, *Int. J. Mech. Tool Des. Res.*, **25**, 15.

Toth, L.S. Jonas, J.J., Gilormini, P. and Borcroix, B. (1992), Length Changes During Free-End Torsion: A Rate Sensitive Analysis, *Int. J. Plasticity*, **6**, 83.

Travis, F.W. and Johnson, W. (1962), Experiments in the Dynamic Deformation of Clamped Circular Sheets of Various Metals Subject to an Underwater Explosive Charge, *Sheet Metal Indust.*, **39**, 456.

Tresca, H. (1864), On the Flow of Solid Bodies Subjected to High Pressures, *C.* R. *Acad. Sci. Paris*, **59**, 754.

Triantafyllidis, N., Maker, B. and Samanta, S.K. (1986), An Analysis of Drawbeads in Sheet Metal Forming, Part I: Problem Formulation, *J. Engng. Mater. Technol., Trans. ASME*, **108**, 321.

Tseng, N.T. and Lee, C.C. (1983), Simple Plasticity Model of the Two-Surface Type, *J. Engng. Mech., Trans. ASCE*, **109**, 795.

Tsuta, T. and Yin, Y. (1998), Rigid/Plastic Theory for Porous Materials Including Void Evolution and Mixed Hardening Effects, *Metals Mater.*, **4**, 359.

Tsutsumi, S. and Kato, T. (994), A New Approach for Estimating Die Profile Friction in Deep Drawing, *J. Engng. Indust., Trans. ASME*, **116**, 183.

Tugcu, P. (1991), Plate Buckling in the Plastic Range, *Int. J. Mech. Sei.*, **33**, 1.

Turvey, G.J. (1979), Thickness-Tapered Circular Plates–An Elastic-Plastic Large Deflection Analysis, *J. Struct. Mech., Trans. ASCE*, **7**, 247.

Tvergaard, V. (1981), Influence of Voids on Shear Band Instabilities Under Plane Strain Conditions, *Int. J. Fracture*, **17**, 389.

Tvergaard, V. (1982), Material Failure in Void Coalescence in Localized Shear Bands, *Int. J. Solids Struct.*, **18**, 659.

Tvergaard, V. and Needleman A. (1984), Analysis of Cup-Cone Fracture in a Round Tensile Bar, *Acta Metall.*, **32**, 157.

Unksov, E.P. (1961), *An Engineering Theory of Plasticity*, Butterworth, London.

Valanis, K.C. (1975), On the Foundations of the Endochronic Theory of Plasticity, *Arch. Mech.*, **27**, 857.

Valanis K.C. (1980), Fundamental Consequences of a New Intrinsic Time Measure Plasticity as a Limit of the Endochronic Theory, *Arch. Mech.*, **32**, 171.

Van Rooyen, G.T. and Backofen, W.A. (1960), A Study of Interface Friction in Plastic Compression, *Int. J. Meek Sei.*, **8**, 1.

Vaughan, H. and Florence, A.L. (1970), Plastic Flow Buckling of Cylindrical Shells Due to Impulsive Loading, *J. Appl. Mech.*, **37**, 171.

Venter, R., Johnson, W., and de Malherbe, M.C. (1971), Yield Loci of Anisotropic Aluminium Using a Frictionless Flat Rectangular Punch, *J. Mech. Engng. Sci.*, **13**, 416.

Vertin, K.D. and Majlessi, S.A. (1993), Finite Element Analysis of the Axisymmetric Upsetting Process Using the Deformation Theory of Plasticity, *J. Engng. Indust., Trans. ASME*, **115**, 450.

Vial, C., Hosford, W.F. and Caddell, R.M. (1983), Yield Loci of Anisotropic Sheet Metals, *Int. J. Mech. Sci.*, **25**, 12.

Voce, E. (1948), The Relationship Between Stress and Strain for Homogeneous Deformation, *J. Inst. Metals,* **74**, 537.

von Karman, Th. (1910), Untersuchungen über Knickfestigkeit, *Mitteilungen über Forschungsarbeit, Wer. Deut. Ing.*, Vol. 81, Springer-Verlag, Berlin.

von Karman, Th. (1942), On the Propagation of Plastic Deformation in Solids, NDRC Report No. A-29.

von Karman, Th. and Duwez, P.E. (1950), The Propagation of Plastic Deformation in Solids, *J. Appl. Phys.*, **21**, 987.

Wagoner, R.H. (1980), Measurement and Analysis of Plane Strain Work-Hardening, Metall. Trans., **11A**, 165.

Wang, A.J. (1955), The Permanent Deflection of a Plastic Plate Under Blast Loading, *J. Appl. Mech.*, **22**, 375.

Wang, A.J. and Hopkins, H.G. (1954), On the Plastic Deformation of Built-in Circular Plates Under Impulsive Loading, *J. Mech. Phys. Solids*, **3**, 22.

Wang, C.M. and Aung, T.M. (2007), Plastics Buckling Analysis of Thick Plates Using p-Ritz Method, *Int. J. Solids Struct.*, **44**, 6239.

Wang, C.M., Xiang, Y. and Chakrabarty, J. (2001), Elastic/Plastics Buckling of Thick Plates, *Int. J. Solids Struct.*, **39**, 8617.

Wang, C.T., Kinzel, G. and Altan, T. (1993), Mathematical Modeling of Plane Strain Bending of Plate and Sheet, *J. Mater. Process. Technol.*, **39**, 279.

Wang, N.M. (1970), Large Plastic Deformation of a Circular Sheet Caused by Punch Stretching, *J. Appl. Mech.*, **37**, 431.

Wang, N.M. (1982), A Mathematical Model of Drawbead Forces in Sheet Metal Forming, *J. Appl. Metalworking*, **2**, 193.

Wang, N.M. (1984), A Rigid/Plastic Rate Sensitive Finite Element Method for Modeling Sheet Metal Forming Processes, in Numerical Analysis of Forming Processes (eds., J.F.T. Pittman et al.), Wiley, New York, p. 117.

Wang, N.M. and Budiansky, B. (1978), Analysis of Sheet Metal Stampings by the Finite Element Method, *J. Appl. Mech.*, **45**, 73.

Wang, N.M. and Shammamy, M.R. (1969), On the Plastic Bulging of Metal Diaphragms Under Hydrostatic Pressure, *J. Mech. Phys. Solids*, **17**, 43.

Wang, X. and Lee, L.H.N. (1989), Wrinkling of an Unevenly Stretched Sheet Metal,*I Engng. Mater. Technol., Trans. ASME*, **111**, 235.
Wasti, S.T. (1970), The Plastic Bending of Transversely Isotropic Circular Plates, *Int. J. Mech. Sci.*, **12**, 109.
Weil, N.A. and Newmark, N.M. (1955), Large Plastic Deformation of Circular Membranes, *J. Appl. Mech.*, **22**, 533.
Weiss, H.J., Prager, W. and Hodge, PG. (1952), Limit Design of a Full Reinforcement for a Circular Cutout in a Uniform Slab, *J. Appl. Mech.*, **19**, 307.
Whiffen, A.C. (1948), The Use of Flat-Ended Projectiles for Determining Dynamic Yield Stress, II, *Proc. Roy. Soc. London Ser. A*, **194**, 300.
White, M.P. and Griffis, Le Van (1947), The Permanent Strain in a Uniform Bar due to Longitudinal Impact, *J. Appl. Mech., Trans. ASME*, **69**, 33.
Whiteley, R.L. (1960), The Importance of Directionality in Deep Drawing Quality Sheet Steel, *Trans. Amer. Soc. Metals*, **52**, 154.
Wierzbicki, T. and Abramowicz, W. (1983), On the Crushing Mechanism of Thin-Walled Structures, *J. Appl. Mech.*, **50**, 727.
Wierzbicki, T. and Jones, N. (1989), *Structural Failure* (eds.), Wiley, New Delhi.
Wierzbicki, T. and Nurick, G.N. (1996), Large Deformation of Thin Plates under Localized Impulsive Loading, *Int. J. Impact Engng.*, **18**, 800.
Wifi, S.A. (1976), An Incremental Complete Solution of the Stretch Forming and Deep Drawing of a Circular Blank Using a Hemispherical Punch, *Int. J. Mech. Sei.*, **18**, 23.
Wilcox, R.J. and Whitton, P.W. (1958), The Cold Extrusion of Metals Using Lubrication at Slow Speeds, *J. Instn. Metals*, **87**, 289.
Wilson, D.V. (1966), Plastic Anisotropy in Sheet Metals, J. Inst. Metals, 94, 84.
Wilson, D.V., Roberts, W.T. and Rodrigues, P.M.B (1981), Metall. Trans., 12A, 1595.
Wilson, W.R.D. and Hector, L.G. (1991), Hydrodynamic Lubrication in Axisymmetric Stretch Forming, Part I: Theoretical Analysis, Part II: Experimental Investigation, *J. Tribology Trans. ASME*, **113**, 659. 667.
Wistreich, J.G. (1955), Investigation of the Mechanics of Wire Drawing, *Proc. Inst. Meek Engrs.*, **169**, 123.
Wistreich, J.G. (1958), Fundamentals of Wire Drawing, *Metall. Rev.* **3**, 97.
Woo, D.M. (1964), The Analysis of Axisymmetric Forming of Sheet Metal and the Hydrostatic Bulging Processes, *Int. J. Mech. Sci*, **6**, 303.
Woo, D.M. (1968), On the Complete Solution of the Deep Drawing Problem, *Int. J. Mech. Sci.*, **10**, 83.
Woodthorpe, J. and Pearce, R. (1969), The Effects of r and n Upon the Forming Limit Diagram of Sheet Metal, *Sheet Metal Indust.*, **46**, 1061.
Woodthorpe, J. and Pearce, R. (1970), The Anomalous Behaviour of Aluminium Sheet Under Balanced Biaxial Tension, *Int. J. Mech. Sci.*, **12**, 341.
Wu, H.C. (1996), The Torsion Test and its Role in Constitutive Equations for Metals, *Chinese J. Mechanics*, **12**, 121.
Wu, H.C. (2002), Anisotropic Plasticity of Sheet Metals Using the Concept of Combined Isotropic-Kinematic Hardening, *Int., J. Plasticity*, **18**, 1661.
Wu, H.C. and Lin, H.C. (1974), Combined Plastic Wave in a Thin-Walled Tube, *Int. J. Solids Struct.*, **10**, 903.
Wu, H.C, Hong, H.K. and Lu, J.K. (1995), An Endochronic Theory Accounted for Deformation Induced Anisotropy, *Int. J. Plasticity*, **11**, 145.
Wu, H.C., Hong, H.K. and Shiao, Y.P. (1998), Anisotropic Plasticity with Application to Sheet Metals, *Int. J. Mech. Sci.*, **41**.
Wu, H.C., Xu, Z.Y. and Wang, P.T. (1997), Determination of Shear Stress-Strain Curve from Torsion Tests for Loading, Unloading and Cyclic Loading, *J. Eng. Mat. Technol., Trans. ASME*, **119**, 113.

Wu, H.C., Xu, Z.Y. and Wang, P.T. (1998), Torsion Test of Aluminum in the Large Plastic Range, *Int. J. Plasticity*, **13**, 873.

Xu, B.Y, Liu, Y.H. and Cen, Z.Z. (1998), Some Developments in Limit Analysis Solutions of Structures, *Metals and Materials*, **4**, 329.

Yamada, Y. and Aoki, I. (1966), On the Tensile Plastic Instability in Axisymmetric Deformation of Sheet Metal, *Int. J. Mech. Sci.*, **8**, 665.

Yamada, Y. and Koide M. (1968), Analysis of the Bore Expanding Test by the Incremental Theory of Plasticity, *Int. J. Mech. Sci.*, **10**, 1.

Yamaguchi, K. and Mellor, P.B. (1976), Thickness and Grain Size Dependence of Limit Strains in Sheet Metal, *Int. J. Mech. Sci.*, **18**, 85.

Yang, D.Y and Han, C.H. (1987), A New Formulation of Generalized Velocity Field for Axisymmetric Forward Extrusion Through Arbitrary Curved Dies, *J. Engng. Indust., Trans. ASME*, **109**, 161.

Yang, D.Y. and Kim, J.H. (1986a), An Analysis for Three-Dimensional Upset Forging of Elliptical Disks, *Int. J. Mack Tool Des. Res.*, **26**, 147.

Yang, D.Y. and Kim, J.H. (1986b), A Rigid/Plastic Finite Element Formulation for the Analysis of the General Deformation of Planar Anisotropic Sheet Metals and its Applications, *Int. J. Mech. Sci.* **28**, 825.

Yang, D.Y. and Kim, Y.J. (1987), Analysis of Hydrostatic Bulging of Anisotropic Rectangular Diaphragms by the Rigid/Plastic Finite Element Method, *J. Eng. Indust., Trans. ASME*, **109**, 149.

Yang, D.Y. and Lee, C.H. (1978), Analysis of Three-Dimensional Extrusion of Sections Through Curved Dies by Conformal Transformation, *Int. J. Meek Sei.*, **20**, 541.

Yang, D.Y. and Ryoo, J.S. (1987), n Investigation into the Relationship Between Torque and Load in Ring Rolling, *J. Engng. Indust., Trans. ASME*, **109**, 190.

Yeom, D.J. and Robinson, M. (1996), Limit Analysis of a Spherical Shell Under Axial Loading on Central Boss, *J. Pressure Vessel Technol.*, **118**, 454.

Yoshida, K. and Miyauchi, K. (1978), Experimental Studies of Material Behavior as Related to Sheet Metal Forming, in Mechanics of Sheet Metal Forming (eds., D.P. Koistinen and N.M. Wang), Plenum Press, New York, p. 10

Yoshida, F. and Uemori, T, (2003), A Model for Large Strain Cyclic Plasticity and its Application to Springback Simulation, *Int. J. Mech. Sci.*, **45**, 1687.

Youngdahl, C.K. (1971), Influence of Pulse Shape on the Final Plastic Deformation of Circular Plates, *Int. J. Solids Struct.*, **7**, 1127.

Youngdahl, C.K. (1972), Dynamic Plastic Deformation of Cylindrical Shells, *J. Appl. Mech.*, **39**, 746.

Yu, T.X. and Chen, F.L. (1992), The Large Deflection Dynamic Plastic Response of Rect Plates, *Int. J. Impact Engng.*, **12**, 603.

Yu, T.X. and Johnson, W. (1982), The Large Elastic-Plastic Deflection with Springback of a Circular Plate Subjected to Circumferential Moments, *J. Appl. Mech.*, **49**, 507.

Yu, T.X. and Stronge, W.J. (1985), Wrinkling of a Circular Plate Stamped by a Spherical Punch, *Int. J. Solid Struct.*, **21**, 995.

Yu, T.X. and Jones, N. (1989), Numerical Simulation of a Clamped Beam Under Impact Loading, *Comput Struct.*, **32**, 281.

Yu, T.X., Symonds, P.S., and Johnson, W. (1985), A Quadrental Circular Beam Subjected to Radial Impact in its own Plane at its Tip by a Rigid Mass, *Proc. Roy. Soc. London Ser. A*, **400**, 19.

Yu, T.X. and Zhang, L.C. (1996), *Plastic Bending Theory and Applications*, World Scientific, Singapore.

Zaid, M. (1958), On the Carrying Capacity of Plates of Arbitrary Shape and Variable Fixity Under a Concentrated Load, *J. Appl. Mech.*, **25**, 598.

Zaid, M. and Paul, B. (1958), Normal Perforation of a Thin Plate by Truncated Projectiles, *J. Franklin Inst.*, **265**, 317.

Zhang, L.C. and Yu, T.X. (1987), An Investigation of the Brazier Effect of a Cylindrical Tube Under Pure Elastic-Plastic Bending, *Int. J. Press. Vess. Piping*, **30**, 77.

Zhang, L.C. and Yu, T.X. (1991), An Experimental Investigation on Stamping of Elastic-Plastic Circular Plates, *J. Mater. Process. Technol.*, **28**, 321.

Zhao, L., Sowerby, R. and Sklad, M.P. (1996), A Theoretical and Experimental Investigation of Limit Strains in Sheet Metal Forming, *Int. J. Mech. Sci.*, **38**, 1307.

Zhu, L. (1996), Transient Deformation Modes of Square Plates Subjected to Explosive Loading, *Int. J. Solids Struct.*, **33**, 301.

Ziegler, H. (1958), Kuppeln Gleicher Festigkeit, *Ingenieur-Archiv*, **26**, 378.

Ziegler, H. (1959), A Modification of Prager's Hardening Rule, *Quart. Appl. Math.* **17**, 55.

Ziekiecwicz, O.C. (1977), *The Finite Element Method*, Second Edition, McGraw-Hill, London.

Zienkiewicz, O.C., Jain, PC and Onate, E. (1978), Flow of Solids During Forming and Extrusion: Some Aspects of Numerical Solutions, *Int. J. Solids Struct.*, **14**, 15.

Zukas, J.A. (1982), Penetration and Perforation of Solids, in *Impact Dynamics* (eds., J.A. Zukas et al.), Wiley, New York, p. 155.

Zukas, J.A. and Gaskill, B. (1996), Ricochet of Deforming Projectiles from Deforming Plates, *Int. J. Impact. Engng.*, **18**, 60.

Name Index

Z

Subject Index

Mechanical Engineering Series *(continued from page ii)*

D. Gross and T. Seelig, **Fracture Mechanics with Introduction to Micro-mechanics**

K.C. Gupta, **Mechanics and Control of Robots**

R. A. Howland, **Intermediate Dynamics: A Linear Algebraic Approach**

D. G. Hull, **Optimal Control Theory for Applications**

J. Ida and J.P.A. Bastos, **Electromagnetics and Calculations of Fields**

M. Kaviany, **Principles of Convective Heat Transfer, 2nd ed.**

M. Kaviany, **Principles of Heat Transfer in Porous Media, 2nd ed.**

E.N. Kuznetsov, **Underconstrained Structural Systems**

P. Ladevèze, **Nonlinear Computational Structural Mechanics: New Approaches and Non-Incremental Methods of Calculation**

P. Ladevèze and J.-P. Pelle, **Mastering Calculations in Linear and Nonlinear Mechanics**

A. Lawrence, **Modern Inertial Technology: Navigation, Guidance, and Control, 2nd ed.**

R.A. Layton, **Principles of Analytical System Dynamics**

F.F. Ling, W.M. Lai, D.A. Lucca, **Fundamentals of Surface Mechanics: With Applications, 2nd ed.**

C.V. Madhusudana, **Thermal Contact Conductance**

D.P. Miannay, **Fracture Mechanics**

D.P. Miannay, **Time-Dependent Fracture Mechanics**

D.K. Miu, **Mechatronics: Electromechanics and Contromechanics**

D. Post, B. Han, and P. Ifju, **High Sensitivity and Moiré: Experimental Analysis for Mechanics and Materials**

R. Rajamani, **Vehicle Dynamics and Control**

F.P. Rimrott, **Introductory Attitude Dynamics**

S.S. Sadhal, P.S. Ayyaswamy, and J.N. Chung, **Transport Phenomena with Drops and Bubbles**

A.A. Shabana, **Theory of Vibration: An Introduction, 2nd ed.**

A.A. Shabana, **Theory of Vibration: Discrete and Continuous Systems, 2nd ed.**

Y. Tseytlin, **Structural Synthesis in Precision Elasticity**

Breinigsville, PA USA
20 November 2009
227695BV00004BB/33/P

9 780387 776736